Community Ecology

Community Ecology

Edited by

Jared Diamond
University of California Medical School, Los Angeles

Ted J. Case
University of California at San Diego

HARPER & ROW, PUBLISHERS, New York
Cambridge, Philadelphia, San Francisco,
London, Mexico City, São Paulo, Singapore, Sydney

This volume is dedicated to G. Evelyn Hutchinson, in admiration for his pioneering contributions to community ecology.

Sponsoring Editor: Claudia M. Wilson
Project Editor: Holly D. Gordon
Cover Design: Mary Archondes
Production: Debra Forrest Bochner
Compositor: York Graphic Services, Inc.
Printer and Binder: R. R. Donnelley & Sons Company

Community Ecology

For information address Harper & Row, Publishers, Inc., 10 East 53d Street, New York, NY 10022.

Library of Congress Cataloging-in-Publication Data
Main entry under title:

Community ecology.

Bibliography: p.
Includes index.
1. Biotic communities. 2. Ecology. I. Diamond, Jared M. II. Case, Ted J.
QH541.C644 1986 574.5′24 85-8477
ISBN 0-06-041202-X

85 86 87 88 9 8 7 6 5 4 3 2 1

Contents

Preface

This book presents a pluralistic approach to community ecology. In this preface we explain briefly what we mean by community ecology, why we find this field interesting and important, and what we mean by a pluralistic approach.

The choices of scale on which to study whole living organisms form a broad spectrum, ranging from one individual organism to all the organisms on earth. Along this spectrum the "community" comprises the populations of some or all species coexisting at a site or in a region. We intend this definition to be flexible (some would say vague) in three respects. First, the site size may vary from an entire biogeographical region to a one-pint laboratory culture bottle. Second, the number of species may vary from a whole local biota to a tiny fraction of it. Finally, the fraction of the biota may be delineated in various ways: a taxonomic set of species, a life form set, a group of mutualists and their resources, a single trophic level, two interacting trophic levels, or a whole food web. The definition of a community is so flexible because, as the following chapters will illustrate, different definitions are suited for asking different questions.

Why study community ecology? Because it is intellectually fascinating, rich with major unsolved questions, important to other fields of biology, and important in preserving our quality of life. Ecologists are challenged by the quest to explain observed patterns of community organization. For example, what limits the species composition of a community? What determines the distributions and abundances of those species that do compose the community? What is the pattern of connections among a community's species? What permits species to coexist? How do communities track environmental change, and how do they evolve? The answers to these questions are important in many other fields of biology besides community ecology, such as parasitology, paleontology, evolutionary biology, and sociobiology. Furthermore, these questions are not only intellectually challenging. They are also crucial to solving the enormous practical problems posed by human impact on natural communities, and posed by the synthetic communities of crops, domestic animals, and timber trees that we have created to support our existence.

In this volume we emphasize a pluralistic approach, because we believe that it is essential to the productive study of community ecology and also that its importance is underappreciated. As applied to scientific research, pluralism means using a diversity of methodologies to obtain data, and a diversity of models to interpret data. For instance, geology and astronomy are pluralistic in their methodology compared to mathematics or organic chemistry; psychology and medicine are pluralistic in their models compared to classical mechanics or molecular biology. Pluralism fascinates some scientists and repels others. Until recently, philosophy of science focused on relatively homogeneous fields such as classical mechanics. As a result, many scientists have been trained to regard pluralistic approaches as soft, unrigorous, unscientific, and indicative of a retarded field. Even scientists who work in pluralistic fields tend to view how science "should" be pursued in ways that are mismatched to their field's special needs. A thoughtful discussion of these problems will be found in Ernst Mayr's (1982) book, *The Growth of Biological Thought*.

Practical constraints require ecologists to use a great variety of methodologies, to work with mathematical models as well as with living animals, and to operate on a great variety of spatial and temporal scales. Methods in ecology include laboratory experiments, observations outdoors, manipulative experiments outdoors, "natural experiments" involving comparative observations at several outdoor sites, and monitoring the results of natural or accidental perturbations such as epidemics or volcanic eruptions. The species that one chooses to study, and the questions that one asks, determine what methods are possible or useful. For instance, laboratory experiments are good for studying wasp predation on beetles, but not for studying killer whale predation on seals. As Chapters 2, 5, 7, 23, and 32 of this book will illustrate, ecologists gain complementary insights into competition from studying growth of fruit flies in laboratory bottles, measurements of cactuses in Mexican deserts, SCUBA-based manipulations of northern New England seaweeds, historical records of introduced birds in Hawaii, and counts of fossil plants over the whole world. In contrast, molecular biologists were able to unravel the entire genetic code solely through short-term experiments in small laboratories.

The answers to interesting ecological questions require a correspondingly diverse array of models. For instance, the importance of any given factor in limiting population numbers varies among species and among sites: Predation often limits meadow mice but not polar bears, and it limits moose on Isle Royale but not in Maine. Suppose for comparison that the genetic code, instead of being determined solely by DNA, was codetermined by seven classes of macromolecules, whose relative role in a given species varied with age, season, weather conditions, and time since the last glaciation and also tended to differ between large and small species, ectotherms and endotherms, and herbivores and carnivores. If this were true, we would surely not have our present complete understanding of the code. Yet this is exactly the problem that ecologists face in trying to understand how species abundances are codetermined by competition, predation, herbivory, disease, parasitism, mutualism, and weather. The answers to general ecological questions are rarely universal laws, like those of physics. Instead, the answers are conditional statements such as "For a community of species with properties A_1 and A_2 in habitat B at latitude C, limiting factors X_2 and X_5 are likely to predominate."

Fields of science dependent on diverse methodologies and conditional models simultaneously derive fascination, strength, and weakness from this fact. The weakness comes from the inherent complexities, the consequent difficulties in obtaining convincing answers, and the risk of blind preference for one methodology or model even where it is inappropriate. The strength arises because conclusions tested in diverse ways become more robust and credible. The fascination stems from the challenges posed by complexity and conditionality, for those who are comfortable with such challenges; many scientists are not and instead prefer fields with more uniform techniques and more clear-cut answers.

Even within sciences dependent on diverse methodologies, there is variation in the degree to which the methodologies have achieved a comfortable coexistence. For example, compare the routine acceptance of methodological diversity in physics or medical science with the disagreements common in psychology or ecology. Physicists interested in atomic spectra understand that comparative observations of stars yield information unattainable through laboratory experiments and vice versa, and that both types of information are needed for full understanding of spectroscopy. One does not hear laboratory physicists branding stellar observation as an inferior technique because astrophysicists cannot visit stars and carry out rigorously replicated experiments there. In contrast, clinical psychologists, experimental psychologists, and psychopharmacologists have been prone to regard one approach to psychology as superior, and to dismiss another approach as unrigorous or bypassing the really interesting questions. Similar claims of methodological superiority have been advanced in much the same words by ecologists pursuing laboratory, field, or natural experiments.

The maturation of community ecology as a discipline has been marked by recurring arguments stemming ultimately from unwillingness to accept this pluralism of methods and results. Such arguments usually boil down to one of four types of assertions: method A is in principle powerful (superior, rigorous); method B is in principle weak (inferior, shoddy); organizing force C is strong or underappreciated; organizing force D is weak or overrated. These assertions are usually inappropriate, virtually meaningless, because useful answers in ecology generally have to be in the form of conditional statements.

Given ecology's complexity, it is tempting to retreat into either of two extreme positions: descriptive particularism, which stresses that every system is unique and complex and that no generalizations other than superficial ones exist; or an oversimplifying search for universal statements and for The Best Method. The goal of this book is to steer between these extremes, and to seek conditional statements that are faithful to the diversity of life. The book specifically asks these questions:

For what species and what problems can each methodology be most profitably applied?

Given that observations can be carried out on various spatial and temporal scales, how does the choice of most appropriate scale depend on the species studied and the question asked?

How do effects of different organizing forces vary among communities?

Given that it is surely impossible to devise one model applicable to all ecological communities, can one at least partition communities among a modest number of types and devise a model for each type?

The following chapters cannot provide tidy solutions to these questions. We feel that they do provide extensive guidelines to the first two questions, tentative partial answers to the third, a rudimentary start toward answering the fourth, and a status report on all four to guide future research.

Jared Diamond
Ted J. Case

Outline of the Contents

This book has been written for three audiences. First, it aims to serve as an introduction to community ecology for readers who are already familiar with general ecology. Second, it seeks to portray the present state of the field, and to formulate empirical and theoretical problems in need of solution, for practising ecologists. Since major issues in community ecology remain unresolved, different chapters of this volume present differing methods, models, or perspectives. For example, the contributors include advocates and deniers of the importance of competition, advocates of equilibrium and nonequilibrium thinking, students of food-limited and space-limited communities, and authors who do or do not find strong pairwise species interactions. Third, we hope to stimulate scientists in other areas of biology to help solve outstanding problems of community ecology, as well as to make more use of what is now known. In particular, we have included four chapters (Chapters 7, 16, 17, and 18) based partly or entirely on fossil communities, since students of modern communities have tended unjustly to neglect the unique perspectives offered by paleoecology.

The book is organized into six sections focusing on experimental methodologies, species introductions and extinctions in the widest sense, spatial and temporal scale, models of communities, forces structuring communities, and diversity of community types. Each section consists of an introductory chapter summarizing the broad issues and two to seven chapters on specific issues or particular communities. In practice, it is impossible to discuss methodology, scale, ecological forces, models, and community types as separate subjects. Thus, most chapters contain much material relevant to sections other than the one in which that chapter is placed. Extensive cross-referencing between chapters is provided.

Some readers will be more interested in particular taxa or in particular ecological forces than in the subjects on which the six sections focus. To guide these readers to the material that interests them, we provide here three outlines of the book: a summary by section and chapter according to the actual sequence of the volume; a brief overview stating where in the book particular taxa receive extended discussion; and a similarly brief overview for particular ecological forces.

SUMMARY BY SEQUENCE OF CHAPTERS

Section One: Experimental Methods in Ecology Ecologists are forced to rely on many types of experiments. Hence Chapter 1, by Jared Diamond, compares the merits of laboratory experiments, field experiments, and natural experiments along seven axes and calls attention to opportunities for more powerful application of each type than is usual at present. It turns out that, for each type, there are certain classes of species or problems for which it offers the best, or even the only, approach.

In Chapter 2 Michael Gilpin, M. Patricia Carpenter, and Mark Pomerantz illustrate how laboratory methods can be extended to the assembly of communities much more complex than those hitherto studied in the laboratory, in this case 28 species of *Drosophila* studied singly, in all pairwise combinations, and in a set of 10. This approach, which may be termed community reconstitution, allows one to test one's understanding of a community by reassembling it from its parts. The *Drosophila* system is sufficiently realistic to address problems posed by actual species-rich competition communities, such as the issue of multiple domains of attraction. Chapter 3, by James Brown, Diane Davidson, James Munger, and Richard Inouye, illustrates the power of long-term, intensive, field experiments, in this case as applied to a desert granivore system of rodents, ants, birds, and seeds. Experimental perturbations reveal a wealth of direct and indirect species interactions and are uniquely suited to tracing out the links within this community.

Among the remaining chapters, laboratory experiments are utilized in Chapters 21 and 31; field experiments, in Chapters 11, 12, 21, 22, 24–26, and 30–33; natural experiments, in Chapters 5, 6, 10, 16, 21, 23, 24, 30, and 33.

Section Two: Species Introductions and Extinctions Laboratory and field experiments permit one to study community effects of adding and removing species under controlled conditions, but only in certain cases and on a tiny scale. To understand such effects operating on a large spatial scale over long periods of time, one must study additions and removals of species by natural events or by accidental or deliberate acts of humans other than experimental ecologists. Jared Diamond and Ted Case survey these events in Chapter 4 and offer a framework for asking questions about their consequences. In essence, a host of experimental perturbations that ecologists could never carry out by themselves has actually been effected and offers insights unattainable otherwise.

The next chapters illustrate this approach by analyzing three such perturbations on progressively bigger spatial and temporal scales. In Chapter 5, by Michael Moulton and Stuart Pimm, the perturbation is human introductions of exotic bird species to Hawaii, creating a wholly synthetic avifauna in the lowlands within the past 125 years. Records of success or failure of over 100 introductions permit calculations of how a species' risk of extinction depends on the number and morphological closeness of other species in the community. In Chapter 6, by Jared Diamond, the perturbation is speciation itself, injecting new species into the montane avifauna of New Guinea over evolutionary time. From comparisons of species pairs at different stages in the speciation process, one can reconstruct how niche differences between related species evolve, and whether present niche differences may reasonably be attributed to the ghost of past competition. Both Chapters 5 and 6 also illuminate the problem of body size differences and Hutchinsonian ratios. Chapter 7, by Andrew Knoll, is again concerned with evolutionary introductions of species, but for plants and on a geological time scale. Comparisons of 391 fossil floras distributed over the past 400 million years let one examine the buildup of plant species diversity with time, its dependence on evolutionary advances in reproductive and vegetative biology, and the relation between introductions of new species and extinctions of old species.

Section Three: Space and Time The amount and distribution of available space and the amount of available time are key elements in almost any ecological discussion. Chapter 8, by John Wiens, John Addicott, Ted Case, and Jared Diamond, provides methodological background by comparing what can be learned from studies on various spatial or temporal scales. It becomes apparent that different scales yield different types of information, so that the choice of appropriate scale depends on the question asked. This conclusion is illustrated by John Wiens in Chapter 9, which demonstrates that patterns of habitat selection by shrubsteppe birds change as the spatial scale of study is progressively reduced from western North America to small census plots.

Temporal variation lies at the heart of the variable-environment hypothesis, in which Wiens (1977) argued that resource limitation, severe interspecific competition, and selection occur only infrequently. Wiens reappraises this model in Chapter 9 and notes a series of time lags between resource inputs and bird population responses in the variable shrubsteppe environment. Another test of the model for a variable environment is provided by Chapter 10, in which Peter Grant examines fluctuations in seed-eating Galápagos finches and their food supply over a decade that included a severe drought. Food limitation and interspecific competition for food occur in most dry seasons, hence at least once a generation.

Spatial patchiness is the theme of Chapter 11 by Peter Kareiva and Chapter 12 by Peter Grubb. Chapter 11 examines effects of patchiness and dispersal on herbivorous insects and their predators. To apply competition theory and predator-prey theory to this system, it is essential to incorporate the patchy distributions of resources, consumers, and predators. Chapter 12, focusing on sparse and patchily distributed plants in European chalk grassland, asks what keeps sparse species sparse, what enables them to persist at all, and whether they are too sparse to affect each other.

Section Four: Equilibrium and Nonequilibrium Communities Peter Chesson and Ted Case introduce this subject in Chapter 13 by providing clear statements of classical competition theory, its extensions incorporating predation and spatial variation, and four newer and more radical theoretical developments. These theories differ from each other in such basic respects as whether they view a community as at equilibrium, as stable, or as subject to effects of history and whether they consider environmental fluctuations important and the environmental mean constant. As Chesson explains in more detail in Chapter 14, environmental variation is not just noise, not just a destabilizing factor, and not just something that delays competitive exclusion. It can actually alter trends, as illustrated by models of variable recruitment, and it can create stochastic mechanisms of coexistence that share many properties with deterministic ones. In Chapter 15 Donald Strong then contrasts the familiar, embattled concepts of density dependence or independence with density vagueness; that is, little or no influence of population density on demographic parameters at intermediate population sizes. The concept of density vagueness finds much support in the real world, and its ecological consequences are explored.

It is paradoxical that students of modern communities regularly debate whether communities are at equilibrium, but rarely consult the abundant fossil evidence relevant to this question. Chapters 16 to 18 offer a fresh perspective by injecting some basic paleoecological facts into the debate. Climate records summarized by Margaret Davis in Chapter 16 emphasize that on any time scale from a decade to 100,000 years climate does not fluctuate about a mean value, but exhibits long-term trends. Thus, it becomes essential to explore how species differ in the rate at which they track a changing environment. For many communities these rates are too slow for the community to remain in equilibrium with the environment. In Chapter 17 Thomas Van Devender then shows how plant and animal remains in radiocarbon-dated packrat middens enable one to reconstruct the detailed vegetational history of the Chihuahuan Desert through the climate fluctuations of the last 35,000 years. Theoretical constructs such as refugia, community constancy, and speciation mechanisms can be tested against the evidence of the middens. The abundant fossil remains of North American small mammals, summarized by Russell Graham in Chapter 18, dismantle the conventional view that species, communities, and climate zones marched together south and then north again during the late Pleistocene and Holocene. Actually, different species moved in opposite directions; communities were massively reshuffled; most late Pleistocene communities had more, not fewer, small mammal species than do modern communities; and Pleistocene habitats and climates often lacked any modern equivalent.

The final chapter of this section, Chapter 19, by Stephen Hubbell and Robin Foster, also finds the notion of stable equilibrium inappropriate—in this case, to describe the results of a study in which all free-standing woody plants over 1 cm in diameter have been mapped and identified in a 0.5-km^2 study plot on Barro Colorado Island. Rather than having highly differentiated niches and finely tuned coevolutionary adjustments to each other, rain forest trees are viewed as forming large guilds of competitively nearly equal species, with rates of competitive exclusion so slow as to be comparable to rates of immigration and speciation.

Section Five: Forces Structuring Communities What is community structure, and what forces produce it? In offering an overview of this much debated question, Jonathan Roughgarden and Jared Diamond suggest in Chapter 20 that the concept of limited membership may unify the diverse patterns proposed as evidence of structure. The chapter examines how limited membership depends on adaptations to the physical environment, dispersal, and species interactions and how these forces interact with each other. The following seven chapters consider some of these forces in communities of fish, plants, mites, mutualists, carrion feeders, and parasites.

Earl Werner describes in Chapter 21 how competition and predation act and interact in communities of freshwater fish. The community ecology of fish differs from that of birds and mammals, but resembles that of many or most other species, in that ontogenetic niche shifts play a central role. Chapter 22, by David Tilman, explores the consequences of a simple theory suggesting that each plant species should be a superior competitor at a particular point along a soil resource : light gradient. This theory illuminates a wide range of phenomena: plant distribution by soil type and resource availability, results of fertilization experiments, primary succession, and plant evolution. In Chapter 23 Martin Cody looks at plant life forms as under opposing selection from two forces: abiotic forces promoting convergence among species sharing a habitat, and biotic forces leading to divergent strategies for capturing light, water, and nutrients. These themes are examined for desert shrubs, desert cacti, and plants of Mediterranean scrub.

The remaining four chapters of this section deal with communities where species live in close association with each other: commensals, mutualists, and parasites. Robert Colwell, in Chapter 24, describes the dispersal of hummingbird flower mites in the bills of hummingbirds. The opposing forces acting on each mite species' choice of host flower species include resource variability, which broadens the flower repertoire, and sexual selection, which narrows it. In effect, the choice of flower is a species-specific secondary sex characteristic, a neglected factor that may contribute to habitat segregation in other animal groups as well. In Chapter 25 John Addicott broadly surveys the population consequences of mutualism, an interaction that has been studied more at the individual level than at the population level. The chapter considers the effects of mutualism on population growth rates and population densities, and the many factors inherent to mutualism that may limit net benefits obtained by participants. Chapter 26, by David Wilson, turns to indirect effects; that is, interactions between species that are mediated by their effects on other species. Do species evolve to increase their fitness through indirect effects and to modify a community in ways that loop back beneficially to themselves? This question is illustrated by experiments on carrion-feeding arthropods and microbes, but such adaptive indirect effects deserve evaluation in many other communities with appropriate population structure. In Chapter 27 Catherine Toft provides a reexamination of parasite communities and a reclassification of parasites. Communities of parasites, parasitoids, and predators are compared as to stability and diversity. Parasite-host relationships can be classified into at least 11 types expected to differ in their population dynamics and levels of diversity.

Section Six: Kinds of Communities Few people would now claim that one model will suffice to describe all communities. Does this mean that every community is unique, or will it turn out that a modest number of models can encompass descriptions of most communities? Chapter 28, by Thomas Schoener, is an initial assault on this difficult problem, which is likely to consume much of the energy of community ecologists in the near future. Schoener proposes a set of primitive axes, 6 related to properties of organisms and 6 to properties of the environment, plus 10 derived axes of community properties. He then explores the relations among these axes for 14 actual communities, ranging from ephemeral intertidal algae to desert granivorous rodents.

The next two chapters compare different kinds of competition communities. In Chapter 29 Peter Yodzis contrasts consumptive competition with competition for space and further divides the latter into dominance control and founder control. In view of the conflicting results of tests of the intermediate disturbance hypothesis, Yodzis presents a more finely differentiated hypothesis, with predictions that depend on two variables: the selectivity of the mortality factor and the kind of competition operating in the community. Chapter 30, by Jonathan Roughgarden, then compares three actual communities: Caribbean *Anolis* lizards, which behave as a food-limited competition community; intertidal barnacles, which behave as a space-limited competition community; and coral reef fish, which share properties of both. A two-way classification of competition communities is proposed, based on whether resources can be partitioned and on whether population structure is open or closed.

The effects of competition and predation on communities of sessile marine species are the subject of Chapter 31 by Leo Buss and Chapter 32 by Jane Lubchenco. Buss examines marine invertebrates, which are especially suited to elucidating the costs and morphological bases of competition and the rankings that it produces. Predation often precludes competition in these communities, but there are at least five conditions permitting competition to persist in the face of predation. Lubchenco's field experiments quantitatively assess the relative effects of competition, predation, and wave action on seaweed colonization of mid-zone rocks of the New England coast. Several biological differences between ephemeral and perennial algae cause these two plant groups to yield differing conclusions for this reassessment.

There has been a long-standing polarization in community ecology between two views.

One view is that communities have strong patterns and that interspecific competition is a dominant force. The other is that communities do not have strong patterns and that predation and physical factors are dominant forces. Suspiciously, proponents of the first view tend to work on terrestrial vertebrates, while proponents of the other tend to work on terrestrial arthropods. In Chapter 33 Thomas Schoener confirms the suspicion that each view may be appropriate for different taxa. Schoener reaches this conclusion by comparing 11 types of patterning for birds, lizards, and spiders on the same Bahamian islands and by synthesizing published results for other localities.

OUTLINE BY TAXA

Chapters with extended discussion of particular groups of taxa are the following.

Terrestrial plants Chapters 3, 7, 12, 16, 17, 19, 22, 23, 25

Marine plants Chapter 32

Terrestrial invertebrates Chapters 2, 3, 11, 15, 24, 25, 26, 27, 33

Marine invertebrates Chapters 30, 31

Fish Chapters 21, 30

Reptiles Chapters 17, 30, 33

Birds Chapters 5, 6, 9, 10, 33

Mammals Chapters 3, 18

Our main gaps in taxonomic coverage are unicellular organisms and freshwater taxa other than fish.

OUTLINE BY FORCES

Chapters with extended discussion of particular ecological forces are the following.

Climate and the physical environment Chapters 2, 6, 9, 10, 11, 14, 16, 17, 18, 20, 22, 23, 32, 33

Competition Chapters 2, 3, 4, 5, 6, 7, 10, 11, 12, 20, 21, 22, 23, 29, 30, 31, 32

Predation Chapters 4, 8, 11, 20, 21, 27, 31, 33

Herbivory Chapters 3, 4, 10, 11, 12, 20, 24, 32

Mutualism Chapters 25, 26

Parasitism Chapters 4, 27

Disease Chapters 3, 4

Sexual selection Chapter 24

ACKNOWLEDGMENTS

This book originated as a set of manuscripts that the authors circulated to each other, met to discuss on April 12–16, 1984, at Calamigos Ranch near Los Angeles, and subsequently revised. We editors acknowledge with pleasure our debt to many individuals and organizations:

University of California Los Angeles, San Diego, and Davis campuses, for financial support;

Carl Biehl, William Bond, Jerri Columpus, Lori Obinger, and Mark Taper, for their help in running the conference and preparing the book;

Sam McNaughton and John Smiley, for their contributions to the conference;

Claudia Wilson and Holly Gordon, our editors at Harper & Row, and Stewart Janes, for their collaboration in production of the book;

Martin Cody, Michael Gilpin, Jonathan Roughgarden, and Thomas Schoener, who shared the responsibilities of this book from initial conception through editing;

And, most of all, to our contributors, for their patience in responding to our seemingly endless series of requests to criticize other chapters and to revise their own chapters.

Contributors

JOHN F. ADDICOTT, Department of Zoology, University of Alberta, Edmonton, Alberta T6G 2E9, Canada, and Rocky Mountain Biological Laboratory, Crested Butte, Colorado 81224

JAMES H. BROWN, Department of Ecology and Evolutionary Biology, University of Arizona, Tucson, Arizona 85721

LEO W. BUSS, Biology Department, Yale University, New Haven, Connecticut 06511

M. PATRICIA CARPENTER, Department of Biology (C-016), University of California at San Diego, La Jolla, California 92093

TED J. CASE, Department of Biology (C-016), University of California at San Diego, La Jolla, California 92093

PETER L. CHESSON, Department of Zoology, Ohio State University, 1735 Neil Avenue, Columbus, Ohio 43210

MARTIN L. CODY, Department of Biology, University of California, Los Angeles, California 90024

ROBERT K. COLWELL, Department of Zoology, University of California, Berkeley, California 94720

DIANE W. DAVIDSON, Department of Biology, University of Utah, Salt Lake City, Utah 84112

MARGARET BRYAN DAVIS, Department of Ecology and Behavioral Biology, University of Minnesota, Minneapolis, Minnesota 55455

JARED DIAMOND, Department of Physiology, University of California Medical School, Los Angeles, California 90024

ROBIN B. FOSTER, Department of Botany, Field Museum of Natural History, Chicago, Illinois 60605, and Smithsonian Tropical Research Institute, Box 2072, Balboa, Republic of Panama

MICHAEL E. GILPIN, Department of Biology (C-016), University of California at San Diego, La Jolla, California 92093

RUSSELL W. GRAHAM, Quaternary Studies Center, Illinois State Museum, Springfield, Illinois 62706

P. R. GRANT, Division of Biological Sciences, University of Michigan, Ann Arbor, Michigan 48109-1048

PETER J. GRUBB, Botany School, University of Cambridge, Cambridge CB2 3EA, England

STEPHEN P. HUBBELL, Department of Biology, University of Iowa, Iowa City, Iowa 52242, and Smithsonian Tropical Research Institute, Box 2072, Balboa, Republic of Panama

RICHARD S. INOUYE, Department of Ecology and Behavioral Biology, University of Minnesota, Minneapolis, Minnesota 55455

PETER KAREIVA, Department of Zoology NJ-15, University of Washington, Seattle, Washington 98195

ANDREW H. KNOLL, Department of Organismic and Evolutionary Biology, Harvard University, Cambridge, Massachusetts 02138

JANE LUBCHENCO, Department of Zoology, Oregon State University, Corvallis, Oregon 97331

MICHAEL P. MOULTON, Department of Zoology, University of Tennessee, Knoxville, Tennessee 37996-0810

JAMES C. MUNGER, Department of Zoology, University of Wisconsin, Madison, Wisconsin 53706

STUART L. PIMM, Graduate Program in Ecology, University of Tennessee, Knoxville, Tennessee 37996, and Australian Environmental Studies, Griffith University, Nathan, Queensland 4111, Australia

MARK J. POMERANTZ, Department of Biology (C-016), University of California at San Diego, La Jolla, California 92093

JONATHAN ROUGHGARDEN, Department of Biological Sciences, Stanford University, Stanford, California 94305

THOMAS W. SCHOENER, Department of Zoology, University of California, Davis, California 95616

DONALD R. STRONG, Department of Biological Science, Florida State University, Tallahassee, Florida 32306

DAVID TILMAN, Department of Ecology and Behavioral Biology, University of Minnesota, Minneapolis, Minnesota 55455

CATHERINE A. TOFT, Department of Zoology, University of California, Davis, California 95616

THOMAS R. VAN DEVENDER, Arizona-Sonora Desert Museum, Route 9, Box 900, Tucson, Arizona 85743

EARL E. WERNER, Kellogg Biological Station and Department of Zoology, Michigan State University, Hickory Corners, Michigan 49060

JOHN A. WIENS, Department of Biology, University of New Mexico, Albuquerque, New Mexico 87131

DAVID SLOAN WILSON, Kellogg Biological Station, Hickory Corners, Michigan 49060

PETER YODZIS, Department of Zoology, University of Guelph, Guelph, Ontario N1G 2W1, Canada

one

EXPERIMENTAL METHODS IN ECOLOGY

chapter 1

Overview: Laboratory Experiments, Field Experiments, and Natural Experiments

Jared Diamond

INTRODUCTION

Community ecologists seek to understand species abundances and distributions. These dependent variables are ultimately controlled by two sets of independent variables: physical environmental factors and other species.

The effects of both sets of factors on individuals of a given species can either be witnessed directly (e.g., as physiological stress, acts of predation, and acts of interference competition) or inferred from other observations (e.g., inferences of exploitation competition from observations of resource utilization). While such observations of individual organisms reveal mechanisms by which the independent variables act on a species, they cannot reveal whether a species' abundance and distribution are thereby affected. To answer this population-level question, one must measure the change in abundance and distribution associated with changes in the independent variables. Experiments of various sorts are a major tool for this purpose.

Community ecologists, like astronomers and geologists, must rely on more types of experiments than do chemists or molecular biologists. Three types are traditionally distinguished: laboratory experiments (LEs), or perturbations produced by the experimenter in the laboratory; field experiments (FEs), or perturbations produced by the experimenter in the field; and natural experiments (NEs), or natural perturbations occurring in the field. In practice, LEs, FEs, and NEs form a continuum.

These three types of experiments differ greatly in ways that set what species can be used and what problems can be addressed. Some species and problems can be conveniently studied by one type of experiment but cannot be studied at all by another type. For instance, in my own research on birds I use laboratory experiments to study food processing by avian intestine, field experiments to study bower decoration by bowerbirds, and natural experiments to study bird distributions in New Guinea national parks. While in these cases the reasons dictating the type of experiment are obvious, the trade-offs are not always as clear and are often overlooked, misunderstood, or controversial. Understanding these trade-offs is essential to appropriate experimental design in community ecology and hence forms a logical starting point for this book.

This chapter begins by considering eight axes along which the types of experiments differ in their relative merits. I then discuss each type of experiment, illustrating its merits and limitations and range of utility in more detail and mentioning possible methodological improvements. The final section then synthesizes this discussion to reach a simple answer to the question, Which is the best type of experiment in community ecology? and to examine why acceptance of this obvious answer has been delayed. The relative lengths of the sections on each type should not be construed as proportional to its virtues. In particular, the strengths and weaknesses of LEs, and the strengths of FEs and NEs, are straightforward and easily described. Much more space has to be devoted to the complicated problem of limitations on FEs and NEs, and what to do about them.

EIGHT AXES FOR TRADE-OFFS

Table 1.1 summarizes, for community ecology, how the three types of experiments differ in their relative merits along each of the eight axes. (Especially laboratory experiments have different merits in other areas of ecology.) For this purpose I distinguish two subtypes of natural experiments. (a) Natural trajectory experiments (NTEs) are comparisons of the same community at various times before, during, and after a witnessed perturbation by nature or by humans other than ecologists. Examples include a storm, volcanic eruption, or the introduction or local extermination of a species. This subtype of natural experiment is discussed at length in Chapter 4. (b) Natural snapshot experiments (NSEs) are comparisons of communities assumed to have reached a quasi-steady state with respect to the perturbing variable (e.g., islands with and without a certain predator, each island having been in that state from the time of earliest observations).

In Table 1.1 I am concerned with intrinsic practical limitations of each type of experiment, rather than with common but curable deficiencies in experimental design. The assignments of Table 1.1 are sketched briefly in this section and in more detail in succeeding sections. The eight axes are as follows.

1. Regulation of independent variables Ecologists regulate actively (often, hold constant) many or all significant independent variables in LEs, one or a few variables in FEs, none in NEs.

2. Matching of sites To compensate for any lack of regulation of independent variables, ecologists attempt to minimize intersite differences in unregulated variables by three methods: replication of sites, selection of sites so as initially to have the same values of unregulated variables (except for the naturally perturbed variable in an NE), and randomization and interspersion of control and experimental sites in FEs and LEs. This matching is likely to be least successful in NSEs, where sites may be distant from each other, and where the status of sites as ''experimental'' or ''control'' (e.g., as having or lacking a certain predator) is given by nature rather than assigned by the experimenter. Matching is likely to be more successful in NTEs, where the same site

Table 1.1 COMPARISON OF THE STRENGTHS AND WEAKNESSES OF DIFFERENT TYPES OF EXPERIMENTS IN ECOLOGY

Axis	Type of experiment			
	LE	FE	NTE	NSE
1. Regulation of independent variables	Highest	Medium/low	None	None
2. Site matching	Highest	Medium	Medium/low	Lowest
3. Ability to follow trajectory	Yes	Yes	Yes	No
4. Maximum temporal scale	Lowest	Lowest	Highest	Highest
5. Maximum spatial scale	Lowest	Low	Highest	Highest
6. Scope (range of manipulations)	Lowest	Medium/low	Medium/high	Highest
7. Realism	None/low	High	Highest	Highest
8. Generality	None	Low	High	High

LE = laboratory experiment; FE = field experiment; NTE = natural trajectory experiment; NSE = natural snapshot experiment.

may be studied before and during a perturbation; still more successful in FEs, where sites may be randomly assigned and interspersed, as well as close or abutting; and most successful in LEs, where the "sites" may be uniform bottles studied at the same lab bench on the same day.

3. Ability to follow response trajectories resulting from perturbations LEs, FEs, and NTEs routinely follow response trajectories after experimental perturbations. NSEs by definition deal with snapshots or quasi-steady states and do not follow trajectories.

4. Maximum temporal scale The experimenter must exert on-going effort to keep LEs and FEs running correctly. Hence LEs and FEs are usually restricted to durations of a few years. NTEs documented by historical or fossil evidence afford the sole opportunity to follow longer response trajectories, including ones lasting millions of years (Chapter 7). NSEs permit one to examine a snapshot of a trajectory after a long time, though not the trajectory itself.

5. Maximum spatial scale Practical problems limit the spatial scale of LEs to around 0.01 hectare (the size of a constant-temperature laboratory room), and FEs usually to less than 1 ha (usually much less, occasionally more). NEs are unlimited as to spatial scale and often use large islands or continents.

6. Range of species and manipulations that can be studied As discussed in Chapter 8, one's choice of temporal and spatial scales determines what one can study. The just-mentioned extreme limitations of temporal and spatial scales, and numerous other practical problems of maintaining plant and animal populations in the laboratory, restrict LE studies of community ecology to only a tiny fraction of the species and processes existing in nature. Similar but not quite so severe restrictions due to scale, plus other practical problems and severe legal and moral constraints, knock FEs out of consideration for a large range of species manipulations, processes, and sites. NEs escape scale problems, escape legal and moral restrictions by studying an already existing situation, and are limited in practice only by the experimenter's ability to find suitable natural situations for comparison.

7. Realism By realism is meant, Are there any natural community and any natural perturbation, even a single one, to which the results of the experiment apply or can be readily extrapolated? NEs are completely realistic: no extrapolation is needed, because they already study natural communities and usually study natural perturbations. The realism of FEs is high but not complete: they too study a natural community, but the experimental perturbation may or may not mimic a natural one. LEs are usually unrealistic by intent: they mostly utilize a highly simplified and regulated community unlike any natural one. There is thus usually an acute question whether the result of an LE applies to any real community at all. (I stress again that this statement applies to community ecology; LEs are often much more realistic in physiological or behavioral ecology.)

8. Generality A result possesses realism if it is immediately known to apply to at least one community; generality, if it is immediately known to apply to many communities. One's ultimate goal is a conclusion with at least some generality, rather than one that applies to just one site in one year. LEs, because they lack realism, also lack generality. NEs and FEs both aim to compare multiple real experimental and control communities. However, NEs obtain their replicates by explicitly sampling natural variation among real communities, while FEs explicitly seek to minimize this variation and to sample adjacent, preferably identical, replicates at a single site. There is thus an acute question concerning the extent to which the result of an FE possesses generality and applies beyond that single site in that year. In effect, FEs compared to NEs gain confidence in the conclusions at one site (axes 1 and 2 above), at the cost of generality.

LABORATORY EXPERIMENTS (LEs)

Examples of LEs

LEs are the experimental approach familiar to chemists, who are often initially puzzled as to why ecologists ever resort to any other approach.

In the laboratory the experimenter can regulate the whole abiotic environment, especially the ecologically critical variables of light, temperature, water, substrate, and nutrients that vary outdoors somewhat unpredictably in space and time. The experimenter can also regulate the biological environment, and in practice most LEs are done on synthetic communities composed of a few (usually two) chosen species. The LE technique is often employed on microorganisms or small arthropods kept in bottles, hence ecologists' somewhat derisive term ''bottle experiment.'' However, similar experiments are also often conducted on plants in greenhouses. As will be discussed in the next section, the dichotomy between LEs using indoor synthetic communities and FEs using outdoor natural communities is artificial; LEs and FEs form a continuum.

An early bottle experiment (*sensu stricto*) was by Pearl and Parker (1922), who used *Drosophila* grown in bottles to demonstrate the decline in *per capita* reproduction with increasing population density. Perhaps the most frequently cited bottle experiments are Gause's (1934) demonstrations of exploitation competition, interference competition, and predator-prey dynamics, using cultures of yeast and protozoans. Other well-known bottle experiments have studied competing beetle species (Park 1962), predator and prey mites (Huffaker 1958), and competing predatory wasps and prey beetles (Utida 1957). LEs on a larger scale are the numerous studies of plants in greenhouses (de Wit 1960, Harper 1977), such as Donald's (1958) growing of two plant species in pots with aerial or soil partitions in order to separate the contributions of root and shoot interactions to competitive depression of growth.

LEs on caged vertebrates have mostly studied behavioral and physiological responses of individual animals, not the dynamics of reproducing populations. These responses have community ecological significance, but the studies fall within the fields of behavioral or physiological ecology rather than community ecology. Examples include Sheppard's (1971) and Heller's (1971) studies of aggressive interactions between two chipmunk species that field observations had suggested to be competitors and Klopfer's (1973) use of a gymnasium filled with oak and pine boughs to examine habitat selection in birds.

The present volume includes LEs on *Drosophila,* fish, and marine invertebrates (Chapters 2, 21, and 31).

Advantages of LEs

LEs are obviously by far the best type of ecological experiment in regard to regulation of independent variables (axis 1 of Table 1.1) and site matching (axis 2, creation of uniform sites adjacent to each other on the same lab bench). Like mathematical theory, LEs take specified simple starting conditions and reveal a range of possible outcomes, which field biologists can then evaluate for the LEs' relevance to actual communities. Famous examples include the central role of bottle experiments by Pearl, Nicholson, and others in our understanding of population growth and regulation; the role of Gause's LEs in stimulating Lack and subsequent field biologists to evaluate competitive exclusion; the influence of Gause's LEs also on the theories and field studies of predator-prey dynamics; and the role of Huffaker's LEs in stimulating field biologists to evaluate the importance of environmental patchiness for predator-prey dynamics (Chapter 11). These instances suffice to make clear the enormous influence that LEs have had for coalescing the fundamental principles of ecology.

Limitations of LEs

Obviously, in most other respects LEs are equally the procedure with the most serious drawbacks, which may be summarized as extreme unrealism and extremely restricted scope. In regard to realism, real communities depend on the direct and indirect effects of a large set of species whose boundaries are hard to predict (Chapter 3). Real communities also depend on physical environmental parameters whose spatial and temporal variances are not mere noise but critical factors in community organization (Chapters 13 and 14). By sweeping away these complexities in order to study a particular process, LEs renounce the goal of illuminating particular actual communities. That is, they are unrealistic and devoid of generality (axes 7 and 8 of Table 1.1). We often have no idea whether the results of an LE apply to any natural community. For instance, the bottle experiments of Gause and his successors told nothing as to whether or not competition is important in natural populations of the yeasts and protozoans studied. One does not even know what variables are worth studying in LEs until their importance in nature has been studied by other means.

In regard to scope, the size of laboratories (axis 5), the modest number of years one can afford to run one LE (axis 4), and the difficulty or impossibility of keeping populations of most species alive and reproducing for many generations in the laboratory all act to restrict LEs to studying community ecology on only a tiny fraction of the world's species (axis 6). (It is difficult to imagine an LE-based paper entitled "Competition between populations of vultures, hyenas, and lions: a 10-generation cage experiment.") Thus, LEs must treat small-sized, short-lived organisms as models of larger ones, despite the systematic ecological differences existing between small and large organisms (Chapter 28).

For these reasons, currently LEs are perhaps the least used of the experimental traditions in community ecology.

Suggested Improvements in LEs

There are at least four ways in which LEs' overwhelming advantage in regulation of independent variables and in site uniformity could be more widely utilized.

1. LEs could be immediately applied to more species. Ecologists have not bothered to do extended laboratory population studies of most suitable species.
2. LEs could be run on a larger spatial scale than is traditional. Greenhouses exist for studying plants; they could be used to maintain populations of small animals not currently studied. Construction of large indoor facilities is expensive, but could in some cases be at least as illuminating as expensive outdoor projects.
3. LEs have almost exclusively studied synthetic communities. There are important natural communities that would lend themselves to the LE approach, such as communities of microorganisms in soil brought into the laboratory.
4. The synthetic communities studied by most LEs have consisted of two species. Perhaps the most important expansion of LEs will consist of creating synthetic communities far more complex and hence far more realistic than those constructed to date. Steps in this direction include Neill's (1975) studies of four crustacean species in all possible combinations; Vandermeer's (1969) experiments on four protozoan species alone and in combinations; Stiven's (1971) studies of 2-, 3-, and 5-species communities of hydras and an amoeba pathogen; and Huffaker's (1958) analysis of predator-prey coexistence in physically complex containers. Gilpin and his colleagues (Chapter 2) take this approach (which they term community reconstitution) a major step further by measuring growth rates of 28 *Drosophila* species singly, in all pairwise combinations, and in a 10-species set and by varying temperature, food quantity, and substrate texture. Such LEs could play a leading role in refining the questions asked by field ecologists, sensitizing ecologists to the kinds of complex outcomes worth looking for in the field, and guiding the design of better FEs and NEs.

FIELD EXPERIMENTS (FEs)

Examples of FEs

FEs differ from LEs in that they are conducted outdoors, and typically also in that they operate on natural rather than synthetic communities. The experimenter usually manipulates only one or a few independent variables. The most common manipulations consist of locally eliminating a species, locally introducing a species (e.g., to a small island), or erecting a fence or cage. The experimenter effectively selects initial values of other independent variables through site selection but does not hold them constant or regulate their trajectories thereafter, in contrast to LEs. There is, of course, a risk that any difference observed in the dependent variable's trajectory between the manipulated and unmanipulated plots is not due to the manipulation but to the natural differences expected between any two field plots. To reduce this risk, the experimenter often arranges replicate manipulated and unmanipulated plots in a randomized, interspersed spatial array.

FEs and LEs intergrade along a continuum, depending on the degrees to which the physical environment is regulated and to which species community composition is synthetic or natural (cf. Schoener 1983b, p. 242). Hybrid studies in which an experimenter creates identical adjacent synthetic communities outdoors instead of seeking similar adjacent natural communities include agricultural plots, numerous other outdoor studies of plant species grown in particular combinations (Chapter 12), and studies of artificial ponds stocked with particular fish species (Chapter 21). Wilbur and Travis (1984) created artificial ponds and allowed them to be colonized naturally by species. It becomes arbitrary where to draw the line among such experiments between LEs and FEs. FEs and NTEs also intergrade: the same manipulation may be launched by an experimental ecologist, by someone else, or by accident (p. 13).

One of the first FEs was performed by Darwin (1859), who demonstrated that mowing or the introduction of grazing animals increases plant species diversity on a lawn (by preventing some species from outcompeting others). Well-known modern FEs have removed limpets to demonstrate their grazing impact on algae (Jones 1948), removed or fenced-out starfish or predatory snails to test their impact on barnacle or mussel prey (Connell 1961a, Paine 1966), removed territory-holding birds to reveal the existence of nonbreeding surplus birds (Hensley and Cope 1951), fumigated mangrove trees to monitor recolonization by arthropods (Wilson and Simberloff 1969), and fenced field mice and noted effects of dispersal on population density (Krebs et al. 1969). The most extensive introduction experiment is by Schoener and Schoener (1983c; see also Chapter 33), who placed lizards on 30 Bahamian islets without natural lizard populations and thereby demonstrated that lizards could reproduce successfully on islets much smaller than those supporting lizards naturally. The present volume includes FEs on terrestrial arthropods (Chapters 3, 11, 24, 25, 26, and 33), rodents (Chapter 3), lizards (Chapters 30 and 33), and plants (Chapters 12, 22, and 25); fresh-water and marine fish (Chapter 21); and marine plants (Chapter 32) and invertebrates (Chapters 30 and 31).

Advantages of FEs

In their merits and drawbacks FEs are intermediate between LEs and NEs along almost all of the eight axes (Table 1.1). There is no single axis along which FEs are the most advantageous type, and also no axis along which they are the least advantageous type. Broadly speaking, FEs purchase gains in realism, spatial scale, and scope over LEs at the cost of losses in regulatory control and site matching; they purchase gains in regulatory control and site matching over NEs at the cost of losses in temporal and spatial scale, scope, and generality. The concluding section of Chapter 3 assesses the advantages and limitations of FEs in the light of a long-term project on desert granivores.

In more detail, compared to LEs, FEs acquire realism by working with actual communities. FEs bring a modest gain in maximum spatial scale, from about 0.01 to 1 ha. (However, there are some larger FEs; perhaps the largest is the experimental isolation of a 100-ha Amazonian forest patch by Lovejoy et al. (1984).) Because of the gains in use of actual communities and larger spatial scale, FEs can study far more species than can LEs. For example, FEs regularly use not just herbs but also trees, and not just yeasts but also rodents.

Vis-à-vis NEs that compare sites naturally having or lacking a particular species, FEs that intentionally introduce or eliminate the species on randomly chosen plots have a major advantage: the species' presence or absence could not be due to some preexisting difference between the sites. Other small-scale manipulations of independent variables can be clearly achieved by FEs when and where one wishes them, including manipulations for which there is no close NE equivalent. For instance, I cannot think of an NE that would have revealed the role of dispersal in microtine population cycles as did the fences of

Krebs et al. (1969), or an NE that would have removed territory-holding birds as cleanly as did the shotguns of Hensley and Cope (1951). NEs and FEs both choose replicate sites, but FEs have the advantage that the sites are likely to be much closer together as well as randomized and interspersed, hence much better matched for independent variables other than the one being intentionally varied.

Limitations of FEs

Limitations Compared to LEs A major sacrifice that FEs make for these gains vis-à-vis LEs is in the matching of sites (axis 2 of Table 1.1). Two types of problems are involved. First, a curable problem: ecologists have started to make widespread use of FEs only recently, with the result that FEs are often still plagued by pseudoreplication of sites (Hurlbert 1984). Second, even in the best designed FEs, there are practical limits on site replication and selection. Sites in spatially heterogenous environments are likely to be poorly matched, and replication is often impossible due to limited resources or because large sites are being studied. For example, in Tinkle's (1982) pioneering study of competition among Arizonan lizards in riparian woodland, he and his field assistants were able to establish one control site of 0.67 ha and one adjacent experimental site of 0.96 ha, to perform repeated removals of two lizard species on the latter site, and to mark 1,244 individual lizards and make many ecological and morphological measurements on them. The control site had more trees, fewer large trees, more rocks, greater structural diversity, and initially denser lizard populations than the experimental site. One of the largest and most important FEs is the Minimum Critical Size project of the World Wildlife Fund and the Brazilian government; currently there is one experimental plot of 100 ha, three of 10 ha, and three of 1 ha in the Amazonian rainforest and there will be others, including one of 10,000 ha (Lovejoy et al. 1984, Lewin 1984). The siting of such large plots in a landscape being cleared for cattle ranching involves many practical considerations other than ones of site matching. It should be emphasized that the FEs of Tinkle (1982) and Lovejoy et al. (1984) are exceptionally well executed and in no way exemplify curable problems of poor design. Rather, they illustrate inevitable practical constraints on large, expensive, labor-intensive field experiments. It is obvious that LEs escape these difficulties of site nonuniformity, and also that FEs are not always able to reap the benefits of their potential advantage in site matching over NEs. In particular, some NTEs can provide better matched experimental and control sites than can large FEs (as will be discussed in the section on p. 13).

The other inherent limitation of FEs compared to LEs is that the experimenter cannot regulate most independent variables, including temperature, rainfall, light, wind, and abundances of many other species. Hence the outcome of an FE may vary with year, season, or geographical location, because the outcome depends on unregulated variables affecting both experimental and control sites. For instance, the removal of desert lizards (Dunham 1980, Smith 1981) and rodents (Morris and Grant 1972) either succeeded or failed completely to affect abundances of competing species, apparently depending on rainfall and hence food availability in the particular year of the experiment. Schoener's (1983b) review of FEs found 11 other studies showing year-to-year variation in the effect of competition. The outcome of removal experiments on fresh-water crustacea varied among seasons of the same year for unknown reasons (Lynch 1978). The frequency of such variable outcomes in FEs is unknown, because FEs are rarely run for enough generations of the species studied even to test for the possibility of such variation (Schoener 1983b). Outcomes may also vary geographically: reciprocal effects of removing ants and rodents on each others' abundance were clear at one Sonoran Desert site and one Chihuahuan Desert site, but unclear or undetectable at another Chihuahuan Desert site (Chapter 3).

In short, because most independent variables cannot be regulated in FEs, it is harder for FEs than for LEs to obtain a reproducible result or to identify the explanation for a varying result. Note that this fact does not mean that FEs are categorically inferior to LEs, but simply that these two methods are useful for different purposes. Variable outcomes due to variation in nature are a *realistic* finding of FEs; the finding does apply to that site at that time.

Limitations Compared to NEs Vis-à-vis NEs, FEs retain four disadvantages of LEs: the same practical limitations of temporal scale, only a modest expansion of spatial scale, restrictions (though less severe ones) on the range of manipulations that can be studied, and problems of generality that will be discussed later in the section on NEs (p. 14).

Consider first the problems of scale. Schoener's (1983b, Fig. 1) depiction of the duration of 164 FEs showed only four studies that lasted over 5 years. Median durations were around 2, 8, 8, and 16 months for FEs on freshwater species, marine species, terrestrial animals, and terrestrial plants, respectively. FEs on a spatial scale exceeding 1 ha are rare. By being condemned to operate on these tiny scales, FEs are blind to whole classes of phenomena detectable by NTEs or NSEs: for instance, genetic changes (evolutionary responses); phenomena that emerge on large spatial scales and that arise from such factors as population patchiness, immigration-extinction dynamics, properties of closed as opposed to open systems, and biogeographic tests of evolutionary potential; population responses in species with long generation times; and chains of indirect effects extending beyond a decade. These and other consequences of temporal and spatial scale are discussed at length in Chapter 8.

A further consequence of the limited temporal and spatial scale of FEs is in rendering them less able than NEs, for two reasons, to follow a perturbation to its equilibrial or steady-state result. First, most FEs examine a small site that is simply fenced off or otherwise distinguished from the community in which it is imbedded. Few FEs study an arena sufficiently large or isolated to make the question of equilibrium even meaningful. Instead, the best that one can hope to attain is a quasi-steady state between internal processes and boundary processes. Unusual in this respect are a few FEs that introduced lizards, spiders, or mammals to islands sufficiently isolated and large to support closed, self-sustaining populations (Chapter 33; Crowell 1983, Schoener and Schoener 1983c). Second, major sustained perturbations to a community are likely to produce a chain of indirect effects lasting too long for an FE to follow to completion. For instance, enclosures from which Brown and colleagues fenced out desert rodents had not reached a steady state after 7 years (Chapter 3). As another example, if interspecific competition is compatible with coexistence in most years but eliminates one species in the rare year of low resource levels, the result in a closed system after a long time would be competitive exclusion. This result would be detectable in an NSE but only rarely in a closed FE, and in the usual open FE immigration might prevent the result from occurring at all.

Practical restrictions on the scope of FEs are of at least eight sorts.

(1) One cannot introduce mobile species that would simply leave, such as birds, other volant species, and small-bodied species impossible to fence or cage. (2) One cannot introduce species requiring a large territory, such as top carnivores, and obtain a self-sustaining population or even a realistically foraging individual in a small enclosure. (3) Species removals in the field have yet to be reported and may be virtually impossible for many types of species (e.g., internal parasites, abundant small-bodied species, subterranean species, hard-to-catch volant species) and for many habitats (e.g., forest canopy). Removal experiments on terrestrial animals are virtually confined to low habitats and strata, such as forest floor and understory, desert, and grassland. FEs in the world's most species-rich communities, the tropical rainforests, are few. (4) Many species would be illegal or immoral to introduce (many predators, pathogens, parasites) or to remove (many birds and mammals). (5) The techniques available for fencing species in or out select entire groups of species on the basis of characteristics such as size (cf. the rodent fences used in Chapter 3) or inability to dig, crawl, or jump past fences. Fence techniques for selecting any arbitrary species or combinations of species are unavailable. (6) FEs cannot realistically simulate important natural disturbances, such as

hurricanes, droughts, and frosts. (7) Few long-term FEs requiring on-going attention have been executed at sites far from biologists' homes. In particular, nearly all FEs are done in the temperate zones and on continents, and very few in the tropics or on islands. (8) FEs are usually forbidden in nature reserves and national parks, such as the sites for the studies of New Guinea and Galápagos birds described elsewhere in this book (Chapters 6 and 10). This limitation on FEs will grow in importance as more and more biological communities become restricted to reserves.

Thus, for many (probably most) individual species and most manipulations in many places, FEs are scarcely possible for practical, legal, or moral reasons.

Suggested Improvements in FEs

This section on FEs concludes with 10 suggestions for expanding their scope and power.

1. There is still much scope for ingenuity in broadening the range of species and problems studied by FEs. Just one example: most FEs have introduced or removed species singly, while study of the interesting questions posed by varying combinations of species has remained the exclusive domain of LEs and NEs (e.g., Chapter 2; Schoener 1975, Diamond 1975). Perhaps some of the practical restrictions just outlined in the preceding section could be overcome through ingenuity.

2. There is especially scope for designing LE-FE hybrids that employ prepared outdoor sites, or synthetic outdoor communities, or partial regulation of environmental variables such as water and light. For some purposes such hybrids will be more illuminating than pure LEs or pure FEs.

3. FEs often still employ poor experimental designs. For instance, an important paper by Hurlbert (1984) discusses at length the curable problems of site replication. Another example is the underutilization of partial factorial designs. As illustrated by Table 3.1 in Chapter 3, such designs permit one to increase the effective number of replicates of experimental plots and to reduce the required number of control plots.

4. Ecologists are still confused about the distinction between "pulse experiments," in which one briefly applies a perturbation and then watches the system relax, and the much commoner "press experiments," in which one applies a sustained perturbation. These two types of FEs have very different interpretations (Bender et al. 1984). For example, many of the species interactions revealed by press experiments are not direct effects, as the experimenter often believes, but chains of indirect effects (Chapters 3 and 20; Bender et al. 1984); additional experiments of the pulse type may have to be run to distinguish these types of effects.

5. Experimental interventions risk introducing artifacts. That is, an experimenter making an intervention presumes to know what salient feature of the intervention causes the observed end result. However, the salient feature may instead be some unforeseen or undetected feature or proximal consequence of the intervention. This risk is modest when a particular species is manipulated by hand, but is large when a fence or cage is erected to keep a particular species in or out, because fences can produce many other effects and can influence many other species (p. 15). In FE-based papers that employ fences, the evidence provided to discriminate among possible causes of an observed fence effect is often disappointingly incomplete. Some of the "epidemiological" criteria for clarifying a problem of causation in NSEs (p. 16) may also be applicable to this problem of causation in FEs.

6. Many FEs could be profitably run for longer than is currently usual, both to reveal the chains of indirect effects, to test for year-to-year variation in the outcome, and to search for outcomes (such as competitive exclusion) likely to be consummated only in an exceptional year.

7. More FEs could be run on closed natural communities, such as small islands, to study equilibria little perturbed by boundary processes.

8. One approach to the problem of how to generalize from an FE at a single site was employed by Lubchenco (Chapter 32). She identified different types of sites in an area, used FEs to assess species interactions at each site type, and then weighted the results by the overall frequency of each type.

9. Another approach to the problem of generality would be to integrate FEs much more closely with NEs. For example, an NE might indicate that the habitat range of species A is restricted by a competing species B present at site 1 but not at site 2; further evidence could be obtained from an FE (a local removal of species A) at site 1. As noted by Connell (1980), integration of FEs with NEs may provide an excellent means to study postulated evolutionary changes in strengths of competition (see Chapter 6 for discussion); the FE-NSE by Abramsky and Sellah (1982) represents a promising start.

10. Many ecologists feel defensive about whether ecology is a rigorous science and are determined to avoid anything smacking of nonrigor. In particular, it is often assumed that rigorous science consists exclusively of experiments carried out to test à priori hypotheses, while experiments carried out simply to see how the system responds are intellectually inferior, and nonexperimental observations unspeakably worse. This attitude grew out of a healthy reaction to ecology's earlier paucity of hypothesis testing and domination by descriptive studies. However, the most profitable approach in science depends on one's state of knowledge. With a little known system one has to begin by describing it, examining its natural variation, and poking it to see how it behaves before one can hope to frame more detailed hypotheses intelligently. Ecological communities are complicated, and our understanding of their connectivity is as rudimentary as is our experience of FEs. Thus, FEs of the "poke-it" type on even the best studied communities may still yield major surprises. (An example discussed for desert granivore communities in Chapter 3 is the change in fungal infections of plants resulting from manipulations of rodent densities.) I do not wish to be misconstrued as implying that hypothesis testing has no place in FEs at present. However, obsession with hypothesis testing should not blind us to the possible value as well of some manipulations unconstrained by prior hypotheses.

NATURAL EXPERIMENTS (NEs)

Examples of NEs

Natural experiments differ from FEs and LEs in that the experimenter does not establish the perturbation but instead selects sites where the perturbation is already running or has run. The perturbation may have been initiated naturally or by humans other than an experimental ecologist. Along with the experimental sites, the investigator selects control sites so that the two types of sites differ in the presence or absence of the perturbation but are as similar as possible in other respects.

NEs fall into two categories: natural snapshot experiments (NSEs), in which one observes only a final steady state or other snapshot of an old system, but not the trajectory that led to it, and natural trajectory experiments (NTEs), in which one observes the trajectory or can reconstruct it from historical or fossil records. Typical NTEs examine the trajectory following a natural disturbance (volcanic explosion, freeze, drought), a human-made disturbance (fire, eutrophication), or an invasion or extinction of a species. Chapter 4 will discuss and illustrate NTEs in detail. In this volume they are utilized in studies of phytophagous mites, trees, birds, and fish (Chapters 5, 10, 16, 21, and 24).

Typical NSEs are studies comparing the abundance, morphology, and habitat range of species A on multiple islands or at multiple sites, some of which have and others of which lack a competing or predator species B. For instance, Schoener and Toft (1983a; see also Chapter 33) found that spiders averaged about eleven times more abundant on 48 Bahamian islands without lizards than on 26 islands with lizards, because lizards prey on and also compete with spiders. Where char and trout occur sympatrically in a lake, each species has a narrower diet and occupies a narrower range of habitats (char in deeper water than trout) than in allopatry. These niche shifts, which are maximal at late-summer times of low resource levels, have been observed in dozens of Scandinavian and North American lakes with two different trout species and two char species (Chapter 21). In this volume NSEs are utilized in studies of desert and Mediterranean scrub plants, spiders, fish, lizards, and birds (Chapters 6, 10, 21, 23, 30, and 33).

NSEs and NTEs intergrade, depending on whether the initial state before the perturbation was observed or could be reconstructed, and on how well the trajectory from initial to present state is documented. NTEs and FEs also intergrade: a species may be introduced or locally removed experimentally by an ecologist, accidentally or intentionally by another human, by some indirect effect of humans (e.g., an introduced disease), or by a natural cause. At first one might expect a sharp distinction between FEs and NTEs, in that FEs have clean layout of experimental and control plots (randomized interspersed design), while NTEs do not. In practice, for large-scale FEs it may be impossible to have good matching of experimental and control plots, or to replicate the experimental plots, or to have a control plot at all. By contrast, site matching in some NTEs may be much cleaner. For example, when the peacock bass, an exotic piscivorous fish, reached Panama's Gatun Lake after introduction into a tributary river by a local businessman, it spread throughout the lake during the course of 5 years as a slow wave with a leading edge of subadults. Zaret and Paine (1973) were able to map the bass's advance from year to year and to deduce its effects from two types of comparisons: approximately bimonthly samples of community composition at several stations for 4 years before and 2 years after arrival of the bass front; and simultaneous comparison of sites already reached or not yet reached by bass, and matched in habitat. Both comparisons revealed similar dramatic effects of bass on prey fish, on bird and fish predators of the prey fish, and on insect and zooplankton and fish prey of the prey fish. It is doubtful that an FE (experimentally motivated introduction) in a lake as large as Lake Gatun would have achieved a cleaner or different design.

Advantages of NEs

In their merits and drawbacks NEs are at the opposite pole from LEs. NEs are the most advantageous experimental procedure in all five respects in which LEs are the most seriously flawed: temporal scale, spatial scale, scope, realism, and generality. Conversely, NEs are the worst procedure in the two respects in which LEs are the most advantageous: regulation of independent variables and site matching. NEs show most of these same merits and drawbacks in comparison with FEs, but to a less marked degree.

The major advantage of NEs is that they provide the sole means to study many perturbations and questions. This arises from NEs' combined advantages of temporal scale, spatial scale, and scope. NEs not only use tiny islands similar in size to the typical FE, but also routinely operate on much larger spatial scales than are possible for FEs or LEs: large islands and even continents. NTEs are the sole technique for following the trajectory of a perturbation beyond a few decades; even durations of one decade are costly and hence very rare for FEs. For example, historical or fossil evidence lets one reconstruct successional or evolutionary trajectories of about 50, 70, 2000, 3 million, and 70 million years resulting from the following perturbations, respectively: elimination of American Chestnut by chestnut blight, insularization of Barro Colorado (Willis and Eisenmann 1979), the sudden crash of hemlock in the northeastern United States 4800 B.P. (Davis 1981a), the closing of the Panama Seaway that launched the Great American Interchange (Marshall et al. 1982), and the extinction of all large terrestrial vertebrates at the Cretaceous-Tertiary boundary. While NSEs do not follow trajectories at all, they do let one examine snapshots reached after long times. For instance, NSEs comparing existing forest fragments of various sizes enabled Willis (1980) and Whitcomb et al. (1981) to deduce how a century of forest fragmentation affected bird distributions, while an FE studying the same problems in Amazonia has only just begun (Lovejoy et al. 1984). The modern mammal fauna of Tasmania and neighboring islands is a snapshot produced by 10,000 years of differential extinction following the sundering of late-Pleistocene land bridges (Hope 1973, Diamond 1984a).

As discussed in Chapter 8, the NEs' expanded spatial and temporal scales open up for study a whole range of problems (including evolutionary ones) that are inaccessible to FEs and LEs. Because of these scale effects and the other practical constraints on FEs previously listed, NEs are the sole way to examine perturbations that cannot, may not, or should not be created deliberately. Chapter 4 describes the species introductions and extinctions available for study through NTEs.

NEs have the further virtue of generality, because they sample a wider range of natural variation among sites than do FEs (LEs do not sample natural variation at all). To set up and maintain an FE involves effort, as in building and maintaining enclosures or removing species. The experimenter deliberately concentrates that effort on immediately adjacent sites (so that control and experimental sites will be well matched), and deliberately samples only a tiny fraction of natural variation among sites. This strategy increases confidence that the conclusion obtained does apply to that study site, but reduces confidence that the conclusion applies to other sites, as the study site might be atypical. NEs instead put effort into studying geographically scattered sites and thereby sample a greater range of natural variation. Any difference that emerges between sites with and without some particular factor is therefore more likely to be valid for a considerable range of sites. For example, Toft and Schoener (1983) were able in 22 days to census five spider species on 116 islands with and without lizards, and to establish that spiders were on the average an order of magnitude more abundant on islands without than with lizards. The sampled islands covered a range of areas, distances, and vegetation heights and complexities, which Toft and Schoener measured in order to disentangle the effect of lizard presence or absence from concurrent effects of these other variables. In the same field time they would have been able to remove most (not all) individual lizards on only 2 islands supporting them, or to introduce a saturating number of lizards to only a few islands lacking them; they would still have had to wait up to several years for spider densities to reach new equilibrium values on the manipulated islands; and, without an NE, they might have been unlucky and picked exceptional islands where lizards did not happen to limit spiders.

Limitations of NEs

Set against these major advantages over FEs and LEs (plus the advantage of realism over LEs), NEs suffer to varying degrees from several major disadvantages.

First, NSEs by definition are NEs where one knows only a single snapshot of a trajectory, not the whole trajectory. As stressed in Chapter 3, obviously this constitutes an enormous loss of information compared to NTEs, FEs, and LEs. The lost information includes not only length of time between causes and effects, but also event sequences that are often decisive in tracing out cause-effect chains. As Brown and colleagues (Chapter 3) put it, "the temporal sequence of events reveals a great deal about the processes by which species affect each other and about the patterns of connectance that link the fates of species."

Second, like FEs, NEs try to minimize differences between experimental and control sites with respect to variables other than the one whose effect is to be examined, by comparing many sites of each type matched for those other variables. However, this matching is likely to be less close for NEs than for FEs, because sites are given by nature rather than selected, and geographically scattered rather than adjacent. This consideration is more of a problem for NSEs than for NTEs: with NTEs one can compare the same site before and after perturbation, as well as perturbed and unperturbed sites at the same time. Thus, the sequence with regard to closeness of site matching tends to be LE > FE > NTE > NSE.

The remaining drawback of NSEs is the one most troublesome to many ecologists: the perturbation is only inferred, not observed or experimentally created. In FEs the experimenter applies the perturbation and randomly assigns sites to be perturbed or not. In NTEs the perturbation is also generally some unequivocally identified event, such as an eruption or the arrival of a disease. In the case of NSEs, however, nature provided the two sets of sites (e.g., those with and without species A). If one observes that species B is more abundant in the absence than in the presence of species A, that could mean that species A itself somehow reduces species B's abundance (e.g., by competing with or preying on it). A necessary (but not sufficient) condition for the correctness of that interpretation is that A's presence or absence at each site must be due to factors of no direct relevance to B's abundance, such as local historical accidents of A's arrival and disappearance. However, it may also be that the two sets of sites differed in some other factor that favored A's presence *and* B's rareness; or (less plausibly in most actual cases) that it was B's abundance that caused A's absence. "There is no certainty that the only difference between the two mountains is the absence of one species. A predator may have been absent, or an essential food organism, soil nutrient, etc., may have been present beyond the boundary on the second mountain" (Connell 1975, p. 462).

In statistical terms, an NSE yields a correlation between two observed differences that distinguish two sets of sites. Other evidence must be examined to decide what is cause and what is effect. This particular burden of extra evidence does not arise with LEs, FEs, and NTEs, where the manipulation itself is known to be somehow the cause of the differences between the two types of sites. Hence there is no doubt that, all other things being equal, a controlled perturbation provides clearer evidence than a preexisting difference between sites.

In practice, I see this extra burden of NSEs as a matter of degree, rather than as the absolute distinction that the contrasting words "cause" and "correlation" suggest. Often, the manipulation employed in an FE does not mimic a natural perturbation but is instead a means used to produce a proximate effect mimicking a natural perturbation. However, the manipulation is actually likely to produce a cluster of proximate effects. While the unnatural manipulation itself surely somehow led to the observed ultimate effects, that fact is not biologically interesting. The interesting question is which proximate effect produced the ultimate effect. Answering this question involves a burden of extra evidence in some but not all FEs. For instance, one of the commonest manipulations in an FE is to fence or cage the experimental sites. Fences and cages produce at least nine types of effects, including preventing the monitored species inside (the primary dependent variable) from emigrating, preventing some other species inside from emigrating, preventing a predator outside from entering, preventing a competitor from entering, preventing a herbivore or a species in some other trophic relationship from entering, providing a perch for hawks or other animals, providing a surface for attachment, providing some protection against wave action in the intertidal, and providing shade. Among six FE-based studies employing fences or cages Chapters 3, 11, 31, Connell (1961a), Paine (1966), and Krebs et al. (1969) attributed the results mainly to the fourth, third, fifth, third, third, and first of these factors respectively, but the decision in each case poses a burden of evidence and may not be easy.

Suggested Improvements in NTEs

NTEs offer a large and underutilized body of clean-cut ecological data (Chapter 4). While field experimentalists are laboriously manipulating species abundances on tiny plots, analogous on-going manipulations on a gigantic scale are receiving little attention (e.g., expansions of parasitic cowbirds in North America and the West Indies, decimations of native fish by introduced piscivores, and on-going elimination of American elms by disease). In Chapter 3, Figs. 3.1 to 3.8 illustrate how FEs involving experimental reductions in populations of selected mammalian herbivores are labor-intensive but illuminating as regards resulting changes in ants, plants, other mammals, and even pathogens. Did any ecologist seize the opportunity that an NTE provided to measure such changes while myxomatosis was decimating Australia's rabbits?

Suggested Improvements in NSEs

Because NSEs provide the sole way to study many types of manipulations, it is not possible to renounce them because of the increased uncertainty and extra burden of evidence involved. How can this burden be met more convincingly than is now usually the case?

Useful guidance may be obtained from epidemiology, a science that for a long time has had formalized, successful procedures for dealing with an analogous problem. "Epidemiology is concerned with the patterns of disease occurrence in human populations and of the factors that influence these patterns. . . . The major epidemiologic problem in evaluating a statistical relationship is to determine whether or not the association is indirect or of etiological significance" (Lilienfeld and Lilienfeld 1980, pp. 3 and 296). These goals parallel the goals of the ecologist who employs NSEs to understand the factors underlying species abundances and distributions, and for whom the main practical problem is to distinguish indirect associations from causal (etiologically significant) associations. Epidemiology texts illustrate how the practicing scientist's use of the word "cause" must differ from that employed by the pure statistician or logician. In effect, one's decision between a causal association and an indirect association depends on details of the association and on auxiliary evidence. As summarized by Lilienfeld and Lilienfeld (1980) in their chapter entitled "The Derivation of Biological Inferences from Epidemiologic Studies," this evidence is of the following sorts:

randomized experimental interventions on human populations, but these are usually not feasible or ethical (no. 1; numbers refer to the corresponding evidence that ecologists can employ with NSEs, as explained below);

"a natural experiment [that] closely simulates the conditions of a randomized, controlled study and thus offers a unique opportunity to establish a causal inference" (Lilienfeld and Lilienfeld 1980, p. 315; no. 2 below);

consistency and strength of the association over various populations (nos. 3 and 4);

comparisons of different but related types of populations (no. 5);

existence of a dose-response relationship between the putative causal factor and the effect (no. 6);

specificity of the association (nos. 7 and 8);

biological plausibility of the putative causal association (no. 9); and

determination of the causal sequence connecting the putative ultimate cause and the effect (nos. 9 and 10).

Let us now examine the corresponding evidence that ecologists can employ with NSEs, as illustrated by some cases where such evidence actually has been employed. Several of these lines of reasoning may also be used in FEs involving an unnatural perturbation (e.g., fencing) for which it is unclear what proximate consequence is responsible for the ultimate outcome.

1. Carry out an FE Just as FEs can be generalized by NSEs, it may be feasible in some cases to strengthen NSEs by FEs. For example, an NSE showed that the vole *Microtus pennsylvanicus* has lower abundance, incidence, and range of occupied habitats on islands with than without a predator, the shrew *Blarina brevicauda* (Lomolino 1984). This suggests reduction of vole populations by shrew predation. Lomolino strengthened this interpretation by an FE: he introduced shrews to three islands initially harboring voles but no shrews, and found that voles declined in abundance or became extinct. Similarly, Hixon (1980) used FEs to test inferences from NSEs about marine fish communities, while Roughgarden (Chapter 30) and Schoener (Chapter 33) did so for *Anolis* communities. Chapter 3 summarizes the extensive FEs that Brown and colleagues have performed on desert rodent communities, stimulated by Brown's earlier NSEs with these communities.

2. Carry out an NTE The clearest examples come from fish communities of lakes (Chapter 21). Comparisons of lakes with and without fish, and of lakes containing a certain fish species with and without a competing or predator fish species, often reveal striking shifts in mean size of zooplankton and in the abundance, diet, body size, and habitat preference of fish. These same types of changes have been witnessed directly in numerous NTEs following the intentional or accidental stocking of lakes with fish.

3. Is the association consistent over many sites? NSEs that compare only a single experimental site and single control site are as dissatisfying as the correspondingly flawed FEs. If outcomes are confirmed for many experimental sites and many control sites, one can be more confident that the cause of the outcome is either the identified difference between the sites or else something closely correlated with it. For example, Toft et al. (1982) were able to examine duck distributions on 236 ponds within a small area, censused for four consecutive years, in their analysis of how five duck species affected each other's presence and abundance. Diamond (Chapter 3) examined how the New Guinea warblers *Sericornis virgatus* and *S. nouhuysi* affected each other's altitudinal range by comparing 9 mountains with both species, 14 with *S. virgatus* alone, and 23 with *S. nouhuysi* alone.

4. Is the association consistent over related populations that are expected to behave similarly with respect to the putative causal link? Chipmunks of the *Eutamias quadrivittatus* group and *E. dorsalis* segregate by altitude where they coexist, but each species expands to occupy the other's altitudinal range on at least 18 mountains supporting only one of the two species. Not only is this pattern consistent over many sites, but it also is exhibited by two different species of the *quadrivittatus* group, *E. quadrivittatus* itself and *E. umbrinus* (Brown and Gibson 1983). Similarly, pairs of New Guinea honey-eater species of the *Melidectes belfordi* group segregate by altitude where they co-occur, while each species expands its niche by several thousand feet to occupy the whole altitudinal transect where it occurs alone (Chapter 6). This pattern is exhibited by all three pairwise combinations of the three species (*belfordi*/*ochromelas*, *ochromelas*/[*rufocrissalis*], [*rufocrissalis*]/*belfordi*), as well as by all three allospecies of the [*rufocrissalis*] superspecies (*rufocrissalis*, *leucostephes*, *foersteri*). Schoener's (1975) study of habitat shifts in the three most widespread West Indian anoles encompassed 20 sites offering nearly all existing species combinations involving these species.

5. Does the association vary predictably over related populations that are expected to behave differently with respect to the putative causal link? Schoener and Toft (1983a; Toft and Schoener 1983) found that abundances of several diurnal spider species were about 10 times lower on islands with than without diurnal lizards. The authors attributed this difference to lizard predation on and competition with spiders. In support of this interpretation, they found little or no effect of lizard presence or absence on nocturnal spiders that would escape predation and competition, nor on a spiny, brightly (perhaps aposematically) colored spider likely to escape predation, and only an inconsistent effect on a spider species whose adults are nocturnal and whose diurnal juveniles may be too small compared to lizards to suffer competition. In Schoener's (1975) study of site-to-site habitat shifts among lizard species identified as potential competitors, the shifts were greater for species pairs with the same than different climatic preferences, greater for similar- than dissimilar-sized species and size classes, and greater for the effect of large lizards on small ones than vice versa. These variations in outcome are consistent with an interpretation of competition but would otherwise be hard to explain. Among New Guinea montane bird species pairs, altitudinal shifts are often observed in the niche of one species between sites differing in the presence and absence of a related species. These shifts decrease in frequency with the size ratio, taxonomic distance, and differences in diet and foraging technique between the two species, suggesting that the shifts arise from competition (Chapter 6).

6. Does the effect size increase with the magnitude of the putative causal factor ("dose-response relationship")? Among 30 sites, Yeaton and Cody (1974) found song sparrow territory size to vary nearly 20-fold, from 0.06 to about 1.15 ha. One observation suggesting varying interspecific competition as the explanation of this variation is that territory size increased regularly with variations in the number of competing species from 0 to 13 (Spearman Rank Correlation Coefficient, 0.93). Negative associations (suggestive of competition) between distributions of duck species increased with interspecific niche overlap along spatial and temporal axes (Toft et al. 1982).

7. Specificity of association: does the association between two variables persist over a wide range of variation in other variables? If so, it becomes unlikely that the other variables are causally involved in the association. In effect, one uses NSEs to explore the natural range of variation over which the association applies. For example, in the Bismarck Archipelago the whistler *Pachycephala melanura* disperses much more readily than the very similar *P. pectoralis* but is absent as a resident on every island occupied by *P. pectoralis*. On islands lacking *P. pectoralis, P. melanura* occupies all available light levels from zones of bright sunlight to nearly permanent cloud cover, all available moisture levels from zones with a 4-month annual drought to constant mist, all available elevations from sea level to the highest summit at 4,650 ft, all available vegetation heights from gardens and coral scrub to the canopy of tall forest, all available habitat densities from open savanna to dense subalpine shrubbery, and all available island sizes from 1 ha to 150 mi^2. This range of variation compatible with residence by *P. melanura* increases the likelihood that its absence from all *pectoralis*-occupied islands is due to the presence of *pectoralis* rather than to some habitat or climatic factor (Diamond 1975).

8. Specificity of association: effects of other variables Another approach to the question of specificity of association is to measure the effects of site differences other than the one initially suspected of providing the causal explanation. For example, if one observes niche shifts in species A associated with the presence or absence of species B, one can measure effects of habitat structure, predation, resource levels, and island area or isolation on species A's niche, in order to test whether these variables rather than B's presence are the ultimate cause of the niche shifts. In their study of island variation in song sparrow territory size, Yeaton and Cody (1974) measured insect abundance and amount of foliage at each site. The former factor proved to bear no relation to territory size, while the latter factor varied in a direction opposite to that required to explain variations in territory size. Lomolino (1984) found that neither island area nor habitat characteristics explained the observed negative relation between vole abundance and shrew presence. Toft and Schoener (1983) found that island spider numbers increased with island area and vegetation height and decreased with island distance, but that the large effect of lizard presence or absence persisted over the range of vegetation heights, over a range of distances up to at least 2.3 km, and over a range of area from 50 to at least 10,000 m^2. Schoener (1975) identified the habitat parameters of greatest predictive value for lizard distribution, measured them at each of 20 sites for each size class of each of 10 lizard species studied, and used the measurements in several different ways to calculate what the habitat niche of a given species would be if other sympatric species were without effect. The calculation confirmed the existence of many site differences in a given species' niche related to the presence or absence of other species and unrelated to habitat variables. Depending on the particular case, the effects of other species became variously more, less, or equally conspicuous after correction for habitat parameters. Critics of NSEs often implicitly assume that, if habitat and resource variables were measured, effects of species interactions inferred from NSEs could only turn out to be weaker than claimed or else nonexistent. Schoener's (1975) result, and that of Yeaton and Cody, illustrate the obvious point that the effects may also turn out to be stronger.

9. Is there a biologically plausible mechanism for the relative causal relationship? In support of his inference that presence of shrews caused reduced abundance of voles on islands, Lomolino (1984) obtained evidence making predation a plausible mechanism: most shrew scats contained vole fur, shrews preferentially eat juvenile voles, and juveniles practically disappeared from vole populations following introductions of shrews. In support of their inference that niche shifts in altitudinally segregating chipmunk species arise from interspecific competition, Brown (1971a), Heller (1971), and Sheppard (1971) quantified interspecific aggression at the transition altitude and in laboratory cages. In support of their inference that island variation in song sparrow territory size was caused by varying numbers of competing species, Yeaton and Cody (1974) measured niche overlaps between the competitors and song sparrows along three niche axes, and measured contraction of song sparrow vertical foraging range from an island with 3 competing species to an island with 11 competing species.

10. Can the observed effect be predicted quantitatively from the putative causal relationship and its observed mechanism? On the assumption that song sparrows progressively expand their niche with decreases in number of competing species, Yeaton and Cody (1974) were able to predict the varying values of island song sparrow territory size by considering the species present on each island and their niche overlap with song sparrow.

Conclusion Uniquely among the types of ecological experiments, the manipulation in an NSE is not directly observed but is instead inferred from site differences. Such an inference bears the obvious burden of demonstrating that the claimed difference rather than some other difference between the sites really caused the outcome. One can never be certain. However, the preceding section has outlined a 10-point strategy, paralleling that used by epidemiologists for testing the putative explanation as well as competing explanations. Most NSEs employ only a small fraction of this strategy, generally point 9, occasionally point 3 as well. I have frequently cited the NSEs by Yeaton and Cody (1974), Schoener (1974), Toft et al. (1982), Toft and Schoener (1983), and Lomolino (1984) because they went to exceptional effort to test alternative causal explanations. Schoener's (1975) test of habitat variables that might confound NSE-based conclusions about lizard competition was so detailed that it has never been attacked by skeptics of competition, but it has also never been imitated by believers in competition. Wider adoption of such methods could reduce the skepticism that ecologists accustomed to LEs and FEs feel toward NSEs.

WHICH IS THE BEST TYPE OF EXPERIMENT?

It will now be clear that this is a silly question, because the answer varies with the species, process, and site studied. NEs tend to have advantages over LEs for studying species interactions among whales; the advantage is reversed for yeasts. If one wants to understand the effects of introducing predator A on rodent B, FEs are often the method of choice to study the changes in abundance after 1 year; NEs are the sole method capable of studying the genetic changes after 10,000 years. FEs are certain to result in expulsion or incarceration for scientists working in national parks or in certain countries; NEs are not. Thus, a pluralistic approach is essential if one wishes to address the range of questions arising in community ecology.

It is therefore puzzling that the ecological literature contains few discussions of the relative merits of all three types of experiments and frequent claims of principled superiority for one type. The failure to utilize the full array of appropriate methods represents a serious impoverishment for ecology today, and its origin warrants careful examination.

Certainly, an ultimate explanation is that ecologists rarely receive equal training in the three approaches. In the past many gifted natural historians lacked experience in laboratory techniques, field experimental methods, or statistics. Today, in part as a backlash, good training in field experiments is infrequently accompanied by equal experience in natural history or the laboratory. My impression is that principled claims of superiority for one type of experiment are most likely to be made by ecologists familiar mainly or solely with that type. However, we still must consider the arguments used to buttress the claims. Most of them seem to me to fall into four categories.

*1. **Focusing on one consideration*** One of the commonest arguments is to cite correctly the inherent advantages of method A, the inherent disadvantages of method B, and the particular shortcomings of a certain study employing method B, while ignoring the inherent disadvantages of method A, the inherent advantages of B, and so on. For instance, field experimentalists wishing to make a general case for FEs over NEs often cite the real advantage of FEs in manipulative control and site matching, while ignoring the real advantages of NEs in maximum spatial and temporal scales and in the range of manipulations that can be studied. Similarly, critics of LEs and FEs often cite their unrealism and limited scale, respectively, while ignoring their compensating advantages.

*2. **Pattern, effect, process, and mechanism*** There is much confusion about these terms. It is sometimes claimed that FEs reveal only "effects," while LEs alone can reveal "mechanisms"; or that NEs or nonexperimental observations yield only "patterns," while FEs alone can identify underlying "mechanisms" or "processes" (e.g., Wilbur and Travis 1984, pp. 113–114).

Actually, what any community ecological experiment attempts to get at is more accurately described as a treatment effect or a cause-effect sequence: the change in X (usually a species abundance or distribution) caused by a change in the physical environment or in the abundance of some other species. In such a case the expression "underlying mechanism or process" would apply to the physiological response or species interactions underlying this experimentally demonstrated change in abundance and distribution. As noted by Schoener (1974a, p. 28; 1983b, p. 275) for FEs (his comment is equally valid for LEs and NEs), the experiment itself rarely investigates the mechanism of the change, although most experimenters nevertheless proceed to offer an opinion about mechanism. Supplementary data, especially direct physiological measurements or else observations of behavior, will always be required to identify the mechanism: for example, to show whether species B reduces species A's abundance by preying on, infecting, parasitizing, driving off, eating the food of, or eating the leaves of species A. Whether the population consequences of this observed behavior or physiological response are then best revealed by LEs, FEs, or NEs depends on the considerations of Table 1.1 applied to each particular case.

*3. **Popperphilia*** In the past decade, a position in the philosophy of science—the falsificationist criterion espoused by Popper—has been abused by some ecologists to support their position in various controversies, such as what type of experiment possesses inherent superiority. The first problem is that, as already noted, this question is not worth posing in such a general form. The second problem is that, even if one espouses a falsificationist criterion, that does not guide one to a preference among LEs, FEs, NTEs, or NSEs. Finally, while Popper's philosophy in general and the falsificationist criterion in particular were formerly considered to be among the significant views within the philosophy of science, most professional philosophers other than Popper's disciples abandoned these views by about 20 years ago. For example, in the comprehensive presentation of major modern philosophies of science edited by Suppe (1977), the falsificationist criterion is cited only briefly in a few places to explain why particular modern philosophers discarded it. Similarly, Popper's overall philosophy is not among those presented at length and still given serious consideration, but is instead summarized briefly as background for understanding its shortcomings and its influence on modern philosophers such as Feyerabend and Lakatos. The recent explosion of Popperphilia among ecologists exemplifies what the philosopher Suppe (1977, p. 19) had in mind when he wrote, "It seems to be characteristic, but unfortunate, of science to continue holding philosophical positions long after they are discredited."

*4. **Models from other sciences*** A recurrent point of view among ecologists is that ecology would progress faster by imitating the rigorous approaches of hard sciences like physics and molecular biology (cf. Strong, Simberloff, Abele, and Thistle 1984, p. viii). In fact, different sciences face different methodological problems, and the successes of physics and molecular biology were made possible by those fields having found appropriate solutions to their particular problems. Ecology must cope with its own distinctive problems. Insofar as ecologists can profitably be guided by other sciences, the best models would be sciences facing problems similar to those of ecology. Molecular biology and some areas of physics are methodologically at the opposite pole from ecology and offer poor guidance. Better models are astronomy, geology, medical science, neuroethology, or vulcanology, which share with ecology practical limitations on experimental intervention, concern with historical or evolutionary problems, and conditionality of results. In those fields the recognition that laboratory studies, field studies, and the comparative approach have a place is no longer controversial, as it still is in ecology. Probably the single most important methodological advance contributing to the recent successes of neuroethology has been the intimate feedback among these approaches: the same investigator *regularly* applies laboratory and field approaches to the same problem (cf. Roeder 1967). For instance, the contributions of Hopkins (1980; Hopkins and Bass 1981) to electric communication in fish have been based on comparing the ecology and electric discharges of 23 fish species (NE), carrying out field playback experiments with discharge patterns (FE), and doing electrophysiological studies on electric organs in the laboratory (LE).

FUTURE PROSPECTS

Prospects for Each Methodology

All three experimental methods could be applied in ecology with more power than they usually are at present. To reiterate briefly the points made in this chapter:

- Laboratory experiments could be applied to a much wider range of species, could be run on a larger spatial scale, could utilize certain natural communities brought indoors, could utilize more complex and realistic synthetic communities, and could be extended to more LE-FE hybrids studied outdoors.
- Field experiments could be applied to more species and problems, could more consistently employ good design, could be run for longer times, could use closed natural communities, and could be integrated with NEs to assess generality. The differences between pulse and press experiments, and between direct and indirect effects, need wider appreciation.
- Natural snapshot experiments (NSEs) could use a whole series of techniques, paralleling those in use in epidemiology, for increasing the strength of inferences about cause. Far more perturbations launched by nature or by humans (NTEs) merit analysis.

Effect Size

Most FEs and NEs still ask only *whether* or not an effect is detectable: for example, whether the presence of species A alters species B's abundance and distribution at the $p < 0.05$ level. It is astonishing how few FEs and NEs go on to study effect size: that is, by *how much* various densities of species A alter B's abundance (Toft and Shea 1983). Such information is essential before ecologists can start to construct explanatory, semiquantitative theories of ecological communities.

Prospects for an Integrated Research Strategy

Ecologists, like scientists in many other fields, can profit by applying different methodologies to the same system. Examples to date include several studies of competitive exclusion in which the same investigators combined two or three approaches. For Caribbean lizards, Roughgarden and colleagues (Chapter 30) combined FEs and NSEs, as did Abramsky and Sellah (1982) for Mediterranean rodents, Grant (1969a) for north temperate rodents, Brown and colleagues (Chapter 3) for desert rodents, and Hixon (1980) for marine fish. LEs were combined with FEs in studies of marine invertebrates by Buss (Chapter 31), and with NSEs in studies of chipmunks by Sheppard (1971) and Heller (1971). Harger (1972) used LEs, FEs, and NSEs to analyze mussel distributions. Evidence from both NSEs and NTEs as well as LEs and FEs was used by Werner (Chapter 21) and Colwell (Chapter 24) to understand communities of fresh-water fish and hummingbird mites, respectively. Schoener and colleagues (Chapter 33; Schoener and Schoener 1984) combined NEs, FEs, and LEs to elucidate why Bahamian lizards require a minimum island area; they compared natural lizard communities on islands of different areas, introduced lizards to lizard-free islets, and studied lizards' overwater dispersal abilities in the laboratory.

These combined studies are exceptionally convincing, because conclusions tested by different methodologies become more robust. These studies also provide a much more complete understanding of their systems than a single-method study could, because each methodology yields some information inaccessible to the others. Finally, the combined studies accelerate progress, as questions raised by results from one methodology lead to further experiments by a different methodology. Thus, the greater use of multiple methodologies in an integrated research strategy may prove valuable in community ecology.

ACKNOWLEDGMENTS

This chapter profited greatly from the criticisms of James Brown, Stuart Hurlbert, Peter Kareiva, and Thomas Schoener.

chapter 2

The Assembly of a Laboratory Community: Multispecies Competition in *Drosophila*

Michael E. Gilpin, M. Patricia Carpenter, and Mark J. Pomerantz

LABORATORY RECONSTITUTION OF COMMUNITIES

For the last eight years we have been studying laboratory systems assembled from about 30 species of *Drosophila* flies. Our motivation for this effort has been as follows. While the ultimate goal of community ecology is to understand actual communities, the direct study of these communities outdoors labors under crippling disadvantages. In the field, environmental parameters vary uncontrollably in time and in space, inordinate numbers of plant and animal species are present, the individual life histories of most of these species and their interactions in pairs or in larger sets are poorly known, and experimental manipulations may be infeasible or doomed to yield ambiguous results (Bender et al. 1984). In the laboratory, one can specify and hold constant the values of environmental parameters, limit the number and identities of species present, and carry out experimental manipulations with few constraints. These facts make it self-evident that it will be difficult or impossible to understand field communities until one has mastered the easier task of understanding laboratory communities.

Unfortunately, much of the previous work on laboratory communities has used systems too grossly simplified to have relevance to the real world. There have been too few species with too little interesting behavior (typically two species, often microorganisms), and too few attempts to mimic environmental variation. The purpose of this chapter is to illustrate how realistically complex communities can be synthesized in the laboratory and how they can be used to address one set of issues currently facing field ecologists: the questions posed by multispecies competition.

Our approach is community reconstitution: the construction of communities from individual parts. We believe that the acid test of one's understanding of any complex system—whether it be a clock, a virus, or an ecological community—comes in trying to reassemble the system out of its pieces. Reconstruction is *the* procedure that accounts for many of molecular biology's successes, for instance, reaching an understanding of mitochondrial function by isolating individual enzymes and soluble factors and by recombining them until the electron transfer ability of the whole mitochondrion has been recreated. Besides offering in our view an essential route to understanding ecosystems, reconstitution is also of growing practical importance to field ecologists faced with the task of restoring damaged ecosystems (Gilpin 1983).

QUESTIONS OF STRUCTURE IN MULTISPECIES COMPETITION

In this chapter we address five problems concerning multispecies competition.

1. Transitivity of interspecies competition Recently there has been extended discussion of pairwise competitive relationships in field systems. Most investigators have found so-called transitive dominance relationships in various phyla, i.e., species A beats species B, B beats C, *and* A beats C (Kato et al. 1963, Hairston et al. 1968, Vandermeer 1969, Lang 1973, Connell 1978, Luckinbill 1979, Goodman 1979). Undisturbed intertidal communities (Connell 1961a, Dayton 1971, Paine 1974) along with many plant successions (Horn 1981, Connell 1978) have a single dominant competitor, which also implies transitivity. Nonetheless, there are theoretical reasons, based on interference competition relationships (Gilpin 1975), to expect competitive intransitivities such that A beats B, B beats C, *but* C beats A. Indeed, Buss (Chapter 31; see also Jackson and Buss 1975, Buss and Jackson 1979) has found intransitive relationships in subtidal communities. How can one rationalize these varying outcomes?

2. Mechanisms of competition Ecologists have often categorized competition as being of two types: the dichotomy of "r" and "K" competition (MacArthur and Wilson 1967) or the related dichotomy of "scramble" and "contest" competition. Both of these dichotomies correspond roughly to Miller's (1967) distinction between the rapid exploitation of resources in a competitively unsaturated environment and the efficient utilization of or control over resources in a saturated environment. We prefer to discuss the somewhat related dichotomy of "exploitation" and "interference" competition (Case and Gilpin 1974). The former term involves the flow of resources from a lower trophic level. The latter term involves direct behavioral interference (by aggression, allelochemicals, or usurping space) against an exploitation competitor, thereby increasing the flow of resources to the species that successfully interferes.

In a multispecies system of competitors, a further dichotomy regarding interference competition must be considered (Case et al. 1979): "generalized" and "specific" interference. We define generalized interference as an action that affects all (or all other) species in a relatively uniform manner, such as habitat destruction or release of a nonspecific poison. Generalized interference may be mathematically represented as the introduction of a negative resource: the more of it exists, the worse all other species do. A general lowering of equilibrium densities of all other species following introduction of a new species would suggest operation of generalized interference. Specific interference, on the other hand, is a negative interaction that affects only one or a subset of the competing species, such as a specific poison, aggression, predation on juveniles, or rape directed at a particular species.

The presence or absence of interference competition of either form affects many of the other questions that we are asking. For instance, specific interference favors the creation of intransitive competition relationships (see discussion below); this makes the reconstitution approach difficult, since each instance of specific interference has to be treated on an ad hoc basis.

3. Niche dimensionality The idea of niche dimensionality is an old one that has at least three different meanings in ecology. One use, which we shall not employ in this chapter, is that of Cohen (1978), who applied the term "dimensionality" to certain properties of food webs. A second, older use is the dimensionality of the fundamental niche, which involves the number of factors limiting the growth of the species population and which has been linked to questions of coexistence among species (Hutchinson 1978). Finally, we shall use the word dimensionality to refer to the number of independent measures of single-species population growth needed to assess interspecific competitive ability.

4. Assembly rules and alternative domains of attraction From empirical patterns of bird species co-occurrences on Southwest Pacific islands, Diamond (1975) induced a set of so-called assembly rules that specified which species combinations were forbidden and which were permitted to occur in nature. These empirical patterns imply that alternative stable communities can be assembled from a certain species pool under a given set of environmental conditions. Gilpin and Case (1976), working with an algorithmic model of interspecies competition, explored theoretically how the expected occurrence of such alternative stable communities should depend on total pool size and on the mean strength of the competition coefficient (alpha). Those theoretical results are consistent with Diamond's empirical observations and suggest interspecies competition as a possible mechanism, although Diamond discussed other contributing factors as well. The theoretical literature refers to such alternative stable communities as ''multiple stable points'' or ''multiple domains of attraction.''

The existence and interpretation of assembly rules have been debated on two grounds. First, Connor and Simberloff (1979) questioned Diamond's empirical conclusions for statistical reasons. However, Diamond and Gilpin (1982; Gilpin and Diamond 1984) showed that these statistical objections were invalid and that a proper statistical analysis confirmed the reality of permitted and forbidden combinations. Second, Connell and Sousa (1983) dismissed evidence for alternative community structure derived from observation of natural communities. They called instead for evidence from experimental perturbation studies. For instance, one could defaunate replicated habitat patches and observe whether alternative communities developed. Our laboratory *Drosophila* system permits us to carry out such an experimental test for alternative communities, but with a decisive advantage over field tests. If alternative communities did develop in the field, this might merely reflect undocumented differences in the identities of the colonizing species arriving at different experimental plots. In the laboratory we have complete control over the ''colonizing'' species, permitting us to separate the contributions of colonization and post-colonization events to the creation of alternative communities.

5. Prediction Predicting the structure of an ecological community obviously involves more complications than do predictions in some other areas of science, such as predicting the next return date of Halley's comet. We believe that two related issues are involved.

First, there is a quantitative issue. Physicists pretend that the only real prediction is a quantitative one, although in reality they expect much higher precision for predicting the absorption spectrum of a hydrogen atom than the location, date, and Richter scale value of future earthquakes in California. Paine (1980), implicitly using the standard of hydrogen atoms rather than earthquakes, concluded that long-term predictions of population sizes are almost certain to be impossible for ecologists. In our view, ecological predictions suffer under a curious but understandable schizophrenia caused by differences between theories and field observations, and ecologists should learn from what physicists do rather than from what they pretend to do. Theoretical models of populations are almost always cast in terms of absolute population densities, while field observations are normally in far more relative terms such as presence-absence or rank order of abundance. The precision of quantitative prediction possible for our laboratory system will suggest limits to realistic expectations for field systems.

Second, and more interesting to us, are questions about whether knowledge of the parts predicts properties of the whole. For instance, can one predict community structure from studies of pairwise interactions? Can one predict the outcome of pairwise interactions from studies of separate species? If the answer is yes in either case, how many parameters are needed? Theory and some laboratory studies suggest that measures of fundamental niche properties (i.e., of the niches of species studied in isolation) *might* yield predictions of the realized niche attained in the presence of competing species. For instance, the theoretician may assume a one-dimensional resource space and draw single-species utilization curves whence derive all predictions of community structure: e.g., species packing depends on the width of the resource utilization niche (see examples in Roughgarden 1979). Lending support to this hope, laboratory studies of microorganisms limited by a single nutrient have shown that a species' half-saturation concentration for nutrient utilization suffices to predict that species' position in the competitive hierarchy (Hsu et al. 1978). These considerations suggest that there might be a single key to predicting community structure. On the other hand, interference competition permits an infinite variety of special cases and intransitivities, while even exploitation competition is unlikely to reveal a single limiting factor over which species assort.

Thus, only empirical studies, beginning with laboratory studies, can reveal the quality and precision of predictions attainable in ecology.

THE *DROSOPHILA* MODEL

The Species Table 2.1 gives the taxonomic names, abbreviations, geographical distribution, and climate preference of the 28 species that we studied.

Since our laboratory *Drosophila* community is to be used as a model for field communities, we

Table 2.1 THE 28 SPECIES USED IN THIS STUDY

Genus and subgenus	Group	Subgroup	Species	Abbreviation	Distribution
Drosophila					
Drosophila	Annulimana		*gibberosa*	Gib	NW TR
	Funebris		*funebris*	Fun	C TE
	Immigrans		*immigrans*	Imm	C TE
			quadrilinecta	Quad	WPI TE
	Mesophragmatica		*gaucha*	Gau	SAM TE
	Pallidipennis		*pallidipennis*	Pallid	NW TR
	Repleta	Hydei	*hydei*	Hyd	C TE
		Mercatorum	*mercatorum*	Merc	NW TE
	Robusta		*robusta*	Rob	NW TE
	Virilis		*virilis*	Vir	OW TE
Sophophora	Melanogaster	Ananassae	*ananassae*	AnB	C TR
		Eugracilis	*eugracilis*	Eug	SEA TR
		Melanogaster	*melanogaster*	MelS	C TT
			simulans	SimA	C TT
		Montium	*birchii*	Bir	ANG TR
			serrata	Ser	ANG TT
	Obscura		*persimilis*	PerO	NAM TE
			pseudoobscura	Psd	NAM TE
	Saltans		*prosaltans*	Pro	NW TR
			sturtevanti	Stv	NW TR
	Willistoni		*equinoxialis*	Eqx	NW TR
			insularis	Inw	NW TR
			paulistorum	Paul	NW TR
			tropicalis	Trop	NW TR
			willistoni	W(+)	NW TR
			willistoni	WW	NW TR
			nebulosa	Neb	NW TR
Zaprionus			*vittiger*	Zap	WAF TR

Genus (italicized) and subgenus (not italicized) names are given in column 1, group and subgroup names in columns 2 and 3, species names (italicized) in column 4, abbreviated names used in this chapter in column 5. WW and W(+) are the same species, but WW is the white-eye mutant.

The first characters in the distributional code give location: ANG = Australia and New Guinea; C = cosmopolitan; NAM = North America; NW = New World; OW = Old World; SAM = South America; SEA = South East Asia; WAF = West Africa; and WPI = West Pacific Islands. The second part of the code gives the climate preference of the species: TR = tropical; TE = temperate; and TT = tropical and temperate.

must initially consider what mappings are possible from bottle to field. Clearly, this depends on the range of behaviors of our *Drosophila* system. *Drosophila* do compete under both r and K conditions. They exhibit mechanisms of exploitation competition, generalized interference, and specific interference. Among these mechanisms are adult food preferences (Sang 1949; Merrell 1951; Buzzati-Traverso 1949; da Cunha et al. 1951, 1957; Cooper 1960), oviposition site preferences (Moore 1952, Del Solar and Palomino 1966, Del Solar 1968, Barker 1971), pupation site preferences (Barker 1971, de Sousa et al. 1968, Sameoto and Miller 1968), interspecific ''rape'' leading to sterile eggs (Narise 1965), poisonous larval metabolites (Weisbrot 1966, Dawood and Strickberger 1969, Budnik and Brncic 1976), larval alterations of mechanical properties of the medium that affect survivorship, and differences in larval food preferences and feeding rates (Lindsay 1958, Cooper 1960, Bakker 1961, Gilpin 1974). All these mechanisms permit both high niche dimensionality and competitive intransitivities. Nothing about *Drosophila* excludes assembly rules and multiple domains of attraction. Very little is either ruled out or necessarily ruled in. Thus, a finding, say, that increased resource heterogeneity permits greater coexistence is unlikely to be based in a simple laboratory artifact. In short, we know of no other laboratory system that realistically addresses so many ecological problems for so little time, energy, and expense.

The Environment The bottles that delimited the laboratory environment of the flies were ¼-pint cream bottles. Most often, the food was a 40-ml plug (about 1 cm deep) of the standard mixture of cornstarch and molasses, laced with proprionic acid and tegosept for mold and bacteria control. We term this system ''thick food.'' However, in some of the work reported here we used different food structures and amendments. A second system, which we dubbed ''thin food,'' consisted of only 3 ml of standard medium atop a 10-ml plug of agar. This food layer was only about 1 mm deep, less than the length of a third instar larva and was, in most cases, totally consumed prior to the onset of pupation. Generation times were thus shorter and these systems were more sensitive to the slight temporal variability in bottle humidity, microflora, etc. In thick-food systems the actual limitation on population sizes could not be the energy in the food, as most of this was still remaining after fly production ceased (Gilpin 1974); hence competition for space, interference competition, or availability of protein may instead be limiting. In thin-food systems the food itself is definitely limiting; whether it is the sole limiting factor is uncertain.

These thick-food and thin-food systems were each run at both 25°C (tropical) and 19°C (temperate) temperatures. These temperatures represent physiological limits for some of the fly species, as mating behavior and egg laying are both quite temperature-sensitive in *Drosophila.* For instance, species Psd (*D. pseudoobscura*) is torpid at 25°C but is fourth in the dominance hierarchy in the thick-food system at 19°C.

A somewhat different system, which we dubbed the ''augmented system,'' had two tablespoons of puffed millet upon which was poured 30 ml of the standard medium. These millet balls perhaps supplied a different nutrient. More importantly, however, they altered the medium's surface texture and its ''soupiness'' under larval activity, thereby possibly influencing the interference interaction of larval drowning.

By virtue of more experimental effort and replications, the 25°C thick-food system can be considered the baseline, with the other experimental treatments seen as perturbations (sensitivity studies) away from it.

Types of Experiments In all experiments adult flies were anesthetized, sexed, and counted into culture bottles. The adults were always removed from the bottle before any progeny (also called recruits or F1s) from the next generation started to emerge (eclose) from their pupal casings. Normally, this removal was done at the seventh day following the introduction. Recruits were then removed at later days or, more typically, weeks. Bottles normally produce recruits for five weeks, during which time the food hardens and excrement accumulates on the inside of the glass bottle. Thus, each bottle ''ages'' over the course of an experiment.

The experiments reported below were of two general types: short-term and long-term. Short-term experiments lasted less than a single generation. Adults were added to a bottle, and all surviving adults and recruits in that bottle were sacrificed and counted at later times. (Ayala et al. [1973] refer to these as input-output or type II experiments.) The three experiments described below to measure single-species parameters (larval production, life history parameters, and population dynamics) used this technique and were carried out solely with thick-food systems.

Long-term experiments involved multiple generations and were carried out with the serial transfer technique (Ayala 1967). A system begins with a single bottle to which adults are added. After one week, a new bottle is added and the surviving adults are transferred to it, usually without anesthetization or counting. After the second week, a third bottle is added to the system, and the surviving adults in the one-week-old bottle (the second bottle) and the recruits in the two-week-old bottle (the first bottle) are transferred to it. Similarly, after the third or fourth week a new fourth or fifth bottle is added, and all adults or recruits in the system are transferred to it. Since bottles cease to produce new recruits after the fifth week, a steady state is reached then: a new bottle is added, but the first bottle can be discarded. Thus, from the end of the fifth week onward the system consists of five bottles of different ages, ranging from one to five weeks. (Ayala et al. [1973] refer to these as type I experiments.) These long-term experiments usually ran for 19 or 24 weeks or, in some cases, until the conclusion was obvious. This technique was used for the two experiments on pairwise competition and on multiple domains of attraction.

We now describe each of these individual experiments.

SHORT-TERM EXPERIMENTS

Larval Production Rate In these experiments we started 25 replicate bottles of thick food, each with 100 adults of a single species. At day 6 we removed the adults, liquified the media by boiling the bottles in a microwave oven, and separated the larvae in a 10% sucrose solution. We then dried this larval population in a vacuum oven and obtained its total dried weight, which we took as a measure of larval production. It represents an integration of the egg laying rate times the growth rate of an individual larva and was determined for each of the 28 species at 25°C and at 19°C. We refer to this measure as the variable LARVPROD.

Life History Parameters Twenty-five replicate adult populations of 13 pairs of a single species were placed in glass shell vials 25 mm in diameter at 25°C. They were given four days to oviposit, after which the adults were removed and the vials monitored on a twice-daily basis for appearance of first, second, and third instar larvae and for pupae. At each monitoring the newly emerged adults were removed and counted, to preclude a second generation. The monitorings and removals continued until after all larval activity ceased (about 6 weeks). Fig. 2.1 illustrates the results of these experiments for the species Psd and Zap. From the emergence curve, which is the average number of newly emergent adults per day, five variables describing life history can be read off: REPLLH, a measure of the replacement (the area under the emergence curve divided by the number of starting adults); RLH, a measure of the exponential rate of increase based on the replacement and the generation time; FSTEMGLH, the day of first emergence; PKEMGLH, the day of peak emergence; and MAXEMGLH, the maximum daily rate of emergence.

Estimates of Growth Parameters Replicate bottles of thick food were initialized with adults of one species over a span of densities designed to run from roughly 10% to 150% of the estimated carrying capacity. This starting population size is denoted N_t. At the end of one week the surviving adults were counted, while the recruited F1s were removed and counted at the end of the second, third, fourth, and fifth weeks by the techniques described by Ayala et al. (1973). The sum of the surviving adults at week 1 and the recruits at weeks 2 to 5 gives an estimate of the total population size after one generation, denoted N_{t+1}. This experiment was carried out for each of the 28 species at 19°C and at 25°C.

The parameters of discrete growth equations

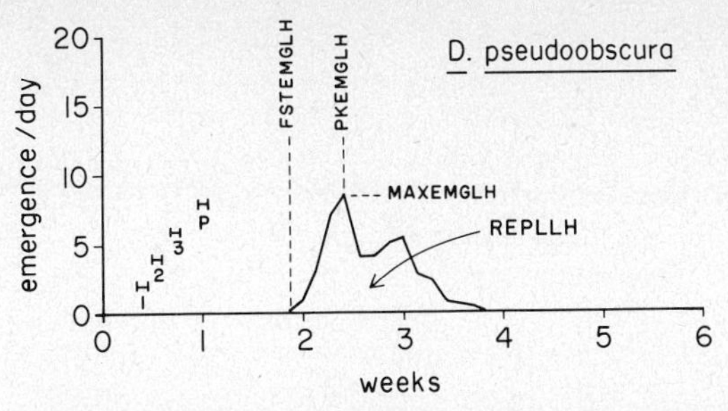

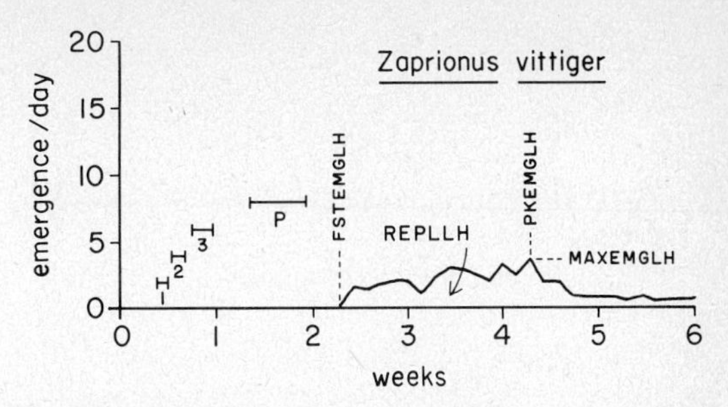

Fig. 2.1 Sample of results from short-term life history experiments. Thirteen pairs of adults of each species were placed in a vial and were removed after laying eggs for four days. Vials were examined twice daily for the presence of first (1), second (2), and third (3) instar larvae and for pupae (P); the horizontal bars in the lower left corner of each graph give the 65% confidence intervals for the presence of these four life stages. Newly emerged adults were removed each day for the first 42 days, and the average emergence of new adults per day is plotted on the ordinate. From these emergence curves one can read off the day of the first emergence (FSTEMGLH), the day of peak emergence (PKEMGLH), the maximum daily rate of emergence (MAXEMGLH), and the replacement (REPLLH: area under curve ÷ number of initial adults). The emergence curve for *D. pseudoobscura* at 19°C is "peaked," while that for *Z. vittiger* at 19°C has a "plateau" shape.

can be extracted by fitting these values of N_t and N_{t+1} to the equation

$$N_{t+1} = N_t + rN_t[1 - (N_t/K)^\theta] \qquad (2.1)$$

where K is the single-species equilibrium density or the carrying capacity of the environment, r is the rate of growth at low density, and θ is a measure of the asymmetry of the growth curve (Pomerantz et al. 1980, Thomas et al. 1980). There is a possible degeneracy in the simultaneous estimation of r and θ, but their product $r\theta$, which is the stability parameter for this system (see appendix C in Pomerantz et al. 1980), is free of this regression pathology. In the analysis and tables below we use KFIT, RFIT, and RTFIT to stand, respectively, for these extracted values of K, r, and $r\theta$.

There were two additional measurements that we made on each species. First, in order to be able to correct for size differences among species, we reared flies under moderate crowding and obtained the average dry weight of one individual (the variable DRYWT). Second, in long-term experiments we counted the number of individuals of a victorious species at the end of a pairwise competition trial (after at least 15 weeks; see below for a more detailed explanation) to obtain an alternative measure of carrying capacity for that species. For species that were usually eliminated in pairwise competition, we obtained this alternative measure (termed K-15) from serial transfer systems that we ran for 15 weeks before counting the number of flies.

LONG-TERM EXPERIMENTS

In the long-term experiments for both single- and multispecies systems, surviving adults and recruits from older bottles were transferred without counting to a fresh bottle of medium. Experiments were generally run for 19 weeks at 25°C and 24 weeks at 19°C. These times amount to roughly 10 generations and were selected because exclusions usually occurred in considerably shorter times and coexistences were maintained for considerably longer times.

Pairwise Competition Serial transfer systems were started with 13 male-female pairs of each of two species. This was done for all possible pairings of all 28 species (28 × 27/2 = 378 pairings) at 19 and 25°C for the thick-food system; for all pairings of 20 species (20 × 19/2 = 190 pairings) at 19 and 25°C for the thin-food system; and for all pairings of 10 species (10 × 9/2 = 45 pairings) at 25°C for the augmented system. Each pairing was replicated two or three times. The flies were not counted at the weekly transfers. Because the different species could be distinguished visually, the possible absence of one of the competitors was noted; after five weeks of such a species's absence the system was terminated and the adult flies counted.

Multiple Domains The remaining long-term experiment was done with 30 different serial transfer systems, using thick food at 25°C and initialized with varying densities of 10 different species: Merc, Ser, Zap, Paul, Neb, Psd, WW, Imm, Eug, and Gib. To increase the total size of the mixed species population, we used three bottles connected by plastic tubing in place of each single bottle normally used in the serial transfer experiments. That is, in the steady state after five weeks there were fifteen bottles in the system: three each of one-week-old, two-week-old, three-week-old, four-week-old and five-week-old bottles. The starting population was mixed and divided into thirds to begin the experiment, and at each transfer the flies were again mixed and divided into thirds. The systems were run for 35 weeks. This experiment is analogous to the relaxation of a supersaturated fauna on a land-bridge island, in that the system is initially stocked with more species than can be retained, with the result that species go extinct.

The initial conditions for the 30 different systems were as follows. In 10 systems the starting numbers of the 10 species were 910, 10, 10, . . . , 10 individuals, with each of the 10 species started once at 910 individuals (91% frequency) in order to explore all "corners" of the system's state space. For the other 20 systems, the initial densities were chosen randomly according to the scheme used by Gilpin and Case (1976) in their computer work on multiple domains of attraction in competition communities. This scheme was in effect a broken stick model, with the line segment (0,1000) broken at nine randomly chosen points and the lengths of the 10 resulting segments taken as the starting densities. In no case were fewer than 5 females and 5 males of a species used in an initialization.

BEHAVIOR OF THE MODEL

Single-Species Studies

Studies at 25°C Table 2.2 summarizes the results of all of our single-species work at 25°C, the temperature selected for intensive study because of faster development times. Table 2.3 gives the correlations between these single-species parameters. (The last two columns of Tables 2.2 and 2.3 are derived from pairwise competition studies and are discussed in the next section.)

Five parameters measure adult production: MAXEMGLH, the maximum rate of emergence of F1s; the replacement REPLLH; the rate of increase RLH; and K-15 and KFIT, the two alternative measures of carrying capacity. (The carrying capacities are heavily influenced by adult production because adult survival is less than 20% per week, so that a high production rate is needed to sustain a high carrying capacity.) There is close concordance among these five measures of adult production: the 10 pairwise correlation coefficients among them all fall between 0.49 and 0.88 and average 0.74 (Table 2.3). Not surprisingly, the correlation between the two measures of carrying capacity is especially high (0.87).

Larval production (LARVPROD) also correlates positively with these five measures of adult production, but more weakly: the five coefficients range from 0.23 to 0.38 and average 0.31. One reason why larval and adult production correlate only weakly is that larvae, though numerous, may not reach the critical weight threshold for pupation (cf. Gilpin's and McClelland's [1979] model for the pupation threshold of *Aedes aegypti*). A second reason is that high larval production may actually interfere with pupation: many larvae pupate on the surface of the medium and are drowned by subsequent larvae. Both reasons mean that larval production is not guaranteed to translate one-for-one into adult production.

The variables FSTEMGLH and MAXEMGLH both correlate negatively with all five measures of adult production as well as with larval production. Small values of these variables (day of first and peak emergence, respectively) mean early emergence. Thus, early emergence favors high larval and adult production.

Studies at 19°C Similar data were collected at 19°C for all species except W(+). Qualitatively, the 19°C data are similar to those at 25°C. Quantitatively, development (e.g., FSTEMGLH and PKEMGLH) was slower at 19°C for 25 of the 27

Table 2.2 SINGLE-SPECIES DATA AND COMPETITIVE RANKS FOR ALL 28 SPECIES ON THICK FOOD AT 25°C

Species	KFIT (flies)	RFIT (flies/wk)	RTFIT	RLH (flies/d)	REPLLH	FSTEMGLH (day)	PKEMGLH (day)	MAXEMGLH (flies/day)	LARVPROD (mg)	K-15 (flies)	DRYWT (0.1 mg)	COMP	RANK
AnB	556	4.5	1.22	.044	1.67	10	11	12	28.3	685	1.60	.497	15
Bir	422	7.9	.95	.025	1.52	12	14	5	12.6	411	1.71	.160	24
Eqx	704	4.1	1.30	.051	1.92	11	13	14	19.7	560	1.20	.698	10
Eug	481	1.8	1.26	.016	1.21	9	11	4	35.4	472	1.75	.276	20
Fun	248	21.2	1.1	.030	1.79	14	15	5	6.0	325	5.29	.537	14
Gau	312	1.4	.92	.000	.98	16	19	7	1.7	312	4.89	.014	28
Gib	136	1.0	.77	.009	1.21	18	20	3	60.3	110	8.20	.275	21
Hyd	295	5.9	.89	.048	3.43	16	17	6	19.2	382	5.50	.964	2
Imm	191	829.7	.83	.017	1.39	13	14	4	5.9	216	5.59	.064	27
Ins	645	2.6	.98	.071	2.84	12	14	17.5	27.4	635	1.53	.713	8
MelS	737	6.7	1.34	.114	4.76	10	14	23	63.1	894	2.53	.997	1
Merc	441	840.9	.85	.028	1.79	13	15	6	26.8	399	4.06	.749	6
Neb	459	13.9	.97	.076	3.72	11	13	10	55.4	199	2.29	.526	12
Pallid	79	771.0	.77	.014	1.13	14	17	3	2.0	171	4.84	.137	25
Paul	485	26.5	1.06	.079	3.35	10	12	10	26.6	576	1.61	.696	11
PerO	124	872.1	.87	.031	1.86	16	21	7.5	10.0	30	3.20	.065	26
Pro	273	2.0	.93	.041	2.54	14	22	7	13.4	248	2.10	.355	19
Psd	261	2.0	.89	.068	3.27	13	16	14	8.9	301	3.17	.484	16
Quad	320	3.3	.99	.000	.98	11	13	5	12.7	243	4.06	.232	22
Rob	238	1.1	1.26	.011	1.27	15	20	3	9.6	160	6.41	.209	23
Ser	592	809.3	.81	.079	4.14	12	14	17	18.7	402	1.89	.421	17
SimA	593	6.6	1.13	.078	2.70	9	11	12	52.7	497	2.15	.898	3
Stv	233	1087.4	1.09	.038	2.98	18	25	5	58.7	285	2.49	.797	5
Trop	979	9.2	1.02	.080	3.12	10	13	18	38.8	607	2.09	.762	7
Vir	621	10.9	1.09	.088	6.63	13	16	15	15.8	645	4.68	.904	4
W(+)	576	7.5	1.13	.049	2.01	10	12	11	30.7	620	2.27	.759	9
WW	430	986.9	.99	.053	1.98	10	12	10	17.7	442	1.58	.419	18
Zap	264	6.0	.84	.017	1.45	12	14	2.5	46.4	207	3.88	.559	13

Names of variables and units (where applicable) are given in the column heads. The variables are as follows. KFIT = carrying capacity, extracted from equation 2.1. RFIT = rate of population growth (r) at low density, extracted from equation 2.1. RTFIT = value of $r\theta$ extracted from equation 2.1, where θ measures the asymmetry of the growth curve. RLH = the rate of increase based on the replacement rate and generation time. REPLLH = the replacement rate, measured from emergence curves (see Fig. 2.1). FSTEMGLH = day of first emergence (see Fig. 2.1). PKEMGLH = day of peak emergence (see Fig. 2.1). LARVPROD = larval production. K-15 = carrying capacity, measured as the number of flies in a single-species system after at least 15 weeks. DRYWT = average dry weight of one fly. COMP = average frequency at the end of pairwise competition with each of the 27 other species; a measure of competitive ability. RANK = competitive rank in Fig. 2.2A (1 = the best competitor, 28 = the worst).

Table 2.3 CORRELATION MATRIX FOR THE PARAMETERS OF TABLE 2.2

	RFIT	RTFIT	RLH	REPLLH	FSTEMGLH	PKEMGLH	MAXEMGLH	LARVPROD	K-15	COMP	RANK
KFIT	−.32	.14	.70	.49	−.22	−.59	.81	.33	.87	.61	−.60
RFIT		−.12	−.13	−.09	−.07	.29	−.18	−.13	−.36	−.21	.20
RTFIT			.00	−.08	−.06	−.19	.05	.04	.19	.16	−.16
RLH				.86	−.14	−.32	.88	.38	.75	.72	−.70
REPLLH					−.11	−.05	.69	.26	.56	.66	−.66
FSTEMGLH						.08	−.20	−.22	−.15	−.01	.01
PKEMGLH							−.36	−.09	−.70	−.24	.21
MAXEMGLH								.23	.78	.56	−.55
LARVPROD									.35	.53	−.55
K-15										.68	−.67
COMP											−.99

species studied at both temperatures. The two exceptions were Fun and Gib, which emerged earlier at 19°C, due apparently to a density-dependent effect: greater egg laying leading to denser larval populations, making the medium softer and more digestible. Despite slower development at 19°C, carrying capacities averaged 23 flies higher than at 25°C, though there was much variation among species. For instance, Neb and Paul, both of them tropical species, were respectively about 70% less and 70% more abundant at 19°C than at 25°C. In fact, for every life history measure some species did better at 19°C and some did worse. Thus, the net effect of temperature on *Drosophila* biology is complex.

Pairwise Competition

Results Pairwise competition was studied in all or subsets of our 28 species under five different conditions of temperature and food type (Fig. 2.2). The 10- or 20-species subsets used in Figs. 2.2C–E were chosen from among the middle-ranking competitors of the 28-species thick-food systems (Figs. 2.2A, B).

A first, and important, conclusion is that our results are highly reproducible insofar as the equilibrium state is concerned. It may take more or less time, but a "strong" competitor invariably defeats a "weak" competitor. The overall measure of our concordance between replicates is 88%.

Mathematically, and with our flies in practice, exclusion is an asymptotic process. In some cases we noted that, although no recruits of the "losing" species were being added to the system, two or three of its adults were still present at the end of an experiment (19 weeks), probably because maximal adult survival times in bottles can be on the order of months. We scored such cases as competitive exclusion by the numerically dominant species.

Transitivity The presentation of results in Fig. 2.2 sorts the species such that the most successful tournament competitor (the highest ranking) is listed in the first row and column and so on to the least successful. Clearly, the triangular form of these matrices suggests a dominance hierarchy for all five systems.

There are various ways to quantify hierarchy. The classical method (Landau 1951) was employed in peck order studies of birds. More recently, Petratis (1979) has investigated such measures in an ecological context. However, both of these methods for quantifying hierarchies are based on a binary win-loss outcome. We feel that our results are more properly trinary, win-tie-loss, since there are doubtless true coexistences between pairs. More to the point, we feel it highly artificial to call a numerical 60:40% coexistence as being a win by the species present at 60% frequency, as a binary win-loss scoring would require (cf. Goodman 1979). Arguments against automatically calling the 60% species the winner include the facts that adults of the less abundant species might weigh more; the actual competitive impact on a species is better measured by the depression of its adult numbers below carrying capacity; adult densities are not neces-

A

```
          111111111122222222
 1234567890123456789012345678

 111111111111111111111111111   1   MelS
0 11111111111111111111111111   2    Hyd
00 *1111111111111111111111111  3   SimA
00* 111111111111111111111111   4    Vir
0000 1*11*1111*1111111111111   5    Stv
00000 ***1111111111111111111   6   Merc
0000** ***1*11*1111111111111   7   Trop
00000** ***11*11111111111111   8    Ins
00000*** **111111*1111111111   9   W(+)
0000*0*** ****11111111111111  10    Eqx
0000000*** *1*1*111111111111  11   Paul
000000*00** 10**111111111111  12    Neb
00000000*00 1*1111111111111   13    Zap
0000000*0**10 01*1011111111   14    Fun
0000*0*0000**1 0**1*1*111111  15    AnB
0000000000**001 *1*111*11111  16    Psd
0000000000000*** *1*11111111  17    Ser
00000000*00000*0* 1111111111  18     WW
000000000000010*00 11111*11*  19    Pro
00000000000000*0*00 01111111  20    Eug
00000000000000000001 *1*1111  21    Gib
00000000000000*00000* 11*111  22   Quad
000000000000000*000000 11111  23    Rob
00000000000000000000*00 1111  24    Bir
000000000000000000*00*00 *11  25 Pallid
000000000000000000000000* *1  26   PerO
0000000000000000000000000* 1  27    Imm
000000000000000000*00000000   28    Gau
```

B

```
          111111111122222222
 1234567890123456789012345678

 111111111111111111111111111   1   MelS
0 11111111111111111111111111   2    Hyd
00 1111111111111111111111111   3   SimA
000 **1111111111111111111111   4    Psd
000* *0111111111111111111111   5    Imm
000** *1*1111*11111111111111   6    Vir
00001* *1111*111111111111111   7    Stv
000000* ******11111111111111   8   Trop
00000*0* ****11*111111111111   9   W(+)
0000000** ****11111111111111  10    Eqx
0000000*** ***11111111111111  11   Paul
0000000**** 1**1*111111*1111  12    Ins
000000*****0 **1111**1111111  13   Merc
00000*0*0**** *111**1*111*11  14    Fun
00000000000*** 111*1*1111111  15 Pallid
00000000*000000 0**111111111  16     WW
00000000000*0001 1*11*101111  17    Eug
000000000000000*0 11*1111111  18    Zap
0000000000000****0 *1**110*1  19    Gib
000000000000**0000* ***11111  20    Ser
000000000000*0*00*0* ****111  21    Rob
000000000000*00*0*** ****11   22    Gau
000000000000000000**** 1*111  23    Bir
00000000000*00001000**0 *111  24   PerO
00000000000000000000**** 1*1  25    Neb
0000000000000*0000100*000 11  26    Pro
000000000000000000*00000*0 *  27   Quad
00000000000000000000000000*   28    AnB
```

C

```
          1111111112
 12345678901234567890

 *1**11*111111111111   1   SimA
* 01*1111**111111111   2   Paul
01 *011111111111111    3    Hyd
*0* 1**1*11111*11111   4    AnB
**10 **111*111111111   5    Zap
000** *1*10*1**11111   6    Neb
000*** 0******111111   7    Eug
*000001 ***1*1111111   8     WW
000*0*** ***11*11111   9    Ser
0*0000*** 11**111111  10    Vir
0*00*1***0 1**011111  11    Stv
00000**0*00 *1*11111  12    Pro
000000**0*** *111111  13   Merc
00000**00**0* *11111  14    Bir
000*0*00*01*0* 11111  15   Quad
000000000000000 *111  16    Rob
0000000000000000* 1** 17 Pallid
00000000000000000 11  18    Fun
0000000000000000*0 1  19    Gib
0000000000000000*00   20    Psd
```

D

```
          1111111112
 12345678901234567890

 **111*111111*111111   1    Stv
* 1*1*1*11*111111111   2   Paul
*0 10**1*11111111111   3    Zap
0*0 0**11**111111111   4    Vir
0011 011010***111111   5   SimA
0***1 *01**011101111   6    Hyd
*0**0* 11****1**1111   7    Bir
0*00010 1*11*1011111   8    Pro
00*01000 111*1****11   9 Pallid
000*0***0 **1**11111  10     WW
0*0*1**00* 1*1*1**11  11    Neb
0000*1*00*0 **111111  12   Merc
0000*0***0** *1**111  13    Eug
*000*0000*0** 1*0111  14    Psd
000000*1***000 01011  15    Ser
000001*0*000**1 ***1  16    Fun
00000000*0*0*10* *01  17   Quad
00000000*0*0001** *1  18    Gib
000000000000000*1* 1  19    Rob
0000000000000000000   20    AnB
```

E

```
          1
 1234567890

 111111111   1    AnB
0 11111111   2   Paul
00 1111111   3    Zap
000 *11*11   4    Neb
000* ****1   5     WW
0000* **11   6    Ser
0000** 1**   7    Pro
000***0 **   8    Gib
0000*0** *   9    Psd
000000***   10    Eug
```

Fig. 2.2 Results from two-species competition trials. A and B are for all 28 species on thick food; C and D are for a subset of 20 species on thin food; E is for 10 species on the augmented system. A, C, and E are at 25°C, while B and D are at 19°C. The abbreviation for the species is given in the right column. Immediately to the left is given the competitive rank of the species (its row number, which is also its column number), from 1 (the best competitor) through 28, 20, or 10 (the worst competitor). A 1 in row "*i*" and column "*j*" indicates that the "*i*th" species excluded the "*j*th" species; a 0 indicates the opposite. For example, in A the 1 in the third column of the second row indicates that Hyd excludes SimA. An asterisk (*) indicates a coexistence or a reversal of outcomes in the replicated trials.

sarily related to larval densities; and the stability of the system may depend on many properties of life history, such that the less frequent species might be the more resilient to disturbance.

Hence, we have decided to employ as a measure of hierarchy one that only counts intransitive exclusions, i.e., the number of 0's that remain in the upper right of the outcome matrix after it has been optimally sorted. For the five systems, the resulting measures of transitivity are:

25° thick food	98%
19° thick food	98%
25° thin food	97%
19° thin food	94%
25° augmented	100%

Thus, as Fig. 2.2 indicates, the degree of competitive transitivity in a given environment is overwhelming.

Shifts of Rank with Environmental Conditions Table 2.4 gives the rank order of the 20 species common to the first four systems, i.e., the thick and thin systems at 25°C and 19°C. The important feature to note about this table is how the rank of some species shifts strikingly with environmental conditions. For instance, with thin-food species AnB is the worst species (rank 20) at 19°C but fourth best at 25°C. Psd has the opposite behavior, going from third best on thick food at 19°C to worst on thin food at 25°C. However, species Quad and SimA retain roughly the same competitive ability (miserable and mighty, respectively) in all four environments. Gause (1934, chap. V) similarly noted shifts in competitive rank of protozoan species with food conditions.

The shifts most susceptible to interpretation are those with temperature. There are 12 cases of species improving and 10 cases of species slipping back by four or more in competitive rank with a shift from 19°C to 25°C (within either the thick-food or thin-food systems). The species that do better at 25°C are drawn disproportionately from the species whose native habitat is the tropics: the pool of 28 species contains 15 tropi-

Table 2.4 COMPETITIVE RANKS OF 20 SPECIES STUDIED UNDER FOUR COMBINATIONS OF TEMPERATURE AND FOOD THICKNESS (FIGS. 2.2A–2.2D)

Species	25°C, thick	19°C, thick	25°C, thin	19°C, thin
AnB	10	20	4	20
Bir	19	16	14	7
Eug	15	11	7	13
Fun	9	8	18	16
Gib	16	13	19	18
Hyd	1	1	3	6
Merc	5	7	13	12
Neb	7	17	6	11
Pallid	20	9	17	9
Paul	6	6	2	2
Pro	14	18	12	8
Psd	11	3	20	14
Quad	17	19	15	17
Rob	18	15	16	19
Ser	12	14	9	15
SimA	2	2	1	5
Stv	4	5	11	1
Vir	3	4	10	4
WW	13	10	8	10
Zap	8	12	5	3

1 = the best competitor; 20 = the worst.

cal and 9 temperate species, but 8 tropical and only 1 temperate species improve at 25°C. Similarly, the species that do better at 19°C are drawn disproportionately from temperate species: 5 temperate and 5 tropical species improve at 19°C.

Also readily interpretable are the shifts in competitive rank between the thick- and thin-food systems at a given temperature. When these shifts were correlated against single-species traits, FSTEMGLH had the highest predictive value: species with later first emergence did comparatively worse in the thin-food system ($p < 0.01$). This is plausible on the grounds that such fly species, when competing against a faster developing species, are more likely to be excluded because they find the limited food supply exhausted before they reach the weight and development limit necessary for pupation.

Prediction of Pairwise Competitive Ability The last two columns of Table 2.2 give two measures of pairwise competitive ability for each species: RANK, its ranked order of competitive ability in Fig. 2.2A (i.e., on thick food at 25°C), and COMP, its average frequency at the end of pairwise competition with each of the 27 other species. RANK ranges from 1 for the best competitor to 28 for the worst; COMP from 99.7% for the best to 1.4% for the worst. Thus, the correlation coefficient between RANK and COMP is negative (−0.99; Table 2.3).

The signs of the correlation coefficients linking COMP or RANK to the single-species measures indicate that high production and early production promote competitive success (Table 2.3). Thus, larval production (LARVPROD) and all five measures of adult production (MAXEMGLH, REPLLH, RLH, K-15, and KFIT) have positive correlations of 0.53 to 0.72 with COMP. PKEMGLH has a negative correlation with COMP: low values of PKEMGLH promote success because low values mean early production (early peak emergence). Since the best competitor has the lowest value of RANK, coefficients for RANK are virtually the same as those for COMP but with reversed sign.

The origins of competitive ability can be explored further by multiple regression analysis. We multiplied KFIT, K-15, and MAXEMGLH by DRYWT (the weight of an individual fly), thereby obtaining 11 independent variables against which to regress competitive rank. The multiple regression explains 85% of the variance in RANK at 25°C and 65% at 19°C; four independent variables enter with a significant *t*-value at 25°C, but none does at 19°C. The lower significance of the regression at 19°C may be because fewer replicates were used.

At 25°C four variables, each with a *t*-value corresponding to $p < 0.01$ or better, explain 80% of the variance in RANK (the other seven variables add only 5%). These variables are the adult production measures RLH, K-15, and MAXEMGLH and the larval production measure LARVPROD. The multiple regression equation is:

$$\text{RANK} = 29.6 - 180\ \text{RLH} - 0.009\ \text{K-15} - 0.14\ \text{LARVPROD} + 0.29\ \text{MAXEMGLH} \quad (2.2)$$

Thus, if a new fly species were added to the tournament, we could accurately predict its rank in pairwise competition on the basis of studying it in isolation. Gause (1934, p. 89) was similarly able to predict the outcome of pairwise competition among yeast species from single-species growth studies.

One detail about equation 2.2 requires comment: the sign for MAXEMGLH. RLH, KFIT, and LARVPROD enter the equation with negative signs and have negative correlations with RANK in Table 2.3 (i.e., large values of these variables tend to yield a RANK nearer to 1 than to 28). However, MAXEMGLH enters the equation with a positive sign even though it too has a negative correlation in Table 2.3. The explanation is suggested by comparing the graphs in Fig. 2.1; species Zap has a low, plateau emergence curve compared to the steeply peaked curve of Psd, and Zap has a correspondingly much lower value of the peak emergence rate MAXEMGLH, yet Zap is a better competitor than Psd. Apparently, the multiple regression is picking up a few such cases and adding their effect as a correction factor to the three other regression variables. A stepwise regression confirms this: MAXEMGLH enters fourth in explaining variance. A mechanistic explanation is that a low MAXEMGLH coupled with a high LARVPROD might indicate that larval activity makes it difficult for pupation to take place; this effect could easily carry over to other species.

Pairwise Coexistence Only 12% of our species pairs (46 out of 378) were able to coexist on thick food at 25°C. When we put 10 species together, we never observed more than 3 species to persist. This result is very different from what happens in nature, where hundreds of *Drosophila* species can coexist in a region. The most important reason for this difference involves the great spatial and temporal heterogeneity of microhabitats in nature, compared to the single such habitat offered in our experiments.

What is it that does permit laboratory coexistence in 12% of our species pairs? There is no tendency for high *K* and low *K* species to coexist, for *K* is highly correlated with competitive ability and the coexisting pairs tend to have very similar competitive abilities, as reflected in the fact that the asterisks [= coexistence] in Fig. 2.2 are clustered about the main diagonal of the outcome matrix. However, when emergence curves (Fig. 2.1) are somewhat arbitrarily classified by shape as either peaked or plateau-shaped, a two-by-two contingency table shows that the coexistences are significantly more common between pairs differing in their emergence pattern. This is quite reasonable and consistent with ideas of production. Species with peaked emergence are better able to exploit the relatively new food in the system, while species with plateau emergence are also able to utilize the food in the older bottles. Thus, coexistence seems to be based on segregation along the niche axis of bottle age.

Assembly Rules

Table 2.5 summarizes the final frequencies (at week 35) of the 30 different trials in which 10 different species were initialized at 30 different sets of frequencies. Examination of Table 2.5 yields five obvious conclusions.

First, the 10-species system always relaxes to a much smaller system: 2 species in 21 trials, 3 species in 7 trials, 1 species in 2 trials, never more than 3 species.

Second, the 10 species differ consistently in how they fared under the 30 different initial conditions. Five species—Ser, WW, Imm, Eug, and Gib—always went extinct regardless of their initial density, which was as high as 91% in 1 trial for each species. Conversely, Paul and Merc almost always survived (in 27 and 25 trials, respectively). Intermediate performers are Neb, Zap, and Psd, which survived in 10, 2, and 1 trials, respectively.

Third, there is close concordance between competitive ability in the 10-species scramble and in the pairwise tournament. From Table 2.5, the descending sequence of survival frequency in the 10-species scramble is Paul > Merc > Neb > Zap > Psd > Ser, WW, Gib, Imm. From Fig. 2.2A and Table 2.2, the sequence in the pairwise tournament is nearly the same: Merc > Paul > Neb > Zap > Psd > Ser > WW > Eug > Gib > Imm.

Fourth, there are assembly rules. Among 10 species, there are $10 \times 9/2 = 45$ possible pairwise combinations, but only 3 of the 45 theoretically possible pairs (Merc-Paul, Merc-Neb, Paul-Neb) ever appeared as a surviving species combination. Of the 21 trials that relaxed to a pair of species, Merc-Paul was the surviving pair in 18 trials. Similarly, there are $10 \times 9 \times 8/6 = 120$ possible species trios, but only 3 of those theoretically possible trios ever appeared as a surviving species combination. Of the 7 trials that relaxed to a trio of species, Merc-Paul-Neb was the trio in 4 trials. Only 2 trials relaxed to a single species, and it was the same species in both cases: Merc. Thus, there are a few permitted or consistently favored species combinations and a much greater number of forbidden combinations. The same pattern was observed for birds of the Bismarck Archipelago and led Diamond (1975) to formulate the concept of assembly rules.

Fifth, these alternative outcomes can in some cases be understood in terms of the starting conditions. Psd, which survived in but one trial (and which probably would eventually have gone extinct there), survived in that trial in which it was initialized at the 91% frequency. It is probably the long survival rates of its adults that accounted for this single case of survival. Zap also survived at a relatively high frequency in the system in

Table 2.5 RESULTS OF COMPETITION AMONG 10 SPECIES INTRODUCED TOGETHER

	Merc	Paul	Neb	Psd	Zap
Corner initializations					
Merc	55	45	0	0	0
Ser	89	0	11	0	0
Zap	0	8	56	0	36
Paul	0	35	65	0	0
Neb	0	5	95	0	0
Ped	0	71	10	19	0
WW	36	64	0	0	0
Imm	66	34	0	0	0
Eug	16	84	0	0	0
Gib	83	17	0	0	0
Broken stick initializations					
(30,15,1,6,5,8,7,13,4,11)	31	69	0	0	0
(8,10,7,6,28,6,1,8,12,14)	74	26	0	0	0
(13,22,9,12,4,7,9,8,11,5)	32	68	0	0	0
(2,7,9,6,3,13,15,22,15,8)	35	65	0	0	0
(4,12,17,11,5,8,3,1,18,21)	55	39	6	0	0
(14,3,8,2,28,4,4,32,4,1)	54	24	22	0	0
(14,2,19,20,3,8,1,2,21,11)	51	49	0	0	0
(33,3,4,3,25,24,2,4,1,1)	100	0	0	0	0
(9,1,28,12,10,4,4,17,5,10)	68	32	0	0	0
(19,4,10,3,3,12,7,13,16,13)	44	44	12	0	0
(8,2,16,6,15,28,20,1,1,3)	24	76	0	0	0
(10,3,36,2,1,20,10,6,6,6)	72	4	24	0	0
(26,11,7,16,7,4,21,3,2,3)	80	20	0	0	0
(20,2,9,11,11,3,10,3,30,1)	0	71	17	0	12
(29,11,12,5,4,6,6,17,8,2)	39	61	0	0	0
(11,26,2,20,4,10,19,4,1,3)	24	76	0	0	0
(2,4,20,1,11,11,7,16,11,17)	21	79	0	0	0
(3,1,5,9,41,16,2,6,12,5)	24	76	0	0	0
(2,1,3,8,37,12,13,9,5,19)	23	77	0	0	0
(9,7,14,8,2,7,22,3,11,17)	100	0	0	0	0

Ten species were introduced simultaneously to bottles of thick food at 25°C. Thirty trials were run, each with a different set of initial frequencies. The trials of the first 10 rows were corner initializations; one species, named at the left of the row (in the first column), was started with 910 individuals, while the other nine species were started with 10 individuals each. The trials of the remaining 20 rows were started with randomly chosen broken stick initializations, given as a vector of percentages at the left of the row, with the 10 species listed in the sequence Merc, Paul, Neb, Psd, Zap, Ser, WW, Imm, Eug, Gib. The numbers in the body of the table are the percentages of the species named in the column heading in the system at the thirty-fifth week. Columns for species Ser, WW, Imm, Eug, and Gib are omitted because those species always went extinct.

which it was started at 91% frequency; it survived in one additional system at 12% frequency and occurred at below 1% in several other systems. Clearly, it is also being excluded; it is simply that exclusion takes longer than the 35 weeks of our experiment. Another instructive example is Merc, one of the two highest-ranking species, which survived in 25 out of 30 trials. In the nine trials where Merc began at 1% and some other species began at 91%, Merc went extinct when the species initialized at 91% was Paul, Neb, Zap, or Psd (the four other highest-ranking species), and Merc survived when the species initialized at 91% was Ser, WW, Imm, Eug, or Gib (the five lowest-ranking species).

CONNECTIONS BETWEEN THE LABORATORY AND THE WORLD

Let us now consider how our laboratory observations relate to the first four issues that we posed at the outset: transitivity of competition, mechanisms of competition, dimensionality, and assembly rules. We shall then conclude with a discussion of the fifth issue, prediction.

Transitivity of Competition

On the issue of transitivity, our laboratory work yielded clear, simple, and illuminating results.

In a homogeneous habitat in the absence of environmental variation, competitive fitness in our 28 *Drosophila* species is highly transitive (94–100% transitive).

At the same time, modest changes in temperature of only 6°C and changes in the thickness or composition of food caused large shifts in the competitive hierarchy (Table 2.4).

Had we been doing our work in the field, against a background of temperature and food type that we could not control, these shifts would have led us to record ''intransitivity,'' but as an artifact. For instance, with thick food but with temperature uncontrolled, Fig. 2.2 yields

Merc beating Paul (at 25°C)

Paul beating Psd (at 25°C)

but Psd beating Merc (at 19°C)

Similarly, at 25°C but with food uncontrolled, Fig. 2.2 yields

Paul beating Hyd (on thin food)

Hyd beating AnB (on thick food)

but AnB beating Paul (on augmented food)

Such shifts of rank due to uncontrolled variables in a spatially heterogeneous environment could lead to field observations of intransitivity. Buss (Chapter 31) documents intransitivity among competing species of sessile marine organisms. However, ours is not the only possible explanation for intransitivity; specific interference is another, and still another applies to the system discussed by Buss.

Mechanisms of Competition

We did not monitor disappearance of resources nor buildup of waste products and other possible allelochemicals with time. We did not systematically study the behavior of individuals. Thus, our evidence about mechanisms underlying the competitive hierarchy that we observed must be inferential. Some indications of the relative importance of exploitation competition, generalized interference, and specific interference are as follows.

Exploitation competition along a single niche dimension would tend to yield a transitive hierarchy of competitive ability. So would generalized interference. Specific interference, however, would tend to yield intransitivity: the outcome of each pairwise contest would depend on the details of the interference that species A happens to practise against species B, rather than on some underlying principle. Since our observed outcomes were so hierarchical, this inclines us to discount specific interference and to favor exploitation and/or generalized interference.

The relative importance of the latter two mechanisms might differ between the thick-food and thin-food systems. In thick-food experiments the food was never used up, hence calories could not have been limiting, but protein might have been.

There is anecdotal evidence for generalized interference under at least some circumstances. The larvae in some cases crowd the entire surface of the food, leaving insufficient space for larvae to breathe or rest. Since larvae do not burrow into the food, only the top layer of food is immediately accessible. After one week the food becomes liquified, and many larvae and eggs drown and are eaten by other larvae. The species Vir (and also Zap) ''destroys the habitat'' by making the medium soupy and then turning it into a hard, asphaltlike surface. Vir's large and active larval population makes it difficult for larvae of other species to survive. Thus, generalized interference directed at larvae rather than at adults might be important.

On the other hand, 85% of the variance in competitive rank at 25°C could be explained by single-species measures of adult and larval production and carrying capacities. This result is certainly most simply explained by exploitation competition. If competitive rank were instead due mainly to generalized interference such as habitat degradation, one might have expected superior competitors to have low carrying capacities—the opposite of reality.

Thus, there is suggestive evidence of both exploitation competition and of generalized interference, as Gause (1934, chap. V) also found for competition among protozoan species. Their relative importance requires more study.

Dimensionality

The pattern of pairwise competition that we observed under any given environmental setting was one-dimensional in the sense of Goodman (1979): species' competitive abilities can be arrayed in a single sequence. When competition involved sets of more than two species, only part of this hierarchical structure was preserved, and the outcome became contingent on starting densities. Even in the case of pairwise competitive rank at 25°C, variation in rank was distributed over at least four dimensions: the independent variables of RLH, K-15, MAXEMGLH, and LARVPROD. Thus, community ecologists using the term ''dimensionality'' need to define their intended use of it carefully.

How can one account for the shifts in competitive rank as the niche dimensions of temperature, food thickness, and food type change (Fig. 2.2)? Under niche theory, each species has a utilization function defined over the dimensions of food thickness, temperature, food type, etc. Normally, competition between species is based on a convolution of these utilizations integrated over the independent axes of the niche space. Such systems permit coexistence where there is sufficient niche separation (the limiting similarity problem). With our system, however, all of the species are evaluated at a single point in the niche space of temperature and food thickness. Thus, competition must be based on the relative rankings of the species at a *point* on the two-dimensional temperature and food-thickness continuum. It should be clear that the relative heights of these utilization surfaces for different species can differ from point to point in such a niche space, yielding changes in competitive rank consistent with these ideas. The cases in which we can most readily interpret these shifts are the improved competitive rank of tropical species at 25°C compared to 19°C and the improved competitive rank of species with early first emergence on thin food compared to thick food. In addition, our analysis of species coexistence indicates still another niche dimension, bottle age; it is along this dimension that coexisting species pairs segregate.

Assembly Rules

Our laboratory results for *Drosophila* species confirm Diamond's (1975) field results for Bismarck bird species: combinations of species drawn from a pool are statistically forbidden or permitted to coexist. The results of both our pairwise systems and our 10-species system fit this pattern. In addition, Diamond found that the presence of a third species could modify or override pairwise rules. This observation opens the door for multiple domains of attraction, since the fate of a species pair depends on the initial composition of the system. Our results also confirm this observation, in that outcome depends on initial frequency (Table 2.5). For instance, the otherwise strong Merc could not succeed at low initial densities against high densities of Paul, Neb, Zap, or Psd.

There are 2^{10} (= 1,024) different possible combinations of 10 species. Our study found 2 to 7 different combinations of species, depending on which states are considered to be transitional and which stable. That is, had the experiment been carried out longer than 35 weeks, it is likely that both Psd and Zap would have gone extinct. The three-species system Merc-Paul-Neb may be unstable and may decay into Merc, Merc-Paul, or Paul-Neb. It is questionable whether Merc-Paul is stable. Regardless, Merc and Paul-Neb are alternative configurations and constitute multiple domains of attraction.

A clue to why outcome depends on initial frequencies is provided by the fate of mighty Merc, highest ranking of the 10 species tested in Table 2.5. Merc survived in 25 out of 30 trials, but went extinct in 5 trials where Merc started at low density (~1%). In these 5 cases Merc yielded to Paul-Neb (plus in 3 cases Psd or Zap, which were probably in the process of disappearing). It appears that low frequency reduces the rate at which females encounter conspecific males, thereby delaying female insemination, shifting the emergence curve to a later time, and lowering the effective growth rate of the population. Thus, the per capita impact of competitors is not linearly related to density, as the Lotka-Volterra competition equations assume.

OUTLOOK FOR PREDICTIONS

Our goal was to explore the potential power of the community reconstitution approach by synthesizing complex laboratory systems of *Drosophila* flies. We used these systems to study problems of multispecies competition. Our experience yields one type of bad news and two types of good news.

The bad news is that it is difficult to understand the structure even of laboratory communities in which one creates and controls a simple homogenous environment, chooses species, and adds those species singly, pairwise, or in higher combinations at will. After eight years of work we still have not established the relative importance of various proximate mechanisms of competition. We do not have detailed interpretations for why competitive rank shifts with food type. Our understanding of what produces the observed assembly rules is rudimentary. If these tasks are difficult in the laboratory, think how much more difficult they will be in the field, where there is an uncontrolled and heterogeneous environment, dozens or hundreds of relevant but little known species, and no opportunity for studying those species in isolation or in pairs.

One type of good news is that it has proved feasible and rewarding to study a complex laboratory system in steps. Life history parameters of single species can be measured as a function of environmental temperature and food supply; pairwise competition can be reconstructed from those single-species parameters; and the outcome of competition within sets of 10 species is illuminated by the outcome of the pairwise contests. This approach tests whether we really have identified the significant components of a higher system, just as does the approach of a biochemist attempting to reconstitute the mitochondrial electron-transfer system from its components.

The other type of good news is that the laboratory system succeeded in capturing the essence of many phenomena important in field ecology. We were able to confirm unequivocally the existence of assembly rules, competitive exclusion, species coexistence by niche partitioning, competitive transitivity, and one mechanism of competitive intransitivity (i.e., environmental heterogeneity). We were able to predict competitive rank and to interpret some shifts in rank with temperature and food thickness. The richness or multidimensionality of single-species behavior observed in the laboratory accords with *Drosophila* lore (cf. Ayala's [1969] article on the variability of intrinsic growth rate in different environments).

Of the three traditions of experimental ecology—natural, field, and laboratory experiments—the laboratory tradition is the one currently being least exploited. We hope that we have demonstrated the potential value of community reconstitution studies pursued in the laboratory.

ACKNOWLEDGMENTS

We thank Frank Correll, Roger Langsford, Roger Stoklas, Joyce Meissinger, Will Thomas, Lisa Brooks, and Tom Philippi for help with the flies. We are especially grateful to Jared Diamond and Ted Case, who continually pressed us to connect our work to ideas generated by field ecologists. This work was supported by National Science Foundation Grants DEB 77-06060 and DEB 79-08085.

chapter 3

Experimental Community Ecology: The Desert Granivore System

James H. Brown, Diane W. Davidson, James C. Munger, and Richard S. Inouye

*I am tempted to give one more instance showing how plants and animals, remote in the scale of nature, are bound together by a web of complex relations. . . . I find from experiments that humble-bees are almost indispensable to the fertilisation of the heartsease (*Viola tricolor*), for other bees do not visit this flower. I have also found that the visits of bees are necessary for the fertilisation of some kinds of clover; . . . Hence we may infer as highly probable that, if the whole genus of humble-bees became extinct or very rare in England, the heartsease and red clover would become very rare, or wholly disappear. The number of humble-bees in any district depends in a great measure upon the number of field-mice, which destroy their combs and nests; and Col. Newman, who has long attended to the habits of humble-bees, believes that "more than two-thirds of them are thus destroyed all over England". Now the number of mice is largely dependent, as every one knows, on the number of cats; and Col. Newman says, "Near villages and small towns I have found the nests of humble-bees more numerous than elsewhere, which I attribute to the number of cats that destroy the mice." Hence it is quite credible that the presence of a feline animal in large numbers in a district might determine, through the intervention first of mice and then of bees, the frequency of certain flowers in that district!*

DARWIN, 1859

INTRODUCTION

Despite the recent emphasis on experimental approaches to ecology, there have been few long-term, intensive experimental studies of terrestrial communities. Such investigations are important because they can provide rigorous independent tests of the inferences obtained from the numerous comparative and observational studies that have produced most of the data and ideas on community structure and function. In addition, because experimental manipulations are perturbations of a kind and magnitude that are usually difficult or impossible to observe without human intervention, they may reveal important patterns and processes that have not been detected by other methods.

For more than a decade, we and our collaborators have been using controlled field experiments to analyze the interactions among desert seed-eating animals and between these granivores and their primary food resources, the seeds of desert annual plants. One purpose of the present paper is to summarize, synthesize, and discuss the current status of these continuing experiments. Many of the results have already been published or are now in press (Brown and Davidson 1977; Brown et al. 1979a, 1979b; Reichman 1979; Davidson et al. 1980, 1984, in press; Inouye 1980, 1981; Inouye et al. 1979; Munger and Brown 1982; Brown and Munger in press). We shall summarize and synthesize the results of these earlier studies in order to document the diverse effects of the different kinds of animals and plants on the organization of this important part of desert ecosystems.

The other goal of this paper is to consider the general implications of these results for contemporary ecological theory and practice. Although our results are generally consistent with both current community theory and with nonexperimental empirical studies of this and other systems, the kinds and magnitudes of the responses to our manipulations reveal a diversity of interactions and a complexity of community organization that is not easily characterized either theoretically or empirically. Long-term experimental studies provide a unique perspective on the organization of communities, because the sustained perturbations set in motion a complex sequence of dynamic behaviors as many different kinds of organisms are affected through both direct and indirect pathways. Perhaps most importantly, our experiments demonstrate several kinds of strong indirect interactions, in which species influence each other through intermediary species. The theoretical importance of such indirect interactions for community organization has been emphasized (e.g., Chapters 20, 26, and 32; Levins 1974, 1975; Levine 1976; Holt 1977; Lawlor 1979; Vandermeer 1980; Patten and Auble 1981; Schaffer 1981; Patten 1982; Bender et al. 1984), but there have been few rigorous field studies to show how these indirect pathways actually operate in natural ecosystems.

THE SYSTEM

The Organisms

The system of seeds and granivores is an important component of desert ecosystems. Primary production in deserts is limited by the availability of water, and the ephemeral or annual plants account for a large fraction of the productivity, because they are able to complete the entire vegetative part of their life cycles during the brief, unpredictable periods when sufficient soil moisture is available following precipitation. These plants produce large crops of seeds, some of which survive buried in the soil for the long intervals between rains. In most desert habitats annuals are taxonomically diverse and account for 85 to 95% of the total seed production. The various species differ in the timing of their life cycles and in the size, shape, and chemical composition of their seeds.

These seeds comprise the primary food resources of a major, taxonomically diverse group of consumers, the granivores. In the Chihuahuan and Sonoran deserts of southwestern North America where we have performed our experiments, three classes of animals have representatives that are specialized to varying degrees to feed on dry seeds.

The first of these classes is the rodents, which include highly granivorous kangaroo rats (*Dipodomys*) and pocket mice (*Perognathus*), as well as more omnivorous deer mice (*Peromyscus*) and harvest mice (*Reithrodontomys*). These small mammals are nocturnal and resident throughout the year (although most pocket mice hibernate during the coldest months). Most species collect large quantities of seeds when they are available and store them underground for use in times of food shortage.

The second major class of desert granivores is the harvester ants, which include the specialized seed-eaters in the genera *Pogonomyrmex, Veromessor,* and *Pheidole* as well as more omnivorous representatives of the genera *Novomessor* and *Solenopsis*. These ants live in colonies that may survive for many years and contain as many as tens to thousands of foraging workers. During limited periods of warm temperature, high humidity, and food availability these ants collect seeds and store them in granaries within their underground galleries.

The third important class of desert granivores is birds, in particular the sparrows that invade the desert in flocks in winter, and doves and quail that may be present, either singly or in flocks, throughout the year. These birds differ conspicuously from both rodents and ants in that they do not store seeds; instead they respond to variation in local food availability by traveling over large distances to exploit abundant seed crops.

This system of seeds and seed-eaters has several advantages for long-term experimental studies. First, the system is a fairly discrete part of relatively simple desert ecosystems. Thus, we can focus our investigations on one closely interacting group of organisms. Second, the system is an important part of the entire desert ecosystem. Annual plants and granivores account for a substantial proportion of the species diversity and biomass of primary producers and consumers, respectively, in deserts. Third, there is sufficient taxonomic and ecological diversity among both the seed-eaters and their food plants to make possible many kinds of interesting interactions. The diversity is sufficiently low, however, that it is possible to analyze experimentally the roles of either individual species or entire groups of species. Finally and perhaps most importantly, the system lends itself, perhaps more readily than most other terrestrial systems, to carefully controlled experimental manipulation. It is both possible and practical to set up numerous replicated plots of biologically realistic size, to remove selected kinds of granivores and plants with almost surgical precision, to manipulate the availability of food for granivores, to census most important species in the community with sufficient precision to quantify their responses to the manipulations, and to maintain the experiments for sufficiently long periods to record the dynamics of the diverse responses. Although major artificial manipulations can be performed on a sufficiently large scale to produce informative results, the effect on the local biota is sufficiently small so as to be politically feasible and morally justifiable.

The Experiments

The basic procedure is straightforward: Find an area of appropriate desert habitat that is as homogeneous as possible, establish replicated plots of adequate size, assign treatments at random, perform manipulations, and census the rodents, ants, birds, and plants at regular intervals to quantify their responses.

The present paper describes the results of two sets of experiments. The first was begun in the Sonoran Desert northwest of Tucson, Arizona, in 1973 and continued until 1977. There were eight circular plots, each 0.10 ha in area. Two replicates of each of the following manipulations were performed to test for the effects of competition between rodents and ants and for the influence of predation by these two classes of granivores on the annual plants: (1) rodents excluded by fencing, (2) ants removed by poisoning, (3) both rodents and ants removed by both fencing and poisoning, and (4) control (both rodents and ants present and neither fenced nor poisoned). It was usually possible to maintain virtually complete exclusion of the taxon designated for removal (see Fig. 3.1, for example). Standardized methods were used to census the rodents, ants, seeds in the soil, and annual plants. Additional, smaller-scale experiments were conducted on these plots in 1977 in order to assess the role of competition within and between species of annual plants as well as the interaction between such competition and seed predation by the granivores. Additional details of methodology are given in Brown et al. (1979a), Inouye et al. (1980), and Davidson et al. (1984).

A second, much more elaborate set of experiments was begun in 1977 and is still continuing. On a 20-ha study area in the Chihuahuan Desert of extreme southeastern Arizona, we set up 24 plots, each 0.25 ha in area. All of these are fenced similarly, except that some have gates (holes) of particular sizes in the fences to allow free passage of selected rodent species. The complex, partial factorial design is summarized in Table 3.1. Basically, there are three main classes of manipulations: (1) exclusion of some or all rodent species by means of different-sized gates in the fences, (2) exclusion of some or all ant

Table 3.1 OUTLINE OF THE 12 EXPERIMENTAL TREATMENTS, INCLUDING CONTROL, IN THE CHIHUAHUAN DESERT

Control		Seed addition		Rodent removal		Ant removal	
Plots 11, 14:	unmanipulated	Plots 6, 13:	large seeds, constant rate	Plots 5, 24:	*Dipodomys spectabilis*	Plots 8, 12:	*Pogonomyrmex rugosus*
		Plots 2, 22:	small seeds, constant rate	Plots 15, 21:	all *Dipodomys* species	Plots 3, 19:	*Pogonomyrmex rugosus* and all *Dipodomys* species
		Plots 9, 20:	mixed sizes, constant rate	Plots 3, 19:	all *Dipodomys* species and *Pogonomyrmex rugosus*	Plots 4, 17:	all seed-eating ants
		Plots 1; 18:	mixed sizes, temporal pulse	Plots 7, 16:	all seed-eating rodents	Plots 10, 23:	all seed-eating ants and rodents
				Plots 10, 23:	all seed-eating rodents and ants		

Note that some of the rodent and ant removal experiments have a factorial design, and duplicate treatments are listed under both headings.

species by means of poisoning the appropriate colonies, and (3) addition of 96 kg of millet seed per year in different-sized particles and in different temporal patterns. We use standardized techniques to census rodents, ants, birds, and annual plants on the plots at regular intervals. Additional methodological details are given by Davidson et al. (in press) and Brown and Munger (in press). In addition, supplemental experiments in which birds or certain plants are selectively excluded have been performed (Inouye 1980, 1981; Davidson et al. 1984) or are in progress.

The Questions

Collectively, these experiments are designed to assess the roles of interspecific interactions among granivores and between granivores and plants. We are especially concerned with the effects of competition within and among the three classes of seed-eaters, predation by granivores on the annual plants, competition among these plants, and indirect interactions that are the result of a series of two or more direct interactions involving intermediary species.

Even this relatively simple system is too complex and we already have too much data to attempt to summarize our current understanding in the limited space available here. The present paper will focus on the interactions involving only the rodents, but these are sufficient to document a rich variety of direct and indirect effects, some of which profoundly influence the structure and function of the entire desert ecosystem. We shall concentrate on the answers that the experiments can provide to the following questions.

1. To what extent do different species of granivorous rodents compete with each other for limited food resources?
2. To what extent do these rodents also compete with the other major classes of seed-eating animals, especially ants?
3. What is the impact of these rodents as seed predators upon the desert plants, and how does this predation interact with competition among the plant species to affect the composition of the flora?
4. What are some of the important indirect effects of rodents on other organisms in the ecosystem, and through what pathways are these interactions effected?

Thus the approach will be to begin by considering direct competitive interactions between closely related, ecologically similar rodent species and then gradually to expand the perspective to include other kinds of organisms and interactions.

RESULTS OF THE EXPERIMENTS

Competition Among Rodent Species

The experiments at the Chihuahuan Desert site include treatments designed to test the hypothesis that seed-eating rodent species compete for limited food resources (see Munger and Brown 1981, Brown and Munger in press, for more details on methods and results). The most direct tests are provided by semipermeable exclosure experiments that use different-sized gates in the fences to exclude selected species. Small gates allow only species smaller than some threshold body size to enter experimental plots, whereas larger gates give all species free access to control plots. In the absence of complicating indirect effects, the competition hypothesis predicts that the smaller species should increase and maintain higher densities on the experimental plots from which larger rodents have been excluded than on control plots where larger species are present.

We have conducted two tests of this hypothesis. The first documents the response of five species of small granivorous rodents to the exclusion of three larger species of kangaroo rats (*Dipodomys*) on four experimental and four control plots (Fig. 3.1). After a lag of approximately nine months following initiation of the manipulation in October 1977, four of the five species of small rodents (*Perognathus flavus, Peromyscus maniculatus, Pm. eremicus,* and *Reithrodontomys megalotis*) increased dramatically on the plots where all *Dipodomys* species had been removed. The fifth species (*Pg. penicillatus*) has shown a tendency to increase as well, but the

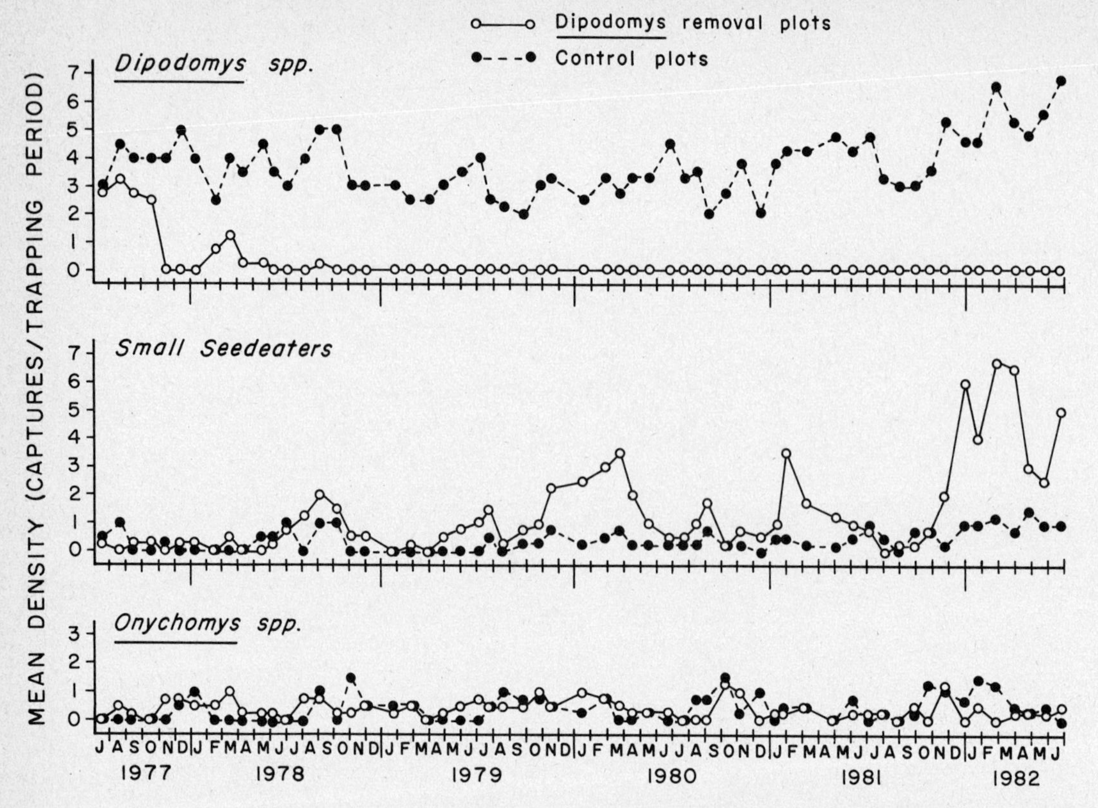

Fig. 3.1 Changes in the densities of three groups of rodents at the Chihuahuan Desert site on four experimental plots where all three *Dipodomys* species were removed beginning in October 1977 (solid lines) compared to the densities on four control plots (dashed lines). (Above) Effectiveness of removal of *Dipodomys*. (Middle) Compensatory increase in total densities of five species of small granivorous rodents. (Below) Lack of effect of *Dipodomys* on the combined densities of two species of insectivorous *Onychomys* species. (From Brown and Munger in press.)

response is not statistically significant. Collectively, the increase of the five small species is highly significant; their combined densities on the experimental plots average 2.2 times higher than on the control plots where *Dipodomys* are present. This result strongly supports the competition hypothesis. That this reflects the effects of competition for shared food resources is additionally indicated by the fact that two species of insectivorous rodents (grasshopper mice, *Onychomys* spp.) that could travel through the small gates showed absolutely no differences between experimental and control plots.

The second semipermeable exclosure experiment involves the removal of only the largest of the granivorous rodent species (*Dipodomys spectabilis,* body weight 120 g). This treatment, begun only in 1980 and consisting of just two replicates, has so far produced one striking result. In the absence of *D. spectabilis,* five of the seven granivorous rodent species shift their foraging behavior to use microhabitats different from the ones used on the control and most of the other experimental plots (Bowers et al. in preparation). *D. spectabilis* is aggressively dominant over smaller species. The patterns of habitat shifts suggest that this aggression is directed primarily toward *D. merriami* (body weight 45 g; see Frye 1983), the most abundant species and one of the next two largest after *D. spectabilis.*

D. spectabilis could affect the other rodent species either directly, by interfering with their foraging, or indirectly, through its effect on *D. merriami*. Surprisingly, despite the pronounced changes in microhabitat use, there are as yet no significant differences in population density of any of these rodents between *D. spectabilis* removal and control plots. Of course, the short duration of this experiment and the small number of replicates would contribute to the difficulty of detecting statistical differences. *D. merriami* and *D. ordii* have increased in the absence of *D. spectabilis,* but these two species also have increased and *D. spectabilis* has decreased on the control plots (Fig. 3.2).

These reciprocal density shifts between *D. spectabilis,* on the one hand, and *D. merriami* and *D. ordii,* on the other, provide additional evidence of competition between these species, especially when their responses to other treatments are also considered (Fig. 3.2). There is a strong tendency for *D. spectabilis* to increase and *D. merriami* and *D. ordii* to decrease on plots to which supplemental millet seeds are added. The reverse trend occurs on the other plots, including the controls, perhaps because the fences interfere with the foraging of *D. spectabilis*. These reciprocal density shifts induced by experimental perturbations provide evidence not only that the three kangaroo rat species compete for seeds, but also that the outcome of this competition is asymmetrical, with the interactions dominated by *D. spectabilis* whenever food resources are sufficiently abundant for it to maintain high densities.

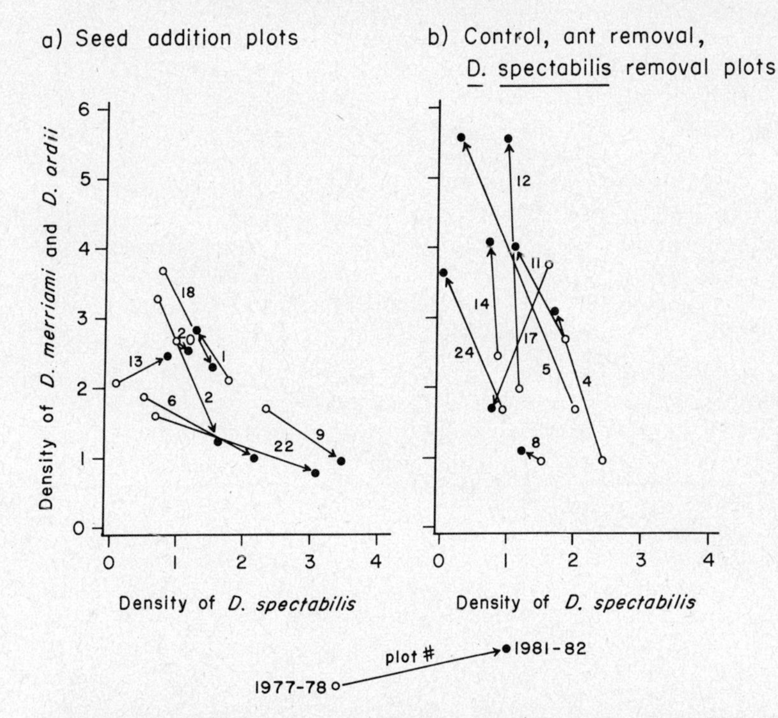

Fig. 3.2 Changes in the total mean density of *Dipodomys merriami* and *D. ordii* relative to changes in the density of *D. spectabilis* in response to various experimental manipulations at the Chihuahuan Desert site. Each line represents changes on one plot between the first year of the study (1977–1978) and the last two years (1980–1982). Note the pronounced reciprocal density shifts, with *D. spectabilis* increasing from 1977–1978 to 1980–1982 at the expense of its smaller congeners in response to seed addition and the reverse trend on other plots. The reciprocal pattern is highly significant (Fisher's exact test, $p = 0.0055$). (Data from Brown and Munger in press.)

This fits well with the biogeographical pattern that *D. spectabilis* is found only in the most productive desert habitats, and it usually occurs in the absence of other *Dipodomys* species in even more productive arid grasslands.

Three aspects of the response of desert rodents to experimentally induced changes in food supply and density of other rodent species warrant special comment. First, all of the changes in population density occurred long after the manipulations were initiated. As discussed in more detail in Brown and Munger (in press), the most likely explanation for these long time lags seems to be either the failure of the rodents to perceive a significant change in the availability of resources (perhaps because a new seed crop is required) or their inability to respond quickly to a detected change (perhaps because of the seasonality of reproduction and dispersal). Second, the magnitudes of the responses to our perturbations, while highly significant statistically, are nevertheless much less than would be expected if those food resources made available by our manipulations were completely utilized by those rodents that potentially had access to them. In fact, although population densities of particular species were greatly affected by our treatments, the almost negligible compensation in consuming biomass (Table 3.2) indicates that most of the seeds made available were not consumed by rodents. Explanations for this phenomenon (see also Brown and Munger in press) include the possibilities that intraspecific interactions, predators, or constraints on foraging behavior or habitat use precluded a more complete response by the rodents and that other classes of granivores consumed a large proportion of the available seeds (see below). Third, the interactions among the rodent species appear to be highly asymmetrical, because aggressive domination of smaller species by larger ones is an important mechanism of competition. We have not tested directly for these asymmetries by removing small species and measuring the response of large ones. We predict that if we did these experiments, however, the magnitude of density compensation would often be statistically undetectable and would always be less than we observe in the reciprocal experiments, in which the larger species are removed.

The results of our experiments, then, strongly support the hypothesis that competition for limited food resources plays a major role in determining the absolute and relative abundances of the rodent species that comprise this community. However, the interactions among the rodents are not so simple as we had naively assumed when we began the manipulations. Indeed, the results reveal such interesting complications as long

Table 3.2 ENERGETIC COMPENSATION (MEASURED IN UNITS OF CONSUMING BIOMASS PER 0.25-HA PLOT) BY DESERT RODENTS TO SUPPLEMENTAL SEEDS AND TO REMOVAL OF SELECTED RODENT SPECIES

	Experimental treatment		
	Addition of metabolizable millet	**Removal of *D. spectabilis***	**Removal of all (3) *Dipodomys* species**
Energy made available (KJ/day)	3060	201.4	439.7
Energetic response (KJ/day)	91	49.4	33.2
Response by what species	All 8 species of granivorous rodents	7 species of smaller granivorous rodents	5 species of small granivorous rodents
Percent compensation	2.9	33.8	9.5

Note that rodent compensation for the additional food made available by either adding seed or removing other rodent species was always low: never more than 33%.

Percent compensation is calculated as: [(1977–1978 consuming biomass minus 1978–1982 consuming biomass for the average of the removal plots) minus (1977–1978 consuming biomass minus 1978–1982 consuming biomass for the average of the control plots)] divided by metabolized energy of added seeds or consuming biomass of the rodents removed.

From Brown and Munger in press.

Table 3.3 SUMMARY OF THE RESULTS OF EXPERIMENTS AT THE SONORAN DESERT SITE IN WHICH ANTS OR RODENTS WERE ELIMINATED FROM PLOTS AND THE UNMANIPULATED TAXON WAS REPEATEDLY CENSUSED

	Rodents removed	Ants removed	Control	Percent increase relative to control	Fraction of comparisons, experimental > control
Ant colonies	543	—	318	70.8	9/10
Rodent individuals	—	151	126	19.8	17/27 (5 equal)
Rodent biomass (kg)	—	5.41	4.21	28.5	17/27 (3 equal)

Values in the first three columns are totals of all censuses. Ants were censused 5 times and there were 2 replicates of each treatment for a total of 10 comparisons. There were 27 comparisons for rodents, 14 censuses of the first replicate (established in August 1973) and 13 censuses of second set (established in December 1973).

Rodent biomass is based on average body weights on the study area of 41.1 g for *Dipodomys merriami*, 28.1 g for *Perognathus baileyi*, 16.9 g for *P. penicillatus*, and 11.4 g for *P. amplus*.

From Brown and Davidson 1977.

time delays, low biomass compensation, and asymmetrical interference interactions.

Competition Between Rodents and Other Classes of Granivores

The first experiments, begun in the Sonoran Desert site in 1973, were designed primarily to test the hypothesis that rodents and ants compete for limited seed supplies. The simple competition hypothesis predicts that each of these taxa should increase in overall population density in response to experimental exclusion of the other. The results support this prediction (see Brown and Davidson 1977, Brown et al. 1979a, for details). Compared to control plots, numbers of ant colonies almost doubled on plots where rodents had been removed, and rodent biomass and censused numbers of individual rodents increased respectively by about 29% and 20% where ants had been removed (Table 3.3).

The second set of experiments was designed in part to test for the repeatability of these reciprocal density changes at the Chihuahuan Desert site. The results, however, are not so distinct as those of the earlier experiment at the Sonoran Desert site. Only *Pheidole xerophila,* one of the smallest of the 10 granivorous ant species, increased in response to removal of rodents (see Davidson et al. in press, for details). Censuses of foraging workers of this species over three years, 1980 to 1982, documented a consistent increase, with densities attaining levels almost 10 times higher on rodent exclusion plots than on controls in the last year. The situation is further complicated, however, because *Pogonomyrmex desertorum,* a somewhat larger species, declined slightly on the rodent removal plots. Thus, it is questionable whether there was any significant compensation in total ant biomass for the missing rodents, despite the dramatic increase in *Pheidole*. We can detect no evidence of significant increases in rodent populations in response to exclusion of ants (Brown and Munger in press).

Clearly, either the results for the Sonoran Desert site are in error, or the competitive relationships between rodents and ants are quite different at the Chihuahuan Desert site. We favor the latter explanation for three reasons. First, Bryant et al. (1976) excluded rodents and ants in habitat similar in vegetation and productivity to the Sonoran Desert site (even though it was in the Chihuahuan Desert of New Mexico) and obtained similar results; the same ant species showed quantitatively similar increases in colony densities in response to removal of rodents. Second, overlap in the diets of the rodents and ants is much greater at the Sonoran than at the Chihuahuan site, indicating that the potential competition for seeds is much greater at the site where the largest and most consistent density compensation was observed (Davidson and Cole, unpublished data). Third, the Sonoran and Chihuahuan sites differ considerably in climate and productivity (especially in the seasonality of seed production in relation to the activity of the different classes of granivores), as well as in the composition of the rodent, ant, and plant species. Since the relationship between rodents and ants apparently is not just simple, direct competition, but is complicated by interactions between these granivores and other organisms (see below), it is not unreasonable to expect that the intensity of competition (at least as revealed by this kind of simple exclusion experiments) might differ between the two sites. A result common to the experiments at both the Sonoran and Chihuahuan Desert sites is that rodents had substantially greater effects on ants than the converse. This is not surprising, given the apparently greater ability of rodents to find and collect seeds (Brown et al. 1975) and the much greater effect of climate on ant than on rodent activity (Davidson et al. in press).

We also have some evidence that rodents compete for seeds with birds. Avian foraging at the Sonoran Desert site was much greater on plots where both rodents and ants had been excluded and large quantities of seeds had accumulated, than on plots where either rodents or ants or both were present and standing crops of seeds were much lower (Brown et al. 1979a). At the Chihuahuan Desert site avian foraging for seeds is significantly greater on plots where supplemental millet seeds have been added than on control plots. Since rodents also increase their foraging in response to seed addition, this suggests that rodents and birds compete for the supplemental millet and perhaps for native seeds as well. Unfortunately, we did not begin to census avian foraging intensively until 1982, about five years after the treatments were initiated. In data collected since then, it appears that birds actually forage less on plots where rodents have been removed than on control plots where rodents are present. We are not yet certain how to account for this result, but it may well represent the outcome of a long-term indirect mutualistic interaction mediated through the direct effects of rodents and birds as selective predators on different but competing species of annual plants. Such an interpretation does not by any means deny the possibility of substantial short-term competition between rodents and birds.

Thus, our experiments indicate that rodents compete for seeds, not only with other closely related rodent species, but also with other distantly related taxa of granivores such as ants and birds. These interactions among distantly related taxa must be considered when attempting to interpret the results of field experiments designed to test for competition between closely related species. When the effects of other organisms that might respond to the same manipulations are not controlled for, they potentially can have a profound influence on the results. For example, the fact that birds consume a significant proportion of the seeds presumably accounts at least in part for the failure of rodents to compensate completely in consuming biomass for the seeds made available by either removing certain rodent species or by adding supplemental seeds.

Predation by Rodents on Annual Plants

Granivores are predators. They kill and eat seeds, which are immature plants. This predation might be hypothesized to have two effects on the prey. First, it should tend to reduce the overall density, biomass, and productivity of plants. Second, to the extent that predation is selective on certain plant species, it should also affect the species composition and pattern of dominance in the plant community. Both of these hypotheses can be tested by comparing the abundances of the various plant species on control plots, where rodents are present, with those on experimental plots where some or all rodent species have been removed. We have such data for both the Sonoran and Chihuahuan sites. Here we shall focus primarily on the effects on annual plants of removing all rodents at the Sonoran Desert site.

Our experiments clearly demonstrate that the influence of rodents as predators on plants is at least as important as their effect as competitors on other seed-eating animals. Fig. 3.3 shows that the density and biomass of both seeds and adults of annual plants increase significantly when either rodents or ants are excluded and especially when both classes of granivores are removed (for additional details see Reichman 1979, Inouye et al. 1980). Shortly after the start of the manipulations there were no significant differences be-

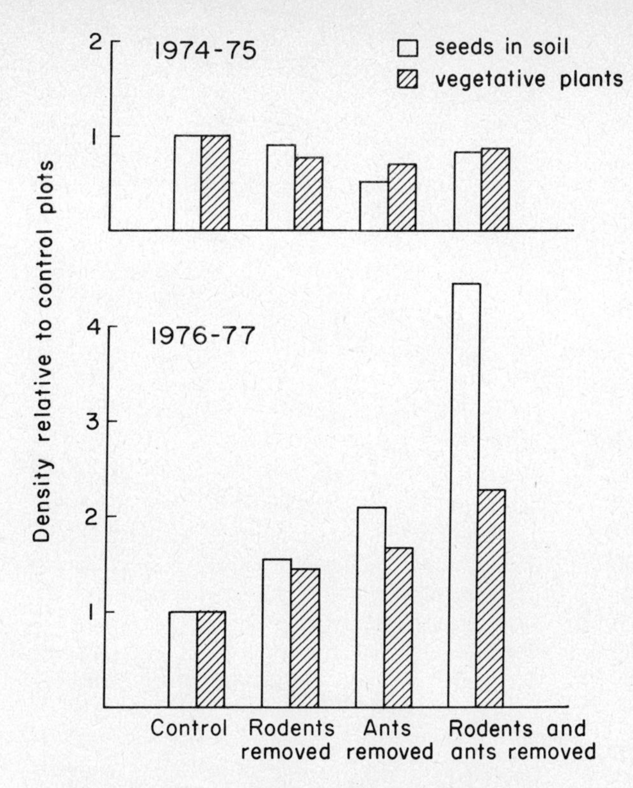

Fig. 3.3 Effects of experimental exclusion of rodents and ants on the densities of seeds in the soil and of vegetative annual plants at the Sonoran Desert site. In 1974–1975, approximately one year after initiation of the granivore removals, there were no significant differences among any of the treatments. Two years later there were significantly more seeds and plants on plots where rodents or ants or both had been removed.

tween experimental and control plots (upper half of Fig. 3.3), but after three years there were substantially more seeds and mature vegetative plants on the experimental plots where one or both taxa of granivores had been removed (lower half of Fig. 3.3). The patterns for seeds and mature plants are qualitatively similar; the apparently greater effect of granivores on the seeds than on vegetative plants can probably be attributed to density-dependent inhibition of germination (Inouye 1981).

Effects of rodent exclusion on composition of the annual flora are equally great. A consistent pattern is for those plant species with relatively large seeds (seed mass > 1 mg) to increase dramatically in density to dominate the annual plant community on plots where rodents have been removed (Fig. 3.4). Thus, individual large-seeded species were 1.5 to 8.2 times more dense on rodent removal plots than on controls. Two of these large-seeded species, *Erodium cicutarium* and *E. texanum,* accounted for over 60% of the annual plant biomass on rodent exclusion plots

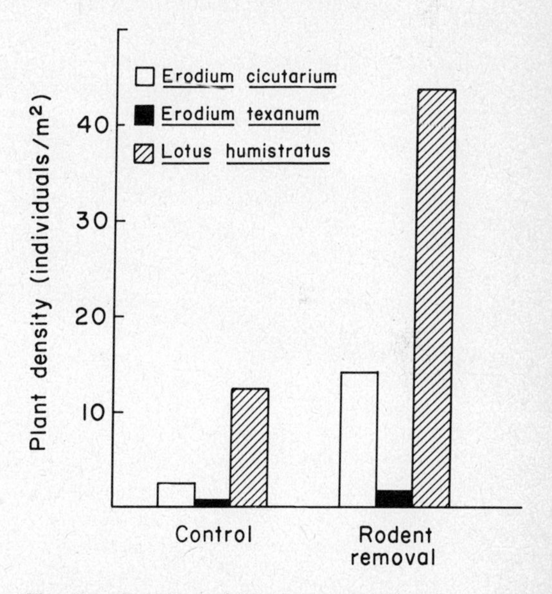

Fig. 3.4 Densities of the three most abundant species of large-seeded annual plants on rodent removal and control plots at the Sonoran Desert site.

compared to less than 30% on plots where rodents were present (see Inouye et al. 1980 for additional details).

The explanation for these results appears to be straightforward. Because of their large body size and the energetic costs of temperature regulation and year-round activity, rodents are constrained to forage selectively for large seeds. This is consistent with other data on the sizes of seeds in rodent diets (e.g. Brown and Davidson 1977, Brown et al. 1979b). The effects of this selectivity on the plant community are profound. Not only do rodents suppress populations of their preferred large-seeded prey, but this suppression is sufficient to prevent these large-seeded plants from completely dominating the annual plant community. Furthermore, this phenomenon appears to be very general, because we can document it at other sites where the habitat and flora are quite different. Thus, large-seeded species dominated the annual plant community in an 18-year-old rodent exclosure constructed by R. M. Turner in the Sonoran Desert east of Tucson, Arizona, but the dominant species in this rocky hillside habitat were completely different from those at our Sonoran Desert site a few kilometers away (Table 3.4). Similarly, preliminary analyses of the plant responses to the manipulations at our Chihuahuan Desert site indicate that large-seeded species are still changing in density to comprise an increasingly large share of the annual plant biomass on plots from which rodents have been excluded compared to plots where rodents are present (Table 3.4; see also Davidson et al. in press). Thus, all of these experiments indicate that rodent predation has a major consistent effect in preventing the domination of desert annual plant communities by a few large-seeded species.

Indirect Effects of Rodents on Other Species

From the magnitudes of the direct effects of rodents, especially as predators on selected plants, it is apparent that they can potentially have important indirect effects on other species that are mediated through these direct interactions. We have tested for only a few of the many possible kinds of such indirect effects, but we can show that some of these can easily be documented. Furthermore, these (and very likely others that we have not yet investigated) probably are extremely important in the organization of desert ecosystems. Here we shall focus on three of the

Table 3.4 RELATIVE INCREASE OF LARGE-SEEDED PLANTS ON PLOTS WHERE RODENTS HAVE BEEN EXCLUDED AT THREE DIFFERENT SITES IN SOUTHEASTERN ARIZONA

Site and species	Family	Seed mass (mg)	Relative increase (density on rodent removal/ density on control)
Sonoran Desert			
Erodium cicutarium	Geraniaceae	1.62	9.2**
E. texanum	Geraniaceae	1.60	1.5*
Lotus humistratus	Fabaceae	1.50	3.57**
Saguaro National Monument			
Astragalus nuttallianus	Fabaceae	1.36	157.3***
Lupinus sparsiflorus	Fabaceae	1.51	8.2***
Chihuahuan Desert			
Erodium cicutarium	Geraniaceae	1.62	7,936.4*
E. texanum	Geraniaceae	1.60	1,000***
Lesquerella gordonii	Brassicaceae	0.94	9.7*
Astragalus nuttallianus	Fabaceae	1.36	4.3***

Asterisks denote significance levels: * = $p < 0.05$; ** = $p < 0.01$; *** = $p < 0.005$.

Note that different plant species and families increased at different sites, but that all had relatively large seeds (mass > 1 mg).

Data are from Inouye et al. (1980), Sonoran Desert; Kurzius and Brown (unpublished), Saguaro National Monument; and Samson, Thompson, Davidson, Kurzius, and Brown (unpublished), Chihuahuan Desert.

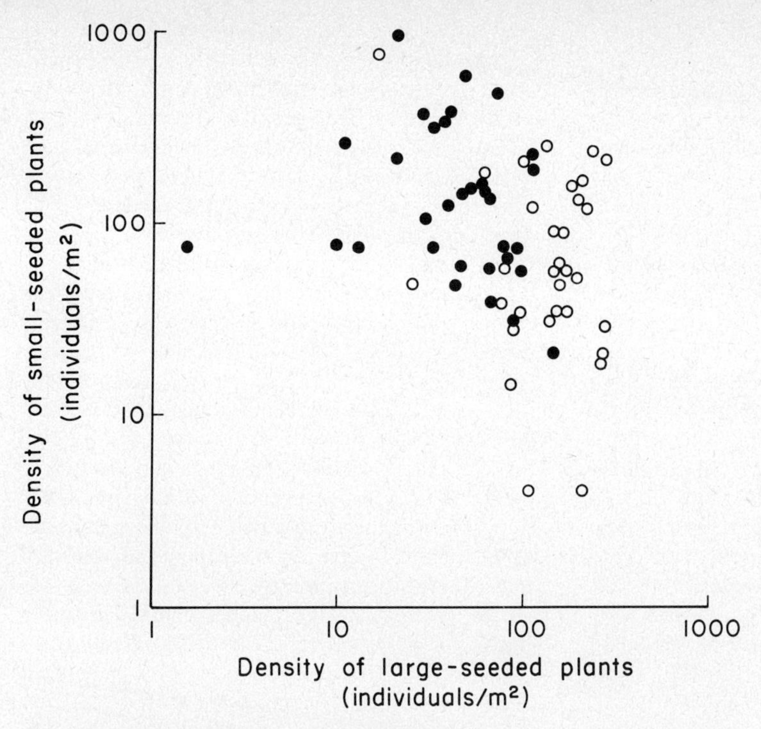

Fig. 3.5 Reciprocal density relationships between large- and small-seeded annual plants on rodent removal (unshaded circles) and control plots (shaded circles) at the Sonoran Desert site. When rodents were removed, large-seeded plants increased greatly in density; this was accompanied by a decrease in the density of small-seeded species as a result of competition.

best documented indirect effects of rodents, all of which are mediated through their direct effects as selective predators on large-seeded annual plants.

The first indirect effect concerns the interaction between rodents and small-seeded plants. Because rodent predation suppresses populations of large-seeded plants, which increase to dominate the annual community when rodents are excluded, we hypothesize that rodents have an indirect beneficial effect on small-seeded plants, provided that small- and large-seeded species compete for limited resources. Experimental evidence strongly supports this hypothesis. The assumption that small- and large-seeded plants compete (in this case probably primarily for limited water, but perhaps also for nutrients) is confirmed by selective thinning experiments, which demonstrate a highly asymmetrical interaction: large-seeded species have a substantial negative effect on small-seeded ones, but any reciprocal effect is too small to detect (Davidson et al. 1984; see also Inouye et al. 1980). Presumably this asymmetry is owing to the fact that the large-seeded species, by virtue of their large seed reserves, have larger seedlings and never relinquish the competitive superiority conferred by this initial advantage.

Thus, when the effects of rodents are also considered, the result is an important interaction between competition and predation that affects the organization of the entire plant community and can clearly be seen in the results of our experiments. On the Sonoran Desert site on plots where rodents were excluded, large-seeded plants increased to dominate the community, and this was accompanied by a substantial decline in the density of small-seeded species, even though there was also a significant increase in total annual plant biomass (Fig. 3.5). On the Chihuahuan Desert site plants are still changing in response to the granivore removal experiments begun in 1977. Preliminary results suggest that the increase of large-seeded winter annuals on rodent removal plots has been accompanied by a decrease in the density of seedlings of the abundant small-seeded species, *Eriogonum abertianum,* but only on plots where ants have not also been removed (i.e., there is a significant rodent-ant interaction effect). This unusual species germinates in winter, but survives as a vegetative rosette until it is able to complete its life cycle with the moisture made available by the summer rains. Because *E. abertianum* begins the summer as an established vegetative plant, rather than as a seedling, it capitalizes on this initial size advantage to grow rapidly and competitively dominate the summer annual community. Its seeds are avidly consumed by ants, and it increases substantially on plots where ants have been removed. Thus rodents, even though their direct effects are primarily on the winter annuals, interact with ants to have an important influence on the summer annuals by limiting the density of the dominant species through a combination of indirect and direct pathways (see Davidson et al. in press).

We can also document an important indirect effect of rodents on the fungus *Synchytrium pallatum,* which is a specific pathogen that infects vegetative parts of the large-seeded plant *Erodium cicutarium* (see Inouye 1981 for details). Because its ability to infect host plants is highly density-dependent, *S. pallatum* in effect competes with seed-eating rodents, even though these competitors attack different life history stages of the plant. This is apparent from the results of our experiments. Exclusion of rodents resulted in a dramatic increase in the densities of both the large-seeded host plant and its fungal pathogen (Fig. 3.6). Although we have no experimental evidence to document it, the reciprocal effect of the fungus on the rodents almost certainly is significant. Because *E. cicutarium* is one of the dominant large-seeded annuals in many desert habitats and in some years infection by

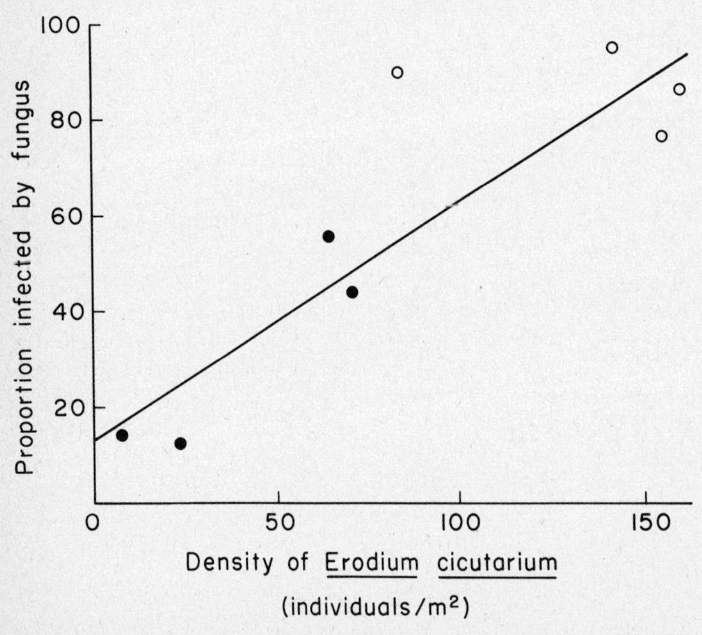

Fig. 3.6 Proportion of individual plants of *Erodium cicutarium* infected by the fungus *Synchytrium pallatum* as a function of both host plant density and rodent predation at the Sonoran Desert site. Note that on plots where seed-eating rodents were removed (unshaded circles), *E. cicutarium* attained a higher density of individuals which then suffered a higher incidence of fungal infection than plants on control plots (shaded circles). (Data from Inouye 1981.)

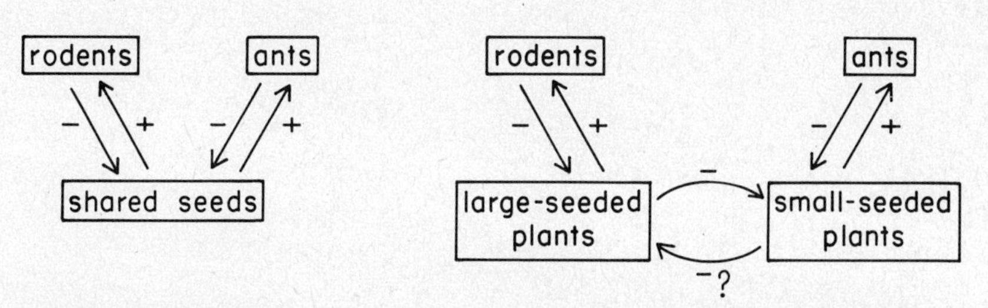

Fig. 3.7 Resource-mediated interactions between granivorous rodents and ants. In the short term the two taxa compete if they overlap in their feeding on limited seeds, but in the longer term the two taxa can have indirect mutualistic effects on each other if they feed differentially on different plant species that also compete with each other.

S. pallatum may cause sufficient mortality to drastically reduce seed production, the fungus probably has a substantial competitive effect on the rodents.

The third documented indirect effect of rodents is on ants. The short-term interaction between granivorous rodents and ants is, if anything, competitive. Presumably this competition results from overlap between the two taxa in the seed resources that they exploit. If, however, the overlap in the diets is not complete, then over the long term the interaction could actually be mutualistic because of the importance of indirect pathways. For such indirect mutualism to occur, it is necessary for the rodents and ants to prey selectively on seeds of different plant species, and for these plants to compete with each other for resources during the vegetative phase of their life cycle (Fig. 3.7). Our data not only show that these conditions are met, but also provide evidence of the expected long-term, indirect, beneficial effect of rodents on ants. Rodents and ants differ significantly in diet, with rodents specializing on larger seeds than most harvester ant species, especially the numerically dominant but tiny ants of the genus *Pheidole* (Brown and Davidson 1977, Brown et al. 1979a). Large- and small-seeded plant species compete for resources (see above and Inouye et al. 1980, Inouye 1982). If plant-mediated indirect interactions between the two taxa are important, we would expect the initial increase in ants in response to rodent exclusion to be followed by a decrease that coincides with appropriate compositional changes in the plant community. This hypothesized pattern is exactly what is observed at the Sonoran Desert site (Fig. 3.8; see Davidson et al. 1984). Note that the abundance of ants on plots where rodents were removed did not ever fall below the initial, premanipulation levels. However, such a result, indicative of a net positive indirect effect of rodents on ants, would not necessarily be expected as long as there is also some direct competition between the two taxa. The magnitude of the indirect effect should depend on such factors as the extent of overlap in the diets of the two taxa and the intensity and symmetry of competition between the two classes of plant species. Our failure to detect any reciprocal positive long-term effect of ants on rodents is probably due to the pronounced asymmetrical competition between small- and large-seeded plants (see above and Davidson et al. 1984).

IMPLICATIONS OF THE EXPERIMENTS

Interspecific Interactions and Community Organization

Recently there has been much discussion of the effect of interspecific interactions on the organization of communities. Some investigators have cast doubt on studies purporting to demonstrate the importance of these interactions (e.g., Connor and Simberloff 1979, Strong et al. 1979,

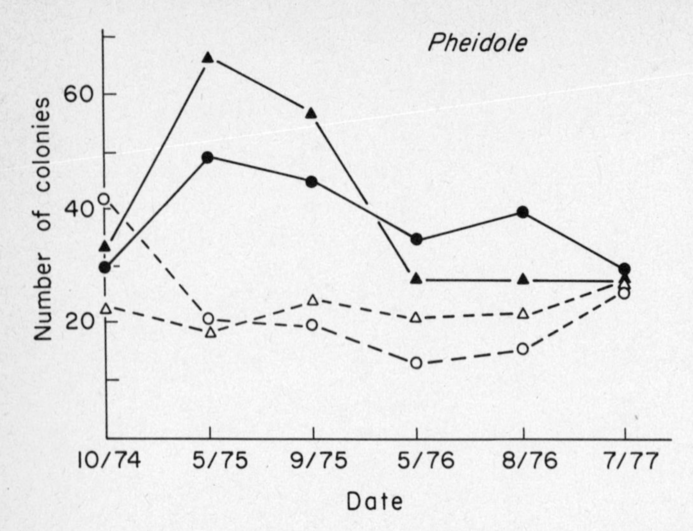

Fig. 3.8 Temporal trends in the densities of colonies of *Pheidole* ants, primarily *P. xerophila,* on rodent removal (shaded symbols) and control plots (unshaded symbols) at the Sonoran Desert site. Circles and triangles refer to different individual plots. In 1974, about a year after initiation of the treatments, there were no significant differences. Ant densities then increased on rodent removal plots (presumably as a result of release from competition) and then decreased again. We suggest that this last decrease is the result of indirect mutualism, mediated primarily through the effect of rodent predation on large-seeded plants that competitively dominate the small-seeded plants that produce the small seeds preferred by these small ants.

Simberloff and Boecklen 1981). Others have suggested that because the physical environment is so variable, the kind of resource limitation that would cause strong interactions rarely occurs in nature (e.g., Wiens 1977; Strong 1983, 1984b). Still others have searched for generalizations by trying to determine whether the organization of particular systems is determined primarily by competition, predation, mutualism, or temporal or spatial variation in the physical environment (e.g., Connell 1975, 1980).

The paucity of rigorous, long-term experimental studies in community ecology has contributed to the failure to resolve these issues. Although there has been an increasing number of experimental manipulations to test for the effects of particular kinds of interactions (e.g., Colwell and Fuentes 1975, Schoener 1983b, Connell 1983), there have been very few detailed, experimental investigations designed specifically to assess the diverse kinds of possible interactions among the species within a community and between the organisms and their nonliving environment (but see, for example, Connell 1961a, Lubchenco and Menge 1978, Paine 1980).

Although our experiments by no means constitute a complete investigation of the desert granivore system, they are sufficient to provide insights into the issues raised above. We can document the occurrence of: (1) competition among rodent species, (2) competition among ant species (not mentioned above, but see Davidson 1985), (3) competition among rodents, ants, and birds, (4) competition among annual plant species, (5) predation by rodents, ants, and fungus on plants, and (6) several indirect effects of rodents and ants on other species in the community that are mediated through their direct effects as predators on plants. Perhaps the most important thing about these interactions is not that they can be shown to occur within the same community, but that most of them are sufficiently strong to have major effects on community organization. Experimental exclusion of rodents, for example, results in substantial changes in the absolute and relative abundances of ants, birds, annual plants, and a fungus on our experimental plots.

We can test simultaneously for the effects of both competition and predation on the annual plants, and we find that they interact with each other in such a way that it would be misleading to suggest that either is more important than the other. Furthermore, other relationships that we have not emphasized in these experiments may be equally important. Thus, predation by carnivorous vertebrates may have as much impact on rodents as the rodents have on plants (e.g., see Kotler 1984). Fluctuations in the physical environment have important influences on all desert communities; in particular, variation in the quantity and timing of precipitation affects primary productivity, competitive relationships among plant species, seed production, and granivore populations (e.g., Brown et al. 1979b, MacMahon 1979). The dramatic increase in small seed-eating rodents in the winter of 1981–1982 shown in Fig. 3.1 is probably a direct consequence of heavy precipitation and high seed production during the preceding months.

Thus, we conclude that it is hazardous to attribute the organization of any reasonably complex natural community to the overriding influence of any single kind of interaction, especially when other kinds of interactions and the relationships between them have not been studied by rigorous experimentation. Although we are cautious in generalizing from the granivore system to other communities, we suspect that similarly complex and interacting relationships among the species and between them and their physical environment will often be found. There is a precedent for this in careful experimental studies in other habitats, such as the intertidal (Connell 1961a, Lubchenco and Menge 1978, Paine 1980). In addition, this complexity is to be expected on conceptual grounds for several reasons.

1. Each species has the capacity to increase at an exponential rate, but continued population growth is eventually checked.
2. Competition occurs whenever different species overlap in their requirements for essential resources that are in short supply.
3. Predation is ubiquitous, because almost every species either eats or is fed upon by other organisms.
4. Mutualistic associations, in which species are dependent on each other for essential resources or other benefits, are common.
5. Each species potentially can limit many others through indirect, as well as direct pathways.
6. All species are limited by spatial and temporal variation in their abiotic environment as well as by intra- and interspecific interactions.

Thus, in order to understand the complex organization of natural communities, it seems necessary to adopt a pluralistic approach that can take into account the diverse, interacting influences of all of these processes.

Theoretical Implications

In a general way the results of our experiments are consistent with current theories that interspecific interactions have important influences on community organization. Our manipulations reveal effects of competition and predation on desert animals and plants that are reminiscent of those described by MacArthur (1958, 1972a), Brooks and Dodson (1965), Paine (1966, 1980), and Lubchenco and Menge (1978) for other systems. In many cases the mechanisms as well as the outcomes are similar: a combination of exploitative and interference competition, selective predation on competitively dominant consumer species, and so on.

We began our experimental research program in 1973, at a time when mathematical models of pairwise population interactions seemed to offer a simple and rigorous conceptual framework for interpreting the patterns of abundance and distribution of closely related species revealed by natural experiments (e.g., MacArthur and Levins 1965, Levins 1968, MacArthur 1972a, May 1973a, Cody and Diamond 1975, Roughgarden 1979). When we set out to test these ideas experimentally, we fully expected that the results would support both the theory and our interpretation of geographical patterns in desert granivore associations (e.g., Brown and Lieberman 1973, J. H. Brown 1975, Davidson 1977).

The experiments do confirm that interspecific competition plays a major role in the organization of the desert granivore system, but they have also revealed a degree of complexity that neither the simple pairwise models nor the geographical comparisons had prepared us to expect. We were surprised to observe the long time lags, highly asymmetrical relationships, slight compensation for absent species, and substantial competition between distantly related taxa that our manipulations have so clearly demonstrated. In retrospect, it is easy to come up with realistic hypotheses to explain these results, but these just emphasize how naive and unrealistic our initial ideas were.

Consider just two examples: the asymmetries and the slight compensation among rodent species. The Chihuahuan Desert study site was one of the localities used by J. H. Brown (1975) in his geographical comparisons of granivorous rodent guilds. Based on measurements of seed and microhabitat utilization at this and other sites, Brown calculated overlaps between species in overall resource utilization using standard techniques. Values of these overlaps, which are supposed to indicate the degree of interspecific competition (but see Case and Gilpin 1974), are uniformly high and symmetrical. Other slightly different methods could have been employed, but the results, while not necessarily exactly symmetrical, would have been similar. These large measured overlaps in resource utilization would lead to the following predictions: (1) the remaining rodent species should increase in density and biomass to compensate to a large extent for experimental removal of selected rodent species; (2) all species that can use millet seeds should increase in response to food addition treatments, and collectively the rodents should consume most of the supplemental food.

Clearly, the results of the experiments support neither of these predictions, and any good biologist who knows rodents and deserts should have no difficulty in coming up with reasonable explanations for the discrepancies. The calculations of overlap in resource utilization did not take into account: (1) pronounced differences between species in both body size and population density that influence resource requirements and habitat utilization; (2) morphological, physiological, and behavioral constraints that affect the interactions between species and their physical environment; (3) effects of interspecific aggressive interference; (4) mechanisms of intraspecific competition, such as territoriality, that could maintain population sizes below the limit set by the availability of food per se; (5) use of seed resources by other animals, such as birds and ants; (6) effects of predators on the resource utilization and population densities of the rodents; and (7) indirect interactions, through which a variety of other organisms could potentially have a wide range of effects. We now have evidence (much of it cited above) that all of these factors are important in determining the roles of different rodent species in the organization of desert communities.

A fundamental problem with much of traditional community theory is that it treats species as if they affect each other in a simple pairwise fashion within a closed system. It ignores the fact that species are imbedded in a complex biotic and physical environment that affects almost every aspect of their ecology, including their interspecific interactions. Although theoretical ecologists have recognized the importance of this imbedding problem (e.g., Levins 1974, 1975; Schaffer 1981; Bender et al. 1984), the additional complexity that it introduces seems to be as difficult to characterize mathematically as it is empirically.

Considerable progress has been made, however, in developing a body of theory that explores the consequences of possible kinds of indirect interactions among species (e.g., Levine 1976, Holt 1977, Lawlor 1979, Vandermeer 1980). Our experimental results show not only that these kinds of indirect interactions occur in natural systems, but also that relationships mediated through these indirect pathways have major effects on the organization of the entire desert community. A simple perturbation, such as the removal of rodents, sets in motion a complex series of changes that ripple through the community, affecting an increasing number of species. Eventually there must be a limit to these changes and the system should approach a new state, but it is a testimony to the importance of these indirect effects that we are still observing pronounced changes in plants and other organisms at least seven years after exclusion of rodents began.

Our experiments suggest a view of community organization in which virtually all species affect each other through a complex web of direct and indirect interactions (see concluding section of Chapter 20 for further discussion). These relationships are highly asymmetrical, nonlinear, and influenced importantly by the physical environment as well as by other species. They vary from site to site, even among superficially similar habitats, and their dynamics are sufficiently slow and complex that many years are required to observe their full effect.

We still have a long way to go, both empirically and theoretically, before we have a really satisfactory understanding of these structural and dynamic properties of communities. Eventually, it would be nice to know enough about the networks of interaction so that we could predict the diverse ramifications of experimental manipulations and identify the frequency-dependent negative feedback loops that must account for the resiliency of the system to natural and experimental perturbations. At present there have been so few comprehensive empirical studies of complex, multispecies systems that it is difficult to know what properties are specific to a particular system and what ones are of general importance. Without an adequate data base to reveal a clear phenomenology of patterns and processes, it is difficult for theoreticians to produce realistic models. Nevertheless, recent progress, such as that in the study of food webs (Cohen 1978, Yodzis 1982, Pimm 1982, DeAngelis et al. 1983), suggests that even the most complex communities may reluctantly reveal their secrets in the face of a combined theoretical and empirical assault.

Advantages and Limitations of Field Experiments

In recent years controlled, manipulative experiments conducted in the field have played an increasingly influential role in making contemporary ecology a rigorous, quantitative science. Several advantages of field experimental methods are frequently mentioned: (1) the precise control of variables that can be achieved by human intervention, (2) the statistical rigor that can be attained as a result of replication and use of powerful experimental designs, and (3) the strength of inference that can be achieved by applying the full logical force of the hypothetico-deductive method. The results of our experiments confirm the importance of all these advantages. For example, they provide evidence for the importance of competition among species of terrestrial vertebrates that not only is difficult to dispute, but also is consistent with the results of other manipulative studies (e.g., Hairston 1980, Pacala and Roughgarden 1982).

Our work emphasizes yet another advantage of field experimental methods, however, and one that is rarely mentioned. This is the ability to monitor the trajectories of the complex dynamic behaviors set in motion by a single controlled and sustained perturbation. Many features of community organization can be analyzed as the system adjusts to the altered state through interactions of varying length, strength, and time constants. Because indirect pathways are chains of direct interactions, the temporal sequence of events reveals a great deal about the processes by which species affect each other and about the patterns of connectance that link the fates of species in different taxonomic groups and trophic levels. Our experiments show that the exclusion of particular species or groups of species may have unanticipated effects on many other species that may take many years to be resolved. Although it may be better hypothetico-deductive science to perform perturbations to test specific hypotheses, rather than just to see what happens, nevertheless these kinds of unexpected results may tell us more about the organization of communities than those predicted by our naive and simplistic hypotheses.

These kinds of insights provided by field experimentation cannot usually be obtained by recording the results of "natural steady state experiments," as discussed in Chapter 1. Comparative observations of unmanipulated systems usually provide only a snapshot that records the more-or-less steady state response of the system to a particular set of environmental conditions. A controlled, replicated perturbation permits precise, quantitative documentation not only of final response of the community to the altered conditions, but also of the complex sequence of dynamic processes by which the changes take place. Terrestrial ecologists, in contrast to some of their colleagues working in the more controlled, replicated environment of lakes (e.g., Brooks and Dodson 1965 and many subsequent studies), have been slow to appreciate the importance of indirect interactions, through which many taxonomically unrelated species can have major effects on each other. We suspect that this is not because such indirect interactions are unimportant in most terrestrial habitats, but rather because the kinds of comparative, nonmanipulative studies that have traditionally been performed did not provide sufficiently precise data to document their effects. For example, imagine trying to convince anyone that the higher density of a fungal plant pathogen on an island, compared to a nearby mainland, was owing to the absence of seed-eating rodents from the island. But we can show experimentally that the density of *Synchytrium pallatum* increased on plots where nothing else was done except to remove rodents.

Although there are those zealous experimentalists who would imply that purely observational studies have no place in modern ecology, this extreme perspective is misguided and at variance with our own experience. Our earlier nonexperimental studies of desert granivores (e.g., Brown and Lieberman 1973, J. H. Brown 1975, Davidson 1977) not only suggested hypotheses about interspecific competition that called for rigorous experimental test, but they also revealed enough about probable mechanisms of interaction to design appropriate experiments and procedures for monitoring the results. Furthermore, although the manipulations have yielded some surprises, they have also confirmed many of the inferences drawn from purely observational studies. For example, comparisons of the abundance, distribution, diet, and foraging behavior of rodents within local habitats and among geographically separated sites clearly suggested important influences of limited food availability, body size differences, seed size selection, microhabitat selection, and competition with ants, and we have subsequently documented these phenomena experimentally. Our experimental program has required a major commitment of time, effort, and resources. To have attempted this project without the background provided by several years of nonexperimental studies would have been not only difficult to justify, but also inefficient, and even likely to fail because of poor design.

Despite the great advantages of the experimental method for dissecting ecological communities and revealing the complexities of their structure and function, we must recognize that many of the important questions cannot be answered by experimental manipulations alone. The experimental method has several limitations and disadvantages. First, the price of the human intervention necessary to manipulate and control variables is the creation of possible artifacts. Although we have tried to minimize the possibilities of such artifacts, it is important to emphasize that the fencing, trapping, poisoning, seed addition, and other manipulations may have unexpected and undetected effects on the community. For example, fencing might differentially exclude predators and folivores, and these could conceivably have caused some of the results we have attributed to granivores. Second, the spatial and temporal scale on which it is practical (and legally and morally permissible) to manipulate ecological systems is necessarily limited. As discussed in Chapter 8, there are good reasons to believe that interactions over geographical spatial scales and evolutionary time scales may affect the organization of local communities. That we might obtain very different results if we were able to perform our manipulations over thousands of square kilometers and maintain the treatments for thousands of years does not mean that the present results are any less important. However, it should encourage us to be cautious in generalizing from such experimental studies and to use ''natural experiments'' to provide the essential geographical and evolutionary perspectives.

Finally, experiments such as ours represent extreme examples of what we call the microscopic approach to community ecology. This approach has its own limitations. Although a detailed analysis of the interrelationships among species and between species and their abiotic environment within one local community can reveal the complexity of patterns and processes that characterize that particular system, it can provide little insight into which of these are specific to that system and which can be generalized to other communities. Furthermore, if there are general rules that govern the organization of communities, it may be impossible or impractical to adduce all of these from microscopic studies of the interactions of individual species (Brown 1981). For example, the statistical distributions of population densities, body sizes, rates of energy use, and areas of geographical ranges among the many species that comprise local communities may elucidate patterns and processes that cannot be discovered from microscopic experimental studies, which must for practical reasons be focused on only a small proportion of the species present. Large-scale geographical studies, especially comparisons of communities inhabiting similar physical environments in widely separated regions inhabited by different taxa of organisms, should continue to provide a valuable perspective.

The challenge of community ecology is to understand the structure and function of some of the most complex natural systems. Those who accept this challenge have no excuse for complaining about its difficulty, but they have every reason for keeping an open mind and encouraging different approaches by a diversity of ecologists.

SUMMARY

Experimental removal of species or groups of species of granivorous animals and experimental addition of seeds have profound effects on desert communities. In the present paper we focus on the response of seed-eating rodents to these manipulations and on the effect of manipulating rodents on other species. These results show that: (1) rodents compete for a limited food supply, not only with other seed-eating rodent species, but also with granivorous animals such as ants and birds; (2) rodents forage selectively for the seeds of certain large-seeded plant species, and as predators they have major impact on these plant species; and (3) as a consequence of their direct interactions with these large-seeded plants, rodents have important indirect effects on small-seeded plants, a parasitic fungus, and granivorous ants.

The diverse and often indirect interrelationships among the species within this relatively simple system have important implications for our understanding of community organization. For the most part, our results provide strong support for some of the conceptual generalizations of community ecology, such as the importance of competition, predation, abiotic factors, and indirect interactions. On the other hand, the specific, quantitative responses to our perturbations reveal a complexity of community organization that remains poorly understood, both empirically and theoretically. Species interact, not in isolated pairs as simple theories originally assumed, but within a complex matrix of other species and the abiotic environment. Species in different taxonomic groups and trophic levels affect each other directly and indirectly, through pathways of varying length, strength, and time constants. Long-term experimental studies have important advantages for investigating the organization of complex ecological systems, but a diversity of theoretical and empirical approaches focused on different spatial, temporal, and organizational scales will almost certainly be necessary to develop a viable general theory.

ACKNOWLEDGMENTS

So many people have helped with the research described here that it is impossible to thank them individually. However, this does not diminish our gratitude for their assistance; without it we could never have undertaken a project of this magnitude. This chapter has benefited enormously from the helpful comments of T. J. Case, J. M. Diamond, P. J. Grubb, and P. Yodzis. The National Science Foundation has generously supported the research from the outset, most recently with Grants BSR 80-21535 to J.H.B. and BSR 80-21537 to D.W.D.

two

SPECIES INTRODUCTIONS AND EXTINCTIONS

chapter 4

Overview: Introductions, Extinctions, Exterminations, and Invasions

Jared Diamond and Ted J. Case

SOME MODEST PROPOSALS

Laboratory and field experiments that introduce or remove species offer obvious advantages for studying community organization. The manipulations occur exactly when and where we ecologists want them, involve those species that we choose and no others, and can be designed to incorporate well-matched controls. Nevertheless, the manipulations attempted to date barely explore the potential scope of these methods. Many other introductions or removals, offering much more profound insights, have yet to be attempted by ecologists. For instance, suppose that we could introduce a lethal disease of eucalyptus into Australia, exterminate selected Galapagos finches, poison the cichlid fishes of Lake Victoria, reintroduce lions to the Great Plains, briefly freeze Britain during the winter of 1989, or destroy all large terrestrial vertebrates on Earth. We could then study community trajectories never revealed by the modest local manipulations that ecologists prefer.

In fact, problems of scale, other practical difficulties, laws, and scruples of conscience restrict ecological experiments to a tiny fraction of those imaginable. But community perturbations, including some of those just mentioned plus others identical in principle, are common. Diseases are currently destroying American elms and eliminated chestnuts 50 years ago; selected birds have been exterminated on Hawaii and other islands; most endemic fishes of Lakes Atitlán and Lanao have been destroyed, and the cichlids of Lake Victoria are now being destroyed; large cats were introduced to South America three million years ago; Britain was frozen during the winter of 1962–1963; and almost all large terrestrial vertebrates disappeared at the Cretaceous-Tertiary boundary. Although the consequences of such events often appall us and their experimental design is usually imperfect, much can nevertheless be learned through their study. Sometimes they even permit good experimental design. Consider, for instance, the elegant spatial and temporal arrangement of controls used by Zaret and Paine (1973) while studying the invasion of a piscivorous fish, or the surgical precision with which chestnut blight eliminated *Castanea dentata* and no other tree.

These natural trajectory experiments, to use the term defined in Chapter 1, enormously expand the scale on which we can trace interactions among a community's species. We recognize four types of unplanned manipulations.

1. Perturbations caused by humans
2. Natural invasions and extinctions that have occurred in recent time and have been witnessed
3. Natural invasions inferred from modern distributional "snapshots"
4. Invasions and extinctions documented in the fossil record.

We will survey these four types in the first half of this chapter.

To help us identify the distinctive questions posed by community trajectories following introductions and extinctions, consider the flow diagrams of Fig. 4.1. This figure raises four questions, which we shall discuss in the second half of this chapter. The same questions arise for planned manipulations, but we are rarely able to consider questions 3 and 4 in planned situations.

1. In the case of an introduction (or invasion), what determines whether it succeeds or fails?
2. What direct and indirect effects follow the introduction or extinction within ecological time?
3. What features of the host community determine whether the impact of a successful introduction is sufficient to cause extinctions?
4. What adjustments follow in evolutionary time on the part of surviving community members if the initial event was an extinc-

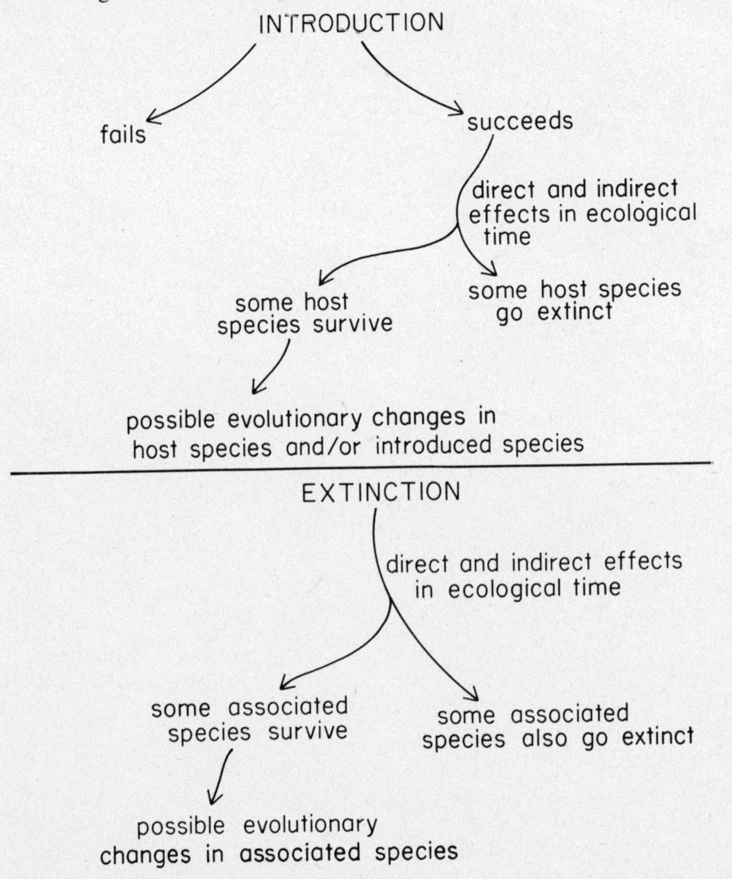

Fig. 4.1 Branching possibilities for the effects of an introduction (or invasion) or an extermination (or extinction) on a community.

tion, or on the part of a successful invader and surviving host species if the initial event was an introduction?

Four other chapters in this volume consider some of these questions in more detail for particular systems: Chapter 5, for birds introduced into Hawaii by humans; Chapter 6, for congeneric birds injected into the mountains of New Guinea by speciation *in situ;* Chapter 7, for the successive groups of large terrestrial plants produced by evolution since the Silurian period; and Chapter 21, for fish introduced into lakes and rivers.

INTRODUCTIONS

Introductions by Humans

Humans have introduced many species to areas where they are not native, either intentionally or by accidental transport of commensals. Introductions affect all six main types of species interaction: predation, competition, herbivory, disease, parasitism, and mutualism.

General summaries of species introductions are given by Elton (1958) and Diamond (1984a). Summaries are also available for particular taxa or regions: birds (Long 1981) and fish (Fisher et al. 1969, Anon. 1981) throughout the world; plants and animals in New Zealand (Thomson 1922; *Proceedings of the New Zealand Ecological Society* 8, no. 6); plants and animals in the Australian state of Victoria (*Victorian Naturalist* 98, January–February); mammals in Hawaii (Kramer 1971) and Micronesia (Barbehenn 1974); and insects in North America (Sailer 1978).

Cases in which introduced species had important effects as *predators* include rats, cats, weasels, pigs, and mongooses introduced to many islands; piscivorous freshwater fish and lampreys, introduced to countless lakes and rivers (Chapter 21; Fisher et al. 1969); ants introduced to Hawaii, the Madeiras, and the Canaries; and predatory rails introduced to several Pacific islands (Diamond 1984a).

The impact of introduced species as *competitors* of native species has been especially striking for fish (Chapter 21; Fisher et al. 1969, Anon. 1981, Crowder et al. 1981). Other cases involve effects of introduced urban birds from Europe (house sparrow, starling) on eastern North American birds (e.g., purple finch, Eastern bluebird) and effects of the introduced ant *Iridomyrmex humilis* on native North American ants. Competitive effects of introduced species on each other will be discussed later in this chapter.

Impact as *herbivores* is important for introduced goats, deer, pigs, Australian possums, rabbits, and other browsing mammals on islands whose floras evolved in the near-absence of native browsers, such as New Zealand and Hawaii; for herbivorous insects introduced from North America to Europe (e.g., the aphid *Phylloxera vitifolii* that eliminated European grape vines in the nineteenth century); and for herbivorous insects introduced from Eurasia to North America (e.g., gypsy moth, spruce sawfly, Japanese beetle).

Notorious cases of impact of species as *pathogens* have included the effect of Dutch elm disease, introduced from Europe to North America, on American elm; of rinderpest, introduced from Asia to Africa, on native African ungulates; of myxomatosis, introduced from Europe (ultimately from Brazil), on Australia's introduced rabbits; and of avian malaria and diseases of introduced poultry and parrots on birds of Hawaii and other islands.

Purposely introduced *parasites* have been one of the main tools used by humans in attempting to control accidentally introduced herbivorous insects. Examples are the role of parasitic wasps in controlling scale insects introduced from Australia to California (DeBach 1966), and of a tachinid fly and ichneumonid wasp in controlling winter moth in Nova Scotia (Embree 1979).

Witnessed Natural Invasions

There are many cases of witnessed self-introductions of species within modern times. What causes these self-introductions is often unknown, but they can sometimes be linked to climate changes, genetic changes, the overcoming of a major geographical barrier by lucky pioneers, human-related habitat changes, and other effects of human activities. Nowak (1971) provides a valuable summary of range expansions of 28 species of mammals, birds, insects, crustaceans, and mollusks in Europe. Summaries of bird range expansions in response to changes in temperature or rainfall are given by Kalela (1949), Merikallio (1951), Gudmundsson (1951), Winstanley et al. (1974), Williamson (1975), Diamond (1984a), and Davis (Chapter 16). Modern self-introductions across broad water barriers include the colonization of New Zealand from Australia by the silvereye *Zosterops lateralis* and other bird species, and the transatlantic colonization of the New World from Africa by the cattle egret.

One particularly interesting self-introduction was the spread of the azure tit (*Parus cyanus*) westward from the Urals to the Baltic in the 1870s and 1880s (Pleske 1912, Vaurie 1957). This expansion took the azure tit into the range of the closely related blue tit (*Parus caeruleus*), with which it initially hybridized and overlapped broadly in habitat. Within several decades these two species achieved reproductive isolation and segregated ecologically by habitat. Thus, this case illustrates the crucial stage in speciation, often inferred but rarely witnessed directly, in which formerly allopatric taxa first achieve sympatry and then develop spatial separation by habitat (presumably because of competition).

Invasions Inferred from Distributional "Snapshots"

This category provides a bridge between witnessed self-introductions and invasions inferred from the fossil record. Suppose that one has a biota for which fossil distributional evidence is lacking and whose species have maintained fixed distributions within historic times. Thus, for any single species considered in isolation, evidence of invasions is absent. However, if one considers together the maps defined by individual distributions of all species, it may turn out that the maps fall into a natural sequence implying an invasion route of which each modern distribution offers a different "snapshot." Whether it is actually feasible to reconstruct invasions in any given biota by this method depends on whether the biota is sufficiently rich in species, the regional geography is simple, invasions occur by only one or a few routes, and species' niches vary consistently with the stage of invasion.

One example considered in this volume is Roughgarden's discussion of how the *Anolis* lizard fauna of species-poor Antillean islands was assembled (Chapter 30). Consistent body size patterns are observed and can be plausibly interpreted by a coevolutionary model driven by interisland invasions and interspecific competition. Williams (1983) provided a similar reconstruction, based on intraisland invasion and competition leading to segregation both by size and by habitat, for assembly of the species-rich *Anolis* faunas of the Greater Antilles.

A second example from this volume is Diamond's reconstruction of speciation in New Guinea montane birds (Chapter 6). The New Guinea cordillera has a simple east-west configuration, making it easy to recognize stages in eastward or westward invasions that led initially to competitors' segregating by altitude and later by size or diet or foraging technique.

The biogeographical literature contains many other examples of invasion routes reconstructed by this method. For instance, several such studies have compared the competitive abilities of invading species derived from biotas of differing species richnesses or from land masses of different areas (Mayr 1944, Darlington 1957, Diamond and Marshall 1976).

Invasions Inferred from Fossil Evidence

A classic natural experiment in self-introduction revealed by the fossil record is the Great American Interchange (Wallace 1876, Simpson 1980, Marshall et al. 1982). The emergence of the Panama land bridge 3 million years ago ended South America's long isolation, during which that continent had evolved endemic families of mammals. Over the bridge, South American immigrant mammals streamed into North America and vice versa, and as a consequence some taxa went extinct and others radiated. When the evolutionary dust finally settled, each continent supported roughly as many mammal families as before the bridge's formation. The quantitative details of the immigrations, extinctions, and radiations constitute a test of the MacArthur-Wilson equilibrium theory on a gigantic scale.

One of the most detailed fossil records of invasions is for the colonization of North America by Asian microtine rodents. They arrived in at least eight waves, dated at 5.3, 4.8, 3.7, 2.5, 1.8, 1.2, 0.47, and 0.17 M.Y.B.P. (Repenning 1980).

Chapters 16, 17, and 18 in this volume discuss invasions deduced from fossil evidence for the Pleistocene. One additional Pleistocene example may be cited here to illustrate how fossil evidence can illuminate central problems of contemporary ecology. A current issue concerns the formation of alternative communities (''multiple domains of attraction'') from a species pool. Chapter 2 explores this problem by putting sets of 10 *Drosophila* species into 30 different bottles and following their fates for 19 weeks. A larger-scale natural exploration of this problem is the repeated elimination of deciduous forest from Europe north of the Alps during glacial periods and its reinvasion of the area during interglacials. Thus, nature put sets of deciduous tree species into northern Europe about 20 consecutive times, and paleontologists (e.g., Davis 1976, 1983a) have reconstructed the fate of each set of immigrants for 10,000 to 20,000 years (the duration of interglacials). The conclusion: ''Differences in the relative abundance of species in the deciduous forest communities of Europe during successive interglacials suggest that the same constellation of species formed alternative communities'' (Davis 1976, p. 17).

Fossil evidence can also be used to reconstruct the impact of species introduced by humans in the distant past. For instance, Australia's largest marsupial carnivore, the thylacine (Tasmanian wolf), persisted on New Guinea and the Australian mainland until the early Holocene, when introduction of its placental equivalent, the dog, and the resulting establishment of wild dingo populations eliminated the thylacine except in the dingo-free refuge of Tasmania. Similarly, Polynesian settlers of New Zealand and Hawaii brought rats, land snails, and lizards whose replacement of native vertebrates and invertebrates is attested by the fossil record.

On the much longer time scale of the last 400 million years Knoll (Chapter 7) traces how the evolution of each successive group of higher plants resulted in the extinction or contraction of the previous dominant group.

EXTINCTIONS

Exterminations by Humans

Since 1600 humans have exterminated about 63 of the world's 4,200 mammal species, 83 of the 8,500 bird species, and numerous cold-blooded vertebrates, invertebrates, and plants. Some of these exterminations resulted from introductions by humans discussed in the previous section, while others had other causes. Diamond (1984a) reviews the six major mechanisms, which are briefly as follows.

Overkill is responsible for most extinctions of large mammals (e.g., Steller's sea cow), many birds (great auk and passenger pigeon), and some trees logged for wood or sap (the wine palm *Pseudophoenix ekmanii*).

Habitat destruction has resulted from fire, wetland drainage, grazing and browsing animals, and deforestation for timber, agriculture, or stock. The diverse victims include all endemic birds of Cebu Island and most native plants of St. Helena Island.

Introduced pathogens offer especially valuable tools to the student of unplanned manipulations because of their high degree of specificity. Dramatic cases besides the examples of plant pathogens mentioned earlier include the extermination of the formerly abundant marsupial carnivore *Dasyurus viverrinus* in Australia and the elimination of Swayne's hartebeest in Africa by rinderpest.

Exterminations by introduced predators occasionally approach the effects of introduced pathogens in their specificity. Effects of introduced piscivores on native freshwater fish offer especially clean models (see the rich literature cited in Chapter 21). Examples include the effects of lampreys on trout and ciscos in the Great Lakes, of largemouth bass in Lake Atitlán, and of peacock bass in Gatun Lake (Zaret and Paine 1973). Nearly as striking were the rapid eliminations of several songbird species by rats that reached Lord Howe and Big South Cape Islands in 1918 and 1964, respectively. There are hundreds of other classic cases, ranging from Hawaiian achatinellid land snails to Australian marsupials.

Effects of introduced competitors on native species are again especially well illustrated by freshwater fish (Chapter 21). Among the multitude of terrestrial examples may be mentioned the retreat of the native squirrel *Sciurus vulgaris* in southern Britain in the face of the introduced *S. carolinensis* from North America; and the extermination of one of the most abundant marsupials on the Australian mainland, the burrowing kangaroo *Bettongia lesueur,* by rabbits competing for burrows and food.

Finally, chemical pollutants have become an important human-related agent of extinction in recent decades. Examples include the recent decimations of the barn owl in Britain and Malaysia by the ''super poisons'' brodifacoum and difenacoum; of most native forest bird species on Guam by DDT (Jenkins 1983); and of lake and stream fish populations by acid rain (Magnuson et al. 1984).

Witnessed Natural Extinctions

Fluctuations in temperature, rainfall, or other natural factors occasionally eliminate local populations, permitting one to study effects on surviving populations (Diamond 1984a).

An example discussed by Colwell in this volume is the local elimination in Trinidad of the hummingbird flower mite *Rhinoseius bisacculatus,* due to an unusually rainy dry season eliminating the flowers on which it depends (Chapter 24). Another mite, *Proctolaelaps certator,* previously recorded only as a rare visitor to that flower species, soon established itself. The return of *R. bisacculatus* to the *Proctolaelaps*-occupied flowers is permitting Colwell to study what factors are responsible for the normal nonoverlapping flower occupancy of these two mite species.

There is a large literature on populations of birds eliminated or decimated by warm weather (little auk in Iceland, Gudmundsson 1951), cold weather (stonechat in much of Wales and Scotland in 1947, Magee 1965), dry weather (effects of the Sahel drought on British populations of the whitethroat, Winstanley et al. 1974), or wet weather (breeding seabirds on Christmas Island, Schreiber and Schreiber 1984). Examples of similar effects on mammals are the local extermination by cold winter weather of the mole *Talpa europaea* in Germany (Stein 1951), and the local extermination by dry weather of the mole *T. caucasica* and vole *Microtus socialis* in the Caucasus (Vereschchagin and Baryshnikov 1984).

One interesting fact learned from these climatic fluctuations is that they may constitute limiting factors for a species only at certain times of the year. For instance, Kalela's (1949) analysis shows that unusual temperatures in a particular month have selective effects on those bird species migrating or breeding during that month. Another important insight resulting from study of climatic fluctuations concerns the role of temperature in controlling the boundary between ranges of northern and southern species that replace each other geographically (Merikallio 1951, Williamson 1975, Järvinen and Väisänen 1979).

Extinctions Inferred from Fossil Evidence

The natural extinctions witnessed in modern times are modest compared to those deduced from the fossil record. For instance, accumulating evidence suggests that an asteroid impact at the Cretaceous-Tertiary boundary extinguished whole sets of species (including the dominant large terrestrial animals, the dinosaurs), thereby permitting the radiation of surviving species (including mammals, which had arisen over 100 million years earlier but had apparently remained competitively suppressed).

An impressive extinction wave documented by the more recent fossil record is the disappearance of about 80% of the large mammal species of North and South America around 10,000 years ago and of a similar proportion of Australia's large mammal species somewhat earlier. Some authors consider these extinctions to be a case of human overkill, while others (including Graham in Chapter 18) relate them to climatic changes (Martin and Klein 1984). There is no doubt, however, that island megafaunas that disappeared soon after humans reached their island were victims of overkill, such as the moas of New Zealand, the elephant birds and large lemurs of Madagascar, and the flightless geese of Hawaii (Martin and Klein 1984).

As an example of a still more selective decimation in the fossil record, pollen analysis shows that 4,800 years ago hemlock (*Tsuga canadensis*) suddenly declined (within 50 years or less) throughout its entire geographical range in eastern North America to one-tenth of its previous abundance (Davis 1981). In some of those areas hemlock never recovered its former abundance, while in others it did but only after 2,000 years. The most likely explanation of the crash is the arrival of hemlock looper or a similar pest.

WHAT DETERMINES WHETHER AN INTRODUCTION WILL SUCCEED?

Theoretical models of communities are centrally concerned with specifying the conditions that make a certain set of species invasible or not by a certain other species (e.g., MacArthur 1972a). The fates of introductions, whether by humans or nature and whether witnessed or inferred from past records, offer extensive tests of such theories.

Demographic and Abiotic Factors

Obvious roles are played by demographic and abiotic factors. Demographic factors include the number and sex ratio of introduced individuals, their mating system, their genetic constitution, and whether or not the species is social (Baker and Stebbins 1965). As for abiotic factors, introduced species are most likely to succeed if they come from a climate zone similar to that in the land of introduction. For instance, the introduction of musk ox from the arctic into Vermont failed because of too warm summers; the introduction of the European pond tortoise into Britain failed because of too cool, cloudy summers; and the northern limit to the spread of the armadillo *Dasypus novemcinctus* in the United States is set by cold winters (Graham 1979, Stuart 1979).

Patterns of Success

In the remainder of this section we discuss the equally important but more complicated problem of how success is influenced by species interactions. A convenient starting point is Elton's (1958) observation that species-rich communities are more resistant to introductions and invasions than are species-poor communities. Elton explained this empirical pattern by the then-current belief that diversity begets stability. This interpretation lost favor following May's (1973a) theoretical demonstration that diversity reduces rather than begets stability in randomly constructed model communities. Even Elton's empirical observation received little further attention. In fact, however, Elton and May were using the word ''stability'' to refer to very different problems. May was concerned with the local stability of a system to small perturbations in such state variables as species densities, while Elton's resistance to invasion involved global stability vis-à-vis incorporation of new state variables such as invading species.

A worldwide review of species introductions to islands and continents (Case, unpublished) confirms Elton's observation and detects three types of patterns. For birds the three patterns are as follows.

First, invasion success (defined as the number of successfully introduced species divided by the number of attempted introductions) declines steeply with species richness (or number of species, S) of the extant native avifauna. For instance, among tropical and subtropical islands, invasion success declines from 90% for Ascension ($S = 0$), to 60 to 80% for Lord Howe, Bermuda, Rodriguez, and Seychelles ($S = 3$ to 17), to 0% for Borneo ($S =$ ca. 420) and New Guinea ($S = 513$).

Second, the penetration of native forest by successfully introduced species also declines with species richness of the extant native avifauna. For instance, on Viti Levu, the largest island of the Fiji archipelago ($S = 48$), the introduced species are strictly confined to human-modified open habitats. On Oahu ($S = 10$) numerous introduced species are abundant in native forest, while Kauai is intermediate between Oahu and Viti Levu both in S (19) and in penetration of forest by exotic birds. Specific examples involving comparisons of the same species among several islands are that the introduced collared dove (*Streptopelia chinensis*) penetrates native forest on Kauai and Molokai but not on Viti Levu, and that the introduced European blackbird (*Turdus merula*) penetrates native forest on species-poor Norfolk Island and New Zealand but not on the species-rich Australian mainland.

Third, there is a strong correlation between number of successful introductions and number of extinctions of recent and subfossil species. For instance, New Zealand has had dozens of each; New Guinea, none of either. The slope of this relationship does not differ significantly from 1.0. As both introductions (Long 1981) and discoveries of extinct subfossil species (Olson and James 1982a) are continuing, we are unsure whether this relationship will prove to be a colossal coincidence (see below).

Data on other introduced vertebrates also obey the first two of these patterns noted for birds. For mammals and lizards, invasion success (or number of successful introductions) and penetration of native forest both decline with number of extant native species. For example, New Zealand, with no native flightless land mammal, has 33 successfully introduced mammal species, of which none is restricted to human-made habitats and many reach epidemic proportions in native forest (Gibb and Flux 1973). Introduced *Rattus rattus* and *R. norvegicus* are ubiquitous pests in forests of tropical Pacific islands lacking native flightless mammals. In contrast, New Guinea, with about 200 native mammals, has few introduced species, *Rattus norvegicus* is very rare, and *R. rattus* is mainly commensal with humans.

Comparison among vertebrate classes extends patterns 1 and 2. Number of native insular species decreases in the sequence birds, lizards, mammals, freshwater fish. Invasion success and penetration of native habitats appear to increase in the same sequence (Case, unpublished; Pernetta and Watling 1978).

Native habitats are the most species-rich habitats for native insular species. Thus, the patterns for each vertebrate class and the comparisons among vertebrate classes all confirm Elton's conclusion: species-rich biotas resist invaders.

Interpretation of the Patterns

If it were not for species interactions, one would expect invasion success to be greater on large islands than on small islands (because of more diverse habitats and higher carrying capacities), and greater in native forest than in open habitats (because of greater structural complexity and more niches). Thus, the cause of lower invasion success on large islands and in native forests must be that their rich biotas of native species resist invaders through predation, competition, disease, parasitism, or (in the case of plant invaders) herbivory. Humans furnish the clearest example of the effects of disease on invasion success: malaria, yellow fever, sleeping sickness, and other tropical diseases were the most important obstacles to colonization of the tropics by Europeans. Most of our information for animals involves competition and predation.

Competition from Native Species Comparison of mainland New Zealand with the nearby island of Little Barrier illustrates the role of competition from native species in excluding invaders (Diamond and Veitch 1981). Weasels and two Eurasian rat species have decimated the native forest avifauna of the mainland but have not yet reached Little Barrier, where the native avifauna consequently remains intact. Numerous introduced bird species that have become abundant in New Zealand mainland forest (e.g., the Australian silvereye and the European chaffinch, blackbird, song thrush, and dunnock) have reached Little Barrier, and some breed there in human-made open habitats virtually devoid of native species. However, the exotic species are almost completely absent from Little Barrier forest, despite its lack of mammalian predators, and doubtless because of its diversity and abundance of native avian competitors. Varying competition from native bird species is probably the main reason why invasion success and penetration of native forest by exotic birds decreases with native species number on islands.

Competition from Introduced Species The most detailed study of how success of introduction varies with competition among the introduced species themselves is for introduced Hawaiian birds (Chapter 5). As that chapter shows, the probability that an introduction will fail increases supralinearly with the number of introduced species already established. Probability of failure is specifically increased by the presence of a morphologically similar congener and rises with the degree of similarity (see Fig. 5.6). The same pattern applies to species introduced by natural invasions or speciation. Case, Faaborg, and Sidell (1983) showed for West Indian birds, Case and Sidell (1983) and Grant and Schluter (1984) for Galápagos finches, and Diamond (Chapter 6) for New Guinea birds that selection of immigrants on the basis of size contributes to the greater-than-random size spacing of sympatric species in these avifaunas.

Lizards, mammals, and parasitic wasps also illustrate competition among introduced species. The gecko *Hemidactylus frenatus,* introduced to Hawaii in the 1940s, quickly displaced two previously introduced geckos (*H. garnoti* and *Lepidodactylus lugubris*) from urban areas (McKeown 1978) and has had exactly the same effect on the same two species in Fiji following its introduction there since 1978 (Watling, personal communication). Species of parasitic wasps introduced sequentially to Hawaii or California each successively replaced the congener introduced previously (Bess et al. 1961, DeBach 1966). Other examples are the competitive effect, documented by field experiments, of an anole (*Anolis cristatellus*) introduced to the Miami area around 1975 on habitat use by an anole (*A. sagrei*) introduced around 1953 (Salzburg 1984), and the displacement by *Rattus rattus* of its earlier arriving congener *R. norvegicus* in Hawaii (Barnum 1930).

We have already argued that species-rich native faunas have severe competitive effects on introduced species. Does the above-mentioned pattern 3 (apparent equality between number of exotic species established and number of native species gone extinct) mean a severe competitive impact of introduced species on native species as well? In fact, examples of such an effect are more limited than the converse effect (cf. Simberloff 1981). Extinctions of native species usually precede successful introductions and are due mainly to habitat destruction and introduced predators and diseases (Diamond 1984a). Insofar as there is any causal relationship between the introduction of exotics and the extinctions of natives, the extinctions are more likely to be among the causes of the introductions than vice versa.

It may seem paradoxical to argue that competition between native and introduced species is asymmetrical. The explanation is that in most cases the extinctions involve initially abundant populations of native species, but only tiny "bridgeheads" of introduced species, whose risk of extinction from demographic stochasticity increases greatly if competition lowers population growth rates. Intact species-rich native faunas must contribute significantly to these bridgehead extinctions. However, that effect is now negligible on islands like Hawaii, where native species are absent in open lowland habitats (Chapter 5).

Predation from Native Species Predation rates are generally higher on adult mammals than on adult birds. Thus, the epidemic success of introduced mammals on many islands may reflect escape from continental predators as well as from continental competitors. The feral goats and pigs that plague Pacific islands could hardly have survived if Captain Cook had also introduced lions and wolves.

Predation from Introduced Species The clearest examples come from the attempts of the New Zealand Wildlife Service (NZWS) to introduce endangered native species to offshore islands now lacking those species. As summarized in the annual reports of the NZWS (*Wildlife—a Review;* see also Gibb and Flux 1973), the absence of particular introduced predators on an island is critical to the success of these introductions. For instance, the tuatara (*Sphenodon punctatus,* a reptile) cannot coexist with *Rattus exulans,* nor the robin *Petroica australis* with *Rattus rattus,* nor ground-nesting birds with *Rattus norvegicus,* nor the saddleback *Philesturnus carunculatus* with cats.

Extrapolation to Other Types of Introductions

In this section we have mainly discussed the fates of species introduced by humans. Similar analyses could be made for witnessed natural invasions, invasions inferred from distributional "snapshots," and invasions inferred from fossil evidence. For example, the outcome of past mixings and evolutionary replacements of faunas is often interpreted in terms of competition; for example, the relative competitive abilities of North and South American mammals may have sealed the outcome of the Great American Interchange. Similarly, the integrity of the world's modern biogeographical regions is often interpreted in terms of competition; for example, the distinctness of Australia's avifauna from that of the Oriental region may be maintained by competitive exclusion of immigrants.

However, the contributions of other interspecific interactions to these outcomes also need evaluation. For instance, a paleontologist in the year A.D. 10,000 who notes that native Hawaiian and New Zealand birds were largely replaced by exotic birds in the years A.D. 1800–2000 might conclude that competition from the introduced exotics exterminated the natives. In fact, as Moulton and Pimm (Chapter 5) stress for Hawaii and as is also true for New Zealand, the natives were mainly exterminated by introduced predators and diseases and habitat destruction before the exotics became numerous. Similarly, the extinction of caribou in Nova Scotia and New Brunswick following the introduction of white-tailed deer was due to a nematode parasite transferred from deer to caribou, not due to competition as one might at first have guessed (Anderson 1965, Embree 1979).

DIRECT AND INDIRECT EFFECTS OF INTRODUCTIONS AND EXTINCTIONS WITHIN ECOLOGICAL TIME

The direct effects of unplanned introductions or extinctions may be followed by a whole chain of indirect effects, just as in the case of experimentally planned introductions or extinctions (cf. Chapter 3). Introductions can cause extinctions of some species and can permit colonization by others, while extinctions can cause further extinctions (Pimm 1979) as well as boosting populations of surviving species. In this section we shall give a couple of examples of these effects for each of the six major types of species interactions.

Predation Introduced predators have been among the commonest causes of extinctions of native prey species on islands (Diamond 1984a). As for indirect effects, extermination of prey species (large herbivorous mammals and flightless birds) by prehistoric humans apparently produced secondary extinctions of many large predatory birds and mammals and carrion-feeding birds known as fossils, such as the giant owls, eagles, and vultures of the West Indies, the eagles of New Zealand and Hawaii, and the late Pleistocene teratorns, eagles, vultures, and mammalian carnivores of North America (Olson 1978, Olson and James 1982a, Steadman and Martin 1984). Extinction of the largest predators (jaguar, puma, harpy eagle) on Barro Colorado Island following insularization led to a population explosion of smaller mammals that served as their prey (monkeys, peccary, coatimundi, opossum), leading in turn to extinctions of ground-nesting birds on which peccaries and coatimundies prey (Terborgh and Winter 1980).

A chain of effects resulting from introduction of a predator has been worked out in the case of *Cichla ocellaris,* a strictly piscivorous fish. Introduced into a tributary of Panama's Lake Gatun around 1967, it proceeded to eliminate 8 of the 11 principal native fish species and to reduce 3 others by 75 to 90%. The secondary effects included an increase in abundance of a ninth (and possibly a tenth) species, whose juveniles had served as prey for species eliminated by *Cichla;* a shift in the zooplankton to forms that had been preferred as prey by the eliminated planktivorous fish species; possibly an increase in abundance of mosquitoes, which had also served as prey for eliminated fish species; and a decrease in herons, terns, kingfishers, and tarpons that fed on the eliminated fish (Zaret and Paine 1973).

Competition When two introduced planktivorous fish species (alewife and rainbow smelt) proliferated in Lake Michigan, three native fish species having negligible food overlap with these exotics remained common, while the seven native species that had feeding habits similar to those of the exotics declined most drastically (Crowder et al. 1981). Whalers in the southern oceans have progressively decimated whale species in descending order of size: blue, fin, sei, and minke whale. Each smaller species has increased its reproductive capacity upon elimination of its larger competitors, and the minke whale, the smallest species, is now more abundant than in the days before large-scale whaling (May et al. 1979). An example of competitive effects between birds and mammals (see also Chapter 3) is the decline of takahe, a flightless gallinule that browses on the same tussock grasses preferred by red deer, after the introduction of the deer to New Zealand (Mills and Mark 1977).

Herbivory Introduced goats converted the forests of St. Helena to grassland, and converted arborescent chaparral on some western North American islands to stands of an unpalatable "goat-proof" plant, tree tobacco *Nicotiana glauca* (Coblentz 1980). The forests of Three Kings Island off New Zealand were similarly converted by goats into monospecific stands of the goat-proof tree *Leptospermum ericoides*, which is yielding again to mixed forest following extermination of the goats by the New Zealand Wildlife Service (Turbott 1963).

Parasitism During this century the brown-headed cowbird (*Molothrus ater*), a brood parasite, invaded large areas of North America to which it was not native and proceeded to decimate numerous songbird populations (Goldwasser et al. 1980, Walkinshaw 1983, Diamond 1984b). The shiny cowbird (*M. bonariensis*), a recent immigrant to Puerto Rico from South America, has similarly driven the yellow-shouldered blackbird nearly to extinction in Puerto Rico (Post and Wiley 1977).

Disease The near-extermination of American chestnut by chestnut blight had the further indirect effects that birch and then oak increased in abundance (Davis 1981). The elimination of elm in Illinois forests by Dutch elm disease triggered a sequence of changes in bird populations, as bird species of edge habitats moved into forest openings created by death of elms and then vanished as the canopy closed over again (Kendeigh 1982). An instance from the fossil record is that the decimation of hemlock 4,800 years ago in eastern North America led to successions of other tree species: for instance, at Mirror Lake, New Hampshire, a rapid transient rise in birch, followed after a century by beech, sugar maple, and hornbeam (Davis 1981).

Mutualism The failure of the tree *Calvaria major* of Mauritius to establish seedlings in recent centuries, despite producing numerous seeds, has been attributed to its adaptation for seed dispersal by the now-extinct dodo (Temple 1977). (Seeds of *Calvaria* and many other tropical trees must pass through birds' intestines to germinate.) All five species of the endemic Hawaiian plant genus *Hibiscadelphus* are extinct or nearly so, due to the human-caused extinction of their pollinators, Hawaiian honeycreepers, whose long, curved bills match the plants' narrow, tubular, curved flowers. The large neotropical mammals that became extinct at the end of the Pleistocene included elephantlike gomphotheres and other frugivores, whose extinction must have affected the plants for which they acted as fruit dispersal agents and must thereby have affected in turn other dispersal agents and seed predators of these plants (Janzen and Martin 1982).

HOW DOES THE IMPACT OF AN INTRODUCED SPECIES DEPEND ON THE RECIPIENT COMMUNITY'S NAIVETÉ?

The impact of an introduced species on the recipient community varies enormously from case to case. Much of this variation can be explained by the community's "naiveté," that is, the extent of its prior experience with functionally similar species. There are clear examples involving predation, herbivory, parasitism, and disease.

Predation Three examples will illustrate the role of prey naiveté.

Introduced mammalian predators have decimated many native mammals in Australia, which had only five native cursorial carnivores of housecat size or larger. In contrast, the impact of introduced mammalian predators has been negligible in South America, which has a rich mammalian carnivore fauna of its own.

It is not surprising that introduced rats have had negligible effects on avifaunas of land masses supporting native rats (Solomon Islands, Christmas Island, Galápagos, North and South America, and Australia). Among those islands lacking native rats, introduced rats quickly decimated the birds of some islands (Hawaii, Midway, Lord Howe, and Big South Cape), but had little or no effect on the birds of other islands (Figi, Tonga, Samoa, Marquesas, Rennell, and Aldabra). Why this difference? Atkinson (1985) noted that the islands without native rats but on which the avifauna is ''immune'' to the effect of introduced rats all have (and the decimated islands lack) native land crabs, tree-climbing scavengers that are the invertebrate equivalent of rats. Thus, long exposure to either native rats or land crabs inoculates birds behaviorally against predation by introduced rats.

A worldwide puzzle in understanding extinction is that North and South America and Australia lost most of their large mammals at the end of the Pleistocene, but Europe, Asia, and Africa did not. Martin's (1984) interpretation is that mammals of the latter but not the former continents had long coexisted with man the hunter, who was able to decimate naive prey when he finally reached the former three continents.

Herbivory Many plant species that have evolved with mammalian herbivores possess antiherbivore defenses, such as spines or unpalatable chemicals. Island plants, lacking any coevolutionary history with mammalian herbivores, lack effective antiherbivore defenses, and this is probably the main reason for the devastation caused by introduced mammalian herbivores on oceanic islands.

Parasitism Long-established populations of brown-headed cowbirds use dozens of bird species as foster hosts. However, there are at least two instances in which cowbirds invading previously unoccupied areas of North America concentrated on naive victims that lacked experience of cowbirds. After cowbirds invaded Michigan, 93% of all cowbird eggs in jack-pine habitat were laid in Kirtland's warbler nests, and the warbler population declined 60% before the U.S. Fish and Wildlife Service began local eradication of cowbirds (Walkinshaw 1983). Cowbirds invading California similarly concentrated on and decimated the California race of the Bell's vireo (Goldwasser et al. 1980).

Disease The major mechanism by which invading human populations have decimated resident human populations has been through diseases to which the invaders had gained immunity through exposure but the residents had not. Classic examples are the death of 40,000 out of Fiji's population of 150,000 when measles first arrived in 1875, the decimation of Canadian Indians by tuberculosis in the nineteenth century, and the death of about half the population of Aztec Mexico (including Montezuma's successor Cuitlahuac) and of Inca Peru (including the ruler Huayna Capac) in the smallpox epidemics of 1520 and 1525, thereby making possible the conquests by Cortes and Pizarro (Zinsser 1935, McNeill 1976).

Similar cases for plants and animals are the elimination of the American chestnut *Castanea dentata* by the introduced Asian chestnut blight, which does not harm the Chinese chestnut *C. mollissima;* and the confinement of native Hawaiian birds today to the malaria-free zone above 3,000 feet, while introduced bird species that are resistant to introduced avian malaria thrive in the malaria zone at low elevations (Warner 1968, Atkinson 1977).

EVOLUTIONARY EFFECTS OF INTRODUCTIONS AND EXTINCTIONS

An introduction does not necessarily lead to evolutionary changes in the introduced species and the recipient community (see Fig. 4.1). First, the introduced species may be unable to establish itself (see section ''What Determines Whether an Introduction Will Succeed?''). Second, if the introduction does succeed, some naive species in the recipient community will be quickly exterminated before they can evolve defenses, for reasons discussed in the preceding section. Third, it may happen that the introduced species and recipient community species were preadapted to coexist, and no evolutionary adjustment is necessary.

In other cases, however, continued interaction of the introduced species and the recipient community leads to evolutionary adjustments of either or both. Occasionally, these genetic changes are rapid enough to be monitored within an ecologist's lifetime; occasionally, they can be documented by historical records; but, most often, they must be inferred by study of ''snapshot'' patterns or by comparisons of allopatric and sympatric populations. Examples are available for evolution (or coevolution) driven by disease, predation, parasitism, and competition.

Disease The history of infectious diseases provides the best studied examples of all steps in Fig. 4.1, including evolutionary and coevolutionary responses understood at the molecular level.

Diseases offer abundant examples of ''introductions that fail,'' when a newly arriving disease either killed or immunized so much of the population that the disease could not propagate itself and hence died out. Examples include the arrival and disappearance of measles in the Faeroe Islands in 1846, of poliomyelitus among Alaskan Eskimos in the 1920s, of O'nyong nyong fever in Uganda in 1959, and of many acutely infectious human diseases without animal vectors (e.g., measles, mumps, rubella, smallpox) in any small and isolated human community (Black 1975, McNeill 1976). Our familiar modern diseases were winnowed from a much longer list of diseases, most of which disappeared after a few epidemics and are known only from historical records, for example, the long forgotten ''English sweating sickness'' of 1485 to 1552 and the ''Picardy Sweat'' of the eighteenth century (Zinsser 1935).

''New diseases'' whose successful introduction to particular human populations or to humans in general was witnessed (Zinsser 1935) include tularemia (transmitted to man around 1900), epidemic infantile paralysis (first recorded around 1840), and syphilis (first European epidemic in 1495).

Influenza virus illustrates an evolutionary response with a well-understood molecular basis by which an introduced species continually evolves to stay ahead of its host species. The virus (especially influenza A) frequently changes two of its antigenic proteins, the hemagglutinin and the neuraminidase, thereby circumventing human antibodies and producing a new epidemic every year or two (Douglas and Betts 1979). Sleeping sickness trypanosomes change their antigens even more frequently, roughly every week.

Numerous initially lethal infectious diseases maintain themselves by evolving decreased virulence: the disease ceases to kill the host and merely causes the host to become sick in ways that spread the disease organism, such as coughing, sneezing, or diarrhea. A famous example is the spectacular transformation of syphilis between 1495 and 1546, from an often quickly lethal condition to the slow illness lasting several decades as we know it today (Zinsser 1935).

Evolution of host resistance is exemplified by the sickle-cell hemoglobin gene, which reduces the impact of malaria on human heterozygotes in West Africa, and the complete loss, in some areas of East Africa, of the Duffy blood group antigen, which in other human populations serves as the receptor for *Plasmodium vivax* malaria.

Finally, the decreasing impact of myxomatosis on Australian rabbits following its introduction in 1950 was caused both by evolution of decreased virulence on the virus's part and of resistance on the rabbit's part (Fenner and Ratcliffe 1966, May and Anderson 1983a).

Predation Lake trout (*Salvelinus namaycush*) have long coexisted with a landlocked population of the sea lamprey (*Petromyzon marinus*) in New York's Finger Lakes. Trout of the upper Great Lakes were isolated from lampreys by Niagara Falls, until construction of the Welland Canal let lampreys bypass the falls and reach Lake Erie in 1921 and Lake Michigan in 1936. The naive trout of Lakes Erie and Michigan were virtually exterminated within a short time, whereas the Finger Lakes trout remain common, perhaps because of having evolved behavioral avoidance of predator attack.

Parasitism In California, where the brown-headed cowbird did not arrive until the 1900s, parasitized Bell's vireos accept cowbird eggs, with the result that hatching and breeding success is low and the vireo population is near extinction (Goldwasser et al. 1980). In the U.S. Midwest, the traditional geographical range of the cowbird, the vireo abandons its nest and builds a new nest if parasitized by cowbirds, with the result that nest parasitism rates are modest, hatching and breeding success is high, and the vireo population is currently expanding (Nolan 1960, Overmire 1962). Evidently, Bell's vireos with long exposure to cowbirds eventually evolved nest-deserting behavior.

Other hosts with long exposure to brood parasites have evolved other forms of antiparasite behavior, such as driving off adult parasites, expelling parasite eggs, or burying the eggs under a new nest floor.

Competition Examples of evolutionary responses to the introduction or extinction of competitors abound. Roughgarden (Chapter 30) discusses evolutionary changes in the size of Caribbean *Anolis* lizards following invasion of one species' range by another; Schoener (1984a) shows for *Accipiter* hawks, and Case, Faaborg, and Sidell (1983) for West Indian birds, how interspecific competition during adaptive radiation of these groups has contributed to the observed pattern of greater-than-random size spacings among sympatric species.

Most of the classic cases of adaptive radiation on islands (Galápagos finches, Hawaiian drosophilid flies and achatinellid land snails, New Zealand moas, Madagascar lemurs) illustrate evolutionary responses of invaders freed of competition by reaching an island with few competing species. The adaptive radiation of mammals in the early Tertiary illustrates an evolutionary response to the extinction of competitors.

Conversely, Knoll (Chapter 7) shows how the evolution of each successive group of vascular plants during the past 400 million years exterminated many or all species of the preceding group. Similarly, of the eight successive waves of microtine rodents that invaded North America from Asia beginning 5.3 million years ago, the later waves eliminated all members of the first two waves, all of the third except for muskrats, and all of the fourth except for bog lemmings (*Synaptomys*) and drove survivors of the fifth wave southwards into mountain refugia (Repenning 1980).

DIRECTIONS FOR FUTURE RESEARCH

The ecological study of invasions, extinctions, exterminations, and introductions has been unjustly neglected. Part of the reason is that their often appalling consequences can make them esthetically unattractive. Another reason is that a properly planned experiment is obviously a much more powerful inferential tool than is an unplanned natural experiment—all else being equal. However, all else is not equal. Zealous empiricism should not prevent us from taking advantage of natural trajectory experiments which, though often lacking rigorous experimental design, nevertheless may involve spatial scales, time scales, types of perturbations, and insights unattainable through exclusion cages and other small-scale manipulations. A few of the many possible directions for future study deserve mention.

The literature on species introduced by humans is extensive. While there obviously are patterns to the varying success and varying impact of introductions, the causes underlying these patterns have been little explored. The quantitative analysis of failed introductions by Moulton and Pimm in this volume (Chapter 5), and Atkinson's (1985) unraveling of the factors governing vulnerability of island birds to rats, hint at the harvest awaiting students of this rich data base.

An equally large and underanalyzed literature is that on invasions. Many species are rapidly expanding their ranges now, such as the great-tailed and boat-tailed grackles in North America and the Syrian woodpecker in Europe. When one of us (J.M.D.) recently visited his boyhood home in a Boston suburb, he was astonished to find that the most abundant bird species now, the house finch, is a recent invader that had been absent during his childhood. What impact had its meteoric rise in numbers had on other bird species? What can be learned by actually observing the local impact of an invasion, or by comparing communities invaded in the recent past, invaded in the more distant past, and not yet invaded? For example, comparisons of North America bird communities exposed for a few decades, a few centuries, and uncounted millenia to brown-headed cowbirds (i.e., the Pacific coast, the Atlantic coast, and the Midwest) offer a unique opportunity to study the evolution of interactions between a brood parasite and its hosts.

The fossil record is another rich data base on past perturbations of communities. Recent notable examples of the quantitative study of these perturbations are the analysis of immigrations, extinctions, and radiations during the Great American Interchange by Marshall et al. (1982); Davis's (1981) tracing of North American tree succession 5,000 years ago, exploiting the abrupt crash of hemlock; and the recognition that successive interglacials let one follow a parade of alternative tree communities (Davis 1976, 1983).

In short, recognizing the right experiment that has already been done may offer insights unattainable through trying to do the right experiment oneself.

ACKNOWLEDGMENTS

We are grateful to Russell Graham, Peter Kareiva, Andrew Knoll, Thomas Schoener, and David Tilman for their suggestions on this chapter.

chapter 5

The Extent of Competition in Shaping an Introduced Avifauna

Michael P. Moulton and Stuart L. Pimm

INTRODUCTION

How important is interspecific competition in determining which species are present in a community? Does competition structure the community in some way? While some ecologists might reply with an emphatic yes to these questions, others are less certain. Too many results once explained by competition can be explained by other hypotheses. Distinguishing the hypotheses has proved to be a difficult task, and one that has generated continuing controversy.

Adding to this confusion is a general lack of agreement, as discussed in Chapters 20 and 28, over what exactly constitutes a community. In looking for community patterns, some authors have examined small sets of species (e.g., Case et al. 1983), whereas other authors have clearly made the distinction between the structure of small sets of species, i.e., guilds, and the structure of large sets of species, i.e., communities (e.g., Bowers and Brown 1982). It is pointless to argue about definitions: if by "community" one means a small set of species, then one must still ask if larger sets show patterns. References to communities in this chapter, following Bowers and Brown, will refer to large-scale taxonomic assemblages in the sense of Jaksic (1981), for example, all songbirds or all pigeons, and not to guilds. This distinction needs to be kept in mind when comparing our conclusions for broadly defined communities with the conclusions that some other chapters in this volume reach for more narrowly defined communities.

What, then, is the mechanism by which competition allegedly modifies communities? For communities structured largely by invasions and extinctions (see Rummel and Roughgarden, 1983, and Figs. 30.1 and 30.2) consider the following. (1) Species that are similar are more likely to compete than those that are not. (2) Competition, by definition, reduces growth rates and population sizes. (3) Small populations are more vulnerable to extinction. (4) With nonrandom extinctions we should observe communities where species are less similar than chance alone dictates. Thus, to answer our questions, we might assemble evidence on (1) growth rates or population sizes with and without the putative competitor(s), (2) extinctions, and (3) the resulting community patterns.

Evidence from Patterns

One way that ecologists have addressed the importance of competition is by seeking certain community patterns such as: (1) reduced taxonomic diversity, for example, a reduced number of congeners on islands (e.g., Grant 1966a), (2) overdispersed morphologies (e.g., Ricklefs and Travis 1980), and (3) checkerboard distributional patterns (Diamond 1975). The term "pattern" implies deviation from what chance alone dictates. And what chance alone dictates is the subject of intense debate (Connor and Simberloff 1979, 1983; Simberloff and Boecklen 1981; Diamond and Gilpin 1982; Gilpin and Diamond 1982; Simberloff and Connor 1982; Simberloff 1983). For example, it is difficult to predict the "ghosts," those species that once belonged, or tried to belong, to the community, but are now extinct. The critical features of the debate are the null hypotheses, and constructing these for natural systems can be tricky (Colwell and Winkler 1984). The null hypotheses attempt to predict the total set of species that should have made it to the communities in question. One predicts the ghosts by subtracting the observed species from this set.

We acknowledge the importance of this debate concerning null hypotheses. It seems that ecologists will forever be prisoners of statistical procedures. But despite the volume of intellectual blood spilled in this debate little headway has been made in answering the primary question of the importance of competition as an organizational force at the community level.

Evidence from Experiments

There is abundant evidence for competition in the form of species' reducing the growth rates or population sizes of similar species. This evidence invariably is based on small-scale experiments (see Schoener 1983b for a review). But competition between small sets of species says little about its extent in an entire community. It is possible that the experimenter has selected those species which, among all species, are most likely to compete. We know that this is the case for our experimental work (Pimm 1978, Pimm et al. 1985). It is also probable that negative results in the search for competition do not find their way into the published literature. And certainly, for some taxa (e.g., insects), demonstrations of competition are rare (Lawton and Hassell 1981).

The critical problem with such experimental demonstrations of competition is that no matter how common they are, competition does not lead inevitably to extinction. In fact, such experiments rarely demonstrate the ability of competition to cause extinctions, and without selective extinctions, competition cannot produce patterns in community membership. Moreover, even if competition does cause extinctions, community patterns could result from agents other than competition. These agents include the tendency for similar species to occupy similar habitats, be closely related phyletically, and have similar dispersal abilities. These agents often affect community structure in ways diametrically opposite to competition, and their effects may predominate. To show that competition shapes communities, one must show that competition is simultaneously *intensive* enough to cause extinctions and *extensive* enough to involve the majority of species, and that the effects of these extinctions *predominate* at the community level.

It is clear that we will never discover whether or not competition shapes communities by simply showing that small sets of species compete or that community patterns exist. These two types of evidence must be linked, and the critical bond between them must come from patterns of extinctions. Thus it is on extinctions that we shall concentrate our efforts. We believe that a consideration of extinctions answers some questions far more clearly than any of the alternatives.

If extinctions could be so useful in answering all our questions, why have they not been considered more often? Species extinctions in natural communities are likely to be infrequent and they may not be where and when we want them to be. A much better strategy would be to create an experimental community and watch some of the species become extinct. This is exactly what Crowell has done with small mammals (Crowell 1973, 1983; Crowell and Pimm 1976).

But these experiments involve few species and are on a small scale. To extend studies of this type, we would need several areas as replicates. These replicates should be isolated from potential immigration and emigration and devoid of addi-

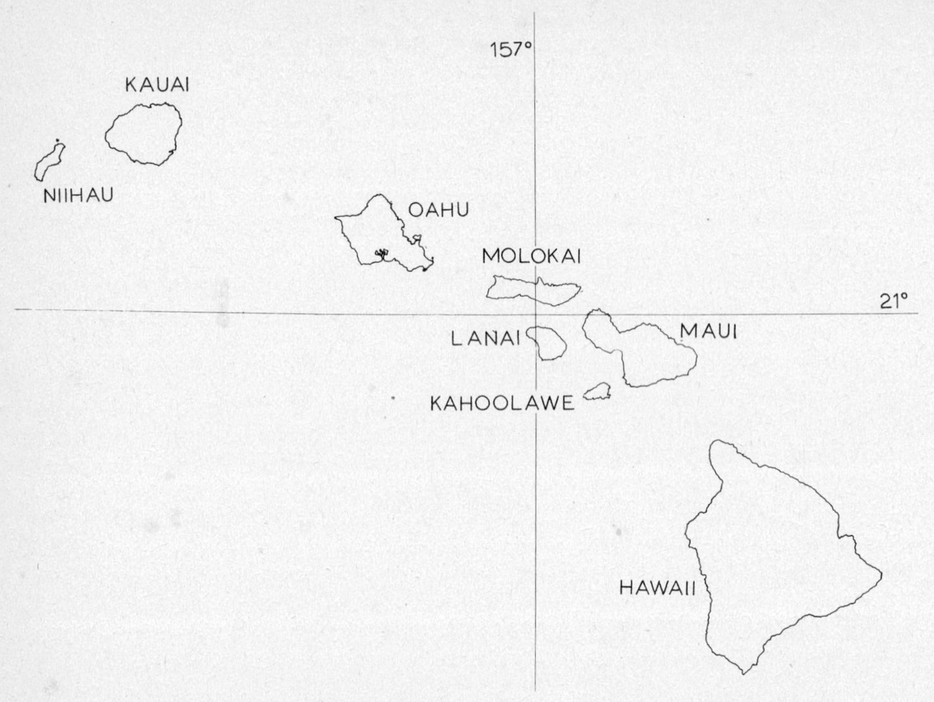

Fig. 5.1 Map of the main Hawaiian Islands.

tional species. We would probably want to watch the system for a century or so, and, for the results to be relevant to some current controversies, it would help if we looked at birds on islands. Impractical as these conditions seem, they are satisfied by the introduced Hawaiian avifauna—an interesting, but not unique case, of humans creating their own animal communities. In the terminology of Chapter 1, the fate of that avifauna constitutes a natural trajectory experiment.

In short, there are many questions involving the role of competition in communities. These questions differ greatly in how important that role should be: the existence of competition does not mean that it is either intensive or extensive; even if it is both, it does not follow that communities are structured by competition. Furthermore the answers to the questions are debatable. On the one hand, small-scale experiments say nothing about large-scale patterns. On the other hand, the analysis of patterns is difficult. There may be right and wrong ways to detect these patterns. But ecologists are so far from agreement that seeking alternative evidence seems, to us, to be the best tactic available. Since one difficulty in detecting patterns stems from predicting the species that have become extinct, it seems reasonable to look at extinctions directly. Such studies are few; the introduced Hawaiian avifauna provides probably the largest source of data.

AN EXPERIMENT IN COMPETITION: THE INTRODUCED HAWAIIAN AVIFAUNA

The remnants of native Hawaiian forests harbor the remains of a large, bizarre endemic fauna and very few exotic species. Below 600 m, however, only the most dedicated observer will find native species in the almost totally exotic forests. Bird species introductions began in the mid-1800s and, through accidental escapes, continue to this day. Different numbers of species have been introduced to each of the six main islands: Kauai, Oahu, Molokai, Lanai, Maui, and Hawaii (Fig. 5.1). Most species have been introduced to Oahu, where most of the people live. Professional ornithologists as well as amateur birdwatchers have kept records over the last century, most actively since the 1940s. This enables us to estimate, at least to the nearest decade, when each species was introduced and if and when it became extinct.

More species of birds have been introduced to the Hawaiian Islands than anywhere else (Long 1981). By the time most of these species were brought to the islands, native habitats below 600 m had been severely altered through grazing by introduced mammals (Tomich 1969); deforestation for timber, particularly sandalwood (*Santalum* sp.; Rock 1913, Kuykendall and Day 1948); and clearing land for cultivation, both by Europeans and the early Hawaiians (Kirch 1982).

In conjunction with this habitat disturbance there was a loss of native species of birds (Henshaw 1902, Greenway 1967, Olson and James 1982b). The native land birds were (and still are) closely associated with native habitats, and in the 1800s these were being replaced by cultivated land, pastureland, and exotic forests (Kuykendall and Day 1948; Schmitt 1977). Introduced species, by contrast, are commonly found in these disturbed habitats; native species are not. Indeed, today it is unusual to find native land birds below 600 m (Berger 1981). Even thirty years ago, only four native passerine species were seen on three of the Honolulu Christmas bird counts (1950, 1954, 1956) organized by the Hawaiian Audubon Society. Of the 8,379 individual passeriforms and columbiforms seen, only 706 were natives (less than 9%). These data strongly suggest that native land birds have played, at most, a minor role in determining which species of introduced birds have succeeded or failed in the islands.

Many of the introductions were made intentionally and with substantial numbers of individuals. For many years there was an organization, the Hui Manu, which existed solely for the purpose of introducing exotic birds into Hawaii. Natural immigration, by contrast, is essentially nil because of the remoteness of the islands. Berger (1981) lists only seven species of passerines as stragglers to the Hawaiian Islands, and these were all recorded on the distant leeward islands and not on the main islands considered here. Since publication of Berger's book there have been reports of starling (*Sturnus vulgaris*) on Hawaii (Elliot 1980) and Oahu (Pyle and Ralph 1981) and great-tailed grackle (*Quiscalus mexicanus*) on Oahu (Bremer 1984), but whether or not these were truly stragglers from the mainland or just cage escapes is unknown. Thus natural immigration of passerines must be considered extraordinarily rare.

In this chapter we consider the introductions from 1860 to the present. During this period the rate of introductions generally increased. The details of our methods are given elsewhere (Moulton and Pimm 1983). We made lists of the dates of introductions and extinctions from a variety of references: Henshaw (1902), Caum (1933), Munro (1944), Berger (1981), Bryan (1958) and the journal *Elepaio* (vol. 1, 1939 to the present). We defined adequate propagule size to consist of at least one male and one female. In practice, propagules were typically larger than this, because of the intentional nature of the introductions. We grouped introductions and extinctions into 10-year periods. Although we could not be confident that an extinction or an introduction took place in the exact year we estimated, we could usually be certain of the time to the nearest decade.

The evidence to evaluate the possible effects of competition comes from four sources: (1) the overall rate of species extinctions, (2) the relationship between the extinctions and morphological similarity, (3) the relationship between the extinctions and taxonomic similarity, and (4) the variations in population densities caused by the different patterns of extinctions on the different islands.

THE RATE OF SPECIES EXTINCTIONS

How should interspecific competition affect the rate of extinctions? If there is no competition, we should expect a linear relationship between extinctions and the number of species present. Equivalently, the chance of each species' becoming extinct is a constant. Consequently, the more species that are introduced, the more there are to become extinct. In contrast, if there is competition, we would expect a nonlinear, accelerating relationship between extinctions and the number of species present. The probability of any species' becoming extinct would rise with increasing numbers of interspecific competitors.

The Pattern of Extinctions

For this analysis, we assembled data on passerine introductions and extinctions from 1860 to 1983.

Fifty species were introduced a total of 125

times to the 6 main islands during the period under consideration. There were 35 extinctions. Dates of the introduction and extinction of species introduced before 1960 are listed in the appendix of Moulton and Pimm (1983). Dates of the introduction and extinction of species introduced after 1960 are available from the authors on request. Forty-two species have been introduced to Oahu, 22 to Kauai, 15 to Maui, 22 to Hawaii, 13 to Molokai, and 11 to Lanai. There were 17 extinctions on Oahu, 7 on Kauai, 2 on Maui, 5 on Hawaii, 0 on Molokai, and 2 on Lanai.

There were no extinctions until the 1920s. Seven species were introduced to all the islands before 1920. In this group there has been only one subsequent extinction and that on only one island: the skylark, *Alauda arvensis,* on Kauai.

These extinctions cannot be viewed as trivial consequences of introducing species to totally inappropriate habitats. There are many examples of species that were present for a reasonably long period of time (10 years or more) and subsequently became extinct (20 of 35, or 57%; see Moulton and Pimm 1983).

We found a significant positive relationship between the extinction rate per decade and the number of species present at the end of each decade, S ($F = 41.2$; $p < 0.0001$). This relationship, however, was significantly improved by the addition to the statistical model of the square of the number of species, S^2 ($F = 7.25$; $p < 0.01$; i.e., the relationship is significantly nonlinear). In contrast the model with S^2 is not improved by the addition of a term in only S ($F = 0.00$; $p = 0.95$). Thus the best model is one with only S^2; i.e., the extinction rate increases faster than the number of species. Equivalently, the per species extinction rate increases linearly with the number of species present. Incidentally, we found that the extinction rate was not significantly related to island size ($F = 0.08$; $p = 0.79$).

We have also attempted to fit our extinction data with the following nonlinear statistical model:

$$Er = \beta_0 (S)^{\beta_1} \qquad (5.1)$$

where β_0 and β_1 are fitted constants, S is the number of species, and Er is the extinction rate. The critical point lies in the magnitude of β_1. Any value significantly greater than unity implies that competition has modified the extinction rate. The larger the value for β_1, the faster the effect of competition intensifies. In effect, it measures the tendency for competition to be a contest or a scramble (Diamond and May 1981). In this nonlinear analysis we estimated $\beta_1 = 1.85 \pm 0.41$. We could fit a wide variety of functional forms to these data (including a piecewise linear model), but there seems to be no escaping the inherently nonlinear nature of these data.

There are many reasons why a species should become extinct. Some species undoubtedly failed to find their appropriate habitats. For others there may have been an insufficient propagule size—although we did attempt to exclude most of such cases from our analyses. Many species probably found suitable habitat in such limited quantities that high population sizes were not possible. But all these possibilities suggest a linear relationship between extinction rate and the number of species introduced. The extinction rate, of course, rises much faster than this (Fig. 5.2), and we find this result to be compelling evidence for interspecific competition. Others (e.g., Williamson 1983) have presented data on rates of extinctions, but they have not always found the data to be of adequate quality to justify tests of curvilinearity.

A nonlinear extinction rate has been claimed by others (Schoener 1976a, Gilpin and Diamond 1976) from indirect methods of fitting various equations to equilibrium numbers of species on islands. The approach most similar to ours in directness is that of Jones and Diamond (personal communication), who used islands censused year after year for breeding birds to obtain series of measured values (Er, S) and found results to fit eq. 5.1 with a β_1 value above 3. We find it encouraging that other authors have reached a similar conclusion using different data sets.

Other Explanations

Could this nonlinear extinction rate be explained by factors other than interspecific competition? There are two possibilities and we discuss why we reject them in detail elsewhere (Moulton and Pimm 1983). These are the arguments in brief.

The first possibility is habitat alteration. Species tend to accumulate through time; there are more species in the later decades than in the early ones on all the islands. Habitat alteration might

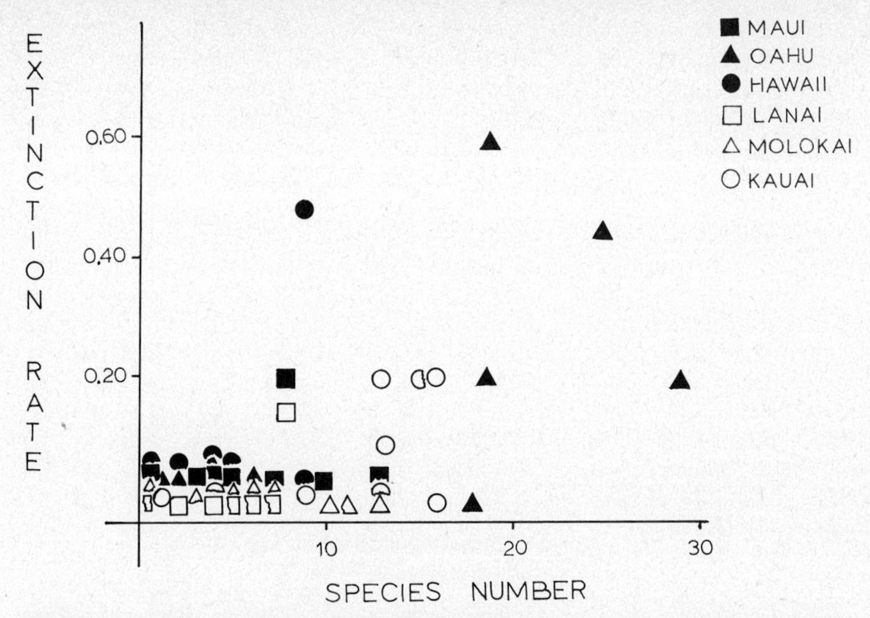

Fig. 5.2 Extinction rate for the period 1860 to 1980 expressed as the number of extinctions per year, averaged over 10-year periods, against the number of passerine species surviving on each island at the end of the 10-year period.

also accelerate through time, with more extensive alterations when there were, coincidentally, more species. As already discussed, the introduced species inhabit disturbed habitats such as cultivated land (which is mainly sugarcane), pastureland, and exotic forests. However, such habitats are probably different from those in which the majority of these species evolved. Moreover, land use data show that these habitats changed far more in the first five decades than the last five; yet there were no extinctions when the habitats were changing the most. Hence this first possibility does not seem a plausible explanation for the nonlinear rate of extinction.

The second possibility deals with factors intrinsic to each species. Some species have been successful nearly everywhere in the world that they have been introduced. Other species, perhaps highly specialized for restricted habitats, probably would fail no matter where they were introduced. The possibility exists that basically successful species were introduced early and species highly prone to failure were introduced later.

There are two arguments against this. First, and perhaps most persuasive, is that several species that are very common in Hawaii today were introduced after 1920, and of four species introduced even after 1960, three are common on Oahu and the fourth on Hawaii, Maui, and Lanai. Second, some species which failed in Hawaii have been very successful elsewhere (Moulton and Pimm 1983).

In short, these two arguments lend little support to the idea that species introduced early were intrinsically more likely to survive than those introduced later. We believe that the nonlinear extinction rate is best explained as an effect of interspecific competition. Thus we conclude that there appears to be competition and that it is seemingly strong enough to effect extinctions.

MORPHOLOGICAL SIMILARITY AND SPECIES EXTINCTIONS

As mentioned in the introduction, statistical problems arise in deciding whether or not surviving species' morphologies are overdispersed in existing communities. The basic problem lies in how one determines which morphologies are missing (i.e., which species did not survive). For Hawaiian birds, this is not a problem; we know which species survived and which species became extinct. Thus we are forced to resist the temptation to tinker with null hypotheses until we reject one. Furthermore, it permits us to look at how morphological patterns develop. We can look for evidence of competition in causing extinctions and we can see whether competition causes communitywide patterns in species' morphologies. Most directly of all, we can look at the relationship between the chance of extinction and morphological similarity to see how far-reaching are the effects of competition.

To do all this, we shall first consider the patterns of coexistence in several pairs of introduced congeners. It is for these species that we are most likely to find evidence for competitively mediated extinctions, if it exists. Second, we shall consider the pattern of species' morphologies for all the introduced species. Third, we determine the relationship between the probability of species extinction and morphological similarity.

Coexistence of Congeners

Moulton (1985) compiled a list of congeneric species pairs using Berger (1981) with some modifications described by Moulton and Pimm (1983). Some species or species pairs had to be excluded. Sometimes there were difficulties in field identification that rendered sight observations suspect. Some congeners were introduced to different islands or widely separated areas of the same island. And, finally, some species were probably not truly wild and others did not occur on the same island at the same time. Details of these exclusions are given in Moulton (1985).

Methods for determining extinction dates prior to 1960 are given in Moulton and Pimm (1983). After 1960 Moulton (1985) considered a species to be extinct when a two-year period passed with no observations reported in the journal *Elepaio*. All these extinctions occurred on Oahu in areas that are frequented by observers.

In these tests bill length was used as a measure of morphological similarity. Grant (1966a, 1968, 1969b) and Schoener (1965) also used bill length. For species already so similar as to be considered congeners this one measure of difference is probably adequate. The measure of morphological similarity involved calculating percent difference (Grant 1969b), here defined to be

$$PD = 100\left(\frac{\text{length of longer bill}}{\text{length of shorter bill}} - 1\right)$$

Of the 18 pairs of congeners included in the analysis, there were 3 cases where both species became extinct, 9 cases where 1 species became extinct (we call these group A), and 6 cases where neither species has yet become extinct (group B). If interspecific competition has caused some of these extinctions, we might expect the morphological differences to be smaller in group A than group B. They are. In group A the differences varied from 1 to 29%, 6 of the 9 pairs differed by less than 10%, and the mean was 9%. In group B the differences varied from 2 to 38%, only 1 of the 6 pairs differed by less than 10%, and the mean was 22%. These 15 cases are illustrated in Fig. 5.3. The difference between the two means is significant ($F = 5.14$; $p < 0.05$).

Communitywide Analyses of Morphological Patterns

Is the overall pattern of the morphologies of the successfully introduced birds overdispersed? In an absolute sense, species' morphologies are undoubtedly not overdispersed, because the evolutionary process produces clumps of similar species. Thus we must search for an overdispersed pattern among successful species *relative* to some larger set of species (i.e., the species pool). As argued above, our previous evidence for competition causing extinctions and, indeed, causing extinctions more commonly in morphologically similar species do not mean that the morphologies will be overdispersed in this relative context. Other factors may predominate and thus swamp the competitive effect.

In seeking such patterns, Grant and Abbott (1980) and Colwell and Winkler (1984) have noted that statistical problems (i.e., an increased probability of not detecting competition when it is present) arise when taxonomically (and hence morphologically) distant species are grouped in these analyses. In fact, Grant and Abbott, following Darwin (1859) and Lack (1947), claimed that even species from different genera should not be included in the same species pool when looking for nonrandom morphologies. And Colwell and Winkler have shown by means of computer simulations that species on average appear more different from one another if drawn from a species pool that includes phylogenetically (=morphologically) distant species.

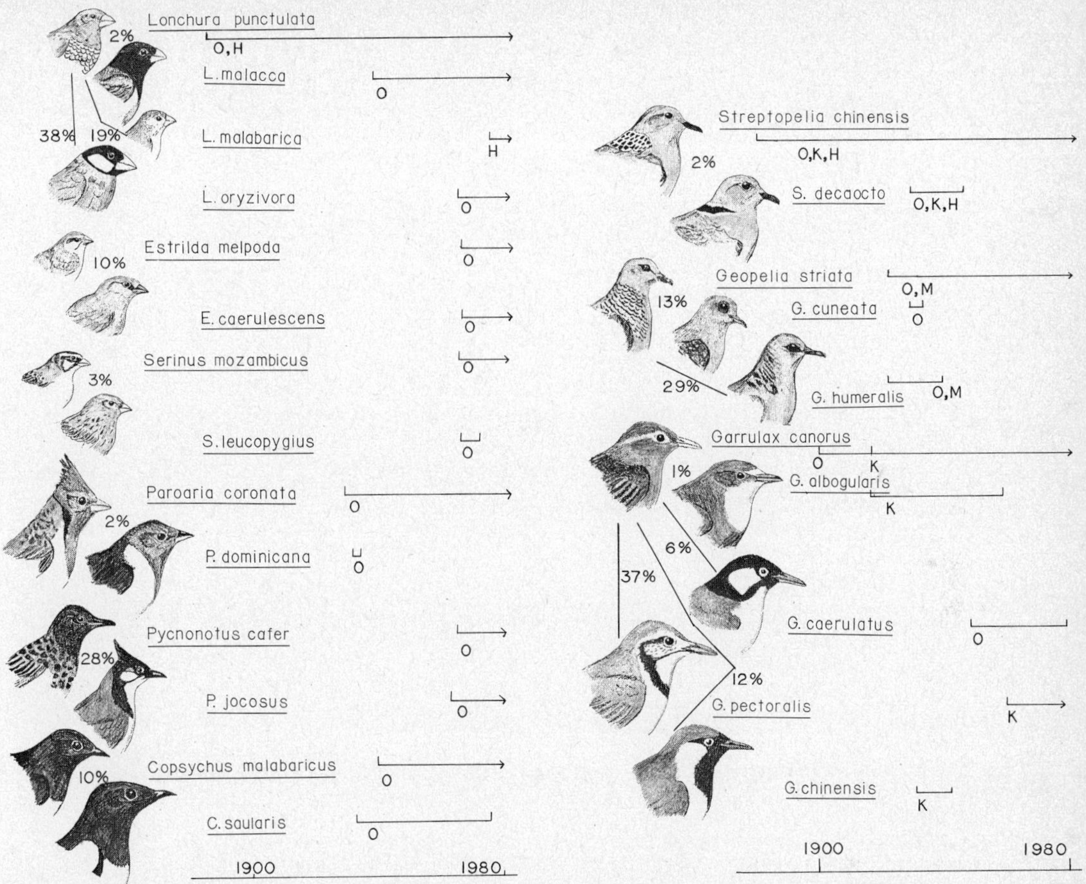

Fig. 5.3 Morphological similarity of congeneric species pairs and their survivorship on various islands: Oahu (O), Kauai (K), Hawaii (H), and Maui (M). Percentages give relative difference in bill length (calculated as defined in the text). Vertical lines indicate approximate years of introduction and extinction, and arrows indicate that the population was still extant in 1980. For example, *Garrulax canorus* was introduced to Oahu around 1910 and to Kauai around 1920 and still survives on both islands. Its congener *G. albogularis,* which differs from it in bill length by only 1%, was introduced to Kauai around 1920 and went extinct around 1960.

We emphasize that our aim here is not merely to detect competition, but to see if its effects predominate at the community level. To do this we must examine large sets of species. It is clear that there are several ways in which we could define such sets. In particular, we could either analyze the passerine species separately from the nonpasserines or treat them together. There are examples in the literature for both choices; Ricklefs and Travis (1980) treated the groups separately, whereas Case et al. (1983) treated them together. We performed our analysis both ways and found that it makes no difference to our final conclusions. Either way, our approach to seeking nonrandom morphologies was similar to others in the literature.

For all the other analyses in this paper we have used data for the period 1861 until the present. This analysis, however, is limited to the century prior to 1960. The fate of some post-1960 introductions was not clear when we started our analyses. Two years of field work have eliminated much of this uncertainty. Unfortunately, we did not record morphological measurements for all of the recent introductions.

We measured six morphological characteris-

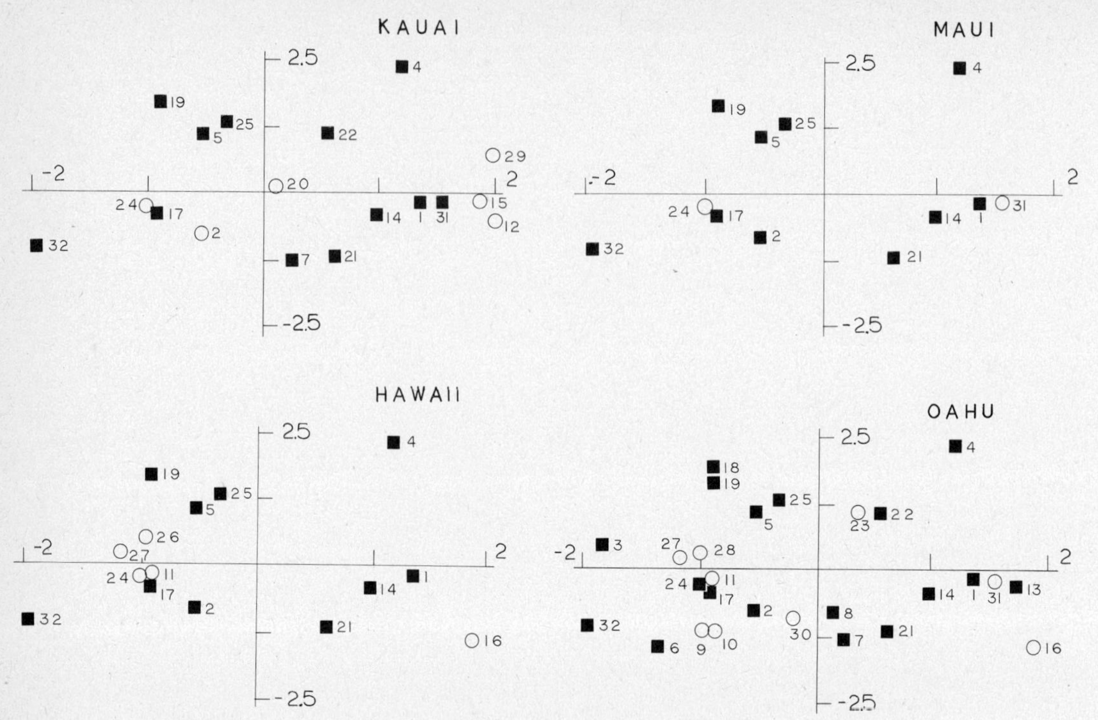

Fig. 5.4 Principal component scores of surviving (shaded squares) and extinct (unshaded circles) species of passeriforms on the main islands. Key: 1 = *Acridotheres tristis*, 2 = *Alauda arvensis*, 3 = *Amandava amandava*, 4 = *Cardinalis cardinalis*, 5 = *Carpodacus mexicanus*, 6 = *Cettia diphone*, 7 = *Copsychus malabaricus*, 8 = *Copsychus saularis*, 9 = *Luscinia akahige*, 10 = *Luscinia komadori*, 11 = *Muscicapa cyanomelana*, 12 = *Garrulax albogularis*, 13 = *Garrulax caerulatus*, 14 = *Garrulax canorus*, 15 = *Garrulax chinensis*, 16 = *Grallina cyanoleuca*, 17 = *Leiothrix lutea*, 18 = *Lonchura malacca*, 19 = *Lonchura punctulata*, 20 = *Melanocoryphora mongolica*, 21 = *Mimus polyglottos*, 22 = *Paroaria coronata*, 23 = *Paroaria dominicana*, 24 = *Parus varius*, 25 = *Passer domesticus*, 26 = *Passerina ciris*, 27 = *Passerina cyanea*, 28 = *Passerina leclancherii*, 29 = *Pezites militaris*, 30 = *Rhipidura leucophys*, 31 = *Sturnella neglecta*, 32 = *Zosterops japonicus*.

tics of each species: wing (chord), tail, tarsus, bill depth, bill length, and bill width. Why use six variables, when, in the previous analysis we used only one? For these analyses, involving a much larger group of species than for the previous analysis, we felt that several morphological characteristics might be needed to adequately define the ecological differences among the species. Other authors have made this same judgment (e.g., Findley 1973, 1976; Karr and James 1975; Ricklefs and Travis 1980; Ricklefs et al. 1981). We are not convinced that such ecological differences can be defined at this level of analysis by a single dimension such as body weight (see Case et al. 1983). There is, of course, no problem in calculating morphological distances between any number of variables. But by including a large number of variables we run the risk of measuring the same, underlying features over and over again, with added statistical "noise" each time. So we reduced the number of variables used to calculate distances between species by using principal components analysis, which produces composite variables by taking linear combinations of the original six variables. We chose only the most important two composite variables because these, alone, accounted for more than an average amount of the data's variability.

The results of this statistical witchcraft are displayed in Figs. 5.4 and 5.5. Each species on each island is represented by a point whose coordi-

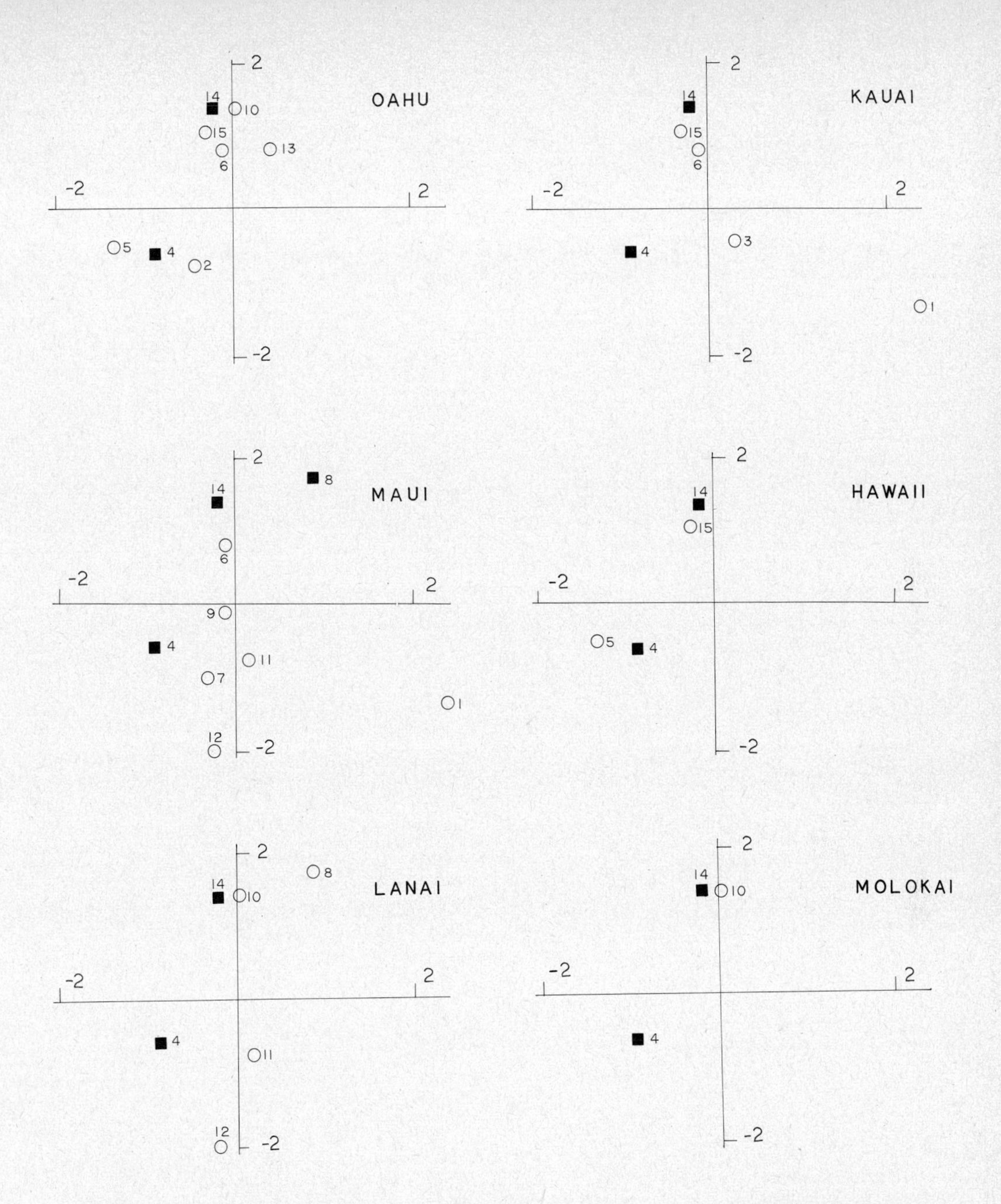

Fig. 5.5 Principal component scores of surviving (shaded squares) and extinct (unshaded circles) species of columbiforms on the main islands. Key: 1 = *Caloenas nicobarica*, 2 = *Chalcophaps indica*, 3 = *Gallicolumba luzonica*, 4 = *Geopelia striata*, 5 = *Geopelia cuneata*, 6 = *Geopelia humeralis*, 7 = *Geotrygon montana*, 8 = *Leucosarcia melanoleuca*, 9 = *Leptotila verreauxi*, 10 = *Ocyphaps lophotes*, 11 = *Petrophassa smithii*, 12 = *Petrophassa plumifera*, 13 = *Phaps chalcoptera*, 14 = *Streptopelia chinensis*, 15 = *Streptopelia decaocto*.

nates are the values of the two composite variables (principal components). The first two principal components accounted for 87% of the total variance in the passeriform analysis, 89% of the total variance in the columbiform analysis, and 81% in the combined analysis.

In the case of the passerines (Fig. 5.4) the first principal component (factor 1) can be interpreted as an overall measure of the bird's size. It correlates highly with length of tail, culmen, wing, and tarsus and to a lesser extent with bill width and depth. The second principal component (factor 2) is highly correlated with bill width and depth. This variable reflects the shape of the bill. Passeriform species with high values of factor 2 have large bill depths for their overall size (these species are granivores). Conversely, species with low values of factor 2 have shallow bills and tend to be insectivores.

In the columbiforms (Fig. 5.5) the first principal component (factor 1) again provides a measure of overall size. However, high values for the second principal component (factor 2) reflect a long tail. We interpret this variable to be an indicator of where the dove feeds: arboreal doves tend to have longer tails than ground-feeding species.

The results of our combined analysis were qualitatively similar to those of the passeriform analysis. Factor 1 was correlated positively with all the original variables, whereas factor 2 had its highest correlations with depth and width of bill. However, in this analysis factor 2 also was correlated (negatively) to a greater extent than in the passeriforms with length of wing and tail.

The remaining stages in our recipe differ from those of previous studies. We consider each island in turn. For a typical analysis, suppose that we have m species in total, that n survived, and that $(m - n)$ became extinct. We can measure the dispersion, in our two-dimensional morphological space, of the n surviving species by a statistic θ_{obs}. The distribution of θ is likely to be complicated. It will depend, for example, on n: the more species that survive, the closer on average they will be to each other. While we cannot look up the distribution of θ in a table, we can generate it using Monte Carlo simulations. There are $(m!/(m - n)!n!)$ distinct ways in which the extinctions can occur: there will be a value of θ, namely θ_i, for each.

Next consider the quantity P, the proportion of the θ_i that equal or exceed θ_{obs}. Clearly if P is sufficiently small (typically, < 0.05), then we would reject the null hypothesis that the morphologies of the surviving species are random. Finally, we can generate several values of P. We can generate not quite 12 values, 6 islands × 2 analyses (passeriforms and columbiforms), because for some combinations there were no extinctions. (There were no extinctions of introduced passeriform species on the islands of Molokai and Lanai between 1860 and 1960.) Individually, the values may not be small enough to reject the null hypothesis, but together they might—in just the same way that a coin's coming up heads once is not suspicious, but coming up heads 10 times in succession would be. We can combine the values of P using a result of Fisher's (1950) which involves converting the proportions to chi-square (χ^2) values (examples in Pimm 1980). These values are additive and give an overall test.

The only omission from our recipe is the choice of the statistic θ that measures dispersion. There are several possible statistics and this number of possibilities raises a problem. Some may be better than others, but there is no guide to which statistic is best; elsewhere in the realm of statistics there are means of analysis to show which statistic is "best" or "sufficient" (Hogg and Craig 1978). We could have run our analyses on all the possible statistics, but this would hardly be fair. Our experience suggests that repeated analyses seem to give "significant" results far more often than they should. Instead we decided to pick one statistic and accept our results. We chose the length of the minimum spanning tree (*MST*) for the morphologies of the surviving species. There is a large number of spanning trees: sets of $(n - 1)$ interspecies distances that connect all the n species. Of these, the *MST* has the shortest combined length and, incidentally, the smallest mean distance between connected species (since that mean = $MST/[n - 1]$).

Our results are shown in Table 5.1. According to our recipe, the surviving species' morpholo-

Table 5.1 RESULTS OF MONTE CARLO SIMULATIONS TO TEST WHETHER THE MORPHOLOGIES OF SUCCESSFULLY INTRODUCED SPECIES WERE OVERDISPERSED RELATIVE TO ALL POSSIBLE SETS OF INTRODUCED SPECIES

Island	I	S	MST	R > O	P_i	χ^2
Columbiforms						
Hawaii	4	2	1.90	2/6	0.33	2.21
Kauai	6	2	1.90	5/15	0.33	2.21
Lanai	6	2	1.90	7/15	0.47	1.51
Maui	9	3	3.21	102/200	0.51	1.35
Molokai	3	2	1.90	2/3	0.67	0.80
Oahu	8	2	1.90	7/28	0.25	2.77
						10.85 NS ($0.9 < p < 0.5$)
Passeriforms						
Hawaii	15	10	7.88	55/200	0.27	2.62
Kauai	18	12	8.41	104/200	0.52	1.31
Maui	12	10	7.88	26/66	0.39	1.88
Oahu	27	18	10.39	32/200	0.16	3.67
						9.48 NS ($0.5 < p < 0.1$)
Combined						
Hawaii	19	12	9.27	50/200	0.25	2.77
Kauai	24	14	9.72	78/200	0.39	1.88
Lanai	13	9	8.85	30/200	0.15	3.79
Maui	21	13	10.29	25/200	0.12	4.24
Molokai	13	12	9.27	4/13	0.31	2.34
Oahu	35	20	11.38	38/200	0.19	3.32
						18.34 NS ($0.5 < p < 0.1$)

I = number of species introduced; S = number of surviving species; MST = length of observed minimum spanning tree for the morphologies of surviving species; R > O = number of random MSTs greater than the observed value divided by number of random MSTs examined; P_i = R > O expressed as a proportion; χ^2 = chi-square values obtained from the proportions.

The number at the bottom of the chi-square column is the sum of the chi-square values, with its probability level in parentheses.

gies are not more dispersed than we would expect by chance. Indeed, they are not even close to rejecting our null hypothesis. Even the combined chi-square test does not reject the null hypothesis and it assumes independence of the *P* values—an assumption we would have trouble defending.

Of course, we could reanalyze our data with different recipes and a host of different statistics. And if we did, we might eventually find a ''significant'' result. But more likely than not, we would be misled and we would certainly avoid facing up to our null hypothesis and its consequences. Simply, in communitywide morphologies, structure is not detectable in this system. Why not?

There could be several reasons why a morphological effect was found in the analyses of congeners (Fig. 5.3), and not for the entire community. We did, after all use a different statistical approach in the second analyses. But if obvious patterns had been present in the communitywide analyses, we think our technique would have detected it. To understand this result, it is important to recall our broad definition of a community. We suggest that the difference between the analyses of congeners and the entire community stems from the level at which the analyses were carried out: the former between morphologically similar species that share an evolutionary history, the latter spanning a much greater range of morphological variation among species that probably do not share an evolutionary history. Species in the former combinations are likely to compete strongly, to enhance significantly the risk of each other's extinction, and to have overdispersed morphologies. Species in the latter combinations are likely to compete weakly, if at all; to contribute little to each other's risk of extinction; and to have nearly randomly distributed morphologies. Hence, when one dilutes the 18 populations of congeneric pairs of Fig. 5.3, which provide evidence for morphological overdispersion, in the much larger set of all 107 populations in Figs. 5.4 and 5.5, most of which are too dissimilar to affect each other, the overdispersion signal becomes submerged. This dilution effect is a frequent problem in searches for nonrandom patterns and has been discussed by Grant and Abbott (1980), Bowers and Brown (1982), Diamond and Gilpin (1982), Harvey et al. (1983), and Colwell and Winkler (1984).

What, then, is the quantitative relationship between chance of extinction and morphological difference? How morphologically similar must two species be to enhance each other's risk of extinction? Our data allow us to examine the relationship directly.

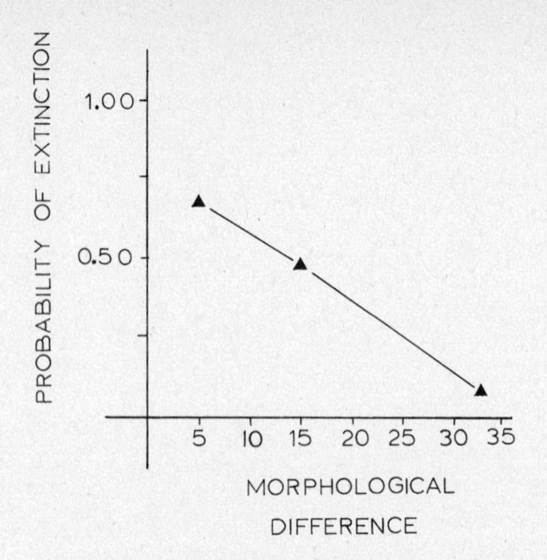

Fig. 5.6 Probability of extinction of a species, plotted against the morphological difference between the species and its most similar congener. Morphological difference is expressed as percentage difference in bill length.

Probability of Extinction Related to Morphological Difference

It is a simple matter to record the morphological difference between congeners and to note whether one or neither of the species became extinct. (We should exclude pairs where both became extinct, since this is not an outcome compatible with pairwise competition.) We can then sort the differences into several groups, calculate the overall proportion of extinctions in each, and so determine its functional relationship to morphological difference. Fig. 5.6 shows that the chance of extinction drops quickly as differences (in this case, in bill length) increase.

If we extrapolate the function to higher values of morphological distance, the function must be nonlinear, because the chance of extinction cannot be less than zero. The function must either become zero at about 35% difference (when we get a step function) or else approach zero more slowly (when the function is certainly curved). This result is almost embarrassing. It suggests that congeners differing in bill ratios by less than 1.35 are likely to be many fewer than we would expect—an old idea in ecology (Hutchinson 1959). Of course, although significant, the confidence intervals on the relationship are very wide. But most importantly, this relationship can be tested on many other bird groups. We only need a pair of calipers, some days in a good museum, and the extensive data compilations of Long (1981) to see if this function is general. A similar relationship for congeners "introduced" by speciation rather than by humans is illustrated by Diamond (see Fig. 6.4).

In summary, we find a relationship between morphological similarity and extinction. But, although powerful, it is a short-range relationship. At the community level, species' morphologies are essentially randomly distributed. The effects of competition may impose some minimum distance between species on this overall pattern. The effects of competition are not sufficient for us to detect any structuring of species' morphologies at the community level.

SPECIES-TO-GENUS RATIO

Our data permit a particularly simple test of whether competition reduces the species-to-genus ratio by increasing the extinction rates among congeners. Because we can observe the extinctions, we do not need to reconstruct them using techniques that are encumbered by the difficulties described by Graves and Gotelli (1983). Forty-three introductions involved species without a congener on the island of introduction. A further 32 introductions involved multiple introductions per genus on the island of introduction: they represent 12 genera. (These data are listed in Table 5.2.)

In compiling these lists, species that survived on one island but became extinct on another were counted as two separate introductions (e.g., *Alauda arvensis* became extinct on Kauai but survived on all other islands). Also, species that were introduced to one island with a congener and another island without a congener were treated as separate introductions (e.g., *Streptopelia chinensis* had a congener on Oahu but not on Maui). Species that survived (or failed) on more than one island were only counted once per category. Finally, we ignored the genus *Uraeginthus* on Oahu, because of confusion in field identification. (As many as three species could have been introduced to Oahu; see Berger 1981.)

In the first group there were 20 extinctions and 23 survivals. In the second group the extinction rate was only slightly higher, but not significantly so as predicted by competition: 16 species survived, 16 became extinct. Simply, we have no evidence to suggest that species belonging to the same genus are more prone to extinction than those that are not. We interpret this result to mean that within a genus factors other than competition dominate the extinction process.

On first reflection, this result appears to contradict the conclusion drawn from Figs. 5.3 and 5.6, namely, that congeners contribute to each other's risk of extinction in proportion to their morphological similarity. The explanation for this paradox is that the analysis based on Table 5.2 asks only whether sharing an island with *any* congener significantly enhances a species' risk of extinction. The answer to this question is negative because there are many congeners that are insufficiently similar to enhance each other's risk of extinction significantly compared to effects of other factors.

We might add, parenthetically, that comparable data to test extinction rates are common. Long (1981), for birds, provides hundreds of examples. We have yet to analyze them, but doing so would test the generality of our conclusions.

SPECIES ABUNDANCE PATTERNS

For our final evidence of the role of competition, we turn to the abundances of the species currently surviving on the islands. Since the intermediate stage between competition and extinction is reduced density, we can look for density differences on the various islands. While such evidence is peripheral to our main thrust of considering extinctions, we find that it extends and supports the results discussed so far.

The data we shall present are experimental, yet they permit density comparisons on a scale many orders of magnitude larger than typical field experiments on competition. Indeed, they are on a scale comparable to many studies that infer competition by looking at natural variations in densities and habitat use (e.g., Lack and Southern's [1949] early observations on the ranges of *Fringilla* species in the Canary Islands; see also Lack 1976).

But while natural variations in density may be tentatively explained by the presence or absence of a competitor, there is always the possibility that we have overlooked some confounding habitat factor that is responsible for these differences. The point of experiments is that it is unlikely that the arrangement of treatments accidentally correlates with any important confounding factors. In this sense, the Hawaiian faunas are experimental: which species went where was determined by the experimenters—the Hawaiians—and not by the environment. We do not expect that Oahu, the island with the most introduced species, is intrinsically capable of supporting more species than the other islands. Indeed, the reverse may be

Table 5.2 SPECIES INTRODUCED WITH AND WITHOUT CONGENERS AND THEIR SUBSEQUENT STATUS

Species	Island	Status	Species	Island	Status
Species without a cogener on the island of introduction			**Species with at least one congener on the island of introduction**		
Streptopelia chinensis	M, L, Mo	S	*Streptopelia chinensis*	All	S
Geopelia striata	L, Mo	S	*Streptopelia decaocto*	K, O, H	E
Chalcophaps indica	O	E	*Geopelia striata*	All	S
Phaps chalcoptera	O	E	*Geopelia cuneata*	O, H	E
Ocyphaps lophotes	O, L, Mo	E	*Geopelia humeralis*	O, K, M	E
Leucosarcia melanoleuca	M, L	E	*Petrophassa plumifera*	M, L	E
Leptotila verreauxi	M	E	*Petrophassa smithii*	M, L	E
Geotrygon montana	M	E	*Pycnonotus cafer*	O	S
Caloenas nicobarica	M, K	E	*Pycnonotus jocosus*	O	S
Gallicolumba luzonica	K	E	*Luscinia akahige*	O	E
Alauda arvensis	O, H, M, Mo, L	S	*Luscinia komadori*	O	E
Alauda arvensis	K	E	*Copsychus saularis*	O	E
Melanocorypha mongolica	K	E	*Copsychus malabaricus*	O	S
Grallina cyanoleuca	H, O	E	*Garrulax pectoralis*	K	S
Parus varius	H, K, O, M	E	*Garrulax caerulatus*	O	E
Cettia diphone	O, Mo	S	*Garrulax canorus*	All	S
Muscicapa cyanomelana	O	E	*Garrulax albogularis*	K	E
Rhipidura leucophrys	O	E	*Garrulax chinensis*	K	E
Garrulax canorus	H, M, Mo	S	*Passerina cyanea*	H	E
Leiothrix lutea	O, M, K, H, Mo	S	*Passerina ciris*	H	E
Mimus polyglottos	All	S	*Paroaria coronata*	K, O	S
Acridotheres tristis	All	S	*Paroaria dominicana*	O	E
Gracula religiosa	O	S	*Serinus mozambicus*	O	S
Zosterops japonicus	All	S	*Serinus leucopygius*	O	E
Cardinalis cardinalis	All	S	*Estrilda caerulescens*	O	S
Passerina cyanea[a]	O	E	*Estrilda melpoda*	O	S
Passerina leclancherii[a]	O	E	*Estrilda troglodytes*	O	E
Paroaria coronata	K, M, Mo, L, H	S	*Estrilda astrild*	O	S
Paroaria capitata	H	S	*Lonchura malabarica*	M, H	S
Tiaris olivacea	O	S	*Lonchura punctulata*	All	S
Sicalis flaveola	O, H	S	*Lonchura malacca*	O, K	S
Sturnella neglecta	K	S	*Lonchura oryzivora*[b]	O	S
Sturnella neglecta	O, M	E			
Pezites militaris	K	E			
Carpodacus mexicanus	All	S			
Serinus mozambicus	H	S			
Passer domesticus	All	S			
Lagonistica senegala	O	E			
Uraeginthus bengalus	H	S			
Estrilda caerulescens	H	S			
Estrilda melpoda	K	S			
Amandava amandava	O	S			
Vidua macroura	O	E			

Islands: O = Oahu; M = Maui; K = Kauai; H = Hawaii; Mo = Molakai; L = Lanai.
Status: S = survivor; E = extinct.
[a]*Passerina cyanea* and *P.leclancherii* did not exist contemporaneously on Oahu, hence we have counted each species separately.
[b]See Moulton 1985.

true; Maui and Hawaii have much greater elevational ranges and larger areas.

Before making our comparisons, we must discuss briefly the range of habitats on the islands. Because the islands are volcanic, they are composed of several conical mountains in various stages of being eroded. Kauai, the oldest and most northwesterly of the six main islands, is the most eroded. Hawaii, the youngest and most southeasterly, is the highest island and its volcanoes are still active. The prevailing winds are from the northeast and drop considerable amounts of water on the eastern slopes of the islands. Mount Waialeale on Kauai, for example, has recorded over 10 m of rain in a year. The mountains, however, provide an effective shield so that the western slopes are often very dry. Water from the eastern sides is often piped across to the western sides and used for irrigating the principal crops: sugarcane and pineapples. Finally, the western sides have mild oceanic climates. Thus we can recognize a number of lowland habitats:

Wet habitats:	wet exotic forest
Dry habitats:	dry forest (kiawe forest and kiawe woodland)
	cropland
	suburban gardens and parks

The sugarcane fields are not easy to census, but access roads through them have grassy verges that support various species of estrildid finches. We shall consider their density variations in another paper.

The suburban areas are widespread on Oahu and are getting that way in parts of Maui. It is sometimes difficult to census birds in suburban gardens and after some preliminary attempts in 1981, we abandoned these censuses.

What we can census easily are the wet exotic and dry forests. The latter we split into two habitats following J. T. Emlen (personal communication). Both dry forest types are dominated by the introduced leguminous tree, kiawe (*Prosopis pallidus*). Kiawe forest is dense with a closed canopy. Kiawe woodland occurs in drier upland areas and has an open canopy.

Our censuses in 1981 and 1983 covered Kauai, Oahu, Maui, Lanai, and Hawaii. In 1983 we concentrated on kiawe habitats and wet exotic forest on Oahu, wet exotic forest and kiawe woodland on Maui, and kiawe forest on Lanai (just across the channel from Maui). We counted birds from June to August using a circular plot technique (Reynolds et al. 1980). Walking through a habitat, we stopped every 100 m and recorded all the birds seen or heard within a fixed time period (8 minutes in wet exotic forest, 5 minutes in kiawe habitats) and our horizontal distance to them. From the distance data, we could estimate what proportion of each species was missed, correct the crude census data, and thus obtain an estimate of true density. Detectabilities, however, vary most across species and habitats, but we would expect them to be similar for a given species in a given habitat, albeit on a different island. Since it is within species and habitats that we wish to make comparisons, we simply present our results (Table 5.3) as birds seen per 100 stops.

There are at least 9 species common to all the main islands; the number of additional species ranges from as few as 1 on Lanai to at least 16 on Oahu (Table 5.4). Clearly, if competition is important, we should expect densities of the 9 ''core'' species to be lower on Oahu than on Lanai and Maui.

Two of the species that occur on more than one island were excluded from our analysis. *Alauda arvensis* is rare on Oahu and lives in a habitat (grassland) that we did not census. *Leiothrix lutea* occurs on both Maui and Oahu, but it is extremely rare on Oahu, and we did not observe it there. Moreover, it typically lives at higher elevations on Maui, and we did not encounter it there during our censuses. Thus we have no abundance comparisons involving these two species.

Not all the remaining species occur in all the habitats we counted. We are left with 20 density comparisons. In 16 comparisons the densities are higher on Lanai and Maui than Oahu. In only 4 comparisons are the densities higher on Oahu. This is significantly fewer than one would expect by chance (binomial test: $p = 0.006$). Two of these involve the sparrow *Passer domesticus* in the two kiawe habitats. But this species is a com-

Table 5.3 RESULTS OF CENSUSES TAKEN DURING 1983

Species	Oahu	Lanai and Maui
Kiawe forest		
Pycnonotus cafer	107	N.P.
Copsychus malabaricus	100	N.P.
Mimus polyglottos	0	25
Acridotheres tristis	5	22
Zosterops japonicus	344	363
Cardinalis cardinalis	263	397
Paroaria coronata[a]	158	25
Carpodacus mexicanus	2	19
Passer domesticus	44	0
Kiawe woodland	**Oahu**	**Maui**
Pycnonotus cafer	123	N.P.
Copsychus malabaricus	27	N.P.
Mimus polyglottos	0	13
Acridotheres tristis	4	33
Zosterops japonicus	219	318
Cardinalis cardinalis	165	277
Paroaria coronata	108	N.P.
Carpodacus mexicanus	19	74
Passer domesticus	31	3
Lonchura punctulata	0	15
Wet exotic forest	**Oahu**	**Maui**
Pycnonotus cafer	69	N.P.
Pycnonotus jocosus	120	N.P.
Cettia diphone	28	N.P.
Copsychus malabaricus	151	N.P.
Garrulax canorus	0	118
Acridotheres tristis	0	88
Zosterops japonicus	287	578
Cardinalis cardinalis	90	60
Paroaria coronata	3	N.P.
Carpodacus mexicanus	51	60
Lonchura punctulata	0	90

Numbers represent the number of observations per 100 census stations.

N.P. = not present on the island.

[a]*Paroaria coronata* has only recently colonized Lanai, where Berger (1981) listed observations for 1976. However, Emlen (personal communication) apparently did not see this species during censuses taken in 1978 and thus assumed that this species was not present on Lanai.

Table 5.4 PASSERINE SPECIES INTRODUCED TO LANAI, MAUI, AND OAHU

Core species	
Alauda arvensis	Eurasian skylark
Mimus polyglottos	Northern mockingbird
Acridotheres tristis	Common myna
Zosterops japonicus	Japanese white-eye
Cardinalis cardinalis	Northern cardinal
Paroaria coronata	Red-crested cardinal
Carpodacus mexicanus	House finch
Passer domesticus	House sparrow
Lonchura punctulata	Nutmeg mannikin
Additional species	
On Lanai	
Lonchura malabarica	Warbling silverbill
On Maui	
Garrulax canorus	Melodious laughing-thrush
Leiothrix lutea	Red-billed leiothrix
Lonchura malabarica	Warbling silverbill
On Oahu	
Pycnonotus cafer	Red-vented bulbul
Pycnonotus jocosus	Red-whiskered bulbul
Cettia diphone	Japanese bush-warbler
Copsychus malabaricus	White-rumped shama
Garrulax canorus	Melodious laughing-thrush
Leiothrix lutea	Red-billed leiothrix
Paroaria coronata	Red-crested cardinal
Tiaris olivacea	Yellow-faced grassquit
Sicalis flaveola	Saffron finch
Serinus mozambicus	Yellow-fronted canary
Estrilda caerulescens	Lavender finch
Estrilda melpoda	Orange-cheeked waxbill
Estrilda astrild	Common waxbill
Amandava amandava	Red avadavat
Lonchura malacca	Chestnut mannikin
Lonchura oryzivora	Java sparrow
On Kauai	
Parus varius	Varied tit
Copsychus malabaricus	White-rumped shama
Garrulax pectoralis	Greater necklaced laughing-thrush
Sturnella neglecta	Western meadowlark
Lonchura malacca	Chestnut mannikin

Core species are defined as species common to all the main islands, additional species as species present on only certain of the main islands.

mensal of man, and it is hard to find kiawe on densely populated Oahu far from human habitation. Thus a case could be made for excluding these counterresults.

In short, we find considerable evidence for competition by examining patterns of species' abundance. In almost all cases, species' abundances are lower on Oahu, where there are more species and hence potential competitors.

CONCLUSIONS

To summarize, we do find evidences for competition. Moreover, we find evidence for competitively mediated extinctions: there is a nonlinear extinction rate, and extinctions are commoner among morphologically more similar species.

We find evidence that competition is extensive. Of the 18 density comparisons (for species away from human habitations), the great majority find densities lower when there are more potential competitors.

We have not estimated what proportion of the extinctions are competitively mediated. Making various assumptions, it is possible to do this from the nonlinear extinction rate and the relationship between morphology and extinctions. But the confidence intervals of such estimates are large. The role of competition in causing extinctions is certainly not trivial (or else we would not have enough data to detect its effects).

But despite these results, at the community level the distribution of species' morphologies is indistinguishable from random. Furthermore, extinction rates are no higher for species introduced with congeners than for species introduced with no close taxonomic relatives.

We feel that these results have two implications for studies of communities in general.

First, extinctions provide evidence for the effects of competition intermediate between changes in densities and resultant community patterns. We argue that no amount of evidence for competition affecting densities of small numbers of species necessarily says anything about the likelihood of large-scale patterns at the community level. It becomes critical, therefore, to see how intensive and extensive are the effects of competition. While we understand that others may interpret any evidence of competitively mediated extinctions as evidence for communitywide structure, we do not. We can detect the effects of extinctions up to the community level, but not at it. Despite selective extinctions, we do not find communitywide morphologies or species-to-genus ratios to be any different from what we would expect by chance. We have explained the reason for this apparent paradox in connection with Figs. 5.4 and 5.5 and Table 5.2, namely, the dilution effect that makes patterns produced by competition much more obscure at the level of a whole community (using our broad definition of a community) than in a narrowly defined guild.

Second, some questions can be answered with much greater simplicity by considering extinctions directly than by piecing together their supposed effects through the difficult and indirect methods of inferring those extinctions. We consider the relationships of extinction to the number of species, morphological similarity, and taxonomic similarity to be particularly pleasing results. This is not because the ideas are new, but because we test these ideas with a minimum of statistical witchcraft.

ACKNOWLEDGMENTS

This chapter benefited greatly from the comments of T. J. Case, J. M. Diamond, P. R. Grant, and D. S. Simberloff. Our field work in Hawaii would not have been successful without the hospitality of Drs. C. J. and C. P. Ralph and Dr. Alan C. Ziegler. B. A. Moulton typed large portions of the manuscript and provided technical expertise. K. Pimm kindly drew Figure 5.1. Our research was partially funded by the Hawaii Audubon Society, the National Geographic Society, and by National Science Foundation Grant DEB 81-03487 to S. L. Pimm.

chapter 6

Evolution of Ecological Segregation in the New Guinea Montane Avifauna

Jared Diamond

INTRODUCTION

Sympatric species are found to occupy different niches, that is, to differ in their exploitation of significant resources.

An ecologist studying local communities today sees these niche differences as finished products of evolution and community assembly, but in fact, they have a history. They must arise at some stage during the course of speciation, as an ancestral population becomes transformed into sympatric daughter populations. Completion of speciation requires the daughter populations to achieve not only reproductive isolation but also ecological segregation (different niches). In this chapter I use the montane avifauna of New Guinea to address four unresolved questions about the evolution of ecological segregation.

Question 1 Do different types of niche differences tend to evolve in a particular sequence? For instance, species may segregate by habitat, body size, diet, or foraging technique. Does one of these modes tend to evolve first and others later?

Question 2 At what stage in speciation does ecological segregation develop? On the one hand, if (as generally believed) the first stage in speciation is the geographical isolation of initially conspecific populations, niche differences might develop fully at that stage. The geographical isolates might diverge ecologically then because they are exposed to different physical environments or biotas in allopatry, and no further ecological adjustment might be necessary when the isolates later encounter each other. On the other hand, the niche differences might instead develop or become magnified during sympatry, perhaps in part as a result of interactions between the species. These two alternatives correspond to what Grant and Grant (1983) termed the ''complete allopatric model'' and the ''partial allopatric model,'' respectively. Lack (1947) and Grant and Grant (1983) favored the partial model; Bowman (1961), Grant (1975), and Connell (1980) favored the complete model.

Question 3 Do sympatric competing species tend to coevolve in such a way as to reduce the competition between them? From a theory of optimal habitat selection Rosenzweig (1979a, 1979b, 1981) argued that competing species may often evolve to adopt nonoverlapping habitat use. Present-day competition would then no longer be detectable, that is, field experiments would reveal no response of either species to removal of the other. Instead, the niche differences would reflect the "ghost of competition past." Connell (1980) doubted that this postulated effect occurs commonly.

Question 4 Does an evolutionary perspective cast light on the debated question of whether sympatric species tend to differ in size, by Hutchinsonian ratios or otherwise?

A major reason why these four questions remain unresolved is a practical problem: the whole course of speciation generally requires far longer than a human lifetime. Thus, the field biologist cannot watch niche differences develop in a particular case and thereby test competing theories. As Rosenzweig (1981) and Connell (1980) noted, this problem makes it hard either to falsify or to support the hypothesis that existing niche differences developed because of past competition.

This chapter offers a method to get around this problem. The method involves identifying species pairs or sets at particular stages in the process of speciation. This approach works well on New Guinea's montane avifauna, for two reasons: New Guinea's geometry makes it exceptionally easy to identify species pairs at various intermediate stages in speciation, and the avifauna's richness assures that each intermediate stage is exemplified by many species pairs. Hence I begin by describing how to identify congeneric species pairs at intermediate stages of speciation. After presenting the data base, I then restate each of the above questions in more detail and discuss it in light of the data base.

Two points about the scope of this chapter should be made explicit at the outset. First, my analysis is only for the New Guinea montane avifauna. The answers to the four questions are unlikely to be universal, but to vary among taxa and regions, as several examples will illustrate. However, the approach used in this chapter may prove useful for addressing these questions in other biotas. Second, even for the New Guinea montane avifauna this chapter does not examine species interactions in general, but only the niche relationships of closely related congeners. Thus, the data base is suitable for posing questions about the evolution of competition, but not of predation or other forces.

HOW TO IDENTIFY STAGES IN SPECIATION

Speciation in birds is believed to be allopatric rather than sympatric.[1] That is, conspecific populations in different areas diverge, become distinct enough to be reproductively isolated, and may subsequently reinvade each others' geographical ranges (Mayr 1942).

If we could observe a particular pair of taxa going through this whole process of speciation, we could obtain unequivocal answers to questions about the evolution of ecological segregation. In a few cases, as in the nineteenth-century expansion of the azure tit (*Parus cyanus*) into the range of the blue tit (*P. caeruleus*), scientists happened to be at the right place at the right time while speciation was proceeding quickly (Pleske 1912, Vaurie 1957). In New Guinea the grey shrike-thrush (*Colluricincla harmonica*) has advanced several hundred miles along an otherwise inferred invasion route (the "counterclockwise circle": see route L, Fig. 6.1) during the 20 years that I have been working in New Guinea.

Usually, however, bird distributions appear fairly static over our lifetimes. To study speciation, we must resort to the method used by scientists concerned with other slow historical processes: the snapshot method. One examines all pairs of species and asks whether each appears to represent a snapshot of an intermediate stage in speciation, and whether the whole process can be reconstructed from all such snapshots. This is the principle of the method by which astronomers deduced the evolution of stars, by which Darwin worked out the formation of coral reefs, and by which geologists came to understand the evolu-

[1]Field studies have shown that the few pairs of New Guinea bird taxa formerly believed to exemplify early stages in sympatric speciation are actually sibling species that have completed speciation (*Peltops blainvillii* and *P. montanus*, *Halcyon torotoro* and *H. megarhyncha*, *Meliphaga analoga flavida* and *M. orientalis citreola;* cf. Diamond 1969a).

tion of landscapes. In essence, our task is like that of a film restorer who is given an old film cut into separate frames and who has to arrange them in proper sequence to reconstruct the film. Naturally, for many biotas, such as the Galápagos avifauna (Grant and Grant 1983, p. 79), the region's geometry and the biota's poverty may make it difficult to reconstruct speciation in detail by this snapshot method.

In New Guinea, however, this reconstruction is relatively straightforward, because New Guinea has a simple geometry: an east-west mountain backbone on the equator, surrounded by a ring of coastal lowlands, with one chain of islands extending to the east (the Bismarcks and Solomons), another to the west (Indonesia), and a long arm of Australia (the Cape York Peninsula) pointing north from the body of Australia toward south New Guinea. New Guinea has a rich avifauna of over 500 breeding superspecies or species. Many of these consist of linear sets of taxa strung out over putative invasion routes and evidently exemplifying snapshots of a speciation process. Examination of these snapshots (e.g., Diamond 1972a, 1985) permits one to identify the seven mechanisms of speciation depicted in Fig. 6.1.

To illustrate the reconstruction process, Fig. 6.2 on page 102 depicts actual distributions exemplifying snapshots of the mechanism of speciation labeled B in Fig. 6.1: formation of eastern and western isolates along the Central Range, leading to the production of montane species. Seven snapshots are shown in Fig. 6.2.

- **(a)** A species is represented from the east to the west end of the Central Range by a single subspecies that does not vary geographically.
- **(b)** The chain of initially uniform populations differentiates, yielding a chain of subspecies from east to west.
- **(c)** Local extinctions due to Pleistocene climatic changes or other reasons produce a distributional gap between eastern and western populations, which are still sufficiently similar that they are assumed capable of interbreeding and are therefore considered subspecies.
- **(d)** With time the degree of difference between eastern and western populations separated by a distributional gap approaches the degree of difference between sympatric species in that genus. Hence the taxa are assumed to be reproductively isolated and are considered allospecies (allopatric species that are still members of a superspecies).
- **(e)** The eastern and western taxa expand westward and eastward respectively until they again abut each other. They continue to expand further until they overlap for a distance of 20 miles (birds of paradise *Astrapia mayeri* and *A. stephaniae*), 100 miles (birds of paradise *Parotia carolae* and *P. lawesi;* Fig. 6.2e), 150 miles (robins *Pachycephalopsis hattamensis* and *P. poliosoma*), 450 miles (rails *Rallicula rubra* and *R. forbesi*), or 700 miles (parrots *Psittacella modesta* and *P. madaraszi*) in the center of New Guinea. However, the west end of the Central Range is still occupied by one taxon alone, while the east end is still occupied by the other taxon alone.
- **(f)** One of the two taxa reaches the opposite end of the Central Range, completely overrunning the range of its sister taxon while still living alone by itself at the other end of the range. Western taxa can overrun an eastern sister (bowerbirds *Amblyornis macgregoriae* and *A. subalaris,* honeyeaters *Ptiloprora perstriata* and *P. guisei;* Fig. 6.2f), or eastern taxa can overrun a western sister (parrots *Vini papou* and *V. josefinae,* robins *Peneothello cyanus* and *P. cryptoleucus*).
- **(g)** The remaining taxon in turn completes its expansion to the opposite end of the Central Range, overrunning its sister and yielding two montane species that occur sympatrically from the east to the west end of the range.

Each of the other speciation mechanisms of Fig. 6.1 is similarly traced out by many pairs of taxa providing snapshots of intermediate stages in speciation. Readers viewing Fig. 6.2 may wonder how a distributional gap along the Cen-

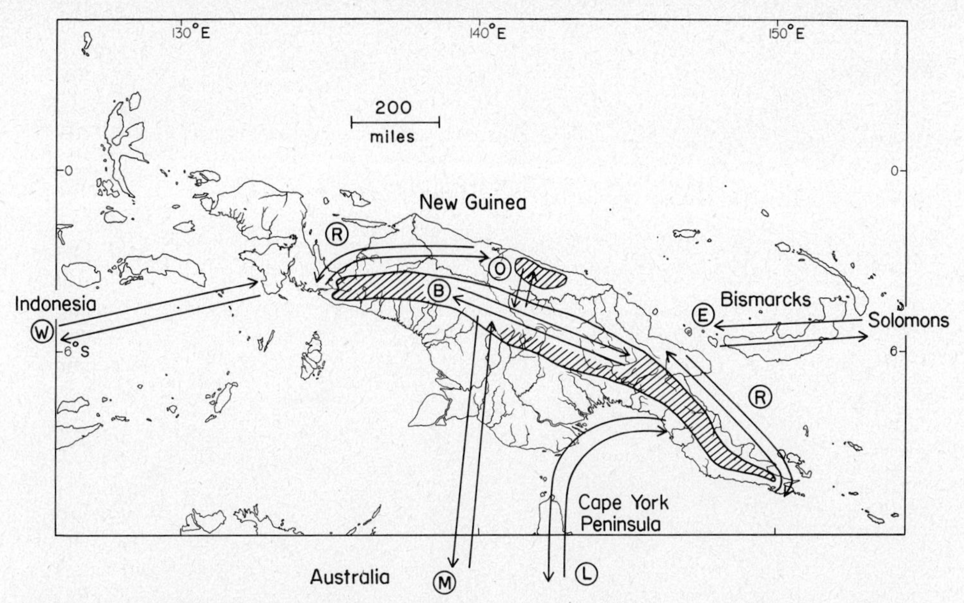

Fig. 6.1 The seven major mechanisms of speciation for New Guinea birds: formation of eastern and western isolates along the mountain backbone of the Central Range (B); formation of a chain of isolates around the ring of lowlands surrounding the Central Range (R); formation of isolates on outlying coastal mountains separated from the Central Range by lowlands (O); invasion from the west via the Indonesian island chain (W); invasion from the east via the Bismarck-Solomon island chain (E); spread of tropical Australian taxa up the Cape York Peninsula and then in a counter-clockwise circle around the New Guinea lowlands (L); direct colonization of the New Guinea mountains by temperate Australian taxa (M). Arrows indicate direction of spread of allopatric sister taxa in the process of achieving sympatry.

tral Range can persist long enough for reproductive isolation to evolve. The probable explanation is that many New Guinea bird species are extremely sedentary. For instance, for most taxa endemic to local areas not a single individual has been recorded in other parts of New Guinea during the history of ornithological exploration.

The process depicted in Fig. 6.2 need not be an inexorable path, with every isolate guaranteed to progress all the way to reproductive isolation and full sympatry. Instead, many isolates may expand and reencounter each other without achieving reproductive isolation, as suggested by the frequent existence of stepped clines in subspecific traits within a single species. Other taxa may become extinct somewhere along the path of Fig. 6.2.

THE DATA BASE

The New Guinea avifauna includes 154 closely related pairs of congeners of which at least one is montane (i.e., confined in New Guinea to the mountains and absent at sea level). The appendix summarizes seven relevant characteristics of each of these 154 pairs, as follows.

1. *Taxonomic closeness* of the two members of the pair, expressed by an index ranging from 1 to 4 on the basis of classical morphological, behavioral, and vocal criteria. Index 1 applies to species whose divergence is little greater than that of strongly marked subspecies; indexes 2 and 3 are progressively more distinct: and index 4 applies to congeners just sufficiently simi-

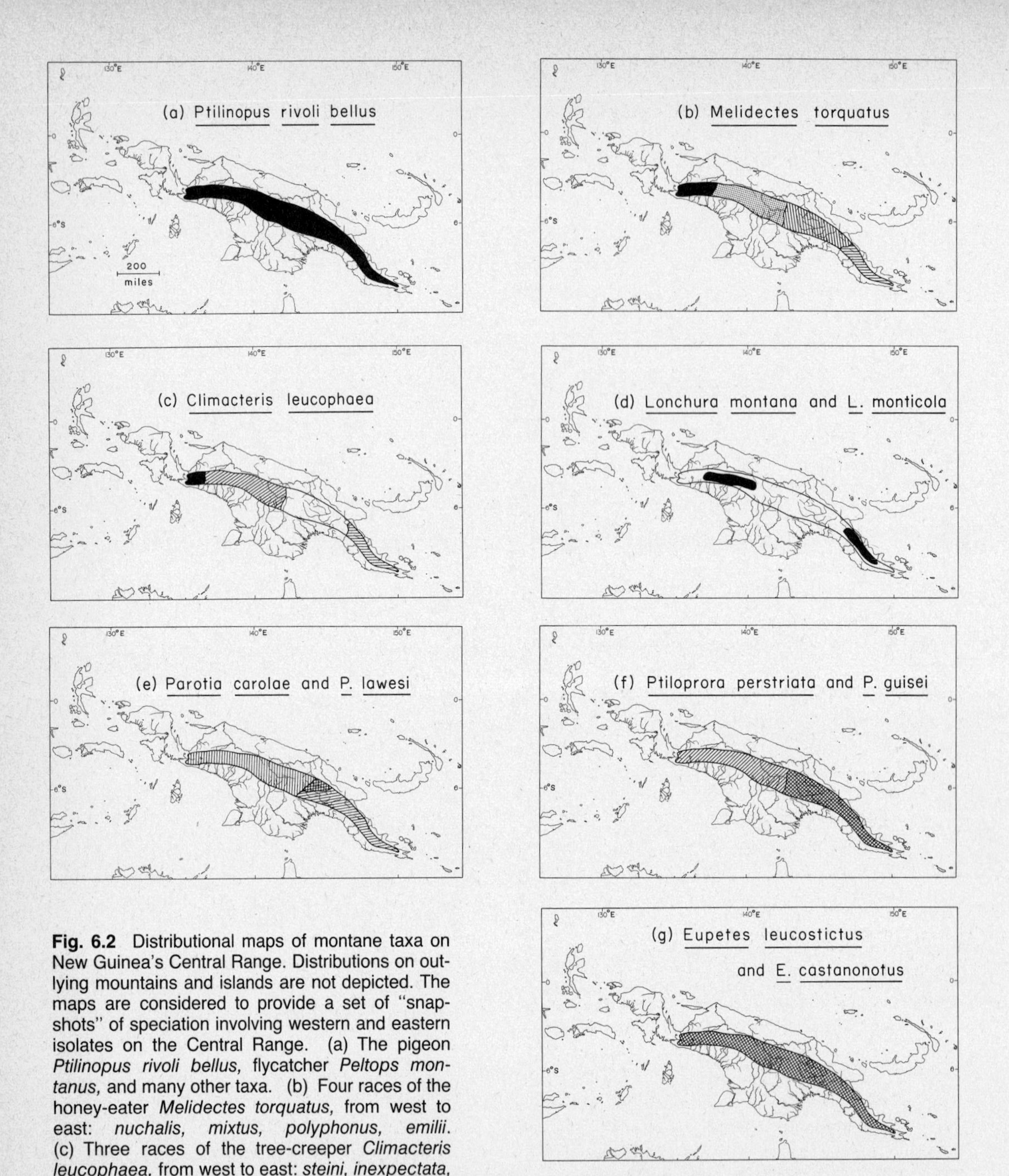

Fig. 6.2 Distributional maps of montane taxa on New Guinea's Central Range. Distributions on outlying mountains and islands are not depicted. The maps are considered to provide a set of "snapshots" of speciation involving western and eastern isolates on the Central Range. (a) The pigeon *Ptilinopus rivoli bellus,* flycatcher *Peltops montanus,* and many other taxa. (b) Four races of the honey-eater *Melidectes torquatus,* from west to east: *nuchalis, mixtus, polyphonus, emilii.* (c) Three races of the tree-creeper *Climacteris leucophaea,* from west to east: *steini, inexpectata, meridionalis.* Note the large area of the Central Range (no shading or hatching) from which this species is absent. (d) The finch allospecies *Lonchura montana* in the west and *L. monticola* in the east. (e) The birds of paradise *Parotia carolae* in the west and *P. lawesi* in the east, with a small area of sympatry. (f) The honey-eaters *Ptiloprora perstriata* (diagonal hatching from lower left to upper right) over the whole Central Range and *P. guisei* (hatching from lower right to upper left) in the eastern half of the Central Range. The birds of paradise *Cnemophilus loriae* and *C. macgregorii* have the same distribution. (g) The log-runners *Eupetes leucostictus* and *E. castanonotus* and many other taxon pairs, sympatric over the length of the Central Range.

lar to be assigned to the same subgenus or subgroup if the genus is diverse enough to be divided. Congeners more divergent than index 4 are not considered because thousands of pairs are involved, but I mention some qualitative conclusions for them below.

2. *Weight ratio* between the two species.
3. *Geographical overlap,* expressed by an index ranging from 1 (complete allopatry) to 5 (full sympatry).
4. *Speciation mechanism,* inferred from distributional evidence and listed as one of seven mechanisms depicted in Fig. 6.1.
5. *Niche differences,* or major modes of ecological segregation in the zone of sympatry, if the two species do occur sympatrically. The modes of ecological segregation between congeneric New Guinea bird species may be grouped into three broad, partially distinct classes: (a) spatial segregation by differences in altitude, habitat, or vertical stratum, by occurrence in a local geographical checkerboard, or as a result of one species occupying the New Guinea mainland while the other occupies offshore islets. (b) Food-related: by foraging technique, or by differences in type of food. (c) Size-related segregation by a greater-than-75% difference in body weight (the reason for choosing the 75% figure is discussed below in connection with Fig. 6.5). Body size affects numerous niche parameters such as size of food consumed, strength of perch necessary for support, and position in interspecific dominance hierarchy (cf. Terborgh 1983).
6. *Niche shift* or *elasticity*. For those species pairs that are at least partly sympatric, I compared the niche of each species at multiple sites in the presence and absence of its congener in order to assess whether niche limits are elastic or rigid. The methods for making these comparisons, and their relation to the question of the ghost of competition past, are discussed below.
7. *Niche differences in allopatry*. For those species pairs for which the speciation process is still incomplete and distributions are still wholly or partly allopatric, the appendix notes whether the two allopatric populations differ ecologically to a conspicuous degree in any of the modes of ecological segregation listed for the zone of sympatry. The purpose of this tabulation is to assess the extent to which ecological segregation in the zone of sympatry involves ecological differences that had already arisen in allopatry (Question 2 of the introduction).

QUESTION 1. DO DIFFERENT TYPES OF NICHE DIFFERENCES TEND TO EVOLVE IN A PARTICULAR SEQUENCE?

Two Examples

The types of niche differences among New Guinea birds are illustrated by Tables 6.1 and 6.2 (see pp. 104 and 105), which categorize the niches of all New Guinea species in two of the most species-rich New Guinea bird genera. Species of *Accipiter* hawks differ in altitude, habitat, body size, foraging technique, and prey type. Species of *Rhipidura* flycatchers differ in altitude, habitat, body size, foraging technique, and vertical stratum.

One's first impression from Tables 6.1 and 6.2 is that randomly chosen pairs of congeners tend to segregate across several niche axes simultaneously. However, when one compares Tables 6.1 and 6.2 with the dendrograms of taxonomic affinities in Figs. 6.3 and 6.4, simpler patterns emerge. Within taxonomic subgroups of accipiters (*buergersi/meyerianus* or *fasciatus/melanochlamys/novaehollandiae*) the species agree in body size, prey, and foraging technique and segregate spatially by altitude or habitat type. The same is true for the subgroups of flycatchers (*albolimbata/hyperythra, brachyrhyncha/rufifrons/rufidorsa, leucothorax/threnothorax/maculipectus*). Thus, combination of taxonomic and ecological information suggests that spatial modes of segregation tended to evolve before other modes among New Guinea species of *Accipiter* and *Rhipidura*.

Table 6.1 NICHE SEGREGATION AMONG NEW GUINEA SPECIES OF *ACCIPITER* HAWKS

Species	Altitude (feet)	Habitat	Weight (grams)	Prey preference	Foraging technique
buergersi	2,000–4,000	Forest	575	Large birds	Soaring goshawk
cirrhocephalus	0–7,000	Forest, savanna	126	Mainly birds	Sparrowhawk
fasciatus	0–3,000	Savanna	220	Generalized	Still-hunter
melanochlamys	5,000–10,000	Forest	235	Generalized	Still-hunter
meyerianus	4,000–9,000	Forest	530	Large birds	Soaring goshawk
novaehollandiae	0–5,000	Forest and edge	247	Generalized	Still-hunter
poliocephalus	0–4,000	Forest and edge	205	Reptiles	Still-hunter

Weights are of adult males, since female weights are lacking for two species. Generalized prey preference means insects, frogs, reptiles, birds, and mammals. The predominant foraging technique of the accipiters termed sparrowhawk is to use the long, slender legs and toes to capture birds on the wing.

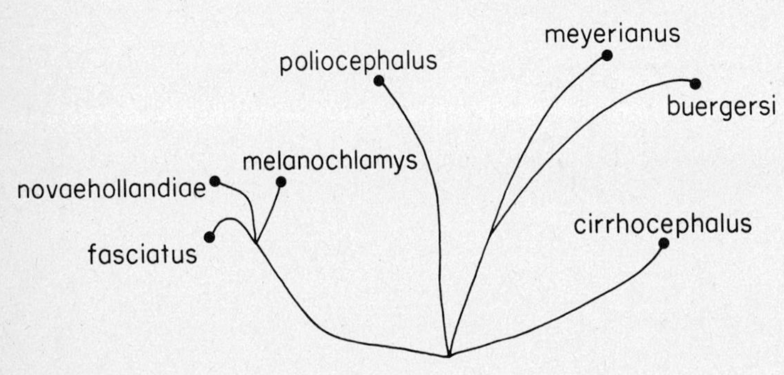

Fig. 6.3 Dendrogram of taxonomic relationships among New Guinea species of *Accipiter* hawks.

Stages in Speciation

To test the generality of this conclusion suggested by two genera, I divided the 154 species pairs of the appendix into 6 groups likely to represent different stages in speciation:

1. Allospecies (relatedness index 1 or 2 in appendix column 2) with no geographical overlap (overlap index 1 in appendix column 4); 27 pairs
2. Closely related species (relatedness index 1 or 2) that are almost entirely allopatric but have marginal sympatry (overlap index 2); 9 pairs
3. Closely related species (relatedness index 1 or 2) with extensive sympatry, but with each species still occupying an area of allopatry (overlap index 3); 10 pairs
4. Closely related species (relatedness index 1 or 2), of which one has completely overrun the range of the second but retains an area of allopatry itself (overlap index 4); 27 pairs
5. Closely related species (relatedness index 1 or 2) that are completely sympatric (overlap index 5); 30 pairs
6. More distantly related species (relatedness index 3 or 4), subdivided into those pairs that are fully sympatric (33 pairs) and those that occur in sympatry but with one or both members also occupying an area of allopatry (18 pairs)[2]

[2]A priori, these 18 cases might represent either a middle stage of speciation comparable to groups 3 and 4 but in which morphological divergence has proceeded exceptionally rapidly, or a late stage of speciation comparable to the other 33 species pairs of group 6. Details of these 18 cases show clearly that the latter explanation is almost always correct.

Table 6.2 NICHE SEGREGATION AMONG NEW GUINEA SPECIES OF *RHIPIDURA* FLYCATCHERS

Species	Altitude (feet)	Habitat	Vertical stratum	Weight (grams)	Foraging technique
albolimbata	4,500–12,000	Forest	ms, us	10.1	Short sallies
atra	3,000–7,000	Forest	ls, ms	11.6	Short sallies, hover-glean
brachyrhyncha	5,000–12,000	Forest	ls, ms	9.3	Beater
hyperythra	0–4,500	Forest	ms, us	11.9	Short sallies
leucophrys	0–6,000	Open	gr	28:4	Glean, pursue
leucothorax	0–3,500	Second growth	ls	17.2	Glean, hover-glean
maculipectus	~0	Swamp forest	ls	19.2	Glean, hover-glean
phasiana	0	Mangrove	ls, ms, us	7.0	Long sallies, glean
rufidorsa	0–3,000	Forest	ls, ms	9.7	Beater
rufifrons	0	Mangrove	ls, ms	11.0	Beater
rufiventris	0–4,000	Forest, second growth	ls, ms, us	15.5	Long sallies
threnothorax	0–3,500	Forest	gr, ls	17.3	Glean, hover-glean

Abbreviations for vertical stratum: gr = ground; ls = lower story; ms = middle story; us = upper story.

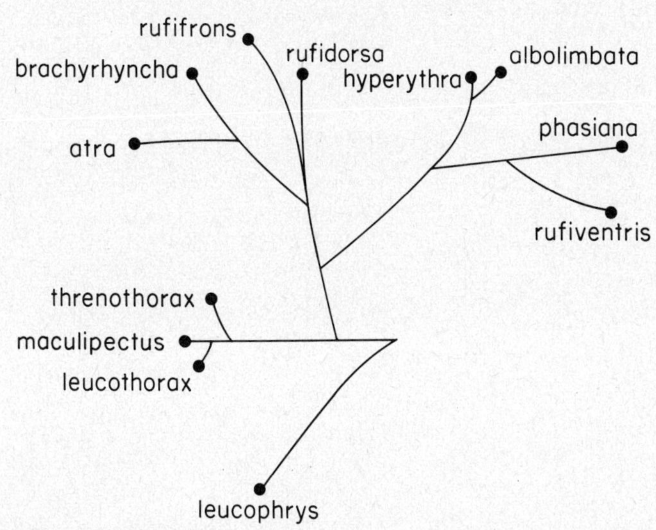

Fig. 6.4 Dendrogram of taxonomic relationships among New Guinea species of *Rhipidura* flycatchers.

Table 6.3 summarizes the modes of segregation in the zone of sympatry for five of these six groups, omitting group 1 (for which there is no zone of sympatry) plus eight species of the other groups for which available information is inadequate. This leaves 119 species pairs that can be analyzed.

Modes of Segregation in New Guinea Montane Birds

Table 6.3 yields four conclusions about modes of segregation corresponding to the six groups.

First, spatial segregation is the dominant mode of ecological segregation through at least

Table 6.3 MODES OF ECOLOGICAL SEGREGATION FOR SPECIES PAIRS AT EACH STAGE OF SPECIATION

Spatial segregation:	Complete				Partial			Negligible		
Size or food difference:	No	Food	Size	Food and size	Food	Size	Food and size	Food	Size	Food and size
Stage of speciation										
2. Marginal sympatry; closely related species (index 1 or 2)	6									
3. Extensive sympatry but each species has an area of allopatry; closely related species (index 1 or 2)	9									
4. Full sympatry for one taxon, partial for other; closely related species (index 1 or 2)	23							1	1	
5. Full sympatry for both taxa; closely related species (index 1 or 2)	18		1		5	3	2			1
6a. Full sympatry for both taxa; less closely related species (index 3 or 4)	17	3	1	1	6	2			3	
6b. Partial sympatry; less closely related species (index 3 or 4)	15		1							

Complete spatial segregation = essentially no spatial overlap; negligible spatial segregation = essentially complete spatial overlap. "Food" includes segregation by type of diet and foraging technique. The numbers are the number of species pairs at the given stage of speciation that segregate by the given mode.

stage 6 of speciation. It is the sole mode of segregation in 88 of the 119 cases, a contributing mode in 25 further cases, and is nonexistent in only 6 cases.

Second, all species pairs analyzed segregate only spatially, as long as each species retains a zone of allopatry (stages 2 and 3). Only when one species overruns the other (stage 4) do the first cases of food-related segregation (by foraging technique or type of diet) and size segregation appear, and these modes do not become at all common until achievement of full sympatry (stage 5).

Third, through stage 4 segregation of a given species pair is only by a single mode, whether spatial, food-related, or size-related. Only with the achievement of full sympatry in stage 5 do species begin to segregate along two or more niche axes, mainly by adding food- or size-related segregation or both to spatial segregation.

Fourth, among submodes of spatial segregation the most prevalent is altitude, acting alone in 63 cases and acting in combination with other modes and spatial submodes in 48 cases. Often but not always, the altitudinal border between the two species is sharp: the highest individual of the low-altitude species abuts and is interspecifically territorial with the lowest individual of the high-altitude species. For instance, on Mt. Karimui the territorial, sedentary warbler *Crateroscelis murina* ranged from sea level to 5,390 feet; at that altitude the highest individual of *C. murina* abutted the lowest individual of *C. robusta,* which occurred from there up to the summit of the mountain (Diamond 1972a). As Terborgh (1971, Terborgh and Weske 1975) also observed in the mountains of Peru, these sharp transitions are usually not close to vegetational ecotones. The sharpness of these transitions may instead reflect habitat selection for purposes of mate finding as discussed by Colwell (Chapter 24), superimposed on species differences in adaptation to altitude or to a correlate of altitude.

The next most important submode of spatial segregation is habitat, acting alone in 3 cases and in combination in 14 cases. Less frequent is segregation by vertical stratum (11 cases), occupation of offshore islands as opposed to the New Guinea mainland (5 cases), and checkerboard exclusion (3 cases).

There are at least two reasons for the dominance of altitudinal segregation in the evolution of New Guinea's montane avifauna. First, the mountains of New Guinea are 16,500 ft high, permitting the development of sequences of up to four or five congeners segregating by altitude (Diamond 1972a, 1973). Second, most of New Guinea is forested, and the little area of nonforest habitat that exists is concentrated in the lowlands. Thus, opportunities for segregation of montane species by habitat are slight.

To express these results colloquially, the easiest way for newly speciated pairs of species to become sympatric in the mountains of New Guinea is by "doing the same thing" at different altitudes. Thus, the species do not yet become syntopic (i.e., they still exploit different microhabitats), but they do become sympatric (they divide the altitudinal transect on countless mountains within the zone of geographical overlap). Later, additional niche differences evolve. Eventually, after the species evolve beyond stage 6 and become different enough to be assigned to different subgenera, segregation along several niche axes simultaneously becomes the rule (cf. Tables 6.1 and 6.2).

Modes of Segregation in Other Faunas

It is worth reemphasizing that these results are based on pairs of New Guinea bird species of which at least one is montane. Analysis of other taxa and other avifaunas will surely yield different patterns. Three examples offer a glimpse at what may lie ahead.

Ernst Mayr and I analyzed bird speciation in the Bismarck and Solomon islands. These archipelagoes consist of hundreds of islands of various sizes, some mountainous but none as high as New Guinea. In the Bismarcks and Solomons altitudinal segregation during speciation is important as in New Guinea. However, spatial segregation by island size (one taxon on large islands, the other on small or remote islands) is equally important, whereas it is relatively much less significant for the New Guinea region.

I have made a preliminary analysis for exchanges of New Guinea and Australian lowland birds. The dominant means of achieving sympatry is by habitat: the New Guinea taxon occupying rain forest, the Australian taxon occupying savanna or open woodland. The lines of segregation of course reflect the relative dominance of each habitat in the land of origin.

E. E. Williams (1972) analyzed the 10 Puerto Rican species of *Anolis* lizards, whose phylogeny has been reconstructed from a combination of morphological, karyotype, biochemical, and ecological evidence. He concluded that in this case the evolution of segregation by size *preceded* segregation by habitat—the opposite of the result for New Guinea birds.

Thus, the answer to question 1 probably depends on the range of habitats in the geographical region considered and also varies among taxa.

QUESTION 2: TO WHAT EXTENT DO ADAPTATIONS PERMITTING ECOLOGICAL SEGREGATION DEVELOP IN ALLOPATRY, BEFORE SYMPATRY HAS BEEN ACHIEVED?

Possible Answers to Question 2

In the previous section we considered the modes by which taxa that have recently achieved sympatry segregate ecologically in the zone of sympatry. We now ask: at what stage in speciation did these niche differences develop?

The possible answers to this question fall between two extremes. At one extreme, called the complete allopatric model, the differing physical or biotic environments to which the taxa were exposed while in allopatry may have led to the development of large niche differences that required no further amplification for the taxa to become sympatric. For instance, one taxon could have evolved in allopatry to live at 0 to 4,000 ft, or to live in savanna, or to eat mainly fruit, or to weigh 80 grams, while the other taxon evolved to live at 5,000 to 8,000 ft, or to live in rain forest, or to eat mainly insects, or to weigh 40 grams. At the other extreme, called the partial allopatric model, the two taxa might have occupied very similar niches in allopatry, and the niche differences permitting coexistence could have developed (or been greatly amplified) through divergence in sympatry, because of interactions between the taxa or for other reasons.

Connell (1980, p. 137) espoused the complete allopatric model: "Instead, it is more likely that they [the two taxa] diverged as they evolved separately so that, when they later came together, they coexisted because they had already become adapted to different resources or parts of the habitat." Grant (1975) formerly also took this position, as did Bowman (1961) for Darwin's finches. On the other hand, Lack (1945, 1947) believed that morphological divergence of Darwin's finches was initiated in allopatry and reinforced in sympatry (the partial allopatric model). Grant (Grant and Grant 1983) came around to this same conclusion as a result of his fieldwork on Darwin's finches. None of these authors, however, based their conclusions on systematic comparisons of niches in allopatry for a whole fauna. Grant and Grant examined one species pair, while Connell merely offered his opinion without analyzing any particular case. Hence this question is in need of an empirical test, by examining for a fauna what fraction of species pairs is closer to each opposite extreme view.

Tests of the Question

The question can be tested with 46 of the 154 pairs of New Guinea bird taxa considered in the appendix. The geographical ranges of these 46 pairs are still partly (19 cases, stages 2 and 3) or wholly (27 cases, stage 1) allopatric. Thus, the niches of the two taxa in allopatry can be compared in all 46 cases, and in the former 19 cases one can additionally compare the niche of each taxon in allopatry with its niche in sympatry (see discussion of question 3 in the next section). Column 8 of the appendix indicates whether there is a major niche difference in allopatry. The details now follow for segregation by size, food, and space, in turn.

Evolution of Segregation by Size Of the 46 cases, there are only 3 in which the weight ratio for the allopatric taxa exceeds 1.75, the minimum value required for segregation by size among syntopic New Guinea birds (to be discussed under question 4). (In 3 other cases the weight ratio falls between 1.59 and 1.63, in 6 cases between 1.26 and 1.42, and in the remaining 34 cases between 1.00 and 1.25.) Whereas only 7% (3/46) of the allopatric taxa have weight ratios above 1.75, 21% (17/81) of the sympatric taxa at stages 5 and 6 of speciation do.

Thus, marked size divergence in allopatry appears to be infrequent in the New Guinea montane avifauna; most size divergence evolves later.

Of those few cases of size divergence in allopatry, there is one in which I can discern a difference in the biotic environments that might have caused the divergence. This is the case of the two bird of paradise allospecies *Diphyllodes magnificus* (95 g) and *D. respublica* (58 g). The former lives on mainland New Guinea and Salawati Island, where it coexists with the congeneric bird of paradise *D. regius* (53 g), while *D. respublica* is confined to Batanta and Waigeu Islands, where *D. regius* is lacking. Perhaps competition with *D. regius* caused *D. magnificus* to evolve larger size.

Evolution of Segregation by Diet or Foraging Technique Among the 46 species pairs that are wholly or partly allopatric, there is no instance of a clear difference in diet or foraging technique between the two allopatric populations, nor is there a case in which such a difference contributes significantly to ecological segregation in sympatry. As discussed under question 1, such differences first appear only after the range of one taxon has been completely overrun by its sister taxon.

Evolution of Segregation by Spatial Niche Of the 27 pairs of purely allopatric taxa, the members of 24 pairs have similar altitudinal ranges and preferences for habitat type and vertical stratum. Of the 19 pairs of partly allopatric taxa, members of 11 pairs occupy similar spatial niches in allopatry, though at least 7 of them segregate by altitude in the zone of sympatry (see discussion of question 3 below). Thus, for New Guinea montane birds it appears that spatial niche differences develop or are magnified in sympatry more often than they develop full-blown in allopatry.

The reason for spatial niche differences' being magnified in sympatry seems often to be competition between the sympatric congeners, which divide between them a spatial gradient that each occupied in allopatry. For example, the taxonomically and ecologically similar robins *Pachycephalopsis hattamensis* and *P. poliosoma* occupy similar altitudinal ranges in allopatry (2,700–4,800 vs. 2,800–5,500 ft, respectively), but in sympatry they divide the altitudinal transect at around 3,200 ft, *hattamensis* living above *poliosoma*.

In a minority of cases the spatial niche differences develop in allopatry. Members of 3 of the 27 wholly allopatric pairs differ in altitude, and members of 1 of these differ also in vertical stratum. Members of 8 of the 19 partly allopatric pairs occupy significantly different altitudinal ranges even in allopatry, and in 2 of these cases there are habitat differences as well. Biotic forces promoting spatial divergence in allopatry are sometimes apparent. For instance, the warbler *Sericornis virgatus* forages on trunks and in the understory at 2,000 to 4,500 ft in north and west New Guinea, while *S. beccarii* of the same superspecies forages on the ground and in the understory at 0 to 2,700 ft in south New Guinea and tropical Australia. *S. virgatus* shares its range with ground-feeding *Crateroscelis* warblers, which are absent from most of the range of *S. beccarii*. Competition from *Crateroscelis* may have caused *S. virgatus* to shift its preferred vertical stratum from the ground to the foliage. As another example, the honey-eater allospecies *Melidectes foersteri* of the Huon Peninsula extends to much higher altitudes (6,200–12,000 ft) than its sister allospecies *M. rufocrissalis* of the Central Range (5,000–7,500 ft). The cause of this difference may be that *M. rufocrissalis* is excluded from high elevations by a high-altitude congener *M. belfordi* (7,000–12,000 ft) absent from the Huon Peninsula, while *M. foersteri* is excluded from low elevations by a low-altitude congener *M. ochromelas* (4,500–6,000 ft) absent from most of the range of *M. rufocrissalis*.

Conclusions About Question 2

In short, it appears exceptional for allopatric populations of New Guinea montane birds to evolve differences in size, diet, foraging technique, or spatial niche parameters already sufficient to permit sympatry. Instead, spatial niche differences most often develop or are magnified when sister taxa come into contact, probably by exaggeration of slight differences in allopatry that cannot be detected with confidence. Nonspatial niche differences generally develop even later. (One may justifiably object that the observed relative numbers of taxon pairs with and without obvious niche differences in allopatry need not be directly proportional to the relative frequencies of niche differentiation in allopatry and in sympatry. For instance, taxa that develop niche differences in allopatry might achieve complete sympatry faster than taxa that do not, so that the former taxa would tend to be underrepresented among existing allopatric pairs. However, it seems unlikely that the apparently much greater frequency of operation of the partial allopatric model is qualitatively incorrect.)

In hindsight, the modest importance of completely allopatric niche differentiation for New Guinea montane birds is not surprising. Most cases of speciation in New Guinea montane birds proceed via allopatric taxon pairs along the Central Range, which is fairly homogeneous in habitat and biota from the western to the eastern end. Strong forces that could produce divergence in allopatry are thus usually lacking, though I have noted a few cases in the preceding discussion.

Niche divergence in sympatry either might constitute coevolution due to competitive interactions of the two species, or might reflect the passage of time and effects of factors other than the competing species. For divergence in size, diet, and foraging technique my data base is inadequate to answer this question. For spatial divergence, coevolution clearly plays a major role: at least 10 of the 19 species pairs at stages 2 and 3 of speciation have altitudinal ranges that differ much more markedly in sympatry than in allopatry, and at least 8 pairs segregate with sharply abutting altitudinal ranges in sympatry.

Thus, pairwise coevolutionary shaping of competitors' altitudinal niches occurs frequently in New Guinea's rich montane avifauna. This conclusion contrasts with Connell's (1980) argument that such coevolution is likely only in species-poor communities, and with the conclusion of Futuyma and Slatkin (1983) that there are surprisingly few unequivocal cases of coevolutionary character displacement among competing species. In particular, Connell reasoned, one might expect each species in a species-rich fauna to encounter so many competing species that close coevolution with any single competitor becomes unlikely. Whether or not this argument is valid in other cases, there is a simple reason why it is invalid for speciation in an area like the mountains of New Guinea. When two allospecies that have been isolated in regions with similar habitats and biotas reencounter each other, they will frequently be very similar ecologically (competition coefficient α approaching 1)—much more similar to each other than either is to other species in the community. Only as a result of mutual adjustment of their niches can the two taxa become sympatric. Thus, during the early stages of speciation in an area without marked geographical gradients in habitat, coevolutionary shaping of competitors' niches is not only probable but almost a necessity.

Comparison with Another Fauna

A preliminary analysis suggests that niche differentiation in allopatry is much more important for New Guinea lowland birds than for montane birds. In the south New Guinea lowlands there are many species pairs consisting of a rain forest species of New Guinea origin and a savanna woodland species of Australian origin (e.g., the kookaburras *Dacelo gaudichaud* and *D. leachii* and the butcherbirds *Cracticus cassicus* and *C. mentalis*). The two taxa segregate in south New Guinea by each occupying the same habitat that it occupies allopatrically in the land of origin. Thus, the habitat differences between wet New Guinea and arid Australia produced the niche differentiation already full-blown in allopatry. From my experiences of avifaunas elsewhere in the world, I would guess that niche differences tend to arise full-blown in allopatry (as assumed by Bowman and Connell) if the source areas differ greatly in habitat, but tend to arise or be amplified in sympatry (as assumed by Lack) if the source areas are similar in habitat.

QUESTION 3. THE GHOST OF COMPETITION PAST: DO SYMPATRIC COMPETING SPECIES TEND TO EVOLVE IN SUCH A WAY AS TO REDUCE COMPETITION BETWEEN THEM?

The Ghost Hypothesis

My analysis of question 1 showed that the commonest mode of ecological segregation among sympatric New Guinea bird species that have speciated recently is spatial, especially segregation by altitude.

My analysis of question 2 showed that those divergent spatial preferences usually develop, or at least become much more marked, during early stages of sympatry.

Based on these conclusions, one might then formulate the following hypothesis, based on a theory of habitat selection developed by Rosenzweig (1979a, 1979b, 1981). When two closely related species with similar diets and foraging techniques first come into contact, the resulting spatial segregation is likely to be phenotypic, i.e., based on reversible choices of habitat made during the lifetime of each individual. For instance, individuals of species 1 might exclude individuals of species 2 from higher altitudes where species 1 has a slight competitive superiority; and species 2 might similarly exclude species 1 from lower altitudes. If species 1 is removed at this stage, some individuals of species 2 will simply reoccupy territories at higher altitudes. After many generations of sympatry, species 2 will tend to lose the genetic adaptations permitting it to live at higher altitudes, since these adaptations are likely to involve costs and since individuals possessing such adaptations would not have the opportunity to leave offspring. Thus, with time the spatial segregation will evolve from phenotypic to genotypic, and field experiments will reveal that species 2 is no longer able to reoccupy high altitudes on removal of its competitors. Rosenzweig coined an apt expression, "the ghost of competition past," for this niche divergence and postulated genetic loss of niche elasticity. This phrase was also used by Connell (1980) in the title of a paper discussing the postulated phenomenon.

Previous Tests of the Ghost Hypothesis

Abramsky and Sellah (1982) carried out a field test of the ghost hypothesis, using two gerbilline rodents in Israel. South of Mt. Carmel the rodents segregate strictly by habitat: *Gerbillus allenbyi* in sand dunes, *Meriones tristrami* in nonsandy habitats. North of Mt. Carmel *G. allenbyi* is absent and *M. tristrami* occurs both in sandy and nonsandy habitats, suggesting that competition with *G. allenbyi* is the ultimate reason for *M. tristrami's* absence from sand dunes in the zone of sympatry. However, removal of *G. allenbyi* by trapping from an experimental plot in the zone of sympatry failed to permit *M. tristrami* to occupy the plot. Abramsky and Sellah concluded that the failure was due to the ghost of competition, that is, that avoidance of sandy habitats had become genetically fixed in *M. tristrami* in the zone of sympatry. Schroder and Rosenzweig (1975), in their studies of two ecologically similar North American rodent species, also observed no response of either presumed competitor to experimental manipulations of the other's density.

While such field experiments are compatible with the ghost hypothesis, they fall short of supporting it. They examine only one pair of taxa, exemplifying a single stage in speciation. They do show that competition is not operating in present time, but they do not show that this outcome was an evolutionary consequence of past competition. Rosenzweig (1981) noted that a single such experiment cannot falsify the ghost hypothesis, since both possible outcomes are compatible with the hypothesis. A positive response (niche expansion of species 2 on removal of 1) would be interpreted by the ghost hypothesis to mean that competition is still operating, while a negative response would be interpreted to mean that past competition had already caused genetic fixation of habitat selection. An alternative interpretation of the negative response is to consider past as well as present competition unimportant. Connell (1980) thus concluded that the ghost hypothesis is unpersuasive.

An explicit test of the ghost hypothesis requires assessing present-day competition in many different pairs of taxa exemplifying various stages of speciation. The later the stage in speciation, the more time is likely to have been available for coevolution in sympatry to produce genetic fixation of habitat selection. Hence a decrease in present competition for taxa at progressively later stages of speciation would support the ghost hypothesis, while no correlation between present competition and stage of speciation is expected if the hypothesis is wrong. The New Guinea montane avifauna, with 154 pairs of taxa distributed over six stages of speciation from complete allopatry to complete sympatry, provides a good data base for the test.

How to Test the Ghost Hypothesis on the New Guinea Montane Avifauna

The test procedure for New Guinea birds cannot be modeled on the field experiments recommended by Connell (1980) and employed by Abramsky and Sellah (1982). For birds capable of flight, it would be difficult to eradicate one species at a local site in the zone of sympatry, or to introduce another species and convince it to stay at one's carefully chosen site. Furthermore, my New Guinea fieldwork is carried out in national parks, and the government would incarcerate or expel any ecologist attempting such experimental eradications and introductions. Thus, the test of ghost effects on spatial niches of New Guinea birds must be performed by means of natural experiments.

Fortunately, these are available in abundance and are of two types. The first type utilizes species pairs at an early stage of speciation (stages 2 and 3, full sympatry not yet achieved) and compares the spatial niche of each species in allopatry and sympatry. The second type utilizes 11 New Guinea mountains disjunct from the Central Range (Diamond 1985), or the numerous satellite islands of the New Guinea region. Since the outlying mountains and islands support varying subsets of the Central Range avifauna, one can identify sites that happen to lack one or the other member of a species pair.

Comparisons of New Guinea sites are not bedeviled by marked variation in climate and habitat: the dominant habitat of the whole New Guinea region is rain forest, except for a small belt of savanna on the south coast. Furthermore, one can compare the niche of a given species at dozens of sites in the presence and absence of a close relative. These considerations reduce, although they do not eliminate, the risk that any niche shift or lack thereof is due to site differences in features other than presence or absence of the congener. In addition, while such confounding factors would introduce noise into the test, they would operate on taxa at any stage of speciation and should not create a relation between frequency of niche shifts and stage of speciation if none exists.

Examples of the Test

To illustrate the range of procedures and outcomes, I give details of seven examples.

1. The closely related warblers *Sericornis nouhuysi* and *S. virgatus* coexist as the most abundant montane trunk-gleaners over the western two-thirds of the Central Range plus two outlying ranges. *S. nouhuysi* occurs without *S. virgatus* on the eastern one-third of the Central Range plus two outliers. *S. virgatus* occurs without *S. nouhuysi* on five outliers plus one mountainous island. I have studied these warblers myself at 15 sites, and others have published studies at 31 more sites. Where the species co-occur (9 sites), *S. nouhuysi* is at higher altitudes (up to 12,000 ft), *S. virgatus* at lower altitudes (down to 2,000 ft), and the two species replace each other at a transition altitude that varies locally from 4,200 to 5,200 ft and averages 4,800 ft. In the absence of *S. nouhuysi* (14 sites) the ceiling of *S. virgatus* varies from 4,300 to 5,100 ft and averages 4,700 ft. In the absence of *S. virgatus* (23 sites) the floor of *S. nouhuysi* varies from 4,500 to 5,300 ft and averages 4,700 ft. Evidently, *S. virgatus* does not expand upward in the absence of *S. nouhuysi,* nor does the latter expand downward in the absence of the former.

2. Pygmy parrots are an ecologically distinctive subfamily restricted to the New Guinea region and composed of two very similar species or superspecies, the montane *Micropsitta bruijnii* and the lowland *M.* [*pusio*].[3] They co-occur on all mountainous islands, replacing each other around 3,000 ft. *M. bruijnii* descends to approximately the same floor on the Moluccan islands of Buru and Ceram, where *M.* [*pusio*] is lacking, while *M.* [*pusio*] ascends to approximately the same ceiling on the high islands of Karkar, Umboi, and Tolokiwa, where *M. bruijnii* is lacking. Thus, there is no evidence of a niche shift in either species.

3. The closely related fruit pigeons *Ducula rufigaster* and *D. chalconota* are endemic to New Guinea, the former living in the lowlands up to

[3]Brackets indicate a superspecies.

3,000 to 4,000 ft, the latter in the mountains down to 4,000 to 6,000 ft. *D. rufigaster* is ubiquitous in the lowlands, so that one cannot examine *D. chalconota's* floor in *D. rufigaster's* absence. However, *D. chalconota* is lacking from all satellite islands and 7 of the 11 outlying ranges, where the ceiling of *D. rufigaster* still does not exceed 3,000 ft. Thus, *D. rufigaster* does not range higher in the absence of *D. chalconota*.

4. The honey-eaters *Toxorhamphus novaeguineae* of west New Guinea and *T. poliopterus* of east New Guinea are sympatric over a distance of 500 miles in central New Guinea. *T. novaeguineae* occurs from sea level up to a ceiling averaging 4,200 ft in allopatry, 4,000 ft in sympatry. *T. poliopterus* extends from a ceiling of 5,000 to 7,000 ft down to a floor averaging 4,300 ft in sympatry (mutually exclusive of the altitudinal range of *T. novaeguineae*), 1,700 ft in allopatry. Thus, *T. poliopterus* greatly expands its altitudinal range in the absence of *T. novaeguineae,* but not vice versa.

5. The honey-eater *Ptiloprora perstriata* occupies one outlier (Wandammen) plus the Central Range from latitude 136°E to its eastern end (latitude 150°E). The related *P. guisei* occupies a different outlier (Huon) plus the Central Range from the eastern end west to latitude 142°E. Throughout the 500-mile-wide zone of sympatry on the Central Range the two species replace each other at a transition altitude averaging 8,800 ft, with *P. perstriata* ranging from this limit up to 12,000 ft and *P. guisei* from this limit down to 4,500 ft. *P. guisei's* ceiling in allopatry on the Huon outlier rises to 11,500 ft, while *P. perstriata's* floor in allopatry on the Wandammen outlier and on the Central Range west of latitude 142°E drops to an average of 5,800 ft. Thus, each species expands several thousand feet in the other's absence (Gilliard and LeCroy 1961, plus subsequent studies).

6. The robin *Pachycephalopsis hattamensis* lives on the western Central Range and four outliers, while its sole congener, *P. poliosoma,* lives on the eastern Central Range and four outliers. In allopatry their average floors are quite similar (2,700 vs. 2,800 ft), as are their ceilings (4,800 vs. 5,500 ft). Their sole sympatry is for a stretch of 150 miles at latitudes 136° to 138°E on the Central Range, where they replace each other at 3,200 ft, *P. hattamensis* living at higher elevation and *P. poliosoma* being crammed into a narrow altitudinal band at lower elevation. Thus, *P. hattamensis* nearly halves *P. poliosoma's* ceiling, while *P. poliosoma* slightly raises *P. hattamensis's* floor.

7. The final example involves a shift in habitat preference rather than in altitude. The robin *Poecilodryas hypoleuca* occupies all New Guinea and three islands at low elevations, while the related *P. brachyura* is confined to an 800-mile stretch of north New Guinea. Everywhere outside of *P. brachyura's* range (nine sites studied by me, many other sites studied by other authors), *P. hypoleuca* occurs not only at the forest edge but throughout the forest interior miles from the nearest edge. I have studied these species at four sites in the zone of sympatry, where I found *P. hypoleuca* confined to second growth and the forest edge, *P. brachyura* in the forest interior. Stein (1936) and Ripley (1964) made the same observation elsewhere within *P. brachyura's* range. Thus, *P. brachyura* excludes *P. hypoleuca* from the forest interior in the zone of sympatry.

Results of the Test

The preceding examples illustrate how natural experiments may be used to test whether a species' spatial niche responds to the absence of a related species from which the given species segregates spatially in the zone of sympatry. Niche expansion in the relative's absence suggests that niche limits in sympatry are determined behaviorally; failure to expand the niche suggests instead that niche limits have become fixed genetically through the ghost of competition.

Table 6.4 summarizes the results of all such experiments for 96 species, categorized according to stage of speciation. At the earliest stage of speciation, when two formerly allopatric taxa have just achieved marginal sympatry, the frequency of niche expansions is 100%. As sympatry becomes more extensive or is achieved for longer times, this frequency decreases until it reaches only 7% for long sympatric, more dis-

Table 6.4 EVOLUTIONARY LOSS OF NICHE ELASTICITY

	Does spatial niche expand in absence of the related taxon?		
Stage of speciation	**Yes**	**No**	**Percent yes's**
1. Allopatry	Untestable		—
2. Marginal sympatry	8	0	100
3. Extensive sympatry	10	4	71
4. Full sympatry for one taxon, partial for other	17	11	61
5. Full sympatry for both taxa; relatedness index 1 or 2	9	10	47
6. Full or partial sympatry; relatedness index 3 or 4	2	25	7

Niche elasticity refers to the following question: For two species that segregate spatially where sympatric, have the niche limits of each species become fixed genetically, or are they elastic in the sense that one species can expand its niche when the other is removed? This question is tested by comparing the niche of a species at multiple sites in the presence and absence of its relative. The number of species yielding yes or no answers to the test of niche expansion are tabulated. Stages of speciation are the same as in Table 6.3.

tantly related pairs of taxa. Thus, the ghost hypothesis is valid for most (apparently, nearly all) speciations involving New Guinea montane bird species.

The ghost hypothesis explicitly assumes that the evolution of habitat selection by sympatric taxa proceeds or is reinforced through competitive interactions during sympatry. Recall from our discussion of question 2 that this is indeed the main course of evolution for New Guinea montane birds. However, the opposite conclusion applies to speciation involving exchanges of lowland New Guinea and Australian birds: habitat selection evolved there mainly because of the contrasting habitats present during allopatry. In that case, the ghost hypothesis would be invalid for exactly the reason stated by Connell (1980): the different habitats occupied by sympatric taxa would not reflect the ghost of competition past. Thus, the answers to questions 2 and 3 are probably linked. The ghost hypothesis may prove valid where the partial allopatric model of niche divergence applies, and invalid where the complete allopatric model applies.

Connell (1980) nevertheless believed the ghost hypothesis to be plausible only for low-diversity communities. This conclusion is incorrect, since the hypothesis is supported for the species-rich New Guinea montane avifauna. What led Connell to this incorrect conclusion? A reading of his paper suggests that he was misled by a logical fallacy and an erroneous belief.

The fallacy is his claim, ''Co-occurrence is the first requirement for coevolution'' (Connell 1980, p. 132). It is surely true that predators must coexist with their prey, and parasites with their hosts, while competitors can segregate spatially. Hence the claim leads one to expect coevolution in predator-prey or parasite-host systems, but not among competitors. However, the claim is illogical: nowhere is it written that the mechanism of coevolution *always* involves the sharing of space. When two species exclude each other from abutting habitat types and each species thereby loses its adaptation for the barred habitat, it is perfectly appropriate to consider that process as an example of coevolution.

Connell's erroneous belief is the one already quoted, namely, that evolution of exclusive habitat preferences is likely *in general* to have preceded achievement of sympatry. As we have seen, this belief is incorrect for most speciations of New Guinea montane birds, though it may prove correct for some other situations.

I also disagree (for the reasons discussed in Chapter 1) with Connell's insistence (1980, p. 135) that a very complicated set of six field experiments is ''both necessary and sufficient'' to test the ghost hypothesis. The New Guinea montane avifauna exemplifies a case in which a test by natural experiments is feasible and permissible, while a test by field experiments is neither.

QUESTION 4. DO SIZE DIFFERENCES PROVIDE A MEANS OF ECOLOGICAL SEGREGATION?

Size Segregation in New Guinea's Montane Avifauna

Hutchinson (1959), Lack (1971), and numerous other field naturalists familiar with local faunas have reported that species with similar diets and foraging techniques, living in the same habitat, often prove to be of rather different body size. Hutchinson further noted that a series of such species pairs clustered around a ratio of 1.3 in linear body dimensions, which would correspond approximately to a ratio of $(1.3)^3 = 2.2$ in weight. The correctness of these claims has been the subject of much recent discussion.

In my discussion of question 1 I noted that marked differences in size, defined arbitrarily as a weight ratio exceeding 1.75, tend to appear late in speciation for New Guinea montane birds. Let us now consider this phenomenon in more detail, in order to gain a fresh perspective on the issue of size segregation.

Fig. 6.5 plots weight ratios for all 148 pairs in Table 6.1 for which ecological information is available. Points are coded according to whether the species segregate by diet (or foraging technique) and whether they overlap fully, partly, or negligibly in spatial niche. It can be seen that species with weight ratios of 1.00 to 1.67 never coexist spatially, unless they have different diets or foraging techniques. The overwhelming majority of species pairs in this range of weight ratios (118 out of 129 pairs, or 91%) exhibits negligible spatial overlap, usually due to segregation by altitude. The other 11 pairs overlap spatially but differ in diet or foraging technique. Not until a weight ratio of 1.79 to 1.83 is achieved does one encounter the first instance of species with the same diet or foraging technique, overlapping fully or partly in space.

At weight ratios from 1.79 to 2.51, the maximum observed in my data set, most species pairs (12 out of 19) do overlap wholly or extensively in space, and 9 of these 12 pairs have similar diets and foraging techniques. (Of the 7 pairs with negligible spatial overlap, 3 are allopatric pairs for which one does not know whether they will actually overlap spatially when they come into geographical contact.) These 9 pairs with spatial overlap and similar diets and foraging techniques may be considered as segregating in ways related to size. The average weight ratio for these pairs is 2.13 ± 0.26 (mean $\pm$ SD).

There are several dozen other pairs of con-

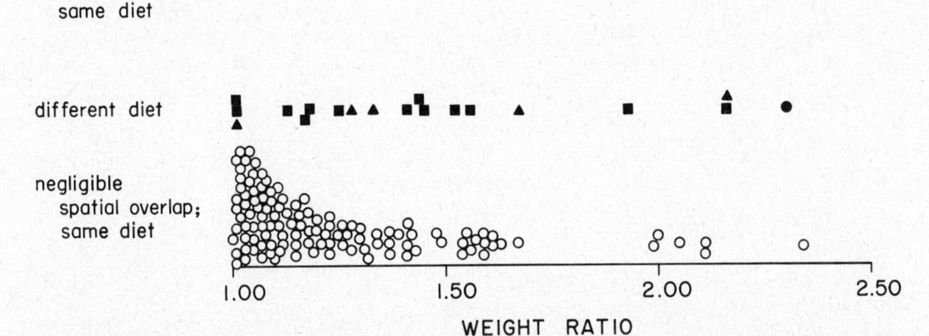

Fig. 6.5 Mode of ecological segregation between members of a closely related species pair plotted against the ratio of their weights. Each point represents one pair of species. Modes of segregation: negligible spatial overlap but same diet and foraging technique (unshaded circle); different diet or foraging technique, with spatial overlap negligible (shaded triangle), extensive (shaded square), or complete (shaded circle); same diet and foraging technique, with spatial overlap extensive (×) or complete (+).

generic New Guinea bird species that also have similar diets, foraging techniques, and broad spatial overlap, but that were not included in the appendix, either because their relationship within their genus is more distant than level 4 or because both members live in the lowlands. Examples include the pigeons *Ducula zoeae* and *D. rufigaster*, frogmouths *Podargus papuensis* and *P. ocellatus*, and the honey-eaters *Myzomela eques* and *M. nigrita*, *Oedistoma iliophum* and *O. pygmaeum*, and *Philemon novaeguineae* and *P. meyeri*. These pairs also tend to have weight differences around a factor of 2.

The frequency of weight ratios much larger than 1 increases with the stage in speciation. Weight ratios of 1.00 to 1.20 characterize 72% of species pairs at stages 1 to 3 (allopatry through partial sympatry), but only 44% of pairs at stage 6. Ratios of 1.79 to 2.51 increase in frequency from 7% at stages 1 to 3 to 16% at stage 6.

The interpretation that I place on these results for New Guinea montane birds is similar to that advanced by other authors for analogous findings with other faunas. Body size has diverse ecological consequences, such as determining size of prey that can be captured, strength of perch necessary for support, density of habitat compatible with maneuvering, concentration and nutritional qualities of foods that can be economically harvested, metabolic requirements, home range size, antipredator strategy, and position on a dominance hierarchy (Terborgh 1983). Thus, species that harvest qualitatively similar food items in similar ways can overlap spatially if they differ sufficiently in size. At the same time the overall significance of size segregation for the whole avifauna should not be exaggerated. Fig. 6.5 and Table 6.3 show that size segregation is about seven times less frequent than spatial segregation for closely related New Guinea montane birds.

The Controversy over Size Segregation

Many early discussions of size segregation were deficient in that they tended to focus on selected cases without explicitly examining the frequency of such cases in the whole fauna or asking whether that frequency was greater than expected by chance. Hence Strong, Szyska, and Simberloff (1979) and Simberloff and Boecklen (1981) compared observed size ratios for various species sets against values generated by null hypotheses based on randomly drawing species from a pool. The observed and predicted ratios were reported usually not to differ significantly, casting doubt on the reality of size segregation. Other authors then noted that several types of statistical errors (discussed by Grant and Abbott 1980, Hendrickson 1981, Case and Sidell 1983, Harvey et al. 1983) and carelessness in assembling data (discussed in Hendrickson 1981, Appendix 1) tended to predispose the analyses by Strong et al. and by Simberloff and Boecklen toward unwarranted acceptance of a null hypothesis. After correction of these errors, analyses of communities of Galápagos finches (Grant and Abbott 1980, Case and Sidell 1983), desert rodents (Bowers and Brown 1982), West Indian birds (Case, Faaborg, and Sidell 1983), *Accipiter* hawks (Schoener 1984a), and introduced Hawaiian birds (Chapter 5) do exhibit significant size segregation.

There is, however, another pair of issues distinct from these questions of statistical procedures and illuminated by Fig. 6.5. These are the related issues of what constitutes a reasonable hypothesis of size segregation, and what data set is therefore appropriate for testing the hypothesis. As I see it, numerous tests of size segregation implicitly assume an unreasonable hypothesis, in that they employ data bases which scramble together data relevant to size segregation and data irrelevant to it. Such scrambling inevitably diminishes the possibility of detecting size segregation.

The problem of data scrambling can be understood as follows. A reasonable hypothesis of size segregation to test would be: "Do species (or related, or confamilial, or congeneric species) that harvest similar foods by similar methods in the same space tend to differ in size?" An unreasonable hypothesis would be: "Do all species (or all related species, etc.) in a local fauna tend to differ in size?" The second formulation is absurd, because one already knows that there are other, more important modes of ecological segregation. As Table 6.3 and Fig. 6.5 showed, size segregation is far less important than segregation based on diet and foraging technique for the New Guinea montane avifauna. A local avifauna in the mountains of New Guinea contains over 100 pairs of closely related species that share similar body sizes but segregate spatially. Fig. 6.5 shows clearly how the second formulation of the hypothesis would therefore defeat the purpose of the proposed test: the 9 ratios of 1.79 to 2.51 that would be analyzed under the first formulation would be swamped by 139 other ratios, of which 129 are much lower and 78 are only 1.20 or less.

A test of size segregation based on all species pairs in Fig. 6.5 would correspond to what Strong et al. (1979) termed "communitywide character displacement," that is, a test based on a whole fauna or a large fraction of one. For instance, Strong et al. (1979), Hendrickson (1981), and Simberloff (1984a) analyzed all confamilial bird species of the Tres Marías Islands, while Strong et al. analyzed all confamilial bird species of the California Channel Islands. Other studies have analyzed congeneric species, but still without attempting to sort them by habitat, diet, or foraging technique. These analyses have generally failed to reveal size segregation convincingly or to reveal it at all, because of three types of data scrambling.

First, faunas of different habitats are scrambled together. Fig. 6.6 illustrates this effect by comparing the weights of all New Guinea fruit pigeon species with those of the widespread species of wet lowland rain forest. The latter are obviously spaced by size. The former are not; they may even be significantly clumped, because of similar-sized species replacing each other in different habitats.

Second, the communitywide tests scramble species with different diets and foraging techniques together, the so-called owl and hummingbird problem. Yet there is no reason to expect such species to have difficulty in coexisting, even if they are of the same size. The disastrous consequences of this type of scrambling for tests of community patterning have been discussed by Bowers and Brown (1982), Diamond and Gilpin (1982), Colwell and Winkler (1984), and Moulton and Pimm (Chapter 5).

Third, some types of resources lend themselves to segregation by consumer size, while others do not. Thus, even if a communitywide test began by separating species into guilds, tested each guild separately, and finally summed the results over all guilds, the test would still be scrambling size-segregated guilds with guilds segregated by other means. For instance, Bell (1984) noted for a New Guinea lowland bird community that obligate herbivores sort well by weight, but carnivores and omnivores do not, because insect prey can be harvested in many different ways unrelated to consumer size, but fruit cannot. My data base supports Bell's reasoning: seven of the nine pairs of New Guinea montane birds that sort by size but not by diet or space are herbivores. The careful analysis of West Indian bird communities by Case, Faaborg, and Sidell (1983) also revealed guild differences in size segregation.

While Simberloff (1984b) and Simberloff and Boecklen (1981) criticized Hutchinson severely for his failure to test size segregation statistically,

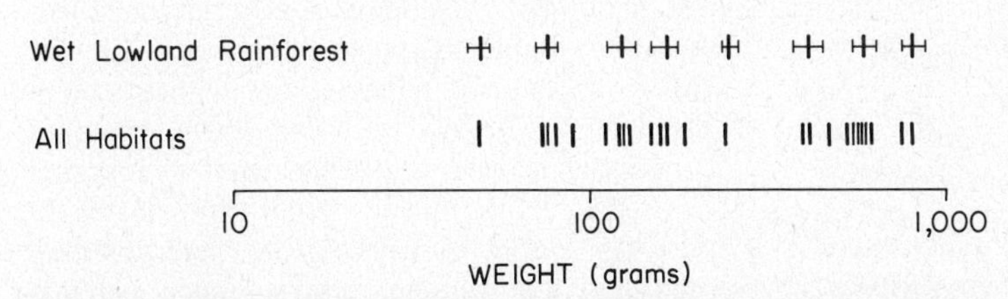

Fig. 6.6 Mean weights of New Guinea canopy fruit pigeons (genera *Ptilinopus* and *Ducula*) on a logarithmic scale. Data above are for widespread species of wet lowland rain forest, data below for all New Guinea species regardless of habitat or geographic range. Horizontal bars indicate standard deviation of weight.

Hutchinson (1959) nevertheless took a precaution that many of his successors did not. He analyzed congeneric species of corixid water-bugs with which he was familiar and which he personally observed to be living abundantly together in the same pool. Numerous subsequent tests of size segregation, such as those by Strong, Szyska, and Simberloff (1979) on Tres Marías and California Channel Islands birds, have been made by authors unfamiliar with the faunas analyzed and therefore unable to sort them into guilds and sets of syntopic species. For the California Channel Islands Strong et al. analyzed many combinations of species that do not even occupy the same island, but are on different islands of the archipelago (see Hendrickson 1981, Appendix 1). The inevitable result has been scrambled data bases inappropriate for testing Hutchinson's idea. Fig. 6.5 shows that clean tests of size segregation will require focusing on sets of species known to have similar diets, foraging techniques, and spatial niches. At least for New Guinea montane birds, Hutchinson was qualitatively correct that such species tend to segregate by size.

Somewhat separate from the question of the reality of size segregation is the question of whether there is a tendency toward a certain size ratio. Hutchinson's series of examples had a length ratio clustering around 1.3, corresponding to a weight ratio of $(1.3)^3 = 2.2$. The average weight ratio for purely size-segregated species of New Guinea montane birds, 2.13 ± 0.26, is close to this value. However, both the generality of this conclusion and its theoretical interpretation remain uncertain (Chapter 30 and Schoener 1985b). What seems to me the first task for reaching a unified understanding of size segregation is to test it on more biotas sorted by habitat, foraging technique, and diet as in Fig. 6.5. Preferably, these should be species-rich biotas to ensure an adequate data base.

CONCLUSIONS

I examined four questions about the evolution of niche differences. These questions were tested by examining closely related, congeneric pairs of species exemplifying snapshots of six stages in speciation. The data base consists of all 154 such pairs of New Guinea bird species of which at least one is montane.

Question 1 Do different types of niche differences tend to evolve in a particular sequence? In the analyzed avifauna, the answer is yes. Spatial segregation (especially by altitude) usually evolves first, while segregation by diet, foraging technique, or size appears later.

Question 2 To what extent do these niche differences evolve in allopatry vs. in sympatry? For New Guinea montane birds, the differences mainly evolve or are amplified in sympatry, with coevolution of abutting habitat preferences playing an important role. However, New Guinea and Australian lowland birds apparently yield the opposite answer. Hence I conjecture that evolution of niche differences in allopatry may be proportional to the habitat differences in allopatry.

Question 3 Do sympatric competing species tend to evolve in such a way as to reduce competition between them? To test Rosenzweig's ghost of competition hypothesis, I used natural experiments to detect niche shifts on removal of a close relative for 96 species at various stages of speciation. The frequency of niche shifts decreased from 100% early in speciation to 7% later, that is, the ghost hypothesis is true for most New Guinea montane birds. For other faunas I conjecture that it will tend to be true or untrue, depending on the answer to question 2 (whether niche differences evolve mainly in sympatry or in allopatry).

Question 4 Do size differences provide a means of ecological segregation for New Guinea montane birds? The answer is yes, but much less often than do spatial niche differences. Closely related, congeneric species pairs with a weight ratio of ≤ 1.67 do not coexist spatially, except in a few cases involving different diets or foraging techniques. Pairs with a weight ratio of ≥ 1.79 often do overlap in space, diet, and foraging technique; the average ratio for such pairs is 2.1 ± 0.3. Methodological problems have often confused analyses of size segregation. It is essential to test for size segregation on biotas sorted by habitat, diet, and foraging technique rather than on scrambled biotas.

ACKNOWLEDGMENTS

It is a pleasure to acknowledge my debt to Bruce Beehler, Brian Finch, Ernst Mayr, Melinda Pruett-Jones, and Stephen Pruett-Jones for discussions of the congeneric bird species analyzed in the Appendix, and to James Brown, Robert Colwell, Ernst Mayr, and Thomas Van Devender for criticism of the manuscript.

APPENDIX

Column 1 The appendix lists all 154 pairs of closely related, congeneric pairs of resident New Guinea bird species, of which at least one member of the pair is montane (i.e., confined in New Guinea to the mountains and absent at sea level, unless in the area of south New Guinea where some otherwise montane species descend to sea level). If several allospecies or sibling species replace each other sequentially along a geographical transect (e.g., the allospecies of *Astrapia* or *Parotia*) or along an altitudinal transect (e.g., the species of *Eupetes, Crateroscelis,* or *Melanocharis*), I analyze only pairs of geographically or altitudinally adjacent species. In each pair of species the one named first is the high-altitude species, if they segregate by altitude, or is otherwise the heavier species. The family name is in roman type. Names in brackets are superspecies.

Column 2 The closeness of taxonomic relationship between the two species is expressed by an index ranging from 1 (divergence barely greater than for strongly marked subspecies) to 4 (congeners just sufficiently similar to be assigned to the same subgenus or subgroup if the genus is diverse enough to be divided). The assessments are based on the taxonomic literature plus my field studies and museum studies (e.g., Diamond 1972a, Diamond and LeCroy 1979). For those familiar with New Guinea bird taxonomy, three examples will serve to illustrate these indexes. The parrot genus *Psittacella* consists of four species, all montane. Following my taxonomic revision (Diamond 1972a), I consider the two most similar species, *P. modesta* and *P. madaraszi,* to be related to each other at index 1; *P. picta* and *P. brehmii* to be related at index 2; and *P. picta* or *P. brehmii* to be related to *P. modesta* or *P. madaraszi* at index 3. Following Koopman's (1957) and subsequent revisions of the honey-eater genus *Myzomela,* I consider the montane *M. adolphinae* and the lowland *M. erythrocephala* to be related at index 2; the montane *M. cruentata* and the lowland *M. nigrita* to be related to each other and to *M. erythrocephala* at index 4; the remaining montane myzomelid (*M. rosenbergii*) and the four remaining lowland myzomelids to be more distant from the aforementioned species than index 4; and *M. rosenbergii* to be more distant than index 4 from all these congeners. Following Goodwin's (1983) revision of the pigeon genus *Ducula,* I consider the montane *D. chalconota* and the lowland *D. rufigaster* to be related at index 2; *D. chalconota* to be related to the lowland *D. myristicivora, D. pistrinaria,* and *D. pacifica* at index 4; and relations of New Guinea's five other lowland *Ducula* species to *D. chalconota* to be more distant than index 4. Naturally, degree of morphological difference corresponding to a given index of relatedness is gauged by morphological differences among subspecies or species of the same genus, for example, striking morphological divergence among congeneric birds of paradise and slight divergence among *Collocalia* swiftlets or *Meliphaga* honey-eaters.

Column 3 Ratio of weights for the two species is given. Average male weight and average female weight were combined to obtain an average weight for each species. In a few cases where weights were unavailable (identified by values in parentheses), I approximated the weight ratio by the cube of the ratio of wing lengths.

Column 4 The degree of geographical overlap between the two species is denoted by an index ranging from 1 to 5: 1 = complete allopatry; 2 = marginal sympatry (zone of sympatry less than one-quarter of the whole geographical range of either species); 3 = extensive sympatry (zone of sympatry more than one-quarter of the whole geographical range for at least one of the species, but each species still has an area of allopatry); 4 = full sympatry for one taxon and partial sympatry for the other (one species shares its whole range with its sister species, but the latter still has a zone of allopatry); 5 = full sympatry (essentially coincident ranges for the two species).

Column 5 The inferred speciation mechanism is indicated in cases where it can be inferred with confidence from distributional evidence. The seven mechanisms are abbreviated by letters (B, R, O, W, E, L, M) explained in the legend to Fig. 6.1.

Column 6 Niche differences in sympatry are indicated by the following abbreviations. A = nonoverlapping altitudinal ranges, pA = only partly overlapping altitudinal ranges. (In the case of social itinerant frugivores or flowering-tree specialists, such as lories, white-eyes, dicaeids, and certain honey-eaters, broader altitudinal overlap was required for a pA designation than in the case of sedentary territorial species.) H = differing habitats (usually, forest vs. second growth or open habitats). V = differing vertical strata (e.g., understory vs. canopy). I = one species occupies New Guinea mainland, the other occupies offshore islets. C = the species replace each other in a local geographical checkerboard. D = differing diets (e.g., fruit vs. insects). F = conspicuously different foraging technique for obtaining the same type of diet (e.g., sallying, leaf-gleaning, trunk-gleaning). M = a mixture of V, D, and F. O = no ecological information available for zone of sympatry. pA, O = same as O except that altitudinal ranges in zone of sympatry are known to be only partly overlapping. S = size difference (body weight ratio in column 3 greater than 1.75; see discussion of Fig. 6.5 in text). G = allopatric species with clean geographical separation.

Column 7 Niche shift, or niche elasticity, is discussed in text section on question 3. The two entries separated by a comma refer to the first- and second-named species of the pair. A, H, I, C = the species expands its niche in the absence of its congener with respect to the indicated niche variable (see codes of niche variables listed under column 6). – = the species does not expand its niche. O = it is not known whether the species expands its niche. See text section on Question 3 for seven examples.

Column 8 For species at an early stage of speciation, such that each species still occurs wholly or partly in allopatry, do the two allopatric populations differ conspicuously in any of the niche variables of column 6? A, H, V, S = yes (see codes of niche variables listed under column 6); – = no; no entry = not known. See text section on Question 2 for examples.

Species pair	Taxonomic closeness	Weight ratio	Geographical overlap	Speciation mechanism	Niche differences in sympatry	Niche shift	Niche differences in allopatry
Casuariidae (cassowaries)							
Casuarius bennettii/[*casuarius*]	2	1.86	5		A, S	A,O	
Accipitridae (hawks)							
Accipiter meyerianus/buergersi	3	1.08	4		A		
A. melanochlamys/novaehollandiae	3	1.12	4		A	O,–	
A. melanochlamys/fasciatus	3	1.07	3	L	A, H	–,–	
Megapodiidae (mound-builders)							
Talegalla cuvieri/jobiensis	1	1.03	1	R	G		–
T. cuvieri/fuscirostris	1	(1.11)	2	R	A	A,O	–
T. jobiensis/fuscirostris	1	(1.10)	1	R	G		–
Aepypodius arfakianus/bruijnii	1	(1.27)	1	O	G		
Rallidae (rails)							
Rallicula rubra/forbesi	2	1.15	3	B	A	A,A	
R. rubra/leucospila	2	1.32	3	B	O		–
R. leucospila/forbesi	1	1.15	1	B	G		–
R. forbesi/mayri	1	1.39	1	O	G		–
Columbidae (pigeons)							
Ptilinopus rivoli/solomonensis	1	1.42	3	E	I	I,I	–
Ducula chalconota/rufigaster	2	1.59	5		A	O,–	
D. chalconota/[myristicivora]	4	1.24	4	W, E	A, I	O,–	
D. chalconota/[pistrinaria]	4	1.41	4	W, E	A, I	O,–	
D. chalconota/pacifica	4	1.61	2	E	A, I	O,–	
Gallicolumba beccarii/jobiensis	3	2.09	5		pA, S		
Psittacidae (parrots)							
Vini papou/josefinae	1	1.26	4	B	A	–,–	
V. josefinae/pulchella	2	1.83	4	B	S		
V. papou/pulchella	2	2.31	5	B	pA, S		
V. rubronotata/placentis	1	1.26	4	R	A	A,A	
Neopsittacus pullicauda/musschenbroekii	1	1.41	5	B	pA, D		
Micropsitta bruijnii/[pusio]	1	1.11	4		A	–,–	
Geoffroyus simplex/geoffroyi	2	1.10	4		A	O,–	
Psittacella picta/brehmii	2	1.83	5	B	pA, S	O,A	
P. modesta/madaraszi	1	1.10	3	B	A	A,A	–
P. picta/modesta, madaraszi	3	1.37	5	B	A		
P. brehmii/modesta, madaraszi	3	2.51	5	B	S		

(Continues)

Species pair	Taxonomic closeness	Weight ratio	Geographical overlap	Speciation mechanism	Niche differences in sympatry	Niche shift	Niche differences in allopatry
Cuculidae (cuckoos)							
Cacomantis pyrrhophanus/castaneiventris	2	1.34	4	M	A		
C. castaneiventris/variolosus	3	1.00	4		H	O,H	
C. pyrrhophanus/variolosus	3	1.34	4		A, H		
Chrysococcyx ruficollis/meyerii	3	1.16	5	B	A	O,–	
C. meyerii/russatus	3	1.03	4		A		
Aegothelidae (owlet-nightjars)							
Aegotheles archboldi/albertisii	1	1.03	2	B	A		A
A. albertisii, archboldi/wallacii	3	1.54	5	B	A		
A. wallacii/bennettii	3	1.09	4	R	A		
A. insignis/albertisii, archboldi	4	2.34	5	B	S		
Caprimulgidae (nightjars)							
Eurostopodus archboldi/papuensis	3	(1.16)	5		A	O,–	
Apodidae (swifts)							
Collocalia hirundinacea/vanikorensis	1	1.15	4		A		
Alcedinidae (kingfishers)							
Halcyon megarhyncha/vanikorensis	1	1.30	5		A	O,–	
Campephagidae (cuckoo-shrikes)							
Coracina montana/melaena	3	1.28	5		A, D	O,–	
C. montana/schisticeps	2	1.37	5	R	A		
C. morio/schisticeps	3	1.13	4		pA, D		
C. morio/tenuirostris	2	1.14	3		pA, H	O,H	
Turdidae (thrushes)							
Amalocichla sclateriana/incerta	2	(2.20)	5	B	pA, S	O,–	
Orthonychidae (log-runners)							
Eupetes leucostictus/castanonotus	2	1.49	5	B	A	O,–	
E. castanonotus/caerulescens	2	1.23	5	R	A	–,O	
Sylviidae (Old World warblers)							
Megalurus gramineus/timoriensis	3	(2.03)	2	M	pA, S, O		
M. gramineus/albolimbatus	2	1.02	1	M	A		

(Continues)

Species pair	Taxonomic closeness	Weight ratio	Geographical overlap	Speciation mechanism	Niche differences in sympatry	Niche shift	Niche differences in allopatry
Maluridae (Australian warblers)							
Crateroscelis robusta/nigrorufa	2	1.13	5	B	A		
C. nigrorufa/murina	1	1.07	5	R	A		
Sericornis nouhuysi/virgatus	1	1.22	4		A	–,–	
S. virgatus/beccarii	1	1.01	1	R, M	G, A, V		A, V
S. rufescens/perspicillatus	1	1.02	1	B	G		–
S. papuensis/perspicillatus	2	1.18	5	B	pA, V, F	O,–	
S. perspicillatus/arfakianus	2	1.01	5	B	A	A,A	
S. arfakianus/spilodera	3	1.31	5	R	A	O,–	
Muscicapidae (flycatchers)							
Peltops montanus/blainvillii	1	1.08	5	R	A		
Rhipidura albolimbata/hyperythra	2	1.18	5	R	A		
R. brachyrhyncha/rufidorsa	4	1.04	5		A	O,–	
R. brachyrhyncha/atra	4	1.25	5	B	pA, V, F		
R. brachyrhyncha/rufifrons	4	1.04	3		A	–,–	
Monarcha frater/cinerascens	1	1.18	3	W, E	A, I	–;A, I	A
Machaerirhynchus nigripectus/flaviventer	2	1.01	5	R	A	O,–	
Microeca papuana/griseoceps	3	1.05	5	B	A		
M. griseoceps/flavovirescens	3	1.21	5	R	A		
Poecilodryas brachyura/hypoleuca	1	1.32	4	R	H	O,H	
P. albispecularis/hypoleuca	4	1.67	5		A, V	O,–	
Peneothello sigillatus/cyanus	3	1.20	5	B	A		
P. cryptoleucus/cyanus	1	1.37	4	B	A	O,A	
Pachycephalopsis hattamensis/poliosoma	2	1.03	2	B	A	A,A	–
Pachycephalidae (whistlers)							
Pachycepala soror/meyeri	2	1.32	4	B	O		
P. lorentzi/schlegelii	1	1.19	4	B	pA, O		
P. schlegelii/soror	2	1.09	5	B	pA, V	A,O	
P. soror/melanura	1	(1.03)	2	M	C	C,C	
P. schlegelii/melanura	2	(1.11)	2	M	A, C	A, C	
P. soror/simplex	3	1.09	4		pA, V	O,–	
P. soror/aurea	2	1.19	5		A, H		
Pitohui dichrous/kirhocephalus	1	1.15	5	R	pA, M		
Artamidae (wood-swallows)							
Artamus maximus/leucorhynchus	1	1.41	4		A	O,A	
Grallinidae (magpie-larks)							
Grallina bruijnii/cyanoleuca	3	2.11	2	L	A, H, S	–,O	A, H, S

(Continues)

Species pair	Taxonomic closeness	Weight ratio	Geographical overlap	Speciation mechanism	Niche differences in sympatry	Niche shift	Niche differences in allopatry
Paradisaeidae (birds of paradise)							
Manucodia chalybatus/jobiensis	1	1.03	4	R	A		
M. chalybatus/ater	3	1.07	5	R	pA, H		
M. chalybatus/comrii	1	(2.34)	1	O	G, S		S
M. keraudrenii/chalybatus	3	1.44	5	R	pA, M		
Cnemophilus macgregorii/loriae	2	1.02	4	B	A		
Paradigalla carunculata/brevicauda	1	1.05	2	B	O		
Epimachus meyeri/fastosus	1	1.20	3	B	A	A,O	A
E. meyeri, fastosus/albertisii	2	1.93	5	B	pA, D, S		
E. albertisii/bruijnii	1	1.39	4	B, R	A	–,–	
Astrapia nigra/splendissima	1	(2.05)	1	B	G, S		S
A. mayeri/splendissima	1	(1.59)	1	B	G		–
A. mayeri/stephaniae	1	(1.18)	2	B	A	O,A	
A. rothschildi/stephaniae	1	1.25	1	O	G		–
Parotia sefilata/carolae	1	1.05	1	B	G		–
P. carolae/lawesi	1	1.08	2	B	O		–
P. lawesi/wahnesi	1	1.01	1	O	G		–
P. [carolae]/"Lophorina" superba	2	2.30	5	B	D, S	–,O	
P. [carolae]/"Ptiloris" magnificus	2	1.01	5	R, L	pA, D	O,–	
P. [carolae]/"Seleucidis" melanoleuca	3	1.01	5	R	A, H, D	O,–	
"Lophorina" superba/"Ptiloris" magnificus	2	2.16	5	R, L	pA, D, S	O,–	
"L." superba/"Seleucidis" melanoleuca	3	2.16	5	R	A, H, D, S	O,–	
Diphyllodes magnificus/respublica	1	1.63	1	O	G		–
D. magnificus/regius	3	1.79	5	R	pA, S	–,O	
Paradisaea rudolphi/[raggiana]	2	1.41	4	R	A	O,–	
P. guilielmi/[raggiana]	1	1.02	4	R	A	O,A	
Ptilonorhynchidae (bowerbirds)							
Amblyornis inornatus/macgregoriae	1	1.05	1	B	G		–
A. macgregoriae/flavifrons	1	(1.07)	1	O	G		–
A. macgregoriae/subalaris	1	(1.28)	4	B	A	A,O	
Sericulus bakeri/aureus	2	1.01	1	O	G		–
Ailuroedus crassirostris/buccoides	2	1.61	4	R, M	A	A,O	
Neosittidae (Australian nuthatches)							
Daphoenositta miranda/papuensis	4	1.06	5	B, M	A		
Meliphagidae (honey-eaters)							
Timeliopsis fulvigula/griseigula	3	1.99	5	R	A, S	O,–	
Myzomela cruentata/nigrita	4	1.15	4		A	A,O	
M. adolphinae/cruentata	4	1.07	4		pA, H		
M. adolphinae/erythrocephala	2	1.07	3		A, H	–,O	

(Continues)

Species pair	Taxonomic closeness	Weight ratio	Geographical overlap	Speciation mechanism	Niche differences in sympatry	Niche shift	Niche differences in allopatry
Toxorhamphus poliopterus/novaeguineae	1	1.05	3	B, R	A	A,–	A
Melipotes gymnops/fumigatus	1	1.05	1	B	G		–
M. ater/fumigatus	2	2.11	1	O	G, S		S
Melidectes nouhuysi/princeps	1	(1.59)	1	B	G		–
M. [princeps]/fuscus	3	(1.67)	5	B	H, D, F		
M. leucostephes/rufocrissalis	1	1.07	1	B	G		–
M. foersteri/rufocrissalis	1	1.23	1	O	G		A
M. belfordi/rufocrissalis	2	1.01	4	B	A	A,A	
M. ochromelas/leucostephes	2	1.29	5	B	A	A,A	
M. foersteri/ochromelas	2	1.48	5	O	A	A,A	
M. belfordi/ochromelas	2	1.19	4	B	A	A,A	
M. rufocrissalis/torquatus	3	1.56	5	B	H		
Meliphaga subfrenata/obscura	3	1.12	5	B, M	A, V	–,–	
M. flaviventer/polygramma	3	2.23	5		S		
M. orientalis/analoga	1	1.20	5	R	A		
M. montana/mimikae	1	1.04	1	R	G		–
M. orientalis/gracilis	2	1.17	3	L	A, H	–,O	A, H
M. orientalis/flavirictus	2	1.09	5	R	pA, F, D		
M. orientalis/albonotata	2	1.55	4	R	A, H	–,O	
M. orientalis/aruensis	3	1.45	5		pA, D		
M. orientalis/[montana]	4	1.52	5		pA, D		
M. [montana]/aruensis	4	1.01	5		pA, D		
Ptiloprora perstriata/mayri	1	1.30	1	O	G		A
P. perstriata/guisei	1	1.23	4	B	A	A,A	
P. perstriata/erythropleura	2	1.29	2	B	pA, O		–
P. [erythropleura]/plumbea	4	1.54	5	B	pA, V, H, F		
P. [erythropleura]/meekiana	3	1.24	5	B	pA, V		
Pycnopygius cinereus/ixoides	4	1.57	5	R	A	O,–	
P. cinereus/stictocephalus	4	1.02	5	R	A, V	O,–	
Dicaeidae (flower-peckers)							
Melanocharis versteri/longicauda	1	1.05	5	B	A		
M. longicauda/nigra	1	1.09	5	R	A		
Zosteropidae (white-eyes)							
Zosterops fuscicapilla/atrifrons	2	1.08	4		A	O,A	
Estrildidae (mannikins)							
Erythrura papuana/trichroa	1	1.33	4		D		
Lonchura montana/monticola	1	(1.03)	1	B	G		–
Lonchura teerinki/castaneothorax	2	(1.12)	4	R, L	C	C,C	
L. [monticola]/spectabilis	3	1.54	4	R, E	A	O,–	
L. vana/castaneothorax	2	1.08	1	R	G	O,C	

chapter 7

Patterns of Change in Plant Communities Through Geological Time

Andrew H. Knoll

On the face of it, the paleoecologist is a misfortunate ecologist.

KRASSILOV, 1975

IS THERE A PALEOBOTANICAL PERSPECTIVE ON COMMUNITY ECOLOGY?

Two myths about the fossil record enjoy widespread currency among biologists. The first is that paleontology has little to contribute toward the solution of biological problems, that it is appropriately the "handmaiden of stratigraphy," best practiced from the basements of geological museums. The opposite view, that given sufficient scrutiny the fossil record can provide answers to all questions of biological history, is equally persistent and equally extreme. Because it documents biological evolution within the context of the earth's physical development, the fossil record can and does provide a valuable perspective on a number of evolutionary issues. However, questions amenable to paleontological investigation are strictly limited by the scales of time resolution available in the geological record and by the postmortem information losses that accompany fossilization.

The limitations of the geological record certainly constrain attempts to reconstruct ancient plant communities from paleobotanical data. As the opening quotation from Krassilov (1975) makes clear, paleontologists cannot hope to make the detailed observations that ecologists use in studies of extant communities; they can resolve only major features of ancient vegetation patterns. On the other hand, paleontologists have the singular advantage of being able to compare patterns across more than 400 million years of vascular plant history. The recognition of general patterns that develop, persist, or change over long time intervals may be the unique contribution that paleobotany can make to community ecology. Potentially, the fossil record can tell us which of the many processes that control community structure in ecological time are important in the development of vegetational patterns on a geological time scale. It may, as well, suggest the long-term evolutionary significance of events that are unlikely to be observed in the course of ecological studies.

In this paper, I will discuss the pre-Pleistocene fossil record of vascular plants, concentrating in particular on large-scale patterns in the composition and taxonomic richness of floras through time.

THE DATA OF THE PALEOBOTANICAL RECORD

In evaluating the structure of living plant communities, ecologists need to know how many and what species are present, and how individuals are distributed within taxa. The anatomy, morphology, and physiological capabilities of populations are important, as are phenological and other aspects of reproductive biology. Interactions among individuals both within and between populations—including potential or actual competition, herbivory and defenses against it, animal-mediated pollination and seed dispersal, and other symbiotic associations—are observed or inferred. All of this information is interpreted within the framework of the physical environment, whose biologically important variables include climate, geomorphology, water availability, nutrient and soil conditions, and the frequency, intensity, and predictability of disturbances. All of these features contribute to the structure of a given community; the question is which of them one might reasonably expect to discern in the fossil record.

As was clear to Darwin (1859) and to Lyell (1830–1833) before him, the fossil record illuminates only selected moments from the history of life. As discussed for fossil faunas in Chapter 18, the preserved record for the most part reveals organisms not as they lived but as they were buried, often after some ill-defined interval of transportation and decomposition. Fortunately, the burial sites of plants, like those of Etruscans, often permit significant inferences to be drawn about the life habits of the organisms preserved. Krassilov (1975) has discussed the practice and problems of plant paleoecology in some detail, and his monograph is usefully augmented in recent reviews by Scott (1977, 1978) and Phillips and colleagues (1974, 1977; Phillips and DiMichele 1981). Discussions of the nature of the paleobotanical record are also available (e.g., Knoll and Rothwell 1981), so no general review is attempted here. In the context of this paper, however, it is necessary to consider the plant fossil record from the particular perspective of the community ecologist.

Inferences About the Plants Themselves

Most obviously, the paleobotanical record provides data on the morphology of ancient plants. The information preserved in individual fossilized organs can be quite detailed. For example, it is often possible to know not only the size and shape of leaves, but also their thickness, cuticle thickness, and stomatal distribution. Stem anatomy can be well preserved, as can that of roots, although for most fossil plants aerial organs are better known than underground portions. Such preservation provides the paleobotanist with an important advantage over invertebrate paleontologists in that the plant fossil record is not characterized by the absence of soft parts. A second advantage of plant paleontology is that although details of physiology are not preservable, it is clear how most ancient plants made their living. One has a reasonable idea of how changes in, say, tracheid or vessel diameter or in leaf surface area would have affected a plant's ability to grow in different environments. A principal disadvantage of the paleobotanical record is that plants tend to become disarticulated and individual organs selectively preserved. Thus, reconstruction of ancient plants is often based on incomplete knowledge, and in many instances only generalized information on life form will be available. This is particularly true of the pre-Pleistocene record, which consists mainly of extinct taxa.

Information on reproductive biology and life cycles comes from preserved sporangia, fruits, seeds, pollen, spores, and, less often, gametophytes, embryos, and even seedlings, as well as from the comparative biology of living plants. One can infer much about the reproduction of Devonian lycopods or Jurassic cycads, although, as in the case of vegetative morphology, such information will often be rather general in nature. Problems arise when dealing with extinct groups such as the Mesozoic cycadeoids, or with groups like the Ginkgoales whose living members are only remnants—and not necessarily representative ones—of once diverse taxa (Hughes 1976). However, even with these plants reasonable constraints on reproduction and life cycles can be made by a combination of paleobotanical and comparative biological insights (e.g., Crepet 1974 on *Cycadeoidea*). Data on reproductive phenology are not likely to be preserved.

Inferences About Interactions

Interactions between organisms must be inferred from the morphologies of preserved fossils. In plants relevant information may be preserved in the morphologies of flowers (e.g., Crepet 1979), pollen or spores (Kevan et al. 1975, Taylor and Millay 1979), fruits or other disseminules (Tiffney 1977), or in possible morphological defenses against predation such as spines or glandular hairs. Animals capable of influencing plant community patterns can be inferred from associated, or at least contemporaneous, faunas. Other clues to plant-animal interactions can be gleaned from coprolites, wounds preserved in plant fossils, or such rarities as insects preserved with adhering pollen (Smart and Hughes 1972, Kevan et al. 1975, Scott and Taylor 1983). Competition must be inferred from long-term patterns of distribution and replacement in the fossil record (Knoll 1984). In general, of all the controls on plant community structure, those involving organismic interactions are likely to be seen most dimly, yet even here general outlines can be discerned.

Sampling Problems

Assemblages of fossil plants can be censused to arrive at some measure of floral composition and taxonomic richness. It is important to recognize what this information does and does not represent, and to recognize both the problems and the potential for relating fossil assemblages to ancient communities (cf. Graham's discussion in Chapter 18 of comparable problems in vertebrate paleoecology). The major problems are two: plants inevitably undergo some amount of decomposition prior to fossilization, and most also are transported to their site of burial. Both of these processes are variable and selective. Differential preservation and transportation will leave some community members unrepresented (or underrepresented) in the fossil record, while augmenting the apparent importance of others. Transportation can also mix elements from originally distinct communities. However, in spite of the distorting influences of these processes, fossil floras still reflect, however imperfectly, the communities from which they were derived, and major temporal changes in assemblage composition or diversity can be used to document corresponding changes in plant communities.

Quantitative studies of abundance and distribution are hampered by the fact that most bedding plane exposures are limited in lateral extent and, hence, are likely to record very local vegetation patterns. Also, sampling is done on disarticulated plant parts, predominantly leaves, rather than on whole plants. If one can accept Harper's (1977) definition of the individual as a single modular unit in plants characterized by modular growth, then the second problem is not as severe as has commonly been supposed. The former difficulty can be surmounted only by the integration of abundance counts from numerous horizons. As such integrations will almost invariably involve a stratigraphic series of bedding planes, fossil plant assemblages will be time-averaged samples of ancient plant associations. Quantitative figures for taxonomic richness and equitability are useful only in a comparative sense and cannot be related directly to equivalent figures for living vegetation. Discussions of techniques in quantitative paleobotany are presented by Phillips and DiMichele (1981), Scott (1977, 1978), and Spicer and Hill (1979). Scott (1978) also discusses the recognition and evaluation of transported elements in fossil plant assemblages.

Inferences About the Physical Environment

What of the physical environment? The composition, texture, geometry, and structures of sedimentary rocks collectively allow estimation of geomorphology, climate, and drainage (e.g., Parker 1977, Retallack 1977, Scott 1978, Hickey 1980), and further inferences can be made on the basis of paleogeographical reconstructions, paleosol (fossil soil) distribution, and structural attributes of the plants themselves (leaf morphology, growth rings, and other features). The intensity and frequency of disturbance are more difficult to gauge. Although evidence can be found in the pre-Pleistocene rock record for floods, fires, and storms, estimates of frequency and intensity must come from records for comparable modern environments (but see Taggart et al. 1982 on volcanic disturbance in Miocene landscapes).

In summary, a combination of observations on fossil assemblages and the rocks that encompass them, and reasonable inference based on the biology of living plants, allows much to be learned about the general structure of ancient plant communities, although it may not permit strong, independent paleobotanical insights to be drawn concerning specific coevolutionary or disturbance-mediated controls on particular ancient community patterns.

Temporal Resolution

Time scale is also an important consideration in evaluations of paleobotanical data. National Science Foundation grant and dissertation research in ecology operates on a time scale of 10^0 to 10^1 years; in the best of circumstances historical records and pattern-based inference may extend the ecological purview to a few hundred years. Pleistocene, particularly late glacial and Holocene, studies provide observations on a longer time scale; resolution can be as fine as 10^2 years or less (Davis 1981a, Webb 1982), and patterns of change occurring on a scale of 10^3 to 10^4 (and, more spottily, 10^5) years are readily observed (Chapters 16 and 17; Davis 1976, 1981b). Pleistocene data are subject to many of the same limitations as those of the pre-Pleistocene record, but have the advantage that extant taxa constitute most of the plant fossils. It might be expected that the Pleistocene record could provide a bridge between the short-term observations of ecology and the large-scale patterns seen in pre-Pleistocene rocks, but this expectation is chastened by the recognition that temporal patterns in Pleistocene floral assemblages are strongly controlled by large oscillations in climate and migrational responses to these changes.

Schindel (1980) and Sadler (1981) have demonstrated that rates of sedimentary rock accumulation are a strong inverse function of the time scale on which observations are made. This obtains because sedimentation is often rapid but discontinuous, a characteristic that is particularly true of terrestrial depositional patterns. Wing and Hickey (1982; see also Behrensmeyer and Schindel 1983) estimated that early Tertiary lignite-bearing beds in the western United States that accumulated in ponds and abandoned oxbows each record a few hundred years of time; periods of nondeposition between fossiliferous units represent 10^4- to 10^5-year intervals. Estimates of accumulation rates for Carboniferous coal-bearing sequences are comparable in magnitude (Duff et al. 1967). This means that, in favorable circumstances, paleobotanists working on pre-Pleistocene rocks may be able to resolve local patterns that develop on an ecological time scale; however, most fossil floras will be time-averaged assemblages representing some hundreds or thousands of years' accumulation. Under these conditions it is principally general patterns that develop on time scales of 10^6 years or longer that can be resolved and interpreted.

FLORAL COMPOSITION AND TAXONOMIC RICHNESS THROUGH GEOLOGICAL TIME

The Nature of the Data Set

In an effort to recognize large-scale patterns of floral change through time, I examined the published records of 391 fossil floras ranging in age from latest Silurian (ca. 410 million years ago) to Pliocene (ending 1.6 million years ago). The floras all consist of compression or compaction remains of macrofossils (sporophytes and the detached organs of sporophytes) that accumulated in lowland deltaic and floodplain environments, mostly under mesic subtropical to tropical climates. Fossil assemblages from known temperate environments are noted in the analysis. Macrofloral remains were chosen for this study because they provide necessary information about life form that cannot always be inferred from fossil spores and pollen. Also, palynofloras (i.e., assemblages based on microfossils) may more readily include foreign elements transported relatively long distances to the site of burial. Still, the paleoecological value of plant microfossils is high (e.g., Chaloner 1968, Phillips et al. 1974, Fredericksen 1980), and a comparative study based on palynological data would be instructive.

By considering only assemblages of macrofossils preserved as compressions and compactions, one minimizes variation introduced by differences in preservational type. For example, petrifaction taxa are often recognized by suites of characters different from those used to identify compressions, and this can influence observed diversity. Equally important, many floras preserved in mineralized peats come from environments that are recognizably different from the mesic floodplains where compressions formed most commonly (e.g., DiMichele et al. in press). As nearly as possible, then, the floras examined in this study represent a comparable range of physical habitats and preservational conditions. It is important to keep the environmental restrictions of this study firmly in mind. Arid and montane habitats are excluded from consideration because their floras are poorly known; swamps are also excluded because, while their constituent plant communities are often well preserved, they can be differentiated from those of mesic floodplains and, hence, eliminated from consideration in an attempt to ensure some measure of environmental homogeneity.

The unit chosen for study is the "flora," and one must understand what this means in paleobotanical parlance and how it translates into units recognized by ecologists. There is no standard formula for determining what constitutes a flora, but in practice almost all students consider a flora to be the total assemblage of fossil plants recovered from a reasonably homogeneous package of rocks, usually a formation, but not infrequently a finer lithostratigraphic unit such as a member. In environmental terms, this means that a flora consists of plants from a mixture of similar and contiguous habitats occurring over an area a few to several tens of kilometers in linear dimension. Individual bedding planes (which only rarely can be traced horizontally for more than 10 to 100 meters) may be the result of a single flood or a brief interval of sedimentation in a lake, but floras are invariably described from a series of bedding planes and thus represent accumulation over a period of several thousand, and perhaps as much as a few million, years. In the terminology of Whittaker (1977), the paleoflora does not represent strictly alpha diversity, i.e., diversity within a community, but rather a time averaging of gamma diversity, i.e., the landscape diversity that includes both alpha and beta diversity (beta diversity being a measure of community heterogeneity within the area under consideration).

As noted in the preceding section, the variable influences of taphonomy, transportation, and diagenesis must be taken into account in evaluations of fossil floras. These factors introduce nonbiological variation into the data. Histories of research also introduce nonbiological variation. Some floras have been studied more extensively than others, and this contributes to their relatively high diversities. On the other hand, it is also true that taxonomically rich floras are often the ones selected for continued study. While one cannot reasonably deny that these factors affect quantitative analyses of fossil plant assemblages, I believe that if one heeds the warning to make only relative comparisons and search only for large-scale patterns, the noise inherent in the system is insufficient to obscure biological signals of ecological and evolutionary interest.

Temporal Patterns of Floral Change

Trends in Species Diversity The mean and standard deviations of within-floral species and genus richness for successive intervals of Phanerozoic time are listed in Table 7.1. To minimize systematic effects of plant disarticulation and the consequent multiplication of organ taxa, a "minimum species" (or "minimum genus") concept was adopted for tabulations (Knoll et al. 1979). The number of species (genera) recorded for each higher taxon was determined as the minimum number of species (genera) sufficient to produce all of the organ species (genera) described from the flora. For example, a flora containing three lycopod stem species, four leaf species, three strobilar species, and one "root" species was considered to contain four species, not eleven. Maximum recorded species and genus numbers for a flora within each period are also listed; in most cases the unusually high diversities represented by these "maximum" figures represent

Table 7.1 TAXONOMIC RICHNESS OF PRE-PLEISTOCENE TERRESTRIAL FLORAS, GROUPED BY TIME INTERVAL

Time interval	Number of floras	Mean species richness (x)	SD_x	Maximum species richness	Mean genus richness (y)	SD_y	Maximum genus richness
Latest Silurian (Pridolian)	5	3.6	2.0	6	2.4	1.1	4
Early Devonian (Gedinnian)	8	4.0	1.4	6	3.7	1.1	5
Early Devonian (Siegenian)	12	7.6	3.1	14	6.8	2.1	10
Early Devonian (Emsian)	7	10.4	4.1	16	8.3	3.1	13
Middle Devonian	11	10.5	2.6	17	9.2	2.0	14
Late Devonian	7	10.7	4.1	16	7.9	4.1	14
Early Mississippian	15	11.9	3.5	19	7.9	2.0	13
Late Mississippian	8	23.3	9.6	34	12.1	3.4	16
Early and Middle Pennsylvanian	50	27.2	15.2	96	12.5	4.6	34
Late Pennsylvanian	28	23.4	7.9	36	12.8	3.3	19
Early Permian	26	29.1	14.1	69	13.6	4.9	22
Early and Middle Triassic	10	26.0	12.8	56	13.3	4.3	22
Late Triassic	10	30.6	23.0	86	17.2	5.8	28
Early Jurassic	20	33.2	27.0	118	18.0	11.3	55
Middle Jurassic	13	30.9	13.7	51	17.6	5.5	26
Late Jurassic	6	30.8	12.3	54	18.5	7.2	33
Early Cretaceous	22	30.2	13.2	70	19.3	8.0	37
Late Cretaceous	17	54.4	29.9	116	38.2	21.1	85
Paleogene	32	70.1	42.9	231	50.2	22.7	113
Neogene	66	66.3	38.8	187	47.6	26.3	135
Permian of Gondwana (cool temperate)	18	16.8	9.2	37	8.3	4.4	19

Species and genera were tabulated as described in the text. Data are from original sources, a list of which is available on request. The columns SD_x and SD_y give the standard deviations of species (x) or genus (y) richness.

deltaic or delta front marine depositional environments with a high degree of mixing of plant remains from different communities and landscapes. These aside, one of the most striking aspects of the data is how closely different authors agree in estimates of within-floral taxonomic richness for any given interval.

Several trends evident in the data are clearly visible in Fig. 7.1, a plot of mean within-floral species richness through time.

First, as might be expected, the oldest known macrofloral assemblages contain few differentiable forms. (Spores in the same rocks indicate a somewhat higher diversity, but one that is comparably low relative to succeeding floras [Knoll et al. 1984].) Within-floral diversity rose rapidly to reach a plateau in late early Devonian (Emsian) times. This plateau persisted into the Mississippian Period.

Second, during the Mississippian within-floral diversity rapidly rose to a new plateau more than double the previous value. This new plateau was then maintained for about 220 million years, or about half of the entire known history of vascular plants.

Third, in middle Cretaceous times within-floral diversity began to increase again, and in this case rapid diversification was not immediately followed by the reestablishment of a diversity plateau. An apparent leveling off is suggested by the Neogene data, but most floras recorded for this interval come from North America, Europe, the Soviet Union, and Japan, and this slight drop from Paleogene diversities must be interpreted in

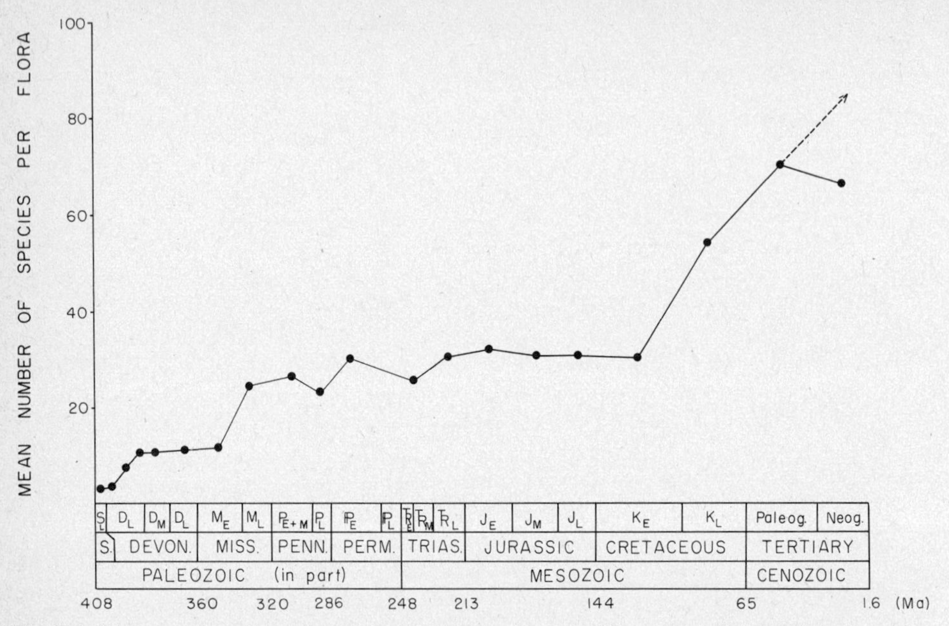

Fig. 7.1 Mean number of species per flora (an imprecise proxy for community diversity) plotted by epoch for the Phanerozoic Eon. Time scale is that of Harland et al. (1982, data from Table 1). The dashed arrow at the Recent end of the graph indicates the estimated Neogene trajectory for moist subtropical floras. Neogene floras used in the compilation reflect increasingly temperate climates during the later Tertiary Period. Older floras used in the construction of the figure are thought to represent mesic subtropical to tropical communities from lowland floodplain, lake margin, and deltaic environments.

light of increasing Tertiary climatic deterioration (Wolfe and Barghoorn 1961, Cousminer 1961). These Neogene floras, then, reflect the differentiation of temperate zone communities. The surprising feature is that Neogene taxonomic richness values decline only slightly; this suggests that contemporary tropical and subtropical diversities (very incompletely known from the fossil record) continued to increase. Indeed, it appears that species numbers within subtropical to tropical communities have been rising continually since the Cretaceous and that a plateau has yet to be established.

The overall pattern of temporal change in within-floral diversity thus consists of a few relatively brief periods of rapid increase bounded by much longer intervals of more or less unchanging taxonomic richness. Although one might have predicted that within-floral diversity would increase continually through geological time, such a pattern characterizes only the last 100 million years, and it stands in strong contrast to the long lasting plateaus of preceding periods.

Sequences of Dominant Taxa Some insight into the controls of within-floral diversity may be obtained by examining the taxonomic composition of floras through time. Fig. 7.2 illustrates trends in the mean percentage composition of floras. The same data are exploded in Fig. 7.3 to facilitate the tracing of any single group's history. It should be emphasized that these figures do not necessarily reflect numerical abundance or relative contribution to biomass. For example, some Mesozoic assemblages are numerically dominated by a few conifer species, while the greatest species richness lies in the less abundant ferns. Relative abundances have infrequently been quantified, but qualitative descriptions of dominance distribution are common.

As shown in Fig. 7.2 and particularly Fig. 7.3, Silurian and Devonian floras consisted of a

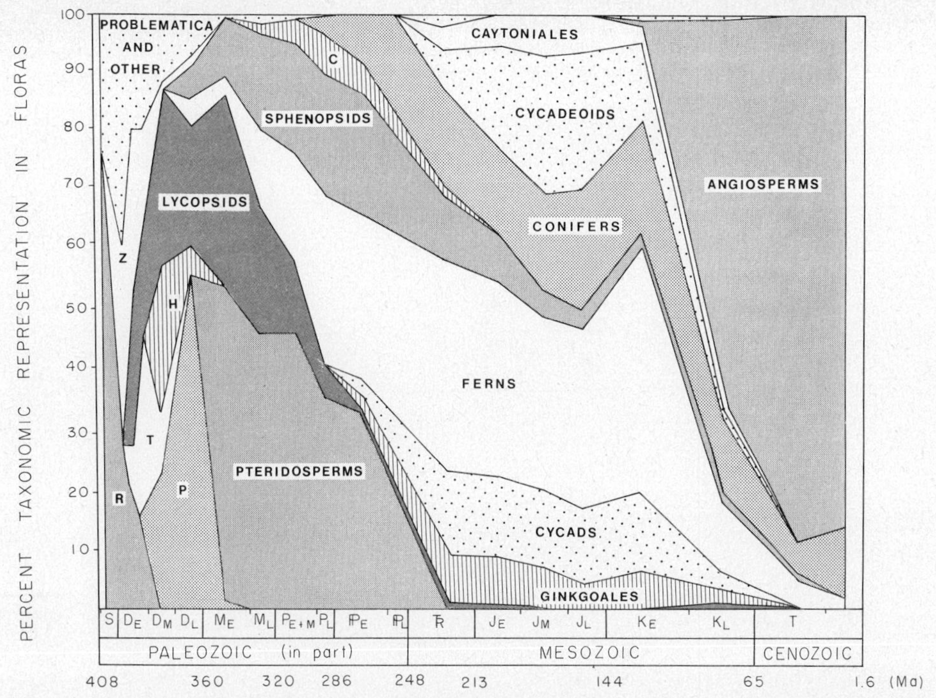

Fig. 7.2 The taxonomic composition of terrestrial floras throughout the Phanerozoic eon (time scale of Harland et al. 1982). As in Fig. 7.1, fossil assemblages reflect principally mesic subtropical to tropical lowland environments. Compositions plotted are mean percentage of species per major group for floras of successive intervals. The category Problematica and Other includes problematic plants as well as several taxa belonging to groups other than the major taxa listed (e.g., the Barinophytales). Silurian and Devonian problematics include nonvascular plants (e.g., *Prototaxites*).

parade of successive dominant groups of spore-bearing plants. The principal contributors to within-floral diversity passed sequentially from the rhyniophytes to the zosterophylls and lycopsids, to the trimerophytes, then to hyenialian (*sensu lato*) plants and early progymnosperms, and finally to advanced progymnosperms such as *Archaeopteris* and its relatives. Available information on relative abundance in these floras indicates a comparable pattern of sequential replacement.

From the Mississippian Period until the mid-Cretaceous floral compositions (and numerical abundances) were dominated by gymnospermous plants. Paleozoic tropical to subtropical floras tend to be dominated by pteridosperms, with marattialean ferns making an important contribution from the latter part of the Middle Pennsylvanian onward (Pfefferkorn and Thomson 1982). A major shift in floral compositions occurred near the end of the Paleozoic Era, with Paleozoic pteridosperm families declining drastically in importance (and eventually dwindling to extinction) concurrent with the expansion of the conifers, which existed in relatively dry "upland" environments in the Late Pennsylvanian Period, and other gymnospermous groups. This floral transition is strongly time transgressive when viewed globally and corresponds to a major period of paleogeographical, tectonic, and, in part related, climatic change (Knoll 1984). A final first-order change in the composition of floras occurred in the Cretaceous Period, with the radiation and rise to ecological dominance of angiosperms.

Steps and Plateaus in Species Diversity Comparison of the data sets on taxonomic composi-

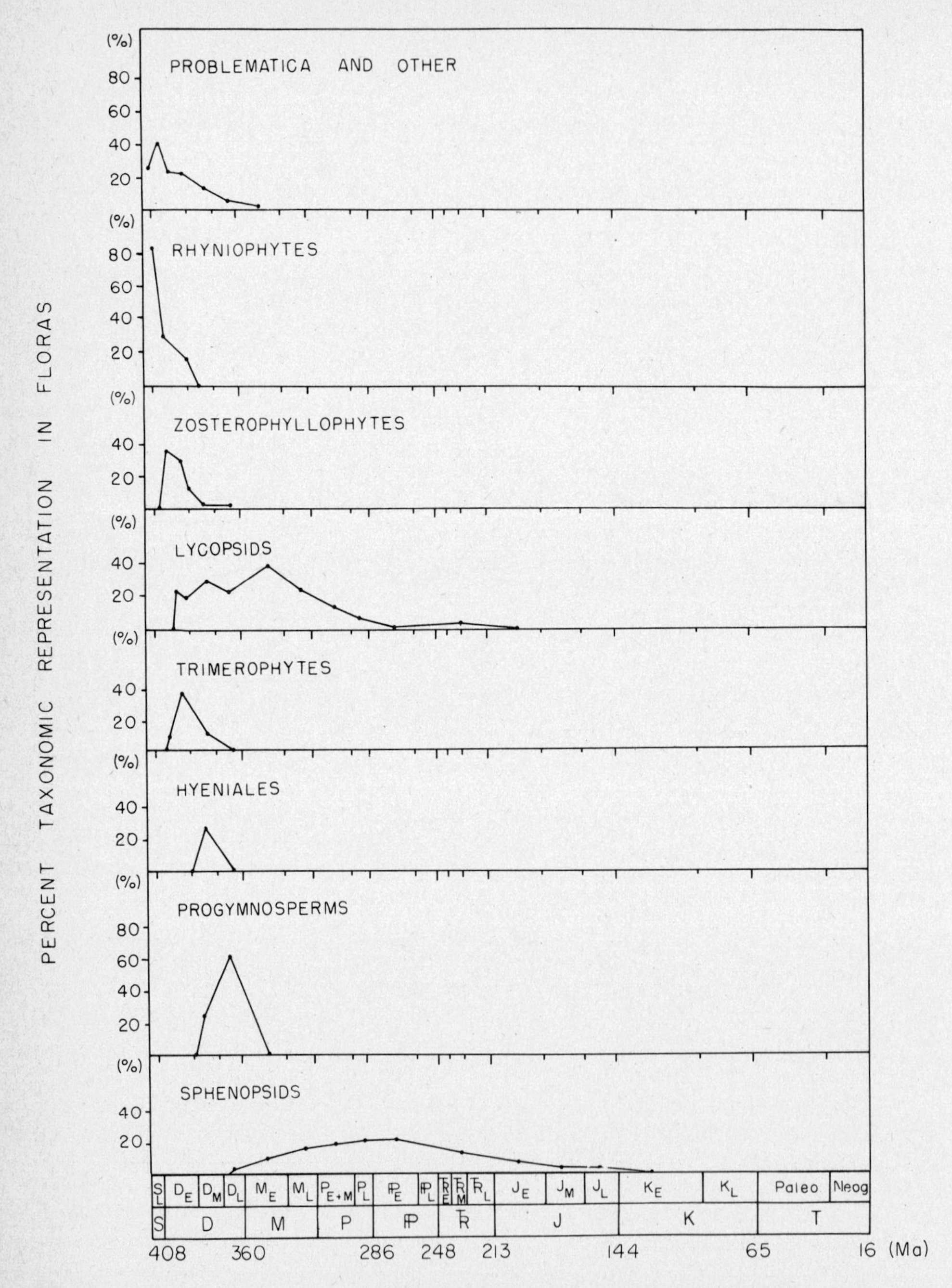

Fig. 7.3 Same data as Fig. 7.2, exploded to clarify temporal trends for individual major groups of plants: Rhyniophytes (R); Zosterophyllophytes (Z); Trimerophytes (T); Hyeniales, including Ibykales, *Pseudosporochnus,* and *Calamophyton* (H); Progymnosperms (P); Cordaitales (C).

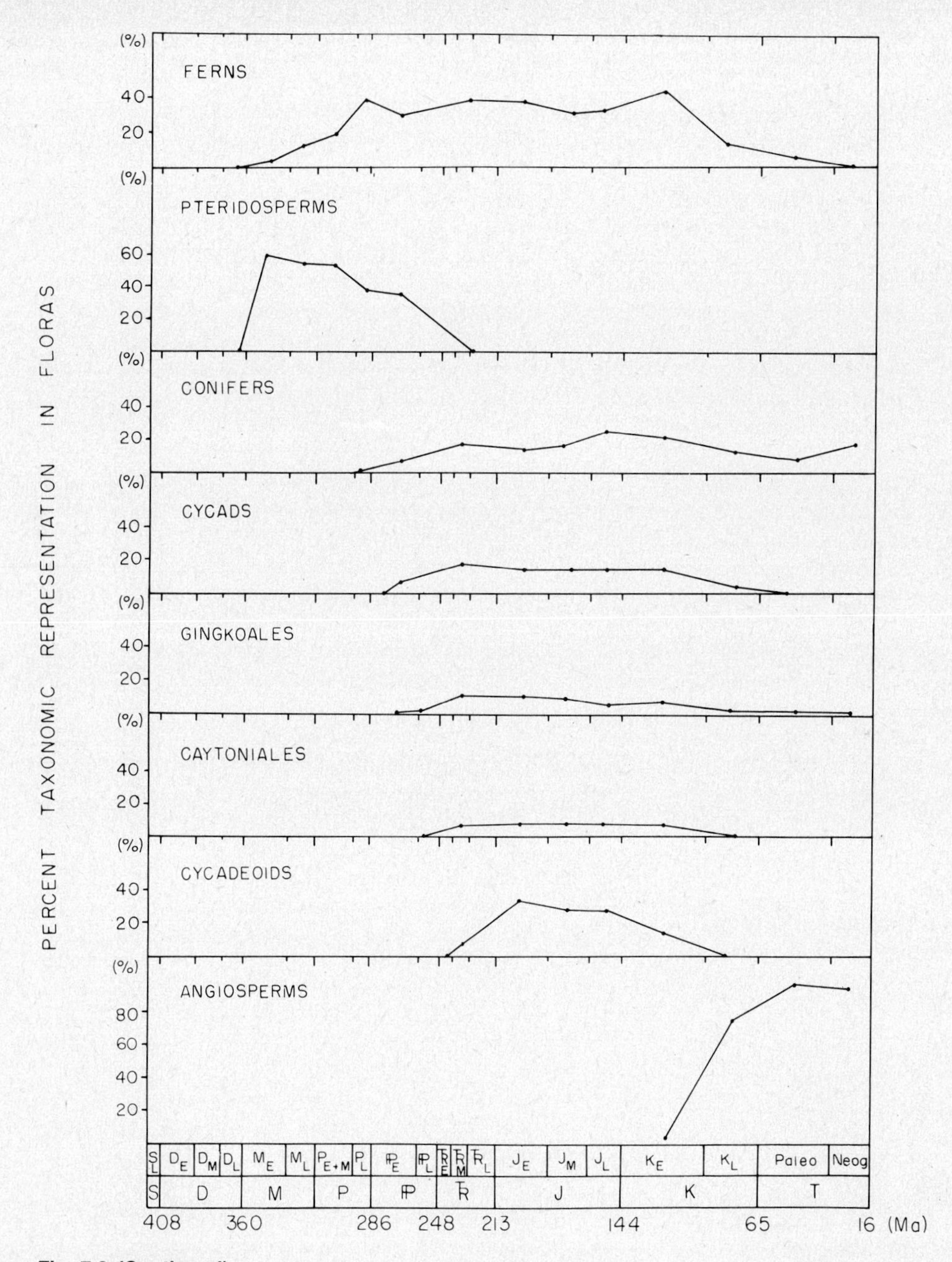

Fig. 7.3 ***(Continued)***

tion and richness shows that the stepwise increases in within-floral diversity accompanied major evolutionary innovations in reproductive biology (Knoll et al. 1979, Niklas et al. 1980). The first rise corresponds to the initial diversification of spore-bearing tracheophytes, the second to the radiation of gymnosperms, and the third to angiosperm evolution. In no case is the diversity increase instantaneous. For example, early vascular plants appeared in the late Silurian or earlier, but species richness did not plateau until some 20 million years or more later, and the latest Devonian origin of gymnosperms was followed by a new diversity plateau only in mid to late Mississippian times. This lag may reflect the kinetics of diversification as modeled by Sepkoski (1979; see also Knoll et al. 1984 for a discussion of early tracheophyte radiation).

The two diversity plateaus do not correspond to periods of taxonomic invariance at the family or ordinal level. On the contrary, major shifts in the composition of floodplain floras occur within intervals of approximately steady-state diversity. Pteridosperm and associated pteridophyte forests of the Pennsylvanian, caytonialean (corystosperm)-dominated communities of the Gondwana Triassic, and Mesozoic gymnosperm-fern floras from the Northern Hemisphere all have similar magnitudes of species richness.

DISCUSSION

The Taxonomic Richness of Pre-Pleistocene Floras

Many ecologists have discussed the *maintenance* of high community diversity in the face of competition, citing the importance of phenomena such as disturbance, predation, life-form and phenological differentiation, and differing capacities for regeneration (e.g., Chapters 12 and 19; Janzen 1970, Grubb 1977, Connell 1978, Hubbell 1979). The *generation* of diversity on an evolutionary time scale has been examined less frequently, but it is this question that is more amenable to paleontological investigation. Among those who have considered the generation of diversity, Whittaker (1977) stressed the role of habitat niche differentiation, while Grubb (1977 and Chapter 12) has emphasized what he terms the regeneration niche.

Mosaic evolution is a prominent feature of vascular plant history. Significant evolutionary changes in vegetative morphology—the features most likely to affect potentials for habitat niche differentiation—have not always coincided with innovations in reproductive biology, the stuff of regeneration niche differentiation. As noted above, significant increases in diversity have consistently accompanied major evolutionary changes in reproductive systems rather than introductions of new vegetative characters. For example, the evolution of trees took place during the *middle* of the Devonian diversity plateau. The absence of significant change in within-floral taxonomic richness during the long interval from the Late Mississippian until the mid-Cretaceous Period, in spite of virtually complete turnover in the vegetative morphologies of the plants that composed these floras, also argues for reproductive control of long-term patterns in community species diversity. Note that the argument is not that periods of diversification are unmarked by morphological differentiation of leaves, stems, and other nonreproductive organs. Since diversity in fossil assemblages is recognized largely by the evaluation of these organs, such differentiation by definition occurred. However, this morphological differentiation may well be a consequence of speciation, rather than its cause.

It is of particular interest to ask why the long late Paleozoic and Mesozoic period of gymnosperm dominance stands out as a conspicuous diversity plateau, while the succeeding era of angiosperm-dominated forests shows continually increasing diversity. Most, if not all, gymnosperms are wind pollinated. Wind distributes pollen nonselectively and over relatively short distances. Also, the presence of foreign pollen grains in the ovules of at least some seed plants inhibits fertilization and germination (Kanchan and Jayachandra 1980, Thomson et al. 1981). Therefore, gymnosperm species in forests may be constrained to live at relatively high population densities, thereby limiting the number of species that can coexist in a given area. Elsewhere in this volume (Chapter 12), Grubb suggests that common species in a community encounter one another frequently and therefore compete strongly, often becoming distributed along an environmental gradient. Under such conditions the success of a new species in a late

Paleozoic or Mesozoic gymnosperm-fern forest would probably have brought about (or been made possible by) the extinction of a preexisting species, and diversity would remain at some approximate steady-state level.

In contrast, angiosperm trees in moist subtropical to tropical forests are pollinated by animals. Animals distribute pollen selectively to target plants over relatively long distances, and this permits populations to persist at low population densities. The result is that more species can be accommodated within a habitat. This comparison has long been invoked to explain the high diversity of tree species in modern tropical forests (Müller 1877, Regal 1977, Raven 1977, Burger 1981), and it draws strong support from the fossil record. The comparison also suggests an explanation for the absence of a diversity plateau in the 100-million-year hegemony of the flowering plants. Able to disperse pollen and seeds over long distances (cf. Regal 1977), angiosperms can establish and maintain sparse populations (see Chapters 12 and 19). Thus, as flowering plants rose to ecological dominance, communities could accumulate higher and higher numbers of rarer and rarer species, with no obvious ceiling on species richness. Insofar as phenomena such as chance disturbance and seed predation may be most effective in maintaining high diversity when populations exist at low densities, their importance in determining community structure should have increased through Cretaceous and Tertiary time.

By stressing the importance of reproductive biology in the generation of increasingly high levels of flowering plant diversity, I do not mean to imply that the tremendous adaptive flexibility of angiosperm vegetative sporophytes has been inconsequential. What I do wish to convey is that in the generation of *within-community diversity* of forest trees and shrubs, the unique reproductive system of the angiosperms has been a principal, and perhaps *the* principal, enabling factor. The effects of morphological and physiological variability are more fully seen in considerations of *total*, or *global, diversity*. The rise in within-floral species richness documented in Fig. 7.1 cannot account for the entire increase in total diversity evident over the past 100 million years (e.g., Hughes 1976, Knoll 1984). Fossil assemblages of angiosperms show much greater degrees of differentiation among habitats and provinces than do those of earlier periods, and this is the additional factor that has contributed to the Cretaceous-Tertiary rise in total fossil plant diversity. Although it cannot be documented with any precision from the fossil record, flowering plants have also diversified dramatically in arid and alpine environments in which gymnosperms and spore-bearing plants appear to have enjoyed only limited success. These components of total diversity certainly reflect the morphological and physiological variability of angiosperms.

Some Further Community Consequences of Reproductive Evolution in Vascular Plants

In addition to its effects on the species richness of floodplain forest trees, the evolution of reproductive systems has influenced temporal patterns of plant community structure in other, less direct ways. For example, the reproductive cycles of extant (and, presumably, extinct) gymnosperms are slow, often taking two or more years to go from pollination to seed germination. Also, with its multicellular gametophyte the gymnosperm ovule requires a sizable investment of resources. Thus, it is not surprising that gymnospermous plants are constrained to be perennial trees and shrubs. In the absence of angiosperms one would expect that occupation of the herbaceous understory would fall to pteridophytes, and, indeed, Triassic, Jurassic, and early Cretaceous forests appear to have been structured along taxonomic lines, with woody gymnosperms rising above a ground layer of ferns and lycopsids. Flowering plants with their rapid life cycles can exist as annual herbs or canopy trees, and, as is clear from Figs. 7.2 and 7.3, they expanded ecologically during the late Cretaceous at the expense of both gymnosperms and understory ferns.

Vascular epiphytes are important elements of many tropical forests, but because taphonomic biases often favor the preservation of woody plants, the history of canopy occupation is difficult to document in the fossil record. As a conspicuous example, fully 25% of living fern species are epiphytic (Madison 1977, Benzing 1983), but the radiation of polypodiaceous *(sensu lato)* ferns that accompanied the angiosperms' rise to dominance is not evident in Figs. 7.1–7.3. The impressive radiation of angiospermous epiphytes, particularly the orchids, is equally poorly documented in the fossil record. Nonetheless, considerations of reproductive biology suggest that the evolution of angiospermy must have had profound effects on the occupation of forest canopies. Benzing (1984) has suggested that the primary requirements for epiphyte hosts are bark texture and stability and openness of the canopy to light, precipitation, and seed, spore, and pollen movement, with nutrient leaching in colonization sites constituting a further factor of significance. In terms of these attributes angiosperm trees often constitute better hosts than gymnosperms. Some living conifers do support vascular epiphyte populations, but the abundance and diversity of canopy dwellers in gymnosperms is low relative to angiosperms (Benzing, personal communication).

Epiphytes reach germination sites by being blown or carried by animals; thus, reproductive systems limit the candidates for epiphytic existence. Gymnosperm seeds are ill suited for such transport, and it is likely that gymnospermous epiphytes were as uncommon in Mesozoic forests as they are today. Spores, on the other hand, are easily blown from site to site and so constitute ideal disseminules for canopy colonization. Many lycopsid species are epiphytes today, and insofar as appropriate sites were available, they probably inhabited canopy environments in the past. Ferns also reproduce by spores, and it is interesting to note that none of the several fern families that were conspicuous elements of Mesozoic moist floodplain communities have living members that are true canopy rooting epiphytes. On the other hand, the polypodiaceous ferns, which have no pre-Cretaceus fossil record, have radiated dramatically in the epiphytic habitat. Perhaps early polypods lived in relatively arid areas unlikely to be well represented in the fossil record; when angiosperms became the dominant forest trees, these ferns would then have been physiologically well equipped to exploit the altered canopy habitat. Mesozoic floodplain ferns, in contrast, may not have been able to adapt to the relatively xeric conditions of canopy environments.

Angiosperms, by virtue of tiny seeds or animal dissemination, are also able to disperse in canopies, and certain flowering plant families have diversified tremendously in the epiphytic habitat. Overall, in terms of canopy-forming trees and canopy dwellers, it seems indisputable that the evolution of angiospermy must have restructured the uppermost reaches of tropical forest communities.

Changes in the Taxonomic Composition of Floodplain Floras: a Competitive Replacement Hypothesis

Plants virtually all have the same general requirements for growth: light, CO_2, H_2O, and simple nutrients. Patterns of moisture and light availability, temperature, and substrate composition are quite variable on the earth's surface, and plants have evolved numerous combinations of morphological, physiological, and life history characters that collectively enable individual populations to exploit different habitat ranges successfully. Nonetheless, because of their similar metabolic needs, it has often been suggested that plants should compete strongly with their neighbors, a hypothesis that has received much support from field and laboratory experiments (e.g., Harper 1977). There are several ways in which the effects of competition can be minimized or retarded in plant communities (e.g., Connell 1978), but it is not clear that competition can be avoided indefinitely. Therefore, given geologically significant lengths of time, one might expect competition to be important in establishing temporal patterns of floodplain community composition.

I have argued elsewhere that some temporal patterns in floral composition are indeed consistent with, and probably best explained by, a hypothesis of competitive replacement (Knoll 1984). This is particularly true for the succession of floras that characterizes early (Silurian and Devonian) vascular plant evolution, and I can attempt to summarize briefly the pertinent observations and arguments.

Early tracheophyte communities consisted of herbaceous plants that colonized open sites via spores and then expanded clonally to form dense monospecific clusters arrayed in a mosaic pattern. Analysis of stratigraphic series of bedding planes indicates that several taxa in succession first appeared as minor elements of communities, rapidly expanded to taxonomic and ecological dominance, and then dwindled in importance, eventually disappearing entirely. The important point is that the stratigraphic distributions of successive dominants overlap (e.g., Fig. 7.3), so that the rise to dominance of one group coincides with the diminution of the previously dominant taxon. Morphological comparisons of successive dominants show clearly a series of changes that must have provided members of each successively dominant group with increased capacities for gathering light and absorbing and transporting water and nutrients (Chaloner and Sheerin 1979, Knoll 1984). Given what is known about competition among living plants (e.g., Harper 1977), the outcome of competitive interaction between a rhyniophyte and a trimerophyte, or between a trimerophyte and a progymnosperm, in a low diversity Devonian floodplain community seems clear. Tilman develops a similar argument in the section of Chapter 22 on evolution of land plants. Thus, the early history of vascular plants consists of a series of species introduction experiments on a gigantic time scale.

Mass extinction followed by rediversification can be eliminated as a possible alternative to competitive replacement on the basis of stratigraphic distributions. Similarly, there is no evidence to support a hypothesis that changes in climate drove the sequence of plant replacements shown in Fig. 7.3. As for a hypothesis based on evolutionary changes among herbivores, a few Devonian fossils preserve evidence of probable arthropod damage (Kevan et al. 1975), but as neither the insects nor land vertebrates radiated until after this interval, it appears unlikely that herbivory could have been the driving force behind changes in community composition.

Like the natural experiment approach to ecological research, the inference of process from paleobotanical pattern is difficult because arguments must be based on the strengths or weaknesses of expected correlations rather than on the rigorous experimental elimination of alternative hypotheses. Therefore, it is prudent to bear the uncertainties of the method clearly in mind. Nonetheless, the hypothesis of sequential competitive replacement in Devonian vascular plant communities is at the very least consistent with the known facts of the fossil record and fulfills all the retrospective predictions that can be made on the basis of the ecology of living plants. No comparable claim can be made for any alternative hypothesis that I have been able to articulate.

A hypothesis of competitive replacement can also be sustained for the transition from Mesozoic gymnosperm-fern vegetation to angiosperm-dominated communities (Knoll 1984). However, the major late Paleozoic compositional change evident in Fig. 7.2 requires another explanation, namely, a botanical response to global climatic change. One agent of community change whose effects are notably weak in the plant fossil record is mass extinction; the paleobotanical record contains no evidence to suggest that globally synchronous mass extinctions played a major role in generating temporal patterns of community composition or diversity at any time prior to the present era of rain forest destruction (Knoll 1984). Individual species may or may not have followed essentially random walks to extinction through geological time (see Hubbell, Chapter 19); the fossil record is not sufficiently complete to address this question.

A Brief Word About Animals

The invertebrate paleontological record provides a rich documentation of animal life in the world's oceans over the past 600 million years, and for every observation I have made about temporal patterns in vascular plant communities there is a corresponding set of data for fossil marine invertebrates. Space does not permit a detailed discussion of the paleozoological record, but a few comparisons between the histories of vascular plant and shelf invertebrate communities are worth noting here.

Compilations of within-fauna and total diversity for skeletonized marine invertebrates reveal a temporal pattern of increase broadly similar to that of plants (Sepkoski 1979, Sepkoski et al. 1981, Bambach 1983). An initial latest Precambrian to Cambrian period of diversification culminated in a short-lived diversity plateau. This plateau was terminated by renewed diversification during the Ordovician Period, following which a second plateau was established. Although marked by temporary decreases associated with episodes of mass extinction, this plateau extended to the end of the Permian Period, when a massive extinction event lowered diversities significantly. Diversity levels recovered by the Jurassic Period, and the ensuing Cretaceous-Tertiary interval has been marked by a sustained increase in taxonomic richness, punctuated every 26 million years or so by extinction events (Raup and Sepkoski 1984). (The incompleteness of the terrestrial animal record makes it difficult to quantify temporal patterns of diversity and composition with any precision, but it is evident that the Cretaceous and Tertiary radiations of insects and mammals have resulted in land communities whose animal components are more diverse than those of any preceding eras. It is also clear that mass extinctions have strongly influenced temporal patterns in the composition of tetrapod guilds.)

This pattern suggests both similarities and differences in the historical development of plant and marine invertebrate communities. On the one hand, both records show prolonged periods of more or less steady-state diversity preceded by intervals of marked diversification. The significance of this similarity is difficult to assess because the controls on diversity trends remain poorly understood for marine invertebrate communities. It is interesting that Bambach (1983) has argued strongly that diversity increases within marine invertebrate communities have occurred through the addition of new guilds to community structure, rather than simply by means of increased packing within existing guilds. If true, this stands in strong contrast to the plant record, where much of the observable increase in within-floral diversity occurred within existing tree, shrub, and herb guilds. The second major difference in the generation of temporal pattern concerns the role of mass extinction. Patterns of vascular plant evolution can be explained satisfactorily in terms of processes observable or inferrable by plant ecologists; however, one cannot hope to understand the evolutionary history of marine invertebrates (or land vertebrates) without taking into account mass extinctions. Much remains to be learned about the fossil records of both plants and animals, but it is clear that comparative studies of plant and animal community histories can provide much information of interest to community ecologists and other evolutionary biologists.

SUMMARY

Because they are subject to the variable influences of postmortem decomposition, transportation, and diagenesis, fossil plant assemblages provide an imperfect reflection of the communities from which they were derived. Nonetheless, major temporal trends in the taxonomic makeup and diversity of fossil floras document underlying patterns of change in the composition of plant communities through time.

Analysis of 391 vascular plant floras ranging in age from latest Silurian (ca. 410 million years ago) to Pliocene (ending 1.6 million years ago) indicates that the history of diversity within floras from subtropical to tropical mesic floodplains is marked by several periods of rapid increase separated by extended periods of more or less unchanging taxonomic richness. The intervals of pronounced diversification coincide with periods of significant evolutionary innovation in plant reproductive biology. The within-floral diversity increase associated with the evolution of angiospermy differs from earlier rises in that it did not soon culminate in the reestablishment of a plateau, but rather has continued throughout the past 100 million years.

It is hypothesized that the pollination and dispersal systems of trees and shrubs found in mesic floodplain communities place constraints on allowable population densities, and that this, in turn, strongly influences the number of species that potentially can coexist in a given area. This hypothesis can explain the long lasting diversity plateau associated with the ecological dominance of successive gymnosperm groups, as well as the continuing rise in within-floral diversity that has marked the period of angiosperm domination.

Although fossil floras are taphonomically biased in favor of woody plants, comparative biological evidence suggests that the evolution of angiosperms also brought about important changes in the occupation of the epiphytic habitat.

Patterns of change in the taxonomic composition of floras through time suggest that, in the long term, plant community structures are more sensitive to the effects of competition and climate than they are to perturbations by impacting bolides or other agents of ecological catastrophe. Evidence for synchronous global mass extinctions of plants is notably weak.

ACKNOWLEDGMENTS

I thank the following persons for helpful comments and discussions: Peter Ashton, David Benzing, Margaret Davis, Jared Diamond, Peter Grubb, Stephen Hubbell, Karl Niklas, David Tilman, Rolla Tryon, and Thomas Van Devender. Elizabeth Burkhardt drafted the figures for this chapter.

three

SPACE AND TIME

chapter 8

Overview: The Importance of Spatial and Temporal Scale in Ecological Investigations

John A. Wiens, John F. Addicott, Ted J. Case, and Jared Diamond

INTRODUCTION

Aspects of scale have major and obvious consequences in ecological investigations. The amount of space available and its grain, the geometric relationships between patches, and the length of time available figure importantly in many ecological discussions (e.g., MacArthur and Wilson 1967, Segel and Jackson 1972, Horn and MacArthur 1972, Slatkin 1974, Levin 1974, Hastings 1977, Casten and Case 1979, Hanski 1983). In a sense, the entire field of island biogeography explicitly focuses on effects of scale. Chapters 10 to 12 of this volume deal with such effects.

In this chapter our concern is not with the familiar biological consequences of scale, but with its less familiar methodological significance. The size of a study area and the duration of an investigation limit what one can see of a community (Dayton and Tegner 1984). As a result, studying ecology resembles what it would be like to conduct a chemistry experiment if the chemist were only a few angstroms long and lived for only a few microseconds. Such a chemist would have difficulty deciphering the overall course of chemical reactions from the random collisions of molecules. The ecological phenomena that we study often operate on spatial scales far larger than our bodies and on temporal scales longer than our own existence and certainly longer than the time span of a research grant. We get only a brief and often dim glimpse of the relevant processes. To use another analogy, we must attempt to reconstruct an entire motion picture from a few consecutive frames of one film or from single frames of many films that we hope are similar. Unless we are aware of the constraints imposed by our scale of operation, and unless we use sound logic in extrapolating from our few frames of observation to a film of ecological and evolutionary processes, we risk either drawing hasty and incorrect generalizations or becoming so mesmerized by proximate details that we fail to discern the broader patterns of nature.

Some of the most vociferous disagreements among ecologists arise from differences in their choice of scale. Those who study the long-term dynamics of a particular community at a single site often reach very different conclusions from those reached by individuals conducting short-term studies of similar communities at many different sites distributed over a large area. Conclusions about the degree of community structure, the importance of noise, the effect of disturbance, the roles of various limiting factors, and the identities of the dominant interactions are likely to differ at these two scales. All too often ecologists accustomed to working on a certain scale react to these differences by disparaging the value of work at another scale, rather than by attempting to reconcile the differences or to benefit from the insights of that work. Short-term, broad-scale studies are often decried as superficial, addressing only patterns rather than processes, producing artifactual findings, and impotent to discern crucial factor X, while long-term studies on a narrow scale are decried as overly particularistic, sensitive to sampling error, devoid of perspective, and unable to discern some other crucial factor Y.

We begin this chapter by discussing the consequences of various choices of spatial scale, ranging from the territory of a single individual to a biogeographical region. We follow with a corresponding discussion of various choices of temporal scale, ranging from a single foraging bout to geological times sufficient for evolutionary change. Finally, these consequences are illustrated by a hypothetical example examining how choice of scale affects what one can learn about predation by coyotes on rabbits. The subsequent chapter by Wiens (Chapter 9) carries out a similar examination in detail for an actual community, the birds of North American shrubsteppe. While our immediate concern in this chapter is with the importance of scale considerations for studies of communities, most of our points are equally applicable to studies of single species.

CHOICES OF SPATIAL SCALE

Although the possible spatial scales for investigations form a continuum, it is convenient to identify five points along the continuum: (1) a space occupied by a single individual sessile organism or in which a mobile individual spends its life, (2) a local patch occupied by many individuals, (3) a region large enough to include many local patches or populations linked by dispersal, (4) a space large enough to contain a closed system (if any exists), that is, a system essentially closed to emigration and immigration, and (5) the biogeographical scale, which is large enough to encompass different climates and constellations of syntopic species.

The Area Used by a Single Individual

Many biological questions can be answered only by prolonged study of a single individual or group. Questions of physiological ecology, sociobiology, foraging ecology, and reproductive biology are often of this sort. The appropriate choice of scale then varies with the species and is defined by the cruising range of the individual or group: a handful of dirt for an annual weed or soil bacterium, a few hectares for a typical resident songbird, dozens or hundreds of square kilometers for a tiger or a band of wild dogs, and the entire continent of Australia for nomadic ducks of ephemeral desert ponds.

Special problems arise when an individual shifts its foraging base with season. For example, there is growing concern that ecologists who study long-distance migrants at summer breeding grounds conveniently near the ecologists' temperate-zone universities may be missing an essential part of the picture. One is doomed to drawing misleading conclusions if the population dynamics of a migratory neotropical bird are studied at its breeding territory in North Dakota, but the main effects of limiting factors operate at its wintering grounds in Venezuela (Chapter 9; Grant and Schluter 1984). Similarly, for monarch butterflies the Mexican forest in which they winter may be as crucial as the California coast where they summer. For whales that commute between the Antarctic and equator or for shearwaters that breed on islands near Australia and then perform a circum-Pacific migration, one can hardly hope to understand even the life of a single individual if one's choice of spatial scale is smaller than the whole Pacific basin.

The Scale of a Local Population or Patch

Observations of one or a few individuals can be very detailed, but they do not provide a secure basis for generalizations. To reduce sampling errors and minimize the influence of individual idiosyncracies, we want a study area large enough that it will contain a sufficient number of individuals to permit statistically secure inferences about the local population or habitat patch. Here, too, the appropriate choice of scale varies with species and depends on population density, social structure, individual cruising range, and the ecological process one wishes to study. Collectively, these factors define an ecological "neighborhood" (Antonovics and Levin 1980). For species composed of sedentary individuals with mutually exclusive territories, for example, the area required for sampling is much larger than the area used by a single individual. For species whose individuals are nomadic or have broadly overlapping ranges, the scale for sampling may be the same as an individual's range. Investigation of predation or herbivory processes within a patch of some plant species may require a different scale from studies of the pollination biology of those plants. In this volume examples of studies based on detailed sampling of local populations appear in Chapters 3, 9, 10, 19, 21, 26, 30, 31, and 32.

The Regional Scale

Standard textbook formulations of competition and predation give the instantaneous growth of each species as a function of its own density and the densities of species with which it interacts. There are usually no terms for the spatial coordinates where these interactions take place nor for the immigration and emigration of individuals across the coordinates. Space is assumed to be homogeneous and the system is assumed to be "well stirred," so that the added complexities of dispersal can be ignored.

Even in homogeneous space, however, violation of the well-stirred assumption produces remarkable consequences. Species that cannot coexist (except briefly) in a small, isolated spatial cell may coexist for long periods in a system composed of many such cells coupled by dispersal. This result does not demand that space be patchy or heterogeneous, although this is often the case. Rather, distributional patchiness can arise as "standing waves," whose perpetuation depends on the continued arrival and departure of individuals and on the interactions between individuals within spatial coordinates. These patches are the so-called dissipative structures of reaction diffusion theory (Turing 1952, Segal and Jackson 1972, Casten and Case 1979). The "waves" or patches cease to exist if removed from the "ocean" that supports them. In addition, of course, distributional patchiness can be caused by patchiness of space itself.

As emphasized in many chapters in this volume, the populations of many species are not distributed continuously over the landscape, but are concentrated in patches for the two types of reasons just described. Qualitatively new phenomena that are invisible at the scale of an individual's cruising range or a single patch may become apparent if one chooses a scale large enough to include many patches coupled by dispersal (Chapters 11, 12, 14, 24, 26, and 30). For instance, the ability of herbivorous aphids to escape predatory ladybugs and develop outbreaks is contingent on environmental patchiness (Chapter 11). Populations of many marine intertidal and coral reef organisms can be viewed as sets of spatially defined patches linked by their contributions to (and recruitment from) the pool of planktonic larvae (Chapter 30). The scale for a population with long-distance dispersal will differ from that of an otherwise similar population with short-distance dispersal. Where superimposed on a heterogeneous habitat or influenced by spatially varying demographic processes, populations may express the sort of hierarchical, metapopulation structure described by Addicott (1978) or Gill (1978). Discerning these patterns requires investigation at both the regional and local patch scales.

A clear example of how choice of scale drastically affects one's conclusions is provided by studies of coral reef fish communities. Sale and Dybdahl (1975), using study sites the size of a single coral head, were struck by the role of stochastic elements and the lack of a role of adult habitat preference in determining year-to-year variation in the fish species composition on a particular coral head. When the study site spans a variety of habitats, however, it becomes clear that different species have different habitat preferences, and the species composition of large reefs appears to be relatively constant from year to year (Anderson et al. 1981, but see Sale and Williams 1982, Sale et al. 1985). These discrepant conclusions arise at least in part from differences in scale: a large stochastic element in the settling rates of larval fish at individual sites obscures the patterns that emerge when many sites are examined, and, alternatively, the importance of such stochastic effects and the limits to the precision of habitat selection are not detected at the broader scale (Sale 1984). The same discrepancies are expected in studies of any species exhibiting much stochastic temporal variation in occupancy of individual sites (e.g., small islands; Haila 1983).

The Scale of the Closed System

Roughgarden (Chapter 30) emphasizes the distinction between closed systems, in which the major processes determining behavior of the system are confined within the system's boundaries, and open systems, in which immigration from outside is an important influence. For example, within any study site the population dynamics of a species will be determined by two components: the local birth and death processes, and the movement of individuals into and out of the site. If our main concern is with questions involving local birth and death processes, then we must choose an area of sufficient size that the overall dynamics are dominated by those processes.

The scale on which one must operate to encompass a closed system varies greatly with the dispersal ability of the species studied. For sedentary species on islands or habitat patches, the island or patch itself offers a natural closed system (e.g., the Galápagos finches discussed in Chapter 10). In the case of the rain forest tree community of Barro Colorado Island, however, Hubbell and Foster (Chapter 19) argue that the island does not constitute a closed system. Instead, many tree populations there are subsidized by immigration from the outside. Thus, it is impossible to understand the Barro Colorado tree community through studies of the island alone; the appropriate unit for understanding it is a much larger one, constituting at least all of Panama. Even more extreme are species such as tardigrades and ferns that disperse worldwide through the aerial plankton and for which the fully closed system is the entire earth.

Whether a system can be considered open or closed also varies with the question asked. For example, suppose that one is concerned with fluctuations in reproductive output of a group of sedentary individuals. That output is likely to depend on resource levels, but an important resource for the sedentary individuals may be mobile organisms whose abundance is determined by processes at a distant site. Also crucial in selecting the spatial scale for an effectively closed system is the temporal scale of the question asked, for example, whether one is concerned with year-to-year population dynamics or with the survival of a species over evolutionary time. Even if one chooses a scale large enough that immigration is too low to affect year-to-year dynamics significantly, immigration may be essential for the population to avoid extinction over long time periods. Thus, the ''scale of the closed system'' does not fall neatly between the regional scale and the biogeographical scale, but can occupy various positions along the continuum of spatial scales.

The Biogeographical Scale

The largest scale is the biogeographical, a spatial scale large enough that at different sites a species will encounter substantially different climates and constellations of syntopic species. In addition, if geographical barriers to dispersal are sufficient, different sites may have genetically differentiated populations of the same species. The biogeographical scale is blind to some phenomena and affords unique insights into others. Because properties of local patches and conditions at different sites are averaged out, much of the detail that can be learned from studies on a smaller scale may be lost (Chapter 9). On the other hand, the biogeographical scale may provide the best opportunity to understand a species' climatic limits, its response to different sets of predators or competitors, or its evolutionary potential. Two examples may illustrate the insights afforded by this largest spatial scale (Chapters 6 and 23 provide additional, detailed examples).

A biologist asked to name the most salient properties of bats would surely mention that they are nocturnal. Some physiologists make their living by studying the sensory mechanisms that enable bats to forage at night. Anyone who had observed *Pteropus* flying foxes in Asia, Indonesia, or New Guinea might be forgiven for assuming these animals to be obligately nocturnal. However, on remote, species-poor Pacific islands such as Fiji, flying foxes forage by day as well as by night. It seems likely that flying foxes are constrained elsewhere to be nocturnal either by the presence of predatory diurnal eagles or by competing diurnal avian and mammalian frugivores. It is unknown whether the diurnal behavior of Fijian flying foxes is a purely phenotypic, behavioral feature or has developed as a slow evolutionary response requiring genetic changes in sensory mechanisms. The point is that no cleverness or long number of years spent in Asia or New Guinea could permit a field biologist there to appreciate a feature of *Pteropus* biology that would become obvious within a few minutes on Fiji.

Our second example involves the island thrush *Turdus poliocephalus,* which is distributed throughout the Pacific from Indonesia to Samoa and is similar to the American robin (*T. migratorius*) and European blackbird (*T. merula*). On New Guinea this thrush is confined to the bitterly cold subalpine forest and alpine grassland at elevations between 2,700 and 4,300 m. In New Guinea's history of ornithological exploration there has never been a single individual recorded below 2,400 m, even as a vagrant, and a long-term field study confined to New Guinea might interpret this fact as a fixed feature of the thrush's biology. On other Pacific islands, however, this thrush's altitudinal floor varies from island to island, and on Rennell, an uplifted coral atoll only 110 m high, it can be seen in abundance hopping around on the hot coral at sea level. Analysis of interisland variation in the thrush's altitudinal range prompts the hypothesis that this variation arises from variations in the presence of other species, more possible competitors being present on large islands and thus confining the thrush there to high altitudes (Diamond 1975). The main point again is that only studies at the biogeographical scale can reveal such patterns and thus suggest something of the evolutionary potential of this species.

CHOICES OF TEMPORAL SCALE

Just as with spatial scale, the appropriate choice of temporal scale depends on the phenomenon and the species to be studied. For any given species one would choose a short time scale to study behavioral responses, a longer time scale to understand population dynamics, and a still longer scale to study evolution and genetic change. Davis (Chapter 16) discusses how the ability to track (i.e., respond to) an environmental perturbation varies drastically among species. If a species' tracking rate is slow compared to the rate of environmental change, the entire concept of ecological equilibrium loses meaning. The interspecies differences that Davis discusses for tracking rate also apply to behavioral responses or population dynamic responses or genetic responses.

The time axis, like the space axis, is a continuum that can be arbitrarily divided into scales of different lengths. For convenience we shall discuss four temporal scales: (1) the time required for a behavioral response, (2) the lifetime of an individual, (3) a span of several generations, and (4) the time required for evolutionary change.

Scale for Behavioral Response

The shortest time scale is that appropriate for studying behavioral responses, such as individual bouts of feeding. Bout frequency differs among species. With appropriate instrumentation one could in a single day gather all the necessary data to publish a paper on chemosensory behavior or feeding strategies of bacteria. Several years would be required to complete the same study on a large constrictor snake, which captures only a few prey individuals in the course of a year.

The Scale Determined by an Individual's Lifetime

Many studies of life history or reproductive biology are best carried out on the time scale of the lifetime of one individual. Naturally, this scale varies with species. The mites studied by Colwell (Chapter 24) have a generation time of about a week, so that a flowering failure of this duration sufficed to exterminate a local population of the mite *Rhinoseius bisacculatus* on Trinidad. For small songbirds with a life span of a few years, Wiens (Chapter 9) points out that there may be a time lag of a year or more in response to habitat change, as a result of site fidelity by adults. The generation times of large vertebrates are such that one may completely miss a critical factor regulating abundance if one's study lasts less than several decades. For instance, major El Niño events occur only a few times per century, but these events may be crucial for understanding the biology of large seabirds such as frigatebirds and boobies, whose life span is many decades (Schreiber and Schreiber 1984). Finally, large trees may have life spans of a few centuries to a millenium or more, with the result that adult populations can coast through an unfavorable period as long as the Little Ice Age without reproducing successfully (Chapter 16)!

The Multigenerational Scale

Questions about population dynamics or demography require study on a time scale lasting at least several generations. Because the generation time for a *Drosophila* fly is on the order of a few weeks, multigenerational studies (such as that of Gilpin et al. in Chapter 2) can be completed in the laboratory. In contrast, Grant's study of Galápagos finches (Chapter 10) has been running for over a decade, but barely qualifies as a multigenerational study. In 1980 a banding study of shearwaters begun in 1947 was not yet close to outliving one generation of the birds (Serventy and Curry 1980).

Three chapters in this volume (Chapters 3, 9, and 10) illustrate the point that the appropriate time scale for a study of population dynamics depends not only on the generation time of the particular species of chief interest, but also on the generation times of other species with which it may interact. This is because environmental or experimental perturbations affect a species both directly and indirectly through chains of effects on other species. For instance, the recovery of Darwin's finch populations from a drought took much longer than the time for rainfall to return to normal or for the finches to produce a brood under normal circumstances; several years were required for the plants on whose seeds the finches depend to return to their former abundance (Chapter 10). Similarly, Brown and colleagues (Chapter 3) experimentally perturbed communities of desert seed-eating animals and noted that the community was still changing 7 years after the perturbation because of chains of indirect effects. In particular, rodent removal enhanced ant abundance in the short term (apparently due to relief from competition), but did not affect ant abundance in the long term! Brown and his associates suggest that rodent-ant competition was eventually offset by an indirect rodent-ant mutualism, mediated by differential feeding on competing plant species.

The Scale for Evolutionary Change

The longest time scale is that required for study of evolutionary problems, which depend on genetic changes and therefore demand many generations. For most organisms this condition makes it impossible to observe the genetic trajectory caused by species interactions directly from laboratory and field experiments. In bacteria or molds, it is true, one can select mutants for resistance to a pathogen or chemical within a few days and complete an experimental study within the funding period of a research grant. Medical problems resulting from natural selection of chloroquine-resistant malaria strains became significant within a few decades of chloroquine's introduction—an inconvenient delay for investigators subject to the hazards of grant renewals, but at least something that a biologist can experience within the life span from tenure to retirement. Ongoing studies of natural selection and evolutionary change in organisms with appreciably longer generation lengths are rare, and they do not exist for animals such as large mammals or seabirds. The many chapters in this volume that deal explicitly with evolutionary problems use either fossil evidence (Chapters 7 and 17) or inferential evidence (Chapters 6, 10, 19, 22–27, and 30).

COYOTES AND JACKRABBITS

To illustrate how the scale of an ecological investigation determines the possible conclusions, let us consider a predator-prey system—coyotes and jackrabbits. Suppose we wish to know whether the abundance of rabbits is significantly controlled or regulated by coyotes and whether coyote abundance is determined by rabbit abundance. We consider six approaches to this question, differing in their spatial and temporal scale. Our treatment is hypothetical, but instructive.

Approaches at Different Scales

1. Short-term behavioral study at one site The biologist chooses a study site large enough to encompass the home range of one pair of coyotes and many rabbits. All available field time is devoted to detailed behavioral observations. At the end of a few months it is established that within the period of observation 52% of the diet of that pair of coyotes consisted of rabbits, and 71% of the juvenile rabbits and 23% of the adults initially present at the site were eaten by coyotes. Thus, rabbits are a major prey for coyotes, and coyotes are a major predator on rabbits.

But this unequivocal result applies only to that one site for that period of a few months. Had the study been carried out at a different season, when mice or fruit were abundant and rabbits hard to capture, the rabbit-coyote interaction might have been found to be unimportant for both species. Or it might happen that the study was carried out at a location with relatively few coyotes, or with rabbits that were few and limited by something else, or with abundant alternative prey for coyotes. Again, the rabbit-coyote interaction would be judged unimportant. Thus, until the study has been repeated many times at many sites, the biologist has no idea about the range of applicability of the study's conclusions.

2. Short-term field experiment To test the coyote-rabbit relationship experimentally, the biologist constructs six fenced, 100-ha enclosures in terrain occupied by coyotes and rabbits. Two enclosures are left as controls containing both species. In two enclosures the coyotes are trapped out, and within half a year the rabbit population has risen significantly. In the other two enclosures the rabbit populations are eliminated, and the coyote pups starve to death and the adults lose weight.

Conclusion: the rabbit-coyote interaction is important to both species. Objection: as with the previous study, the conclusion applies only to a single, short time period at one site. With additional grant support the biologist may extend the experiment to test the conclusion at one other site and retest it at a different season at the old site, but such short-term extensions improve one's confidence in the generality of the conclusions only slightly.

3. Medium-term field experiment After the short-term perturbations just described, the biologist remains at the old site and monitors it for 10 years. Whether the short-term effects are sustained or reversed in the long run depends on time lags and on the chain of indirect effects.

In the coyote-removal enclosure the rabbit numbers initially increase, but to the extent that coyotes normally selectively remove rabbits that are sick from diseases or parasites, the long-term result might be a greater frequency of infected rabbits and ultimately a lower population density of rabbits. In the rabbit-removal enclosure another herbivorous mammal formerly suppressed by competition from rabbits, such as voles, might prosper. Thus, if coyotes could survive the initial lean period, their numbers might ultimately rise to a level even higher than that existing before removal of rabbits.

The conclusion from observations over a period of time would be that rabbits and coyotes happen to determine each other's abundance where they coexist, but that different factors are unmasked as limiting each species when the other species is removed. We have developed greater insight into the causal factors determining the dynamics of the system by expanding our temporal scale, but this conclusion still suffers from the disadvantage that it is based on a single site.

4. Medium-term censuses at one site Eschewing experimental perturbations that are expensive to carry out and might introduce artifacts, the biologist instead makes careful repeated censuses of rabbit and coyote numbers at a single site for 10 years. The result: to a first approximation, the two species seem uncoupled, and the densities of each are more closely predicted by climatic changes. A statistical analysis might reveal a positive or negative association between the densities of the two species over time, at least for some time lag values in cross correlations.

There are two reasons why this study yields a different result from the previous two. First, the temporal fluctuations in this study do not result from experimental manipulation of a single variable initially affecting only one species, but result at least in part from environmental fluctuations directly affecting both species. It is not the case that a climatic change in one year directly affects only one species, with the other changing as a result (e.g., high rain leading to more abundant rabbits, with coyotes increasing after a lag; high rain depressing coyote abundance, with rabbits increasing after a lag). Instead, the climatic change acts directly on both species, and whether the direct effects of the climatic change on the two species are parallel or opposite affects whether rabbit and coyote abundances are positively or negatively associated. A second reason is that environmental fluctuations affect many other species besides rabbits and coyotes, each species' population responding at different rates and with different time lags, causing further changes in other species that eventually filter back to affect rabbits and coyotes. Before the ripples from one environmental fluctuation have disappeared, a new environmental change may be setting off new waves of response.

Thus, interpretation of the temporal fluctuations is complicated by environmental factors acting simultaneously on coyotes and rabbits and by chains of indirect effects. And, even if information about the rabbit-coyote interaction could be teased out, the conclusion still applies only to a single site.

5. Short-term regional census To cure the problem of limited spatial scale in the previous studies, the biologist greatly expands the spatial scale, visits many sites that differ in numbers of rabbits and coyotes and in various other ways, and looks for spatial correlations in abundances of the two species. Each site is visited only once.

The outcome depends in part on whether rabbits and coyotes can disperse among the sites. If sites are contiguous, allowing dispersal, coyotes or rabbits can move to any temporarily more desirable location. Consequently, the pattern that emerges may be due to background environmental variation, which can affect the two species similarly. For example, when one finds a site with lots of rabbits, one will probably also find many coyotes there, not a few coyotes.

It is also possible that in selecting the sites to visit, the biologist inadvertently favors locations more suitable for rabbits than for coyotes (close to roads or adjacent to agricultural fields, for example), producing negative distributional and abundance correlations between the species. The conclusion might be that high predator densities lead to reduced prey densities, but this is not necessarily the correct explanation.

Further, because each site is sampled only once, temporal variations in rabbits, coyotes, and other environmental factors at each of the sites are not known. Abundances of the two may in fact be closely coupled, with changes in predator numbers lagging behind changes in prey numbers in a textbook predator-prey fashion. If different sites are surveyed at different stages of localized prey-predator oscillations, the absence of a clear pattern when they are combined in correlational tests leads to the conclusion that the species are not interdependent. Thus, the conclusions are no longer restricted to a single site, but they become sensitive to just which sites are sampled and when.

6. Biogeographical study The biologist repeats the above study, but on a biogeographical scale, such that sites are far apart or separated by barriers to dispersal. For example, the biologist may compare islands of similar topography and climate, all of which have had rabbits for a long time, but only some of which have coyotes.

This natural experiment resembles the short-term or medium-term coyote-removal field experiments (study 2 or 3). The field experiments, however, measure rabbits' short-term or medium-term response to removal of coyotes (pulse experiment, in the terminology of Bender et al. 1984). In contrast, the biogeographical comparison of islands with different faunas is measuring steady-state differences in the communities after long-term adjustments have resulted in the configuration of a community (press experiments, in the terminology of Bender et al. 1984). For ancient islands we have the further complexity of evolutionary adjustments as well as behavioral and demographic ones. Each island population may no longer be a genetic replicate of its mainland relatives.

Further, unlike the field removal experiments,

we are uncertain that the isolated locations do not differ in ways that are important to rabbits other than the presence or absence of coyotes. The influences of such factors must be assumed to be averaged out across the sites being compared.

Comparative Assessment of These Rabbit-Coyote Studies

None of these approaches yielded the same answer as any other approach. At best, if biologists had been lucky enough to perform studies 1 and 2 at the ''right'' site in the ''right'' year, studies 1, 2, and 6 might have yielded similar conclusions, while studies 3, 4, and 5 yielded different conclusions. The probable outcome would be the type of discussion that permeates the ecological literature: the authors of each study preferring the conclusion obtained by that study, stressing confounding factors particular to and obfuscating the other five studies, and offering principled arguments why their method is inherently superior to the other five methods.

All of these approaches have strengths and weaknesses. Which approach is the ''best'' one? The answer depends on the original question asked by the study. If the question is to test whether interactions between rabbits and coyotes are directly and immediately important, studies 1 and 2 would be potentially illuminating, studies 5 and 6 tangential, and studies 3 and 4 downright confusing. If the question is whether the rabbit-coyote interaction was evolutionarily important, study 6 would be the most useful. If instead one is interested in the rabbit-coyote interaction within the broader context of other factors limiting each species, studies 3 and 4 would be more useful.

Clearly, there are sound reasons why different choices of scale are likely to yield different types of information and, in turn, prompt different interpretations and conclusions. There is nothing inherently wrong with careful, intense studies conducted at a single site or with well-structured comparisons over a biogeographical area. Each can provide insights into ecological patterns and suggest how those patterns may have been formed by various processes. The inadequacy of the narrow-scale approach becomes apparent when one wishes to generalize from specific findings, and the limitations of the broad-scale approach are most evident when one desires to understand proximate cause-effect relationships. The ideal solution would be to explore a wide spectrum of spatial and temporal scales, because no single choice of scale yields a complete understanding. A full understanding of the rabbit-coyote interaction, for example, would require studies spanning the continuum from the species' behavioral interactions with each other and their physiological relationships with the local environment to the species' long-term history over their entire geographical ranges. In practice, of course, limited time and resources may prevent one from studying the entire spectrum of scales and may force one to choose one or two. The choice must be made carefully, remembering that the scale will dictate the methods used, the results obtained, and the interpretations that can be made of them using sound logic. The most useful choice of scale will depend on the species studied, the question asked, and logistical constraints.

chapter 9

Spatial Scale and Temporal Variation in Studies of Shrubsteppe Birds

John A. Wiens

INTRODUCTION

The challenge of community ecology is to discover the patterns of natural assemblages of organisms and to explain them in terms of controlling processes. During the past several years this has become a much more difficult endeavor than was envisioned during the halcyon days of 1960 to 1975, when many ecologists discerned clear patterns and the processes producing them with apparent ease. Several factors have contributed to this growing disquietude in community ecology (see Schoener 1982, Lewin 1983, Simberloff 1983, Wiens 1983a), but prime among them has been a developing awareness of environmental variability and its effects.

Time and space are not constants in ecological systems. Environments vary, but they do so with varying amplitudes, periodicities, and degrees of stochasticity in time, and the pattern and extent of these variations change in space as well. Several contributions to this volume (Chapters 10–12, 16–19, 30, and 32) provide especially clear documentations of the complex spatiotemporal dynamics of environments on a wide array of scales. This environmental variability forms the background against which species occupy habitats and communities are assembled. However, species differ from one another in their responses to environments and to these variations in environmental conditions, because species differ in their sensitivities to changes in the complex of other species and resources with which they may interact, and because species also differ in their ''tracking inertia,'' or responsiveness to environmental changes (Wiens 1984). Communities of organisms therefore vary in space and, at any given point in space, in time. As discussed in Chapter 8, the scale of space and time on which ecological systems are viewed thus makes a difference in the patterns that are detected or the processes that are proposed to account for them.

These insights are by no means new. They were a cornerstone of the view of ecology developed by Andrewartha and Birch (1954) three decades ago, and MacArthur (1972a) called attention to the importance of both temporal and spatial variation at a time when confidence in seemingly unequivocal documentations of community patterns and processes was at a peak. MacArthur noted, for example, that food might be scarce enough to cause severe competition between species in only 1 year in 20, but this might be enough to eliminate the inferior species from the area during the interim. In a habitat where both could normally coexist, one might thus witness severe competition only occasionally. MacArthur (1972a, p. 21) concluded from this that "the ecologist watching the populations may well not see them competing severely although the biogeographer has strong evidence that competition must sometimes occur." This suggested that a biogeographical scale might be most appropriate for discerning competitive patterns in communities. Elsewhere in his book, however, MacArthur (1972a, p. 186) considered the difficulty of determining the most appropriate scale for studying diversity patterns, noting that a very large (biogeographical) area would be subject to the complicating effects of speciation and history, while an area that was too small would hold an inadequate sample of species and individuals. MacArthur suggested that a relatively homogeneous habitat of a size just large enough to hold an adequate sample of species might be most appropriate, as the traces of history might have little effect at that scale.

MacArthur's observations hinted at the difficulties of dealing with variations and scaling of time and space in community studies, but it was several years before the impact of these difficulties began to be appreciated. In order to provide some perspective on these difficulties, I will explore in this chapter some of the effects of varying the spatial scale on which communities are studied and consider some aspects of temporal variation in environments and communities. I will do this by synthesizing portions of our recent research on bird communities and their habitat relationships in semiarid shrubsteppe systems. I will conclude by discussing the broader implications and limitations of this work and by commenting in particular on current views regarding temporal variability in environments.

GENERAL BACKGROUND

The viewpoint on spatial scale and temporal variation in ecological systems that I develop in this chapter grew out of studies that John Rotenberry and I have conducted on breeding bird communities in grassland and shrubsteppe habitats in North America over the past decade. The grassland work was carried out as part of the International Biological Program and was centered on several sites in tall-, mixed-, and short-grass prairies (Wiens 1973a). The more recent shrubsteppe studies were conducted in the northern portions of the Great Basin and adjacent areas (Wiens and Rotenberry 1981a). All of these areas are characterized by cold winters and hot summers, with rainfall occurring predominately in the summer in the grasslands and during the winter in the shrubsteppe. These seasonal patterns are more pronounced and predictable in the grasslands than in the shrubsteppe, where precipitation shows little within-year autocorrelation. Variation in annual rainfall is greatest in short-grass prairies, intermediate in shrubsteppe, and least in tall-grass prairies (Wiens 1974).

Regionally, these areas contain a moderate diversity of breeding birds, but within local habitats diversity is generally low, on the order of two to seven species. Tall-grass avifaunas are dominated by Eastern meadowlarks (*Sturnella magna*), dickcissels (*Spiza americana*), and grasshopper sparrows (*Ammodramus savannarum*), short-grass communities by horned larks (*Eremophila alpestris*) and lark buntings (*Calamospiza melanocorys*), and shrubsteppe by sage thrashers (*Oreoscoptes montanus*), Brewer's sparrows (*Spizella breweri*), and sage sparrows (*Amphispiza belli*). Western meadowlarks (*Sturnella neglecta*) are widely distributed throughout short-grass and shrubsteppe ecosystems. All of these species are generally similar ecologically, and locally co-occurring species occupy broadly overlapping territories and rarely interact behaviorally with individuals of other species.

We surveyed these bird communities and measured aspects of their habitats using carefully standardized methods (Rotenberry and Wiens 1980a, Wiens and Rotenberry 1981a). At most of the sites we established 9-ha gridded plots in areas of representative habitat surrounded by large areas of similar habitat, and then determined densities of species on the plots by mapping the territories of breeding males. One subset of the shrubsteppe data was gathered using

610-m linear transects, on which bird densities were estimated following a modification of Emlen's (1971, 1977) census procedures. Habitat structure was measured by recording the incidence of various physiognomic categories of vegetation at vertical point samples arrayed in a stratified random design in each plot or transect area. Features of the height stratification of vegetation were also recorded at these points, and measures of vertical and horizontal heterogeneity in habitat structure were derived from these point samples (see Rotenberry and Wiens 1980a). In the shrubsteppe locations coverages of shrub species were also determined from incidences at point samples.

Other aspects of our studies of grassland and shrubsteppe bird communities have included analyses of morphological patterns (Wiens and Rotenberry 1980a, 1981b; Wiens 1982), trophic relationships (Wiens and Rotenberry 1979, Rotenberry 1980), and behavioral patterns of habitat use (Wiens 1985a; Wiens, Van Horne, and Rotenberry in prep.; Wiens, Rotenberry, and Van Horne submitted). These analyses produce results that parallel in many ways those of the habitat studies, so they will not be considered here.

SCALE-DEPENDENT PATTERNS OF HABITAT OCCUPANCY

Features of habitats occupied by birds have long figured importantly in discussions of their niche relationships (e.g., Lack 1933; Svärdson 1949; Wiens 1969; Cody 1974a, 1978; Schoener 1974a), following the argument that habitat is either a resource of direct relevance to birds or is tightly associated with other resources that are in limited supply. Natural selection should thus be expected to tune the patterns of habitat selection of species so that they are optimally adjusted to the habitat responses of other potential competitors in the community (Cody 1974b, 1981). Evidence bearing on this proposition has been gathered on a wide range of spatial scales, from detailed observations of individuals in quite local settings to broad comparisons on a biogeographical scale. It seems to have been implicitly assumed that the patterns that emerge are not strongly affected by the differences in scale and that observations obtained at any scale of investigation bear equally on documenting niche partitioning and, thus, community structuring.

How valid are these assumptions? Our investigations were conducted at geographical scales ranging from comparisons among sites over half of North America to comparisons of areas occupied or not occupied by birds within single 9-ha plots (Fig. 9.1). This spectrum can be used to explore possible scale-dependency in the patterns that emerged. Detailed analyses of much of this information are presented elsewhere (Rotenberry and Wiens 1980a, Wiens and Rotenberry 1981a, Wiens 1985a), so I will only summarize the patterns here.

Biogeographical Scale

At the broadest scale we may ask how the breeding densities of birds on study plots correlate with habitat measures for those plots by comparing plots distributed over a spectrum from tall-grass prairies through short-grass prairies to shrubsteppe (Rotenberry and Wiens 1980a). At this scale (Fig. 9.1A) most of the numerically dominant species exhibited strong correlations with individual features of habitat composition or structure. Eastern meadowlarks, dickcissels, and grasshopper sparrows, for example, occupied sites with considerable grass coverage and deep litter, but little bare, unvegetated ground; overall, the vegetation was tall, dense, and distributed in a relatively even, unbroken sward. Sage and Brewer's sparrows and sage thrashers exhibited quite different patterns, inhabiting sites with low coverage of grasses, considerable bare ground, and high coverage of shrubs. Sage sparrows and sage thrashers also showed strong correlations with several measures of vegetation patchiness. Densities of horned larks varied inversely with the overall height of the vegetation.

Because many features of these habitats covary, we used principal components analysis (PCA) to explore these patterns further. The first dimension derived by this procedure (PC I) accounted for 41% of the variation in the original set of habitat measures and reflected a gradient of increasing horizontal patchiness within habitats, while the second component (PC II) accounted for 22% of the variation and represented increas-

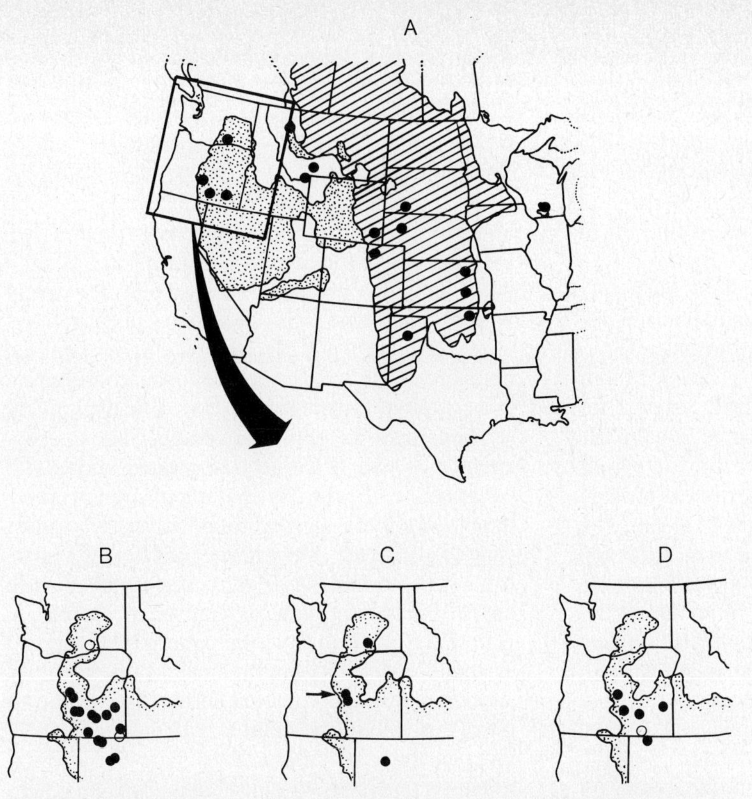

Fig. 9.1 Spatial scales at which analyses of bird-habitat relationships were conducted. (A) Biogeographical scale. Shaded circles indicate study sites, hatched area grassland, and stippled area shrubsteppe. (B) Locations of the 14 sites used in a regional, within-shrubsteppe analysis. Unshaded circles indicate the Owyhee and ALE sites used in a nonlinear analysis of the effects of changing the geographical scale. (C) Locations of 4 sites used in a second regional-scale analysis. Arrow indicates the site at which local-scale investigations were conducted. (D) Locations of 5 sites used in an analysis of temporal variation in bird-habitat relationships. Unshaded circle indicates Guano Valley, where an "experimental" perturbation was conducted.

ing vertical structuring and heterogeneity. All of the above bird species varied significantly in relation to these derived dimensions of habitat structure: Eastern meadowlarks, dickcissels, and grasshopper sparrows loaded negatively with respect to PC I and positively on PC II (horizontally homogeneous but vertically structured habitats), while sage and Brewer's sparrows and sage thrashers loaded positively on both components (high horizontal and vertical patchiness). Horned larks were negatively related to PC II, reflecting their general occupancy of low-stature habitats. At this scale of resolution, then, several clearly defined suites of species emerged that covaried in distribution and abundance in association with variation in features of habitat structure and that were distributionally largely exclusive of other suites of species.

Regional Scale

What happens when we reduce the spatial scale of investigation and focus on bird-habitat associations within the shrubsteppe portion of this tallgrass–shrubsteppe biogeographical spectrum? If the habitat selection of a species is relatively constant over large portions of its range, as Beals (1960) and Noon et al. (1980) have suggested, we might anticipate little change in the broad-scale patterns of habitat associations. Because a reduction in scale leads one to examine habitat relationships *within* suites of distributionally covarying species more closely, one might even expect to see subtle variations between species on the general theme of habitat response revealed in the broader analysis, reflecting more fine-tuned habitat partitioning.

Linear Analyses Our correlational analyses for a series of locations within the shrubsteppe region of the Pacific Northwest (Fig. 9.1B; Wiens and Rotenberry 1981a) provide little support for either of these expectations. At this regional scale none of the associations between bird densities and habitat variables noted for the three dominant shrubsteppe species (Sage and Brewer's sparrows and sage thrashers) at the broader scale emerged as statistically significant, nor were the birds clearly associated with variation in habitat structure as synthesized in PCA components. Horned larks also failed to exhibit the correlations with measures indexing low vegetation stature that were apparent in the broad-scale analysis. On the other hand, Western meadowlarks, which showed no significant habitat correlations in the broader analysis, varied strongly with several features of habitat structure at this regional scale, generally displaying a pattern of habitat response that paralleled that of Eastern meadowlarks in the broad-scale analysis. In general, then, most of the species in this region appeared to vary in distribution and abundance largely independently of variation in features of habitat structure, and this was especially true of the most characteristic shrubsteppe species.

At this regional scale we also measured the coverages of different shrub species on the sites, permitting an analysis of relationships between bird abundances and habitat floristics as well as structure. Sage sparrows and Western meadowlarks were positively associated with sagebrush (*Artemisia tridentata*), while sage thrashers and Brewer's sparrows varied independently of sagebrush coverage, but exhibited negative relationships with hopsage (*Atriplex spinosa*) and budsage (*Artemisia spinescens*). Horned larks were positively associated with gray rabbitbrush (*Chrysothamnus nauseousus*) coverage.

Another analysis conducted at this same regional, within-shrubsteppe scale provides some important perspective on the foregoing results. This analysis used measures of bird densities and habitat features from plots at four sites distributed from southeastern Washington through north-central Nevada (Fig. 9.1C) that were surveyed for two to seven years (Wiens, Rotenberry, and Van Horne submitted). On the basis of this data set sage sparrow densities varied positively with grass and bare ground coverage and negatively with shrub coverage, hopsage and budsage coverage, an index of horizontal patchiness, and the diversity of physiognomic life forms contributing to vegetative coverage; there was no significant relationship to sagebrush coverage. Sage thrasher abundance was negatively correlated with shrub coverage (and with its major constituent sagebrush coverage), but exhibited no other significant correlations. Abundances of Brewer's sparrows and horned larks were uncorrelated with any of the habitat features that we measured.

Finally, in contrast to the pattern that emerged at the broader biogeographical scale, the bird species within this regional shrubsteppe environment varied in distribution and abundance largely independently of one another. Pairwise correlation tests revealed few significant positive or negative associations between species, a pattern shown as well by multivariate tests. If habitat partitioning by structural features occurred among most of the birds breeding in this shrubsteppe habitat type, it was generally too subtle to be revealed by statistical analyses.

Nonlinear Analyses All of these analyses used linear correlational or multivariate procedures, as has been customary in avian habitat studies. There is little reason, however, to expect this assumption of underlying linearity in the responses of birds to habitat variation to hold in nature (Meents et al. 1983). To explore nonlinear aspects of habitat responses, we applied response-surface models (Myers 1971) to data gathered at the regional scale, using information from 14 sites surveyed over a 3-year period (Fig. 9.1B; Rotenberry and Wiens submitted). Generally, the nonlinear model provided a considerably better fit when bird densities were regressed on PCA components derived from habitat structure and habitat floristics data sets. The basic patterns that were expressed in the linear analysis were unchanged but strengthened by the consideration of nonlinear effects. This is not especially surprising, as there are more terms included in the nonlinear model, but it nonetheless points to the importance of nonlinearities in the habitat responses of birds. Partial correlation analyses indicated no significant interactions among bird species with habitat variation factored out, again suggesting that the bird species responded to habitat variation largely independently of the presence or the abundance of other species.

These response-surface analyses can be used to examine the sensitivity of bird-habitat association patterns to small changes in scale at this regional level. These analyses were conducted on data gathered from a network of sites located in central and southeastern Oregon and extreme northern Nevada, and the correlations thus represent patterns over this series of sites. Do the same patterns remain if the scale is altered to include additional sites located within the shrubsteppe habitat type but beyond the geographical boundaries of the original set of sites?

The fit of a data set to a regression may be expressed in terms of mean residuals, the average departure of observed site-specific values from those predicted by the regression equation. The mean residuals for most of the response-surface regressions conducted on the original data set were relatively low (Table 9.1), confirming the generally significant fit of the relations. One of the additional sites, Owyhee, was located in extreme southeastern Oregon, within 16 km of sites used in the original analysis (Fig. 9.1B). When values for this site were related to the original response-surface regression, residuals for Western meadowlarks and sage thrashers remained relatively low, those for sage sparrows increased somewhat, and the fit of the model to the values for horned larks and Brewer's sparrows was quite poor (Table 9.1). The other site (ALE) was in the Columbia River basin of southeastern Washington, over 400 km from the closest site used in the original analysis. Inclusion of this site in the analysis reduced the fit for Brewer's and sage sparrows and Western meadowlarks substantially, while horned lark residuals increased somewhat and those of sage thrashers only slightly.

Collectively, these comparisons, like those of the four-site linear analysis reviewed above, suggest that there is a strong component of geographical variation in the relationship between avian densities and habitat variables and that the magnitude of this effect is different for different species. Developing a regression model that provides a significant fit to bird-habitat data within one portion of the shrubsteppe provides no assurance that the same relationship will hold if the spatial scale is altered, even within the same general habitat type.

Local Scale

At either the biogeographical or regional scales patterns are defined by averaging bird densities and habitat measures for entire study plots and then using statistical procedures to derive correla-

Table 9.1 MEAN RESIDUAL VALUES FOR SHRUBSTEPPE BIRD SPECIES IN RELATION TO RESPONSE-SURFACE REGRESSIONS BASED ON HABITAT STRUCTURE AND FLORISTICS DATA SETS

		Mean residual values		
Species	Data set	Original sites	Owyhee	ALE
Horned lark	Structure	15.3	131.9	49.8
	Floristics	8.4	131.2	29.7
Western meadowlark	Structure	9.2	6.8	51.8
	Floristics	13.0	9.1	54.2
Sage sparrow	Structure	25.8	40.8	115.7
	Floristics	20.3	53.0	59.0
Brewer's sparrow	Structure	69.0	204.6	253.9
	Floristics	71.1	143.3	290.6
Sage thrasher	Structure	5.8	8.1	11.1
	Floristics	4.2	2.0	12.7

Mean residuals for the original sites are based on the regressions for 14 sites; residuals for the Owyhee and ALE sites are in relation to those regression models.

tions through between-plot comparisons. A third scale of analysis may involve comparisons *within* local study plots, between areas actually occupied by breeding individuals and unoccupied parts of the plot. This scale perhaps most accurately mirrors the actual patterns of habitat selection by individuals of different species and would thus seem most likely to be sensitive to species interactions (Wiens 1969, 1985a).

We analyzed such within-plot habitat associations by mapping the territories of breeding individuals of each species within plots and then calculating values for habitat measures from samples taken in areas included in and outside of territorial boundaries of each species. This research is in a preliminary stage of analysis (Wiens, Rotenberry, and Van Horne submitted), so only one example will be presented here.

We calculated values of habitat features for occupied and unoccupied areas on three plots in central Oregon shrubsteppe (Fig. 9.1C) over a three-year period for all of the species breeding on these plots. With respect to grass and shrub (largely sagebrush) coverage, sage sparrows, Brewer's sparrows, and sage thrashers each occupied portions of the plots that agreed rather closely in coverage values with the values for the entire plots (the latter two species less closely than sage sparrows), while horned larks occupied habitats in only a small portion of the range of habitat conditions potentially available to them, especially with respect to shrub coverage (Fig. 9.2). At this scale the first three species appeared not to be particularly selective of habitat conditions within plots, while horned larks clearly were. Values for unoccupied areas in the plots, however, suggest that sage sparrows may have actively avoided some portions of the plots (as indicated by the wide variance of unoccupied data points in Fig. 9.2), while areas in plots not occupied by Brewer's sparrows or sage thrashers were, for the most part, not clearly different from occupied areas. In general, values of habitat features for areas not occupied by the species within a plot were more variable than measures from occupied areas. From this pattern one might possibly infer that the species occupying a plot tend to converge in their habitat-occupancy patterns. Sample sizes for the unoccupied data sets for each plot are generally small, however, and the greater variance of these values may reflect inadequate sampling instead.

Such "presence-absence" patterns provide information on active habitat selection by birds as they establish territories. Can additional insights into these patterns be gained by looking at the sequence of territory establishment among individuals within a population? We lack appropriate information to answer this question in the shrubsteppe systems, but an analysis of a single grassland bird community (Wiens 1973b) is instructive. There, breeding territories were mapped frequently from the initial arrival of the birds in the spring until the peak of the breeding season, so the habitat characteristics of the first territories to be established could be compared with those of territories established later in the season, when options for territory placement in the area were increasingly constrained. Areas initially occupied by grasshopper sparrows contained less grass cover, shallower litter, and greater density and average height of forbs than the area as a whole, while Savannah sparrows (*Passerculus sandwichensis*) initially occupied portions of the area with opposite characteristics. As the populations of each species built up, however, more and more of the area was occupied, and differences between occupied and unoccupied areas and between the habitat occupancy patterns of the two species diminished to the point of being indistinguishable. Evidence of habitat selection at this within-plot scale of resolution, therefore, is sensitive to the degree to which densities are high and the plot is saturated with breeding individuals.

TEMPORAL VARIATIONS IN SHRUBSTEPPE BIRD COMMUNITIES

Much of the ecological theory that applies to communities assumes for mathematical convenience that natural systems are at or close to resource-determined equilibria (see Chapter 13), but the real world is often impressive in its variability (Wiens 1977, 1985b). For a variety of reasons communities change over time in the presence and absence of species, in the density levels of these species, and in their densities relative to one another. The stability and predictability of other aspects of ecological relationships among

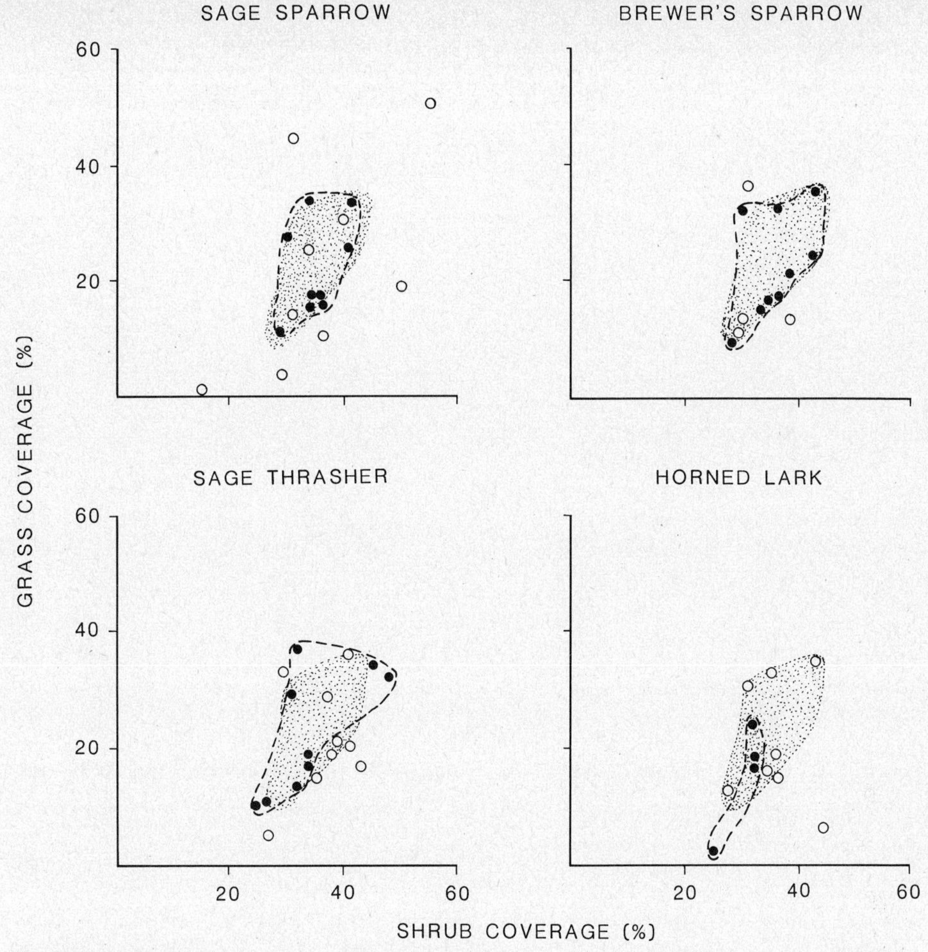

Fig. 9.2 Distributions of percent coverages of grass and shrubs (sagebrush) in areas occupied (shaded circles) and not occupied (unshaded circles) by breeding birds of four species in three plots at a site in central Oregon, 1981 to 1983. The stippled area represents the range of average values for the entire plots.

the species hinge on the relative constancy of these patterns. Variations in time as well as in space thus compromise our ability to detect and interpret patterns in communities.

How variable are the bird populations and communities in the habitats we have studied? Consider, as an example, the temporal dynamics in one of our local shrubsteppe plots, which was surveyed by the same procedures over an 8-year period (Fig. 9.3). Several features are apparent in this example, and these generally characterize all of our shrubsteppe and grassland study areas. First, the number of breeding species present in the community varied substantially, from a low of three species in 1976, 1979, and 1981 to a high of six or seven species in 1978, 1982, and 1983. The "core" species (Hanski 1982), that is, those that are generally widespread in the region and that tend to be relatively abundant where they occur, varied in density, but were always present. Brewer's sparrows clearly dominated this community numerically, contributing 43 to 75% of the individuals present. Their densities, however, were actually somewhat more stable through time than those of the other two core species (Brewer's sparrow CV breeding density =

19.5%, sage thrasher = 23.6%, sage sparrow = 33.2%). Collectively, this set of three species accounted for 70 to 100% of the individuals present in the plot. The other species varied substantially in their presence or absence from the plot and in their densities when they were present. Gray flycatchers (*Empidonax wrightii*), for example, were not recorded on the plot until 1981, even though they were breeding nearby, but in 1982 they contributed 13% of the total density of breeding birds in the community. None of these density variations showed any clear relationship with variations in annual growing-season (October–April) precipitation (Fig. 9.3). At this local scale, then, the community has some aspects of stability (the persistence of the three core species), but a substantial amount of variability (in species composition and relative abundances of species) as well. Cody (1983) found similar patterns in the South African fynbos bird communities he surveyed.

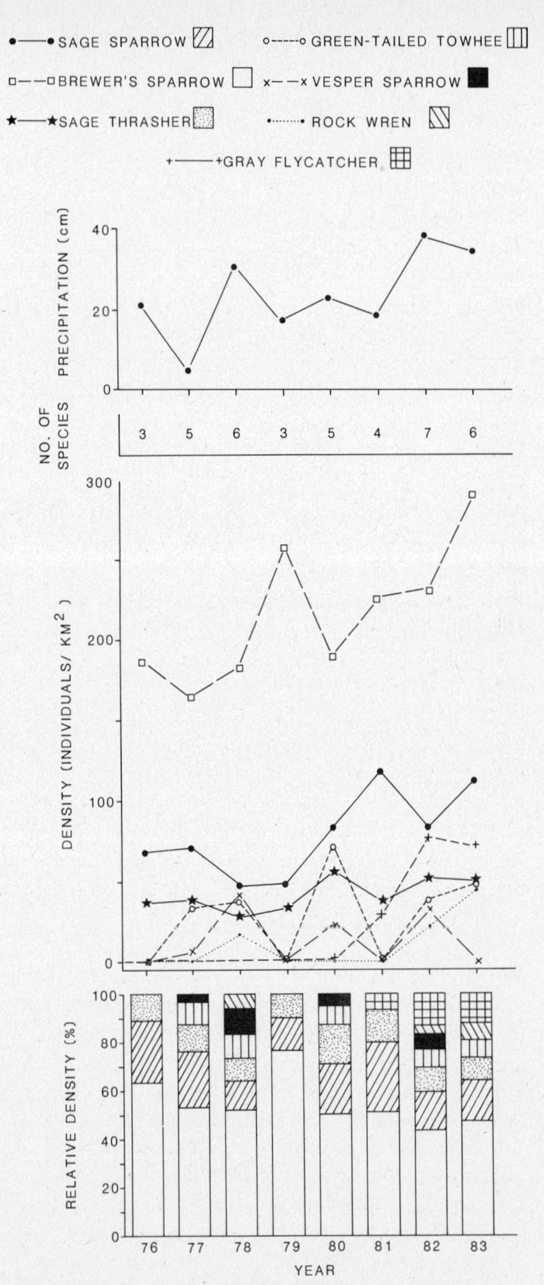

Fig. 9.3 Annual variation in growing-season (October–April) precipitation, the number of breeding species present, population densities of those species, and their relative densities in a sagebrush-dominated study plot in central Oregon, 1976 to 1983.

Correlational Approaches to Temporal Variation

In an attempt to deal with such patterns of variation in a more general manner and to relate them to underlying variation in habitat features, John Rotenberry and I analyzed data gathered from 14 transect study areas in southeastern Oregon over a 3-year period (Fig. 9.1B; Rotenberry and Wiens 1980b). This period began with an extremely dry year (27% of normal growing-season precipitation), followed by two wet years (136% and 126% of normal), so we could assess how the different components of the system—habitat structure, floristic composition, and bird densities—responded to this climatic variation.

Changes in habitat structure were clear in several of the variables we measured. In the first wet year following the dry year, vegetation height and coverage increased significantly and several measures of horizontal patchiness decreased, largely as a consequence of increased coverage of grasses and annual and perennial forbs over the array of sites. The same patterns generally persisted in the following year as well. Shrub species coverages, on the other hand, did not change significantly over the region during the 3-year period. This is not surprising because, given the woody, perennial growth form and great longevity of these shrubs, their coverage at any particular location is not likely to change much between years. There were also no consistent yearly changes in densities of bird species over the study sites, however, even though densities at each location did vary considerably between years. Variations in bird densities on any given site were largely independent of variations on other, nearby sites, and thus no general temporal trend to the variation was apparent. As a consequence, density variations were not associated with the consistent patterns of temporal change in measures of habitat structure.

In a similar study K. G. Smith (1982) monitored the responses of birds to drought-induced environmental changes in several montane habitats. The impact of the drought on the birds differed between habitats—communities in deciduous forests were strongly affected, while those in alpine meadows were relatively unaffected. Smith found the greatest differences between predrought and drought years; unfortunately, we were unable to address this comparison in our shrubsteppe study.

We have recently extended our earlier analysis by considering the temporal dynamics of a subset of five of these locations (Fig. 9.1D) over a 7-year period (Wiens and Rotenberry in prep.). We used PCA to derive synthetic axes of variation among measures of habitat structure and (separately) of shrub species coverages and composition and examined the patterns of correlations of the distribution and abundances of bird species with these PCA axes. The results generally paralleled those of our earlier study (Rotenberry and Wiens 1980b). The PCA dimensions associated with increasing coverage of grasses and litter and increasing vegetation height and with a gradient among the sites in coverages of forbs and various annuals varied significantly in time. Both of these components of vegetation structure are responsive to variations in annual precipitation, and the significant yearly variation most likely was a consequence of variations in the regional climate between years. At a central Oregon site, for example, growing-season precipitation varied by a total of nearly 700% during the period 1976 to 1983 (Wiens, Rotenberry, and Van Horne submitted). Some of the variation in grass coverage, however, also reflected an increase in coverage of cheatgrass (*Bromus tectorum*) from 1977 through 1983, a trend that has been general over much of the West (Mack 1984). None of the PCA factors based on shrub species exhibited any significant yearly component of variation. As before, the densities of bird species varied substantially, although most of this variation was attributable to between-site differences and relatively little was associated with consistent yearly variation across the series of sites. There was no apparent response or tracking of the variations in habitat structure by concordant variation in bird densities.

We have also used nonlinear response-surface analyses to examine temporal variations in the habitat relationships of these shrubsteppe bird communities. In the analyses described earlier in this chapter, we derived regression relationships between bird densities and structural and floristic PCA dimensions from 1977–1979 data. These regressions may be used to predict the bird densities expected on five of the same sites (Fig. 9.1D) over the following 4-year period on the basis of their habitat structure or floristics during that time. If the habitat at a site were to change between the two time periods, the position of the site with respect to the PCA dimensions would change, placing it at a different position on the nonlinear regression surface and altering the bird densities predicted. As before, we gauged the predictiveness of the nonlinear structural and floristic models by calculating mean residual values for each bird species for the two time blocks (Table 9.2). In general, the 1980–1983 residuals were about the same as those for the 1977–1979 period from which the regressions were generated; only the residuals for sage sparrows and horned larks on the floristic model were significantly greater in the 1980–1983 time period. Across this set of locations, then, the relationships between these bird species and structurally or floristically defined PCA dimensions that were apparent in analyses of a 3-year data set generally persisted over the subsequent 4-year time block.

It should be noted, however, that in this analysis relationships were averaged over a region (the five sites) and over several years (the two time blocks). Thus, although analyses that consider the detailed year-to-year dynamics of specific sites fail to reveal a close tracking of temporal variations in habitat features at either local or regional scales, averaging across several years at a regional scale preserves a general pattern of bird-habitat relationships. Here, then, the temporal as well as the spatial scale on which the patterns are

Table 9.2 COMPARISON OF 1977–1979 AND 1980–1983 MEAN RESIDUAL VALUES FOR SHRUBSTEPPE BIRDS IN RELATION TO RESPONSE-SURFACE REGRESSIONS BASED ON HABITAT STRUCTURE AND FLORISTICS DATA SETS

Species	Data set	Mean residual values	
		1977–1979	1980–1983
Horned lark	Structure	15.3	23.7
	Floristics	8.4	17.7*
Western meadowlark	Structure	9.2	13.1
	Floristics	13.0	16.0
Sage sparrow	Structure	25.8	28.7
	Floristics	20.3	37.2*
Brewer's sparrow	Structure	69.0	85.0
	Floristics	71.1	65.9
Sage thrasher	Structure	5.8	6.7
	Floristics	4.2	6.4

Values for 1977–1979 are the residuals in relation to the regression models generated from 14 sites for those 3 years. Values for 1980–1983 are from a subset of 5 of these sites in relation to the models generated from the 1977–1979 data sets.
* = $p < 0.05$.

viewed affects their clarity and their interpretation.

An "Experimental" Perturbation: the Effects of Time Lags

The habitat changes that occurred during these studies did not affect shrub species coverages to a detectable degree. These are seemingly important, however, in the habitat relations of at least some of these shrubsteppe birds. We may take advantage of an "experimental" perturbation that occurred during our studies to examine how the birds might respond to a more massive temporal habitat change involving shrub coverages as well as features of habitat structure (Wiens 1985a). In the spring of 1980 the Bureau of Land Management and the Oregon State Land Board aerially sprayed our study plot at Guano Valley, in south-central Oregon (Fig. 9.1D), with the herbicide 2-4D, as part of a large-scale "range improvement" program. The following fall the dead sagebrush was broken down and removed and crested wheatgrass (*Agropyron cristatum*) was planted. We had initiated our studies on this plot in 1977 and thus could record the changes in habitat that resulted from these manipulations and from the responses of the bird populations to those changes.

The manipulation indeed dramatically altered both the structure and the composition of the habitat. In the year following the manipulation the site changed from one with substantial shrub coverage and open ground with high horizontal heterogeneity in the distribution of vegetation to a configuration of reduced shrub coverage, increased coverage of grasses and litter, and low heterogeneity. Sagebrush coverage was reduced from 27% to 4%. In 1982 and 1983 sagebrush coverage increased somewhat (because the spraying was not entirely effective), and grass coverage increased from 4 to 5% to nearly 60% as the crested wheatgrass became established and as cheatgrass invaded the site following the soil disturbance.

The bird populations, however, exhibited no clear responses to these habitat changes. Densities of sage sparrows, for example, remained virtually unchanged in the year following the manipulation, while Brewer's sparrows continued a decline that had begun in the previous (premanipulation) year. During 1982 and 1983 Brewer's sparrow densities increased dramatically and then decreased to low levels again, while sage sparrow densities gradually diminished.

We believe that the absence of a clear, immediate response by the birds to the habitat changes may reflect the fidelity of territorial males to locations in which they have bred in previous years. Adults of some species may return to previous breeding locations despite major habitat changes (Hildén 1965), and our observations of marked birds at other shrubsteppe sites indicate that some individuals may return to the same territorial location over as many as six successive years. If individuals at Guano Valley behaved similarly, some of the 1980 breeders might have returned and established territories in subsequent years, and the expression of a clear population response to the habitat manipulation might take several years, depending on the turnover rate of individuals in the population.

DISCUSSION

Spatial Scale

Our work has clearly indicated that the spatial scale on which patterns are viewed *does* make a difference in the patterns that are detected. Shrubsteppe bird species clearly exhibit different relationships to one another and to underlying habitat features when the data are considered at a biogeographical, regional, or local scale, and the processes contributing to those patterns probably vary with scale as well (this contrasts with Tilman's view in Chapter 22).

Consider, for example, sage sparrows. At the broadest scale this species was clearly associated with several features of habitat structure, varying positively with shrub and bare ground coverage and negatively with grass cover. At a regional scale including a series of 14 locations within the shrubsteppe, however, these associations were no longer evident, although a clear statistical relationship with sagebrush coverage emerged. In another regional analysis, however, sage sparrow densities were not significantly associated with sagebrush coverage and in fact were negatively related to overall shrub coverage. There they were positively associated with both grass and bare ground coverage, features that were significantly related to sage sparrow density variations at the biogeographical scale, but in different manners from each other. Finally, at a more local scale of resolution sage sparrows seemed to be generally unresponsive to variations in shrub or grass coverage within 9-ha study plots, although often portions of the plots not occupied by individuals had unusually high or low shrub or grass coverage.

Patterns of habitat occupancy of the other species were similarly sensitive to changes in spatial scale. There appeared to be rather little consistency in the quantitative patterns of habitat associations of these birds over the spectrum of scales we considered. Why might this be?

One possibility, of course, is that the birds are attuned to features of the habitat that we have not measured or have measured at the wrong scale. Given the variety of features that we have measured, however, and the various statistical approaches that we have taken in their analysis, this seems unlikely. Another possibility is related to the spatial scale of sampling in our investigations. Our data were derived from plots or transects that ranged in size from 0.1 to 0.4 km^2. These are relatively large areas for bird surveys, yet they may have been small enough that they failed to sample the actual community composition and densities accurately. In particular, if not all of the available habitat is fully occupied by breeding individuals (as is usually the case in shrubsteppe communities), differences in densities between plots (or variations in plot densities between years) may not reflect actual population patterns but rather may be due to sampling error. This is because vacancies that occur in an unsaturated habitat may or may not be filled largely by chance, with the result that the number encountered on small plots may vary from place to place or year to year even if the population at a larger scale is invariant (Rotenberry and Wiens 1980b, Wiens 1981). These effects will be more severe if relatively few individuals occupy a plot, but they are likely to pose problems at all densities short of complete saturation of the habitat. The habitat patterns revealed by comparing occupied with unoccupied areas within plots may likewise be sensitive to such chance variations. Thus, at a local scale at least some of the variations and patterns that occur may be quite sensitive to the size of the sample areas used. Connell and Sousa (1983) have likewise noted the potential effects of small study areas on attempts to assess the stability of ecological systems.

Most community patterns are discerned by comparing values obtained from a number of such local sample areas, and the patterns are thus not only sensitive to the sizes of the sample areas themselves but to the geographical area over which the comparisons are made. Our documentation of different patterns of bird-habitat relationships at a biogeographical versus a regional scale, for example, is clearly influenced by the total spectrum of habitat conditions included at the two scales (Wiens and Rotenberry 1981a). When one is regressing y on x, a correlation that is significant when one has a broad range of values of x may disappear when the range of x is more restricted. Thus, when we consider the biogeographical spectrum from shrubsteppe through tall-grass prairies, species such as sage sparrows display significant correlations with the habitat gradient (x) because they are present only in one portion of it; the correlation is essentially a presence-absence analysis. Within the shrubsteppe region the habitat gradient is restricted to various kinds of shrubsteppe, however, and sage sparrows are present over the entire gradient; correlations depend on more subtle variations in breeding densities of the birds in relation to habitat variables. Given the many sources of variation in local densities (including the sampling effects noted above), clear habitat associations may fail to emerge unless they are extremely strong. Additional complications are introduced if in fact the habitat affinities of species change geographically (Collins 1983, Karr and Freemark 1983; see examples for New Guinea montane birds in Chapter 6). Averaging across a habitat spectrum within which a species changes its habitat responses may fail to reveal the actual responses. Obviously, any statements about the habitat relationships of bird species in this system have meaning only in the context of the scale on which they are derived.

To some extent, these inconsistencies may be a consequence of the lack of closure of the system on at least some of the scales of our investigation. Certainly at a local scale these shrubsteppe bird communities are open systems, similar in a way to the marine systems described by Roughgarden (Chapter 30). All of the species breeding on our shrubsteppe plots are migratory, wintering in deserts of the southwestern United States and northern Mexico. The young birds that are produced from local plots emigrate following the breeding season and join a "pelagic pool" of sorts, from which new recruits to the local plots may be drawn in the following year as the birds arrive again in spring. Despite intensive banding efforts, we have never observed any of the young birds produced on the plots to return to the plots or nearby areas in successive years; recruits thus come from other local areas.

On an annual basis, then, shrubsteppe breeding assemblages at local and probably regional scales of analysis are open systems. This means that the patterns that we do see or the failure of expected patterns to materialize at these scales may be consequences of events occurring outside of the system as we have defined it. Dayton and Tegner (1984) have called attention to the effects that large-scale events or processes (e.g., El Niño) can have on what is observed at a smaller scale in marine systems, and similar relationships undoubtedly occur in terrestrial systems as well. In the shrubsteppe system, for example, it is possible that the populations we have studied on the breeding areas are limited by conditions in their wintering areas, as we (Wiens 1977, Wiens and Rotenberry 1980a, Rotenberry and Wiens 1980a) and others (Fretwell 1972, Pulliam and Parker 1979, Dunning and Brown 1982) have suggested. If populations of different species belonging to a given local breeding assemblage (or populations of the same species breeding in nearby local sampling areas) winter in different areas and are thus subjected to differing resource levels and events during the winter, their breeding densities may well vary independently of one another and fail to match proximate habitat conditions very well. If recruits to local populations are drawn from some pool of an unknown spatial extent and derivation, this will produce further distortion of local community patterns. When these local community samples are combined over some arbitrarily determined regional scale, the patterns that emerge may thus be epiphenomena, artifacts of sampling rather than consequences of real biological processes such as competition, predation, colonization, habitat selection, survivorship, and the like.

These problems of spatial scale and the flux of individuals are simplified if one deals with a closed system, such as the *Anolis* assemblages studied by Roughgarden (Chapter 30), the birds, lizards, and spiders investigated on small islands by Schoener (Chapter 33), or the finches on Galápagos islands examined by Grant and his colleagues (Chapter 10). Much of the value of island systems as research settings is in fact due to the clear closure that they physically impose on many population and community processes. Few terrestrial systems on continents or large islands are so clearly closed, however; even sharply bounded and isolated habitat patches (e.g., Whitcomb et al. 1981) or mountain peaks (e.g., Chapter 6) are subjected to flows of dispersing individuals, which may affect local community patterns (May 1981c). Because it is generally impossible to expand the spatial scale of such systems until they are in fact closed, one must instead study them by sampling from them, and the scale on which samples are drawn and compared thus becomes critical. The type of information acquired varies with the scale studied (Chapter 8).

There are thus several reasons for variation in community features and habitat associations with changes in spatial scale. In our studies we have identified the following (although this listing by no means exhausts the possibilities for this or other systems).

1. Habitat selection within local plots is more clearly expressed at low densities and will therefore differ between areas with different densities.
2. The occurrence of vacancies in a territorial population and their subsequent reoccupancy may be strongly influenced by chance effects in an unsaturated habitat, leading to sampling errors.
3. At a regional or biogeographical scale patterns are detected by comparing over a series of local samples, and if those samples are subject to the effects noted in (1) and (2), such comparisons will be sensitive to just which local areas are included.
4. Changing the geographical scale of analysis will alter the range of habitat conditions included, influencing the likelihood that statistical correlations will emerge.
5. Geographical variation in the habitat preferences of the same bird species may blur analyses on a biogeographical scale.
6. The system being studied may not be closed, and events on wintering areas (for example) may influence breeding-area dynamics. If different breeding populations winter in different areas, direct comparisons among them may not be appropriate.

Temporal Variations and Lags

If natural environments and the communities they contain are in a steady-state equilibrium, the patterns they express will be relatively stable (subject only to sampling error), and studies of relatively short duration will provide a reasonably accurate image of the basic patterns. As natural environments appear to vary on virtually any scale on which they are viewed (Chapters 16–18; see also Dayton and Tegner 1984), however, expectations of such an equilibrium are probably in vain. If the communities are not in equilibrium, or if the equilibrium is a dynamic one in which attributes of individuals, populations, and the community change rapidly to track environmental variations, a short-term approach will provide only a misleading snapshot of a changing community frozen at a point in time. To see the true dynamics, a longer time scale is required.

Features of environments and habitats in the shrubsteppe *do* vary in time, often dramatically, and bird populations and communities vary as well—witness the fluctuations apparent in Fig. 9.3. Nonetheless, in this system the habitat variations and the avifaunal variations are not closely associated, at least on the scale of year-to-year changes viewed at local or regional levels. To some degree, this may be a consequence of the same sampling problems that affect the analysis of spatial patterns. If sampling is inadequate or biased by an inappropriate scale, the dynamics of the populations may give the appearance of "density vagueness" (Strong 1984b; Chapter 15). Bird dynamics may also fail to match habitat dynamics, however, because of time lags, a mismatching of the time scales of response of different components of the system to environmental variations. These lags distort patterns of resource tracking that otherwise might be detected through suitable sampling and correlational procedures, complicating attempts to derive and interpret community patterns.

Some indication of the complexity of time lags (or compounded delays in the expression of indirect effects) in the relationships between environmental driving variables, resources, and community components is apparent in the simplified conceptualization of Fig. 9.4. Variations in a climatic factor such as precipitation input to a system should influence both habitat and food resources, but such effects are not likely to be direct. Our shrubsteppe studies, for example, clearly show that some components of the vegetation (grasses, annuals, and some perennial forbs) respond within a growing season to rainfall variations, while other life forms (shrubs) are unresponsive to variations on this time scale and may react only on a time scale of decades to longer-term climatic variations. The degree to which habitat structure or floristic composition changes with climatic variations is thus a function of its growth-form composition; one would expect a more immediate response in a habitat dominated by grasses or desert annuals than in one dominated by woody perennials.

Food resources such as arthropod populations may also be influenced by variations in precipitation, but with time lags reflecting their characteristic life history or seasonal phenology traits. As the arthropod component of a system is either directly or indirectly dependent on the vegetation present, time lags in responses may also be mediated through the vegetation component. Thus, not only may there be a lag in the growth response of some life forms of the vegetation to a climatic change, but the translation of this response into a response in the arthropod component is delayed by the processes of feeding, individual growth and reproduction, and emergence phenology among the arthropods.

Additional time lags are introduced in the conversion of these changes in resource states into variations in densities of bird populations. As the ''experiment'' at Guano Valley showed, there may be a considerable time lag in the response of bird populations to habitat changes, probably as a consequence of site tenacity. The translation of increased or decreased abundances of arthropods in a habitat into changes in bird density is expressed through changes in reproductive output or survivorship in the bird populations, and time lags are also involved in these responses. If the dynamics of the bird populations are also influenced by density-dependent factors (e.g., territo-

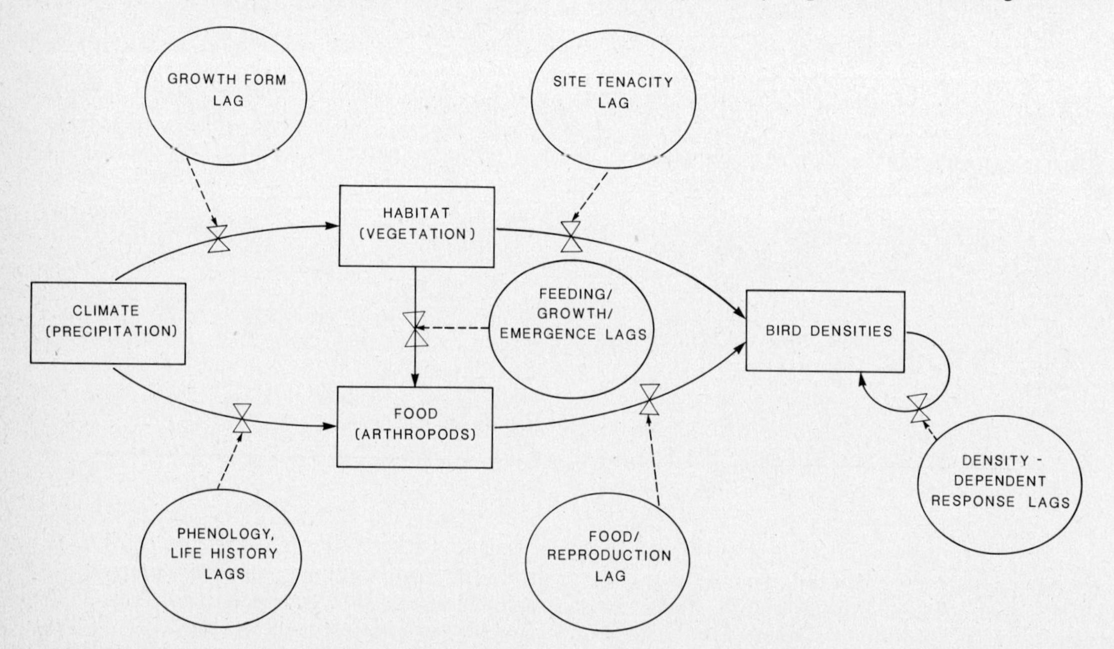

Fig. 9.4 Possible influences of variations in climatic factors (in this case, precipitation) on bird population densities, showing avenues of effect and sources of time lags.

rial spacing, social dominance effects on breeding systems or emigration), additional time lags are introduced. Finally, the extent and phasing of such time lags may vary between individuals of different sexes or age classes, between species in a community, and between habitats, producing a bewildering heterogeneity of temporal dynamics in the community as a whole.

If this sort of scenario is at all common in nature, it seems unlikely that the sort of close tracking of resource variations envisioned by Cody (1981) will occur very often. The appearance of variability and unpredictability in the dynamics of local bird communities may thus reflect the complexity introduced by these time lags rather than our failure to measure varying resource levels tracked by a changing set of consumers from year to year, as Cody (1981, p. 107) suggested. If environmental variations are extreme, however, they may affect some of the intermediate stages shown in Fig. 9.4 immediately or bypass these transitional steps and act directly on bird densities. In either case time lags are reduced and bird densities or community properties will track the environmental changes more closely. An elegant documentation of such tracking is provided by Grant's studies (Chapter 10) of the responses of finch populations and their seed resources to a severe drought episode in the Galápagos islands. Thus, close resource tracking may well occur if environmental variations are extreme (although some responses to extreme environmental variations such as El Niño may still display a substantial lag; Barber and Chavez 1983). More often, however, its expression is likely to be obscured by a great many factors, especially variable time lags in responses. These complicate both observational and experimental studies (e.g., Chapter 3), and make a long-term perspective essential.

Variable Environments Revisited

These considerations relate to a hypothesis of the effects of environmental variability on the role of competition in structuring communities that I proposed several years ago (Wiens 1974, 1977). Schoener (1982) has recently commented on this view, and Grant (Chapter 10) has carefully evaluated several aspects of it using data from the Galápagos finch system, so it is appropriate to reconsider it briefly here.

The hypothesis that I presented in 1977 must be understood in its temporal context. It was intended to stand in contrast to what I perceived to be the prevailing view in community ecology at that time (e.g., Cody and Diamond 1975). From the literature of 1960 to 1975 it seemed clear that many ecologists believed that natural systems were at or close to equilibria determined by resource limitation, that selection on resource-utilization traits was more or less incessant, and that competition between species was the primary component of this selection and therefore played the major role in structuring communities. The existence of competition as a process in communities was generally inferred on the basis of the apparent agreement of observations with the patterns predicted by theory (see Wiens 1983a). I suggested instead that in variable environments resource levels would vary between superabundance and scarcity and that in many systems the times of scarcity ("ecological crunches") might be relatively infrequent. It would be primarily during such crunches, however, that competition would be intense and act as a major selective force on the attributes of species and communities. Some of the time, then, the system might not be resource-limited, selection intensity would be reduced, and populations and communities might depart from the clear patterns expected in a competitive community and predicted by existing theory. Competition was thus viewed as intermittent in its occurrence and effects, and as a consequence the occurrence of competition could not simply be inferred from community patterns, but required more rigorous documentation. Grant (Chapter 10) has presented the basic structure of this view in greater detail.

In a more contemporary context (Schoener 1982) the basic arguments boil down to a contrast between three views: that (1) environments are relatively stable and contain consumer communities that are at equilibrium with resource levels (the prevailing view in the 1960s and early 1970s), (2) environments and resources vary, but populations and communities track these variations closely and maintain a dynamic equilibrium (Cody's 1981 view), or (3) environments vary such that a resource-limitation threshold is crossed only intermittently, competition is not necessarily frequent, and populations and communities are often nonequilibrial (my 1977 view). These three views correspond to the ones labeled in Chapter 13 as classical competition theory, direction 1, and direction 2, respectively. The discussions between Schoener and myself (Schoener 1982, 1983a; Wiens 1983b) center on view 3, and in particular on how frequent or infrequent periods of resource limitation or ecological crunches should be in order to mesh with view 3 or with Schoener's modification of view 1.

To determine the frequency with which some resource-limitation threshold is passed requires a clear conception of the resource dynamics and of resource limitation (Wiens 1984b); one must know what represents a limiting level of a given resource for a given population size. Simply charting the temporal dynamics of resource abundance as recorded by some sampling procedure or as measured indirectly, while interesting, does not say anything about variation in resource limitation in the absence of additional information on what represents a limiting level of resource abundance. Likewise, direct information on these dynamics cannot be obtained by recording temporal variations in resource-use patterns by organisms in the absence of information on resource levels, because resource use does not necessarily index resource availability (Wiens 1984b). Individuals may respond to changes in the composition and abundance of a resource pool by expanding or contracting their use of resources, and the different species in a community may thus vary in the extent of niche overlap among them as resource levels fluctuate, as Schoener (1982) has documented. A reduction in resource levels might lead to a reduction in niche overlap between species as they change from converging on a single superabundant resource to diverging to follow their own species-specific specialities (e.g., Smith et al. 1978, Schluter 1982a). Such shifts may occur (1) in nonlimiting situations, in which the resource-limitation threshold is not crossed and competition may not occur; (2) as the threshold is crossed, in which case competition is intermittent; or (3) in continuously limiting situations, in which resources change from limited to even more scarce, intensifying competition. Clearly, without some knowledge of the location of a resource-limitation threshold, such niche shifts and changes in niche overlap cannot be used as documentations of either resource limitation or competition. Documenting resource limitation in nature is difficult, of course (Newton 1980), but with careful measurements in the right sort of situation it can be done (e.g., Jones and Ward 1979, Heinemann 1984; Chapter 10). Assessing the impact of environmental variations on population and community attributes and the frequency with which ecological crunches are interposed among relatively benign, nonlimiting conditions requires that we carefully consider what "resources" actually are, at what levels (relative to population densities) they become limiting, and how individuals and populations respond to such limitation.

As a hypothesis, the "variable environment" view as originally framed is unfortunately untestable in principle, as it predicts only that not all times will be characterized by limited resources and competition. Thus, for example, the studies of Dunham (1980), Smith (1981), and Tinkle (1982) on lizards, of Grant and his colleagues (Chapter 10; Smith et al. 1978; Grant and Grant 1980b, 1983) on Galápagos finches, of Pulliam and Parker (1979), Pulliam (1983), and Dunning and Brown (1982) on wintering sparrows, of Smith (1982) on montane birds, of Eriksson (1979) on fish and waterfowl, or of Karr and Freemark (1983) on tropical birds may be regarded as at least partially consistent with a variable environment view, as they demonstrate intermittency in resource limitation and competition. In their reviews of field experiments on competition, both Connell (1983) and Schoener (1983b) found that competition was of intermittent occurrence in roughly half of the multiyear experiments they considered. Schoener (1983b) nonetheless concluded that "variability in the existence of competition is rare," while Connell (1983), noting that he differed with Schoener, stated that "apparently it is not rare for competition to vary in either time or space." In a broader frame environmental variations such as infrequent tidal extremes (Bleakney 1972) or El Niño (Barber and Chavez 1983) represent clear ecological crunches with long-lasting consequences.

All of these examples tend to corroborate the

hypothesis. One could always argue, however, that a failure to record periods of nonlimitation would not falsify the hypothesis, but would simply mean that one has not studied the system long enough (this parallels the argument that a failure to document niche differences between coexisting species does not falsify the niche partitioning hypothesis, as other unmeasured niche dimensions on which the species might segregate always exist). The hypothesis would be more amenable to testing if it made quantitative predictions regarding how different frequencies of limiting periods might influence population dynamics, niche overlap patterns, community structure, and the intensity and direction of natural selection on species' attributes. This is an area in which theoretical efforts would be especially rewarding.

As it now stands, the variable environment view does contain some testable assumptions (Chapter 10), but primarily it served to call attention to aspects of community ecology other than those expressed in the conventional wisdom of the 1960s and early 1970s (Wiens 1983b). Chapters 13 and 14 of this volume summarize more recent theoretical discussions of variable environments and their effects on competition. Admitting that environments vary and that populations may not always be resource-limited does not lead to the conclusion that competition does not exist or that it is unimportant as a process structuring communities. It is clearly important in some communities, perhaps especially in relatively closed systems such as islands or relatively sedentary resident populations (e.g., some *Parus* assemblages; Alatalo 1982) or in guilds specialized upon a restricted resource base (e.g., nectarivores). Resolution of the differences in Schoener's, Cody's, and my views requires clearer definition of the effects of varying frequencies and severities of ecological crunches on communities and better empirical data on populations *and* resources.

Conclusions

The studies we have conducted in grassland and shrubsteppe environments are seemingly plagued by difficulties of both spatial and temporal scaling, and these complicate our attempts to detect clear patterns in the relationships of bird species and communities to their habitats. It is now becoming apparent that these complications derive at least in part from the temporal variation of the environments and time lags in biotic responses, the mobility of the organisms and the openness of the system, and the consequences that these impose on measurement and sampling. These difficulties have become apparent in our studies because we have made the effort to examine the dynamics of the system on several different spatial scales and over a reasonably long time period. They are likely to occur in many ecological systems, but they are rarely detected for several reasons. First, ecologists have generally been inattentive to scaling problems, perhaps because of the attractiveness of equilibrium-based theory and deep-rooted preconceptions about the "balance of nature." These have led to the expectations that natural systems will express equilibrial configurations in most situations and that spatial scale and temporal variations are not likely to have major effects in distorting anticipated patterns. Second, ecological systems have not often been investigated in a way that would reveal the sensitivity of patterns to spatial scales or time lags. Long-term studies conducted over a range of spatial scales are necessary. Such studies may require considerable time and effort, however, and are not likely to appeal to all ecologists, especially those with "wanderlust" (Diamond 1978). Until rather recently, there has been little compelling reason to conduct such studies, but they are now clearly needed. Not all ecology need be done this way, of course. Important insights emerge from a pluralism of approaches, guiding philosophies, and scales of investigation. I would suggest, however, that we can no longer afford to ignore the implications of spatial scale and temporal variation in the conduct and interpretation of ecological research, for they are profound.

SUMMARY

Environmental variability in space and time has major effects on community patterns. Because these effects differ at different scales, the patterns detected in ecological studies and the processes proposed to account for them are critically dependent upon the scale of investigation.

I consider some aspects of scale-dependency of community patterns using information on relationships between breeding birds and features of their habitats in grassland and shrubsteppe environments of North America. Clear relationships are apparent at a biogeographical scale spanning the spectrum from tall-grass prairies through shrubsteppe, and in particular a suite of shrubsteppe bird species displays significant correlations with several aspects of habitat structure. When these same species are considered at a regional, within-shrubsteppe scale, however, these patterns of habitat associations vanish, although some relationships with variations in coverages of certain shrub species are clear. If the same analyses are performed using a different set of samples from the same shrubsteppe region, different patterns emerge. At a local scale, within 10-ha study plots, some species appear to select breeding habitat, while other species occupy areas in the plots that are indistinguishable from unoccupied areas. Thus, demonstrating a relationship of variation in bird densities or community composition to habitat variables at one scale of resolution provides no assurance that the same relationships will hold if the scale is changed, even if one remains within the same general habitat type. Over all of these scales there is little evidence of clearly defined habitat partitioning among coexisting bird species.

Shrubsteppe bird communities also vary from year to year in the presence and absence of species and in their relative densities, and thus the ecological relationships among locally co-occurring species are clearly in a dynamic state of flux. Several structural attributes of shrubsteppe habitats change through time in response to environmental variations, but there are no apparent regionwide responses of breeding bird populations to these changes. There is thus no evidence from this system that the birds track habitat changes at all closely. An ''experimental'' perturbation of habitat structure and composition in one of our study plots suggests that site fidelity of breeding birds may produce substantial time lags in responses to habitat changes.

Several factors may contribute to such scale-sensitivity in these and other systems. Local samples of communities may be biased by the scale of sampling. When broader-scale pattern analyses are conducted by combining or comparing sets of such samples, the results may be strongly influenced by just which samples are included. Combining different sets of samples over the same scale or combining samples on different scales may produce different patterns, at least partly as a consequence of the variation that exists at the local scale. This is especially likely if local areas are not fully saturated with individuals or species, or if the system is not closed, but is open to the movement of individuals or the effects of events occurring over a larger area or elsewhere in the system. Time lags in responses to changes in habitats or other resources may be complex and may differ between different species in a community, and these also influence the detection of ecological patterns. The appearance of variability and unpredictability in the dynamics of local communities may thus reflect the complexity introduced by these time lags rather than a failure to discern varying resource levels that are closely tracked by consumers from year to year. Such time lags may also contribute to the intermittency of resource limitation and competition that is hypothesized to occur in temporally variable environments.

ACKNOWLEDGMENTS

Our grassland and shrubsteppe research has been supported by the National Science Foundation, most recently through Grants DEB 80-17445 and BSR 83-07583. John Rotenberry and Beatrice Van Horne assisted in several phases of the research. They and Peter Grant, Jared Diamond, Martin Cody, and Jonathan Roughgarden read an initial draft of this chapter, and their comments were tremendously helpful in clarifying many points. Yevonn Ramsey prepared the figures.

chapter 10

Interspecific Competition in Fluctuating Environments

P. R. Grant

INTRODUCTION

The publication of the proceedings of a symposium on community ecology to honor the memory of Robert MacArthur (Cody and Diamond 1975) marked the final flourish of an era of unbridled enthusiasm for the notion that interspecific competition is important in conferring structure upon communities of organisms. Within the following 2 years three serious challenges were mounted. First, Caswell (1976) introduced a null model for community structure, as an analogy to the neutral model in population genetics. He compared the performance of this model with a more traditional one incorporating competitive effects and found the null model to be sometimes superior in accounting for patterns in nature. Second, Rathcke (1976) employed a random model to detect competitive effects in the distribution of stem-boring insects and found very little evidence for competition. Third, Wiens (1977) subjected the assumptions of competition theory to critical analysis and concluded that often the theory does not apply to the real world, largely because the underlying assumptions are not met.

From the first two challenges there has developed a vigorous debate on the appropriateness and suitability of null models in ecology (Harvey et al. 1983). The third challenge has been influential in the emergence of a more circumspect view of the role of competition in vertebrate communities (e.g., Dunham 1980), especially among birds (Wiens 1983a, 1984a, 1984b). Rather surprisingly, it has attracted less critical attention than have the null models. This is unfortunate because the paper by Wiens (1977) presents cogent discussion of several important processes in communities, whereas the null model papers are basically concerned with methods of analysis, hypothesis testing, and statistical inference, in other words, with pattern analysis. Yet everyone would like to understand processes, since patterns are simply their product.

A Challenge to Competition Theory

When theory is developed in the language of algebra it is referred to as formal theory; informal theory is verbal. Wiens (1977) questioned formal competition theory, that which has been developed in terms of mathematical equations such as Lotka-Volterra equations and used by MacArthur and Levins (1967), Cody (1974a), Schoener (1974b), and many others. Competition is defined as a negative effect of one species upon the population size of another arising from their joint exploitation of environmental resources (see Tilman [1982] and Wiens [1984b] for discussion of what constitutes a resource).

Wiens (1977) focused on four critical assumptions that are frequently left unstated or are set aside.

1. Populations, guilds, or communities are at an equilibrium determined by resource limitation.
2. Selection on the system of resource-exploiting attributes considered by theory is continuous and intense.
3. Competition is the major selective force acting upon the resource-utilization traits or determining the distribution of species.
4. The predicted optimal state is available to the population in its evolutionary development.

He then asked whether these assumptions are met. A review of temporal variation in several terrestrial environments led him to conclude that assumption 1 is frequently not met. This being so, assumptions 2 and 3 are probably not often met, and assumption 4 is also in doubt. As a consequence, the theory is not often applicable to the real world. He drew three further, cautious conclusions of an empirical nature: (1) competition may be less pervasive in its effects than is usually assumed, (2) what we witness in nature may at times represent merely a coarse fit to the optimal states predicted by theory, and (3) documentation of the competition process in nature may be extremely difficult if environments vary through time (see also Wiens 1983a, 1984a, 1984b).

Responses to the Challenge

The overwhelming response has been one of approval. Using Citation Indexes to identify sources (about 150), then restricting attention arbitrarily to articles published from 1978 to 1983 in *Ecology* (32), *American Naturalist* (15), *Evolution* (2), and *Auk* (12), I found the three most cited conclusions to be the idea that competition occurs only rarely and at times of extreme environmental stress (cited 27 times), the suggestion that communities are not in equilibrium (11), and the view that most of the evidence for competition is questionable (11). These are all empirical matters that are not dependent upon the correctness or failure of formal theory. Evolutionary concerns, such as how frequently selection occurs and its relationship to competition, have been cited rarely. Only one set of authors (Sih and Dixon 1983) endorsed the view, without further discussion, that assumptions of theory should not be accepted uncritically, and only one of the 61 papers registered anything that could be described as criticism. Thomson (1980) objected that setting up a contrast between equilibrium and nonequilibrium views of communities is partly artificial (see also Wiens 1983b). For understanding the evolution of properties of species, he suggested, species composition of a community is more important than fluctuations in population sizes.

Beyond the limits of this survey I am aware of two other authors who have registered mild disapproval. Schoener (1982, 1983b) interpreted the large body of experimental evidence for interspecific competition to show the importance and frequency of the process in nature, arguing that neither competition theory nor optimal foraging theory is incompatible with natural and experimental evidence of fluctuating population sizes and environmental conditions (see also Schoener 1983a, Wiens 1983b, and Chapter 9). Cody (1981, p. 107) addressed the challenge when he wrote, "Recently complaints that bird community structure appears variable and unpredictable from year to year (cf. Wiens 1977) may be more a reflection on our failure to measure varying resource levels tracked by a changing set of consumers from year to year rather than a failure in the consumer community to match resources that we wrongly assume to be constant." He then continued, "But in some years and some habitats food resources may be superabundant; I recognize, but will not discuss further, that this possibility would lift selective pressure for habitat selection and community structure would degenerate." With this extra acknowledgment he seems to be agreeing with Wiens that the occasional suspension of food limitation would eliminate the effects of interspecific competition on community structure.

A Reassessment

It is important to reassess this challenge for three reasons. First, no author since then has articulated as comprehensive a critique of the applicability of competition theory to nature. Wiens' 1977 paper stands as the clearest expression of a critical viewpoint. Second, the conclusions drawn in that paper have attracted more attention than have the detailed arguments by which they were reached. There is a need to examine the arguments. Third, the assumptions of competition theory were rejected or seriously doubted on the basis of general considerations; data contributed to those considerations but were not used specifically to test any of the assumptions.

Pertinent data on populations, resource limitation, competition, and natural selection are available from a study of Darwin's finches on the Galápagos islands. I shall now consider the assumptions one at a time to determine how much they are upheld or contradicted by those data. I shall restrict attention to contemporary processes and will not discuss how competition in the past might have determined the characteristics of sympatric assemblages of these species. The evidence for competition as a historical process that has determined the nonrandom co-occurrences of species and nonrandom ecological characteristics of sympatric species has been discussed and summarized in several papers (e.g., Grant 1981a, 1984; Grant and Schluter 1984; Schluter and Grant 1984a).

POPULATIONS, GUILDS, OR COMMUNITIES ARE AT AN EQUILIBRIUM DETERMINED BY RESOURCE LIMITATIONS

In formal competition theory, as summarized in Chapter 13, carrying capacities are fixed and equilibrium occurs when they are reached. In the real world carrying capacities are not fixed, and in this sense the assumption is violated (Wiens 1977). However, the important element of the assumption is resource limitation. This will be the focus of my discussion. I will use the concept of equilibrium to mean ''state of balance,'' not constancy. This is the Oxford Dictionary meaning. Balance, in the present context, implies a fixed relationship between numbers or biomass of consumers and numbers or biomass of resources; it does not imply fixed numbers of consumers and resources. Others discuss equilibria in terms of stability (e.g., Connell and Sousa 1983), and this corresponds to my usage only to the extent that the relationship between consumers and resources is stable.

I shall ask whether and how often consumer abundance is limited by resources, not in communities as a whole, but in a single guild of seed-eating finches on several of the Galápagos islands. The answer is provided by the results of an analysis of data on finch numbers and their food supply obtained painstakingly over 10 years by Abbott et al. (1977), Smith et al. (1978), Schluter (1982a, 1982b), and Schluter and Grant (1984b). But first a few remarks are needed about a typical annual cycle of events on the Galápagos to place the analyses and their results in perspective.

The Galápagos have a seasonal climate, despite being on the equator. A wet season in the first few months of the year is followed by a longer dry season (Grant and Boag 1980). The rise and fall of food supply is summarized in Fig. 10.1. Typically it rises in January following the first heavy rains as a result of the production of flowers, seeds, fruits, and arthropods, chiefly caterpillars. It falls after April at the end of the wet season at a time when only a few new seeds and fruits are being produced, and it rises a little in November as a result of the flowering of *Opuntia* cactus. The demand placed upon the food supply is also shown to rise and fall. First it rises as a result of population increase due to breeding in the wet season; then it falls as a result of population decrease due to mortality, and possibly emigration, in the dry season. Consumption by finches is the main reason for the decline in food supply from May to November. The latter part of this period is the time of potential food limitation for finches. It is illustrated in a hypothetical annual cycle in Fig. 10.2.

All six species of ground finches in the genus *Geospiza* are members of a seed-eating guild.

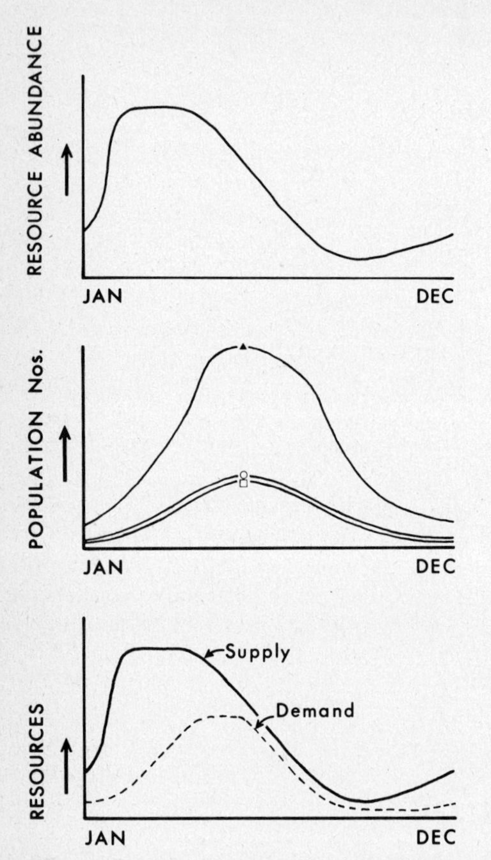

Fig. 10.1 Some suggested relationships between finches and their resources from a study of three species on Genovesa in the breeding season from January to May and in the nonbreeding season in November 1978: *Geospiza difficilis* (triangle); *Geospiza magnirostris* (circle); *Geospiza conirostris* (square). The graphs summarize the patterns observed on other islands and in other years (e.g., Smith et al. 1978). (After Grant and Grant 1980a.)

To varying degrees they feed on animal and other plant matter (Abbott et al. 1977; Smith et al. 1978; Grant and Grant 1980a, 1981; Schluter 1982b; Schluter and Grant 1984b), but toward the end of the dry season there are few such alternative foods available. Other members of the seed-eating guild are doves (*Zenaida galapagoensis*), mockingbirds (*Nesomimus* species), native rats (*Nesoryzomys narboroughi* and *Oryzomys bauri*), and introduced black rats (*Rattus rattus*). Doves and mockingbirds occur on almost all of the islands. Doves tend to concentrate on large seeds, including two large and common types avoided by finches, *Merremia aegyptica* and *Ipomoea linearifolia,* both members of the Convolvulaceae (P. R. Grant and K. T. Grant 1979). Mockingbirds feed extensively on the fleshy fruits of *Cordia lutea* and *Opuntia* species, but regurgitate the undigested seeds, which the larger finches then eat. Some seeds are also consumed by native and black rats (Clark 1982). Among the islands included in the following analyses, native rats occur only on Fernandina and black rats occur only on Santa Cruz. There are no seed-eating ant species. Since most of the seeds consumed by finches are not exploited by the other, nonfinch, members of the guild, the other members will be ignored in the analyses.

Seven islands have been studied in the months October to December and before the *Opuntia* flowering season. Total finch biomass is strongly and positively correlated with seed biomass among these islands ($r = 0.962$, $p < 0.01$; Fig. 10.3). This is the balance, or fixed relationship, to be expected if at this time of the year the guild is at an equilibrium determined by resource limitation (Fig. 10.2). The analysis, with more extensive data, confirms the previous findings of Smith et al. (1978), Grant and Grant (1980a), and Schluter (1982a); see also Schluter and Grant (1984a). In contrast to these results, there was no demonstrable relationship between consumers and seed resources in the wet season (Abbott et al. 1977, Smith et al. 1978), although invertebrates should have been censused and added to the seed resources to produce a more comprehensive estimate of total food supply at this time.

Six sites at different elevations on the island of Pinta were studied in November 1979 by Schluter (1982b). The correlation between finch biomass and seed biomass among sites was positive and strong at this time ($r = 0.966$, $p < 0.01$; Fig. 10.4). Positive correlations were also found between densities of some individual species and their food supplies (Schluter 1982b).

The evidence is consistent with a hypothesis of food limitation, but does not demonstrate it. On the island of Pinta, for example, birds might distribute themselves more commonly in the food-rich sites than in the poorer sites and thus give rise to the observed correlation even when population sizes are not limited by food (Schluter

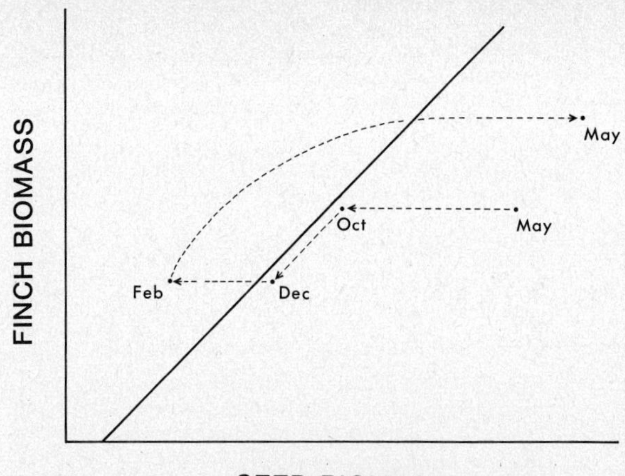

Fig. 10.2 Hypothetical relationship between finch biomass and seed biomass in an annual cycle. The solid line represents the carrying capacity of finch populations determined by seed biomass alone. Finches are shown to be limited by the seed supply from October to December. This period of food limitation ends when nectar and pollen in *Opuntia* flowers become available. Finch numbers and biomass increase through breeding from January or February to April or May in a typical year. Biomasses in successive Mays are shown to be different to illustrate annual variation in both finch and seed production.

1982b). Birds do not redistribute themselves under changing food conditions among islands, however, yet the same relationship between consumers and resources is seen among islands as within the island of Pinta. Indeed, the interisland regression line in Fig. 10.3 provides an adequate fit to the intraisland scatter of points in Fig. 10.4. The possibility of food limitation is further indicated by declines in the sizes of all populations during the dry season prior to this time, declines that match the declines in food supply (Schluter 1982b).

The direct way to investigate food limitation is to follow populations of known size through

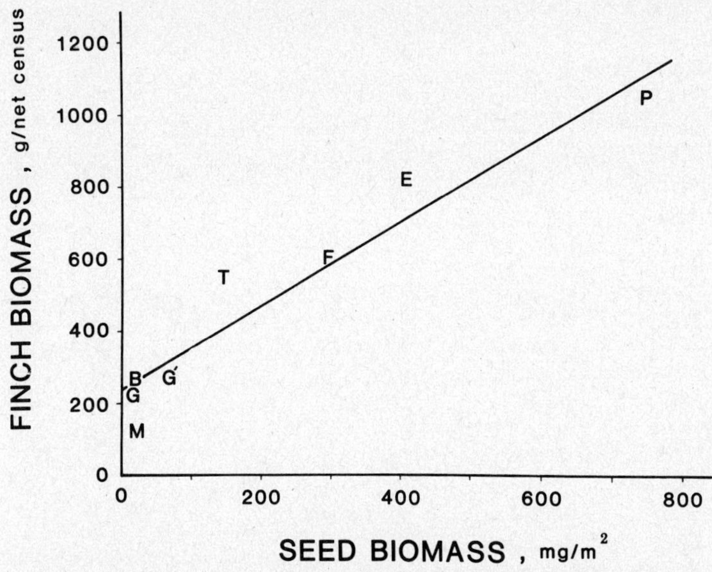

Fig. 10.3 Biomass of ground finches, *Geospiza* species, in relation to seed biomass in the late dry season on seven islands: Marchena (M), October 1979; Genovesa (G), November 1973; Bahía Borrero on Santa Cruz (B), December 1973; Tortuga (T), November 1981; Fernandina (F), October 1981; Española (E), November 1981; Pinta (P), November 1979. Genovesa was surveyed a second time (G′), but the point was not included in the estimation of the line of best fit by least squares regression. (Data from Smith et al. 1978 and Schluter and Grant 1984b.)

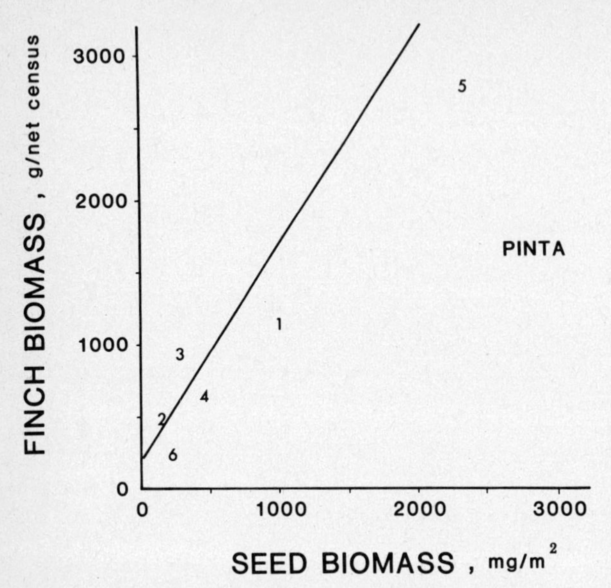

Fig. 10.4 Biomass of ground finches, *Geospiza* species, in relation to seed biomass at six sites on Pinta in November 1979. The sites are numbered from 1 to 6 from low to high elevation. The line has been estimated by least squares regression from the points in Fig. 10.3, excluding the data from Pinta and the second survey on Genovesa. The line fits the points quite well, although the slope is a little high. (Data from Schluter 1982a and personal communication.)

time. From our studies this can be done only on the small island of Daphne. The rains failed almost completely on this island in 1977 (Grant and Grant 1980b), and to a rough approximation it may be said that the period May 1976 to January 1978 was one long dry season. During this period there was a precipitous decline in both total finch biomass and seed biomass, and the two biomasses were positively correlated with each other ($r = 0.921$, df 3, $p < 0.05$; Fig. 10.5). Each of the resident species treated separately showed a significant correlation: for *G. fortis* $r = 0.914$ ($p < 0.05$) and for *G. scandens* $r = 0.946$ ($p < 0.05$). The same results are obtained when finch numbers are substituted for finch biomass, and numbers of each species are then correlated with the biomasses of the seeds that each species exploits ($p < 0.05$).

Short of experimental manipulations of finch numbers or food supply on the appropriate scale and with appropriate controls, which are not permitted by the National Parks regulations governing scientific work on the Galápagos, these results provide as strong a case as can be made that food is sometimes limiting.

On some islands and in some years food is not

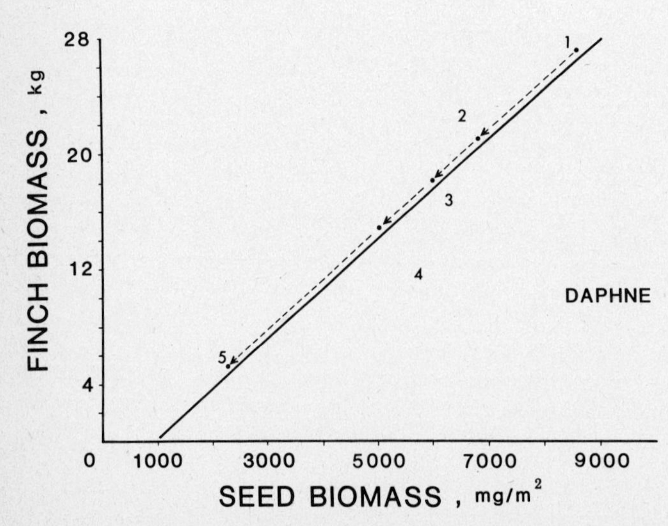

Fig. 10.5 Biomass of ground finches *Geospiza* species, in relation to seed biomass on the island of Daphne through an extended dry period. Censuses were taken in May 1976 (1), December 1976 (2), March 1977 (3), June 1977 (4), and December 1977 (5). The arrows illustrate the progressive decline in finch and seed biomasses. The line of best fit (solid line) was calculated by least squares regression. (Data from Boag and Grant 1981, 1984a.)

limiting in the dry season, and at such times there is little mortality from any cause. This happens when the rains are extensive, seed production is large, and seeds are not severely depleted by the time that *Opuntia* flowering begins; flowers offer an additional source of food in the form of pollen and nectar. All of this happened in 1973 on the island of Daphne. Survival from April to December was 90% in *G. fortis* and 95% in *G. scandens* (Grant et al. 1975, and subsequent revision unpublished), and seed supplies remained high (Smith et al. 1978).

Food limitation occurs, but how often it occurs is difficult to determine. I suspect it occurs in most years. The high survival of finches on Daphne in 1973 was exceptional. Usually there is almost no adult mortality in the breeding season, but moderate to substantial mortality in the nonbreeding (dry) season, especially of birds born that year (Boag and Grant 1984a, Price and Grant 1984).

Fig. 10.6 shows how population sizes have changed drastically on Daphne over a 10-year period. It would appear from the figure that populations might have been limited by food prior to 1977, but that thereafter they were not, as shown by the fact that they remained at postdrought low levels for several years. Such reasoning implies an inability of the population to increase rapidly through reproduction. This is partly correct; given the amount of reproduction that subsequently occurred, the population could not have reached predrought levels in less than two years, even without juvenile mortality. But the rate of increase of the population and hence its recovery time were governed by food supply recovery in

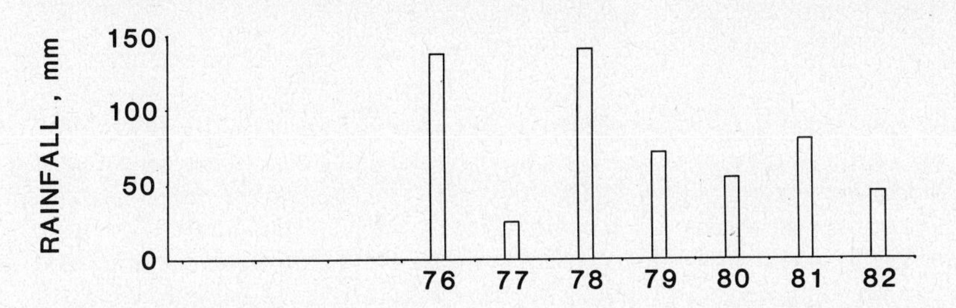

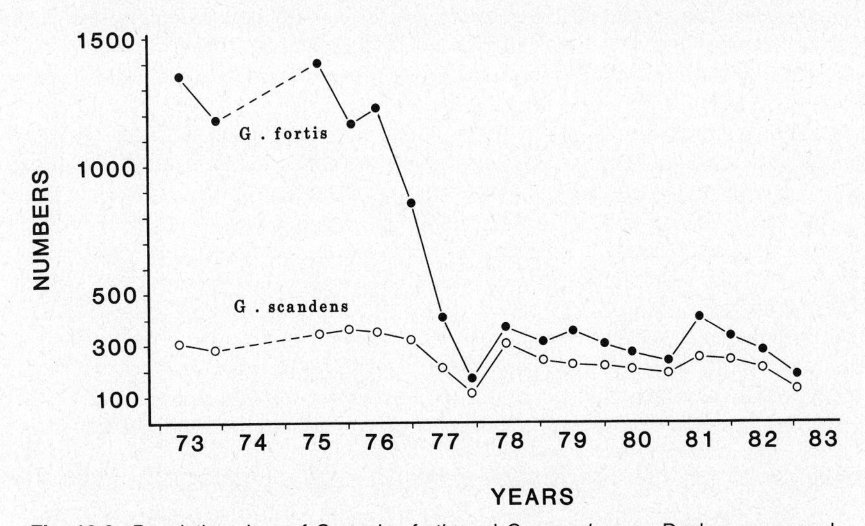

Fig. 10.6 Population sizes of *Geospiza fortis* and *G. scandens* on Daphne, censused twice a year except in 1974. Note the population crash caused by the drought of 1977; see Fig. 10.7 for the crash in seed biomass. (Data from Boag and Grant 1981, 1984a, and unpublished.)

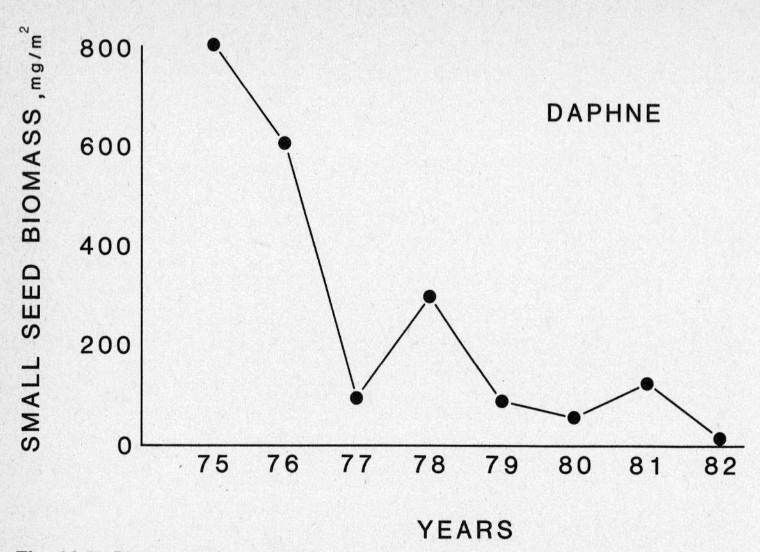

Fig. 10.7 Biomass of small seeds on Daphne at the beginning of the dry season (June–July), 1976 to 1982. The seeds do not include *Tribulus cistoides, Opuntia echios, Cenchrus platyacantha,* and *Bursera malacophylla.* (Data from Boag and Grant 1984a and unpublished.)

two ways. In the breeding season the supply of caterpillars and pollen restricted the period of reproduction to 1 to 4 months in each of the years 1978 to 1982 (Boag and Grant 1984b, Millington and Grant 1984), and in the nonbreeding season the reduction in seed biomass caused high mortality, especially among young birds. The second effect appears paradoxical in view of the low population density, but the paradox is removed by correcting the implicit assumption of a rapid return of the carrying capacity to the predrought level. Many seed-bearing plants died in the drought of 1977, especially *Heliotropium angiospermum* (Grant and Grant 1980b, Boag and Grant 1984a), and they were not immediately replaced. In fact *Heliotropium* became common again only in the extraordinarily wet year of 1983. The seed-determined carrying capacity for finches was depressed for several years after the drought of 1977. The reduction in biomass of small seeds is shown in Fig. 10.7. Small seeds are crucial to the survival of birds in their first year (Price and Grant 1984).

I conclude that the guild of seed-eating finches and its component populations are limited by food resources in the dry season of many, perhaps most, years. Finches can live for more than 10 years. Droughts have occurred recently once every 10 years, and less severe dry conditions accompanied by high juvenile mortality have occurred once in every 2 to 3 years (Grant and Boag 1980, Grant 1985). Generation times of finch populations are on the order of 3 to 4 years. Therefore, the effects of food limitation may be experienced as frequently as once a generation, and strong effects may be experienced once a decade.

In reaching this conclusion I have ignored other potentially limiting factors such as predation and disease. The effects of predators (hawks, owls, egrets) on finch populations seem to be generally small where they have been studied in detail (Boag and Grant 1984a), although owl predation reduces breeding success in some species in some environments (Grant and Grant 1980a). Disease is possibly important but undetected. Symptoms similar to those of avian pox are occasionally seen, but certainly some birds recover. These two factors and others may interact with finch food supply in some unknown way to determine finch numbers, but I cannot think of a simple biological explanation that would render spurious the demonstrated correlations between finches and their food supply.

SELECTION ON THE SYSTEM OF ATTRIBUTES CONSIDERED IS ASSUMED TO BE CONTINUOUS AND INTENSE

The strongest statement of this assumption is the following: "Conventional competition theory thus envisions selection as incessantly acting against all but the best adapted or optimal phenotypes in populations" (Wiens 1977, p. 591). This clearly implies intense stabilizing selection on the phenotypes or "attributes," which elsewhere are indicated to be morphological traits, such as bill dimensions of birds or behavioral traits that are employed in exploiting resources. Wiens continues, "If this were not so, deviations from the optimum would be expected to arise, and we should witness considerable variation about the predicted optimum." In other words, under relaxed selection phenotypic distributions should broaden considerably. Fig. 10.8 provides an illustration, and Wiens argued that this occurs in fluctuating environments.

I do not believe that formal competition theory makes the assumption as stated; therefore its violation, should this occur, does not damage the theory. Much of the theory does not deal explicitly with phenotypic traits at all (e.g., see Schoener 1982), and in that which does the variance is often ignored. Variance is ignored in nonevolutionary models because the mean is an adequate index of a phenotypic distribution for purposes of relating that distribution to the distri-

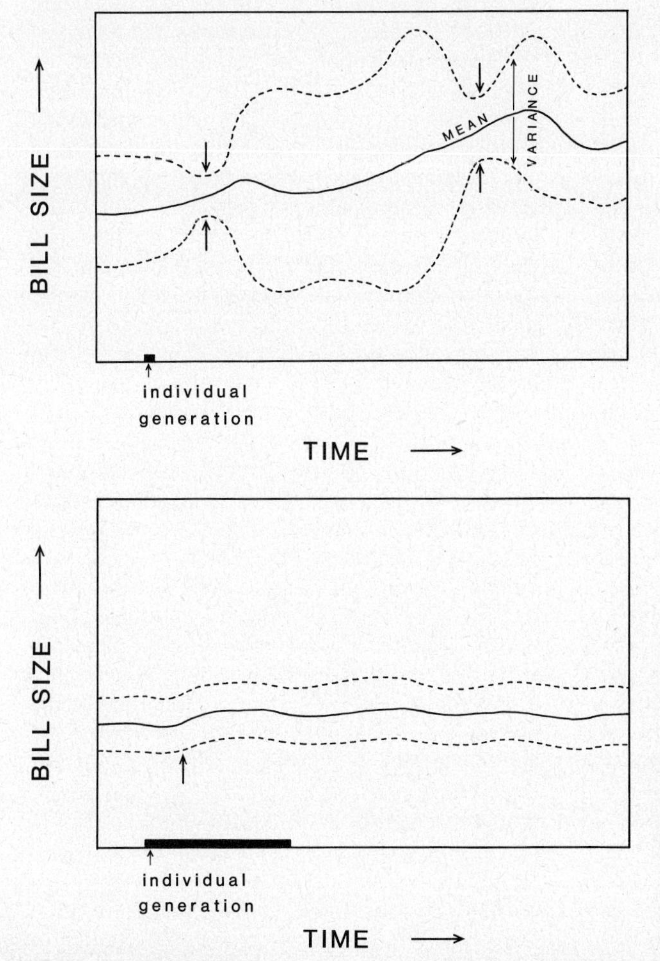

Fig. 10.8 Changes in the mean and variance of bill size in a fluctuating environment. The upper figure is hypothetical and shows intense stabilizing selection (arrows) at two times of severe environmental stress over a period of about 70 generations. At other times stabilizing selection is relaxed and variance in bill size and resource utilization (not shown) increases considerably. (After Wiens 1977.) The lower figure shows small fluctuating changes in mean bill size but little change in variance of *Geospiza fortis* on Daphne over a period of three to four generations. (Based on data in Grant et al. 1976; Boag and Grant 1981, 1984b, and Price and Grant 1984.)

bution of resources exploited. On the other hand, phenotypic and genetic variance are important in evolutionary models (e.g, see Roughgarden 1979, Slatkin 1980). For these, stabilizing selection is required to counteract the effects of drift on the mean. The influence of stabilizing selection on the variance has also been modeled (Lande 1976, Bulmer 1980). Indeed, the prevalence of stabilizing selection is a major tenet of quantitative genetics theory applied to continuously varying phenotypic traits in natural populations (e.g., Lande 1979, 1980a). The question, then, is whether stabilizing selection is supposed to act continuously and intensely or just occasionally and weakly. Studies of *G. fortis* and *G. scandens* on Daphne show that selection does occur but neither continuously, in a literal sense, nor (usually) intensely (Boag and Grant 1981; Grant and Price 1981; Price, Grant, Gibbs, and Boag 1984).

The important attributes are bill dimensions. These, particularly depth, determine the size and hardness of seeds that can be cracked. This has been shown in comparisons of individuals of the same species (Bowman 1961, Grant et al. 1976, Grant 1981b) and of different species (Abbott et al. 1977, Grant 1981b, Schluter 1982a, Schluter and Grant 1984a). Therefore, it can no longer be objected that the critical link between overlap in feeding structure of different species and overlap in their utilization of limiting resources has not been satisfactorily demonstrated (Wiens 1977). At least for Darwin's finches it has (Abbott et al. 1977, Schluter and Grant 1984a).

Stabilizing selection on the bill dimensions of *G. scandens* has been demonstrated (Grant and Price 1981; Price, Grant, and Boag 1984; Boag and Grant 1984): variances were reduced at a time of stress. Since then (1977) we have seen nothing like the massive increase in variance depicted in Fig. 10.8 which Wiens suggested occurred under conditions of relaxed selection. In the period 1978 to 1982 annual adult mortality has not been very high, and changes in variance have not been detected. I know of no study where a large increase in variation has been observed, and it is not expected from quantitative genetics theory (e.g., Falconer 1981). Admittedly a true test of the variance-release hypothesis would require observations made over several generations (see Fig. 10.8), whereas the study on Daphne has spanned only two or three generations depending on how generation length is defined. But variances are approximately stable over longer periods of time. Variances estimated from samples of current populations are similar to those estimated from samples of the same populations collected mainly at the turn of the century and now in museums (e.g., B. R. Grant and P. R. Grant 1979, Boag and Grant 1984b, Grant et al. 1985).

Thus the obverse of the assumption as stated by Wiens (1977), that "considerable variation about the predicted optimum" should occur under benign conditions, is not correct for bill dimensions, and it has not been observed. But on the other hand, failure to observe a massive increase in variation does not mean that stabilizing selection is necessarily continuous and intense.

Intense selection on the other species on Daphne, *G. fortis,* has been observed, but it was directional and not stabilizing (Boag and Grant 1981). Selection occurred during the drought of 1977 when food was limiting and the supply of seeds was nonrandomly depleted. Birds with large bills were favored, apparently because they were able to crack the large and hard seeds that remained in moderate abundance when smaller seeds had become scarce. Two similar but less pronounced selection episodes have been documented since then, and estimates of the direct effects of selection upon bill dimensions have been obtained (Price, Grant, Gibbs, and Boag 1984). But although population means increased (cf. Fig. 10.8), there were no detectable changes in variances (Boag and Grant 1981, Grant and Price 1981).

The effects of selection were transmitted to the next generation; in other words, microevolutionary changes occurred (Boag 1983). They occurred because bill variation is largely heritable. The heritability of a trait is the proportion of the phenotypic variance that can be attributed to the additive effects of genes. Heritabilities of the three bill dimensions, length, depth, and width, have been estimated by the offspring-midparent regression technique (Boag and Grant 1978, Boag 1983). All exceed 0.65. These high values could be inflated if genotypes are distributed nonrandomly among environments (habitats or patches) on the island, but there is little evidence of genotype-environment correlations from our studies (Boag 1983, Grant 1983a), and experimental studies have failed to detect effects of the rearing environment on final adult size in other species of passerines (Smith and Dhondt 1980, Dhondt 1982; but see James 1983 for geographical effects). Thus, there is a large amount of additive genetic variance for bill traits in the population of *G. fortis* on Daphne. Boag (1983) confirmed the evolutionary potential of the bill size variation by showing that the calculated response to selection occurring in 1977 agreed closely with that predicted from heritabilities and selection differentials. High heritabilities have also been estimated for bill traits in the *G. scandens* population on Daphne (Price, Grant, and Boag 1984), as well as for bill traits of *G. conirostris* on the island of Genovesa (Grant 1983a).

To summarize, it seems likely that *G. scandens* is subject occasionally to varying intensities of stabilizing selection, and *G. fortis,* a more generalized species in its diet (Boag and Grant 1984b), is subject to fluctuating directional selection (Grant et al. 1976; Price, Grant, Gibbs, and Boag 1984). Directional selection may operate in different directions either within (Price and Grant 1984) or between generations. The net effect over a period as long as a generation may be weak, overall stabilizing selection. On the scale of years selection is intermittent; in some years there is no selection. On the scale of generations it may be continuous, but even on this scale the intensity of stabilizing selection seems to vary. There is certainly no indication of long-term relaxation or absence of stabilizing selection, in contrast to the implication of Fig. 10.8 (upper). These observations are consistent with the requirements of theory.

The Connection Between Assumptions 1 and 2

The link was made by Wiens (1977) in the following way. In variable environments the variances in resource utilization traits become severely restricted at times of intense environmental stress and strong stabilizing selection. Populations of low-fecundity organisms such as birds make a slow recovery from crashes, whereas recovery of their resources is faster, and this results in population sizes frequently being below equilibrial sizes for several generations, hence not limited by food. When populations do recover, the phenotypic variances may become broad, but the broadening process is initially slow because genetic variance was lost due to founder effects and inbreeding when the populations were small. Broad phenotypic distributions indicate lack of selective constraints and thus the continuing absence of food limitation of population sizes for many more generations (see Fig. 10.8, upper). Populations are therefore food-limited only at the relatively infrequent times of environmental stress.

Data from Darwin's finches do not support the two major ideas of rare food limitation and strongly fluctuating variances. The finches live in variable environments and are stressed by the failure of their food supply to regenerate in years of low rainfall. Their population sizes fall, in parallel with their seed-determined carrying capacities. Finch populations have the potential to recover from crashes in two years. Food supply sets a limit to the actual rate of recovery. Therefore these populations are often, but not always, at fluctuating equilibrial sizes.

Phenotypic variances of Darwin's finches do not show the major fluctuations depicted in Fig. 10.8 (upper). Genetic variation was not depleted by the single intense selection event associated with high mortality in *G. fortis* (Boag 1983, Price et al. 1984). There has been no opportunity for founder effects to arise during the course of the Darwin's finch study, since these effects follow from "the establishment of a *new* population by a few original founders . . . which carry only a small fraction of the total genetic variation of the parental population" (Mayr 1963, pp. 211–212; Lande 1980; italics mine). In any event the idea that founder effects are important in contexts such as this is based on a misunderstanding of the nature of genetic variation underlying phenotypic variation in continuously varying traits (see Lewontin 1965, Lande 1980, for details). Only rare alleles are likely to be lost in a given generation in the sampling of small numbers (Lande 1980a). Populations did not remain small on Daphne for several generations following the drought of 1977 (they increased greatly in 1983), so loss of genetic variance arising from continued inbreeding did not occur. More generally, the opportunity for random effects in Darwin's finch populations is small, given the overriding importance of selection (Price, Grant, and Boag 1984).

COMPETITION IS THE MAJOR SELECTIVE FORCE ACTING UPON THE RESOURCE-UTILIZATION TRAITS

Wiens (1977) pointed out two ways in which this assumption may not be correct. The first is that other factors might prevent populations from reaching sizes at which competition is intense. An intrinsic factor (low fecundity) has been mentioned above, but extrinsic factors such as predation should also be considered, and indeed they must be invoked to account for the long absence of food limitation and the weakness of stabilizing selection over the large number of generations shown in Fig. 10.8 (upper). This is a variant of an argument about the importance of predation made by G. C. Varley to David Lack when the latter was preparing his monograph on Darwin's finches (Lack 1947, Grant and Grant 1980a). It has been substantiated by the several experiments conducted since then that have shown how predators can keep prey populations below the sizes at which competition for food has detectable effects on growth and survival (e.g., Slobodkin 1964, Paine 1966, and many others subsequently). However, food limitation and natural selection associated with such limitation have occurred sufficiently frequently in our studies of Darwin's finches to make this objection inapplicable to those organisms.

The second objection is more complex. It starts with the supposition that only at times of population crashes, through restriction of variances, is overlap between species in resource-utilization traits eliminated (Fig. 10.9), the assumption being that there is a close correspondence between morphological variances and the breadth of feeding niches. Wiens (1977, p. 594) then argues, "These differences may indeed be the products of competitive interactions during the period of resource limitation, but it is also possible that they portray differential responses of the populations to a common selective regime (given populations with differing genetic compositions to begin with) or responses to different selective agents associated with the ecological crunch. Thus, even at this time, species differences cannot be interpreted solely in terms of competition."

With regard to Darwin's finches, the starting supposition is incorrect. Overlap in bill dimensions of sympatric species does not become eliminated at times of environmental stress (Fig. 10.9). What does happen is that foraging behavior and diets change, and overlap in feeding niches diminishes (Fig. 10.10). This has been observed to occur more than once on several islands toward the end of the dry season when seed supplies are generally low (Smith et al. 1978, Grant and Grant 1980a, Schluter 1982b, Boag and Grant 1984a; note that Wiens 1984b has incorrectly represented these results as showing the opposite). The small changes in bill dimensions that are brought about by selection may arise in part because of interspecific competition for food (Grant 1983b), but they do not result in the elimination of overlap.

On Daphne the feeding changes that cause interspecific niche overlap to diminish under stringent conditions are not the same in the two species present, in contrast to what might be expected from the scheme in Fig. 10.9 (upper). Diets of the specialist *G. scandens* become narrower, diets of the generalist *G. fortis* become broader (Fig. 10.10). There is no detectable between-phenotype component to the niche width of *G. scandens,* but even in *G. fortis,* which does have such a component, broadening of the diet seems to be largely a within-phenotype phenomenon, i.e., all individuals broaden their diets (Boag and Grant 1984a). There are only two discrete differences between the species: *G. fortis,* unlike *G. scandens,* crack the large and hard mericarps of *Tribulus cistoides* and stones of *Bursera malacophylla,* and extract and consume the seeds. Other differences between the species are matters of proportions. Both species feed on the several other food types available, but *G. scandens* largely specializes on the seeds and other products of *Opuntia echios,* and *G. fortis* feeds on the full variety of seeds, including those of *O. echios,* in approximate proportion to their availability (Boag and Grant 1984a).

So the argument should be rephrased in terms

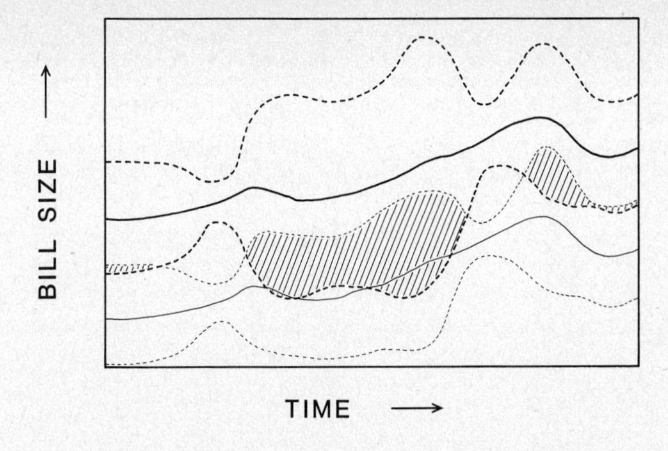

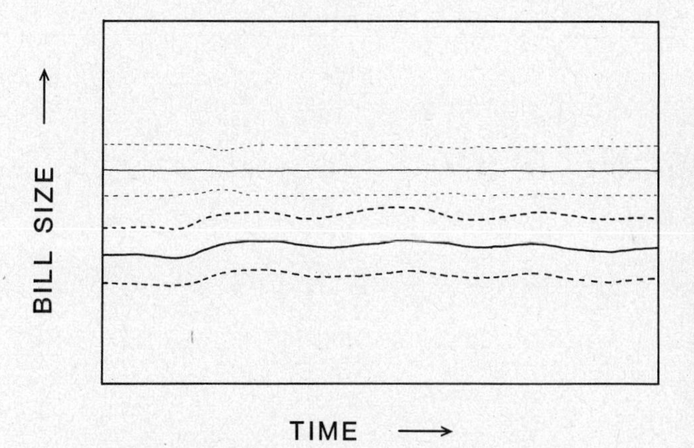

Fig. 10.9 Changes in overlap of bill size frequency distributions of two species living in a fluctuating environment. The upper figure is hypothetical and shows the overlap (hatched) to be eliminated twice at times of severe environmental stress. (After Wiens 1977.) The lower figure shows permanent nonoverlap between *Geospiza fortis* (thick lines), taken from Fig. 10.8, and *G. scandens* (thin lines). The trait is bill length; in bill depth the two species are almost identical. Note that *G. scandens* shows a reduction in variance once, but no change in the mean. (Based on Grant and Price 1981 and Boag and Grant 1984b.)

of niche overlap and not bill dimension overlap. As such, it is similar to Bowman's (1961) objections to Lack's (1947) interpretation of bill size differences between species. Lack suggested that interspecific competition for food caused species to diverge in bill size under natural selection. Bowman argued that bill sizes were adapted to diets independent of any influence from other species (see Grant 1981a, 1984; Grant and Schluter 1984; and Schluter and Grant 1984a for recent discussion). The rephrased Wiens' argument is that reduction in niche overlap can be the product of independent responses of the species to diminishing food conditions.

The counterargument is that if food is limiting to two sympatric species whose diets overlap, competition between them occurs. *G. fortis* and *G. scandens* populations on Daphne were food-limited in 1977, their diets overlapped (Fig. 10.10), and therefore they competed for food. Their diets diverged and the overlap diminished; as a result, direct competition was possibly reduced but certainly not eliminated because there was dietary overlap at all times, as shown in Fig. 10.10 and discussed above. To show this argument for competition to be wrong, it would be necessary to show that the seeds jointly exploited by the two species had no influence on the survival of members of either. I doubt if it could be done, especially as the feeding niche of *G. scandens* is included in the feeding niche of *G. fortis* (Grant and Grant 1980b). Obviously, a controlled experiment (with how many replicates?) would be needed to show cause and effect beyond all reasonable doubt. In the absence of such an experiment, we can perform a thought experiment. Imagine that all *G. scandens* left the island in 1976. *G. fortis* numbers would have declined

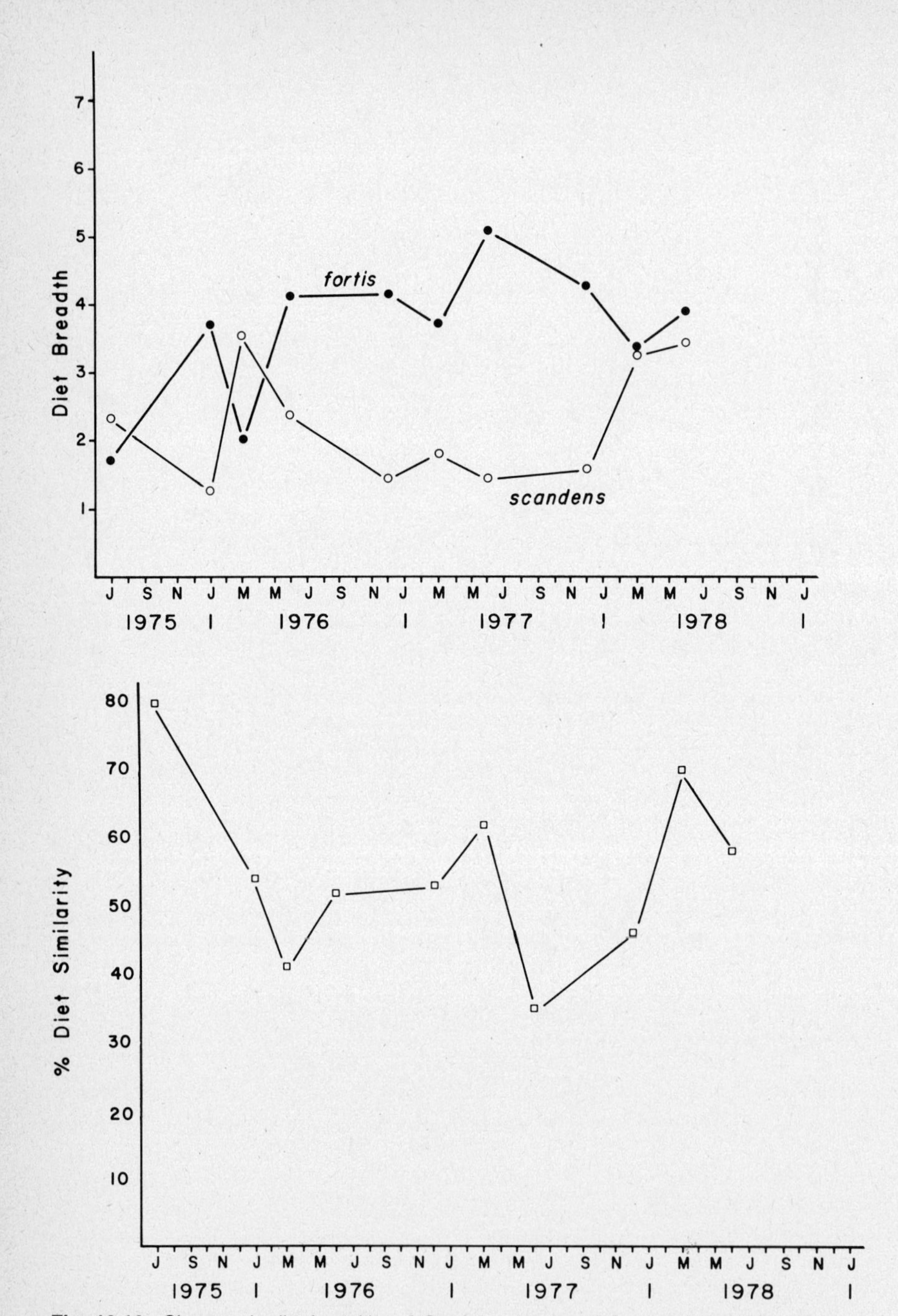

Fig. 10.10 Changes in diet breadths of *Geospiza fortis* and *G. scandens* (above), and changes in diet similarity on Daphne (below). When a drought occurred in early 1977, *G. scandens* decreased and *G. fortis* increased its diet breadth; these changes were reversed after rainfall in early 1978. (Taken from Boag and Grant 1984b.)

more slowly in 1977, sustained in part by the food made available by the convenient departure of *G. scandens*. The reciprocal thought experiment yields a reciprocal result. I am therefore arguing that niche divergence at times of food limitation is not solely the product of independent responses of the two species to their diminishing food supplies, but is driven in part by the consumption of jointly exploited foods (see also Smith et al. 1978; Grant and Grant 1980a, 1980b; Schluter 1982b).

How else might we recognize interactive effects? They can be detected statistically by the use of partial correlation and regression techniques (Minot 1981, Dhondt 1977, Dhondt and Eyckerman 1980a). In the future it will be worth applying these techniques to data on the population sizes of *G. fortis* and *G. scandens* on Daphne, changes (losses) in numbers in the dry season, and seed abundance. At the moment these quantities are known for only the seven years of continuous study, 1976 to 1982, and the data are scarcely sufficient for analysis.

I have discussed this critique as if the premise is correct; that theory assumes competition to be *the* major selective force acting upon the resource-utilization traits. But does theory depend upon this assumption? With regard to bill sizes of birds I doubt it, although discussions of competition in the 1960s and 1970s gave the impression of such a dependence (J. A. Wiens personal communication). The bill of a bird is an instrument for gathering and dealing with food, so the major selective forces acting on bill size variation arise from properties of the foods. Competition can have only an indirect influence through its effect upon food supply. Theory only requires competition to be an important component of the selective forces acting upon variation in resource-utilization traits, not the major selective force itself.

THE PREDICTED OPTIMAL STATE IS AVAILABLE TO THE POPULATION IN ITS EVOLUTIONARY DEVELOPMENT

If this assumption is not true, "selection will maintain suboptimal states because the intermediate stages necessary to reach the predicted optimum confer reduced fitness" (Wiens 1977, p. 592). This is another way of saying that a higher peak in an adaptive landscape cannot be reached by a population currently on a nearby lower peak because the peaks are separated by a valley. Wiens (1977) raised this as a possible objection to competition theory, but did not discuss it.

Two points need to be made. The first is partly one of semantics. In applications of optimization theory to the performance of organisms, such as foraging, an optimum is specified in a system that is subject to constraints (Maynard Smith 1978). If, in the present context, an adaptive valley is an identified constraint, then the peak beyond the valley is not part of the system, and the population has already achieved its optimal state by virtue of its position on the lesser peak. However, identifying and assessing constraints are often a major problem. A valley in an adaptive landscape is not necessarily an insuperable barrier. Lande (1976) has modeled the dynamics of valley transgression by genetic drift. The adaptive topography and drift determine the likelihood that a barrier will be crossed (see also Wright 1982).

Second, regardless of the finer points of barrier transgression, the evidence from Darwin's finches shows that populations occupy adaptive peaks (Schluter and Grant 1984a). The observed phenotypes of sympatric species correspond closely to those expected from food characteristics of islands. For members of the seed-eating guild, no empty niches (unoccupied peaks) have been identified. The correspondence between bill sizes and food availability is not surprising in view of the high heritabilities of bill dimensions and the occasionally strong directional selection acting on bill variation. Phenotypic traits of a population are evolutionarily plastic.

Thus, for Darwin's finches the assumption is correct.

OVERVIEW

This discussion leads me to a different view of the validity of the four assumptions from the one held by Wiens (1977). First, I have argued that assumptions 2 and 3 are not as restrictive as they appear to be in their formulation by Wiens (1977). Second, evidence from nonexperimental field studies indicates that all four assumptions are met to some extent by a seed-eating guild of

Darwin's finches. Therefore members of the guild are not usually below an equilibrium level set by resources in their variable environment and hence rarely subject to interspecific competition. On the contrary, within each generation food limitation and interspecific competition occur in one or more dry seasons, and populations are subject to pervasive but probably weak stabilizing selection which holds them at or close to their optimal state.

GENERALIZING FROM A SINGLE CASE STUDY

Formal theory purports to be general, whereas empirical studies are circumscribed. It is desirable to establish the degree of applicability of theory by finding out the extent to which empirical studies yield results that are consistent among themselves and with the assumptions of theory. I therefore ask, to what extent are the characteristics of Darwin's finch guilds shared by other birds in other parts of the world? Specifically I ask three questions: (1) Are populations frequently limited by resources? (2) Do species compete for those resources? (3) Does selection act on those traits involved in the exploitation of those resources?

Regarding the first question, there is consistent but indirect evidence that emberizine finches elsewhere are sometimes limited by the amount of food available in the nonbreeding season. Winter densities of seed-eating finch (sparrow) species, either singly or collectively, are positively correlated with the amount of rainfall in the preceding summer (Dunning and Brown 1982). This implies local food limitation, because the amount of summer precipitation determines the amount of seed production (Pulliam 1975, Pulliam and Brand 1975, Pulliam and Parker 1979). Pulliam and Parker (1979) showed that numbers of the migrant sparrows at one site in Arizona were positively correlated with grass seed production in the years 1972 to 1976. Depletion of seeds was virtually complete in years of low seed production but not in years of high production. Their conclusion that local sizes of migrant finch populations are regulated by food supply in some years (the years of low production) is similar to my conclusion regarding the food limitation of Darwin's finch populations. Capurro and Bucher (1982) have similar correlational evidence for the limiting effect of food supply on seed-eating bird populations in an arid part of Argentina. Similar but less quantitative arguments have been made for birds in the arid parts of Africa (Morel and Morel 1974, Maclean 1976) and Australia (Serventy 1971, Davies 1976).

In less variable but seasonal environments food has a similar limiting effect on bird populations in the nonbreeding season. To take one example from the many detailed studies of bird populations in Europe, van Balen (1980) has shown that the size of the crop of beech nuts, an important winter food, has a large effect on the winter survival of juvenile great tits, *Parus major;* and the experimental provision of supplemental food enhances overwinter survival of all birds (see also Klomp 1980, Newton 1980, Janssen et al. 1981). In addition to these limiting effects, food supply can also determine the rate of increase of a population by governing the number of broods that are raised and the overall breeding success (e.g., Newton 1980, Boag and Grant 1984a, Millington and Grant 1984).

Therefore, I suggest that food limitation of bird population sizes in the nonbreeding season is a widespread if intermittent phenomenon in seasonally varying environments (see also Schoener 1982).

The second question is whether or not species compete for food. To demonstrate that species compete for food, it is necessary to show that one species consumes some of the foods that limit the population size of another species. Controlled experiments are the means by which such demonstration is made, but under many circumstances experiments are infeasible or forbidden. As an alternative, competition is inferred from the demonstration of statistical (negative) effects of the population size of one species on the population size of another when each exploits at least some of the resources exploited by the other (Dhondt 1977, Dhondt and Eyckerman 1980b, Minot 1981). Less directly it can be inferred from patterns of coexistence observed and predicted from competition theory (Pulliam 1975), but not without difficulty in rejecting alternative hypotheses

(Pulliam 1983). Competition for food, space, or roosting places has been directly demonstrated in seven experimental studies (Davis 1973; Slagsvold 1978; Williams and Batzli 1979; Högstedt 1980; Dhondt and Eyckerman 1980a, 1980b; Minot 1981; Garcia 1983; see also Alatalo 1982). While negative interactive effects on individuals were demonstrated, not all of these studies showed that one or more populations suffered from the effects. In contrast to these results, Thompson and Lawton (1983) found no evidence of interspecific competition for extra food supplied in winter in a short field experiment with unconstrained birds in England. But the experiment was not well suited to the hypothesis under test, and the results are not easy to interpret for several reasons, some of which are carefully discussed by the authors. Results of experiments with captive birds feeding on non-natural foods (e.g., Grant 1966b, Alatalo and Lundberg 1983) are even more difficult to interpret because confinement makes the test situation so artificial.

In general, then, the experimental evidence for interspecific competition for food among birds is not extensive. It is richer for other groups of vertebrates (e.g., see Connell 1983, Schoener 1983b, and discussions for mammals and fish in Chapters 3 and 21), probably because birds, regardless of their diets, are not good subjects for experiments that require confinement of populations. It was for this reason that years ago I chose small mammals rather than birds for experimental investigations of interspecific competition (Grant 1972).

The third question, concerning selection on bill dimensions, cannot be answered clearly because there are no published studies comparable to ours. There is heritable variation in the bill dimensions of some other species (Smith and Zach 1979, Smith and Dhondt 1980), but selection on that variation has not been shown. Selection on a species introduced to North America, *Passer domesticus,* has been shown (e.g., Lande and Arnold 1983): selection may be a recurring process (Rising 1973, Johnston and Fleischer 1981, Fleischer and Johnston 1982), but the evidence is indirect (Lande and Arnold 1983). Selection may have also occurred on bill size in another species, *Carpodacus mexicanus,* introduced to eastern North America (Aldrich 1982). Heritabilities of morphological traits have not been determined in these two species.

From this survey it appears that processes revealed by a study of Darwin's finches are exhibited at least sometimes by other species of seed-eating birds elsewhere. The evidence is clearly not uniformly good, and elsewhere other factors make it difficult to discern those same processes. The question for empiricists to answer is whether interspecific competition for food is rare and of little consequence since other processes such as predation predominate, or whether it is common and important but for many practical reasons its real role in community dynamics is exceedingly difficult to demonstrate. For example, a general problem with all of these continental studies is that the "communities" are open systems allowing ingress and egress. Migrant finch (sparrow) populations, like sunbirds in Africa (Gill and Wolf 1979), may be locally limited by food supply, but their members then fly somewhere else and are lost from the study site. Since the populations are not shown to suffer, competitive effects are not known to occur, even though individuals must compensate for food losses by changing their activity patterns (Gill and Wolf 1979). In temporally varying environments like arid regions the problem of individual mobility is especially acute because the incidence of migration and nomadism outside the breeding season is high (Serventy 1971). And in the breeding season food density may not limit the breeding bird density except only occasionally. Studies performed at this stage of the annual cycle have yielded little evidence of interspecific competition (Rotenberry 1980; Wiens and Rotenberry 1980a, 1980b, 1981a; Rotenberry and Wiens 1980; but see Dhondt and Eyckerman 1980a, 1980b; Högstedt 1980; Minot 1981). Therefore, the time at which food limitation and competition are most likely to occur is the time when they are least easy to investigate. A similar problem confronts the student of seabirds. It can be avoided only by studying sedentary populations (see also Wiens 1984b). Most of these are found on islands or at latitudes on continents closer to the equator than are the residences of most biologists.

These complications aside, it is tempting to generalize beyond birds and declare the findings from studies of seed-eating birds to be applicable to all animals. Certain consistencies encourage this generalization. For example, in taxa as different as insects, fish, frogs, and birds, diets become restricted and niche overlap diminishes as food abundance declines, suggesting that competitive effects are similar in such diverse organisms (Smith et al. 1978, Schluter 1981, Schoener 1982). Moreover extensive, and taxonomically unrestricted, surveys of the literature on experimental studies of interspecific competition have led to broad agreement that competition occurs widely in nature although not at all times and not uniformly in all taxa (Connell 1983, Schoener 1983b). And if the findings are applicable to many animals, why not to plants as well? There is a solid experimental foundation to the view that competition is an important process in plant communities (Chapters 22, 32; Harper 1977, Jackson 1981, Connell 1983, Schoener 1983b), and Chapter 32 specifically addresses the frequency of competition in a plant community. Genetic variance governing some of the properties that enable plants to exploit the environment has also been well established (e.g., Bradshaw 1965, Antonovics and Primack 1982), although field data on the appropriateness of assumptions 2 and 3 are scanty.

Despite these consistencies it has yet to be determined how far we can generalize about competition. A necessary step is to strengthen the link between theory and empirical knowledge. Competition theory should be elaborated to accommodate some complexities encountered in empirical studies. Carrying capacities fluctuate. Consumers and resources recover from disturbances at different rates, and those rates vary according to the disturbing and subsequent circumstances. Resource frequency distributions are often asymmetrical and partly discontinuous (Slatkin 1980, Schluter and Grant 1984a). Theory that accommodates these complexities should be tested with a view to establishing (1) how frequently competition occurs, (2) between which organisms, (3) under what environmental conditions, (4) with what duration and intensity, and (5) with what direct and indirect consequences of an ecological and evolutionary nature.

The importance of competition in relation to the many other processes in communities will be understood more clearly when answers to these questions are obtained. Here I think experiments have a less than absolute role, and attempts to answer one or more of these questions from experimental results alone (e.g., Schoener 1982) will always be insufficient (see Chapters 1 and 8 for discussion). Since adequately controlled and replicated experiments are, to some extent, artificial, they will not by themselves give us an unambiguous answer to these questions. Furthermore, complex field experiments are difficult to control adequately, and their results are correspondingly difficult to interpret (Bender et al. 1984). Experiments need to be conducted in conjunction with simultaneous long-term nonexperimental studies of the same organisms in the same environment to establish the link between natural and experimental processes (e.g., Tinkle 1982). A variety of studies of different organisms is needed, because a variety of answers to these questions is to be expected—which brings me to my last point.

The Cody and Diamond (1975) volume concluded with some important remarks by G. Evelyn Hutchinson on the subject of generalization. They bear repeating. He recalled MacArthur's (1972b) awareness that excessive generalization from theory without a check from empirical studies could be misleading. But Hutchinson (1975, p. 515) cautioned that gaining a broad understanding of nature through empirical studies "will require even greater detailed knowledge than ordinarily goes into ecological studies today, since such a program, in essence, is comparative." The need for such a program is the subtheme of this chapter.

SUMMARY AND CONCLUSIONS

Mathematical equations of the Lotka-Volterra type have been used to explore theoretically the consequences of competition between species under various conditions. Wiens (1977) has argued that the theory often does not apply to the real world because its assumptions are not met. A consideration of populations living in temporally varying environments led him to reject, or at best seriously doubt, the applicability of the following four assumptions:

1. Populations, guilds or communities are at an equilibrium determined by resource limitation.
2. Selection on the system of resource-exploiting attributes considered by theory is continuous and intense.
3. Competition is the major selective force acting upon the resource-utilization traits.
4. The predicted optimal state is available to the population in its evolutionary development.

I have used data on Darwin's ground finches (genus *Geospiza*) on the Galápagos islands to assess the validity of both the assumptions and the challenge to them. The data consist of abundances of finches and of seeds (their principal dry season food), both of which were estimated on several islands, and these quantities plus diets and morphological changes observed in populations of two species on Daphne Island over a 10-year period. In the many situations such as these where experiments are infeasible or forbidden, interspecific competition can be recognized when two conditions are met: (1) population sizes are limited by resources and (2) one species consumes some of the resources that limit the population size of another.

I reach the following five conclusions. First, guilds of seed-eating finches appear to be subject to resource limitation in the dry season of most years, hence at least once a generation, as shown by strong positive correlations in time and in space between finch biomass and seed biomass. Second, competition between species for food occurs at these times as a result of overlap in their diets. Third, I suggest that theory does not require selection on resource-exploitation traits such as bill size of finches to be continuous and intense; in fact both stabilizing and directional selection have been demonstrated, but they are not continuous and rarely intense. Fourth, it is probably incorrect to insist that competition is necessarily the major selective force acting upon bill size for competition theory to apply, since the trait is used for dealing with food (the direct "force"), and competitive effects can have only an indirect influence through their direct effects on food. Fifth, predicted optimal bill sizes of Darwin's finches are available to, and possessed by, the ground finch populations.

Thus, the assumptions are largely met by Darwin's ground finches. This says nothing about the correctness of the theory; other, more mechanistic theories (e.g., Tilman 1982) may be more powerful and hence preferable. It does say that the theoretical arguments derived from the use of Lotka-Volterra models are not rendered untenable by unrealistic assumptions. While theory has license to continue, it should be elaborated to accommodate both fluctuating carrying capacities and asymmetrical, partly discontinuous, resource distributions. At the same time the need for detailed empirical studies to test theory is urgent. For example, species of seed-eating finches living in continental regions show evidence of food limitation and possible competition, but their often migratory or nomadic habits have prevented detailed examination of the factors that affect population sizes. As yet, too little of the evidence is experimental, and we are far from being able to generalize confidently about the frequency and intensity of competition for food among these birds, or indeed among any animals.

ACKNOWLEDGMENTS

I thank J. M. Diamond, J. Fry, T. Getty, D. Goldberg, B. R. Grant, S. L. Pimm, T. D. Price, B. J. Rathcke, D. Schluter, T. W. Schoener, J. A. Wiens, and M. Zuk for helpful discussion and comments on the manuscript. Research on Darwin's Finches has been supported by the Dirección General de Desarrollo Forestal, Quito, the Galapagos National Parks Service, the Charles Darwin Research Station, and N.S.F. (U.S.A.) and N.S.E.R.C. (Canada).

chapter 11

Patchiness, Dispersal, and Species Interactions: Consequences for Communities of Herbivorous Insects

Peter Kareiva

INTRODUCTION

Although much theorizing in community ecology assumes an even distribution of organisms in a homogeneous environment, nature rarely satisfies this assumption. Instead, communities are an irregular patchwork of plants and animals. Patchiness begins at the level of the physical environment: landscapes vary in topography, soils vary in nutrient or mineral content, and habitats occur in patches. Patchiness is also created by chance disturbances such as fires and windthrows in forests, or bashing logs in the rocky intertidal. Not only do organisms reflect these environmental heterogeneities, they also create their own spatial patterning through group living, limited dispersal of offspring, and nonuniform consumption of resources. Even when the environment is continuous and homogeneous, certain types of species interactions will generate patchiness in population densities (for an example, see Levin and Segel's 1976 discussion of ''diffusive instability'' in predator-prey systems).

Thus, patchiness is an inescapable feature of biological populations. Plants, for example, almost always exhibit clumped spatial dispersions (Kershaw 1973). Aggregation is so prominent among insects that their tendency to form aggregations is often treated as a key life history trait (Taylor 1971); there is, in fact, a statistical law called Taylor's Power Law that relates the degree of clustering in natural populations of insects to their local density (Taylor 1965; Southwood 1978, pp. 42–45). Patchy distributions are also commonplace among birds (Diamond 1980), mammals (Wiens 1976), and virtually every other taxonomic group that has been examined.

To date, the influence of patchiness on ecological communities has been investigated primarily with mathematical models, most of which have been published in the last decade (Slatkin 1974, Levin 1974, Hastings 1977, Shigesada et al. 1979, Atkinson and Shorrocks 1981). These models reveal that fundamental properties of multispecies systems can be altered whenever variation in space is a factor in species interactions, that is, when dynamics are ''spatially distributed'' (see Levin 1978 for a review of this theory). For example, Levin (1974) has shown that subdivision of a habitat into patches and interpatch migration may permit coexistence among competing species that would otherwise exclude each other. Additional models demonstrate that predator-prey or host-parasite interactions can be stabilized if the patterns of enemy attack are sufficiently aggregated (Hassell and May 1985; but see Murdoch et al. 1984 for counterexamples).

Because dispersal determines the coupling between patches, the degree of ''mixing'' in the environment, and the rates at which predators are capable of aggregating at prey patches, dispersal is always a key parameter (though sometimes only implicitly) in models of patchy systems. While I focus on the consequences of dispersal for interactions in patchy *heterogeneous* environments, Yodzis (Chapter 29) shows how the simple addition of dispersal in continuous *homogeneous* space profoundly changes the conditions for coexistence among competitors. In either a homogeneous or heterogeneous environment, theory provides a compelling case for the importance of dispersal in spatially distributed species interactions. When organisms move about and interact in space, entirely new possibilities emerge for species interactions, possibilities that cannot be represented adequately by traditional Lotka-Volterra theory.

Although mathematical ecologists were not the first to appreciate spatial heterogeneity (see the prescient paper by Elton in 1949), ''patchiness'' became a common theme in community ecology only after modelers had formalized its significance. In spite of the fact that concepts from patch theory are now regularly used to interpret data on ecological communities, experimental tests of patch models (e.g., Paine and Levin 1981; Hanski and Ranta 1983) are still few. Since the theory of spatially distributed species interactions has such great explanatory potential, we clearly need experiments to scrutinize its applicability to natural communities.

The goal of this chapter is to motivate such experiments and to indicate the sorts of insights that might be gained by studies examining the community-level consequences of patchiness and dispersal. I will focus on interactions involving herbivorous insects, especially monophagous insect species. In my own research I have chosen to test patch theory with these plant-insect systems because they provide well-defined and manipulatable patches (see Kareiva 1984, 1985), because rates of insect movement between patches can be unambiguously measured (e.g., Kareiva 1982a), and because several lines of evidence pinpoint patchiness and dispersal as important features of insect population biology (Kareiva 1981). Insect populations often have been described as mosaics of extinctions and recolonizations, with dispersal the source of recolonists (Andrewartha and Birch 1954, Ehrlich 1983). For herbivorous insects in particular, there is overwhelming evidence that the spatial arrangement and patchiness of food plants affect the numbers of insects per plant (Cromartie 1981, Stanton 1983), with some suggestion that dispersal underlies these effects (Kareiva 1983a). Finally, dispersal seems to provide insects with a key mechanism by which they can respond to the spatial and temporal unpredictabilities in their lives (Southwood 1977). Taken together, these biological observations indicate that patchiness is influential in plant-insect associations.

This chapter begins by considering factors that cause herbivorous insects to be patchily distributed: patchy distribution of host plants, spatial variation in plant quality, and temporal fluctuations due to weather or effects of other species. I then review biological consequences of patchiness and describe field experimental studies of how patchiness affects the impact of predatory ladybugs on their aphid prey. Finally, I discuss how patchiness complicates studies of competition and predation involving herbivorous insects.

CAUSES AND CONSEQUENCES OF PATCHINESS IN PLANT-INSECT SYSTEMS

Causes of Patchiness

Herbivorous insects frequently make their living on archipelagos of food plants surrounded by a sea of unpalatable vegetation. In these situations the dispersion of food plants represents the most obvious patchiness with which insects must deal. A further templet of heterogeneity is impressed onto patches of conspecific food plants by nutritional quality, which usually varies from one plant to the next (Jones 1983, Alstad and Edmunds 1983) and even from one leaf to the next on the same plant (Whitham 1983). Such intraspecific variation in the quality of plants as food for herbivores is caused by variations in soil nitrogen (McNeill and Southwood 1978), by environmental stresses (Lewis 1984), by genetic differences among plants (Moran 1981, Wainhouse and Howell 1983, Service 1984), or by changes in foliage that are induced by herbivore attack (Haukioja and Niemala 1979, Carroll and Hoffman 1980, Schultz and Baldwin 1982; but see Fowler and Lawton 1984 for a contrary view). The enormous patchiness characteristic of herbivore populations is not surprising in light of the variation in these ''confusingly large number of host plant influences'' (Strong, Lawton, and Southwood 1984, p. 188).

Temporal fluctuations in weather often vary out of phase between locations and thus create spatial patchiness in plant-insect systems. For example, the thermal environment, which potentially limits opportunities for reproduction in butterflies, varies on temporal scales ranging from hours to years, and does so heterogeneously in space (Kingsolver 1983a, 1983b; Cappuccino and Kareiva 1985). Dempster (1983) has argued that fluctuations in weather are responsible for most intergeneration variation in lepidopteran natality. Less directly, weather can disrupt the timing of insect development relative to food plant phenology so that herbivores face an absolute absence of food, or a food whose quality has deteriorated due to factors such as plant senescence or the accumulation of antiherbivore defenses (Feeny 1970, Dixon 1976, Cappuccino and Kareiva 1985). Such phenological disruptions have been implicated as the primary causes underlying local extinctions of butterflies (see Ehrlich 1983).

Besides weather, a further cause of fluctuations in herbivore populations may be involvement in coupled-oscillator systems, such as host-disease interactions (Anderson and May 1980), predator-prey interactions (Hassell 1978), or plant-herbivore interactions (Crawley 1983). Herbivorous insects that participate in such interactions and possess high potential rates of reproduction and nonoverlapping generations are especially likely candidates for oscillatory dynamics (May 1974a).

Although there are thus many reasons why insect populations should fluctuate, it is rarely possible to attribute the oscillatory behavior of any particular population unambiguously to a specific cause. Regardless of cause, however, temporal fluctuations in the abundance of phytophagous insects seems to be the rule (see Chapters 13 and 14 for a discussion of fluctuations in general). Overall, insect-plant systems are thus best represented as checkerboards of subpopulations fluctuating (sometimes to extinction) through time. This is essentially the mosaic view of population dynamics that Andrewartha and Birch (1954) and den Boer (1970) advocated as a challenge to the simple density-dependence of logistic equations.

Some Consequences of Patchy Dispersion of Food Plants

Related Effects of Plant, Density, Search Prowess, Weather, and Predation It is well known that volatile plant compounds serve as powerful attractants for herbivorous insects (Finch 1980). An appreciation of the existence of such highly refined chemical interactions often prompts the conclusion that specialized herbivores are well-equipped to find their patchily distributed food plants rapidly and efficiently. The contrary situation, in which specialized herbivores experience difficulty locating seemingly abundant food plants, is, however, surprisingly common. Although their survival depends on finding a suitable host, many herbivores bumblingly search in a haphazard fashion and must literally bump into a host plant before they recognize it as food (Dethier 1959, Messina, 1982, Cappuccino and Kareiva 1985).

The ineffectiveness of searching herbivores means that we cannot trust our casual impressions of food availability. For example, in a field of apparently abundant food (2.4 food plants/m^2), as many as 80% of the resident *Melitaea harrissi* (Lepidoptera: Nymphalidae) died before they could locate a host plant (Dethier 1959)! These problems of host finding are important to understanding how herbivores can be food-limited despite a surplus of uneaten foliage (Crawley 1983). Those herbivores that are inept at finding food plants may consequently remain so scarce that they have no noticeable depleting effect on their food. Nonetheless, such species could be considered food-limited in the sense that their populations would probably increase if there were more food plants, thus making it easier for individual herbivores to locate a food plant.

Beyond the direct effects of food limitation, patchy resources and difficulties of host-finding create a scenario in which weather and predation also play complicating roles in plant-insect dynamics. Even inept herbivores in search of patchy and scarce plants could be successful if they had enough time. Because of poor weather, however, herbivorous insects often have only limited opportunities for searching (Cappuccino and Kareiva 1985). In such situations herbivore populations might seem to be governed by weather, since weather limits the availability of these searching opportunities. Yet it is the rarity of food plants that sets the stage for weather to have this influence: if there were a homogeneous mat of food plants, herbivores would be guaranteed success in their search, and time and the availability of suitable weather would probably be of little consequence.

Predation can also become a key agent in herbivore population dynamics as a result of the problems that herbivores have in finding food plants. Instead of starving, herbivorous insects frequently become victims of predators, a risk that seems to be greater during the process of active searching for food (e.g., Dethier 1959). In such systems the toll of predation on herbivores depends on the difficulties of host-finding and thus ultimately on host plant density.

One conclusion from this discussion is that it is dangerous to attribute the limitation of herbivorous insects to any single, simple factor. Because of their limited searching prowess, the population dynamics of herbivorous insects are subject to the intertangled influences of plant density (Crook et al. 1979, Kemp and Simmons 1979), weather (Hayes 1981), and predation (Clark and Holling 1979).

Differential Responses of Herbivores to Contrasting Patterns of Plant Patchiness Numerous experiments document the varied effects of plant spacing on different herbivore species (Kareiva 1983a; Stanton 1983; Strong, Lawton, and Southwood 1984). For example, some herbivore species, like flea beetles on collards (Cromartie 1975), attain their greatest numbers per plant where their hosts grow in large, dense, monospecific patches. Other species, like cabbage worms on collards (Root and Kareiva 1984), attain their highest numbers per plant where their hosts are scattered widely in small isolated patches. Hence different herbivores may prevail on the same food plant species depending on the plant's spatial patterning (Cromartie 1975). Since complex spatial mosaics are characteristic of plant populations, herbivore coexistence may thus depend partly on variation in plant dispersion (Kareiva 1983a).

Recent models by Cain (1985) indicate that randomly searching herbivores should have greater difficulty finding food plants that are clumped than those that are evenly distributed in space. The escape predicted for plants as a result of clumped dispersion does not rest upon overlapping zones of detectability (regions within which herbivores can detect the presence of the plant) and consequently smaller total target areas. Instead, the effect of clumping in Cain's models is to limit the success of herbivore searching to a degree that depends on the mobility of herbivores and the total amount of time they are allowed to search. Cain predicts that as herbivore mobility increases, the negative effect of plant clumping on herbivore searching success will decline asymptotically to zero. Field experiments using cabbage worms and varying dispersions of collards have verified Cain's predictions (Cain et al. 1985). Early instars of experimentally released caterpillars found collards growing in clumps less frequently than uniformly spaced collards, but clumping had no effect on searching success in the more mobile late instars (see Fig. 11.1).

These models and experiments suggest that plants may generally gain some escape from herbivores by growing in aggregations, the degree of that escape decreasing with the mobility of the searching herbivores. In addition, Cain's models provide one concrete mechanism by which mobility can dictate the species-specific relationships between herbivores and plant dispersion that were discussed above.

Some Consequences of Spatial Heterogeneity in Plant Quality

Features of plants such as nitrogen content, secondary compounds, and leaf toughness which

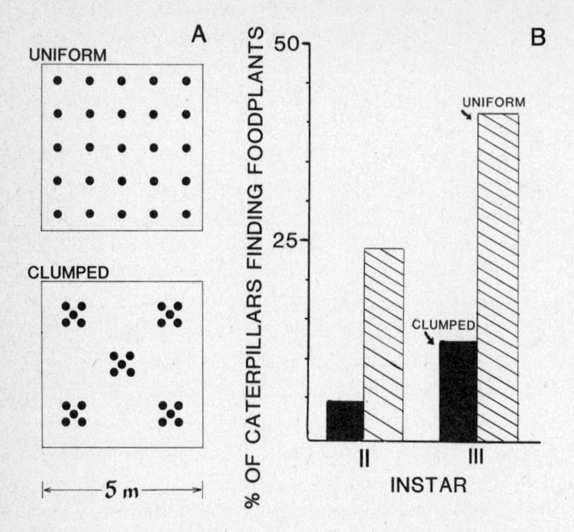

Fig. 11.1 Differences in larval *Pieris rapae* searching success (percent of caterpillars that located a host plant within 60 min) on clumped vs. uniformly dispersed collards. Both experimental plots were 5 × 5 m with collard densities of 1 plant/m^2, but with different spatial arrangements, as shown at the left in A. Differences were significant for instar II and III larvae (B), but not for the more mobile instars IV and V (not shown).

collectively might be referred to as plant quality, affect the feeding behavior, growth rate, and reproduction of herbivores. Conversely, herbivore attack often alters these same plant characteristics. Although herbivores rarely kill their food plants, they may nonetheless compete with one another by reducing plant quality. In such instances a "good" food plant to one herbivore species (because it is rich in nitrogen or low in defensive compounds) will probably represent a "good" plant to other herbivores. Thus, instead of the negative correlations between spatial distributions of species that are often sought as evidence of competition (Diamond 1979), competing herbivores may show significant *positive* associations; Toft (Chapter 27) notes the same phenomenon for parasite communities. For example, the two phytophagous beetles *Phyllotreta cruciferae* and *Phyllotreta striolata,* which feed primarily on *Brassica* plants, consistently are positively correlated in their plant-to-plant distributions, yet sometimes compete as has been shown by removal experiments (Kareiva 1982b). These results caution against using spatial distributions to study competition among herbivorous insects.

Plant quality can, of course, vary for reasons other than patterns of herbivore attack. Of particular interest are the numerous examples of genetic differences among plants with respect to their suitability as food (Moran 1981, Wainhouse and Howell 1983, Service 1984). This genetic heterogeneity in plant quality raises the possibility that insects might evolve into distinct subpopulations adapted to different plant genotypes. In such cases it would be misleading to study plant-insect systems without reference to the mosaic of genotype-genotype interactions. Although it is clear that herbivores can respond evolutionarily to features of their food plants (Gould 1983), only a handful of studies document significant interactions between insect genotypes and plant genotypes within natural populations (e.g., Via 1984, Service 1984). Nonetheless, it is becoming increasingly evident that students of plant-insect associations cannot automatically ignore the consequences of genotype-genotype interactions (Antonovics 1976), which could profoundly influence the magnitude of population fluctuations or the stability of species interactions (e.g., Levin and Udovic 1977, Roughgarden 1979).

Spatial variation in plant quality, whatever its underlying causes, places special importance on patterns and rates of insect movement. For example, dispersal is the process by which herbivores select plants of higher quality, and dispersal rates constrain the effectiveness with which herbivore populations can concentrate on the better food plants (Kareiva 1982a, Parker 1984). If herbivores differ in their mobilities or if habitats differ in the dispersal that they allow, these differences will influence the responsiveness of herbivores to plant quality or to the effects of competitors on plant quality (Kareiva 1982b).

Dispersal will also be key to the interplay between genetic variation in plant quality and the evolution of specialized herbivore genotypes. Herbivores can differentiate genetically with respect to particular plant genotypes only if dispersal is such that gene flow does not swamp the action of selection (Mitter and Futuyma 1983).

Some Consequences of Fluctuations in Weather and Temporal Heterogeneities

By influencing survival (Ehrlich et al. 1972), natality (Hayes 1981), activity (Kingsolver 1983a, 1983b), phenology (Cappuccino and Kareiva 1985), and plant quality (White 1969), fluctuations in weather often produce dramatic fluctuations in the abundances of insects. These fluctuations can create an impression of disorder in insect assemblages, especially if species oscillate independently of one another. For example, over the course of a 3-year study of the insects associated with *Solidago canadensis* (a perennial goldenrod), a different species has been the dominant herbivore each year, and each species that has been a dominant herbivore has also been virtually absent from the community during at least one other year (personal observation). In other herbivore communities, such as the phytophagous insects associated with bracken (*Pteridium aquilinum*) in England, there is remarkably constant structure despite fluctuations in abundance of individual species (Lawton 1983). It would be profitable to try to understand why some herbivore communities exhibit relatively stable structures, whereas other communities vary enormously from year to year.

Some herbivorous insects are characterized by infrequent outbreaks, the causes of which remain a matter of debate (e.g., the gypsy moth and the spruce budworm). These outbreak dynamics can provide the basis for dramatic and surprising species interactions. In the northeastern United States I observed such an interaction while studying the butterfly *Pieris virginiensis,* whose caterpillars feed primarily on the woodland herb *Dentaria diphylla*. Because *Dentaria* is rarely defoliated or even attacked by other herbivores, *P. virginiensis* would seem to be immune from competition (although several different insects are specialized to feed on *Dentaria*). In 1981, however, vast stands of *Dentaria* were consumed, leaving a large percentage of *P. virginiensis* caterpillars to starve to death (personal observation). The insect that consumed the *Dentaria* was not part of the plant's normal fauna—the culprit was the gypsy moth, which after stripping the forest canopy dropped to the floor and ate any herbs it encountered. However rare such gypsy moth outbreaks are, their drastic consequences imply that the gypsy moth could be *Pieris virginiensis*'s most important competitor. Herbivores that go through boom-and-bust population cycles may, during their boom phase, unexpectedly reduce the populations of a wide variety of phytophagous insects (Strong, Lawton, and Southwood 1984).

So far I have discussed how fluctuations, both in weather and in populations themselves, can create what appear to be unpredictable population dynamics. Another aspect of these fluctuations is that demographic rates (birth and death schedules) vary markedly through time. This means that the timing of events such as emergence or diapause in a herbivore's life cycle will typically determine that herbivore's relative success. Phenology may thus be a key feature in the adaptation, specialization, and niche partitioning of herbivorous insects. Whereas the importance of synchrony between plants and their herbivores is commonly cited (e.g., Crawley 1983, pp. 338–339), the importance of synchrony between herbivorous insects and their predators is less appreciated. But the same idea applies: Phenological synchrony can govern the effectiveness of predators at harvesting their herbivore prey. For example, the predaceous stinkbug *Perillus circumcinctus* specializes on chrysomelids in the genus *Trirhabda* primarily by timing its oviposition such that its nymphs mature during the brief period when *Trirhabda* larvae are most easily captured (Evans 1982a).

When the foraging success of predaceous insects depends in this manner on developmental timing, fluctuations in weather may interfere with predator control of herbivore populations. This idea was first promulgated by Connell (1970), who suggested not only that predators are often more vulnerable to weather fluctuations than are their prey, but also that short periods of unfavorable weather may enable prey to escape predation by growing to large sizes while their predators remain relatively inactive. Several authors (e.g., Lord and MacPhee 1953, Clausen 1958) provide examples of entomophagous insects that experience proportionately greater mortality during severe weather conditions than their prey. Recently, Evans (1982b) showed that unusually cool spring temperatures suppressed the activity of predatory stinkbugs, but not that of their prey, tent caterpillars. As a result of a cold snap in 1977, tent caterpillars grew to large sizes while stinkbugs were inactive, and effective predation by stinkbugs was considerably reduced. If it proves generally true that weather deters predaceous insects more than herbivorous insects, this factor could be responsible for geographical variation in the impact of natural enemies on insect communities (Connell 1970).

In short, numerous significant consequences of patchiness for pairwise interactions involving insect herbivores (plant-herbivore, herbivore-predator) are now established. In the next section I consider all three trophic levels at once and examine how the effects of plant patchiness are passed on to influence the coupling between herbivores and their predators.

HOW PLANT PATCHINESS AFFECTS THE IMPACT OF PREDATORS ON HERBIVORE POPULATIONS

Laboratory Experiments and Theory

Working with a predatory mite and a prey mite that feeds on oranges, Huffaker (1958) manipulated spatial heterogeneity by varying the numbers of oranges, rubber balls, vaseline barriers, and wooden poles in a laboratory microcosm. Whereas in simple systems (that is, systems composed of only a few oranges) both mite species usually went extinct, Huffaker attained sustained prey-predator oscillations when he increased the spatial complexity by using 120 oranges intermixed with vaseline barriers and poles. Apparently, the inclusion of sufficient spatial heterogeneity stabilized the predator-prey interaction so that coexistence became possible (Hassell 1978).

Huffaker's classic experiments have inspired numerous models of spatially distributed dynamics in which predator and prey interact and disperse among patches (Maynard Smith 1974, Hastings 1977, McMortrie 1978). These general models all suggest that spatial patchiness can contribute stability to predator-prey interactions. Moreover, in a great variety of difference equation models that are explicitly tailored to insects, the stability of predator-prey (or host-parasite) systems is increased when distributions of attacks become more aggregated (see excellent review by Hassell and May 1985). This effect occurs because clumped distributions of attacks allow some prey patches partially to escape attack. To further examine these ideas about patchiness and the stability of predator-prey interactions, I have been manipulating plant patchiness and following the consequences of these manipulations for the dynamics of an interaction between predatory coccinellid beetles (ladybugs) and the herbivorous aphids on which they prey. The experiments I describe in the following section are, in a sense, Huffaker's laboratory experiments moved to the field.

Test System and Experimental Design

Goldenrods (*Solidago*) are a remarkably successful plant group, with over 100 species native to North America (Fernald 1950). Some species can form essentially monospecific stands that dominate fields for several years. Goldenrods harbor a diverse fauna of relatively specialized herbivores that range in diet from those favoring a single *Solidago* species to those accepting numerous representatives of the genus and perhaps related genera (Messina 1978, 1982; Messina and Root 1980).

My experiments in goldenrod communities have taken place at Brown University's field station in Bristol, RI, where *Solidago canadensis* is the dominant species in open field habitats. To remove vegetation diversity as a complicating factor, I have restricted my manipulations to selected portions of fields in which *S. canadensis* is effectively a monoculture. Although four species of herbivores and two predators are common in my experimental fields, only the aphid *Uroleucon nigrotuberculatus* and its coccinellid predator *Coccinella septempunctata* seem to be strongly coupled (Kareiva 1984). Here I will focus on the predator-prey interaction repre-

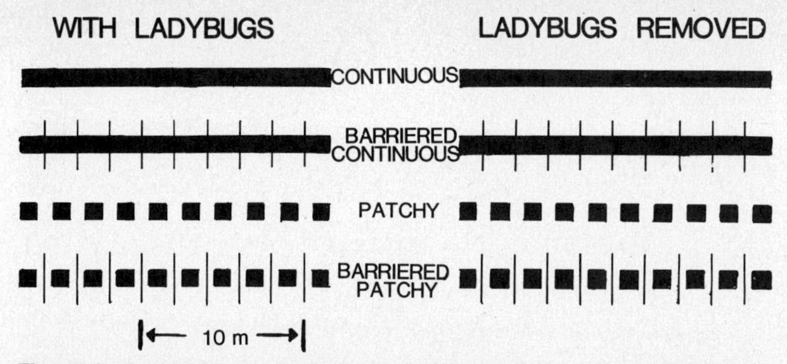

Fig. 11.2 Arrangements of patches and dispersal barriers used in goldenrod experiments. Actual linear strips or arrays of patches often curved slightly to conform to preexisting stands of goldenrod monoculture. Arrays within fields were separated by a minimum of 3 m. Although there was some movement by insects between arrays or strips, most movement was along the linear axis of each array or strip.

sented by these two species. *U. nigrotuberculatus* specializes on *S. canadensis;* it is almost never found on plants other than goldenrods and ventures only infrequently onto goldenrods other than *S. canadensis*. Although *Coccinella* is a generalist predator, *Uroleucon* comprises about 90% of its diet at the Brown field station. Details on the natural history of these species can be found in Kareiva (1984).

To manipulate patchiness, I mowed goldenrod fields into pairs or triplets of linear arrays. The arrays were a 20 × 1 m continuous strip, a 20-m row of 1-m^2 patches, or one of the preceding treatments plus intervening curtain barriers (see Fig. 11.2). These manipulations alter the rates at which aphids and ladybugs move along the arrays and thus affect the mixing of populations within each array (Kareiva 1984). For some experiments I also added a ladybug removal treatment.

All subsequent comparisons or graphical analyses concern only data from within the same goldenrod field. The basic data were obtained by repeatedly censusing 8 to 10 m^2 of goldenrod vegetation in each array of goldenrod. These square meters were either patches (if a row of patches was being sampled) or permanently staked quadrats (if a continuous strip of goldenrod was being sampled). In each square meter I regularly counted all aphids and ladybugs (recording life stages as well) on 10 randomly selected stems in 1982 and on 20 randomly selected stems in 1983. These 1-m^2 sampling units contained an average of more than 50 goldenrod stems. In patches of this size I have observed as many as 50,000 aphids and 30 ladybugs persisting for weeks. Because 1 m^2 of goldenrod can maintain such substantial populations of ladybugs and aphids, I will discuss my data in terms of within-patch (or within 1 m^2) population interactions, coupled by between-patch dispersal.

Results of Manipulating Goldenrod Patchiness

Patchiness had an enormous effect on aphid densities (Figs. 11.3, 11.5). For instance, in the 1982 experiment *Uroleucon* was 10 times more abundant in arrays of patches with barriers than in continuous strips of goldenrod (Fig. 11.3). The frequency of stems occupied by *Uroleucon,* however, did not differ between treatments (Kareiva 1984). The density differences evident in Fig. 11.3 reflect localized outbreaks of aphids to thousands per stem. Such outbreaks are common in patchy and barriered treatments, but rare in continuous strips of goldenrod.

That this effect of patchiness involves the predator-prey interaction was demonstrated by three further types of experiments carried out in 1983 (Figs. 11.4, 11.5). First, experimental removal of ladybugs increases aphid abundance, whether the field is continuous, patchy, or barriered (Fig. 11.4). Second, after removal of ladybugs the aphid density is the same in continuous

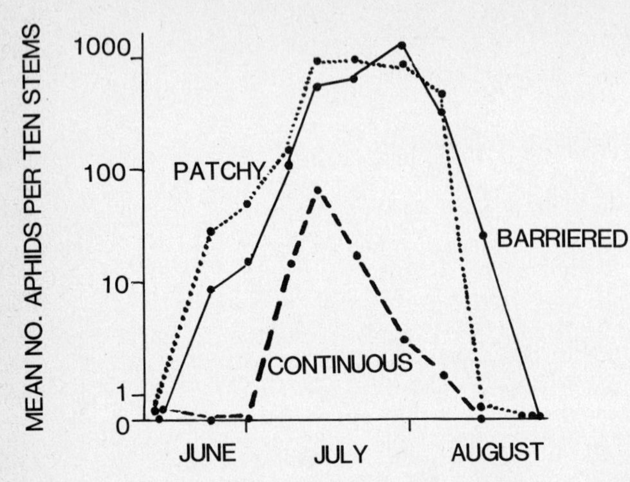

Fig. 11.3 The influence of patchiness manipulations on the density of the aphid *Uroleucon nigrotuberculatus* in 1982. Each point represents a mean from censusing 10 patches. Similar results were obtained in two other fields, although the mean densities differed between fields.

as in patchy fields (Fig. 11.5b), in contrast to the situation in the presence of ladybugs (Fig. 11.5a). Finally, barriers cause aphid densities in the presence of ladybugs to be the same in continuous as in patchy fields (Fig. 11.5c).

Thus, the effect of patchiness is mediated by predation and arises from the effect of patchiness on insect movement (which is equivalent in patchy and barriered arrays) rather than from any unforeseen experimental artifact. Taken together, Figs. 11.4 and 11.5 indicate that with increasing patchiness of vegetation the ladybug-aphid interaction changes so that aphid populations increase, occasionally to outbreak densities. Importantly, because *Uroleucon* does not switch host plants (it overwinters in goldenrod fields), the patterns that I observe all develop from within goldenrod fields.

Why Aphids Escape Predation More Successfully When Their Food Plants are Patchy

A mechanistic understanding of *Coccinella* foraging behavior and of *Uroleucon* demography suggests why patchiness frees *Uroleucon* from predation. One key phenomenon is that per capita reproduction in *Uroleucon* increases as the numbers of *Uroleucon* feeding together on goldenrod stems increase (Fig. 11.6). This represents a form of positive density-dependence (or autocatalysis) that may be common among aphids (Kareiva 1984). Since *Coccinella* forages nonrandomly, it tends to aggregate at patches of high *Uroleucon* density (Kareiva 1984). As long as *Coccinella* can aggregate at incipient aphid outbreaks, it may prevent aphids from reaching densities at which the autocatalytic effect overrides mortality due to predation. Patchiness is important because it interferes with ladybug movement, reducing dispersal rates to one half of what they are in continuous goldenrod strips (Kareiva 1984). This in turn decreases the effectiveness of *Coccinella's* aggregating behavior and makes it more likely that aphid colonies can escape ladybugs for prolonged periods. By flagging and following the fates of numerous aphid colonies, I observed that colonies in patchy goldenrod were attacked less frequently by *Coccinella* and attained larger maximum sizes (Fig. 11.7). It remains to be seen whether the preceding verbal description can be mimicked by a formal mathematical model representing the interplay of dispersal, aphid autocatalysis, and ladybug aggregation.

Whereas theory consistently suggests that patchiness stabilizes predator-prey interactions, I found that goldenrod patchiness increases the likelihood of aphid outbreaks. This "contradiction" emphasizes that the existing theory is largely phenomenological and that the actual effects of patchiness will depend on the mechanistic details of demography and foraging behavior. Difference equation models (see review by Hassell and May 1985) do not really contain explicit representations of habitat patches or movement

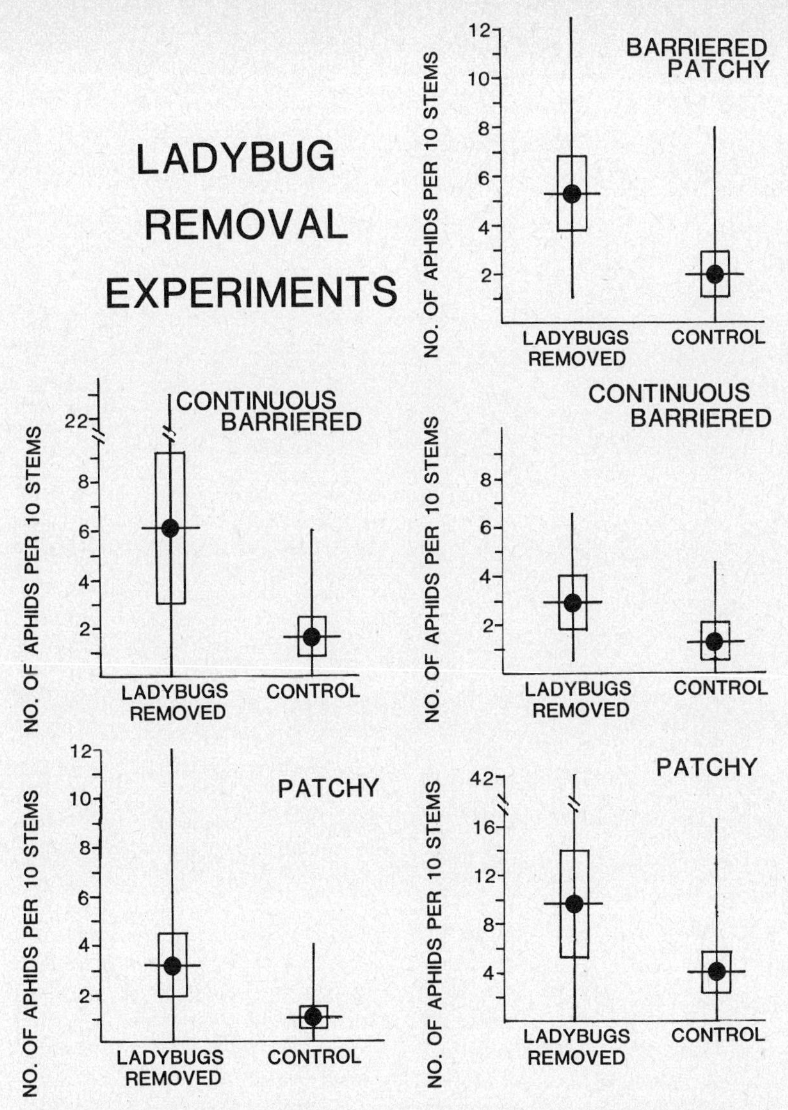

Fig. 11.4 The increase in *Uroleucon nigrotuberculatus* density with the removal of ladybugs in five pairs (ladybug removal and control) of goldenrod strips: two patchy, two continuous barriered, and one barriered patchy. The data shown are the mean, standard deviation, and range of the peak density of aphids during the summer of 1983 in each patch or permanently staked quadrat.

between patches. However, the prediction of these models about the consequences of aggregated attacks is borne out by my results; in particular, the more stable situation (continuous goldenrod strips) is the system in which ladybugs can aggregate most effectively. More general patch models, such as those analyzed by Hastings (1977) or Maynard Smith (1974), explicitly include movement, but do not provide for aggregating predators. We need a more sophisticated theory to describe the interplay of patchiness and nonrandom foraging.

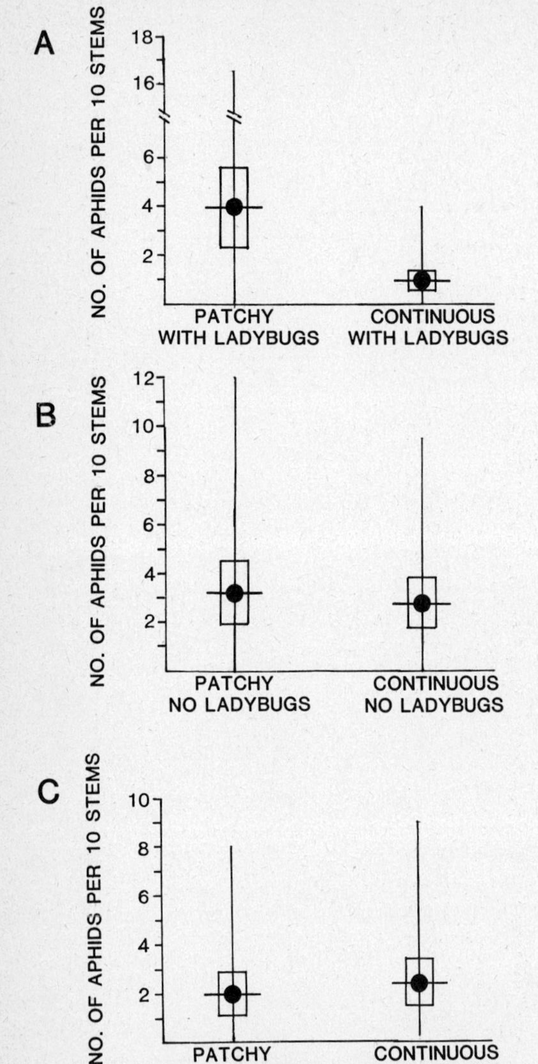

Fig. 11.5 Tests of how the effect of patchiness on aphid densities depends on dispersal barriers and predation. Data are plotted as in Fig. 11.4.

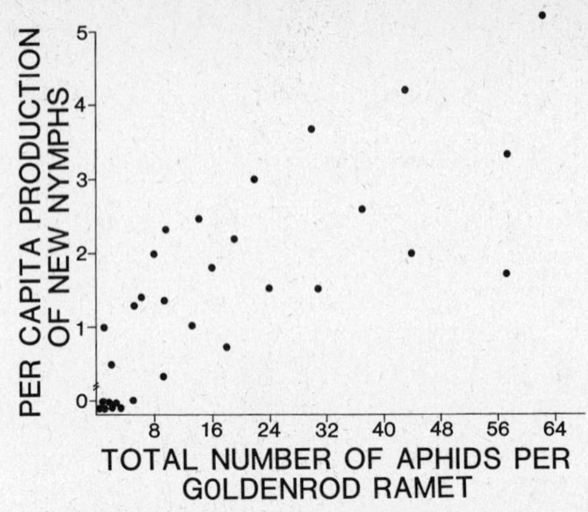

Fig. 11.6 The relationship between per capita reproduction in *Uroleucon nigrotuberculatus* and the number of aphids per stem. Reproductive rate was calculated by dividing the net increase in nymphs observed over three days by the total number of reproductive aphids on the stem at the beginning of the experiment.

CONSEQUENCES OF PATCHINESS FOR ECOLOGICAL THEORY AND THE TESTING OF THAT THEORY

Ecologists studying insect populations have been especially critical of traditional ecological theory, such as logistic equations and Lotka-Volterra equilibrium models of species coexistence (Andrewartha and Birch 1954, Price 1980, Strong 1984a). Many field observations suggest that insect populations rarely reach levels at which competition plays a major role (den Boer 1970). Temporal fluctuations in populations are typically so unpredictable that it is hard to find clear evidence of density-dependent predation (Chapter 15; Dempster 1983). Schoener (Chapter 33) shows that arthropod community patterns are notably less predictable than are vertebrate community patterns. Insects in general, and herbivorous insects in particular, clearly do not fit simple Lotka-Volterra community theory.

One key to understanding the erratic behavior of insect communities is patchiness. Difficulties in making sense of insect communities arise because (1) our theories of species interactions have only just begun to include spatial and temporal heterogeneity and (2) our field approaches and interpretations of field data too often neglect the spatial dimension.

In the next two sections I discuss the problems posed by patchiness for the application of competition and predator-prey theory to assemblages of herbivorous insects. These are problems that limit our understanding of ecological communities in general, not just of plant-insect associations. The consequences of patchiness are, however, more conspicuous for interactions among

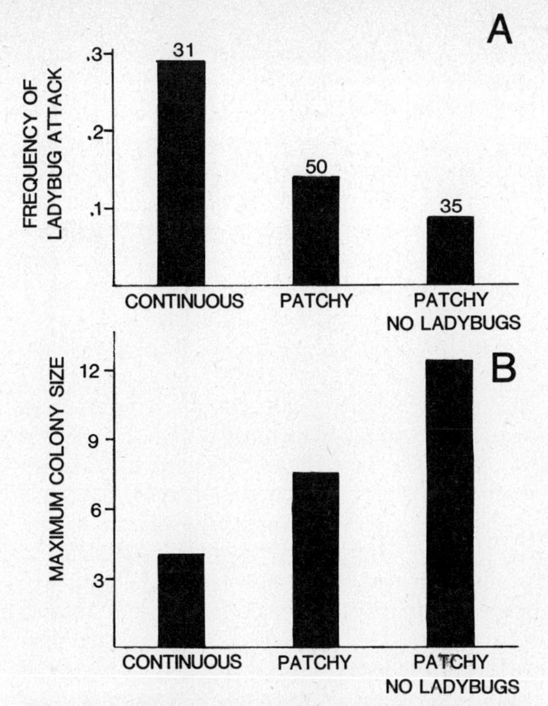

Fig. 11.7 Demographic data on the fate of randomly selected aphid colonies within continuous, patchy, and patchy with no ladybugs (ladybug removal) treatments. The colonies were censused three to five times weekly throughout the summer of 1983 or until the colony went extinct, that is, until all the aphids on a goldenrod ramet disappeared. (A) Proportion of colonies that, at any time during their existence, were discovered being attacked by ladybugs. Differences between treatments are significant. The attack rate is not zero for the ladybug removal treatment because, in fact, it was impossible to remove and keep removed *all* ladybugs. Numbers above bars represent the numbers of aphid colonies that were followed in each treatment. (B) Maximum size attained by colonies in the various treatments. All differences are statistically significant. The numbers of colonies observed were the same as in A.

insect species because the patchiness in these systems occurs at scales that are especially easy to observe (e.g., plants or clumps of plants).

Applying Competition Theory to Patchy Plant-Insect Systems

If herbivores compete, they probably do so through their effects on plant quality. There is little hope of understanding competition in insect herbivore communities without first understanding how herbivore populations depend on plant quality, which varies temporally and spatially. The same problems arise for the dependence of mammalian herbivores on plant quality, as discussed in Chapter 18.

One problem is that herbivory alters plant quality in a variety of nonlinear and devious ways. In some cases damage induces antiherbivore defenses in plants only after a threshold is crossed (as opposed to a graded response). These defenses then persist for varying lengths of time, with a variable time lag between the actual damage and the appearance of the defenses (Rhoades 1985). Time lags and thresholds complicate studies of competition in insects, just as they do in birds and desert granivores (Chapters 3, 9, and 10). For example, plants may change their defensive profiles as much as a year after herbivore damage (e.g., Haukioja and Niemala 1977, Haukioja 1980), and the effects of competition will be correspondingly delayed. In these cases competitor removal experiments lasting less than a year would fail to detect any effects of competition that might involve such induced defenses.

Plant quality also varies independently of herbivory, as a function of plant age, growing season, and physical stress. Thus, plants may move through brief periods of vulnerability during which the possibility of competitive interactions greatly increases (Parker 1985). The key role of plant quality in limiting herbivore success may explain the asymmetry that seems typical of interspecific competition among phytophagous insects (see Strong, Lawton, and Southwood 1984; Crawley 1983). Such asymmetry is expected when the effect of herbivory on plant quality depends on the timing of the damage, with the result that one species disproportionately reduces plant quality because it attacks the plant at a critically vulnerable stage.

Not all herbivory reduces plant quality. Some herbivores feed more efficiently, not less efficiently, as their density per plant increases, for example, aphids (Way and Banks 1967) and caterpillars (Tsubaki and Shiotsu 1982). Such autocatalysis or positive density-dependence in herbivore feeding could create especially complex competitive dynamics, including multiple steady states and rapid transitions between different equilibria. Two herbivores might act mutualistically at low densities and competitively at higher densities. For example, two aphid species could in mixed colonies improve each other's feeding efficiency, yet could compete when population densities grew to levels at which autocatalysis was no longer a factor. The dynamics to be expected from competitive interactions among herbivores are diverse because plants respond to herbivory in such a variety of ways.

The third trophic level of predators or parasites contributes yet another complication to the interplay of herbivory and plant quality, especially when the effectiveness of plant defenses at curtailing herbivore population growth depends on the activities of "natural enemies" (Lawton and McNeill 1979, Price et al. 1980). This phenomenon may be common because a major effect of plant defenses often is to increase the period of time during which a developing herbivore is exposed to predators, parasites, or diseases (by slowing development or by requiring longer daily bouts of feeding). Thus, certain plants will be resistant to herbivores only when natural enemies are present to take their toll (Starks et al. 1972). The possibility of herbivory's reducing plant quality to an extent that would result in competition may also depend on the activities of natural enemies. Without natural enemies, deteriorating food may be of little consequence to herbivore population growth. Thus, whereas we normally think of predation as preventing competition, it could, in some cases, enhance competition among herbivorous insects, just as among freshwater fish (Chapter 21) and sessile-marine invertebrates (Chapter 31). Even without mentioning patchiness or dispersal, it is clear that the intricacies of the feedback loop between plant quality and herbivory are too complex to be captured by Lotka-Volterra competition theory. Considerations of patchiness and dispersal further emphasize the need for a fresh theoretical approach.

Competition among insect herbivores depends in several ways upon plant patchiness and insect movement. As mentioned earlier, limited mobility and searching ineptness often produce small herbivore populations that leave large proportions of food plant resources undiscovered and unused. With so few plants attacked by herbivores, there will be even fewer plants jointly attacked by more than one species of herbivore. Although the infrequency of these joint attacks suggests that the effects of competition will not be dramatic, this does not mean that competition can be ignored. In analyses of competition between herbivorous insects the relevant parameter is not the percentage of plants without herbivores, but rather the percentage of those plants supporting herbivores that are occupied by more than one species. Competition will necessarily be subtle, almost "secretive," if it does occur among those herbivores that are inept searchers.

For all herbivores, whether inept or skilled searchers, mobility constrains their response to spatial variation in plant quality. Hence, mobility will influence the types of competitive dynamics that are possible. When herbivore mobility is unlimited, the effects of competition should immediately be visible as behavioral avoidance of plants damaged by competitors. Alternatively, when mobility is minimal, herbivorous insects may be forced to accept damaged plants and thus suffer reduced survival or reproduction due to the activities of their competitors. When competition occurs between species that occupy ephemeral patches (e.g., patches of food plants), dispersal can allow competitively inferior species to persist in the system (Atkinson and Shorrocks 1981). Finally, stochastic variation in patchy environments may promote coexistence among species (Chapter 14).

Predator-Prey Theory for Patchy Environments

Unlike the situation with competition theory, in which there is a shortage of models suited to insects, several predator-prey and parasite-host models seem to apply well to insect population systems (Hassell 1978, 1985; Hassell and May 1974, 1985). Importantly, predator-prey theory includes models that address environmental patchiness, nonrandom predator search, and predator aggregation. A common and crucial simplification in these models, however, is that spatial patterns and animal movements enter equations phenomenologically rather than mechanistically. Patterns of distribution in space (e.g., of "attacks") are assumed and then analyzed. By collapsing complicated processes such as nonrandom predator search into simple equations, much insight into predator-prey dynamics has been gained (Hassell and May 1985). For example, the central conclusion that clumped population distributions enhance population stability has been repeatedly attained (Hassell and May 1985).

To supplement this relatively successful theory of predator-prey dynamics, we now need models that explicitly involve the spatial dimension and that derive patterns of aggregation and species interactions from assumptions about movement processes. It would be especially useful to be able to translate data on individual foraging movements, such as turning behavior following a prey encounter (i.e., ''area-restricted search,'' Murdie and Hassell 1973), into a macroscopic population term for predator aggregation. By grounding population models on the details of behavior, such an approach may allow us to predict long-term population dynamics from short-term observations of individual foraging and movement episodes. Thus, instead of simply knowing that area-restricted search is stabilizing, we might be able to compute the percent reduction in prey populations that a predator could accomplish, given a particular feeding rate, dispersal rate, and aggregating response. Another goal would be to incorporate optimal foraging theory, which has been remarkably fruitful, into models of predator-prey population dynamics.

Since theory emphasizes the importance of spatially density-dependent mortality in arthropod predator-prey stems (Hassell 1978), field studies of predation need to employ spatially stratified sampling, as Wiens (Chapter 9) also points out for studies of habitat selection in birds. Without appropriate spatial stratification, significant density-dependence in rates of predation or parasitism may well be obscured because data lump together opposing trends (Hassell 1985). For example, using nested quadrat sampling of parasitism rates in the holly leaf miner (*Phytomyza ilicis*), Heads and Lawton (1983) found highly significant density-dependent parasitism at the smallest spatial scale (0.03 m^2), but no density-dependence at all in samples of 1 m^2. When dealing with predators that leave behind no evidence of their activities, even spatially stratified sampling may be inadequate. If predators aggregate quickly at prey patches and then deplete the prey rapidly, the whole density-dependent process may be so transient that it cannot be detected without following through time individual predators and cohorts of prey. For example, by creating experimental patches of aphid prey and releasing marked ladybugs, I found that the *Coccinella* ladybugs aggregate within hours at prey patches in goldenrod fields (Kareiva 1984). Normal sampling typically fails to detect this aggregating response because ladybugs arrive, consume the aphids, and disappear far too rapidly.

Standard entomological approaches for analyzing population dynamics, such as key factor analysis of life table data, are not effective for analyzing spatially heterogeneous and variable predator-prey interactions (Hassell 1985). Improved statistical approaches and perturbation experiments will be required to tease out such regulating factors in natural populations of insects.

SUMMARY

Largely because of the ubiquitous effects of temporal and spatial heterogeneity, dispersal plays an especially important role in the population ecology of herbivorous insects.

1. The mobility and searching prowess of a herbivorous insect determine its sensitivity to food plant dispersion and also determine the likelihood that it will be food-limited in a world lush with foliage.
2. By providing recolonists, dispersal governs each insect population's vulnerability to extinction in a hostile, capricious environment.
3. Dispersal is what permits insects to sample their environment and to select the better food from a pool of plants heterogeneous in quality. When there is a premium on selecting the better food, mobility may well be the major determinant of competitive success in herbivorous insects.
4. In predator-prey systems dispersal affects both the stability of the interaction and the ability of predators to control prey populations at low levels. For example, interfering with the movement of predatory ladybugs interfered with their ability to

aggregate at patches of aphid prey and thereby to control aphid population growth.

5. Dispersal, through its influence on gene flow, constrains opportunities for differentiation among herbivore subpopulations with respect to genotypes of plant defenses that vary in space.

Because dispersal can be manipulated and measured in plant-insect systems, these many influences are experimentally tractable. Perhaps the more formidable challenge is the theoretical one of combining dispersal, not necessarily simple diffusion (see Kareiva 1983b), and patchy plant-insect dynamics in models of species interactions. Whatever its difficulties, theory can no longer neglect the fantastic variability that characterizes insect populations in space and time and the importance of dispersal in moderating this heterogeneity. For plant-insect systems, it matters a lot that the world is not homogeneous.

ACKNOWLEDGMENTS

I am grateful for support provided by the Science and Engineering Council of Great Britain (through Grant GR/C/63595) and by the National Science Foundation (Grant DEB 82-07117), and especially for the hospitality shown by the Centre for Mathematical Biology at the University of Oxford. For comments and discussions, I thank J. Bergelson, P. Bierzychudek, M. Hassell, J. Lawton, S. Levin, R. May, R. Root, D. Strong, and H. Surowiec. R. Paine and J. Diamond showed special courage and patience in reading *two* versions of the manuscript, each time making valuable suggestions. *Coccinella* experiments were supported by a GSRF grant from the University of Washington.

chapter 12

Problems Posed by Sparse and Patchily Distributed Species in Species-Rich Plant Communities

Peter J. Grubb

INTRODUCTION

Most of those writing on the maintenance of species richness in plant communities have ignored the problems posed by sparse and patchily distributed species. Yet, in all communities that are extremely species-rich, most species are likely to be sparse, that is, have "consistently low populations" (Rabinowitz 1981). In addition, there is abundant evidence that most species in species-rich communities are patchily distributed. As soon as these simple points are appreciated, a series of major questions arises.

1. Do the same species remain sparse, year after year? If so, what keeps them sparse? Conversely, what keeps them from going extinct?
2. Do the distributional patches of a species remain fixed in space from year to year, or do they move around?
3. To what extent do sparse or patchily distributed species encounter and affect each other?
4. If sparse or patchily distributed species encounter and affect each other only slightly, does niche differentiation play any significant role in maintaining their coexistence?

When I reviewed in 1977 the many mechanisms that may contribute to the maintenance of species richness in plant communities, I was as guilty of ignoring the implications of sparsity and patchiness as most other writers on the subject. I have set out my revised approach rather briefly in a recent review intended to cover a wide range of subjects (Grubb 1984), and my chief purpose here is to consider critically a larger body of evidence. What I have to say is very much in agreement with the views arrived at independently by Benzing (1981) for the special case of epiphytes.

The first part of this chapter is a brief recapitulation of the kinds of niche differentiation that can be found in plant communities. The second part presents an analysis of some recent research that addresses the series of questions just posed for a moderately species-rich community, chalk grassland. The third and final part is a discussion of the results from chalk grassland first in relation to the ideas expressed by other authors in this volume, and second in terms of their implications for our understanding of the world's most species-rich plant communities.

RECAPITULATION OF TYPES OF NICHE DIFFERENTIATION IN PLANT COMMUNITIES

In my earlier review (Grubb 1977) I wrote, "The niche of a plant is taken here to be the definition of its total relationship with its environment, both physicochemical and biotic; such a definition necessarily includes a statement of the role played by the plant as well as a statement of its tolerance." Four component niches were recognized: the habitat niche, the life form niche, the phenological niche, and the regeneration niche. Each of these is considered briefly here. These niche concepts are also utilized by Hubbell and Foster, Tilman, and Cody (Chapters 19, 22, and 23) in their discussions of plant communities.

Habitat Niche

The extent to which differentiation in the habitat niche is important in making possible coexistence of species in a community probably varies a great deal. For example, variation in microhabitat at the scale of an adult herb is generally greater in a forest than in a grassland, not only because of differential shading effects by the trees, but also because of the relict effects of tip-up mounds, rotting tree trunks, the way in which large roots raise the soil level locally, and the way in which certain trees acidify or alkalinize the soil under them.

Life Form Niche

It is generally appreciated that plants representing different scales of organization may fairly readily coexist, for example, trees and herbs or trees and shrubs. It may also be that similar but more subtle differences in life form among herbs, shrubs, or trees are important in promoting their coexistence. Cody (Chapter 23) argues the case in respect to semidesert shrubs, and the issue arises later in this chapter in respect to grassland herbs.

Phenological Niche

The phenology of a plant is its seasonal pattern of development. For two plant species to coexist as a result of phenological separation, they must capture light, water, and mineral nutrients at different times of year. In the simplest case uptake by one species at one time of year would have no impact on what was available for the other species. In practice, species are only partially complementary in this respect. What matters is that one species of a pair is not able to use enough of the resources available during the growing season of the other to eliminate it. Thus, perennial shrubs in semidesert can markedly inhibit the growth of ephemerals after rain (Friedman et al. 1977), but cannot take up enough resources to eliminate the ephemerals. Similarly, the later-growing species in tall-grass prairie can never eliminate the earlier-growing species, although they overtop them for much of the summer (Curtis 1959). The soil supply of water and nutrients is renewed by the next spring, and the early species are able to grow above the litter of the previous year's later and taller plants and so gain access to light.

There is plenty of evidence for phenological separation in vegetative functioning in forest-floor communities, grasslands, and semideserts. Most of the relevant work has been done on the capture of light and water by herbs and shrubs, but there is emerging evidence in respect to nutrient uptake (Verosoglou and Fitter 1984) and trees (Rogers and Westman 1979). Phenological niche differentiation in plants provides one of the clearest examples of coexistence mediated by temporal variation (Chapter 14).

Regeneration Niche

Despite the considerable scope for niche differentiation in the adult's microhabitat tolerance, life form, and phenology, there are many groups of coexisting species that have widely overlapping habitat tolerances and very similar life forms and phenologies. I argued earlier that, for such groups of species, differentiation in requirements for regeneration is likely to be of paramount importance. The word "niche" has unfortunate connotations in this context, and it seems that some ecologists have misunderstood my concept of "regeneration niche" as a narrowly spatial one, taking it to refer only to the sort of gap to be filled in respect to size or surface features. In fact, as originally defined (Grubb 1977), the regeneration niche explicitly includes the requirements for effective seed set, characteristics of dispersal in space and time, and requirements for germination, establishment, and onward growth that have to do not only with gap shape and size, but also with weather, pests, and diseases. Examples were given to illustrate all these points, and in particular year-to-year variation in seed production and seedling establishment.

Some people have misinterpreted the regeneration niche as having to do chiefly with the position of a species in time during an "internal succession" or "serule" following a disturbance, but I emphasized in my earlier review that this is only one of many ways in which species may be differentiated in their requirements for regeneration. I was more concerned to explain the coexistence of several to many species at any one stage in a "serule" (Grubb 1977, p. 121).

The argument for the importance of the regeneration niche was put forward originally on an intuitive basis, although reference was made to the models of Skellam (1951) and Botkin, Janak, and Wallis (1972) which illustrated the effectiveness of differences in dispersability of seeds and in tolerance of seedlings and saplings for maintaining coexistence. More recently Chesson and Warner (1981) have demonstrated by means of a model the way in which year-to-year variation in recruitment of juveniles can promote coexistence, and this matter is taken up at length by Chesson in Chapter 14. Chesson's models show that variation in recruitment through time can maintain species-richness only if there are overlapping generations. These are provided by vegetative plants in the case of perennials and by seeds in the seed bank of the soil in the case of many annuals and biennials.

In the following section clear examples of differentiation in the regeneration niche appear. However, the important new issue that arises and that is also central to Chapter 19 by Hubbell and Foster is the extent to which differences in niche are necessary at all for maintained coexistence.

RESULTS OF RECENT RESEARCH ON CHALK GRASSLAND DEMONSTRATING THE EFFECTS OF SPARSITY AND PATCHINESS

The research to be described concerns one of the best known species-rich plant communities in northern and central Europe. Grassland on freely draining, infertile calcareous soil has a striking floristic uniformity over a large area, stretching from southern France and southern Germany to the British Isles and southern Scandinavia (Willems 1983). It is developed on soils that supply only small amounts of available nitrogen and phosphorus. Wherever it is actively managed by grazing, cutting, or burning and no fertilizer is added, it may be found to contain 30 to 40 species of vascular plants per square meter, sometimes even more. While most species are perennials, short-lived plants can be conspicuous and include annuals (all hemiparasitic Scrophulariaceae-Rhinanthoideae), biennials (mostly Gentianaceae but also *Linum catharticum*), and pauciennials, that is, species that take two to four years to flower and flower one to three times before dying (mainly Compositae and Leguminosae).

The structure of the community, in its most species-rich form, is a low, rather dense turf 2 to 10 cm tall. If management ceases, it quickly develops into a tussocky sward some 30 to 50 cm high, in which fewer species survive and which shrubs and trees invade. The taller form resembles the meadow steppe of the southern USSR, with which it shares many species. Chalk grassland is a community brought into existence and maintained by human activity. It has replaced mixed deciduous forest, and very probably species have been recruited from a number of natural communities (cf. Grubb 1976). It is a fine-grained community. After seeing 30 to 40 species in a square meter of the short-turf form, one might find on a whole hillside covering several hectares only 50 species that could be said to be found most often (these days) in chalk grassland. Almost certainly many more species would be present on such a slope, especially at disturbed microsites, but these could be assigned primarily to other communities, for example, those of scrub edges, rabbit warrens, localized wet patches, grasslands on more fertile soil, and even arable fields.

The perennials of chalk grassland vary in life form, including small tussocks, rosettes, and creeping subshrubs. Collectively they form a "matrix," in the "interstices" of which the short-lived plants come and go (cf. Gay, Grubb, and Hudson 1982). Most of the interstitial gaps have an area in the range 1 to 10 cm^2. The marine algal communities discussed by Lubchenco in Chapter 32 are surprisingly similar in structure, as emphasized in Tables 28.2 and 28.3, consisting of large perennial and small ephemeral species.

Three relevant studies of chalk grassland are reviewed here. The first study concerns the matrix-forming perennials in the short-turf form of chalk grassland. In this community certain species prove to be sparse year after year and at site after site, raising the following questions: What keeps these species sparse? What enables them to persist at all? Are they too sparse to affect each other? The second study examines populations of interstitial short-lived plants in the same short turf. In this community the rank order of species abundance varies considerably from year to year, and the sites of maximum density for a given species tend to shift from year to year as "drifting clouds of abundance." Here, too, one has to ask whether these species are too sparse for the operation of intra- and interspecific density-dependent effects, and how sparse species manage to persist. The third study turns to longer-lived perennials in the tall-turf form of chalk grassland for evidence about how seed output varies at the ultimate patch level, that defined by the individual plant.

Matrix-forming Perennials in Short Turf

Sequences of Relative Abundance Let us begin our discussion of matrix-forming perennials by examining how species differ in abundance, and how constant these relative abundances are from year to year or from site to site.

Typical dominance-diversity curves, of the type introduced by Whittaker (1965), are shown for the perennials at a site in southern England in Fig. 12.1; they are based on the work of Mitchley (1983). The site is grazed quite heavily by cattle in winter and very lightly by rabbits in summer. The turf is mostly 2 to 10 cm tall in August.

The results for 1981 in Fig. 12.1 show that the total cover increased appreciably during the growing season, but even at the end of the growing season only 17 species had a repetitive cover of more than 4%, while for 16 species the value was less than 4%, that is, there are many sparse species.

It is clear from Fig. 12.1 that species of matrix-forming perennials differ greatly in relative abundance. However, Table 12.1 shows that there is remarkable year-to-year constancy in relative abundance; the sparse species of one year are also the sparse species of the next. Moreover, when sites 150 km apart but with comparable aspect and management are compared, there is a highly significant constancy about the hierarchy ($p < 0.01$ for most comparisons; Grubb, Kelly, and Mitchley 1982). There is also evidence for a similar hierarchy in data collected over a period of 20 years at a wide range of chalk sites in England (Mitchley 1983).

Two questions arise immediately: How do some species become more abundant than others? How is it that the sparse species do not disappear? After providing tentative answers to these questions, we can then consider a third question: To what extent do the sparse species affect each other; that is, do they influence each other's abundance, and are there any constraints on their coexistence?

Experimental Evidence for Interference Competition As all the grasses and sedges involved, and most of the dicotyledons, seem to be long-lived, and as the plants of many species are intimately mixed in the sward, it seemed likely to Mitchley and me that the rather strict hierarchy must be maintained through interference (*sensu* Harper 1961) between adult plants. An experiment was therefore set up on a natural chalk grassland profile, from which the top 5 cm of soil had been removed, to test whether or not a regular hierarchy of interference effects could be found under appropriate growing conditions, and, if so, whether it corresponded with the hierarchy in spontaneous grassland. Six species were planted in all possible combinations, two typically abundant species, two typically less abun-

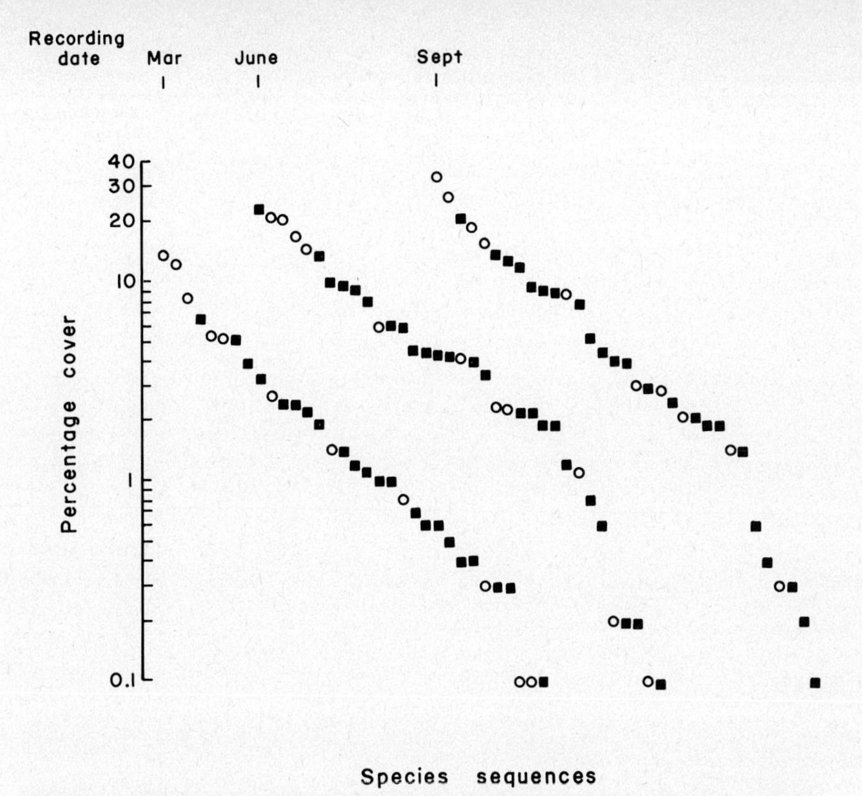

Fig. 12.1 Dominance-diversity curves, based on repetitive cover values, for the perennials in chalk grassland on a south-facing slope at Castle Hill National Nature Reserve in southern England in March, June, and September 1981: monocotyledons (circles); dicotyledons (squares). For the nature of the data base, see Table 12.1. The figure plots percentage cover for each species, arranged from left to right in descending sequence of percentage cover. (From Mitchley 1983.)

dant, and two typically sparse. Well-spaced seedlings were planted out in June, percentage establishment was mostly high, and the species interacted strongly in the following year. A fully transitive sequence of interference effects emerged after 15 months (Grubb 1984, Fig. 3). With one exception this hierarchy agreed well with that found in spontaneous turf, suggesting that interference competition is what produces the abundance hierarchy of Fig. 12.1 and Table 12.1.

In the experiment, which was carried out in an exclosure, and therefore free of grazing by rabbits, sheep, or cattle, the tallest plants grew to a height of about 15 cm, and the interference effects appeared to be a function of the height at which each species held its leaves. This possibility had not seemed at all obvious in the spontaneous turf grazed to a height of 2 to 10 cm, but analysis of relative leaf position in large numbers of point quadrats in spontaneous turf showed a highly significant correlation with abundance ($p < 0.001$; Grubb 1984, Fig. 4). All the sparse plants have their leaves relatively low in the canopy, while most of the abundant species have their leaves relatively high in the canopy. The leaf area index, that is, area of leaf per unit area of ground, is about 3.5 $m^2 \, m^{-2}$ in June, and the mean leaf angle is about 50° from the horizontal. If each leaf layer absorbs about 70% of the incident photosynthetically active radiation, shading effects could be considerable (cf. Kira, Shinozaki, and Hozumi 1969). There is clearly a *prima facie* case for the importance of competi-

Table 12.1 RANK ORDER AND COVER VALUES FOR 29 SPECIES OF MATRIX-FORMING PERENNIALS IN CHALK GRASSLAND AT ONE SITE IN THREE SUCCESSIVE YEARS

		Rank order			Repetitive cover values (%)	
		1980	1981	1982	1981	1982
	Sanguisorba minor	1	1	1	22.7	24.9
G	*Brachypodium pinnatum*	2	3	3	20.1	20.7
G	*Bromus erectus*	3	4	5	16.3	19.4
G	*Festuca ovina/rubra*	4	2	2	20.9	22.1
	Cirsium acaule	5	6	7	13.3	12.1
	Thymus praecox	6	9	9 =	9.1	7.8
	Leontodon hispidus	7	7	6	9.8	12.3
	Asperula cynanchica	8	│ 13	12	5.8	5.9
	Lotus corniculatus	9	10	13	7.9	5.5
G	*Briza media*	10	11 =	8	5.9	8.3
	Hippocrepis comosa	11	8	9 =	9.5	7.8
	Hieracium pilosella	12	14	15 =	4.5	3.4
G	*Koeleria macrantha*	13	│ 18	│ 14	4.1	3.6
	Carex flacca	14 =	│ 5	4	14.2	19.8
	Succisa pratensis	14 =	17	15 =	4.2	3.4
	Centaurea nigra	16 =	15	│ 19	4.4	3.2
	Pimpinella saxifraga	16 =	│ 20	22 =	3.4	2.7
	Filipendula vulgaris	18 =	16	│ 20 =	4.3	2.9
	Phyteuma orbiculare	18 =	19	18	3.9	3.3
	Plantago lanceolata	20	│ 11 =	11	5.9	6.0
G	*Avenula pratensis*	21	21 =	│ 15 =	2.3	3.4
	Polygala vulgaris	22	25 =	26	1.9	1.8
	Ranunculus bulbosus	23 =	25 =	28	1.9	0.7
G	*Danthonia decumbens*	23 =	21 =	22 =	2.3	2.7
	Plantago media	25	23 =	│ 27	2.2	1.7
	Scabiosa columbaria	26	23 =	24	2.2	2.6
	Carex caryophyllea	27	28	│ 20 =	1.1	2.9
	Viola hirta	28	27	25	1.2	1.9
	Prunella vulgaris	29	29	29	0.8	0.4

Vertical bars indicate changes of 4 or more in rank between successive years. G denotes grass species. Equal signs mean that cover values were the same. The table is based on the results of 4,800 point quadrats on each sampling occasion, 240 of them being distributed at random in each of 20 randomly sited stands of 0.48 m^2, each one randomly sited in an area of 16 × 20 m.

In 1980 "nonrepetitive cover" for each species was determined from the percentage of the points at which it occurred, irrespective of the number of leaves hit at any one point. In 1981 and 1982 repetitive cover was determined from the total number of hits on each species (often more than one at a given point) divided by the number of points, and multiplied by 100. Equals signs beside rank order indicate a tie (cover values the same within two significant figures).

Data are based on Mitchley's (1983) study at a site in southern England in June, the same site used for Fig. 12.1.

tion among shoots for light in setting up the hierarchy among perennials in spontaneous turf.

However, Mitchley (1983) also reported an experiment carried out in the Cambridge Botanic Garden using shallow pans of the same kind of soil and in which there appeared to be minimal overlap of shoots. He tested 13 species against 2 species which are normally abundant and found a hierarchy of interference effects which again paralleled the hierarchy in spontaneous turf, with a few notable exceptions ($p < 0.01$). There is thus a *prima facie* case for the role of below-ground interference among the roots, as well as for above-ground interference among the shoots.

Reasons for Persistence of Sparse Species Interference between adults, whether mainly for light or mainly below ground, can explain the fact that there is a hierarchy and that this is rather strictly maintained. How are the sparse species prevented from disappearing?

It seems that there are two answers. The first is grazing, and this is shown clearly by the fact that most of the sparse, lower-growing species are lost when grazing is stopped. There is no evidence that the sparse species are more shade-tolerant and thus comparable with the herb layer in a continuous forest stand. With few exceptions, the sparse species do not, for example, go further into shade than the abundant species at the edges of woods (cf. light-requirement values given by Ellenberg 1982 for the species in Table 12.1). Between episodes of grazing the taller-growing species can suppress the lower-growing species, but do not have time to oust them; during grazing, especially the nonselective grazing of cattle, the taller-growing species lose more "capital."

There is evidence that a second kind of explanation is also important. Among the dicotyledons several of the sparse species have shorter-lived individuals and are more effective at becoming established from seed than the abundant species with longer-lived individuals (Mitchley 1983). The persistence of the species with shorter-lived individuals is probably promoted by small-scale disturbances in the turf, such as hoof prints of grazing animals.

Interactions Among Sparse Species The issue central to the theme of this paper is whether or not the sparse species affect each other materially. Is their abundance determined primarily by their contacts with the abundant species or with other sparse species? Ideally, we want to know how often a sparse species meets another sparse species as opposed to an abundant species.

Mitchley's point quadrat data are not suitable for an analysis of association, since the small size of the sample point tends to lead to a predominance of negative associations as a result of exclusion effects (cf. de Jong et al. 1983). However, it is clear that most sparse species are overtopped to a considerable degree by the more abundant species, and few of the relations between sparse species are other than random. For example, when the species recorded in June 1981 were divided into the 15 more abundant and 14 less abundant, there were only 3 significant interspecific relations between less abundant species in terms of relative leaf position, but 37 significant relations between less abundant and more abundant species (Mitchley 1983). Looking at the problem in another way, the total cover values were such that the chance of a "sparse" species (defined as one with cover less than 3% in June) encountering an "abundant" species (one with cover more than 3% in June) would have been about 10 times greater than its chance of encountering another sparse species, if the distributions of the species had been random.

Very probably the distributions of the species, even in the intimately mixed short turf, are not random. Nevertheless, there seems to be a strong case for thinking that the abundances of the sparse species are determined by their relations with the abundant species rather than with each other. Whether any sparse species is ever in much danger of being ousted by another sparse species is doubtful, although there is the possibility that any one sparse species might be ousted by a substantial number of the other sparse species acting in concert.

Interstitial Short-Lived Plants in Short Turf

We now turn from the matrix-forming perennials of short turf, a community in which we have seen that relative abundances vary little from year to year, to a community with drastically fluctuating relative abundances: the interstitial short-lived plants that share the same habitat.

Inconstancy of Relative Abundances With a reasonably constant management regime, the numbers of flowering individuals of interstitial short-lived plants per unit area are found to fluctuate markedly from year to year, especially in the case of annuals and biennials. These fluctuations have been documented over six years on two transects of 50 × 0.5 m (one east-facing and one south-facing) at the site in southern England already discussed in relation to the matrix-forming perennials. Some of the results obtained after four and five years respectively have been published already (Grubb et al. 1982; Grubb 1984). The results for the east-facing transect after six years are summarized in Fig. 12.2, from which one can conclude that the rank order of abundance varies greatly from year to year. These fluctuations in the relative numbers of short-lived plants contrast strongly with the very constant rank order of the perennials over three years in the same period (Table 12.1).

With the exception of 1981, when the turf was taller following inadequate grazing in winter, the gross structure of the turf has appeared similar in

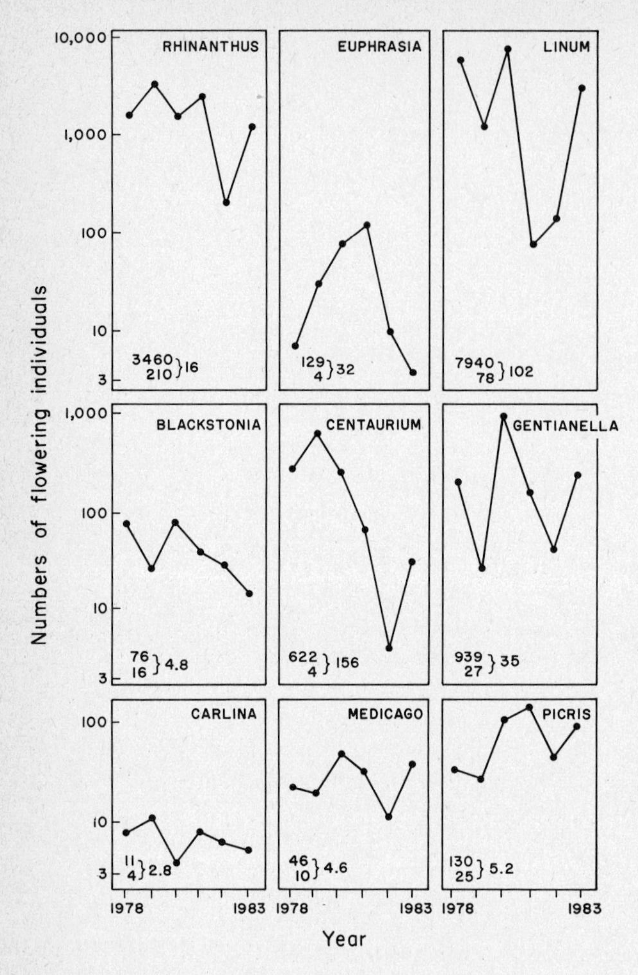

Fig. 12.2 Numbers of flowering individuals of nine species of short-lived plants (logarithmic scale) in six successive years (1978 to 1983) on a transect of 50 × 0.5 m on an east-facing slope of short turf. The transect was within 200 m of the stands used for Fig. 12.1. The highest and lowest numbers recorded, together with their quotient, are shown in the lower left corner of each graph. There are two annuals (*Rhinanthus minor* and *Euphrasia officinalis*), four biennials (*Linum catharticum, Blackstonia perfoliata, Centaurium erythraea,* and *Gentianella amarella*), and three pauciennials (*Carlina vulgaris, Medicago lupulina,* and *Picris hieracioides*). Note that the patterns of fluctuation with time differ among species, so that the rank order of species differs from year to year.

each year. For the more abundant species a rational explanation of at least the general trends in abundance can be given. The different patterns of change through time reflect differences in the regeneration niche in respect to sensitivity to turf height, drought, and grazing by molluscs, and the extent of the seed bank in the soil (Grubb 1984, Tables 1, 2). For the less abundant species, that is, the pauciennials, chance factors are likely to confuse any explanation of changes in abundance at the level of 25 m^2. Indeed, the numbers of flowering plants of *Carlina* on the east-facing transect (50 × 0.5 m) have been found to correlate very poorly with the number in a parallel sample of 50 × 5 m immediately upslope (from 1980 to 1983: 4, 8, 6, and 5 versus 26, 30, 40, and 38).

The results from the two transects on year-to-year fluctuation have generally been in agreement in a qualitative sense, even for the sparser species. However, for some species there has been disagreement in the scale of fluctuation, and for others there has been qualitative disagreement in trend when at low density (cf. Grubb et al. 1982, Table 4).

To summarize, there are marked fluctuations in abundance with time, and many of these fluctuations can be explained, but at low densities there seems to be increasing noise resulting from random effects.

Clouds of Abundance We now have to consider the extent of patchiness in any one year, and the extent to which the areas of greatest density

Table 12.2 SUMMARY OF HOW THE DISTRIBUTIONS OF THREE PAUCIENNIAL SPECIES CHANGED OVER FIVE SEASONS IN QUADRATS OF 0.5 × 0.5 m

	East-facing transect				South-facing transect			
	(a) Occupied in 1979 but not 1983	(b) Occupied in 1983 but not 1979	(c) Occupied in both 1979 and 1983	(a) + (b) / (c)	(a)	(b)	(c)	(a) + (b) / (c)
Carlina vulgaris	15	9	11	2.2	1	13	4	3.5
Medicago lupulina	9	9	22	0.8	10	6	14	1.1
Picris hieracioides	11	22	9	3.7	11	11	5	4.4

Data are based on the east-facing transect of 50 × 0.5 m used for Fig. 12.2 plus a south-facing transect. Numbers of individuals of three pauciennial plant species (all individuals of *Medicago* and *Carlina*, only flowering individuals of *Picris*) were recorded on 0.5-m quadrats. For *Picris hieracioides* the years compared are 1979 and 1982, not 1979 and 1983, because 1979 and 1982 represent years of lesser abundance overall (see text).

Numbers refer to the number of quadrats.

change from year to year. Table 12.2, based on the distributions of three pauciennial species along two transects over six years, summarizes the extent of change on both transects.

Some details relevant to the interpretation of Table 12.2 are as follows. In the case of *Picris* I have deliberately compared two years with relatively low totals (1979 and 1982) to show that the species does not retreat to the same refugial patches each time the numbers dip down. For *Carlina* it is practicable to record individuals of all ages, and the species is known to have a negligible amount of dormant seed in the soil (Schenkeveld and Verkaar 1984); when it is marked absent, it is wholly absent. For *Picris* it is not practicable to plot immature plants because they are indistinguishable from plants of *Leontodon hispidus,* which is also common in the turf, and this difference in treatment undoubtedly contributes to the apparently higher spatial turnover in *Picris* compared with *Carlina*. However, *Medicago* can be mapped at all ages after germination and really does change position less quickly than *Carlina,* probably for at least two reasons. First, the fruits of *Medicago* are held only 1 to 3 cm above the ground (cf. 10 to 15 cm in *Carlina*) and have no obvious means of dispersal (cf. pappus in *Carlina*). Second, most individuals are polycarpic (cf. *Carlina* strictly monocarpic), and so the species has more than one chance of perpetuating itself in a given quadrat of 0.25 m^2. There is evidence of a modest amount of dormant seed for both *Medicago* and *Picris,* and thus these plants are not necessarily wholly absent when marked absent in terms of vegetative and flowering individuals.

Table 12.2 shows that the distributions of *Carlina* and *Picris* tend to shift with time, as "drifting clouds of abundance." For both *Carlina* and *Picris* the number of quadrats occupied both in the years 1979 and 1983 (or 1982) was much less than the number occupied in only one of those two years. For *Medicago,* however, the "clouds of abundance" are relatively fixed in space: as many quadrats were occupied both in 1979 and 1983 as in one of those years alone.

For many of the more abundant species, where we are less concerned with presence or absence than with levels of abundance, there is clear evidence that the "clouds" of greatest abundance move from year to year. Results for *Rhinanthus* have been published already (Grubb 1984). Kelly (1982) analyzed the correlation between the number of plants flowering in a quadrat in one year and the number flowering in previous years. As an example of Kelly's results, consider the contrasting patterns for *Gentianella* and *Blackstonia* (Table 12.3). *Gentianella,* which is confined to particularly short turf, showed a marked constancy in pattern, while *Blackstonia,* which is not confined to particularly short turf, showed much less constancy.

To summarize, some species exhibit constancy in patch distribution related to (1) small differences in turf height, as in *Gentianella,* and (2) limited dispersal of seeds and polycarpy, as in *Medicago*. However, there is clear evidence the "clouds of abundance" of most species (e.g.,

Table 12.3 TEST OF THE MOBILITY OF THE "CLOUDS OF ABUNDANCE" IN TWO SPECIES OF BIENNIAL

	Correlation coefficients					Numbers of plants				
	$n-1$	$n-2$	$n-3$	$n-4$	$n-5$	$n-1$	$n-2$	$n-3$	$n-4$	$n-5$
	***Gentianella amarella* (east-facing transect)**									
1979	0.459***					27/208				
1980	0.252*	0.425***				939/27	939/208			
1981	0.311**	0.084	0.319**			159/939	150/27	159/208		
1982	0.230*	−0.103	0.116	0.206*		42/159	42/939	42/27	42/208	
1983	0.372***	0.241*	0.108	−0.094	−0.036	249/42	249/159	249/939	249/27	249/208
	***Blackstonia perfoliata* (east-facing transect)**									
1979	−0.180					28/76				
1980	0.188	0.215*				76/28	76/76			
1981	−0.048	0.245*	−0.016			40/76	40/28	40/76		
1982	0.101	−0.094	−0.007	0.070		31/40	31/76	31/28	31/76	
1983	0.339**	0.049	−0.043	−0.079	−0.041	16/31	16/40	16/76	16/28	16/76

Columns 2 to 6 give the correlation coefficient between the number of flowering plants per quadrat in one year and the number in the same quadrat 1, 2, 3, 4 or 5 years previously. Columns 7 to 11 give the total numbers of flowering plants involved. For comparisons involving 1979 and 1983 there were data from 100 quadrats of 0.5×0.5 m on each transect. For comparisons involving 1978 the numbers in years 1979 to 1983 in quadrats of 0.5×0.5 m were correlated with half the number in the corresponding 1.0×0.5 m quadrat, which was the unit of recording in 1978. Note that patches of *Gentianella*, but not of *Blackstonia*, tend to occupy the same quadrats in successive years. Analysis by D. Kelly of data collected by Kelly and the author.

Asterisks indicate significance levels: $* = p < 0.05$; $** = p < 0.01$; $*** = p < 0.001$.

Carlina, Picris, Blackstonia) do move about over the slope through time. We have no critical evidence on whether the inconstancy in patch distribution results from purely random effects of dispersal and of slight differences in the height of grazing and in the distributions of predators and pests, or whether once-favorable patches tend systematically to become unfavorable as the result of a buildup of specific predators and pests.

Significance of Density-Dependent Effects
The huge fluctuations in numbers of flowering individuals in four of the species (*Blackstonia, Centaurium, Gentianella,* and *Linum*) and the considerable fluctuations in numbers in three other species (*Euphrasia, Rhinanthus,* and *Picris*), taken in conjunction with the fixity of the apparent structure of the perennial matrix in most years, strongly suggest that most species spend much of the time well below saturating densities. For example, there are probably enough suitable microsites in the turf every summer for adults of the very slender one- or two-stemmed *Linum* to reach the order of 10^4 to 10^5 individuals per 25 m^2, but in fact the numbers are only about 10^3 or even 10^2 in most years. The species normally cannot fill anything like all the physical gaps available to it, because the regenerating stages are just too demanding; the seedlings are too susceptible to turf that is a little too tall, to molluscan predators, and to drought. In addition, seed output is modest (about 25 per plant on average), and the process of dispersal in space is not very effective.

This analysis leads one to pose the following question. To what extent are the numbers of flowering individuals of plants of this type controlled directly by the environment and to what extent by intraspecific or interspecific density-dependent effects? If the single species rarely reach densities such that intraspecific density-dependent effects are marked, what likelihood is there that the combined densities of pairs of species will be great enough for systematic interference effects between species to occur?

In order to investigate density-dependent effects, a detailed study of survival and performance has been made for *Rhinanthus* over five years on a 2-m^2 stretch (4 × 0.5 m) of the east-facing transect used for Fig. 12.2. Except for *Linum* in its years of greatest abundance, *Rhinanthus* is the most abundant of the species of short-lived plants and therefore the one in which we might expect the greatest incidence of density-dependent effects. Because the "clouds of abundance" move through time, any long-term study in a fixed sample area will sometimes be of plants at a higher-than-average density compared with the whole transect (or slope) and sometimes at an average or lower density. In the first two years the mean density of flowering individuals of *Rhinanthus* in the study area of 2 m^2 was about twice that for the transect as a whole (× 1.6 and × 2.1); in later years it declined to × 1.4, × 0.06, and × 1.2 (Fig. 12.3). Almost all seedlings emerge in March and early April. Over five years the number of seedlings present in each of the 200 subquadrats of 10 × 10 cm in the sample plot has been recorded on or about May 1; since the third year individual seedlings have been mapped to improve the accuracy of recording. The numbers of flowering individuals, and the numbers of fruits and flowers borne by them, have been recorded in early August.

The survival of the seedlings is shown in Fig. 12.3. Leaving aside the exceptionally dry summer of 1982 when survival to flowering was only 0.7% (only 117 mm of rain in April–July compared with 213 to 292 mm in the four preceding years), survival has been in the range 13 to 40%. In Fig. 12.4 the survival to flowering is shown in relation to density on or about May 1. Only in 1979 was there a marked density-dependent effect; the seedlings had a high survival in the wet spring, but those in dense patches grew very little and many of them died during the dry summer months (Fig. 12.3). In 1980 the spring was dry; density-independent mortality was high, and the densities during the later part of the plants' growing season were too low for appreciable density-dependent effects to occur. In 1981 the spring was wet, but the turf was exceptionally tall and it seems that this was the cause of the relatively high mortality early on; the plants that did survive produced exceptionally large numbers of fruits (Fig. 12.3). In 1982 severe desiccation obliterated almost all the seedlings. In 1983 the number of seedlings was low on May 1 (almost all from dormant seed), and survival was moderate.

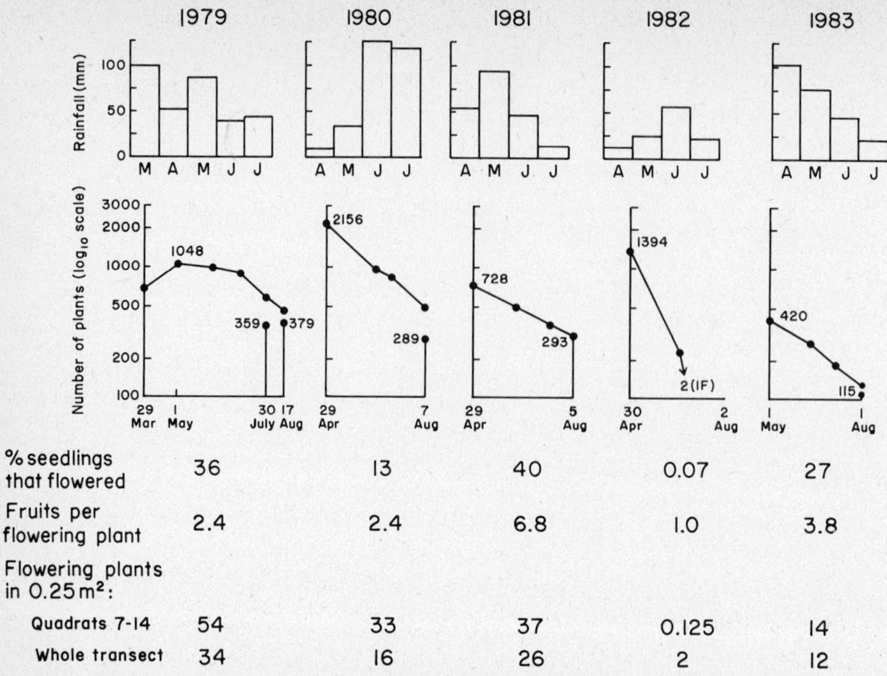

	1979	1980	1981	1982	1983
% seedlings that flowered	36	13	40	0.07	27
Fruits per flowering plant	2.4	2.4	6.8	1.0	3.8
Flowering plants in 0.25 m²:					
Quadrats 7-14	54	33	37	0.125	14
Whole transect	34	16	26	2	12

Fig. 12.3 Fates of seedlings of the hemiparasitic annual *Rhinanthus minor* in an area of 4 × 0.5 m and rainfall in five successive years. The area studied was within the transect used for Fig. 12.2, and the rainfall values are for a comparable site about 18 km to the east. The number above each curve of number of plants is the peak number of seedlings for that year; the one or two numbers and vertical bars at the right are the numbers of flowering plants late in the season. Data below the graphs are the percentage of seedlings flowering on or about August 1, the number of fruits per flowering plant, and the mean density of flowering plants in this small area compared with that on the transect as a whole.

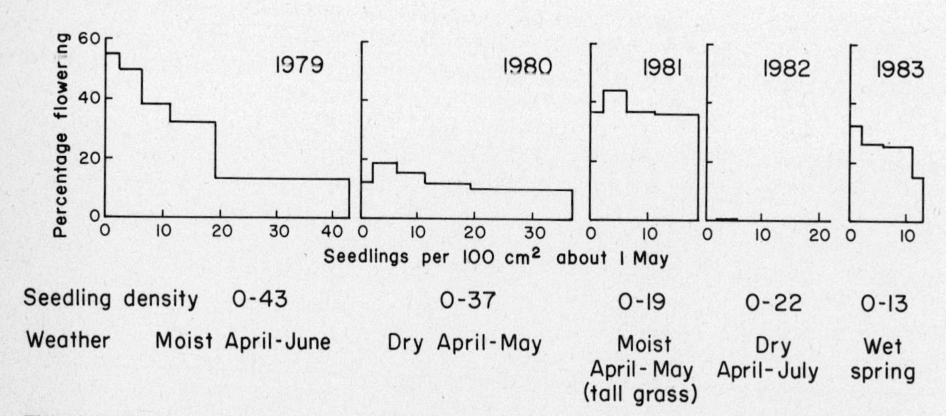

Fig. 12.4 Percentage of seedlings of *Rhinanthus minor* present on or about May 1 that flowered on or about August 1 in relation to the density of seedlings about May 1. Data below the graphs give the range of seedling density and a summary of the weather for each year. Data are based on 200 subquadrats of 10 × 10 cm within the study area of 4 × 0.5 m used for Fig. 12.3. The low percentage flowering value for the highest density in 1983 is based on a single subquadrat of 100 cm^2. Note that density-dependence (a decline in percentage flowering with density) was present only in 1979.

Thus, in only one of five years, 1979, was an appreciable density-dependent effect seen on survival to flowering. In that year there was also an appreciable effect on fruit output (Table 12.4). In other years there was only a slight density-dependent effect on performance or none at all. In 1979 appreciable density-dependent effects occurred with more than 6 seedlings per subquadrat of 100 cm^2 (Fig. 12.4, Table 12.4), and such subquadrats occupied 25% of the studied area. Since this area had about twice the density of flowering plants found on the whole transect, one might estimate that density-dependent effects occurred on about one-eighth of the slope in the one year in five when such effects were important.

Kelly (1982) reported the results of more sophisticated studies on the density-dependence of survival and performance of three species over three years in similar grassland near Cambridge. His results were based on the survival and performance of plants in squares of 5 × 5 cm. He found for *Euphrasia pseudokerneri, Gentianella amarella,* and *Linum catharticum* a few instances of density-dependence, but they occurred over less than 5% of the study area and mostly only one year in three. The one stage in the life cycle where density-dependent effects might occur, and which has not been tested by Kelly or me, is the survival of seeds from the time of liberation to the time of germination.

Using the approach that he had adopted in searching for density-dependent effects within species, Kelly (1982) went on to look for evidence whether a given species was affected by the density of either of the other two species at the site near Cambridge. He found no such evidence at naturally occurring densities over three years. He also added seed (collected from nearby) at the density of 100 per 25 cm^2 and looked for interspecific effects with these unnaturally high densities. Only one of the six possible effects (E → G, G → E, E → L, L → E, G → L, L → G) was significant: *Euphrasia* at high density reduced the performance of *Linum*. Whether this was a parasitic effect or not is unknown.

Thus, even those short-lived plants that are common overall and intermingled are still sufficiently sparse not to affect each other very much. If we turn to the species that are sparse overall, it seems that interspecific effects are probably even rarer. The seedlings are nearly always concentrated near the flowering individuals of the previous year, and as these are often in separate patches, the seedlings of different species are generally nowhere near each other. Clearly, this result is dependent on three properties of the species concerned: modest seed output (about 50 to 100 per flowering plant), poor survival from seed formation to seedling formation, and limited dispersal.

Overview of Sparse Species in Short-Turf Chalk Grassland

Let us now summarize the answers obtained to the three questions that I initially posed about sparse species in chalk grassland.

1. What keeps the sparse species sparse? The main factor is interference competition from abundant matrix-forming perennial species. Certain matrix-forming perennial species remain

Table 12.4 FRUITS PRODUCED PER FLOWERING INDIVIDUAL OF *RHINANTHUS* AS A FUNCTION OF PLANT DENSITY AS SEEDLINGS ON MAY 1 AND AS FLOWERING PLANTS ON AUGUST 3

Number of seedlings per 100 cm^2	Number of flowering plants per 100 cm^2				
	1–2	3–4	5–6	7–10	(Mean)
1–6	2.8	2.7	2.6	—	(2.7)
7–11	1.4	2.6	2.3	2.7	(2.4)
12–43	1.9	1.6	1.6	1.4	(1.8)
(Mean)	(2.6)	(2.3)	(2.2)	(2.0)	

Results are based on 404 flowering plants in 1979. Note that the effect of density is greater at the seedling stage than at the adult stage.

sparse year after year, because these species are low in the dominance hierarchy of shoot and root competition among perennials. For the short-lived interstitial species, all of which are patchily distributed, sensitivity to drought and predation also contributes to sparseness. Unlike the sparse matrix-forming perennials, the interstitial species vary in relative abundance from year to year and have shifting "clouds of abundance." Their abundance each year is partly determined by various factors affecting regeneration and partly represents random variation.

2. What enables the sparse species to persist at all? The main factor is herbivory by grazing animals; the sparse species largely disappear in the absence of the grazing. Grazing favors the sparse matrix-forming perennials because they are shorter and lose less capital than the abundant taller matrix-formers. Also, some of the sparse matrix-formers are maintained by small-scale disturbance, because they are shorter-lived and more effective at becoming established from seed than the abundant species. Among the interstitial species the pauciennials are consistently sparse, but persist by outcompeting any of the commoner and shorter-lived interstitial species should they meet them; the pauciennials have larger seeds (1.0 to 1.3 mg) than all the annuals and biennials (0.01 to 0.22 mg) except *Rhinanthus* (1.6 mg). On the other hand, the pauciennials can never become abundant because of their modest seed output (about 50 to 100) and relatively long time to flower (2 to 4 years) coupled with losses to predators, disease, and shading.

3. To what extent do the sparse species affect each other? They have little effect on one another. Their abundance is mainly determined by their relationships with the abundant species, not with each other. Among the interstitial species intraspecific density-dependence is rare; direct tests for interspecific effects have yielded negative results except at artificially increased densities; and interspecific effects are likely to be even rarer in the species that are sparser overall and that are consequently our best models for the sparse species in extremely species-rich communities.

The conclusions reached by Kelly and me for the interstitial species are closely compatible with those reached independently by During et al. (1985) and Schenkeveld and Verkaar (1984), studying many of these same species in chalk grassland in South Limburg, the Netherlands.

Seed Output of Individuals of Selected Perennials in Tall-Turf Chalk Grassland

The ultimate level of patchiness is the individual plant. For a study of how seed output varies among individual plants (Dickie 1977), we turn from short-turf to tall-turf chalk grassland near Cambridge. In this community the matrix is formed almost exclusively by the tussock grass *Bromus erectus,* and the interstitial plants are mostly perennial dicotyledons of varying life length. Annuals and strict biennials are absent, and only two species of pauciennial are present (*Daucus carota* and *Picris hieracioides*).

Dickie (1977) reported the estimated seed output for mapped individuals of six species at two sites in 1974, 1975, and 1976. His results demonstrated year-to-year variation, with no two species showing the same pattern of variation through time (Fig. 12.5). These variations through time could be related to (1) rooting depth and the response of shoots to drought, especially important in the great drought of 1976, (2) sensitivity to cool cloudy periods in one species (*Cirsium acaule*) with inflorescences near the ground, (3) sensitivity to grazing by rabbits in another species (*Pimpinella saxifraga*), and (4) sensitivity to infestation with a specific insect pest in another (*Centaurea scabiosa*). The six species certainly have different regeneration niches.

The finding of greatest interest in the context of this chapter was the large amount of plant-to-plant variation. This is illustrated in Fig. 12.6. The implication of this variation is that, if one compares different gaps that become available for colonization in the same year, the seeds of any two species may arrive in very different proportions at two gaps only a few meters apart or at two gaps at different sites along a slope where the overall performance of different species can be related well to environmental factors. For instance, consider seed arriving from immediately neighboring plants at points A–D in Fig. 12.6.

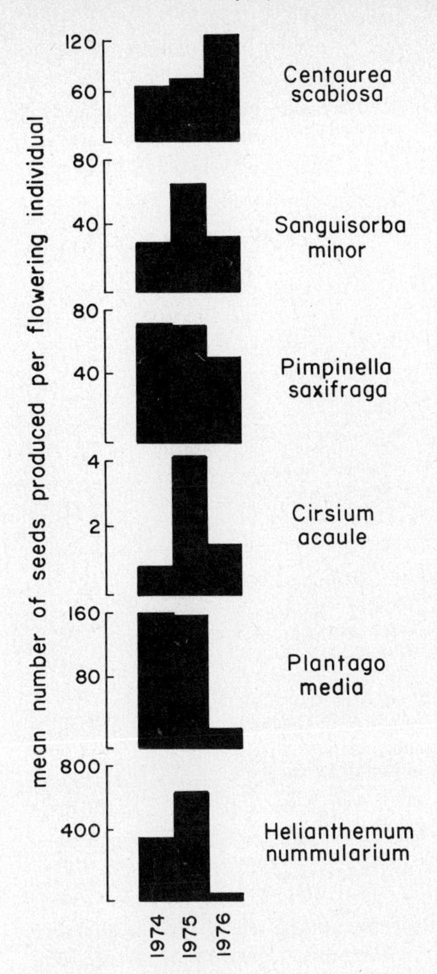

Fig. 12.5 Estimated mean number of seeds produced per flowering individual for five species of perennial, and the number per rosette for *Cirsium,* in three successive years at the Devil's Ditch Local Nature Reserve in eastern England. The figure is based on 100 mapped individuals of 5 species and 50 of *Helianthemum.* Note that seed production varies from year to year in each species and that no two species show the same pattern of variation with time. (Based on Dickie 1977.)

At point A two species (*Plantago* and *Scabiosa*) formed seeds in similar proportions in the two years, but a third species (*Pimpinella*) failed in the second year. At point B two species (*Pimpinella* and *Plantago*) failed in the second year. At point C the ratio between two species (*Plantago* and *Scabiosa*) was quite different in the two years, while at point D it was similar. Clearly, these findings matter only if seed dispersal occurs for the most part over only short distances and if most species are represented little or not at all in the seed bank in the soil. Both these conditions apply to most perennial dicotyledons in chalk grassland.

The origin of this plant-to-plant variation is not known, but presumably includes genetic variation as well as differences in age and neighbors. It is a fact that in many species most adult individuals do not flower every year, and as a result it is easy for individuals of one species to be out of step with each other. Whatever the origin of interplant variation in seed output, it clearly adds a great deal of "noise" to the system and might delay considerably any trend to competitive exclusion.

DISCUSSION

Chalk Grassland in Relation to the Approaches of Other Authors

Relation to Yodzis' Community Types The two major functional groups of plant species in chalk grassland provide excellent samples of the community types designated by Yodzis (Chapter 29) as undergoing "spatial competition" rather than "consumptive competition." The matrix-forming perennials in the short-turf form of the community show "dominance control," while the interstitial short-lived plants show "founder control." We may expect, therefore, that the two groups of species within one community will have different properties. For example, they should respond differently to predation, as set out in Yodzis' Table 29.1.

Relation to Chesson's Discussion of Variable Recruitment The matrix-forming perennials are especially interesting in relation to Chesson's (Chapter 14) models of maintenance of species richness through variable recruitment. Dickie's (1977) work has shown that seed production by perennials in the tall-turf form of chalk grassland is indeed variable from year to year, and it is likely that this is also true for perennials in the short-turf form. Records for one small area on the transect used for Fig. 12.2 certainly show year-to-year fluctuation in recruitment of seedlings of perennials. Chesson's models predict that main-

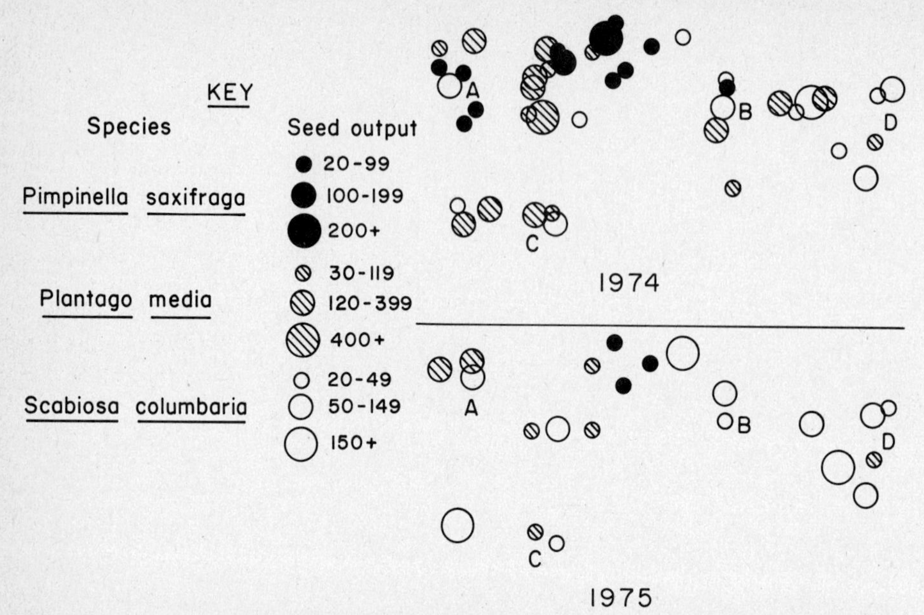

Fig. 12.6 Seed production of individuals of particular species of perennial dicotyledons mapped in two successive years on a 10-m stretch of the Fleam Dyke in eastern England. For each year the figure shows the positions of mapped individuals and indicates the seed output of those individuals by the code shown. For the significance of the four imaginary gaps at A, B, C, and D see text. The main conclusion to be drawn from the figure is that the relative abundances of species in the local seed supply change from year to year in different ways at different points in the turf. (From Dickie 1977.)

tenance of species richness by variable recruitment is compatible with a strong hierarchy of abundance among the adults, and this is what seems to be happening among the matrix-forming perennials.

Intraspecific and Interspecific Density-Dependence The limited incidence of density-dependent effects in the control of population size in the short-lived plants illustrates well the "density-vague" type of control discussed by Strong (Chapter 15). It is not that density-dependent effects have no part to play. In a favorable year, especially in patches that happen to be densely populated at that time, marked density-dependent effects can be found, and these put an ill-defined ceiling on population size. The important point is that in most years, over most of the ground, the plants concerned are far below saturating densities and are not subject to appreciable density-dependent effects.

By analogy it is argued that the sparse species are likely to affect each other little, whether they be interstitial short-lived plants of chalk grassland, or insects among the 200-odd species known on oak trees (cf. Strong, Lawton, and Southwood 1984). It is not that they will never have any impact on each other. For example, *Euphrasia officinalis* in its year of peak abundance on our transect in Sussex (Fig. 12.2) *was* found to have a deleterious effect on *Rhinanthus minor* in those few quadrats where it was most abundant (Kelly 1982). The point is that the impact will occur so rarely that even species with extremely similar niches may coexist for a long time. The system envisaged is like that modeled by Levins and Culver (1971) in one way and by Shmida and Ellner (1984) in another.

In mechanistic terms, as emphasized by Benzing (1981), there are many more suitable microsites for adults than are filled in practice. The two chief reasons why many remain empty are that the chance of establishment from seed in any given site is small and the chance of arrival at a suitable site is small. It is not realistic to try to determine "limiting similarity" with models which assume that all species invade all gaps, or even that all species invade gaps in relation to their abundance, irrespective of distance from gap.

Relation of Niche Differentiation to Coexistence

The hierarchy among the perennials is related to life form, the lowest-growing on the whole being the least abundant. It might be argued that they avoid competition in part by being complementary in this respect: either potentially tall and able to suppress others between episodes of grazing or short and able to avoid losing capital during grazing. As emphasized by Cody (Chapter 23), the four aspects of the niche characterized earlier in this chapter should not be regarded as wholly independent. In chalk grassland the plants of lower-growing life form tend to be shorter-lived and able to regenerate more readily from seed. Most perennial species in chalk grassland seem to overlap widely in their phenology of vegetative development, but one sparse species (*Ranunculus bulbosus*) avoids competition with the more abundant species by maximum canopy development in the spring.

The point to be made totally explicit is that we should always expect niche differentiation between any two species that are generally abundant in a habitat, but we should not necessarily expect any niche differentiation between any two species that are always sparse in the same habitat. Ideally, one would like to see those who make models of realistic systems express the results in terms of time to extinction. In practice, it seems that, with the exception of very small populations, the relevant models usually predict either rapid extinction or a very long time to extinction (Chapter 19; Chesson 1982). The "very long time" should be seen against the background of continuously fluctuating climate emphasized by Davis (Chapter 16) and the chance that one species might be marginally favored for one millenium and another for the next.

In reality, as made plain by Varley (1949), it is most improbable that any two taxonomically "good" species will have exactly the same niche. Instead, the question is whether or not this inevitable niche differentiation has any significant role in maintaining coexistence. For pairs of regularly sparse species this seems doubtful, although it must be emphasized that a lot more critical field observations and experiments are needed to establish the point one way or the other. One can be easily misled. For example, when first studying diminutive winter annuals on a fixed sand dune system, on which they normally cover 1 to 10% of the ground at maturity, I expected negligible interference between species. Nevertheless, when a careful demographic study of four coexisting species was made by Kelly, a hierarchy of interference effects was found along with clear evidence of density-dependent effects within species (Grubb et al. 1982).

Implications of Results from Chalk Grassland for Other Types of Plant Community

Other Communities Posing Similar Problems

Chalk grassland is not at all exceptional in being divisible into matrix-forming and interstitial species. Natural alpine grasslands of both Northern and Southern Hemispheres have essentially the same structure, and indeed the same floristic groups often form the interstitial species (Rhinanthoideae as annuals and Gentianaceae as biennials). Probably many mires and savannas fit this scheme, too, and so do semideserts with their matrix-forming shrubs and interstitial ephemerals. The same idea, in the broadest sense, might be extended even to rain forests, if we treat the trees as forming the matrix and the epiphytes as a special type of interstitial species.

Chalk grassland probably is exceptional in having such a fixed hierarchy of abundance among the matrix-forming perennials, but a considerable degree of constancy is probably to be found in many communities, especially if records are stratified for "time after last disturbance" in those communities subject to periodic sheet destruction, such as mediterranean-climate shrublands, boreal forests, and many natural grasslands. In other words, many communities probably contain an appreciable number of species that arc usually sparse, and the possibility that any niche differentiation between them has little to do with their continued coexistence must be taken seriously. Of course, they must have niches that are different from those of the abundant species, or they would indeed be lost.

The communities generally regarded as most species-rich and most in need of explanation in the face of classical niche theory are the heathlands of South Africa's Cape region (Chapter 23; Kruger 1979) and of southwestern Australia (with about 30 to 120 species of vascular plant in any one closely defined habitat); the montane rain forests of the tropics (with up to 150 species of trees and shrubs in one habitat type in one area and as many vascular epiphytes; Grubb and Stevens 1985); and above all the lowland rain forests of the tropics (with hundreds of tree species in what may appear to be one habitat type; Chapter 19). What is significantly different about the rain forest trees, compared with chalk grassland perennials, is that they show no hierarchy of abundance among the species in any one forest type if this is sampled at sites several kilometers apart, and in this respect they resemble the interstitial short-lived plants of chalk grassland with their founder control (*sensu* Yodzis). One major reason for the lack of a clear hierarchy is presumably the extreme similarity in life form between the tree species of one tolerance type, for example, strongly light-demanding or strongly shade-tolerant. Another reason could be a lack of consistent selection at the seed, seedling, and sapling stages by predators and pests.

Variable Recruitment Hubbell points out that there seem to be rather few different niches in lowland rain forest in terms of the gap size that is most favorable for regeneration, and this conclusion parallels that of During et al. (1985) for interstitial short-lived plants in chalk grassland. What the studies on chalk grassland show is that it is another aspect of the regeneration niche—variable recruitment—that distinguishes the species from each other. In the case of tropical lowland rain forest there is still little known about the extent of year-to-year variation in recruitment. The few available studies certainly suggest that there is a lot of variation in flowering and seed fall (Medway 1972; Foster 1982b).

However, care is needed in defining "variable recruitment." Strictly, we need to follow *effective* recruitment all through its lifetime. It is not enough to know year-to-year variation in seed fall. As a compromise with the search for ideal data, it would be better to find whether or not there is appreciable decade-to-decade variation in a given area in the population of saplings of, say, 1 to 5 m height, since these have a much greater chance of becoming trees than any crop of seeds or newly germinated seedlings.

It may well be that variable recruitment will help explain species richness in tropical lowland rain forest, as discussed by Chesson and by Hubbell and Foster in this volume, and that the variation will prove to involve (as in chalk grassland perennials) not only systematic differences between species (as in Fig. 12.5) but also variation among individuals of the same species (as in Fig. 12.6). It seems improbable however, that hundreds of tree species have significantly different niches, even in terms of variable recruitment, and I suggest that in the long run sparsity and patchiness are likely to emerge as being important in the maintenance of species richness in tropical lowland rain forest.

Historical Origins of Species Richness The study of the maintenance of species richness should never be divorced from a study of its origin. Grasslands on calcareous soils in Europe are much richer in species than grasslands of similar productivity on strongly acidic soils. It seems that the most likely explanation is a historical one, best expressed in terms of "apparency," that is, the number of suitable sites available, integrated through space and time (cf. Feeny 1976). During the Quaternary period in Europe (and possibly for longer) there has been a greater apparency of calcareous grassland than acidic grassland; there has always been a substantial amount of steppe on the predominantly loessic soils of southwest Asia and southeast Europe, disturbed by natural fire and grazing animals, but there has been no comparable treeless or savanna-covered area on acidic soils. In contrast, in North America the Atlantic coastal plain has a long history of extensive fire-disturbed savanna type vegetation on acidic soils, and in that area now we find man-made annually burnt grasslands at pH 4 with as a great density of species as chalk grassland in Europe (Walker and Peet 1984). Historical factors are at least as important for tropical lowland rain forest, though in this case the apparency may have been less crucial than relative constancy of climate which could have been all-important in enabling populations of patchily distributed sparse species to survive.

Conclusions and Suggestions for Further Research

In the past most ecologists interested in the maintenance of species richness in plant communities have assumed that most species are encountering and affecting most other species in any given community over most of its area for most of the time. This chapter challenges that view. It is suggested that sparsity and patchiness have important roles in the maintenance of species richness because they insure that in the most species-rich communities most species are not encountering and affecting each other most of the time. At the mechanistic level emphasis is placed on the importance of the spatial dynamics of the pattern and on the limited dispersability of most plants.

So far, the positive evidence for the revised viewpoint is limited to critical studies on a few species in a community that is only moderately species-rich by world standards. Three kinds of study are needed now for a wide variety of species-rich communities: (1) critical observations on the fates of individuals of various sparse species in relation to the local densities of other sparse species in spontaneous stands of vegetation; (2) experiments in which single sparse species are removed and the effects on other sparse species are monitored; and (3) experiments in which the densities of individuals of single sparse species are increased and the effects on other sparse species are monitored. Results are bound to accrue most quickly from studies on annuals, biennials, and pauciennials like those reported in this chapter, but long-term studies are needed on long-lived perennials, for example, in rain forest and mediterranean-climate heathland. The question of the extent to which single sparse species are materially affected by collections of other sparse species also needs to be tackled.

ACKNOWLEDGMENTS

I am indebted to John Dickie, David Kelly, and Jonathan Mitchley for their imaginative and enthusiastic contributions to the study of chalk grassland, to the Nature Conservancy Council and Cambridgeshire Naturalists' Trust for permission to work on their reserves, to Alan Bowley for logistic support, to Richard West for making me see the possible importance of Quaternary history in relation to species richness of chalk grassland, and to Peter Chesson, Martin Cody, and Peter Yodzis for criticism of my draft paper.

four

EQUILIBRIUM AND NONEQUILIBRIUM COMMUNITIES

chapter 13

Overview: Nonequilibrium Community Theories: Chance, Variability, History, and Coexistence

Peter L. Chesson and Ted J. Case

INTRODUCTION

To what extent are the attributes of natural communities predictable? The development of much ecological theory has proceeded under the assumption that natural communities can be described by models with stable equilibria. The stability of the equilibrium means that historical effects, chance factors, and occasional environmental perturbations play a small role. Because the system heads toward equilibrium, the effect of history disappears, environmental perturbations have no lasting effect, and chance is limited to a role in migration as it affects the arrival of species at a particular locality. Such models have been used to suggest that natural communities do indeed have a highly predictable structure. However, theories based on stable equilibria have been questioned on the grounds that (1) in natural systems the environment is continually changing, often with pronounced effects on populations, and (2) the species in many communities do not appear to have the attributes necessary for stable equilibria in models.

These findings demand that we enquire about the roles of environmental variability and unstable population dynamics. Does environmental variability make the properties of communities unpredictable, or does it simply lead to different predictions? If a community has nothing like a stable equilibrium point, can it possibly have predictable properties, or will its structure be dominated by chance factors and historical effects? More generally, what sort of theory can community ecology hope to have? These are difficult questions, yet some interesting theoretical and empirical progress has been made, much of which is discussed in this book. It is the purpose of this chapter to review and synthesize this progress.

We begin with a discussion of the assumptions and conclusions of classical competition theory and its extensions. Key assumptions involve the notions of equilibrium and stability, which we define. We then discuss four different theoretical and empirical approaches to the questions above. Some of these approaches yield results that are similar to those of the classical theory and its extensions, yet emphasize rather different features of species such as life history traits and responses to environmental change. Other approaches give quite different results and suggest that history, chance factors, migration rates, speciation rates, and climatic change all are important influences on community structure.

EQUILIBRIUM THEORIES OF COMMUNITY STRUCTURE

There are three major equilibrium theories of community structure: classical competition theory and two modifications of the classical theory.

Classical Competition Theory

Hutchinson (1959) argued that competition is the predominant process tending to limit species diversity and that competition leads to pattern in community structure. Although the mathematical modeling of competition had begun much earlier, Hutchinson's ideas led to a tremendous interest in the process of competition and to the development of a sophisticated theory of community structure that may be called classical competition theory. Armstrong and McGehee (1980) and Schoener (1982) review the key developments in the evolution of the theory, and Roughgarden (Chapter 30) presents a formulation of the theory.

The essential assumptions of classical competition theory are:

1. The life history characteristics of species can be adequately summarized by the population's per capita growth rate.
2. Deterministic equations can be used to model population growth; in particular, environmental fluctuations can be ignored.
3. The environment is spatially homogeneous, and migration is unimportant.
4. Competition is the only important biological interaction.
5. Coexistence requires a stable equilibrium point.

An early and key prediction of the theory is that at least n limiting resources are required for the coexistence of n species. The presence of n limiting resources is necessary, but not sufficient for the coexistence. Sufficient conditions involve the notion of *limiting similarity:* to coexist, the n species must be sufficiently dissimilar in their use of the available n or more resources, that is, they must use the available resources in sufficiently dissimilar proportions. Dissimilarity in resource use of animals, when the resources are all types of food, is often expected to be reflected in body size differences, with larger animals concentrating on larger food.

This classical theory naturally became concerned with evolutionary notions and postulated that species in a community would evolve in response to interspecific competitive pressures. Such coevolution, coupled with repeated invasions of new species and extinctions of residents by competitive exclusion, was postulated to yield communities in which the theoretical limits to similarity are approximately achieved, endowing real-world communities with highly predictable properties. Schoener (1982) discusses these results in detail. Roughgarden (Chapter 30) argues that this approach is applicable to communities of Caribbean *Anolis* lizards.

Equilibrium Predation

If a predator is added to the equations of the classical theory, relaxing assumption 4, some of the predictions are changed. Broader limits to similarity of resource use may be possible (Roughgarden and Feldman 1975), and n species may be able to coexist on fewer than n resources. In this case predators may take the place of one or more resources and so may represent limiting factors (Levin 1970). The new theory then predicts that at least n limiting factors are required for the coexistence at equilibrium of n species. The precise predictions depend on the complexity of the predator's behavior (Chapter 29), including the possibility that a single predator may represent more than one limiting factor. Grubb, Buss, and Lubchenco apply this approach to communities of grassland plants and marine sessile invertebrates and plants (Chapters 12, 31, and 32).

Equilibrium Spatial Variation

If species compete for a single resource, but the environment favors different species in different patches, then it is possible for n species to coexist in a system consisting of at least n patches (Chapter 30). Although a patchy environment, this is still an equilibrium situation: there are n different equilibria for each of the n patches, and these equilibria are stable. Variations on this theme are discussed by Levin (1974), Goh (1980), and Tilman (1982 and Chapter 22). An alternative to spatial variation in the environment is the existence of multiple stable points (Levin 1974), which provide spatially varying equilibria and permit coexistence under circumstances denied by the classical theory.

DEFINITIONS OF NONEQUILIBRIUM AND STABILITY

Equilibrium theories are currently under challenge, but as early as 1961 Hutchinson had proposed the beginnings of an alternative theory. His basic idea was that lack of equilibrium could be an explanation for species diversity. Recognizing that many more phytoplankton species coexist in lakes than can possibly be explained by the classical theory, he applied the logical contrapositive: If equilibrium implies that there can be no more species than limiting resources, the observation of more species than limiting resources implies that the hypothesis of equilibrium must be wrong. Hutchinson then went on to explain the diversity of phytoplankton communities in terms of intermediate-frequency temporal variation.

Hutchinson's idea was not given a great deal of attention, perhaps partly because other equilibrium explanations dominated the intellectual scene and perhaps partly because the mathematical theory of nonequilibrial situations was slow to develop. However, the observation that none of niche differentiation, predation, nor equilibrium spatial variation seems to be an adequate explanation of coexistence in some communities (Chapter 19; Sale 1977; Wiens 1977; Connell 1978; Hubbell 1979, 1980) and the finding of substantial temporal variation in densities, environmental variables, or population parameters (Chapters 9–12, 15–19, 30, 32, 33; Grubb 1977; Wiens 1977; Sale 1977, 1980; Hubbell 1980; Butler and Keough 1981; Keough 1983; Caffey 1985; Swarbrick 1984; Underwood and Denley 1984) argue for a close examination of Hutchinson's suggestion that the structure of some communities might best be explained by nonequilibrium ideas.

What Is a Nonequilibrium Explanation?

Although we have referred to "equilibrium" rather loosely until now, it is essential that we be more precise. In ecology there is no unanimity about the definition of equilibrium. There can be different kinds of equilibria depending on the way a system is modeled and the nature of the solution. A limit cycle can be regarded as an equilibrium and so can the strange attractors associated with chaotic population dynamics. For stochastic models solutions are often sought in terms of an equilibrium probability distribution. In multivariable systems we might expect some variables or composite variables to reach an equilibrium, while others drift about indeterminately. The equilibrium theory of island biogeography is a good example: species number is expected to reach an equilibrium point (or an equilibrium density function, depending on the formulation), while the set of species present is continually changing.

Nevertheless, the classical theory and its extensions involving predation and spatial variation are based almost exclusively on point equilibria at which species abundances remain constant over time. The second extension involves the idea that different spatial locations may have different point equilibria. Thus, at equilibrium there is variation in space in the densities of the species in the community, but the population density at each spatial location remains constant over time. We shall call *nonequilibrium* any situation where species densities do not remain constant over time at each spatial location. Clearly, questions of scale can arise with this definition because, as emphasized by Connell and Sousa (1983) and Murdoch et al. (1985), in the real world fluctuations on very small spatial scales necessarily occur, and such fluctuations on a small spatial scale may average out to yield relatively constant population densities on a larger spatial scale. However, from the point of view of developing theory, one asks whether the fluctuations are an explanation of a community phenomenon of interest or are incidental to the explanation. If fluctuations or changes in population densities on some spatial scale are an essential part of a theory of some community phenomenon, then we shall refer to the theory as a nonequilibrium theory. From this perspective, "equilibrium island biogeography" is a nonequilibrium theory.

Some of the nonequilibrium theories discussed below depend on a patchy environment in which fluctuations occur on a local spatial scale, but populations may show constancy on a larger spatial scale. Other theories depend on fluctuations or changes in population densities on the largest spatial scales included in the model.

What Is Stability?

All of the equilibrium theories discussed above involve stability of the equilibrium, and in many cases the equilibrium is globally stable. Global stability of the equilibrium means that the system will return to equilibrium following any displacement. Such stability has four important consequences for community theory.

1. *Community conservation*. The community will show little tendency to lose species with time. Indeed, global stability implies that in the absence of external perturbations no loss of species will ever occur.
2. *Community recovery*. The community can recover from events that drive any of the species to low density.
3. *Community assembly*. The community can be built up by immigration of species from outside the system, for combinations of species that are capable of coexisting will increase to their equilibrium values.
4. *Irrelevance of history*. Because the community approaches equilibrium, the effects of past abundances of the species disappear. Note, however, that aspects of history such as the order of arrival of species will generally affect community structure unless species arrive over a time span that is so short that no extinctions have had time to occur. For example, a globally stable predator-prey system will nevertheless fail to form if the predator arrives first.

These aspects of stability may also be shared by the models of nonequilibrium theories. For example, systems that are nonequilibrial locally in space may still have an equilibrium for the total community as a sum of all the local communities, and this equilibrium may be globally stable, carrying with it the four properties listed above. More generally, for many nonequilibrium models the definition of coexistence is *invasibility* (see Chapter 14), which means that each species can increase from low density when all species are present in the community. The idea of invasibility applies to both deterministic and stochastic models of fluctuating populations. It is essentially the property of community recovery, and it implies community conservation and community assembly also. Invasibility often leads to the fourth property, irrelevance of history, because it often implies that the system will approach an equilibrium probability distribution for the abundances of the species in the system. Thus the system "forgets" previous abundances.

Properties 1 to 4 define what we shall call a *stable community*. As we have seen, the concept of a stable community is independent of the concept of equilibrium as defined here in terms of point equilibria. Many of the nonequilibrium theories below are indeed stable community theories. We now consider the different directions that have been taken in the approaches to nonequilibrium theory that are discussed in this book.

NEW THEORETICAL DIRECTIONS

Direction 1: Fluctuations and Continuous Competition

The simplest deviation from the assumptions of the classical theory is relaxation of the requirement of a point equilibrium, while retaining the idea that competition is important and occurs continuously. However, this simple deviation invalidates one of the key predictions of the classical theory. For instance, Armstrong and McGehee (1980) showed that if population dynamics lead intrinsically to limit cycles, then it is possible for many species to coexist on a single limiting resource.

In Armstrong's and McGehee's model the

environment does not vary in time. Instead, fluctuations in population densities and resource levels derive from instability of the model's point equilibrium. A fluctuating environment is more in line with Hutchinson's (1961) ideas about the effects of deviations from equilibrium. Indeed, a fluctuating environment can lead to predictions that differ from the classical theory, in particular, to the prediction that many species can coexist on a single limiting resource (Chapter 14). It is not necessary for fluctuations in resources or environmental variables to reduce the intensity of competition for these deviations from the classical theory to occur. What is important is that fluctuations occur in the competitive rankings of the species. Generally, these fluctuations are assumed to result from stochastic variation in the environment from year to year, but can also result from regular (for example, seasonal) environmental variation.

Temporal variation is the driving force in direction 1 theories. However, it has long been recognized that local populations and local environments may experience fluctuations that are out of phase from place to place. Elaborate community theories have been built on this premise, again with predictions differing substantially from those of the classical theories (Chapter 29; Atkinson and Shorrocks 1981; Chesson 1984, 1985; Comins and Noble 1985). These theories are compatible with continuous and intense interspecific competition, but competitive rankings of the species vary through time and space. The fluctuations in competitive rankings occur in two distinct ways. Fluctuations in migration rates into particular patches may occur, causing fluctuations in the numerical advantage that a species has in a particular patch. Alternatively, the competitive ability of individuals present in a patch may be environmentally dependent and therefore may fluctuate with the changes in the local environment.

Although these theories differ from the equilibrium theories by the absence of point equilibria and although many involve stochastic processes, they are all stable community theories as defined here. Thus, they have the properties of conservation, recovery, assembly, and irrelevance of history that the equilibrium theories possess.

In addition to stability properties, the theories of direction 1 have a number of other features in common with the equilibrium theories. For instance, none of these theories predicts coexistence of identical species. Although the species may not differ in resource use, they will usually differ in some other sense. They may have different functional responses (different changes in resource capture rates with changes in resource density), as in Armstrong and McGehee (1980) and Hsu et al. (1978), or they may respond differently to temporal variation in the environment. In this sense these new theories may be considered as an enlargement of the classical theory rather than a strict alternative.

Despite similarities to the classical theory, these new theories suggest profound differences in the way communities should be studied. Clearly, the new theories emphasize that dynamic rather than purely static features are extremely important, and they also emphasize such species characteristics as functional responses and life history traits, including dispersal abilities. Indeed, life history characteristics can have profound effects in the presence of environmental fluctuations. For instance, Chesson (Chapter 14) shows that life history characteristics that tend to buffer a species against unfavorable environmental events (for example, long life and iteroparity) will also promote coexistence of competing species in a fluctuating environment. Other life history traits may promote competitive exclusion or have no effect on coexistence. Thus, it is not just the fluctuating environment that is important for coexistence, but the combination of a fluctuating environment and the possession of certain kinds of life history traits.

In the theoretical models of this direction, all members of a guild are assumed to have the same overall kind of life history (for example, are all long-lived and iteroparous). However, in the real world fundamental differences in life histories may well lead to important differences in responses to the environment. For example, they may lead to different time lags for responses to an event that two species find equally favorable or unfavorable. Thus, differences in life history traits, in addition to possession of particular kinds of life history traits, may be important factors in the coexistence of competitors in a variable environment.

These ideas suggest an explanation of coexistence of strongly competing species from very different taxonomic groups, for example, harvester ants, finches, and rodents (Chapter 3). Differences in life history, physiology, and behavior among these different organisms may lead to very different responses to the environment, so that a fluctuating environment promotes coexistence. On the other hand, congeners may have fewer opportunities for differential responses to the environment and therefore may more frequently show differences that lessen competition, for example, the differences in morphology and habitat distributions discussed in Chapters 5, 6, 10, 23, and 30 for closely related species of birds, lizards, or arid-zone plants.

Direction 2: Fluctuations and Discontinuous Density-Dependence or Competition

A second direction of deviation from classical theory emphasizes fluctuations in density or environmental variables as dominant processes. These fluctuations take place on an ecological time scale, and population dynamics may be density-independent much of the time. Strong, Kareiva, Grant, Grubb, and Wiens (Chapters 9–12 and 15) discuss the field evidence that many species fluctuate greatly in abundance, and that these fluctuations are often strongly related to environmental factors and only weakly related to population density or interactions with other species. Strong suggests that density-dependence may become more important at high and possibly very low densities, but that much of the time the dynamics of many species, especially insect species, are little influenced by density and are thus "density-vague." In this view density-dependence sets the range over which population fluctuations occur, but most of the changes in population density occur in a density-independent manner.

Studies of populations with density-vague dynamics are unlikely to reveal any clear density signals except at the extremes of densities observed in nature. However, if one wishes to explain a population's mean density, when sampled over time, a study of density-dependence at the population extremes will be necessary. Indeed, density-dependence and density-independent fluctuations will interact to produce this mean density, as commonly observed in stochastic population models (e.g., May 1973a).

The community consequences of intermittent density-dependence and interspecific competition have been explored in models by a number of authors (Koch 1974, Huston 1979, Chesson 1983). In all cases fluctuations in environmental factors reduce the densities of several potentially competing species to levels where competition is weak and population growth is for a time insensitive to density. Thus both intra- and interspecific competition fluctuate in intensity with time. In all cases coexistence is promoted by these fluctuations. While these models do not explicitly address evolution, Wiens's verbal model (discussed in Chapters 9 and 10) considers the effects of environmental fluctuations on the intensity of selection and the likelihood of character displacement.

Other community models in which environmental fluctuations can lead to periodic reductions in densities, reducing competition and promoting coexistence, involve a patchy environment (Slatkin 1974, Caswell 1978, Hastings 1980, Hanski 1983). In these models (also reviewed in Chapter 14) local extinctions wrought by environmental factors or predators open up space and may permit species to colonize and grow for a time without the influence of interspecific competition. Potential real-world examples involving forests and sessile marine organisms are discussed by Connell (1978), Fox (1979), and Paine and Vadas (1969).

It was pointed out above that the mean density in a population with density-vague dynamics results from the interaction between density-dependence and density-independent fluctuations. Similarly, properties of communities subject to intermittent competition may result from an interaction between density-independent fluctuations and competitive effects. This is especially clear in the lottery model with vacant space (Chesson 1983), where occasional density-dependence and environmental variation both have profound effects on the relative abundances of the two species. If space limitation is completely eliminated from this model, even slight differences between the two species can lead to the domination of a single species. However, the joint action of occasional space limitation and environmental fluctuations can prevent this result.

The predictions of the nonequilibrium models of this second direction share many similarities with theories based on the absence of point equilibria, (direction 1) and, indeed, much of Chesson's discussion (Chapter 14) deals simultaneously with these two directions. Both directions can be looked upon as enlarging the standard theory. For instance, the patchy environment examples involve organisms doing different things from one another: Some are good dispersers, while others are good competitors. Moreover, in most cases coexistence occurs in the sense of invasibility, as discussed above, and so these theories are mostly stable community theories. Huston's theory is an exception, for coexistence in his theory does not depend on differences between species and it is not a stable community theory. It is discussed in detail under direction 4 below.

Direction 3: Changing Environmental Mean

Directions 1 and 2 consider fluctuations in the environment about some mean value, and generally the mean and the variance of these environmental fluctuations are assumed to be constant over time. However, Davis (Chapter 16) emphasizes that the mean of climatic environmental fluctuations cannot be considered to remain constant over any time scale that may reasonably be considered ecological time. Moreover, both Davis (Chapter 16) and Van Devender (Chapter 17) document community changes in response to changes in mean values of the year-to-year climatic fluctuations. Neither classical competition theory, nor its extensions based on point equilibria and a constant environment, nor the new stable community theories of directions 1 and 2 address this possibility. Does this mean that these theories must fail?

The answer to this question depends on the rate of change of the frequency distribution of the year-to-year environmental fluctuations (the mean and variance of these fluctuations) relative to the speed of community dynamics. Slow changes in this distribution imply that the community can track the predictions of the stable community theories (assuming that the models on which they are based apply in the short term). For instance, if a stable community theory predicts the mean abundances of a number of species in a community as a function of the mean of the environmental fluctuations, then the actual community means can be expected to follow closely these predicted mean abundances as they change with the mean environment. In particular, present-day populations and communities can be explained on the basis of the environmental fluctuations that are observed at present without recourse to changes in the mean (and variance) of these fluctuations that have occurred historically. A number of different authors have studied such tracking behavior explicitly in simple models (Hubbell 1973; Roughgarden 1975b, 1979; Nisbet and Gurney 1982).

Davis suggests that short-lived organisms may well fit this picture since their population dynamics are fast relative to the rate of change of the mean environment. However, both Davis and Van Devender note that long-lived organisms, such as forest trees, respond to changes in the mean environment with considerable time lag; indeed, Davis notes that some present-day forests appear to be genetically maladapted to current conditions. For such communities, appeal to tracking cannot salvage the stable community theories.

In principle, it is simple to modify the models of the stable community theories—just make the mean of the environmental fluctuations change with time. The analysis of such models is not quite so simple, but more important, the amount of information necessary for a prediction increases enormously because stability property 4 (irrelevance of history) will no longer be true. Since the community is constantly adjusting to new conditions, but never completes the adjustment before conditions change again, past abundances of the species in the community remain relevant to the present abundances and community structure. In particular, the present community cannot be explained simply by studying it today.

Although clearly complex and likely to proceed largely as an empirical endeavor for some time, the study of communities from this direction appears rich and rewarding. For instance the studies of Davis, Graham, and Van Devender (Chapters 16–18) show that the relationships among different species can change dramatically with climatic change. Life history characteristics and dispersal abilities have a profound influence on the tendency of species to track environmental change. During the Pleistocene forest trees were much poorer trackers of environmental change than herbs or mammals. Individual trees could withstand very long periods of unfavorable weather, sometimes only as below-ground biomass with the ability to recover with the return of more favorable climate. Such species therefore may persist at a locality longer than others, but also may reappear more quickly. Because of such differences in the tracking tendencies of different species, with each successive advance and retreat of the glaciers communities did not simply shift geographically back and forth. Rather, the entire community composition changed. These results suggest that the evolutionary histories of many species are likely to be quite varied.

Direction 4: Slow Competitive Displacement

Hubbell and Foster (Chapter 19) argue that many tropical forest tree species are essentially ecologically identical, having identical resource requirements and responding to the environment in identical ways. If this is true, then only reproductive incompatibility provides an ecological distinction between individuals of different species, and it follows that at any given time the difference in the densities of the two species is just as likely to increase as to decrease, quite independently of the environmental conditions or the population densities of the species. Fluctuations in numbers result simply from the uncertainty in the lives of individuals (within-individual variability; Chesson 1978), and these fluctuations lead to a random walk in population densities with no stabilizing tendencies at all. (Theoretical mean population growth rates are always zero.) Stable community theories clearly do not apply to this scenario. However, Hubbell and Foster show that in any reasonably large forest the time for competitive elimination of a species is long, and this is their explanation of coexistence. For the maintenance of diversity on long time scales, appeal to speciation and migration is necessary, as discussed below.

Slow competitive elimination has been suggested as an explanation of coexistence in several other settings. Shmida and Ellner (1984) put forward the slow dynamics hypothesis in a deterministic model of species with similar competitive abilities. The similarity in competitive ability and the near equality of intra- and interspecific competition mean that competitive elimination will be slow. In the Hubbell-Foster model there is complete competitive equality in the community, and elimination occurs only as a result of random drift in numbers. It is implicit in both models that competition occurs continuously. On the other hand, Huston's (1979) theory relies on intermittent competition: Periodic reductions in population density reduce the frequency of intense competition and hence the speed at which species are competitively eliminated from the system. Overall similarity in growth rates enhances the opportunities for coexistence.

These theories of long times for extinction and slow competitive elimination satisfy only stability property 1 (community conservation). Thus, they are theories of conservation of the existing species pool and do not address the question of how that pool comes about in the first place. In this sense they are less complete than the stable community theories, which usually explain how a community can be assembled by the invasion of new species and coevolutionary adjustments among residents (Roughgarden 1979 and Chapter 30). In models with zero or near-zero mean growth rates there is no predictable community assembly. Only by chance population fluctuations will a potential invader increase in numbers and become part of the community. Indeed, it is most likely that a rare species will be eliminated quickly, and no species can be regarded as having any degree of permanence.

Communities obeying such slow elimination models may work rather like the neutral gene model in population genetics (see Roughgarden 1979). Given an ecological interpretation, this model says that newly arriving species increase and become part of a community purely by chance. The net number of species in the community is determined by the arrival rate relative to the population size. Species enter the community as a result of speciation. Alternatively, or in addition, species arrive by migration from surrounding areas, but the ultimate answer is still, of course, speciation. Consistent with these ideas is the finding of Hubbell and Foster that the time for competitive elimination is comparable to the time for speciation.

For the sorts of communities discussed under direction 4, chance and history may be major factors shaping community structure. The time scale of slow elimination and slow input of species via speciation and migration will be on the order of the geological time scale. Past environments or conditions can play an important role in setting up communities in which slow but inevitable elimination occurs. For instance, it may be that the present tropical forests are derived by amalgamation of communities that were isolated during glacial periods (Haffer 1969, Vuilleumier 1971, Livingstone 1975). This isolation presumably led to genetic divergence and allopatric speciation, and so the present diversity of tropical forests may in some measure be a reflection of elevated speciation rates that occurred historically. Indeed, in such systems the actual number of species present will be critically dependent on speciation rates and historical changes in these rates, including climatic and geological events that divide or amalgamate communities. The structure of such systems therefore may well have a strong historical imprint.

CONCLUSIONS

Table 13.1 summarizes the general assumptions and predictions of equilibrium and nonequilibrium theories. It is necessarily an oversimplification and presents only the major emphases of these theories. The table makes no attempt to cover extensions of the theories to incorporate evolution and invasion.

The nonequilibrium theories of directions 1 and 2 differ in the way density-dependence and competition are perceived, but they produce similar results. The two theories can be considered enlargements or generalizations of classical competition theory and its equilibrium extensions, both in the nature of coexistence and in the predictions of the theories. These nonequilibrium theories are stable community theories, as we have defined them, for coexistence involves the ability to recover from events that take a species to low density. This in turn implies a number of other properties generally associated with stability in equilibrium theories.

Like the equilibrium theories, directions 1 and 2 require that species must differ from one another if they are to coexist. However, the focus of these differences is not on how the species use resources, but on other factors such as fluctuating density-independent mortality. The species may be identical in their use of limiting resources, yet may still coexist because they have different responses to a fluctuating environment. Moreover, these two new directions emphasize that the life history characteristics of species are critically important. Coexistence thus results from the combined effects of fluctuations and the possession of certain kinds of life history traits.

Directions 3 and 4 are different from directions 1 and 2. Direction 3 puts great emphasis on historical factors and casts doubt on other directions, especially directions 1 and 2. Unless communities can be regarded as closely tracking gradual climatic change, we have at present no theory that can adequately handle the concerns of direction 3. Indeed, these concerns suggest that vastly different approaches to community ecology may be necessary. However, the empirical paleoecological approaches of this direction are revealing, and it is clear that they have much to offer to the understanding of communities that exist today. For instance, the suite of species present at a locality today may have a historical imprint that can be uncovered by looking back in time.

Although the major concern of direction 4 is still with coexistence, it is not a stable community theory, and it emphasizes overall similarity of species for their long-term coexistence, not the presence of differences. Species composition of a community is conserved because competitive exclusion is slow, but species composition will show no tendency to recover following a perturbation. Diversity of a community depends critically on rates at which similar species may competitively eliminate one another, relative to migration rates and speciation rates. Indeed, since this direction deals with events taking place on a long time scale, the history of climatic and

Table 13.1 EMPHASES OF DIFFERENT COMMUNITY THEORIES

	Equilibrium theories			New directions			
	Classical competition theory	Predation extension	Spatial variation extension	(1) Fluctuations and continuous competition	(2) Fluctuations and discontinuous competition	(3) Changing environmental mean	(4) Slow competitive displacement
Assumptions							
No environmental fluctuations	Yes	Yes	Yes	No	No	—	Yes/no
Constant mean environment	Yes	Yes	Yes	Yes	Yes	No	Yes
Spatial homogeneity	Yes	Yes	No	Yes/no	Yes/no	—	Yes
Life history traits adequately summarized by growth rates	Yes	Yes	Yes	No	No	No	Yes
Continuous competition	Yes	Yes	Yes	Yes	No	—	Yes/no
Predictions: Factors Involved in Coexistence of *n* species							
At least *n* resources, limiting factors, or patches	Yes	Yes	Yes	No	No	—	No
Limits to similarity of resource use	Yes	Yes/no	No	No	No	—	No
Differential responses to fluctuating environmental variables or resources	No	No	No	Yes	Yes	—	No
Particular kinds of life history traits	No	No	No	Yes	Yes	Yes	—
Shapes of functional responses	No	Yes	No	Yes/no	—	—	No
stability	Yes	Yes	Yes	Yes	Yes	No	No
Overall similarity of species	No	No	No	No	No	—	Yes
History	No	No	No	No	No	Yes	Yes

A dash indicates that no particular emphasis has yet emerged.
Yes/no means that the theory suggests different answers in different circumstances.

geological change may well be very important to its predictions.

In dealing with chance, variability, and history, none of these approaches denies that predictions can be made about the properties of communities. The approaches all suggest a much more complicated world than classical competition theory, but even the largely historical approach of direction 3 allows us to relate community change in a changing environment to the life histories of the constituent species. As emphasized by Hubbell and Foster (Chapter 19), these new approaches to ecological theory may lead to predictions of a different nature, but they do not leave us without predictions.

These theories have already revealed a variety of new patterns that may be expected in guilds of competing species. While competition may be common in nature, as shown by field experiments (Connell 1983, Schoener 1983b), the new theories imply that classical patterns of resource partitioning, habitat segregation, or limitation by predators should not necessarily be expected. There has been much debate over whether patterns in nature do support classical competition theory (Schoener 1984, Simberloff 1984a). However, competition can still be important to community structure, as suggested by the new theories, without producing any of the patterns predicted by the classical theory.

Although our discussion of new theoretical directions has dealt mainly with a single trophic level, the results clearly have important ramifications for entire food webs and cast doubt on the deterministic modeling effort in that area. Indeed, some well-studied food webs have revealed a preponderance of nonequilibrium phenomena (e.g., Dayton 1975a).

Finally, we close with a plea for a pluralistic approach to the problem of diversity and species coexistence. It is quite reasonable to suppose that two species coexist in part because they have some differences in resource use, in part because they have different responses to the environment, and in part because the average net advantage that one species has over another is small, leading to a long competitive elimination time. Techniques that permit a quantitative partitioning of the components of an explanation are important for the study of systems that involve several determining factors, which undoubtedly constitute most of the ecological world.

ACKNOWLEDGMENTS

We are grateful for helpful comments on the manuscript by Joseph Connell, Theresa Dahlem, and Jared Diamond.

chapter 14

Environmental Variation and the Coexistence of Species

Peter L. Chesson

INTRODUCTION

Variation is a striking feature of many real populations and communities. Variation in both time and space occurs in the environment, in species densities, and in relative densities. The widespread occurrence of variation raises a number of important questions. Of what significance is variation to the dynamics of populations and communities? What note should ecologists take of variation? Are the effects of predictable variation, such as seasonal environmental variation, different from the effects of unpredictable variation, such as yearly rainfall or temperature? How is it possible to tell if the structure of a community depends on environmental variation?

Historically, ecologists' attitudes to variation have themselves been quite varied. Some, such as Andrewartha and Birch (1954), have focused on variation or fluctuations in population numbers and have sought the causes of such variation. Community ecologists usually have rather different questions in mind and view variation mainly in terms of how it affects community structure or how it affects the study of community structure. Four different views of variation are set out below with a focus on stochastic variation, that is, variation that is not predictable with high confidence from one time to the next. The weather gives us many examples of stochastic environmental variation, but it must be remembered that stochastic variation still has its predictable aspects. For instance, yearly rainfall may be highly stochastic, while the average rainfall over a sufficient number of years is highly predictable.

Among the views of variation discussed here, view D is the one that is most strongly supported by models. Models of coexistence by means of recruitment variation are presented as a principal illustration of this view. Other models have not necessarily supported the idea that environmental variation promotes coexistence, but this chapter will show how the seemingly conflicting results of a variety of models can be understood within the context of a single general model. This general model suggests how certain kinds of life history traits may be generally favorable to coexistence in a temporally varying environment. If the environment varies in space and time, then life history traits such as dispersal may also be important, and models incorporating dispersal in a spatially varying environment have strong similarities to models of recruitment variation in time.

It is often assumed that stochastic mechanisms of coexistence, such as those discussed here, will lead to population dynamics that are fundamentally different from those associated with deterministic mechanisms of coexistence. However, I shall argue that these two different sorts of mechanisms of coexistence may yield quite similar population dynamics in the real world.

FOUR VIEWS OF VARIATION

View A: Variation is noise, and it tends to obscure what is really happening in a system. This view makes no distinction between variation and sampling error. Sampling error is a common problem in ecological research. For instance, there are various methods one can use to try to determine the size of a small mammal population in a particular locality at a particular time (Seber 1982), but all entail sampling errors that can make it difficult to get an accurate result. However, there is nevertheless a true value for the population size for the particular place and time, and this value is potentially knowable.

Now consider a fish population. One may wish to know the number of juvenile fish maturing to adulthood, that is, the number of new recruits, in a particular year. This too has a true, potentially knowable value, but fisheries biologists often want more; they want to know the relationship between the size of the adult stock and the number of new recruits. One can measure the number of new recruits for different values of the adult stock, for the adult stock varies in time. Getting the relationship will certainly involve sampling error, but something else is going on also. Although two different years may have the same adult stock sizes, the actual numbers of recruits for these two years may be quite different (Cushing 1982). This has nothing to do with sampling error, for the numbers of recruits for the two different years are truly different, as would be seen if every individual in the population were counted. Thus, if sampling error could be eliminated, there would still not be any precise relationship between stock and recruitment. Nevertheless, there will be a mean value for recruitment over many years with the same stock size, and view A treats the relationship between this mean value of recruitment and the stock size as the "true" relationship. Variation from this relationship is regarded as noise, or error, that need not be distinguished from errors resulting from the difficulties of sampling.

The remaining three views do make a distinction between sampling error and variation in the true population size and assume that variation in population size, and in environmental factors, is important to the workings of a population or the organization of a community.

View B: Variation is present and important. If the variation is stochastic, it is harmful in the sense that it increases the probability of extinction of a population and decreases the diversity of a community. This view pervades deterministic approaches to modeling. A stable equilibrium point, or a stable cycle, is regarded as necessary for community persistence. Stochastic variation causes displacements from equilibrium, and if sufficiently strong, it may cause some species to become extinct. For a community to persist in the face of strong stochastic variation, there must be strong deterministic forces opposing the displacements from equilibrium that are caused by the stochastic variation.

Many ideas and theories have developed in response to this view, including ideas on limiting similarity, multiple stable points, resilience, and vulnerability (May and MacArthur 1972, Holling 1973, May 1974b, Goh 1975, 1976, Leigh 1975, Beddington et al. 1976). As emphasized, this view mostly concerns stochastic variation because of its unpredictable properties.

View C: Variation leads to interruptions or reversals of biological interactions. It thereby slows down competitive exclusion and may prevent it from occurring on an ecologically relevant time scale. Hutchinson (1961) gave the first detailed discussion of this view. He suggested that competing species may coexist as a result of changes in the environment that reverse the order of competitive superiority among the species. Hutchinson argued that if such environmental changes occur with a period roughly equal to the time for competitive exclusion of a species, then competitive exclusion would be prevented and a rich community might result. However, if the period between environmental changes is quite different from the time for competitive exclusion (either longer or shorter), then competitive exclusion would still occur.

Ideas on disturbance (Paine and Vadas 1969, Connell 1978, Caswell 1978, Fox 1979, Abugov 1982) usually involve the notion that environmental events or predators interrupt biological processes and prevent monopolization of resources (usually space) by a single species. Intuitively, interruption of biological processes, especially competition, should delay competitive exclusion. Theoretical support for this idea comes from Huston's (1979) model, in which disturbance leads to a reduction in the population sizes of all species and consequently a temporary reduction in competition. On the other hand, Caswell's (1978) analysis of a different model of disturbance assumes at the outset that the degree of delay of competitive exclusion is the appropriate focus of the analysis.

That environmental variation may primarily slow the rate at which competitive exclusion occurs is not an uncommon view, and although not stated explicitly in Hutchinson's original discussion, it is implicit in his arguments. In this view environmental variation is very different from mechanisms like resource partitioning that lead to stable equilibria in models and thereby eliminate trends toward extinction of one or more species. Variation is not seen as eliminating these trends; it merely slows them down. Often associated with this view is the idea that population dynamics should appear very unstable with large fluctuations in species relative abundances (Grossman 1982, Grossman et al. 1982).

View D: Variation not only slows processes like competitive exclusion, but changes trends as well. Species that cannot coexist in a constant environment may in the presence of variation all show positive average growth rates at low density. Each species would thus have an upward trend whenever its density is low. Moreover, in this view variation can lead to the creation of a central tendency for population fluctuations, analogous to the existence of a stable equilibrium point in a deterministic model. Variation is not greatly distinguished in its action from other forces in a community like competition or predation. Like these forces, variation may change average growth rates in a positive or a negative direction depending on the circumstances.

Levins (1979) gives perhaps the best expression to date of this view. He suggests that when a limiting resource varies in abundance, species may partition different aspects of resource variation. For example, some species may act like consumers of the resource variance, while others are consumers of the resource mean. By so partitioning resource variation, several species may coexist on a single limiting resource.

View D is also expressed implicitly in a number of models of disturbance in a patchy environment (Slatkin 1974, Crowley 1979, Hastings 1980). Disturbances are assumed to occur asynchronously in the different patches of a large (effectively infinite) habitat. These models all produce the result that disturbance can lead to an indefinite, stable coexistence, in the total habitat, of two or more strong competitors. Moreover, the coexisting species all have positive growth rates at low density. Crowley (1979) and Chesson (1981) argue that the sort of coexistence found in these models of infinite habitats is a reflection of tendencies that are present in more realistic models of finite habitats, but are more difficult to see there.

According to view D, it does not greatly matter whether environmental variation is regular (e.g., seasonal) or stochastic. With both kinds of variation the environment goes through a number of states, and both give a predictable frequency of different environmental states at least in the long term. These long-run frequencies of the different states are seen as one of the most important aspects of the environmental variation. Naturally, a regularly varying environment is predictable in the short term as well as in the long term, but this is not of overriding importance.

A considerable amount of evidence from stochastic models has accumulated in support of view D, and the rest of this chapter is an elaboration of this view.

COEXISTENCE MEDIATED BY TEMPORAL VARIATION

An important role of environmental variation in population dynamics is suggested by the relative fragility of the juveniles of many organisms. Juvenile mortality rates are high and sensitive to environmental conditions. Thus, if the environment varies, juvenile survival varies widely. In contrast, adults may have higher survival rates that are relatively insensitive to environmental conditions. However, reproductive rates, like juvenile survival, may vary substantially with environmental factors. Indeed, since there is believed to be a trade-off between reproduction and adult survival (Murdoch 1966, Goodman 1974, Nichols et al. 1976, Schaffer 1979), some organisms may maintain high adult survival by varying their reproductive effort in response to the environment. Deferring reproduction in response to unfavorable environmental conditions may also permit deferring the use of resources that are allocated to reproduction (Harper 1977, Nichols et al. 1976, Tyler and Dunn 1976). A likely outcome of this deferral of resources is both a greater mean magnitude and a higher level of variability in recruitment. Extreme reproductive variation in some long-lived organisms, for example, some perennial plants, is believed to be not so much a response to environmental conditions, but a mechanism that thwarts predators (Silvertown 1982).

The characteristics of high and relatively unvarying adult survival coupled with highly variable juvenile survival or highly variable reproductive rates are not uncommon species properties, but are especially well documented for fishes (Gulland 1982, Cushing 1982) and perennial plants (Grubb 1977 and Chapter 12, Harper 1977, Hubbell 1980). These characteristics are also becoming increasingly apparent for sessile marine organisms (Butler and Keough 1981, Keough 1983, Underwood and Denley 1984, Caffey 1985). I now present models of communities having these features, for they provide an especially valuable demonstration of the validity of view D.

Models of Communities

Consider a community of n species, each having the characteristics discussed above. An equation for the dynamics of the system is

$$X_i(t+1) = (1-\delta_i)X_i(t) + R_i(t)X_i(t) \quad (14.1)$$

where $X_i(t)$ is the population of the ith species at time t, δ_i is the adult death rate, and $R_i(t)$ is the recruitment rate (the per capita number of new individuals entering the adult population). Equation 14.1 expresses the new population at time $t+1$ as the sum of adult survival and recruitment.

While adult survival is assumed not to vary in time, the recruitment rate varies as a function of both the environment and species densities. The recruitment rate is the product of the reproductive rate and the juvenile survival rate and reflects the effects of interactions within and between species, in addition to the effects of a varying environment. Thus, the recruitment rate can be expressed in the form

$$R_i(t) = f_i[E(t), X_1(t), \cdots, X_n(t)] \quad (14.2)$$

where f_i is some function, $E(t)$ is a multidimensional variable representing the state of the environment at time t, and n is the number of species. The different dimensions of the environment can be things like temperature and moisture and can also include disturbances of different kinds. The involvement of a fluctuating environment in the recruitment rate means that regulation of recruitment may be density-vague (Chapter 15).

Because population growth is fundamentally a multiplicative process, we can best understand the consequences of equation 14.1 by taking logs. Then we see that the change in log population size is

$$\ln X_i(t+1) - \ln X_i(t) = \ln[1 - \delta_i + R_i(t)] \quad (14.3)$$

This change in log population over one time interval can be thought of as the instantaneous per capita growth rate applicable for that time period.

Since an adult lives an expected life time of

$(\delta_i)^{-1}$ time units (breeding seasons), it is helpful to divide this growth rate by δ_i to obtain a growth rate measured on a per generation time scale, a "scaled growth rate." This facilitates the comparison of species having different longevities. Expressing the recruitment rate on this same time scale as $\rho_i(t) = R_i(t)/\delta_i$, the "scaled recruitment rate," we obtain

$$\begin{aligned}\delta_i^{-1}\,[\ln X_i(t+1) - \ln X_i(t)] \\ &= \delta_i^{-1}\ln[1 - \delta_i + R_i(t)] \\ &= \delta_i^{-1}\ln\{1 + \delta_i[\rho_i(t) - 1]\} \quad (14.4)\end{aligned}$$

Notice that the population increases from one time to the next if $\rho_i(t) > 1$ and also that the change in log population from any time t_1 to some other time t_2 $(\ln X_i(t_2) - \ln X_i(t_1))$ can be obtained by summing equation 14.4 over the values $t = t_1$ to $t = t_2 - 1$, and multiplying by δ_i. In particular, the sign of the sum of the scaled growth rate over any period indicates whether the population has increased over that period.

The scaled growth rate is plotted as a function of scaled recruitment in Fig. 14.1. The ρ axis has a log scale, which means that the scaled growth rate for $\delta_i = 1$ plots as a straight line. The curve for $\delta_i = 0$ is the limit of the scaled growth rate as death rates are made small. The most important feature of these curves is that, except when $\delta_i = 1$ (nonoverlapping generations), the scaled growth rate has a lower bound equal to

$$\delta_i^{-1}\ln(1 - \delta_i) \quad (14.5)$$

which is achieved when recruitment fails completely ($\rho_i(t) = 0$). For small δ_i ($\delta_i \leq 1/3$), the lower bound given by equation 14.5 is reasonably approximated by -1. Thus, when generations are overlapping, the scaled growth rate can never be less than the value of equation 14.5, which is the value determined by adult survival from the previous period. When $\delta_i = 1$, there is no adult survival, and it follows that when recruitment fails, the scaled growth rate is $-\infty$. Note that in no case is there an upper bound on the growth rate as a function of the recruitment rate; if a large recruitment occurs, the growth rate will be large. Notice also that the scaled growth rate curves up more strongly (is more convex in the mathematical sense) for smaller adult death rates.

These properties of the scaled growth rate are explored in Fig. 14.2, where it is assumed that scaled recruitment fluctuates between the two values 1/5 and 5. If the scaled recruitment rate takes these values with equal frequency, then the average growth rate can be found as the midpoint of the line joining the two values of the scaled growth rate, for example, the point C joining A and B in Fig. 14.2. Although the figure illustrates only the case $\delta_i = 0.1$, this average value is positive for all values of δ_i except for $\delta_i = 1$, indicating that the population will show a net increase over all periods in which these two values for the scaled recruitment are equally frequent. The actual change in log population size is found by multiplying the average growth rate by the time period involved.

Values of the average growth rate for cases when the two values of the recruitment rate are not equally frequent are found in a similar manner. The points D and E are placed at 1/4 and 3/4 of the distance from $\rho = 1/5$ to $\rho = 5$, corresponding to cases where poor recruitment has a relative frequency of 3/4 and 1/4. The intersections of the vertical lines through D and E with the straight lines joining the values of the growth rates give the average growth rates with these recruitment frequencies.

With nonoverlapping generations the average growth rate is always negative whenever the poor recruitment is more frequent. However, with overlapping generations it is possible to have a positive average growth rate even though recruitment is poor much of the time. The smaller the adult death rate, the more likely it is that this will be so, for then the scaled growth rate is more strongly curved as a function of scaled recruitment. When adult death rates are small, survival from periods of strong recruitment can more than compensate for periods of poor recruitment, even when periods of poor recruitment are more frequent (provided such poor recruitments are not too frequent). This is reflected in the lower limit (equation 14.5) to the growth rate. Moreover, the average growth rate is relatively insensitive to the actual magnitude of poor recruitments (Fig. 14.3). Indeed, positive average growth is still possible in the face of complete recruitment failure ($\ln \rho = -\infty$), provided strong recruitments occur at other times.

These ideas extend beyond the case where re-

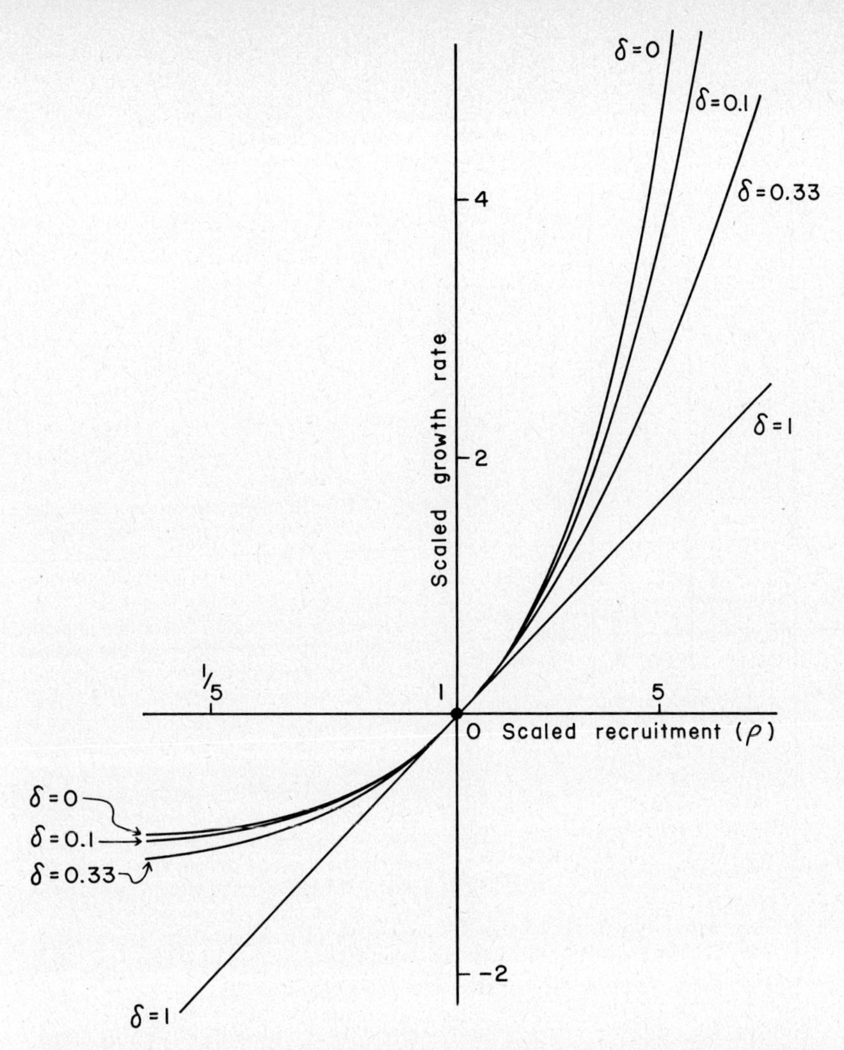

Fig. 14.1 Relationship between scaled growth rate (per capita growth rate scaled by the generation time) and scaled recruitment (per capita recruitment scaled by the generation time), for different values of the adult death rate, δ. Scaled recruitment is represented on a log scale. For $\delta = 1$ the relationship is linear, but for $\delta < 1$, the relationship is curved, with stronger curvature for smaller δ. Note that for all cases of $\delta < 1$ the scaled growth rate approaches a finite constant as scaled recruitment decreases.

cruitment takes on just two values to cases of arbitrary variation in recruitment (Chesson 1983). In particular, the simple idea that positive average growth is possible on the basis of periods of strong recruitments alone, when generations are overlapping, applies generally and has been termed the "storage effect" (Chesson 1983).

The interest here in the effects of variable recruitment and overlapping generations is not so much for what they have to say about a single species, for that has been discussed at length by others (Murphy 1968, Schaffer and Gadgil 1975, Hastings and Caswell 1979, Goodman 1984), but for what they have to say about coexistence of a set of interacting species. For example, species in strong competition may depress each other's recruitment rates. The results above show that a species may still be able to have a positive aver-

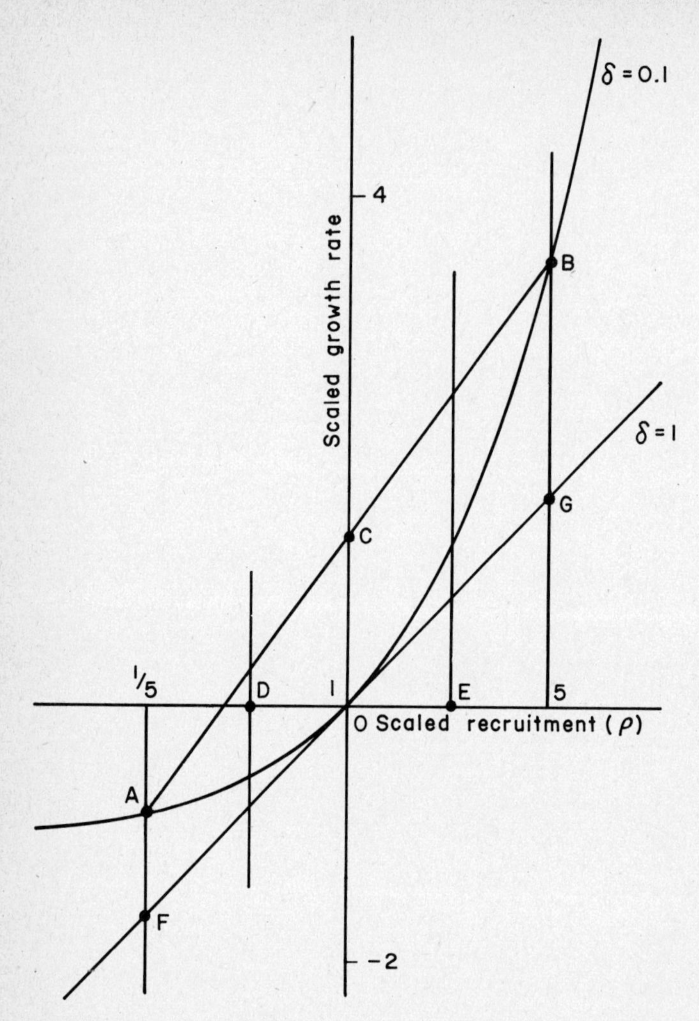

Fig. 14.2 Average growth rates in a fluctuating environment. If the scaled recruitment, ρ, fluctuates between 5 and 1/5, then the scaled growth rate fluctuates between the values A and B ($\delta = 0.1$) or F and G ($\delta = 1$). The average growth rate lies on the straight line joining the two values of the scaled growth rate. For example, if ρ takes its two values equally frequently, then the average growth rate is at C ($\delta = 0.1$) or 0 ($\delta = 1$). If $\rho = 1/5$ occurs three times more often than $\rho = 5$, then the vertical line through D intersects lines AB and FG at the average growth rates. If $\rho = 5$ occurs three times more often than $\rho = 1/5$, then the line through E intersects at the average growth rates. Note that with $\delta = 0.1$ the average growth rate is positive in all of these cases, while for $\delta = 1$ (nonoverlapping generations) the average growth rate is positive only when the value $\rho = 5$ is more frequent.

age growth rate when faced with strong competition, provided it still has periods when it is able to recruit well. Moreover, it can do this even if much of the time it is at a disadvantage to other species and often recruits poorly. As suggested by Hutchinson (1961), variation in the environment may vary the relative competitive abilities of species, leading to strong recruitments for different species at different times.

A positive average growth rate at low density is important for persistence and, indeed, is a useful stochastic persistence criterion (Turelli 1978, 1981; Chesson 1982). At low density intraspecific competition will be minimal. Consequently, if the environment sometimes gives a species the edge in terms of interspecific competition, the per capita number of recruits, $R_i(t)$, may be large. If such good recruitments are sufficiently strong, given their frequency, then the species will be able to recover from low density and hence persist in the system; its average growth at low density will be positive. In this way a number of negatively interacting species may be able to coexist as a result of varying recruitment rates.

The Lottery Model

The idea that varying recruitment can promote the coexistence of competing species is simply illustrated by the lottery model of competition (Chesson and Warner 1981). In this model competition is assumed to be for space. Although

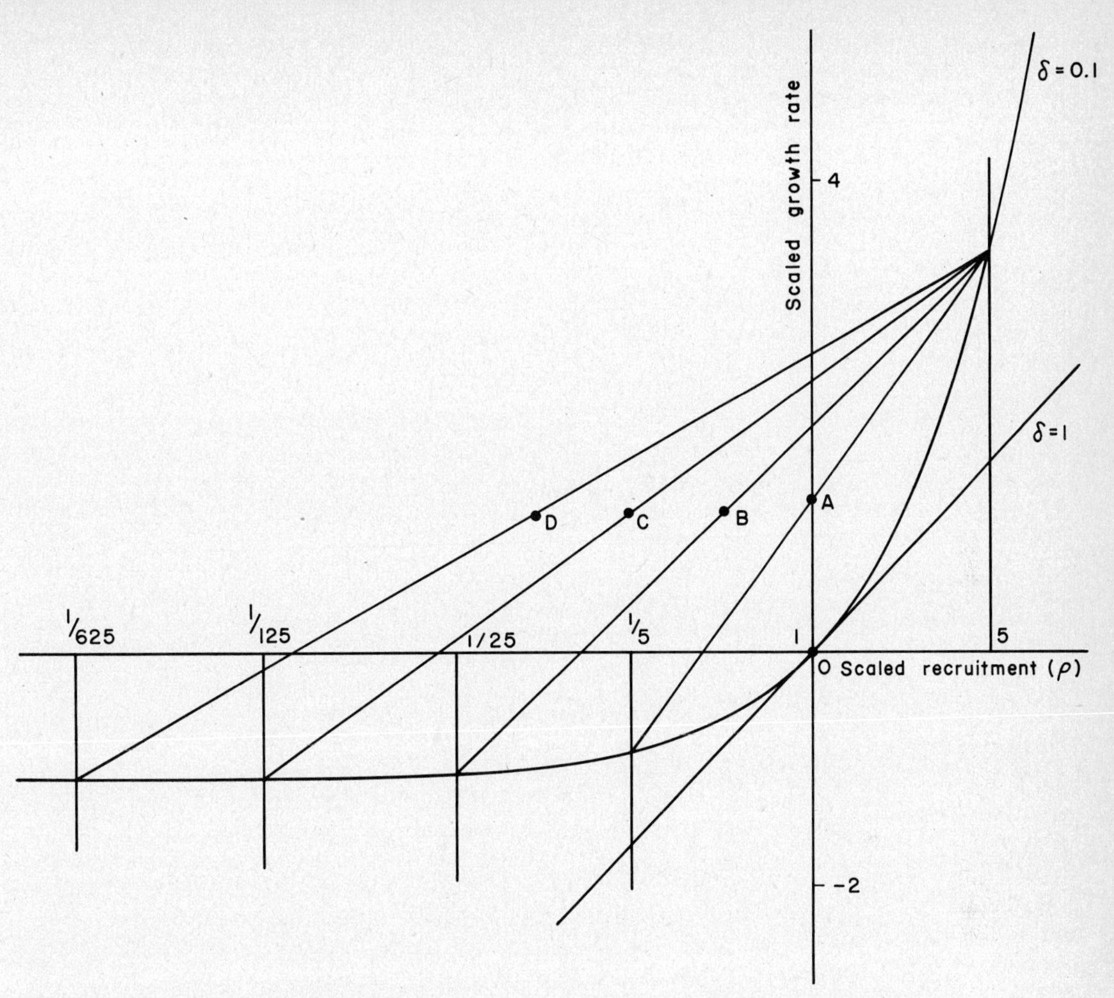

Fig. 14.3 Insensitivity of average growth rates to the magnitude of poor recruitment. The two values of recruitment are assumed equally frequent. The points A through D give the average growth rates when poor recruitment has the values 1/5 through 1/625, for $\delta = 0.1$. (A through D are the midpoints of the line segments.) For nonoverlapping generations ($\delta = 1$) the average growth rates are strongly negative when the poor recruitment is 1/25 or less.

originally formulated to apply to territorial reef fishes, as discussed by Sale (1977), it is a useful model for other space-holding organisms as well. In particular, Comins and Noble (1985) have applied the model to forest trees, and Shmida and Ellner (1985) have suggested modifications of the model for other plant communities. Woodin and Yorke (1975) discussed a similar model for soft-bottomed benthic invertebrates, and the model is also closely related to Hubbell's (1979) model for tropical forests as discussed below.

In the lottery model potential recruits become adults by acquiring units of space, and space becomes available by death of adults. Thus, for a two-species system the amount of space becoming available is

$$\delta_1 X_1(t) + \delta_2 X_2(t) \qquad (14.6)$$

because this is the number of adults that have died. The number of juveniles of species i seeking to settle in this space is assumed to be $\beta_i(t)X_i(t)$, where $\beta_i(t)$ is the product of the per capita birth rate and the presettlement juvenile survival rate and is assumed to be a function of the environment, $E(t)$. In the forest example $\beta_i(t)$ might represent the average number of viable seeds produced by a tree in year t, and it may also include any density-independent mortality of seedlings. It can be thought of as representing the number of adult trees that would result from the average seed crop of a tree of species i, in year t, if the seedlings were grown in the absence of competition from other juveniles or adults. In the reef fish example $\beta_i(t)$ represents the product of the brood size of a female for year t and juvenile survival in the plankton for that year.

The lottery model assumes that space is allocated at random (by ''lottery'') to the juveniles of the two species. Thus, the fraction of the available space allocated to juveniles of species i is

$$\frac{\beta_i(t)X_i(t)}{\beta_1(t)X_1(t) + \beta_2(t)X_2(t)} \qquad (14.7)$$

Multiplying equations 14.6 and 14.7 to get the number of new recruits and adding in adult survival, we find that the adult population at time $t + 1$ is

$$X_i(t+1) = (1 - \delta_i)X_i(t) + [\delta_1 X_1(t) + \delta_2 X_2(t)] \frac{\beta_i(t)X_i(t)}{\beta_1(t)X_1(t) + \beta_2(t)X_2(t)} \qquad (14.8)$$

The formula for the recruitment rate is

$$R_i(t) = [\delta_1 X_1(t) + \delta_2 X_2(t)] \frac{\beta_i(t)}{\beta_1(t)X_1(t) + \beta_2(t)X_2(t)} \qquad (14.9)$$

It is important to note that in these equations the species densities sum to the total amount of space, so that when one species is at low density, the other species is at high density.

As noted above, the important values of the scaled recruitment rate are those occurring at low density. Setting $X_i(t) = 0$ in equation 14.9, these low-density values are found to be

$$\rho_1(t) = \frac{\beta_1(t)/\delta_1}{\beta_2(t)/\delta_2}$$

$$\rho_2(t) = \frac{1}{\rho_1(t)}$$

These expressions compare the ratio of β and δ for one species with the ratio for the other species. Although calculated for low density, a little algebra shows that these values for the $\rho_i(t)$ determine for all densities whether a species will be increasing. Thus, if $\rho_1(t) > 1$, then $\rho_2(t) < 1$, and species 1 increases at the expense of species 2. Since the $\rho_i(t)$ are determined entirely by the environment (in this system), there is no possibility of coexistence if the environment does not vary. (The best that can be achieved is a neutral equilibrium when $\rho_1(t) \equiv \rho_2(t) \equiv 1$.)

The reason for the failure of coexistence in a constant environment can be understood from the fact that among juveniles interspecific and intraspecific competition is equally intense, for only in this way can allocation of space be random. It follows that whichever species is at an advantage either reproductively or in adult survival will be able to displace the other species. However, when the environment varies, the advantage may shift between the two species. If generations are overlapping, this variation can lead to coexistence in the sense that each species shows upward trends from low density.

To see how coexistence occurs in variable environments, consider first a case where the species are equal on average, but the environment fluctuates so that the high value of ρ in Fig. 14.2 occurs half the time and the low value occurs the rest of the time. Note that when one species has the high value of ρ, the other has the low value of ρ. The average low-density growth rates for both species are then obtained at the intersection of the vertical axis with the line joining the values of the growth rates. Thus, the mean growth at low density is positive for both species. This is illustrated in Fig. 14.2 for the case $\delta_i = 0.1$, but positive average growth rates for both species occur in all cases except the case of nonoverlapping generations ($\delta_i = 1$). When $\delta_i = 1$, the average growth rates are 0, and consequently a random walk occurs with one species eventually approaching extinction (Chesson and Warner 1981).

If the conditions are changed so that one species has an advantage over the other (for example, has three times as many favorable periods as the other), then the vertical lines D and E of Fig. 14.2 are used, instead of the vertical axis, to get the average values of the growth rates. The average growth rate of the advantaged species is given by line E, while that of the disadvantaged species is given by line D. Note again that when generations are overlapping, both species can have positive average growth rates at low density and thus will tend to increase whenever their densities become low. With nonoverlapping generations, however, only one species can have a positive average growth rate at low density, and the other species therefore must become extinct. Overlapping generations have introduced an asymmetry between the effects of good periods and bad periods; poor recruitments can be made arbitrarily poor to the point of complete recruitment failure without a large change in the average growth rates.

Thus, when the environment varies in such a way that each species has times when it can have strong recruitments, the net effect is to favor positive growth rates at low density for both species. The fact that both species tend to increase from low density means that they will coexist in the system. More detail on the nature of the coexistence is given later, but the important thing to note at this stage is that both species show upward trends at low density. Without variation in the environment it is only possible to have one species with an upward trend at low density. Thus, the stochastic environment has changed the overall trends in population growth. It has not merely slowed down or delayed competitive exclusion.

GENERALIZATIONS OF THE LOTTERY MODEL

There are two critical requirements for coexistence in the lottery model, and these two requirements are found to lead to coexistence of competing species quite generally. First, environmental variation should permit each species to have periods when it has strong recruitment rates at low density. Second, generations must be overlapping, with adult death rates not greatly affected by competition. The second requirement places a lower bound on the growth rate and ensures that strong recruitments will have a greater effect on the average growth rate than poor recruitments.

These two requirements together favor positive average growth rates and promote coexistence, given a broad variety of forms for the recruitment rate as a function of both the environment and species densities (Chesson 1983, 1984; Warner and Chesson 1985). For coexistence it is not necessary that recruitments of the two species should be strictly negatively related. The negative relationship in the lottery model is a result of the assumption that all space is filled and thus is an outcome of strong competition, not a requirement for coexistence. However, variation in the ratios of recruitment rates for different species in the system does seem to be a general requirement.

These ideas are not restricted to coexistence of just two species, but can apply to coexistence of an arbitrary number of species. Moreover, the variation in recruitment that is required for coexistence of two or many species does not have to be great if the species are similar to each other in average recruitment rates, mortality rates, and competitive abilities (Chesson 1984). Large variation in recruitment will be necessary for coexistence if there are big differences in these average properties of the species.

Overlapping generations are an essential part of coexistence achieved by temporal fluctuations in recruitment. However, overlapping generations can be achieved in ways other than through adult survival. For instance, annual plants with a seed bank have overlapping generations. Indeed, Ellner (1984) has discussed models that rely on variable recruitment to a seed bank and has demonstrated that coexistence results from variation and from overlapping generations. Both means of achieving coexistence are examples of the storage effect in action (Chesson 1983, 1984; Warner and Chesson 1985). The idea is that recruitments from different breeding periods become added together in the adult population or the seed bank; essentially, they are stored there subject to a rate of attrition equal to their death rates. These stored recruits are used every breeding season, but their use is most important during favorable periods, because then strong recruitments can replenish the store. That an adult population can be composed mostly of individuals from a few strong recruitments is well known for marine fishes, where good data are available (Gulland 1982, Cushing 1982). Such strong recruitments must be the major contributors to future population growth. However, the main point here is that coexistence of competing species is promoted by such storage of recruitment. A species that normally recruits poorly because of competition with other species can nevertheless persist if it has occasional good periods when it recruits well and replenishes its store.

One can think of recruitment also in terms of acquisition of resources by the individuals in a population, especially if future reproduction is closely related to the total amount of resource that the population holds (Abrams 1984a, Warner and Chesson 1985). The storage of resource provides protection against periods when resource acquisition is poor. If the times that are favorable for acquisition of resources are different for different species, coexistence will be promoted.

Since fishes often have such obvious storage of recruitment in the adult population, they are candidates for coexistence by the storage effect, if competition can be shown to affect recruitment. Grubb (Chapter 12) discusses an assemblage of long-lived perennial herbs that have fluctuating recruitment rates. Although Grubb identifies a number of other possible explanations of coexistence, it seems likely that the storage effect at least contributes to coexistence. Other likely areas of application are perennial plants in general, long-lived benthic marine organisms, and annual plants with a seed bank (Shmida and Ellner 1984, Warner and Chesson 1985).

The lottery model and the generalizations discussed here all assume that the population is large enough for population size to be treated as a continuous variable. Moreover, demographic stochasticity (May 1973a) or within-individual variability (Chesson 1978) is ignored. This can be done only for large populations and only under certain conditions (Chesson 1981). These models, together with the majority of deterministic models in ecology, have been termed "infinite population models" (Chesson 1982) to correspond to the terminology of population genetics. Models that recognize within-individual variability and have discrete changes in population size are called "finite population models." As discussed by Chesson (1982), the results of most finite population models converge on those of their infinite population counterparts as population size is increased, and this is known to be true for the finite version of the lottery model (Chesson 1982). Thus, such infinite population models can be validly taken to represent the properties of large, yet still finite populations.

GENERAL CONDITIONS FOR COEXISTENCE IN A TEMPORALLY VARYING ENVIRONMENT

We have seen how a temporally varying environment promotes the coexistence of competing species. However, such variation need not always be favorable to coexistence. In fact, depending on the circumstances, a temporally varying environment may do nothing or may even promote competitive exclusion (Turelli and Gillespie 1980, Turelli 1981, Chesson and Warner 1981). These different results can be understood in terms of how the effects of the environment on different species in the system interact.

To discuss the interaction of environmental effects, we first note that each species will have a response to the environment that will be expressed in quantities like the birth rate or death rate, which we normally use as parameters in a model. We shall now assume that just a single parameter varies with the environment, and this parameter will be denoted by E' for species 1 and E'' for species 2. For instance, in the lottery model $E' = \beta_1(t)$, the parameter giving the per capita rate of production of juveniles of species 1, and $E'' = \beta_2(t)$. We have seen previously that the instantaneous growth rate of a species at low density, which we shall call g here, is critical to persistence. This growth rate will often depend on both E' and E''. For example, in the lottery model the value of g for species 1 is

$$g = \ln\left[(1 - \delta_1) + \frac{\delta_2 E'}{E''}\right] \qquad (14.10)$$

where $1 - \delta_1$ represents adult survival and $\delta_2 E'/E''$ is per capita recruitment.

A general symmetrical model of two competing species, incorporating both E' and E'' in the growth rate of each, is analyzed in the appendix to this chapter. The symmetry of the model implies that the species have the same sorts of population dynamics, affect each other symmetrically, and experience the same frequencies of environmental states, but usually at any one time E' will not equal E''. Also, interspecific and intraspecific competition are equal in this model. In spite of these simplifications the model is in other respects very general, and it seems to indicate what is likely to occur in broad circumstances.

The symmetry properties of the model imply that in a constant environment a species at low density will have a 0 growth rate, that is, a 0 value of g, and will not be able to recover from that low density. However, when the environment varies, the average value of g can be different from 0, and this new value depends on the interaction between E' and E''. The interaction between E' and E'' is said to be negative or antagonistic if a simultaneous increase in both E' and E'' leads to a change in g less than that accounted for by the sum of the changes due to each variable considered separately. This is the situation for g in the lottery model (equation 14.10), and in the general symmetrical model it leads to a positive average value of g for both species.

Thus, when E' and E'' are antagonistic, the two species will coexist; this result will apply regardless of the magnitude of the environmental variation, subject only to the condition that E' and E'' are not always equal, that is, the effects of the environment on the two species must sometimes be different.

Conversely, if E' and E'' are synergistic, the average values of g will be negative, and the stochastic environment will promote competitive exclusion.

Finally, if E' and E'' have just additive effects on g, the stochastic environment will yield 0 average values for g and will promote neither coexistence nor competitive exclusion.

Specific examples of antagonism between E' and E'', and hence coexistence in a stochastic environment, include not only the lottery model but also the consumer-resource models of Abrams (1984a) and the seed bank model of Ellner (1984) with randomly varying germination rates. These examples include asymmetrical cases, suggesting that antagonism and a variable environment will generally promote coexistence in such cases as well. However, in asymmetrical cases g will not be 0 in a constant environment. Whether the average value of g is made positive for both species in a stochastic environment will depend on both the degree of antagonism and on the magnitude of the environmental variability.

In the consumer resource models of Abrams (1984a) E' and E'' are the resource consumption rates of the two species. From these models Abrams notes that the storage effect is not the only way that coexistence can result from temporal variation in the environment. Indeed, the results above show that antagonism between E' and E'' is the key. However, this does not alter the fact that the storage effect is necessary in a number of specific models (Chesson 1983) and is operative in most of Abrams' specific examples. Moreover, the overlapping generations feature provides a general mechanism by which antagonism can occur. By giving a lower bound to the growth rate, overlap in generations means that a low value of E' and a high value of E'' will not be as harmful as suggested by the sum of their separate effects. Thus, it will lead to antagonism.

The failure of temporal variation to have interesting effects in the stochastic version of Lotka-Volterra competition analyzed by Turelli and Gillespie (1980) can be traced directly to considerations of this section. In their model the environmental variation is included in the per capita growth rate in an additive manner. Most important, the instantaneous growth rate of a species involves only the environmentally varying parameters of that species, that is, only E' is included in the growth rate of species 1 and only E'' in the growth rate of species 2. Thus, in the symmetrical cases studied by Turelli and Gillespie (1980) the environmental variation has no effect on coexistence. Asymmetrical cases studied by them give mixed results, but their essence is that environmental variation neither broadens nor narrows the conditions for coexistence. The discrete versions of stochastic Lotka-Volterra competition studied in some detail by Turelli (1981) similarly have each species' instantaneous growth rate depending only on its own environmentally varying parameters, and so these models also show minimal effects of environmental variation.

It is possible to build models involving the storage effect that do not include both E' and E'' in a species' growth rate, and then the storage effect does not promote coexistence. However, when both E' and E'' are included, the storage effect is likely to promote coexistence as explained above. If environmental variation affects birth rates and juveniles compete, then both E' and E'' are necessarily included in the growth rates of both species. All published models of the storage effect have this feature.

The considerations of this section also allow us to examine cases of the lottery model in which death rates, rather than juvenile production rates (birth rates), vary. It is then easily checked that E' and E'' are synergistic and as a consequence promote negative growth rates at low density for both species, as discussed by Chesson and Warner (1981). The lottery model with variation in both birth rates and death rates involves a combination of two opposing effects. However, if death rates are generally small, birth rate variation is more important than death rate variation and continues to promote coexistence, but the quantitative details are modified by death rate variation and its correlation with birth rate variation.

COEXISTENCE MEDIATED BY VARIATION IN SPACE

That spatial variation in the environment or in species abundances can promote coexistence is an old idea, discussed in detail by numerous authors (Levin 1974, 1979; Yodzis 1978; Goh 1980; Dale 1978; Hastings 1980). The purpose of this section is to point out how spatial variation in recruitment, in particular, can promote coexistence in a manner that is closely related to coexistence promoted by temporal recruitment fluctuations. In a patchy environment recruitment can vary in space in two ways. First, there may be a fixed spatial pattern of heavy and light recruitment, that is, some places may always be superior sites for recruitment of a particular species. Second, the spatial pattern of heavy and light recruitment may change with time; that is, sites of heavy recruitment one year may be sites of light recruitment the next year and vice versa. If offspring disperse among patches, then coexistence of competing species is promoted by either kind of spatial variation in recruitment (Chapter 30; Chesson 1984, 1985; Comins and Noble 1985). The mechanism of coexistence is essentially the same as that for temporal variation, with dispersal taking the place of overlapping generations.

That coexistence can result from spatial variation in recruitment is important, because such spatial variation is found commonly in marine systems of space-holding organisms (see reviews in Chapter 30 and by Caffey 1984). Grubb (Fig. 12.6 and Table 12.5) demonstrates the same conclusion for a terrestrial plant community. Moreover, Hubbell and Foster (Chapter 19) point out that a spatially varying pattern of recruitment is to be expected in a tropical forest, because seedlings may remain in a suppressed state beneath the canopy until a canopy gap occurs above a particular group of seedlings. This effect reduces temporal variation in recruitment of a species for the forest as a whole, but introduces a temporally varying spatial pattern of recruitment, and this sort of recruitment variation can also promote coexistence.

Hubbell and Foster (Chapter 19) favor a different hypothesis for the diversity of tropical forests, suggesting that species abundances will be subject to drift in an area of forest through independent random processes of birth and death for individual trees (demographic stochasticity or within-individual variability) and that local diversity is maintained by immigration from neighboring areas. Hubbell's (1979) model for drift at a locality corresponds to the finite population version of the lottery model (Chesson 1982), but incorporates no environmental variability and therefore has no tendency to stabilize species numbers.

Yodzis (1978 and Chapter 29) also focuses on local community dynamics where migration is an important process in maintaining the community in an environmentally homogeneous setting. Unlike Hubbell, however, Yodzis assumes that deterministic processes are responsible for species extinction at a locality.

THE NATURE OF COEXISTENCE

The definition of persistence that has been used above is essentially the idea that a species should be regarded as persisting if it tends to increase from low density. We have seen how, under certain circumstances, environmental variation can change the trends in population growth so that all species tend to increase at low density, and we have regarded this as meaning that the species coexist. This definition of coexistence is usually called "invasibility," and it is also used with deterministic models (e.g., MacArthur and Levins 1967, Armstrong and McGehee 1980).

It is now common to view coexistence in a deterministic model in terms of what it ought to mean when environmental variability is imposed. In this setting invasibility captures the essence of coexistence, for in the presence of environmental variability it is unreasonable to expect a community to remain at equilibrium, and stability of the equilibrium is then often seen as providing restoring forces back toward equilibrium and away from extinction of any of the species (Holling 1973; Goh 1975, 1976; Beddington et al. 1976). Thus, stability of the equilibrium creates trends in the direction of increase from population densities that are low relative to the equilibrium density. Of course, invasibility in a stochastic model involves a fluctuating increase about a mean trend. However, global stability of the equilibrium of a deterministic model is surely not to be viewed any differently when translated into its meaning for a community in the stochastic real world.

Although invasibility involves some key properties of coexistence and persistence, it is not regarded as a complete criterion for coexistence in either deterministic models or stochastic models. The reason is that although it guarantees positive trends at low density, it still does not eliminate the possibility of long periods of time being spent at low values (Armstrong and McGehee 1980, Chesson 1982). To prevent this, additional criteria are needed, and these can often be shown to apply to coexistence in stochastic models (Turelli and Gillespie 1980; Chesson 1982, 1983; Ellner 1984).

In some cases it is possible to get a detailed idea of the population fluctuations that are produced in a stochastic model. For instance, in the lottery model one can determine the probability distribution of population size when adult death rates are small. This probability distribution is approximately normal, with variance proportional to the adult death rate. This means that when adult death rates are small, population fluctuations will be small, regardless of the time scale on which the fluctuations are observed. Thus, if one used a time scale commensurate with the generation time, as recommended by Connell and Sousa (1983), one would judge a system of long-lived species that coexist according to the lottery model to be a very stable assemblage.

These results appear to apply quite generally to models of recruitment variation Chesson (1984). They can be understood from the fact that in a long-lived species the adult population will be a sum over many seasons of recruitment, and thus the fluctuations at the level of recruitment will average out in the adult population. This phenomenon has been observed for real populations as well as for models (Chapter 12; Cushing 1982), and in models it permits the dynamics of the adult population to be described approximately, but not explained, by deterministic equations (Chesson 1984). However, the averaging out of recruitment fluctuations certainly does not mean that they are unimportant, for in some models (e.g., the lottery model) these fluctuations are necessary for coexistence, even though they do not cause large fluctuations in adult population densities. The resolution of this paradox involves the fact that the average of a nonlinear function is generally different from the nonlinear function of the average, $\overline{f(X)} \neq f(\overline{X})$. The recruitment rate is generally a nonlinear function of more basic quantities like the β_i, and even though the variation in recruitment tends to average out, the average that is achieved reflects variation in the β_i. That is, it depends on the variances of the β_i, not just their means. Moreover, since the function that converts the β_i into recruitment rates depends on population densities, the effect of variation in the β_i on the average recruitment is density-dependent and has much to do with coexistence in models of recruitment variation.

The overall conclusion from this section is that the kind of coexistence that results from environmental variation is not fundamentally different from that found in deterministic models.

DETERMINISTIC VERSUS STOCHASTIC MECHANISMS OF COEXISTENCE

What is a stochastic mechanism of coexistence? The lottery model has the property that only one species can persist in the system in a constant environment. However, if the environment varies, coexistence of two or more species is possible. Thus, it seems appropriate to regard variation as the mechanism of coexistence. But if the organisms are long-lived, little fluctuation in population density may be present, even on a time scale that is long compared to the generation times of the species. Thus, stochastic mechanisms of coexistence do not have to be associated with obvious stochastic variation in population densities.

As discussed previously, Turelli's and Gillespie's (1980) study of a stochastic version of the Lotka-Volterra model revealed essentially no overall effect of temporal variability on coexistence. Species may coexist in this model, but they do so because of factors that are unrelated to stochastic variation. For example, one such factor is resource partitioning; increased overlap in resource use will prevent coexistence. Such mechanisms that do not involve stochastic processes may be referred to as "deterministic." However, stochastic variation may cause large population fluctuations in this stochastic version of Lotka-Volterra competition, even though it has little effect on coexistence.

The mechanism of coexistence is determined by the effect that its addition or removal has on coexistence. It cannot generally be recognized by presence or absence of population fluctuations, for both deterministic and stochastic mechanisms are compatible with population fluctuations of any magnitude. Methods for determining the mechanism of coexistence on the basis of population fluctuations (Grossman 1982, Grossman et al. 1982) are therefore unreliable. The hypothesis that the mechanism of coexistence is stochastic is too general to be tested by crude population dynamics alone. However, specific forms of the hypothesis can be tested. For example, if the hypothesis is coexistence by the storage effect, then variation in recruitment rates should be observed. Moreover, the amount of recruitment variation can be compared with the contribution from other factors to see if it is a sufficient explanation of coexistence (Warner and Chesson 1985).

CONCLUSION

Stochastic models have revealed an enormous potential for stochastic mechanisms of coexistence. These mechanisms often involve the interaction of variation with the life history characteristics of organisms. For example, the interaction of recruitment variation and overlapping generations leads to stochastic coexistence by the storage effect. In the case of the storage effect overlapping generations largely function in a manner that moderates the effect of unfavorable periods. An attractive hypothesis is that natural selection will often act in a manner that leads to life history traits with the effect of moderating unfavorable conditions. The results discussed above for the general symmetrical model of competition in a stochastic environment suggest a corollary to this hypothesis: Natural selection will often lead to life history traits that promote coexistence in a stochastic environment.

Coexistence under a stochastic mechanism can be just as strong and stable as that occurring under a deterministic mechanism. All species can show positive average growth rates at low density and may even give the appearance of tending toward a stable equilibrium point. Indeed, deterministic approximations can be found to some stochastic models that depend heavily for their properties on the presence of the stochastic components. Environmental variation can change the trends in population growth so that a positive central tendency, much like a stable equilibrium point, appears. Environmental variation need not always change the trends in a direction that favors coexistence; sometimes it may instead favor competitive exclusion. The important point is that the potential effects of variation are not just to increase population fluctuations or to slow down competitive exclusion, but to create new trends as well. In this sense variation is not very different from other sorts of factors affecting population and community dynamics.

SUMMARY

Ecologists have viewed stochastic temporal variation in a number of different ways: as noise that obscures the important features of a system, as a destabilizing factor that decreases diversity, and as a factor that interrupts or reverses the process of competitive exclusion and thus promotes diversity by slowing the trends to competitive exclusion. Models of stochastic environments have led to a fourth view: variation is a factor that changes trends by altering average population growth rates. In particular, a stochastic environment can convert competitive exclusion into a situation in which all species have positive average growth rates at low density. Thus, variation reverses the trends toward competitive exclusion; it does not merely slow these trends down.

Models of recruitment variation illustrate this last view. These models are intended primarily for organisms, such as perennial plants, fishes, and sessile marine organisms, in which recruitment to the adult population is much more variable than adult survival. However, modifications of these models can apply to other communities, such as annual plants with seed banks.

A variable environment need not always change trends in a direction that promotes coexistence; trends in the direction of competitive exclusion can sometimes be created. The different circumstances for these different effects can be deduced from a general symmetrical model of competition in a variable environment.

Stochastic mechanisms of coexistence, such as recruitment variation, share many properties with traditional deterministic mechanisms of coexistence, such as resource partitioning or frequency-dependent predation. For both stochastic and deterministic mechanisms, coexistence involves positive average growth rates at low density for all species, and the actual population dynamics observed for the two sorts of mechanisms can be quite similar.

ACKNOWLEDGMENTS

Discussions with many people have helped me put this chapter together, but I am especially grateful to Ted Case for suggesting its overall structure and for guiding me through the various stages of its preparation. Theresa Dahlem, Jared Diamond, Peter Grubb, Stephen Hubbell, Donald Strong, and Peter Yodzis provided helpful comments on the manuscript.

APPENDIX

A general model of two-species competition in a stochastic environment is given by the following equation:

$$X_i(t+1) = X_i(t)G[E_i(t); E_1(t), X_1(t); E_2(t), X_2(t)] \quad (14.A1)$$

where the function G of five variables is the finite rate of increase of species i, and i can be either 1 or 2. $E_i(t)$ is a parameter giving the effect of the environment on species i, and it occurs in two places in equation 14.A1. Its function in the first place, before the first semicolon, is assumed to relate to the favorability of the environment of species i in the absence of competition. If the environment is favorable to species i, then it may contribute more competition to the system, for example, by producing more juveniles to compete, as in the lottery model. The second place that $E_i(t)$ occurs, as either $E_1(t)$ or $E_2(t)$, is assumed to relate to competition by species i. G is thus independent of the value of $E_i(t)$, in this second term, when $X_i(t)$ is equal to 0 (when species i cannot contribute any competition to the system).

Two symmetry assumptions are imposed on this model. The first is that

$$G[E_i(t); E_1(t), X_1(t); E_2(t), X_2(t)] = G[E_i(t); E_2(t), X_2(t); E_1(t), X_1(t)]$$

This assumption can be interpreted as meaning that the contribution of a species to the level of competition in the system does not depend on the species' identity. The second symmetry assumption is that the bivariate distribution of $[E_1(t), E_2(t)]$ is the same as that of $[E_2(t), E_1(t)]$. Thus, the two species will have the same frequencies of favorable periods, and the frequency of times when it is good for species 1 but bad for species 2 will equal the frequency of times when it is good for species 2 but bad for species 1.

Two additional assumptions have nothing to do with symmetry. First, the values of the environment are independent from one time to the next. (This is the standard random environment assumption.) Second, $E_1(t)$ is sometimes different from $E_2(t)$.

The analysis of this model is a standard invasibility analysis (Turelli 1978). If species 1 is absent from the system, then the resident species (species 2) is assumed to have an equilibrium probability distribution, and its mean instantaneous growth rate with respect to this distribution is 0, that is,

$$\mathbf{E}g[E_2(t); 0, 0; E_2(t), X_2(t)] = 0 \quad (14.A2)$$

where $g = \ln G$, and $\mathbf{E}$ means "theoretical mean" or "expected" value. The mean growth rate of species 1 (the invader) while at low density will be

$$\mathbf{E}g[E_1(t); 0, 0; E_2(t), X_2(t)] \quad (14.A3)$$

and species 1 will be able to persist in the sense of invasibility if expression 14.A3 is positive. To identify the situations when this will be so, I define the environmentally dependent average growth rate, $h(E', E'')$, according to the formula

$$h(E', E'') = \mathbf{E}g[E'; 0, 0; E'', X_2(t)] \quad (14.A4)$$

where E' and E'' are treated as nonrandom values for the purpose of taking the average. Thus, the expected value in equation 14.A4 averages over the possible values of the resident density with the environment held fixed.

In terms of h, equations 14.A2 and 14.A3 become

$$\mathbf{E}h[E_2(t), E_2(t)] = 0 \quad (14.A5)$$

and

$$\mathbf{E}h[E_1(t), E_2(t)] \quad (14.A6)$$

The symmetry assumption on the environmental values now implies that

$$\mathbf{E}h[E_1(t), E_1(t)] = \mathbf{E}h[E_2(t), E_2(t)] \quad (14.A7)$$

and

$$\mathbf{E}h[E_1(t), E_2(t)] = \mathbf{E}h[E_2(t), E_1(t)] \quad (14.A8)$$

Symmetry thus implies that the question of whether species 1 can increase from low density in the presence of species 2 is the same as the question of whether species 2 can increase from low density in the presence of species 1. Using equations 14.A7 and 14.A8, it is clear that both species will be able to increase from low density if

$$\begin{aligned} &\mathbf{E}h[E_1(t), E_2(t)] + \mathbf{E}h[E_2(t), E_1(t)] \\ &> \\ &\mathbf{E}h[E_1(t), E_1(t)] + \mathbf{E}h[E_2(t), E_2(t)] \end{aligned} \quad (14.A9)$$

Inequality 14.A9 will be true in general if the two arguments E' and E'' of the function $h(E', E'')$ are antagonistic, and then the two species will coexist. On the other hand, if E' and E'' are synergistic, both species will have negative mean instantaneous growth rates as invaders, and so competitive exclusion will be favored.

In the text antagonism and synergism are discussed for g rather than h, but since h is derived by averaging over g, antagonism or synergism for g implies the same for h. The analysis using h is, however, more general.

chapter 15

Density Vagueness: Abiding the Variance in the Demography of Real Populations

Donald R. Strong

Since all models are wrong, the scientist must be alert to what is importantly wrong. It is inappropriate to be concerned about mice when there are tigers abroad.

Box, 1976

INTRODUCTION

Every view of communities and ecosystems assumes some form of population model. Most population models in ecology are deterministic and emphasize negative feedback that is continuous over the entire spectrum of density. In this chapter I argue for more liberal models of population change, with looser relationships between density and change, especially at intermediate densities, where many populations spend most time. Substantial variance occurs routinely in the density relationships of natural populations; it is a ''tiger abroad'' in Box's sense. Other contributions to this volume that stress related aspects of natural variability are Chesson on ''views C and D,'' Grant on the demography of Galápagos birds, Schoener on island spiders, Colwell on hummingbird mites in flowers, Grubb and Hubbell and Foster on plant demography in diverse communities, Kareiva on goldenrod insects, and Wiens on shrubland migratory birds (Chapters 9–12, 14, 19, 24, 33).

Antecedents

Abiding the variance in density relationships is a venerable, if somewhat bohemian, tradition in ecology, and I believe that some of our precious conceptual antiques are displayed to good effect in this light. For example, the old row over density-independence and density-dependence can be profitably viewed as a disagreement over the significance of demographic variance. The outline of what I term ''density-vague'' demography was adumbrated by Milne's (1957) ''imperfect'' relationships between density and fluctuations of population, and much of my ''liberal population regulation'' was anticipated by Ito's (1961) diagrams of high suppressive ceilings to population transit, below which change is largely stochastic. Ehrlich et al. (1972) proposed a scheme similar to liberal regulation and called it "envelopist" population dynamics, and Southwood and Comins (1976) proposed similar possibilities within their "synoptic" population model.

Attention to the community implications of natural variability in concrete problems was paid by a few ecologists as early as the 1920s (Thompson 1956) and has developed substantially since then (e.g., Andrewartha and Birch 1954). Crucial conceptual advances from the Dutch school of ecology, den Boer's "Spreading the Risk . . ." and Reddingus' "Gambling for Existence" (Reddingus and den Boer 1970), and from others (Murphy 1968) elevated the study of natural variability from the status of an esoteric crotchet, as many had come to regard it during the 1950s and 1960s, to that of an exciting idea. Currently, the understanding of so many concrete processes in ecology is infused and anastomosed with natural variability that it has become a major theme of research (Roff 1974, Simberloff 1978, Chesson 1978, Underwood 1978, Hubbell 1979, Paine and Levin 1981, Chesson and Warner 1981, Sale 1982, Keough 1983, Keough and Butler 1983, Steele and Henderson 1984, Thompson 1984, Cappuchino and Kareiva 1985).

My thoughts on density vagueness have been greatly influenced by two graphical models that wed density-dependence (DD) to density-independence (DI): the models by Horn (1968) and Enright (1976). Horn treated DD and DI as separate entities and graphed a line of dN/dt as a function of N to represent DD, transected by an ever-increasing line representing DI. The result is fluctuations of N through time that combine effects of DD with DI. The two functions can change positions through time, hence can be thought of as having variance.

Where Horn's model might be considered a closet approach to natural variability in demography, Enright's model is in the open and plots birth and death rates as a function of N, both with and without variance. Either birth or death rate can be made DI if its mean value is drawn to have a slope of zero. The height of a DI rate on the graph relative to the other rate contributes to "equilibrium" population sizes, and in this way Enright showed that DI factors can influence the geographical limits of species.

DENSITY VAGUENESS

Dispersion, inconsistency, weakness, and indistinctness in the key demographic functions are what I call density vagueness: variance in relationships between density and birth, death, and migration rates. Density vagueness grades from being slight where there is little scatter about a relationship, through intermediate where the relationship is present but highly variable, to great when an expected density relationship can be inferred only by means of devout Malthusian faith. Density vagueness means more than contingent variation, which can be known deterministically and factored out by knowledge of another or other variables; density vagueness means inherent stochasticity.

Slight and tidy density vagueness is illustrated by a drawing of intersecting birth and death rates with variance (Fig. 15.1). In this example density vagueness is the variance about regression lines. Often, combined variances, such as in the difference between per capita birth and death in Fig. 15.1b, are broader than individual variances of the component functions (Travis 1982). So, we have an intermediate range of densities within which population change is not distinctly related to density; even this slight form of density vagueness precludes point equilibrium in population size. The slope of density relationships is also important. Shallow slopes, even though statistically significant, will have the weakest influence on population dynamics. Of course, higher variance about a regression line means a lower slope estimate, *ceteris paribus* (Pielou 1974, p. 56).

In nature, relationships between density and rates of birth, death, migration, and net population change are likely to be less neat than in Fig. 15.1b (e.g., Fowler 1981) and to take any of several forms, from indistinct "fuzzy thresholds" to "spreading scatters" of points with correlated mean and variance (Fig. 15.2). The differences between explicit and vague density relationships are like those between a smooth ramp and a rugged back-switching trail up a mountain. The richness and variety of a trail are not included in a

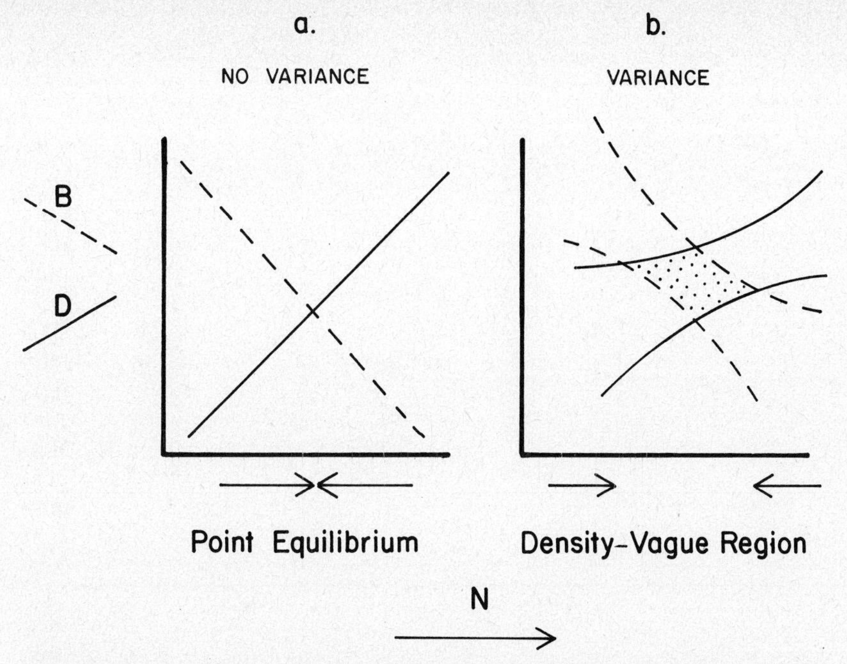

Fig. 15.1 Per capita birth (B) and death (D) rates as a function of population size (N). (a) With no variances. Point equilibrium in N occurs at the intersection of the two functions (assuming no migration). (b) With variances. The paired lines indicate prediction intervals around regression lines that create a density-vague region where the functions cross. Within the density-vague region change in N is stochastic.

ramp, and smoothed averages do not well describe variable density relationships.

Things are often not as neat as Fig. 15.2. Components of birth, death, or migration in any particular data set may be completely unrelated to density under the conditions and over the range observed (e.g., Strong, Lawton, and Southwood 1984, Table 5.2). The lack of a density correlation usually means not a constant density-independent rate with no variance, but rather a highly variable set of points with no trend in relation to density.

CONTINGENCY AND RESIDUAL VARIANCE

Operationally, observed variance in density relationships is composed of "contingent" variance and of residual or inherent variance. Contingent variance (contingency) is deterministic variation in the relationship accounted for by (an)other variable(s). For example, a density relationship in survivorship of mussels might be affected by moisture and temperature, which are greatly influenced by inundation. So, in turn, tide is another variable that would explain some of the variance in mussel survivorship during several weeks or months; the density relationship is contingent upon tidal inundation. But, even after tide and other variables with consistent influences are taken into account, still other influences will cause residual variance in the relationships. For our mussels, differences in the aspect and composition of particular rock faces or pilings, the happenstance of small population sizes, weather, waves, or predators, and even genetic and developmental differences among individuals all contribute to the residual variance that we find in natural density relationships.

This perspective of contingent and residual stochasticity is analogous to analysis of covariance, with contingent variables being covariates and residual stochasticity being variance about

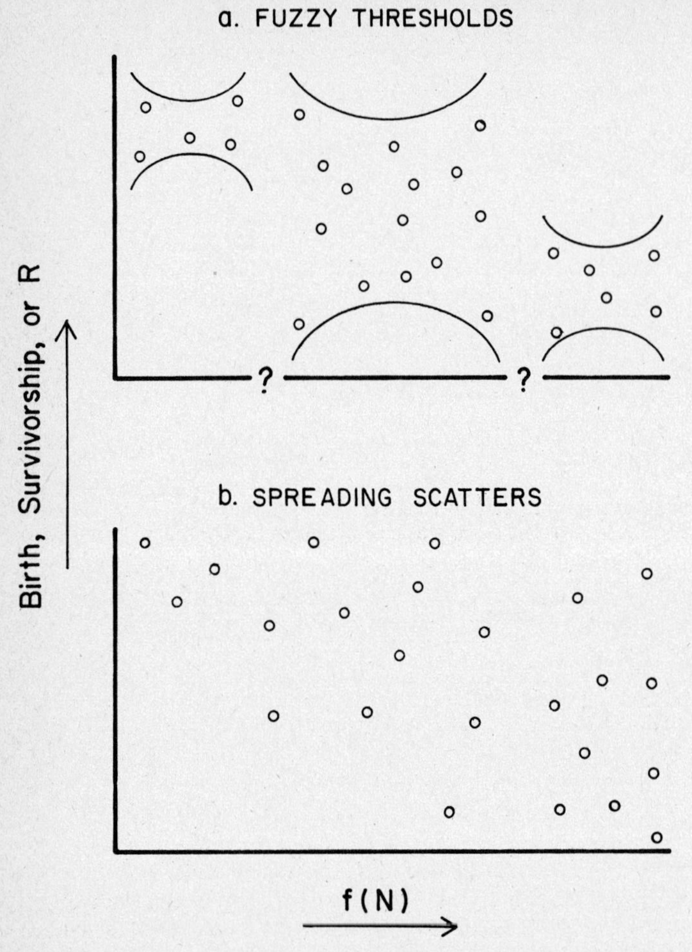

Fig. 15.2 Two moderately vague density relationships of the sort likely to emerge from actual observations of natural populations. (a) Fuzzy thresholds. These are indistinct transitions between levels of response, with different levels appearing to have different mean and variance. I have broken the abscissa subjectively at the question marks to amplify the effect of a threshold. (b) Spreading scatters, without even fuzzy thresholds. Points indicate individual observations, and lines indicate ranges.

the regression line that remains after the influence of covariates has been accounted for.

It is unlikely that all variance can be reduced to contingency in concrete density studies, because many ecological variables have markedly inconsistent, haphazard, unpredictable—inherently stochastic—effects on population change. Substantial residual variance will usually, if not always, remain in natural density relationships if for no other reason than that its cause is environmental stochasticity.

Fig. 15.3 illustrates some possible forms of contingent and residual stochasticity in density relationships. One effect of contingency is to reduce variance about a density relationship, as illustrated by the diagonal arrows indicating effects of variables C_1 and C_2 in the figure.

Contingencies can also influence density rela-

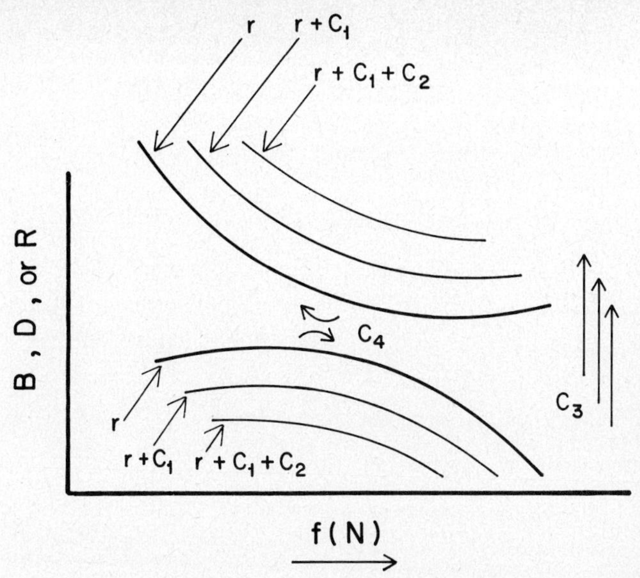

Fig. 15.3 Forms of contingent and residual stochasticity in density relationships. Contingency can be accounted for by knowledge of additional variables, as in analysis of covariance. Residual variability (r) remains after contingency has been subtracted. C_1 and C_2 are contingent variables that reduce the variance and leave the mean and slope of the relationship unchanged. C_3 is a contingent variable that increases the mean value of the relationship, and C_4 is a contingent variable that changes its slope. Contingent variables may interact to resolve or dissolve a density relationship. Contingency on small or short scales is likely to become residual stochasticity on larger and longer scales. Lines indicate prediction intervals.

tionships more profoundly by changing the mean or slope, as shown by the vertical (C_3) and rotational (C_4) arrows in Fig. 15.3. For our mussels, knowledge of inundation might reduce variance in our survivorship function by revealing several different relationships for different average tidal heights, each with a different mean. It might also reveal that different inundations cause different slopes in density relationships for mussel survival. If slope changes are distinct enough, contingent variables might even interact statistically and have the fascinating effect of resolving or dissolving a density relationship. For example, a significant relationship between survivorship and density at low temperature may disappear at high temperature. The statistical interactions among variables are likely to be complex and creative in density relationships. Two recent results illustrate this point. Fowler and Antonovics (1981) for plants and Caffey (1984) for barnacles found quite high contingency at very small spatial scales, which propagates into residual variance at larger spatial scales.

DENSITY-INDEPENDENCE AND DENSITY-DEPENDENCE

Perhaps the most important accomplishment of the old debates over population dynamics is the understanding that, while much population change is potentially affected by density, virtually no natural population changes solely as a function of density (Solomon 1957, Klomp 1962, Bakker 1964). This fact and residual stochasticity mean that a simple dichotomy between density-dependent and density-independent is inadequate to describe the rich character of density relationships in nature. For example, Figs. 15.2a and b show some relationships between density and population change, so both are density-dependent, but they are different from one another and very different from tight relationships such as Fig. 15.1a and even Fig. 15.1b. Both DI and DD variously influence the character of density relationships.

The inadequacy of the alternative "DI or DD" becomes clear from the current ecological literature. The term density-dependence has a whole series of usages: continuously strong influence as in Fig. 15.1, accelerating influence as in the mean of Fig. 15.2, or some sort of upper density limit to population transit similar to the ceiling of Fig. 15.4. Density-independence variously indicates constant unvarying influence, a regularly varying one, or a randomly varying one with either high or low variance; this is a tremendous range of possibilities! I shy away from arguing that any population or community relation-

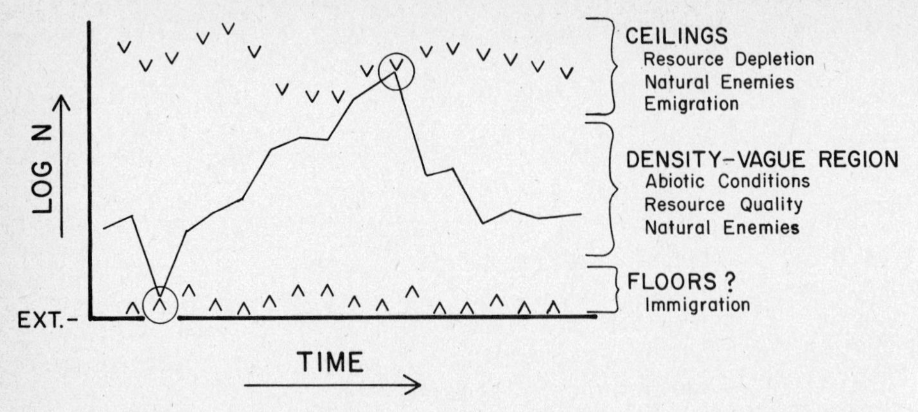

Fig. 15.4 Population change under liberal population regulation. Liberal population regulation means that change in abundance is only vaguely related to density at intermediate densities, between ceilings (arrows pointing downward) and floors (arrows pointing upward). "Ext." means extinction. At the upper circle the population falls as a result of encountering a ceiling. At the lower circle the population has been rescued from extinction by encountering a floor. Most change in this population is unrelated to density.

ship is inherently independent of density because of the operational difficulty of demonstrating the absence of an effect and because of the possibility of indirect effects (DeBach 1958). An unmeasured variable might resolve a density relationship in the data, new data might indicate one among already measured variables, and variables can interact. The productive questions concern the forms of density relationships and their range, strength, and consistency.

LIBERAL POPULATION REGULATION

Liberal population regulation is a major implication of density vagueness in demographic functions (Fig. 15.4). I mean by liberal regulation something qualitatively distinct from the regulation of the logistic and Lotka-Volterra equations, from models with set points that are converged upon from above and below, and from continuously strong negative feedback over the full range of density. The under- and overcorrections of time lagging, cycling, chaos, and cascading bifurcations of density-explicit dynamics (May 1981a) are not liberal regulation. Rather, under liberal regulation "control," "regulation," and "negative feedback" of population (mechanisms of return to intermediate densities) occur largely outside of fairly widely separated upper and (perhaps) lower bounds to population change. These bounds are likely to be "soft" and to increase their effect gradually, to cause gentle return toward intermediate densities. As well, neither biotically nor abiotically created bounds are likely to be static in position or in quality.

Without the closely apposed bounds and constant correction of density-explicit regulation, much population change need not be much affected by density; many populations will spend most time in intermediate regions away from density bounds. Perhaps the most important implication of this sort of regulation is that it cannot be well understood without experiments (Murdoch 1970, Bender et al. 1984). The statistical difficulties of identifying true population regulation by means of just post hoc statistics are well known, especially where density relationships have much variance (St. Amant 1970, Maelzer 1970, Pielou 1974, Slade 1977).

CEILINGS AND FLOORS FOR POPULATION

The notion of liberal population regulation calls attention to distinctions among mechanisms of population change acting in different regions of density. "Ceilings" and "floors" are the two bounds to population transit, beyond which most negative feedback occurs, and these two density regions have different mechanisms of density change and different demographic conditions from the central density-vague region.

Ceilings for insect populations caused by resource depletion have been treated by Pollard (1981). Some other likely candidates for the causes of population ceilings are heavy losses to natural enemies and emigration (Fig. 15.4). For brevity, here I use resource depletion broadly to include both exploitative and interference competition, but understand the ambiguity that this shortcut can generate (Sale 1982). An interesting question is why some populations rapidly and consistently climb to ceilings, while others frequently remain below any solid impediment to increase. Ecologists have directed much attention toward populations near ceilings (Chapters 10, 33).

Floors to density are more poorly understood than ceilings simply because sparse populations are more difficult to find and study. Standard deterministic models such as the logistic and Lotka-Volterra have floors of sorts, where per capita growth grades continuously to a maximum as population declines to extinction. The Allee effect is actually a hole in the floor, with decreasing net growth rates as the lowest possible densities are approached.

Hard density floors would appear as distinctly increasing net population growth rate as extinction is approached. The lack of a floor would appear as unchanging net growth rate as extinction is approached (Fig. 15.5).

One mechanism that could create floors for local populations is background immigration, which, per capita, is low relative to resident population except at very low density. This is an interesting unexplored area of research that can be approached by experiments that bar immigrants, especially during population lows.

Identifying ceilings and floors for natural populations requires special caution and great attention to actual demographic mechanisms. Decline from high and increase from low population does not necessarily mean that some regulatory force has come into play. Populations return toward intermediate densities from both highs and lows because of factors unrelated to density. The weather can change for the worse to cause decline from high density, and it can change for the better to allow increase from low density. In populations brought down by the end of a favorable season, "pseudoregulation" occurs as a matter of course (St. Amant 1970). Understanding ceilings and floors requires substantial knowledge of natural history and, in many cases, experiments. It cannot be done just by correlations of crude density with density change.

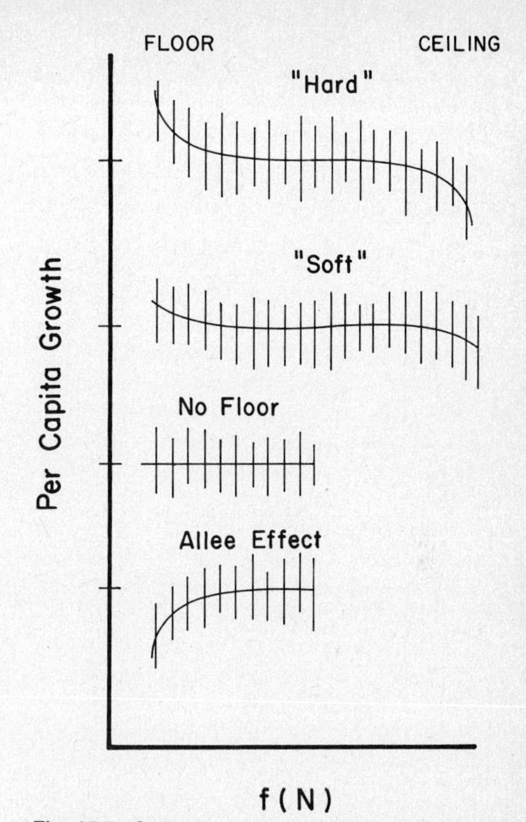

Fig. 15.5 Ceilings and floors in liberal population regulation as inflections in per capita growth, downward and upward respectively, at the extremes of the density range. Sharp inflection indicates a hard ceiling or floor, gentle inflection a soft ceiling or floor. Horizontal lines indicate means, and vertical lines indicate standard deviations. The ticks on the vertical axis indicate growth rates that give no net change in N.

Many populations may have no endogenous mechanistic floor, no negative feedback at very low density created by consistent relationships among birth, death, and migration. Rather, one alternative is an environmental, exogenously imposed, floor of "stochastic boundedness," which can prevent extinction (Chesson 1978). In this case favorable environments are frequent enough that sparse populations normally increase rather than wander to extinction. The density-vague region in liberal regulation makes stochastic boundedness more likely because sparse populations would have little DD, which resists population increase; thus only marginally favorable environmental conditions might suffice to rescue a sparse population.

THE OLD RANDOM WALK BUGBEAR

It is important to distinguish between unlimited and limited, bounded, random walks in population dynamics. Some ecologists who do not deal much with the dynamics of natural populations still believe that change unrelated to density is logically impossible, on the authority of the old argument concerning unlimited random walks. This argument said that populations with no negative feedback in density will ultimately either increase to infinity or decrease to extinction; ergo, population change must be density-dependent. As far as it goes, the argument is correct, but it is not very useful in ecology because it does not tell us the kind of density-dependence, its intensity, variance, and limits. Density vagueness and its manifestation in liberal regulation, which operates like a random walk at intermediate densities, predispose neither to extinction nor to increases to enormous density. Nisbet and Gurney (1982) discuss some mathematics of limited random walks for population.

Reddingus (1971) made a telling mathematical case against the utility in ecology of arguments about unbounded random walks. Their power and generality derive only from the ''ultimate'' outcome after an infinite amount of time has passed: ''It can never be observed that animal populations persist *forever*'' (my italics, Reddingus 1971, p. 37). What happens at the mathematical limits after an infinite amount of time can be of little value in dealing with real populations in finite time (see also Cain, Eccleston, and Kareiva 1985). Reddingus went on to show that some simple models with no density-dependence can yield population dynamics that avoid both enormous densities and extinction for quite some time.

APPARENT DENSITY VAGUENESS

Neither demographic variation nor erratic population dynamics need be caused by density vagueness. Completely deterministic dynamics with variance-free density relationships can produce the gamut of different dynamics, from constancy, through cycles with regular and increasing amplitude, to ''chaos'' (May 1981a), even in populations with overlapping generations (May 1980).

Apparently, density-vague relationships such as those in Figs. 15.1b and 15.2 might just be due to simple contingency and might be resolved into tight density relationships with knowledge of an additional variable or two. Table 15.1 applies the term ''veiled density-dependence'' to contingency, with complicated versions of logistic-type mechanisms determining dynamics.

From this neat extreme we move to ''slight density vagueness,'' the grey area of mixed deterministic and stochastic influences at intermediate densities (Table 15.1, Fig. 15.1b).

The other extreme is dominant density vagueness at intermediate densities. In this case intermediate densities could have a large number of causes; some are listed in Table 15.1. Soft factors have relatively small influences, but none overwhelms demography. Hard factors have an overwhelming effect, such as a well-timed drought that kills almost all of a year class of seedlings or an exceedingly abundant plankton crop that allows spectacular recruitment of larval fish. In the density vague region neither soft or hard factors

Table 15.1 REASONS FOR APPARENT DENSITY VAGUENESS

I. Veiled, but strong and continuous density-dependence; little residual density vagueness
   1. Fluctuating K or r in logistic dynamics
   2. Nonlinear density-dependence
   3. Changing rate of density-dependence in space or by age
   4. Time lags in density-dependence

II. Slight density vagueness (Fig. 15.1b)
   1. Linear
   2. Nonlinear; thresholds

III. Dominant density vagueness; little density-dependence at intermediate densities
   1. Little effective resource depletion
   2. Multiple soft factors
   3. Erratic hard factors
   4. Dominance by autecology, phenology, or seasons

are likely to have any but casual or inconsistent relationships with population density.

EXAMPLES

Insects

Life table studies of insects provide many excellent examples of density vagueness (Varley et al. 1973; Podoler and Rogers 1975; Southwood and Reader 1976; Hurd and Eisenberg 1984; Royama 1984; Strong, Lawton, and Southwood 1984; Strong 1984b). Above the density-vague regions, some populations have a fairly distinct ceiling caused by resource exhaustion (Dempster and Lakhani 1979). In some cases the environment is so continually fine that populations and ceilings meet frequently (Dempster and Pollard 1981). In other cases, such as tsetse fly (Rogers 1979), viburnum whitefly (Southwood and Reader 1976), and the classical winter moth at Whytham Wood in England, ceilings are frequently higher than populations and are soft, variable, or indistinct.

Density floors, per capita increases that are distinctly greater at the lowest densities, may occur for insect populations, but this question has not been critically investigated. Local extinctions do occur fairly frequently in some species (Ehrlich et al. 1972, Simberloff 1978, Singer and Ehrlich 1979), and continual reimmigration does create ''iterative'' extinction in other cases (Rey and Strong 1983). But, it is not known how often background immigration or some other factor *prevents* local extinction and thus creates a density floor.

Parasitoids Insect parasitoids raise one of the most intriguing topics in density relationships. They have long been modeled as causing continuous and strong density-dependent mortality in populations of their host insects (Chapter 27; Nicholson 1954, Hassell 1978) and equally long have been considered by others to cause mortality that is vaguely, at best, related to host density (Thompson 1956; Dempster 1975, 1983). Parasitoids undoubtedly lower the running average of host density in some circumstances, even greatly as demonstrated by successes of biological control (DeBach 1979). In natural populations, however, the rate of mortality caused in host insect populations by parasitoids is not often strongly density-dependent in space or time; some correlations between density and mortality caused by parasitoids are even negatively related to density, but with high variance (Morrison and Strong 1980, Stiling and Strong 1982)!

Most intriguing are reexaminations of two cases of successful biological control by parasitoids; the winter moth in Nova Scotia and the olive scale in California. In Nova Scotia the parasitoid fly *Cyzenis* was introduced and, perhaps with the aid of another introduced parasitoid, probably caused the decline in winter moth density over a decade or so (Hassell 1980). But close examination of the rate of mortality caused by *Cyzenis* does not reveal patent density-dependence; host mortality did not increase with density in any consistent way, at least as far as we can tell retrospectively from the census data (Dempster 1983).

Similarly, the olive scale in California was much reduced in density after the successful introduction of two parasitoid species. The authors' original interpretation was that some of the data hinted at density-dependence in scale mortality caused by the parasitoids, but these hints are ''all inconclusive, possibly because of the many interrelated and confusing variables involved'' (Huffaker and Kennett 1966, p. 336).

I have suggested a model of slight density vagueness for olive scale and its parasitoids (Strong 1984b). My idea was that the parasitoid introductions had the effect of moving a statistical regression of combined olive scale mortality to a point lower on the density scale, such that per capita birth and death rates were roughly equal at a much lower density after than before the parasitoid introductions. The variance about the regression would produce the observed variance in scale density both before and after introduction of the parasitoids.

Recently, Murdoch, Reeve, Huffaker, and Kennett (1984) have critically reappraised the data on mortality caused to the olive scale by its pair of parasitoids in California. They came to the conclusion that there is even more density vagueness than I had inferred! I was guilty of liberal theology in drawing slight density vagueness with continuous slope when evidence even in the original publication (Huffaker and Kennett

1966) suggested the correctness of an agnostical view for this system, with a relationship at least as vague as in Fig. 15.2.

Very Fuzzy Thresholds? The possibilities for density relationships of olive scale and its parasitoids are now three: very low and very fuzzy thresholds, no density-dependence at all, and very slight and inconsistent density-dependence. The first and third of these alternatives include DD and so would have to operate at such low densities or so inconsistently as to be transparent to the rigorous statistical tests of Murdoch et al. (1984).

The functional significance of the first alternative is importantly different from that of the second and third, because it implies a consistent and functionally significant cybernetic quality, a negative feedback, among the three species, while the second and third alternatives do not. In favor of the first alternative, we know from Huffaker and Kennett (1966) that the parasitoids can cause high mortality at low olive scale densities; even in the region of any threshold, mortalities could be as high as at any density. With some imagination (faith?) we might see a tendency for the mortality rates to be lower at the very lowest scale densities (in the lower left corners of Figs. 4 and 5 in Murdoch et al. [1984]).

Immigration that reverses local extinctions (of any combination of the host and its pair of parasitoids) might play a role in creating a systemwide fuzzy threshold. Murdoch et al. (1984) suggest that some local extinctions may actually occur in the system. However, the published evidence suggests to me that, contrary to the coexistence of predator and prey mites on oranges in Huffaker's (1958) beautiful lab experiments, these parasitoids do not drive local olive scale populations to extinction as a matter of course; mortality rates are not consistently high enough. Infrequent local extinction followed by reimmigration could contribute to a fuzzy threshold and vague density-dependence, but it is not the lion's share of the explanation.

Alternatives two and three—no or at best trivial DD—are difficult to rule out for the olive scale system. It is consistent with the evidence that the parasitoids kill olive scale in a manner unrelated to its density, imposing a quite variable, but density-independent mortality. Killing scale in this manner would bring down average densities as observed.

Crude Versus Refined Density On the other hand, contingent density-dependence in mortality from parasitoids cannot be ruled out for the olive scale system either. Such contingencies as live scale protected from parasitoids by overlying dead scale, climatic succession and fluctuation between years, and the interactions of the two parasitoid species (Huffaker and Kennett 1966) make one wonder whether density under one set of conditions is functionally equivalent to density under another. Does this contingency mask density-dependence? In a similar system one crucially contingent variable for density relationships has been found by Luck and Podoler (in press) for parasitoids of California red scale on citrus. Parasitism rate depends greatly upon scale size, and this size relationship varies among parasitoid species. ''Crude'' density, just a head count of scale, gives a misleading picture of parasitism rate functions. ''Refined'' density, contingent upon the size distribution of the host population and upon the particular parasitoid species, is necessary for understanding these density relationships.

Organisms Other Than Insects

The stock recruitment relationships used in fisheries are often spectacularly density-vague. The deterministic ideal is a convex line of recruitment that rises and then falls with increasing parental stock (density). In nature the only facets of this ideal that are consistently realized are that very low and sometimes very high stocks produce low recruitment; intermediate stock sizes produce recruitments that range from very low to very high (e.g., Sale 1982, Fig. 7). The variance at intermediate densities probably results from the fact that recruitment is extremely contingent upon abiotoic and synecological factors (Everhart and Young 1981, Laurence 1981, Cushing 1982).

Marine invertebrate populations are likely to have vague density relationships and quite liberal population dynamics (Frank 1965, Loosanoff 1966, Sutherland 1974, Speight 1975, Black

1977, Underwood 1978, W. H. Wilson 1983, Petraitis in press, Caffey 1984, Wethy et al. unpublished observations).

Plants give some interesting cases of density vagueness. The limited random walks of density change for all but the most common trees in Hubbell's and Foster's tropical forest on Barro Colorado Island (Chapter 19) are a good example, as are the dynamics of the small plants in Grubb's chalk grassland (Chapter 12) and of sparse species in some North American prairies (Rabinowitz 1981). Density vagueness is described by Shaw (1983), who found in experiments with the herb *Salvia lyrata* ''weak or conflicting responses'' at intermediate densities and ''much mortality and stunting'' at abnormally high densities. Though diseases are sometimes contracted by plants in ways roughly related to density, relationships have a great deal of variance especially at low densities and where many species mix together in natural communities (Burdon and Chilvers 1982, Augspurger 1983).

CODA: IMPLICATIONS FOR COMMUNITIES

Density vagueness is a form of natural variability. Rather than noise, natural variability is an essence of ecological music (Simberloff 1980). In Chesson's (Chapter 14) views C and D, natural variability can reverse and interrupt determinism and even create novel processes. Sale (1982, p. 139) hits the heart of the implications of density vagueness for communities: Deterministic theory is built upon the assumption that ''rate of population growth is strongly dependent upon current population size,'' that density has a ''direct'' and ''immediate'' influence upon population change. Sale's lottery model (Sale 1977, Chesson and Warner 1981) is an attempt at interspecies competition theory with attenuated influence of density, with density-vague demography. The implication of the lottery model is a nature that works in ways very different from that treated by most theory, in which persistence is more an element of autecology, of accommodation for real changing environments, than of machinelike synecological cybernetics in an environmental vacuum.

''Gleasonian persistence'' as opposed to mathematical stability is the main implication for communities of density-vague demography, as illustrated by the complexities of parasitoid influence upon host insect population dynamics, discussed above. Mathematical ''stability'' relies solely upon reciprocal dominance of population dynamics between species, while Gleasonian persistence relies upon a larger set of influences that includes both autecology and synecology. In Gleasonian persistence species may affect one another, but normally will not reciprocally dominate dynamics and existence. Without continuous, intense density-dependence, mathematical stability is wide of the mark as an explanation of why sets of species persist. More than delicate synecological coactions, autecologies may be the key to communities.

SUMMARY

1. Variance and dispersion complete the skeleton of density relationships, of which average tendencies are only a rudiment in natural populations. Density vagueness is variance in relationships between density and rates of birth, survivorship, migration, and net population growth. This variance grades from slight, where the density signal is strong and continuous, to great, where there is little if any measurable influence of density upon a demographic parameter over a substantial portion of the abundance spectrum.

2. The dichotomy of ''density-dependent or density-independent'' is inadequate for the rich array of real demographic relationships. Much population change has the potential to be affected by density, and there is no population that changes solely as a function of density. The strength, variability, range, and consistency of density relationships determine population dynamics.

3. Operationally, observed variation in density relationships can be either contingent or residual variance. Contingency (contingent variance) is deterministic variation that is resolved with knowledge of additional variables. Variation in density relationships caused by age structure or time lags is contingency, as is some environmental variation caused by simple differences in time and space. Residual variance remains after all contingency has been removed from density relationships, is due to environmental and demographic stochasticity, and is often substantial in natural populations.

4. Liberal population regulation results from density vagueness in demographic functions. It is population dynamics without much, if any, influence of density at intermediate population sizes. Bounded or limited random walks are analogous to liberal regulation. Ceilings to population transit, where density-dependence is more pronounced, exist above the intermediate, density-vague regions in liberal regulation. Floors to population, with increased per capita rates of population growth at lowest densities, may also exist in some cases, but little is known about the demography of sparse natural populations. Floors could be caused by background immigration.

An interesting possibility for endogenous floors, which are caused by interactions among birth, death, and migration rates, is ''stochastic boundedness'' (Chapter 14), in which favorable environmental conditions with concomitant population increases are so frequent that sparse populations are usually rescued before wandering to extinction. Density vagueness increases the possibility of stochastic boundedness because of the lack of much density-dependent resistence to increase at low densities.

5. The literature is full of examples of density vagueness and liberal regulation. Life table studies of insects provide many examples, and parasitoid influence upon host population dynamics is especially interesting in this light. Recent reexaminations of classic cases of biological control have found that mortality caused by parasitoids is much less clearly related to density of host insect populations than many models have assumed.

Stock recruitment relationships for fish are often quite density vague, as is population change in diverse plant communities (rain forest trees, grasslands, and prairies).

6. Density-vague demography implies a community theory different from that based upon deterministic demography. The lottery theory for vertebrates and coexistence of parasitoids with host insect populations not based upon tight synecology are two examples. When density vagueness is substantial, mathematical stability of populations due to delicate cybernetics of species interactions may be less important than is Gleasonian persistence due to a combination of autecological and synecological forces in accounting for species coexistence in communities.

ACKNOWLEDGMENTS

I thank Mike Blouin, Ted Case, Peter Chesson, Rob Colwell, Mike Gilpin, Nick Gotelli, Conrad Istock, Peter Kareiva, Mick Keough, Peter Petraitis, Stuart Pimm, Ann Thistle, Joc Travis, Joel Trexler, and Tony Underwood for ideas and vigorous argument about density vagueness and natural variability.

chapter 16

Climatic Instability, Time Lags, and Community Disequilibrium

Margaret Bryan Davis

INTRODUCTION

Climatic variability is an important exogenous factor affecting community structure. Most discussions assume that perturbations occur with a characteristic probability that remains constant through time (Whittaker 1975, Leigh 1975, May 1981a). It is also usually assumed that variations in temperature or rainfall are bounded within limits and distributed symmetrically about some mean value (Botkin and Sobel 1975).

Climatic parameters change in a directional way through time, however, and vary about a trajectory of values, rather than around a constant mean. No matter what time scale is considered within the last 2 million years—decades, centuries, or millennia—climatic parameters are continually changing (Fig. 16.1), moving upward or downward in an erratic fashion. For the last 50 years or 500 or 1,000—as long as anyone would claim for ''ecological time''—there has never been an interval when temperature was in a steady state with symmetrical fluctuations about a mean. Instead, there has always been directional change (Mitchell 1977). Only on the longest time scale, 100,000 years, is there a tendency toward cyclical variation, and the cycles are asymmetrical, with a mean much different from today.

The thesis of this paper is that directional changes of climate have profound effects on the structure of biological communities. This is the third of the nonequilibrium community theories discussed in Chapter 13. The community effects arise because species vary in their responses to climate. Some organisms track climate closely, reacting to conditions each year, while others respond so slowly that only long-term climatic trends have any observable impact. Thus, both the magnitude and the timing of responses to the same climatic trend are different for each species. Communities do not respond as units to directional climatic change; instead, the variations in rates of change among component species cause faunal and floral disequilibrium and change the patterns of species abundances resulting from competition and predation. The complex patterns of distribution and abundance that result are difficult to decipher unless climatic variations in both space and time are taken into account.

The first part of this chapter describes time lags in the responses of plant and animal species

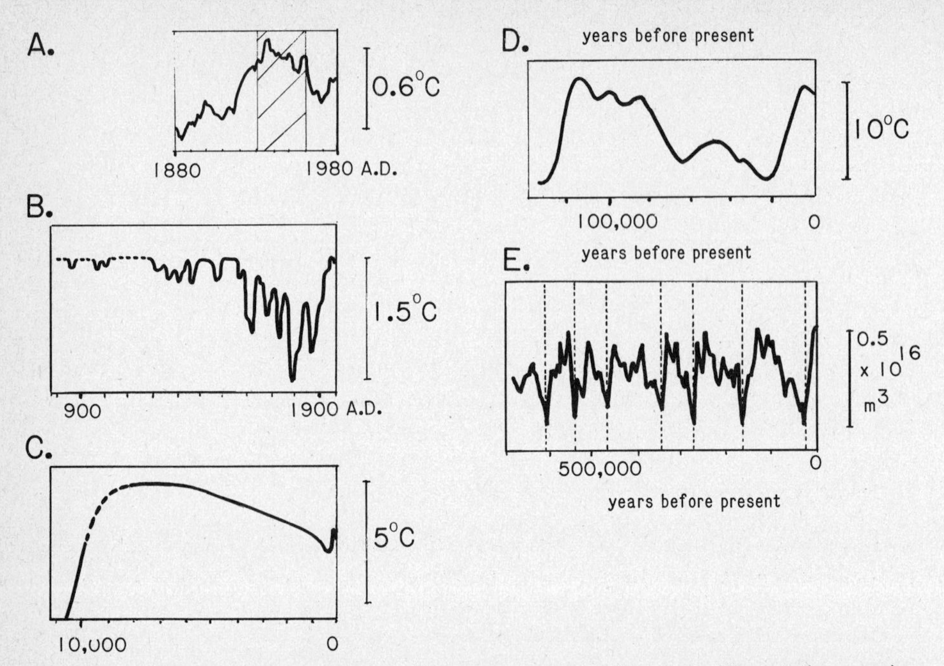

Fig. 16.1 Temperature changes in the Northern Hemisphere at different time scales. (A) Instrumental record of annual temperature over the last 100 years at latitudes 23.6°N to 90°N. Hatched area is the 30-year period 1931–1960 used to calculate so-called climatic means. (From Hansen et al. 1981.) (B) Reconstructed air temperature over the last 1,000 years in the North Atlantic, based on accounts of sea ice. (From Lamb 1977.) (C) Annual temperature over the last 10,000 years in northeastern United States, based largely on fossil vegetation records. (From Davis et al. 1980.) (D) Reconstructed air temperature over the last 100,000 years in Europe, based on vegetation records, record of eustatic sea level, and fossil and geochemical records in deep-sea cores. (Modified from Mitchell 1977.) (E) Global ice volume during the last seven glacial-interglacial cycles, based on oxygen isotopes in deep-sea cores. (Modified from Mitchell 1977, Johnson 1982.)

to changes in climate during the past century and discusses factors responsible for delayed responses. The second part discusses the impact of climate change on plant and animal communities at several different time scales. The third part considers spatial differences in the amplitude and direction of climatic changes, because such differences complicate the interpretation of community patterns. The final part is a case study of a community of interacting species, each of which responds at a different rate to climatic change.

TIME LAGS IN BIOTIC RESPONSES TO CLIMATIC CHANGE

A striking change in climate has occurred in many parts of the world during the last century. Between 1880 and 1950 the mean annual temperature of the Northern Hemisphere rose 0.6°C, and since 1950 the temperature has declined 0.3°C (Mitchell 1977, Hansen et al. 1981) (Fig. 16.1A). The responses of plants and animals to these changes serve to document differences among species in sensitivity to climate. For many species responses at the population level have been delayed years or decades after the climatic event.

Animal Population Responses to Climatic Change

The most rapid responses to climatic change occur among animals. Population density or geographical range may change the same year or within a few years of the onset of climatic warming or cooling. Extreme temperatures or unusually severe storms can destroy food supplies and wipe out a local population, facilitating colonization by a better adapted species. The effects of storms and hard winters are documented for hummingbirds (Gass and Lertzman 1980), redshanks (*Tringa totanus*) and oystercatchers (*Haematopsus ostralegus;* Davidson and Evans 1982), Belding's ground squirrels (Morton and Sherman 1978), and butterflies (Ehrlich 1983).

Barnacles Changes in densities occur rapidly when climatic change alters the relationship between competing species within a community. The climatic warming since 1880 changed water and air temperatures along the British coast, and the northern barnacle *Balanus* was replaced over a wide area by the southern barnacle *Chthamalus* (Fig. 16.2; Southward and Crisp 1954). Both species were already present in the community, but one species gained at the expense of the other (Connell 1961a). A two- to three-year time lag resulted from the life span of individual barnacles, because the sessile adults are replaced after their natural life span by juveniles of the better adapted species (Southward and Crisp 1954). Here, as in many other examples, climate affects the mortality of young life stages more than the mortality of adults.

Birds Bird species have varied in the speed of their responses to recent climatic changes. Rapid changes in geographical range are possible. Species that are present every year as visitors may begin to nest as soon as conditions become more favorable. For example, the black-headed and herring gulls (*Larus ridibundus* and *L. argentatus*) began to breed in Iceland early in the twentieth century (Gudmundsson 1951). Rapid responses also occur for birds like the fieldfare (*Turdus pilaris*), which was carried by changed wind patterns to a new site, a change in geographical range that was necessarily simultaneous with the onset of a new climatic regime (Salomonsen 1951, Williamson 1975). Other bird species in the North Atlantic region have responded more slowly to the climatic warming since 1880, gradually gaining "momentum" in their northward expansion. Thus, green woodpeckers (*Picus viridis*) and blackbirds (*Turdus merula*) are still moving northward into Scotland, while simultaneously ospreys (*Pandion haliaetus*) and snowy owls (*Nyctea scandiaca*), which track

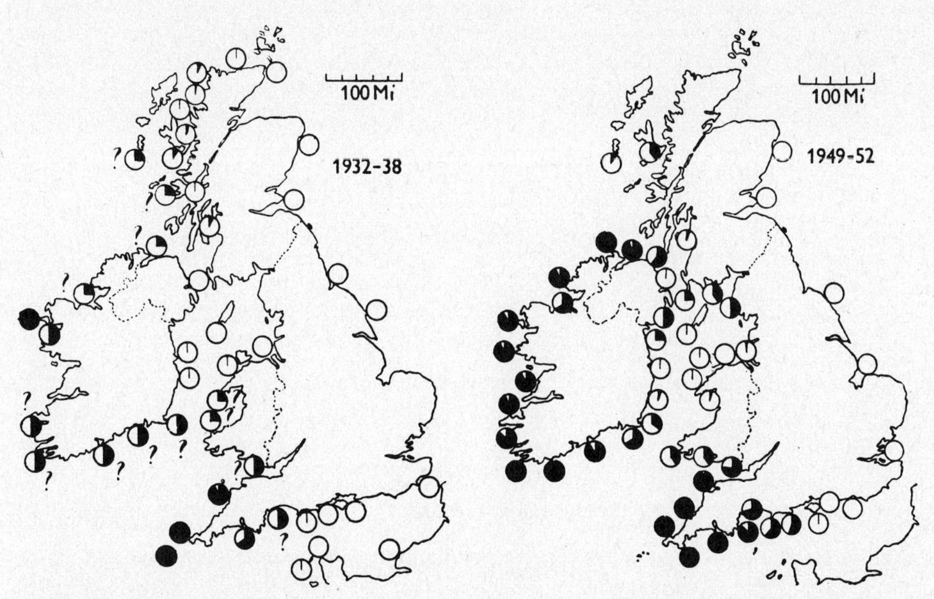

Fig. 16.2 Changes in relative abundances of the barnacles *Chthamalus* (black) and *Balanus* (white) at a number of stations around the British Isles. *Chthamalus* increased at the expense of *Balanus* at many stations during this 15-year interval of rising temperatures. (From Southwood and Crisp 1954.)

climate more closely, are already moving southward in the same region in response to the temperature decline since 1950 (Williamson 1975).

The sage sparrow (*Amphispiza belli*) displays site fidelity, continuing to attempt to nest for several years after a vegetation change has occurred (Chapter 9). Species with similar behavior will inevitably lag habitat and climatic changes by several years. Järvinen and Ulfstrand (1980) argue for the importance of habitat in shifts of Scandinavian bird distributions, suggesting that changes brought about by human activities have had greater influence than climate. For a few species, however, climate appears more important than habitat (Salomonsen 1951). There are also many well-documented examples of northward range expansions of birds in Iceland and Greenland (Gudmundsson 1951), where habitats have changed relatively little during the present century.

Insects Twenty-one species of Lepidoptera have been added to the British fauna since 1920, as southern species have expanded their ranges northward and northwestward into the British Isles in response to the climatic warming. Several native species have expanded northward within Britain, for example, the white admiral (*Limenitis camilla*). For those species that track climate most closely, northward expansion ceased after 1950, but others with longer time lags are continuing to expand northward with the momentum gained during the pre-1950 climatic amelioration (Burton 1975).

Mammals The climatic warming of 1880 to 1950 was accompanied by drought periods in the dust-bowl region of the United States. Several prairie mammal species with short life spans showed rapid density changes in response to decade-long droughts. Near Hays, Kansas, jackrabbits (*Lepus californicus melanotis*) increased threefold, white-footed mice (*Peromyscus maniculatus*) remained unaffected, and prairie voles (*Microtus ochrogaster*) virtually disappeared by the second year of the 1932 to 1939 drought. Voles did not reappear until several years after the drought ended (Tomanek and Hulett 1970). A competing species, the cotton rat (*Sigmodon hispidus*), has been expanding northward since 1900 in response to higher temperature. Between 1932 and 1947 the cotton rat extended its range in Kansas at the rate of 10 kilometers per year (Hoffmann and Jones 1970, Cockrum 1948).

Martin (1956) and Glass and Slade (1980) have studied the interactions between cotton rats and prairie voles in the region where the two species have recently come into contact. Cotton rats have aggressive interactions with voles and are displacing them from a portion of their habitat. These rapid behavioral adjustments indicate that small mammal populations can track climatic changes closely, that is, within several years. One can suppose that interactions and adjustments similar to this occur continuously around the borders of a species range, as the interacting species expand and contract from one decade to the next in response to changing climate. Hoffmann and Jones (1970) suggest that numerous larger-scale adjustments of mammal ranges have occurred over the past several thousand years in response to climate.

Animals with longer life spans respond more slowly to environmental change. In east Africa years of high rainfall correlate with years when elephants reproduce successfully; populations in the Wankie game preserve are made up of cohorts dating from years of high rainfall. The variability of rainfall and the time lag of response are such that populations never even reach a stationary age distribution. Elephants live about 60 years, and are not reproductive before age 10 (Wu and Botkin 1980); time lags may be an order of magnitude longer than for small mammals.

Plant Population Responses to Climatic Change

Plants show much longer time lags than animals. Lags are lengthened by the tenacity of resident plants, which continue to compete successfully for space even in the face of changed climate. Adult plants can survive many catastrophic weather events, in extreme cases dying down to the ground but sprouting back up again when the bad weather is over. Iversen (1944) observed severe damage in Denmark to holly (*Ilex*) and to ivy (*Hedera*) growing on trees during the severe winters of 1939 to 1942. However, both species recovered. Ivy at ground level was protected by snow; many of the holly trees, although killed above ground, resprouted from the roots.

Marginal populations frequently survive for long time periods under climatic conditions where flowering and seed set seldom occurs (Moore 1976, Iversen 1944). For example, in the North American arctic north of tree line black spruce (*Picea mariana*) krummholz reproduces by layering and only rarely produces seed. When destroyed by fire, it regenerates poorly (Payette 1983). In case of species that bank seeds in the soil, occasional germination can keep species resident in the community for many years. Mountain ash (*Sorbus aucuparia*) near tree line in Sweden originates from seed carried by birds from lower elevations, where the tree can flower and set fruit successfully (Kullman 1983). These mechanisms all tend to delay the extinction of plant populations even when local climate is unfavorable.

Colonization of newly suitable habitats is limited for some plants by competition from resident species. For other plant species colonization is delayed by ineffective dispersal of seeds, and a third group of species may be hindered by low intrinsic rates of increase. Brown et al. (Chapter 3) report that populations of annual desert plants in experimental exclosures are still increasing seven years after release from seed predation.

Prairie Communities Because productivity and percent cover of crop plants respond immediately to unusual weather, people generally have the impression that plants are sensitive to climate. Changes in natural plant communities, however, such as a switch in the dominant species, are slow to occur. The change is frequently delayed for years or even for decades. Fig. 16.3 shows that decrease in cover in a two-species short-grass prairie community was progressive during the drought years 1932 to 1939. By 1940, after eight years of drought, 80% of a typical quadrat was bare ground. That space was rapidly colonized during the 1940s by buffalo grass. When rainfall reached high values in the early 1950s, blue grama replaced buffalo grass. After the 1952 to 1957 drought, in contrast to the earlier drought, blue grama recovered abundance more

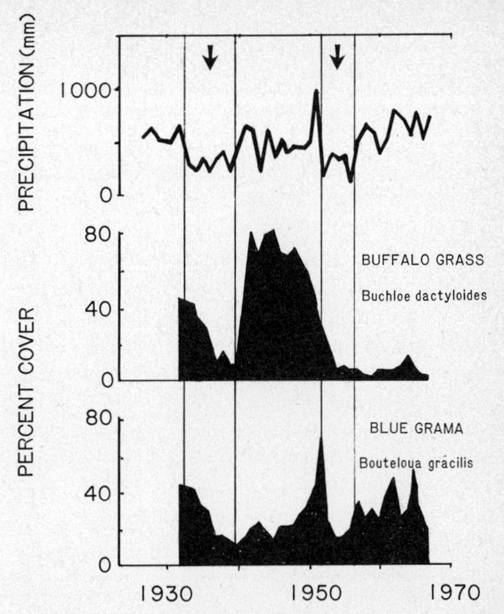

Fig. 16.3 (Top) Rainfall record near Hays, Kansas, over a 40-year interval. Arrows indicate two drought periods, 1932–1939 and 1952–1957. (Bottom) Percent cover within a two-species short-grass prairie community from 1932 to 1967. The two species were equally abundant in 1932, but blue grama was dominant in 1967. (After Tomanek and Hulett 1970.)

quickly. Although the two species were equally abundant in 1932, buffalo grass was dominant in 1945, and blue grama was dominant in 1967. Prairie communities with high diversity showed more complex reactions to the two droughts and still longer delays in response (Tomanek and Hulett 1970).

Forests Forest communities display longer lags than herbaceous communities in responding to climatic change. Again, individual trees will show growth responses to climate within a few years' time, but community-level changes may be delayed for many decades.

In the White Mountains of California bristlecone pines (*Pinus longaeva*) show a narrowing of annual rings during the seventeenth to nineteenth centuries, indicating that the lowered temperatures (Fig. 16.1B) slowed the growth of trees near tree line. No new seedlings were established for 200 years between 1700 and 1900. However, the tree line remained at the same elevation because the mortality of adult trees was not affected. Since 1900 conditions have been more favorable (Fig. 16.1A). Growth rates of trees have increased, and numerous seedlings have become established at and above tree line (LaMarche 1973). Advances of tree line have also occurred in Finland (Erkamo 1952) and in Sweden (Kullman 1983).

In Alaska the climatic record from tree rings suggests that summer temperatures were 3°C colder than today between 1820 and 1860. During this cold period few if any spruce were established near tree line in the Brooks Range. However, many seedlings and small trees are now growing above tree line that were established in the warmer climate that has prevailed since 1860 (Brubaker and Cook 1983). Advances of tree line like these apparently occur more rapidly than community changes within forests, because adult trees are not present to compete with seedlings as they occupy new habitat.

Modeling Community-level responses to climate have been simulated in a computer model of forest growth (Davis and Botkin 1985). In the model the growth rates of trees are assumed to be affected by shading from taller trees in 100-m^2 plots, as well as by species-specific responses to temperature, precipitation, soil and life history parameters. Thus the model can predict the outcomes of competition under different climatic conditions. In a series of modeling experiments a forest growing under conditions that caused sugar maple (*Acer saccharum*) to be dominant was subjected to a 200-year-long cooling. The 2°C drop in temperature, imposed in year 800 of the simulation, was sufficient to cause a change in the dominant trees, and spruce (*Picea rubens*) replaced sugar maple. When the change was reversed in year 1000 and the original temperature was restored, sugar maple became dominant again (Fig. 16.4). For our present discussion, the most interesting result is the time lag of the response. Changed basal area was not obvious until 50 years after the climatic change, and spruce did not reach a constant abundance level for 200 years. In fact, because of continued growth of young trees, basal area for spruce continued to increase for 50 years after the climate change was

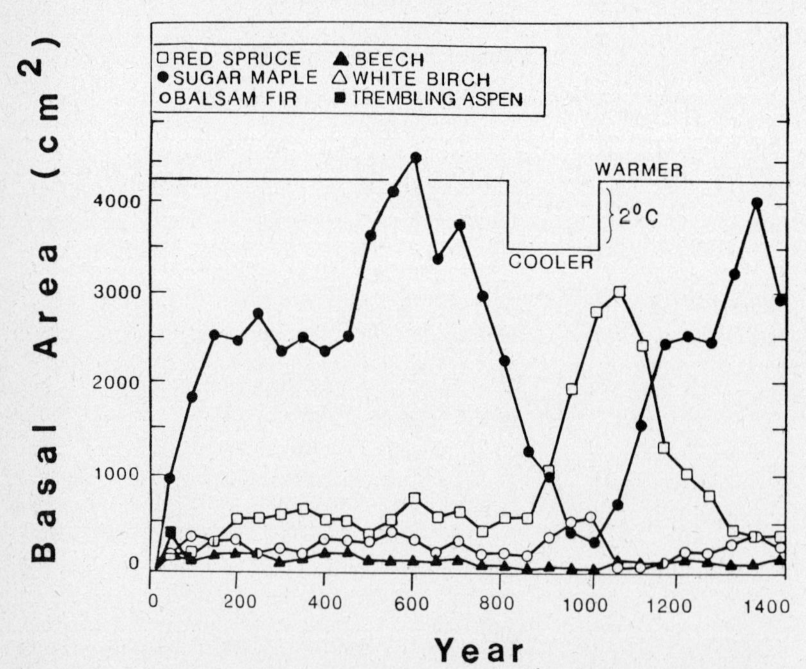

Fig. 16.4 Simulated growth of a forest subjected to a 200-year episode of cool climate when temperatures were lowered 2°C. Basal area was averaged from 20 simulated forest plots of 100 m^2 each. Temperature changes were fed in at years 800 and 1000 of the simulation. The dominant species changed in response, but with a long time lag. (From Davis and Botkin 1985.)

reversed in year 1000. In the simulations the replacement of the dominant species lagged the temperature change by 100 to 200 years even though seedlings were available for all species throughout the experiment (Davis and Botkin 1985).

The delayed community response in the simulations came about because the temperature change caused the large maple trees to grow more slowly, but had little effect on longevity. These individuals continued to dominate the plots; spruce entering the plots grew slowly because of shade from maple, even though the new temperature regime was closer to the optimum assumed for this species. As the canopy maple trees aged and died, they were gradually replaced by spruce, a slow process because of the long life span (400 years) assumed in the model for both species. The model illustrates some of the mechanisms that result in time lags in the responses of plant communities to climatic change.

Additional modeling experiments showed further that the forest community response to temperature change was sufficiently slow that simulations of climatic episodes of short duration (but of the same magnitude) caused only small changes in the relative abundances of the two dominant trees, and very short-term events (25 years) caused no detectable changes. Periodic disturbances were simulated in additional experiments; these had the effect of speeding up the response, although a time lag of a century was still observed (Davis and Botkin 1985).

These simulations suggest that climatic changes that persist for only a decade or two will not change most forest communities, and century-long directional temperature trends will not result in noticeable changes among canopy trees until almost a century after the event. This result may explain the absence of high-frequency climatic events from the temperature curve shown as Fig. 16.1C; that curve is based almost entirely on fossil records of changes in the forests of northeastern United States (Davis et al. 1980).

In contrast to the records of birds, insects, mammals, and tree line communities, closed forest communities provide few examples of changes that can be attributed to climate during the present century. If changes have been set in motion, they may be reflected only in the age structure of tree species; these changes will not become apparent among the canopy trees until the twenty-first century (Davis and Botkin 1985). Of course, it is difficult to distinguish the effects of climatic change from similar effects created by human disturbance. There is debate, for example, whether damage to mycorrhizae that resulted in widespread birch dieback in Nova Scotia in the 1930s was caused by warmer climate or by higher soil temperatures following logging (Hawboldt 1952). However, the paucity of examples of forest responses to recent climatic events reinforces the impression that trees are slow to adjust, at least relative to most animals. As the climate changes, plant communities may frequently be in nonequilibrium situations, lagging far behind in their adjustment to the changes.

Evolutionary Response

The many changes in geographical distributions that have occurred during the Quaternary suggest that local populations are rarely able to evolve physiological adaptations to changing climate. Apparently, the rate of climatic change overwhelms the evolutionary response of most species. Instead of adapting to new conditions, local populations die away and populations build up in distant localities where conditions are closer to the species optimum.

The speed with which adaptations to changing climates can occur is not known. Vaartaja (1959) reported that seed for forest plantations in Europe was generally collected about 100 km south of the plantation site, because the resulting trees grew faster than trees from local seed. They were cued by day length to enter dormancy on a later date and as a result utilized a longer growing season. Vaartaja suggested that the mature seed-producing trees became established as seedlings under the climatic regime of the early years of the century, when the growing season was shorter. Therefore ecotypes 100 km to the south were better adapted than local populations to the warmer climate of the 1950s. This argument invokes a lag in evolutionary adaptation to changing climate.

Animals with shorter life spans may be able to adapt more quickly. Potts (1983) suggests that short-lived and long-lived reef organisms have been affected differently by the Quaternary; he argues that the extremely long-lived corals have been in evolutionary disequilibrium throughout. Sea level has changed frequently during the Quaternary; levels are never maintained for more than about 3,000 years within the range that permits continuous growth of shallow-water reefs. This length of time is only a few generations for a colonial organism like a coral, insufficient time for reproductive isolation to develop before changed sea level causes extinction of local populations and reestablishment of corals at lower or higher levels from propagules coming from distant sites. This process has produced extensive intraspecific variation, but few new species. Short-lived reef-dwelling organisms, in contrast, have been able to speciate during the short intervals of relatively stable sea level, differentiating into clusters of endemic species (Potts 1983).

CLIMATIC CHANGE AND COMMUNITY RESPONSES

Let us now consider change and response on four time scales: the 100,000-year scale of the glacial-interglacial cycles during the Quaternary, the 1,000-year scale of the Holocene, the 100-year scale of the "Little Ice Age," and the 10-year scale of the present century.

100,000-Year Time Scale

The Quaternary period (the last 2 million years) has been characterized by large, periodic fluctuations of climate. Colder intervals saw the growth of large continental glaciers (Fig. 16.1E). The subtraction of water from the ocean caused sea levels to fluctuate. Interglacial climates, as warm as those prevailing today, recurred 18 to 20 times, roughly once every 100,000 years. These interglacials are brief events on the time scale of the Quaternary, lasting 10,000 to 20,000 years; they are interruptions in what is basically a cold, glacial period. During the last glacial maximum, 18,000 years ago, ice sheets covered 30% of the land area of the continents, and sea level fell 100 m, exposing the continental shelves. These changes affected the reef faunas of the tropics (see above). Aridity was widespread at low latitudes (causing contraction or displacement of rain forests), temperatures were lower (causing montane glaciers to form even at the equator), and the Gulf Stream was deflected away from Europe, where tundra replaced forests across all of Europe north of the Alps, and steppe replaced woodland in the Mediterranean region (Peterson et al. 1979).

Temperate Forest Communities The glacial-interglacial cycles caused major disruptions of temperate vegetation. Many temperate deciduous tree species were displaced entirely from the areas of their modern ranges. As yet these glacial populations are poorly known, because few sedimentary deposits of full-glacial age contain fossil evidence from deciduous trees. The inference from the record, and from the patterns of northward expansion of these trees as the climate warmed, is that temperate forest species survived in small populations at scattered sites both in North America and in Europe (Davis 1976, 1981, 1983b; Huntley and Birks 1983).

As the ice melted, beginning 16,000 years ago, the northward expansion of plants lagged behind the retreat of the ice. In North America oak (*Quercus*) and maple (*Acer*) advanced northeastward from the Mississippi Valley region relatively rapidly, but hickory (*Carya*) expanded more slowly, reaching Michigan 10,000 years ago and Connecticut 5,000 years ago. Chestnut (*Castanea*) was the slowest deciduous tree species to extend its range northward, reaching its northern range limit in New England just 2,000 years ago. Hemlock (*Tsuga canadensis*) and white pine (*Pinus strobus*), conifers that are characteristic of temperate, predominantly deciduous forest communities, moved northwestward rather than northeastward, spreading from presumed refuges along the Atlantic coast (Fig. 16.5).

The different rates of expansion and the crisscrossing migration paths suggest that northward expansion was not a response to simple warming or to a northward movement of climatic zones similar to those that can be recognized in eastern North America today. Instead, biotic factors such as seed dispersal, establishment on immature soils, or competition from already established vegetation may have limited the expansion of many species, at least part of the time. If so, the northern range limits were not in equilibrium with climate for several thousands of years (Davis 1981).

An alternative point of view is presented by

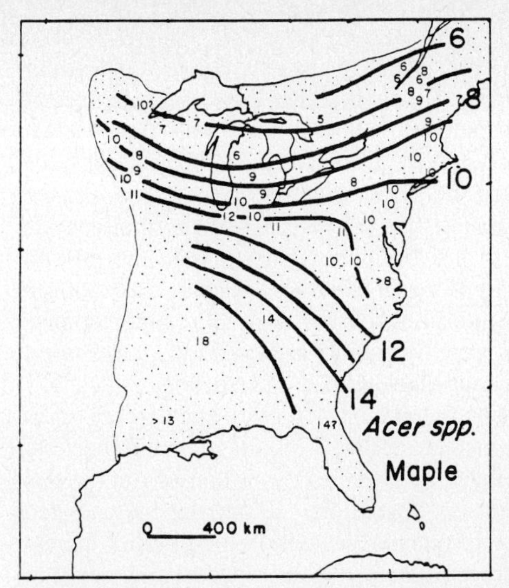

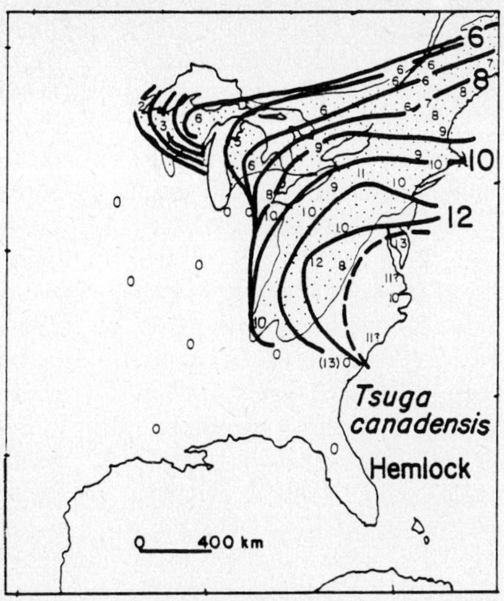

Fig. 16.5 Map of northward range extension of maple and hemlock during the Holocene. Solid lines represent the estimated position of the advancing frontier at 1,000-year intervals. Large numbers are thousands of years before the present. Small numbers represent radiocarbon ages (in thousands of years) of local arrivals of maple and hemlock, as deduced from a sharp increase in the numbers of fossil pollen grains in lake sediments. Stippled area is the present range. Maple, like many deciduous trees, advanced to the north and to the northeast from refuges in the southern Mississippi valley region. In contrast, hemlock moved northward and westward from a refuge area somewhere near the Atlantic coast. (From Davis 1981.)

Webb (1984). He believes that different climatic factors limit each species and that these parameters can vary independently; the criss-crossing patterns represent to him a complicated record of changing climatic parameters. As Graham also argues in Chapter 18, the climatic regimes 5,000 to 10,000 years ago may have no modern analog, which Webb believes is adequate explanation for the lack of modern equivalents for the low-diversity forest communities of the early Holocene.

Regardless of causality, the patterns of northward movement resulted in unique forest communities that changed from one millennium to the next as new species arrived and invaded the forest. Even widespread forest communities that today are similar to one another, such as northern hardwood forests (beech, birch, sugar maple, and hemlock), have had different histories in different regions. In New Hampshire maple arrived 9,000 years ago, hemlock 7,500, and beech 6,500 years ago. In east-central Wisconsin maple arrived at least 7,000 years ago, and hemlock and beech only 2,000 years ago (Davis, unpublished data). Beech arrived before hemlock in central Michigan, whereas hemlock preceded beech in northern Michigan. Oak-hickory communities in southern Michigan have been in existence 5,000 years longer than apparently similar oak-hickory forests in Connecticut (Davis 1981).

Seed Dispersal Large-scale delays in the adjustment of species ranges to climate can be attributed to problems of seed dispersal. For example, after beech had become established in southern Michigan, there was a delay of 1,000 to 2,000 years in the appearance of beech on the far side of Lake Michigan. A similar delay occurred after beech populations had expanded to the Straits of Mackinac (Davis et al. 1984). Dispersal was presumably by birds, which crossed physical barriers (water or prairie) 30 to 100 km wide.

Dispersal of seeds by wind seems to be more effective, because hemlock moved into the same region more rapidly. Hemlock has also established several dozen outlying colonies 50 to 150 km beyond the main, continuous population. The fossil record shows that these outliers are not relics from a previous, more extensive population, but instead new colonies established ahead of the advancing species front (Davis, unpublished data).

The rapid rates of northward expansion for all of these species can be explained only if seed dispersal occurred over greater distances, and with greater frequency, than has been generally supposed. Although expansion apparently could not keep up with climate change, it still was rapid (100 to 400 m per year) relative to expectations of ecologists (van der Pijl 1969, Davis 1981). Note, however, that 100 m per year is two orders of magnitude slower than the expansion rate of cotton rats in the course of the twentieth century (Hoffmann and Jones 1970).

Alternate Communities A series of interglacial deposits in England shows that alternate forest communities developed there during the various interglacials (West 1970, 1980). To some extent these were influenced by stochastic events that affected which species arrived in Great Britain before it was isolated by rising sea level. But evolutionary trends influenced the success of particular species. Spruce (*Picea*), which appears early in the sequence in older interglacials, migrated to the British Isles from refuges in western Europe. In the mid-Quaternary these western populations apparently became extinct. After that time spruce arrived in Britain late during each interglacial, expanding across Europe from refuges in the east, just as it has during the Holocene. During the course of the Quaternary, hazel (*Corylus*) has become more and more successful. In each successive British interglacial it appears earlier in the sequence and reaches greater abundance. In the early Holocene it was the dominant tree until the arrival of taller, shade-tolerant trees (West 1980).

Linkages with Animal Communities Mammalian assemblages in North America during the glacial maximum include extinct genera and combinations of extant species that are unknown today (Chapter 18). The lag in northward movement of particular species of trees, however, does not seem to have affected the dispersal of mammals. Mammals, like many birds, are apparently more sensitive to vegetation structure than to species composition (Chapter 18). There are, of course, bird species like the Kirtland's warbler (*Dendroica kirtlandii*) that breed only in forests of jack pine (*Pinus banksiana;* Van Tyne 1951). In this case the dispersal of warblers must have followed the dispersal of the tree. However, the fieldfare (*Turdus pilaris*), which normally nests in trees, has been found nesting on the ground above tree line in Swedish Lapland. Salomonsen (1951) interprets this behavior as a rapid response to the recent climatic amelioration, a response more rapid than the rise in tree line.

Fossil assemblages in lake and stream sediments provide a good record of water beetles. These insects moved northward rapidly at the end of the last glaciation, colonizing deglaciated areas with pioneer vegetation long before the establishment of forest. These species now occur in regions of boreal forest and are apparently limited by temperature, but not by the presence of trees (Morgan et al. 1983). Herbivorous insects often feed on a restricted assemblage of plant species and are therefore more likely to be affected by delays in dispersal of plants. However, many highly specialized taxa can and do use alternate hosts or in fact are generalists in another part of their geographical range (Fox and Morrow 1981).

The apparent ease with which temperate and boreal plant communities were invaded by trees, and the decoupling in the past of various components of these ecosystems, suggest that forests and the animals associated with them are not tightly organized as communities, perhaps as a result of Quaternary history. The Quaternary with its glacial-interglacial cycles is a period of unstable climate within which dispersal ability and the ability to invade established vegetation have been important for survival.

1,000-Year Time Scale

Climatic conditions have been more or less comparable to modern conditions during the last 10,000 years, the Holocene (Fig. 16.1C). During the first 5,000 years of this interglacial, however, changed orbital parameters caused changes in the insolation reaching high latitudes, resulting in warmer summers and cooler winters in the Northern Hemisphere and intensifying monsoonal climates in tropical regions (Kutzbach 1981). Plants and animals with southern affinities spread northward. For example, the European swamp turtle (*Emys orbicularis*) spread into Sweden; it now reaches its northern limit in Germany. These changes suggest that the climate in Europe was warmer, at least during the summer months (Degerbøl and Krog 1951).

Changes in the position of the prairie-forest border in Minnesota suggest lessened precipitation during the early Holocene (Webb et al. 1983). There is evidence for greater rainfall in central Africa (Kutzbach 1981) and higher temperatures in the mountains of Papua New Guinea (Hope 1976). In northeastern United States temperatures rose the equivalent of 2°C (Davis et al. 1980). The precise dates differ from region to region, but the highest temperatures during the early Holocene are generally recorded between 5,000 to 8,000 years ago (Fig. 16.1C). In northwestern Canada, however, where residual ice sheets had less influence, maximum summer warmth occurred earlier, about 10,000 years ago (Ritchie et al. 1983).

The climatic changes of the early Holocene were sufficiently large and rapid that plant species could not change geographical range quickly enough to occupy all the habitat that became suitable for them. Time lags are suggested by the history of two coniferous trees, white pine (*Pinus strobus*) and hemlock (*Tsuga canadensis*). In the White Mountains of New Hampshire both are now confined to low elevations. In the early Holocene both species grew at elevations as much as 350 m above their present limit. If they had moved into this region immediately as the climate warmed and were limited by climate rather than dispersal, we could expect fossils earliest at low elevation sites. Then as the climate warmed, fossils would appear at successively later dates at higher elevations, much as Kullman (1983) has observed for the recent rise in tree line in the Handölan valley (Fig. 16.7). Instead, fossil evidence for each species appears *simultaneously* at all elevations as soon as any fossils appear in the region, white pine 9,000 years ago and hemlock 7,500 years ago. Apparently, the climate had warmed to conditions similar (at least during summer) to today 10,000 years ago, when spruce began to grow abundantly at and above the elevations of present-day tree line (Spear 1985). Although conditions apparently favored its growth, white pine was delayed 1,000 years in colonizing the region, and hemlock was delayed 2,500 years (Davis et al. 1980). The differences in lag time suggest differences in the ability of the two species to disperse seed and to become established.

100-Year Time Scale

The "Little Ice Age" is well documented from historical and instrumental records. This cold period began around A.D. 1250, became more intense around 1450, and culminated in the seventeenth and eighteenth centuries, when montane glaciers advanced in many regions (Fig. 16.1B). Low temperatures, especially in winter, affected the geographical distributions of many mammals, which disappeared from the fauna of northern Europe, to reappear after 1850 in response to climatic warming (Ahlmann 1953, Hustich 1952). Kalela (1952) documents several examples in Finland.

There is relatively little evidence that frequencies of forest trees changed during the Little Ice Age. Among the few examples, hemlock became more abundant in northwestern lower Michigan (Bernabo 1981), and beech (*Fagus*) extended its limit westward a few tens of kilometers in upper Michigan (Woods and Davis 1982). A major change in forest composition occurred 400 years ago in the Big Woods region of Minnesota, where the direct effect of climatic change on the vegetation was amplified by a decrease in wildfire frequency (Grimm 1983). Here oak scrub was replaced by maple-basswood forest. The Big Woods forest community described in detail by Daubenmire (1936) was only one or two generations old.

10-Year Time Scale

Superimposed on these longer events are variations from one decade to the next, short-term trends that reverse themselves after 10 or 20 years. The downturn in temperature in the Northern Hemisphere since the 1950s is an example (Fig. 16.1A). Given directional trends at several time scales, what significance is there to "mean annual temperature''? The U.S. Weather Service averages over a 30-year period, traditionally 1931 to 1960, which is in fact the warmest 30-year interval we have experienced in the last 500 years! Recently, more extensive tables have been issued that show averages for 1940 to 1970. In most regions of the United States the 1940–1970 temperatures are lower than the 1931–1960 values, reflecting the cooling trend in the Northern Hemisphere since 1950. There is no ''typical'' 30-year interval that would be appropriate as a standard (Brubaker 1981). Ecologists need to be aware of the arbitrariness of the term ''mean annual temperature'' and to recognize the long-term changes that occur in the system they are studying.

SPATIAL VARIATION

Regional Differences Within the United States

Local changes in climate can result from shifts in the positioning of the general circulation of the atmosphere. Other local changes result from changes in the amount of heat received from the sun; these changes are translated into changes in circulation patterns transferring heat from one part of the globe to another. In either case a change at one spot on the earth's surface is linked with the worldwide circulation and will be correlated with a series of changes elsewhere on the globe. The changes that take place elsewhere may be opposite in sense or may involve quite different climatic parameters. This point is illustrated by maps of differences between nineteenth-century and twentieth-century climate in the United States, based on instrumental measurements (Wahl and Lawson 1970). Two maps of deviations in climatic parameters are shown in Fig. 16.6. In the Great Lakes region and along the East coast annual temperatures were 2°F lower in 1860 than in 1931–1960, while in the Great Basin they were 2°F higher. Annual precipitation also shows a distinct pattern; the Southwest received 20% more precipitation than at present, while there was little change along either coast.

The between-region differences are as large as the anomalies we have been discussing throughout this paper, that is, they are large enough to affect the distribution and abundance of species and therefore to affect community patterns in space.

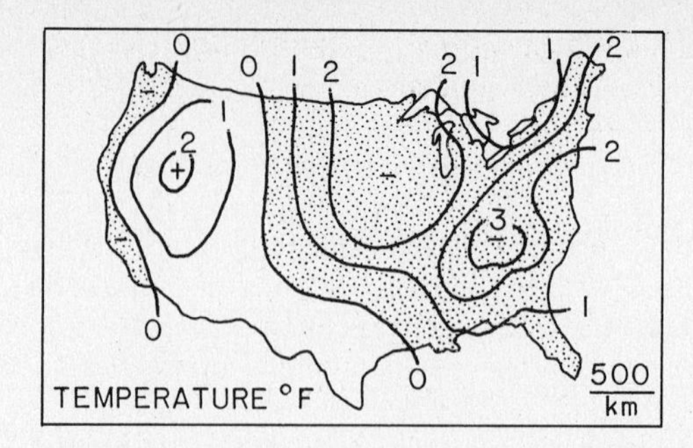

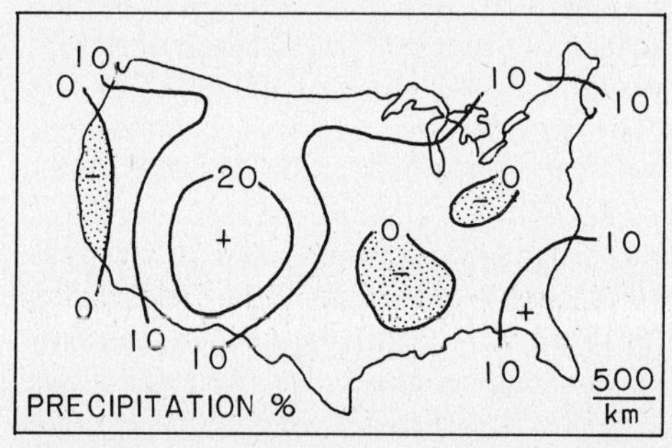

Fig. 16.6 Maps illustrating how temperature (top) and rainfall (bottom) in 1860 differed from so-called climatic mean values, that is, those for 1931–1960. Stippled areas were colder or drier in 1860, while unstippled areas were warmer or wetter. Notice that in 1860 the eastern and Great Lakes regions of the United States were colder, Florida was wetter, and the western region (except for the Pacific coast) was warmer and wetter than in 1931–1960. (After Wahl and Lawson 1970.)

Spatial patterns in the amount of change and even in the direction of change add to the complexity of climatic effects on community structure. The influence of exogenous factors may be difficult to assess unless both time and space dimensions are taken into account (Chapters 8, 9). For example, changes in the bird fauna of Iceland since the end of the last century are interpreted as an adjustment to climatic change (Gudmundsson 1951), while changes in the bird fauna of the Channel Islands of California during the same interval are used to measure turnover rate assuming faunal equilibrium (Diamond 1969b). Both interpretations are reasonable, because the north Atlantic region has experienced a large temperature rise since 1880, whereas the climate of southern California has been relatively stable (Fig. 16.6).

The Tropics

Changes in the tropics are of particular interest to ecologists, because one explanation for the diversity of tropical communities assumes a long period of climatic stability. Abundant evidence exists to show that the tropics experienced changes in precipitation and temperature during the last glacial interval and during the early Holocene. Direct fossil evidence from areas of lowland tropical rain forest is available from northern Australia, where rain forest was replaced by sclerophyll forest (Kershaw 1976). Paleoecological studies from mountainous areas and from areas of dry forest corroborate the impression of lowered temperatures and changed precipitation patterns throughout the tropics (Kendall 1969, Walker 1970, Bradbury et al. 1981, Singh et al. 1981). The geographical displacements of forest species may have been more modest than in the temperate zone, but all the evidence that we have so far argues against survival of rain forest communities intact.

Tropical regions have also experienced recent climatic changes. A long-term instrumental record from Indonesia shows the familiar upward trend of January and July temperatures since 1880 (Lamb 1977). When records are averaged for different sectors of the globe, the temperature rise since 1880 at low latitudes (23.6°S to 23.6°N) has only half the amplitude of the rise at high latitudes of the Northern Hemisphere (23.6°N to 90°N). Temperatures continued to rise at low latitudes after 1950, instead of reversing as in high northern latitudes (Hansen et al. 1981).

Short-term weather variation occurs in many regions of the tropics (Leigh 1975). Droughts in tropical Brazil are correlated with large-scale anomalies in oceanic and atmospheric circulation (Hastenrath 1984). The latter show short-term changes similar to El Niño. Given lag effects in the response of vegetation, a change in the frequency of episodes of unusual weather could have major effects on a tropical forest community. In a tropical rain forest in Queensland tree species now present in the canopy are not present as juveniles (Connell 1978). The same is true in tropical dry forest in Costa Rica (Hubbell 1979). This age structure is not easily explained without invoking climatic variation or disturbances that change conditions for the establishment of seedlings.

A CASE STUDY: FORESTS OF THE HANDÖLAN VALLEY, SWEDEN

At the community level climate change is important to the degree that response lags differ among the components of the community. If all species responded in the same way and at the same rate, a community could migrate intact from one latitude to another, maintaining the same climatic environment. Chaney (1944) appears to have visualized Tertiary ''geofloras'' maintaining their integrity in space and time in much this manner. However, in this paper I have emphasized differences among species in the timing of their responses. These differences preclude the possibility of geographical displacement of intact communities. Still, many discussions of community structure consider only one guild of organisms or one related group of animals or plants, rather than several trophic levels. To what extent do guild members differ in speed of response to climatic change?

A case study makes it clear that even minor differences in autecology and life history can have major effects on the responses of individual species. The forests near the tree limit in the Handölan valley in central Sweden have been studied extensively by Kullman (1979, 1981a, 1981b, 1983). Each of the dominant trees in the forest—birch (*Betula pubescens*), pine (*Pinus sylvestris*), and spruce (*Picea abies*)—responds to climate with a different time lag and consequently is affected by climatic change on a different time scale. Birch, which forms the alpine forest limit, tracks climatic changes most closely. It becomes established at sites where shallow snow accumulations melt early in the season. The climatic amelioration of this century with warm summers has permitted birch to advance upslope by as much as 100 m in elevation. Age measurements of birch populations at tree line (Fig. 16.7) show

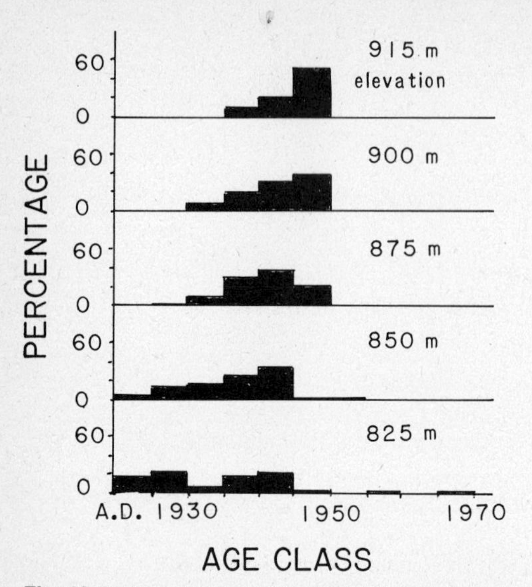

Fig. 16.7 Age structure of birch forest at various elevations near tree line in the Handölan valley of Sweden. The age structure reflects the rise in tree line that occurred in response to warm summers during the first half of the century, with establishment of trees at progressively higher elevations from 1915 to 1940. Establishment ceased at the higher elevation sites in 1950, when the climate became cooler again. (After Kullman 1983.)

that birch became established at 825 m as early as 1915, while colonization began progressively later at higher elevations. Establishment of seedlings ceased in 1950, when the summers became cooler again. Seedlings that were already established at 915 m continued to grow, an example of the inertia illustrated by spruce in the computer simulation (Fig. 16.4). In the case of birch, seed germination and seedling establishment appear more sensitive to summer temperature than are survival and growth of already established trees (Kullman 1983).

Spruce is infrequent near tree line, but in places there are spruce growing as high on the slopes as birch, with which spruce competes strongly. Where the snowfall is heavy, spruce krummholz occur. Since 1915, when the Handölan forests were first studied in detail, no new spruce have become established at high elevations. However, previously established spruce krummholz began to grow vigorously as the climate became warmer, increasing in height, at least until 1950. All of the spruce trees and krummholz shrubs are less than 200 years old, postdating fires which apparently destroyed any preexisting stands of spruce in the early eighteenth century. Reestablishment of spruce has apparently been inhibited by competition from birch, which grows abundantly at high elevations. Fossil studies indicate that spruce first colonized the Handölan valley only 1,300 years ago, when it reached this part of Sweden in the course of its expansion westward from glacial refuges.

Pine in the Handölan valley is even more sensitive than birch to snow cover, but unlike birch, the upper elevation for pine trees has not risen in response to the warmer climate of the present century. Pine grew well above present tree line in the early Holocene and has been retreating since then (Fig. 16.1C). Kullman found evidence that a fire in 1711 destroyed stands of pine near the present upper tree limit for the species; these stands have failed to regenerate. It appears that pine trees growing now at high elevations became established during the centuries of warm climate immediately preceding the Little Ice Age (Fig. 16.1B).

As the climate became colder in the seventeenth and eighteenth centuries, adult trees were able to persist, but seedlings did not survive. When fires in the early eighteenth century destroyed these stands, which were essentially relict from a previous climatic regime, they did not regenerate. The few pine trees that still persist at high elevations are several hundred years old (Kullman 1983). Thus, the elevation of the upper *tree* limit for pine is lagging the climate by about 400 years, although fires did speed up the retreat of the *forest* limit of pine to lower elevations.

While birch is tracking decadal climatic changes, pine is lagging by several hundred years. Spruce, which can tolerate heavy snow cover, has been affected less by the recent changes in climate; its abundance seems to be a function of fire frequency and competition from birch. In fact, the geographical range of spruce is still affected by glacial events more than 15,000 years ago.

This case study makes it clear that species frequencies in the forest community of Handölan cannot be explained as equilibrium abundances resulting from competition or predation or other factors traditionally used to explain community structure. Although competition, herbivory, and disturbance are all operating on the system, response to climatic change is also important.

CONCLUSIONS

Both Hutchinson (1961) and Connell (1978) have discussed the role of environmental change in the maintenance of diversity, arguing that if environmental changes occur sufficiently rapidly, competitive exclusion cannot occur before the advantage of the winning species is reduced by the changed environment. Hutchinson was considering seasonal changes in lakes, but Connell discussed climatic changes, debating whether directional trends could occur rapidly enough to affect forest species. The Handölan Valley example answers Connell's query, showing that climatic changes on several different time scales affect this forest community. In systems where species lifetimes are shorter than for trees and in systems for which there is no long-term record as often exists for forests, the impact of climatic change may be more difficult to document (Connell and Sousa 1983). However, the processes described in the Handölan Valley will operate for birds, mammals, or insects, although on a shorter time scale. It seems likely that one or more components of all communities will not be in equilibrium with the prevailing climate, that is, will not have achieved the distribution and abundance that would be predicted as the outcome of biotic interactions under that particular climatic regime.

It would be valuable for theoreticians to address the impact of long-term environmental trends, modeling systems where competing species respond to change with time lags of different length. Shugart et al. (1981), for example, have used a simulation model to study forest responses to long-term cooling and warming, demonstrating that alternate communities result under the same climate during continuous warming versus during continuous cooling. The asymmetry is caused by time lags similar to those discussed in this paper.

The view of community structure that I have presented contrasts with models that strive to explain species abundances and the maintenance of species diversity in forest communities under equilibrium or climax conditions (cf. Horn 1981, Woods 1979). The climatic-instability view emphasizes the dynamic nature of biotic communities, with species frequencies (and sometimes species composition) changing continually, even during the lifetimes of individual organisms. Species frequencies and age structures at a single point in time have little predictive value for the future, except where inertial effects (due to growth of already established individuals) continue existing trends despite changes in the physical environment.

SUMMARY

Climate is inherently unstable, changing continually on all time scales. Directional climatic changes affect the distribution and abundance of organisms. There are many examples of changes in population size and geographical distribution that have been caused by decade- or century-long directional climatic trends.

In the Northern Hemisphere temperatures rose 0.6°C between 1880 and 1950. Many species of birds, mammals, and insects extended their ranges northward in response. Those animals, especially insects and some species of birds that track climate closely, responded within a year or two of the onset of climatic change. Other species have responded more slowly, with a time lag of several years or even one or two decades. These slower species are still expanding northward, while others with faster responses are already expanding southward again in response to the climatic cooling that has occurred in the Northern Hemisphere since 1950.

Time lags of plants are generally longer than animals. Although biomass and cover respond quickly, the importance of species within the community may lag the climatic change by several decades or, in the case of forests, by a century or more. Dispersal of plants to newly suitable habitats can occur at an average rate of several hundred meters per year, but this is several orders of magnitude slower than for animals, which can expand at several kilometers per year (cotton rat) or several hundred kilometers per year (fieldfare).

Time lags vary by as much as an order of magnitude from one species to another, because they depend on behavior, life span, life history traits, intrinsic rate of increase, and dispersal. Competition from resident species is important, especially for plants, because resident plants are tenacious, often surviving unfavorable weather for decades even when seed production or seedling survival are seriously impaired.

Time lags affect community structure, because they determine the timing of each species' response to a given climatic change. The differences in the timing of responses disrupt community patterns and prevent migration of the intact community to a favorable environment. Furthermore, climatic changes occur in complex spatial patterns, and consequently the distortions of community patterns occur in both space and time.

A case study of a forest community shows that each of the three dominant trees tracks a different climatic signal and on a different time scale. Although competition and occasional fires affect the abundance and distribution of these species, they are also influenced by the climatic changes occurring from one decade to the next in the present century, from one century to the next over the past 500 years, and the major change that occurred when the interglacial began 10,000 years ago. Pine abundances at high elevations are higher than one would expect under equilibrium conditions, due to a lag in adjustment of pine to the cooler climates of the eighteenth and nineteenth centuries. Spruce abundance is less than expected because spruce was destroyed by fires in the eighteenth century. Birch is more abundant than expected because it became established at high elevations during the climatic warming before 1950. Although spruce and birch compete, it is difficult to deduce from their present frequencies what equilibrium abundances should be expected.

Given the slow responses of many species to climatic change and given the instability of the climate at all time scales, many plant and animal communities, or at least components of those communities, will be in disequilibrium, continually adjusting to climate and continually lagging behind and failing to achieve equilibrium before the onset of a new climatic trend. The species composition and the abundances of species within a community cannot be interpreted on the basis of biotic interactions without also considering responses to directional climatic trends.

ACKNOWLEDGMENTS

This work has been supported by the National Science Foundation and by the University of Minnesota Graduate School.

chapter 17

Climatic Cadences and the Composition of Chihuahuan Desert Communities: The Late Pleistocene Packrat Midden Record

Thomas R. Van Devender

INTRODUCTION

Fossil communities can be recognized by studying fossil assemblages of various kinds. Remains of plants may be preserved in sedimentary deposits as macroscopic leaves and fruit (Chapter 7) and as microscopic pollen and spores (Chapter 16). Bones of vertebrate animals may be preserved in open sedimentary sites or in cave deposits (Chapter 18). All of these fossil accumulations are "death assemblages" that were transported to the site of deposition. The paleocommunities captured in these assemblages are in many cases mixtures of organisms that lived many miles apart or hundreds to thousands of years apart in time.

Rich assemblages of well-preserved plant macrofossils in ancient packrat middens provide detailed records of local plant communities for the last 35,000 years along a latitudinal gradient in the Chihuahuan Desert in New Mexico, Texas, and northern Mexico. Packrat middens have the great advantage that the fossils provide a time capsule of a fossil plant community gathered from a small area in a relatively short time period.

Packrat Ecology and Midden Methodology

Packrats, or wood rats, are medium-sized rodents in the genus *Neotoma*. About 20 species of *Neotoma* are found from sea level to 3,350 m elevation in many habitats coast to coast and from British Columbia to Guatemala. *Neotoma albigula* (white-throated packrat) is the species that lives in most habitats in the Chihuahuan Desert, while *N. mexicana* (Mexican woodrat) is restricted to more mesic woodlands and forests in the mountain islands. Most of the fossil middens from the Chihuahuan Desert were constructed by *N. albigula,* although a few teeth of *N. mexicana* have been recovered as well.

Packrats carry various objects to their houses for use as construction materials or food (Finley 1958). Their waste piles or middens usually contain rich assortments of plant fragments collected within 30 to 50 m of the house. When packrats live in rock shelters, houses are not well developed, but the middens often serve as urination perches and become very hard, dark, and shiny. The plant remains incorporated into the hardened middens represent a random sampling of 50 to

100 years of packrat collections from a local plant community. This is long enough to provide excellent sampling of annuals and short-lived perennials, but is usually less than the standard deviation on the associated radiocarbon date. The remains of each plant species in a midden assemblage can be ranked internally using a relative abundance scale from 1 to 5 (rare to abundant). Comparison of modern midden assemblages with line transect data demonstrates that the middens provide an excellent reflection of the local plant community within 30 m of the rock shelter. Any biases introduced through the packrats' collecting preferences are present throughout the data base.

These ancient middens can be preserved as long as they are dry. The oldest packrat midden discovered to date is more than 50,000 years old (Spaulding et al. 1983). The plant remains or packrat fecal pellets are excellent for radiocarbon dating, allowing the assemblages to be placed in a time framework. Sequential series of packrat midden assemblages from a single area provide excellent records of the *in situ* development of the local plant communities through time.

Packrat Midden Analyses

Beginning in the 1960s the analyses of well-preserved plant macrofossil assemblages from radiocarbon-dated packrat middens have provided a wealth of information on the history of vegetation and climate in the arid areas of North America (Van Devender and Spaulding 1979, Spaulding et al. 1983). The macrofossil assemblages have documented widespread expansion of woodland and forest communities into most North American deserts during the last glacial period, the Wisconsin.

In recent years fossil packrat middens have been found at 24 Chihuahuan Desert sites (Fig. 17.1) ranging from the San Andres (latitude 33°11′N) and Sacramento (32°50′N) mountains in south-central New Mexico, south into the Big Bend of Texas (29°12′N) (Lanner and Van Devender 1981), and to the Bolson de Mapimi in Coahuila and Durango, Mexico (26°N; Van Devender in press, Van Devender and Burgess in press). About 160 midden assemblages have been analyzed with 180 radiocarbon dates on materials from them. At least 400 species of plants, including most of the dominants in modern woodland, desert-grassland, and desertscrub communities, have been identified from the samples. The current emphasis is to build detailed local vegetation chronologies along a latitudinal gradient through the Chihuahuan Desert. Differences in composition and the timing of change can help infer atmospheric circulation patterns through time and the role of dispersal in the development of the vegetation.

Goals of This Chapter

I will use the historical evidence from Chihuahuan Desert packrat middens to address five problems of interest to ecologists: (1) the age of modern communities, (2) the notion of equilibrium as applied to communities, (3) the evidence for refugia, (4) the differing histories of faunas and floras, (5) the origin of the latitudinal gradient of species richness. As background, I will briefly summarize the vegetational history of the Chihuahuan Desert from the Miocene until the first dated packrat midden around 50,000 B.P.

THE MIOCENE REVOLUTION

In the early Tertiary the plant communities of the western United States were mixtures of plants found in conifer forests (*Pinus*), sclerophyll woodlands (*Quercus, Cercocarpus*), and broadleaf deciduous forests (*Magnolia*) with plants from dry tropical forest including cuajilote (*Bombax*), elephant tree (*Bursera*), palms (*Brahea, Sabal*), figs (*Ficus*), apes' earrings (*Pithecellobium*) and tree ferns (Axelrod 1979). Some of the subtropical species lingered into the middle Miocene (Mint Canyon local flora, 10 to 13 million years ago) of California as Sierra Madrean woodland and forest plants increased in importance.

The uplift of the Rocky Mountains, the Sierra Madre Oriental, and the Sierra Madre Occidental beginning in middle-late Miocene gradually caused a revolution in the biota. Eventually, the mountains were high enough to have major effects on the upper circulation of the atmosphere. Relatively modern climatic regimes developed with more regionalization of climate, general drying and relatively arid conditions in many areas, especially in rainshadows, and greater seasonal extremes in temperature and precipitation. The biotic results included the segregation of the flora into new relatively modern communities including tundra, boreal forest, grassland, and desertscrub. Broadleaf deciduous forests were displaced to the southeastern United States and the mountains of northeastern Mexico, and dry tropical forest was displaced to lower latitudes. This trend continued until the modern deserts were formed between 5 to 8 million years ago in the latest Miocene (formerly the middle Pliocene; Axelrod 1979). At the same time a major evolutionary radiation in the biota heralded the rise of many groups that dominate the landscape today, including grasses, composites, toads (Bufonidae), frogs (Ranidae), lizards (Iguanidae), snakes (Colubridae, Elapidae, Crotalidae), passerine birds, and cricetid rodents. Reasonably close structural and functional ancestors of modern species, if not the modern species themselves, have been in existence at least 5 million years in these groups. The origins of middle Tertiary broadleaf deciduous or dry tropical forest plant species could be much older.

ICE AGES: CYCLES OF CHANGE

The advent of glacial climates in the Pleistocene had profound effects on the distribution of organisms and communities over the entire globe, and these effects decreased from the poles to the equator. In glaciated areas the entire biota immigrated into the area in each subsequent interglacial period. In central New Mexico (latitude 36°N), hundreds of miles south of the continental glaciers, the vegetation changed from mixed-conifer forest to pinyon-juniper-oak woodland or grassland (Betancourt and Van Devender 1981). In southern New Mexico and Trans-Pecos Texas, pinyon-juniper-oak woodland grew in areas that now support Chihuahuan desertscrub (Wells 1966; Van Devender, Betancourt, and Wimberly 1984; Van Devender in press). Entire vegetation zones were absent during glacials and developed only during interglacials. Today, ponderosa pine (*Pinus ponderosa*) is an important, widespread dominant in the pine forest zone between mixed-conifer forest and pinyon-juniper woodland from Mexico to British Columbia. In the last glacial mixed-conifer forest abutted juniper or pinyon-juniper-oak woodlands directly. Not only was the range of ponderosa pine greatly reduced to the south, but it was apparently an uncommon tree in south-central New Mexico and southern Arizona. Similar interglacial range expansions probably occurred in many plants.

South of latitude 30°N the effects of glacial climates were less catastrophic, but still very important. Changes in the general circulation of the atmosphere under glacial temperature regimes caused major shifts in the amounts and seasonal distributions of rainfall. In the northern subtropics (ca. 20°N to 30°N) increased annual rainfall and lower summer temperatures favored mixed communities composed of woodland, desertscrub, and subtropical thornscrub plants (Van Devender and Burgess in press). Farther south the glacials appear to have brought drier conditions to the highlands of central Mexico (Watts and Bradbury 1982) and to the tropics where the rain forests of Amazonia and Venezuela gave way to tropical savannahs (Campbell 1982). However, the extremely arid coastal deserts in Ecuador and Peru appear to have been much wetter than today due to a local accentuation of the El Niño equatorial current (Campbell 1982). Indirect effects of continental glaciers were probably reflected at these latitudes by changes in composition, structure, and distribution of communities as precipitation regimes changed. Indeed, the few places on the globe where the ice ages did not significantly alter the plant communities are probably still under ice or water.

Traditionally, four glacial periods were recognized in the Pleistocene and widely correlated between Europe, North America, and South America despite the fact that deposits beyond the range of the radiocarbon dating technique (ca. 40,000 years) are difficult to date. However, with the development of global correlation techniques using the paleomagnetism of fine-grained sediments and of paleoclimatic analyses using planktonic fauna and oxygen isotope ratios in deep-sea cores, the traditional four-period terrestrial glacial sequence has been questioned. It now appears that there have been as many as 15 to 20 glacial periods in the 1.8 million years of the

Pleistocene (see Fig. 16.1E; Imbrie and Imbrie 1979). Mounting evidence supports Milankovich's astronomical explanation of the cyclic patterns of glaciations in the Pleistocene, with ice ages in times of insolation minima.

Moreover, the glacials were roughly 100,000 years in length and interglacials only 10,000 years. Thus, interglacials, like the present Holocene, spanned as little as 10% of the Pleistocene. As the Chihuahuan Desert packrat middens discussed in the following section demonstrate, the modern climatic and biotic regimes only developed 4,000 years ago in the late Holocene. Assuming that previous interglacials were similar, the modern regime probably characterized less than 4% of the Pleistocene. The modern plant communities are really rather ephemeral, modified versions of latest Miocene communities!

VEGETATIONAL HISTORY OF THE CHIHUAHUAN DESERT

The Chihuahuan Desert (Fig. 17.1) is unique among the deserts of the American Southwest because it is an interior continental desert in the rainshadows of the Sierra Madre Occidental to the west, the Sierra Madre Oriental to the east,

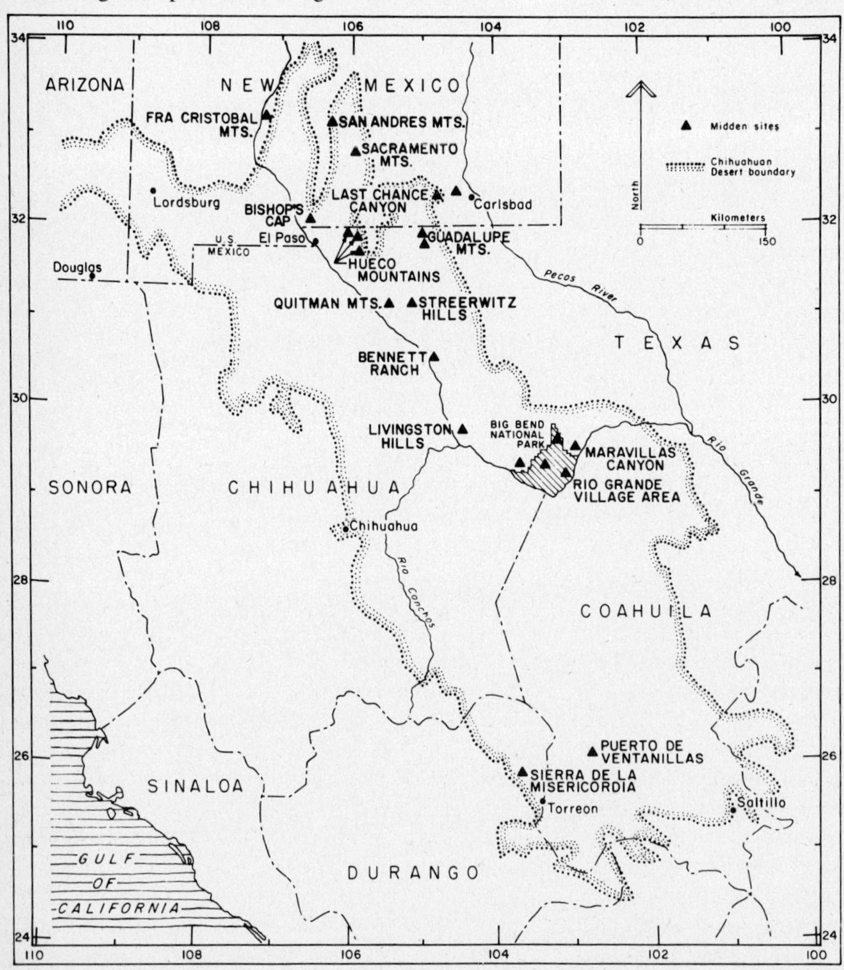

Fig. 17.1 The Chihuahuan Desert, United States and Mexico. Triangles mark fossil packrat midden localities. (Boundary of the Chihuahuan Desert after Schmidt 1979.)

the highlands of the Mexican plateau to the south, and the Rocky Mountains to the north (Johnston 1977). Portions of the Chihuahuan Desert occur in the states of Arizona, New Mexico, Texas, Chihuahua, Durango, Coahuila, Zacatecas, Nuevo Leon, and San Luis Potosi along its 1,600-km northwest-to-southeast axis (latitude 22° to 33°N).

The vegetation gradients reflect general precipitation and temperature regimes, with desertscrub communities in the hot, dry lowlands of the Big Bend of Texas (600–1,675 m) and the Bolson de Mapimi (1,075–2,000 m) gradually replaced by desert-grassland to the north, west, south, and southeast (Schmidt 1979, Morafka 1977, Johnston 1977). A relatively mesic mesquite scrub community is found on the coastal plain of the Gulf of Mexico and contiguous low areas in Texas, Coahuila, Tamaulipas, and Nuevo Leon (Muller 1939, 1947). The vegetation of the southeastern portion is much more subtropical than other areas of the Chihuahuan Desert. In desertscrub communities there is a gradual increase to the south in the number of species and the importance of succulents and other frost-sensitive species. Pine-oak woodland and pine forest are present at higher elevations on montane islands in the Chihuahuan Desert and on the surrounding mountain ranges.

Midden Chronologies

Packrat midden chronologies have been completed in the Chihuahuan Desert from the Sacramento Mountains (Van Devender, Betancourt, and Wimberly 1984) and the San Andres Mountains (Van Devender and Toolin 1983) of New Mexico, and from the Hueco Mountains (latitude 32°N; Fig. 17.2), Maravillas Canyon (29°33′N), and Rio Grande Village area (29°12′N; Fig. 17.3; Van Devender in press) of Texas. Similar chronologies from Chaco Canyon, northwestern New Mexico (36°02′N; Betancourt and Van Devender 1981; Betancourt, Martin, and Van Devender 1983) and the Abajo Mountains, southeastern Utah (37°26′N; Betancourt 1984) extend the latitudinal gradient northward onto the Colorado plateaus.

All of the Chihuahuan Desert vegetation sequences are similar in showing stepwise changes in the dominant woody perennials and succulents with the most mesic in the late Wisconsin and the most xeric in the last 4,000 years in the late Holocene (Fig. 17.2). In all areas the end of the late Wisconsin is marked by a reduction or loss of the more mesic plants including *Pinus edulis* (New Mexico pinyon), *P. remota* (Texas pinyon), and *Juniperus scopulorum* (Rocky Mountain juniper) about 11,000 years ago (Lanner and Van Devender 1981). With the exception of the Rio Grande Village sequence, the early Holocene vegetation was a transitional xeric oak-juniper woodland (Van Devender 1977). The middle Holocene was a desert-grassland period lacking both woodland and many important Chihuahuan desertscrub plants. The development of the modern communities was in all cases a late Holocene event. The vegetation changes that mark the boundaries between the early, middle, and late Holocene have some temporal variability, but were completed by 8,000 and 4,000 years ago, respectively. The Rio Grande Village chronology (Fig. 17.3) differs from the others in the importance of Chihuahuan species in the late Wisconsin pinyon-juniper-oak woodland and the early Holocene development of desert-grassland (Van Devender in press).

The only human impacts detected on vegetation in packrat midden records are prehistoric cutting of pinyon for fuel in Chaco Canyon about 1,000 years ago (Betancourt and Van Devender 1981; Betancourt, Martin, and Van Devender 1983) and shifts in the composition of Chihuahuan desertscrub communities due to grazing by goats and cattle in the Sacramento Mountains in the last few centuries (Van Devender, Betancourt, and Wimberly 1984). Thus, the packrat midden remains show that the Chihuahuan Desert vegetation changed in the late Pleistocene from pinyon-juniper-oak woodland to oak-juniper woodland to desert-grassland, ending in the present Chihuahuan desertscrub in the last 4,000 years.

Refugia

The concept of a refugium as a geographical area where organisms can survive unfavorable cli-

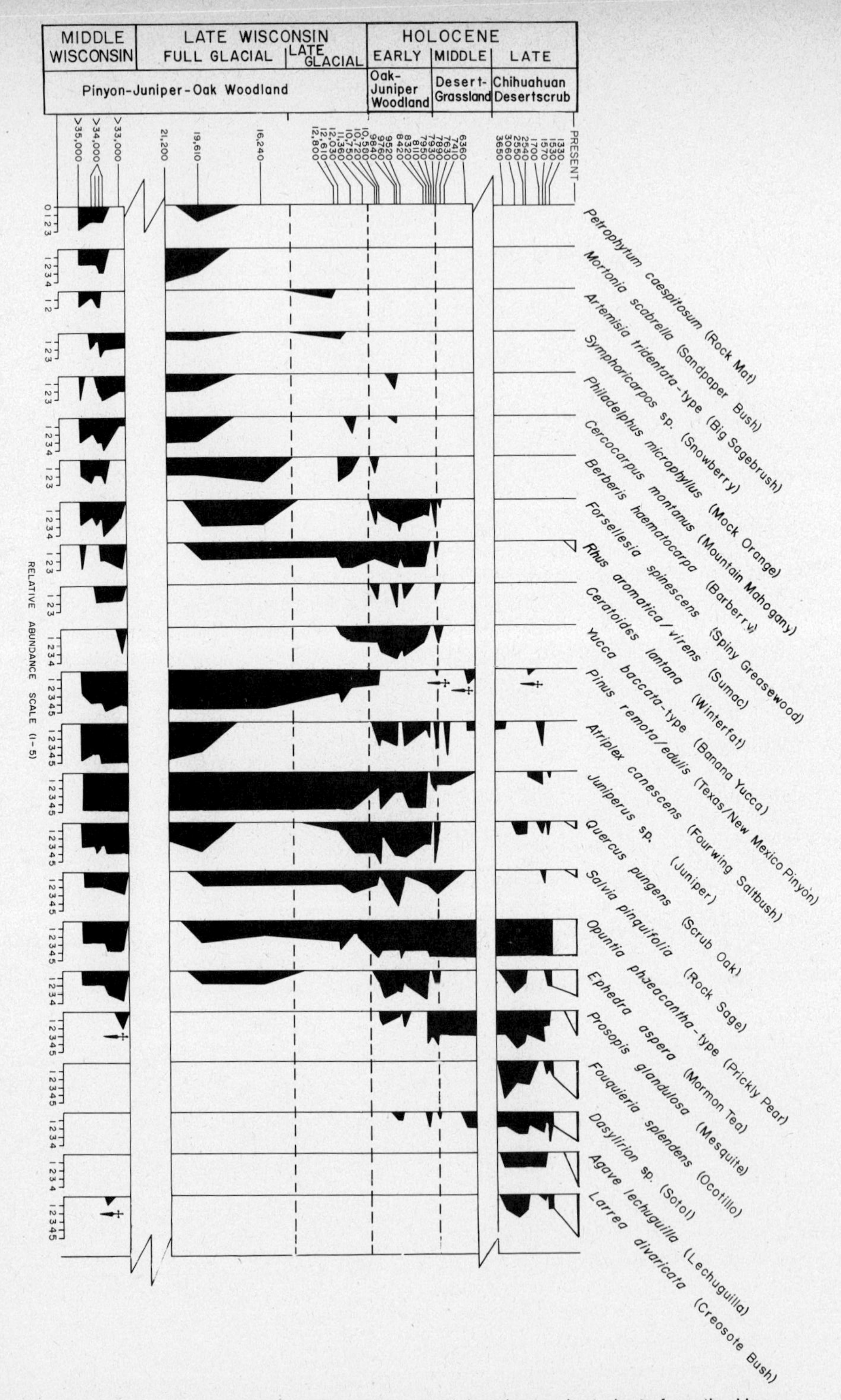

Fig. 17.2 Chronological sequence of selected woody perennial and succulent plants from the Hueco Mountains, El Paso County, Texas. The plants illustrate *in situ* vegetational development from a late Pleistocene pinyon-juniper-oak woodland to postglacial oak-juniper woodland, desert-grassland, and modern Chihuahuan desertscrub. Relative abundance scale: 1 = rare; 2 = uncommon; 3 = common; 4 = very common; 5 = abundant. Vertical scale is in years before present. Ages are radiocarbon dates on individual packrat midden assemblages. Daggers mark suspected contaminants.

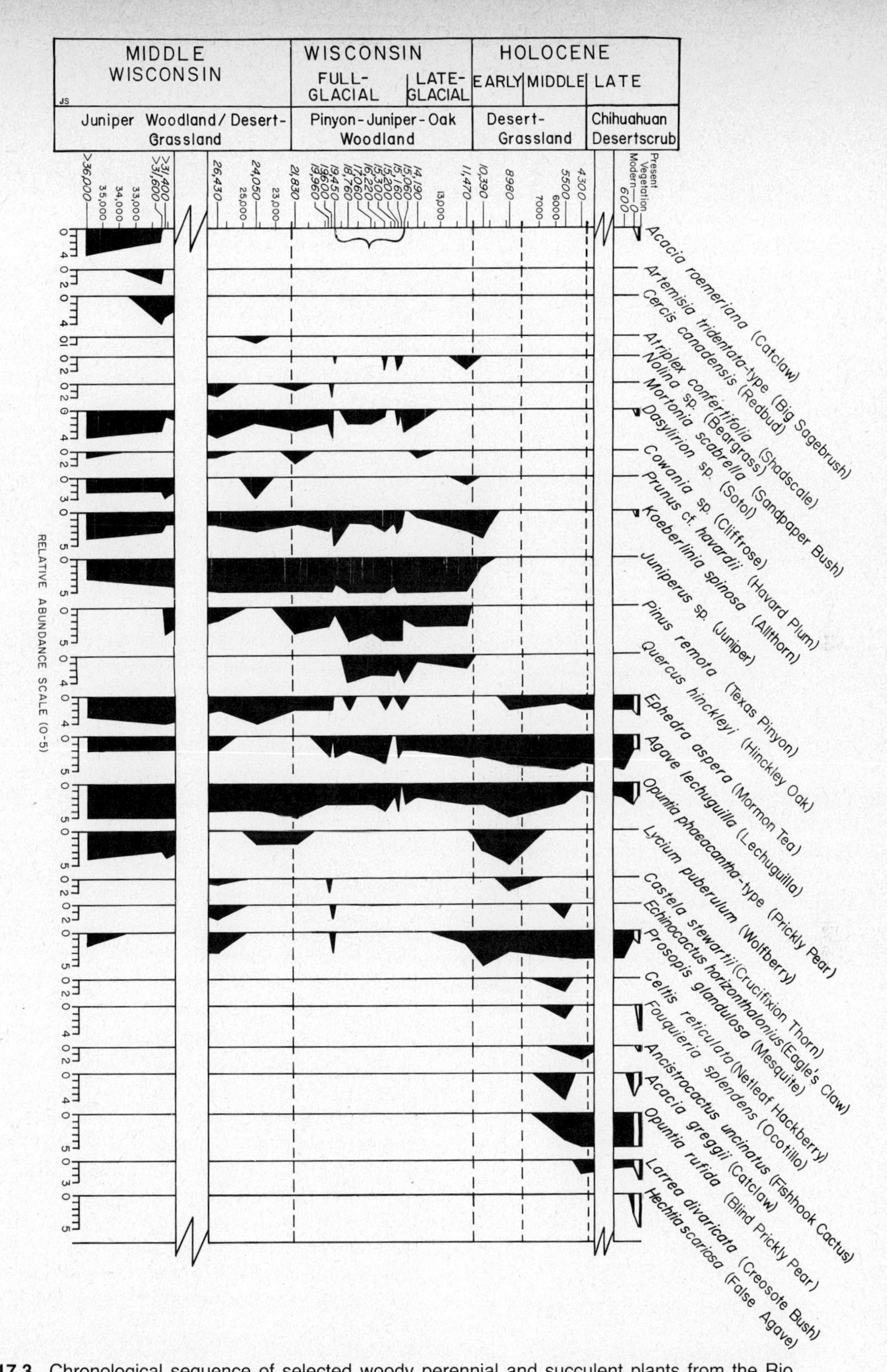

Fig. 17.3 Chronological sequence of selected woody perennial and succulent plants from the Rio Grande Village area, Big Bend National Park, Brewster County, Texas. The plants illustrate *in situ* vegetational development from Pleistocene juniper woodland–desert-grassland and pinyon-juniper-oak woodland to postglacial desert-grassland and modern Chihuahuan desertscrub. Relative abundance scale: 1 = rare; 2 = uncommon; 3 = common; 4 = very common; 5 = abundant. Vertical scale is in years before present. Ages are radiocarbon dates on individual packrat midden assemblages. Bracket marks the relatively stable period in the full glacial discussed in the text.

matic regimes has long been important in biogeography (Brown and Gibson 1983). The midden sequences provide excellent opportunities to evaluate the role of refugia and dispersals in Chihuahuan Desert plant communities.

During the last full glacial, woodland communities dominated by *Pinus edulis* were restricted to elevations of 1,220 to 1,830 m in a relatively small area between latitude 32° and 34°N in south-central New Mexico and adjacent Trans-Pecos Texas (Fig. 17.4). In the Holocene *Pinus edulis* expanded its range as far north as the Wyoming border (40°N; Van Devender, Betancourt, and Wimberly 1984).

The Chihuahuan desertscrub plants in the same area showed several different responses. *Atriplex canescens* (fourwing saltbush) remained in place as a member of the Wisconsin pinyon-juniper-oak woodland. In the San Andres Mountains it was living in a full-glacial mixed-conifer forest with *Pseudotsuga menziesii* (Douglas fir), *Pinus ponderosa,* and *Picea pungens* (blue spruce). The upper limits of its range today are well below the lowest blue spruce. Other dominants in the modern desertscrub community were displaced southward in the Wisconsin and dispersed into the area as middle Holocene desert-grasslands (honey mesquite, sotol) and late Holocene Chihuahuan desertscrub (lechuguilla, creosotebush, ocotillo) communities developed.

The Rio Grande Village (2° to 3° to the south) midden sequence shows a similar pattern with some of the modern desertscrub plants persisting in late Wisconsin woodlands (lechuguilla, honey mesquite, allthorn) and plants near their northern limits today (blind prickly pear, candelilla, false agave, resurrection plant) arriving in the middle or late Holocene.

In turn, two late Wisconsin midden samples from Durango and Coahuila (4° to the south) record expansions of juniper and Texas pinyon into the lowlands of the Bolson de Mapimi, where succulents including *Opuntia rufida* (blind prickly pear) continued to be important. These assemblages contained a mixture of plants that dispersed from many areas including the south (Fig. 17.5; Van Devender and Burgess in press). Consideration of the reptile and amphibian communities led Morafka (1977) to conclude that the Bolson de Mapimi was a relatively unmodified desertscrub refugium in the late Pleistocene, but these plant results show that the situation was more complex. Meyer's (1973) pollen record from the marshes of Cuatro Cienegas, Coahuila, suggested little vegetation change in the last 20,000 years.

Thus, none of the late Wisconsin midden assemblages from the Chihuahuan Desert can be viewed as unmodified desertscrub communities. Rather, some species occurred in the late Wisconsin to the south of their present geographical range. Other species occurred in the late Wisconsin in the same geographical area as they do at present, but associated with a different habitat and set of species than today. In each study area there were plants in late Wisconsin samples that dispersed to more northern areas in the Holocene. Most of them are now desertscrub plants that were then living in equable, mixed woodland communities. In the traditional sense none of these areas were refugia. While portions of the Chihuahuan Desert could be viewed as a refugium for some species of desert plants, reptiles, and amphibians, the plant communities were substantially restructured.

Community Stability

General sequences of plant communities for the last 30,000 years are easily seen in the biochronological diagrams in Figs. 17.2 and 17.3. These sequences resemble the Miocene developmental sequence, where a woodland developed as subtropical and broadleaf deciduous trees withdrew, and continued drying with the uplift of the mountains gave first more xeric woodlands, then grasslands, and finally Chihuahuan desertscrub. Similar developmental sequences probably occurred with each of the 15 to 20 glacial-interglacial cycles in the last 1.8 million years. Each time that this occurred, the communities were probably slightly different because of variations in the relative lengths and climatic severity of glacials and interglacials, and the evolution, immigration, or extinction of important plants and herbivores. Certainly the Pleistocene dispersal of *Larrea divaricata* (creosotebush) from South America into the Chihuahuan Desert and later across the Continen-

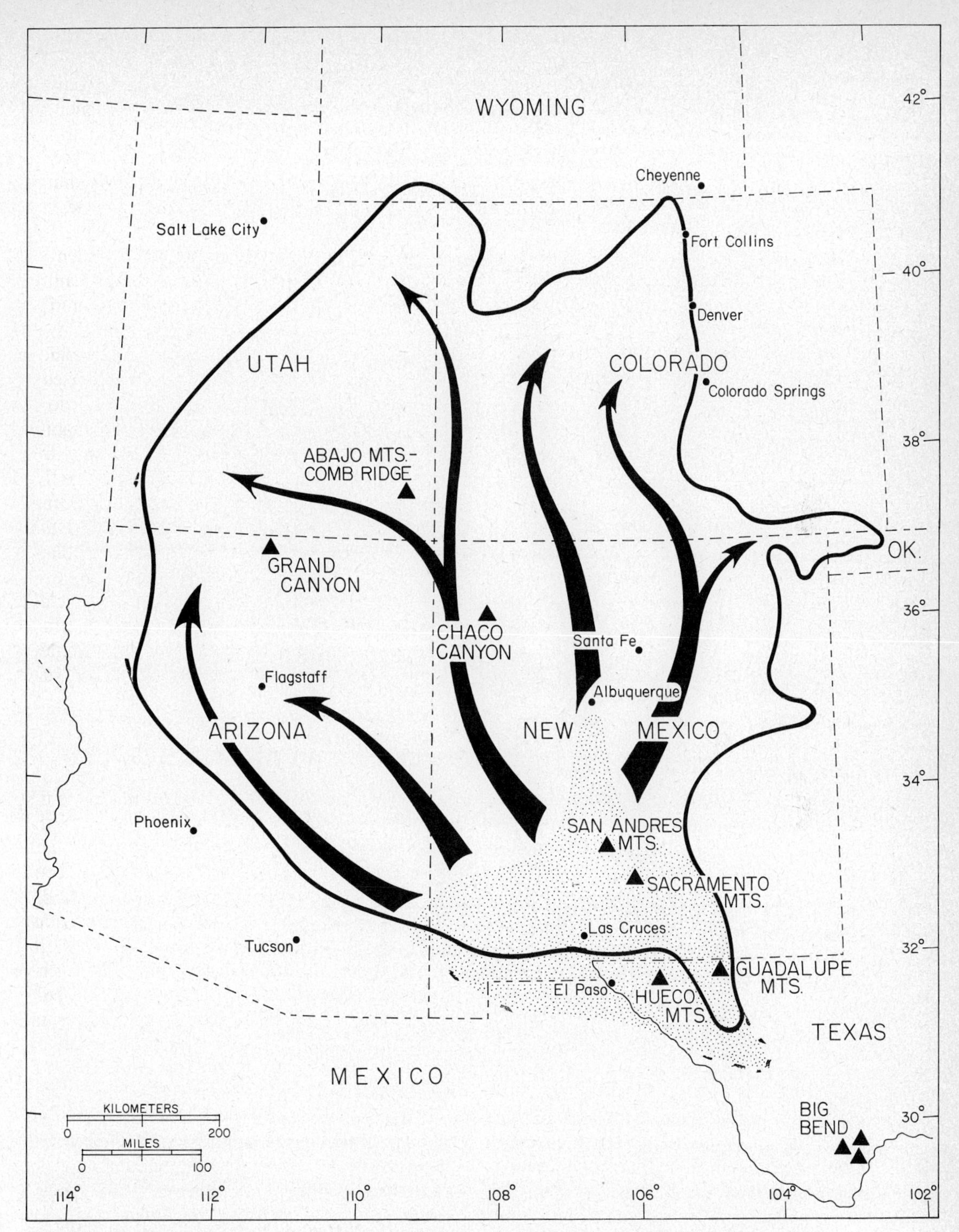

Fig. 17.4 Generalized modern distribution (area bounded by solid line), suggested late Wisconsin range (stippled), and inferred postglacial dispersal of *Pinus edulis* (New Mexico pinyon). Triangles mark packrat midden sites. (From Van Devender, Betancourt and Wimberly 1984.)

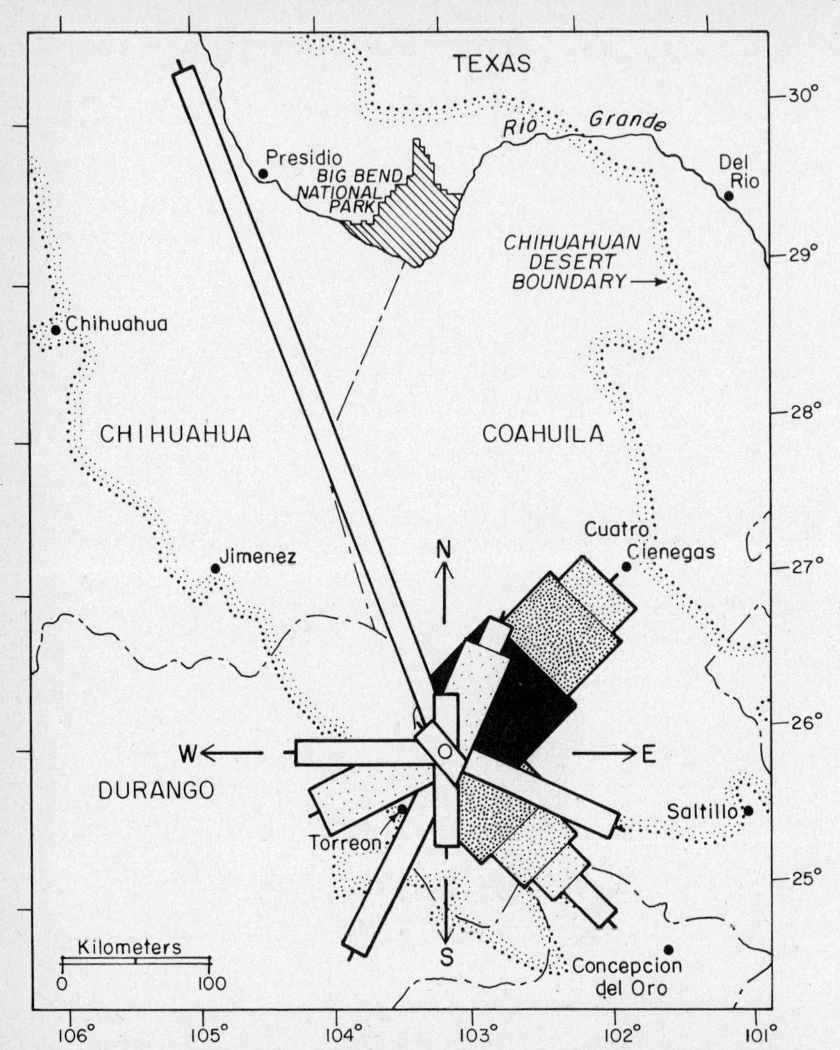

Fig. 17.5 Radial diagram summarizing dispersal events from source areas reflected by extralocal plants in late Wisconsin packrat middens from the Bolson de Mapimi. Bar orientation is direction of nearest population, length is distance to nearest population (24–580 km), and width and stippling density reflect number of plant dispersals (1–6). Records from two samples are combined, and multiple distances for individual species are included. Midpoint of diagram is between the packrat midden sites. (Boundary of the Chihuahuan Desert after Schmidt 1979.)

tal Divide to the Sonoran and Mohave deserts (Wells and Hunziker 1976) changed the compositions of plant communities in most of the interglacial warm deserts of North America. The biotic landscape changed a great deal with the evolution of *Agave lechuguilla* (lechuguilla), a widespread, abundant succulent in most of the Chihuahuan Desert today.

To assess the question of community stability through time, let us examine the Rio Grande Village profile in detail. The vegetation sequence presented in Fig. 17.3 illustrates the relative abundances of 27 out of a total of 114 species. These are the larger, more prominent, and longer lived trees, shrubs, and succulents. Grasses, perennial herbs, and annuals are not included. A general stepwise vegetation sequence is readily visible in the diagram: juniper woodland–desert-grassland (middle Wisconsin), pinyon-juniper-oak (late Wisconsin), desert-grassland (early Holocene), and Chihuahuan desertscrub (late Holocene). Similarity indices between adjacent samples (SI = $2W/A + B$, where A and B = the number of species in samples A and B, W = the number of shared species) point out several interesting facets of stability in the communities. For a period of at least 4,000 years in the Wisconsin full glacial (19,450 to 15,060 B.P.), the woody perennials in the assemblages were very stable (SI = .75–.89, $\bar{x}$ = .80). This involves a total of seven plants with *Pinus remota* (Texas pinyon), *Juniperus* (juniper), and *Quercus hinckleyi* (Hinckley oak) present in similar relative abundances in all seven samples. *Koeberlinia spinosa* (allthorn) was in six samples and *Ephedra aspera* (Mormon tea) in five. These are the plants responsible for the general physiognomic structure of the vegetation. The stability is even more impressive because the eight midden samples were collected at elevations from 680 to 835 m from five separate rock shelters within a 4-km radius.

However, the perennial succulents are much more variable throughout the sequence. The intersample similarity indices in the same full-glacial series were .44 to .80 ($\bar{x}$ = .56). The records of the shorter lived plants in the assemblages are even more variable, although some of the variability could be due to sampling. Thus, it appears that the plant communities can be very stable for a long period when the structural dominants alone are considered. As more plants are included in the community, the variability through time increases. Stability appears to depend on which types of plants are included in the community, that is, how the limits of the community are defined. The differences in stability could well be due to different response times of different species in modestly fluctuating climates (see discussion in Chapter 16).

FAUNAL HISTORY OF THE CHIHUAHUAN DESERT

In this section I will discuss the probability that the plant-animal associations we see today are not characteristic of the Pleistocene as a whole. Although bones are usually rare in packrat middens, they are directly associated with radiocarbon-dated plant assemblages. The occasional bone preserved in a midden may be collected by the packrat or be the remains of another inhabitant of the rock shelter. However, most of the bones appear to be transported to the site by raptorial birds (see Chapter 18) or by small predatory mammals such as *Bassariscus astutus* (ringtail) (Mead and Van Devender 1981), because the bones are isolated, disarticulated elements that often appear digested.

Mohave and Sonoran Desert Midden Fauna

The amphibians, reptiles, birds, and small mammals from Mohave and Sonoran desert middens from Arizona and California have been published (Van Devender, Phillips, and Mead 1977; Van Devender and Mead 1978; Mead and Phillips 1981; Cole and Mead 1981; Mead, Van Devender, and Cole 1983). These studies concluded that the fauna was much more conservative than the flora, with most animals remaining near the sites in late Wisconsin and early Holocene woodlands. Numerous reptile species that are now totally or mostly restricted to desertscrub habitats were living in woodlands as little as 8,000 years ago, for example, *Gopherus agassizi* (desert tortoise), *Coleonyx variegatus* (banded gecko), *Sauromalus obesus* (chuckwalla), *Uta stansburiana* (side-blotched lizard), *Chionactis occipitalis* (shovel-nosed snake), *Crotaphytus wislizeni* (leopard lizard), and *Lichanura trivirgata* (rosy boa).

Chihuahuan Desert Midden Fauna

Although the Chihuahuan Desert midden faunas are still under analysis, some of the more interesting results can be summarized here. The only records of an extinct animal were tooth fragments of *Mammuthus* sp. (mammoth) from two middle Wisconsin pinyon-juniper-oak middens from the Hueco Mountains. Most of the amphibians, reptiles, birds, and small mammals identified from middens of all ages still live near the sites. Common Chihuahuan Desert animals found in middle or late Wisconsin pinyon-juniper-oak woodlands include *Crotaphytus collaris* (collared lizard), *Holbrookia maculata* (lesser earless lizard), *Urosaurus ornatus* (tree lizard), *Uta stansburiana* (side-blotched lizard), *Arizona elegans* (glossy snake), *Crotalus lepidus* (rock rattlesnake), and *Hypsiglena torquata* (night snake). The differences between the modern and Pleistocene habitats occupied by these animals are not striking because the modern ranges of most of the animals extend well to the north of the Chihuahuan Desert into the pinyon-juniper-oak woodlands of north-central New Mexico and in some cases onto the Colorado plateaus and into the Great Basin.

Gopherus agassizi today is restricted to desertscrub habitats below juniper woodlands in the Sonoran and Mohave deserts west of Tucson, Arizona (longitude 111°W). In the late Wisconsin its range extended as far east as El Paso, Texas, and Carlsbad, New Mexico (longitude 104°27′E), where it was living in pinyon-juniper-oak woodlands (Van Devender, Moodie, and Harris 1976; Moodie and Van Devender 1979). A record of *Neotoma mexicana* (Mexican wood rat), a mountain forest dweller today, in the early Holocene juniper woodlands of Bishop's Cap, south-central New Mexico, represents a more modest range change. Teeth of *Microtus* sp. (voles), rodents of moist mountain forests and meadows, were found in three Wisconsin middens from the Hueco Mountains, Texas. The nearest modern vole populations are at least 650 m higher in elevation in the massive Sacramento Mountains (120 km to the north-northeast).

Phrynosoma douglassi (mountain horn lizard) is a widespread inhabitant of forests and grassland from British Columbia to northern Mexico. In Texas it is restricted to isolated populations in the upper elevations in the Guadalupe and Davis mountains. Bones from two late Wisconsin middens from the intermediate Streeruwitz Hills document a full-glacial connection between the modern relict populations. *Gerrhonotus liocephalus* (Texas alligator lizard) is a large Mexican lizard whose range extends northward into central Texas. A full-glacial midden record for it from near the ghost town of Terlingua suggests that it was more widespread in the lowlands in the late Wisconsin woodlands and that the isolation of the relict population in the Chisos Mountains in Big Bend National Park occurred in the Holocene. Milstead (1960) discussed the distributions of 14 additional species that have similar relict populations in the Chihuahuan Desert biotic province (Blair 1950).

Bones from two colorful king snakes, *Lampropeltis mexicana* (gray-banded king snake) and *L. triangulum* (milk snake), were found in a full-glacial pinyon-juniper-oak sample from the Rio Grande Village area. *Lampropeltis mexicana* is a secretive snake found in habitats ranging from mountain woodlands to Chihuahuan desertscrub. Milk snakes are widespread animals found in forest and woodlands in much of North America and northern South America. *Agkistrodon contortrix pictogaster* (Trans-Pecos copperhead) is a western subspecies of the copperhead of the eastern United States that is restricted to Trans-Pecos Texas in habitats ranging from woodland to Chihuahuan desertscrub. All of these animals became adapted to desertscrub habitats in the Big Bend of Texas in the Holocene as Wisconsin woodlands disappeared. Indeed, this is the only portion of the ranges of milk snake or copperhead where they live in nonriparian habitats in hot, dry desertscrub communities. *Sceloporus jarrovii* (Yarrow's spiny lizard), a widespread forest and woodland animal in the Sierra Madre, provides a similar example of an animal that was stranded in, and became adapted to, Chihuahuan desertscrub in the Bolson de Mapimi in Durango and Coahuila in the Holocene. These are examples of local populations of animals that have adapted and survived in habitats that are unusual for the species as a whole today or in the past.

Regional Faunal Records

Rich faunas have been studied from many areas in the central United States as far south as central Texas (Chapter 18; Lundelius 1967, Lundelius et al. 1983) and Nuevo Leon, Mexico. The faunas document many significant southward range changes in the late Wisconsin for microtine rodents, shrews, etc. Dry Cave in southeastern New Mexico produced several rich late Wisconsin faunas from within the Chihuahuan biotic province (Harris 1970, 1977). Several voles (*Microtus longicaudus, M. mexicanus, M. ochrogaster*), the dwarf shrew (*Sorex nanus*), and the yellow-bellied marmot (*Marmota flaviventris*) moved into the area from higher elevations in the Rocky Mountains to the north. Least shrew (*Cryptotis parva*) extended its range westward from central Texas. Sagebrush vole (*Lagurus curtatus*) and sage grouse (*Centrocercus urophasianus*) extended their ranges to the southeast from the Great Basin.

In comparison, the late Wisconsin range changes in animals recorded in packrat middens from the core of the climatic Chihuahuan Desert (Schmidt 1979) are few in number and usually do not involve great distances. The only vole records are from near the northern edge of the Chihuahuan Desert, and remains of shrews have not been found at all. I view this as a regional difference that is directly related to differences in late Wisconsin environments. In areas to the north and east and at higher elevations where the paleoclimates had much greater rainfall or much cooler summers, there was a greater change in the fauna. Although in the Chihuahuan Desert the late Wisconsin rainfall was greater and summers were cooler than today, summers were warm enough for a modest development of summer monsoon and the spring-summer drought (Van Devender, Betancourt, and Wimberly 1984). The late Wisconsin pinyon-juniper-oak woodlands in the lowlands of the Chihuahuan Desert were probably not mesic enough to support many microtine rodents or shrews. The cricetid and heteromyid rodents in the modern faunas have broad ecological tolerances and often range up into open woodlands.

In summary, Chihuahuan Desert packrat middens provide numerous examples of northern species displaced to the south in the late Pleistocene. In some cases the middens trace the origins of modern range disjunctions from more continuous ranges in the past. In addition, and surprising to modern ecologists, the middens show that many animal species occurred during the late Pleistocene in the same geographical area that they inhabit today, but were living in different habitats. Thus, animal and plant species did not simply shift geographically during the Pleistocene as whole communities; the communities were extensively reshuffled. Similar conclusions are reached by Graham and Davis for other communities elsewhere in this volume (Chapters 16 and 18).

LATITUDINAL GRADIENTS IN SPECIES RICHNESS

In equal-sized sample plots in the Chihuahuan Desert the total number of species and often the number of species in a single genus increase to the south. A similar trend occurs in the approximately 1,000 endemic plants, with greater numbers to the south (Johnston 1977). The majority of Chihuahuan endemics have affinities with the Sierra Madre Oriental to the southeast.

Similar latitudinal gradients in species richness are found in many organisms in many areas on the globe. In recent years explanations of latitudinal species gradients have focused on the biotic interactions involved in competition, species packing, coevolution, and resource partitioning (Pianka 1966, MacArthur 1972a, Brown and Gibson 1983). Although Fischer (1960) presented arguments that the species richness gradient was controlled by latitudinal climatic gradients, the role of past climates has not been adequately considered. The detailed paleoecological record from the last glacial-interglacial cycle of the Chihuahuan Desert suggests a strong mechanism to explain the accumulation of species at lower latitudes.

The Ghost of Extinction Past

Speciation is most likely to occur when a population becomes geographically isolated. The frequencies of populations' becoming isolated in fluctuating environments may be similar at different latitudes if the landscapes are similar. However, the differences between modern and glacial climates and the probability of extinction of isolated populations at the end of either an interglacial or glacial increases to the north with latitude. In the northern Chihuahuan Desert the packrat middens provide evidence that local extinctions of many displaced plants were rapid and thorough at the end of the late Wisconsin. Speciation has occurred only in specialized habitats such as cliff faces, cold air sinks, and gypsum deposits. To the south, as the difference between modern and glacial climates becomes less, the probability of extinction decreased, hence that of speciation increased. The climate to the south has been stable only in the sense that the environmental extremes of cold, heat, and drought have never been great enough to cause major extinctions. The repeated climatic fluctuations in the Pleistocene and the resulting changes in the distributions of organisms have provided a great many opportunities for allopatric speciation. Gypsum beds and playa surfaces have been repeatedly exposed, and limestone sierras have repeatedly become islands as pluvial lakes dried and filled. Repetition of similar climatic sequences in each climatic cycle could easily result in the independent evolution of similar adaptations in closely related organisms.

The "ghost of competition past" model (see discussion by Diamond in Chapter 6), with speciation in allopatry and with intense competition for a short time after sympatry is established leading to niche separation and little subsequent competition, is an appealing mechanism to explain the sympatry of closely related species. In latitudes where the environment has fluctuated repeatedly, but where extremes have never been harsh enough for extinction to simplify the biota or hinder evolution, there has been a gradual accumulation of new and old species. Thus, historical factors may be a major cause of the latitudinal gradient in species richness.

SUMMARY

Fossil packrat middens are hard, dark organic deposits preserved in dry rock shelters that contain well-preserved plant and animal remains collected from within 30 to 50 m. Radiocarbon-dated series of packrat middens from local geographical areas are providing detailed sequences of the development of vegetation through the last 35,000 years. Midden sequences from a number of areas along a latitudinal gradient in the Chihuahuan Desert provide examples of the behavior of plant communities through the last glacial-interglacial (Wisconsin-Holocene) cycle.

The general vegetation sequence from pinyon-juniper-oak woodland to oak-juniper woodland to desert-grassland to Chihuahuan desertscrub is probably similar to the developmental sequence that led to the formation of the Chihuahuan Desert 5 to 10 million years ago in the late Miocene. This sequence has probably been repeated, albeit with some variation, in each of the glacial-interglacial climatic cycles in the Pleistocene. Recent work suggests that there have been 15 to 20 glacial cycles, not the traditional 4, in the last 1.8 million years. Glacial periods were about 100,000 years long and interglacials only 10,000 to 20,000 years. The climate and vegetation of the modern Chihuahuan Desert are only about 4,000 years old and characterize only about 4% of the Pleistocene.

The fossil packrat midden assemblages do not provide any evidence of Pleistocene refugia in the sense of a geographical area where the plant community compositions were unmodified by ice age climates. There are many records of plants persisting in the southern portion of their range and expanding widely to the north in the Holocene. However, many of these were modern desertscrub plants that were living in mixed woodland communities in the late Wisconsin.

Detailed analyses of the packrat midden chronologies demonstrate that the woody perennial dominants that structure the physiognomy of the plant communities can be very persistent. Near Rio Grande Village in the Big Bend of Texas, Texas pinyon, juniper, and Hinckley oak were very stable for at least 4,000 years during the late Wisconsin full glacial. However, the perennial succulents, grasses, and herbs in the same series were much more variable with time, so that the overall stability of the community depends on which plants are considered to be in the community. Differences in longevity and life history undoubtedly cause different species to have different response times to moderately fluctuating climates.

A "ghost of extinction past" model is presented to explain the latitudinal gradient in species richness. The difference between glacial and interglacial climates increased to the north. The probability of extinction of an isolated population at the end of either a glacial or an interglacial was likewise greatest in the north. The extinction of isolated populations was less likely to the south, and the opportunities for speciation in allopatry consequently increased. Considering the 15 to 20 glacial-interglacial fluctuations in the Pleistocene as well as earlier fluctuations, there have been many opportunities for speciation. At lower latitudes the climatic extremes of cold, heat, and drought have never been great enough to cause wholesale extinctions of species. The ''ghost of competition past'' model, with speciation in allopatry and with intense competition after sympatry is established leading to niche separation and little subsequent competition, probably best explains the sympatry of closely related species.

ACKNOWLEDGMENTS

Discussions with Tony L. Burgess, Paul S. Martin, Julio L. Betancourt, Robert S. Thompson, Rebecca K. Van Devender, Laurence J. Toolin, Charles H. Lowe, David Morafka, A. Michael Powell, Russell W. Graham, Margaret B. Davis, Andrew H. Knoll, and Jared Diamond helped in the formation of the ideas in this paper. The Chihuahuan Desert packrat midden research has been supported by National Science Foundation Grants 76-19784 and 80-22773 to T. R. Van Devender. Additional funds were provided to study the Hueco Mountains Holocene middens by Ft. Bliss Army Base contract DABT 51-82-M-3459EA through G. de Garmo and K. von Finger. Jean Morgan and Cindy Turner typed the manuscript. Jacqui Soule and Chuck Sternberg drafted the figures.

chapter 18

Response of Mammalian Communities to Environmental Changes During the Late Quaternary

Russell W. Graham

INTRODUCTION

During the last glacial maximum (~18,000 years ago) in North America, glacial ice covered most of Canada and much of the upper midwestern United States (Fig. 18.1). The biotas that previously inhabited these areas were physically displaced by the glacial ice sheets, and the attendant glacial climates and environments affected the distributions of organisms in unglaciated areas as well. Shortly after the glacial maximum continental ice sheets began to melt and rapidly withdraw to the north. By ~8,000 B.P. the continental glaciers occupied only a small portion of arctic Canada. The ablation of the continental glaciers opened vast areas to biotic recolonization.

This pattern has been repeated many times during the Pleistocene. For example, analysis of oxygen isotopes ($^{18}O/^{16}O$) in the shells of fossil foraminifera from deep-sea cores provides proxy measures of paleotemperatures and ice volumes for the Pleistocene (Fig. 16.1E). These data suggest that there have been at least 22 alternating stages of high and low ice volumes in the Northern Hemisphere during the last million years (Shackleton and Opdyke 1973).

Oscillations in the physical environment would have caused repeated fluctuations in the biota. How ancient, then, are the ecological communities that we see today? Also, given the environmental fluctuations of the Quaternary, is it likely, as ecologists often implicitly assume, that species that coexist today have had a long continued history of opportunity for coevolutionary adjustments to each other?

The answers to these questions depend on how species ranges shifted during the Pleistocene. If we focus on middle and high latitudes in the Northern Hemisphere, one possibility is that communities moved as a whole from north to south during glacial periods and from south to north during interglacials (Blair 1958, 1965; Dansereau 1957; Martin 1958). If this were the case, then modern communities and guilds may have had their present species composition for a long time, and there would have been substantial intervals of time for coevolution.

If, however, various species shifted their ranges for different distances, in different direc-

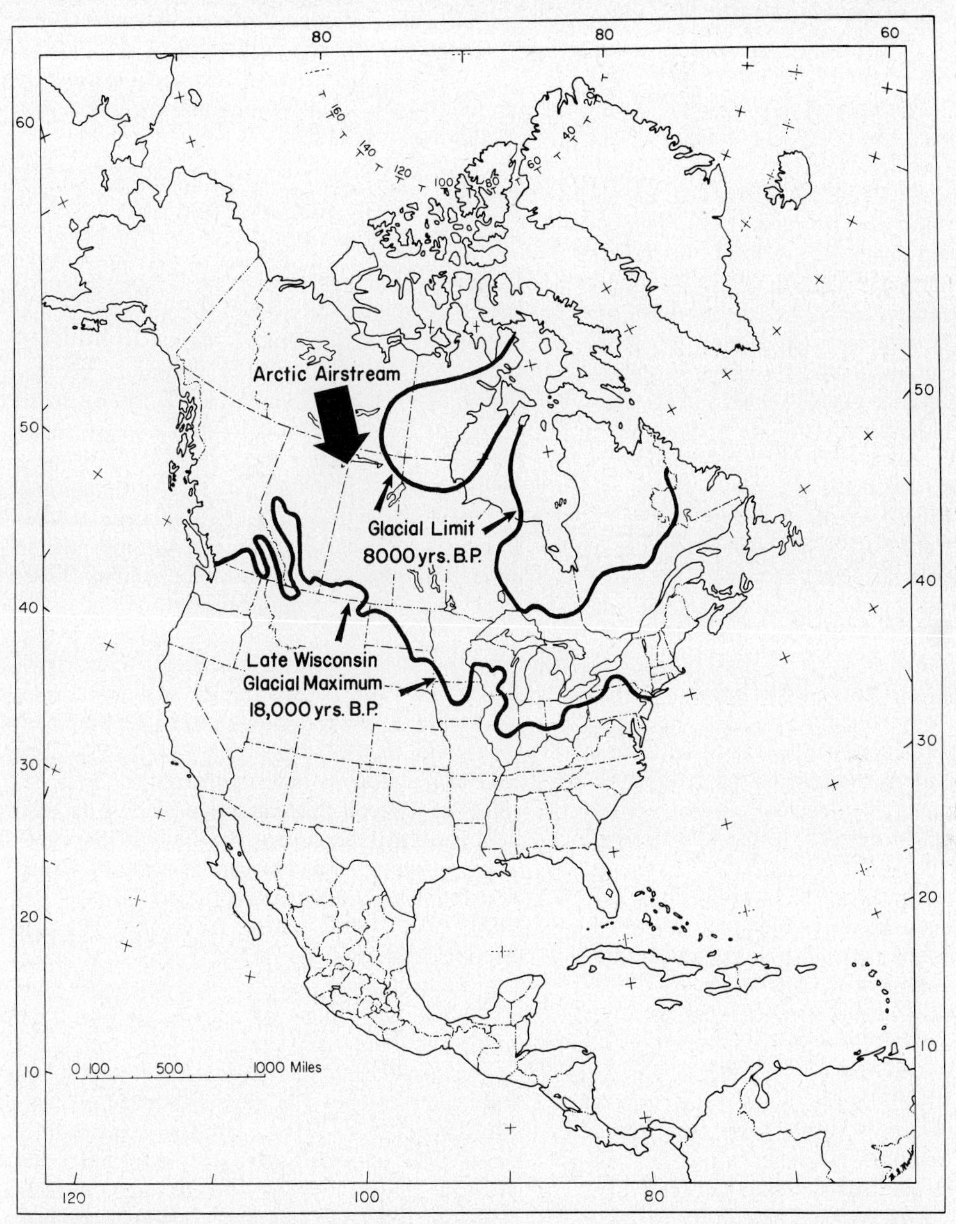

Fig. 18.1 Southern margin of continental ice masses in North America at 18,000 B.P. and 8,000 B.P. Retreat of the glacial ice northward during the late Pleistocene and early Holocene not only exposed new terrains for biotic recolonization, but it also altered climatic patterns by allowing arctic air masses to migrate southward in the early Holocene.

tions, at different times, or with different response rates, then species sets in the temperate zones would have been subjected to numerous reshufflings during the Pleistocene. Any given species could have belonged to a whole series of clusters of sympatric species through time (Cushing 1965, 1967; Davis 1976; Graham 1976, 1979; Livingstone 1967, 1975; Wright 1981). If this were the case, then coevolutionary adjustments would have had to have been rapid or would have been nonexistent (Hoffman 1979).

Species migrations during the Quaternary would have also affected species diversity for any given area. The simplest expectation would be that any site in the temperate zone supported fewer species during glacials than during interglacials. The basis for this simple assumption is that today species diversity generally increases from the poles toward the equator, and that climate zones shifted poleward during the interglacial and toward the equator during the glacial periods. However, if species responded individually to environmental fluctuations, then one might expect different diversity patterns in the Pleistocene.

All of these alternative hypotheses about biotic responses to fluctuating environments during the Quaternary can be tested by the fossil record. I shall discuss these alternatives for the Quaternary small mammal fauna of eastern North America. Small mammals, especially insectivores and rodents, are well suited for this study because of their inability to migrate large distances in a short period of time, their relatively easy and consistent identification as fossils, their occurrence in substantial numbers in fossil deposits, and their adaptations to local environments. Also, most of the late Pleistocene species are extant today and their environmental adaptations are fairly well known.

This chapter begins with a brief summary of the evidence available for reconstructing vanished communities from fossil evidence. I then compare the late Pleistocene and Holocene with respect to geographical ranges of individual small mammal species, integrity of faunas, habitat heterogeneity, and species diversity. I consider the reasons for the differences between the late Pleistocene and Holocene that emerge from these comparisons. Finally, I use these results to reexamine the extinction of the Pleistocene megafauna, the phenomenon that will first come to the minds of many readers when thinking about faunal changes at the end of the Pleistocene, but one that was actually accompanied by the many other changes in small mammals discussed in this chapter.

METHODS FOR RECONSTRUCTING PALEOCOMMUNITIES

Fossil floras and faunas are not complete, unbiased ''snapshots'' of biological communities that lived in the past. Instead, as discussed for fossil floras in Chapter 7, paleocommunities are derived from the accumulation and burial of fossil organisms, and these collections contain many inherent biases. One method to control these sampling biases is to compare fossil communities at different times (e.g., Pleistocene versus Holocene), since all fossil communities would have been subject to similar accumulation processes and similar sampling biases.

In the analysis of paleocommunities it is also important to demonstrate that they are not artificial samples that result from the spatial and temporal mixing of past communities. Therefore, it is essential to consider in detail how collections of fossil organisms have been formed. It is beyond the scope of this paper to provide a detailed discussion of taphonomy, the study of the processes by which living organisms are converted into fossil assemblages (see Behrensmeyer and Hill 1980). I shall instead confine myself to a brief discussion of four primary taphonomic factors (depositional environments, agents of bone accumulation, rates of sedimentation, and postdepositional disturbances).

The depositional environments of different sedimentary bodies can affect the types and size fractions of the fossils preserved. For example, in high-energy environments (such as beaches and channels of fast-running streams) only durable materials, like teeth and dense bone, will be preserved. These environments also tend to wash away the remains of smaller animals and to concentrate the fossils of larger ones. Also, high-energy depositional environments may transport fossils for some distance. Low-energy depositional environments (such as ponds, lakes, overbank stream deposits, and wind-blown deposits) generally establish a more complete sample of all components of the fossil communities, and they are also more representative of local environments. In this study I have used faunas only from low-energy environments or caves.

Caves frequently contain rich fossil deposits because they serve as focal points for bone accumulation. Bone deposits in caves are usually not the result of long-distance transport, but reflect local environments. The catchment, or area sampled for cave faunas, is partially dependent upon the size of the internal drainage system, which is usually quite small compared to open fluvial (river) systems. Also, the catchment for cave faunas is dependent upon the home ranges of predators or scavengers who serve as vectors for bone accumulation. Owls are frequently the primary vectors for fossil microvertebrate faunas, especially small mammals. Most owl species retain prey items in their digestive systems for only one day (Reed and Reed 1928), and they usually forage within a few kilometers of their roost. Thus, studying cave faunas that result from owl pellet accumulations can help minimize the likelihood of mixing communities from broad geographical areas.

Cave bone deposits may contain an overrepresentation of the remains of the predators and their prey. With regard to the prey species, these samples may be skewed by all the variables involved in prey selection by different species of predators. Bones can also be contributed to cave deposits by resident species (such as bats and snakes). In the case of pit caves (vertical shafts), the random victims of these natural traps may be still another significant component of the bone accumulation. Therefore, paleocommunities do not represent a complete spectrum of the past biota, but instead a mixture produced by several types of sampling.

For these reasons I have attempted to compensate for many of the biases by analyzing fossil faunas with similar taphonomic histories. Even though the samples may be biased, the biasing processes should be similar and the comparisons should be valid. Furthermore, because most speleological microvertebrate samples are the result of owl predation and because most owl species prefer rodents and insectivores as prey, I have restricted my comprehensive analyses of species densities to these mammalian taxa.

Depositional environments with rapid rates of sedimentation provide the greatest temporal separation of events; sites with slow rates of deposition yield the lowest resolution. For most late Quaternary sites radiocarbon dating can be used to establish rates of sedimentation. Often the only datable materials in cave deposits are bones, which generally do not give reliable dates (Land et al. 1980). However, bone dates can be accepted as minimum ages (Lundelius 1967).

Rapid deposition is frequently manifest at intact kill sites, that is, archeological sites where human hunters have killed various species of prey. If these sites are not buried rapidly, then natural processes such as decay and erosion will destroy the integrity of the association of artifacts and prey. Thus, when one finds intact kill sites, one can be confident that they represent material accumulated over short intervals of time, perhaps less than 10 years, and that paleocommunities from these sites are not the result of extensive time averaging processes.

Bone accumulations can also be altered by postdepositional processes such as chemical breakdown in soils, bioturbation or burrowing by organisms, and redeposition. Processes that cause mixing of deposits of different ages will produce artificial fossil assemblages that do not represent biological communities. These processes can frequently be identified in the field during excavation by differences in preservation of bones or by chemical analyses of bones. Obviously, mixed fossil assemblages must be excluded from paleocommunity analyses. In short, although the fossil record is not a "snapshot" of the past, it can be used with care to contribute an invaluable temporal perspective to community ecology and evolution.

LATE QUATERNARY SPECIES MIGRATIONS

One of the most common responses for mammalian species to Pleistocene environmental fluctuations was the southward displacement of northern species and the displacement of montane species to lower elevations. For example, the collared lemming (*Dicrostonyx*) today resides in the arctic tundra of extreme northern Canada (Fig. 18.2a). During the late Pleistocene it was widely distributed in the northern half of the United States, where it has been found as a fossil at elevations below 1,500 m and was not restricted to mountain tops (Kurten and Anderson 1980).

Many other boreal and montane species, such as the yellow-cheeked vole (*Microtus xanthognathus*), northern bog lemming (*Synaptomys borealis*), and arctic shrew (*Sorex arcticus*), similarly expanded their distribution to lower latitudes and altitudes. These range shifts were partly the result of an obvious effect: General boreomontane climate zones shifted southward and to lower elevations during the Pleistocene. However, they also resulted from a subtler effect: Seasonal patterns of late Pleistocene climates were quite different from modern ones, especially during the summer. Thus, boreal species whose southern or low-altitude limit is set today by warm summer temperatures were especially likely to expand their ranges during the Pleistocene, when climates were cooler and moister.

Northward migrations of southern species during the late Pleistocene are not as apparent as the southward migrations of northern species. However, species of wood rats (*Neotoma*), ground squirrels (*Spermophilus*), jaguars (*Felis onca*), and jaguarundi (*Felis yagouaroundi*) did extend their distributions further north during the late Pleistocene. These range extensions suggest that late Pleistocene winters did not suffer the intense cold extremes characteristic of modern winters. It is also interesting to note that many extinct species with southern counterparts today—tapirs (*Tapirus*), capybaras (*Hydrochoerus*), and armadillos (*Dasypus*)—ranged into the southern and central parts of the United States during the late Pleistocene.

Species that are today restricted to the deciduous and coniferous forests of the eastern United States expanded their distributions westward during the late Pleistocene (Fig. 18.2c). Likewise, species that currently inhabit the dry grassland environments of the Great Plains ranged much further eastward (Fig. 18.2d). More taxa migrated westward than eastward (Graham in press a). These westward and eastward shifts are related to the east-west moisture gradient in North America east of the Rocky Mountains. As with the temperature gradient, relaxation of the moisture gradient would allow eastern and western species to inhabit different microenvironments within the same area. In some cases the shifts in eastern and western species occurred at different times (Foley 1984, Rhodes 1984).

INDIVIDUALISTIC RESPONSE

Distributional shifts of species during the Quaternary were not mass movements of groups of species (communities). Instead, each species responded to environmental changes individually. Species migrated in different directions, at different times, and for different distances, as emphasized by Davis (Chapter 16) for plant species. As a striking example of individualistic responses, consider the collared lemming (Fig. 18.2a) and brown lemming (Fig. 18.2b). Today the brown lemming extends further south than the collared lemming, but in the late Pleistocene the collared extended at least 1,000 km further south than the brown. Even though only one Pleistocene site with brown lemming is known south of its present range, our knowledge of its southern limit during the Pleistocene is unlikely to be an artifact of sampling, since it should have been recovered from the numerous fossil sites in the northern United States if it had had a wider distribution in the late Pleistocene.

The differential migrations of these two lemmings illustrate that factors other than climatic ones were involved in some late Pleistocene range shifts. If climate was the sole limiting factor, then one would expect a direct relationship between the range limits today and those in the Pleistocene. In other words, because the brown lemming has a more southern distribution than the collared lemming today, it should have spread further south than the collared lemming during the late Pleistocene. Instead, the opposite pattern is apparent in the fossil record. Therefore, other factors, such as vegetation, disease, and species interactions, may have overridden any direct effects of climatic change.

INTERMINGLED PLEISTOCENE BIOTAS

The southern, eastern, and western migrations of species did not cause the wholesale exclusion of other species groups to isolated refugia. Instead,

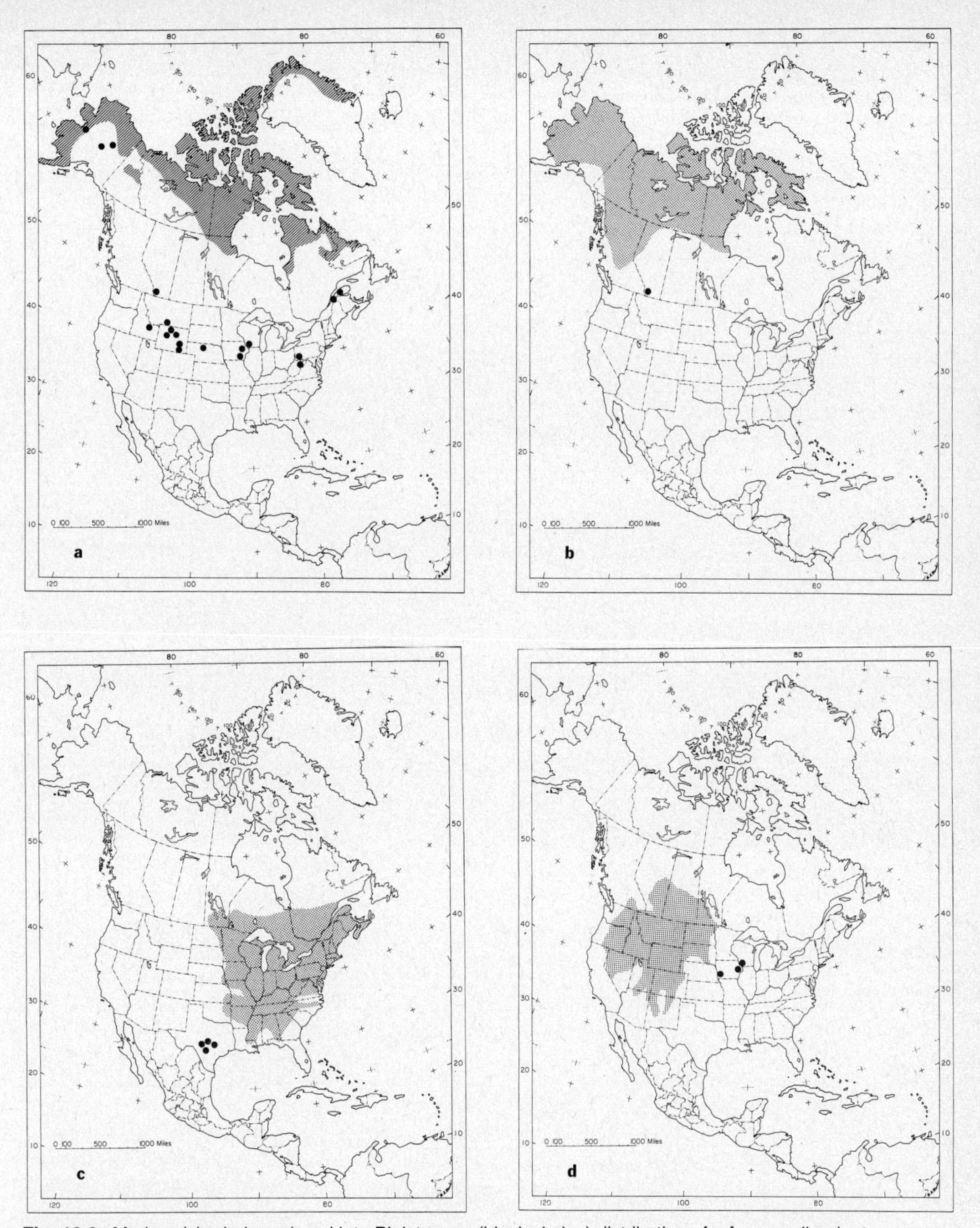

Fig. 18.2 Modern (shaded area) and late Pleistocene (black circles) distributions for four small rodent species. Compared to the modern distribution, the late Pleistocene distribution lies (a) to the south for the collared lemming (*Dicrostonyx*), (b) to the south for the brown lemming (*Lemmus*), (c) to the west for the eastern chipmunk (*Tamias striatus*), and (d) to the east for the northern pocket gopher (*Thomomys talpoides*). Note that *Dicrostonyx* is a more northerly species than *Lemmus* today but was a more southerly species in the Pleistocene.

arctic and boreal species were amalgamated with biotas from more temperate latitudes. Eastern and western species were also intermingled. The result of these individualistic responses is that Pleistocene biotas were composed of species that today live in separate habitats and remote geographical areas (Fig. 18.3). This intermingling is a crucial, worldwide characteristic of Pleistocene biotas (Fig. 18.4) that is little appreciated by ecologists studying modern communities.

The late Pleistocene biotas have been referred to as ''disharmonious'' because their species associations are unlike any modern assemblages (Lundelius et al. 1983, Semken 1974). This choice of term is somewhat misleading, because the late Pleistocene biotas were undoubtedly in harmony with prevailing environments. Therefore, I use the term ''intermingled'' to refer to such Pleistocene biotas without modern analogues.

Mixed fossil assemblages similar to the intermingled biotas from the late Pleistocene could result artifactually from numerous processes that do not reflect actual biological communities. For instance, redeposition of older fossils in younger sediments, mixing by burrowing organisms, and slow depositional rates with rapid environmental fluctuations (temporal compression) can produce intermingled patterns. These mixtures can be differentiated by careful fieldwork and laboratory procedures. However, mixed assemblages may also result from geographical and habitat mingling during deposition of the fossil deposits.

Do the intermingled Pleistocene biotas represent real communities, or did they arise as artifacts of taphonomic processes? Many types of evidence support the former conclusion.

The most compelling evidence for intermingled biotas as reflecting actual community patterns is the lack of intermingled associations in most Holocene fossil assemblages. Presumably, the taphonomic processes of specific depositional systems (such as caves, streams, and lakes) were the same in the Pleistocene and Holocene. Therefore, if the intermingled nature of the late Pleistocene biotas was a result of artificially mixed assemblages, then intermingled biotas should also appear in Holocene fossil assemblages. Semken's (1983) review of 68 Holocene fossil sites from the eastern United States ranging in age from 10,000 years ago to historic times clearly demonstrates the lack of intermingled associations in Holocene faunas. In contrast, a review of 178 late Pleistocene faunas revealed a predominance of intermingled associations (Lundelius et al. 1983).

Intermingled faunas have also been found with Pleistocene paleoindian kill sites (Johnson and Holliday 1980, Slaughter 1975). As already discussed, these archeological sites require rapid burial to maintain the integrity of the association of the human artifacts with the bones of prey. Sedimentary deposits at these sites may accumulate in tens of years. These intermingled faunas are therefore not the consequence of the temporal compression of numerous events.

Floral studies exhibit the same intermingled pattern. Late Pleistocene floras contained intermingled associations unlike any modern assemblages (Cushing 1965, 1967; Davis 1976; West 1964; Wright 1981), and these species associations had disappeared by the early Holocene in the eastern United States (Webb et al. 1983). These intermingled patterns are apparent in both the pollen and the plant megafossil (needles, cones, nuts, etc.) records. A contribution of long-distance transport to the megafossil record can generally be eliminated. Hence, these intermingled floras were not produced by mixing of materials from distant sites. Analyses of later Quaternary floras also allow a high-resolution dating and calculation of sedimentation and pollen influx rates, thereby eliminating the mixing of materials from different time levels as a cause for the intermingled biotas.

A further type of evidence is that intermingled biotas are not restricted to a certain geographical area, but appear to have a global distribution during the Pleistocene (Graham and Lundelius 1984) (see Fig. 18.4). Intermingled biotas are not limited to particular depositional environments, and they are not confined to specific taxonomic groups. Assemblages of late Pleistocene terrestrial mollusks (Miller 1976), beetles (Morgan and Morgan 1980), amphibians and reptiles (Holman 1976), and birds (Guilday et al. 1977, Lundelius et al. 1983) as well as mammals exhibit intermingled associations.

All these types of evidence leave no doubt that the late Pleistocene intermingled biotas represent communities and environments without modern analogues.

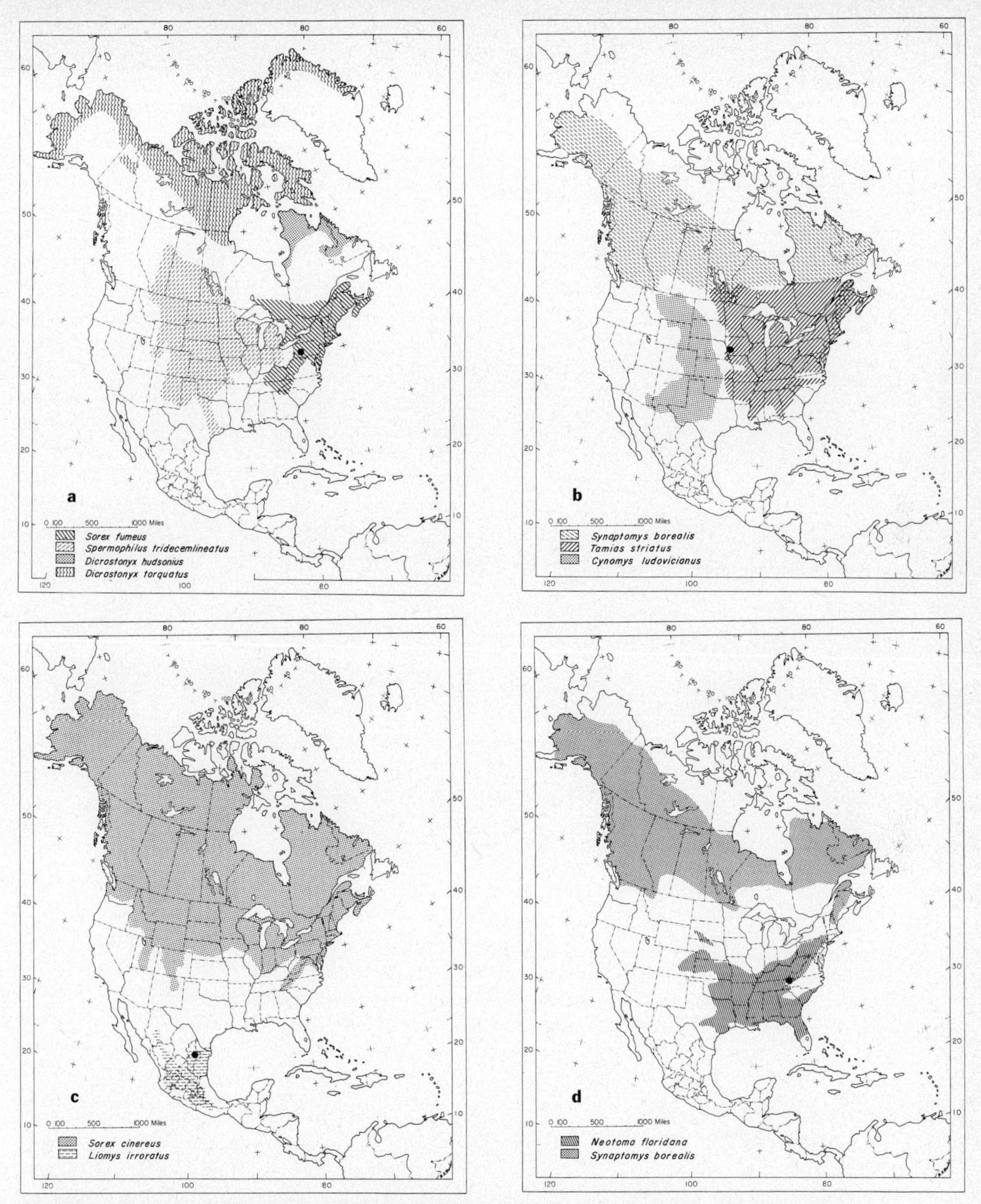

Fig. 18.3 Four examples of intermingled late Pleistocene biotas in North America. In each map the black circle shows the location of a representative site where the named species co-occur as fossils in a primary context (not mixed artificially), and the shaded areas are the modern distributions. Note that in each case the modern ranges are largely or completely disjunct, so that co-occurrence at the same site would be unlikely or completely impossible today. (a) New Paris No. 4 (11,000 B.P.), Bedford County, Pennsylvania, United States. Only one species of *Dicrostonyx* occurs in this fauna. (Guilday et al. 1964.) (b) Craigmile (23,000 B.P.), Mills County, Iowa, United States. (Rhodes 1984.) (c) San Josecito Cave (no date, late Pleistocene), Nuevo Leon, Mexico. (Jakway 1958.) (d) Baker Bluff Cave (~10,500–19,000 B.P.), Sullivan County, Tennessee, United States (Guilday et al. 1978.)

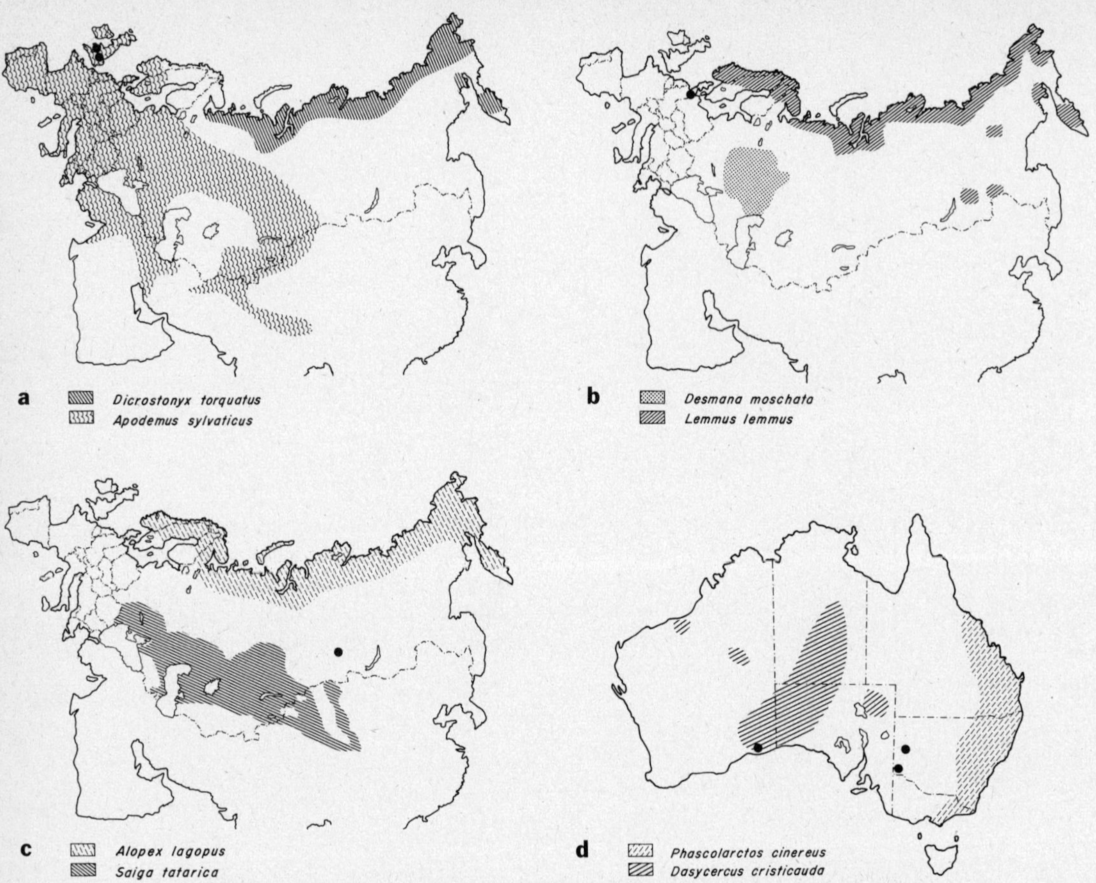

Fig. 18.4 Four examples of intermingled late Pleistocene biotas in Great Britain, Eurasia, and Australia. As in Fig. 18.3, circles mark representative fossil sites where species with largely or completely disjunct distributions today (shaded areas) co-occurred during the late Pleistocene. (a) Cat Hole, South Wales (upper Paleolithic, 15,000–10,000 B.P.) and Robin Hood's Cave, Derbyshire (upper Paleolithic, 28,500–10,600 B.P.), Great Britain. (Campbell 1977.) (b) Meiendorf (older Dryas, 12,000–11,800 B.P.) and Stellmoor (older Dryas, 12,000–11,800 B.P. and younger Dryas, 11,000–10,000 B.P.), Holstein, Germany. (Degerbøl 1964.) (c) Middle Yenisei River sites (early and late Sartan, 30,000–10,000 B.P.), Siberia, USSR. (Klein 1971.) (d) Lake Menindee, New South Wales (26,000–18,000 B.P.), Lake Victoria, New South Wales (>18,000 B.P.), and Madura Cave, Western Australia (37,880–15,600 B.P.), Australia. (Lundelius 1983.)

THE FINE-GRAINED MOSAIC OF PLEISTOCENE HABITATS

The intermingled biotas of the late Pleistocene were probably, in part, the result of greater heterogeneity in habitat types than exist today. The heterogeneous environments of the late Pleistocene were composed of a fine-grained mosaic of habitats that today occur far apart. In fact, some of the late Pleistocene habitats have no modern equivalents at all. McNaughton (1983a) has shown that spatial heterogeneity of species distribution is one of the most important features contributing to the maintenance of the Serengeti grassland ecosystem.

The biota from the mastodon bone bed at Christensen Bog (Graham et al. 1983) in central Indiana provides an excellent example. The vertebrate fauna and associated flora are from deposits that date from 13,220 B.P. to 12,060 B.P.; the age of the fauna is closer to the 13,220-year-old date at the base of the deposit. The fossils were encapsulated by a fairly rapid sedimentation rate of 10 cm/100 years.

Floral remains (pollen and macrofossils) from these deposits indicate a much more diversified forest than exists in the area today. The flora contained mixtures of spruce (*Picea*), fir (*Abies*), birch (*Betula*), larch (*Larix*), and a variety of other woody taxa such as oak (*Quercus*), elm (*Ulmus*), hazel (*Corylus*), ash (*Fraxinus*), and hickory (*Carya*) (Whitehead et al. 1982). There was also a diverse aquatic flora in the kettle lake. The fossil vertebrate fauna from the same stratigraphic horizon and temporal period also reflects the diversity of habitats. The fauna was composed, in part, of mastodon (*Mammut*), giant beaver (*Castoroides*), caribou (*Rangifer*), turtles (*Chrysemys, Trionyx, Chelydra*), ducks (cf. *Anas*), and turkey (*Meleagris*). Today, caribou inhabit boreal forest and tundra environments, whereas wild turkey and many of the turtles reside primarily in deciduous forests.

The coexistence of these diverse floral and faunal elements at Christensen Bog during the late Pleistocene was facilitated by the variety of habitats present in a fine-grained vegetational mosaic. Diverse habitat associations extend well back into the late Pleistocene and occur in many different geographical areas and physiographical settings (Graham in press a, Guthrie 1982, Lundelius et al. 1983, Rhodes 1984).

SPECIES NUMBERS

The Holocene mammalian fauna of North America has been regarded as "depauperate" or "impoverished" in comparison to the Pleistocene (Semken 1983). While the most glaring example involves the megafauna that disappeared at the end of the Pleistocene, the higher species diversity of late Pleistocene mammalian communities is also reflected in the small mammal guilds of which most of the species are extant today (Fig. 18.5). Application of conventional diversity measures, such as the Shannon-Weaver index, to fossil samples is often inappropriate because of the many inherent biases in the fossil sample. Hence my analysis instead uses species number (Graham 1976). I focused on soricid (shrew) and arvicolid (microtine rodent) species because their fossil remains are readily identifiable and because they are fairly abundant in fossil deposits. These taxa are also the common prey of many owl species that sample relatively local environments and retain prey items for a short time. Fig. 18.5 shows that local numbers of shrew and microtine species in the late Pleistocene in the midcontinent were approximately double those in the same area today.

Could these apparently higher species numbers in Pleistocene faunas be an artifact from mixing of spatially or temporally distinct sites? No, because such artifacts would presumably apply at any time in the past, yet species numbers demonstrate a decline from Pleistocene to Holocene fossil sites (Fig. 18.5). This pattern seems to be independent of depositional environments, since caves and alluvial sites exhibit the same pattern, although caves tend to have a higher species numbers (points ● vs. ▲ in Fig. 18.5). In addition, modern owl pellet samples (◆ in Fig. 18.5), the presumed source of micromammal fossil deposits, exhibit the same species number patterns as the Holocene fossil sites.

In mountainous areas, such as the western United States and the Appalachians, topographical relief probably contributed importantly to species diversity in the Pleistocene as it does today (Simpson 1969). However, these complications can be eliminated by examining flat areas. The reduction in species number from the Pleistocene to the Holocene illustrated in Fig. 18.5 is also exhibited by a chronological sequence of fossil faunas from Iowa, a state in which the highest and lowest point differ by only 363 m, and from other low-relief areas of the Great Plains (Graham in press a).

The higher species numbers for shrews and microtines are, in part, a result of the intermingling of arctic and boreal biotas with temperate biotas. In essence, the displacement of arctic and boreal biotas by glacial climates did not cause the relocation of more southern communities; instead, these biotas were amalgamated to form new communities in the temperate zone.

Higher species numbers in the late Pleistocene are not restricted to shrews and microtines, but have also been documented for other rodent fami-

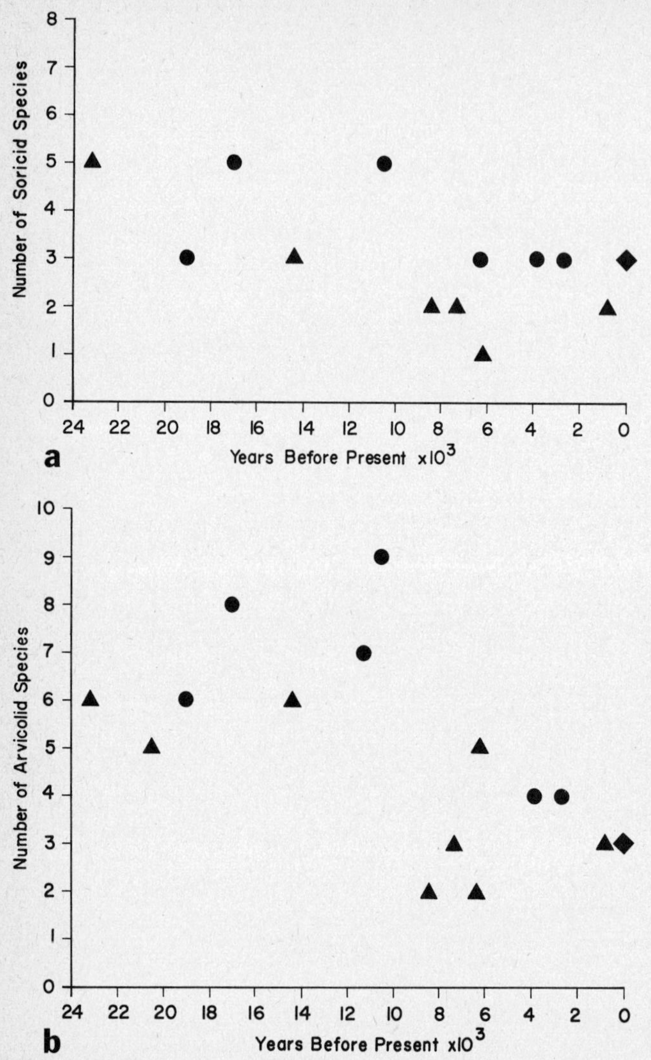

Fig. 18.5 Ordinate: number of soricid (shrew) and arvicolid (microtine rodent) species at individual fossil sites in the eastern United States. Abscissa: absolute age of the site (the Pleistocene-Holocene transition is around 10,000 B.P. Triangle = open sites; circle = cave sites; diamond = modern owl pellet sample from central Illinois (see text for explanation of site types). Sites at approximately the same absolute latitude were selected, since shrew and microtine species numbers in this area increase with latitude during the late Pleistocene as they do today (Graham 1976). In calculating arvicolid species number I lumped *Microtus ochrogaster* and *M. pinetorum* because of difficulties in identification, and I excluded *Ondatra zibethica* from all counts. Note that species numbers of both shrews and microtines decrease from late Pleistocene to Holocene fossil sites.

lies (Lundelius et al. 1983). Because of sampling and identification problems, it is uncertain whether these increased species numbers occurred universally in all terrestrial vertebrate communities. Late Pleistocene amphibian and reptilian communities also exhibit the intermingling of northern and southern biotas (Holman 1976). However, in contrast to the mammalian communities, the intermingling of herpetological communities occurs at more southerly latitudes. The reason for this difference may be that there are no herpetological equivalents for the northern mammalian communities of the tundra. Therefore, the southward displacement of northern herpetological taxa can only be detected at more southern latitudes than that of the mammalian fauna.

MAINTENANCE OF PLEISTOCENE DIVERSITY

At least three primary factors may have been involved in maintaining the intermingled biotas and higher species numbers during the late Pleistocene: equable climate, habitat diversity, and plant quality.

The nature of the late Pleistocene climate may have been one critical factor. Although the climates of glacial periods are generally characterized as being colder than at present, the actual configuration of the glacial climates was more complicated. Glacial summers were cooler and did not experience the extremes of heat characteristic of modern climates in North America. Also, winters during the glacial period exhibited less extreme cold than modern winter seasons, probably as a direct result of the glacial ice mass in North America that would have formed an orographic barrier to the southward movement of the arctic air mass during the winter (Fig. 18.1). In addition, air descending from an ice mass more than 2 km thick would have been adiabatically warmed.

Late Pleistocene climates south of the ice margin would therefore have been more equable than modern climates for these areas (Hibbard 1960, 1970). The equable climatic model is supported by independent evidence from the study of isotopic compositions of fossil wood (Yapp and Epstein 1977) and bone (Land et al. 1980). Limits of distribution for many species are controlled by seasonal climatic extremes, the southern limits of boreal species being set by hot summer temperatures, the northern limit of temperate species being set by cold winter temperatures. Hence, the relaxation of these seasonal temperature extremes during the late Pleistocene would have led naturally to higher species numbers. These patterns are consistent with MacArthur's (1975) predictions for higher species diversity in areas with reduced seasonal climatic extremes.

Higher habitat diversity in the Pleistocene would have been a second factor contributing to higher mammalian species diversity. As previously discussed, Pleistocene environments were more heterogeneous than modern ones. Greater habitat diversity would have led to more mammal species in Pleistocene sites compared to modern ones, just as it does in comparisons of different sites in the Great Plains and Rocky Mountains today (Graham in press a).

Finally, the diversity of herbivores may be a function of the effective nutritional value of the vegetation, just as Kareiva (Chapter 11) discusses for insect herbivores. Some plant species, such as black spruce (*Picea mariana*) and alder (*Alnus*), grow slowly in "poor sites" and defend their photosynthetic capital from herbivores throughout their life. Therefore, they are poor forage for herbivores. Other plant species, such as quaking aspen (*Populus tremuloides*), balsam poplar (*Populus balsamifera*), and paper birch (*Betula papyrifera*), live in better sites, grow more rapidly, tolerate more herbivory, and are relatively undefended in their mature growth form (Bryant and Kuropat 1980). Thus, they provide a more utilizable resource for herbivores.

The equable late Pleistocene environments may have selected for a better mix of palatable plant species than exists today and may thereby have afforded a greater availability of nutritious forage to mammalian herbivores. For example, the late Pleistocene spruce forests of the eastern United States contained a wider variety of palatable plant species than does the modern spruce forest (Wright 1981). Also, the late Quaternary vegetational succession in central Alaska exhibits a higher effective nutritional value in the Pleistocene than in the Holocene (Graham in press b). Indirect consequences of effective nutritional value may have also been important. Guthrie (1982) believes that reduced snowfall in unglaciated parts of Alaska during the Pleistocene may have increased effective nutritional value by allowing the herbivores to forage on frost-hardy herbs for a larger period of time in the winter.

HOLOCENE CHANGES

A rapid warming of the global climate at the beginning of the Holocene (~10,000 B.P.) significantly altered the physical and biotic environments of North America. Melting glaciers exposed new landscapes in previously glaciated terrains and eliminated old landscapes by a eustatic rise in sea level which inundated pre-Holocene terrestrial environments on the continental shelves. Retreat of the continental glaciers in response to global warming also changed climatic patterns by allowing arctic air masses to extend further southward, especially during the winter. These changes in North America created a more continental Holocene climate, marked by colder winters and hotter summers.

These climatic changes from the Pleistocene to the Holocene caused individualistic species migrations along environmental gradients. The colder winters and hotter summers of the Holocene limited the ranges of species distributions along these gradients. The ranges of many formerly coexisting species were drawn far apart from each other. Hence the fine-grained habitat mosaic of the Pleistocene was pulled apart into simpler communities that became widely separated geographically. Northward shifts of many species' ranges into deglaciated terrains lowered the diversity of small mammal communities at middle latitudes. Thus, modern mammalian communities are geographically young, less than 10,000 years old.

Throughout the Holocene, climatic fluctuations have continued to cause shifts in the ranges of individual mammal species (Hoffman and Jones 1970, Semken 1983). Also, Holocene mammalian communities have been altered by local extinctions in isolated environments like those on mountain tops (Brown 1971b, 1978). However, the magnitude of these changes appears to be small compared to the Pleistocene-Holocene transition.

EXTINCTION OF THE PLEISTOCENE MEGAFAUNA

At the end of the Pleistocene there was a major extinction of terrestrial vertebrate species, especially among birds and mammals. Although this was a global extinction event, the heaviest losses were among medium- to large-sized mammalian herbivores and carnivores in North and South America and in Australia. This extinction of the Pleistocene megafauna is the most dramatic of the changes in mammalian communities during the late Quaternary.

North America, for instance, supported during the Pleistocene a suite of large mammalian herbivores and carnivores that are now extinct. The structure of these communities in North America was probably similar, although not identical, to the modern mammalian communities in Africa. For example, it is not uncommon to find the remains of camels (*Camelops*), horses (*Equus*), mammoths (*Mammuthus*), peccaries (*Platygonus*), antelopes (*Antilocapra*), and bison (*Bison*) in Pleistocene deposits of the Great Plains. Pleistocene mammalian communities of the eastern United States frequently contain extinct large browsers such as American mastodon (*Mammut*), giant beaver (*Castoroides*), woodland musk ox (*Symbos*), and stag-moose (*Cervalces*). Large carnivores that are now either extinct or have more restricted distributions, such as short-faced bear (*Arctodus*), dire wolf (*Canis dirus*), cheetah (*Acinonyx trumani*), and lion (*Felis atrox*), were probably the primary predators on the Pleistocene large herbivores of North America.

The cause of the megafaunal extinction has been the subject of a long debate, of which both sides are presented in a recent book by Martin and Klein (1984). According to one well-known view, the first appearance of humans in North and South America and Australia was the major cause of the megafaunal extinction. According to this view humans' ability as efficient predators permitted them to overexploit the "big game" and cause its extinction.

An alternative view, however, notes that the megafaunal extinction at the end of the Pleistocene coincides with major environmental changes throughout the world (Graham and Lundelius 1984). The diverse large herbivores of the Pleistocene may have been supported by the habitat mosaic of the Pleistocene. Herbivore species could have partitioned plant resources by habitat or else by selective feeding on different plant species, plant parts, or plant growth stages (Guthrie 1982, Martin 1982). The development of continental climates at the end of the Pleistocene produced several environmental changes that would have profoundly affected the large herbivores and may have caused their extinction: destruction of Pleistocene habitats, disruption of herbivore feeding strategies (Graham and Lundelius 1984), reduction in effective nutritional value of the vegetation (Guthrie 1982, Graham in press b), and shortening of the growing season (Guthrie 1982). In this view, as in the opposite view that stresses humans as the agent of extinction, extinction of the large carnivores was a direct result of extinction of their diet, the large herbivores.

The environmental-change model of megafaunal extinction begs the question why extinction events were not associated with previous glacial-interglacial cycles. Supporters of this model must assume that the general patterns of change in the physical and biological environments were repeated many times in the past, but that the specific details of each fluctuation varied significantly. This assumption now appears to be supported. For instance, the vegetational succession of the preceding glacial-interglacial cycle (Illinoian-Sangamonian) in the Midwest consisted of a series of stages similar to those of the Wisconsinan-Holocene. However, the species compositions of the Illinoian-Sangamonian biotas were quite different from those of the Wisconsinan-Holocene, and the Illinoian biotas suggest less severe climatic conditions than in the late Wisconsinan (King and Saunders 1984). Furthermore, there is substantial evidence from the terrestrial biota throughout the world to suggest that the last interglacial had less cold winters and more equable climates than the Holocene (Graham in press a, Stuart 1982, Webb 1974).

Thus, the Holocene cannot be used as a direct analogue for the last interglacial. It is instead a somewhat individual interval in earth history, as probably was each glacial-interglacial episode. Hence the diversity patterns and species compositions for Wisconsinan and Holocene mammalian communities may not be directly applicable to previous glacial-interglacial periods. Understanding the particular qualities of the late Pleistocene-Holocene environmental change is fundamental to comprehending the late Pleistocene extinction and the evolution of our modern communities.

CONCLUSION

A myopic view of community evolution from the late Pleistocene to the present might reason as follows. With the advance of ice sheets, species retreated from north to south as whole communities and then spread north again in synchrony when ice sheets melted. Pleistocene habitats and species communities were for the most part similar to those existing today at some higher latitude. By analogy to the latitudinal gradient in species diversity that can be observed today, species number at a given site rose from the late Pleistocene to the Holocene. The only major qualitative difference between Pleistocene and Holocene communities is that the Pleistocene megafauna is now extinct, and this may have been an idiosyncratic consequence of human hunters rather than of general ecological considerations.

Reality proves to be quite different from this simple picture. Different species shifted north, south, east, or west over different distances and at different times. There are even species pairs in which the more northerly species in the late Pleistocene became the more southerly species in the Holocene. As a result, communities have been massively and repeatedly reshuffled. These facts need to be integrated into models of coevolution, which often make the implicit assumption that communities or species associations have remained intact throughout a long history. Except for land under or at the edge of the ice sheets, late Pleistocene communities had higher rather than lower species diversity of small mammals, and higher local habitat diversity, than do modern communities. Some Pleistocene habitats were entirely without modern equivalents.

No one questions that these dramatic changes in small mammal communities and habitats at the end of the Pleistocene were due to environmental changes rather than to human intervention. I therefore suspect that these same environmental changes may have also caused the extinction of the Pleistocene megafauna. Whether or not this suspicion is correct, the evidence from North American small mammals makes clear that present configurations of ecosystems are quite unlike those of much of the Pleistocene.

ACKNOWLEDGMENTS

I thank Jared Diamond for his efforts and suggestions in helping improve this paper. James Brown, Thomas Van Devender, and Andrew Knoll also provided valuable and constructive comments on an earlier draft. Neal Woodman assisted with editorial comments. I am also indebted to Mary Ann Graham for her assistance in preparation of the manuscript. Julianne Snider is thanked for drawing all of the figures. This is contribution No. 78 of the Quaternary Studies Program of the Illinois State Museum.

chapter 19

Biology, Chance, and History and the Structure of Tropical Rain Forest Tree Communities

Stephen P. Hubbell and Robin B. Foster

INTRODUCTION

Theoreticians abhor unique events. Perhaps it is not therefore surprising that theoretical community ecology has paid little heed to historical explanations for ecological phenomena unless compelled to by incontrovertible evidence. It is more difficult to account for disregard of the idea that many of the macroscopic properties of ecological communities are statistical descriptions of underlying random processes. Fears that community ecology might cease to exist if chance and history are important to the structure of natural communities are of course unwarranted: it simply changes the nature of the questions asked. Quantum mechanics did not put physics out of business, even though it overthrew classical determinism, but it did alter the direction of physical inquiry and led to new testable predictions. Similarly, chance and history as ecological mechanisms do not preclude predictability of large-scale properties of ecological communities (Chapter 13).

This chapter concerns the structure and dynamics of tropical rain forest tree communities, in particular, why chance and history might be expected to play an especially important role in structuring these communities. We present the fourth of the nonequilibrium community theories discussed in Chapter 13. Chapters 12, 22, and 23 by Grubb, Tilman, and Cody stress the evidence for niche differentiation and strong biotic interactions in other plant communities. Evidence for niche differentiation and biotic interaction among tropical tree species is also strong, and some of it is discussed here. However, chance and history are emphasized in this paper because their roles have been inadequately explored in the literature, and their significance is not as widely appreciated. We will draw heavily on the data from a study of the structure and dynamics of the tropical forest on Barro Colorado Island (BCI), Panama, where we have mapped and identified all free-standing woody plants over 1 cm in diameter in a 0.5-km^2 plot (Hubbell and Foster 1983). Our current evidence on the role of chance and history is circumstantial and the arguments speculative, but such is the state of our knowledge.

CHANCE AND UNCERTAINTY IN TROPICAL TREE COMMUNITIES

One may question whether chance is a proper causal mechanism or simply an excuse for ignorance of true causality, but this philosophical issue is not the present concern. The thesis here is that chance and biological uncertainty are inherent properties of species-rich tropical rain forests and are of profound significance to the evolutionary, population, and community ecology of tropical tree species. The observations made here may be appropriate only for high-diversity plant communities. However, biological complexity and uncertainty are not unique to the plant world, but also characterize complex animal communities as well (Glesener and Tilman 1978, Howe 1985).

Consider the case of the seasonal moist forest of BCI. This forest is only moderately rich in tree species by tropical wet forest standards. Nevertheless there are more than 300 woody plants with a stem diameter of at least 1 cm dbh (diameter at breast height) in the 50-ha plot, not counting lianas, and nearly 200 species which achieve diameters of 20 cm dbh or larger (Hubbell and Foster 1983).

How are we to account for the maintenance of so many tree species in sympatry or close peripatry? The view of tropical forests from the perspective of traditional ecological mechanics (the classical competition theory outlined in Chapter 13) is that they are stable, competitively coevolved sets of niche-differentiated tree species, each of which has a means of exploiting limiting resources that is sufficiently different to permit its continuing persistence in the community. The forest is in taxonomically stable equilibrium or near-equilibrium, such that each species limits its own growth more than that of other species. Because of the great age of the tropics and its benign and stable climate, there has been a slow, steady accumulation of species. Coevolution of all these species under competition in a stable environment has led to finer and finer partitioning of limiting resources, leading in turn to greater and greater degrees of specialization (Ashton 1969). Many equilibrium hypotheses have been proposed for the maintenance of high tree species richness in tropical forests, and evidence bearing on some of these for BCI will be reviewed shortly. Before doing so, however, we consider the hypothesis that chance plays a major role in structuring tropical tree communities.

Although formalized mathematically only recently, the idea that chance might play an important role in the structure and dynamics of tropical forests has a long history in the literature (Corner 1954). That the extraordinary species richness defies easy explanation in equilibrium and selectionist terms has been realized for some time (Federov 1966, Poore 1968, van Steenis 1969). The richest forests have guilds of very similar species, often including many apparent ecological and evolutionary homologues in suites of many congeneric species (Federov 1966). Moreover, the dominant role of wind in disturbing the canopy (Webb 1958; Cousens 1965; Whitmore 1974, 1975, 1978; Brokaw 1982), thereby opening up gaps in unpredictable spots, and the importance of gaps for the regeneration of most if not all species (Hartshorn 1978), suggested that being in the right place at the right time might be more important to a tree's success in reaching the canopy than the taxon to which the tree belongs (Baur 1968, van der Pijl 1969, Strong 1977). Indeed, if the probability of success is essentially the same on a per capita basis for all mature phase species, then the composition and relative abundance of tree species in the forest would tend to drift in a random walk.

These points give rise to the observation that the biotic environments shaping the life histories of the trees in the rain forest are inherently statistical in nature. For example, gap formation is fundamentally and inherently a random process. It is theoretically impossible for humans or trees to predict precisely which tree will next be killed by lightning or felled by wind, although some trees may be more likely to fall than others (Hubbell and Foster 1985a). This might be of little consequence to predictability in biotically simple communities, but in species-rich plant communities biotic central tendencies are only very weakly developed. For a given rain forest tree, a rooted organism, the selective environment that determines its reproductive success is to a large extent its immediate biotic neighborhood. If this neighborhood is predictable and simple, that is, composed of relatively few species over evolutionary time, the potential for competitive niche differentiation is greater than if the neighborhood is unpredictable and complex.

The importance of this observation depends critically on the biotic unpredictability of tree neighborhoods in the rain forest. Of particular concern is the uncertainty of neighborhoods of individual trees and the degree of similarity of these neighborhoods from one tree to the next. With the detailed map data on the 0.5-km^2 rain forest plot on BCI, we have the opportunity to examine the local neighborhoods of tens of thousands of canopy trees and their saplings. For canopy adults (trees over 30 cm dbh) and their 20 nearest neighbor trees in the canopy, the results reveal substantial levels of local diversity.

The distribution of the species means of neighborhood species richness around individual trees in 98 species is shown in the top graph of Fig. 19.1. Among the first 20 neighbors the average number of different species is 14.1; over two-thirds of the 20 neighboring trees are of different species. The bottom graph shows the mean neighborhood species similarity for any 2 conspecific trees of the same 98 species. Only about 30% of the 20 nearest neighbors of any 2 adults of a given species can be matched by species. If we consider more neighborhoods than for just 2 trees, the species diversity increases rapidly with number of tree neighborhoods (Fig. 19.2). In the immediate neighborhoods of just 10 conspecific trees, for example, we can expect to encounter nearly 60 canopy tree species. The curve above the points is what would be expected if all trees were randomly intermixed. The observed cumulative species richnesses (mean ± 1 SD) are slightly lower due to imperfect mixing of species in the forest, but departures from random mixing are not great.

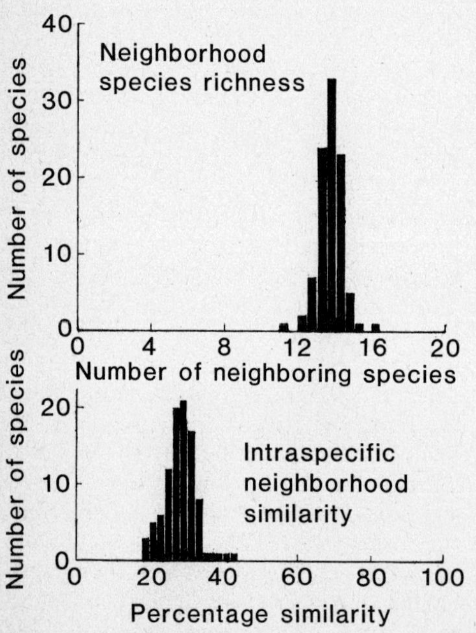

Fig. 19.1 Neighborhood diversity analysis for 98 BCI canopy tree species for trees more than 30 cm dbh. Analysis performed on first 20 nearest neighbors more than 30 cm dbh. (Top) Distribution of species means for the number of tree species among the 20 nearest neighbor trees. (Bottom) Distribution of species means for the percentage of trees among the 20 nearest neighbors that can be matched to species in the neighborhoods of two conspecific trees.

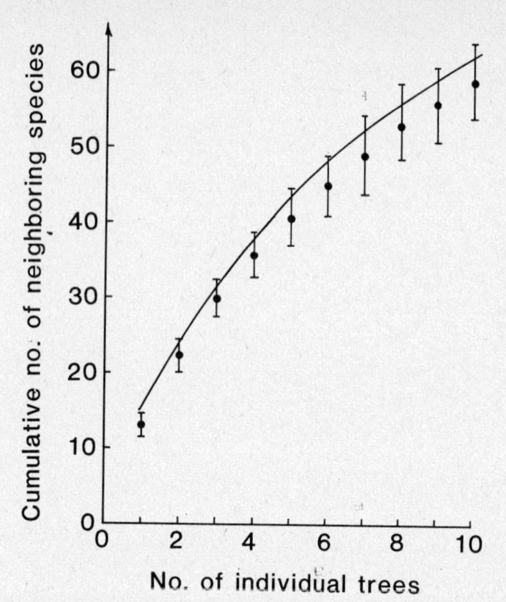

Fig. 19.2 Species-neighborhood curves for BCI canopy trees, showing the rapid rise in species richness with the addition of just a few 20-tree neighborhoods. Curve is the expected curve for randomly intermingled species at their observed total abundance in the BCI plot. Points are observed mean values for the cumulative curve. Error bars are 1 SD.

Neighborhood species richness is potentially only a small fraction of the total biotic uncertainty within the immediate neighborhoods of individual trees. As the trees grow through different size classes, they encounter a continually shifting mix of competitors (Hubbell and Foster unpublished). There is unknown genetic diversity on top of this species diversity as well (Loveless and Hamrick 1985), so that classifying neighbors only to species underestimates the genotypic uncertainty of tree neighborhoods in the rain forest. The impact of unpredictability and dissimilarity of biotic neighborhoods among individual adults on the nature of selection acting on rain forest tree species is potentially profound. What is selectively favored in each local biotic environment may differ from adult to adult, such that there is little constancy from tree to tree in the direction of selection. The postulated result is that evolution will reflect the temporal and spatial average of the selective conditions created by a suite of ever-changing and diffuse competitors. No two tree species encounter each other frequently or consistently enough in evolutionary time to evolve pairwise character displacement. A similar argument has been made to explain the origin of diffuse mutualisms in the case of plant-pollinator and plant-disperser systems (Howe 1985).

This view is very different from the traditional one that rain forest tree taxa are highly niche differentiated in terms of their physical and biotic environmental requirements. Ashton (1969) argued that under the predictable and favorable climatic regimes of the tropics, natural selection would be dominated by biotic interactions, thereby leading to ever-narrowing ecological niche breadths of the constituent species. However, specialization may not occur even in the presence of very intense competition. Implicit in this view is the assumption that fairly precise and specific coevolution of competitors occurs, which requires that competing species pairs be predictably associated with each other through ecological and evolutionary time (Connell 1980, Howe 1985). If, on the other hand, a species has unpredictable neighbors during its evolutionary and biogeographical history, there may be little constancy in the direction of selection to lead to the expected specialization.

Neighborhood biotic uncertainty may instead often lead to convergent generalization in response to the time average of the diffuse selective forces imposed by the sum of all competitors. As an example, tree species competing for light collectively cast shade on their own seedlings and saplings and those of other species. This diffuse competition should result in selection for generalized shade tolerance in the young stages of all these species, not in selection for specialized shade tolerance for the degree or quality of shade cast by particular species (Connell personal communication). Selection should lead to a convergence to similar generalized abilities to tolerate the shade conditions in the forest because (1) shade competition is diffuse and (2) the species casting the shade at any particular time and place are unpredictable. Convergent evolution for shade tolerance could be an explanation for the observed high degree of morphological similarity among the seedlings and saplings of many rain forest tree species.

Similar hypotheses can be made about competition for limiting soil nutrients. There is little evidence of host specialization in vesicular-arbuscular mycorrhizal species found in association with tropical trees, at least in the neotropics. Janos (1982) has argued that the dependence of neotropical trees on the same few mycorrhizal fungus species for mineral nutrient uptake limits their ability to exclude one another competitively. Indeed, shared mycorrhizal species may be a major competitive equalizing mechanism, and selection may then strongly favor infectable tree genotypes in any species not yet having a mycorrhizal association, that is, lead to competitive convergence in nutrient competitive ability.

GUILDS OF TREE SPECIES IN TROPICAL FORESTS

Any theory that biotic uncertainty promotes convergent generalization in rain forest tree communities must be reconciled with the undeniable existence of life history differences among tropical trees. Whether these differences cluster into biologically distinct guilds is more controversial. Nevertheless, some real guilds seem easily recognizable, such as a group of pioneer species specialized for rapid colonization and growth in large tree-fall gaps. As a group these species tend to share characteristics such as seed dormancy, shade intolerance, rapid elongation of a monopodial shoot, low wood density, small size at first reproduction, bird or bat dispersal, and short life spans. In the BCI forest there also appear to be guilds of edaphic and topographical specialists. Some tree species are largely restricted to slopes, whereas others are predominant on flat ground or in the seasonal swamp in the plot (Hubbell and Foster 1983). Shade-tolerant shrubs and understory trees are also recognizable guilds. Finally, there are gap-edge regeneration specialists that apparently do best in the moderate light environments of gap edges, which may be less stressful sites than gap centers.

The prevalent view in the literature is that there is a near continuum of adaptive strategies for tropical tree species. Many authors have speculated that high degrees of specialization occur in tropical trees for conditions of regeneration (e.g., Ricklefs 1977, Grubb 1977, Denslow 1980), but the evidence is largely anecdotal at present. We do not yet see clear evidence for such a high degree of specialization for regeneration niches in the BCI forest. Maybe this evidence will be forthcoming as we learn more about the life histories of the tree species in the forest, and in particular whether recruitment fluctuations are the mechanism of coexistence of many tropical tree species (see below). At present, however, we see evidence for only a dozen or so guilds, each of which is composed of many morphologically and phenologically similar tree species. It is within these major life history modes of existence in the rain forest that we suggest adaptive convergence and generalization occurs. According to this view, there are major adaptive discontinuities and trade-offs between guilds, such that there are only a few basic strategies for making a living as a tropical tree. The concluding section of the next chapter (Chapter 20) contrasts this view of community organization with several alternative views.

In a detailed study of gaps and tree species distributions in relation to gaps, gap edges, and forest interior in the BCI forest, we did find regeneration guilds, but only a few (Hubbell and Foster 1985a, 1985b). For example, there is a guild of heliophilic trees, which constitute about a quarter of all BCI canopy species. These species are never common in the BCI forest. Conversely, common BCI trees are disproportionately represented among the group of species that are statistically indifferent to the microsite distinctions between gap, gap edge, and forest interior in the BCI forest.

We suggested two hypotheses. First, obligate heliophiles that require large gaps to reach maturity will necessarily be rare because large gaps occur infrequently in the BCI forest both in space and time. Thus, as a class this group of species should be gap-limited. A prediction of this hypothesis is that the percentage of heliophiles should increase if the disturbance rate increases. When the canopy species compositions of hectares are compared, hectares with more open canopy architecture at 30 m, suggesting recent history of greater disturbance, do exhibit higher proportions of heliophilic species in their canopy layer (Hubbell and Foster 1985a). Second, a necessary condition to becoming an abundant tree species in the BCI forest is having the ability to mature in both gap and nongap areas under the average disturbance regime characteristic of the BCI forest. The gap disturbance regime on BCI is one that produces mostly small gaps; of 638 recorded gaps in the 0.5-km^2 plot, 58.6% (374) were of the smallest size recorded (25 m^2), whereas only 18 (2.8%) were 400 m^2 or larger.

The distribution of small saplings (1 to 5 cm dbh) was analyzed in 81 canopy tree species in gap, edge, and nongap microsites. Of these 81 species, 14 species (17.3%) showed statistically significant gap association in their small saplings, 4 (4.9%) were significantly associated with edge microsites, and 6 species (7.4%) showed significant negative association with gaps in their sapling distributions. The remaining 70% of the BCI species showed no apparent specialization in their regeneration requirements, exhibiting the number of small saplings in gap, gap edge, and forest interior expected by chance (Hubbell and Foster 1985b). We agree with Hartshorn (1978) that approximately three-quarters of the species probably do require a gap at some stage of their ontogeny, but we suggest that the regeneration requirements of many of these species are, for all practical purposes, functionally identical.

In spite of such clear evidence for specialization into broad regeneration guilds, there is little evidence in support of finer levels of regeneration niche specialization among the BCI trees. Obviously, each tree species can be distinguished from all others in the plot on the basis of morphological differences, but these character differences among plant species should not be accepted as *prima facie* evidence for niche differentiation.

The question, as discussed by Yodzis (Chapter 29) and Grubb (Chapter 12), is which if any of these differences are necessary and sufficient for competitive coexistence. Unlike Fedorov (1966), we do not make the argument that natural selection is unimportant in producing the observed differences between ecologically similar species in guilds. The argument is simply that if such differences do not arise from interspecific competition, then they do not represent character displacement and are irrelevant to supposed competitive niche differentiation. Of course, it may be that the differences are still important to the stability of the species association of the guild if they reflect a past filtering of species able to coexist by virtue of differences established prior to contact. Alternatively, their differences could simply reflect slightly different ways of solving the same autecological problem starting from historically different parent genetic material.

Summarizing the argument, interspecific competition does not invariably lead to specialization and may more often lead to greater generalization (and similarity) among guild members, particularly if interspecific competition is diffuse and the direction of selection is the result of smoothing of erratic and unpredictable local competitive effects. Thus, in species-rich tree communities the dominant evolutionary consequence of unpredictability in the biotic environment may be a kind of diffuse coevolution resulting in large guilds of competitively coequal or nearly coequal species, rather than in competitive niche differentiation (character displacement). Under such circumstances chance and history (that is, the initial local and biogeographical distribution of species) are likely to come into greater prominence, compared to biotic effects, in determining the relative abundance of tropical tree species in particular forest stands.

MAINTENANCE OF TREE SPECIES RICHNESS

According to this model, many tropical tree species coexist not in spite of, but because of being functionally equivalent generalists. Once they have solved the basic physiological problems of survival within the broad adaptive zone of their life history guild, there are few ecological forces that can lead to their systematic elimination from the community.

There are at least three arguments that support this line of reasoning. The first argument is based on a theoretical result that comes from studies of the persistence times of identical species in model communities. The second argument is genetic and comes from consideration of the effect of intrapopulation genetic variability on processes of competitive exclusion. The final argument (discussed in the next section) concerns the nonequilibrium status of tropical forests.

Drifting Relative Abundance

Theoretical studies of the community drift model (Hubbell 1979, Wright and Hubbell 1983), in which species abundances are in a multinomial random walk with equal per capita probabilities of birth and death, reveal that times to local extinction of a tree species can be long indeed. Populations of a few thousand adults can be viewed as essentially immune to extinction over geologically significant time spans, long enough for speciation to become an important process (Hubbell unpublished). This observation is consistent with Levinton's (1979) theory of diversity equilibrium. If there is a limit to the total tree biomass in a region, then a regional equilibrium tree species richness can arise without invoking diversity control through niche subdivision of resources. As more species are added to the region, there will be a decrease in mean population size of resident tree species, and a stochastic equilibrium in diversity will be established when average population sizes have decreased to the point where extinctions are occurring at the same rate as new species appear (Hubbell 1979).

Thus, Gause's (1934) ''axiom'' that no two species with identical niche requirements can coexist for long is false—if by ''for long'' we mean hundreds of generations. We can illustrate this conclusion with some simple calculations for the community drift model. Consider the case of an island forest cut off from immigration, which holds *K* trees regardless of species. Let one tree be killed at random across species, and let it be replaced by a tree of a species drawn at random from the surviving species with probabilities determined by current relative species abundances. Although this is a ''null'' model, it should be noted that biotic interactions via diffuse competition are extremely intense in the model: No species can increase except at the expense of another species, since there is a finite limit on the total individuals of all species in the community. The equilibrium fate of a species in such a community is either extinction or complete dominance. Extinction is the more probable outcome for any species having a starting abundance of less than half the total individuals in the community.

Suppose there are N individuals of a species in a community of K total individuals of all species. The time, t_N, measured in terms of total number of tree births or deaths in the community, that it will take this species to drift to extinction ($N = 0$) or to complete dominance ($N = K$) from a starting abundance of N individuals, $0 < N < K$, can be found analytically:

$$t_N = (K-1)\left[(K-N)\sum_{j=1}^{N}(K-j)^{-1} + N\sum_{j=N+1}^{K-1} j^{-1}\right]$$

The first term of the sum is the mean time that the species will spend at abundances below N; the second term is the mean time that it will spend at abundances above N, before extinction or complete dominance. When N is much smaller than K, $t_N \approx N(K-1)(1+\ln K)$. We can illustrate the dramatic increases in time to extinction or total dominance that occur by increasing the size of the community (K) and holding the relative abundance of the given species (N) constant at a fixed fraction of K (Fig. 19.3). For example, more than 1 million trees will turn over in a forest of K = 2,000 trees before a species starting with $N = 200 = 0.1K$ trees will go locally extinct. Times to extinction are shortened if more than one tree dies at a time ($D > 1$), but they are still long unless a large fraction (e.g., a quarter of the forest is killed in a succession of major disturbances.

Obviously, such models are oversimplifications of reality, but they make an important point that added realism seems unlikely to change, namely, that ecologically equivalent species can coexist in drifting relative abundance for long periods of time with no refuge from intense diffuse competition. We can therefore no longer

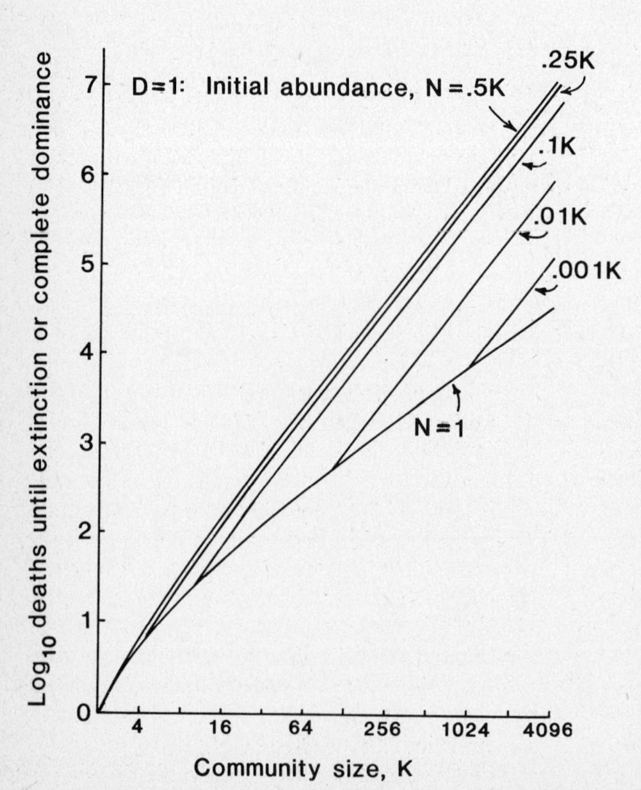

Fig. 19.3 Times to extinction or complete dominance for a species beginning with an abundance of N individuals in the community drift model. The graph is a representation of the equation in the text relating extinction or dominance time to community size, the total number of individuals of all species in the community (K), and the initial population size (N) of the focal species. The death rate (D) is one individual per unit time. The bottom boundary curve is for N fixed at one individual for arbitrary K. The diagonal lines slanting upward from the curve for $N = 1$ give times to extinction or complete dominance for N as various fixed proportions of K, that is, constant relative abundance. Note that time to extinction or dominance increases with K and with N.

argue that species persistence is sufficient evidence by itself that a community is competitively stabilized by niche differentiation. However, one would expect greater species turnover among rare species and less similar relative species abundance patterns in nonequilibrium communities governed solely by drift. The empirical challenge is to discriminate between these models of community organization on the basis of their quantitative rather than qualitative predictions.

Intrapopulation Genetic Variability

The second argument for the maintenance of many ecologically near-equivalent tree species in tropical forests is a genetic one. Genetic variability among the individuals of a species should make competitive exclusion less likely to go to completion than if all individuals were identical specialists (Buckley 1983). Turkington and Harper (1979) showed that divided *Trifolium* clones were better competitors when replanted in that part of the pasture from which they originally came than when they were transplanted elsewhere in the pasture. Diversity in locally adapted genotypes should promote greater equalization of competitive ability between species if species are more difficult to displace locally because of such local adaptation.

In line with previous arguments, the adaptive generalization of rain forest tree species can result from two processes. On the one hand, species move closer together in their mean phenotype in response to the long-term average of similar diffuse competitive pressures. On the other hand, soft disruptive selection in diverse local biotic environments results in the maintenance of high levels of intrademic genetic variability in tree species populations. Recent evidence on the genetic architecture of rain forest tree populations supports this argument. Loveless and Hamrick (1985) have found more intrapopulation than interpopulation genetic variation in several BCI tree and shrub species, a result that was not expected from previous studies of temperate zone tree populations (Loveless and Hamrick 1984). This view also fits with the observation that dioecy is common and outcrossing the rule in most neotropical trees (Bawa and Opler 1975, Bawa 1979). Indeed, we may argue that pervasive sexual reproduction in tropical trees in species-rich forests is in fact due to the high biotic uncertainty of local environments, which in turn is partly due to the sexual reproduction by one's competitors (Levin 1975, Glesener and Tilman 1978).

We need to develop theory for the extent of and limits to the coevolution of competitors in species-rich communities. Most existing models of coevolution of competitors assume that species can be ordered along a gradient and that the identity of neighboring species on the gradient is predictable over evolutionary time. In such forced encounters, it is perhaps to be expected that when competitive functions are symmetrical, niche differentiation evolves, whereas when they are asymmetrical, competitive elimination usually results (Roughgarden 1983). Recent simulation studies reveal, however, that it is possible to construct models of coevolution under conditions of diffuse competition from biotically uncertain competitors which do *not* result either in competitive niche differentiation or in competitive exclusion (Hubbell and Hoffmaster unpublished).

TROPICAL FORESTS: EQUILIBRIUM OR NONEQUILIBRIUM COMMUNITIES?

The current controversy over whether tropical forests are equilibrium or nonequilibrium communities is of central concern to the question of the role of chance and history in structuring these tree communities. The terms "equilibrium" and "nonequilibrium" are now commonly used in discussions of tropical forest organization and dynamics (Connell 1978, 1981; Huston 1979; Hubbell 1979), but each author means something slightly different by them. Here we use an "equilibrium community" to mean a community in which there is stability as well as persistence of a particular taxonomic assemblage of species in a defined area (also see discussion in Chapter 13). Equilibrium communities do not necessarily have sharp community boundaries (Whittaker 1956, 1967) or precisely constant patterns of relative species abundance, but they are systems in which the requirements of each species are met in a manner that does not reduce (and may even enhance, as an incidental effect) the stability of the assemblage.

Many mechanisms for stabilizing species assemblages in plant communities have been proposed. These include separate niches for each species (e.g., Richards 1952, Grime 1977, Grubb 1977), frequency dependence due to an inability of tree species to regenerate beneath or near themselves (Aubreville 1938, 1971; Gillette 1970; Janzen 1970; Connell 1971; Horn 1975), frequency dependence due to uncorrelated recruitment fluctuations in long-lived trees (Chesson and Warner 1981, Warner and Chesson 1984), the intermediate disturbance hypothesis (Connell 1978), the competitive nonsaturation hypothesis (Huston 1979), nutrient ratio theory (Tilman 1980, 1982), and nontransitive competitive hierarchies (Jackson and Buss 1975, Yodzis 1978).

An early example of what today would be classified as an extreme equilibrium view was proposed by Clements (1916), who argued that plant communities were analogous to organisms, with each species functioning as an essential part or ''organ'' for the whole, and that communities matured through a developmental process (succession) analogous to organismic ontogeny. These ideas soon came under attack by Tansley (1935) because they were not easily reconciled with natural selection and by Gleason (1926) and later by Whittaker (1956) because they were inconsistent with the data on biogeographical and local patterns of plant species distribution. More recently, the classical paradigm of succession (Odum 1969) as an ontogenetic, self-generating process has come under attack by a number of authors (e.g., Drury and Nisbet 1973, Connell and Slatyer 1977), who prefer to describe succession not so much as a community phenomenon, but as the collective unfolding of the individualistic life histories of species specialized along points of a regeneration or stress gradient.

This autecological and evolutionary perspective is implicitly or explicitly defended in the writings of many modern plant ecologists (e.g., Decker 1959, Yarranton 1967, Pickett 1976, Grime 1977, Harper 1977, Peet and Christensen 1980, Runkle 1981). A related view is of succession as a set of processes involving nontransitive, sometimes cyclic, competitive hierarchies (Watt 1947; Aubreville 1971; Horn 1975, 1981; Strong 1977; Yodzis 1978; Acevedo 1981) that produce fixed-point or cyclic species equilibria. The notion that there might be more ''regeneration niches'' in tropical forests has been explicitly proposed as an explanation for the higher species diversity of tropical rain forests (e.g., Budowski 1961, 1970; Grubb 1977; Ricklefs 1977; Strong 1977; Hartshorn 1978).

In recent years there has been some confusion over the use of the term ''equilibrium'' when applied to forests subject to differing levels of natural disturbance. Connell (1978) has suggested that intermediate levels of disturbance promote higher species richness in tropical forests, and he applied the term ''nonequilibrium'' to such forests because the disturbances acted to prevent complete space monopoly by a few competitively codominant species. If, however, the disturbance regime is a relatively predictable component of the environment, we would classify Connell's intermediate disturbance hypothesis as an equilibrium hypothesis. Because disturbance is an ever-present and natural feature of the real world to which species adapt, specializing to particular transient but predictably recurrent environmental conditions, this seems to be a perfectly valid example of the equilibrium competitive niche differentiation model.

The critical distinction between equilibrium and nonequilibrium communities, to our way of thinking, is whether a particular species assemblage is stabilized by intrinsic biotic interactions under the given environmental regime of the site, and whether, if the species mixture is altered, the community tends to return to its former composition. The nonequilibrium view is that there are no (or only weak) intrinsic stabilizing factors in the community tending to maintain the integrity of the current taxonomic assemblage through time. Indeed, there may not be any destabilizing factors either, so that the patterns of relative species abundance should continually change or ''drift'' as species come and go from the community (Hubbell 1979). As discussed previously, this is not to deny the existence of broad guilds of species specialized to occupy gap or mature phases of the forest, or the understory or canopy at maturity, but to argue that on uniform substrates and within guilds the composition of tree species is determined largely by historical accident, and the number of species is determined by the balance between immigration and extinction (Hubbell 1984). This view has its origins in the "equilibrium" theory of island biogeography (MacArthur and Wilson 1967). In the present context calling this an equilibrium theory is a misnomer because the basic theory predicts a taxonomic nonequilibrium in island communities with continual species turnover, even though the species richness of the island communities may reach a steady state.

EVIDENCE FOR DENSITY-DEPENDENCE

Stability in virtually all equilibrium community models, especially those which assume competitive niche differentiation, requires that each species in the equilibrium community limit its own growth to a greater extent than that of its competitors. Implicit in this requirement is that there be negative density-dependence, such that the per capita reproductive performance of adults should decline with increasing adult density. We therefore performed a regression of the number of juveniles (J) per hectare on the number of conspecific adults per hectare (A) using the following model:

$$J = a_0 + a_1A + a_2A^2$$

The precise relationship is unknown, but this model is the simplest density-dependent model possible, with a_0 added as an intercept to explain juvenile immigration from surrounding hectares. If the second-order coefficient a_2 is significant and negative while a_1 is positive, or if the first-order coefficient a_1 itself is negative, then we have evidence for density-dependence. In this case we expect juvenile density to be an increasing concave (decelerating) function of adult density or else a decreasing function. If there is no density-dependence, then there should be no change in the per capita production of juveniles with increasing adult density, a_1 would be positive, and a_2 would not be significantly different from zero. This would produce a linear and positive relationship between juvenile and adult density, or no consistent relationship if processes are density-vague (*sensu* Chapter 15 and Strong 1984b).

Although the analogy of the regression model above to the model of logistic growth, $dN/dt = rN - (r/K)N^2$, should be apparent, we need to exercise caution in interpreting the results. In general, we cannot make inferences about dN/dt from static, single-census measurements. We are limited to statements about whether the per capita production of juveniles shows density-dependence through to the production of the size class of juveniles observed, not to the production of new adults.

When graphs are made of the number of juveniles per hectare versus the number of adults per hectare for BCI tree species (Fig. 19.4), about half of the species show no evidence for density-dependence. Of the 48 most common species, 52% (25) have no significant second-order density effect. Of the remaining 23 species, 21 display negative density-dependence and 2 display positive density-dependence. With one exception, the negative density-dependence in 21 species seems likely to be too weak to regulate the numbers of adults of these species; the significant second-order coefficients are small numbers, so that the curvilinearity they produce in the regression is slight. This suggests that many of the tree species may be far below the adult densities per hectare at which strong density effects would be observed. However, the most common canopy tree species, *Trichilia tuberculata* (Meliaceae) with more than 1,100 individuals over 20 cm dbh, exhibits strong first-order negative density-dependence (Fig. 19.4).

The frequent lack of density-dependence might be the result of a spatial scale problem (discussed in Chapter 8), namely, that a hectare is inappropriately large (or small) for density effects to be observed. We checked for scale dependency of the results by repeating the analysis at scales of 10 × 10 m, 20 × 20 m, 50 × 50 m, and 200 × 200 m. For large canopy tree species, the analysis cannot be done effectively at 10 × 10 m because there are rarely more than one and never more than two adults in a single 10 × 10 m subquadrat. At 20 × 20 m and 50 × 50 m the results showed fewer species with density-dependence than at a plot size of 1 hectare (100 × 100 m). However, this could be attributed to a loss in power of the test due to a reduction in the range of adult numbers per quadrat, coupled with an increasingly important edge (immigration) effect. At a plot size of 200 × 200 m the results were similar to the 1-hectare results.

The effect of the numbers of adults of other

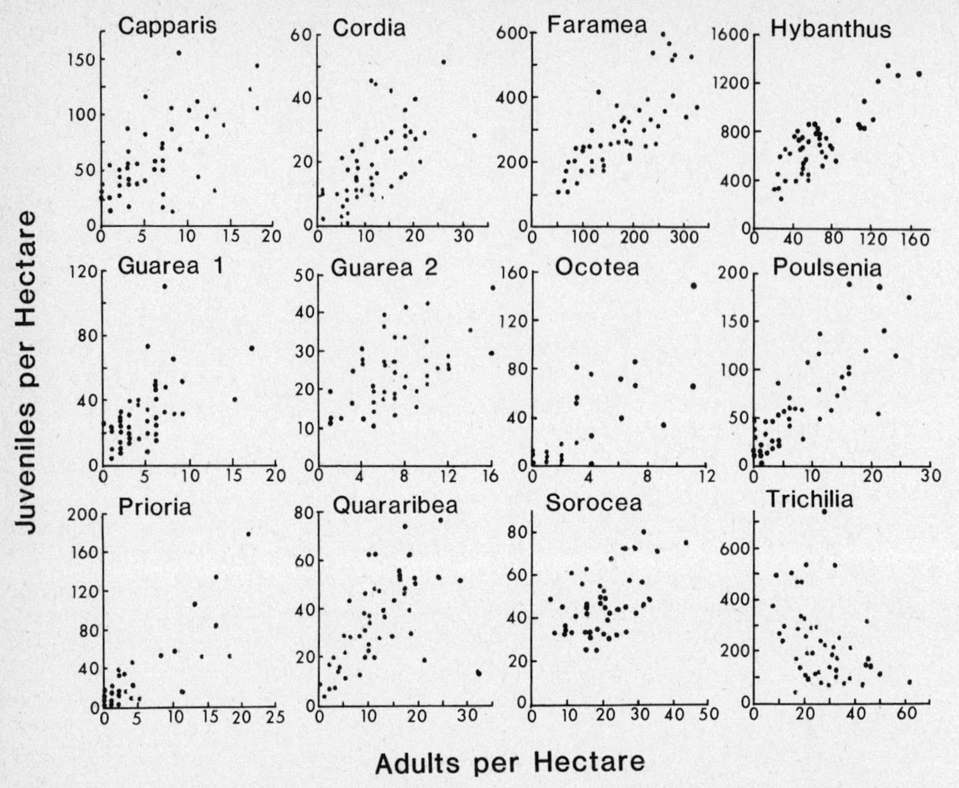

Fig. 19.4 Graphs showing the relationship between number of juveniles per hectare and the number of adults per hectare for *Trichilia tuberculata,* the most common canopy tree in the BCI plot, and for 11 other BCI species chosen haphazardly. Of these species, only *Trichilia* exhibits significant negative density-dependence. *Hybanthus* and *Faramea,* two of the most common species in the plot, and the remaining species in the figure exhibit no significant second-order effects of adult densities on juvenile densities. Species are: *Capparis frondosa, Cordia lasiocalyx, Faramea occidentalis, Hybanthus prunifolius, Guarea guidonia* (Guarea 1), *Guarea* sp. nov. (Guarea 2), *Ocotea skutchii, Poulsenia armata, Prioria copaifera, Quararibea asterolepis, Sorocea affinis,* and *Trichilia tuberculata.*

species on the number of juveniles of the focal species per hectare must also be considered. In equilibrium models it is the relative magnitude of interspecific versus intraspecific density effects that is important to the stability of the species assemblage. In a separate set of regressions the number of adults of all other tree species in the hectare was included as a variable; in another set the product of the number of adults of the conspecific and of heterospecific species was added. In none of the 48 most common species (all species examined) were these variables significant. This is perhaps not too surprising in view of the fact that the number of trees of all species in the BCI forest is almost constant from one hectare to the next: the mean number of trees over 20 cm dbh per hectare is 151.9 with a coefficient of variation of only 0.093.

An equilibrium argument can be erected to explain these data, namely, that the taxonomic assemblage of each hectare is a tree community in local equilibrium. If this were true, however, then one either has to postulate that each hectare has its own unique community matrix, since abundances of the common tree species are often radically different among hectares, or one needs to suppose that the rare tree species are "keystone competitors" that dictate the patterns of relative species abundance among the common tree species. The nonequilibrium interpretation also has its weaknesses, one of which is the assumption that the different adult densities reflect different stages in the growth of local populations of a given tree species.

The equilibrium view is supported by strong density-dependence in *Trichilia*, the most common canopy tree; 1 in 7.7 canopy trees is a *Trichilia*, and the crowns of many of these individuals are touching the crown of at least one other *Trichilia*. At such densities common species may be prevented from further monopolization of space by intense intraspecific competition. Our current hypothesis, however, is that a majority of the tree species, unlike *Trichilia*, are nowhere near the densities at which strong density-dependence would be detectable. If populations of most BCI tree species are far below intraspecific saturation of the local habitat, then population fluctuations in these species may resemble a density-independent random walk.

REPRODUCTIVE PERFORMANCE AND TREE ABUNDANCE

Other data relevant to the equilibrium-nonequilibrium status of the BCI tree community come from patterns of senescence of tree populations in relation to abundance. By ''senescence'' we refer to the proportion of adults out of the total population of a given species. If a population of a species is composed largely or exclusively of adults, then either the population is declining or else reproduction in the species is episodic in nature (Hubbell 1979). Data on population abundance and senescence are interesting in light of an ingenious equilibrium hypothesis recently developed mathematically by Chesson (Chapter 14) and Chesson and Warner (1981) from verbal models of Sale (1977) and Grubb (1977). They proved in a stochastic two-species model that coexistence between two species was possible if adults were long-lived and fluctuations in recruitment occurred. In multispecies systems the condition for coexistence is that each species have a period in which its recruitment success is greater than the average recruitment success in all other species, or that it have a positive growth rate when rare (Warner and Chesson 1984). To take tree communities as an example, we may suppose that new canopy gaps are created at a finite rate per year. If tree species have recruitment pulses at irregular intervals, then a rarer species has a potential frequency-dependent advantage over the more common species if it reproduces more in years when the common species reproduce less. This mechanism is appealing because reproduction in a number of BCI trees is infrequent, and even though most species fruit every year, they may show considerable annual variation (Foster 1982, Howe 1983).

Patterns at the community level are also expected. If rare tropical tree species are being maintained in the rain forest primarily by stochastic frequency-dependent recruitment fluctuations, then there ought to be a general tendency for the proportion of adults of species to decline with increasing rarity of species. This would be indicative of greater average per capita reproductive performance when species are rare. A plot of data for 169 species on population senescence (percentage adults) versus adult population size reveals a great deal of scatter; however, if there is a trend, it is in the wrong direction from that predicted (Fig. 19.5). When species are divided into three classes of adult abundance—rare (1–9), occasional (10–99), and common (100 and up)—the percentage of species with more than 20% adults declines from 54.0% in rare species, to 42.5% in occasional species, and further to 35.9% in common species.

One can raise several immediate objections to this analysis. It assumes that all tree species have the same potential maximum population density in the forest. It also does not take account of differing survivorship patterns in these tropical trees. A 10-year study of sapling and tree mortality on BCI does show that patterns of survivorship were variable from species to species, particularly in sapling and subadult stages (Lang and Knight 1983).

A potential problem for the theory may arise if common species reproduce every year but rare species do not: the theory requires that each species have a period when its recruitment success is greater than the average recruitment success of the other species. For some hypothetical calculations, we can use the method suggested by Warner and Chesson (1984) to test whether there is sufficient recruitment variability for 2-species coexistence to occur. Consider a case of two species, one common and one rare, that have equal death rates. Suppose that in all years except one

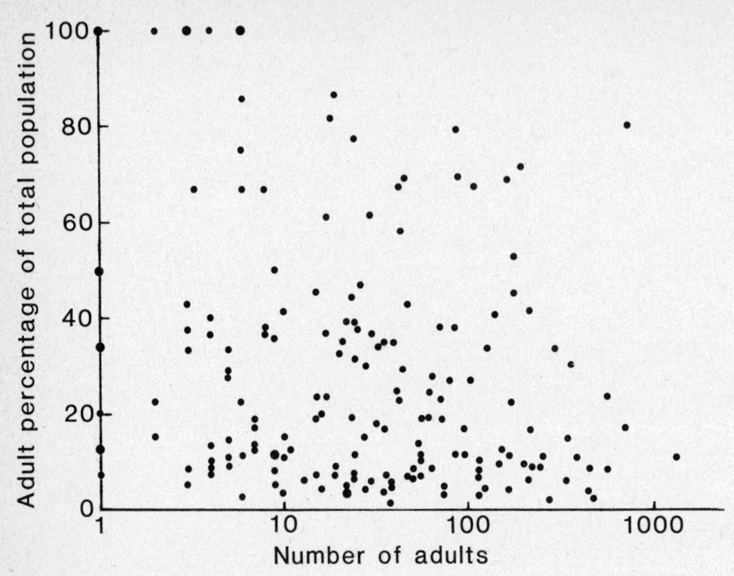

Fig. 19.5 Scatter plot showing a weak negative relationship between the percentage of adults of the total population and abundance (measured by number of adults) for 169 BCI tree species. See text for interpretation.

out of n years the common species recruits better on a per capita basis than the rare species. The common species will persist if its average recruitment success exceeds that of the rare species in the $n - 1$ years by the factor $n/(n - 1)$. However, the rare species will only persist provided that, in the year it reproduces, its per capita recruitment exceeds that in the common species by a factor greater than n. For a species that reproduces every four to six years, such as *Tachigalia,* this factor must be greater than 4 to 6. For a senescent and rare species whose population consists solely of adults no younger than a century, this factor must be greater than two orders of magnitude.

It seems likely that the recruitment fluctuation hypothesis or "storage effect" contributes to the maintenance of species richness in tropical tree communities, but it remains to be seen how quantitatively important the effect is. There is some question, for example, whether obligately outcrossed tropical trees are "stochastically bounded from below" (Chesson and Warner 1981). Tree species can be driven to local extinction when rare because of reproductive Allee effects. In obligately outcrossed species there almost certainly exist densities below which mate availability becomes a problem. We believe that perhaps as many as a quarter of the rarest species in the 50-ha plot are not self-maintaining, but are sustained solely by long-distance immigration from population centers outside the plot. This may explain why there is a significant difference between the pattern of species abundance in dioecious and hermaphroditic species in the BCI forest. Fifty percent of 191 hermaphroditic species have less than 20 adults, whereas only 20% of 50 dioecious species are so rare (Hubbell and Foster 1985b). Perhaps normally outcrossed hermaphroditic species fare better than dioecious species do if hermaphrodites are capable of facultative selfing under conditions of extreme rarity.

INVASIONS AND EXTINCTIONS

A nonequilibrium condition of the BCI forest is also suggested by circumstantial evidence for tree species invasions and extinctions in the 50-ha plot. Probable extinctions of four rare shrub species were observed in 1983 in an El Niño year. In contrast to the increased rainfall in that year in the Galápagos (Chapter 10), BCI experienced the driest dry season in 60 years (D. Windsor, personal communication). These four extinctions happened to species characteristic of, and more common in, wetter forest than that on BCI.

The case for species invasions is based on inferences from contrasting patterns of spatial dispersion of juvenile and adult size classes in certain species (Hubbell and Foster 1983). Many species exhibit spatially well-mixed populations of adults and juveniles, but some do not; and some of the latter species show clear signs of invasion. For example, large adults (>60 cm dbh) of *Beilschmiedia pendula* (Lauraceae) are concentrated in the northwest corner of the plot (Fig. 19.6, top). Small adults (30 to 60 cm dbh) appear in two new patches (Fig. 19.6, middle); whereas 1 to 4 cm dbh saplings are spread much more widely, especially following the steeper slopes (Fig. 19.6, bottom). The pattern of sapling invasion is uncorrelated with either the fine or coarse pattern of gaps, so we conclude that it is not driven by gap successional dynamics. Indeed, *Beilschmiedia* is one of the species that is statistically indifferent to the microsite distinctions of gap, gap edge, or forest interior in the plot.

CONCLUSIONS AND SPECULATIONS

In this paper we have endeavored to present some of the theoretical reasons why chance and history are likely to play a significant role in structuring tropical tree communities. At the present writing the case for taxonomic nonequilibrium of the BCI tree community is circumstantial. Both equilibrium and nonequilibrium hypotheses can be erected to explain most of the current BCI results. In many cases, however, we suggest that the equilibrium hypotheses are less parsimonious as explanations than nonequilibrium hypotheses. It is our general conclusion that in species-rich tropical tree communities the influences of biotic uncertainty and locally unpredictable disturbances may often dominate over the effects of pairwise and predictable biotic interactions. We do not argue against any role for biotic factors in these communities; we maintain only that, whatever stabilizing influences they may exert, their community-level effects are considerably weaker than the dominating effects of immigration, extinction, and habitat insularity. The picture may be very different in bird communities, where high mobility coupled with habitat selection elevates the predictability and importance of biotic interactions.

We conclude with some brave speculations about what is really going on in tropical tree communities. It is our current belief that, within broad climate regions of the neotropics, there is probably little besides distance, various barriers to dispersal, and maybe some minor genetic tuning to prevent a host of additional tree species from residing in the BCI forest. According to this view, biotic interactions between species in tropical tree communities may be intense, but they are not very effective in stabilizing particular taxonomic assemblages of tree species, in causing competitive exclusion, or in preventing invasion of additional species. Biotic interactions are important as selective agents shaping the major life history guilds for tree species. However, these guilds appear to be broad adaptive zones rather than highly specialized niches, created by selection operating under a regime of diffuse competition in a biotically unpredictable world. We see no evidence that these large tree guilds are in any sense saturated with species or noninvasible. Therefore, we think that the principal control of tree species richness in most local tropical forests does not depend heavily on the community ecology and biotic interactions of the current species residents. Instead we suspect that the maintenance of high tree species diversity in many tropical forests has much more to do with regional tree species richness and availability of potential immigrants, which in turn are dictated by the interaction of climate, the historical dynamic biogeography of particular tree taxa, local dispersal, and speciation processes on a regional and subcontinental scale.

SUMMARY

The spatial structure and dynamics of species-rich tropical forests suggest that chance and biological uncertainty may play a major role in shaping the population biology and community ecology of tropical tree communities. This paper considers some potential effects of biological uncertainty on the nature and extent of possible competitive niche differentiation among tropical tree species. It is suggested that a common outcome of spatially and temporally uncertain com-

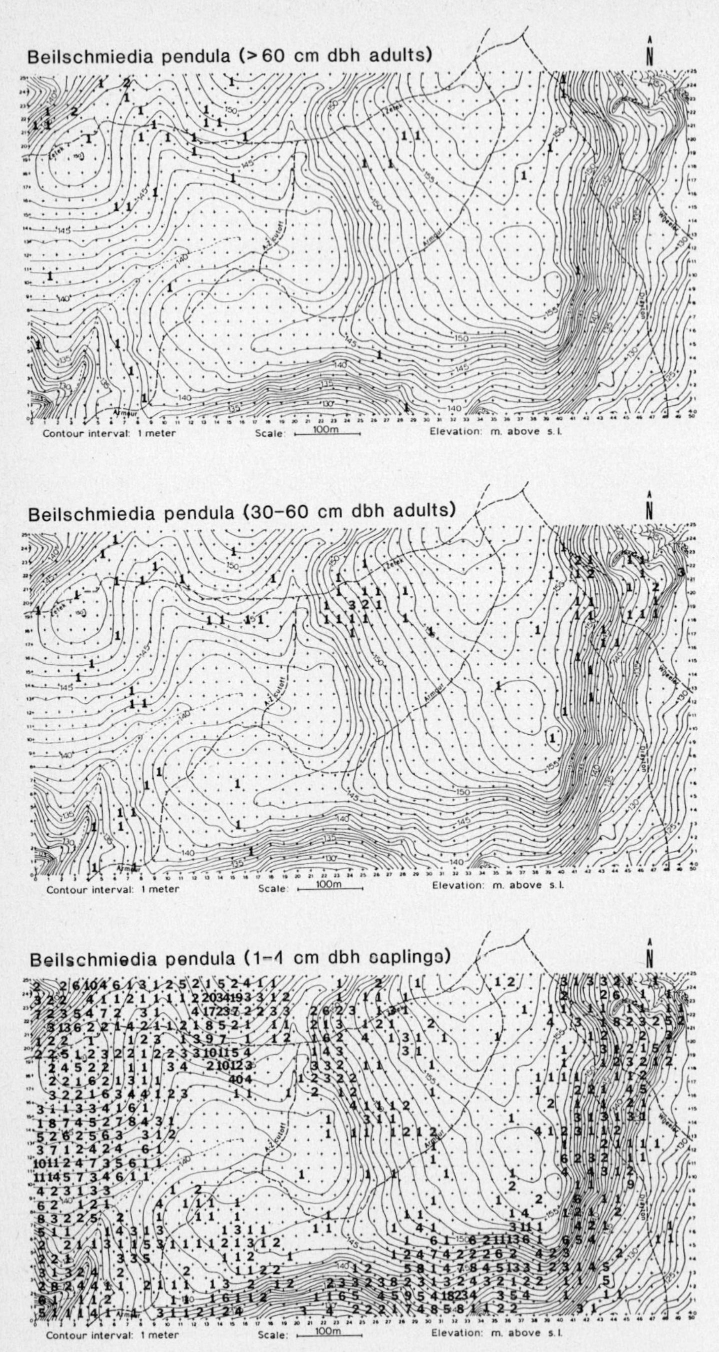

Fig. 19.6 Topographical maps of the 0.5-km^2 BCI plot showing the dispersion of individuals of the tree *Beilschmiedia pendula.* (Top) Large adults (>60 cm dbh). (Middle) Small adults (30–60 cm dbh). (Bottom) Small saplings (1–4 cm dbh). The numbers on the maps represent the number of trees of the given size in the designated 20 × 20 m quadrat.

petitors is likely to be a diffuse coevolution of generalist tree species within a few major life history guilds, rather than the pairwise coevolution of specialists in competitive equipoise. As a result, tropical forests may accumulate tree species in part because forests of generalists may not be particularly resistant to invasion, and in part because there may be few forces besides drift that can lead to the systematic elimination of generalist tree species once established. These suggestions are considered against a backdrop of data on the composition and spatial structure of a 0.5-km^2 plot of tropical forest recently mapped on Barro Colorado Island, Panama.

ACKNOWLEDGMENTS

We thank the National Science Foundation, the World Wildlife Fund, and many private donors for financial support of the tropical forest research on Barro Colorado Island, Panama. We appreciate the continuing interest and support of the Smithsonian Tropical Research Institute in the project. We also thank the more than 50 field assistants who have helped with the fieldwork over the past four years, especially the many excellent University of Panama students who have worked with us. Finally, we thank Ted Case, Peter Chesson, Jared Diamond, Peter Grubb, Henry Howe, and Leslie Johnson for reading this manuscript and making many helpful suggestions for its improvement.

five

FORCES STRUCTURING COMMUNITIES

chapter 20

Overview: The Role of Species Interactions in Community Ecology

Jonathan Roughgarden and Jared Diamond

INTRODUCTION

The objective of community ecology is to explain the variety and abundance of organisms at any place and time. No spot on earth is identical to another, and no species is identical to another. Yet this natural variety does not imply that each community must be considered as one of a kind or that a comparison of communities will not reveal common themes in their composition and in the processes that occur within them.

This chapter begins by considering what communities are, why they are worth studying, and what their ''structure'' (if any exists) might consist of. We suggest ''limited membership'' as a unifying theme underlying various phenomena often proposed as evidence of community structure. Accordingly, we consider the causes of limited membership, of which species interactions are one. We then review the major types of species interactions and the complexities involved in representing a population interaction. Finally, we pose what seem to us to be the leading unsolved questions about the role of species interactions in community ecology.

WHAT IS A COMMUNITY?

Since the turn of the century biologists have grappled with the problem of how to approach levels of organization above the single individual. Indeed, the phrase ''level of organization'' might be construed as presumptuous, for it is not obvious, a priori, that there is *any* organization above the individual. Yet scientists generally accept the single-species population as a natural object (or system) of investigation; it has properties, including its geographical distribution, abundance, and composition in terms of the ages and genotypes of its members, that are important and require explanation.

If it had emerged that populations occurred in rigidly defined groups, whose component species were no more substitutable than are the parts of a Chevrolet automobile engine for a Chrysler engine, then one would have immediately defined such a group as a community. Such a view of the community as a superorganism was formerly entertained seriously (e.g., Clements 1936), but has proved not to correspond to reality. Species associations vary from site to site (Whittaker 1956,

Bray and Curtis 1957, Fager 1957). Furthermore, all three chapters of this volume that discuss Pleistocene communities (Chapters 16, 17, and 18) emphasize that species occurring together today were not rigidly associated in the past.

Given the fact that a natural unambiguous definition of communities does not exist, ecologists use the word community in different ways, and there is a whole spectrum of inclusiveness, as illustrated within this volume. The most inclusive definition of a community is all the organisms in a prescribed area. This definition is usually restricted in at least four ways—spatial, trophic, taxonomic, and life form—to yield less inclusive communities (see especially Chapter 33). A spatial definition might include all the species in a single habitat (e.g., a pond) or in a particular stratum of a habitat (the forest canopy, or hard surfaces in the intertidal zone; Chapter 31). A trophic definition includes all the species at one trophic level (plant community, parasite community, or carrion community; Chapters 7, 22, 26, 27) or a pair of trophic levels (herbivorous insects and their predators; Chapter 11) or the guild of all species using the same category of resources (all desert seed-eaters, whether insects, rodents, or birds; Chapter 3). A taxonomic definition consists of all species of some higher taxon (fish, spiders, *Drosophila*, introduced birds, *Anolis* lizards, small mammals, hummingbird mites, congeneric New Guinea birds; Chapters 2, 5, 6, 18, 21, 24, 30, 33). A life form definition comprises all species of the same life form (desert shrubs, seaweeds, rain forest trees, or grassland matrix-forming perennials; Chapters 12, 19, 23, 32). Mixed definitions are also common: seed-eating Galápagos finches, coral reef fish, shrub-steppe birds (Chapters 9, 10, 30).

Thus, as the chapters of this volume illustrate, the community is a set of species defined in various ways whose relative virtues depend on the question asked.

WHY STUDY COMMUNITIES AND SPECIES INTERACTIONS?

Merely defining communities does not establish that they are worth studying. Is it possible that community ecology could be reduced to the simultaneous, but separate and independent, study of single-species populations, as though by separate investigators who never talk to one another except to coordinate shopping trips for field supplies?

The widespread existence of phenomena such as mimicry, plant secondary compounds, symbiosis, resource partitioning, and animal pollination and fruit dispersal, none of which can even be discussed without reference to species interactions, suffices to show that the independent study of separate populations does not work in general. Every heterotroph preys on some other species, whether plant (herbivory) or animal (classical predation). All species use nutrients and space also used by other species, hence every species is potentially involved in interspecific competition. Almost all species are attacked by parasites and pathogens, and many species are involved in mutualism. Thus, population interactions must always be considered as possibly important even in studies whose primary focus is on a single population.

While the ubiquity of population interactions is obvious, it is not obvious, a priori, whether the interactions have far-reaching consequences at the community level. Do the combined effects of all the interactions in a community impart "structure" to it?

WHAT IS COMMUNITY "STRUCTURE"?

The search for community "structure" has been controversial among ecologists for over 50 years, partly because of disagreement about the specific object of the search: How would we recognize "structure" if we found it?

Many properties of communities have been considered as possible evidence of community structure. These include species abundance relations (May 1975b), correlations between body size and abundance (Schoener and Janzen 1968), food web patterns (Cohen 1978, Yodzis 1980, Pimm 1982, Sugihara 1983, Briand 1983), distributions of species in ecomorphological space (Karr and James 1975), body size differences (Chapters 3, 5, 6, 23), relations between α, β, and γ diversity (Cody 1975), and geographical trends such as latitudinal gradients in life history traits and species diversity (Fischer 1960, Pianka 1966, Simpson 1969).

This list could be extended, but its length and diversity already beg the question of deciding what counts as a structural pattern and whether all patterns are equally significant indications of structure. If *any* pattern counts as evidence of structure, the term begins to lose meaning. For instance, even a set of randomly chosen numbers exhibits a pattern (the predictable frequency distribution of the numbers).

It may ultimately turn out that there is no unifying theme among the community properties suggested as evidence of structure. However, it seems worth trying to find a unifying theme, and the best candidate seems to us to be Elton's (1927) phrase ''limited membership'': Why is it that what *does* occur together constitutes a limited subset of what *might* occur together? More specifically, why does any regional species pool constitute only a subset of all species on earth? Why do the ecomorphologies of species in any actual pool constitute only a subset of those ecomorphologies that seem physiologically possible? Why does the set of species coexisting at any point in space and time constitute only a subset of the regional species pool?

CAUSES OF LIMITED MEMBERSHIP

We think that causes of limited membership fall into three categories: lack of traits that would permit a species to function under the physical conditions of the environment, limitations on dispersal, and interactions among species.

Physical Environment

The first criterion to be met by a potential entrant to a community is that it be able to persist under the physical conditions where the community occurs. These physiological conditions include the seasonal climatic pattern of temperature, rainfall, light, and wind, plus altitude, soil texture, water salinity and depth, and availability of inorganic and organic nutrients. Lack of adaptations to local conditions prevents many species in a regional pool from occurring at some places and times within that region. Many chapters in this volume discuss distributional limits imposed by various aspects of the physical environment: rainfall (Chapters 9–11, 18, 24), temperature (Chapters 2, 11, 16–18), light (Chapters 22, 23), altitude (Chapters 6, 24), soil water (Chapter 23), soil nutrients (Chapter 22), storms (Chapters 11, 33), water depth (Chapter 21), and wave action (Chapters 30, 32).

Since a community's physical conditions must be faced by all its members, the members will inevitably share some adaptive features. These shared responses to the physical environment also provide the basis for the schemes of community classification that rely on correlations between life forms and environmental conditions. These shared responses also provide the basis for the belief that there is convergent evolution of species living under similar environmental conditions, independent of phyletic origin (Cody and Mooney 1978, Orians and Paine 1983).

Some ecologists choose to consider these adaptations to a common physical environment as a limit to community membership and hence as a determinant of community structure. Others prefer to regard these adaptations as determining the pool of eligible species, rather than the membership obtained from that pool.

Dispersal

Difficulties of dispersal often prevent species from reaching areas on earth where the physical environment is one for which they possess the necessary adaptations. This failure to reach potentially suitable environments partly explains the world's distinct biogeographical regions, for example, why species that evolved in the Mediterranean zone of Europe do not also occur in the corresponding climatic zone of southwest Australia and vice versa. Dispersal difficulties also prevent many species from occupying nearby islands with suitable habitats (e.g., the Bahamian spiders discussed in Chapter 33), or from occupying continental areas as soon as environmental conditions permit (e.g., the slow recolonization of deglaciated areas of North America by trees; Chapter 16), or from occupying patches of habitat except as a population flickering in and out of existence (e.g., plants in chalk grassland and tropical rain forest; Chapters 12 and 19). Dispersal difficulties are even crucial in understanding why herbivorous insects often do not reach suitable food plants, and why predatory insects do not reach those herbivores, only a few meters away (Chapter 11). The frequent success of human introductions of species to sites where they were formerly absent (Chapters 4, 5, 21, 24) proves that limited membership in such cases was due to dispersal barriers, not to lack of adaptation to the physical environment.

Dispersal has positive as well as negative effects on limited membership. That is, while inability to disperse to a site may lower membership by preventing arrival of some species adapted to the local physical environment, dispersal may also boost membership at a site above that expected from the properties of the site in isolation. For instance, if a site is surrounded by a variety of habitats, dispersal will cause the site to have more species than would occur if the site lay in a stretch of homogeneous habitat. This feature of dispersal contributes to explaining why local species diversity at continental sites usually exceeds that on islands: Continents tend to have a larger variety of habitats than islands within the dispersal distance of the organism involved.

The weakest possible community structure arises when dispersal and the physical environment are the sole determinants of membership, that is, when a community contains all species whose individuals arrive in sufficient numbers and can exist under the physical conditions of the site. In fact, some scientists might prefer to say that such a community has no structure at all, except for those two limitations on membership.

Species Interactions

Interactions, the subject of the remainder of this chapter, may influence community membership negatively or positively. Competition and predation are the interactions most often suggested as restricting community membership, but there are also examples of limitation by herbivory, disease, and parasitism. Conversely, interactions can boost membership in several ways. First, some species, especially sessile animals and plants, enhance the structural complexity of the habitat and thereby permit the addition of other species. Second, the presence of some species depends on mutualistic, predatory, parasitic, or herbivorous interactions with other species. Finally, both predation and competition can reduce the abundance of some species to levels permitting the entry of other species.

Links Between Causes of Limited Membership

While we have introduced the physical environment, dispersal, and species interactions as distinct factors influencing community membership, in reality they act in concert. Dispersal and competition often interact in that local success is decided between two competing species in favor of the one that arrives first and then excludes the other (Chapters 12, 29, 31, 32). The interplay between dispersal and predation is the key to understanding patchy distributions of herbivorous insects feeding on goldenrod (Chapter 11). We shall also discuss later how the major classes of species interactions (predation, competition, symbiosis, herbivory, etc.) interact with one another.

Moreover, many purported cases of limitations imposed by the physical environment are not at all "pure" examples of autecological limitations, as often assumed, but are instead environmental limits codetermined by species interactions. Examples discussed in this volume include the distribution of competing insect (Chapter 2) and bird (Chapter 16) species along temperature gradients, the distribution of plant species along light gradients (Chapters 22, 23), the distribution of fish species along depth gradients (Chapter 21), the distribution of bird species along altitudinal gradients (Chapter 6), the distribution of plant species along soil water (Chapter 23) or soil nutrient (Chapter 22) gradients, and the interaction of predation and weather to limit herbivorous insects (Chapter 11). Laboratory, field, and natural experiments designed to study these mixed examples show that the environmental limit to species A's distribution in the presence of species B often does not represent the limit of species A's fundamental niche, but instead the temperature (light level, altitude, etc.) at which species A is eliminated by a competitor (predator, pathogen, etc.).

Ecologists often focus on one limiting factor and speak of a population as being resource-limited *or* predation-limited *or* parasite-limited *or* recruitment-limited. These are clearly only extremes. In general, a population is simultaneously limited by all these factors in the sense that it will show a short-term increase upon the addition of food, the provision of additional living space, the removal of predators and parasites, or the arrival of more recruits. While there are circumstances in which one factor is quantitatively more important than others, one should be wary of classifying a species as being limited by only one of the several factors that affect its abundance and distribution.

SPECIES INTERACTIONS

We shall not attempt here to summarize the kinds of species interactions that have been described. A survey of many of these has appeared in Futuyma and Slatkin (1983), and another will appear shortly in Ehrlich and Roughgarden (in press). However, it seems useful to review briefly some of the major interactions, in order to indicate where they are discussed in this book and also to mention what has not received extensive coverage in this book.

Herbivory, Predation, and Parasitism

The most obvious interactions are these three, in which one organism eats another or part of another. Herbivory is discussed in this volume in Chapters 3, 4, 11, and 32; predation in Chapters 4, 11, 21, and 23; and parasitism in Chapters 4 and 27.

It is an unsettled question whether herbivory on certain plant parts (e.g., seeds, pollen, and nectar; Inouye 1978) is more likely to involve resource limitation than herbivory on other parts (e.g., leaves). The experiments of Brown et al. (Chapter 3) show limitation of granivorous desert rodents and ants by seeds. Strong (1984a) argues that leaf-eating hispine beetles are not limited by food but more likely by predation, parasitoids, and weather. The long-term study of the butterfly *Euphydryas editha* by Ehrlich et al. (1975) found pronounced differences among populations: severe competition for food in some populations, but not on Jasper Ridge. This study also showed that the quality of the plant material consumed is critical and that there is great spatial variation in the accessibility of plants to herbivores. The variation in accessibility reflects variation in the temperature and other aspects of the microclimate that allow herbivores to survive on the plant and that allow or prevent predators on those herbivores to succeed. Kareiva (Chapter 11) argues that abundance of uneaten food is not incompatible with resource limitation of herbivorous insects, when one considers their limited food-searching abilities.

The converse question, whether leaf-eating herbivores limit their food plants, yields equally variable results. Some plants are regularly defoliated by insect herbivores, while other plants are consistently only lightly damaged. Long-term consequences of herbivory include the evolution of defense traits, such as the defensive toxins that fill long-lived leaves (Gilbert and Raven 1975).

There has long been a dispute about whether predation influences the abundance of prey species. Errington (1956) argued that predators catch only those prey individuals weakened by disease or age. However, there is now a wealth of field experimental evidence from many terrestrial and marine systems showing that predation can affect prey abundance (Holmes et al. 1980, Menge and Lubchenco 1981, Pacala and Roughgarden 1984). Examples discussed in this volume are the control of herbivorous insect outbreaks by predatory ladybugs (Chapter 11) and the reduction of prey fish population densities by predatory fish (Chapter 21). Two further classes of evidence for prey control by predators are the high abundances reached by hare, reindeer, elk, and rats on predator-free islands (Scheffer 1951, Troyer 1960, Schnell 1968, Klein 1968, Windberg and Keith 1976) and the reductions in prey populations often achieved by introduced predators (Chapter 4).

Toft (Chapter 27) discusses the corresponding question of how parasites affect host population size, and the same question arises for pathogens. The myxomatosis virus clearly affected rabbit abundance in Australia (Fenner and Myers 1978), human populations have often been devastated by plagues, and Chapter 4 gives other examples of host populations reduced by introduced parasites or pathogens. Nonetheless, endemic parasites have not been experimentally removed from natural populations to see if the host's population size increases, although the possibility of a positive result is suggested by the improvements in human survivorship that often result from medical advances against parasites and pathogens. Thus, we do not know what effect a "typical" parasite has on the host population.

Competition

Competition is discussed in Chapters 2, 3, 5–7, 10, 11, 21–23, and 29–32. Some general issues that emerge include alternative mechanisms of competition, the occurrence of competition among taxonomically remote species, competition's evolutionary consequences, its frequency, its relation to multiple stable states of communities, and the possible relation between habitat selection and intraspecific competition for mates.

Chapters 2, 29, and 30 provide extensive discussion of the alternative mechanisms underlying competition. Some basic dichotomies that these chapters set out are those between interference and exploitation competition, food-limited and space-limited systems, and competition for partitionable and nonpartitionable resources. Interference competition involves one individual's driving off a competing individual of the same or another species or, in the ultimate form, killing and possibly eating it. For instance, *Drosophila* larvae drown and eat each other (Chapter 2), colonial marine invertebrates kill cells of neighboring colonies which they contact (Chapter 31), larger fish eat smaller fish of their own and competing species (Chapter 21), and larvaphagy occurs among filter feeders in soft-sediment marine communities (Woodin 1976).

Alternatively, competition may involve the exploitation of resources important to another individual species. Examples of resources in short supply include: food for seed-eating mammals, ants, and birds (Chapters 3, 10), for *Anolis* lizards (Chapter 30), for *Drosophila* (Chapter 2), and for fish that eat plankton or other fish (Chapter 21); space for sessile marine invertebrates (Chapters 30, 31), for terrestrial and marine plants (Chapters 12, 32), and for larval *Drosophila* (Chapter 2); light for terrestrial and marine plants (Chapters 12, 22, 23, 32); water for desert plants (Chapter 23); and soil nutrients for plants generally (Chapter 22).

While most studies of competition have focused on groups of species that are fairly close taxonomically, there is growing evidence for strong competitive effects between members of taxonomically diverse groups. Examples include seed-eating rodents, ants, and birds (Chapter 3); herbivorous large mammals and grasshoppers (Sinclair 1975); carrion-feeding microbes, beetles, and flies (Chapter 26); plankton-feeding fish and shrimp (Chapter 21); plankton-feeding fish and birds (Hurlbert et al. 1985); piscivorous seabirds and humans (Mills 1981); and nectarivorous birds and insects (Carpenter 1979). The mechanism of competition in most of these cases is exploitation competition, but aggressive interference has also been noted, such as between hummingbirds, moths, and bees feeding at the same flowers (Carpenter 1979, Brown et al. 1981). In addition, the rotting of animal carrion and fallen fruit constitutes a case of interference competition, namely, chemical interference by microbes with vertebrates and invertebrates that compete with them for the carrion or fruit.

Evolutionary consequences of competition are discussed in Chapters 6, 7, 22, and 30. Communities in which sympatric species partition resources may be pictured as being built up either by selection of invaders that already differ sufficiently from existing community members or by divergent evolution of two species initially showing only minor differences. In reality, these two formulations represent opposite extremes of a continuum; selection of invaders seems more important for *Anolis* lizards of the eastern Caribbean (Chapter 30), while divergent evolution seems more important for New Guinea montane birds (Chapter 6). Chapters 7 and 22 discuss the role of competition in producing the evolutionary sequence of dominant plant taxa seen in the fossil record.

The frequency and intensity of competition may vary with time and in space. On this variation depend the answers to questions such as whether natural selection is continuous or episodic, whether coevolutionary adaptation of a species to particular competing species is possible, whether population regulation is density-dependent, and how competition compares in importance with other factors. These issues are discussed in Chapters 9, 10, 12, 13, 15, and 32.

Theoretical studies (Chapters 29 and 30; Gilpin and Case 1976) and empirical studies (Chapters 2, 31, and 32; Diamond 1975) suggest that competition interacting with dispersal may lead to multiple stable states, that is, alternative species sets reaching dominance at different places or times under similar environmental conditions. The most detailed exploration in this volume is by Gilpin et al. (Chapter 2), who proved the existence of multiple stable states in laboratory communities of *Drosophila*.

Segregation of closely related species by habitat is often attributed to inter*er*specific competition for resources, each species being restricted to that habitat in which it is superior. In the case of hummingbird mites, however, Colwell (Chapter 24) argues that habitat segregation instead arises largely from intr*a*specific competition for mates. Selection of a particular host plant (the mites' habitat) is seen as a secondary sexual characteristic facilitating not only the search for conspecific mates, but also reproductive isolation from closely related species. Competitive mate searching deserves consideration as an agent that fine tunes habitat use by vertebrates.

Symbiotic Interactions

The interactions that have received the least attention from community ecologists are those involved in symbiosis, where one organism inhabits or otherwise lives in close association with another. Symbiotic interactions include parasitism, commensalism, mutualism, and disease. This volume devotes four chapters to these interactions: the discussion of parasitism by Toft (Chapter 27), of hummingbird mite transport on hummingbirds by Colwell (Chapter 24), and of mutualism by Addicott and Wilson (Chapters 25 and 26). Recent reviews are those by Feinsinger (1983) on pollination, by Vermeij (1983) on symbiosis in marine habitats, and by May and Anderson (1983b) on disease's ecological and evolutionary consequences.

INTERACTIONS AMONG INTERACTIONS

Interspecific interactions do not occur in isolation from each other; they affect each other strongly. The following brief comments introduce a complex subject involving diverse phenomena whose formal representation will be deferred to the next section.

Addicott (Chapter 25) emphasizes that mutualism often arises when one species affects the strength of the interaction between two other species. Mutualism may involve one species protecting another against predators (anemonefish and their anemones, coral crabs and their corals), or against parasites (cleaner wrasses and the fish that they clean), or against herbivores (ants and their myrmecochorous host plants), or against competitors (mites that pierce the eggs of flies competing with the mites' phoretic beetles; Chapter 26).

Interactions between predation and competition are discussed in Chapters 15, 21, 29, 31, and 32. Predation may increase or decrease competition, and competition may increase predation. The best known of these interactions is that predators may reduce densities of competing species below the level where competition would be important, as discussed in this volume for seaweeds, fish, and herbivorous insects (Chapters 15, 21, and 32). Less familiar is the converse effect of enhancement of competition by predation, when risk of predation compels prey species to feed together in the same habitat (Chapter 21) or to compete for a limited number of refuges (Chapter 31). A major consequence of interspecific competition among fish is to enhance the risk of predation by reducing growth rates (smaller fish are more at risk from predators than are larger fish; Chapter 21).

Predation and disease interact in that many studies have shown sick animals to be more likely to succumb to predators (Errington 1956).

Predation and parasitism interact in the same way as do predation and disease: Parasitized animals are especially subject to predation. In addition, the mechanism by which a parasite with a complex life cycle gets from one host to the next host often involves predation: host no. $X + 1$ eats host no. X (e.g., humans, pigs, and the roundworm causing trichinosis; Chapter 27).

Parasitism or disease may tip the outcome of a competitive encounter by weakening a participant that would win in the absence of the parasite or pathogen or that would survive the parasite or pathogen in the absence of the competitor.

The interaction between herbivory and competition is illustrated by Darwin's (1859) classic experiment, showing that grazing animals increase plant species diversity by reducing a dominant competitor. As Grubb and Lubchenco (Chapters 12 and 32) discuss, this interaction is important in maintaining the species diversity of chalk grassland plant communities and seaweed communities. One mechanism of competition among herbivorous insects involves the changes in plant quality caused by herbivory (Chapter 11).

Predation affects herbivory by forcing herbivores to change their diets in order to reduce the risk of predation.

Thus, of the 15 pairwise interactions between the 6 major classes of interspecific interactions ($6 \times 5/2 = 15$), we are familiar with important examples for 11. Undoubtedly, readers will be able to find examples of the remaining four pairs (mutualism, herbivory, or parasitism interacting with disease; herbivory interacting with parasitism). It should be clear that the separate consideration of any one type of interspecific interaction oversimplifies reality.

THE REPRESENTATION OF INTERACTIONS

The simplest representation of an interaction between two species is that the strength of the interaction, that is, its impact on population dynamics, is directly proportional to the product of the abundances of the participating species. For example, if the left-hand side of an equation for the population dynamics of species i is dN_i/dt, then the interaction of this species with species j would be represented on the right-hand side as the term aN_iN_j, where the a is a constant of proportionality specific to the particular type of interaction being modeled in the environment where the interaction is happening. This simple approach may be unsatisfactory for two reasons: higher-order interactions and distributed interactions.

Higher-order Interactions

If the interactions must be represented by a more complicated expression than aN_iN_j, the interaction is said to be a "higher-order interaction." One example would be a term like $aN_i^cN_j^d$, where c and d are not equal to 1 and would typically exceed 1. (If c and d are both equal to 1, as is often assumed in models, then the interaction is said to be a second-order interaction.) Similarly, a three-way interaction, like $aN_iN_jN_k$, is a higher-order interaction. Examples of population dynamic models with higher-order interactions include those of Ayala et al. (1973) and Schoener (1973).

There has long been interest in whether interactions are only "simple" second-order interactions or whether they must be represented as higher-order interactions. Wilbur (1972) detected statistically that higher-order interactions were present among larvae of ambystomid salamanders in Michigan ponds, but later studies like that of Morin (1983a) have failed to confirm this result. Pomerantz (1981) and Case and Bender (1981) point out serious errors that have been made in such studies purporting to detect higher-order interactions. A particularly important point is that much of the apparent nonlinearity is removed by using a higher-order model for the *intraspecific* density-dependence, leaving the interspecific interactions represented by a residual that is satisfactorily described as a second-order interaction.

The best evidence for higher-order interactions comes from a known mechanistic basis. Clear examples are the frequent cases in which mutualism arises as a result of one species affecting the strength of predation, parasitism, competition, or herbivory between two other species (Chapter 25). Further examples are that the intensity of competition between different species for refuges increases with predation (Chapter 31), that disease increases an individual's risk of predation (Errington 1956), and that competition between herbivorous insects is influenced by herbivory-induced changes in plant quality (Chapter 11).

Of the ''interactions among interactions'' discussed in the preceding section, not all need to be represented as higher-order effects. In particular, many of those cases constitute what may be called the ''coaction'' of two or more second-order interactions, mediated by a change in abundance. To understand this distinction, consider again the simple representation of an interaction between species i and j as aN_iN_j. If the coefficient a itself depends on the abundance of a third species N_k, then this constitutes a higher-order interaction, in which the third species qualitatively alters the interaction. But if the presence of a third species only affects N_i or N_j or both, that constitutes a coaction; the third species affects the overall impact of the interaction on a species simply by affecting the frequency with which the interaction occurs. For example, Darwin's (1859) classic experiment showed that herbivores increase plant diversity by reducing a dominant competitor; Chapters 12 and 32 and Morin (1983a) offer further examples. This phenomenon is usually just the net outcome of the combined population dynamics of competition and predation. The herbivore does not change the nature of the competitive interaction between the harvested species; it simply lowers the density of consumed individuals and thereby reduces the number of competitive encounters.

Distributed Interactions, Including Ontogenetic Niche Shifts

An interaction between populations is the summation of many interactions between individuals. It is the population effect of those individual contacts, which have varied outcomes depending on the age and state of the participants and on where the contact occurs (Chapter 31). How do we determine what population interaction corresponds to the interaction that one may observe between two individuals? At present the best information on population dynamic parameters comes from population-level experiments, which lump large numbers of individuals in a large area. For improved theoretical models, however, the population's structure will have to be subdivided into parts within which the interactions can be treated as, on the average, homogeneous. We use the expression ''distributed interaction'' to refer to such models of population interactions based on subdivided population structure.

One of the clearest examples of distributed interactions involves ontogenetic niche shifts. Here the variation in interactions is a function of life history phases. The importance of this subject has been overlooked by students of higher vertebrates, because juveniles of most bird and mammal species do not begin to forage independently of their parents until they are already close to adults in body size, diet, and foraging behavior. However, as Werner (Chapter 21) emphasizes, birds and mammals are exceptional in this respect. Most invertebrates, lower vertebrates, and plants undergo major changes in size and niche with age. Body size often varies with age over many orders of magnitude. The niche differences between larval and adult invertebrates or lower vertebrates of the same species usually dwarf the niche differences among adults of related species of these taxa. The metamorphosis of many amphibia from aquatic juveniles to terrestrial adults is one example. Thus, representations of interspecific interactions may have to specify the age or size class of each species involved.

INDIRECT CONSEQUENCES OF SPECIES INTERACTIONS

If a species is involved in more than one direct pairwise interaction, then the chains of pairwise interactions connecting species produce what have come to be known as ''indirect effects.'' In this volume indirect effects receive particular attention in Chapters 3, 4, 25, 26, and 32. The simplest example is a variation of the adage, ''The enemy of my enemy is my friend.'' If A competes strongly with B, and B competes strongly with C, then an increase in C tends to lead to an increase in A because an increase in C produces a decrease in B, which leads to an increase in A.

Indirect effects are receiving increasing attention for theoretical and practical reasons. First, the great increase in field perturbation experiments is leading to the frequent detection of indirect effects (Chapters 3, 21). Second, indirect effects are a potentially huge source of bias leading to an overestimate of the frequency and strength of direct interactions in a community. In the example above, an experiment that lowers the abundance of species C for a long time will result in an increase in species A. A common error is to conclude that C has a direct positive effect on A with a time lag, reflecting perhaps differences in the generation time between the species. An experimenter insensitive to the possibility of indirect effects would fill the pairwise interaction matrix for a community with numbers supposedly representing the direct effect of each species on every other species. In fact, in the example there are only two direct interactions (A with B and B with C), not three.

In a recent important paper Bender et al. (1984) discuss methods to separate direct effects from the confounding indirect effects. They distinguish press experiments, where the abundance of a species is maintained at a different level (including species removal) throughout the experiment, from pulse experiments, where the abundance of one species is initially perturbed but not controlled thereafter. Bender et al. show that pulse experiments reveal direct effects, but that press experiments, which are by far the most common field experiments, confound direct with indirect effects.

In evolutionary time it is theoretically possible for indirect effects to lead a species to evolve itself to extinction, and to interfere with the evolution of coadaptation between interacting species (Levins 1974, Roughgarden 1977). Moreover, Wilson (Chapter 26) shows for a system of carrion-feeding species that selection at the community level can lead to coordinated functions among component species of the community.

WHEN CAN SPECIES INTERACTIONS BE NEGLECTED IN COMMUNITY ECOLOGY?

Although we have focused on the role of interactions in community ecology, not all of the interactions need be taken into account all of the time. The theoretical issue involved is called "information hiding" in computer science (Wirth 1984). When focusing on a single species or on one or two strongly interacting species, the question is whether one can safely merge the interactions of those species with all the other components of the community into the general "environment," that is, into the general circumstance surrounding the system of interest. Success in autecological studies involves a judicious choice of what species interactions to omit. When can all the prey species for a consumer be lumped together as "food"? When can the space that the consumer lives in be lumped as "habitat"? When can the predation on a species be considered as density-independent mortality and merged into the survivorship curve of the species? Similarly, the success of a synecological study involves the judicious omission of lower-level considerations, especially those involving mating behavior and genetics.

The search for working methods of information hiding is a trial-and-error process. In computer science a large program is constructed from modules, each written by a team of programmers. Each team knows little about the others' work. In ecology we, in effect, inherit a large program of nature, and our challenge is to determine its natural modules.

UNSOLVED QUESTIONS

Here we venture to suggest what seem to us to be four leading unsolved questions that species interactions pose for community ecology today.

1. How should we represent a community's state, or, to use the terminology of computer science (Wirth 1976, Winston and Horn 1984), what data structure can we use to represent a community? To go beyond a concept of community structure based only on the concept of limited membership, we must describe for a community something like "what is going on where."

Two important considerations are that the spatial position and functional role of each species must both be represented; and that species must be grouped into hierarchies of interacting species with regard to both spatial position and functional roles. If we begin with the most inclusive definition of a community—all the organisms within a prescribed area—then a community may be compared to a township in human affairs. A township has neighborhoods, trades, and industries. The neighborhoods are defined by physical proximity, which leads to certain interactions. Members of each trade are mobile, share a distinctive activity, and may compete because of similarity in activity. Industries are composed of interdependent trades. Neighborhoods are somewhat analogous to habitats or local areas, trades to ecological guilds.

The analysis of graphs based on food webs, as pioneered by Cohen (1978), has offered a fine start. It is ultimately based on a two-state classification of interaction strength, simply as absent (0) or present (1). This simplification is of debated value (Pimm 1982). However, we feel that more important omissions in the food web data structure are of spatial relationships and hierarchy.

2. How can we increase our understanding of the number and strength of connections among a community's species? For example, communities have some sections where membership is limited by species interactions and other sections where membership is limited only by dispersal and by the size of the pool of appropriate species. Do the sections of a community interact? Hubbell and Foster (Chapter 19) and Grubb (Chapter 12) argue that plant communities consist of species sets such that resource partitioning exists between sets, while the species within each set are virtually interchangeable and have random dynamics. Are communities composed instead of species most of which have strong direct interactions with each other? May's (1973a) models of interacting populations suggest that this should be unlikely. Are communities composed of clusters, each consisting of a few species interacting strongly with each other, but weakly with all other species? Such a community would exhibit both few direct and few indirect effects among species. Are there a few threads of strong interaction running throughout the community and ultimately coupling each member to the other, such that there are direct interactions among few species pairs, but strong indirect interactions among many? This latter picture may correspond to the concept of the desert granivore community arrived at by Brown et al. (Chapter 3).

3. How can we accumulate and synthesize information on the relative contributions of the physical environment, dispersal, and the major species interactions in determining community composition? The preliminary syntheses of Hairston et al. (1960) and Menge and Sutherland (1976) need to be evaluated and extended, as in Chapters 28, 32, and 33, where the relative importance of more than one type of species interaction is assessed.

4. How can we integrate community ecology with studies at the next higher and lower levels: the ecosystem and the individual, respectively? Combination of the community and ecosystem approaches is needed to understand how species interactions affect ecosystem attributes like nutrient flow and productivity. By combining community and individual approaches, ecologists can develop submodels of population interactions in terms of the interactions between individuals, and can assess what consequences the observed behavior of individuals has at the population level.

chapter 21

Species Interactions in Freshwater Fish Communities

Earl E. Werner

INTRODUCTION

The unique characteristics of different taxa constrain the ways in which species communities can be organized. For example, the relative importance of different types of interspecific interactions, barriers to dispersal, and limiting environmental factors vary among taxa (Chapters 20, 28). Community ecology has traditionally emphasized the study of birds and mammals (and, more recently, lizards), and this has naturally resulted in a certain myopia. Among the respects in which birds and mammals prove exceptional is that juveniles of most bird and mammal species resemble adults ecologically and morphologically and are nearly as large as adults by the time that they are old enough to forage for themselves. In contrast, conspecific individuals of most other taxa vary greatly in body size and often undergo drastic changes in morphology and ecology as they grow. This fact has major consequences for community structure and species interactions.

Considerations of such ontogenetic size changes loom large in understanding freshwater fish communities, the subject of this chapter. My presentation has two sets of goals. First, I summarize the varied types of evidence for the role of species interactions in structuring freshwater fish communities. This evidence includes patterns of resource partitioning, indications that food resources are limiting, natural experiments documenting niche shifts between sympatry and allopatry, short-term and long-term field experiments, and effects of species introductions. I then examine in more detail how competition and predation interact in fish through habitat selection, growth rates, and ontogenetic niche shifts. The ontogenetic shifts greatly complicate study of species interactions and often cause species to influence each other in opposite ways at different life history stages. I offer a framework for predicting the occurrence of these shifts. Finally, I suggest how some of the lessons learned from study of species interactions in fish are relevant to a wide range of taxa and must be incorporated into a theory of species interactions.

PATTERNS IN RESOURCE PARTITIONING

A traditional means of examining community structure has been to quantify patterns in resource partitioning among species. Such patterns have often been used to infer that competition is, or has been, an important force in communities, though there are also other reasons why species might differ in resource use. In his review of the literature up to 1973, Schoener (1974a) listed only four papers on resource partitioning in fish, but such studies have recently increased greatly.

Ross (1985) surveyed over 230 studies through 1983 where three or more species of fish were examined. He concluded that fish exhibit substantial resource partitioning along the traditional niche axes of food, habitat, and time. In contrast with Schoener's finding for most other animals, trophic separation in fish appeared more important than habitat separation. However, this conclusion must be treated with caution, as fish have typically been collected rather than observed, with the result that niche differentiation by food use (deduced from stomach contents) rather than by habitat use has been stressed. Furthermore, the investigator can generally subdivide resources operationally much more finely in the case of food than of habitat, thereby giving the impression that separation is greater on the former dimension.

Ross's survey showed that temporal separation was much less important than trophic or habitat separation, but it still appeared more important for fish than for other groups. Seasonal separation (especially in breeding times) and diel separation were strongly associated with family or ordinal lines, suggesting phylogenetic constraints.

These studies may suggest competition to be important in structuring fish communities, but the evidence is rarely compelling. Other reasons for species differences in resource use are not excluded, and resource use often varies greatly with season and with stage of life cycle. Thus, traditional analyses of patterns of resource use provide at best only a superficial understanding of the potential interactions in fish communities. Hence I now consider more direct evidence that resource limitation is important in fish communities.

ARE RESOURCES EVER LIMITING IN FISH COMMUNITIES?

Evidence from Individual Growth Rates

If the just mentioned patterns in resource partitioning among fish represent effects of exploitative competition, one should be able to demonstrate that resource limitation is important in these communities. As has been the tradition in studies of resource partitioning for other taxa, studies with fish rarely test whether the demand for resources exceeds the supply. Fortunately, for fish a large but diffuse (and somewhat independent) literature indicates that resource limitation often occurs. The major reasons why such data exist are that individual growth in fish is extremely flexible and thus a sensitive indicator of resource levels, and that fish conveniently carry a record of their growth history in scales and otoliths. Furthermore, growth has been a central concern of the vast literature on applied problems of fish management. This literature shows that growth rates often vary with the density of both conspecific and heterospecific individuals.

In freshwater ponds and lakes extensive experimental studies clearly illustrate the sensitivity of individual growth rates to population density (e.g., Hall et al. 1970, Hepher 1978, Backiel and Le Cren 1978). Transplant experiments further demonstrate an immediate increase in growth rates and an increase in ultimate size when fish are transferred to a richer environment (Alm 1946, Martin 1966, Burnet 1970). Studies in which populations are thinned out have often shown intra- and interspecific density-dependence of growth rates (Beckman 1943, Alm 1946, Johnson 1977, Healey 1980, Persson 1985) or growth and recruitment rates directly governed by abundance of major prey species (e.g., Forney 1977). Weatherley (1972) has reviewed many studies showing the effects of resource limitation on fish growth, and I review several such studies in subsequent sections.

In streams convincing evidence also exists for intraspecific density-dependence of growth and recruitment rates, especially among juvenile salmonids (Johnson 1965, Le Cren 1973, McFadden 1969, Gee et al. 1978, Elliott 1984). Clear dome-shaped stock recruitment curves have been demonstrated in several cases (e.g., Gee et al. 1978, Elliott 1984), with intraspecific territoriality being strongly implicated as the mechanism responsible, especially during the important period when feeding territories are first being set up by juveniles.

Effects of Fish on Resources

Another type of evidence for food limitation in lake fish communities is the demonstrable effect of fish on the abundance and composition of their food. There is now a large literature showing how size-selective predation by fish can dramatically lower the mean size and alter the species composition of zooplankton in communities ranging in size from small ponds and lakes (Hall et al. 1970, Nilsson and Pejler 1973) to Lake Michigan (Wells 1969). The increase in food intake rates associated with selection of larger prey (Werner et al. 1983b, Mittelbach 1983) and the positive relation observed between food particle size and fish growth rates (Paloheimo and Dickie 1966, Wankowski and Thorpe 1979) both argue compellingly that the decrease in plankton size resulting from fish predation tends to increase food limitation for the fish (Kerr and Martin 1970, Mittelbach 1983). In more complex habitats, such as sediments and vegetation, evidence as to whether fish significantly affect their prey communities is inconsistent (Ball and Hayne 1952, Thorpe and Bergey 1981, Allan 1982, Crowder and Cooper 1982).

The above evidence indicates that resource limitation and intraspecific density-dependence often, though surely not universally, occur in fish communities. However, such data rarely indicate to what extent this resource limitation can be affected by other species. Hence the following section examines evidence for the importance of interspecific competition. I first review comparisons of species in sympatry and allopatry, then survey short- and long-term experimental work, and finally examine evidence provided by the introduction of exotic species.

EVIDENCE FOR SPECIES INTERACTIONS

Niche Shifts in Sympatry and Allopatry

Comparisons of species' niches in sympatry and allopatry, especially among islands or between islands and the mainland, have provided evidence for species interactions in many taxa (natural snapshot experiments in the terminology of Chapter 1). Lakes obviously possess many of the properties of islands, and such comparisons of lake fish communities have also suggested strong interactions between species. Clearly, the more species-rich the communities are, the more difficult it is to ascribe shifts in resource use by one species to the presence or absence of another species. Fortunately, both Scandinavia and North America have regions with many depauperate lakes that provide excellent opportunities to examine simple fish communities with differing species compositions.

A classic example involves the brown trout and arctic char in high-altitude Scandinavian lakes (Svardson 1976).[1] Nilsson (1963, 1965) compared allopatric and sympatric populations of these species in many lakes. In allopatry both species have similar size-frequency distributions and food habits. In sympatry, however, important niche shifts occur; char feed much more extensively on zooplankton, while trout feed on caddisfly larvae and terrestrial insects. These tendencies are stronger later in the season when food levels are lowest (Nilsson 1960), resulting in considerable habitat segregation: trout in the shallow littoral regions, char in the deeper pelagic regions. When both species coexist in a lake, the char are more abundant than the trout, and total fish biomass is greater than in lakes with trout alone. A long history of char introductions into trout lakes has verified over and over again the detrimental effect of char on the abundance and individual growth rates of trout. References to this effect can be found as early as the mid-nineteenth century, and cases have been reported where trout went extinct locally (Svardson 1976).

Remarkably parallel interactions occur among the salmonids of small western North American lakes. For example, cutthroat trout and Dolly Varden char coexist in many coastal lakes of British Columbia, but can also be found allopatri-

[1]Throughout the text I use common names of species. Scientific names are listed in the appendix at the end of the chapter.

cally. Each species utilizes fewer zones of the lake in sympatry than in allopatry (Andrusak and Northcote 1971). In sympatry each species exhibits a narrower diet than in allopatry, and differences are again maximal in late summer, when resource levels are lowest. In sympatry Dolly Varden concentrate on bottom fauna, cutthroat on surface foods, and in the laboratory Schultz and Northcote (1972) demonstrated that Dolly Varden are indeed more efficient at using bottom organisms, while cutthroat capture surface prey more rapidly. Nilsson and Northcote (1981) also examine cutthroat trout in allopatry (17 populations) and sympatry (10 populations) with rainbow trout and described similar sorts of niche shifts. Body size also shifted dramatically; rainbows are larger than cutthroats in allopatry, but cutthroats are consistently larger than rainbows in the same lake. Examples of such niche shifts in sympatry and allopatry for other environments can be found in Fraser (1978), Hixon (1980), Magnan and FitzGerald (1982), and Finger (1982).

A pattern noted in both the Scandinavian and North American lakes was that more littoral or benthic species were depressed in the presence of more pelagic species. The studies of salmonids in Sweden suggested a hierarchy of competitive dominance that was related to a species' ability to utilize open-water plankton; pelagic planktivores are dominant over littoral species (Svardson 1976). As early as 1910 Ekman (1910) pointed out that char populations were greatly decreased or exterminated in lakes following the introduction of whitefish (*Coregonus* spp.); which are more effective planktivores. Filipsson and Svardson (1976) extended this finding to 30 Swedish lakes into which whitefish are known to have been introduced, causing a great decrease in char populations (see also Nilsson 1963). Depression of more littoral species by more pelagic species has now been noted in Europe with arctic char and brown trout (Nilsson 1963), normal and dwarf char (Nilsson and Filipsson 1971), cisco and smelt (Svardson 1976), and whitefish and perch (Svardson 1976) and in North America with cutthroat trout and Dolly Varden char (Andrusak and Northcote 1970, 1971) and with cutthroat and rainbow trout (Nilsson and Northcote 1981).

The reason for this competitive dominance of pelagic over littoral species is that pelagic species are more efficient planktivores and presumably cause greater reductions in abundance or size of zooplankton. Nilsson and Pejler (1973) showed that the zooplankton community changes to smaller and smaller forms as more efficient planktivores are added to the fish community. This reduction in zooplankton size and abundance suppresses use of open water by more littoral species sympatric with more pelagic species, and this may be especially critical when the littoral species are very small.

If competition is important in these communities, one expects to observe density compensation. That is, species in species-poor lakes should compensate in numbers or biomass for missing species confined to more species-rich lakes of equivalent productivity. Tonn (1985) described a striking case of compensation, and the possible interactions that can cause it, in five similar northern Wisconsin lakes containing from one to three species in a nested pattern (mudminnows alone, with yellow perch, or with yellow perch and golden shiner). In the presence of yellow perch (with or without golden shiner, which contributes little to total biomass) the mudminnow accounts for only 10% of the total biomass. When mudminnows are found alone, however, their biomass equals the whole biomass of the two-species or three-species communities, meaning that density compensation is complete.

The observations of niche shifts have often been only qualitative and do not constitute controlled experiments. Such natural experiments are always frought with the possibility of uncontrolled differences that could contribute to the patterns seen (see Chapter 1). What makes the foregoing examples reasonably clear is that the systems are simple, there are many replicates of allopatric and sympatric populations, and there have been numerous species introduction "experiments," with consistent consequences. In several cases the differences in foraging efficiency or behavior of the species have been verified in the laboratory (Nilsson 1963, Schultz and Northcote 1972). Taken as a whole, these studies provide compelling evidence of competitive interactions that can markedly affect species population sizes, individual growth rates, and patterns of resource use.

We now turn to evidence from controlled field experiments.

Experimental Studies of Niche Shifts

Experimental studies of fish communities confirm that competitive interactions can be responsible for the niche shifts observed in the sympatric-allopatric comparisons. Hixon (1980) noted that two surfperch species found on inshore marine reefs segregated by depth in sympatry, but occupied all reef habitats in allopatry. With this natural experiment as background, he then performed a controlled reciprocal removal experiment. The two species were similar morphologically and overlapped extensively in diet. Removal of the shallow-water species was followed by a significant movement of the deeper-water species into shallow water. In the reverse experiment the shallow-water species did not move deeper, and Hixon was able to show that the shallow-water zone has higher food densities and was preferred by both species. Hixon (1980) concluded that interference competition was the mechanism responsible for the shift in sympatry.

Larson (1980) performed a similar experiment in the same environment with two rockfish species that also segregated by depth. Reciprocal removals again resulted in a marked movement of the deeper-water species into shallow water and a slight movement of the shallow-water species to deeper water. Interspecific territoriality was inferred to be the mechanism underlying this competitive interaction.

Fausch and White (1981) have also provided experimental evidence that interference competition is important in maintaining niche differences between stream salmonid species. Brown trout have been introduced into North America from Europe and have often been reported to affect native North American brook trout populations. When brown trout were removed from a section of stream, brook trout shifted to using more advantageous feeding and resting positions formerly occupied by the brown trout.

The field experiments just described were on systems that yielded evidence for niche shifts due to interference competition. My own studies with sunfishes (*Lepomis* spp.) provide a contrasting case where exploitation competition appears responsible for niche shifts. When we used SCUBA gear to quantify abundances of sunfishes, we observed distinct habitat partitioning among species (Werner et al. 1977, Hall and Werner 1977). Food studies indicated that this habitat separation contributes to striking overall resource partitioning, at least among the larger size classes of fish (Keast 1978, Mittelbach 1984). In addition, growth rates in the field often are far below those permitted by a species' physiological capacity (Werner and Hall unpublished; see also Beckman 1948, Carlander 1977, Wiener and Hanneman 1982). These facts suggest the occurrence of competition in the field.

In order to investigate the mechanisms underlying resource partitioning among sunfishes, we introduced species into experimental ponds 30 m in diameter, where factors such as fish density, individual size, presence or absence of congeners, and habitat structure could be manipulated. The ponds contained natural stands of emergent and submersed vegetation with prey populations similar to those of natural littoral zones. For example, a large series of experiments with three congeners of the genus *Lepomis* (bluegill, pumpkinseed, and green sunfish) demonstrated competitive effects of these species on each other (Werner and Hall 1976, 1977, 1979). Each species, when stocked in ponds alone, preferred the vegetation zone where larger prey were found and hence where higher foraging rates were possible. In the presence of congeners the green sunfish remained in the vegetation, but the bluegill and pumpkinseed exhibited marked niche shifts to the open water and sediment habitats, respectively (Werner and Hall 1976), and growth rates of all species were reduced. Habitat use in the presence of congeners in these experiments closely paralleled that actually seen in the field (Werner et al. 1977). This attempt to reconstitute a community by adding competing species in various combinations is similar in spirit to the laboratory reconstitution of *Drosophila* communities by Gilpin et al. (Chapter 2).

Several subsequent experiments were performed to test the inference that these niche shifts resulted from species differences in ability to harvest resources from these habitats. By ranking species ordinally according to their relative foraging efficiencies in the different habitats (Werner and Hall 1977, 1979), we correctly predicted the temporal order of habitat shifts in a pond as exploitation by the fish caused resources to decline in the preferred habitat (Werner and Hall 1979). There is evidence for similar seasonal shifts of habitat in natural lakes, as peak spring resource levels decline through the summer (Nilsson 1960, Seaburg and Moyle 1964, Laughlin and Werner 1980). When we built foraging models to predict these habitat shifts and tested model predictions in a natural lake (Mittelbach 1981) as well as in the experimental ponds (Werner et al. 1983b), there was good correspondence in both cases between the predicted and actual habitat use of the fish. Thus, the fish appear to utilize habitats so as to maximize energy return, and the habitat shifts in the presence of congeners presumably reflect the differential depletion of resources in particular habitats by the congeners.

Long-Term Studies

The experimental studies presented above have been relatively short-term studies, where the effects of competition were expressed in niche shifts and, in some cases, in growth rates as well. Presumably, these short-term responses lead to longer-term effects on population dynamics that influence species coexistence and relative abundances. Such effects can be sought in studies involving longer-term manipulations of selected species, though most such studies have lacked controls.

For instance, Johnson (1977) removed 85% of the estimated standing crop of a dominant species, the white sucker, from a Minnesota lake and followed the changes in fish community structure for seven years. Yellow perch densities increased by an order of magnitude over the preremoval levels, and growth rates of individual perch increased markedly as well. Walleyes responded more slowly, but their growth rates and population size also increased. Juvenile suckers increased eightfold. Since suckers overlap in diet with at least some of the life history stages of yellow perch and walleye, it seems clear that reduction of the sucker population allowed considerable competitive release.

Persson (1985) experimentally reduced the roach population of a Swedish lake to 27% of its original numbers. The lake contained eight fish species, of which roach and perch were the most abundant. Following roach removal, individual growth rates increased in both the roach and perch populations, and young-of-the-year perch increased nearly fourfold. The zooplankton community eventually shifted to larger forms. As in all such studies, there was no control for year-to-year differences, but Persson was able to discount the effects of the most probable confounding factors (e.g., year-to-year temperature differences). Similar effects of experimental reductions of populations of particular species were reported by Ricker and Gottschalk (1941) and by Rose and Moen (1952).

The converse of species removals is species introduction. Introductions of exotic species, though again not a carefully controlled experiment, have often provided evidence for the role of competition (see Chapter 4). In some cases introductions have evidently led to species replacements. For instance, the brook silverside was the only silverside found in Lake Texoma, Oklahoma, from 1944 to 1953. Inland silversides were discovered in the lake in 1953 and by 1955 had completely replaced brook silversides (McComas and Drenner 1982). The replacement was able to proceed rapidly because both species have annual life cycles. Brook silversides are still found in the tributaries to the lake, but are excluded from the lake itself. Laboratory studies of feeding mechanics and efficiencies of the two species showed that the inland silverside is much more effective on evasive prey such as copepods and is superior in competition experiments in small pools (McComas and Drenner 1982).

Fraser (1978) studied three salmonid populations for six years prior to and following the introduction of yellow perch into a small Ontario lake. After introduction of the yellow perch, growth rates of all salmonid species declined by more than 50%, and each salmonid species exhibited drastic shifts in food habits.

Of all natural experiments with fish, perhaps those on the grandest scale are provided by the fish communities of the Great Lakes after invasion of the rainbow smelt and then construction of the Welland Canal permitted invasion of the alewife and sea lamprey (S. H. Smith 1968, Stewart et al. 1981). The consequences included decimation of the primary native piscivore by lampreys and overfishing, permitting a rise in alewife and rainbow smelt abundance, resulting in turn in extreme reductions or local extinctions of at least 11 native planktivore species (due to competition or else predation on eggs and larvae). All this was accompanied by profound changes in size structure of the zooplankton community.

These various short-term and long-term experimental studies indicate that competition can often be an important force in structuring fish communities. I now turn to the evidence that predation can also be important, a conclusion for which the just cited chain of events in the upper Great Lakes already provides some support.

Effects of Predators

The introduction of top carnivores illustrates the effects of predation on community structure in fish. Zaret and Paine (1973) documented effects of the introduction of the peacock bass, a piscivore native to the Amazon River, into Gatun Lake in the Panama Canal Zone. The actual progress of the species through the lake was followed, and thus areas that the predator had not yet reached provided controls assessing community change in areas reached by the predator. Pre- and postinvasion censuses around Barro Colorado Island indicated the possible elimination of six out of eight common native fishes and drastic reduction of the seventh. These effects, of course, do not represent the final new equilibrium, but rather the short-term devastation resulting from the introduction. Effects of the peacock bass were also documented on other trophic levels, such as zooplankton and other vertebrates. Such effects are well enough documented in lake systems that research is now in progress to explore the potential of such ''biomanipulation'' to relieve eutrophic algal blooms in lakes (e.g., Stenson et al. 1978, Shapiro and Wright 1984).

A long-term study of the yellow perch–walleye interaction in Oneida Lake, New York, also illustrates important effects of a predator on a fish population. Yellow perch account for 70 to 98% of the stomach contents of the walleye (Forney 1977). Good evidence exists from the period 1954 to 1973 that growth of walleye was directly correlated (Forney 1977) and mortality inversely correlated (Forney 1976) with the abundance of young yellow perch. The perch did not appear to be food limited (Noble 1975), but exhibited large fluctuations in year class strength (the number of individuals in each annual cohort). Mortality increased the variance in year class strength, because walleyes consumed a higher proportion of small than large year classes (see also Nielsen 1980). Other studies providing evidence that predators limit fish population densities and in some cases affect prey relative abundance are those by Hoover (1936), Foerster and Ricker (1941), Carlander (1975), Maclean and Magnuson (1977), and Hall and Ehlinger (1985).

I have focused on the evidence for species interactions among fish, but I do not wish to imply that they are the sole factors structuring fish communities. Other factors include effects of history, such as glacial cycles, and effects of weather, such as floods in the breeding season, droughts, and oxygen depletion in winter (e.g., Horowitz 1978, Moyle and Li 1979, Tonn and Magnuson 1982). I now turn to the important role of growth rates and body size in these species interactions.

HOW DO COMPETITION AND PREDATION INTERACT IN FISH COMMUNITIES?

The above review indicates that both competition and predation can have important effects on fish community structure, at least in certain systems. Further, some of these effects are manifest in familiar ways, such as niche shifts between sympatry and allopatry.

When we consider species interactions in fish more closely, however, it becomes clear that results at the levels presented so far can offer only a superficial understanding. In large part the problem stems from the fact that competition and predation interact in complex ways in fish communities and thereby produce novel community dynamics. This interaction arises mainly in three ways. First, the mere presence of predators can greatly affect habitat use by prey species and hence the prey's interactions with other species.

Second, both growth rates (and/or competitive ability) and predation on a species vary with body size and therefore change during ontogeny. Thus, competition and predation are intricately linked through growth rates. Third, the above factors interact to cause ontogenetic niche shifts in most fish species, and such shifts can even change the sign of the interaction between two species. The next sections briefly explore the first two of these factors, while the discussion of ontogenetic niche shifts follows thereafter at greater length. The interaction of competition with predation is discussed elsewhere in this volume in Chapters 31 and 32.

Indirect Effects of the Risk of Predation

The mere presence of predators may have extensive size-specific indirect effects on community structure, some of them mediated through competition. Different habitats in the aquatic environment, such as vegetation compared to open water, expose prey to different risks of being captured by predators (Savino and Stein 1982, Werner et al. 1983a). As a consequence, small fish are confined to areas of cover such as vegetation or rocky reefs, even though more open habitats often afford much more food (Jackson 1961, Werner et al. 1983a, Mittelbach 1984). In such cases, as also suggested for herbivorous insects and sessile marine invertebrates (Chapters 11 and 31), the predator's presence can actually intensify competition, especially if resources in those habitats that offer refuges from predation are limited (Mittelbach 1984).

This striking effect is opposite to predation's familiar effect of reducing competition among prey by removing prey individuals. However, by confining small prey fish to habitats with cover, predators can also release larger individuals of prey species from intra- or interspecific competition with smaller size classes and species, and can thereby provide these larger individuals with a competitive refuge, enabling them to continue to grow (Mittelbach 1981, Werner et al. 1983a). Such indirect effects of predators greatly affect habitat utilization patterns in lakes, and the consequences for community structure are obviously important but little studied.

Interaction of Competition and Predation Through Growth Rates

Competitive effects in fish are largely mediated through changes in size-specific growth rates, which in turn interact strongly with size-specific predation rates. It is misleading to consider these processes independently, and a realistic appraisal of how they affect community structure must begin with a synthetic view of how they interact.

Mortality rates are generally believed to be inversely related to size in fish (Ware 1975, Shepard and Cushing 1980, Werner and Gilliam 1984, Peterson and Wroblewski 1984), and it is often felt that predators are the actual agents of most of this early life history mortality, though there is rarely unequivocal evidence to this effect. Because the total number of potential predators is usually greater for smaller organisms and because vulnerability to a given predator generally decreases with size, overall risk of predation quickly decreases as an individual grows larger. Hence, increased growth rate reduces the time spent in smaller, more vulnerable size classes and thereby reduces the overall probability of mortality.

Gilliam has shown (1982, Werner et al. 1983a) that if a change in the environment multiplies the growth rate by a factor c ($c < 1$ for growth reduction), then the new survivorship (L') over a certain size interval can be expressed as a power of the original survivorship over the same size interval (L, i.e., survivorship at the original growth rate):

$$L' = L^{(1/c)}$$

Simply stated, if growth rate is reduced by 1/2, the animal is spending twice as long in some size range, so the new survivorship is L^2 instead of L. That is, a given reduction in growth can have an enormous effect on survivorship if survivorship is already low. Hence the strong effect of competition on fish growth rates can regulate mortality rates in a very nonlinear way. These ideas have a long tradition in the fish population dynamics literature going back at least to the 1940s (Ricker and Forester 1948, Beverton and Holt 1957, Ware 1975, Shepard and Cushing 1980). Clearly, growth and the direct and indirect effects of predation will greatly influence ontogenetic niche shifts. Therefore, I next examine such shifts in fish.

ONTOGENETIC NICHE SHIFTS IN FISH

Examples of Ontogenetic Niche Shifts

Because resource utilization abilities and risk of predation vary with body size and because the body weights of conspecific individuals commonly span as much as one to four orders of magnitude, extensive ontogenetic shifts in food or habitat use are nearly universal in fish (Werner and Gilliam 1984). These shifts may be continuous. For instance, the range of prey sizes may increase with body size, while the smallest prey size changes very little (Wilson 1975), so that the niche of all smaller size classes is included in that of all larger classes. Alternatively, the ontogenetic shifts may be relatively abrupt, as is often the case among fish. Such size-specific ontogenetic niche shifts complicate how competition and predation interact. For instance, two species may relate to each other as competitors at one stage and as predator and prey at another stage.

Size-specific shifts in food or habitat type have been documented in many species (Crossman and Larken 1959; Martin 1966; Keast 1977, 1978; Ross 1978; Grossman 1980; Stoner and Livingston 1984; Mittelbach 1984; and natural history accounts such as those in Scott and Grossman 1973). In some cases such shifts are associated with morphological changes (Montgomery 1977, Christensen 1978) or with changes in social behavior, such as the shift from schooling behavior when young to solitary behavior when adult (Helfman 1978).

Piscivorous species commonly undergo three or four rather abrupt shifts. For instance, as largemouth bass grow, they switch from feeding on zooplankton to littoral invertebrates and then to fish (Gilliam 1982). Two species of porgy (family Sparidae) progress through five and six well-ordered stages, respectively (Stoner 1980, Stoner and Livingston 1984). Many cases of a shift from carnivory to herbivory at some stage have been reported (e.g., Gibson 1982). Species such as the gizzard shad and threadfin shad shift at a size of about 25 mm from particulate feeding on zooplankton to filter feeding on phytoplankton,. detritus, and zooplankton (Drenner et al. 1982). Thus, such shifts can be important in structuring species interactions. Let us now try to understand why they occur.

Causes of Ontogenetic Niche Shifts

One reason for many of these ontogenetic shifts seems clear: Many fish prefer larger prey, and when ontogenetic shifts occur, they almost always involve shifts to larger prey. Theoretical and empirical studies indicate that growth rates are positively correlated with food size (Paloheimo and Dickie 1966, Wankowski and Thorpe 1979). Martin (1966) showed among 30 lake trout populations that, in populations for which a switch to piscivory was possible, individuals attained maximum weights at least 2.5 times those in populations that remained planktivorous. This difference involves a phenotypic effect within the life of an individual: Individuals transferred from lakes where planktivory was obligatory to lakes where piscivory was possible were subsequently found to feed on fish, and their growth rates increased accordingly.

Many ontogenetic shifts in habitat, however, also appear to be related to risk of predation. As noted earlier, the juveniles of many species are confined to areas where the risk of predation is lower, and as they grow, more open habitats can be exploited (Jackson 1961, Helfman 1978). This is clearly the case for many of the sunfishes that I study (Werner et al. 1983a; Mittelbach 1981, 1984). In a natural lake Mittelbach (1981) found that large size classes of bluegills (>100 mm) foraged in those habitats that maximized return. Smaller fish, however, did not; they spent nearly all their time in the vegetation, even when the profitability of the open water was much higher. Since the smaller fish are those vulnerable to the major predator in the system (the largemouth bass), it appeared that the size at which the ontogenetic shift to open water occurred was influenced by the risk of predation. We tested this hypothesis in an experimental pond and found that the smaller size classes of bluegills spent significantly more time in the vegetation in the presence than in the absence of predators, even when foraging rates were as much as threefold higher in more open habitats (Werner et al. 1983a). In the absence of predators all size classes foraged

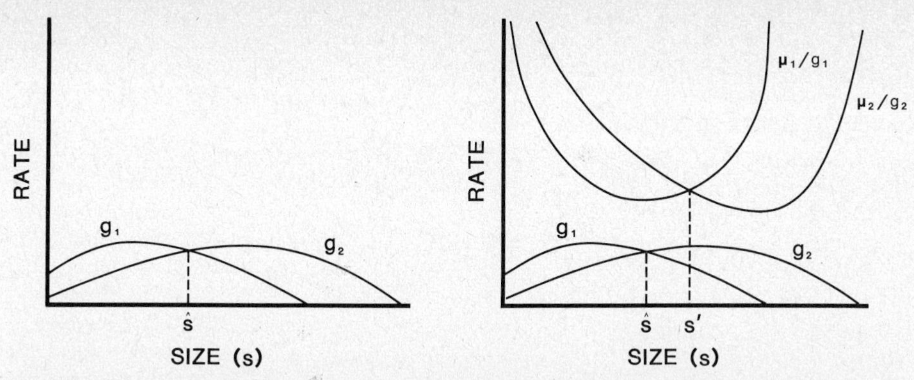

Fig. 21.1 Model for predicting the body size at which ontogenetic niche shifts should occur. (Left) Absolute growth rates (g) as a function of fish body size (s) in two habitats, neglecting habitat differences in the size-specific risk of predation. The ontogenetic switch in habitats should occur at size ŝ in order to maximize growth rates. (Right) Absolute growth rates as above, but taking into consideration habitat differences in the size-specific rates of predation. The curves μ/g display the ratio of mortality rate to growth rate as a function of fish body size in the two habitats. The figure is drawn on the assumption that mortality rates decline with size in each habitat, but are higher in habitat 2 than in habitat l. The ontogenetic switch in habitats should now occur at size s′ (>ŝ) in order to maximize fitness when predation risk as well as growth rates are considered.

in the more profitable habitats. Thus, prediction of ontogenetic niche shifts requires a framework that at least considers the interplay of foraging rates and predation risk.

Predicting Ontogenetic Niche Shifts

The universal expression of ontogenetic niche shifts in fish and their obvious importance for species interactions (see next section) indicate the need for a theory to predict when such shifts should occur. Such a theory is also needed to understand how selective pressures change ontogenetically and mold the particular life history or morphology of a species.

We have attempted to provide such a framework (Gilliam 1982, Werner 1984, Werner and Gilliam 1984). The facts that fecundity and sometimes egg size are a direct function of size in fish (Bagenal 1978, Elliot 1984) and that mortality rate is often an inverse function of size imply a considerable advantage to maximizing growth rates (Werner and Gilliam 1984). Thus, it seems reasonable to hypothesize that, all else being equal, ontogenetic niche shifts occur so as to maximize absolute growth rates or surplus energy at each size. Growth rate curves can then be plotted as a function of body size for each available habitat or for each food type, if this is appropriate, and niche shifts should occur at a size corresponding to the intersections of these curves (Fig. 21.1 left).

It is unlikely, however, that the size-specific risk of predation will be identical in two habitats, and this alters our predictions of the switching size that maximizes fitness. Since growth and mortality rates are expressed in different units, they must be related through the common units of fitness, that is, r in the Euler-Lotka equation. Gilliam (1982) has approached this problem through optimal control theory and size-specific demographic models (vital rates in fish are predominantly size- rather than age-specific; Werner and Gilliam 1984). In equilibrium populations where $r = 0$, the solution for juveniles (among which most ontogenetic shifts occur) is simply to minimize the ratio of mortality rate to growth rate at each size and thus to shift habitats accordingly (Fig. 21.1 right). Intuitively, minimizing the ratio of mortality to growth at each size allows the addition of each increment of size at a minimal cost of mortality and hence maximizes the probability of reaching the next size. The solutions for adults or for a changing population are more complex (see Gilliam 1982 or Werner and Gilliam 1984 for an account).

In general, the problem of ontogenetic niche shifts involves trade-offs of current growth, mortality, and birth rates and is closely related to the problem of optimal life histories (Werner and Gilliam 1984). The above theory is simply a beginning, but it permits us to predict quantitatively the habitat preference of each size class and hence to predict ontogenetic niche shifts. This is a necessary step for evaluating the complex interactions reviewed in the following sections.

IMPLICATIONS OF ONTOGENETIC NICHE SHIFTS FOR SPECIES INTERACTIONS

Juvenile Bottlenecks

How do ontogenetic shifts influence the interplay of competition and predation and thus species interactions? One implication of the relation between food size and body size is that young of larger species often have resource use requirements similar to those of older individuals of smaller species. As a consequence, competition during this stage can cause a significant bottleneck to recruitment to the adult stage of the larger species. Furthermore, many species use similar resources when small because the risk of predation confines them to rocky reefs or beds of vegetation but then can diverge when larger (Jackson 1961, Werner et al. 1983a, Mittelbach 1984). All these factors can combine to produce interspecific interactions with very complicated dynamics.

The juvenile bottleneck can be exacerbated if trade-offs exist in features that adapt species to alternate ontogenetic niches. For instance, I quantified foraging costs for ''piscivorous'' and ''invertebrate-feeding'' body plans in fish feeding on different food types (Werner 1977). Significant trade-offs in feeding efficiency existed. For example, the piscivorous species typically passed through a zooplankton feeding stage when small, yet was burdened at this time with the piscivorous morphology and thus was a poor competitor for zooplankton. Such trade-offs appear to be widespread in fish and may be the basic reason for both the hierarchy of competitive effects in the Scandinavian salmonids discussed earlier and for their ability to coexist by partitioning resources.

Consider first the simplest case: intraspecific size class interactions in species such as planktivores that do not undergo dramatic ontogenetic shifts in diet, but whose diet changes continuously as they grow larger. Because foraging efficiencies are usually size-related, most interactions between age classes will be highly asymmetrical. This asymmetry may operate in either direction: either large or small individuals may be the more efficient competitors, depending on the distribution of resources (Wilson 1975, Mittelbach 1981, Werner and Gilliam 1984, Hamrin and Persson unpublished observations). The resulting interactions can cause wide fluctuations in year class strength or population cycles, as have been demonstrated in many marine and freshwater fishes (e.g., Jonsson and Ostli 1979). For instance, Hamrin and Persson (1985) report a two-year oscillation in vendace in Sweden. The diet of this species is almost entirely pelagic zooplankton; larger individuals can use larger food sizes, but the smallest food size is the same for all individuals. During periods of low resource levels in late summer, when all prey are small, smaller fish are favored as they have higher size-specific growth rates. Thus, in years with strong year classes the competitive advantage of smaller fish severely suppresses the reproductive output of co-occurring adults, thereby releasing the younger year class from competition the following year when they reproduce. Hamrin and Persson (1985) list several examples of vendace and cisco populations where oscillations generally occur with a period correlated with age of maturity.

Typical examples of two interacting species that overlap in resource use only for a period in the life history include juveniles of species that diverge in niche when larger, or juveniles of a large species that interact with a smaller species. Several studies illustrate the effects of planktivores on the early life history stages of other species (Svardson 1976, Li et al. 1976, Kohler and Ney 1981, Fast et al. 1982). Emery (1975) suggested such a scenario for smallmouth bass in a large Ontario lake. Smallmouth bass were censused in the lake from 1936 on, and around 1950 cisco were introduced as a prey fish to improve growth of lake trout. As cisco populations grew, average bass size declined, but it was found that growth rates in 3-to-12-year-old bass remained the same as before cisco introduction.

It appears that cisco compete for plankton with smallmouth bass less than 3 years old, thereby reducing growth of young bass and consequently size at each succeeding age. Other examples of this type of interaction will be mentioned later.

In the sunfishes of small, warm lakes small size classes of nearly all species are found in dense vegetation (Werner et al. 1977, Keast 1978, Mittelbach 1984). Hence resource overlap is high early in the life history, while resource use of larger individuals diversifies as they move into more open areas. Small bluegill and pumpkinseed sunfish (<75 mm), for instance, overlap by about 50% in diet, whereas large individuals (>75 mm) overlap by only 2 to 8% in lakes with a true limnetic zone (Mittelbach 1984). Longear sunfish also overlap by about 50% in diet with the pumpkinseed when small, but less than 20% when large (Laughlin and Werner 1980). Growth of the smaller size classes is lower in the field than in ponds without competing species, suggesting that food limitation is occurring. Growth rates increase when the shift to the "adult" niche is made (Mittelbach 1984, Werner and Hall unpublished). In the particular case of the bluegill and pumpkinseed, laboratory foraging experiments and pond experiments suggest that these species are roughly equal in their abilities to utilize the vegetation and that relative abundances of adults of the two species (ranging from 25:1 to 1:1) are correlated with relative production of "adult" resources across lakes (Mittelbach 1984). Thus, the early life history bottleneck may not differentially affect these two species, and indeed one might expect evolutionary convergence in their abilities to utilize the early life history habitat.

Mixed Competition-Predation Interactions

Yet another level of complexity is added to interactions in size-structured populations when ontogenetic shifts lead to mixed competition-predation interactions. Most piscivorous species, for instance, begin life by feeding for some period on zooplankton or small invertebrates before becoming large enough to make a living as a piscivore. As a consequence, such species often compete when they are small with species that serve as their major prey when they are large. We now have a number of illustrations of the importance of such interactions.

A classical example is found in the work of Larkin and his colleagues on Paul Lake, British Columbia. Paul Lake was fishless until stocked with rainbow trout in 1909. Between 1937 and 1949 the annual catch from the lake had stabilized (Crossman and Larkin 1959). All size classes of trout fed on a mixture of plankton, benthos, and terrestrial insects (Larkin and Smith 1954, Crossman and Larkin 1959). Just prior to 1950 a smaller species, the redside shiner, entered the lake from upstream in the drainage basin. The shiner also fed on plankton and benthos and overlapped in the littoral zone with young trout. After the shiner invasion large trout (>20 cm fork length) fed on shiners and dropped plankton from their diet. As the shiner population increased during the following 10 years, the growth of large trout (>20 cm) increased in response to the availability of shiners as food. In contrast, the growth rate and amount of food in the stomachs of small trout (<20 cm) declined markedly; the trout required effectively a full year longer to reach a given size. Thus, the two life stages were affected in opposite ways by the shiner invasion! Though only catch data were available, it appeared that the overall effect of the shiner was to reduce the trout population in the lake.

The introduction of threadfin shad in California as a prey species to improve largemouth bass growth provides a parallel example. Growth of bass in the piscivorous size classes increased as expected after shad introductions. However, the growth rate of bass at early stages was often just as markedly reduced, presumably due to competition with shad for zooplankton. This competition often led to poor survivorship and smaller year classes of the bass (Fast et al. 1982).

Li et al. (1976) documented analogous effects on growth of black and white crappies in a California lake, following introduction of a planktivorous prey species, the Mississippi silverside. After the introduction of silversides growth of both crappie species when small and planktivorous was significantly depressed. However, most annual growth increments in crappies that were large, piscivorous, and feeding on silversides were significantly greater than before the silverside introduction. Several studies have demonstrated an identical, mixed competition-predation effect of the alewife on its predators (McCaig 1980, Kohler and Ney 1981).

Interestingly, an invertebrate plankton predator often introduced to increase growth of prey fish, the opossum shrimp, has in Lake Tahoe caused the virtual disappearance of planktonic cladocerans (Richards et al. 1975) and the collapse of the prey fish (kokanee) population (Wydoski and Bennett 1981). Thus, this may be a parallel interaction to those above, except that it involves a vertebrate and invertebrate. It is likely that as we become sensitive to these mixed competition-predation interactions, we will find them abundantly represented in most fish communities.

PROSPECTUS

The various types of evidence that I have reviewed, including observations, natural experiments, field experiments, and introductions, indicate that resource limitation and species interactions often play a significant role in structuring freshwater fish communities. The long-term studies suggest that growth depression and niche shifts due to competitors or predators can have long-term effects on the absolute and relative abundances of species. Fish communities offer good potential for field experiments designed to unravel these interactions. Furthermore, fish can often be brought back to the laboratory to study how species interactions depend on factors such as size, morphology, and behavior.

Several conclusions from the study of fish communities are relevant in varying degrees to other taxa. These conclusions revolve around the ways that growth, size, and size-related ecological factors combine to affect species interactions. Studies of fish communities show that, if we are to understand species interactions, we must consider the sequence of stage-specific or size-specific interactions that individuals experience during their ontogeny. Though community ecologists have long focused on *mean* interspecific size differences as a factor in interspecific interactions, the logical extension of these ideas to interactions between size classes of populations has largely been ignored. Yet, with the notable exception of birds and mammals, most taxa exhibit long periods in which juveniles forage independently, increase greatly in size at flexible rates, and undergo large ontogenetic shifts in resource use (Werner and Gilliam 1984).

Such shifts have been widely documented among invertebrates and lower vertebrates. Werner and Gilliam (1984) provide examples from a diverse array of organisms including snakes, lizards, turtles, amphibians, zooplankton, leeches, spiders, gastropods, and aquatic insects. Studies of freshwater zooplankton (e.g., Neill and Peacock 1980) and salamanders (e.g., Morin 1983b) in particular have clearly documented the resulting complex interactions and their effects on community structure. Many of these interactions are of the mixed competition-predation type. Such size-dependent sequences of interactions have the consequence that subtle changes in interaction strength at one particular stage can have cascading effects and produce radical changes in community structure (e.g., Neill and Peacock 1980). Possible results include multiple stable community states; strong possibilities of indirect effects (Werner and Gilliam 1984), of which I have given examples in this chapter; and the possibility that a collection of communities, all structured by the same deterministic processes, may exhibit no pattern in composition and thus may appear randomly assembled.

Though I have stressed the relevance of these ideas to communities of mobile animals, these considerations are relevant to sessile organisms as well, both plants and animals. Interactions among plants are highly size-dependent, resulting in very asymmetrical intra- and interspecific interactions (e.g., between juveniles and adults). The importance of such interactions in plants has been stressed by Ross and Harper (1972), Goldberg and Werner (1983), and Grace (1985). Buss (1980 and Chapter 31) has discussed consequences of the extreme size dependency in interactions among sessile animals, leading to intransitive and frequency-dependent interactions and thus complex community behaviors.

These ideas also bear directly on the problem of the evolution of complex life histories (Istock 1967, Werner and Gilliam 1984). In anurans, for instance, metamorphosis represents a shift from aquatic herbivore to terrestrial carnivore; totally unrelated competitors and enemies are faced in these two stages. The evolution of this life cycle must be considered from the perspective of the interactions at each stage, and the framework used in this chapter may prove useful for predicting body size at metamorphosis.

There is a critical need to incorporate population structure into our models of species interactions. As we have seen above, population structure often maps to major niche changes during ontogeny. The conventional approach of considering species populations monomorphic in size or treating only adults is wholly inadequate for studying community ecology of species undergoing ontogenetic niche shifts—that is, for many or perhaps most species.

ACKNOWLEDGMENTS

I thank Ted Case, Jared Diamond, Mathew Leibold, Gary Mittelbach, Craig Osenberg, and Tom Schoener for insightful comments on an earlier draft of this paper. Mathew Leibold helped with the literature search, and Laura Riley provided indispensable logistic support. This work was supported by National Science Foundation Grant BSR 81-19258. This is contribution No. 543 of the Kellogg Biological Station.

APPENDIX

Common name	Scientific name
Alewife	*Alosa pseudoharengus*
Arctic char	*Salvelinus alpinus*
Black crappie	*Pomoxis nigromaculatus*
Bluegill	*Lepomis macrochirus*
Brook silverside	*Labidesthes sicculus*
Brook trout	*Salvelinus fontinalis*
Brown trout	*Salmo trutta*
Cisco	*Coregonus artedii*
Cutthroat trout	*Salmo clarki*
Dolly Varden char	*Salvelinus malma*
Gizzard shad	*Dorosoma cepedianum*
Golden shiner	*Notemigonus crysoleucas*
Green sunfish	*Lepomis cyanellus*
Inland silverside	*Menidia beryclina*
Kokance	*Oncorhynchus nerka*
Lake trout	*Salvelinus namaycush*
Largemouth bass	*Micropterus salmoides*
Longear sunfish	*Lepomis megalotis*
Mississippi silverside	*Menidia audens*
Mudminnow	*Umbra limi*
Opossum shrimp	*Mysis relicta*
Peacock bass	*Cichla ocellaris*
Perch	*Perca fluviatilis*
Pumpkinseed	*Lepomis gibbosus*
Rainbow smelt	*Osmerus mordax*
Rainbow trout	*Salmo gairdneri*
Redside shiner	*Richardsonius balteatus*
Roach	*Rutilus rutilus*
Rockfish	*Sebasthes spp.*
Sea lamprey	*Petromyzon marinus*
Smallmouth bass	*Micropterus dolomieu*
Smelt	*Osmerus eperlanus*
Surfperch	*Embiotoca spp.*
Threadfin shad	*Dorosoma petenense*
Vendace	*Coregonus albula*
Walleye	*Stizostedion vitreum*
White crappie	*Pomoxis annularis*
White sucker	*Catostomus commersoni*
Yellow perch	*Perca flavescens*

chapter 22

Evolution and Differentiation in Terrestrial Plant Communities: the Importance of the Soil Resource: Light Gradient

David Tilman

INTRODUCTION

Although there are over 300,000 species of terrestrial vascular plants worldwide, these species occur together in communities that are often similarly structured. One striking pattern is the convergence of unrelated species to a common set of morphological, physiological, and life history traits in widely separated but physically similar habitats worldwide (Orians and Paine 1983). This has perhaps been best studied in areas with Mediterranean climates (e.g., Mooney 1977, Cody and Mooney 1978). Another pattern is that, for a wide variety of geographical areas, plant communities have their peak diversity in relatively resource-poor habitats (e.g., Beadle 1966; Dix and Smeins 1967; Whittaker and Niering 1975; Al-Mufti et al. 1977; Connell 1978; Grime 1979; Huston 1979, 1980; Tilman 1982; Shmida et al. 1985). Still another pattern: within a given geographical region much of the variation in the local composition of plant communities is associated with the type of parent material on which the soil formed (e.g., Lindsey 1961; Hole 1976; Rabinovitch-Vin 1979, 1983; Jenny 1980). Moreover, the qualitative patterns observed in many cases of both primary and secondary succession are often similar in a variety of regions. The causes of such similarities are one of the major mysteries facing plant ecologists. Might such similarities imply that a few general underlying processes have greatly influenced terrestrial plant evolution and community structure? Or, must a unique explanation be invoked for each situation?

There are two types of potential explanations for these patterns. First, it may be that most plant species occurring in a locality are differentiated in the traits that determine their responses to major biotic or abiotic constraints. Such differentiation could allow the long-term persistence of the species. If plants experienced similar constraints in a variety of habitats and if there were a limited number of ways to respond to these constraints, this could provide an explanation for the patterns mentioned above. In theory, there are an almost unlimited number of ways in which plants could be differentiated. All that is required is that plant species be inversely ranked in terms of their response to two or more resources or limiting factors (Tilman 1982). The possibilities include (1) specialization on different ratios of various soil resources or light (Tilman 1980, 1982), (2) specialization on certain periods of a seasonally fluctuating climate (the ''phenological niche'' of Grubb, Chapter 12), (3) differentiation in age or size structure leading to multiple stable community equilibria (Chapter 2; Hassell and Comins 1976), (4) specialization along the continuum from exploiting average resource availability to exploiting variance (stochasticity) in resource availability (Chapter 14; Armstrong and McGehee 1980, Levins 1979, Tilman 1982), (5) differentiation in the competitive abilities of plants versus their susceptibility to herbivory or predation (Lubchenco 1978 and Chapter 32; Tilman 1982). This list is by no means exhaustive nor are these processes mutually exclusive. However, if one or a few of these processes proves to be of overriding importance in a variety of habitats, it could explain the similarities mentioned above.

Alternatively, it may be that species are not differentiated with respect to such traits, but are functionally identical. This could be considered to be a ''neutral species'' hypothesis, in many ways comparable to the neutral allele hypothesis of population genetics. If species are not differentiated, the composition of a community would be determined purely by random processes of invasion and extinction (Chapter 19; MacArthur and Wilson 1967). Although this may seemingly explain patterns observed on small spatial scales, it is unlikely to explain the long-term persistence of species within a region. Hubbell (1979) showed that, in the absence of continual species immigration from nearby source areas, a model of forest dynamics based solely on chance colonization and disturbance led to extinction of all but one species. Random processes of colonization and disturbance within a local habitat lead to random walk to an absorbing boundary, extinction. Such extinction is merely slowed by considering larger areas. Although many ecological processes have a large component of randomness, each species must respond differently to variable versus constant conditions if stochasticity is to explain long-term persistence of species.

If species are not identical and are not undergoing random walk to extinction, they must be differentiated in the traits which determine their response to major biotic or abiotic constraints. All plants are consumers of various resources, some of which may be in short supply relative to need. Plants require soil nutrients, water, and light. Their ability to use these is highly dependent on temperature, humidity, pH, and oxygen availability in the soil. Also, all plants are themselves potentially the resources of various herbivores and predators. Thus, the major constraints on plants are competition for limiting resources and herbivory. Both competition and herbivory may be equally important. In this chapter, though, I consider just resource competition among abundant species. I concentrate on abundant species because the patterns they show may more closely reflect the major processes structuring the community. Although there are insights to be gained from considering rare species (e.g., Chapter 12; Rabinowitz and Rapp 1981), rare species may be exploiting uncommon habitat conditions or rare events and thus may not give as much information on the major processes structuring the community.

Most of the theory presented in this chapter is limited to equilibrium conditions. I do this not because I believe that plant communities are at equilibrium, but because it is potentially simpler and thus more consistent with Occam's razor to explore equilibrium explanations before invoking more complex dynamic models. Simple, equilibrium models may be able to explain many of the patterns we observe on large spatial scales (thousands of individual plants) or over relatively long periods of time (at least the generation time). On smaller spatial or temporal scales, demographic stochasticity (May 1973a) and spatial stochasticity (Tilman 1982) are likely to decrease predictive ability. There are, though, dynamic processes not included in equilibrium models which can lead to the long-term persistence of species (Chapter 13; Armstrong and McGehee 1980). Such dynamic processes are not considered in this chapter.

I focus this chapter on what I believe may be a general feature of plant competition for resources. I consider it to be a starting point for a more detailed, mechanistic approach to plant community ecology. All terrestrial plants require various mineral nutrients, water, and light. For most of these species, mineral nutrients and water are obtained below ground from the soil, whereas light is acquired above ground. Because each plant requires both above-ground and below-ground resources, its ability to grow in a habitat will be determined by the availabilities of these resources relative to the plant's morphological and physiological abilities to acquire above-ground and below-ground resources. In this chapter I first present a simple theory of plant competition for nutrients and light (Tilman 1982, 1985). This theory suggests (1) that each species should be a superior competitor for a particular point along a nutrient: light gradient and (2) that changes in the relative availability of these resources, through either space or time, should lead to changes in the composition of community.

I then review six lines of evidence, each of which has major elements consistent with plants' being differentiated in their relative requirements for soil resources versus light. The lines of evidence are (1) soil heterogeneity and spatial distributions of plants, (2) patterns in primary succession, (3) patterns of evolution in the earliest terrestrial plants, (4) results of experimental manipulations of vegetation (5) patterns of resource availability in natural and experimentally manipulated vegetation, and (6) patterns of life history variation. All of these patterns suggest that one of the major axes for the evolution and differentiation of terrestrial plants has been the gradient from habitats with low availabilities of soil resources but high availability of light at the soil surface to habitats with high availabilities of soil resources but low availability of light.

RESOURCE COMPETITION THEORY

The Lotka-Volterra competition equations are the traditional approach to competition theory in ecology. Although these equations are the most general description of interspecific competition (May 1973a), they are more phenomenological than mechanistic. Although less general, a resource-based approach to competition allows the conditions for competitive dominance or coexistence to be expressed in terms of the resource requirements and consumption rates of the species. These parameters are more easily measured and may more accurately reflect the actual mechanisms of interaction than the abstract parameters of the Lotka-Volterra model. However, for a situation in which two species compete for two resources, a resource-based approach needs information on four variables. To abstract this to a two-dimensional plane means that information on resource-dependent isoclines must be combined with information on the rates of supply and consumption of the two resources.

Competition Along a Nutrient: Light Gradient

The basic theoretical construct used for resource competition theory is shown in Figs. 22.1 and 22.2; it has been discussed in depth in Tilman (1980, 1982). Consider, first, a single plant species living in a spatially homogeneous habitat in which there are two limiting resources (Fig. 22.1). Let R_1 and R_2 be the environmental availabilities of each resource, and let S_1 and S_2 be the maximal amounts of R_1 and R_2 that can occur in the habitat in the absence of consumption. I have called the point (S_1,S_2) the resource supply point and have assumed that the rate of resource supply is proportional to the difference between the maximal (S_i) and ambient (R_i) amounts of each resource (Tilman 1982). The growth rate of the plant population will depend on R_1 and R_2. The plant population will be at equilibrium when its growth rate exactly balances its loss rate. An isocline can be drawn to show the availabilities of the two resources for which this would occur. Along this isocline, $dN/dt = 0$. The isocline has a right-angle corner because the two resources are essential (Tilman 1980). The various mineral elements and light required for plant growth are essential because increased availabilities of one of these resources cannot overcome limitation by another resource. Thus, in Fig. 22.1 the plant is limited by either R_1 (above the broken line) or by R_2 (below the broken line). If resource levels are greater than those on the isocline, population size will increase. If they are less, population size will decrease.

Because this model includes both species densities and resource levels, equilibrium requires that $dN_1/dt = 0$ and $dR_j/dt = 0$ for all species i

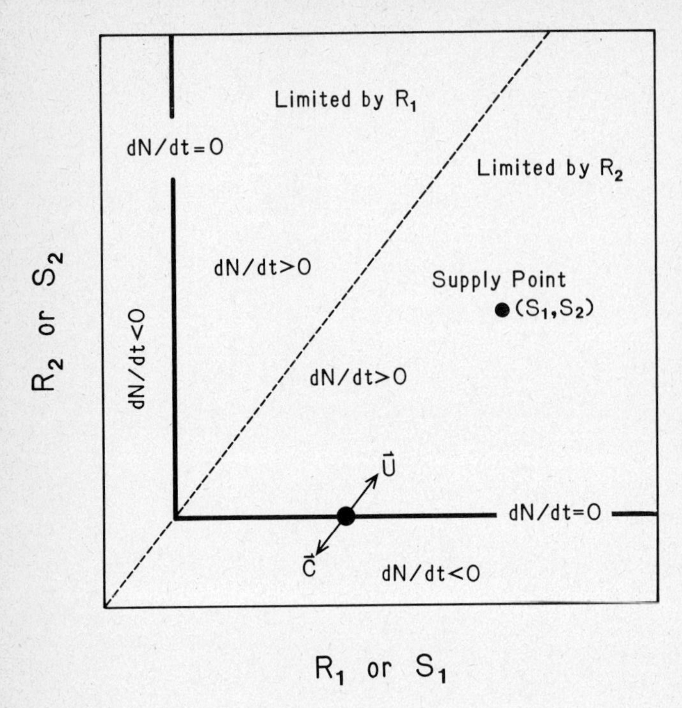

Fig. 22.1 Theoretical construct for resource competition theory. The solid line with a right-angle corner is the resource-dependent growth isocline of a plant. Along this isocline the growth of the plant population equals its loss rate. The broken line shows the proportions of the two resources, R_1 and R_2, for which this population is equally limited by each. Optimal foraging theory predicts that this population would consume these resources in this proportion. Thus, the consumption vector, $\vec{C}$, is parallel to the broken line. The point (S_1,S_2) is the resource supply point. The rate of supply of the resources, in the absence of consumption, is assumed to be $dR_1/dt = a(S_1 - R_1)$ and $dR_2/dt = a(S_2 - R_2)$. S_1 and S_2 are the total amount of all forms of resources 1 and 2 in the habitat. The vector $\vec{U}$ is the resource supply vector. The point shown on the resource-dependent growth isocline is the equilibrium point associated with the supply point shown. At this point birth balances death, and resource supply balances resource consumption. The simple definition of resource supply employed allows each resource supply point to be easily mapped in to its associated equilibrium point. This is further illustrated in Fig. 22.2 and Tilman (1982).

and resources j. For all points on the isocline, $dN/dt = 0$. However, there will only be one point on the isocline and one population density of this species for which $dR_1/dt = dR_2/dt = 0$. This will be the point on the isocline at which the plant consumes the resources (the vector labeled $\vec{C}$) in the same proportion in which they are being supplied (the vector labeled $\vec{U}$). At this point population density will adjust until the total rate of resource consumption by the population equals the total rate of resource supply. For the supply point shown, the species is limited by R_2 at equilibrium and unaffected by changes in R_1.

Now consider two plant species, A and B, each of which requires two essential resources. Because this chapter is most concerned with the effects of competition for soil resources versus light, let the resources of Fig. 22.2 be available soil nitrogen and light available at the soil surface. Species A and B are differentiated such that A is the superior competitor for nitrogen, but B is the superior competitor for light. Because of this differentiation, the isoclines cross. The point at which they cross is a two-species equilibrium point, that is, a point of coexistence of the two species. At this equilibrium point resource levels are such that each species is limited by a different resource. In habitats in which there is a low rate of supply of nitrogen (supply point z), both species will be nitrogen-limited and species A, the superior competitor for nitrogen, should competitively displace species B. The mechanism of displacement is the reduction of available nitrogen levels to the point z′ on the isocline of species A, at which point species B has insufficient nitrogen to survive. Similarly, in habitats in which there is low availability of light (such as supply point x), species B should competitively displace species A because B can reduce light at the soil surface to a level (point x′) below that required for species A to survive. In intermediate habitats, in which both nitrogen and light are limiting (supply point y), consumption by the two species will focus environmental nitrogen and light levels into the two-species equilibrium point (point y′), and the species will coexist.

As shown, the two-species equilibrium point is stable. Stable coexistence requires that each

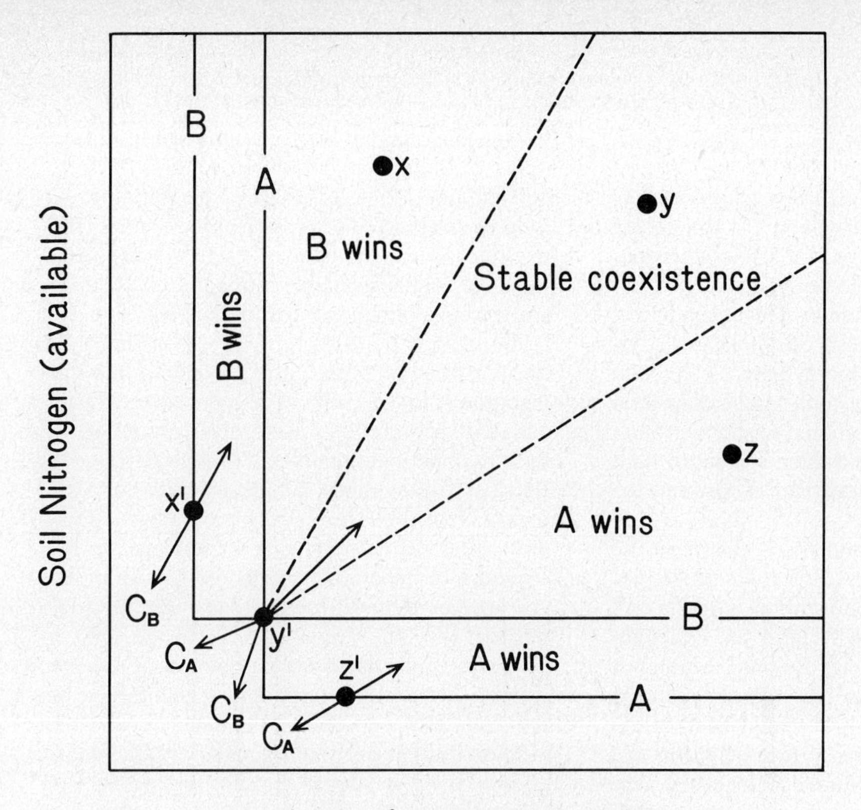

Fig. 22.2 Competition for soil resources versus light. The two solid lines with right-angle corners are light- and nitrogen-dependent growth isoclines for species A and B. Each isocline shows the environmental availabilities of nitrogen and light at the soil surface for which the long-term growth rate of the population equals its long-term loss rate. Species A has the lower requirement for nitrogen and is predicted to displace species B competitively from habitats in which both species are limited by nitrogen, such as the habitat with supply rates represented by point z. The equilibrium levels of nitrogen and light associated with supply point z are the point labeled z′. Thus, the processes of resource consumption and supply map supply point z into equilibrium point z′. Species B has the lower requirement for light and is predicted to displace species A from habitats with low availability of light such as habitat x. Supply point x will be mapped into equilibrium point x′. The point at which the isoclines cross is a two-species equilibrium point. These two species can stably coexist in intermediate habitats that have supply rates of nitrogen and light for which each species is limited by a different resource. All habitats with supply points in the region of stable coexistence will be mapped into the two-species equilibrium point. Thus, supply point y leads to equilibrium point y′. Equilibrium occurs when resource consumption equals resource supply. For each of the three equilibrium points (x′, y′, and z′) illustrated, there are a set of associated vectors. These vectors show consumption by species A and B (C_A and C_B) and resource supply (unlabeled) at each of these points.

species inhibits its own growth more than it inhibits the other species (Tilman 1980, 1982). This will occur if each species, compared to the other, consumes relatively more of the resource that limits it at equilibrium. For the two-species equilibrium point in Fig. 22.2, each species is limited by the resource for which it is the poorer competitor, species A by light and B by nitrogen. Species A is shown to attenuate more light per unit nitrogen consumed than B. This seems likely since species A, being the superior nitrogen competitor, should produce more biomass per unit nitrogen than B. Because light attenuation is at least approximately proportional to above-ground biomass, species A should attenuate more light per unit nitrogen consumed than B, making the equilibrium stable.

In a more complete theory differences in the maximal heights of these species would also influence the stability of the equilibrium point. As will be discussed, plants that are superior light competitors, such as species B, may be taller at maturity than plants that are superior nutrient competitors. At the two-species equilibrium point species B would be nitrogen-limited and species A would be light-limited. Species A, the shorter species and the better competitor for nitrogen, would mainly shade itself and thus would inhibit itself more than it inhibited species B. Species B, the taller species and the better competitor for light, would be nitrogen-limited. However, it could not reduce nitrogen to levels at which species A would be nitrogen-limited. Thus, even when differences in plant height are considered, it seems likely that the two-species equilibrium point would be stable because each species would tend to inhibit itself more than it inhibited the other species. These predictions could be easily tested. If two species are coexisting because of the mechanisms shown in Fig. 22.2, fertilization with nitrogen should favor the taller, overstory species (species B) more than the understory species. Similarly, addition of light should favor the shorter, understory species (species A) more than the overstory species.

Soil nitrogen and light intensity at the soil surface are not independent resources. Although the rate of supply of nitrogen may be considered to be a trait of the soil and local physical conditions, light intensity at the soil surface is determined by plant biomass and architecture and thus depends on nitrogen. To see if this dependence might modify the qualitative predictions of Fig. 22.2, I made a dynamic model of competition for nitrogen and light (Figs. 22.3 and 22.4). Nitrogen and light were assumed to be perfectly essential, with the rate of weight gain of each species determined by either available nitrogen or light at the soil surface, whichever led to the lower rate. For both nitrogen and light, growth was a Michaelis-Menten function of the current availability of that resource (Tilman 1977). The species were differentiated such that they were inversely ranked in their competitive abilities for nitrogen versus light. All individuals within a population were identical in their resource requirements. Thus, the model did not explicitly include age or size structure or genotypic or phenotypic variability. All species experienced the same density-independent loss rate. Light intensity at the soil surface was controlled by plant biomass. Each species required a different amount of nitrogen per unit biomass and each attenuated a different amount of light per unit biomass, with the values chosen to give stable equilibrium points. By using light at the soil surface as a limiting resource, I assumed that seedling growth rate strongly influenced a plant's ability to become an adult, that is, that establishment is a critical stage in its life history (Grubb 1977 and Chapter 12). This is a common assumption, as evidenced by the use of seedling ''shade tolerance'' classes to predict replacement patterns in forests (Decker 1952). It is, though, a potentially critical assumption that deserves further exploration because different size classes of plants do experience different light levels.

Fig. 22.3 illustrates that the qualitative predictions of Fig. 22.2 still hold even with the dependence of light on nitrogen and the other assumptions made above. As in Fig. 22.2, there is a region in which species A displaces species B, a region in which both species stably coexist, and a region in which species B displaces species A. The regions are determined by the supply rate of nitrogen (controlled by total soil nitrogen in the model), light levels (dependent on plant biomass), and the nitrogen and light requirements

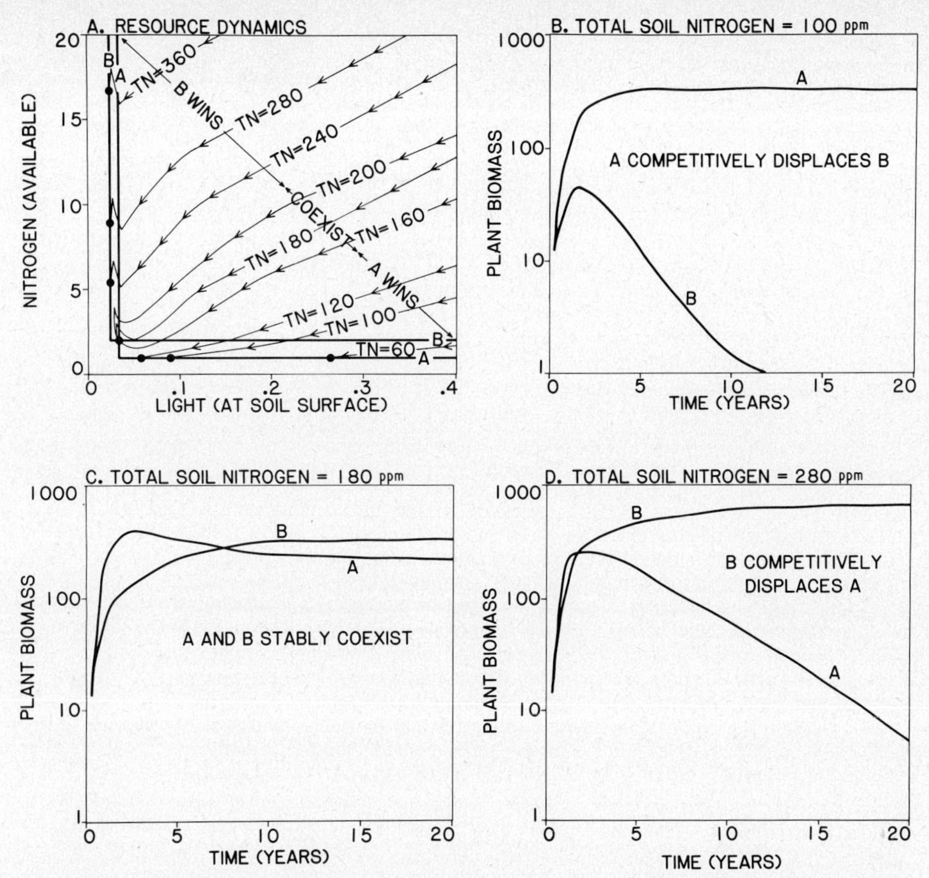

Fig. 22.3 Soil nitrogen, light intensity, and plant biomass. (A) An explicit model of plant competition for nitrogen and light (Tilman 1985) was formulated and solved numerically using a digital computer. The general assumptions of the model are given in the text. The lines with right-angle corners are the light- and nitrogen-dependent isoclines. The points on the isoclines show equilibrium availabilities of nitrogen and light associated with various rates of supply of soil nitrogen. The thin line coming into each of these points shows the trajectory followed by nitrogen and light levels during a particular simulation of competition between species A and B. The simulations have identical parameters except the rate of supply of nitrogen, which is determined by the total nitrogen (TN) level, assuming that 10% of total nitrogen is actually convertible into an available form. All simulations start with full sunlight, available nitrogen equal to 10% of TN, and densities of species A and B of 10. These simulations show the same relationship (i.e., mapping) between resource supply points (here, total nitrogen) as the graphical model in Fig. 22.2. (B) In habitats with low rates of supply of nitrogen, species A competitively displaces species B. The dynamics of this displacement are shown for the case in which TN = 100 ppm. Note that species A has a lower requirement for nitrogen than species B (see isoclines of part A of this figure). (C) In habitats with intermediate supply rates of nitrogen, nitrogen and light levels are reduced, at equilibrium, to the point of stable coexistence. The species stably coexist for TN values between about 150 and 210. A case with TN = 180 ppm is shown. (D) In habitats with high availabilities of soil nitrogen, both species are light-limited. Under such conditions species B, the superior competitor for light, displaces species A, such as is shown for TN = 280 ppm.

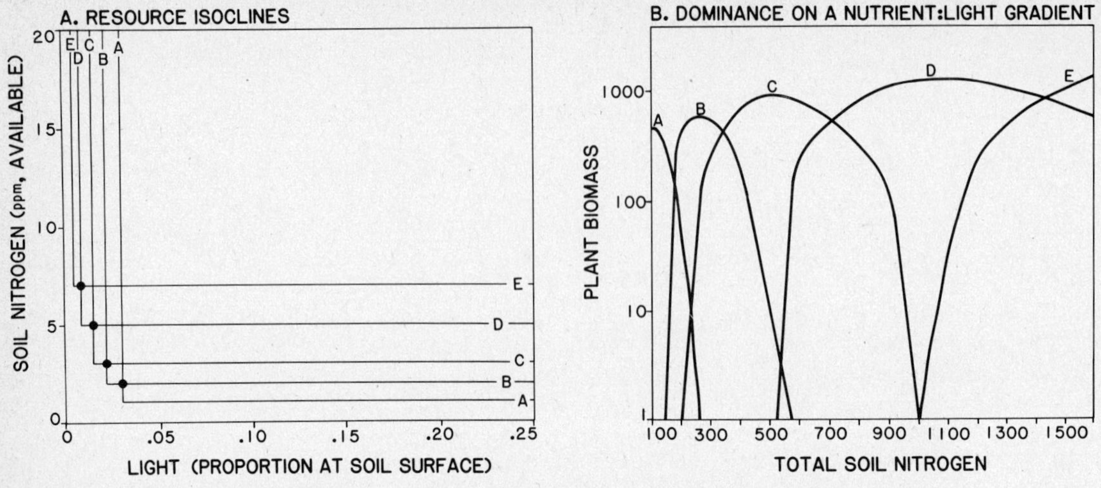

Fig. 22.4 Model of resource competition applied to several species. (A) Nitrogen- and light-dependent isoclines are shown for five species, labeled A to E. Two-species equilibrium points are indicated by dots. The five species are chosen so that they are inversely ranked in terms of their competitive abilities for nitrogen and light. Their consumption rates of nitrogen and light make the four two-species equilibrium points stable. (B) The parameters specified in part A are used for numerous simulations of competition among these five species. Except for having three additional species, the model used is identical to that of Fig. 22.3. For each simulation performed at a particular level of total soil nitrogen, the density of each species is recorded after 50 years of competition. These densities are graphed against total soil nitrogen. They show that these five species are separated along the total soil nitrogen axis in a way consistent with their nitrogen and light requirements. Thus, this can be considered to be a nitrogen : light gradient. Note that species A, which has the lowest requirement for nitrogen, is dominant at total soil nitrogen levels of 100, followed in order by species B, C, D, and E according to their nitrogen and light requirements.

and consumption characteristics of the species. These regions can be considered to be a gradient from low nitrogen but high light habitats to high nitrogen but low light habitats.

All models are logical devices which convert a set of assumptions into predictions. If the predictions of a model are inconsistent with observations, the underlying assumptions may be wrong or the logic of the model may be wrong or incomplete. In this chapter I make many major simplifying assumptions about plant biology and habitat structure. This is done to determine the implications of a few general and ubiquitous aspects of resource competition among terrestrial plants. It is the simplicity of this approach that makes it useful. Only by eliminating many other potentially important processes is it possible to determine the patterns caused by a given process. If a simple approach can explain some of the complexity of nature, then by extending the approach to its limits and beyond we may be able to determine what other processes need to be added to the model to better understand nature.

The qualitative patterns predicted by this model would not be changed significantly if many of these assumptions were made more realistic. I believe that these patterns would hold even if species had genotypic variability, if species experienced different mortality rates, if mortality were density-dependent, if the isoclines had curved rather than right-angle corners, or if there were point-to-point spatial variability in resource supply rates (see Tilman 1982). The predictions made above are critically dependent on the assumption that species are inversely ranked in their competitive abilities for the two limiting resources. If this were not so, one species could competitively displace all others from the habitat (called a "superspecies" in Tilman 1982, but a "supercompetitive species" here).

The patterns may depend on the assumption that all individuals within a population are morphologically identical. If age- or size-dependent resource requirements were added, there could be life history bottlenecks (Lynch 1978) and thus multiple stable equilibrium points (Hassell and Commins 1976). A life history bottleneck could occur, for instance, if a shrub, which was a poorer light competitor, could grow to such a density in the absence of trees as to reduce light at the soil surface to the point at which tree seedlings could not successfully invade a habitat in which they otherwise would grow into the dominant species. There are no conceptual barriers to adding size or age structure to the model. However, this would immensely increase the complexity of the model (e.g., Botkin, Janak, and Wallis 1972; Botkin 1981). One effect of size structure, as mentioned above, may be to stabilize equilibrium points. It may be that plant life history traits are constrained in a manner in which multiple life history bottlenecks are not common. If this is so, it would be possible to explain many of the patterns observed without the need to invoke size structure. However, only further observational, experimental, and theoretical work can answer this question.

The Resource Ratio Hypothesis

The implications for community structure of the mechanisms assumed in Figs. 22.2 and 22.3 are clearer when applied to numerous species (Fig. 22.4). Five species are shown to differ in their requirements for a limiting soil resource (nitrogen is used) and light at the soil surface. Species A is the best competitor for the soil resource, followed by B, C, D, and then E. The species are inversely ranked in their requirements for light, with species E the best competitor for light, followed by D, C, B, and then A. This gives four two-species equilibrium points, each of which has an associated set of habitats in which a pair of species stably coexist. For any given community, it is critically important which soil resources are limiting (Tilman 1982). I discuss Fig. 22.4 in a broad way to stress the generality of limitation by soil resources versus light. I do not wish to imply, though, that all limiting soil resources are identical. Clearly, different species tend to dominate low phosphorus versus low nitrogen versus low water soils (Tilman 1982) and to be dominant at different points along an experimental nitrogen:magnesium gradient (Tilman 1983, 1984).

Different levels of total soil nitrogen create a soil resource:light gradient that ranges from habitats with resource-poor soils but high availability of light at the soil surface to habitats with resource-rich soils but low availability of light at the soil surface. Numerical solution of the model represented in Fig. 22.4A led to separation of these five species along this gradient. The superior competitor for the limiting soil resource is dominant in resource-poor soils and the superior competitor for light is dominant at the other end (Fig. 22.4B). This theory thus predicts that each species should be a superior competitor for a particular range of relative availabilities (or ratios) of the limiting resources and should exist in a habitat only when the ratios of supply rates fall in its range. If any processes lead to changes in the availability of soil resources through time, this theory would predict that there would be comparable changes in the species composition of the plant community. I have termed this the "resource ratio" hypothesis of succession (Tilman 1982, 1985).

If there is spatial heterogeneity in the supply rate of soil resources within a given locality, this theory would predict comparable spatial variability in the composition of the plant community. Such spatial heterogeneity could allow many more species to coexist than there were limiting resources (Tilman 1982). Such coexistence assumes that a species would not be displaced from a favorable microhabitat because of random mortality combined with inundation by propagules from neighboring microhabitats that favor other species. As discussed in Tilman (1982), spatial heterogeneity may also explain the observation that high diversity plant communities are found on resource-poor soils, if the amount of spatial heterogeneity does not increase more than linearly with the average resource level.

THE SIX PATTERNS

Simple theories, if they are to be useful, should provide insights into many different aspects of the structure of a system. For most of the remainder of this chapter I will apply this theory to the six patterns mentioned above in order to determine the generality and limitations of this approach.

Soil Heterogeneity and Plant Dominance

The first point may seem so obvious as to be trivial, but it is not. On broad geographical scales the major variables along which plant species are separated are climate and soil type (e.g., Jenny 1980). On smaller scales climate is much less variable, but soil can be quite heterogeneous because many soil characteristics depend on depositional and erosional patterns and on its slope and aspect. Within relatively homogeneous climatic areas, soils are strongly correlated with plant community composition. Although cause and effect have been frequently debated, soil and vegetation maps of areas show marked similarities. One aspect of soils not influenced by plants is the initial substrate upon which soils formed. Initial substrates are often highly correlated with the dominant vegetation type. For instance, within Wisconsin the four largest areas with soils formed on wind-deposited sand were dominated by pine barrens (jack pine, oak, and prairie grasses) before settlement (Hole 1976). On a smaller spatial scale Lindsey (1961, p. 434) reported that vegetation patterns for seven counties of northern Indiana "definitely tend to follow substrates produced by different modes of glacial action, even though I minimized edaphic differences by mapping forest types as they occurred on only well-drained, median portions of the terrain mosaic." Within the Upper Galilee of Israel Rabinovitch-Vin (1979, 1983) found that "different climax plant communities grow in the same altitudinal belt under similar climatic conditions." These differences, Rabinovitch-Vin felt, were explained by soil differences that were dependent on the type of parent material on which the soil formed. Similar patterns have been reported on spatial scales ranging from meters to kilometers (e.g., Box 1961, Snaydon 1962, Piggott and Taylor 1964, Beals and Cope 1964, Zedler and Zedler 1969, Hanawalt and Whittaker 1976).

Not only are there correlations between soils and the composition of plant communities; these correlations have an underlying pattern. If soils within a locality were ranked from those with the lowest rates of supply of the limiting resource (often nitrogen or water) to those with the highest rate of supply, the dominant vegetation would be seen to change from plants of short stature at maturity to increasingly taller species. Thus, in Wisconsin, once fire frequency is controlled for, the poorest soils are dominated by lichens, forbs, grasses, and scattered pines or oaks. Richer soils are dominated by closed-canopy oak or pine forests, and still richer soils are dominated by maple (Hole 1976). The Wisconsin vegetation patterns could be described by Fig. 22.4 if lichens had the resource requirements of species A, grasses had the requirements of species B, pines and oaks had the requirements of species C and D, and maples had the requirements of species E. The most likely limiting soil resource is nitrogen. This same pattern occurs in the vegetation at the University of Michigan Biological Station near Pellston, Michigan (personal observation). Grime (1979) also reports that species that are taller at maturity tend to be found on richer soils, while shorter species tend to be found on more resource-poor soils or in very stressed environments.

Similar patterns have been reported in all five of the global regions with a Mediterranean-type climate. In four of these regions—California, Chile, Sardinia (Italy), and the Cape of southern Africa—the vegetation changes from low and open scrub to tall and dense scrub along a gradient from low to high availability of water, an important limiting soil resource (Cody and Mooney 1978). The heights at maturity of the dominant plants range along this gradient from less than 30 cm to more than 10 m. The species sequence along such a gradient could be explained by the relations shown in Fig. 22.4 if the best competitors for the limiting soil resource, water, were short at maturity and thus poor competitors for light, and if the best competitors for light had high water requirements. Similar changes from dominance by short to dominance by tall species occur in this vegetation following fire or other disturbances (Cody and Mooney 1978).

In southwestern Australia, the fifth global region with a Mediterranean climate, Beard (1983) reported that the vegetation composition was highly dependent on both the annual rainfall of a region and the parent material on which the soil was formed. Within a given climatic region, such as the Coolgardie Botanical District, the vegetation ranged from 3-m-tall scrub heath on sandy soils to 25-m-tall salmon gum woodland on the most productive soils, the white, kaolinitic clays. When Beard (1983) viewed the vegetation on a broader spatial scale, he found that, holding parent material constant, there was the same type of sequence from 3- to 5-m scrub heath to 25-m forest along a climatic gradient from 200 to 1,200 mm/year of rainfall.

For the New World tropics, Beard (1944, 1955) reported that a major axis of separation of species was from areas receiving sufficient rainfall all year to areas with only seasonal rainfall. Beard called this sequence the "seasonal formation series." The areas with both the highest rainfall amounts and no dry months had tropical rain forest with dominant plant species reaching 40 to 50 m in height. Areas with a few dry months each year had evergreen seasonal forest with maximal heights of 35 m. Along this gradient of increasing seasonal limitation by water, the other four formations Beard recognized had maximal heights of 25 m for the semievergreen seasonal forest, of 15 to 20 m for the deciduous seasonal forest, of 5 to 9 m for the thorn woodland, and of 2 to 4 m for the cactus scrub. He stated that "there is in reality one long unbroken series, in which the formations are artificially delimited stages." Beard (1955) reported four other formations for tropical vegetation. Along each of these gradients the mature, dominant plants ranged in height from a maximum of 25 to 35 m on the most productive end of the gradient to a maximum of 3 to 8 m on the least productive end.

For the native flora of the northeastern United States, Marks (1983) asked, "In the primeval landscape, where were the plants that are abundant today in old fields?" Based on the habitats described in old floras, the plant composition of present forest openings, the plant composition of persistently open habitats, and the seed dispersal abilities of old field plants, Marks concluded that the majority of the native flora found today in old fields were not fugitive species colonizing newly created, high-light openings, but had evolved in persistently open, marginal habitats such as sand plains, gravel bars, eroded stream banks, limestone outcrops, talus slopes, rock crevices, and eroded ridgetops. In contrast, the beech, maple, and other canopy trees probably had evolved in habitats with rich soils and closed canopies. This further supports the view that plants of shorter stature seem to be superior competitors on resource-poor soils, but are displaced from richer soils by taller plants.

Primary Succession

The parent materials in which soils form often contain all of the mineral elements required by plants except nitrogen (Jenny 1980). Several authors have suggested that changes in soil nitrogen during succession may be a major cause of the observed pattern of vegetation change. In Glacier Bay, Alaska, total soil nitrogen increased 6-fold to 10-fold during the first 100 years following glacial recession (Crocker and Major 1955). For the Mendenhall glacier, Alaska, total soil nitrogen increased 6-fold during the period from 10 to 120 years after glacial recession, and it increased a similar amount for the Herbert glacier (Crocker and Dickson 1957). Olson (1958) reported an approximately 15-fold increase in total soil nitrogen during the first 300 years following dune formation along the southern border of Lake Michigan. Possibly because of the evolutionarily unlikely mechanisms for soil change proposed by Clements (1916), and possibly because many recent studies of succession have focused on secondary succession on rich soils (see McIntosh 1981), soil processes have recently been assigned secondary importance in succession (e.g., Connell and Slatyer 1977, Grime 1979). There is, though, no reason to ascribe soil development to altruism in early successional plants. Soil is a mixture of mineral particles from the parent material, refractory organic compounds that soil micro- and macroorganisms could not utilize, and the soil flora and fauna itself. Photosynthetic plants do not form soil; soil organisms do. The formation of soil is an evolutionary indirect effect of the death of photosynthetic plants and the life of decomposers.

The pattern of change in primary succession and in many secondary successions on poor soils is a pattern in which there seems to be a tight correspondence between soil and plants. As soil nitrogen levels increase during succession, total plant biomass increases. Increased plant biomass leads to increased attenuation of light by the plants. At the soil surface, where newly establishing seedlings and shoots grow, light availability decreases steadily as plant biomass increases. Thus, the habitat available for newly establishing plants changes dramatically during succession, from a habitat with low availability of nutrients but high light intensity to a habitat with high availability of nutrients but low availability of light. The approximately 10-fold increase in soil nitrogen during succession is associated with a comparable decrease in light availability at the soil surface. The relative availability of nitrogen to light, which could be expressed as the ratio of total soil nitrogen to light availability at the soil surface, thus changes by about 100-fold during succession.

As the relative availability of these resources changes through time, the composition of the plant community also changes. The actual species involved depend on the geographical location and the local climate and substrate, but the general trends are often the same. The bare mineral substrate is initially dominated by soil algae, mosses, or lichens, followed by taller herbaceous plants (often capable of nitrogen fixation), then by taller woody shrubs, and finally by even taller trees. The "endpoint" of the successional sequences in a given area seems to be highly dependent on the parent material (Olson 1958, Jenny 1980), and very long-term studies suggest that there is no endpoint (Walker et al. 1981). This type of successional sequence could be explained by the changes in the relative availability of nitrogen and light if the dominant plant species are differentiated in their competitive abilities for these two resources as hypothesized in Fig. 22.4. The early successional species (species A and B of Fig. 22.4) would have lower requirements for the limiting soil resource, nitrogen, but higher requirements for light than the later successional species (D and E). If this relationship exists, changes in the relative availability (or ratio) of nitrogen and light would lead to changes in the composition of the plant community. The eventual composition of the plant community would be determined by the eventual ratio of nitrogen to light at the soil surface. This general approach, the resource ratio hypothesis of succession, predicts that succession should be a directional or predictable process only to the extent that the relative supply rates of the limiting resources change in a directional or predictable manner (Tilman 1982, 1985).

One of the longest successional sequences studied is in the sand dunes of eastern Australia, where the existing vegetation may represent a 400,000-year chronosequence. This site shows the type of successional sequence discussed above, followed by a long period of "retrogression" during which progressively shorter plants become dominant. As predicted by the resource ratio hypothesis, the retrogression is associated with a progressive loss of soil nutrients, especially phosphorus and calcium, caused by leaching (Walker et al. 1981).

Evolution of Land Plants

Similar patterns may have occurred during the early evolution of land plants. The first land plants were probably single-celled algae (Stebbins and Hill 1980), which lived on the moister areas of bare mineral substrate, just as many soil algae do now (Bold 1970). Through most of the Precambrian era the terrestrial ecosystem contained cyanobacteria, algae, bacteria, and, later, fungi (Golubic and Campbell 1979, Knoll and Rothwell 1981). The early mineral substrates probably provided most of the elements essential for algal growth, except nitrogen, which does not occur in mineral substrates (Jenny 1980). Just as nitrogen-fixing cyanobacteria (blue-green algae) are common in bare but moist mineral soils (Bold 1970) and in nitrogen-limited lakes (Smith 1983) today, the first major dominant plants of the terrestrial habitat may have been nitrogen-fixing cyanobacteria. Unfortunately, as Stebbins and Hill (1980) lamented, there are no fossil remains from these habitats.

As terrestrial algae grew, the remains of these photosynthetic plants provided energy for various bacteria and, in the late Precambrian, fungi. Refractory organic compounds from the soil cyanobacteria, algae, bacteria, and fungi would have accumulated within the original mineral substrate. That mixture of refractory organic compounds, minerals, and microorganisms was the first soil. As occurs now during primary succession, this soil substrate was probably more nutrient-rich, especially in nitrogen, and probably had better water-holding capacity than the original parent material. Thus, as cyanobacterial and algal growth eventually and indirectly led to higher availabilities of limiting soil resources, plant biomass increased. The soil surface likely was covered with dense mats of cyanobacteria and algae. In these mats light attenuation could have been extreme. Because light is a directional resource supplied from above, algae on top of the mat would obtain more light, but would have lower availability of soil resources.

This spatial separation of essential resources into below-ground resources (nutrients) and above-ground resources (light) may have been a major factor favoring multicellularity in early land plants. A multicellular alga could, if properly oriented, obtain both soil nutrients and light. Because nutrient absorption is dependent on cell surface area, a multicellular alga with some cells in the substrate but some cells above the substrate would have less surface area for nutrient uptake than algae that were completely in the substrate. The cells above the substrate would, though, be in an area with a higher availability of light. Such a morphology could have been favored on a nutrient-rich substrate even though above-ground structures might have been acquired at a cost in the ability of the alga to obtain nutrients. This suggests that a major factor favoring multicellularity in land plants may have been the simultaneous need for a below-ground and an above-ground resource. Multicellularity would have allowed specialization of cell function within an individual. However, such multicellularity would have been a compromise in which a plant sacrificed some of its ability to compete for one limiting resource to gain competitive ability for another.

A finer-scale view of the early evolution of vascular plants suggests that competition for light may have been a major selective force. In floodplain areas during the Devonian (ca. 395 to 345 million years ago), there were dramatic replacements of one major vascular plant group by another (Figs. 7.2 and 7.3). Knoll (1984) attributes these replacements to interspecific competition, mainly for light. The earliest fossil vascular plants, the Rhyniophytes, were short and leafless with dichotomously branching, photosynthetic axes. Some of these gave rise to the Trimerophytes, which were taller and had stronger stems than the Rhyniophytes. As the Trimerophytes became dominant, the Rhyniophytes went extinct (Knoll 1984). In the middle Devonian the Progymnosperms were derived from the Trimerophytes and eventually displaced them. The next groups to dominate were the Lycopsida (e.g., club mosses and quillworts), the Sphenopsida (horsetails), the Pteropsida (ferns), and the Pteridospermales (naked-seeded plants). The transition to these groups from the single-celled algae represents a transition from plants which were one cell "tall" to plants which were as tall as our present trees (Tiffney 1981). The main factor favoring increased plant height is competition for light (Givnish 1982). As early mineral substrates became nutrient-rich soils, plant biomass increased, and light became a more important limiting resource than nitrogen or some other soil resource. Much of the pattern in the early evolution of land plants may have been caused by increasingly strong competition for light favoring taller plants in the habitats in which the parent material and topography allowed the formation of nutrient-rich soils.

As taller forms arose, they could have displaced shorter forms from the resource-rich soils. However, they would have been able to displace existing plants from resource-poor soils only if the new, taller forms were both better competitors for light and soil resources. A species which is a superior competitor for several limiting resources through a broad range of resource availabilities is a "supercompetitive species." Many of the extinctions of early vascular plants (Knoll Chapter 7 and 1984) may have been caused by the evolution of supercompetitive species. As discussed in Tilman (1982), selection within such supercompetitive species is likely to favor specialization to local conditions, and such specialization could lead to speciation within the supercompetitive species. It seems likely, though, that the majority of the species that evolved along the soil resource:light gradient were not supercompetitive species, but had traits that represented trade-offs along this gradient. To the extent that new species were superior competitors for light but inferior competitors for soil resources, the earlier plants could have coexisted with the new species. The present flora suggests that species descended from some of the early plants of short stature, such as soil algae, mosses, lichens, and horsetails, do have a range of availabilities of soil resources and light through which they are dominant or coexist with descendents of the later evolving but taller plants. This is consistent with differential requirements for a soil resource and light such that the shorter species have lower requirements for soil resources but higher requirements for light than the taller species (see Fig. 22.4).

Fertilization Experiments

In mesic, north temperate areas agricultural practice and numerous fertilization experiments performed on natural vegetation have indicated that, of all the mineral elements required by plants, nitrogen is often most limiting (e.g., Lawes, Gilbert, and Masters 1882; Milton 1947; Willis and Yemm 1961; Specht 1963; Thurston 1969; Til-

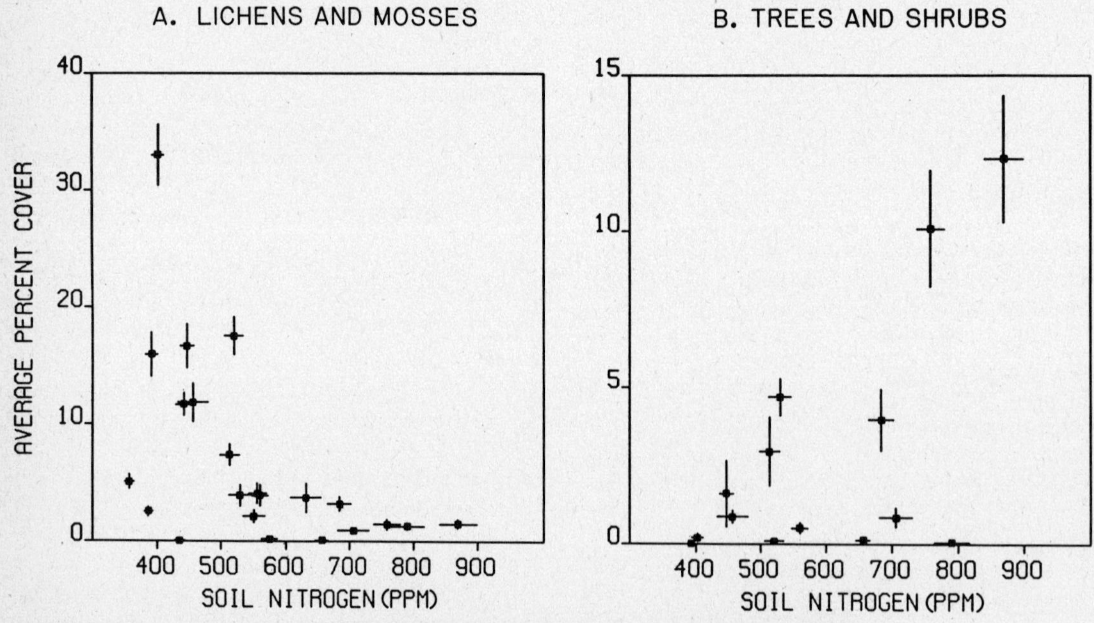

Fig. 22.5 Species dominance and limiting soil resources. Each point in this figure shows the mean percent cover (n = 100) and the mean total soil nitrogen (n = 100) of an old field at Cedar Creek Natural History Area. The lines from each point show the standard error of the mean. Note that lichens and mosses have their peak dominance on poor soils, while trees and shrubs dominate old fields with high (for Cedar Creek) total soil nitrogen levels. Fields with intermediate levels of nitrogen are dominated by herbaceous perennials. These correlational patterns, though, are confounded by the positive correlation between soil nitrogen and age of the old field. (Inouye et al. manuscript in preparation.)

man 1982, 1984). Within fields that we have sampled at the University of Minnesota's 2,300-ha Cedar Creek Natural History Area, which occurs on a glacially deposited sand plain, the most nitrogen-poor soils have total nitrogen levels from 100 to 400 ppm. These soils are dominated by mosses and lichens with areas of encrusting soil algae. Richer soils are dominated by herbaceous vascular plants, and even richer soils by woody shrubs and trees (Fig. 22.5). Following nitrogen fertilization within a field with poor soils, lichens were competitively displaced by vascular herbs within two years (unpublished data). Nitrogen fertilization of plots on newly disturbed soil at Cedar Creek has shown that early successional annual plants are rapidly displaced on plots receiving high levels of nitrogen, but persist on the nitrogen-poor soils of control plots (Tilman 1983, 1984).

In greenhouse studies that I have performed, silica sand substrates watered with all mineral nutrients except nitrogen have invariably become covered with a mat of crust-forming algae. Thicker crusts were formed in pots receiving slightly higher levels of nitrogen, but there was little obvious algal growth in the pots receiving even higher nitrogen levels; those pots instead had a dense growth of herbaceous vascular plants shading the soil surface. None of the annual herbs that I was growing in these pots reproduced at the lowest nitrogen level, but the crust-forming soil algae flourished. Although these are admittedly preliminary data, they are consistent with the hypothesis that soil algae are better competitors for nutrients, but are inferior competitors for light than vascular plants.

The addition of limiting soil resources to terrestrial vegetation has consistently resulted in dramatic changes in plant dominance (Tilman 1982). More than 125 years ago, Lawes and Gilbert (1880) started fertilizing a regularly mowed pasture at the Rothamsted Experimental Station in England with different combinations of mineral nutrients. Their manipulations, still in progress today, led to marked changes in dominance (Brenchley and Warington 1958). The patterns of species dominance that have resulted from these fertilizations are consistent with each species' being a superior competitor for a particular ratio of availabilities of the limiting resources, especially nitrogen and light (Tilman 1982).

Patterns of Resource Availability

The 6- to 15-fold changes reported for soil nitrogen during primary succession may be compared to the local spatial variation in soil nitrogen. The toposequence from Isanti to Sartell soils at Cedar Creek Natural History Area ranges on average from 4,500 ppm total nitrogen to 400 ppm, an 11-fold difference (Grigal et al. 1974). On a smaller spatial scale 216 samples were collected within 30 $\times$ 45 m areas in each of three old fields at Cedar Creek. Total soil nitrogen in the upper 15 cm of soil ranged from 405 to 1,240 ppm in the 14-year-old field, from 123 to 830 ppm in the 25-year-old field, and from 332 to 994 ppm in the 48-year-old field (Tilman 1982). These values represent nitrogen variation of from 3.0- to 6.7-fold within 0.14-ha areas. Thus, the variation in soil nitrogen through time, which several authors have suggested may play a role in succession, is not much greater than the variation observed through space at a given time in a small area. This suggests that if soil change is an important determinant of succession, local spatial variation in soils may be an equally important determinant of vegetation patterns at a given time.

The theory presented in Figs. 22.3 and 22.4 assumes that light and a limiting soil resource should form a natural gradient, that is, that the availability of light at the soil surface should be negatively correlated with the availability of a limiting soil resource. Several studies at Cedar Creek Natural History Area have shown that the major limiting soil resource in both old fields and native oak-savannah vegetation is nitrogen (Tilman 1983, 1984). In the spring of 1982 total soil nitrogen was measured in 258 control plots for experiments on the effects of nitrogen on natural vegetation in fields of different successional ages at Cedar Creek. The control plots did not receive any nitrogen fertilization. In August of that year light attenuation to the soil surface was measured in the plots. Both within each field and among all fields there was a significant negative correlation between total soil nitrogen and light attenuation

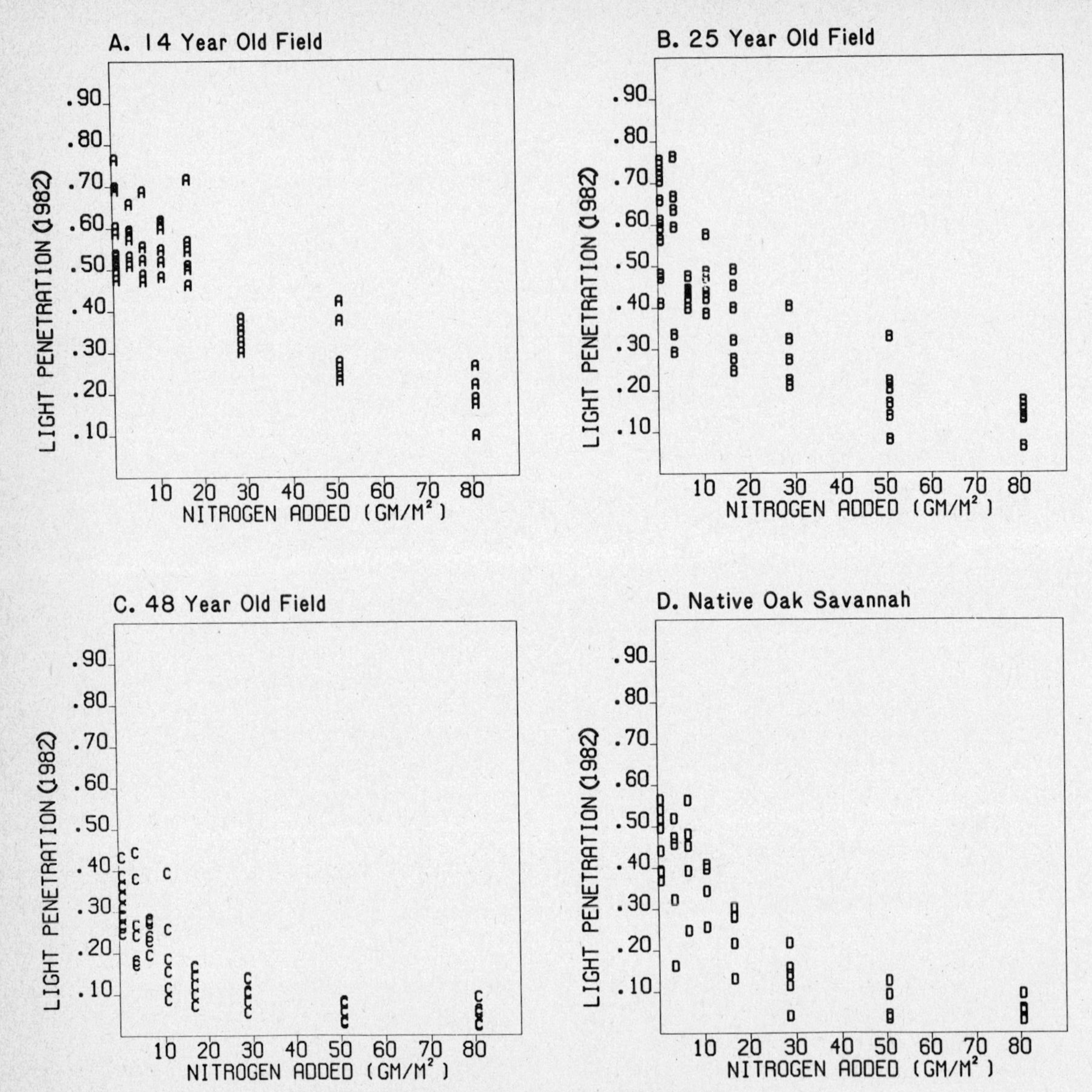

Fig. 22.6 Dependence of light available at the soil surface in August 1982 on the rate of addition of nitrogen (as ammonium nitrate) to four fields that year. Note the highly significant decrease in light available at the soil surface with increasingly high rates of nitrogen supply.

to the soil surface. This is a partial test of the assumption, but it depends on a further assumption: that the supply (mineralization) rate of nitrogen is proportional to total soil nitrogen. Pastor et al. (1984) have shown that the supply rate of nitrogen is also dependent on several other soil factors. A clearer test would come from fertilization experiments in which the supply rate of available nitrogen is controlled. Hence nitrogen fertilization experiments were performed and showed that light penetration to the soil surface decreased consistently with nitrogen addition (Fig. 22.6). Considered together, the control plots and the nitrogen fertilization experiments strongly support the hypothesis that there is a naturally inverse nitrogen: light gradient in and among the fields studied.

Plant Life History Evolution

The traits of a plant species should reflect the selective pressures of the environment in which it evolved. One hypothesis that has been used to explain the evolution of many plant life history characteristics has stressed the differential abilities of plants to grow in disturbed versus undisturbed areas. Differences in plant growth rates, age at first reproduction, seed size, and seed dispersal abilities are often cited as being consistent with the hypothesis that each plant species falls somewhere along a continuum from being an "r-selected, fugitive species" to being a "K-selected, stable habitat species," although the terms r and K are now invoked less frequently. The work reviewed above and the treatment of one case in depth by Marks (1983) suggest that the plants that are now dominant during early succession did not evolve in habitats in which their ability to colonize recently disturbed sites was the major selective force, but rather to great increases in reproductive rate. This being so, it is necessary to explain how plants with delayed reproduction may have evolved from plants that reproduced earlier. How, for instance, could mosses have evolved from soil algae? Reproduction in mosses is delayed because moss spores undergo cell division to form multicelled, relatively tall, moss gametophytes. The moss gametophyte, because of its height, could outcompete soil algae for light. In habitats in which light was a major limiting resource, individuals that delayed reproduction and by so doing became taller were specialized on particular relative availabilities of a limiting soil resource versus light. Other species seem to have evolved at other points along the soil resource:light gradient. Although many processes influence their life history evolution, how may the life histories of plants depend on their position along this gradient?

An individual plant can allocate its potential growth in several different ways: to asexual or sexual reproduction, to increases in height and photosynthetic area, or to increases in absorptive tissues for below-ground resources. When a plant allocates more of its potential growth to one of these functions, there is less potential growth that can be allocated to the other functions. The age at first reproduction is one of the most important determinants of fitness (Cole 1954). Slight decreases in the time to first reproduction can lead and obtained more light could have been favored over those that reproduced sooner.

This pattern of association between height at maturity and delay until first reproduction seems to hold throughout many plant lines. There would be an optimal age at first reproduction that represented a compromise between the increase in future reproduction that could come from greater allocation to vegetative growth versus the "price" of delayed reproduction caused by such allocation. Because light is attenuated through the canopy in a negative exponential manner, plants that allocate their growth in low-light habitats so as to reach the canopy can have immense increases in light capture. In contrast, plants that allocate their growth below ground in low-nutrient habitats probably receive only proportionate increases in nutrient uptake. Thus, delayed reproduction may be more commonly favored in light-limited habitats than in nutrient-limited habitats.

On very nutrient-poor soils, a plant will be limited by nutrients. Plants that allocate more growth to below-ground structures for nutrient acquisition would be favored over those that allocate more to above-ground structural and photosynthetic tissues, since light and a limiting nutrient are essential, not substitutable for each other. Optimal foraging theory for essential resources predicts that plants should "forage" so as to be equally limited by all essential resources (Tilman 1982). The "foraging strategy" of a plant is expressed by its allocation to below-ground structures for nutrient acquisition versus its allocation to above-ground structures for light capture. Both a plant that allocated so much of its potential growth to above-ground photosynthetic structures that it became nutrient-limited and a plant that allocated so much of its potential growth to below-ground structures that it became light-limited would be outcompeted by an individual that allocated its growth such that it was equally limited by both nutrient and light. On any given soil the optimum allocation to nutrient versus light acquisition is the allocation that leads a plant to be equally limited by both resources. Additionally, plants that are good competitors for nutrients are often capable of rapid uptake and storage of nutrients (Marks 1974), which are used for future growth (Chapin 1980).

Tall plants face a major structural problem. Plants are basically poles. Even vertical poles firmly attached at their base are unstable because, once perturbed, their own weight pulls them in the direction of the pertubation. Greenhill (1881) calculated the greatest height consistent with stability in a vertical pole or tree of given dimensions. For taller plants to be stable given even minor pertubations, they must have a stronger stem. Of all the different types of plant stems, the strongest are those of hardwood trees. Hardwood stems are also the most energetically expensive for a plant to make. Thus, there are two costs for a plant living in a low-light habitat when it increases its height at maturity: delayed reproduction and the necessity of producing a more energy-expensive stem (Givnish 1982). These plants, though, are growing in a habitat in which energy is their main limiting resource. For them to produce a stronger stem, they must necessarily grow more slowly. This suggests that competition for light necessitates slower growth and delayed reproduction in order for plants to be taller at maturity. Because these plants reproduce later in life, they may tend to be more long-lived than plants that are superior competitors for nutrients. However, because differential survival of adults and young may be a major selective force for long life span (Charnov and Schaffer 1973), not all plants that are good nutrient competitors need be short lived.

For situations in which nitrogen is the limiting soil resource, there is a further physiological reason why a plant would be a superior competitor for only one region along the nitrogen:light gradient (Chapin 1980). Chapin states that "photosynthetic rate is proportional to leaf nitrogen concentrations over a broad range, because the bulk of the leaf nitrogen is directly involved in photosynthesis as a component of photosynthetic enzymes and chlorophyll." Nitrogen that has been allocated to leaves cannot be used in roots for the production of the enzymes involved in the active transport of nitrogen into cells. Thus, the allocation of nitrogen to uptake versus photosynthesis would influence the growth rate of a plant in a particular habitat.

Seed size is influenced by many factors, including the trade-off between dispersal ability and competitive ability of seeds (Werner and Platt 1976). For plants that can regenerate in undisturbed vegetation, those species that are good competitors for light must be able to grow as seedlings in the shade of other plants. For these species larger seeds might be favored because they would allow a more rapid increase in height (Grime and Jeffrey 1965) and thus in light capture. In contrast, for plants that are similarly dominant under high nutrient but low light conditions but that can regenerate only in openings (disturbances), seed dispersal ability may be more important than seed competitive ability. This suggests that smaller seeds may be favored in such species at higher soil resource:light ratios. Additionally, because disturbances are ephemeral, rapid initial growth may be favored in such species. Thus, either larger or smaller seeds may be favored at high nutrient:light ratios depending on the regeneration requirements of the plants.

Many factors other than a plant's position along the soil resource:light gradient will also influence its life history traits. In total, these considerations of the evolution of plant life histories suggest that, in general, compared to plants that are superior competitors for light, plants that are superior competitors for soil resources should have relatively rapid vegetative growth, should invest a relatively greater proportion of their growth in structures for nutrient acquisition than for photosynthesis, should reproduce earlier in life, and should be shorter at maturity.

All plants are phenotypically variable. Because phenotypic plasticity can partially mask interspecific life history differences, the life history traits of a species should be observed in the habitat in which it is naturally dominant. For in-

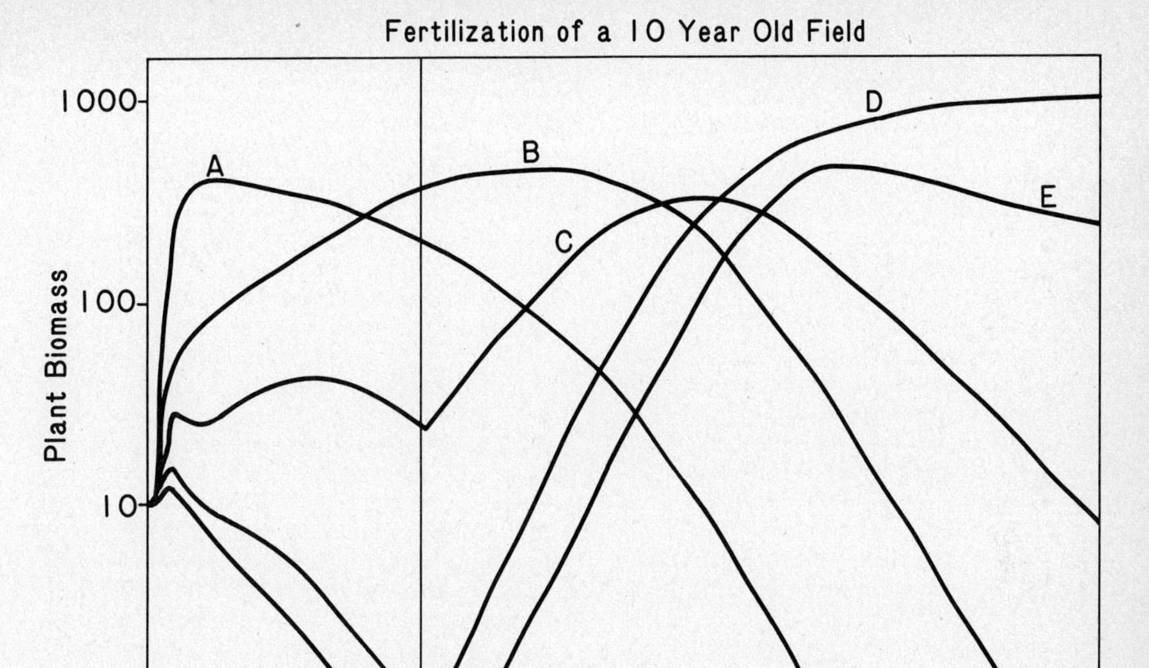

Fig. 22.7 Dynamics of competition among five species for nitrogen and light. The model is the same as in Fig. 22.4. In this case the initial soil has a total nitrogen of 200 ppm, which leads toward coexistence of species A and B, with B more abundant than A. After 10 years of such competition, the supply rate of nitrogen is increased six fold. This leads to the eventual coexistence of species D and E. However, there is a significant transient response. Species C, which is not a superior competitor under either set of conditions, becomes dominant during the transition because it has a higher growth rate than species D and E. This higher growth rate is consistent with the life history predictions in the text. Species C would not become dominant during this transition if all species had the same maximal growth rates. Thus, differences in the maximal growth rates of species may be very important in determining the transient dynamics observed following a pertubation. This transient response took 12 years before damping began, a rather long time in relation to ecological research.

stance, aspen (*Populus grandidentata*), which tend to be dominant on poor soils, are capable of growing as tall as sugar maple (*Acer saccharum*) on rich soils. However, their relatively weak stem causes them to experience much higher loss rates on these soils than maple. The phenotypic plasticity that allows aspen to grow into the canopy on rich soils does not allow them to coexist with maple in these forests, but does slow the rate at which they are displaced.

Life history differences may have a profound effect on the transient dynamics shown by species during their approach to equilibrium. The resource-dependent isoclines in Figs. 22.1–22.4 (and in numerous figures in Tilman 1980, 1982) represent equilibrium conditions. Many different dynamic models can give identical resource-dependent isoclines. Thus, it is possible for there to be marked differences in dynamics away from equilibrium for species that have identical equilibrium resource requirements. For many short-term experiments, such as most experiments performed on terrestrial plants, the transient response may be more easily observed than the eventual equilibrium. The transient response could be easily misinterpreted (see Bender, Case, and Gilpin 1984). If plant species differ in their traits along a nutrient:light gradient as discussed above, especially in maximal growth rates, the initial species to become dominant after a manipulation may not be the best competitor for the experimentally imposed conditions. For instance, Fig. 22.7 shows a simulation of the development of a plant community on a nitrogen-poor soil. After 10 years this model community was fertilized so as to increase the rate of supply of nitrogen 6-fold. This led to a successional sequence in which species D and E were the eventual dominants. However, short-term observations could have led an investigator to conclude that species B or C was the superior competitor for the newly imposed conditions. The transient increase by species B and C would not have occurred if they did not have higher maximal growth rates than species D and E.

CONCLUSIONS

In this chapter I have reviewed relations between terrestrial plants and their resources from a variety of perspectives. Such seemingly unrelated observations have often been ignored in trying to understand the causes of pattern in plant communities. Yet these patterns—patterns that occur over temporal scales from one year to thousands of years and over spatial scales from meters to kilometers—may provide a significant insight. What would be a more parsimonious explanation for the species composition, dynamics, diversity, and physical structure of plant communities than an approach that explained broad- and small-scale spatial patterns and short- and long-scale temporal patterns? Is it not possible that the same processes that led to the differentiation and evolution of early land plants are operative today structuring plant communities?

These various views of plant communities and plant evolution strongly suggest that one of the major axes of plant differentiation may be their competitive abilities for soil resources versus their competitive abilities for light. It seems that terrestrial plants may have trade-offs in their requirements for these limiting resources such that the species are inversely ranked in their competitive abilities for above-ground versus below-ground resources. Plants are further differentiated in their abilities to compete for different below-ground resources (Tilman 1982) and in the dependence of their growth rate on such physical factors as pH and temperature. If plants prove to be specialized on different points along a soil resource:light gradient, it may be better to classify plants as ''low-nutrient but high-light species'' or ''high-nutrient but low-light species'' rather than as ''early successional'' or ''late successional'' or ''fugitive'' or ''climax'' species. Indeed, it would be best to specify the actual point along the gradient at which each species has its optimal competitive ability. The life histories of old-field and prairie plants with which I am familiar seem much more consistent with the view that they are evolved for different points along a soil resource:light gradient than in response to different disturbance rates.

The view presented in this chapter provides one possible explanation for some otherwise difficult-to-explain relations in terrestrial plants. North temperate regions are inhabited by hundreds of species of vascular plants that seem to be able to coexist because they are differentiated with respect to one another. However, paleoecological evidence (Chapter 16 and Davis 1981b) has shown that these species are not part of a long-existing potentially ''coevolved'' community. The present species combinations seem to be unique to the past 10,000 years since glacial recession. How, then, could species that evolved apart for the preceding 100,000 years happen to have traits that allow so many of them to invade and coexist in recently deglaciated habitats? A similar question has been raised by Goldberg and Werner (1983), who point out that plants, because they are bound to a particular locality, interact most with their immediate neighbors. In a multispecies community, how could species have coevolved if they were infrequently associated as neighbors?

Both of these relations might be explained if species are envisioned as being differentiated along one or a few major axes. For instance, if each species is a superior competitor for a particular point along a soil resource:light gradient, it doesn't matter exactly which species a plant has as its neighbors. The neighbors will either be superior competitors for light but inferior competitors for the nutrient than the plant, or they will be superior competitors for the nutrient but inferior competitors for light. Given neighbors with such traits, it is easy to see how the traits of a species could be subject to stabilizing selection. When plants invaded a recently deglaciated region, the other species invading that region would have been similarly differentiated. Even though a particular species had not evolved its traits in response to the group of species with which it interacted in the new area, each had evolved traits that allowed it to be a superior competitor for a particular point along the nutrient:light gradient under some climatic conditions.

The view presented in this chapter may be contrasted with another common view of plant community processes, the gap-replacement perspective. I consider my view to be a statement of the mechanisms by which one individual replaces another, on average. However, it could be claimed that light gaps are an exception to the view presented in this paper, since light gaps have both high-light and high-nutrient availability. I would counter this assertion by noting that nutrient levels are elevated in light gaps for a relatively short period of time and that only a small percent of the mature individuals in a forest die any given year. Thus, the majority of the seedlings and saplings in a forest experience low light levels. Their abilities to survive and grow under these conditions would determine which of them was tallest and thus most likely to become dominant when a gap was created. Too detailed a view of the short-term dynamics of light gaps may obscure the long-term average effect of the interactions among forest plants. The short-term dynamics may more closely reflect life history traits evolved for reasons other than growth in light gaps.

There are many ways that plants may be differentiated, each of which is potentially important in a given system. Any attempt to apply this theory to a particular plant community must consider the actual constraints of that habitat and modify the theory accordingly. Additionally, this theory represents a major simplification of plant interactions and habitat structure. This theory is a framework upon which more complete and realistic models, including dynamics models (Tilman 1985), may be built. However, I want to stress that patterns in terrestrial plant spatial distributions, succession, evolution, and life histories seem generally consistent with the hypothesis that one of the major factors leading to differentiation in terrestrial plants may have been competition along the soil resource:light gradient. If further work tends to support the importance of the soil resource:light gradient, it may have profound implications for our view of the evolution of plant life histories and the processes allowing the long-term persistence of plants in a given region.

SUMMARY

Limiting soil nutrients and light tend to form a natural gradient. Habitats with low availabilities of soil resources have low plant biomass and thus high availability of light at the soil surface. Habitats with rich soils support high plant biomass and thus have low availability of light at the soil surface, where it is required for the growth of seeds and shoots of newly establishing plants. Because plants require both soil resources and light for growth, each plant necessarily has a limited region along a soil resource:light gradient for which it is a superior competitor. Such soil resource:light gradients may have been a major axis for the differentiation and evolution of early land plants and may help explain the life history patterns we see in current plants. Compared to plants that are superior competitors at low nutrient levels, plants that are good competitors for light would be expected to be taller at maturity, be slower growing, have a relatively greater investment in structural and photosynthetic tissue, and reproduce later in life. The patterns of early evolution of land plants, their successional sequence, and the life history traits of various species seem generally consistent with this hypothesis.

ACKNOWLEDGMENTS

This material is based upon work supported by National Science Foundation Grant BSR 81-143202-A02. I thank Edward Cushing, Margaret Davis, Eville Gorham, and Patrice Morrow for numerous discussions we have had about this material. I thank Martin Cody, Jared Diamond, Peter Grubb, Andy Knoll, Robert Sterner, Ted Case, David Wedin, Terry Sharik, Sam Scheiner, and the Naturalist-Ecologist Training Program students at the University of Michigan Biological Station for their comments on this paper.

chapter 23

Structural Niches in Plant Communities

Martin L. Cody

INTRODUCTION

Plant ecology has its origins in the description of vegetation and the classification of plant distributional patterns (Schimper 1903, Raunkiaer 1934). From these beginnings the three main areas or levels of modern studies were derived. (1) The phytosociological approach describes plant associations within habitats and plant changes among habitats (Clements 1916, 1936; Weaver and Clements 1938; Whittaker 1975). (2) The population approach is concerned with the population dynamics of individual species in specific environments (Harper 1977, Grime 1979). (3) The physiological-biophysical approach interprets the life processes of individual plants relative to physical and chemical factors (Nobel 1979, Gates 1980). In contrast, much of animal ecology has operated at a level between the whole association and the single population, namely, at the level of guilds of ecologically related species. Since such guilds are chosen so that their members are most likely to affect each other's density, distribution, and other ecological characteristics, it is largely within such species groups that the significance of ecological and adaptive features can be resolved.

Some studies do exist on the ecology of plant guilds, excellent examples being those by Abrahamson and Gadgil (1973) and Werner (1979) on goldenrods (*Solidago*), by Biehl (1985) on buckwheats (*Eriogonum*), by Harper and McNaughton (1962) on poppies (*Papaver*), by Tilman (1977) on freshwater algae, by Yeaton et al. (1985) on *Yucca* species, and by McNaughton (1984) on Serengeti grasses. However, plant studies at this level have been undertaken much less frequently than for animal groups, and it is at this level that the niche concept is most useful. The axes along which related plant species segregate need first to be identified, so that questions about resource partitioning, niche shifts, and density compensation, similar to those pursued by animal ecologists, can be formulated and answered. In this chapter I shall discuss how the component species of several plant guilds are differentiated with respect to life form, especially in the morphology of their leaves, root systems, and branching patterns.

PLANT LIFE FORMS AND NICHE AXES

A plant's life form includes those aspects of its gross morphology that indicate how it disposes its photosynthetic surfaces to trap light above ground and how it arranges roots for water and nutrient uptake below ground. Differences between species in life form indicate a divergence in strategies of light, water, and nutrient collection; Grubb (1977, Chapter 12) regards life form and three other possibly less fundamental niche axes—habitat differences, phenological differences, and differences in processes associated with regeneration—as critical in the segregation of plant species.

The selective pressures on a plant's gross morphology are many and varied, but they can be sorted into two general sorts: abiotic forces promoting life form convergence and biotic forces promoting life form divergence. Abiotic constraints act similarly on all of the plant species in a habitat or microhabitat and result in the general morphological similarity of coexisting species relative to the plants of different habitats. Thus, microphylly or stem succulence in desert shrubs, broad sclerophylly in Mediterranean-climate chaparral (Schimper 1903), and simple, drip-tipped, evergreen leaves in tropical rain forests (Richards 1964) are characters shared by many species in these habitats. Such convergent characters may change in unison across species with changing environments (see Box 1981). The change in life form from shrubs to graminoids to trees with increasing moisture availability up mountain slopes in the southwestern United States (Whittaker and Niering 1965), the reduction in leaf area with increasing elevation in Philippine dipterocarp forests (Brown 1919) and in the Peruvian Andes (Terborgh personal communication), and the association of broader-stemmed cacti with steeper, rockier slopes and of narrower-stemmed cacti with denser vegetation and shadier sites are further examples of common solutions in many plant species to specific abiotic challenges. There may also be similar responses among coexisting plants to herbivores, such as similar branching patterns that render their foliage less accessible.

Opposed to the abiotic forces promoting life form convergence are biotic forces that favor divergence in life form, by means of which different species utilize different methods of light, water, and nutrient collection and thereby coexist in the same habitat. Within a habitat different life forms reflect different utilization of light or water, or different solutions to problems such as temperature extremes, drought, or herbivory. Plants committed to one particular life form are thereby committed to a particular mode of resource utilization, and their effects on plants of different life form are reduced.

Recently, there has been renewed interest in interpreting plant life forms, especially in forest trees (Halle and Oldeman 1970; Horn 1971; Halle, Oldeman, and Tomlinson 1978; Tomlinson and Zimmermann 1978). However, the difficulties in tackling the life form problem in forest trees are formidable. Trees have extremely complicated life forms that are not easily quantified, that vary within and among environments to the extent that form constants are not readily discerned, and whose adaptive significance is not always obvious (e.g., Oldeman 1979). Such problems can be circumvented in at least four ways. Adaptive morphology can be studied (1) by selecting vegetation with a simple, two-dimensional structure (e.g., Sarukhan 1974, Thomas and Dale 1976, Werner 1979), (2) by selecting vegetation in which plant individuals are isolated in their above-ground portions, as in deserts, (3) by focusing on just one aspect of plant form, such as leaf morphology, or (4) by selecting vegetation with a simple and easily quantified morphology, such as leafless and minimally branched columnar cacti.

This chapter uses all four of these simplifying notions to examine the adaptive significance of plant life forms in two habitats, deserts and Mediterranean-climate chaparral or fynbos. In deserts I discuss two groups of plants: leafy shrubs and stem-succulent cacti. Leafy shrubs are convergently similar in above-ground morphology in open deserts, but differ radically in root systems, with implications for near neighbor preferences. Variation in stem diameters and branching patterns of stem-succulent cacti appears to reflect the relation between photosynthetic area and available moisture. In chaparral and equivalent dense scrublands, I illustrate the divergence in leaf morphology among species and interpret interspecific differences in terms of physiological adaptation to different light, temperature, and humidity microhabitats.

ASPECTS OF LIFE FORM IN DESERT HABITATS

Mojave Desert Shrubs

Apart from the distinct stem- and leaf-succulent species of cacti and yuccas, the Mojave Desert vegetation appears to be a remarkably homogeneous assemblage of 1-m-high bushy shrubs. This superficial uniformity belies a considerable plant species diversity, with 0.01-ha quadrats in level granitic alluvium around 1,200 m elevation supporting 6 to 12 species and 0.1-ha plots producing around 30 species (Cody 1985). Although taxonomically diverse and drawn from many families, many shrubs are similar in stature and overall appearance. One view of this similarity is presented in Fig. 23.1, which shows how leaf area is accumulated horizontally and vertically through the plant. Three species (a–c) have ephemeral leaves, but photosynthetic stems, and four (e–h) are evergreens with various leaf longevity and turnover rates; one leaf succulent (d) is included for comparison.

Analysis of the spatial relationships among species reveals that the vegetation is far from a random mix of plants. There are distinct near-neighbor preferences such that certain species occur in proximity far more often than they might by chance alone, and others are significantly less

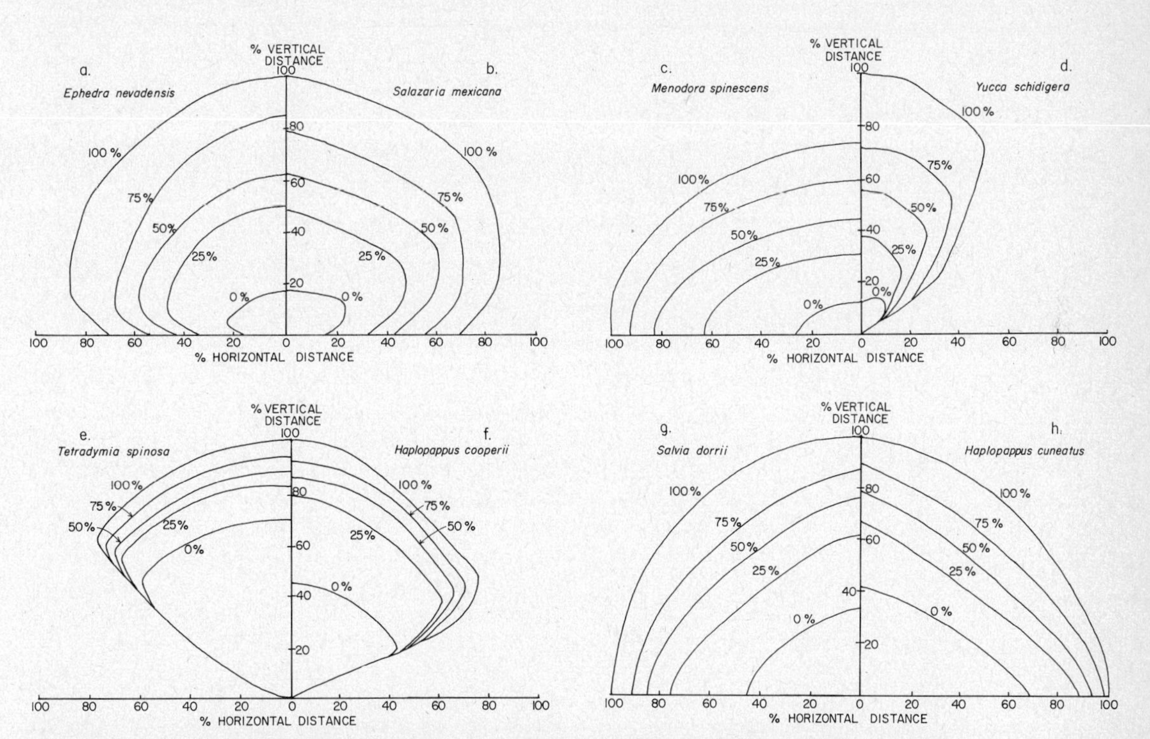

Fig. 23.1 Distribution of photosynthetic surface area within eight species of Mojave Desert shrubs. The horizontal and vertical axes represent horizontal and vertical distances from the shrub stem at ground level, and the contour lines show cumulative percentage of leaf photosynthetic area as a function of these relative distances. Species a–c have photosynthetic stems, d is a leaf succulent, and species e–h are leafy and variously deciduous. Note the strong similarities in growth form between different (and taxonomically unrelated) species within groups a–c and e–h. (After Pahlavan 1977.)

common as neighbors than might be expected (Cody 1985). Some of these neighbor preferences are indicated in Fig. 23.2. At sites in the Granite Mountains, three species of cylindropuntia cacti, *Opuntia ramosissima, O. acanthocarpa,* and *O. echinocarpa,* together with the yucca, *Yucca schidigera,* are the most common larger plants. The first two opuntias are larger and tend to avoid their own and each other's company, and the smaller *O. echinocarpa* also interacts negatively with the more dominant *O. acanthocarpa*. The large yucca also tends to be evenly spaced and avoids conspecifics, but is a significantly preferred neighbor of each of the three cylindropuntia species and reciprocally prefers *O. acanthocarpa* as a neighbor.

At Mojave Desert sites in the Mid Hills two moderately sized shrubs are *Hymenoclea salsola* and *Cassia armata*. Each tends to be separated from conspecifics, but each shows a significant preference for the other as a near-neighbor. At the same location three smaller shrubs, *Haplopappus cooperi, Acamptopappus sphaerocephalus,* and *Salvia dorrii,* are clumped in distribution, apparently due to their interstitial distribution among the larger plants, but each shows an avoidance of the other species in the trio. These near-neighbor preferences are exerted in deep, uniform substrate devoid of topographic irregularities and are apparently produced by differential mortality of young plants that germinate near inappropriate neighbors.

In explanation of these near-neighbor patterns, it appears that an important aspect of life form in these desert shrubs is the structure of the root system. Species with different and comple-

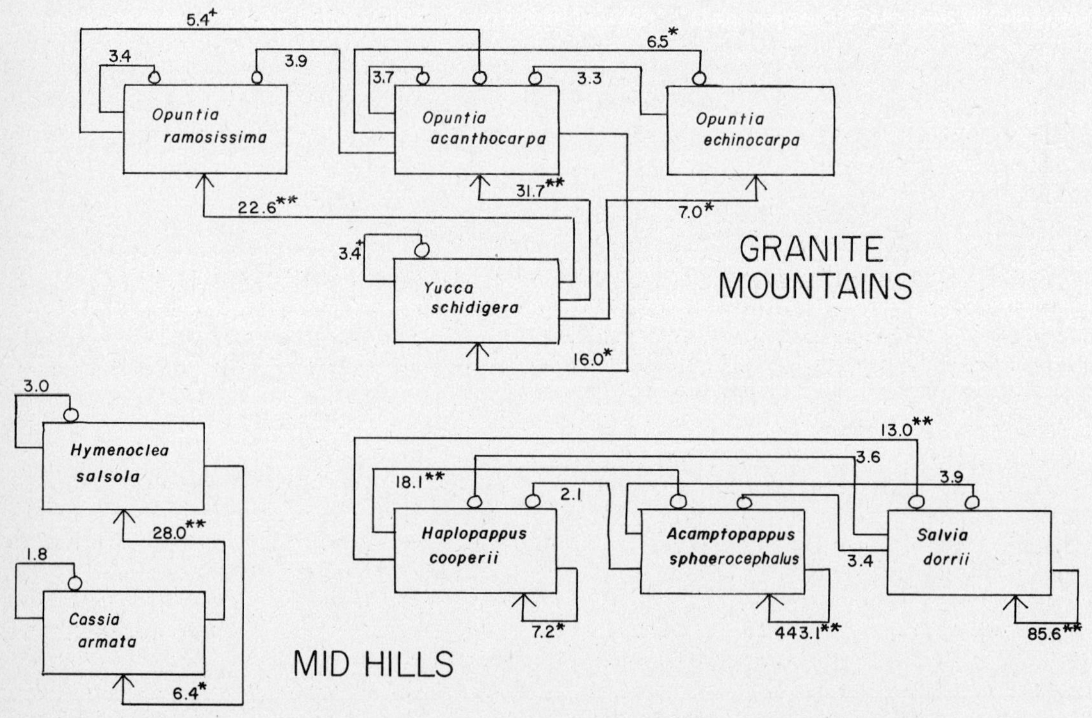

Fig. 23.2 Diagrammatic representation of neighbor preferences of shrubs at two locations in the Mojave Desert. A line from one species to another that terminates in a circle indicates that the first species occurs less often than expected by chance as a neighbor of the second. A line that terminates in an arrow indicates that the first species is a preferred neighbor of the second. Figures adjacent to the lines give chi-square values, which are significant at a level $p < 0.10$ at one site at least (+), significant at $p < 0.05$ at one site (*), or significant at $p < 0.05$ at two sites (**). At the Granite Mountains interactions among four large species of cylindropuntias and yuccas are shown; at the Mid Hills those between a pair of medium-sized species (left) and a trio of smaller species (right) are shown.

mentary root systems will appear more often as near-neighbors, and species with similar and conflicting root systems will be rarer as neighbors, compared to expectations based on pure chance (Yeaton and Cody 1976, Yeaton et al. 1977, Cody unpublished observation). Taking advantage of erosion at the edges of washes after heavy rains, I have excavated several dozens of shrubs, including individuals of most of the more common species. This work reveals that root system morphology is extremely species-specific, far more so than above-ground morphology, and some of the contrasts among species are shown in Fig. 23.3. Some species maintain spreading roots within 15 cm of the surface (*Ferocactus, Echinocereus*). Some are nearly as shallow rooted (*Yucca* spp., the perennial grass *Hilaria,* the mint *Salazaria mexicana*). Other species have deep taproots and take up water and nutrients from more than 2 m below the surface (*Hymenoclea, Cassia*), while still others, including the *Opuntia* cacti, are deeprooted but have spreading root systems rather than taproots. Most species are intermediate, but show obvious interspecific differences.

Given the differences among species in root system morphology, the patterns in near-neighbor preferences make sense. Of the large yuccas and opuntias, the former have spreading and particularly shallow roots, and the latter have deep and spreading roots; both root systems would mitigate against conspecific near-neighbors, but permit proximity of yuccas and opuntias with their complementary root systems. Taprooted species with few laterally spreading roots, such as *Hymenoclea* and *Cassia,* may grow close to each other and often appear as mutual near-neighbors; any interspecific differences in the depth of lateral root spread would favor allospecific over conspecific neighbors. My data on root system morphology in the smaller, interstitial species show extensive similarities among species, especially between *Haplopappus* and *Salvia;* such root systems must compete extensively for moisture uptake (Fig. 23.3). Among these plants conspecifics may be clumped by virtue of numerical advantages in seed and seedling establishment near parent plants.

It would appear that a large part of the explanation for the coexistence of many shrub species with similar above-ground life forms might be attributed to their different root systems and correspondingly different water uptake strategies. These differences can be exercised to a greater extent in coarse, gravelly sands through which water percolates rapidly and is thus available at different depths. In support of this, it is on the coarse granitic alluvial fans that one finds maximum shrub diversity in the Mojave Desert.

Much earlier work on root system morphology (see Cannon 1925 for desert plants, Weaver 1965 for prairie vegetation) allows similar observations on interspecific differences, but the role of these differences in niche segregation and the maintenance of species diversity has not been closely examined. However, Parrish and Bazzaz (1976) have documented segregation of root systems in successional vegetation, and Jenik (1978) has suggested that different species of tropical trees have a diversity of root systems, implying a diversity of water and nutrient uptake modes, that may aid their coexistence. Hubbell's and Foster's data (Chapter 19) on spacing patterns of tropical trees, however, do not indicate neighbor preferences.

Cactus Life Forms

Cactus Geometry, Photosynthetic Area, and Water Storage The stem-succulent Cactaceae are particularly convenient for the study of adaptive morphology and life form ecology because of the geometric simplicity of their stems and branching patterns. Aside from the platyopuntias, these plants are essentially cylinders (aboveground), varying chiefly in the radius (r) and length (l) of the cylinder and in how it is divided into branch segments. Here also the selective advantages of different life forms can be deduced. Since biomass B is simply $\pi r^2 l$ and photosynthetic area A is $2\pi rl$, the ratio of photosynthetic area to biomass scales as r^{-1}. Growth rates should be higher with higher ratios of photosynthetic area to biomass and should increase with decreasing radius r. On the other hand, both water storage ability and water conservation should be proportional to the ratio of biomass to photosynthetic area, increasing with stem radius r (especially in species with pleated or ribbed stems). Thus, there will be a direct trade-off be-

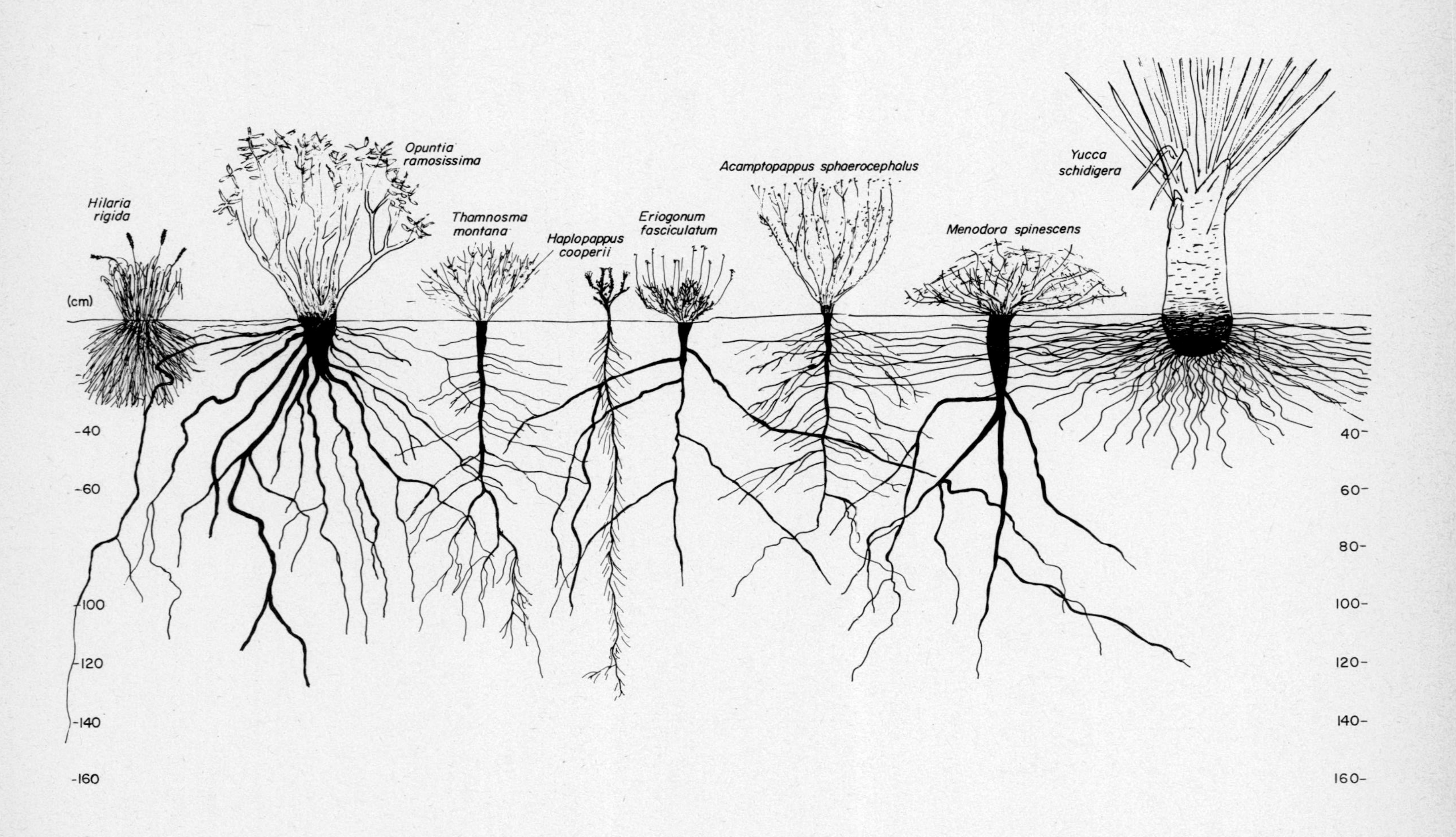
Hilaria rigida
Opuntia ramosissima
Thamnosma montana
Haplopappus cooperii
Eriogonum fasciculatum
Acamptopappus sphaerocephalus
Menodora spinescens
Yucca schidigera
(cm)
-40
-60
-100
-120
-140
-160
40-
60-
80-
100-
120-
140-
160-

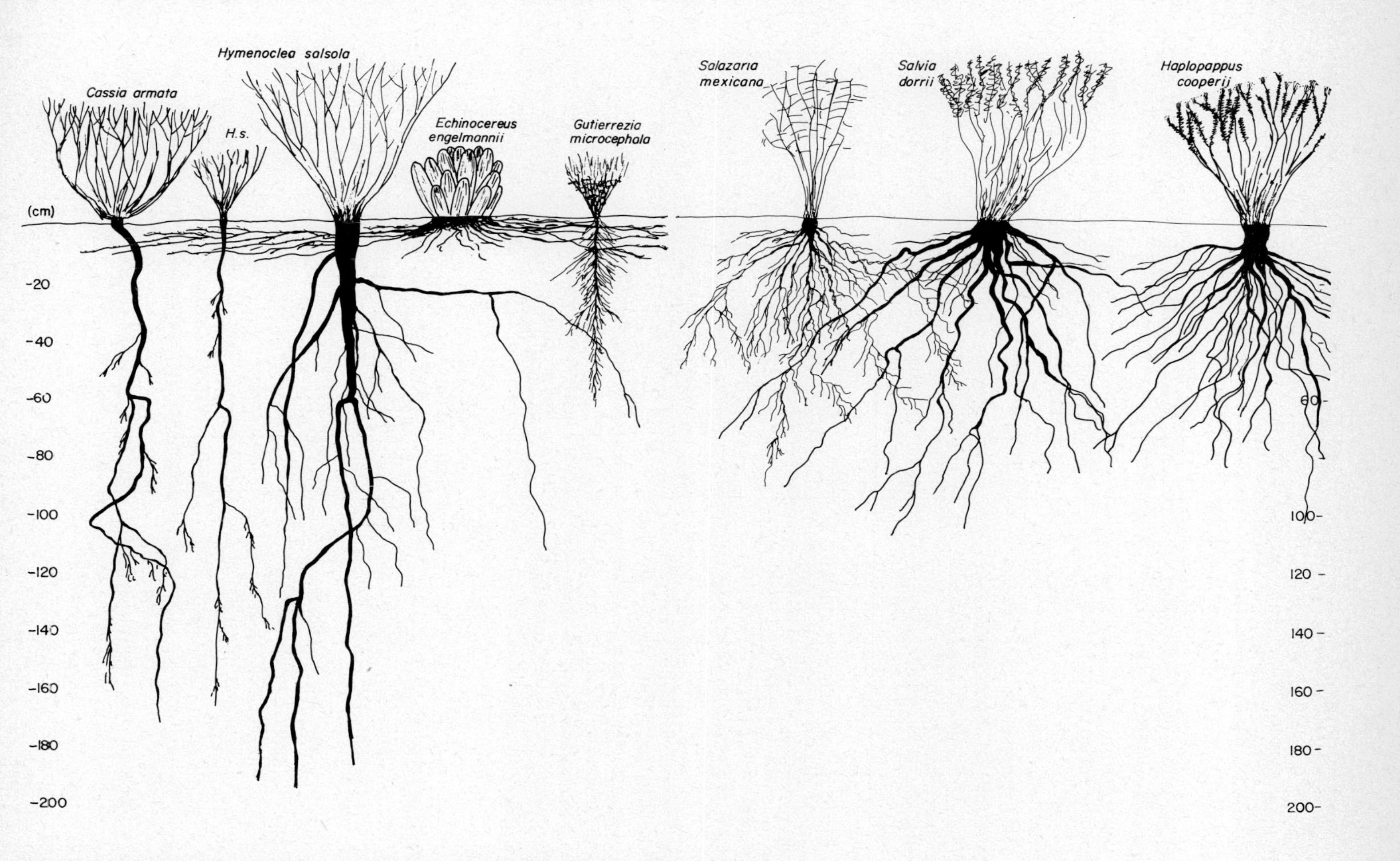

Fig. 23.3 Root systems of Mojave Desert shrubs, drawn from field excavations, diagrams, and photographs. Root system morphology is notably species-specific, but varies widely among species. Species may be shallow surface rooters (*Echinocereus, Hilaria, Yucca*), may be taprooted (*Cassia, Hymenoclea*), or employ more spreading roots at various intermediate depths (*Haplopappus, Menodora, Salazaria, Salvia*). Differences in root system morphology are thought to be the basis of specific near-neighbor preferences among these shrub species.

tween growth rate and water storage and conservation, the compromise being reflected in stem radius and concomitant growth form (Cody 1984).

Besides varying in above-ground life form, cacti also vary considerably in their root systems. Species with wide stem radii generally possess shallow root systems adapted for rapid water uptake from near the substrate surface, whereas narrow-stemmed species have deeper root systems to tap less ephemeral water that percolates down through the substrate (Yeaton et al. 1977, Yeaton and Cody 1979, Cody 1985). Observations confirm that much-branched and narrow-stemmed species (such as *Opuntia ramosissima* in the Mojave Desert and *O. arbuscula* and *O. leptocaulis* in the Sonoran Desert) are restricted to sandy flats, where their deep roots can tap water year-round and their advantage of high *A/B* can be exercised. Wide-stemmed and less-branched species, such as the shallow-rooted barrels *Ferocactus* and the hedgehog cacti *Echinocereus*, predominate on steeper and rockier slopes where their larger-diameter and ribbed stems aid in storing water that would otherwise rapidly run off their slopes and be lost to them. Thus, along habitat gradients from steep, rocky slopes to sandy flats there is a trend in cactus morphology from shallow roots, wide stems, and minimal branching to deep roots, narrow stems, and maximal branching. Species should be limited in their upslope distributions by abiotic factors associated with water shortage and poor water retention, but downslope by biotic factors associated with low *A/B* ratios that become increasingly unfavorable where a more constant water supply is available and higher transpiration and growth are possible.

The Cylindropuntias Although morphology and habitat are generally related in cacti as described above, in fact several species can coexist at a single site of uniform slope and substrate. In these cactus "communities," segregation by stem radius (and other morphological variables associated with it) seems to be of primary importance in species coexistence. This is shown in Fig. 23.4, where more conveniently measured stem circumferences were assessed in species on relatively flat (but variously rocky) ground at the southern end of Isla Carmen, Sea of Cortez

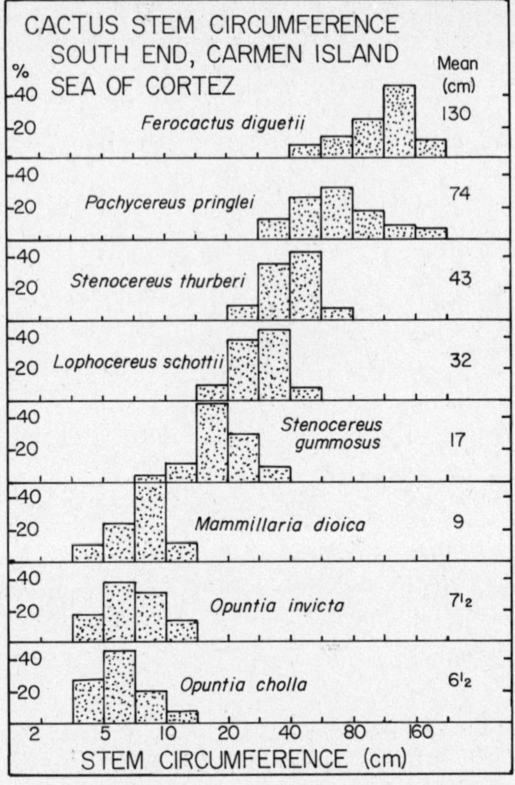

Fig. 23.4 Distribution of stem circumference in the eight cactus species that occur on the southern end of Isla Carmen, Sea of Cortez, showing segregation among species. The three species most similar in stem circumference are very dissimilar in growth form, with *Mammillaria dioica* short (<20 cm) and unbranched, *Opuntia invicta* branching at the substrate surface and forming broad mats, and *O. cholla* erect (to 2 m) and arborescent.

(Cody et al. 1984). Adjacent species differ in circumference by factors of 1.2 to 2.0 (average 1.5); the only species close in stem radii are the two *Opuntia* species, of which one (*invicta*) forms mats at ground level, while the other (*cholla*) is erect and arborescent. This segregation of coexisting plant species by stem radius is analogous to the segregation by size often discussed for animal species (see examples for mammals, birds, and lizards in Chapter 3, Fig. 5.6, Figs. 6.3 and 6.4, and Chapter 30).

There are three common cylindropuntia species in the Mojave Desert. Despite differences in their tolerances for steeper or rockier slopes, habitat overlaps among them are extensive. For example, all three (*Opuntia acanthocarpa*, *O. echinocarpa*, and *O. ramosissima*) co-occur on bajadas and gentle foothill slopes near the Granite Mountains and elsewhere in the general vicinity (Cody 1985). Stem diameter decreases from the first to the last, and their branching patterns are also quite different. *O. acanthocarpa* and *O. ramosissima* are of comparable biomass, whereas *echinocarpa* is smaller (except where it occurs without *acanthocarpa*, such as over much of Joshua Tree National Monument, California). Of the two major species, *ramosissima* is deeper rooted than *acanthocarpa* (Cody 1985, 1986, unpublished observation). The same difference also characterizes *Yucca brevifolia* and *Y. schidigera*, and soil depth might determine which commonly co-occuring species pair, *O. ramosissima–Y. brevifolia* or *O. acanthocarpa–Y. schidigera*, dominates the desert vegetation (see Cody 1985).

The life form of these and other cylindropuntia species can be characterized by their size-specific envelopes, with the size of individuals measured by the number of terminal joints or "termini." Such a representation is shown in Fig. 23.5 for four species, two (*fulgida* and *acanthocarpa*) measured in Organ Pipe Cactus National Monument, southwest Arizona, and two (*bigelovii* and *echinocarpa*) in Joshua Tree National Monument, southern California. Contours of termini with increasing shrub size show the different ontogeny of growth form in different species. Stem diameter (included in the figure as circumference) is largest (5 to 6 cm; circumference 18 cm) in *bigelovii*, the least-branched species that attains only 100 termini; it is least (ca.

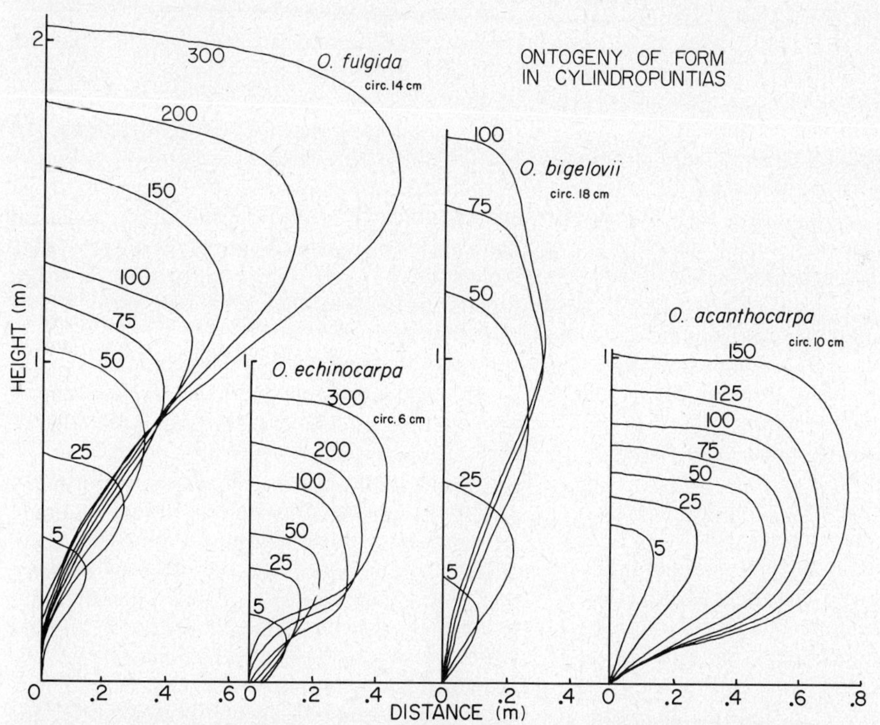

Fig. 23.5 Growth form in cylindropuntia species shown by shrub contours in the plane of horizontal and vertical distance from the shrub base. Shrub size is represented by the number of terminal stems or joints, and numbered contours of terminal joints with increasing shrub size show the ontogeny of growth form in different species. Species differ also in stem circumference (given as "circ. 14 cm," etc.), in their maximum sizes (number of termini), and in their root systems. At some Sonoran Desert sites four to six species co-occur with different growth forms and further segregate over differing slope angle and substrate type.

2 cm) in *echinocarpa,* the small-statured, most-branched of the four species (up to 300 termini). The branching patterns of the species are rather different, with *echinocarpa* showing a rapid ramification terminally, *bigelovii* producing regular laterals from a central trunk, *fulgida* being a more ramified and decentralized version of *bigelovii,* and *acanthocarpa* producing regular sequential branches with branch elongation within its globose form. These four species (albeit with *echinocarpa* populations of smaller-sized individuals) coexist with at least three additional cylindropuntias of yet different growth forms within a few hectares at many Sonoran Desert sites, where an additional component of their ecological segregation is by slope angle and substrate.

The Columnar Cacti Cacti in the tribe Pachycereeae (Gibson and Horak 1978) are the dominant members of their family over much of the Sonoran Desert, although they are unrepresented in the cooler Mojave Desert. These plants include the spectacularly large and arborescent saguaros and cardónes so characteristic of the Sonoran Desert vegetation. Because of thick, upright stems that are less ramified, their branching patterns are more easily quantified than the patterns of the opuntias; they are discussed in detail in Cody (1984).

The life forms of columnar cacti can be represented as plots of numbers of branches versus height above the ground. Locally coexisting desert species range from low and much-branched species of small stem radius to taller and minimally branched species of large stem radius. Besides varying among species at a site, these branching fingerprints also vary within species, indicating that branching morphology is not a species-specific constant, but rather a function of the cactus community and an adaptation to biotic as well as abiotic environments (Cody 1984).

Some of the variation in the branching patterns of columnar cacti within and among species at various sites in the Sonoran Desert and its fringes is shown in Fig. 23.6. Species composition varies among sites, particularly as *Carnegiea gigantea* (saguaro) is absent from Baja California and *Stenocereus gummosus* is essentially absent from Sonora. As many as four species occur where precipitation reaches 150 to 200 mm per annum (La Paz, Puerto Libertad), but this number drops to two or three species with decreasing precipitation in central Baja California (Cataviñá, San Ignacio, Pozo Aleman), drops to one or two species on smaller islands in the Gulf of California (Isla Santa Cruz), and declines with increasing latitude from northern Sonora into southern Arizona (Lukeville).

Fig. 23.6 shows the branching patterns of the columnar cacti at each site and includes a pictoral indication of the life form of the larger but variable cardónes, *Pachycereus pringlei.* This species is generally the tallest and least branched species of the community (e.g., at Cataviñá, San Ignacio). However, at Puerto Libertad, where it encounters saguaros, which are still taller and less-branched, it is shorter and more branched than elsewhere. And where cardónes occur alone on Gulf islands, they are still shorter and more branched, appearing morphologically more like absent species such as senita (*Lophocereus schottii*) or organ-pipe cactus (*Stenocereus thurberi*). On the small land-bridge island of Monserrate, four cactus species are represented as on the adjacent mainland, but the *Stenocereus gummosus* and *Lophocereus schottii* converge in branching pattern there. In all, I have illustrated two mainland or peninsular sites with two species, two with three species, and two with four species. At no two of these sites are species compositions the same, yet the segregation of species by life form appears similar among sites and is clearly affected by numbers of coexisting species.

Where precipitation increases to 400 to 500 mm in eastern Sonora (Ures) and in southeastern Baja California (San Bártolo), the desert is replaced by thorn scrub, in which two columnar cacti persist (Fig. 23.6). *Pachycereus pringlei* is replaced by the congeneric *P. pecten-aboriginum,* which is joined by *Stenocereus thurberi* in this habitat. Both species are similar in peak branch numbers, height above the ground (corresponding to midcanopy height), and stem radius, indicating that branching morphology is severely constrained by growth in the closed-canopy thorn scrub.

In northwestern Baja California (Colonet, Fig. 23.6) precipitation likewise increases toward the

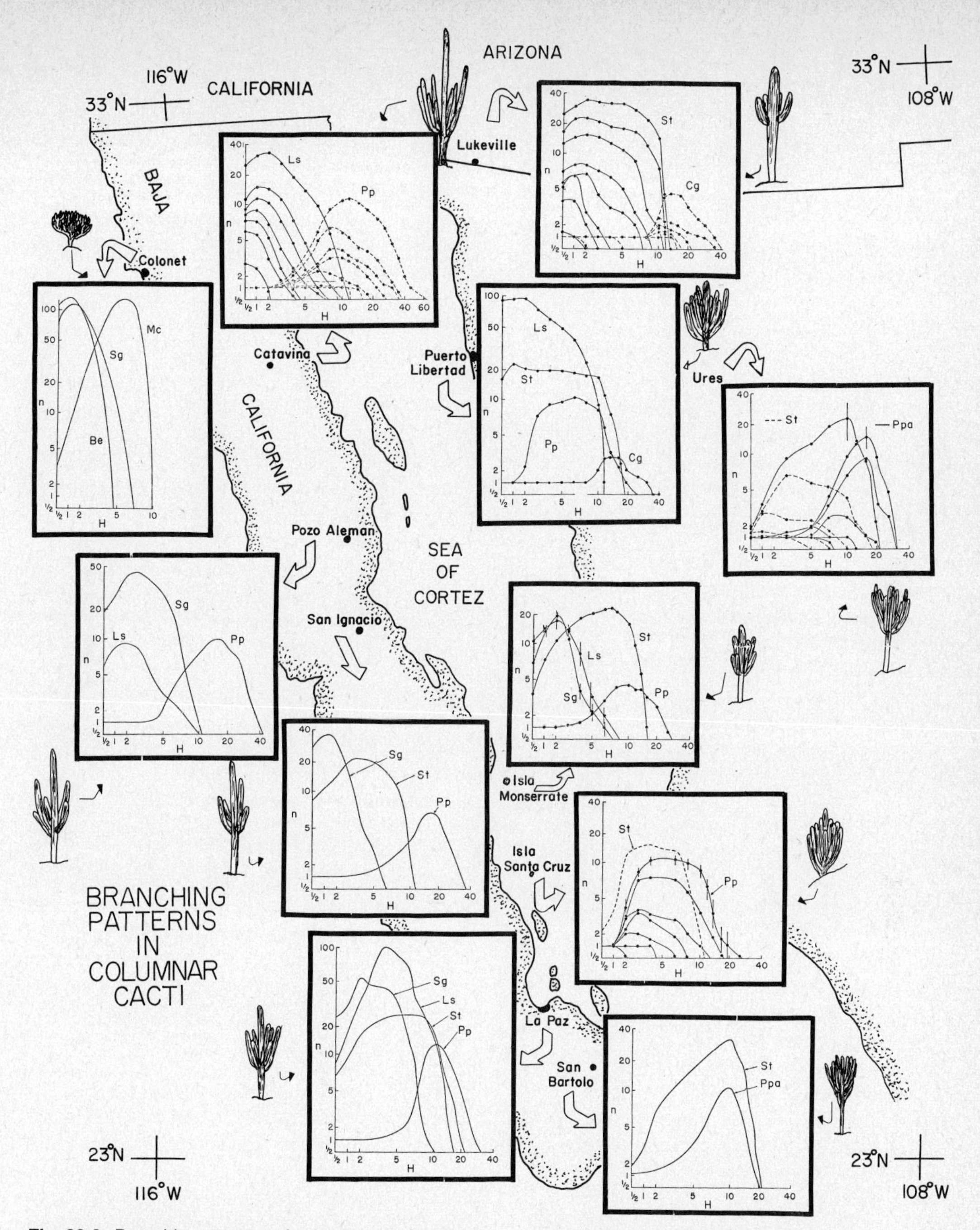

Fig. 23.6 Branching patterns of columnar cacti at sites in and near the Sonoran Desert on both sides of the Sea of Cortez. Branching patterns are represented as plots of numbers of branches (n) versus height above ground in feet (H). Species are shown either as families or curves of different size classes (Cataviñá, Lukeville, Ures, Isla Santa Cruz) or as a single curve, the envelope of the species' curve family (other sites). From one to four species coexist at a site, where their branching patterns are segregated. An ideogrammatic figure of the branching and size of *Pachycereus* at each site is included, except at the Colonet and Lukeville sites, where the figure is of *Myrtillocactus* and *Carnegiea*, respectively. Species abbreviations: Be = *Bergerocactus emoryi;* Cg = *Carnegiea gigantea;* Ls = *Lophocereus schottii;* Mc = *Myrtillocactus cochal;* Pp = *Pachycereus pringlei;* Ppa = *P. pecten-aboriginum;* Sg = *Stenocereus gummosus;* St = *S. thurberi.*

Mediterranean-climate zone. Here two different columnar cacti occur, *Myrtillocactus cochal* and *Bergerocactus emoryi,* and these along with *Stenocereus gummosus* are all narrow-stemmed and much-branched species with high *A*/*B* ratios, clearly foregoing water storage and conservation in favor of high photosynthetic area for competing with the broad-leaved shrubs of this mesic zone.

Thus, Fig. 23.6 illustrates two principles. Species coexisting at a single site differ in morphology, and populations of a single species at different sites differ morphologically in such a way as to provide morphological differences within each particular set of coexisting species at each site.

Columnar Cactus Growth Form, Climate, and Habitat The two most important aspects of aerial growth form in columnar cacti are clearly stem radius and branching pattern. These morphological features appear from fragmentary evidence to correspond to interspecific differences in root systems. They combine into a specific morphological syndrome that ranks species from, narrow-stemmed, high *A*/*B*, deep-rooted, low, sprawling, and highly branched forms to wide-stemmed, low *A*/*B*, shallow-rooted, tall, and minimally branched species that are good water-storers. The Sonoran Desert, with over 200 mm precipitation, generally supports four species along this range of morphotypes, and each apparently can trap light and water in different ways with different schedules.

Variations in climate and habitat affect this coexistence norm, as indicated in Fig. 23.7. At high latitudes, as in the Tucson Mountains, only the thickest-stemmed species can withstand the low temperature (Nobel 1980), and at their northern limits thinner-stemmed species are restricted to south-facing 20° slopes that receive the maximum solar radiation, as at Lukeville, Rcho. Arenoso. On such slopes temperature problems are minimized, and the advantages to high *A*/*B* are exercised to the fullest (Cody 1984). In thorn scrub a substantial stem radius is required to support a plant that can reach the 5- to 8-m canopy, but thicker-stemmed species with less branching and lower *A*/*B* ratios are excluded, for example, at San Bartolo, Ures. Only the shallow-rooted and thickest-stemmed cacti can occupy rocky hillsides, a habitat conducive to *P. pringlei* especially and in which the squat and unbranched barrel cacti *Ferocactus* reach high densities (for example, Santa Cruz, Santa Catalina). In the dense, low, and shrubby chaparral (for example, Colonet), only the narrow-stemmed, high *A*/*B* cacti survive. Further north in San Diego County it is the even more narrow-stemmed and more branched cylindropuntia *Opuntia parryi* that persists as the chaparral vegetation becomes increasingly mesic, tall, and dense.

Columnar Cacti in Subtropical Deciduous Woodland Whereas columnar cactus life forms vary dramatically in desert habitats, they show a considerable convergence in thorn scrub. This point is illustrated to the south in Oaxaca (central Mexico) in subtropical deciduous woodland with about 1,000 mm of precipitation. Here columnar cacti are perhaps more diverse than in any other habitat type. The vegetation is a rather open and airy woodland, with a somewhat discontinuous canopy between 5 and 12 m. A study site of just 5 ha in this habitat produced eight species of columnar cacti, but considerably more species could be found by taking advantage of different aspects and angles of slopes and a limited elevational range.

Data were collected (in April 1983) on branching patterns within the 5-ha site near Teotihuacan in gently rolling topography. Five species were common: *Stenocereus stellatus, Myrtillocactus geometrizans, Pachycereus weberi, Escontria chiotilla* and *Cephalocereus collinsii;* the three less common species at the site were *Pachycereus marginatus, Stenocereus dumortieri,* and a species of *Neobuxbaumia*.

Unlike the desert species, the woodland cacti show relatively minor differences in stem radii; all common species are 13 to 22 cm in diameter. Further, the largest individuals of the more common species are remarkably similar in branching pattern (Fig. 23.8). These individuals constitute the largest sixth (in terms of biomass or total stem length) of the species' samples at the site. However, comparisons among species in the smaller size classes, presumed younger individuals exemplified by those in the fourth and second sixths of the size ranges, reveal that interspecific simi-

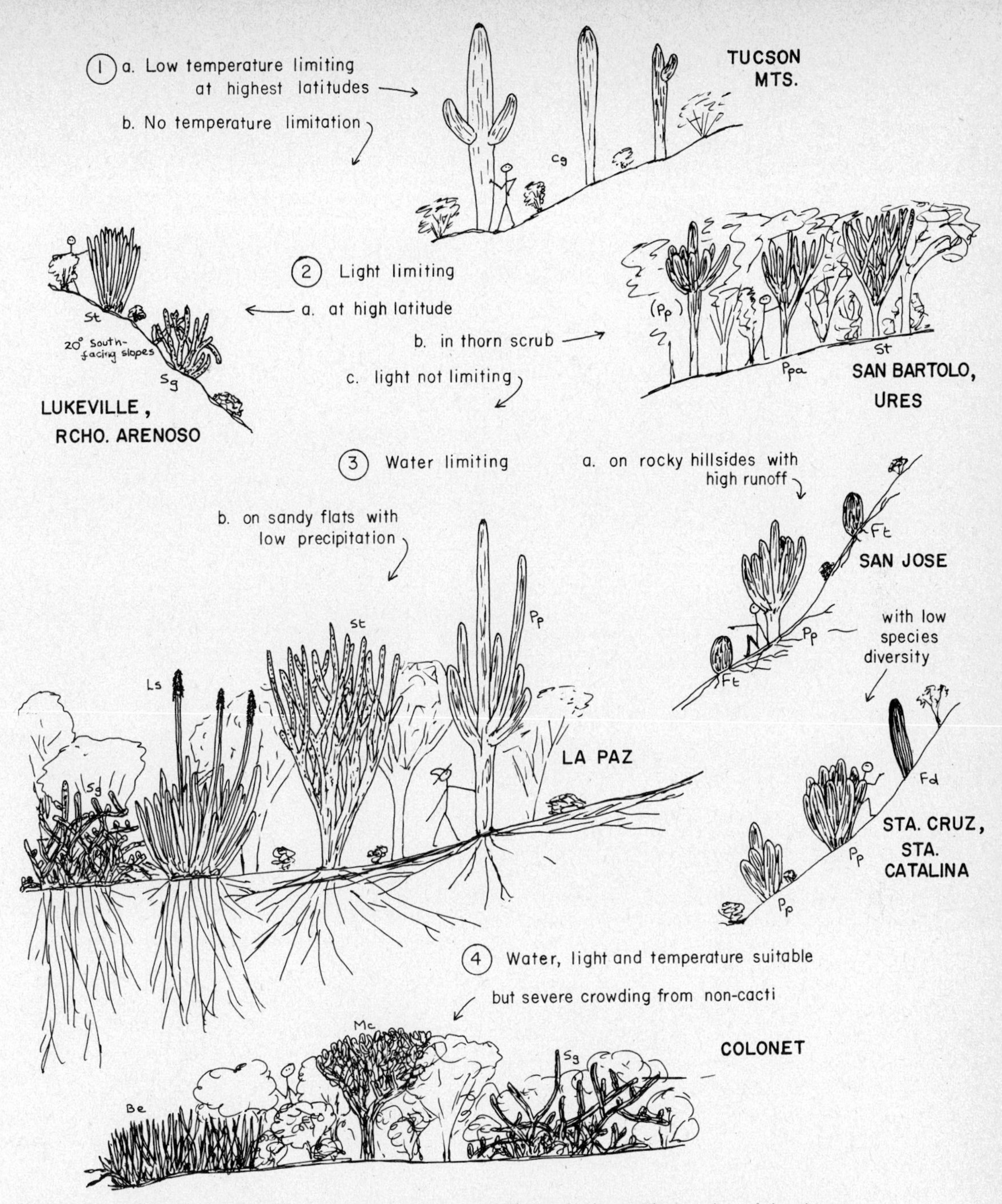

Fig. 23.7 A pictoral algorithm for columnar cactus morphology. At the northern edge of the desert either low temperature (1) or low radiation (2) may be limiting, and light may also limit the incursion of cacti into thorn scrub and deciduous forest, where they are subject to shading from nonsucculent trees. Water is limiting throughout the Sonoran Desert (3), where a diversity of branching patterns and associated root system morphologies permits coexistence of several species at a site. On steep, rocky hillsides, where runoff is fast and water-retaining substrate is shallow, only the wide-stemmed species with shallow roots with good water uptake and storage can persist (San Jose, Sta. Cruz, Sta. Catalina). Cacti begin to be outcompeted in chaparral (4) by evergreen plants with higher photosynthetic areas and no need for water storage mechanisms. Species abbreviations: Fd = *Ferocactus diguetii;* Ft = *F. townsendianus;* other abbreviations as in Fig. 23.6. The sketched human figure indicates scale. (Modified from Cody 1984.)

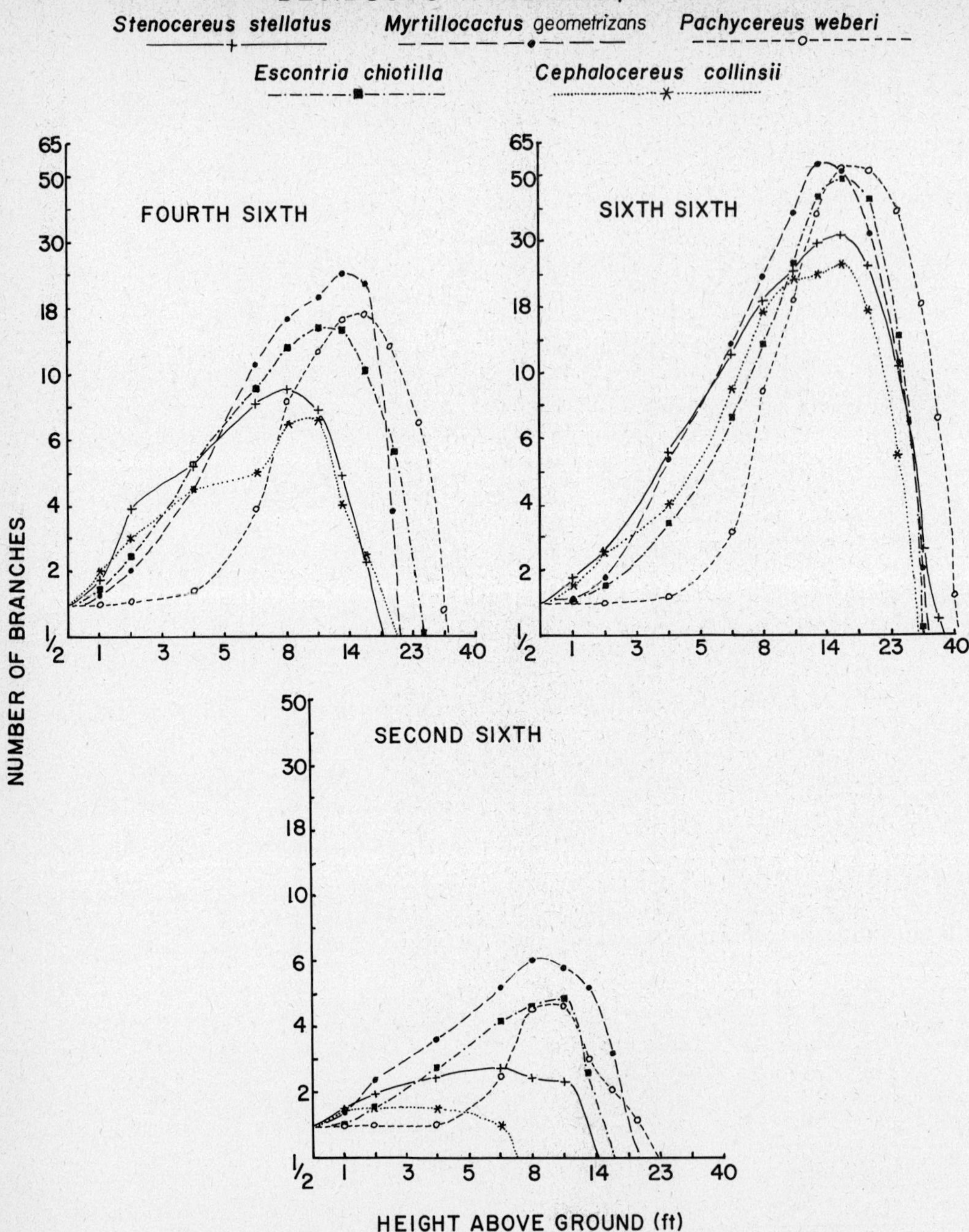

Fig. 23.8 Branching patterns of five species of columnar cacti in tropical deciduous woodland in Oaxaca, south-central Mexico. Individuals are grouped into six size classes (based on total stem length), from the sample of largest (sixth sixth) to intermediate (fourth sixth) to small (second sixth) individuals. These cacti have similar branching patterns as adults, but the juvenile individuals are increasingly divergent with decreasing plant size. This divergence supports the notion of different regeneration strategies for different cactus species in this closed woodland, but convergent light-capturing strategies for the adults in the canopy of the vegetation.

larities in branching patterns are not as great as with the largest plants. Thus, while adult plants that have reached the canopy of the woodland vegetation are similar in growth form, the younger individuals are not. A particular growth form appears to be optimal for carbon-fixing in the canopy, but a variety of growth forms are viable in the variously shady and patchy understory. It may be that different cactus species can grow best in certain sorts of light gaps or in certain sorts of light-shade regimen and that coexistence is determined by regeneration niches for the younger age classes, rather than by different conditions for the adults.

Thus, it appears that segregation by life form in adult cacti is possible only in open desert vegetation where light can be trapped efficiently down to near-ground levels. In taller and denser vegetation, such as thorn scrub and deciduous woodland, adult cacti converge in branching pattern and in other morphological aspects such as stem radii. Interspecific segregation is apparent only in juvenile growth forms, which presumably are adapted to different sorts of light gaps on the woodland floor, though this remains speculative at present.

LEAF MORPHOLOGY IN MEDITERRANEAN-CLIMATE SCRUB

Convergent Evolution in Shrub and Leaf Morphology

Some of the earliest observations on the adaptive nature of plant form were based on the similarities among the broad-leaved evergreen sclerophyllous shrubs of the five Mediterranean-climate regions of the world: "In all of these widely separated countries the vegetation bears essentially the same stamp, in spite of deep-seated differences in the composition of the floras" (Schimper 1903). Detailed studies of these basic similarities have been published (see Mooney and Dunn 1970, DiCastri and Mooney 1973, Mooney 1977, Thrower and Bradbury 1977, Cody and Mooney 1978, Miller 1981, Kruger et al. 1983, and numerous references within and beyond these).

General similarities of life form in this type of vegetation notwithstanding, there are adaptive morphological shifts with local variation in elevation, aspect, and precipitation (Mooney et al. 1974, Miller 1983). Even within narrowly circumscribed sites the vegetation displays considerable morphological variety among species, especially in leaf characteristics (Cody and Mooney 1978, Mitrakos and Christodoulakis 1981). In this section I examine the leaf-form diversity in shrubs of the South African fynbos, the Mediterranean-climate vegetation equivalent to Californian chaparral, Chilean matorral, and Mediterranean macchia. I interpret this diversity as adaptive responses to two pressures: physical factors, such as vegetation microhabitats, and nutrient characteristics of the substrate.

South African Fynbos

This vegetation is dominated by shrubs of the family Proteaceae, usually with an abundance of smaller heaths Ericaceae, restios Restionaceae, and many other perennial plants. Indeed, fynbos is reputed to be one of the world's richest vegetation types in terms of species numbers, and with its many locally restricted endemic species contributes to the classification and distinction of the Cape Floral Kingdom. The proteoid elements of the fynbos date from the Miocene and in fact predate shrub dominance and summer-dry climates in the Cape (Deacon 1983). The vegetation has undergone extensive migrations over the southern half of Africa with the changing climates over geological time (Deacon 1983, Axelrod and Raven 1978) and is generally regarded as a product of long and extensive differentiation throughout shifting climatic conditions over a diverse topographical region, and not particularly adapted to the current summer-dry conditions of the western Cape (Levyns 1964, Specht 1973).

Fynbos grows on soils derived from Table Mountain sandstone, which are extremely nutrient-poor. Nitrogen and phosphorus levels are an order of magnitude lower than in California and Chile (Specht and Moll 1983), a feature shared with parts of southern and southwestern Australia with which there were ancient land connections and where there has been similar but independent radiation of the Proteaceae, presumably a parallel response to the impoverished soils.

Plant species diversity in the fynbos is extremely high; it has several times the α-diversity of chaparral, and many times the β- and γ-diversity (species turnover with habitat and with geographical area respectively: Kruger and Taylor 1979, Bond 1983). High diversity in the fynbos has been attributed to a range of factors including the age and history of the vegetation, the influence of climatic changes over time, and topographical diversity (Tilman 1980, 1982). Proteads are equipped with special roots to aid nutrient uptake in poor soils, and the formation of these roots is induced by nutrient scarcity (Specht 1973). In fact, fynbos biomass is about the same (ca. 5,000 g/m^2) as in chaparral of comparable postfire age, and annual biomass accumulation is generally around 50% higher than in chaparral (Ehleringer and Mooney 1983). However, chaparral maintains its productivity over several decades at least (Schlesinger and Gill 1980), whereas 20-year-old fynbos is already obviously senescent and has $3\frac{1}{2}$ times as much dead material as living biomass (van Wilgen 1982).

Leaf Morphology in Protead Communities In order to investigate the role of life form diversity in the maintenance of the inordinately high plant diversity in fynbos, the patterns of leaf morphology in its dominant shrubs must first be documented and examined. The leaves of proteads are often large and thick; leaf specific weights average nearly twice those of chaparral species (Mooney 1983), and their longevities are estimated to be upwards of a decade in some species (cf. Diamantouglou and Mitrakos [1981] for short-lived leaves on Grecian shrubs). Several to many year classes of leaves may be present on the shrub at any time.

I collected data on leaf morphology in protead communities at four fynbos sites in mountain ranges in Cape Province from October to December 1979. The species measured include all of those present at my study sites in three genera, *Protea, Leucospermum,* and *Leucadendron*. These are the largest and most common shrubs in the family, with largest stature usually in *Protea*. I took 50-leaf samples for each species from several year classes preceding the current crop and from several different individual shrubs. The samples included leaves that spanned the entire leaf-size range. From these measurements I generated species-specific distributions of leaf morphology in the plane of log (leaf length) and log (leaf width/length); these variables measure leaf size and shape respectively and are expressed on logarithmic scales as befits such phenotypic dimensions.

The data from four sites are presented in Fig. 23.9. The sites are located from west to east, along a gradient of increasing summer rainfall and increasing *Leucadendron* diversity; *Leucadendron* species, but not those of *Protea,* have their main growth period during the summer, and their diversity clearly increases in the more favorable eastern sites with wetter summers (Rutherford 1981). In the Cedarberg range in the northwestern Cape (Fig. 23.9a) a level 5-ha site on Pakhuispas produced seven species: *Protea nitida, P. glabra, P. laurifolia, Leucadendron procerum, L. salignum, L. laureolom,* and *Leucospermum calligerum*. The *Leucadendron* species are dioecious, and males and females were measured and scored separately. Fig. 23.9a shows the distribution of these six species on the leaf morphology plane. All species are well segregated, from the long, broad leaves of *nitida* at the upper right to the shorter and much narrower leaves of *salignum* at the bottom. Note that the sexes of *salignum* are almost entirely distinct in leaf morphology, whereas those of the similarly dioecious *procerum* show about a one-third overlap.

Similar leaf morphology portraits of protead communities are shown from Sir Lowry's Pass in the Hottentots Holland range just east of Cape Town (Fig. 23.9b), from Robinson's Pass in the Outeniqua range north of Mossel Baai (Fig. 23.9c), and from Swartbergpas north of Oudtshoorn, just 110 km north of the preceding site yet the home of a rather different set of proteads (Fig. 23.9d). I draw five conclusions from the data of Fig. 23.9.

1. Protead species segregate on the leaf morphology plane with minimal interspecific overlap, yet narrowly adjacent leaf morphologies. The one apparent exception is the relative similarity of *Protea eximia* and *P. nitida* in the Swartberg. Although both species were recorded

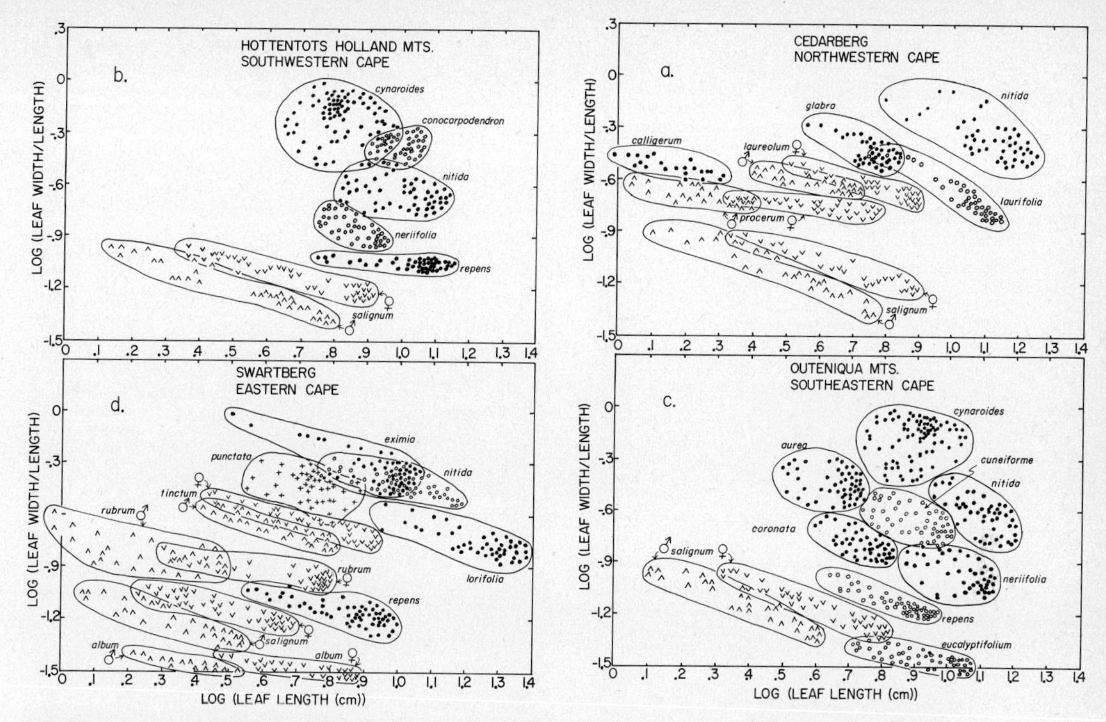

Fig. 23.9 Leaf morphology variation in the species of three dominant genera of Proteaceae—*Protea, Leucospermum,* and *Leucadendron*—at four geographically distinct sites in fynbos vegetation, Cape Province, South Africa. Approximately 50 leaves of each species were sampled, leaf lengths and widths measured, and the species represented as a cluster on the leaf morphology plane of leaf size, or log (leaf length), and leaf shape, or log (leaf width/length). Note that males and females of the dioecious *Leucadendron* species (*laureolum, procerum, salignum, tinctum, rubrum, album*) are displayed separately.

in the same 5-ha site, the former was restricted to westerly slopes and the latter to the more southerly slopes, and the two are habitat-segregated with limited occurrence as neighbors. The segregation patterns strongly suggest that coexisting species sets are selected for compatibility based on leaf morphology.

2. A given leaf morphology niche is represented by different species at different sites. Thus, *Protea glabra* (Cedarberg), *P. aurea* (Outeniqua), and *P. punctata* (Swartberg) are all similar in leaf morphology, as are *P. nitida, P. cynaroides,* and *P. eximia* at the same three sites. Other examples might be drawn from the figure. These species replacements reinforce the suggestion that selection has favored the segregation of species on the plane, and the replacement of species by others with morphologically equivalent leaves from one site to another indicates that leaf morphology "niches" can be filled and preempted.

3. Between sites a given species can shift its leaf morphology, which is thus less a species-specific constant than a function of species composition and available niche space on the leaf morphology plane. *P. nitida,* for example, has shorter and narrower leaves in the Hottentots Holland Mts. and Outeniquas compared to the Cedarberg and Swartberg, and in these sites the larger and broader-leaved *P. cynaroides* was not present. In the Outeniquas with the shorter-leaved *P. coronata* present, *P. neriifolia* has longer leaves than in the Hottentots Holland where *P. coronata* is absent. In the Outeniquas *P. repens* has much shorter leaves, vacating areas on the plane it occupies in the Hottentots Holland that are now occupied by the shifted *P. neriifolia*.

4. *Leucadendron* species are sexually dimorphic in leaf morphology to varying degrees, from 100% overlap in the monomorphic *L. eucalyptifolium* to partial overlap (*L. procerum, L. tinctum*) to near zero overlap (*L. salignum, L. album*). Intersexual differences in leaf morphology supplement and complement interspecific differences, in that the plane is crowded with a diversity of leaf forms, but overlap between adjacent forms is low or nil.

5. Female *Leucadendron* have higher leaf areas than males, but this may be accomplished by longer leaves (as in *L. rubrum, L. procerum,* and *L. album*), wider leaves (as in *L. tinctum*), or a combination of the two (as in *L. salignum*).

The segregation of protead species by leaf morphology suggests the possibility that different species are pursuing different strategies in light capture and photosynthesis and that leaf morphology niches covary with specific microhabitats in the vegetation. This possibility is explored first by examining the spatial structure of the vegetation (at Elandskloof in the Aasvoëlberg) and then by relating leaf morphology to the biophysics of leaf environments and carbon fixing.

Elandskloof: Habitat Segregation and Spacing Patterns Leaf morphology measurements were made in September 1980 at a fifth site, Elandskloof in the Aasvoëlberg near Villiarsdorp, selected because its topography, slopes, and aspects were more varied than at the four sites just discussed. Table 23.1 gives census data from three locations at Elandskloof. On site 1, a shallow 8° southeast-facing slope and *P. repens* was dominant with *L. rubrum* common and *L. salignum* rarer. On site 2, characterized by steeper and more southerly slopes, *P. laurifolia* was the most common species and *salignum* outnumbered *rubrum*. On site 3, with still steeper and more westerly-facing slopes, both *nitida* and *salignum* became common and *rubrum* was absent. The less abundant *P. acuminata* and *L. spissifolium* were both most common at intermediate slope angles, and a fourth *Leucadendron* species, unidentified, was rare.

Leaf morphology data from the three Elandskloof areas are shown in Fig. 23.10, from

Table 23.1 RELATIVE ABUNDANCES OF PROTEAD SPECIES IN THREE SITES AT ELANDSKLOOF

Site	Slope angle	Aspect	Shrub density	Pa	Pl	Pn	Pr	Lr	Ls	Lp
1	8°	135	.59	.034	.051	0	.542	.246	.127	.0
2	13°	210	.37	.054	.513	0	.095	.095	.148	.095
3	20°	240	.27	.038	.038	.321	.019	0	.508	.076
Mean height (meters)				.96	2.31	4	1.94	1.46 (f)	.83 (f)	
								1.17 (m)	.82 (m)	
SD				.15	.82	—	.30	.20 (f)	.18 (m)	
								.10 (m)	.09 (m)	
Mean width (meters)				.60	1.38	4	1.20	.72 (f)	.70 (f)	
								.76 (m)	.92 (m)	
SD				.14	.66	—	.26	.25 (f)	.12 (f)	
								.28 (m)	.18 (m)	

Aspect refers to compass orientation.

Species abbreviations: Pa = *Protea acuminata;* Pl = *P. laurifolia;* Pn = *P. nitida;* Pr = *P. repens;* Lr = *Leucadendron rubrum;* Ls = *L. salignum;* Lp = *L. spissifolium.*

Mean heights and widths are tabulated separately for male (m) and female (f) individuals of *Leucadendron* species.

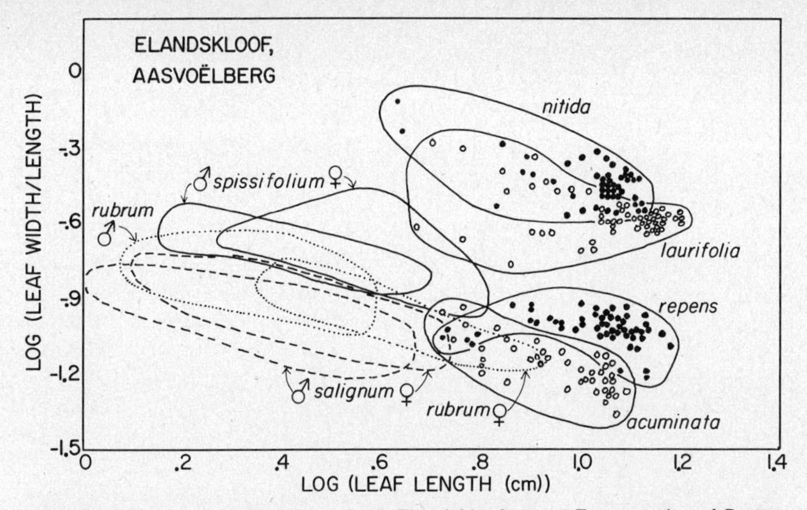

Fig. 23.10 Leaf morphology from three Elandskloof areas. Four species of *Protea* and three of *Leucadendron* occur at Elandskloof, but here the greater overlaps in leaf morphology are correlated with differences in the habitats (slopes of different angle and aspect) and in the positions in the vegetation they occupy. Leaf morphology is depicted as in Fig. 23.9. The *Leucadendron* species most similar in habitat requirements, *salignum* and *spissifolium,* are the two most dissimilar in leaf morphology.

which it is immediately apparent that interspecific overlaps are considerable and much greater than those depicted in Fig. 23.9. There is thus an association between species segregation by habitat (slope angle and aspect) and increased interspecific similarity in leaf morphology. The strong inference is that the former permits the latter.

Within sites at Elandskloof the spatial arrangement of plants is far from random, and this is shown by near-neighbor analysis as for the Mojave Desert shrubs discussed above. At site 1, for instance, *repens* (Pr) occurs as its own near-neighbor just as often (15/30 times) as is expected by chance (16.3), but occurs near female *rubrum* (Lrf) more than twice as often (10/30 times) as is expected by chance (4.6), to the exclusion of other species as potential neighbors (chi-square = 8.36, $p < 0.05$). Lrf, on the other hand, tends to avoid its own company (observed = 1, expected = 4.59) and, quite reasonably, to prefer the company of the males of its species (Lrm: observed = 5, expected = 2.79). In turn, Lrm are weakly associated with *L. salignum* (Lsf + Lsm: observed = 7, expected = 3.81), and Lsm and Lsf avoid Pr and associated preferentially with Lrf (observed = 18, expected = 9.18) and with their own kind, including the opposite sex (observed = 17, expected = 7.62). Thus, the spatial patterning of shrubs at the site deviates from random in the direction of specific neighbor preferences.

In Table 23.1 data on mean height and width of shrubs show the differences in stature among species. At site 1 Pr comprises 67.5% of the protead cover when abundances are corrected for size, Lr second with 11.8%, Pl third at 11.1%, and the remaining species comprise the last 10% of the cover. Note that while the sexes in *rubrum* do not differ significantly from each other in shrub width ($t = 0.33$, $p \gg 0.1$), they do differ in height ($t = 4.08$, $p < 0.001$); on the other hand, in *salignum* the sexes are similar in shrub height ($t = 0.14$, $p \gg 0.10$), but they differ in width ($t = 3.07$, $p = 0.01$). The spatial organization can be described relative to the large and dominant *P. repens* individuals. These have Lrf as most likely neighbors, an average of 0.92 m away, Lrm individuals a further 0.60 m away, Lsm 0.34 m further out into the interstices among the scattered dominants, and lastly Lsf centered in the interstices an average of 0.92 m from the Lsm.

This sort of analysis at the other sites produces similar results. Where *Protea laurifolia* (Pl) is most common, *L. rubrum* preferentially occupies positions closest to this larger shrub, with each sex more likely to have the other sex than the same sex as a near-neighbor. Both *salignum* (Ls) and *spissifolium* (Lp) occur toward the more open habitat of the interstitial spaces among the larger shrubs. Both species show a significant avoidance of the larger Pl and Pr (Ls: observed = 16, expected = 36.4; Lp: observed = 21, expected = 36.4), both show a significant preference for Lr (Ls: observed = 8, expected = 5.7; Lp: observed = 9, expected = 5.7), and both show a preference for their own company (Ls: observed = 15, expected = 3.0; Lp: observed = 4, expected = 1.9), with all trends stronger in Ls. The two interstitial species Ls and Lp themselves use space differently, for Ls is three times more common near Lr and one-third as common near Pl as is Lp.

In site 3 *Protea nitida* (Pn) dominates, and neighbor preferences are much less conspicuous than in the more crowded sites. This may indicate that the selective processes for microsite preference operate only in the more densely occupied areas. The observation is further evidence that the spatial patterning of the shrub species owes its generation to the close packing of the individuals themselves, presumably because denser vegetation creates a mosaic of microhabitats which differentiates among species in terms of their growth and survival. The relative similarities in leaf morphologies at Elandskloof thus correlate with both a segregation of species by habitat and a segregation of species within the vegetation by neighbor position.

In short, comparison of Figs. 23.9 and 23.10 at first suggests that species exhibit much greater morphological overlap at Elandskloof than at the four sites in Fig. 23.9. The explanation is that Elandskloof is more topographically diverse. At Elandskloof species are nonrandomly distributed by slope and aspect, with the result that relative abundances of individual species differ among the three sites of differing slopes and aspects (Table 23.1), and with the further result that species have significant neighbor preferences even within each site. In effect, morphologically similar species segregate by habitat, and Fig. 23.10 mixes results from several habitats much more than does Fig. 23.9.

I know of no other data on protead spacing patterns, but Taylor and van der Meulen (1981) have reported habitat segregation by slope angle among seven protead shrubs in the Rooiberg, southern Cape, and Laidler et al. (1981) reported a similar segregation of species on Table Mountain relative to rock cover.

Sexual Dimorphism in *Leucadendron* There are some 80 species of *Leucadendron* recognized by I. J. M. Williams (1972), all essentially within the southern Cape Province area. Species in this genus illustrate the point that morphologically similar species tend to segregate spatially, either by geographical range or by habitat. Using data supplied by Williams, I have plotted the positions of the largest male leaves (to the left) and female leaves (to the right) in the leaf morphology plane of Fig. 23.11. There are 60 species represented, with three species off the plane (the short- and round-leaved *concavum* at the upper left and the linear-leaved *comosum* and *nobile* at the bottom); *spissifolium* is represented on the plane by its five subspecies (74a–e). The length of the line joining male and female points is an estimate of sexual dimorphism, crude because a large, and even major, component of sexual differences may lie orthogonal to rather than along these lines.

There are 32 species pairs with crossed lines representing morphological overlap. Of these, 21 have nonoverlapping geographical ranges (at the resolution of $\frac{1}{2}$-degree latitude-longitude squares). Of the 9 pairs with range overlaps, 1 pair does not overlap at collection sites, and the remaining 8 show a mean overlap less than 50%. Given that species co-occupy $\frac{1}{2}$-degree squares, they are expected by chance (based on the numbers of collection sites) to co-occur at 23 sites, but in fact do so at just 12 sites (chi-square = 5.28, $p < 0.01$). My conclusion is that morpho-

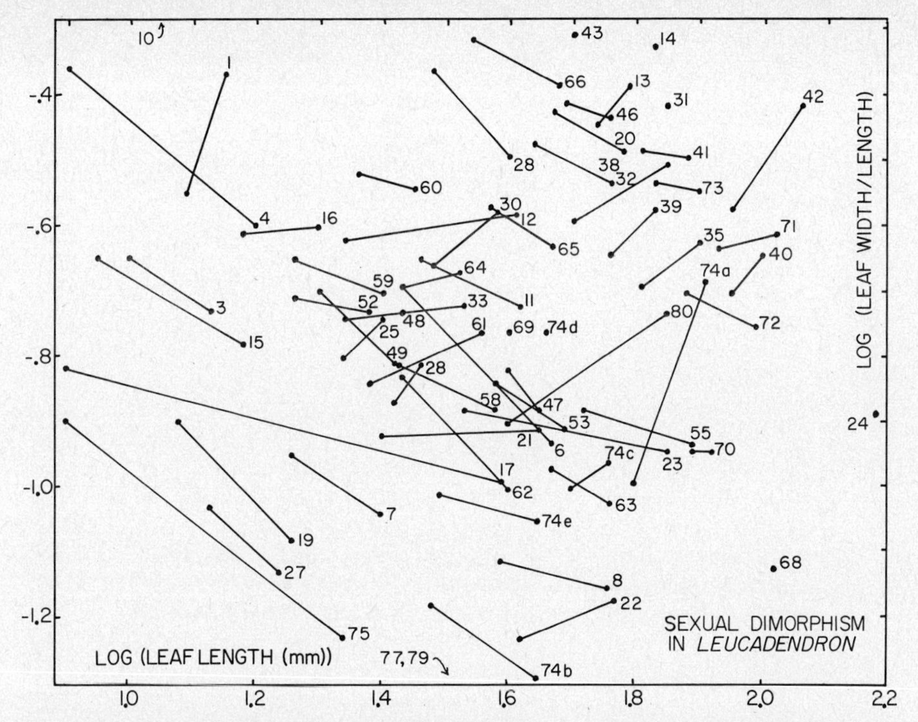

Fig. 23.11 Sexual dimorphism in *Leucadendron* species. Lines connect average male (to the left) and female (to the right) leaves; numbered dots without lines are sexually monomorphic species; and three species (no. 10, upper left; nos. 77 and 79, bottom) are off the figure. Sexual dimorphism decreases generally from left (shorter leaves) to right (longer leaves) and is achieved generally by dint of relatively longer leaves in females of short-leaved species, but relatively wider leaves in females of long-leaved species. (Data from I. J. M. Williams 1972, to whose species the numbers in the figure correspond.)

logically similar species tend either to be allopatric or to occupy different habitats (such as different slopes, aspects, or elevations). A simulation of random scatterings ($n = 20$) of lines with the observed distribution of line lengths (amounts of sexual dimorphism) on the leaf morphology plane produces 53.7 ± 13.8 crossings rather than 32 observed ($p = 0.1$ that this happens by chance).

In this set of *Leucadendron* species leaf length increases significantly with plant height ($r = .417$, $n = 84$; $p \ll 0.01$). Although the significance of interspecific variation in the degree of sexual dimorphism is unknown, sexual dimorphism is greater in narrow-leaved plants of short stature ($F = 4.54$, $p = 0.01$); this is in accord with the notion that the variety of microhabitats that might be exploited differentially between sexes increases lower in the vegetation. However, sexual dimorphism is not related in any simple way to range size, number of overlapping species, or the mean number of coexisting species of *Leucadendron*.

Leaf Morphology, Microhabitats, and Performance

The Biophysics of Leaf Size and Shape Several hypotheses might account for interspecific segregation in leaf morphology of protead shrubs. Apostatic selection from herbivores is one possibility; I know of no evidence for it, and I have observed only minimal damage to protead leaves, as befits their longevity and presumed chemical protection. Leaf morphology might be a specific phenotypic character evolved in past times in other sites, but interspecific variation among sites and the consistency and repeatability of the segregation patterns belie this viewpoint. A more plausible explanation is that leaves of different sizes and shapes occur on shrubs in association with specific microhabitats and that leaf morphology is selected to produce the maximum seasonal photosynthetic rates in those microhabitats. Then, as Oechel et al. (1981) put it, "Within the plant community, there may be several carbon uptake and allocation patterns possessed by different species. The combination of these species possessing different carbon efficiencies may result in maximized total resource use and minimized overlap in resource use."

To evaluate this hypothesis, a good deal of information is required. Climate produces seasonal regimes of precipitation, radiation, and temperatures; substrate (slope, runoff, soil depth) and root systems limit moisture availability; soil limits nutrient levels and availability. The vegetation itself produces a variety of microclimate habitats, since light, temperature, humidity, and radiation vary both horizontally and vertically from the top of the canopy to the ground. A particular leaf morphology, although just one aspect of the plant phenotype, performs in a certain fashion on a certain microhabitat and over the seasons fixes carbon according to a morphology-specific schedule. Of course, this performance is affected by other morphological and physiological factors, such as leaf thickness, stomatal resistance, leaf angle, and density in the shrub.

Although sorting and interrelating these various factors is a job for the plant physiologist, a few salient points will be mentioned here. A number of authors have described ways in which a leaf's morphology is expected theoretically to affect its transpiration rate, heat balance, and the thickness of its boundary layer. (See also Givnish [1978, 1979] and Givnish and Vermeij [1976] for discussions on the adaptive significance of leaf morphology in an ecological contest.) Mooney and Gulman (1979) discuss plots of boundary layer conductance as a function of leaf width for various wind speeds. Gates (1980) writes equations for leaf energy balance as a function of the following variables: radiation absorbed, used in photosynthesis, and released by respiration; transpiration rate; ambient temperature and temperature of the leaf; wind speed; characteristic leaf dimension (length of long axis relative to wind flow); and leaf emissivity. These relations are tabulated by Gates and Papian (1971) for various parameter values. One use to which the tables may be put is shown in Fig. 23.12, which plots transpiration rates versus wind speed for a variety of leaf sizes and shapes under conditions of 0% and 100% relative humidity with radiation level, resistance, and ambient temperature held constant. As might be expected, transpiration rates are greater in larger and narrower leaves, are greater in drier situations, and fall off with wind speed. Note that a smaller leaf lower in the vegetation and sheltered from wind may show the same transpiration rate as a larger leaf higher in the vegetation where wind speeds are greater.

The question of the effect of leaf size and shape on the convective heat transfer of a leaf was approached earlier by Martin (1943), who assessed this rate as proportional to (leaf length)$^{-.3}$ and (leaf width)$^{-.2}$. This relation plotted on the leaf morphology plane of Figs. 23.9 and 23.10 would yield a series of contour lines of slope $(-\frac{1}{2})$, with the line corresponding to the thinnest boundary layer at the lower left, and the line corresponding to the thickest layer at the upper right. If one looks again at Figs. 23.9 and 23.10, points for individual species do cluster about lines of slope $(-\frac{1}{2})$, suggesting that these biophysical considerations contribute importantly to the observed pattern of species segregation by leaf morphology.

Parkhurst and Loucks (1972) discuss optimal leaf size in relation to efficiency of water utilization, that is, the rate of carbon dioxide uptake relative to water loss. The water use efficiency of shaded leaves is higher when they are large, but is higher in unshaded leaves that are smaller.

Taylor (1975) modeled leaf form as a function of temperature, water use, and productivity. Rather than use raw leaf length and width, he transforms leaf dimensions into a characteristic leaf dimension (D) by factors related to leaf shape; D is then related to resistance (the inverse of conductance), temperature, transpiration, and net photosynthesis rate. At 30°C an order of magnitude increase in D produces a 20% increase in net photosynthesis at low leaf resistance, but es-

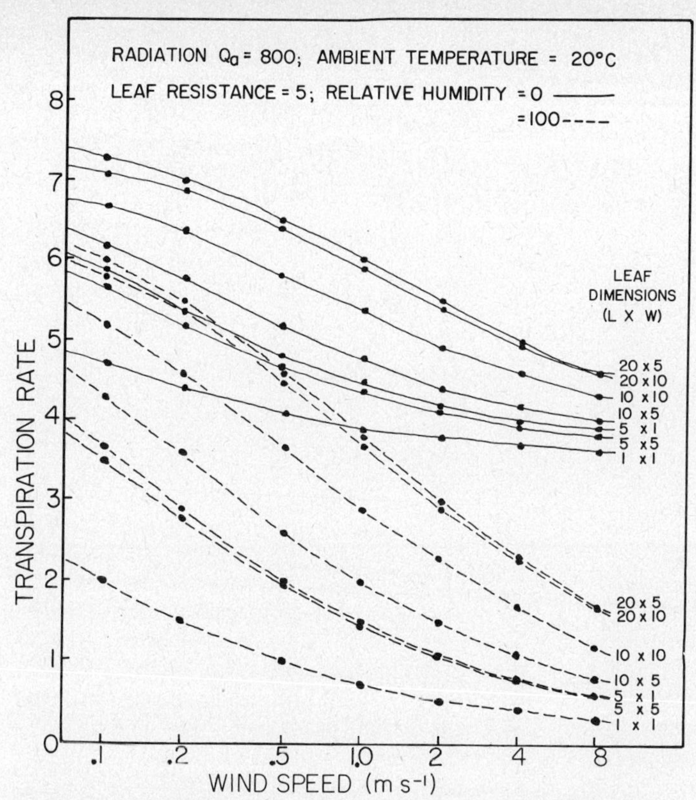

Fig. 23.12 Transpiration rate (g $cm^{-2}s^{-1}$) in leaves as a function of wind speed (m s^{-1}) and leaf size and shape (individual curves) in two humidity regimes (0% and 100%). Smaller leaves have thinner boundary layers that are less affected by wind speeds, but relatively more affected by changing humidity. Ambient temperature, radiation level, and leaf resistance are held constant in the figure, but these variables also determine the performance of leaves with differing morphologies. (Calculated from data presented by Gates and Papian 1971.)

sentially no change at high leaf resistance. The impression is that over a variety of realistic microclimate habitats, different shapes of leaves could all achieve a healthy net photosynthesis.

Empirical Observations of Leaf Performance

Most of the empirical information on leaf performance in Mediterranean-climate habitats comes from the California chaparral and Chilean matorral. Oechel et al. (1981) use such data to draw up "photosynthetic response surfaces" for various species, showing how net photosynthesis varies with temperature, light, and moisture for the given species' leaf morphology. When one simulates effects of changes in leaf parameters, invariably the net photosynthesis rate is predicted to decrease with changes in morphology from the actual morphology, indicating that leaf morphology is indeed selected for maximum photosynthesis and carbon fixing. Leaf performance in spring can be assessed by five variables: optimal temperature for photosynthesis, range of temperatures over which net photosynthesis is within 95% of maximum, percent of maximum photosynthetic rates at 5°C and at 40°C, and maximum photosynthetic rate. Leaf morphology is measured by three variables: leaf width, leaf specific weight, and leaf area.

Values for these eight variables describing leaf morphology and performance have been pub-

Table 23.2 CORRELATION MATRIX FOR THREE LEAF MORPHOLOGY VARIABLES (1–3) AND FIVE LEAF PERFORMANCE VARIABLES (4–8) DERIVED FROM MEASUREMENTS ON FOUR PAIRS OF CALIFORNIAN-CHILEAN CHAPARRAL-MATORRAL SHRUB SPECIES

	1	2	3	4	5	6	7	8
1. Leaf width	1.000	.358	.980	−.431	−.433	−.097	−.478	−.499
2. Leaf specific weight		1.000	.334	−.182	−.218	.003	−.441	.028
3. Leaf area			1.000	−.301	−.399	−.174	−.422	−.605
4. Optimal temperature for photosynthesis				1.000	.301	−.424	.369	−.330
5. Temperature range for 95% of maximum net photosynthesis					1.000	.603	.918	−.241
6. Percent of maximum photosynthesis at 5°C						1.000	.463	.139
7. Percent of maximum photosynthesis at 40°C							1.000	−.360
8. Maximum photosynthetic rate								1.000

lished for four species pairs of California-Chile ecological equivalents (Oechel et al. 1981, Thrower and Bradbury 1977); the correlation matrix is given in Table 23.2. Examination of this matrix yields the following conclusions. (1) Narrower leaves achieve higher maximum photosynthetic rates and a better performance at high temperatures. (2) Leaves with larger areas have lower maximum photosynthetic rates and a lower range of optimal temperatures. (3) Thicker, more sclerophyllous leaves (with higher specific weight) may have high maximum photosynthetic rates, but perform best at lower temperatures and over a narrower temperature range.

In general, chaparral leaves have maximum photosynthetic rates over a broad range of temperatures centered around 20°C. But despite the breadth of temperature range over which these shrubs perform at near-optimal capacity, relatively minor shifts in environment associated with changing elevation alter species performance and in consequence their relative abundances. Oechel et al. (1981) show that the percent cover of five common chaparral species increases with their net seasonal photosynthesis (g CO_2/g/season). This trend holds among species ($r = 0.84$, $p < 0.01$) and also within species; 15/20 within-species increases in seasonal photosynthesis are associated with increases in the percent cover of these species (chi-square = 5.0, $p = 0.025$). Thus, carbon-fixing performance affects plant abundance, and selection for its efficiency should be strong.

These findings support the view that different leaf morphologies can be optimal in the different microhabitats offered by the open fynbos vegetation. Of course, carbon-fixing strategies are expected to vary with other aspects of the plant's ecology, such as broader aspects of habitat (slopes, aspects, soil depth), water uptake strategies, and growth and reproduction phenologies. Fynbos shrubs are known to display a diversity of growth phenologies (Kruger 1981), presumably in line with a diversity of carbon-fixing strategies that are related to their leaf morphology-microhabitat niches. A good deal of further work is required before this hypothesis can be adopted or even tested, but fynbos vegetation can provide a rich source of material for the study of plant life forms and their adaptive significance.

CONCLUSIONS

In this chapter I have developed the notion that differences in life form in plants permit interspecific segregation in strategies of light interception from above and in strategies of water and nutrient uptake from below, and thereby promote species coexistence. Life form diversity, then, is a fundamental niche axis in plant communities and may constitute an important conceptual basis for explanations of species diversity and density.

In leafy desert shrubs that encounter harsh above-ground conditions, an adaptive radiation in life form diversity occurs in the morphology of root systems, which affects species' neighbor preferences. In desert stem-succulent cacti, the trade-offs between large photosynthetic area and a good ability to store and conserve water determine life form, and different species ply their respective trades via compromise life forms wherever conditions of water and light availability permit. Other species with other life forms strongly influence this availability.

In the diverse fynbos vegetation life form diversity is exemplified by the variety of leaf morphologies in the larger shrubs, each leaf morph being segregated from others in the community and borne by a shrub of a particular stature with certain spatial relationships with other neighboring shrubs. It appears that different leaf morphs may be adapted to photosynthesize optimally in different microhabitats, and a wide variety of these microhabitats occurs in the open fynbos vegetation on nutrient-poor soils.

The segregation of plant life forms into different structural niches would seem to be an aspect of community ecology that is as yet poorly developed, but worthy of more attention. The extent to which it might be useful in interpreting the diversity of tropical forest trees (Chapter 19) or of chalk grassland and prairie herbaceous flora (Chapter 12) remains to be seen. Similarly, the extent to which structural niche axes covary with other niche axes, such as phenology and regeneration strategies, remains largely unexplored.

ACKNOWLEDGMENTS

The fieldwork on diversity and morphology in North American arid areas was supported by a grant from Resources for the Future and by several contributions of the University of California Research Committee. Fieldwork in South African fynbos communities was made possible by grants from the John Simon Guggenheim Foundation, the National Geographic Society, and the Percy Fitzpatrick Institute for African Ornithology. I am extremely grateful for the breadth and depth of this support and also to several colleagues, especially Prof. Harry Thompson, for my continuing education in plant ecology and adaptive morphology. For many helpful comments on the text I thank W. Bond, P. J. Grubb, D. Tilman, and the editors.

chapter 24

Community Biology and Sexual Selection: Lessons from Hummingbird Flower Mites

Robert K. Colwell

INTRODUCTION

For most kinds of plants and animals, it is difficult to define many of the most fundamental variables of population and community biology, which makes it hard to answer interesting questions about those variables. What is a "resource"? What is the spatial and temporal pattern of resources from the point of view of the organism? How are closely related species affected by differences in resource patterning? Why are some species more specialized than others in their use of resources, and how does specialization vary among communities? What is the role of reproductive behavior in the evolution of resource use? What is "dispersal," and how do we recognize a disperser? How are dispersal and colonization related to community composition, species interactions, speciation, and biogeography?

Hummingbird flower mites are obligingly rigid in their use of the floral resources of their host plants and are stereotyped in their use of hummingbirds for transportation. They either dine at home on the host plant (on nectar and pollen exudates) or make clear their commitment to dispersal by actively boarding the bills of visiting hummingbirds (Colwell 1973, 1979, 1983, 1985; Dobkin 1985). The mites are diverse and widespread enough to permit comparative studies among species within sites and among communities between sites. They are small enough, prolific enough, and sufficiently oblivious to manipulation to permit field and laboratory experiments on a demographic scale, yet such experiments pose no threat whatever to human welfare, and any effects on the ecosystem are quite temporary.

This chapter begins with an analysis of the organization of local assemblages of hummingbird flower mite species in a biogeographical context. The analysis sets the scene for an evaluation of the conflicting forces that shape patterns of species interactions in these communities. I will argue that a form of sexual selection, hitherto unappreciated as such, plays a leading role in restriction of the host plant repertoires of hummingbird flower mites, and quite possibly in habitat restriction in other kinds of animals as well. The chapter concludes with further synthesis of ecological and evolutionary approaches by taking a look at how behavior, population structure, species interactions, natural and sexual selection, and rare events interact in the extinction and speciation of hummingbird flower mites.[1]

BIOGEOGRAPHY AND LIFE HISTORY

Hummingbirds range from Alaska to Tierra del Fuego, with the highest local diversity in wet tropical forests. The number of coexisting species declines precipitously with latitude, elevation, and isolation (Feinsinger and Colwell 1978). Hummingbird flower mites follow this pattern of species diversity closely, ranging from a single species at the latitude of California or central Chile to some 20 sympatric species in the Arima valley of Trinidad.

This ecologically defined group of mites encompasses all described species of the genus *Rhinoseius,* which ranges from northern California to Chile, and a diverse, tropical lineage within the genus *Proctolaelaps*. (Both genera are in the family Ascidae, suborder Gamasida [=Mesostigmata].) These two ecologically convergent lineages have spawned an impressive adaptive radiation of perhaps 200 species (Colwell 1979). To date, only about 40 species are described, but we have additional specimens from some 50 hummingbird species and from nearly 100 host plant species in 17 plant families, as yet largely unstudied—and the Amazon basin is still unaccounted for.

All known species of hummingbird flower mites apparently have similar life histories. Mated females lay their eggs in flowers (Colwell 1973), bracts (Dobkin 1983), or under nearby leaves of the host plant (eggs are held by the mother until hatching in some cases). The brief egg stage is followed by an active, feeding, six-legged ''larva,'' succeeded by eight-legged protonymph and deutonymph stages, and then by an adult male or female stage. Generation time is about a week (Colwell 1973, Dobkin 1983). All stages feed on nectar, and later stages include pollen exudates in their diet as well. No resting stages are known. Based on studies of related gamasine mites, it is likely that the reproductive strategy of hummingbird flower mites is parahaploidy—all zygotes start out diploid and biparental, but those that become males are haploid after the embryo stage, whereas females remain diploid (Oliver 1983).

Hummingbird flower mites feed and breed within the floral corolla. They move freely on foot between flowers of the same inflorescence or from nearby refuge sites in the inflorescence (bracts or bud clusters), where they may spend a day or two, to newly opened flowers (Dobkin 1983). Thus the mites on a single inflorescence represent an approximately random-mating ''breeding group.'' Density on host plants differs greatly among mite species; the range is from about 5 to more than 1,000 adults per inflorescence.

Movement of mites between inflorescences and between plants is almost exclusively on hummingbirds (Colwell et al. 1974); the mites ride within the nasal cavity of the bird. The mites do not feed on birds and have no detectable effect on their carriers in normal densities (zero to 15 mites per bird).

AFFILIATION OF MITES WITH HOST PLANTS

From our studies in Costa Rica, Mexico, California, Chile, and the West Indies, a coherent pattern of resource use emerges. Locally, each mite species has a well-defined host plant repertoire that comprises from one to several species of hummingbird-pollinated plants. The host plant repertoires of sympatric species are either rigidly nonoverlapping or completely overlapping (Table 24.1). The use of hummingbirds for dispersal, on the other hand, appears to be entirely opportunistic. An individual hummingbird, regardless of species, usually carries a sampling of mites from all the plant species the bird has recently visited, the passenger manifest sometimes includes as many as five mite species. Yet, as I will later discuss in detail, each mite disembarks from the bird only at flowers of its own host plant species.

Monopoly and Overlap

With few exceptions, each mite species monopolizes its resource base, whatever the

[1] I am greatly indebted to David Dobkin, Amy Heyneman, and Shahid Naeem for their collaboration in much of the work I will discuss.

Table 24.1 HUMMINGBIRD FLOWER MITES AND THEIR HOST PLANTS

Mite species[a]	Host plant species
Arima valley, Trinidad, West Indies (elevation 200–300 m)	
Rhinoseius trinitatis	*Heliconia trinidatis* (Heliconiaceae [Musaceae])
R. spinosus	*Heliconia psittacorum* (Heliconiaceae [Musaceae])
R. klepticos and *Proctolaelaps certator*[b]	*Heliconia wagneriana* and *H.* aff. *tortuosa* (Heliconiaceae [Musaceae])
P. certator	Dry season: *Aechmea fendleri* (Bromeliaceae)
R. bisacculatus (until 1976 and after 1980) and *P. certator* (since 1976)	*Costus scaber* (Zingiberaceae)
R. fidelis	*Costus niveo-purpurea* (Zingiberaceae)
P. contentiosus	*Renealmia exaltata* (Zingiberaceae)
P. belemensis	*Monotagma spicatum* (Marantaceae)
R. phoreticus	*Guzmania* sp. (Bromeliaceae) and *Pitcairnia integrifolia* (Bromeliaceae)
R. hirsutus and *P. contumex*	*Cephaelis muscosa* (Rubiaceae)
P. kirmsei	Dry season: *Palicourea crocea* (Rubiaceae) Wet season: *Hamelia patens* (Rubiaceae)
P. jurgatus	*Isertia parviflora* (Rubiaceae)
P. glaucis	*Centropogon surinamensis* (Lobeliaceae)
P. rabulatus	*Mandevilla hirsuta* (Apocynaceae)
P. mermillion	Unknown
Monteverde, Costa Rica (elevation 1,400 m)	
R. nr. *erro* and *P.* nr. *belemensis*	*Heliconia tortuosa* (Heliconiaceae [Musaceae])
R. colwelli	Wet season: *Centropogon cordifolius* (Lobeliaceae) and *Bomarea* sp. (Alstroemeriaceae [Amaryllidaceae]) Late dry season: *Centropogon solanifolius* Late wet season: *Columnea microcalyx* (Gesneriaceae) Sporadic: *Pitcairnia brittonia* (Bromeliaceae)
R. chiriquensis	Early wet season peak: *Hamelia patens* (Rubiaceae) Late wet season: *Cuphea* sp. (Lythraceae) Dry season: *Lobelia laxiflora* (Lobeliaceae)
R. richardsoni	*Cavendishia* sp. (Ericaceae) and *Maclaenia* 2 spp. (Ericaceae)
Cerro de la Muerte, Costa Rica (elevation 3,100 m)	
R. colwelli	All seasons: *Centropogon valerii* and *C. talamancensis* (Lobeliaceae) Seasonal: *Bomarea* 2 spp. (Alstroemeriaceae [Amaryllidaceae])
R. richardsoni	*Cavendishia smithii* (Ericaceae) and *Maclaenia glabra* (Ericaceae)
Volcan Colima, Jalisco, Mexico (elevation 2,000–2,500 m)	
R. epoecus	Dry season: *Lobelia laxiflora* (Lobeliaceae) Wet season: *Castilleja integrifolia* (Scrophulariaceae)
Northern coastal California, United States (elevation 0–200 m)	
R. epoecus	Summer only: *Castilleja affinis, C. latifolia,* and *C. franciscana* (Scrophulariaceae)

[a]Both genera Suborder Gamasida (= Mesostigmata), Family Ascidae.
[b]Both mite species occupy both host plants at all seasons.

number of plant species involved. Put another way, generally only one species of hummingbird flower mite is found on any particular host plant species. The exceptions, however, are significant and intriguing. In every case in which two species of hummingbird flower mite regularly share floral resources, the pair consists of one species of the genus *Rhinoseius* and one of *Proctolaelaps* (Table 24.1). There is not a single case known to us in which congeners regularly share a host plant species. These cases of regular co-occupancy are to be distinguished from "mistakes," the 1 in some 200 individuals that is found in the "wrong" host plant.

We have investigated three cases of "double occupancy" in some detail. Census data and behavioral observations for the mite species pair *R. hirsutus* and *P. contumex,* both monophagous on the host plant *Cephaelis muscosa* in Trinidad, provide no evidence for complementary distributions or complementary densities among inflorescences nor for niche differentiation within inflorescences. Likewise, Dobkin (1983) found little differentiation in the use of the plants *Heliconia wagneriana* and *H.* aff. *tortuosa* by the mites *R. klepticos* and *P. certator* (Table 24.1). There was some evidence for complementary shifts in relative abundance with season, but both mite species use both hosts, and their densities are *positively* correlated among inflorescences within host species. Heyneman (data to appear elsewhere) experimented with the pair of mite species associated with the host plant *Heliconia mathiasi* in Costa Rica. She tested the hypothesis that one species specializes on pollen and the other on nectar, but could find no differential attraction of the two species to these substances.

Monophagous Species

In the case of monophagous mites, such as *Rhinoseius trinitatis* (Dobkin 1984), *R. fidelis,* or *Proctolaelaps glaucis* in Trinidad (Table 24.1), the single host plant species of each mite species flowers all year, although there are often distinct flowering peaks and lulls (Feinsinger et al. 1982). Because there are no resting stages in hummingbird flower mites, monophagous species cannot exist except on such reliable host plant species, which are common only in lowland wet tropical sites.

Polyphagous Species

In contrast, hummingbird flower mites of tropical montane, subtropical, and temperate assemblages are almost exclusively polyphagous. The polyphagous species fall into two classes: sequential specialists (Colwell 1973) and opportunists. Sequential specialists, such as *Proctolaelaps kirmsei* in Trinidad, rely on two or more host plant species, no one of which flowers all year, but which together span all seasons. The mite *R. epoecus,* which travels on California migrant hummingbirds, plays a variation on the sequential specialist theme; it has summer host plants in California, and shifts between winter host plants and summer ones in west-central Mexico (Table 24.1) (Colwell and Naeem 1979).

Opportunists have one or two "home base" host plant species that flower year-round, but also occupy one or more additional host plant species that have brief flowering seasons. Examples include *Proctolaelaps certator* in Trinidad and *R. colwelli* and *R. richardsoni* at Cerro de la Muerte in the Costa Rican highlands (Table 24.1) (Colwell 1973, 1983).

Unifying Features of Host Plant Affiliations

There are three unifying features in these patterns of host affiliation, which are exemplified by the representative communities outlined in Table 24.1.

First, each mite species requires a year-round floral resource base.

Second, in the tropical lowlands the resource base is almost always a "minimal" one, either a single host plant that flowers year-round (monophagous mites) or a series of seasonally flowering species no one of which flowers for enough of the year to support a mite species on its own (sequential specialists). In the tropical lowland assemblages that we have studied, opportunists almost always have only one "home-base" species, that is, one year-round host plant species. (*P. certator* and *R. klepticos* in Trinidad are exceptions.) Moreover, the seasonal species that an opportunist exploits would not collectively support an additional mite species.

Third, polyphagous species dominate the hummingbird flower mite fauna of higher elevations and latitudes. Sequential specialists and opportunists with more than one year-round host are common in these communities. In contrast, monophagous species dominate the fauna in the lowland tropics. In the Arima valley of Trinidad, for example, 11 of the 15 species of hummingbird flower mites for which the host plants are adequately known are monophagous.

EVOLUTION OF HOST PLANT AFFILIATION

My interpretation of patterns of affiliation of hummingbird flower mites with their host plant species is as follows. Two opposing forces are at work in the evolutionary adjustment of host plant repertoires; one force broadens the repertoire, the other restricts it.

To the degree that resources are seasonal or unreliable, selection broadens the repertoire. The floral resources of polyphagous species are either seasonal or unpredictable from year to year (Colwell 1973, 1974). A single atypical year in the flowering pattern of the host plant of a monophagous mite can cause the mite's local extinction, as I will document in a later section.

The broadening force, then, is resource variability. Put another way, resource variability sets the limit to specialization. There is nothing new about this idea in biology (Klopfer and MacArthur 1960), not to mention in microeconomics, but hummingbird flower mites demonstrate the phenomenon with special clarity.

Several candidates contend for the role of restricting force—and I will argue that more than one of them play a part in limiting host plant repertoires. Three facts must be accounted for: (1) the frequent restriction of host repertoires to the minimum permitted by resource variability, often producing true monophagy, (2) the remarkable lack of any overlap in resource use among species of the same genus, and (3) the complete reciprocal overlap between certain allogeneric pairs of species. I will evaluate, in turn, a series of hypotheses that might account for these facts.

The Special Adaptation Hypothesis

Perhaps each mite species is so closely adapted to life on its own host plant species (or to all the host species in its repertoire) that it is unable to survive and reproduce elsewhere. The special adaptations might be morphological, behavioral, or biochemical. In its pure form this hypothesis makes no invocation of interspecific competition; each mite species simply occupies as broad a spectrum of host plants as its adaptations permit.

Morphological Adaptations There is evidence for special morphological adaptation in some species of hummingbird flower mites. *Proctolaelaps certator* in Trinidad occupies flowers surrounded by a moat of water in the bracts of *Heliconia wagneriana* and *H.* aff. *tortuosa*. Among the Trinidad *Proctolaelaps,* this mite is the only species that can walk on water. Other species of mites, placed during transplant experiments into the flowers of *H. wagneriana,* eventually wander out of the flower and drown (Dobkin 1983). On the other hand, having special feet doesn't prevent *Proctolaelaps certator* from also occupying a bromeliad and a ginger (Table 24.1), neither of which has water-filled bracts.

A second morphological example concerns mite body size, which, in general, is roughly correlated with host flower size (Fig. 24.1). The explanation for this correlation is unknown. (Mangan [1982, Fig. 2] reports a strikingly similar correlation between the size of different species of cactophilic *Drosophila* and the diameter of their host cacti, also unexplained.) Whatever the explanation for the correlation in Fig. 24.1, *Proctolaelaps glaucis* (point g) clearly deviates from the pattern; it is "too small" for its flowers. However, if *P. glaucis* were much larger, it would not fit through the narrow openings between the outer corolla tube and the nectar chamber of the flowers of its host, *Centropogon surinamensis*. (Hummingbirds stick their tongues through these openings.) Mites much larger than *P. glaucis* very likely could not survive on this plant, but other small species would have a

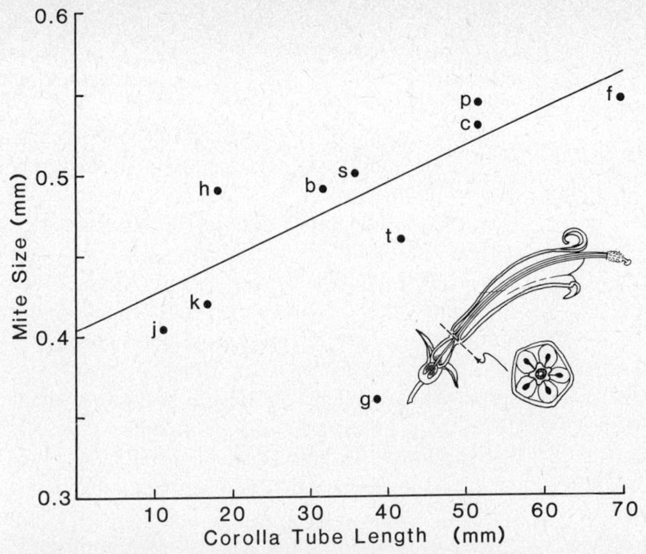

Fig. 24.1 Relationship between mite size and corolla tube length, and an example of special adaptation. Each of the points on the graph represents a different species of hummingbird flower mite from the Arima valley of Trinidad, West Indies. The regression line ($r = .84$) was computed for all points except g, which represents *Proctolaelaps glaucis* in the flowers of its host plant *Centropogon surinamensis,* pictured in the figure. *P. glaucis* is extraordinarily small, given the size of its host flower, and this small size permits the mite to reach the nectar chamber (proximal to the constriction at the broken line) by passing through the five tiny openings (darkened in the cross section) between the bases of the stamens. The openings are about 0.4 mm wide. Corolla tube length is measured inside each flower, from the deepest point to the lower "lip" (if any) of the flower. Mite size represents the length of the dorsal shield, computed as the intersex mean, based on 10 individuals per sex, weighted by field sex ratio estimates. The species and their host plants are listed in Table 24.1. The letters in the figure represent the first letter of the specific epithet of each mite species; the only ambiguities are: k = *Proctolaelaps kirmsei;* p = *Rhinoseius klepticos;* b = *R. bisacculatus;* c = *P. certator.* (Drawing by Shahid Naeem.)

chance. Indeed, *P. kirmsei* (point k), only slightly larger than *P. glaucis,* survived and reproduced in flowers of *C. surinamensis* in transplant experiments, as described in a later section.

Behavioral Adaptations Egg-laying behavior differs among species of hummingbird flower mites as a function of plant phenology and morphology. For example, mites that occupy plants of the genus *Centropogon* (e.g., *R. colwelli* in Costa Rica [Colwell 1973, 1983] or *Proctolaelaps glaucis* in Trinidad) lay their eggs inside the flower, which continues to secrete nectar and attract hummingbirds for a week or more—plenty of time for eggs to hatch. The great majority of lowland tropical hummingbird flowers, on the other hand, last less than a day, which forces female mites to lay elsewhere on the host plant. *Proctolaelaps kirmsei* on *Hamelia patens* lays at leaf-vein intersections on the underside of leaves just below the inflorescence. *Rhinoseius trinitatis* on *Heliconia trinidatis* lays among the bracts subtending the inflorescence (Dobkin 1983), a frequent pattern for many other species. *R. epoecus* on *Castilleja* in California is ovoviviparous, or nearly so (Colwell and Naeem 1979). It is easy to imagine both compatible and incompatible egg-laying behaviors among species in a community.

The one-day flowers of some species actually fall from the plant later the same day they open. A striking example is the plant *Heliconia psittacorum,* home of the mite *Proctolaelaps spinosus* in Trinidad. These mites leave the flowers for the safety of the inflorescence bracts just in the nick of time each day. When Dobkin (1983) introduced the mite *P. certator* into flowers of this plant, they failed to leave the flower and perished. The flowers of the four plant species that *P. certator* normally inhabits (Table 24.1) all remain on the plant, even after pollination and wilting. The behavior of *P. spinosus* permits it to exploit one-day, deciduous flowers, while the behavior of *P. certator* makes sense for the persistent flowers of its own host plant.

Biochemical Adaptations Special physiological adaptation to host plant biochemistry, a phenomenon common among herbivorous and seed-eating arthropods, is another factor that might limit the scope of the host plant repertoire. Perhaps the digestive physiology of each mite species is finely tuned to the nectar or pollen biochemistry of its host plants. Our tentative evaluation of this possibility is based on phylogenetic patterns, both experimental and natural introductions of mites to novel host plants, and biochemical analyses of nectar.

One of the classical sources of evidence for biochemical coevolution of host plants and phytophagous arthropods that consume leaf or stem tissue is the approximate congruence of host plant and arthropod phylogenies, at least at the level of genera or families (Ehrlich and Raven 1964, Mitter and Brooks 1983, Futuyma 1983). For example, each of the 70 or so species of all 11 genera of heliconiine nymphalid butterflies oviposits on plants of the family Passifloraceae (Benson et al. 1975).

Hummingbird flower mites and their host plants show no such pattern. The mite genera *Proctolaelaps* and *Rhinoseius* occupy plants of some 20 plant families, both monocots and dicots. The host repertoire of a single polyphagous species frequently includes plants of very diverse affiliation. The record for catholicity is currently held by *Rhinoseius colwelli* at Monteverde, Costa Rica (Table 24.1). This species occupies a series of five seasonally flowering plants, which span two dicot orders and two monocot orders. Likewise, both mite genera are represented among mites that occupy different host plant species in the same plant family (e.g., Bromeliaceae, Gesneriaceae), different species in the same plant genus (e.g., *Centropogon, Heliconia, Costus*), or even the very same flowers of the same plant species (the cases of double occupancy discussed earlier).

Experimental introductions of alien hummingbird flower mites to adoptive host plants (with their usual occupants absent) are often successful, in the sense that the mites survive and reproduce. (The details of several cases will appear elsewhere.) Inflorescences in bud are isolated with Tanglefoot and allowed to begin flowering inside mesh bags, or inflorescences already in flower are prepared in a similar way after elimination of any natural occupants by several days of flower picking. We introduce adult alien mites into the adoptive flowers with the aid of a 0000 artist's paintbrush and then rebag the inflorescence. Seven to 10 days (about a generation) later, the adoptive inflorescences are examined for mites, which are individually cleared and mounted for positive identification of species and instar. Failures to persist or reproduce may be accounted for by genuine incompatibility with the adoptive host (special adaptation) or inadequate technique. The successes nonetheless establish a surprisingly high level of compatibility, given the rigid patterns of host affiliation found in nature.

The natural introduction of a hummingbird flower mite to a native host plant will be described in the following section on speciation and extinction. We have also documented two natural introductions to exotic plants. The mite *Rhinoseius epoecus* now occupies cultivated species of African *Aloe* and *Kniphofia* (both

Liliaceae) in northern California gardens, in addition to its native host plants of the genus *Castilleja* (Table 24.1). The same mite species wintering in Mexico with migrant hummingbirds occupies a different species of *Castilleja* as well as *Lobelia laxiflora* (Lobeliaceae; Colwell and Naeem 1979). Meanwhile back in Africa, *Aloe* and *Kniphofia* species are occupied by "sunbird flower mites" (undescribed *Proctolaelaps* spp., but a lineage distinct from the hummingbird flower *Proctolaelaps*). To complete the irony, I have found sunbird mites in flowers of *Hamelia patens,* a neotropical hummingbird mite plant, in a garden in Kenya.

The lack of congruence between hummingbird flower mite and host plant systematics contrasts sharply with the taxonomic pattern of "miteless" hummingbird-pollinated plants, which cluster in major lineages. Lack of hummingbird flower mites is largely a family-level characteristic; the hummingbird-pollinated species of Malvaceae, Onagraceae, Labiatae, Convolvulaceae, Solanaceae, Acanthaceae (except for two species of *Aphelandra*), and several smaller families lack hummingbird flower mites.

For the most part, these plants have flowers well within the morphological and phenological range of "mite-occupied" hummingbird-pollinated plants, yet they remain unoccupied in spite of constant exposure to mites on generalist hummingbirds. For example, in Trinidad the common hummingbird *Glaucis hirsuta* often alternates feeding on interspersed plants of the miteless *Pachystachys coccinea* (Acanthaceae) and plants of *Heliconia wagneriana,* occupied by the mites *Proctolaelaps certator* and *Rhinoseius klepticos*. Hummingbirds netted in these mixed patches almost invariably carry mites, yet we have not found a single mite in the hundreds of *Pachystachys* flowers that we have dissected.

Amy Heyneman has done a major analysis of 43 biochemical variables for "mite-supporting" and "miteless" nectars (to be published elsewhere). She reports that while no single biochemical difference consistently separates the two groups, a multivariate distinction is quite clear-cut, using only three or four variables. Behavioral experiments show that in some cases, but not all, mites prefer sugar-water (matched for concentration) to the nectar of "miteless" plant species—a phenomenon that suggests repellent substances.

Special Adaptations: Conclusions Special adaptations clearly set certain upper limits to the breadth of host plant repertoires, but on present evidence do not appear to provide an adequate accounting for the high degree of host specialization seen in these mites. Moreover, the special adaptation hypothesis provides no causal explanation of the orderly patterns of overlap.

The Interspecific Competition Hypothesis

Interference Competition Does interspecific competition account for the narrow host repertoires of hummingbird flower mites and for their patterns of overlap in resource use? The very first pair of sympatric species that I studied, *Rhinoseius richardsoni* and *R. colwelli* at Cerro de la Muerte in Costa Rica, engage in mortal combat when placed together experimentally (Colwell 1973)—the ultimate form of interference competition. (This observation has been repeated by several generations of students in courses given by the Organization for Tropical Studies.) I concluded at the time that the lack of overlap in resource use by these two species could be accounted for by interspecific aggression, in conjunction with the foraging routes of hummingbirds (Colwell 1973, 1979). In short, it looked as if host-specificity (nonoverlap) was simply the most evolutionarily expedient way for individuals of one species to avoid the damaging hostilities of other species.

It is now clear that the violent behavior of these two species is not typical of all hummingbird flower mites. Staged encounters between seven pairs of mite species of the Arima valley in Trinidad failed to turn up a single case of serious physical damage from interspecific fighting. Although males of most, but not all, species did kick, shove, and chase males of the other species in arenas or dissected flowers, the interactions appeared to be largely ritualized and not obviously different from intraspecific interactions among males. On the other hand, allospecific females were not attacked, but were repeatedly "investigated," courted, and sometimes mounted by allospecific males.

Behavioral interactions between species of different genera did not differ in any consistent way from interactions between congeners. However, it appeared that interactions were especially placid between *Rhinoseius hirsutus* and *Proctolaelaps contumex,* which share the host plant *Cephaelis muscosa*. Given our failure to discover any evidence for differences in the way these two species use the plant, one might have thought it especially likely to find interference mechanisms between mites of this pair of species. The lack of any significant negative correlation between their abundances (among inflorescences) and the peaceful coexistence of the two species suggest that their resources are not limiting.

Exploitation Competition If interference competition provides no general explanation for patterns in the use of host plants by hummingbird flower mites, exploitation competition might nonetheless play a leading role. Small differences in the level of special adaptation to particular host plant species could, in principle, leave a different species the victor in the scramble for each reliable resource set. In this scenario competitive exclusion is responsible for the elimination of overlap in resource use, and the narrowness of repertoires results from interspecific competitive pressure. (For the moment, let us ignore the problem of the origin of local diversity—the supply of lean and hungry species ready to take over poorly utilized resources.) This hypothesis, however, is faced with the obvious contradiction of the cases of allogeneric double occupancy discussed in the previous section.

In principle, the exploitation hypothesis could be tested in the field for particular sets of similar species. In Trinidad, for example, we have succeeded in establishing experimental populations of *Proctolaelaps kirmsei* in unoccupied inflorescences of *Centropogon surinamensis,* the usual host of *P. glaucis*. The two mite species could be introduced into bagged, virgin *Centropogon* inflorescences simultaneously and in both sequential orders (the second species after establishment of the first) in replicated experiments. The interspecific competition hypothesis would be disproven, for this pair of species, if the alien species, *Proctolaelaps kirmsei,* were regularly to prevail under all three treatments. Equal fitness of the two species on *Centropogon*—a draw—would also constitute disproof, as long as experimental error were not too suspect and sample size were respectable.

Pure exploitation competition is weak, at best, when resources are not limiting, and in any case is unlikely to produce zero overlap. It is unusual to find a mite-occupied host flower that does not have a standing crop of nectar and pollen, in spite of its inhabitants (Dobkin 1984). For many species only a fraction of the flowers of a given host species are even inhabited (Colwell 1973, Dobkin 1983). The agent or agents responsible for the limitation of mites below the level of their food supply is not known (predation on eggs is my current unsupported best guess), but the lack of food limitation casts a pall on the hypothesis that exploitation competition is widely responsible for the nonoverlap of host repertoires among congeneric hummingbird flower mites.

Interspecific Competition: Conclusions To the extent that special adaptations differ among hummingbird flower mite species associated with different host plant species, exploitation competition has the potential to restrict host plant repertoires more narrowly than the broad absolute limits set by special adaptations themselves. However, the high degree of resource limitation required for complete and stable competitive exclusion (no overlap among species in resource use) is not supported by the evidence, nor can the coexistence of certain pairs of species with identical host repertoires be accounted for. Interference competition may reinforce differences in host plant repertoire in some cases, but the evidence does not support a primary role for this factor.

The Sexual Selection Hypothesis

Sexual selection arises from variance among individuals in mating success, whereas natural selection depends upon variance in survival and fecundity (Darwin 1871, Bateman 1948, Fisher 1958, Arnold 1983). In courtship behavior and morphology, hummingbird flower mites manifest sexually selected traits typical of polygynous social species. Multiple mating by males, in this case, is accentuated by the female-biased sex ratios of these species (Wilson and Colwell 1981; Colwell 1981, 1982). I will argue in this section that the relevance of sexual selection to the community structure of hummingbird flower mites is twofold. (1) Classical sexual selection may play a major role in establishing and maintaining reproductive isolation between incipient species (Fisher 1958, Carson 1978, Templeton 1980, Lande 1981). (2) Restriction of host plant repertoire to the minimum permitted by resource variability may be the result of self-reinforcing host preference based on differential mating success—a second kind of sexual selection.

Classical Sexual Selection Both male-male combat and female choice probably play a role in courtship and mating among these mites. When two hummingbird flower mites meet, they immediately palpate each other, each tapping its forelegs on the other's dorsum. The forelegs are heavily instrumented with tactile and chemosensory setae; the mites have no eyes. If both mites are males, a lunging and shoving match immediately ensues. These scuffles are usually quickly settled by withdrawal of one or both of the participants, but sometimes lead to escalated fights. When two females meet, they usually remain in close contact, continuing to palpate each other. The contact may last for many hours if the mites are inactive, or it may be brief if the mites are actively feeding or if a male arrives. When a male encounters a female, after mutual dorsal palpation, he continues to palpate her, while he moves around to one side, leaving his forelegs resting across her dorsum. Mating (venter-to-venter) may or may not follow, depending on time of day and on the female's response—she may move quickly away. In a densely populated flower these various activities can produce quite a frenzy.

Morphologically, males and females are strikingly distinct in most species of hummingbird flower mites (see the drawing on p. 619 in Janzen 1983). Among *Proctolaelaps* species, males are considerably smaller than females, with much heavier setae in most species, including a highly enlarged setal spur on the fourth pair of legs that is used during male-male combat, much as a cock uses his tarsal spur. Among *Rhinoseius* species, the two sexes are quite similar in size, but the dorsal and leg setae of males are much longer and stouter, and raptorial spines (highly modified setae) appear on the second pair of legs, which are used in male-male combat as well as to hold the female during copulation.

Females of closely related species are often difficult to distinguish externally (although the sclerotized, internal spermathecae are usually quite distinct), whereas the males differ strikingly in such seemingly arbitrary features as the length ratio of adjacent rows of dorsal setae and setal curvature, as well as the length and shape of the intromitive organ, the spermatodactyl. In this regard the mites parallel some genera of their hummingbird carriers, as well as many other lineages considered to be undoubted subjects of sexual selection (Darwin 1871, Gilliard 1969, Mayr 1972, J. L. Brown 1975, Thornhill and Alcock 1983).

Sexual Selection for Host Fidelity The essence of the host fidelity hypothesis is that the most reliable way for a dispersing mite to maximize mating success is to disembark from its hummingbird carrier at the right host plant species. And what is the "right" host species? Like the shape of the bowerbird's bower or the color of a hummingbird's gorget or the dialect of a sparrow, the identity of the "right" host plant is rather arbitrary, within certain limits.

Under this hypothesis, host plant preference is a secondary sex characteristic. Just as female choice can set off a runaway process of sexual selection for largely arbitrary characteristics of males (Fisher 1958, Lande 1980b, Kirkpatrick 1982, Arnold 1983), the evolution of host plant preference on the basis of differential mating success is a self-reinforcing process that leads to monophagy, in the absence of countervailing natural selection for expansion of the host plant repertoire. Before taking up some historical and theoretical aspects of this hypothesis, I will attempt to evaluate the case for its operation among hummingbird flower mites.

As far as we know, mating takes place only on host plants, not during dispersal on hummingbirds. To judge from the fervor that characterizes mating among mites in flowers, it is very unlikely that females disperse unmated, and females taken at random from flowers and isolated in the laboratory rarely fail to lay eggs. However, it is quite probable that females must mate more than once to realize their maximum potential fecundity. Egg-laying by experimentally isolated females ceases after a day or two in the laboratory, whereas the average reproductive period for females is probably at least a week. Males appear to be more or less continually engaged in courtship and mating, although there is a daily cycle of intensity.

Rigorous studies of mites of the allied family Phytoseiidae have shown that once-mated females lay significantly fewer eggs than females allowed to mate multiply (10 references cited by Amano and Chant 1978). Phytoseiid females have a maximum lifetime fecundity of only 50 to 100 eggs, produced over a period of 2 to 4 weeks, depending on species; male phytoseiids, on the other hand, are capable of producing 500 to 1,000 progeny, in the species studied by Amano and Chant (1978). The evidence we have points to similar values for hummingbird flower mites.

Dispersal and colonization are essential features of the lives of hummingbird flower mites. The "extinction" of each and every breeding group of mites is a certainty, because every inflorescence inevitably ceases to flower. Moreover, the completion of flowering in an inflorescence is not a random event, but a programmed one. However high the local fitness of a gene, its glory will be short-lived without representation in the dispersal pool and successful spread to other groups, especially new ones. Regardless of the absolute level of local adaptation to the host plant, as measured by potential fecundity and survival, the global relative fitness of a gene is ultimately a function principally of its representation in the dispersal pool on hummingbirds, from which new and existing inflorescences are colonized.

However, in addition to the representation of a gene in the dispersal pool, the lifetime mating success of its bearers is also a critical element of global fitness. Consider a mite on a bird. Unless every species of host plant the bird visits is equally saturated with potential conspecific mates, the mite will have more descendants if it disembarks at the currently most popular meeting place for its species. Host plant fidelity simply maximizes returns on the risky investment of dispersal.

The heritable basis of any behavior that differentially increases mating success will be favored by sexual selection, although the net effect on gene frequencies may depend on other factors (Fisher 1958, Lande 1980b, Arnold 1983, Halliday 1983, O'Donald 1983). If we are correct that the fecundity of female hummingbird flower mites is significantly increased by multiple matings (which clearly do occur, whatever the benefit), then the mating success of males is a function of continual access to receptive females, whether the latter are previously mated or not. Likewise, the fitness of females depends on the continued presence of males. Although host fidelity in this system differs in some respects from the kinds of characters usually envisioned by the architects of sexual selection theory (see below), there is no theoretical reason to exclude host fidelity from the rubric of sexual selection, and several good reasons to include it, as I will argue below.

Sexual Selection for Host Fidelity: Historical Considerations The role of sexual selection in focusing host plant preferences or, more generally, habitat preference seems not to have been previously recognized. Parker (1978) argued for the evolution of "encounter site conventions" by means of sexual selection (an "accelerating female choice spiral") in insect courtship and mating. Although Parker (1978) and Thornhill and Alcock (1983, pp. 182–207) considered female feeding and oviposition sites as a common focus of mating encounters, particularly among species in which females mate several times, they were concerned primarily with selection on the mate-searching behavior of males, taking as a given that females were already restricted to these sites. In their review of lekking behavior, an undoubted product of sexual selection, Bradbury and Gibson (1983) favor the explanation that lek sites are "hot spots" of female availability. From the point of view of dispersing male hummingbird flower mites, the "correct" host plant is just such a hot spot, at which females in search of mates will also do well to disembark.

The general idea that narrow habitat preference may serve as a means of finding mates is not new, although it has certainly never been accepted into the canon of conventional wisdom as an explanation for resource partitioning, perhaps because of the hegemony of interspecific competition as an explanation for all such patterns. The relationship between habitat preference and mate finding first came to my attention in a paper by Rosenzweig (1979a), who calls the phenomenon "rendezvous habitat selection," and in discussions with C. H. F. Rowell, who had independently proposed the hypothesis to explain host plant fidelity in rain forest grasshoppers (in which, like hummingbird flower mites, only allogeneric species share host plants in sympatry). Prokopy (1968), Zwölfer (1969, 1974), and Labeyrie (1971) made earlier statements of the idea for insects, and Rohde (1977) rediscovered it in his work with the parasites of fish gills, which he claims are restricted to arbitrary microsites by "selection to increase intraspecific contact." Many of these authors also recognized the role that habitat fidelity may play as a reproductive isolating mechanism (see below), a phenomenon that, in principle, is distinct from the idea of mate finding per se.

Sexual Selection for Host Fidelity: Theoretical Considerations From a theoretical point of view, to say that sexual selection based on "female choice" favors the evolution of some heritable feature of males is simply to assert that males exhibiting that feature have a higher-than-average frequency of mating (Halliday 1983). If females vary in some heritable trait that affects the probability that they will mate with different kinds of males, a genetic correlation between the preferred trait in males and the preference trait in females will result from assortative mating (Fisher 1958, Lande 1981, Kirkpatrick 1982, Arnold 1983, O'Donald 1983).

In this framework, fidelity of males to the prevailing modal host plant (for their species) can easily be seen as a sexually selected trait. However, in contrast with orthodox cases the female character that is expected to become correlated with this male feature is precisely the same trait: fidelity to the prevailing modal host plant. The genetic correlation, in this case, seems far more likely to be based on parallel expression of the same genes in the two sexes than on linkage disequilibrium, although sex differences in ploidy and dispersal patterns add complexity. (In any case, O'Donald's [1983] objection to Lande's [1980, 1981] model are probably irrelevant for this trait.)

Male dispersal varies among species of hummingbird flower mites. Species with the largest mating groups have little or no male dispersal, whereas males of species that live in small groups are almost as likely to disperse as female conspecifics. A full discussion and interpretation of this intriguing pattern is beyond the scope of this paper and will appear elsewhere. For present purposes it must be asked if host plant fidelity can be seen as a sexually selected characteristic in species with nondispersing males, who, dying in their natal inflorescence, have no active choice to make regarding host plant.

To the degree that female mating success in itself affects female fitness, I have argued that sexual selection operates directly on the host preferences of dispersing females, who do well to seek the company of conspecific males. (A solitary mated female could mate with her sons, but will have to wait a generation to do so.) For host fidelity as a trait, it would be inconsistent to consider the increased fitness the trait provides to males, through advantage in number of matings, as a fundamentally different phenomenon from the increased fitness the same trait, and probably the same genes, provide to females, through advantage in number of matings. Nonetheless, some will probably insist that any female trait that increases realized fecundity is favored by natural, not sexual, selection (e.g., Mayr 1972). I hope it is obvious that any disagreement on this point is purely semantic.

Finally, if one is willing to consider the mating success of the sons of dispersing females, then sexual selection for female host fidelity is in part based on male mating success even in species in which males do not disperse from their natal inflorescence. In effect, dispersing mothers make a host plant choice for their sons, who guarantee their own access to conspecific females simply by staying put. (This idea differs critically from the "sexy son" hypothesis of Weatherhead and Robertson [1979] in that the sexually selected feature—host plant fidelity—is not detrimental to female fitness.)

Sexual Selection: Host Plant Fidelity, and Reproductive Isolation In an assemblage of hummingbird flower mites in which every habitable, reliable host plant species (or set of species) is already occupied by a different mite species, host plant fidelity has a double payoff in terms of mating success. In addition to increasing the density of potential mates (as argued in the previous section), disembarking from a hummingbird at the right host plant species avoids wasting time, energy, and even gametes on mistaken courtships and matings with alien species. In this way sexual selection, operating on differential mating success, reinforces reproductive isolation through host plant fidelity. Trivers (1972), Mayr (1972), Otte (1979), and Parker (1979) discuss sexual selection in relation to the reinforcement of reproductive isolation in a general sense—a separate issue from the more contentious one of the role of sexual selection in the *origin* of reproductive isolation (see below).

In the present context, the role of host plant fidelity in reproductive isolation provides an appealing hypothesis for the orderly patterns of overlap in resource use among hummingbird flower mites: (1) Selection against losses to hybrid matings or attempted matings keeps congeneric mite species from sharing host plant species. (2) Two species of different genera, on the other hand, are more likely to differ sufficiently in general morphology and courtship behavior that premating reproductive barriers are effective even within the same inflorescence. (3) Under this hypothesis, the precise congruence of host plant repertoires (Table 24.1) within each coexisting *Rhinoseius-Proctolaelaps* pair testifies to a consistent balance between the repertoire-broadening effect of resource unreliability and the repertoire-narrowing effect of sexual selection, given a particular pattern of floral resources and hummingbird foraging routes in space and time. In effect, each pair provides a sample size of two, instead of the usual one, for an evolutionary experiment on the interaction of sexual and natural selection for host plant affiliation.

Sexual Selection: Conclusions To a mite, the fitness advantage of choosing a host plant that is already monopolized by its own species is twofold. First, the potential number of appropriate mates will be greater, from the point of view of dispersing males, dispersing females, or the sons (and daughters) of dispersing females. Second, behavioral interactions with alien species are avoided, including possible sterile matings or even physical harm. Sexual selection for host plant fidelity is inevitably a force that tends to narrow host repertoires. It is countered by natural selection for broader repertoires based on resource variability.

The somewhat arbitrary nature of host plant affiliations makes the sexual selection hypothesis an especially explanatory one for host plant fidelity. Moreover, this hypothesis accounts not only for the narrowness of host repertoires, but also for the lack of overlap in host plant affiliation by congeners and for the complete overlap between certain allogeneric pairs. Congeners may well be sufficiently similar that losses to interspecific mating are significant. Allogeneric species, on the other hand, are more likely than congeners to be so distinct morphologically or behaviorally that they have a negligible effect on each other's courtship and mating activities.

EXTINCTION AND SPECIATION

Local Extinction of *Rhinoseius bisacculatus*

On my first visit to the Arima valley of Trinidad in 1973 the mite *R. bisacculatus* monopolized the host plant *Costus scaber* and was often collected (by aspiration) from hummingbird visitors to this plant. The pattern was unchanged when I returned to Trinidad in 1975 and 1976. During the 1976 visit, however, I witnessed the extinction of this mite throughout (at least) the Arima valley as a result of a flowering failure in the host plant. The rainiest dry season in anyone's memory apparently slowed the maturation of new inflorescences that normally replace the old ones at this time of year.

Only 10 days passed with no open flowers, but when the new inflorescences at last began to bloom, *R. bisacculatus* was nowhere to be found. Instead, the opportunist species *P. certator* soon began to appear in the flowers of *Costus scaber*. Collections from flowers made over the following four years continued to produce only the interloper *P. certator* at our sites in Trinidad.

However, the extinction of *R. bisacculatus* was clearly local. This mite occupies the same host plant on the island of Tobago, some 40 km northeast, and continued to do so during the absence of *R. bisacculatus* from the Arima valley. (Moreover, the type specimen of the species was taken from a hummingbird near Belem, Brazil [Fain et al. 1977].) Thus it was not too surprising, when, in 1980, isolated individuals of *R. bisacculatus* began to appear again on birds at our Trinidad sites and eventually became fairly common again in flowers of the host plant, although still well mixed with *P. certator*.

It is not yet clear from the available data whether *P. certator* is being replaced by the returning *R. bisacculatus* or whether the two species will now coexist like certain other allogeneric pairs. They certainly did not coexist in anything approaching equivalent numbers before the "extinction" event. Looking back at our earliest collections from *Costus scaber*, from 1973 and 1975, we found only about 1 in 200 non-*R. bisacculatus* individuals mixed in with *R. bisacculatus*. However, the most common "mistake-maker" was none other than *P. certator*. By chance, these two species were one of the pairs I studied in 1975 for interspecific behavior. The *P. certator* mites were taken from their principal host, *Heliconia wagneriana*, and the *R. bisacculatus* mites were taken from *Costus scaber*. Males were very aggressive to each other and engaged in extended grappling fights. It will be interesting to reexamine interspecific behavior, especially if a stable coexistence should happen to develop.

Extinction as a Check on Sexual Selection

Although it now appears that the "extinction" of *R. bisacculatus* was local and temporary, the genuine extinction of hummingbird flower mite species must be a fairly common event on the time scale of millennia, or even mere centuries. All that is required to snuff out a monophagous species is a short period of flowering failure throughout the range of the mite, regardless of the long-term health of its host plants. If I am correct that sexual selection continually tends to contract the host repertoire, the ultimate check on this self-reinforcing process may be extinction.

Beginning with Darwin (1871), sexual selection theorists have recognized that natural selection sets limits on the sometimes extravagant products of sexual selection. However, almost without exception, the assumption has been made that these opposing forces act on the same time scale, producing a steady balance (e.g., Fisher 1958). By its very nature, however, sexual selection operates in every generation, at least in a stabilizing mode. In contrast, some of the forces that constrain Fisher's "runaway process" are likely to be quite episodic and perhaps very infrequent.

In the case of hummingbird flower mites one may guess that the frequency of flowering failures is directly proportional to the number of host plants in the repertoire of each mite species. Seasonally flowering species "fail" every year; the mites that use them are polyphagous. At the other extreme, geographically widespread host plants that flower all year have probably been home to the same monophagous mite species for eons. In the middle ground, host plants of restricted geographical distribution that normally flower all year may well support the evolution of a succession of monophagous species, each one ultimately doomed to extinction in some millennial flowering failure.

Host-Finding Behavior and Host Shifts

We have done an extensive series of field-laboratory experiments with T-shaped, capillary-tube "choice chambers" to investigate the cues that hummingbird flower mites use to find their host plants. Experiments with many different mite species were designed to test for discrimination between natural nectars, artificial nectars (sugar solution matched to a particular natural nectar for the concentration of 6- and 12-carbon sugars), floral tissue, and pollen. (Amy Heyneman will publish details of the technique and results elsewhere.)

Our experiments show that mites are attracted to virgin nectar of their own host plant species, in tests against matched artificial nectar or virgin nectar from the host plants of other mite species in the same habitat. On the other hand, mites do not distinguish between the virgin nectar of these other host plants and matched artificial nectar. These results suggest that mites are not repelled by the nectar of host plants of other species of hummingbird flower mites. In contrast, there is some evidence for actual repellence by the nectar of some "miteless" hummingbird-pollinated species (see previous section on biochemical adaptations).

The ability of these mites to discriminate among host plant nectars by chemosensory means and the rarity of "mistakes" by disembarking mites strongly suggest that the choice to enter a particular flower from the bird's bill is made from a distance. The use of the bird's nostrils by dispersing mites may function primarily as a way to exploit olfactory information about the bird's flower visits—information carried in the rapidly reciprocating air stream as the hummingbird respires two or three times per second.

Successful experimental introductions of alien mites into virgin adoptive host plants (see earlier section on biochemical adaptations) suggested the possibility of conditioned host preferences. We carried out a series of experiments with several mite species to investigate the potential for shifts in host preference (details and quantitative results will appear elsewhere). For example, the normal dry season host of the mite *Proctolaelaps kirmsei* in Trinidad is the plant *Palicourea crocea*. We succeeded in establishing *P. kirmsei* in bagged, unoccupied inflorescences of *Centropogon surinamensis*, the home of a different mite.

To our surprise, we were able to show a significant shift (in laboratory choice tests) toward preference for *Centropogon* nectar among survivors and descendants of the introduced mites, after only five days to a week. Similar experiments with other mite species and other adoptive hosts lend further support to a potential for shift in host preference by conditioning, although not all combinations produced significant results. The conditioning effect we have been able to demonstrate is not strong and may not be general. Nevertheless, any degree of conditioning would intensify the focus of host preference and would amplify the effectiveness of other factors (see below) in the establishment of reproductive isolation. Sexual selection for host fidelity may act, in part, simply by increasing "conditionability."

In contrast with the lack of good evidence for "larval conditioning" of adult host preference in phytophagous arthropods (Futuyma and Mayer 1980), the effect of adult experience on subsequent host preference has recently been shown for certain host-specific *Drosophila* (Jaenike 1982), tephritid flies (Prokopy et al. 1982), and beetles (Rauscher 1983). As far as I know, ours is the first evidence for the conditioning of host preference among mites. Unfortunately, the design of our experiments does not reveal at what point in the mite life cycle conditioning occurs.

Reproductive Isolation and Speciation

Taken together, the organismal, population, and community biology of hummingbird flower mites strongly suggests that speciation may sometimes be a very local process. Dispersing mites make "mistakes" in host choice at the rate of about half a percent, based on our identification of more than 12,000 slide-mounted individual mites from the Trinidad assemblage. These errors effectively guarantee rapid colonization of any host plant species whose previous occupants have become extinct or of a potential host plant species that has arrived without mites in a new locality.

There is probably little, if any, inbreeding depression, because the mites inbreed regularly anyway (Haldane 1940) and because deleterious recessive alleles are weeded out in the functionally haploid males (e.g., Hoy 1977). Thus even a solitary mated female can initiate a new breeding group.

Gene flow between mites on the ancestral host plant species and an incipient colony on a novel host plant will depend, in the short run, on four factors: (1) the frequency of hummingbird "traffic" between the old and the new host plant species, relative to traffic between individuals of the new host plant (Colwell 1973, Fig. 5), (2) the effectiveness of rapid host-preference conditioning (see previous section), (3) founder effects on morphological and behavioral isolating mechanisms, and (4) founder effects on genetically specified host preference (if such a preference exists).

If the interaction of these factors produces a level of genetic isolation sufficient to permit genetic divergence of mites on the novel host plant, then, in the longer term speciation may be consolidated by runaway sexual selection for host fidelity, as well as by classical sexual selection for male morphology or courtship behavior and the correlated preference among females (Templeton 1980, Lande 1981). Adaptation to special local conditions may then proceed, unhindered by gene flow from the ancestral population.

I do not intend to argue that all or even most speciation events in these mites take place sympatrically, but the potential seems very clear. The ecological conditions required by models of sympatric speciation based on assortative mating are closely met (Maynard Smith 1966, Bush 1974, Futuyma and Mayer 1980, Littlejohn 1981, Bush and Diehl 1982). Systematic studies by OConnor et al. (1985) on the Trinidad hummingbird flower mites have turned up a possible example: *Proctolaelaps belemensis, P. contumex, P. contentiosus,* and *P. certator* form a tight, local sibling species group. Perhaps it is no accident that *P. certator* was the mite that became rapidly established in *Costus scaber* after the ''extinction'' of *R. bisacculatus*. At the level of clades, lineages that are ''hard-wired'' for host preference and never make mistakes are perhaps less likely to become ancestors of new species, at least by sympatric speciation, than conditionable mites that make mistakes.

The idea that sexual selection plays a role in the evolution of reproductive isolating barriers between incipient species, first suggested by Gulick (1890), is the basis of a large, venerable, and sometimes contentious literature (see Thornhill and Alcock 1983, pp. 408–414, and Littlejohn 1981 for a useful overview). Little evidence exists for reproductive character displacement between closely related species in sympatry, but there is a growing body of theory and empirical evidence for the *incidental* divergence of sexually selected characters as a result of founder effects, changes in the expression of traits in a novel social environment, and alterations in learned components of courtship (Carson 1978, Kaneshiro 1980, Templeton 1980, Lande 1981, Thornhill and Alcock 1983).

Because of their mating system, subdivided population structure, and pattern of dispersal and colonization, hummingbird flower mites seem especially prone to reproductive isolation through founder effects on sexually selected aspects of morphology and courtship behavior. Shifts in host preference through rapid conditioning fit nicely under the rubric of alterations in learned components of mating success. The fact that every case of double occupancy of host plant species involves allogeneric mites suggests not character displacement, but ''differential colonization'' on the basis of ethological distinctness of potential occupants: congeners do better to keep to separate host plants to avoid the cost of courtship confusion and, perhaps, the costs of hybridization.

WHAT IS GENERAL, WHAT IS NOT?

Some of the principles that hummingbird flower mites demonstrate are quite general, though more obvious than in other kinds of organisms. Other aspects of their ecology and evolution are probably restricted to small arthropods that live in temporary microhabitats. However, this category includes most land animals with legs, so the restriction is not severe. In this section I will try to evaluate the generality of the principal forces that structure assemblages of hummingbird flower mites.

Limits to Specialization

The role of resource variability, or reliability, in setting the upper limit on ecological specialization of consumers is a piece of conventional wisdom that, for most organisms, is difficult to evaluate. I believe it to be very general, though often complicated by dispersal or physiological escape (diapause, dormancy). Hummingbird flower mites nicely demonstrate this principle because their need for flowers is simple and constant.

Reasons for Specialization

The legitimacy of what I have called ''special adaptations'' as sufficient explanations of ecological specialization and community patterns is probably highly overrated for many groups. Although adaptation sets the bounds, history and accident create a certain arbitrariness, which takes many forms—host plant affiliations, in the case of the hummingbird flower mites. Whereas not every mite is equipped to live in every host plant, most could survive and reproduce in *some* other sympatric host plants.

This pattern is common among phytophagous arthropods and many explanations have been proposed to account for it (Futuyma 1983). Clearly, the causality is often complex and highly dependent on life history, species interactions, and resource patterning. In the case of hummingbird flower mites, I have argued that special adaptations (mostly morphological and behavioral) and sexual selection for host fidelity are the principal forces that restrict host repertoires, with a sporadic role for interference competition.

Sexual Selection for Host or Habitat Fidelity

When resources are in short supply, the payoff for finding or controlling them is great. In contrast, when resources are not usually limiting, as appears to be the case for hummingbird flower mites, mate finding may become relatively more important as a focus of selection. Sexual selection for "competitive mate-searching," to use Parker's (1978) term, is likely to be strongest among mobile animals (and perhaps among rare, animal-pollinated plants) that live in complex, patchy habitats at low densities. If food is easy to find and mates are scarce, frequency-dependent reinforcement of conventional mating sites centered on arbitrary host plants or food sources may evolve.

Although many arthropods doubtless fit these criteria, there are two well-studied insect systems that have many ecological, behavioral, and evolutionary parallels with hummingbird flower mites: tephritid fruit flies and certain lineages of *Drosophila* (especially the Hawaiian group, but perhaps other groups as well, such as the cactophilic species [Barker and Starmer 1982]). These groups would seem especially likely candidates for the operation of sexual selection for host plant fidelity.

Many species in both these groups of flies share the following characteristics with hummingbird flower mites. (1) They oviposit only on certain host plant species and (2) mate on or very near the host plant (Cavalloro 1983, Prokopy 1983, and Zwölfer 1983 for tephritids; Heed 1971, Spieth 1974, and Parsons 1981 for *Drosophila*). (3) Both males and females mate many times (Spieth 1974 and Markow 1982 for *Drosophila;* Thornhill and Alcock 1983, pp. 463–465, for both groups of flies). (4) Classical sexual selection clearly plays an important role in courtship and mating (Bush 1969 and Prokopy 1980 for tephritids; Spieth 1974, Carson 1978, and Kaneshiro 1980 for *Drosophila*). (5) Conditioning is implicated in oviposition behavior (Roitberg and Prokopy 1981 and Prokopy et al. 1982 for tephritids; Jaenike 1982 for *Drosophila*). (6) There is evidence for sympatric speciation (Bush 1969, 1974 for tephritids; Richardson 1982 for *Drosophila*).

No one would dispute the role of sexual selection in the evolution of sex-attractant pheromones of males. In many arthropods these substances are elaborated from substrates that originate as secondary plant chemicals. For example, certain male danaid butterflies produce a courtship aphrodisiac derived from pyrrolizadines elaborated by the larval food plant (Edgar 1982). It seems likely that the ancestors of these butterflies used the larval food plant as a rendezvous site for courtship and mating (Edgar et al. 1974), a phenomenon that I have claimed is no less a product of sexual selection than are courtship pheromones. It may be that a significant proportion of arthropod courtship pheromone systems originated as host plant rendezvous systems.

It is possible that habitat use by vertebrates may be affected by competitive mate searching as well, at least in the "fine-tuning" of local distributions in complex environments such as tropical forest. For example, Diamond (Chapter 6) gives many examples of sharp altitudinal boundaries between congeneric New Guinea birds, and Terborgh (1971) provides others for Andean birds. These boundaries may have been sharpened by sexual selection for the use of habitat as an encounter site convention. In an early paper Lack (1933) came close to this idea, but later rejected it in favor of interspecific competition for resources (Lack 1944).

A possible example of sexual selection for habitat partitioning in lizards is discussed by Lande (1982) in the context of classical sexual selection. On the other hand, in a precise reversal of this hypothesis Kiester (1979) points out that conspecifics may be used as cues for finding ecologically appropriate habitat. In any case, given the frequency with which species distributions of vertebrates are correlated with habitat discontinuities (Endler 1977, Littlejohn 1981), I would urge vertebrate ecologists to treat sexual selection for habitat partitioning as a serious alternative hypothesis to special adaptation and interspecific competition for resources.

SUMMARY

Hummingbird flower mites (Ascidae: *Rhinoseius* and *Proctolaelaps*) spend most of their lives in the inflorescences of hummingbird-pollinated plants, where they feed on nectar and on pollen substances, mate, and produce young. The mites move between inflorescences by stowing away in the nostrils of hummingbirds. They require a year-round supply of flowers. At temperate latitudes and in the tropical highlands, only one or two species of these mites are found in sympatry, each affiliated with several species of seasonal host plants. By contrast, in tropical lowland forest, nearly 20 species of hummingbird flower mites may coexist, most of them monophagous. In all communities studied, most host plant species support only one mite species, but when two mite species share the same host, the mites are invariably of different genera (one *Rhinoseius* and one *Proctolaelaps*).

On the whole, the host plant repertoire of each mite species appears to be no broader than the minimum required to provide a reliable year-round resource base—a single host species in many cases. However, experiments show that mites can survive and reproduce in sympatric host plants outside their normal repertoire. Several forces might contribute to the extreme restriction of the natural host repertoire. Special morphological and behavioral adaptations are required to cope with floral morphology and phenology in some cases, but the limits are broad. Biochemical factors probably play a role in keeping mites out of "miteless" plant species, but there is no evidence for coevolution of mites and acceptable host plants based on host chemistry.

Although some pairs of species are very aggressive toward each other, most are not, so that interference competition alone cannot account for the pattern of host use among mite species. Competitive exclusion based on exploitation is also unlikely; mite populations are generally well below carrying capacity, and stable coexistence occurs in the cases of double occupancy of host species, with no evidence of resource partitioning or inverse density correlation.

Given the life history characteristics, mating system, and population structure of hummingbird flower mites, I suggest that sexual selection based on differential mating success may be largely responsible for the evolution of host plant fidelity and monophagy: Individual mites that disembark from birds at the "correct" host plant find more mates. Thus affiliation with a particular host plant species is effectively a secondary sex characteristic, expressed in both sexes, that becomes focused by frequency-dependent selection, a special case of Fisher's "runaway process." Mating success is also increased by avoidance of mistaken courtship and mating with congeners, which accounts for the absence of congeneric species that share the same host plant. I suggest that the greater divergence of morphological and behavioral premating isolating barriers at the intergeneric level accounts for the several cases of coexistence of noncongeners in the same host plant.

As with other sexually selected characters, there is a degree of arbitrariness in host plant affiliation and a degree of exaggeration—expressed as monophagy—that can result in extinction if the host species experiences a widespread flowering failure. I contend that such extinctions must be rather frequent and that the conditions are favorable for the replacement of such casualties by sympatric speciation: (1) mating and oviposition both take place on the host plant, (2) local breeding groups persist for many generations before the inflorescence finishes flowering, (3) experimental evidence for rapid shift in host preference suggests that conditioning adds a further element of assortative mating, (4) new breeding groups are founded by small numbers of individuals, followed by rapid increase in group size by reproduction, and

(5) founder effects on courtship and morphology may initiate rapid changes in isolating mechanisms through classical sexual selection.

ACKNOWLEDGMENTS

I am grateful to my friends and colleagues without whom this work would not have been possible: David Dobkin, Amy Heyneman, and Shahid Naeem collaborated with me on many aspects of this work. Each has made individual contributions as well, which I have endeavored to credit in the text. There would be no Monteverde data on hummingbird flower mites without Peter Feinsinger, and no systematics without Barry O'Connor and Shahid Naeem. I also express my thanks to Carol Baird, Jack Bradbury, Robin Chazdon, Martin Cody, Blaine Cole, David and Debbie Clark, Jared Diamond, Bernard Hallet, Lauren Howard, Mark Kirkpatrick, Gary McCracken, Ken Marten, George Powell, Jack Price, Emily Reid, George Roderick, Michael Rosenzweig, Hugh Rowell, Donald Strong, David Sloane Wilson, and James Wolfe. This work was supported by National Science Foundation Grant DEB 78-12038 and the University of California, Berkeley, Committee on Research.

chapter 25

On the Population Consequences of Mutualism

John F. Addicott

INTRODUCTION

Mutualism is a process that results in reciprocal net benefits, at either an individual or population level, for two different species (Boucher et al. 1982). If those individuals of different species that associate with each other achieve higher fitness than individuals that do not associate, then mutualism exists at an individual level (Roughgarden 1975a, Keeler 1981). If two or more species each increase the per capita population growth rate of the other [$\delta(dN_i/N_i dt)/\delta N_j > 0$], then mutualism exists at a population level (May 1981b).

Recent interest in mutualism (Colwell and Fuentes 1975, Risch and Boucher 1976) is leading to a better understanding of the existence, distribution, and importance of beneficial interactions. Mutualism may be facultative or obligate, symbiotic or nonsymbiotic, and tight or diffuse, and benefits may arise through many different mechanisms (see Addicott 1984). Mutualism permits organisms to exploit marginal environments (Harley 1970, Lewis 1973, Davidson and Morton 1981a, Huxley 1982). Hypotheses about indirect or nonobvious mutualism are increasingly common (Owen and Wiegert 1976, 1981; Vandermeer and Boucher 1978; Hutchinson 1978; Lawlor 1979; Petelle 1980; Vandermeer 1980), and symbiotic mutualisms may be ubiquitous (Margulis 1975). For ecologists, two major problems remain: how does mutualism evolve (Roughgarden 1975a, Keeler 1981, Thompson 1982), and what are the population consequences of mutualism?

In this chapter I will examine two problems concerning the population consequences of mutualism. First, under what conditions will processes that are mutually beneficial to individuals also have beneficial effects upon population density or population growth (Gilbert 1977, Templeton and Gilbert 1985)? Second, are there factors inherent in mutualism that constrain the net benefits available to one or both species?

The motivation for the first question arises from the need to reconcile the differences between cost-benefit models (Roughgarden 1975a, Keeler 1981) and population models (Whittaker 1975, May 1981b) of mutualism, and from a desire to understand the effects of mutualism on population regulation. Cost-benefit models explore conditions for the existence of mutualism, but only at an individual level. Population models usually assume that mutualism affects both population growth rates and equilibrium densities (May 1981b, but see Addicott 1981). However, the link between benefits at an individual level and effects at a population level is not clear, and there may not even be any link: Mutualism could exist at an individual level without affecting equilibrium densities or population growth rates. Since there are relatively few studies of mutualism at a population level, it is difficult to assess how frequently mutualism at an individual level leads to effects at the population level. I shall examine the evidence for a link between individual and population effects, and I shall relate this link to the complexity of association among mutualists.

Conflicting views on how mutualism affects population regulation also need to be resolved. Does mutualism decrease (May 1981b) or increase (Addicott 1981) persistence and return time stability, and can mutualism account for numerical constancy of populations (Gilbert 1977)? If mutualism affects population growth rates primarily at low population sizes, this could lead to strong buffering of population dynamics, because mutualistic populations temporarily depressed to low densities would increase more rapidly toward equilibrium densities than nonmutualistic populations.

The second question, the problem of constraints on net benefits in mutualism, arises from differences between theoretical and natural history approaches to mutualism. In population models global stability requires either some arbitrary constraints on net benefits (May 1981b) or application of some limiting factor external to the mutualism (Whittaker 1975). Although limiting factors external to mutualism can certainly be important, there are also factors inherent in mutualism that can limit the extent to which one species benefits another. These inherent limiting factors become apparent, not by considering mutualism as a single, homogeneous process, but by appreciating the natural history of mutualism, particularly the diversity of processes that can lead to reciprocal benefits (Addicott 1984).

I shall explore some of the ways in which such limits arise. For example, costs and benefits may depend on density or frequency; there may be interdependencies between costs and benefits; or intraspecific competition may be an inevitable consequence of the way in which benefits arise. Conflicting selection pressures or population effects may arise and may limit benefits in complex mutualisms.

Thus, the focus of this chapter is on the population consequences of mutualism arising from mutually beneficial interactions at an individual level. By this choice of topics I am specifically excluding from consideration those mutualisms that exist only at the population level, that is, food web mutualisms arising as "nonobvious" or "gratuitous" consequences of complex indirect pathways of interactions within food webs (see Chapters 3, 20, and 26 for discussion; also Lawlor 1979, Davidson 1980, Vandermeer 1980, Bender et al. 1984). For example, species A, B, and C may never associate directly with each other while competing for a set of common resources. If both A and B, and B and C, compete intensively while A and C compete little, then the net effect of A and C on each other can be positive, even though they are direct competitors (Lawlor 1979, Bender et al. 1984). Such food web mutualisms are an interesting phenomenon. However, they have mainly been explored theoretically, and they are mutualisms that exist only at population level. In this sense they are relevant to the general question of the interrelationships between population-level and individual-level mutualism, but they lie beyond the scope of this chapter.

Exclusion of these food web mutualisms does not mean, however, that I am concerned only with simple mutualisms involving the interactions of just two symbiotic species. While I shall consider simple, two-species interactions, such as plant-pollinator interactions, mycorrhizae, nitrogen-fixing bacteria, and endosymbiosis, individuals of two species can also associate and benefit one another in much more complex ways, often depending upon the presence of a third species. For example, the beneficial interactions of crabs and corals (Glynn 1976), ants and seeds (Culver and Beattie 1978), ants and ant-plants

(Inouye and Taylor 1979), and ants and homopterans (Addicott 1979) all involve at least one other species besides the mutualists. As a specific example, the beneficial effects of crabs (*Trapezia* sp.) for pocilloporid corals occurs through the crabs' inhibition of feeding by starfish on corals (Glynn 1976). Just because a third species is involved does not mean that these are food web mutualisms. In food web mutualisms the existence of benefits depends upon the indirect effects of changes in the population density of the third species. This is not the case with the mutualisms that I am considering. There need only be a behavioral effect of deterring or removing the third species. A change in the population density of the third species might occur, but it is not required for the existence of these mutualisms.

DOES MUTUALISM AT THE INDIVIDUAL LEVEL AFFECT POPULATION DENSITY?

In some cases the answer to this question is obviously, but trivially, yes. By definition the complete removal of an obligate mutualist causes the density of the other species to decline to zero. However, decreasing the number of symbionts per host or the proportion of hosts with symbionts in an obligate symbiotic mutualism need not affect the host's population density (Templeton and Gilbert 1985). Similarly, neither the complete nor partial removal of facultative mutualists will necessarily cause the equilibrium level of the other species to decline.

Evidence that individual-level mutualism affects population densities is difficult to obtain. Demonstrations of the existence of mutualism usually involve experimental manipulations, followed by observations of appropriate effects upon individual fecundity or survivorship. Short-term studies on organisms with long generation times rarely permit inferences about the population consequences of individual-level mutualism. Correlations between diversity or density of mutualists in different locations (Ostler and Harper 1978, Beattie 1983) provide only weak evidence. Consequently, much of the following discussion has to rely on inferences about possible density effects, given the structure or complexity of a given mutualistic system and knowledge of other ecological processes.

Proportion of Life History Involved in Mutualism

If individuals of two species associate during only one life history stage, the chances of an effect at the population level are greatly diminished (Gilbert 1977), as the following examples will illustrate.

Pollination mutualisms usually involve just one kind of beneficial interaction for both pollinator and plant, and the benefits may be restricted to a limited part of the life cycle. For example, *Heliconius* butterflies pollinate cucurbit vines, but feed on their nectar and pollen (Gilbert 1977). Neither species affects that life history stage of the other that is likely to limit population density (Gilbert 1977). As with many other plants (Harper 1977), wide variation in natural levels of cucurbit seed production may only weakly affect recruitment of adult plants, and *Heliconius* populations are probably limited by density-dependent intraspecific competition occurring on their larval food plants (Gilbert 1977).

A similar example of pollination mutualism involves seed production in yuccas (*Yucca* spp) that depend for pollination on yucca moths (*Tegeticula* spp). Yucca moths may or may not limit seed production (Udovic 1981, Udovic and Aker 1981, Addicott 1985), but under present conditions yucca recruitment is probably limited most strongly by insufficient rainfall for successful germination and seedling establishment or by intense seed predation and browsing on seedlings (Campbell and Keller 1932).

In some myrmecochorous plants benefits derived from dispersal by ants occur during just one life history stage. For example, the primary benefit to the plant *Datura discolor* from dispersal of its seeds by the ants *Veromessor pergandei* and *Pogonomyrmex californicus* is the removal of seeds from beneath *D. discolor* bushes, where discovery by mammalian seed predators is likely (O'Dowd and Hay 1980). The interaction is beneficial to ants because ants feed on elaiosomes. However, unlike other ant-seed dispersal mutualisms, few adult plants are found associated with ant mounds, as seeds are subsequently dispersed by wind or water. The existence of a population effect for *Datura* depends upon the relative importance of seed pool size and the number of safe sites for germination.

Restriction of benefits to one life history stage decreases, but need not eliminate, the possibility of a density effect. In the remarkable mutualism between inflorescence spiders (*Peucetia viridans*) and flowers of *Haplopappus venetus* (Louda 1982a), spiders use *H. venetus* inflorescences as foraging sites, capturing both pollinators and seed predators. Although spiders decrease seed set because of interference with pollinators, the interaction still benefits *H. venetus,* since seed predation is decreased even more. Mutualism with spiders affects densities of *H. venetus,* because seed production limits recruitment (Louda 1982b).

With more continuous association between mutualists, as in endosymbiotic mutualisms, density effects are more likely, but are still not certain. Zoochlorellae, such as those that associate with *Paramecium bursaria* or *Chlorohydra viridissima,* benefit their hosts either directly by transfer of metabolites or indirectly by deterring predation. Zoochlorellae significantly increase the density of *C. viridissima* in competition with *Hydra littoralis* under laboratory conditions (Slobodkin 1964). The presence of zoochlorellae in *P. bursaria* significantly reduces the rate of predation by *Didinium nasutum* on *P. bursaria* (Berger 1980). This should have a significant effect upon the density of *P. bursaria,* because of the continuity of association and the importance of predation in the dynamics of protozoan populations (Salt 1967).

Mutualism and Limiting Resources

Two mutualisms involving different discrete and limiting resources, carrion and dung, illustrate different possible effects of individual-level mutualism on population density.

Burying beetles (*Necrophorus* spp.) and their phoretic mites (*Poecilochirus necrophori*) are mutualists (Chapter 26; Springett 1968). The larvae of *Necrophorus* feed on carcasses of small mammals. Carcasses are a limiting resource for burying beetles, and the production of carcasses is independent of beetle densities. The larvae of flies (*Calliphora*) utilize the same carcasses, and there is intense competition between *Calliphora* and *Necrophorus* for carrion; in the presence of *Calliphora* no *Necrophorus* larvae successfully complete development unless phoretic mites are also present. Mites pierce fly eggs, effectively eliminating *Calliphora* from the system. In the field the importance of the mites depends on the length of time required by the adult beetles to bury carcasses (see Fig. 26.3). Thus, even though mites cannot affect the production of the resource, they can affect its subsequent availability, and therefore they should increase the population density of beetles.

Another mutualism involves moss and flies. Members of the moss family Splachnaceae grow exclusively upon substrates with high organic content, such as dung and carrion (Bequaert 1921). Flies are attracted to the sporophytes of *Splachnum* spp. and *Tetraplodon* spp. by a secretion produced on the hypophysis, and flies carry the sticky moss spores to fresh dung.In the bogs of northern Alberta all moose dung produced during the summer is immediately colonized by one of four species of *Splachnum* (P. Marino personal communication). Under these conditions dung is a limiting resource for *Splachnum,* as increased amounts of dung should lead to increased numbers of moss populations. Although these mosses require flies for dispersal, under present conditions changes in the number of flies would probably not affect the number of moss populations.

Complexity of Mutualistic Benefits

The importance of individual-level mutualism at the population level may also depend upon the complexity of the mutualistic interaction. In simple mutualisms just one benefit may exist, and just one component of fitness may be affected. However, there are numerous biological mechanisms, direct and indirect, whereby one species can benefit another (Addicott 1984). Direct benefits include nutrient transfer, energy transfer, providing of habitat, dispersal of gametes, and modification of the abiotic environment. Indirect benefits depend on a third species that feeds on predators, deters predation, increases prey availability, decreases niche overlap, deters competition, increases competition, feeds on competitors, or competes with competitors. The more ways in which a mutualist benefits another species, the more likely it is to affect the density of that species.

In ant-homopteran (aphid) mutualisms, ants can have many different effects. Ants deter predation, remove potential competitors, increase aggregation, increase individual fecundity and development rates, and decrease the impact of physical factors such as heavy rainfall (Way 1963, Skinner and Whittaker 1981). Not only do ants have many effects, but they also associate with all life history stages of homopterans (Way 1963). Only when homopterans disperse are they unassociated with ants. However, the more dense the ants are, the sooner will an aphid be found and tended during the early phases of population growth when aphid populations are most vulnerable to destruction (Addicott 1978). The short generation time of aphids has allowed unequivocal demonstrations of how homopteran population density is limited by ants (Way 1963).

Ants also have a multitude of effects in other mutualistic systems. For instance, ants affect the fitness of myrmecochorous forest herbs at several life history stages. Ants remove the herbs' seeds from locations where they are vulnerable to seed predators, increase germination rates, remove seeds from locations of intense intra- or interspecific competition, and place seeds on ant nests, which are favorable locations for germination, growth, and maturation (see Beattie 1983). In the ant-acacia mutualism, ants deter both mammalian and insect herbivores, decrease competition from vines and other plants, and lower fire-caused mortality rates (Janzen 1966, 1967). In the leaf cutter ant–fungus mutualism, leaf cutter ants provide suitable substrate for fungal growth, remove other fungi that could compete for substrate, and deter herbivorous insects that feed on the fungi (Quinlan and Cherret 1978). In all three systems ants influence their mutualists in many ways, increasing the probability that changes in numbers of ants would lead to changes in the densities of myrmecochorous plants, acacias, and fungi.

Thus, we see that the impact of individual-level mutualism on the density of the other species varies with three factors: the complexity of the mutualism; whether an organism associates with its mutualist for a small or large part of its life history; and whether the mutualist is able to increase the availability of a limiting resource. There are many mutualisms in which a density effect is likely to occur, but there are many others in which this is not the case. Now I shall turn from consideration of effects on population density to effects on population growth rates.

DOES MUTUALISM AT THE INDIVIDUAL LEVEL AFFECT POPULATION GROWTH RATES?

Where mutualism at the individual level affects population density, there must also be an effect upon population growth rate. However, partial or complete decoupling between mutualism at the individual and population levels is possible. There might be an effect upon population growth rate, but not upon equilibrium density (Addicott 1981, Wolin and Lawlor 1985). Mutualism could be important at densities below equilibrium levels, affecting the rate at which a population could respond to a perturbation or the rate of growth toward an equilibrium density following colonization of a new habitat. Effects on growth rate and density could also be absent. In this case, effects from the individual level would be seen only in the genetic composition of the population.

In this section I shall examine evidence for the effects of individual-level mutualism on population growth rates as opposed to equilibrium population density. I shall consider an example of mutualism where net benefits decline with increased intraspecific density and an example where mutualism clearly increases growth rates but may have little or no effect upon equilibrium density. A corollary of such patterns is increased resiliency or return time stability (Addicott 1981), and so I shall also consider the importance of mutualism in buffering population fluctuations.

Effects of Mutualism on Population Growth Rates

My own work with aphids and ants on fireweed allows a partial examination of the effects of mutualism at densities well below equilibrium (Addicott 1979). Four species of aphids feed on fireweed (*Epilobium angustifolium*) in the Rocky Mountains of Colorado; only three are tended by ants, and only two (*Aphis varians* and *A. helianthi*) show beneficial responses to tending. Aphids of a given species that occur on single flowering shoots of fireweed constitute a local population. The responses of local populations of both *A. varians* and *A. helianthi* to tending depend on the aphid density in local populations. Small populations show the greatest positive response to tending. As local density increases, responses to tending decrease or reverse, with tended populations more likely to decline than untended populations! Thus, net benefits decline with decreased intraspecific density. I have not yet determined which of the many effects of ants on aphids are the causes of this density-dependence, and I do not know the extent to which equilibrium aphid density (if it exists) increases in response to tending. However, a consequence of this density-dependent mutualism should be an increased population resiliency (see below).

Recruitment of many conifers depends on the caching behavior of corvids (Vander Wall and Balda 1977, Hutchins and Lanner 1982, Tomback 1982). In North America pinyon jays and Clark's nutcrackers remove seeds from cones and bury them in caches of one to a few seeds at depths suitable for germination, in locations removed from most other seed predators, and in sites suitable for growth of juvenile trees. Although there is a considerable cost to this type of seed dispersal, it appears to be effective in maintaining recruitment and in range expansion (as illustrated in Fig. 17.4). In the Great Basin of the western United States there has been a range extension of 500 to 640 km for several corvid-dispersed pines in 8,000 years during the Holocene (Wells 1983). The observed rate of expansion is compatible with corvid dispersal of seeds, whereas wind and gravity dispersal can account for only about 3.2 km of the range expansion (Wells 1983). This establishes a plausible case for the corvid mutualists' limiting population growth rates of pines. However, it is unclear whether equilibrium densities would be as strongly affected.

Stability of Mutualist Populations

If the strength of mutualistic effects is greatest at low densities, then the theoretical instability of obligate mutualisms in variable environments (May 1981b) may be unimportant. May (1981b) hypothesizes the existence of a low density threshold, below which one or both mutualists would inevitably suffer extinction. However, the buffering action of mutualism could significantly decrease the probability of ever approaching this threshold.

Heliconius ethilla populations in Trinidad showed no differences in population sizes either within or between two study areas during a period of two years (Gilbert 1977). This constancy of numbers may reflect three factors, all involving mutualism (Gilbert 1977). *Passiflora* are the larval food plants for the *Heliconius* butterflies. *Passiflora* possess extrafloral nectaries that attract microhymenopteran egg parasites, and these cause high mortality in *Heliconius* eggs. This mutualism between *Passiflora* and egg parasites apparently results in relatively constant levels of early stage parasitism on *Heliconius*. *Heliconius* adults feed on the pollen of cucurbit vines, allowing both constant egg production throughout the year and long reproductive life. All these effects should result in a buffering of population fluctuations.

In aphid-ant or aphid-ant-plant mutualisms there may also be a strong buffering of population dynamics as a result of the mutualism. For instance, there is a complex mutualism involving mountain birch (*Betula pubescens* spp. *tortuosa*), its major herbivore the autumn moth (*Oporinia autumnata*), and aphids (*Symydobius oblongus*) that are tended by ants (*Formica aquilonia*) (Laine and Niemelä 1980). During outbreaks of

Oporinia, there is severe defoliation of birch, but not where ant nests are established and ants routinely tend aphids. Since honeydew forms a major part of the diet of ants, aphid populations on birch make "high, stable ant populations possible during all phases of herbivore fluctuations. When an outbreak of some defoliating caterpillar (e.g., *Oporinia*) occurs, large numbers of foragers continuously patrolling in trees after honeydew can readily switch on caterpillars" (Laine and Niemelä 1980). Thus, although there is a significant cost to birch of maintaining aphids during nonoutbreak conditions, the result is a more stable system. Populations of subterranean aphids being tended by *Lasius flavus* are relatively constant, perhaps because of density-dependent harvesting of the aphids by *Lasius* (Pontin 1978).

Several workers have argued that facultative interactions may lead to buffering of systems (Vance 1978b, Culver and Beattie 1978, Bristow 1981, O'Dowd 1982). The important feature of facultative relationships is their considerable flexibility. For example, any of a number of ant species may be capable of associating with aphids or extrafloral nectaries and of providing adequate defense. Furthermore, ants provide flexibility in ecological terms, because they have many different effects (see above). Another kind of flexibility is shown by the variety of organisms that use extrafloral nectaries (EFNs) and potentially protect the plant from herbivores. For example, Stephenson (1982) found that the EFNs of *Catalpa speciosa* attract ants, coccinellids, and parasitoids, all of which attack the sphinx moth larvae feeding on *Catalpa*.

Consequently, there are numerous reasons to expect that for both obligate and facultative relationships mutualism can potentially buffer the system against population changes and changes in the external environment. In many cases this may be the primary consequence of mutualism at the population level. However, this aspect of mutualism needs further study.

WHAT LIMITS NET BENEFITS IN MUTUALISTIC SYSTEMS?

In this section I shall explore the various mechanisms that constrain the net mutualistic benefits available to individuals of a given species. The idea that the per capita effect of mutualism between species might decrease with increasing intraspecific density (Gause and Witt 1935) is incorporated in models of mutualism with nonlinear isoclines (Whittaker 1975, May 1981b, Christiansen and Fenchel 1977, Vandermeer and Boucher 1978). Whittaker (1975) interprets such nonlinearities as arising from factors external to the mutualism. However, in many systems these changes are integral aspects of the mechanism producing benefits and can be understood by comparing how costs and benefits vary with density. It turns out that benefits may be limited for any of several reasons. (1) Costs increase with density faster than do benefits, (2) there is a ceiling on benefits, (3) costs and benefits are interrelated and constrain each other, (4) mutualism enhances intraspecific competition *among* the mutualists for resources unrelated to the mutualism, (5) there is intraspecific or interspecific competition *for* mutualists, (6) costs and benefits vary in space or time, and (7) complexity constrains benefits. We shall now discuss each of these reasons in turn.

Faster Increase in Costs Than Benefits

As the density of one mutualist increases relative to the other, the relative values of costs and benefits may change. For example, in some pollination systems the costs of maintaining pollinators may increase with pollinator density, even though benefits do not. Yucca moths not only pollinate yuccas, but also oviposit within the yucca pistils, and the yucca moth larvae feed the developing seeds (Riley 1892). The costs of feeding yucca moth larvae can be high: 5 to 100% of seeds in set, nonaborted fruit (Addicott, in preparation). There are consistent differences among species of yucca in the damage caused by yucca moth larvae. These differences are related to characteristics of the plant that affect the success of yucca moth eggs. For example, in *Y. baccata,* no yucca moth larvae mature from eggs deposited at the time of pollination. These eggs are deposited in the pistil, below the style, in a zone where only rudimentary ovules occur. In *Y. elata,* eggs are deposited in the style rather than in the interior of the locule, and few eggs survive.

Differences in damage among populations and years within a species probably reflect the number of adult yucca moths available during the flowering and early fruiting period. High densities of moths per flower, or peak flight seasons later than peak flowering seasons, can lead to increased input of yucca moth eggs. Aker and Udovic (1981) report that some yucca moths oviposit in fruit that is already developing, with the result that the fruit has two cohorts of yucca moth larvae and seed damage is greater than usual. For example, in *Y. baccata* only fruit in which moth eggs are deposited when the fruit is already developing suffer any damage at all. Thus, in the yucca moth system there will be increasing costs associated with increased densities of adult yucca moths relative to yucca flowers.

Ceiling on Benefits

As the density of one mutualist increases, its ability to benefit the other may not continue to increase. This depends upon the way in which benefits arise. If benefits arise through deterring competition or predation (see Rai et al. 1983, Addicott and Freedman 1985), maximum benefits occur when all effects of competition or predation are removed. If the costs of supporting this mutualist are fixed, then the net benefit will increase to some maximum. If the costs of supporting this mutualist are a function of its density or activity, net benefits might increase initially and then decline.

Handel (1978) studied competition among three species of sedges, one of which (*Carex pedunculata*) is myrmecochorous. *Carex pedunculata* is a relatively poor competitor, producing lower dry weight, fewer rosettes, and fewer flowering culms in the presence of either of the other species. However, ants carry seeds of *C. pedunculata* to their nests, many of which are located in fallen logs. These logs act as refugia, lasting up to 20 years, because dispersal of the other species to logs is unlikely. Thus, myrmecochory in this system should effectively lower interspecific competitive interactions resulting from habitat overlap.

Viola nuttallii is myrmecochorous in the Rocky Mountains of Colorado (Turnbull and Culver 1983). Seed predation by nocturnally active rodents is high, but seeds are shed at midday, when ants are most active. Consequently, ants remove about 88% of seeds, leaving only 12% for rodents. Higher densities of ants apparently could have little additional benefit for *V. nuttallii* in terms of escape from predation. Similarly, increasing the reward provided by the elaiosome would not significantly increase the benefits of myrmecochory.

Interrelation of Costs and Benefits

There may be a tight interrelationship between costs and benefits: As costs decline so may benefits, or as benefits increase so may costs. In Louda's (1982a) system of spiders, plants, pollinators, and seed predators (see above), costs and benefits are interrelated temporally. If spiders occupy inflorescences too early, then both benefits and costs are likely to increase. The earlier spiders are present, the more seed predation will decline. But fewer seeds will be produced, because spiders also capture pollinators. Alternatively, spiders occupying the inflorescence later will have little if any detrimental effect upon pollination, but they will also miss many seed predators. Consequently, the costs and benefits in this system are interdependent and constrained by the independent responses of all four components to temporal environmental variability (Louda 1982a). This system is also density-limited. Should spider densities become too great, more pollinators are likely to be captured, seed production will drop, and there will be fewer plants in the next generation to be occupied by the offspring of the current generation of spiders (Louda 1982a).

Although grazing mutualisms are controversial, they do provide examples that intermediate densities of the grazer are beneficial, while higher densities may be detrimental. In grazing mutualisms "plants have the capacity to compensate for herbivory and may, at low levels of herbivory, overcompensate for damage so that fitness may be increased" (McNaughton 1983b). This could be the result of mechanisms intrinsic to the plant involving physiological or developmental changes, or of extrinsic mechanisms involving modification of the environment. Simberloff et al. (1978) present a similar case for possible beneficial effects, based upon isopod root borers on red mangroves. In both these examples the existence of individual-level mutualism is plausible, but not yet demonstrated.

One benefit for plants in pollination mutualisms involves outcrossing, which can lead to either higher reproductive success or more successful offspring. Using hand pollination, Price and Waser (1979) showed maximal seed production in *Delphinium nelsoni* with crosses from parents separated by distances of about 10 m, presumably reflecting a balance between inbreeding depression and outbreeding depression. However, under field conditions average pollen dispersal distances were about 1 m, so that quality of matings was suboptimal. Could plants manipulate their pollinators so as to obtain higher quality crosses? Decreased nectar rewards might force the pollinators to make longer flights between plants (Zimmerman 1982), but at the risk of an overall decline in seed set or the loss of pollinators to other flowers. More fundamentally, if pollinator movements increased, this would simply increase the genetic neighborhood, increasing the zone of inbreeding depression, and shifting the point of optimal outcrossing distance even further away (Waddington 1983). Thus, potential benefits in terms of quality of matings are strictly limited.

Intraspecific Competition Among Mutualists

As the density of a mutualist increases, it is likely to experience increased levels of intraspecific competition for resources unrelated to mutualism. For example, as aphid densities increase on individual host plants, fecundity declines, development times increase, and size decreases (Dixon 1977). As the size of the metapopulation increases, high-quality host plants become scarce, leaving only host plants on which potential population growth is severely limited (Addicott, unpublished). However, these and numerous other examples simply tell us that other factors extrinsic to and unaffected by mutualism must eventually limit populations.

A more interesting situation exists if the mechanism of benefit leads directly to and enhances intraspecific competition. Davidson and Morton (1981a) describe myrmecochory in two Australian arid zone chenopods, *Sclerolaena diacantha* and *Dissocarpus b. biflorus*. In locations where soils are crusty, red, alluvial loams, these plants grow almost exclusively upon the mounds of *Rhytidoponera* ants, which are the dispersal agents. ''Myrmecochory in these chenopods appears to be an adaptation for highly directional dispersal to favorable microsites'' (Davidson and Morton 1981a). However, unlike the situation in the eastern deciduous forests of North America (Culver and Beattie 1980), there is a low turnover rate of nest sites. Consequently, ''competition among seedlings may be intensified as a result of the aggregation of diaspores on nest mounds'' (Davidson and Morton 1981a).

In corvid-conifer systems, caching leads to clumping of seeds. Vander Wall and Balda (1977) found caches of 5 to 10 seeds of limber pine and piñon pine, but only 1 seed from each cache survived, due to intense intraspecific competition. Thus, the caching behavior removes the seeds from sites of high predation and places them in sites favorable to germination, but it also reinforces intraspecific competition.

Intraspecific Competition for Mutualists

If the effects of mutualism on population density are constrained (see above), mutualists may become a limiting resource, and benefits may be a decreasing function of the ratio of mutualist densities. As one species becomes relatively more numerous than the other, some individuals may lack mutualists, or the mutualists may be unable to provide benefits as large as when relative frequencies are more even.

Such effects are seen routinely in ant-EFN and ant-homopteran systems, because ants compete both intra- and interspecifically (e.g., Bradley 1973), thereby limiting their densities. Protection of bean plants by artificial EFNs depends on habitat, with protection occurring only where ant activity levels are high (Bentley 1976). O'Dowd and Catchpole (1983) were not able to demonstrate any protective effect of ants present at the EFNs of *Helichrysum* spp. in Australia. However, the proportion of capitula occupied by ants was never greater than 0.4, and the maximum average number of ants per capitula never exceeded 1.4. These values were very low and would indicate that ant densities are too low, relative to plant densities, to provide adequate protection. Large wild cherry trees experience greater levels of damage from tent caterpillars, because more egg masses are laid on them, but fewer ants tend EFNs per bud (Tilman 1978). An ant "colony's foraging ability may be saturated by a large tree" (Tilman 1978). Fewer populations of aphids on fireweed are tended by ants later in the season, as the number of populations increases (personal observation). In a year of high aphid (*Cinara* spp.) density on redwoods, many populations are untended and there are fewer ants (*Camponotus modoc*) per population than in a year of low aphid density (Tilles and Wood 1982). Treehopper densities can exceed the ability of ants to tend all aggregations, leaving the treehoppers vulnerable to predation (Wood 1982).

Similar examples exist in pollination and seed dispersal systems. In pollination systems seed production may be strongly limited by the availability of pollinators, as demonstrated by increased seed production with hand pollination over natural pollination. This has been demonstrated many times (see Bierzychudek 1981), but it does not occur in all systems (see Stephenson 1981). In the dispersal of nutmeg by birds, the greater the number of trees fruiting simultaneously within 50 m of a given tree, the less fruit is removed from a given tree (Manasse and Howe 1983).

Interspecific Competition for Mutualists

Services or resources provided by one mutualist may be utilized by many different species, providing the potential for interspecific competition (see Waser 1983) and limitation of benefits for a given species. Two systems illustrate some of the complex and potentially limiting effects of competition for mutualists.

Buckley (1983b) studied a system involving a plant (*Acacia decurrens*) that possesses EFNs, a honeydew-secreting membracid insect (the treehopper *Sextius virescens*) that feeds on *A. decurrens*, ants (*Iridomyrmex* sp.) that tend both EFNs and membracids, other herbivores of acacia, and the predators and parasites of membracids. Membracids have no detectable direct negative effect on acacias. Ants are beneficial to membracids, and in the absence of membracids ants are beneficial to acacia. However, since both EFNs and membracids attract ants, this provides the opportunity for competition to occur between the plant and the membracids for the service of ants. Ants could conceivably abandon EFNs entirely without a negative effect upon acacia, because ants could still remove herbivores. However, membracids "reduce the efficacy of plant defense by providing an alternative source of sugar which is more abundant and less dispersed and therefore attracts the ants away from the EFN" (Buckley 1983b).

The second example of interspecific competition for mutualists is provided by Davidson's and Morton's (1981b) study of Australian myrmecochorous plants (chenopods: see above). The chenopod *Sclerolaena diacantha* exists at lower population densities where it coexists with the chenopod *Dissocarpus b. biflorus* than where it occurs alone. This could be the result of either competition for dispersal to ant mounds, or competition among seedlings upon ant mounds, or both (Davidson and Morton 1981b).

It is possible that there could be either simultaneous or sequential competition among each group of mutualists for the other (Heinrich 1976, Culver and Beattie 1978). Plants could compete for pollinators, and pollinators could compete for plants. Or, ants could compete for seeds, and plants could compete for ants. These ideas have never been thoroughly tested, as competition is looked for in just one direction or the other.

Certain species act as mutualists yet provide no benefit, while other species are clearly parasitic on mutualistic systems. This has been inferred for pollination systems in which some species may provide no reward to pollinators, depending on deception by being similar to species providing rewards (Gentry 1974; Watt et al. 1974; Brown and Kodric-Brown 1979; Boyden 1980, 1982) or by being highly variable in floral morphology (Heinrich 1975). In both cases visits are likely to be made by young, inexperienced pollinators (Watt et al. 1974, Heinrich 1975). This could have a negative effect on the plants that produce rewards, as visits might be lost.

Variability of Costs and Benefits in Time and Space

The population definition of mutualism $[\delta(dN_i/N_i dt)/\delta N_j > 0]$ may be interpreted to imply that net benefit must occur at all times, places, and densities. But this definition is overly restrictive, given the change in costs and benefits as a function of density and given the temporal and spatial variability of real systems. A particular mutualism may be limited to certain large geographical areas by problems in the environments (Harley 1970; Losey 1972, 1974, 1978; Beattie and Culver 1981; Davidson and Morton 1981a; Huxley 1982). Alternatively it may be limited on a finer scale (N. G. Smith 1968, 1979; Crush 1975; Warner and Mosse 1982).

More interesting situations are presented by systems in which costs and benefits fluctuate temporally (Tilman 1978, Laine and Niemalä 1980, Sinclair and Norton-Griffiths 1982). During outbreaks of the autumn moth (*Ophorina autumnata*) on birch, ants tending aphids can have large beneficial effects, but during the intervening low-density years ants and aphids may provide no benefit (Laine and Niemalä 1980). Ants foraging at EFNs on wild cherry trees can provide considerable benefit to the trees by removing larvae of tent caterpillars (Tilman 1978), but only for about three weeks when the caterpillars are small, because ants are incapable of attacking the larger instars. Spatial and temporal variation in costs and benefits are thus one additional factor limiting the ability of mutualists to respond to each other.

Constraints Imposed by Complexity

Where the benefit obtained by a mutualist depends on the interactions of a number of species, this complexity may constrain potential benefits. The best example of this kind of complexity is provided by systems composed of plants, homopterans, ants, herbivores, and enemies of homopterans (Room 1972, Skinner and Whittaker 1981, Buckley 1983b, Fritz 1983).

The basic mutualism in these systems involves ants tending homopterans. This mutualism may be beneficial to the host plants if ants deter herbivores as well as the predators of homopterans. However, the complexity of the system allows many ways in which net benefits to the host plants may be limited.

First, the homopteran-ant interaction may be so successful that densities of homopterans are high enough to constitute a significant cost to the host plant (Dixon 1971a, 1971b). Thus, as ant densities increase, effectiveness of herbivore defense might increase, but so would costs.

Second, the potential benefits of the mutualism are a function of herbivore loads, which may vary considerably in time and space (Laine and Niemalä 1980, Messina 1981, Fritz 1983). Costs to the host plants of maintaining homopterans are likely to be relatively constant, and therefore the interaction between host plants and the ant-homopteran mutualism could shift repeatedly from beneficial to detrimental and back.

Third, ants may interfere with other organisms beneficial to the host plant. Ants lower predation on a leaf-mining beetle of black locust. For instance, ants interfere with an hemipteran predator or a leaf-mining beetle on black locust trees. Only when ants are the most important predator of the host tree's herbivores does net benefit to the tree result (Fritz 1983).

Finally, ants may already have been attracted to the host plant by EFNs. If homopterans attract ants differentially and if they are more aggregated than the EFNs on the host plant, then there may be a detrimental effect of the ant-homopteran mutualism for the host plant (Buckley 1983b).

The complexity of these and other mutualisms constrains the extent to which one species may benefit another.

SUMMARY

Despite renewed interest in mutualism, we know little about its population consequences, because most field studies have focused on mutualism at an individual level. This is primarily a matter of practicality, as it is easier, for example, to determine if fecundity increases in the presence of another species than it is to follow the population through an entire generation. Theoreticians have emphasized either the development of population models, including models of food web mutualisms, or cost-benefit models of mutualism at the individual level. At the moment there seem to be few studies that connect work at the population level with work at the individual level.

In this chapter I have addressed the problem of how the structure and mechanism of individual-level mutualism affect the dynamics of mutualistic species. Mutualism does have effects upon population growth and density, but certainly not in all or even most interactions, because many aspects of the structure of mutualistic benefits lead to a lack of density effects. How mutualism influences population growth rates is still poorly understood, but effects on population resiliency may be great.

Whether benefits are felt at a population or individual level, many factors inherent to mutualism limit the magnitude of net benefits. In some cases the absence of an effect of mutualism on the population density or dynamics of one species may make that species a limiting resource for the other. In other cases limitations on benefits are a natural result of the mechanism by which mutualistic benefit arises. A close examination of the structure of a mutualistic system will show whether or not such effects are likely, but ultimately observations must be made on systems at different times, places, and densities to see these effects.

chapter 26

Adaptive Indirect Effects

David Sloan Wilson

INTRODUCTION

The study of most subjects proceeds from the simple to the complex, so it is not surprising that our understanding of biological communities is based largely on two-species interactions. Lately, however, many ecologists have realized that species can be strongly influenced by species with which they do not directly interact (Chapters 3, 20, and 32; Neill 1974, Levins 1974, 1975, Colwell and Fuentes 1976, Levine 1976, Holt 1977, Lawlor 1979, Schaffer 1981). A simple example is shown in Fig. 26.1A, which portrays a consumer species (A), a resource species harvested by the consumer, and a species that competes with the resource species (''competitor''). The consumer and the competitor do not interact directly, but they affect each other through their interactions with the resource. The consumer suppresses the resource (by harvesting it), thereby benefiting the competitor. The competitor also suppresses the resource (by competing with it), thereby harming the consumer. The sign of these ''indirect effects,'' can be obtained by multiplying the signs of the arrows that connect the species to each other.

This paper explores the evolutionary implications of indirect effects. The basic problem is illustrated in Fig. 26.1B, which shows a second type of consumer, B, that differs from A by including the competitor species in its diet. Two sorts of benefits can result from this behavior. First, there is the direct caloric gain from eating the competitor (the positive arrow leading directly from the competitor to the consumer). Second, there is the indirect benefit of suppressing the species that suppresses the resource (multiplying the negative signs on the arrows leading from the consumer to the competitor and from the competitor to the resource yields a positive sign). The first is feeding behavior, while the second is weeding behavior. Optimal foraging theory attempts to predict diet choice purely on the basis of direct caloric gain (e.g., Krebs et al. 1983). Let us assume that the competitor species is a poor optimal foraging choice. Nevertheless, consumer B might still have a higher fitness than consumer A, if the indirect benefit is sufficiently large.

More generally, if indirect effects are important in nature, then to what extent do organisms evolve to increase their fitness via indirect ef-

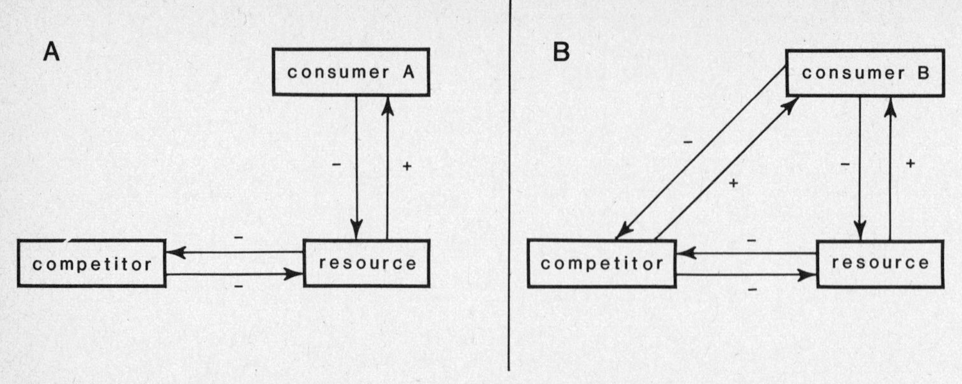

Fig. 26.1 Indirect effects. Consumer A and competitor affect each other only indirectly, through their direct effects on the resource. Consumer B has a beneficial indirect effect on itself by suppressing the competitor that suppresses the resource.

fects? Are species adapted not only to feed efficiently and to protect themselves from predators, but also to modify their community in ways that loop back beneficially to themselves?

This question cannot yet be answered, in part because it is so rarely asked. The concept of adaptive indirect effects is not totally absent from the thinking of community and evolutionary ecologists, but its distribution is extremely spotty. In studies of mutualisms the adaptive evolution of indirect effects is a commonplace idea. Consider, for example, the mutualism between crabs and corals carefully documented by Glynn (1983). A coral colony is inhabited by a single mated pair of crabs, who chase away coral predators, thereby preserving their own domicile. On a local scale the crabs modify their community in a way that loops back beneficially to themselves. Many similar examples could be cited. The benefit of interference competition is also an indirect effect, although it is only rarely stated as such (e.g., Roughgarden 1976, Schoener 1983b). It would be easy to redraw Fig. 26.1 using two consumers on a single resource, rather than a consumer, a resource, and a competitor. On the other hand, I am unaware of a single paper on optimal foraging theory that even acknowledges the possibility that diet choice might have indirect consequences. In my own opinion, adaptive indirect effects tend to be recognized in specific cases where they are obvious (it is hard to interpret the crab mutualism or interference competition any other way), but are ignored otherwise. They certainly have not found their way into our abstract concepts of species interactions and the organization of biological communities.

THE PROBLEM OF SHARED BENEFITS

When an organism modifies its community, it may or may not be the sole recipient of the indirect effects. For example, assume that consumers A and B in Fig. 26.1 each occupy an exclusive territory and that the competitor and resource species do not move between territories. Consumer B is then the only beneficiary of its own weeding activities and will prevail over consumer A if the benefits of weeding outweigh the cost of foraging suboptimally. Now assumc that all four species exist in a single, freely mixing community. In this case consumer A is, in anthropomorphic terms, a freeloader that enjoys all the benefits of consumer B's weeding activities, without expending any efforts of its own. If this single community is isolated from all others (e.g., a small island with negligible immigration), and if there is any cost at all to weeding (i.e., consumer B departs from optimal foraging rules), then consumer A has the highest fitness and will prevail (Wilson 1976, 1980). For this situation it is fully appropriate to ignore adaptation via indirect effects, because there is no way for them to evolve, no matter how beneficial they might be. This is a special case of a more general problem in evolutionary theory, that Hardin (1968) referred to as "the tragedy of the commons."

Exclusive territories and freely mixing communities are, of course, two extremes of a continuum. Most species in nature exist between these extremes in a complex matrix of semiisolated local populations connected by dispersal. The indirect effects of an organism's activities are shared to a degree, but not among the global population. Freeloaders always have a higher fitness than weeders (or other solid citizens) in their immediate vicinity, but if the types are patchily distributed, then patches in which solid citizens are common fare better than patches in which freeloaders are common. The evolution of adaptive indirect effects depends on the relative strengths of these opposing forces, which in turn are governed by a large number of factors loosely referred to as "population structure" (e.g., genetic variation among patches, persistence of patches, local population regulation, details of dispersal among patches, in addition to the costs and benefits of solid citizenry). Wilson (1980) and Uyenoyama and Feldman (1980) provide details.

To summarize, adaptive indirect effects are not expected to occur in all communities, but only in those with the right kind of population structure. With this caveat in mind, let us look at some examples of indirect effects in natural communities and how they might be modified by the evolutionary process. I discuss first a set of examples that I have been studying in carrion communities and then two speculative examples from the rocky intertidal zone and plant-soil communities. I conclude by considering whether these cases constitute evolved adaptations and whether adaptive indirect effects create alliances among species.

INDIRECT EFFECTS IN NATURE

Carrion Communities

My own research concerns a multispecies community that develops on small carcasses, such as dead mice. This community is dominated by a highly specialized genus of beetles (Silphidae, *Nicrophorus*) that bury the carcasses in underground chambers and raise their brood on them. These burying beetles are sufficiently numerous that most small carcasses are found and buried within 24 hours of their death. Several species usually coexist in an area. They have well-defined niches based on phenology and short-term temperature variation, but still often encounter each other on both small and large carcasses when the adult beetles aggregate to feed themselves (Wilson, Knollenberg, and Fudge 1984).

Burying beetles compete not only with each other, but also with flies and microbes. Almost every carcass is visited by flies before the burying beetles arrive, and the carcass is thoroughly innoculated with soil microbes during the burial process. Competition with flies tends to be an all or none process, with a given carcass producing either flies or beetles (D. S. Wilson 1983a, unpublished data). Competition with microbes may have been an important factor in the evolution of parental care; when the parent beetles are removed, a dense mold grows over the surface of the carcass, and brood success is low. Brood success is variable even with the parents present, and microbes very likely contribute to this variability (Wilson and Fudge 1984).

In addition to the beetles, flies, and microbes the community on small carcasses contains a diverse group of mites (at least 14 species from 4 families: Parasitidae, Macrochelidae, Uropodidae, Anoetidae). Mites cannot fly; they disperse by attaching to the bodies of burying beetles, an interaction termed phoresy. Depending on the beetle species and the locality, an average individual beetle can carry over 400 mites from 5 species and 3 families. (There is also at least one species of phoretic nematode about which little is known.)

Now we are in a position to examine the evolution of indirect effects. Fig. 26.2 shows a mite's-eye view of the carrion community. There are four potential resources for the mites (excluding other mites): the carcass, the flies, the microbes, and the beetles. In addition, the beetles have a positive effect on the mites, as agents of transport to other carcasses. As in Fig. 26.1, let us consider two distinct types of mites. Mite A is an optimal forager that maximizes its own fecundity without regard to indirect effects. We do not actually know the optimal diet for any mite species, but there is no reason to expect the indirect consequences of that diet always to be positive.

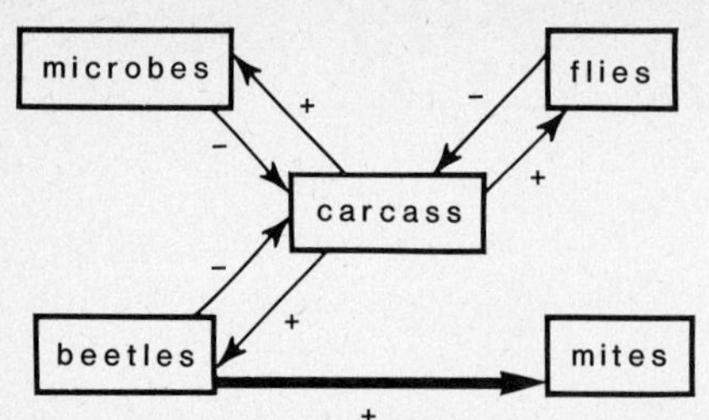

Fig. 26.2 Indirect effects in a carrion community. Flies, microbes, and beetles feed on the carcass. Beetles benefit mites as agents of transport. The mites would indirectly benefit themselves by feeding on flies and microbes, and indirectly harm themselves by feeding on the carcass and beetles.

Mite B attempts to maximize its fitness through indirect effects, by voraciously attacking flies and microbes, while ignoring the carcass and especially the beetles. As with mite A, there is no guarantee that these behaviors will simultaneously maximize personal fecundity. On general principles it would be unusual for two different things to be maximized by the exact same set of behaviors.

The opposing forces of selection that operate in structured populations can easily be visualized for this particular community. On any single carcass the freeloading mite A has more offspring than the solid citizen mite B, and beetles that survive through the efforts of mite B transport both types. On the other hand, carcasses vary in the species composition of their mites and in the genetic composition of single species. Those communities that eliminate flies and microbes because of an excess of mite B will outproduce communities that do not eliminate flies and microbes because of an excess of mite A. *The differential productivity of alternative communities opposes selection for maximum fecundity within communities.*

Which of these forces prevails in the real world? It will take years to answer this question thoroughly, but so far the evidence points toward adaptive indirect effects. I have reported elsewhere on a parasitid mite that pierces fly eggs (*Poecilochirus necrophori* Vitz.; D. S. Wilson 1983a). The egg-piercing behavior dramatically increases beetle brood success in the field, but does not increase the personal fecundity of the mites. The mites do eat the eggs, but if eggs are not provided, they feed equally well on the carcass (D. S. Wilson 1983a, unpublished data). We have since extended these experiments, and an updated story follows.

The basic experiment involves placing pairs of beetles on mice in the field, either with or without the egg-piercing mite *P. necrophori*. Fig. 26.3 shows the combined results of three replicates for the beetle *Nicrophorus tomentosus*, totaling 146 broods. The mite has no effect on brood success when the carcass is deeply buried, but it has a strong positive effect when the carcass is shallowly buried. The reason appears to be that the beetles themselves kill fly eggs deposited before burial by systematically removing the hair of the carcass with their mandibles. If the carcass is deeply buried, such that a layer of soil lies between it and the surface, additional fly oviposition is prevented and the mite is superfluous. But if the carcass is only shallowly buried, such that the top of the chamber is open to the surface, then fly oviposition continues throughout brood development.

Three *Nicrophorous* beetle species in addition to *N. tomentosus* exist at this study site. They differ in body size and therefore in the depth to which they can bury a carcass, as shown in Table 26.1. Based on burial depth, we predict that *N. orbicollis* and *N. sayi* will rarely benefit from *P. necrophori*. This prediction is confirmed for *N. orbicollis* in three replicates totaling 152 broods (unpublished data). Experiments with *N. defodiens* are in progress.

We therefore have an interesting situation. The egg-piercing behavior of *P. necrophori* seems to benefit two of four coexisting beetle species. This mite does not require fly eggs for its own reproduction, however, and should be found in association with all four beetle species. This is not the case. When beetles are collected in pitfall traps, the two smallest species have many more *P. necrophori* than the two largest species (Table 26.1). Furthermore, we have shown that individual mites can distinguish among beetle species and have strong, genetically encoded attachment preferences (Wilson 1982 and unpublished data). Two morphs exist at this location and prefer *N. tomentosus* and *N. defodiens* respectively

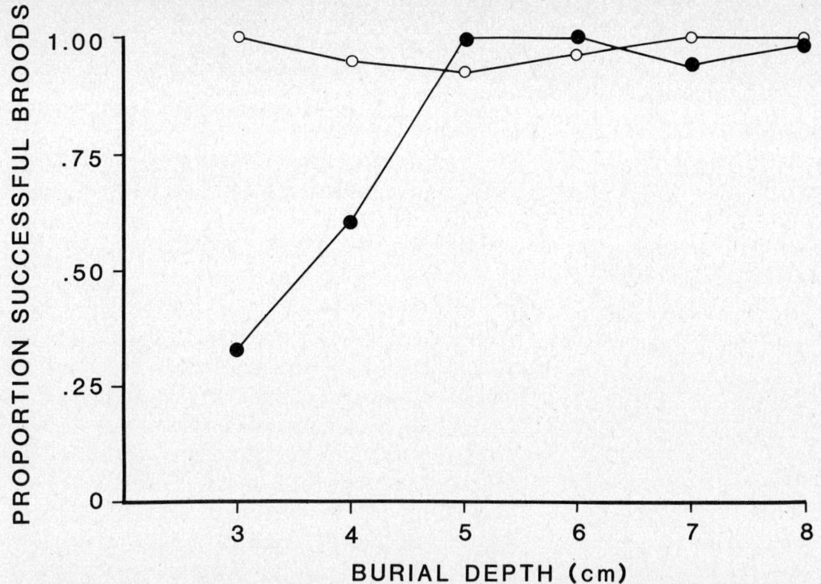

Fig. 26.3 Effect of burial depth on proportion of *N. tomentosus* broods that succeed. Solid circles represent beetles without *P. necrophori;* open circles represent beetles with *P. necrophori*. The differences between treatments for burial depths of 3 cm and 4 cm are statistically significant (chi-square = 7.47, $p < 0.01$ and chi-square = 4.48, $p < 0.05$, respectively).

(Table 26.1). In other words, those few *P. necrophori* that are found on *N. orbicollis* and *N. sayi* will abandon ship when given a choice.

Why is the mite *P. necrophori* associated primarily with the beetles that it benefits, and why does it avoid two other species that appear equally habitable from its own standpoint? It is important to stress that while *N. orbicollis* and *N. sayi* have few *P. necrophori*, these beetles have other mite species that are associated primarily with them. We have recently initiated experiments in which *N. orbicollis* are placed on mice in the field, either with or without their entire phoretic communities. These experimental beetle broods are visited by free-flying burying beetles, whose mites sometimes jump ship. To our surprise, we find that *N. orbicollis* with their own mites removed end up with more *P. necrophori* than beetles with their normal phoretic community (unpublished data, chi-square = 4.59, df = 1, $p < 0.05$). In other words, *P. necrophori* may be actively excluded from *N. orbicollis* by other species of mites.

Do the mites normally associated with *N. orbicollis* increase the fitness of their beetle? So far

Table 26.1 BURYING BEETLES AND MITES

	N. defodiens	*N. tomentosus*	*N. orbicollis*	*N. sayi*
Body size (g)	.044	.072	.099	.099
Mean carcass burial depth (cm)	3.93	4.5	6.02	6.10
Mean number of mites per beetle	10	12	2	3
Mite morph preference	Yes	Yes	No	No

For four species of burying beetles (Silphidae, *Nicrophorus*), the table gives: the beetle's body size; the mean depth at which the beetle buries a carcass; the mean number of mites *P. necrophori* found on a beetle; and whether there is a *P.necrophori* morph preferring that beetle species at my study site.

Data from Wilson, Knollenberg, and Fudge 1984 and unpublished data.

the answer is yes, but the result is only marginally statistically significant. At worst their effect is neutral. We have yet to discover a mite that decreases the fitness of its carrier beetle.

Although these results are not conclusive, perhaps they permit some guarded speculation. Let us assume that the population structure of this community is conducive to adaptation via indirect effects. Then we predict that in addition to feeding and avoiding predators, phoretic mites should form a protective network, a sort of external immune system, around the beetles that will carry them to future carcasses. We might also predict that different beetle species will require different phoretic associates to the extent that they face different problems for their survival. We have already seen how flies can be a problem for one beetle species, but not for another. Finally, we might expect a single beetle species to encounter different problems over its geographical range, with corresponding variation in its phoretic community. If burial depth varies with soil type, for example, then *P. necrophori* might benefit *N. orbicollis* in some regions, but not in others. We have already discovered *P. necrophori* morphs that prefer *N. orbicollis* at some localities (Wilson 1982).

Mytilus californianus Communities

Rocky intertidal zones have been intensively studied for decades, giving rise to a general paradigm of community structure based on competition, predation, and physical disturbance. A major problem for the mussel *Mytilus californianus* is the risk of being dislodged by wave action. Because mussels are attached not only to the substrate but to each other, they often are dislodged not as individuals, but as whole patches (Paine and Levin 1981). Although this is a purely physical process, it can be aggravated by biological factors. Fouling organisms that attach to mussels, such as brown algae and barnacles, greatly increase resistance to flow and therefore the probability that a patch will be dislodged (Witman and Suchanek 1984).

Suchanek (1985a, 1985b) has added a new dimension to the problem by studying the diverse community of mobile organisms that dwell within mussel beds, many of whom graze upon fouling organisms. When Suchanek removed these grazers, the density of fouling organisms increased dramatically. Once again, we have two alternative interpretations of grazing activity. Are grazers merely foraging optimally, with the happy consequence that patches persist longer? Or are grazers departing from optimal foraging rules in order to increase patch persistence? Unlike carrion communities, *Mytilus* communities do not come in discrete units, but they do vary in species composition over appropriate spatial scales (Suchanek personal communication). If these alternative communities vary in their effect on fouling organisms, such that some are torn away at a greater rate than others, then it seems possible that this process of differential patch extinction might influence the evolution of single species and the organization of whole communities.

Plant-Soil Communities

Most plants obtain their nutrients from a structurally and chemically complex medium called soil. In most cases the nutrients must be in a relatively simple and water-soluble form. Nutrients present in large organic or insoluble inorganic molecules are unavailable to the plants. The distribution, transport, and conversion of nutrients are all influenced by the activities of heterotrophic soil organisms (bacteria, fungi, nematodes, arthropods).

The heterotrophs themselves require large organic molecules ultimately derived from plants. This includes the living plants, their discarded parts, their exudates, and the heterotrophs that feed upon them. While killing and eating a living plant may be advantageous over the short term, in another sense it is like killing the goose that lays the golden egg. For example, microbial activity is many times greater within a few millimeters of living roots than in bulk soil. Many soil organisms profit most by living under a healthy, fast-growing plant (Coleman et al. 1983).

Without belaboring the point, we are again faced with the same question as for the above two examples. Do soil heterotrophs just look for food, or are they engaged in increasing their fitness through more subtle, indirect means? Are there subcommunities within the entire soil community that are organized to prevent nutrients from leaving the root zone, to gather nutrients for the plant, to convert nutrients from an unavailable to available form, to make the soil chemically and structurally beneficial for plant growth, and to protect the plant from other heterotrophs that are specialized for the quick kill? We know the answer to this question for mycorrhizal fungi, but what about other components of the soil community? Furthermore, what is the population structure of mycorrhizal fungi? How many spores colonize a single plant? When a single "individual" fungus transports nutrients into a plant, how many other fungi gain?

DO THESE INDIRECT EFFECTS CONSTITUTE EVOLVED ADAPTATIONS?

In evaluating the above three examples, it is important to distinguish between the consequences and the purpose of an organism's activity. Indirect effects are likely to be numerous and important in any collection of interacting species, even those without any previous history of interaction. In addition, a significant fraction of indirect effects are likely to benefit the species that cause them, without having specifically evolved to do so. The mere facts that phoretic mites sometimes benefit their carrier insects, intertidal grazers increase the persistence of mussel beds, and soil organisms sometimes benefit plants do not signify that these indirect effects are evolved adaptations. They may be consequences of nonevolved behaviors or behaviors that evolved for different reasons (for example, feeding efficiency). Of course, the distinction between consequence and purpose is a general problem in evolutionary biology and is not confined to indirect effects. For *any* trait, a positive effect on fitness is only the first step in demonstrating an adaptation.

The above three examples are intended simply to illustrate some ways that organisms can benefit themselves through their effects on other species in their vicinity. We do not know if they are evolved adaptations, and only for carrion communities is a systematic effort being made to find out.

ALLIANCES OF SPECIES

Adaptive indirect effects have the interesting property of appearing mundane in some contexts and controversial in others. Humans are masters at benefiting themselves through their effects on others, and human "ultrasociality" might be better explained by this process than by kin selection (Campbell 1983). Such interactions are by no means restricted to humans, however, for we readily acknowledge them in interference competition and mutualistic associations.

Adaptive indirect effects become controversial not when they exist among nonhumans, but when they exist among whole populations, as opposed to single individuals. We expect a damselfish to protect and nourish its sea anemone, but we do not expect soil communities to protect and nourish plants. More generally, our ideas about species interactions are very different from our ideas about individual interactions.

As previously mentioned, this double standard is partially justified. The old superorganism tradition assumed as a general rule that species behave as individuals (e.g., Emerson 1960), and this is certainly not correct. On the other hand, the modern individualistic tradition assumes as a general rule that species never behave as individuals (e.g., Williams 1966), and this is also incorrect (D. S. Wilson 1983b). Each of these positions corresponds to an extreme of the continuum referred to earlier. What has not been done, until recently, is to explore the middle range of population structures that lie between individual territories and single freely mixing communities. Until this middle range is better understood, we will not know which standard to apply to any particular natural community.

How would adaptive indirect effects alter the organization of biological communities? Even in the absence of well-documented examples, our familiarity with human behavior and individual level interactions makes it fairly easy to answer this question. Harmonious relationships are *not* an inevitable result of adaptive indirect effects; interference competition is purely negative, and even if phoretic mites have evolved to protect burying beetles, they do it by exterminating flies and microbes. Adaptive indirect effects do not eliminate conflicts among species (or individuals), but they might often cause species to form alliances in shared conflicts with other species or the abiotic environment. If one focuses on the organization of the alliances, then cooperation obviously becomes a dominant (although not the only) theme. This is in sharp contrast to the last 20 years of community ecology, with its emphasis on competition, predation, and disturbance.

In addition to creating alliances among species, adaptive indirect effects would increase the sheer diversity of evolved relationships, both positive and negative. Apart from individual mutualisms, we rarely think about evolved relationships between a prey and its refuge, a predator and the species that provide nutrients to the competitor of its prey, and so on. Indeed, adaptive indirect effects are potentially so diverse that a categorization at this point would be premature. It is far more important simply to document their existence above the level of individual interactions. Perhaps this unexplored subject will receive more attention in the future.

SUMMARY

Organisms can increase their fitness not only by feeding efficiently and successfully avoiding predators, but also by modifying their communities in ways that indirectly benefit themselves. Adaptive indirect effects are ubiquitous among humans, but among nonhumans they are recognized only for situations in which individuals are the sole recipients of their own indirect effects. General theories of species interactions and community structure tend to ignore the possibility of adaptive indirect effects. This tendency is justified for single freely mixing communities, but perhaps not for many other communities that are broken up into patches of local interactions. For these communities, we might expect to find a greater diversity of evolved relationships and a higher proportion of cooperative relationships than the last two decades of community ecology would lead us to expect. Examples abound for human communities and individual interactions; the only question is the extent to which they will be found on the population and community levels.

chapter 27

Communities of Species with Parasitic Life-styles

Catherine A. Toft

INTRODUCTION

Ecologists have recently proposed that communities of parasitic species differ systematically from those of free-living species. In this chapter I focus on one property of parasite communities: their species diversity. Two aspects of diversity need to be distinguished: the total number of species on earth or in some large geographical area (termed γ-diversity); and the number of species that coexist locally, for example, that potentially share the same host populations (termed α-diversity). Understanding γ-diversity in parasites is an evolutionary problem that has been treated elsewhere (Rohde 1979, Brooks 1980, Holmes and Price 1980, Price 1980, Holmes 1983). This chapter is concerned entirely with the ecological problem of what determines α-diversity in communities of parasites. Although evolutionary and ecological time scales are difficult to untangle, for parasites more than for predators, isolating and exploring the ecological consequences of parasitic life-styles allow insights about how species of parasites coexist.

The chapter necessarily takes a population dynamics approach. In even phrasing a question about communities of parasites compared to organisms with other life-styles, one is implicitly asking how the interaction *between* trophic levels affects communities on one of the trophic levels. The purpose of this chapter is to make this approach explicit. Hence, I explore how characteristics of parasitic life-styles determine the dynamics of parasite-host interactions in contrast to predator-prey interactions, and how these dynamics might determine the number of coexisting species in different types of parasite communities compared to those of predators. How these dynamics might affect α-diversity of the *lower* trophic level, that is, of the hosts, is a separate question that has been discussed by others (Holmes 1979, Freeland 1983, Holmes and Price 1985), but this chapter also yields some new expectations on the topic.

The chapter is organized as follows. The first section presents some definitions and a general overview of parasitic life-styles. The second and longest section develops insights into population dynamics from simple models, focusing on how the major differences between life-styles determine behaviors of parasite and host populations. This second section initially examines the primary characteristics defining parasitic life-styles, or those that determine the numerical response (the parasites' rate of increase), and then examines secondary characteristics that are associated with but do not define parasitic life-styles, especially those characteristics related to complex life cycles, host specificity, and competition among parasites. The third section presents expectations about the α-diversity of parasite communities that arise from the dynamical characteristics of parasite-host interactions. A final section identifies four characteristics of parasitic life-styles that seem most strongly to affect parasite-host interactions and uses these characteristics as branching criteria for a classification of parasitic life-styles. This classification is intended to identify ecologically meaningful categories of parasites and to help in understanding ecological aspects of parasite communities, primarily α-diversity. I emphasize that this classification is meant to serve a specific purpose, not to exclude or supersede other classifications that serve different purposes. The new classification makes clear how many types of parasite-host systems have yet to be modeled mathematically.

I conclude that characteristics of parasitic life-styles lend stability to parasite-host interactions and that parasites are less likely to regulate the lower trophic level than are predators. In general, more parasite than predator species should be able to coexist under any given circumstances, and some types of parasites should occur in more diverse communities than others.

PARASITES AND PARASITIC LIFE-STYLES

Price (1980, p. 4) presents a useful definition of "parasite"; although from *Webster's Third International Dictionary,* it precisely states the essential characteristics of parasites (numbers are mine): "A parasite is an organism (1) living in or on another living organism, (2) obtaining from it part or all of its organic nutriment, (3) commonly exhibiting some degree of adaptive structural modification, and (4) causing some degree of real damage to its host."

From a population standpoint, the parasites on the upper trophic level increase at the expense of the hosts on the lower (characteristics 2 and 4). These two characteristics define all "+−" two-trophic-level interactions (notably, parasitism and predation), as opposed to "+0" or commensalism and "++" or mutualism, both of which can involve the passage of nutrients from one trophic level to another, and as opposed to "−−" or competition, which occurs on one trophic level.

As symbiotes (characteristic 1), parasites are distinct from predators, which are free-living. Symbiotic organisms are governed by conditions within or on the host, leading often to characteristic 3, while free-living organisms are governed by external conditions. Importantly, if parasites cause the host to die, they may also die before they have completed growth or reproduction. We will develop the dynamical consequences of this basic feature of parasites shortly.

Parasites often have complicated life histories to cope with the dependency on the host required by their symbiotic life-style. Dispersal, including the problems of finding new hosts as well as mates, is a primary constraint. Typical life cycles involve one to several larval stages that are adapted for growth, asexual reproduction, and dispersal, plus an adult stage in which sexual reproduction and perhaps dispersal takes place.

One, several, or all of these stages may be parasitic. Many parasites have free-living stages; such stages may be (1) primarily for dispersal and involve no growth, (2) primarily for growth, or (3) primarily for reproduction. Almost every variation on these themes is exhibited by some species with a parasitic stage sometime in the life cycle (Table 27.1).The stage at which the organism is parasitic determines whether the nutrients gained from the host in a symbiotic relationship are used for growth (parasitic larval stages) or for reproduction (parasitic adult stage). We shall see that these life history details also have important dynamical consequences for parasite-host interactions.

It is helpful to oversimplify this variety by defining two types of life histories, the complex life cycle (sometimes abbreviated CLC) and the direct life cycle. Following Wilbur (1980), a complex life cycle is one "that includes an abrupt ontogenetic change in an individual's morphology, physiology, and behavior, usually associ-

Table 27.1 AN ECOLOGICAL CLASSIFICATION OF +− TWO-TROPHIC-LEVEL LIFE-STYLES, BASED ON A POPULATION DYNAMICS SCHEME, FOR TAXA WITH 100 SPECIES OR MORE AND SMALLER GROUPS OF SPECIAL INTEREST

Type Number	Name	Members (common names in parentheses)	Life-style characteristics: Adult	Immature	Probability of host death	More than one generation?
I	Macroparasites	Platyhelminthes (flukes and tapeworms); Acanthocephala (thorny-headed worms); Nematoda (round worms); Crustacea; Pentastomida; Diptera puparia (louse and bat flies); Strepsiptera; Gasteropoda; parasitic plants, e.g., Loranthaceae (mistletoe), Cuscutaceae (dodder), various Scrophulariaceae (broom-rape)	P	P	2	No?
II*	Microparasites	Protozoa, Fungi, viruses, bacteria	P	P	2	Yes
III*	Ectoparasites and others	Mallophaga and Anoplura (lice), Amphipoda (whale lice), Acarina (mites), Homoptera (aphids, etc.), Hemiptera (plant bugs), Thysanoptera (thrips), Tardigrada, Coleoptera, Nematoda	P	P	2	Yes
IV	Insect ectoparasites	Siphonaptera (fleas), Dermaptera	P	FL2	2	Yes?
V	Miscellaneous	Crustacea, Coleoptera, Nematoda	P	FL1	2	No
VI	Biting arthropods and others	Diptera, all divisions (e.g., mosquitoes, midges, horse and deer flies); Lepidoptera; Coleoptera; Agnatha (lampreys)	FL3	FL1	3	No
VII	Biting arthropods and others	Glossinidae (tsetse), Hemiptera (bed bugs), Acarina (mites and ticks), Hirudinea (leeches), vampire bats, Galapagos finches	FL3	FL2, 3	2–3	No
VIII	Parasitoids	Diptera (e.g., bee flies), Hymenoptera (parasitic wasps), Coleoptera, Lepidoptera	FL4	P	1	No
IX	Myasis-causing Diptera and others	Diptera (e.g., bot flies), Lepidoptera, Coleoptera, Nematomorpha (horsehair worms)	FL4	P	2	No
X	Herbivorous "parasitoids"	Lepidoptera (butterflies and moths), Diptera (e.g., true fruit flies), Hymenoptera (sawflies and gall wasps)	FL4	P	3	Yes?
XI	Miscellaneous	Pycnogonida, Bivalvia (larvae of freshwater clams), Crustacea, brood parasites (mainly birds)	FL1	P	2	No
XII	Predators	Almost all phyla have representatives	FL1	FL1	1	No
XIII	Grazing herbivores and detritivores	Almost all phyla have representatives	FL1	FL1	2–3	No

The last four columns give the branching criteria for this classification, abbreviated as follows. Adults and immatures: P = parasitic, whether ectoparasitic or endoparasitic; FL1 = completely free-living with growth or reproduction; FL2 = free-living larva, with larval nutrition provided by parent from host; FL3 = blood meal required for reproduction or growth; FL4 = free-living stage primarily for dispersal, mating, and oviposition and not using host for nutrition. Probability of host death: 1 = a developmental necessity; 2 = probable with heavy infestation, but not a developmental necessity; 3 = rare or accidental. More than one generation: Yes = more than one life cycle (generation) can be completed in or on a single host individual; No = each generation of parasites must transfer to a new host individual.

*Types II and III are distinguished as follows. In determining population dynamics the number of *infected hosts* is important for type II, and the number of *parasites per host* is important for type III.

ated with a change in habitat.'' Here habitat can mean host. In contrast, in a direct life cycle the individual grows at a gradual rate from birth or hatching through maturity and reproduction and does not generally change its habitat, host, or the general way it makes a living.

Three types of parasitic life-styles (Table 27.1) are customarily distinguished, both from one another and from predators (e.g., Anderson and May 1979; May 1982 Table 1): macroparasites (I), microparasites (II) and parasitoids (VIII). It is important to note here that all categories of life-styles grade into one another and are not the discrete categories that the definitions imply. Microparasites, including the viruses, bacteria, protozoans, and fungi, typically are small and reproduce inside the host, completing at least one life cycle in a single host individual. Macroparasites, such as the parasitic helminths (all eukaryotes), are larger and usually cannot or do not complete one life cycle within or on a single host individual. Both microparasites and macroparasites can have either direct or complex life cycles, and in species with complex cycles typically both growth and reproductive stages are parasitic. Typically, also, in both microparasites and macroparasites death of the host is not a developmental necessity, although some of the more virulent microparasites come close. In most forms death of the host results immediately in death of the parasite. In contrast, parasitoids, nearly all of which are hymenopterans and dipterans, are parasitic only in the larval stage, and death of the host is a developmental necessity in most species. Usually one larva develops from one host individual, typically the immature stage of another insect, which the larva kills by the end of its own development. The free-living adult stage is primarily for dispersal, mating, and oviposition. The adult either does not eat at all or takes nectar and pollen from flowers. Less commonly, females may sip fluids of hosts provisioned for larvae.

Predators (XII in Table 27.1) differ from parasites in three respects: (1) They are free-living in all stages, (2) they kill the prey as a developmental necessity, and (3) they require many, not just one, prey individuals for growth and reproduction. Predators, too, may have either direct or complex life cycles.

So far I have not been specific about the exact position of the lower trophic level; hosts may be heterotrophs (e.g., animals) or autotrophs (e.g., plants). However, nearly all predators, by the above definition, take animal prey except for seed- and fruit-eaters and phytoplanktivores. Whether an organism is accepted as a ''parasite'' often depends on whether the host is a plant or an animal, although the organism otherwise fits the above definition; many organisms with plant hosts are not considered parasites in the traditional view (see below and Price 1980). Rather than make an arbitrary distinction, below I emphasize life-style characteristics that determine population dynamics regardless of the type of prey or host used. By this view, organisms that get nutrients from plants may sometimes be separated from those getting nutrients from animals and sometimes not, depending on the exact characteristics of the life-style.

The three traditional designations for parasitic life-styles—microparasite, macroparasite, and parasitoid—have provided a basis for both the theoretical and empirical work done so far. In the next section an exploration of model systems tailored for these three life-styles reveals how their essential characteristics determine the behavior of parasite-host interactions, which often differs from the behavior of predator-prey systems. Using these characteristics, however, many more categories of parasitic life-styles, at least 11, can be distinguished from the vast diversity of parasitic species. Importantly, each of these categories could potentially exhibit different population behavior and therefore different community characteristics. These are the topics of subsequent sections.

DYNAMICS OF PARASITIC LIFE-STYLES

My aim in this section is to gather as much intuition as possible about the population dynamics of organisms with parasitic life-styles through the examination of simple models of two-trophic-level interactions. Such models are often criticized for their lack of biological detail, yet their parsimony helps us to gain insights and to clarify our thoughts. A model stripped to the essentials can better reveal the relationship between each life-style characteristic and its dynamical consequences than one embellished with detail. Moreover, population processes in general can be understood only with the aid of mathematical models, for these processes occur on a temporal and spatial scale that humans cannot observe (Chapter 8; Roughgarden 1979, p. 13).

The Basic Models

Two types of models, epidemiological and ecological, need to be distinguished; only the second type interests us here. Epidemiological models, which have a somewhat longer history than ecological models (see Bailey 1975, Anderson 1982, and Anderson and May 1985, for synopses of this literature), fix the size of the host population as a simplifying assumption and examine the dynamics of parasite numbers in a host population of constant size. This assumption is not unrealistic for epidemiologists, considering that they are most concerned with human and domestic animal hosts. Ecological models, in contrast, leave the host population free to vary as a result of the interaction with the parasites. May and Anderson (1979, Table 2) give a fuller classification of these models. Because knowing the impact that parasites have on host populations is central to understanding how these populations behave in nature, we must consider ecological models here.

In modeling the interaction between trophic levels, the equations for different levels are coupled by the functional response term. The functional response, which is defined as the relationship between the number of prey (hosts) captured per predator (parasite) and prey (host) density, describes the harm that comes to the lower trophic level. It also determines the numerical response, which in turn describes the rate at which prey (hosts) are converted into new predator (parasite) individuals. The general form for overlapping generations is:

$$dN/dt = R(N) - F(N,P) \cdot P \qquad (27.1)$$
$$dP/dt = P \cdot G[F(N,P)] - \delta P \qquad (27.2)$$

where N is the number of individuals on the lower trophic level and P that on the upper trophic level, the function $R(N)$ describes the rate of increase of the lower trophic level in the absence of the upper, $F(N,P)$ describes the functional response, $G[F(N,P)]$ describes the numerical response, and δ is the death rate of the predator unrelated to the prey (see also Pimm 1982). Analogous equations can be written for nonoverlapping generations (May 1973b, Hassell 1978). Table 27.2 lists some basic forms for the functional and numerical responses of the major life-style patterns that have been modeled.

Two aspects of population behavior are of special interest in considering the interaction between trophic levels: (1) stability and (2) the ability of the upper trophic level to regulate the lower. Because these terms have multiple meanings, it is necessary to define them for our purposes.

An interaction between two populations is *stable* if sizes of both populations are positive at equilibrium and return to this equilibrium after being perturbed from it (Pimm 1984). The upper trophic level *regulates* the lower if the equilibrium size of the lower population, N^*, is smaller when the upper trophic level is present, $P^* > 0$, than when it is not. The *degree* of regulation is determined by how far any particular N^* is depressed below its predator- or parasite-free value.

Simple ecological models deliberately give the host population an exponential, density-independent growth rate, $R(N) = rN$, so that hosts never reach a stable equilibrium in the absence of the parasite. One could provide density-dependent forms of $R(N)$, such as $R(N) = rN(1 - N/K)$, but it is simpler to assume that the host population will grow in the absence of the parasite until it reaches another limitation. This assumption helps to identify the exact impact that parasites have on the host population.

I shall examine three departures from a stable equilibrium in which the upper trophic level regulates the lower. First, the population on the upper trophic level can drive that on the lower below a critical threshold, resulting in extinction of one or both: $N^* = 0$, $P^* = 0$.

Second, the upper population can fail to regulate the lower, which then grows until it reaches a ceiling set by other factors. Depending on the model, the lower population can grow at the rate that it would in the absence of the upper population, or, more realistically, it can grow at a somewhat slower rate, dragging the upper population along with it until it is limited by other factors

Table 27.2 BASIC FORMS OF FUNCTIONAL AND NUMERICAL RESPONSES FOR TWO-TROPHIC-LEVEL LIFE-STYLES THAT HAVE BEEN MODELED

Life-style	Functional response $F(N,P)$	Numerical response $G[F(N,P)]$	Comments	References
Overlapping generations				
Predator (XII)	αN	$b(\alpha N)$		Volterra 1926
Macroparasites (I)	α	$bN - \alpha N \cdot f(P/N)$	Functional response (α): where parasites occur in hosts, P/N instead of P is used in equation 27.1 and Ns cancel. Numerical response: where b subsumes transmission dynamics, and $f(P/N)$ is distribution-dependent	Anderson and May 1978, May and Anderson 1978, Kostitzin 1934
Microparasites (II)	α	$bX - \alpha Y$	Assuming no recovery, and with Y estimating P. Here b = transmission rate	Anderson and May 1979
Nonoverlapping generations				
Predator (XII)	$\exp(-\alpha P_t)$, the likelihood of not being attacked	$c[\{N_t[1 - \exp(-\alpha P_t)]\} - \beta P_t]$	Predators and parasitoids must also differ in the form of α (Table 27.3); for parasitoids, $c = 1$, $\beta = 0$	Beddington et al. 1976, Hassell 1978
Parasitoid (VIII)		$N_t[1 - \exp(-\alpha P_t)]$		

N = number of individuals on lower trophic level; P = number of individuals on upper trophic level; α = capture rate, the rate of parasite-induced host mortality; b = conversion rate of food into progeny; c = number of progeny per prey-host; β = number of prey-hosts needed to produce eggs, where $b = f(c,\beta)$; X = number of susceptibles; Y = number of infected individuals.

See Table 5.1 of May (1981b) for more forms of functional and numerical responses.

(e.g., Anderson and May 1979). Under this possibility, the interaction is nevertheless *persistent,* and this is a form of stability (Pimm 1984) because both species stay in the system: $N > 0$, $P > 0$.

Third, even when a stable equilibrium regulated by the upper trophic level exists, it can have *degrees* of stability, defined by the *resilience,* or rate of return to equilibrium, and by *variability* in population densities over time (Pimm 1984). The most stable equilibrium is one in which population sizes are most constant (a point equilibrium) and to which the system returns quickly and monotonically. A less stable situation is a point equilibrium to which the system returns more slowly or with oscillations (oscillatory damping). A still less stable equilibrium is a stable limit cycle, in which populations fluctuate regularly through time. The greater the amplitude of the stable limit cycle, the greater the possibility that one species will go extinct, crossing some critical threshold at the lower point of the cycle. Ultimately, simple deterministic models can move from stable limit cycles to unstable, chaotic behavior (which is stochastic in appearance) depending on parameter values (May 1975a).

Stability can be evaluated for small perturbations around equilibrium (local stability) or for larger perturbations that move population densities to any feasible value, that is, $N \geq 0$, $P \geq 0$ (global stability). In nonlinear systems local and global stability are often distinct, for example, as with alternative stable equilibria in "state space." Local and global stability can also be used to mean physical, rather than state, space. Models can deal with a closed (spatially homogeneous) system, in which dynamics of immigration and emigration are not modeled, or an open (spatially heterogeneous) system, in which immigration and emigration occur among segments of the population isolated in patches (Caswell 1978). In open systems a global equilibrium can be stable even when the local equilibria are not (e.g., Levin 1974).

Judging stability and the upper population's ability to regulate the lower reveals much about the mechanisms underlying population behaviors. More importantly, these two qualities lead us directly to expectations about community patterns in organisms with different types of life-styles. I next explore model systems tailored for microparasites, macroparasites, parasitoids, and predators, examining how characteristics of these life-styles determine stability of the interaction between trophic levels and the ability of the upper level to regulate the lower. I divide these life-style characteristics into two categories: (1) primary characteristics, or those that determine the numerical response of the upper trophic levels and that reflect the essential (by definition) differences between life-styles, and (2) secondary characteristics (in particular those associated with complex life cycles and the greater specificity of parasites), or those that, regardless of why they evolved, affect population dynamics in ecological time.

Primary Characteristics: Numerical Response

The characteristics distinguishing various types of two-trophic-level interactions determine the numerical response, or rate of increase, of the population on the upper trophic level. All these life-styles by definition involve some harm to organisms on the lower trophic level. Two sorts of harm may be distinguished: death and loss of reproduction.

All life-styles can be ordered according to probability of death of individuals on the lower trophic level. Although this probability varies over a continuum, there is a discontinuity that has important dynamical consequences for two-trophic-level interactions, that is, whether host-prey death is ($p = 1.0$) or is not ($p < 1.0$) a developmental necessity.

Macroparasites and Microparasites For life-styles in which host death is not a developmental necessity, death of the host can result in the untimely death of the parasite, that is, before a sufficient number of new hosts are infected by the next generation of parasites (transmission). Hence a tension exists for all parasites except parasitoids between exploiting the host and causing their own deaths. For parasite and host populations, this tension enhances the stability of the interaction, generally at the expense of the parasite's ability to regulate the host population.

In simple population models the mechanism

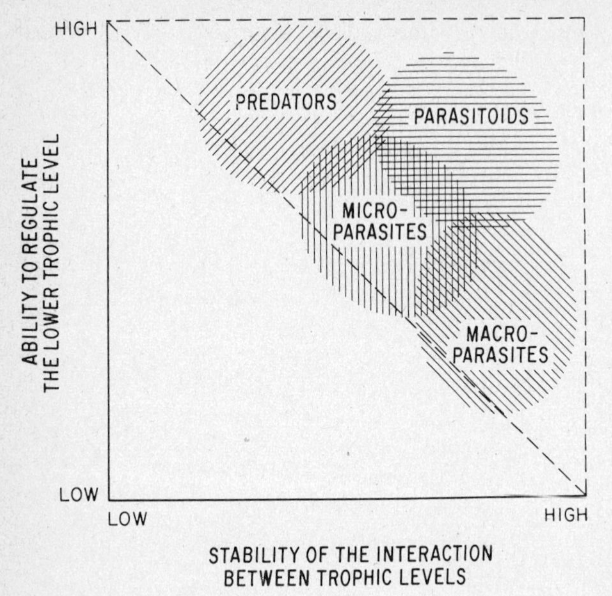

Fig. 27.1 Dynamical properties of the four types of two-trophic-level interactions modeled so far (macroparasites, microparasites, parasitoids, and predators) based on primary life-style characteristics. Stability of the system, whether a parasite-regulated equilibrium exists or not, is evaluated by (1) resilience, (2) persistence, or (3) population variability in time, including that caused by various time lags. Because characteristics of each life-style grade into one another, boundaries around each life-style are arbitrary (and hypothetical!). Low inherent stability and low regulation of the lower trophic level do not arise in +− coupled systems, in part because stability includes persistence: If the upper trophic level has no effect on the lower, it cannot cause extinctions or oscillations.

of this phenomenon can be traced to the self-damping properties of the parasite population. Parasite-induced host mortality, α, enters as a negative term (See Table 27.2) in both equations 27.1 and 27.2 for macro- and microparasites, but not for predators. The negative feedback on both populations counters the natural propensity of coupled +− systems to oscillate (e.g., May 1973a).

This intrinsic self-damping is strengthened in macroparasites by their nearly universal tendency to aggregate in the host population (Crofton 1971, Anderson and May 1978). The result is that most members of the parasite population are contained in disproportionately few hosts and, moreover, in hosts most likely to die from parasite-induced causes. The latter is even more true if α is related to parasite density faster than linearly, as is frequently the case (Anderson and May 1978). Thus, when parasites cause host deaths, these deaths have a stronger effect on the parasite population than on the host's, because the parasites are concentrated in the fraction of the host population that dies. As aggregation of parasites and parasite-induced host mortality increase together (especially with nonlinear forms of α), a smaller range of parameter values will permit a parasite-regulated equilibrium as opposed to the host's escaping regulation by the parasite. At the same time such an equilibrium, when it exists, is more stable in that the system will return more quickly to it after a perturbation (stronger damping) (Anderson and May 1978).

Interactions between microparasites and their hosts tend to have properties intermediate between those of macroparasite and predator interactions (Fig. 27.1). Microparasite-host interactions tend to be less stable than those of macroparasites, and microparasites tend to be more able to regulate the lower trophic level than macroparasites, for four reasons. First, microparasites, like macroparasites and unlike predators, die when the host dies, but the potential for self-damping is lower than in macroparasites. The self-damping effect of aggregation is lost, because the number of microparasite individuals per host is usually not relevant. Instead, a host is either infected or not; it becomes *infected* when an inoculum of threshold size invades it. After the parasite population reaches a certain size in the host, the host becomes *infective,* that is, able to infect new hosts. Until such time, the host is *latent*. Moreover, the concept of "individual" in most microparasites is hazy. For these reasons, dynamics of microparasite populations are described in simple models as the changing numbers of hosts in susceptible, latent, infective, or immune states, depending on the model (e.g., Anderson and May 1979, Anderson 1982). Here, number of parasites is not modeled, nor is it biologically relevant to do so. Thus, the strongly stabilizing aspects of aggregation in the host population do not occur in microparasite-host systems.

Second, transmission is more clearly a function of virulence (α) in microparasites than in macroparasites. While the host must survive until transmission is possible, many microparasites effect transmission by the very symptoms that cause the host distress and that may lead to its death (see May and Anderson 1983a). For example, many microparasites that are transmitted when fecal material is ingested by the next host cause profuse diarrhea in the host; in especially virulent forms such as cholera, death of the host results directly from the dehydrating effects of the diarrhea (e.g., Burrows 1959). The microparasites remaining in the host die when it dies, but the symptoms leading to the host's death are nevertheless their opportunity for transmission. (Again, the concept of individual is unclear in microparasites, and this situation may also be ripe for group selection.) Thus, there is some self-damping, but less than in macroparasites. As a result, microparasites also have greater potential to regulate the host population than do macroparasites (Anderson and May 1979).

Third, reproduction inside a single host can decrease the stability of simple model systems by introducing a region of parameter space in which the parasite can drive the host to extinction (model E of May and Anderson 1978). This happens because generations of parasites inside a single host allow the parasite population to overtake the host population. Depending on the details of the parasite's life history, reproduction inside the host could outweigh the self-damping features of parasites. But here we must consider whether the number of individuals per host is important; some microparasite infections may be better described by May and Anderson's model E and others by the host-state models mentioned above. In either case the likelihood of a stable microparasite-host interaction is reduced.

Fourth, microparasites tend to occur as short-lived infections and to produce long-lasting immunity in the host. These features introduce time lags that increase the tendency of microparasite populations to oscillate either in the form of stable limit cycles or periodic fade-outs and outbreaks (Anderson and May 1979, May and Anderson 1979). A common observation is that microparasites tend more to occur as epidemic infections, macroparasites as endemic infections.

Death of the host is a more conspicuous and universal phenomenon than loss of reproduction in the host, which does not even occur in parasitoids or predators. Loss of reproduction, including parasitic castration, has been primarily reported for invertebrates, which serve as hosts for larval macroparasites and for a variety of microparasites (see discussion in May and Anderson 1978). The degree to which parasites limit reproduction as opposed to causing hosts to die is insufficiently known to estimate its importance for various types of parasites.

We have been evaluating stability from two points of view: whether each feature of parasitic life-styles will affect (1) the range of parameter values allowing a parasite-regulated equilibrium, as opposed to driving the host to extinction, and (2) the dynamical properties of a parasite-regulated equilibrium when it exists. The first point of view is of relatively little use in comparing the population dynamics of different life-styles in nature. We simply do not know what values these parameters take. Very likely, real parasite-host systems are characterized by a far-from-random set of parameters (May and Anderson 1978), as is equally true of predator-prey systems, because the systems we see in nature are those persisting over long periods.

From the second point of view, however, both macro- and microparasitic life-styles reduce the tendency of these populations to oscillate with their hosts, because of their essential self-damping relative to predators (May and Anderson 1979). *The stability of a parasite-regulated equilibrium is enhanced by the same property that reduces the parasite's ability to regulate the host.* In the absence of a parasite-regulated equilibrium the host will grow exponentially until limited by something else, taking the parasites along with it. In such a system both species persist, and because the parasites do not regulate the host, they are no longer able to drive oscillations in the host population. Thus, both kinds of parasites enjoy more stability in interactions with the lower trophic level and are less likely to regulate it than are predators. There is a trade-off here between stability and the ability to regulate, and the three life-styles compared so far can be arranged along a continuum based on these two properties (the diagonal axis in Fig. 27.1). Microparasites are intermediate on this continuum, with the least virulent forms approaching characteristics of macroparasites and the most virulent forms approaching those of predators.

Parasitoids Parasitoids can be compared to equivalent predators if we convert the Volterra-type predator-prey equations to discrete time (May 1973b). It would not be fair to compare continuous- and discrete-time model systems directly because of the inherently less stable properties of the latter. Keep in mind, however, that this is not simply a defect of the model systems, but reflects the basic time lag that is a real consequence of nonoverlapping generations, such as would occur in higher latitude arthropod populations.

In contrast to macro- and microparasites, there is no trade-off between ability to regulate and stability of interaction in parasitoids compared with equivalent predators.

Parasitoids have inherently less efficient functional responses than predators because of the way they "attack" prey. A parasitoid lays an egg on each host and then leaves the host to be reencountered, whereas predators remove prey after an attack (leading to Nicholson's "competition curve" for parasitoids). Ever fancier and more mechanistic functional response terms (Table 27.3) do not alter this trend, with parasitoids remaining more inefficient at attacking hosts than predators (Royama 1971, Rogers 1972, Arditi 1983). In behavioral time parasitoids' inefficiency at high prey densities, when the parasitoids' regulating effect is most "needed," destabilizes interactions with the lower trophic level. Other behaviors can counteract this inefficiency, such as aggregation in patches of high prey-host densities and mutual interference (Comins and Hassell 1979; Hassell 1979, 1982; May 1978a; but see Murdoch et al.

Table 27.3 SPECIFIC FORMS FOR THE FUNCTIONAL RESPONSES OF PREDATORS AND PARASITOIDS WITH NONOVERLAPPING GENERATIONS, WHERE THE GENERAL FORM IS $F(N_T,P_T) = \mathrm{EXP}(-\alpha P_T)$

On the upper trophic level	α
Predator, $T_p = 0$	$a[T - (T_h \cdot N_a/P_t)]$
Nondiscriminating parasitoid, $T_p = T_h$	$\dfrac{aT}{1 + aT_hN_t}$
Discriminating parasitoid, $0 < T_p < T_h$	$\dfrac{a[T - \{(T_h - T_p) \cdot N_a/P_t\}]}{1 + aT_pN_t}$

All forms are mechanistic derivations for a type II, or invertebrate, functional response, in which the predator-parasitoid's attack rate asymptotes when plotted against prey-host density. In these expressions the leveling off is due to the time required for handling prey-hosts, T_h. Unlike predators, all parasitoids waste at least some time handling already parasitized hosts, and this wasted time, T_p, causes their functional response curves to level off sooner than those of predators. Hence, parasitoids are more "inefficient" than predators. The most inefficient parasitoid cannot discriminate between parasitized and unparasitized hosts, i.e., $T_p = T_h$. Functional response for nondiscriminating parasitoid is as in Holling's (1959) disc equation.

N_a = number of prey-hosts attacked = $N_t[1 - \exp(-\alpha P_t)]$; T = total time available for search, i.e., the adult female's lifespan; T_h = time for handling unparasitized hosts and prey; T_p = time for handling parasitized hosts; a = search (encounter) rate $aT = \alpha$ (in the simplest Nicholson-Bailey model).

From Arditi (1983).

Table 27.4 RATE OF INCREASE OF DIFFERENT LIFE-STYLES, BROKEN INTO GROWTH (C) AND REPRODUCTION (β) COMPONENTS

Life-style	c (number of progeny per host-prey), life stage from egg to sexual maturity	β (number of host-prey to produce eggs), life stage from sexual maturity to death
Predator	1 progeny per many prey $0 < c < 1$	Many prey per egg $\beta >> 0$
Macroparasite	Many progeny per host $c > 1$	1 host per many eggs $0 < \beta < 1$
Microparasite	Very many progeny per host $c >> 1$	1 host per many "eggs" $0 < \beta < 1$
Parasitoid	1 progeny per host $c = 1$	0 hosts per egg $\beta = 0$

After Beddington et al. (1976).

1984), but insect predators and parasitoids alike exhibit these traits. (Actually, few data exist to quantify the relative efficiencies of predators versus parasitoids, as evidenced by Tables 3.1 and 5.1 in Hassell [1978]; this would be a ripe area of investigation for biocontrol.) Behavioral aspects of parasitoids' life-style would seem to decrease both their ability to regulate and the stability of their interactions with the lower trophic levels relative to equivalent predators (see also Hassell and Waage 1984).

However, on a time scale exceeding one generation, characteristics of the parasitoid life-style enhance both the ability of parasitoids to regulate the lower trophic level and the stability of the interaction compared to predators. From what we can learn from simple model systems, these characteristics more than overcome the destabilizing inefficiencies of parasitoid behavior (Hassell 1978). Beddington et al. (1976) partitioned the rate of increase (b in Table 27.2) into two components: c, or the "conversion rate," expressed as number of progeny of the predator-parasitoid per prey-host, which reflects how prey-host individuals are turned into predator-parasitoid individuals able to reproduce; and β, or the minimum number of prey-hosts that an adult needs to reproduce, that is, to produce eggs. In more general terms, c describes how many prey-hosts are needed for predator-parasitoid growth from birth to maturity, and β describes how many prey-hosts are needed for predator-parasitoid reproduction from mating to death. Partitioning the term governing the numerical response in this way provides a more general framework useful for comparing across life-styles (Table 27.4) and for considering how complex life cycles affect population dynamics.

The rate of increase, as determined by c and β, of parasitoids is much more "efficient" than that of equivalent predators, in marked contrast to the functional response. On a population turnover, as opposed to a behavioral, time scale, parasitoids can turn hosts into more parasitoid individuals for the next generation than predators can turn prey into new predator individuals. A parasitoid (typically) needs no hosts to produce eggs ($\beta = 0$), and each host it successfully attacks turns into one progeny ($c = 1$); the same number of prey could be converted into only a fraction of a predator. As a result, parasitoids can respond to their host populations more quickly than predators can to prey populations. In effect, parasitoids have a smaller time lag in their interaction with the lower trophic level.

There is another way of viewing this result. Both Nicholson-Bailey and Volterra models of predator-prey interactions yield the classical result that stability of the interaction decreases as prey reproductive rate, ln λ or r, increases compared to the predator's relative ability to depress the prey population, (see Hassell 1978, Fig. 2.7; May 1981b, Fig. 5.2). As prey reproductive rate increases or predators become more efficient in behavioral time (that is, as each predator eats more prey for the number of offspring it produces), populations move from a monotonically damping, stable equilibrium point to increasingly more oscillatory behavior until the system reaches chaos. The mechanism for this trend is the decreasing ability of the upper trophic level to follow the lower in time. The discrepancy in their reproductive rates introduces a time lag that throws the system into oscillations.

Comparison of All Four Life-styles Wollkind et al. (1982) present a model that bridges the various models discussed here. Theirs is an age-structured model describing arthropod predators with overlapping generations; it yields both upper and lower thresholds on the prey density that characterizes stability. Their model illuminates the mechanism. A too efficient functional response, on a behavioral time scale, enhances the upper trophic level's ability to drive the lower to extinction. This result also appears in Rosenzweig's and MacArthur's (1963) graphical model for predators. A too inefficient functional response permits the lower trophic level to escape regulation by the upper and to grow exponentially. The relationship between stability and ability to regulate depends on that between functional and numerical responses; this is reflected in the stability criterion of Wollkind et al. (1982), which is a ratio between numerical and functional response terms. This relation between functional and numerical responses is what we have seen implicitly in our review of the previous models. If the functional response is too efficient relative to the numerical response, as for predators, the upper trophic level can both depress the lower too much and introduce a time lag that causes the populations to oscillate.

Table 27.4 summarizes the differences among four life-styles discussed so far. Ability to respond to the lower trophic level, that is, to reduce time lags, increases with c and decreases with β. Macro- and microparasites have high to very high values for c; one host suffices for many parasite progeny. β is small for these parasites, much smaller than for most predators, but not as small as for parasitoids. Were it not for the property of macro- and microparasites to die when they cause host deaths, they could respond to the lower trophic level more quickly than parasitoids—in fact they could overtake it, as we have seen above. How these factors balance is impossible to say without accurate estimates of the model's parameters from real systems.

Parasitoids, on the other hand, are prevented from overtaking the host populations because of the 1 : 1 conversion of hosts to progeny. This conversion ratio permits them to respond quickly, but not excessively to changes in the host population. With no automatic self-damping properties in either predators or parasitoids—host-prey death being required for successful development—they should be better able to regulate the lower trophic level.

The life-style types studied so far form more of a triangle than a single linear continuum with respect to the two population properties, ability to regulate and stability (Fig. 27.1). Parasitoids occur on one point of the triangle, having a combination of enhanced ability to regulate and enhanced stability in their interaction with the lower trophic level. No wonder that parasitoids can make such effective biocontrol agents!

Secondary Characteristics

Complex Life Cycles A few model systems, particularly those incorporating age structure (see Wilbur 1980), lend some insight into how complex life cycles might affect population dynamics. In sum, the work done so far predicts both stabilizing and destabilizing effects that arise from different mechanisms and produce some unknown net effect. To this extent, the results from model systems are inconclusive. Complex life cycles exert two general effects on population dynamics; they produce time lags, and they produce several types of heterogeneity.

As usual, time lags produce oscillatory behaviors including stable limit cycles. A model of the Anderson-May system for macroparasites can incorporate complex life cycles by relaxing the assumption that transmission is "saturated." That is, the dynamics of immature stages occur on a time scale equivalent to, not much faster than, the dynamics of the adult parasite and host populations and so cannot be assumed to be at equilibrium with respect to the host population. This situation would describe any life cycle in which considerable growth occurs when the immature stages live apart from the adults, whether the immature stages are also parasitic or are free-living. The result in either case is increased time lags, because now three or more population components (e.g., adult parasites, hosts for the adults, parasite larvae, and possibly hosts for the larvae) must interact with one another, instead of only two (adult parasites and hosts).

Any pronounced age-structure effect introduces heterogeneity into the system parameters; this is true in both complex and direct life cycles. One assumption could be that complex life cycles have by definition more pronounced, in the sense of more disjunct, age structures. In general, heterogeneity, resulting from age-specific reactions to variation in both endogenous parameters (parameters of population growth) or exogenous (environmental) parameters, produces oscillatory behavior (Beddington 1974, Oster and Takahashi 1974). Moreover, Oster and Takahashi (1974) show how, in two-population interactions, population waves "excited" by the age profile of one trophic level can feed back and excite similar waves in the other, resulting in "endogenous limit cycles of very different origin than the Lotka-Volterra type." Age-structure heterogeneity can be, in these ways, destabilizing. It is not clear, however, whether these destabilizing effects would be greater or actually less in complex than in direct life cycles. The radical disjunction between immature and adult age classes in complex cycles might "break" the resonance generated by age-profile heterogeneity, because each age class interacts uniformly with a separate host population. Models specifically tailored to investigate properties of complex versus direct life cycles would be useful.

Other age-structure effects can lead to several kinds of heterogeneity that are stabilizing. The aggregation of parasites in the host population is almost surely due to heterogeneity in the dynamics of transmission (Skellam 1952, Feller 1943, Anderson and Gordon 1982), that is, the interplay between adult parasites in hosts, dispersing larvae whether free-living or in their own hosts, and new hosts for them to infect. Recall that this aggregation mediates the strong, stabilizing self-damping of macroparasites. Again, this kind of heterogeneity can arise in both direct and complex life cycles, and more information is needed on whether increasing the complexity of the life cycle increases the heterogeneity. Age-specific heterogeneity in intensity of intraspecific competition is directly stabilizing (Tschumy 1982), and that of interspecific competition introduces nonlinearities that result in multiple stable points and increased opportunity for coexistence among species (Hassell and Comins 1976). In both cases the mechanism is the abrupt ecological difference between immature and adult stages in complex life cycles that reduces intra- and interspecific competition.

Whether the stabilizing or destabilizing influences of complex life cycles win out is not revealed by either existing population models or evolutionary models (Istock 1967, Slade and Wassersug 1975). Nevertheless, available evidence suggests that selection often increases the complexity of life cycles. For example, in many groups, especially the trematodes and the insects, complex cycles occur in the most evolutionarily advanced forms. In all groups that possess complex cycles, whether derived or primitive, larval and adult forms become increasingly more divergent in ecology and morphology, and metamorphosis becomes increasingly more rapid and radical (Brown 1977, Wassersug and Hoff 1982), that is, complex cycles evolve to become more complex. These observations suggest that the net effect of complex life cycles could not be too destabilizing and might even be stabilizing in certain situations.

If we examine complex life cycles in trematodes, a group for which the evolution of life cycles is particularly well understood (Pearson 1972), we see more possibilities for what complex, as opposed to direct, life cycles accomplish and, moreover, for how complex cycles may influence population dynamics. The direction of evolution in this group appears to be from commensals, in the intestine or in the mantle cavity of molluscs, to one-host parasites for which molluscs are the definitive host (no such forms exist today), to two-host life cycles with typically a mollusc or crustacean intermediate host and invariably a vertebrate definitive host. From such a two-host system, more hosts have been added, such that the first intermediate host is still an invertebrate from which transmission occurs by free-living dispersing larvae, and subsequent intermediate hosts are typically small vertebrates that are eaten in line by the next host, until the final vertebrate definitive host. Interestingly, the number of hosts in trematodes reached a maximum of four (with a total of seven different developmental stages counting free-living ones), and there is evidence in several groups that hosts were dropped evolutionarily. In existing trematodes the modal number of hosts is two to three, suggesting that life cycles of intermediate complexity are optimal, in at least some ecological circumstances.

Details of trematodes' life histories suggest that complex life cycles in macro- and microparasites may be designed to maximize simultaneously the likelihood of encountering a host, dispersal, and opportunity for growth. Intermediate hosts are used for asexual reproduction. They are typically small and numerous in their environments; they often exhibit a high prevalence of infection (they are easy for parasite larvae to find) but have a limited capacity for dispersal and parasites' growth (e.g., Crofton 1971, Sankurathri and Holmes 1976). Definitive hosts are large, long lived, and often capable of long-range dispersal. In a typical fluke life cycle, once the first dispersing larva encounters an intermediate host, it magnifies itself many (hundreds of) times asexually to stage an attack on the next intermediate host, which is larger and offers greater opportunity for growth. From there the "individual" (divided into many asexually produced selves) travels up the food chain as a result of each host's being eaten by the next host. This mode of transmission must be more reliable than swimming about to find a host at random. (The parasite often modifies host behavior to increase the probability of its being eaten [Holmes and Bethel 1972].) In fact, a driving force for evolution of complex cycles may have been that the parasite thereby profits rather than suffers from a predator's eating its host: The predator itself becomes a host, and the former definitive host becomes an intermediate host.

We thus get a picture of macroparasites isolated in hosts, which are in turn isolated in favorable habitat (see also Price 1980). Dispersal of parasites from one host to the next and of hosts from one habitat to the next may be difficult, especially for parasites with complex life cycles, because they must disperse to complete even a single life cycle. This sort of temporal-spatial heterogeneity resembles open model systems (defined above) designed to investigate how environmental heterogeneity affects population interactions. Such models (e.g., Slatkin 1974; Levin 1974; Allen 1975; Hastings 1977, 1978; Caswell 1978), which take varying forms and incorporate different interactions (competition, predation, or both), yield a common outcome. The effect of environmental heterogeneity always enhances stability or persistence and therefore coexistence, provided dispersal is not too fast or too slow. Macroparasites, with their prodigious levels of egg production and asexual reproduction, may well achieve intermediate levels of dispersal ability. Intuitively, the odds of finding a host may seem to us to be one in a million, but that is to an order of magnitude the number of eggs that some of these worms produce per individual.

To reiterate, all organisms experience environmental heterogeneity. But parasites confined in hosts, especially those with complex life cycles, must experience relatively more environmental heterogeneity, and, if so, their population dynamics may be more stable, at least globally, as a result. Whatever the complex cycle's effect on population dynamics, it must add complexity to the community structure of organisms that possess it in direct proportion to its degree of complexity (see also Wilbur 1980 and below).

Specificity and Competition Parasites commonly show specific adaptations to their hosts, as might be expected for symbiotic organisms. The causes for their often remarkable specificity have been reviewed and debated for the major groups of parasites (Holmes 1973b, Day 1974, Rohde 1979, Waage 1979, Price 1980, Holmes 1983). Two sorts of coevolution have been proposed: (1) coevolution between trophic levels—the famous "arm's race" or "red queen effect" (Futuyma and Slatkin 1983)—in which adaptations by hosts to counter parasites are matched by adaptations by parasites to counter them and so on; (2) coevolution on the same trophic level between parasites to avoid competition within a host, or character displacement (e.g., Roughgarden 1976, Slatkin 1980). The relative importance of the two mechanisms is the subject of controversy (e.g., Holmes 1973b, Rohde 1979, Brooks 1980, Holmes and Price 1980), but no one questions the greater specificity of parasites relative to predators.

No matter how such specialization arises evolutionarily, it should increase α-diversity by reducing competition in ecological time. Competition, all models agree, is a destabilizing interaction that puts an upper limit on coexistence by causing local extinction (see summaries in May 1973a, Hassell 1978). Coevolution among three trophic levels has been implicated in enhancing predator diversity by this mechanism (review in Toft 1985). However, parasites, with their even greater specificity, may be able to avoid competition even more than predators and thereby coexist in more diverse communities (Rohde 1979, Price 1980).

Laboratory studies show that macroparasites can compete for limited resources in the host (review by Halvörsen 1976); Holmes (1961) has demonstrated that parasites limit the range of host tissues they use when they compete interspecifically. Moreover, both in nature and in the laboratory, parasites often interfere with one another, interspecifically and intraspecifically, by various host-mediated immune responses (Schad 1966) and by directly preying on each other within the host (review by Lim and Heyneman 1972).

But how frequently does such competition occur among macroparasites in nature, and does it affect α-diversity of their communities? The evidence so far is based primarily on statistical studies and is therefore indirect. Nevertheless, many of these studies (Chappell 1969; Holmes 1973a; Hobbs 1980; Bush and Holmes 1983, in preparation) indicate that competition, probably exploitative competition for food, can be important in natural communities of intestinal macroparasites. The main evidence that these studies present for ongoing competition consists of habitat shifts along the intestine by certain species pairs in "concurrent infections" versus "single infections," that is, in sympatry versus allopatry, as Holmes (1961) originally showed experimentally. Another form of evidence is a test of the broken-stick hypothesis, devised by Bush and Holmes (1983), who showed that parasite species are more uniformly distributed along the length of duck intestines than expected by chance. However, where habitat shifts occur, indicating interspecific competition, species are often *more* likely to occur in the same host individual than expected by chance, as Kareiva (Chapter 11) also notes for competing herbivorous insects.

The peculiar life cycles of macroparasites are responsible for this apparent paradox and throw a novel slant on determinants of community structure. Parasites are positively associated because of the paths they travel up food chains (e.g., Esch 1971, Cannon 1972, Wootten 1973, Bush and Holmes in preparation). Suites of parasites share intermediate hosts, and the definitive host will accumulate parasites using the intermediate hosts that it eats (Bush and Holmes in preparation). Thus, by riding up a food chain, a parasite may be forced to associate with species most likely to compete with it. Of course, competitive exclusion could still eliminate species after they arrive in the definitive host, and this process may account for the negative associations that we see. Habitat shifts (and character displacement) in species likely to co-occur may permit the positive associations in the face of potential competition. (Thus, statistical tests for negative interactions, as envisioned by Cohen [1970], may well not elucidate the importance of competition in parasites without independent evidence of interactions.)

Hassell and Waage (1984) have recently reviewed the evidence for interspecific competition in parasitoids and its role in shaping parasitoid communities. Despite the greater specificity of parasitoids relative to insect predators, competition among them is evident in field studies. Even though parasitoids are specific to their hosts, an incredible number of species still share the same host (e.g., Zwölfer 1961, Askew and Shaw 1974). These observations suggest that α-diversity of parasitoid communities may well be enhanced by the high specificity of parasitoids, which allows more species to co-occur until interspecific competition reaches some limiting level.

While evidence is suggestive for intestinal macroparasites and parasitoids (the main groups studied so far), the exact relationship between specificity, ongoing competition, and α-diversity remains to be fully explored. The hypothesis that still needs to be tested is whether the greater specificity of parasites, relative to predators, reduces ongoing competition, which in turn increases the α-diversity of parasite communities.

EXPECTATIONS ABOUT PARASITE COMMUNITIES: α-DIVERSITY

This section summarizes themes about population dynamics from the previous sections and draws together their consequences for the diversity of parasite communities, much as May (1978b) has done for insects.

Stability and the ability to regulate are two characteristics of two-trophic-level interactions about which simple population models can give insight and which also have consequences for community-level patterns.

A trend pervades the results from models of two-trophic-level interactions: characteristics of parasitic life-styles that have been investigated so far lend stability to interactions between parasites and their hosts. In particular, the very characteristics that distinguish the life-styles, the primary characteristics determining the numerical response, most uniformly result in increased stability in parasite-host interactions compared to predator-prey interactions. Secondary characteristics of parasites such as those associated with complex life cycles and host specificity, are more ambiguous in their effects; some increase and some decrease stability, and the net effect is unclear.

Theoretical and empirical confirmations of the stability-enhancing effects of primary characteristics come from studies of parasites in a broader context. In a theoretical study Pimm and Lawton (1978) fitted parasitoid links of the Hassell (1978) type into larger food chains and found that the enhanced stability of parasitoid-host interactions permitted more levels in food chains than possible with other life-styles. Connell and Sousa (1983) reviewed empirical studies of stability in natural populations and found that the five parasite populations fell into their two most stable categories, with four out of five in the most stable category.

Models attribute this increased stability to the parasite's basic numerical response. Parasites can respond more quickly to variations in their host populations than can predators, but at the same time their life-style limits their ability (compared to that of predators) to exploit the lower trophic level. Simple models suggest that parasites will less often regulate their hosts than predators regulate their prey in nature, as Holmes (1982) concluded independently from a review of empirical studies. Because parasites are, first, relatively unlikely to regulate their hosts and, second, more specialized, I speculate that parasite species are less likely to compete in nature than are predator species. Because interspecific competition is a destabilizing interaction, stability of predators is reduced even more (see also May 1973a, Hassell 1978, Pimm 1982).

The same reasoning leads us to expect that parasitoids would be more likely to compete than macroparasites, because the self-damping properties of the latter make them least likely to regulate the lower trophic level. Microparasites may be more able to regulate the lower trophic level than macroparasites, because of their higher virulence and less pronounced self-damping, and they also should be more likely to compete than macroparasites. To some extent competition on the upper trophic level and lower stability between trophic levels go hand-in-hand, thereby exacerbating an already less stable situation. Parasitoids appear to break this trend somewhat, because their life-style combines highly stabilizing features with no self-damping and with host death as a developmental necessity. The greater specificity of parasites (including parasitoids) relative to predators would work to alleviate competition that may remain in ecological time and thus to increase the stability of parasite-host systems even more.

While the relationship between stability and diversity is complicated and controversial, excellent summaries of the issues are bringing out some consistent answers (May 1973a; Pimm 1982, 1984). More diverse communities, all else being equal, tend to be more fragile in important ways: The more species, the fewer interactions there should be between them, if all species are to co-occur stably, and the less resilient the component populations will be (Pimm 1984). However, if the interactions between species are inherently more stable or persistent, more species and more interactions between them will be permitted than in a community with less stable interactions (e.g., Pimm and Lawton 1978). We have seen that the characteristics that define various life-styles cause parasite-host links in food webs to be inherently more stable than predator-prey links. As a result, parasite communities are expected to have more sympatric species than predator communities. Possibly also, parasites might be able to violate the *n*-predators–*n*-prey expectation of various theoretical treatments (reviewed by May 1973a). In other words, parasites can be expected to occur in relatively more top-heavy food webs than predators. Recent discussions of predator-prey ratios (Pimm 1982, Briand and Cohen 1984) have not distinguished "predators" with different life-styles. It would be interesting to separate the parasitoids from free-living predators in these webs (none of the webs includes macroparasites, which of course occur in the organisms in these studies).

In addition, parasitic life-styles should themselves exhibit different levels of ecological diversity in proportion to stability of interactions with hosts and to relative abilities to regulate hosts. These propositions could be tested with data on diversity of actual communities, given comparable methods of measuring diversity for all systems. Preliminary results of a literature survey that I am making indicate that more species of macroparasites and parasitoids coexist than do predator species, relative to the number of their hosts or prey. In contrast, relatively fewer species of microparasites than macroparasites or parasitoids coexist per host species, more on par with predators.

My conclusion that characteristics inherent to parasitic life-styles should lead to greater ecological diversity in parasites is not inconsistent with Price's (1980), although we have arrived at similar conclusions by different reasoning and by invoking perhaps different mechanisms. His view is that parasites (at least macro- and microparasites) live in patchy, isolated environments and that this condition leads to unpredictability and hence to nonequilibrium. He emphasizes that as a result competition should be unimportant. Unlike Price, I do not relate patchiness to instability and nonequilibrium; rather, I conclude from a study of population models that patchiness leads to global stability. Moreover, I emphasize that parasite-host interactions should show higher local stability. However, I am not suggesting that parasite-regulated equilibria are common in these interactions. Nevertheless, Price and I agree that competition should be less important in parasites than in predators.

Finally, because parasites are less likely to regulate the lower trophic level, they may not increase species diversity of host communities in the straightforward way that predators do for prey communities (Paine 1974). However, parasites can have other effects on host coexistence (e.g., Freeland 1983, Holt and Pickering 1985).

CLASSIFICATION OF PARASITES: A POPULATION DYNAMICS SCHEME

Until now, we have recognized the traditional three categories of parasites: macroparasites, microparasites, and parasitoids. However, the above discussion on population dynamics allows us to approach a fuller and more specific classification by identifying four crucial life-style characteristics.

1. Is the organism parasitic or free-living? As symbiotes, parasites are governed by conditions within the host, leading often to specific adaptations to the host. Importantly, if parasites cause the host to die, they also die.
2. In which life stage, adult or immature, is the organism parasitic? Beddington et al. (1976) first distinguished how population dynamics may be sensitive to whether the lower trophic level contributes to the upper's growth (immature stages) or to reproduction (adults). Age-specific survival, which could be determined by different factors in parasitic and free-living stages, also affects population dynamics (review in Oster 1978).
3. What is the probability of death of the host? As mentioned at the outset, there is an important breakpoint between 100% probability and less than that. This breakpoint determines the probability that the upper tropic level will regulate the lower and also determines whether the upper trophic level will self-damp.
4. Does the organism complete a life cycle within a single host individual? May and

Anderson (1978) showed that this trait has the destabilizing result that parasite reproduction can outstrip the host's and drive it to extinction. In this way parasites can overexploit their hosts as can predators.

Once the enormous diversity of parasitic life-styles is appreciated, we can recognize at least 11 categories of parasites that may potentially exhibit different population dynamics and therefore different community characteristics. In Table 27.1 I have used the four life-style characteristics listed to classify all groups with 100 species or more (plus smaller groups of special interest) in which at least one life stage is parasitic as defined above. The three traditional categories are macroparasites (I), microparasites (II), parasitoids (VIII). Two free-living categories are included in Table 27.1 for comparison: predators (XII) and ''grazing herbivore-detritivores'' (XIII). Members of category XIII, while entirely free-living, do not kill prey individuals as do predators, but rather take bits of organic matter from living or already dead organisms. Of the groups traditionally studied as parasites, approximately 1.5% of eukaryotic species worldwide are macroparasites, 3% microparasites (excluding prokaryotes), and 10% parasitoids.

This new, ecological classification of parasites can direct future research in two ways. First, we must recognize that life-style characteristics of these 11 groups could cause their populations to behave in very different ways, yet adequate models are presently lacking for 8 of these groups! Second, the classification invites a search for consistent biological differences between communities of species in these 11 groups.

Much of Table 27.1 is self-explanatory, and I will not discuss each category. Instead, I next highlight features of some of the parasitic life-styles in Table 27.1 and indicate promising directions for further study.

Arthropod Ectoparasites: Lice versus Fleas (Life-styles III and IV)

Models with overlapping generations (Anderson and May 1978, May and Anderson 1978) may be applicable to ectoparasites that are confined to hosts throughout their life cycles, such as lice and some mites (life-style III), provided that their transmission dynamics are reasonably well described by these models. Fleas (life-style IV), however, should differ fundamentally from lice and mites because their larvae occur in the host's nest, where they are nevertheless fed from the host by the parents whose blood-rich feces drop into the nest. The lower evolutionary diversification of fleas relative to lice has long been noted (Rothschild 1952, Traub and Starcke 1980) and attributed to constraints on flea larvae in the nest. For example, fleas seem to be more nest-type-specific than host-specific, and organisms without permanent nests seem to have fewer fleas. In addition, adult fleas are sufficiently mobile that they have a much lower probability of dying when the host dies, so flea populations should exhibit less self-damping than those of lice. Yet fleas are not sufficiently free-living to be put into category VII.

Biting Organisms (Life-Styles VI and VII)

Biting organisms that require blood for both growth (including larviparous forms) and reproduction (life-style VII) have greater potential for coupling to the host than type VI organisms, and in fact life-styles in category VII grade into those of types III and IV. For this reason I have separated them, as parasites, from the free-living grazing herbivore-detritivore category (XII) in which organisms take bits of organic matter from living plants and already dead organisms. Glasgow (1963) makes the interesting case that hematophagous insects may well be food-limited. In some tsetse areas he computes that there is just enough blood available in the host population to support the tsetse densities seen. Yet the population dynamics of organisms living independently of the host should be very different from those of type I or III ectoparasites, because they do not die when hosts die and they are subject to external environmental factors. They are also unlike predators in that host death is not a developmental necessity, and their population dynamics may be more similar to those of grazing herbivores. (Models describing the latter are reviewed by Caughley and Lawton [1981].) Because parasites in life-styles VI and VII do not self-damp and they potentially can use host resources to a limiting extent, they could very well be subject to competition. Indeed, even biting organisms of type VI, such as tabanid flies, may compete as adults for host blood (Waage 1979, Waage and Davies in preparation).

Herbivorous Parasites (Life-Styles III and X)

Price (1980) generated controversy by including organisms interacting with plants as parasites. Yet, such organisms are not truly free-living. I also consider many as indistinguishable from parasites with animal hosts, just as we would not quibble about rusts and nematodes being plant parasites. The sessile Homoptera, such as scales and aphids, should exhibit much the same dynamics as species in type III traditionally considered ectoparasites, such as lice and mites, and I have not separated them.

Other groups are similar to parasitoids of type VIII in that the adults are free-living stages engaged primarily in dispersal and mating; the adults oviposit on host individuals that happen to be plants (e.g., lepidopterans, saw flies, and dipterans such as tephritids and agromyzids). The crucial variable is probability of host death, and this difference is great enough that I have separated the herbivorous ''parasitoids,'' type X, from the insectivorous ones, type VIII. Greatly lowering the probability of host death could decrease the ability of the upper trophic level to regulate the lower and thus reduce interspecific competition. On the other hand, herbivorous parasitoids can produce more than one individual on a host and, without self-damping, might overexploit the host population, resulting in increased competition. In addition, herbivorous parasites (of both types III and X) are lower on the food chain than parasites with animal hosts; this introduces the well-known complication that predation and parasitism regulate herbivorous parasites (Hairston et al. 1960, Lawton and Strong 1981, Schoener 1983b). Such considerations would make their population dynamics and community structure very different from insectivorous parasitoids.

SUMMARY

The dynamics of parasite-host interactions lead us to expect parasitic species to coexist in more diverse assemblages than do predators, in relatively more top-heavy food webs. The very characteristics that define parasitic versus predatory life-styles lend greater stability to parasite-host interactions in model systems.

Parasites are less likely to drive coupled oscillations with their host populations than are predators, because parasites respond to the lower trophic level with reduced time lags.

Parasites are less likely than predators to overexploit and indeed even to regulate the lower trophic level, because of less efficient functional responses and self-damping.

Parasites are less likely to compete with each other for this reason and because of the greater levels of specialization they exhibit.

Based on criteria that affect population dynamics in these ways, one can recognize at least 11 major parasite-host patterns likely to exhibit different population dynamics and hence different levels of α-diversity.

Finally, because parasites are less likely to regulate their hosts than are predators, they probably do not increase diversity of the host communities in the straightforward way that keystone predators do.

ACKNOWLEDGMENTS

I am grateful to J. Diamond and T. Case, who both patiently read the manuscript several times, and to P. Chesson, M. Cody, P. Kareiva, S. Pimm, J. Roughgarden, and D. Strong for valuable comments on various drafts.

six

KINDS OF COMMUNITIES

chapter 28

Overview: Kinds of Ecological Communities—Ecology Becomes Pluralistic

Thomas W. Schoener

INTRODUCTION

A major consensus emerging from this volume is that ecological communities in general, and those studied by the contributors in particular, are very different from one another. This agreement is more than just the trivial recognition that each community has unique features. Rather, with a few exceptions, the *emphasis* is on differences, not similarities.

It has not always been so in community ecology. MacArthur presented the opposite view when he wrote in *Geographical Ecology* (1972a) that "the ecologist and the physical scientist tend to be machinery oriented" and "the machinery person tends to see *similarities* among phenomena" as opposed to *differences*. "Machinery people," if by that one means those who construct mathematical models, are well represented in this volume. Yet even their chapters emphasize differences (e.g., Chapters 14, 29, 30, 33)! Clearly, theoreticians have found that a relatively precise mapping of biological diversity onto mathematical models can be both illuminating and fun.

In looking back, one may wonder whether the emphasis on differences really does constitute such an extreme shift in viewpoint. In Chapter 33 I give some examples of positions I consider more monolithic than those held by most chapter authors in this volume. There I argue that positions on both sides of the competition/density-dependence/regularity issue were as extreme as they were in part because of the taxonomic provincialism of their advocates.

I doubt, however, that this is nearly the whole story. Perhaps as a result of the introduction of a small number of simple mathematical models, those of Volterra and Lotka in particular, ecology developed Cohen's (1971) widely quoted characterization of "physics envy." These models constituted the first attempt to make the ecologists' often vague verbalizings precise, indeed to show (as Robert MacArthur again said [personal communication]) that there exists *some* mathematical model that will give the conclusions claimed from the conditions and processes assumed.

This worthy objective, however, seems sometimes to have been transformed into the optimistic hope, if not the outright assumption, that the models would play a similar role in ecology as in, say, mechanics. The models might not so much quantitatively describe biological reality, but certainly they would at least be qualitatively correct (e.g., Levins 1968). The dual credo was held that small variations in model structure will not affect conclusions and that variations (maybe not so small) in the fit of real organisms to the models will not jeopardize the models as devices for at least qualitative understanding. To quote again MacArthur's *Geographical Ecology:*

> Very likely no population ever grows exactly according to Volterra's equations. Ecologists who use them are following reasoning something like this: The true, correct equations are probably "near to" the Volterra equations, and the behavior of such equations will be "near to" the behavior of the solutions of Volterra's equations.

The practice of employing a small number of simple models to understand ecological reality has, of course, the implicit assumption that this reality can be embodied in a small number of simple "laws."

The dismantling of this simple (now we would say simplistic) position and its replacement by the position described above is a history of false starts and ironies. It has been summarized by Colwell (1984) in a chapter for a symposium titled *A New Ecology* (Price, Slobodchikoff, and Gaud 1984). The bombshell that Colwell dropped on that symposium was the following quotation (which I shorten here).

> I predict there will be erected a two- or three-way classification of organisms and their geometrical and temporal environments, this classification consuming most of the creative energy of ecologists. The future principles of the ecology of coexistence will then be of the form "for organisms of type A, in environments of structure B, such and such relations will hold." This is only a change in emphasis from present ecology. All successful theories, for instance in physics, have initial conditions; with different initial conditions, different things will happen. But I think initial conditions and their classifications in ecology will prove to have vastly more effect on outcomes than they do in physics.

When asked, no one at the conference could identify the author, and I believe this would be true of ecologists generally. In fact, the author of this quotation illustrating the "new" position was (again!) Robert MacArthur, and the quotation was from a paper (MacArthur 1972b) published in the same year as *Geographical Ecology,* from which I have quoted extensively above to illustrate the "old" position.

To compound the irony, unaware of MacArthur (1972b), I wrote the following in a review of *Geographical Ecology* (Schoener 1972) that was intended to be at least mildly critical.

> There is [a] way to match nature's complexity with quantitative models. That is to construct a *family* of analytically tractable models focused upon some major phenomenon (such as predation), but each differing in the limited parameters or processes incorporated and each having a specifically designated applicability. . . . From a historical perspective it seems inevitable that the next years of model building in ecology will result in a gradual buildup of alternative, analytically tractable models with both qualitative and quantitative ends, models that will not all claim or be shown to mimic the same phenomena. This should lead to an embracing of the empiricist's precious detail with mathematical theory. . . .

The "new" ecologists in the above-mentioned symposium came mostly independently to much the same conclusion—Colwell's chapter is subtitled "Community Ecology Discovers Biology." Some of the contributors to the present volume doubtless arrived at the same position again independently of any of these potential (but not actual) antecedents.

All this convergence seems hopeful, but I still have the reservations that I expressed in 1972: The degree to which even very specific models can even qualitatively mimic specific systems *while remaining reasonably simple* is an open question. Moreover, if we need too many models, even though each is of no more than intermediate complexity, we will be lost in a vast encyclopedia of special cases. To be mildly pessimistic on this score, we are now discovering that variations in the structure of mathematical models that, if verbalized, most ecologists would be indifferent to can give *qualitatively* different predictions (e.g., Gilpin and Justice 1972, Turelli 1981; Schoener 1985a reviews such results). Yet all of us who are urging a greater diversity of mathematical and conceptual approaches to community ecology hope, indeed must hope, that the final list of assumptions and outcomes will not be too large. As yet, we are not even close to knowing whether or not this is so: Both model construction and data gathering are too inchoate for any kind of empirical pronouncement, despite what we sometimes hear.

As if in fulfillment of MacArthur's prediction, the following is a crude preliminary attempt to give some of the ingredients that might be contained in a *pluralistic* theory of community ecology. I am sure it could be vastly better, and I would be pleased if it were moderately successful in generating proposals for improvement or even replacement.

TERMINOLOGY

The most general definition of an ecological "community" is a set of species populations that occur in some place. Both the boundaries of the set and the boundaries of the place are in this general definition arbitrary, and there are several alternative specific definitions (Preface, Chapter 20). As is clear from the context of this volume, the implicit definition of community most commonly used in this volume is almost, but not quite, equivalent to "guild," the set of species populations in some place that utilize a particular kind of resource (Root 1967). The difference is that, as is so traditional among ecologists, a taxonomic subclass of species belonging to the same guild is nearly always studied. Possibly the main reason ecologists specialize taxonomically is that expertise with a variety of organisms is difficult to obtain. However, concentration on "taxon-guilds," rather than guilds, does have the important methodological advantage of making it easier to order communities by various properties characterizing the kind of organism composing them, not just by resource use. This removes the implicit emphasis on resource competition and can lead to much greater theoretical understanding (e.g., for life history traits see Chapter 13).

Nevertheless, a great deal of variation in organismic properties can exist within a single taxon-guild. For example, littorine molluscs in New England can have very different dispersal traits: One releases planktonic egg capsules that can disperse kilometers, whereas another broods larvae that scarcely disperse at all (Lubchenco personal communication). Hence, if we wish to stress similarity in a whole set of functional properties of organisms, including but not restricted to how they use resources, we sometimes will need to go to a community unit still smaller than the taxon-guild. While I have considerable reservations about adding a new term to an already jargon-laden biological vernacular, there seems in this case no alternative but to do so, as the concept is entirely distinct. I propose the term "similia-community" (similia = similar things) to describe that set of species occurring in one place that are similar with respect to crucial organismic and environmental traits. (A list of these traits is to be suggested shortly.)

For some purposes restriction to taxon-guilds or similia-communities may lead to greater understanding than guilds or other definitions of communities, but for many other purposes this will not be the case. Isolation of a vertical connection—a consumer and its resource—is often desirable. The same is true for a horizontal connection. For example, Brown et al. (Chapter 3) argue plausibly for considering all members of the desert seed-eating guild—mammals, birds, and ants—together when trying to understand their ecological properties. On the other hand, it would be hard to generalize much about the organismal properties (size, mobility, and so forth) of these three types of seed predators. Thus, similia-community is the most useful definition for my purposes.

The axes along which communities are to be ordered are of two kinds, *primitive* and *derived*.

Primitive axes consist of two kinds, *organismic* and *environmental*. They are united by being near the root of the causal network leading to various derived characteristics. Organismic primitive characteristics are basic properties of the sort of organism whose populations comprise the community. Examples are body size, generation time, and mobility. Such characteristics, if much varied from their present state, would result in a new kind of organism; a tree that flew about like a bird would no longer be a tree. Environmental primitive characteristics are properties of the physical *and* biological environment in which the community occurs. Examples are openness of the resource input and severity of physical factors. Obviously, as discussed below, there will be correlations between the various organismic axes or between the various environmental axes. Moreover, position on certain environmental axes is often modulated by position on organismic axes. For example, the perceived severity of physical environment depends on the homeostatic ability of the organisms involved.

Derived axes order many of the ecological characteristics whose patterns we wish to explain. Examples are the relative importance of physical versus biological processes, the number of species, the rate of species turnover, and the relative importance of history.

PRIMITIVE AXES

To keep things under some control, I have limited myself to six axes each from the classes organismic and environmental. Some of the axes are representative of a particular genre of axis, and I have generally avoided using more than one representative from each genre. By necessity, only those characteristics representable along a single dimension could be used.

Organismic Axes

O1. Body size (small to large) The size of an organism is one of its many ecologically important morphological properties, which include shape and growth form among others.

O2. Recruitment (open to closed) Open communities have populations whose recruits—certain earlier developmental stages or simply adult immigrants—are produced at great distances from the community's place. In contrast, closed communities are self-contained: the recruits come from within. Because the type of recruitment depends heavily on dispersal traits of individual organisms, it is included here rather than as an environmental axis. Roughgarden (Chapter 30) elaborates on the distinction.

O3. Generation time (short to long) Generation time is defined as the average age of reproduction. It it a demographic property weighted by survivorship and fecundity curves. As such, it is correlated (necessarily in some cases) with a variety of other demographic-reproductive characters, such as r and longevity. Other ecologically important characters in this genre are number of reproductive seasons, age at first reproduction, and fecundity per reproductive bout. Reproductive mode (viviparity versus oviparity, for example) might also be included.

O4. Individual motility (sessile to mobile) Some organisms, having grown to maturity, never move (trees), others seldom move (anemones), others can move quite rapidly but do not move far for behavioral reasons (territorial lizards), while others are extremely motile (whales).

O5. Homeostatic ability (low to high) Homeostatic ability measures the degree to which organisms are independent of physiological stresses caused by the environment. For terrestrial organisms, its most important component is generally degree of endothermy; for aquatic ones, homeostasis with respect to salinity may be most important.

O6. Number of life stages (low to high) A variety of organisms (e.g., holometabolous insects, many parasites, certain algae) pass through a number of life stages that are morphologically and often ecologically quite distinct. Such organisms are said to have complex life cycles (e.g., Wilbur 1984). Incredibly, the known maximum number of such stages is seven, occurring in certain trematode parasites (Chapter 27).

On this axis I intend to discriminate organisms with different numbers of discrete life stages, and also organisms that spend much time independently of parental care, at various ecological positions while still within a particular life stage. Thus lizards, which change their habitats and diets with age, lie farther from the low end of this axis than do altricial birds, even though little lizards grow into big lizards and nothing more. This axis embodies the condition referred to by Werner (Chapter 21) as ontogenetic niche shifts. Also included here is Grubb's (1977 and Chapter 12) concept of the regeneration niche, in which not just properties of the adult plant but also properties of early developmental stages (seeds and seedlings) are of major importance in allowing coexistence.

Environmental Axes

E1. Severity of physical factors (high to low) Environments differ in the harshness of physical factors to life in general. Moreover, certain kinds of organisms are more sensitive to given degrees of harshness than others. This axis could instead be scaled to measure the relative importance of physical versus biological processes, though I prefer (see below) to view the latter as a derived axis. Were communities largely physically controlled, a whole host of climatic axes (e.g., various measures of temperature, rainfall, incident solar radiation, severity of wave action) would have a primary role. As this seems minimally the case for the communities considered in this volume, I have left out such specific axes, doubtless to the astonishment of many ecosystem ecologists.

E2. Trophic position (low to high) The locus that the members of a guild occupy in the trophic web (roughly their trophic level) is widely recognized to be of major importance in understanding a variety of derived characteristics of communities, although there is some disagreement over details. To a great extent organismic characters are intimately related to trophic position, but I have listed trophic position as an environmental axis, because it is partly a function of the presence or absence of other species.

E3. Resource input (open to closed) In some ecological communities the populations of interest profoundly affect their resources because the resources grow and reproduce in the place that they are consumed. This is a kind of closed system. Closed systems are the sort modeled by Volterra; a real-world example might be wolves and moose on Isle Royale. An open system of resource input has resources flowing into the community's place from the outside, for example, winged dipterans being blown onto a small island from elsewhere. Mathematically, this kind of open-closed axis and the organismic axis of recruitment (O2) can be identical (Schoener 1976), but they need not be, and in any case they order ecologically distinct characteristics.

E4. Spatial fragmentation (fragmented to continuous) This designation is meant to apply to the type of environment suitable for the kind of species composing the community of interest. Species on islands surrounded by water are in a fragmented state when the islands are considered *in toto*. The same is true for mainland species distributed over habitat patches. Examples are specialist plants of tree-fall sites (Chapter 19), burying beetles on carcasses (Chapter 26), fishes in lakes (Chapter 21), hummingbird mites on inflorescences (Chapter 24), and herbivorous arthropods on individual plants (Chapters 11, 15). Differences in microsites, as for terrestrial herbaceous plants, also constitute a kind of fragmentation (Chapters 12, 22).

This characterization is a tricky one in several regards. First, if we consider only those populations on a single island to be the community, then we no longer have fragmentation (but we may have a lot of openness in recruitment), whereas if we consider the archipelago to provide the spatial bounds of the community, we have much fragmentation (but less openness). Second, species can create patchiness in their distributions independently of any environmental patchiness (e.g., Levin 1974); then we in fact have a derived character. More on such issues can be found in Chapters 8 and 11.

E5. Long-term climatic variation (high to low) This is the only form of temporal variability included as an axis here. Although they certainly could be listed separately, I have lumped within-year and between-year variability together for two reasons. First, the two are often correlated, temperate areas being somewhat less buffered in the moderately long term from climatic variation than tropical ones. Second, the two often have similar effects on derived axes, e.g., average degree of ecological overlap (Schoener 1982).

Seasonal variation occurs as a relatively predictable sequence to which adaptation can become attuned; this is much less true for between-year variation. Ecologists continue to find seasonal variation interesting, as it is variable itself over the earth's surface. Of course, as tropical ecologists are ever fond of pointing out (e.g., papers in Leigh et al. 1982), seasonality is buffered and ameliorated by well-adapted organisms, but, as Janzen (1984) has recently remarked, absolute variation is so much greater in temperate than tropical areas that this has to be ecologically significant. Moreover, many large aquatic areas should ordinarily be less seasonal than terrestrial ones in the same general geographical region.

Temporal variation can also occur with respect to the other environmental axes listed here, and such variation might be characterized by its amount as well as by the degree of autocorrelation. Collapsing time in the way I have done it here is therefore somewhat unsatisfactory. The whole subject is extensively discussed by Chesson in Chapter 14.

E6. Partitionability of resources (low to high) Some resources are intrinsically relatively uniform and thereby nonpartitionable, for example, units of two-dimensional space such as bare rock available for colonization. In contrast, others are intrinsically variable and thereby potentially partitionable, for example, the sizes of arthropod prey and the chemical composition of leaves. Chapters 29 and 30 illustrate the usefulness of this axis. Of course, here again certain types of organisms will perceive or respond to a given degree of variability more than others (Schoener 1974a), but, also again, there is an underlying organism-independent variation.

Correlations Among Primitive Axes

Correlations exist among the six organismic axes and separately among the six environmental axes. These correlations help generate hypotheses about biological phenomena on levels lower than community ecology and thereby ultimately contribute to their understanding.

With respect to the organismic axes, adult body size is correlated to intrinsic rate of increase (Fenchel 1974) and thereby to generation time. Body size, together with degree of endothermy, is strongly related to degree of homeostatic ability. Smaller animals are more likely to have numerous life stages than are larger ones. Causal bases of these correlations contribute to the theory of physiological ecology and developmental biology, among other things (e.g., Bonner 1965, Gates 1980, Bartholomew 1982).

With respect to environmental axes, correlations illuminate the physical workings of climatic factors as well as certain interactions between physical and biological factors. For example, an open system with respect to resource input is more likely to be found on small islands than large islands and under seasonal than nonseasonal regimes.

Finally, evaluating environmental axes for organisms varying along organismic ones contributes toward understanding evolutionary adaptation. Small organisms are less likely to be found in (absolutely) physically extreme environments or at the top of trophic webs (Schoener 1974a, Connell 1975). Generation time limits the kinds of seasonality that can be accommodated (e.g., Lack 1968). Homeostatic ability contributes toward occupancy of physically stressful environments (e.g., Bartholomew 1982).

Correlations between the primitive axes just discussed and the derived axes to be discussed later contribute toward theory for higher levels and in particular toward understanding ecological communities. In the next sections we explore how this is so, with illustrations from the communities studied by the contributors to this volume.

POSITION OF REAL COMMUNITIES ON PRIMITIVE AXES

From the contributions to the present volume, it was possible to select 14 similia-communities for ordination along the axes just described (Table 28.1). More than 14 communities are discussed in the various contributions, but I selected only those communities for which authors (singly or in combination) presented a relatively complete picture of both primitive and derived characteristics, either in their chapters or elsewhere. Chapters concentrating on a single interaction or using particular communities mainly as examples for theoretical points were excluded, except as adding to descriptions based on other chapters. For two chapters (Chapters 12 and 32), I distinguished two similia-communities within a single broad

Table 28.1 COMMUNITIES RANKED IN TABLES 28.2 AND 28.3 ACCORDING TO PRIMITIVE AXES

Abbreviation	Similia-community	Location	Chapter
EAlg	Ephemeral intertidal algae	New England, United States	32
PAlg	Perennial intertidal algae	New England, United States	32
IVas	Interstitial vascular plants	Chalk grassland, Great Britain	12
MVas	Matrix vascular plants	Chalk grassland, Great Britain	12
CTre	Continental forest trees	Temperate, tropical North America	16, 19
HMit	Hummingbird mites	Neotropics	24
THer	Temperate herbivorous insects	North America	esp. 11, also 15
ISpi	Subtropical island orb spiders	Bahamas	33
PFis	Pond and lake fishes	North America	21
CFis	Coral reef fishes	Various shallow tropical waters	21, 30
ILiz	Island lizards (insectivorous iguanids)	West Indies	30, 33
CBir	Shrub-steppe, continental birds (mainly granivorous and insectivorous)	North America	9
IBir	Tropical island birds (mainly granivorous and insectivorous)	esp. Galápagos; also New Guinea, Hawaii, Bahamas	esp. 10; also 5, 6, 33
CHet	Continental desert heteromyids	Southwestern United States	3

taxon, as do the authors, because the two vary in certain primitive axes and because such variation has been related in illuminating ways to positions along derived axes. Table 28.1 is rather specific as to the place where the community is found, and that location (e.g., on islands as opposed to mainlands, temperate as opposed to tropical) contributes to obvious differences in position along certain environmental axes.

Tables 28.2 and 28.3 give the positions of 14 communities along organismic and environmental axes, respectively. I stress that these positions are crude approximations; no significance should be ascribed to sizes of gaps between communities, and the most reliable orderings are between communities treated in the same chapter.

Several points emerge from inspection of Tables 28.2 and 28.3. First, despite some serious gaps in the coverage of this volume (for example, freshwater animals lower than fishes), a great deal of variation is evident across both types of primitive axes. Second, this variation is not entirely haphazard; rather, there are similarities between certain communities (draw a line through the positions for each community and note resemblances). Some of these similarities are not surprising, for example, island lizards (ILiz and island birds (IBir) occupy roughly similar positions for all axes but O2 and O5. Other similarities are perhaps more unexpected, for example, those between ephemeral marine algae (EAlg) and interstitial vascular plants (IVas). Still, it is the variation, rather than the similarity, that is striking. If the primitive axes chosen here act in different ways on community characteristics, then one can begin to comprehend the size of a complete theory of community ecology.

SOME RELATIONS BETWEEN PRIMITIVE AND DERIVED AXES: TOWARD A PLURALISTIC THEORY

I now discuss 10 derived axes and show how variation in the location of the real communities just considered might be understood from variation in their positions along the primitive axes of Tables 28.2 and 28.3. (The label for each derived axis is followed by a list of the primitive axes that are suspected to be important for it.) As some of these derived axes (''higher'' ones) involve other derived axes (''lower'' ones) for their under-

Table 28.2 ORDINATION OF REAL COMMUNITIES ALONG ORGANISMIC AXES

Axis	Ranked Communities											
O1. Body size (small → large)	HMit	THer ISpi		IVas EAlg		ILiz MVas	PFis	PAlg IBir CFis	CHet CBir			CTre
O2. Source of new individuals (closed → open)*		ILiz PFis		CHet		IVas MVas	CTre	IBir†		THer ISpi HMit CBir	PAlg	CFis EAlg
O3. Generation time (small → large)	HMit	THer ISpi EAlg	IVas		CFis PFis ILiz CHet	CBir IBir	PAlg MVas					CTre
O4. Mobility (sessile → mobile)	PAlg EAlg CTre IVas MVas		HMit THer	ISpi		ILiz	PFis CFis	CHet	IBir†		CBir	
O5. Homeostatic ability (low → high)	HMit ISpi THer		EAlg IVas	ILiz	MVas PFis PAlg	CFis CTre					CBir IBir CHet	
O6. Number of life stages (small → large)	CBir IBir CHet				ISpi ILiz		PAlg EAlg CFis CTre PFis IVas MVas		HMit	CHer		

*Islands considered separately.
†Galápagos study.

standing, my discussion to some extent proceeds from the lower to the higher.

D1. Relative importance of physical versus biological processes (O1, O5; E1, E2) This axis is closely correlated with axis E1, and, indeed, in an all-other-things-being-equal situation would be identical to it. However, because the importance of physical factors is considered *relative* to biological ones, dependence is less than total. To illustrate, small organisms (O1) or organisms with lower homeostatic ability (O5) would ordinarily be more sensitive to physical factors. For example, orb spiders in the Bahamas are apparently more affected by low-level storms than are lizards on the same island (Chapter 33 and Schoener in preparation). Physical factors would therefore be more important for the spiders *if* at the same time predation (E2), which also is typically more severe on small organisms, were absent. But predation is absent only on the smaller islands, so it is only on those islands that physical factors predominate. Spiders on larger islands are heavily affected by predation from lizards, whereas those lizards are heavily affected by competition.

D2. Relative importance of predation versus competition (O1, O3, O6; E2) This is one of the several spectra that can be erected between two interspecific interactions. It appears to be the most important one for understanding population sizes and species occurrences, although work on mutualism (Chapter 25) and parasitism (Chapter 27) is much less extensive. Position along this axis can be related to position along a number of primitive axes.

First, trophic position (E2) is related to this

Table 28.3 ORDINATION OF REAL COMMUNITIES ALONG ENVIRONMENTAL AXES

Axis	Ranked Communities										
E1. Severity of physical factors (high → low)		ISpi EAlg			CFis IVas	PAlg MVas CTre	THer	CBir	PFis	CHet IBir ILiz	
E2. Trophic position (low → high)		IVas MVas PAlg EAlg CTre		CHer HMit	ISpi	THet IBir CBir CFis		PFis ILiz			
E3. Resource input (open → closed)*		ISpi	ILiz			CBir CFis	IBir†			CHet PFis	THer CTre EAlg PAlg IVas MVas HMit
E4. Spatial fragmentation (broken → continuous)*		HMit THer	IVas		MVas	EAlg	PAlg		CFis	IBir ISpi PFis ILiz	CHet CTre CBir
E5. Long-term climatic variation‡ (high → low)	CBir CTre (D) THer MVas IVas	CHet	PFis PAlg EAlg				IBir†	ISpi ILiz		HMit CTre (H)	CFis
E6. Partitionability of resource (low → high)	PAlg EAlg		IVas	MVas CTre		THer	HMit CFis		ISpi	CHet IBir CBir ILiz PFis	

D = Davis; H = Hubbell.
*Islands considered separately; adults or life stages similar to adults, e.g., later instars of spiders or hemimetabolous insects, considered only.
†Galápagos study.
‡Seasonal, or over a period of about 10 years.

axis by the Hairston et al. (1960) hypothesis (discussed in Chapter 32) that competition dominates terrestrial carnivore and producer communities while predation dominates terrestrial herbivore communities. The degree to which this logic or related logic (e.g., Menge and Sutherland 1976) should work and does work has recently been debated (e.g., Chapters 32 and 33; Schoener 1982, 1983b, 1985c; Connell 1983, Hairston 1985). So far as this volume is concerned, chapters considering terrestrial herbivores (Chapters 11, 15) do not find competition very important, whereas most of those considering terrestrial carnivores or species deemed similar in the extended sense of Slobodkin et al. (1967) do (birds, lizards, and rodents as discussed in Chapters 3, 5, 6, 30, and 33, but see discussion of birds, mites, and spiders in Chapters 9, 24, and 33). Chapters considering the relative importance of competition versus predation for terrestrial producer communities (Chapters 12, 22, 23) are somewhat split. Wilson's (Chapter 26) work on burying beetles is also consistent with the Hairston et al. hypothesis for decomposers.

Second, generation time (O3) is apparently important. Short-lived ephemeral marine algae are affected less by competition relative to predation (herbivory) than are perennial algae (Chapter 32). Somewhat similarly, interstitial plants of chalk grassland, while affected most strongly by competition from matrix plants, are affected more by herbivory than are the matrix plants; matrix species are perennial, whereas interstitial species are annual, biennial, or pauciennial (Chapter 12). The interstitial plants also tend to be more influenced by physical factors (D1). In both of these examples, organisms with a short generation time seem to be more adapted for colonization of short-lived, unpredictable locations (indeed, logically they must be partly responsible for the character of these locations) and, in the algae at least, do not develop extensive defenses against herbivory.

Third, small organisms (O1) compete less on average than do large ones, at least in part because of vulnerability to predation (Chapter 33; Connell 1983, Schoener 1983b).

Fourth, for organisms with several life stages (O6), at least some of the stages are small. When others are large, as in fishes (Chapter 21), marine algae (Chapter 32 and Lubchenco 1983), and matrix plants (Chapter 12), predation can control one segment of a population and competition can control another, creating possible bottlenecks and confounding any simple characterization of the community as a whole.

D3. Nearness of populations to carrying capacity (D1, D2; O2; E3, E5) This axis is closely related to the axes D1 and D2. If predation or physical factors predominate, populations will spend more time away from carrying capacity. This is illustrated by the ephemeral versus perennial marine algae or the spiders versus lizards, just discussed.

A primitive axis, long-term temporal variation (E5), is also obviously relevant here. Such variation might create lags between population numbers and their resources such that often the former would be below carrying capacity (Wiens 1977). The argument is plausible for the shrub-steppe birds that Wiens (Chapter 9) studies, but Grant's attempt to test it for an insular avifaunal community, the Galápagos finches, failed to find significant lags (Chapter 10). Possibly resource variation was smaller on Grant's tropical islands than in Wiens' temperate continent, but the former variation was quite large by absolute standards.

Alternatively, other primitive axes, those dealing with open versus closed systems (O2, E3), may be involved. Possibly a lack of fit of populations to their resources in continental systems is accentuated by variable input from outside the area of study, perhaps in terms of resources but particularly in terms of recruitment. Because of their openness, small pieces of continents may have a lot of species that do not quite belong where they are and are maintained mainly by input from elsewhere. This picture is consistent with some of Wiens' (Chapter 9) results on patterns and spatial scale; at certain small scales, but not the largest, species abundances do not correlate well with habitat features. In contrast to continents, immigration would be minimal on isolated islands, even more so if they were tropical, as organisms, particularly birds, in low latitudes tend to be sedentary (e.g., Mayr 1966, Diamond 1974).

D4. Outcome of competition (O4; E6) As both Yodzis and Roughgarden (Chapters 29 and 30) have pointed out, the degree to which one or another competitive outcome predominates (along the axis from coexistence through priority effects to exclusion) depends upon several factors, some of which we have designated as primitive axes.

If the resource is relatively nonpartitionable (E6), as for small units of two-dimensional bare space, then the possible competitive outcomes are invariant exclusion of one species by another or priority effects in which the identity of the species that wins depends on initial abundances (Chapter 29). In such a community, dominance relationships will often favor a hierarchy. This in turn will allow the operation of the intermediate disturbance effect, in particular where competitively dominant species are preferred by predators (Chapters 29, 33).

Where the resource is partitionable, as for island *Anolis* lizards, species are less apt to be arranged in a hierarchy. Rather, they are arranged latitudinally along a resource spectrum, and proximity in niche space determines the degree of competition (Chapter 30).

Various other interference mechanisms, if complicated enough, can affect this simple scheme. In particular, if chemical (*sensu* Schoener 1983b) competition occurs, such that each species produces one or more toxins and has variable susceptibilities to the various toxins produced by other species, networks rather than hierarchies are possible (Chapters 2, 31, but see Quinn 1982).

Finally, position on the mobility axis (O4) determines which types of competition are possible. Sessile organisms necessarily occupy a unit of space, a precondition for preemptive (*sensu* Schoener 1983b) competition, although competition in sessile organisms could alternatively occur through other mechanisms—overgrowth, chemical, or even consumptive (Chapter 22). Entirely sessile organisms by definition cannot compete via a territorial or encounter mechanism, whereas entirely mobile organisms cannot compete via a preemptive or overgrowth mechanism.

D5. Species abundances (D1, D2; O1, O5; E1, E2) To make this axis one-dimensional, we could order communities by their average degree of commonness versus rarity, though perhaps a multidimensional description would be better (e.g., May 1975b). Nonetheless, certain taxon-guilds have been characterized as being composed entirely of sparse organisms, for example, interstitial plants of chalk grasslands (Chapter 12). In the case of such plants, as well as sparse matrix species in the same system, rarity is apparently caused mainly by competition with abundant matrix species (D2, E2). Physical factors (drought, disturbance by the feet of grazing mammals) additionally contribute to rarity (D1, E1). In contrast, universally small population sizes in island orb spiders have been tentatively related mainly to predation; physical factors may also be important, and competition may contribute as well (Chapter 33).

Rarity may also be linked to homeostatic ability (O5). Endotherms, such as birds on Bahamian islands, have greater energy requirements and smaller densities than ectotherms (lizards) on the same islands (Chapter 33). Similar relations to body size (O1) also exist.

D6. Temporal variation in population size (D1; O1, O2, O3, O5; E1, E2, E3, E4) Lizards (and probably birds) show far less variation in population size than do terrestrial arthropods, both in the same system (the Bahamas) and among data from various different systems (Chapter 33). Workers on herbivorous insects are particularly fond of stressing such variation (Chapters 11, 15). It can be directly related, through mortality, to the relative importance of physical factors (D1), in turn understandable from the primitive environmental axis of severity of physical factors (E1), itself related to the primitive organismic axes of body size (O1) and homeostatic ability (O5).

Probably no less important, however, is the openness of the system with respect to recruitment (O2). Many arthropods disperse from greater distances than do lizards, for example (Chapters 11 and 33; Kareiva 1983a). The role of openness is tricky here. Were many potential locations for populations unoccupied, then moderate openness could generate variation in population size (Chapter 11; Quinn 1979). On the other hand, a steady flow of immigrants into a particular community can reduce population fluctuations as a nonequilibrium condition (Schoener 1976b). Finally, Kareiva (Chapter 11) has shown experimentally that relatively large variation in population size in certain herbivorous insects occurs in a system which is both patchy (E4) *and* contains predators (E2).

Other primitive axes can also be related to this derived one, at least in theory. First, small generation time (O3), implying a high r, translates directly into increased fluctuations in population numbers (e.g., May 1981a). This effect is consistent with the greater population-size fluctuations of interstitial than matrix plants (Chapter 12). Second, as also shown by population models (e.g., Schoener 1976b, 1978), systems open with respect to resource input (E3) show less population fluctuation than do closed ones.

D7. Species turnover (D1, D2, D7; O1, O2, O3, O5; E1, E2, E3, E4) Degree of turnover shows the same qualitative difference between terrestrial arthropods and vertebrates as does degree of population variation (Chapter 33). Inasmuch as turnover includes extinction rate, there is a necessary relation between this and the previous derived axis (for a mathematical formulation, see Wright and Hubbell 1983). Hence variation along each of the two derived axes is roughly parallel and subject to much the same sort of explanation. For example, generation time (O3) and turnover are inversely (and approximately linearly) related (Schoener 1983c).

The relationship between openness with respect to recruitment (O2) and turnover at a local level is illustrated well with coral reef fishes; despite intense territorial competition, local relative abundances of species are quite variable in time because of open recruitment (Chapter 30; Sale 1984).

Finally, species extinction can be related to the amount of time during which a community's populations are sparse, and this in turn can be related to the relative importance of predation (D2) when predation is indiscriminate (Chapter 33).

D8. Number of species (O1, O2, O3, O5, O6; E1, E2, E3, E4, E5, E6) The number of species in ecological communities is of less interest now than a decade or two ago (e.g., papers in Lowe-McConnell 1969). Nonetheless, the ecological maintenance of species diversity is a theme running through a variety of chapters (e.g., Chapters 12, 19, 22, 24, 27).

Almost all primitive axes listed above have been related to species diversity (see Pianka 1983 for references and summaries, Whittaker 1972).

- O1. Smaller animals can be more specialized, hence more diverse, than large animals (e.g., Schoener 1969, Brown 1981).
- O2. Long-distance dispersal enhances the ecotone effect, allowing species that do not "belong" in an area to persist there and hence increase diversity (Chapter 30).
- O3. The more generations, the greater the rate of speciation and the more species.
- O5. Ectotherms require less energy, so can be more specialized and have more species than endotherms.
- O6. The greater the within-species variance, the fewer species can be packed along a resource axis; the greater the importance of each life stage, the greater the potential for maintaining diversity, for example, the regeneration niche (Chapter 12).
- E1. Physically stressed (*sensu* Sanders 1969) environments allow fewer species.
- E2. Organisms at higher trophic levels receive less energy, so cannot be maximally as diverse as those at lower levels.
- E3. Open systems with respect to resources are more stable and thereby allow more species to coexist (Schoener 1976b).
- E4. Spatial fragmentation affects extinction in predator-prey systems, among others, and can lead to more or fewer species (Chapter 11).
- E5. A variable and unpredictable climate disfavors specialization and thereby reduces diversity. For example, in hummingbird mites lowland tropical species are more host-specific than are montane species or those of higher latitudes (Chapter 24). In contrast, predictably fluctuating environments can allow specialization to different seasons or different types of years, increasing diversity (Grubb 1977).
- E6. The more heterogeneous a resource, the more species can partition it.

One conclusion from this litany is that species diversity is so derived an axis that almost any primitive axis is important for some system or other in its explanation.

D9. Importance of "chance" (D5, D8; O4; E2, E4) Chance is a sticky concept that takes a variety of meanings in ecological discussions, though it is no less protean in other fields. It is argued to play the following roles in the communities discussed in this volume (see also Chapter 13).

First, in tropical lowland rain forest trees (Chapter 19) species are diverse (D8) and sessile (O4). Hence, the environment in which offspring of particular species develop is special (composed of a few trees) but unpredictable (because there are so many possibilities compared to realizations). This system makes it impossible for resource partitioning *or* strong hierarchies to develop; most species persist, at least in the short term, which is actually very long so far as trees are concerned. If the system is losing species, it is doing so slowly. In a system with few strong rules, ''chance'' determines the composition of the community.

A second, rather similar view is Grubb's (Chapter 11) treatment of interstitial plants, in which chance is argued to play a strong role in intraguild interactions—all species are rare (E4, D5) and their location in space unpredictable. (Tilman [Chapter 22] has a rather different view of similar organisms).

A third example, in which ''chance'' is measured by degree of demographic stochasticity, involves Bahamian orb spiders (Chapter 33). Here, it is argued, predation (E2) reduces spider populations to such low levels (D5), or islands colonized by spider propagules are so small (E4), that stochastic extinction becomes of major importance, in turn reducing regularity in certain ecological patterns.

D10. Importance of history (D4; O3; E5) Where generation time (O3) is large and long-term variation (E5) sufficiently high, it may not be easy to interpret communities on the basis of present characteristics alone; some degree of history is necessary. Temperate forest trees are an excellent example (Chapter 16). History is also important, though in a less epochal sense, when priority effects obtain in competition (see discussion of D4 above), or more generally when multiple stable points exist (Chapter 2).

CONCLUSION

The eventual goal of an analysis such as this one is to answer at least two questions. First, do certain attributes of communities (derived axes) tend to be determined by particular characteristics of the organisms and of the environment (primitive axes)? Second, is each community effectively unique, or is there a modest number of ''types'' of communities?

At present I cannot guess how the answers will turn out. The analysis of this chapter obviously represents only the most tentative start. Only 14 communities were considered, with some severe qualitative lacunae, such as that parasites and mutualists are not represented. Some relationships are fuzzy or at least non-monotonic (e.g., E2 with D2), possibly implying that additional primitive axes would be desirable.

With these disclaimers, we may note that in this preliminary survey each of the 12 primitive axes chosen is suggested as important for understanding at least one of the 10 derived axes chosen. Conversely, each derived axis requires for its understanding generally more than one or two, but not nearly all the primitive axes—on the average, five of them. This hints that a theory of community ecology with modest goals would be of large but not overwhelming size. For the derived axes of species diversity and temporal variation in population size, however, even my initial attempt portends a very large theory.

Consideration of more communities will surely suggest additional primitive and derived axes. Will the entire theory then increase greatly in complexity? The answer to this question will depend on whether the average number of primitive axes needed to explain a given amount of variation along individual derived axes increases or stays roughly constant. I am cautiously optimistic.

ACKNOWLEDGMENTS

I thank James Brown, Ted Case, Robert Colwell, Jared Diamond, Peter Grubb, Peter Kareiva, Jane Lubchenco, and Earl Werner for reading a previous draft.

chapter 29

Competition, Mortality, and Community Structure

Peter Yodzis

INTRODUCTION

Interspecific competition has long occupied a preeminent position in the theory of ecological communities. Recent questioning of this emphasis on competition has resulted in a healthy reappraisal of some of our most fundamental ideas. For the most part the reappraisal has considered *when* and *where* there is competition in nature. Less attention has been focused on our whole conception of *how* competition affects community structure in those cases where it does.

What is generally thought of as ''competition theory'' is really the theory of one particular kind of competition: pure exploitation or consumptive competition (*sensu* Schoener 1983). This theory does not apply to five of the six kinds of competition distinguished by Schoener (1983)! In order to assess properly the role of competition in nature, we need a broader theoretical framework.

I am going to argue here that there are at least three essentially different ways in which competition can control community structure. I am also going to discuss how alterations in mortality factors such as predation, grazing, and physical disturbance influence competition-mediated community structure in these three different cases.

CONSUMPTIVE VERSUS SPATIAL COMPETITION

Imagine an area of space that is inhabited by several species we want to study (and by many others besides) and that contains certain resources indispensable for the lives of our study species. Assume that these indispensable resources are distributed evenly enough so that the area seems homogeneous to our study species. Assume further that none of the study species preys on or parasitizes any of the others and that there are no mutualistic interactions among them.

Since the resources are indispensable to our study species, and since the resources are, inevitably, limited in supply, our study species may very well have to compete for those resources. (If the study species never become abundant enough to tax the supply of resources, competition might be so weak as to be, for all practical purposes, nonexistent. However, I am going to consider here situations in which competition does occur.)

I shall call a set of species satisfying these conditions a *competitive community*.

There are basically two ways that an individual organism may go about getting its required share of the available resources. One way is to collect a fraction of the resources from the whole area; the other way is to collect all the resources from a fraction of the area. (There is, of course, a continuum of intermediate strategies between these two extremes.) Adoption of the first alternative by several individuals will lead to *consumptive competition* among them. Adoption of the second alternative will lead to *competition for space*.

In consumptive competition each individual ranges more or less freely through the entire area, collecting resources as it wanders. Different individuals may try to avoid one another as much as possible, and if they are successful in this avoidance, we will have pure consumptive competition: Each individual hinders the others solely by consuming resources that they might otherwise have consumed. Consumptive competition may be accompanied by encounter or chemical competition (*sensu* Schoener 1983b), but that does not affect its basic nature in terms of utilization of space.

In competition for space some essential resource can be obtained only by occupying, more or less exclusively, some portion of space. The essential resource may be space itself, as in attachment sites for sessile filter feeders or nest sites for hole-nesting birds. Most plants need to occupy a certain amount of space below ground in order to obtain water and nutrients, and a certain amount of space above ground in order to obtain light.

On the level of individuals competition for space occurs on a much smaller spatial scale, in relation to the body sizes of the competitors, than does consumptive competition. On the level of populations competition for space can take place on a very large scale. Thus, while individual adult sessile filter feeders only centimeters apart may not be in competition for space, populations with pelagic larvae may compete over regions of tens or hundreds of miles (Chapter 30).

Competition for space could be viewed simply as a special case of consumptive competition, with space being the resource that is consumed. This view has been articulated by Tilman (1982, Chap. 8). I believe, however, that competition for space is so different from what we normally think of as consumptive competition that it makes more sense, and will ultimately prove more fruitful, to think of it as a completely different category of competition. Certainly space is quite different from any other resource. While an individual item of food, once consumed, is gone forever, an individual unit of space, once "consumed," can be recovered at a later time and used by another "consumer" individual. At the same time, recovery or renewal of the "space" resource within an area is closely linked with, if not identical to, mortality in the "consumer" population, while food resources are renewed on a time scale and by processes that have nothing to do with consumer mortality.

Moreover, the nature of coexistence among competitors for space is profoundly different from that of coexistence among consumptive competitors. The basic difference follows from the contrast in spatial utilization at the individual level that I outlined two paragraphs ago. Because of the global nature of consumptive competition, competitive exclusion cannot be avoided in this context if competition is too strong. Thus, coexistence of consumptive competitors *requires* niche differentiation, be it through partitioning of resources or through partitioning of temporal utilization. As I shall show later in this chapter, competitors for space can coexist without niche differentiation. The theory of consumptive competition is well known and I shall not say much more about it. MacArthur (1972a) and Roughgarden (1979) provide excellent summaries (see also Chapters 10 and 30), and Casten and Case (1979) explore the effect of dispersal. Since spatial competition theory is less well known, I shall summarize it in the next five sections (Yodzis 1978 is a somewhat more technical exposition).

PLANTS AND COMPETITION FOR SPACE

I think that nothing I have said will be particularly controversial if it is applied to competition among sessile animals. But I have also mentioned plants as competitors for space. Writers such as de Wit (1960), Harper (1977), and Werner (1979) may find this quite congenial, but other, equally thoughtful, biologists may take exception to the notion that plants should be viewed as competing primarily for space. If plant species compete for space and therefore can coexist without niche differentiation, then what about the careful delineations of the plant niche by Cody, Grubb, and Tilman in Chapters 12, 22, and 23 of this volume?

These are elegant papers, and I do not question the existence of the differences among plant species explicated by these and other authors. The question is what these differences have to do with coexistence. I do not know the answer to this question and I do not believe that anybody else does either.

As Cole (1960) pointed out, "by definition, no two species are identical, so that if one looks closely enough he is bound to find something that can be considered a difference in the ecological niches." It is exceedingly difficult to establish that putative niche differentiation among coexisting species is what permits them to coexist. Within the context of consumptive competition, most ecologists nevertheless have accepted niche theory, partly because of a large body of data which, at the very least, is not inconsistent with it, but mainly because of the lack of a credible alternative paradigm for the coexistence of consumptive competitors.

I think it likely that some of these differences among plant species *do* help promote coexistence, but I think they are best viewed as refinements of a basic explanation in terms of competition for space. Grubb's (1977) regeneration niche is fully consonant with a spatial competition approach. Indeed, reading Grubb's paper helped clarify my own thinking about competition for space. Hubbell's and Foster's approach to plant communities (Chapter 19) is also compatible with mine. I would hope that a consideration of pure competition for space, without any niche differentiation, might similarly provide a background against which to view niche relationships.

UNDERLYING PROCESSES OF COMPETITION FOR SPACE

There are undoubtedly many factors that contribute to the coexistence of competing species in nature. Eventually, one would like to put together a Grand Explanation that combines and balances all these factors in the right way for each system. But first we will have to understand each factor by itself, and then see how the different factors relate to one another. Therefore, in order to isolate features characteristic of spatial competition, I am going to discuss pure competition for space in this section and the following two. There will be no local niche differentiation and no environmental heterogeneity. Then I will devote a section to the role of environmental heterogeneity in spatial competition.

Competition for space often occurs on a seasonal basis. For example, many bird species establish and defend territories during the breeding season, but range freely during the rest of the year. This could have important implications for community structure, if the populations in question are limited by processes that occur during the period of spatial competition. However, I am going to discuss here the more clear-cut situation in which each individual occupies a unit of space for most of its lifetime.

The most obvious examples of this kind of competition for space are sessile organisms. Motile organisms may also sometimes appropriately be thought of in this fashion, if they live in an environment consisting of several habitable "islands" in a "sea" of uninhabitable space.

The basic processes in this kind of competition are dispersal, establishment, growth, and mortality. Most competitors for space have a definite dispersal phase—seeds, spores, and larvae are examples. The scale of dispersal may not be much greater than the size of an adult, or it may be over whole geographical regions. Establishment at a spatial location generally also takes place early in life, but this establishment may only be provisional. In some cases competition is at its severest early in life (probably the case for many plants; Werner 1979), while in others it is closely linked with growth (as in many sessile fauna). Jackson (1977a), Dayton (1983), and Buss (Chapter 31) review competition among sessile fauna, Grubb (1977) and Werner (1979) among plants.

These processes have been studied in patch occupancy models (Cohen 1970, Levins and Culver 1971, Horn and MacArthur 1972, Slatkin 1974, Hastings 1980, Horn 1981, Hanski 1983), reaction-dispersal models (Levin 1974, 1976; Yodzis 1976a, 1977a, 1978; Allen 1983), lotteries (Hubbell 1979, Chesson and Warner 1981), and simulations (Botkin et al. 1972, Moser 1972, Shugart et al. 1973, Maguire and Porter 1977, Karlson 1981, Karlson and Jackson 1981). Different studies have assumed different specific details of the underlying processes, and yet their results are remarkably similar. Therefore, the basic picture of competition for space that has emerged from these studies is probably quite robust. I will sketch this picture using some of my own results from reaction-dispersal theory (Yodzis 1978).

The area of space under consideration is thought of as divided into a number of "cells," with competition occurring *within* each cell and dispersal *among* the cells. (These cells are called "patches" by Yodzis [1978] and other authors, but it is better to reserve that word for certain spatial patterns in species distribution that result from the dynamics, rather than for the more elemental concept that I am calling "cell" here.) The cells may be either discrete, naturally occurring units or artificial divisions of continuous space imposed by an investigator. The size of the cells should be such that the individuals or clones within a cell can reasonably be thought of as competitors. Hence, a cell may be no larger than an adult individual, or it may be of the order of home range size, depending on the biological context.

The extent of the system, that is, the set of all cells, depends partly on the range of dispersal and partly on the viewpoint one wishes to adopt. In many cases it is natural to consider a habitat unit in the usual sense. A reasonable example of this would be a forest; while it is true that there will always be a trickle of long-range dispersal into and out of a forest, most dispersal of tree seeds is over distances at most of the order of what would (nowadays) be regarded as a small forest. However, if the norm is long-range dispersal, as for many pelagic larvae, and one wishes to treat a closed system, then the system may span whole geographical regions. On the other hand, even in the case of long-range dispersal, one could also treat a single habitat unit as an open system with settlement from outside. Roughgarden (Chapter 30) discusses this distinction further.

The rate of dispersal can have important implications for competitive community dynamics. This was first pointed out by Levin (1974) and proved for a rather wide class of models (Levin 1976). If dispersal rates are too high, there is so much "mixing" among the cells that they function as one big homogeneous system; competition then loses its local character and the dynamics look quite different. In this section and the next I will assume that dispersal rates are low enough so that the local character of competition is maintained. I will return to the "complete mixing" case in the section on environmental variability.

Just to get started, we may think of the *local* competition *phase* of the dynamics as being described by the usual Lotka-Volterra equations of consumptive competition, with the competition coefficient α_{ij} being proportional to the per capita effect of local population j on the growth rate of local population i. However, these alphas will seldom have much to do with resource overlaps, since most competition for space involves a great deal of interference—for perfectly good evolutionary reasons (Case and Gilpin 1974).

In some cases this kind of description may have a direct biological interpretation. For example, the replacement of a tree in a forest may be determined by the dynamics of the seedling populations in the local unit of space under its crown (Horn 1981). Hence, this essential element of forest dynamics may not be so different from de Wit's (1960) competition experiments. In other cases, such as coral reefs (Maguire and Porter 1977) or other systems where adult-adult competition is important, the biological interpretation of the alphas may require a much finer partitioning of space or a completely different local dynamic may be called for.

The outcomes of pairwise interactions between species are related to the alphas as follows:

$\alpha_{ij} < K_i/K_j$ and $\alpha_{ji} < K_j/K_i$ ⟶ coexistence

$\alpha_{ij} < K_i/K_j$ and $\alpha_{ji} > K_j/K_i$ ⟶ species i dominates species j

$\alpha_{ij} > K_i/K_j$ and $\alpha_{ji} > K_j/K_i$ ⟶ contingent competition

In the first case the two species can coexist locally. In the second case species i will always win a local competition with species j; species i *functionally dominates* species j. In the third case one species will always win a local competition, but the identity of that species will be *contingent* upon the initial densities of the two species.

In consumptive competition, where there is no real distinction between local and global competition, the first case is the only interesting one. Depending on how important niche differentiation among competitors for space turns out to be, this case may or may not be interesting in competition for space. But certainly with the ground rules adopted here (no local niche differentiation), the first case is the only *un*interesting one.

We can generate model communities of spatial competitors by letting a number of species, chosen at random from some "pool," colonize an area of empty space. Allowing dispersal within that area and local competition within cells, we can wait for the establishment of an equilibrium, then investigate the species composition of the resulting community. As we shall see below, sometimes we might have to wait forever for the establishment of an equilibrium.

The pool of potential colonizers is specified if we give probability distributions for all relevant parameters of those species. In the work that I am going to describe, each parameter was simply chosen at random from some fixed interval of real numbers. It turns out that all of the interesting results depend on the value of just one parameter, which I will denote by C. It is the probability for a competition coefficient α_{ij} to be greater than K_i/K_j. The probability for a randomly chosen pair of species to coexist locally is then $(1 - C)^2$. Since I want to see what happens without this kind of local niche differentiation, I will never choose C smaller than 0.9. Then only one pair of species in a hundred can coexist locally. However, the probability for a randomly chosen pair of species to have a contingent interaction is C^2, which will always be appreciable.

Larger alphas correspond to larger values of C. Therefore, I will call C the *strength of competition*.

SOME COMMUNITY PROPERTIES THAT RESULT FROM PURE COMPETITION FOR SPACE

If $C = 1$, then all pairwise interactions are contingent; there are no dominant interactions. Hence, once a species has become established in a cell, it cannot be eliminated by dispersers from other cells. Community structure is essentially a relict of the original colonization episode, so I call such a community "founder controlled."

If $C < 1$, there will still be some contingent pairwise interactions, but there will also be some functional dominance. Therefore an initial colonizer cannot necessarily maintain its presence in a cell; intercell dispersal or growth of individuals (depending on the biology of a particular system) will bring about a "reshuffle" of cell occupants, with some species being eliminated from the system entirely. Dominance interactions are crucial for this process, and, as we shall see below, the overall degree of dominance in the community as a whole strongly affects the outcome. Therefore, I call such a community "dominance controlled."

Dominance Control

Assume first that $C < 1$. Then the probability for a randomly chosen colonizing species to dominate another colonizing species, chosen at random, is $C(C - 1)$. This degree of dominance is the crucial factor in determining community structure.

After the original colonization episode, there will be a sequence of intercell invasions due to dispersal or to individual growth. This process will often terminate in an equilibrium, but in certain communities an equilibrium is never attained. Rather, in some cells the identity of the locally numerically dominant species repeatedly cycles through a subset of the colonizing species. I call this kind of behavior a "quasi cycle" (Yodzis 1977a, 1978). A similar behavior in Lotka-Volterra systems with immigration has been studied by Gilpin (1975) and May and Leonard (1975).

Quasi cycles result from intransitive dominance relationships. The simplest example occurs if three colonizing species A, B, C are such that A dominates B, which dominates C, which dominates A ("dominance" being functional dominance in the sense of the preceding section). This kind of relationship was first observed by Jackson and Buss (1975) among the epifauna that inhabit the undersurfaces of certain foliaceous corals. Subsequent observations are detailed by Buss

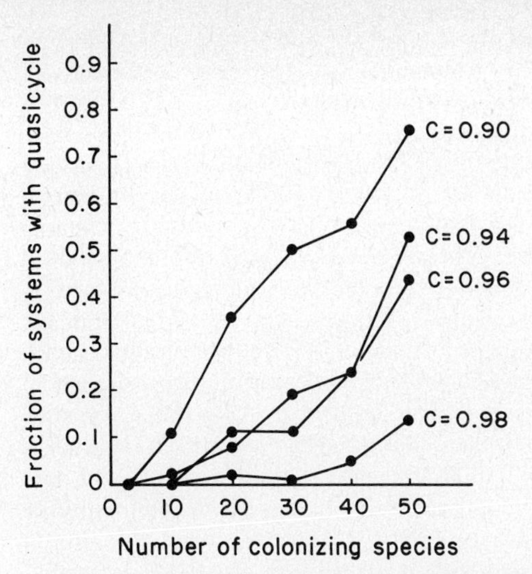

Fig. 29.1 Fraction of randomly assembled systems having a quasi cycle versus number of colonizing species, for several values of the competition strength *C*. (After Yodzis 1978.)

(Chapter 31) and discussed further by Gilpin et al. (Chapter 2).

For the model communities being described here, the fraction of communities with a quasi cycle is plotted against the number of colonizing species for several different values of the competition strength *C* in Fig. 29.1. It will be seen that there is quite a potential for quasi cycles, particularly in species-rich systems with weak competition. However, Fig. 29.1 probably overestimates the likelihood of quasi cycles. I shall discuss this point further in the next section (see also Yodzis 1978, p. 52).

For now, let us consider the communities that do approach an equilibrium. In the approach to equilibrium the number of species in the community will at first increase steadily as more and more species arrive, then decrease steadily to an equilibrium value as subdominant species are eliminated. This behavior is often seen in studies of succession in plant communities. Dayton (1973a) has portrayed such a process in considerable detail for the epifauna in a rocky intertidal zone (Fig. 29.2).

Equilibrium species richness is plotted against the number of colonizing species for a number of different values of the competition strength *C* in Fig. 29.3. As the number of colonizing species

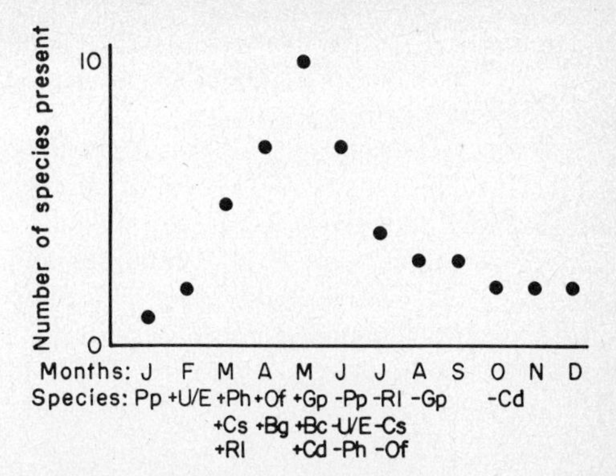

Fig. 29.2 Species present in over half of the treated plots in the intertidal zone of San Juan Island, Washington, United States. + = appearance of a species; − = disappearance of a species. Species abbreviations: Pp = *Porphyra perforata;* U/E = *Ulva* or *Entermorpha;* Ph = *Polysiphonia hendryi;* Cs = *Colopomenia sinuosa;* Rl = *Rhodomela larix;* Of = *Odonthalia floccosa;* Bg = *Balanus glandula;* Gp = *Gigartina papillata;* Bc = *Balanus cariosus;* Cd = *Chthamalus dalli.* (After Dayton 1973a. Copyright 1973 by the University of Chicago.)

increases, equilibrium species richness at first also increases. However, equilibrium species richness eventually levels off at a value determined by the competition strength *C*, and once this saturation value is reached, further increases in the number of species available for colonization have no effect on the number of species actually present in the community.

As long as the supply of potentially colonizing species is adequate, equilibrium species richness is determined by the competition strength *C*, not by the number of colonizing species. If competition is stronger (as expressed here by a larger

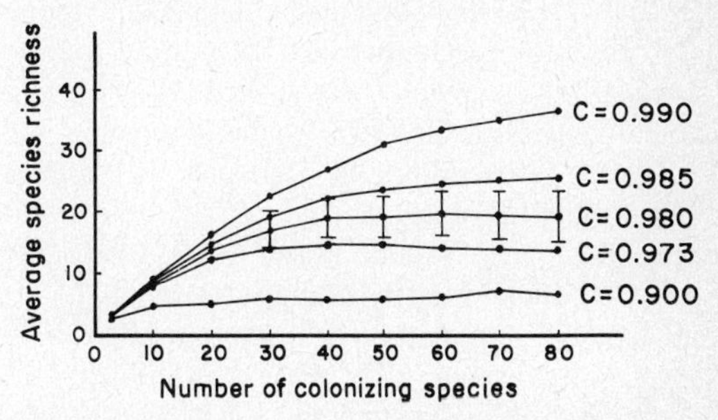

Fig. 29.3 Average species richness of randomly assembled equilibrium communities versus number of colonizing species, for several values of the competition strength *C*. (After Yodzis 1978.)

value of C), contingent interactions are more likely and dominance is less common. Hence, the stronger the competition, the more diverse the community. This statement is true for both the richness and evenness components of diversity (Yodzis 1978, sec. III.4.3). Indeed, abundance-rank curves for these model communities look, superficially at least, very much like the geometric and lognormal abundance-rank curves seen in real plant communities (Yodzis 1978, pp. 61, 63).

Founder Control

If $C = 1$, there is no functional dominance in the pool of colonizing species. All interactions are contingent. In this case community structure is governed entirely by the colonization process rather than by processes intrinsic to the community, so I call the resulting communities founder controlled.

There are no quasi cycles in founder-controlled communities. In the approach to equilibrium species richness simply increases steadily until the entire area is occupied, instead of behaving as in Fig. 29.2. Equilibrium species richness depends on the number of species in the pool of colonizers and the size of the area under consideration. For a given number of potential colonizing species, species-area curves will level off at larger areas for founder-controlled communities than for dominance-controlled communities.

Most natural communities involving competition for space show clear signs of dominance control. But founder control, or something close to it (perhaps dominance control with C very close to 1), is also sometimes observed. Among the best-known examples are the marine fouling communities studied by Sutherland (1974), Sutherland and Karlson (1977), and Schoener and Schoener (1981b). Grubb (Chapter 12) argues that interstitial plants of chalk grassland are another example. Neither Porter (1974) nor Bak and Luckhurst (1980) found any evidence of functional dominance in coral reefs in the Caribbean, though many other reefs do exhibit dominance. Similarly, Sammarco (1982) did not find evidence of dominance in Caribbean macroalgal communities, though other such assemblages have shown strong dominance.

Dominance control becomes manifest only in the later stages of community development. In early successional stages the behavior of all communities is essentially founder controlled.

On a more speculative note, I have suggested (Yodzis 1978) that lowland tropical rain forests may be founder controlled or something close to it. The general features of these forests in relation to temperate zone forests—high species richness, high evenness, species-area curves that show little sign of leveling off at large areas—are exactly the attributes of a founder-controlled community. Some authors (e.g., Dobzhansky 1950, Williams 1964) have claimed that competition is particularly ''keen'' in the tropics, which would lead one to expect founder control. Hubbell and Foster (Chapter 19) also argue for founder control in tropical forests.

DOMINANCE: THE FUNDAMENTAL PARADIGM OF COMPETITION FOR SPACE

We see, then, that communities that result from pure competition for space have a rich structure, and this structure is determined by the nature of dominance relations among the competing species. *In pure competition for space, dominance relationships play the same crucial role in determining community structure that niche relationships play in pure consumptive competition.*

I motivated the discussion by talking about local Lotka-Volterra models and their alphas, but one would obtain the same basic results from a formalism couched in terms of probabilities p_{ij} for species j to dominate species i in local competition (at a site occupied first by species i), at whatever is the appropriate stage in the life cycle of the organisms under consideration.

In some cases these dominance relationships might turn out to be deterministic, with all the p_{ij} equal to 0 or 1. This seems likely especially when competition occurs at the adult stage along with growth. In other cases there may well be a stochastic element. One might expect a significant element of stochasticity when competition occurs at an early stage of individual development and affects the success of initial establishment.

There is a modest literature on these kinds of dominance relationships. Karlson and Jackson

(1981) distinguished "hierarchies," in which all species are ranked in a simple linear fashion, and "networks," which contain intransitive dominance relationships and lead to quasi cycles. Unfortunately, some authors, picking up on this distinction, have written as if these two were the only possibilities for community dominance relationships. This is far from the case. What I have called contingent interactions can produce a rich taxonomy of community dominance relationships, with important implications for community structure.

When competitive ability depends on a single attribute only (say, speed of growth), then one expects that species can be ranked in a hierarchy depending on the degree to which each species possesses that one attribute. More generally, the fewer the factors determining competitive superiority, the most likely are hierarchies. Hence, Fig. 29.1 clearly overestimates the likelihood of quasi cycles in relatively simple situations, such as the more or less pure competition for light in periphyton assemblages (Hoagland et al. 1982). One might expect to find quasi cycles in a complicated system like a temperate zone forest, but in this case the dynamics are so slow that direct observation of a quasi cycle would be difficult.

Empirically, while there is an enormous anecdotal literature on dominance relationships, relatively few studies have systematically gone about "measuring the p_{ij}," or some such, for an entire community. The first study to do this for sessile fauna seems to have been that of Buss and Jackson (1979), who found intransitivities in the dominance relationships among the sessile invertebrates in cryptic coral reef habitats. Quinn (1982), on the other hand, found a hierarchy among the sessile organisms in a low intertidal zone, as did Gilpin et al. (Chapter 2) for *Drosophila* species. There was little stochasticity in either of these studies.

Botanists, largely following the lead of de Wit (1960), have measured a lot of dominance relationships for chosen pairs of species. I am not aware of any work on a whole community along these lines, though studies by Caputa (1948), Bornkamm (1961a, 1961b), Jaquard (1968), Stephens and Waggoner (1970), Botkin et al. (1972), Moser (1972), Shugart et al. (1973), Horn (1975), and Goldsmith (1978) do tend in this direction. For plants, too, most dominance relationships seem to be deterministic, but there is also a significant stochastic element. Indeed, one could regard some of the attempts to explicate a "plant niche" as attempts to reduce this stochasticity to its underlying causes.

ENVIRONMENTAL VARIABILITY AND COMPETITION FOR SPACE

Throughout the preceding three sections I have assumed a homogeneous environment, both temporally and spatially, in order to get at the essential features of competition for space. It turns out that pure competition for space implies by itself a rich community structure, not at all unlike what one observes in a good many natural communities. In this section I shall indicate how environmental variability can be expected to affect that basic picture. The study of environmental variability is an exciting and rapidly developing field; witness Chapter 14 by Chesson. I am merely going to give here my impressions of how some of this work relates to spatial competition.

Founder Control

Thus far I have assumed that dispersal rates are low enough that the local character of competition is maintained. Within this context environmental variability is not likely to have much effect on a founder-controlled community, as long as the contingent nature of pairwise interactions is maintained. If fluctuations are severe enough, they can cause species richness to decline by preventing occupation at any one time of all available space.

In the "complete mixing" domain of high dispersal rates, the local character of competition will break down and the ideas of spatially homogeneous, global competition will apply: Coexistence in a constant environment is, with contingent pairwise interactions, impossible, and all species but one must go extinct. Hubbell's and Foster's community drift model (Chapter 19) and Chesson's lottery model (Chapter 14) are two similar models that, while not expressing dispersal explicitly, look much as if they describe this situation.

Hubbell's and Foster's model incorporates

demographic stochasticity, making it particularly well suited to the study of extinction. Hubbell and Foster find that extinction occurs on a long, indeed evolutionary, time scale. Chesson's model does not take demographic stochasticity into account, but it does incorporate environmental variability. Chesson finds that this variability can enable an arbitrarily large number of species to coexist forever in a founder-controlled fashion.

One way or the other, even with complete mixing, founder-controlled coexistence of an arbitrarily large number of species still seems a viable paradigm.

Dominance Control

Consider first dispersal rates low enough to preserve the local character of competition. In this context environmental variability can enhance the diversity of dominance-controlled communities. If temporal variability is such that different species are dominant at different times, and if the species involved are sufficiently long-lived, then dominance will be averaged out in the long term and diversity will increase—an extreme form of Chesson's storage effect. Even without the storage effect, if environmental variability is associated with an overall deleterious effect on population growth that affects dominant species at least as much as it affects subdominant species, then diversity will increase (Yodzis 1978, secs. IV.3.1, IV.3.2).

Spatial variability can also enhance diversity, if it has the effect of enabling different species to dominate at different locations within a community (Chesson's between-patch variability [Chapter 14]). Grubb (1977) discusses this idea in considerable depth for plant communities. Tilman (1982) elaborates this kind of coexistence in the special case where local competition is purely exploitative.

Consider finally the complete mixing domain of high dispersal rates. Functional dominance could easily be added to the community drift model and the lottery model. It seems obvious that one would get communities whose structure is dominance controlled in exactly the same sense as the dominance-controlled communities discussed elsewhere in this paper.

Indeed, it may prove difficult to distinguish which of the three versions of dominance control (low dispersal, high dispersal with drift, high dispersal with stochastic boundedness) we are witnessing in any given natural system. What I am suggesting is that the basic paradigm of dominance control, like that of founder control, is very robust.

THREE TYPES OF CONTROL OF STRUCTURE BY COMPETITION

I suggest that there are basically three ways in which competition can control community structure.

1. *Niche control.* When there is pure consumptive competition, community structure springs from niche differentiation.
2. *Dominance control.* When there is pure spatial competition, with some species capable of functional dominance, community structure springs from dominance relationships.
3. *Founder control.* When there is pure spatial competition, with no clear functional dominants, community structure springs from the colonization process.

Notice that the three categories are arranged in order of increasing competition strength. One might add a fourth category "no competition," which would, in terms of community structure, be difficult to distinguish from the "strongest" competition, founder control!

I believe that most natural competitive communities fall rather clearly into one of these three categories, but there is certainly a degree of blurring. Thus, founder control can be viewed as a limiting case of dominance control, and communities that are close to, but not quite, founder controlled might more appropriately be thought of as founder controlled than dominance controlled. On the other hand, colonization events will often play some role in the structure of dominance-controlled communities. Consumptive competition sometimes occurs together with some form of interference, which may produce a degree of dominance (and spatial or temporal segregation). And niche differentiation will sometimes enable a degree of local coexistence among competitors for space (for example, vertical stratification in a forest).

These cases of blurring may indicate that a threefold classification is too simplistic, and a finer classification may be needed. However, a *less* finely differentiated view is certainly too simplistic.

Some communities, rather than blurring my distinctions, seem to combine neatly two or more categories. Grubb (Chapter 12) provides a nice example. In his chalk grasslands the perennials, viewed as a community in themselves, are dominance controlled. This community is "perceived" by the annuals as a matrix of openings, which are occupied in a founder-controlled fashion. Each year the matrix (which itself changes from year to year as the perennial community develops) is occupied by a fresh founder-controlled community of annuals. Similar processes operate on other time scales for other short-lived plants.

This sort of combination may well be quite common for plants. The Mojave Desert looks very much like just such a combination of dominance-controlled, matrix-forming perennials and founder-controlled annuals. I do not know of any examples of combined control for epifauna, but it would be surprising if there were none.

Finally, a word should be said about spatial pattern in species distributions in the three cases. In a homogeneous environment after a long enough time, niche control and founder control should yield random distributions of all species, while under dominance control one should see some highly overdispersed distributions. In practice, though, the effect of dominance on spatial pattern is much confounded with other factors. For example, environmental patchiness can produce patchy species distributions in all three kinds of structural control. And, on realistic time scales differing dispersal ranges among different species can affect spatial pattern strongly. Kershaw (1973) provides a stimulating review of spatial pattern in plant communities.

MORTALITY AND COMPETITIVE COMMUNITY STRUCTURE: A PLURALISTIC REPLACEMENT FOR THE INTERMEDIATE DISTURBANCE HYPOTHESIS

Mortality factors such as predation, grazing, and physical disturbance can have a profound effect on community structure, and there is a rich empirical tradition of studying the influence of some single mortality factor on communities of sessile organisms.

For instance, Darwin (1859, chap. 3) allowed a lawn that had long been mowed to grow freely and found that "out of twenty species growing on a little plot of mown turf (three feet by four) nine species perished, from the other species being allowed to grow up freely." In another classic study Paine (1966) found that removal of the top predator from an intertidal community resulted in a decrease in the number of major space-utilizing species in the community from 15 to 8 (see also Paine 1974). Chapters 31 and 32 analyze this problem for sessile marine plants and animals.

Intermediate levels of mortality have often been associated with an increase in species richness. This has lead to the enunciation by a number of authors of an "intermediate disturbance hypothesis" and to a considerable literature of observational tests and theoretical "derivations" of the hypothesis over a period of at least 20 years.

Some studies have contradicted the hypothesis (Table 29.1). Despite this, the intermediate disturbance hypothesis lives on in the pantheon of current ecology, with each deviation from it treated as an isolated special case.

There is a better procedure, and I believe that ecology will not come of age until it embraces this procedure. It is to reject hypotheses that have been falsified, to replace them with other, more finely differentiated and pluralistic hypotheses (which usually will include the "rejected" hypotheses as particular, clearly defined cases), and to test the new hypotheses. In other words, rather than seek grand generalizations, we should try to articulate the precise conditions under which various different outcomes are to be expected.

Partly to illustrate this way of doing ecology and partly because it highlights yet another essential difference between competition for space and consumptive competition, I am going to summarize here the literature on mortality and species richness of competitive communities. The conclusions, summarized in Table 29.1, constitute part of a pluralistic hypothesis to replace the oversimplified intermediate disturbance hypothesis.

Table 29.1 EFFECT ON SPECIES RICHNESS, *S*, OF INCREASING THE INTENSITY OF A PARTICULAR MORTALITY FACTOR (AT INTERMEDIATE LEVELS OF MORTALITY)

Community control	Selectivity (species preference)	Effect on *S*	Theory	Observation
Niche	None	Decreases	1, 37	2, 27
	For dominant	Increases	6, 34, 36	
	For subdominant			
	Switching	Increases	28, 34	
Dominance	None	Increases	4, 14, 30, 38	3, 5, 7–11, 15, 19–24, 26, 31–33
	For dominant	Increases	38	13, 18, 25
	For subdominant	Decreases	38	12, 13, 18
	Switching			
Founder	None	Decreases	39	16, 17, 26, 29, 35
	Switching			

Numbers in the table correspond to the following references: 1 = Abrams 1977; 2 = Addicott 1974; 3 = Ayling 1981; 4 = Caswell 1978; 5 = Connell 1970; 6 = Cramer and May 1972; 7 = Darwin 1859; 8 = Day 1972; 9 = Dayton 1971; 10 = Dayton 1973b; 11 = Dickman and Gochnauer 1978; 12 = Glynn 1976; 13 = Harper 1969; 14 = Hastings 1980; 15 = Heinselman 1973; 16 = Hunter 1980; 17 = Hunter and Russell-Hunter 1983; 18 = Lubchenco 1978; 19 = Loucks 1970; 20 = Menge 1976; 21 = Osman 1977; 22 = Paine 1966; 23 = Paine 1971; 24 = Paine 1974; 25 = Porter 1972; 26 = Porter 1974; 27 = Risch and Carroll 1982; 28 = Roughgarden and Feldman 1975; 29 = Sammarco 1982; 30 = Slatkin 1974; 31 = Slobodkin 1964; 32 = Taylor 1973; 33 = Tomkins and Grant 1977; 34 = Vance 1978a; 35 = Waser and Price 1981; 36 = Yodzis 1976b; 37 = Yodzis 1977b; 38 = Yodzis 1978; 39 = Yodzis unpublished.

At least two distinctions have to be drawn in order to arrive at a satisfactory understanding. First, one needs to distinguish among niche control, dominance control, and founder control of competitive community structure. Second, species selectivity of the mortality factor can play an essential role in determining the outcome. I shall distinguish four forms of selectivity: no preference, preference for one or more functionally dominant species, preference for one or more functional subdominants, and switching (a disproportionate preference on the part of a predator for the most abundant prey species at any time).

Again, I do not claim this is the ultimate scheme of distinctions, but rather the minimally differentiated scheme that we need at our present level of understanding. In particular, some interesting ideas are emerging about the influence of spatiotemporal relationships (involving patchiness of the mortality) on the outcome (Vance 1979, Hanski 1981, Paine and Levin 1981, Miller 1982), but right now it is just too difficult to relate these ideas to the existing empirical literature.

I think that eventually it will be necessary to distinguish as well whether or not there is feedback from the state of the community to the intensity of the mortality factor being studied. For instance, if a predator on the community is limited by this food source, the dynamics will not be the same as when the predator is limited by some other factor. In most instances of physical disturbance there will be little or no feedback. Different kinds of feedback (for instance, different predator functional responses) will also eventually have to be distinguished.

Table 29.1 summarizes a fairly thorough survey of the literature, carried out sporadically over a period of years. In each case the "effect on *S*" (*S* is species richness, the total number of species in the community) is the one predicted by the cited theoretical papers. There was some ambiguity in the assignment of observational papers to these categories, and someone else might well have made other assignments.

A number of published studies were not included in the table for various reasons. For instance, if the mortality rate is high enough, there will be a negative effect on species richness, relative to what I am calling intermediate levels of mortality, no matter what the nature of the competitive community or of the mortality. This seemed to be the explanation for decreases in *S* observed by Paine and Vadas (1969), Dayton

(1975b), Nash (1975), and Duggins (1980). On the other hand, absence of any effect on species richness in studies by Mook (1981), Hoagland (1983), and Morin (1983a) could be due to relatively low mortality rates.

Virnstein (1977) maintained that there was no competition in his community of benthic infauna (he observed an increase in diversity under predation). The nature of the competitive community in Abele's (1976) study of decapod crustaceans inhabiting reef coral heads is not clear to me. Other studies (Hall, Cooper, and Werner 1970; Grigg and Maragos 1974; Nicotri 1977; McCauley and Briand 1979; Lynch 1979) were not included in the table because species selectivity of the mortality did not seem to fit any of my four categories. While there is a rich literature about grazing by rabbits on the English chalk grasslands mentioned above as examples of a neat combination of dominance and founder control (e.g., Tansley and Adamson 1925; Hope-Simpson 1940; Watt 1957, 1960), the rabbits affect both the dominance-controlled perennials and the founder-controlled annuals (with opposite results, Table 29.1). Thus, one gets all kinds of effects, which are extremely difficult to unravel. However, the study of Waser and Price (1981) on grazing by cattle in the Sonoran Desert is included in the table because this grazing seems to affect primarily the annuals.

I have not found any clear conflict between theory and observation to the best of my ability to interpret the observational papers in terms of these categories. But practically all of the observations involve competition for space. There is enormous scope for empirical studies to fill in empty or sparsely populated boxes in the table!

CONCLUSIONS AND SUGGESTIONS FOR FUTURE RESEARCH

1. Pure competition for space implies a rich community structure, without any need for niche differentiation or environmental heterogeneity. It would be interesting to see more empirical studies that adopt this point of view and try to measure whole community dominance relations, especially for plants, if only in order to explicate the shortcomings of this viewpoint.

2. The theory of competition for space should be broadened to include local niche differentiation and environmental heterogeneity. Both of these considerations will augment species diversity. But pure competition for space already allows a great deal of diversity. Hence, the vital, and exceedingly difficult, question is: What are the relative contributions to diversity of these three factors in any given case?

3. Table 29.1 replaces the intermediate disturbance hypothesis with a more finely differentiated complex of hypotheses for the influence of mortality on competitive community diversity. There is a need for empirical studies to test some of the contrasts predicted in the table. Particularly desirable are studies that involve pure consumptive competition.

4. The threefold classification of competitive communities proposed here needs criticism. Are these three categories good ones in the sense that most natural communities fit one category pretty well or at least neatly combine categories? Which are the communities that do not fit tidily into this scheme?

ACKNOWLEDGMENTS

In developing these ideas, I have benefited from discussions with many field biologists, too numerous to name here, who generously shared with me their experiences and impressions of particular systems. I would like especially to single out Hans Burla, who has shown me as much about the human spirit as he has about biology. Carl Biehl, Doug Larson, Denis Lynn, and Sandy Middleton were helpful with particular points that arose in the preparation of this paper. I am grateful as well to Peter Chesson, Peter Grubb, Jon Roughgarden, and, especially, Ted Case for critical remarks on the first draft.

chapter 30

A Comparison of Food-Limited and Space-Limited Animal Competition Communities

Jonathan Roughgarden

INTRODUCTION

During the 1960s and 1970s much of the attention of community ecology was focused on competition, yet this focus was on two kinds of systems with apparently different properties, as discussed in the preceding chapter (Chapter 29). Studies of terrestrial vertebrate communities emphasized the exploitation of limiting food supplies as the mechanism of competition and resource partitioning as the circumstance permitting coexistence between competing species. Studies of sessile marine invertebrate communities living on hard substrate emphasized overgrowth and physical contact as the mechanism of competition and mortality that removed organisms before extensive contact could occur as the circumstance permitting coexistence. The community ecology of other systems was frequently viewed by analogy to these systems. In the literature one sees investigators approaching new and previously unstudied systems with a view either toward finding the ways that resources are partitioned or toward finding the sources of mortality that reopen new space for the organisms to use. In particular, both these perspectives appear in the study of coral reef fish.

This chapter presents a review of the ecology of a food-limited community as exemplified by *Anolis* lizard populations of the eastern Caribbean and of the ecology of a space-limited community as exemplified by the barnacle-dominated community of the high rocky intertidal zone in central California. (Chapter 10 demonstrates that Galápagos finches constitute another food-limited community, while Chapters 31 and 32 discuss space-limited communities of sessile marine invertebrates and plants.) The chapter then analyzes coral reef fish communities and concludes that this system shares features in common with both terrestrial food-limited systems and marine space-limited systems, so that both pictures seem necessary to understand the ecology of coral reef fishes. In order to provide a framework for the possible synthesis of diverse results from many competition systems, I offer a classification of such systems based on two criteria: population structure and whether resources can be partitioned.

A review of findings from case studies of

food-limited and space-limited systems seems timely. Although there has been much research during the last 15 years on Caribbean anoles and on the North American rocky intertidal zone, the theoretical premises of the research, the theoretical predictions from models about these systems, and the actual facts about these systems are often not appreciated. Today, the generality of competition as a major causal factor in determining the structure of natural communities is not clear, nor is it known to what extent those communities in which competition is important are comparable to one another. Nonetheless, a few points in common among seemingly different types of competition communities are now apparent.

FOOD-LIMITED COMMUNITIES

Studies of food-limited communities tend to refer, at least implicitly, to mathematical models for how competition may produce pattern in a community and to use observations of pattern as indicators of the processes that may have produced the patterns. Hence I will concentrate on what is known about the suitability of competition models for food-limited competition communities, on what those models predict, and on whether those predictions are true.

Theoretical Premises: Population Dynamics

Competition models, even if fully validated, license an inference in one direction only. The models have the logical form "if competition and A, then P," where A represents some additional premises and P represents some predictions, usually pertaining to whether coexistence between certain species is possible and, if so, to the degree to which they use the same resources. Current theory does not license the converse inference, that is, inferences of the form "if P, then competition." No competition model has yet predicted a pattern of species coexistence that is logically *diagnostic* of competition.

Nonetheless, patterns that match those predicted to result from competition offer clues that competition may be involved. This is analogous to observing a warning-colored caterpillar; there is a fair chance it is unpalatable, but a feeding experiment must be done to know for sure. Similarly, patterns of species coexistence may make a convincing circumstantial case for competition, but if it is important to know for certain, then the case must be strengthened with experimental studies that perturb the populations or their resource base.

The major premises of models for food-limited competition are:

1. Food is the main limiting resource.
2. The mechanism of competition is mainly exploitative.
3. Food is partitionable.
4. The strength of competition increases as the degree of partitioning decreases.

The second premise may seem surprising. Most theoretical models of food-limited competition have assumed exploitative rather than interference competition. Competition models have been formulated that represent mechanisms other than, or in addition to, exploitative competition. They have not been solved in general for more than two species and typically will not be (Smale 1976). What is sometimes referred to as "community theory" or, more modestly, "competition theory" or "niche theory" is primarily a theory about exploitative competition. Nonetheless, some important special cases of interference competition have been examined (Case and Gilpin 1974, Case and Casten 1979). In particular, if the strength of the interference roughly matches the strength of the exploitative component of competition, then the results of exploitation-only theory and mixed exploitation-interference theory agree.

Additional premises, that can be relaxed to some degree, are:

5. The community dynamics are modeled by the Lotka-Volterra competition equations (that is, quadratic terms with fixed coefficients).
6. The partitioning is described with one or more resource axes.
7. The overlap in resource use between competitors leads to a matrix of competition coefficients that is positive definite. (The competition matrix is the matrix of competition coefficients, each element of which

is usually denoted as $\alpha_{i,j}$. A positive definite matrix is a matrix whose eigenvalues are all positive, as is automatically true of competition matrixes that are symmetrical.)

The seventh premise is a refinement of the second premise concerning exploitative competition. The precise reason why the Lotka-Volterra equations have been analyzed for more than two species in niche theory is that the matrix of competition coefficients is assumed to be symmetrical or, if not symmetrical, transformable to a symmetrical matrix. The origin of this competitive symmetry lies in the concept of overlap, which is inherently symmetrical. If the strength of competition is proportional to overlap, then the competition matrix, so to speak, "acquires" a symmetry as well. This mathematical property is the basis of the main theorems in niche theory for N competing species ($N > 2$) that guarantee the global stability of positive equilibria (see review in Roughgarden 1979, chap. 24). Without this mathematical property, even in the simple Lotka-Volterra equations for three or more species, the system may not approach a steady state and the species may permanently oscillate instead. A mathematical example of a limit cycle in the three-species Lotka-Volterra equations was offered by May and Leonard (1975), based on a very asymmetrical scheme of intransitive competition.

Moreover, the context of the discussion, including the use of Lotka-Volterra models further requires that:

8. The populations are primarily closed in the sense that the production of most recruits is from within the boundaries of the population. Further, the number of recruits produced depends on the amount of food relative to the number of animals in the region where the populations occur.

9. A complete set of interacting populations has been identified, and the principal interaction among them is competition. The interaction with other members of the ecosystem is weak enough to be treated as only a perturbation to a picture that is primarily a competitive interaction among the main actors.

10. Miscellaneous additional requirements include that the environment changes slowly relative to the response time of the populations to those changes and that the populations be continually present in the habitat.

In the future additions to the theory may take account of seasonality, extensive migration among habitats, and time lags. Similarly, both age structure and cross-predation, where the adults of each species prey upon the juveniles of all species, have yet to receive their needed theoretical attention. In general, however, competition theory is not extremely fragile as is, for example, the theory of predator-prey interactions based on the structurally unstable Volterra equations. Nonetheless, the predictions of the theory for food-limited competition need adjustment when other factors are known to be quantitatively significant.

Population Dynamics Premises and *Anolis*

Most of the rather extensive set of premises for the possible applicability of theory for food-limited communities presently appear to be met, more or less, for *Anolis* lizard populations of the Caribbean.

Food as limiting resource Four lines of evidence suggest that food is limiting. (1) Food addition leads to an increase in body growth rates under natural conditions (Licht 1974, Stamps 1977). (2) The abundance of *A. gingivinus* on St. Maarten shows a perfect rank-order correlation with the number of insects caught in tanglefoot traps, for sites where its competitor, *A. wattsi,* is absent (Roughgarden et al. 1983a). (3) Addition of lizards to experimental enclosures constructed in natural habitat leads to a lowering of body growth rates; both interspecific and intraspecific density effects are detected (Pacala and Roughgarden 1982, 1985). (4) Finally, removal of lizards from such enclosures leads to a doubling of the number of insects trapped on the forest floor and to a 10- to 30-fold increase in the number of spiders in the forest understory (Pacala and Roughgarden 1984).

Space not limiting Surprisingly, since adult anoles are territorial and species assume characteristic positions within the vegetation, space is evidently not limiting. On St. Eustatius the two anole species perch at different heights in the vegetation. Yet, when anoles that perch high in the vegetation were experimentally restricted to the low perch heights used by the other species on the island, there was little enhancement of the competitive effect between the species (Rummel and Roughgarden 1985a). Moreover, on St. Maarten, where there was preliminary evidence of interspecific territoriality, studies revealed that the interspecific territoriality was confined to interactions between males of *A. wattsi* and females of *A. gingivinus,* which have the same body size (Bohlen 1983). In general it seems that the use of space is derivative upon the use of food. Large lizards typically perch higher than small lizards as though to acquire a vantage from which to search for the larger prey they characteristically use, and interspecific territoriality is confined to forms of the same body size because such forms tend to consume the same size of prey.

Exploitation and other mechanisms of competition The evidence that food is limiting points to an exploitative mechanism for competition. Recall especially the increase in arthropod abundance when lizards are experimentally removed. However, as noted later, adults prey on juveniles of all species, and this leads to both interspecific predation and cannibalism, similar to what is frequently observed with fish (Chapter 21).

Partitioning of food by size Food is partitioned with respect to prey size; the relation between lizard size and prey size has been established both interspecifically and intraspecifically (Schoener and Gorman 1968, Roughgarden 1974a) and confirmed repeatedly over the last 15 years. Moreover, this partitioning in the eastern Caribbean is a direct reflection of the choice of prey taken from the same place in the forest. Both anoles of St. Eustatius forage primarily on the forest floor, even though the larger species perches above the smaller species (Adolph and Roughgarden 1983). The difference in prey size used by these species is not an artifact of foraging in different microhabitats.

Competition and its strength A series of short-term field experiments over the last five years with the anoles of St. Maarten (where maximum adult body length of the two species differs by 15 mm) and St. Eustatius (where maximum adult body length of the two species differs by 30 mm) has established that there is strong, reciprocal, present-day competition between species whose body sizes are similar, and weak, reciprocal competition between species whose body sizes are greatly different (Pacala and Roughgarden 1982, 1985; Rummel and Roughgarden 1985a, Roughgarden et al. 1983b). The experiments demonstrating competition revealed effects on survivorship, fecundity, rate of body growth, prey size, and prey volume in stomach contents. The experiments also suggested that there is asymmetry in the competition such that any lizard has a higher competitive effect (alpha) against a smaller lizard than vice versa (see also Dunham 1980 for better evidence of asymmetry in iguanid lizards). Moreover, circumstantial evidence for competition and for the dependence of its strength on similarity in body size has existed since 1977 (Roughgarden et al. 1983a, 1984b). On St. Maarten the smaller species is found only in the central hills, while the larger species is found in all habitats, including the central hills. The field experiments confirmed the earlier evidence of competition based on natural experiments, in which comparison of locations that differed in the presence and absence of a competitor revealed shifts in perch height and species abundance. Now there is no doubt that interspecific competition occurs between anoles or that its strength depends on the similarity in body size of the species.

More specific premises are as follows.

Lotka-Volterra model The extent to which the Lotka-Volterra equations faithfully represent the interspecific competition between anoles is unknown. Studies are planned with experimental enclosures in natural habitat in the eastern Caribbean that will examine the suitability of various models for interspecific competition.

Resource axes The partitioning of food is described by the resource axis of prey size. This is the only axis for the anole communities of the northeastern Caribbean, and possibly for the eastern Caribbean generally. In the Greater Antilles, where the anoles do take food from different places in the vegetation, the perch positions are probably not reducible to the prey-size axis, and more axes seem to be operating.

Positive definite competition matrix For two-species communities the matrix of competition coefficients is automatically positive definite. The competition matrix for communities of three or more species is also positive definite if the competition coefficients are correctly predicted by the various multiplicative overlap indexes. Use of such overlap indexes has yet to be validated, however.

The premises pertaining to the context of competition theory for the anoles also seem to be satisfied, as follows.

Closed population The lizard populations on an island are effectively closed populations. However, spatial variation in habitat throughout an island, from montane forest to rainshadow desert, suggests that an island population should often not be modeled as a single homogeneous population, but as a population with spatially varying parameters.

Major actors The anoles on islands of the eastern Caribbean are the major diurnal ground-feeding insectivorous vertebrates. Birds, primarily the pearly-eyed thrasher, *Margarops fuscatus,* and the sparrow hawk, *Falco sparverius,* are predators on anoles; birds are not competitors because there are no insectivorous ground-feeding birds (Adolph and Roughgarden 1983). Of course, other species, including a snake, teiid lizard, and iguana, geckoes, arthropod predators, and intestinal parasites all, may interact with anoles; as experience accumulates a possibly important role for these interactions may be discovered. At present the evidence suggests that it is appropriate to focus on the competitive interaction between the anoles as a major factor in their ecology.

Slow environmental change relative to population dynamics speed Seasonality is a fact in the eastern Caribbean, but its magnitude and timing vary among sites. The most xeric sites, especially locations in the rainshadow of mountains, have the largest seasonal variation (Stamps 1976, Rose 1982). Seasonality in reproductive condition has long been known (Licht and Gorman 1970).

Table 30.1 shows census records for two sites on St. Maarten from July 1977 to August 1984. The abundances have been fairly constant with a tendency for the within-year variation to exceed the between-year variation; the summer abundances appear lower than the winter abundances. Most reproduction occurs in the winter, the period following the rainy season, while the abundance evidently drops by as much as a factor of two during the summer, which follows the dry season.

The pattern of seasonality reveals what I feel are the main technical questions that are currently unsolved for *Anolis* population dynamics, and further studies are in progress. The conditions for egg survivorship, the selectivity of females in choosing egg-laying sites, and the possibility that egg sites are limiting are unexplored. The mechanism for the seasonal decline in abundance is unknown. It possibly represents starvation, but a better hunch is that the decline is caused by predation, both by birds and by lizards themselves. If so, competition may have a strong interference component that, unlike the case for marine space-competition systems, will be reciprocal and not strongly hierarchical.

A possible importance of cross-predation in *Anolis* would find a parallel in observations showing that adults of fish (Chapter 21; Werner, Gilliam, Hall and Mittlebach 1983a; Sissenwine 1984) and marine soft-sediment invertebrates (Wooden 1976) feed on larvae of their own and of other species.

More generally, a better mechanistic understanding of the coefficients that appear in the Lotka-Volterra competition equations, or in whatever population dynamics models are eventually used, is needed. This understanding will probably come from developing a tailor-made model for the dynamics of *Anolis* and then com-

Table 30.1 ABUNDANCE (ESTIMATED NUMBER PER 100 M^2) OF *ANOLIS GINGIVINUS* AND *A. WATTSI* AT TWO SITES ON ST. MAARTEN, NETHERLANDS ANTILLES

	Boundary		Pic du Paradis	
	A. ging.	*A. wattsi*	*A. ging.*	*A. wattsi*
July 1977	50.1 (2.2)	0	8.4 (2.6)	10.5 (0.7)
July 1978	65.0 (3.0)	0	14.2 (5.5)	24.9 (3.3)
July 1979	64.8 (2.8)	0	7.6 (6.2)	21.2 (2.2)
March 1980	129.8 (5.5)	0	14.6 (3.1)	43.9 (5.2)
November 1980	95.4 (3.0)	4.4 (1.8)	39.8 (4.7)	56.8 (2.7)
March 1981	72.1 (3.3)	2.0 (1.0)	13.2 (2.9)	39.4 (2.5)
October 1981	122.3 (6.5)	1.3 (0.7)	15.7 (4.8)	54.3 (6.1)
January 1983	111.7 (5.0)	2.0 (0)	18.6 (2.7)	44.0 (2.2)
June 1983	54.4 (1.9)	0	14.4 (1.8)	30.6 (3.5)
March 1984	114.7 (3.5)	0.5 (0)	14.6 (1.5)	41.8 (1.2)
August 1984	49.3 (2.3)	0	6.4 (1.4)	33.0 (1.3)

Figure in parentheses is the standard error.
Census methods are detailed in Heckel and Roughgarden (1979).

paring that model to the Lotka-Volterra equations, in order to determine an equivalence among the parameters for some ranges of population sizes.

Thus, most of the premises for the use of population dynamics competition theory appear to be met, more or less, in anoles of the eastern Caribbean, and presumably to some extent in other *Anolis* communities as well. The picture of the basic ecology of *Anolis* is gradually filling in. It is still incomplete, and my acceptance of niche theory as a model for this system is still tentative.

Theoretical Premises: Evolution

If, in addition to affecting population dynamics, competition for food is hypothesized to affect the evolution of the competing species, evidence on still more premises is needed.

First, there must be evidence that the populations have evolved together. Geological data should demonstrate that the habitat has existed for, say, a few thousand generations. Systematic or paleontological data should demonstrate that the populations have been together in the habitat together during that time and, while there, have differentiated, at least biochemically, from their ancestral stock.

Second, the initial condition for that evolution should be specified. In particular, if two species arc presumed to have been similar to one another and to have evolved differences in response to competition from each other, then the grounds for their initial coexistence as similar species should be explicated. Did both species simultaneously colonize newly formed habitat, as in the recolonization of the volcano Krakatoa after its eruption or the colonization of the Limfjord by marine mud snails when the land separating salt water from fresh water was destroyed in a storm (Fenchel 1975)? Or, as Diamond (Chapter 6) demonstrates with closely related pairs of bird species from New Guinea, did the species first exist primarily in allopatry and then develop differences as a result of evolution in a zone of overlap that became progressively wider as the species differences increased?

The specification of the initial condition in this way is needed because, if there is competition for food whose strength relates to similarity in body dimensions, then it is problematic to postulate that two initially similar species could coexist to begin with. If a scenario cannot be justified that explains why the initial condition consists of two similar species, then differences between the competing species are more likely the result of the invasion by dissimilar species than the result of divergent evolution by initially similar species.

Evolutionary Premises and *Anolis*

Evolutionary change At least some of the anoles of the eastern Caribbean have coevolved as competitors with one another. There are 23 islands in the eastern Caribbean and 32 populations that are taxonomically distinct (Underwood 1959, Etheridge 1960, Lazell 1972, Williams 1976). Time obscures the ancestral identities, whether from Puerto Rico or Central America, of these 32 populations. Thus, these populations have been in the eastern Caribbean theater for a long time, and where there are more than one species on an island, the species ranges overlap broadly so that the populations are in broad to total contact with each other. The islands themselves vary in geological history and topography. Some have recent volcanic formations, while others are limestone platforms that overlay an older volcanic base (Woodring 1954, Tomblin 1975, Pregill 1981). All the banks in the eastern Caribbean have been emergent since the Pleistocene.

Initial condition Though coevolution has surely occurred among some of the anoles sharing an island, species differences need not be a result of coevolution. The anoles of the northeastern Caribbean comprise a group with a characteristic derived karyotype (Gorman 1973) called the *bimaculatus* group. Its affinities lie with Puerto Rican anoles (Gorman and Kim 1976, Gorman et al. 1980a). The islands of the northeastern Caribbean are too small to have allowed speciation (multiplication of species, *sensu* Mayr 1963). Instead, the closest relative of a species on a northeastern Caribbean island having two species is always a population on a nearby island, and not the other species with which it presently co-occurs (Lazell 1972, Gorman 1973, Gorman and Kim 1976, Williams 1976). Thus, the use of competition theory for the biogeography of *Anolis* in this region must be as concerned with invasion as with coevolution.

The anoles of the southeastern Caribbean comprise the *roquet* group. This group does not have close affinity with Puerto Rican anoles or with the *bimaculatus* group that lies between them and Puerto Rico (Yang et al. 1974). The phylogeny of the *roquet* group is unclear, but possibly derived from the neighboring continent some long time ago.

The source faunas, whether Puerto Rican or continental, presently contain a large diversity of body sizes, including forms larger and smaller than any now seen in the Lesser Antilles. The study here concentrates on the fate of forms that enter the Lesser Antilles and not on maintenance of diversity in the source faunas.

The initial condition for the coevolution between anoles on an island in the northeastern Caribbean probably consists of two different-sized species. The difference in sizes is a prerequisite for a second species to invade an island where a resident is already established. The initial differences between the species were acquired in allopatry and do not represent the result of *in situ* coevolutionary divergence. This picture of the initial condition squares with the known phylogeny in the northeastern Caribbean, the case history of anole colonization on Bermuda as discussed later (Wingate 1965), and the following biogeographical facts. Specifically, there is no instance among *Anolis* anywhere in the eastern Caribbean where a recent invader and an older resident of similar morphology and foraging behavior share an island and segregate by habitat with the recent invader living only on the coast and the older resident only in the interior. Nonetheless, the possibility of coevolution producing *in situ* differentiation according to the scenario envisaged by Diamond (Chapter 6) should be considered for the islands of St. Vincent and Grenada in the southeastern Caribbean which may be sufficiently large and topographically complex to have permitted speciation.

The phylogeny of anoles in the central Caribbean, including the Greater Antilles, is also not clear. The original interpretations of Etheridge (1960) and used by Williams (1976) are based solely on osteological characters. These original interpretations have been challenged by others using karyotypic, electrophoretic, and albumin immunological data (Schochat 1976, Gorman et al. 1980b, Wyles and Gorman 1980). In particular, the distinction between the so-called alpha and beta sections of the genus is probably not valid, and the fauna of Puerto Rico is probably polyphyletic, unlike the scheme depicted in E. E. Williams (1972). Nonetheless, the *cristatellus* group in Puerto Rico is a single radiation as E. E. Williams (1972) claimed, and electrophoretic data indicate that it is older than the *bimaculatus* radiation of the northeastern Caribbean (Gorman et al. 1980a).

The anoles of the Bahamas are not taxonomically distinct from populations on the nearby Greater Antilles and are clearly recent colonists.

Thus, the islands of the eastern Caribbean are systems where evolutionary change has clearly occurred. Their biogeography is a result of both this evolutionary change and, especially in the northeastern Caribbean, the ecology of colonization.

Predictions from Food-Limited Competition Theory

Most of competition theory is irrelevant to explaining the actual differences that exist between coexisting species. This unhappy fact results from early focus on the issue of limiting similarity, that is, the limit to the similarity between two or more species consistent with their coexistence. This theory predicts which species cannot coexist and must therefore occur in some regional pattern consistent with local allopatry, like a checkerboard pattern or a pattern of altitudinal replacement. Limiting similarity theory does not address the issue of the degree of similarity of coexisting species, other than to say that they must differ by at least a certain degree (which itself is zero in some circumstances).

The degree of similarity between species that do coexist may be termed the "realized similarity." So far, two hypotheses that could explain the realized similarity have been explored theoretically (Rummel and Roughgarden 1983, 1985b): invasion and coevolution.

Invasion-Structured Communities These communities are assembled by the sequential addition of species to the fauna. The idea is that competition between the residents and the potential invader influences the likelihood that the invader will succeed. The source fauna is hypothesized to contain a spectrum of body sizes that are potential invaders. By considering the sequential addition of those types of invaders that are most likely to succeed, one gradually assembles a fauna. Figs. 30.1 and 30.2 illustrate the step-by-step assembly of a community according to the Lotka-Volterra competition equations. The figures present "trees" that represent all the path-

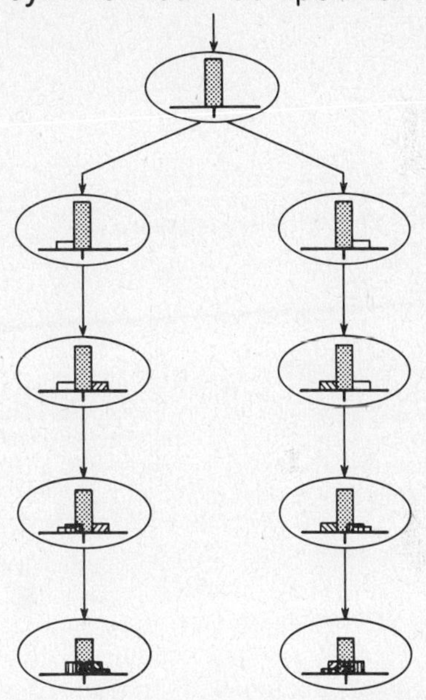

Fig. 30.1 Faunal buildup for an invasion-structured community with symmetrical competition. Each node represents a community. Within each node each vertical bar represents a population. The position of the bar on the horizontal axis indicates the niche position of the population, its height indicates the abundance of the population, and its width indicates the niche width (twice the standard deviation of the population's resource utilization curve). The carrying capacity function is unimodal with its peak at the center of the horizontal axis; its width is indicated by the length of the horizontal axis (π times twice the standard deviation of the carrying capacity function). The diagram illustrates the pathways of faunal buildup as a result of sequential invasion. The island begins with a population whose niche position corresponds with the peak of the carrying capacity function. The island is then invaded either by a smaller form, leading to the left branch, or by a larger form, leading to the right branch. The two branches terminate with the communities illustrated at the bottom of the figure. Notice that the final communities are not symmetrical, although the two final possibilities are mirror images of each other.

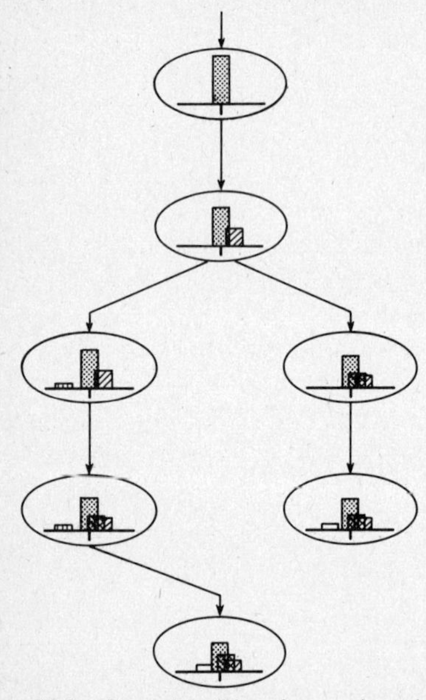

Fig. 30.2 Faunal buildup for an invasion-structured community with asymmetrical competition. Asymmetrical competition means that an individual has a higher competition coefficient against a smaller individual than vice versa. In both pathways the community accumulates large-sized species.

ways by which the fauna can be assembled, starting with one species and ending with the maximum number that can be supported. At each node in the tree the community is represented by a set of bars whose position on the horizontal axis represents the phenotype of the species (its niche position or body size) and whose height represents the population size of the species. There is one bar per species in the community.

Fig. 30.1 shows that even with symmetrical competition it is possible for asymmetrical communities to form. The asymmetrical communities appear as a pair of mirror image possibilities.

A kind of asymmetrical competition can also be studied in the context of these models. Consider the competition between animals of differing body sizes, and, by convention, assume that an advantage lies in larger size. Then this asymmetry can be incorporated into the competition function that specifies the competition coefficient in the Lotka-Volterra equations. This function indicates the degree of competition between two individuals as a function of the difference in their body sizes. If the reciprocal competition coefficients are equal, then the competition function is symmetrical about the vertical axis and is typically modeled as a bell-shaped curve centered on the origin. If an advantage accrues to larger individuals, then the competition function is not symmetrical and is typically modeled as a bell-shaped curve that is shifted upward and to the left (see illustrations in Roughgarden 1979, chap. 24).

Fig. 30.2 shows that asymmetrical competition causes the community to accumulate large species. The niche separation is low among these large species, but the small species are widely spaced from each other and from the large species. Note the high overlap, indicating the presence of strong competition. This strong competition is quite consistent with stable ecological coexistence. Furthermore, although not illustrated, the faunal buildup frequently involves the invading species' replacing one or more residents; the sequence of phenotypes in the community resulting from the progression of invasions and extinctions converges to a specific phenotype and does not result in a cycle of phenotypes.

One may develop alternative models of invasion-structured communities by varying the assumption of who qualifies as an invader and how often invaders arrive. The predictions of Figs. 30.1 and 30.2 come from a model where invaders were presumed to enter one at a time, that is, each invasion was presumed to go to completion before another began. The criterion for the invader's phenotype was that the invader should "see" the largest difference between the carrying capacity and the amount of resources used by the current residents, relative to potential invaders of other phenotypes. The one-at-a-time invasion seems appropriate to the rarity of invasions by anoles among Caribbean islands. Other formulations might be more appropriate to systems where invasions are more frequent or are structured by dispersal behavior.

Coevolution-Structured Communities These communities are assembled by an alternation of invasion with coevolution. A coevolved fauna is invaded, this fauna coevolves, then invasion occurs again, and so forth, until all possible paths of faunal assembly are investigated.

With symmetrical competition the community eventually attains a symmetrical configuration, as Fig. 30.3 illustrates. The buildup to the two-species state from the one-species state (Fig. 30.3, left) illustrates how, in this model, what would normally considered to be character displacement (Brown and Wilson 1956) occurs. In the final two-species state the niche of each species is displaced an equal amount, but in the opposite direction, away from the point occupied by the first species. The usual scenario envisaged as leading to this two-species state (e.g., Connell 1980) is that somehow the community is invaded by a species that is similar to the original resident. The two then evolve away from each other. This scenario is, however, unlikely according to competition theory because it is unlikely for a successful invader to have a phenotype similar to the resident to begin with. Instead, as Fig. 30.3 illustrates, the invader is predicted to enter with a phenotype quite far from the resident. Once established, both species move in the same direction and come to a halt when each is equally displaced from the position originally occupied by the resident.

With symmetrical competition the evolution is always convergent, erasing the asymmetrical niche positions that temporarily exist after an invasion and leading to a single configuration of

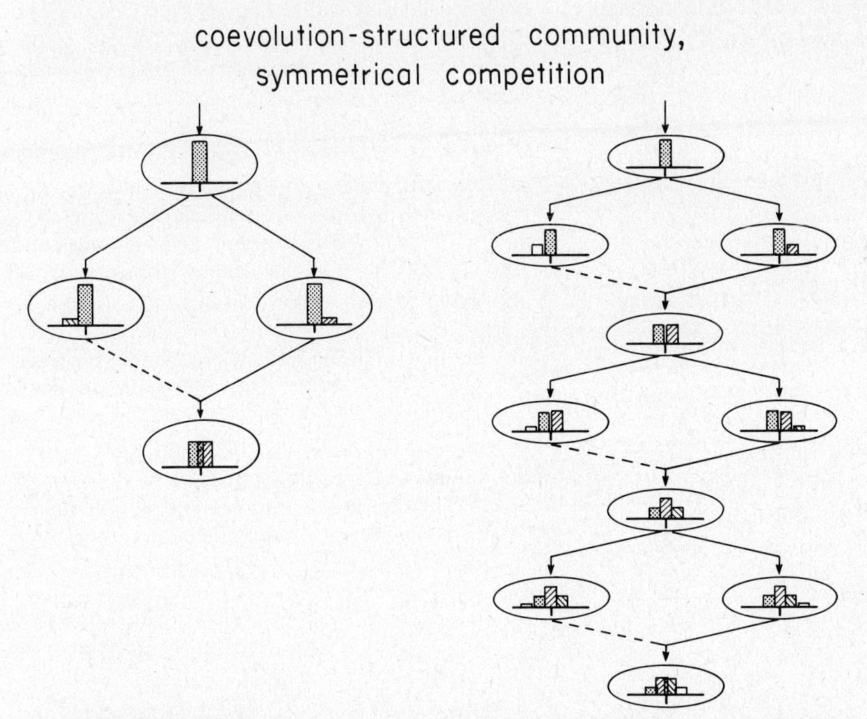

Fig. 30.3 Faunal buildup for a coevolution-structured community with symmetrical competition. The diagram illustrates the pathways of faunal buildup that result from repetitions of an invasion followed by coevolutionary readjustment. In the left panel the island begins with one species. It is invaded either by a smaller form, leading to the left branch, or by a larger form, leading to the right branch. Both possible newly invaded communities converge during coevolution to the community drawn at the bottom. In the right panel the carrying capacity function is wider than in the left panel. (Note that the ellipses surrounding each node in the right panel are narrower than in the left in order to enclose the longer horizontal axis.) The result of all pathways is again a single symmetrical community. It has approximately even spacing among its members.

niches for each level of species diversity (Fig. 30.3, right). The niches in the center are more tightly packed than the niches toward the tails of the carrying capacity function, but this difference in the degree of packing is slight. These symmetrical coevolved model communities seem to be the only instance where competition theory predicts a community with approximately evenly spaced niches—a point relevant to the issue of Hutchinsonian ratios (see next section). (Approximately evenly spaced niches are found rarely for special parameter values with invasion communities, but not consistently as with symmetrical coevolved communities.)

With asymmetrical competition the predictions are quite different from the pictures of character displacement and equally spaced niches that are widely believed to be the principal results of competition. With asymmetry, extinction frequently results during coevolution. In Fig. 30.4 a community with one species builds to a community with two species by the invasion of a species that is larger than the resident. With asymmetry invasion by a smaller species is unlikely. Then the pair shifts together, and the larger species (the invader) "overtakes" the smaller species (original resident), resulting in the extinction of the smaller species. The invader now evolves to assume the position of the original resident.

coevolution-structured community,
asymmetrical competition

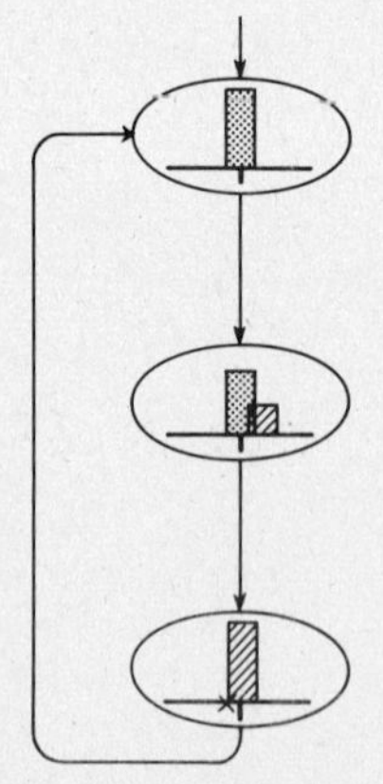

Fig. 30.4 Faunal buildup for a coevolution-structured community with asymmetrical competition. An invasion by a large species is followed by the extinction of the original resident (at the position marked X on the lowest node) during the coevolutionary readjustment phase. The remaining species then converges to the initial condition. This picture appears to apply to the biogeography of *Anolis* lizard populations in the northeastern Caribbean.

The cycle of invasion leading to extinction of the original resident during coevolution with the invader and ending with the remaining resident attaining the phenotype of the original resident is analogous to the "taxon cycle" proposed by Wilson (1961) for the evolutionary biogeography of ants in Melanesia. According to Wilson, a recently established invading species evolutionarily shifts into preferred forest habitat and out of the marginal coastal habitat to which it was originally adapted. This shift causes an extinction of older residents in forest habitat and also vacates marginal coastal habitat that is then once again available to an invader. The extinctions in forest habitat coupled with invasions in coastal habitat combine to keep the overall species number on the island approximately constant.

Extinctions during coevolution are frequently predicted in models of coevolution with asymmetrical competition. For example, one can have cycles between one-species and two-species stages (Fig. 30.4), cycles between two-species and three-species stages, and so on. Clearly, competition theory does *not* predict that evolution among competing species inevitably leads to niche segregation, resource partitioning, and long-term stability.

Finally, the coevolved communities tend to accumulate species that are small or positioned near the peak of the carrying capacity curve. This skew is opposite to that predicted for invasion-structured communities for the same degree of asymmetrical competition. Moreover, the smaller species will tend to be older than the larger species, according to this model.

Hutchinsonian Ratios Hutchinsonian ratios are constant ratios of body weights between adjacent members in a sequence of size-ranked species from a food-limited competition community. These ratios correspond to equally spaced niches where the horizontal axis is taken to be the logarithm of the body size. The existence of a characteristic Hutchinsonian ratio is often sought as evidence of community structure caused by competition (Chapters 5, 6; Schoener 1974a,

Wiens 1982). Yet, according to competition theory, this pattern is produced in a community by competition under the exceptional circumstance of symmetrical coevolution, and even then only approximately so.

Nonetheless, competition theory offers the surprising result that a kind of Hutchinsonian ratio emerges very generally as an "ensemble" property of many competition communities considered together, even though the ratios may be absent as a striking regularity in each of the particular communities of the ensemble. Schoener (1984) discovered statistical evidence of size differences among sympatric bird-eating hawks. For the analysis, Schoener produced a species pool by formally placing all the existing hawk species into an "urn." Then, by drawing species at random from this pool, he produced an "expected" distribution of size differences between species. The expected frequency distribution of size differences monotonically decreases as the difference increases, that is, more similarly sized birds are expected to co-occur than differently sized birds according to random draw. In contrast, the actual distribution of size differences was multimodal, indicating a likelihood of greater size difference between co-occurring forms than would be expected if they were thrown together by random draw. Also, the modes were regularly spaced.

Rummel and Roughgarden (1985b) used the same method to analyze a large set of theoretically constructed competition communities. (This is the set of communities from which the examples of Figs. 30.1–30.4 were selected.) All the species in all the theoretical communities were cast into a common urn and then sampled to provide an expected distribution of niche differences. The actual distribution of niche separations from among all the theoretical communities was also determined. The expected frequency distribution of niche separations monotonically decreases as the separation increases, and the actual distribution is unimodal with most niche separation distances (d/w) lying between 1 and 3. This theoretical result is consistent with Schoener's (1984) finding using the sizes of bird-eating hawks, although in our study the actual distribution tended to be unimodal rather than multimodal as in Schoener's study. Thus, though competition theory does not generally support the existence of regular niche spacing among the species within any particular community, it does suggest that an ensemble of communities will show a statistical regularity in size differences.

Predictions and *Anolis*

The predictions of competition theory for food-limited communities accord with, and appear to explain, the biogeography and evolutionary history of *Anolis* throughout the Caribbean.

Morphology and Coexistence In each of the Greater Antilles the local faunas consist of three to five anoles that coexist in the same habitat. The fauna of Puerto Rico, for example, can be divided into four ecological groups termed "structural niches" (Rand 1964, 1967; Williams 1983): grass anoles, trunk-ground anoles, trunk anoles, and trunk-crown anoles. The criteria for membership in a structural niche rely on functional and morphological characteristics that are independent of geographical distribution. In particular, the body size characteristics of adjacent structural niches markedly differ. The grass anoles are small, the trunk-ground anoles are medium sized, the trunk anoles are small, the trunk-crown anoles are medium sized.

The important observation concerning structural niches is that the Greater Antillean islands of Puerto Rico, Hispaniola, and Cuba each have several species jointly occupying certain of the structural niches. Yet, except at narrow zones (1 km or less) where the ranges of two species overlap, there is no situation where members of the same structural niche coexist. Instead, members of the same structural niche are allopatric, typically replacing one another along altitudinal or other environmental gradients. Thus, throughout the vast area of the Greater Antillean islands, several anoles from different structural niches ubiquitously coexist, while anoles from the same structural niche do not coexist except at narrow zones of overlap at species borders.

The history of introductions of anoles among Caribbean islands reveals that an invader can succeed if it meets residents of a different size and fails if it meets a resident of the same size or structural niche. Wingate (1965) chronicled the spread on anoles on Bermuda. In 1940 a medium-sized anole from Jamaica was introduced to Bermuda for the biological control of fruit flies. During five years it spread throughout the island. Some years later the large anole from Antigua was introduced and became established in woods in the center of the island. When the medium-sized anole from Barbados was introduced, it failed to spread and exists only in small enclaves near the beaches on the north tip of the island.

Other introductions throughout the Caribbean persist as enclaves. For example, in the Dominican Republic *A. porcatus* from Cuba is found in Santo Domingo at the site of a former world trade fair, and *A. cristatellus* from Puerto Rico is found in the sugar refinery town of La Romana. Both these introduced populations have not spread against the native anoles that occupy similar structural niches and that surround these enclaves. Yet anoles are generally excellent colonists if a native analogue is not present. Indeed *A. carolinensis* has spread from Cuba throughout much of the southeastern United States, although the circumstances of its introduction are not known.

Body Size and Evolution The biogeography of the eastern Caribbean anoles appears to substantiate the evolutionary predictions of competition theory with asymmetrical competition (Fig. 30.4, left). Specifically, St. Maarten is an island where the smaller species is competitively excluded from low elevation sites by a larger species. The smaller form is restricted to the central hills in what might be interpreted as a relictual distribution, while the larger species is present in all habitats, including the central hills. The larger species on St. Maarten is itself related to still larger species on nearby islands, inviting the interpretation that it entered St. Maarten as a large anole and evolved a smaller size there and has nearly driven the smaller species to extinction as a result.

Although the smaller form seems eventually destined for extinction, the central hills appear to provide a refuge at this time, as suggested by the stability of its border since 1977. The eventual coup de grace is likely to be either a change in weather making the habitat as xeric as the lowland sites from which the smaller form is presently excluded, or else habitat destruction by people that could also cause a desertification of the habitat. Indeed, *A. wattsi* has already become extinct on Anguilla (on the same bank as St. Maarten) since it was collected there in the 1930s, presumably as a result of habitat destruction (Lazell 1972). Similar habitat destruction on the Antigua bank and the St. Kitts bank, where the interspecific competition is not as strong, has not led to the extinction of the smaller forms, all of which are relatives of the *A. wattsi* on St. Maarten. Thus, a species' susceptibility to extinction as a result of change in climate or habitat destruction appears to depend in *Anolis* on the strength of its interspecific competition with other anoles.

The only island with a single species that has a biogeographically anomalous body size is Marie Galante. Here the size is larger than that of the 15 other species that are solitary residents of islands. This condition may be interpreted as the aftermath on an island where the smaller species has already become extinct, as seems about to occur on St. Maarten. Indeed, if early collection records are to be believed, a smaller species on Marie Galante has become extinct since the turn of the century (Garman 1887).

Perhaps most significantly, the fossil evidence of *Anolis* on the Antigua bank shows that the large anole there was much larger than it now is as recently as 4,000 years ago (Etheridge 1964, G. Pregill personal communication). This precisely accords with the theoretical prediction (Fig. 30.4, left) that an island is most likely to be invaded by a large form which then evolves a smaller body size and causes competitive exclusion of the original resident and also accords with the interpretation of the distribution on St. Maarten discussed above.

The fossil evidence cannot be taken as an unqualified confirmation of the coevolutionary prediction, however, because during the 4,000 years that the evolutionary change occurred other components of the fauna became extinct, and human land use presumably aridified the habitat (Steadman et al. 1984). Certainly European colonization led to wholesale destruction of lowland forests. Yet human land use, though an important consideration, cannot explain the key features of the biogeographical pattern of anoles in the eastern Caribbean. Why, for example, has the larger anole on St. Maarten become so small? Has St. Maarten had a history of more devastation than nearby islands like Antigua, where the larger anole has not become as small as on St. Maarten?

The extensive archeological research in the eastern Caribbean (review in Rouse and Allaire 1978) reveals that St. Maarten has had no more, and probably had less, land use than the islands of the Antigua and St. Kitts banks. Dr. A. Verstees, the leader of the archeological research team from University of Leiden currently working a precolumbian dig on St. Eustatius, writes: "Saladoid Indians have not been found at St. Maarten up to now" (personal communication). Villages of such Indians have been found on Antigua, St. Kitts, Saba, and St. Eustatius. Current evidence suggests that land use by Indians practicing agriculture is more recent than A.D. 1000 on St. Maarten, while it dates to perhaps A.D. 1 on the other nearby islands. Thus, though land use and extinctions of other parts of the fauna are important components of the history of these islands, these considerations do not substitute for the account of evolutionary change predicted by competition theory involving invasion and coevolution.

Paleobiological information tends to strengthen the connection between ecology and evolutionarily biology offered by coevolutionary competition theory. Assertions (Pregill and Olson 1981, Olson 1982) that evolutionary ecologists have not taken historical information into account are clearly incorrect.

The anoles of the Caribbean thus constitute a food-limited competition community with a partitionable resource. Species differences in resource use are the condition for coexistence and for an invader's entry into a community. Evolution may, however, reduce species differences, culminating in an extinction. I now turn to the quite different picture that emerges in a community whose members compete for a nonpartitionable resource, space.

SPACE-LIMITED COMMUNITIES

Studies of space-limited communities, especially of sessile marine invertebrates on hard substrate, tend to rely on induction from the accumulated results of small-scale field experiments. There are few theoretical models for marine systems to guide the synthesis of these experimental results, to aid the design of field studies on a larger spatial scale, and to address evolutionary issues. To strike a new tack, I here begin with simple models for community processes in the intertidal zone, offer evidence that these models are approximately correct, and suggest further directions for theoretical research. This approach leads to modifications of inductive generalizations in marine ecology, including the intermediate disturbance principle, that increase their generality. This approach also leads to a classification of the conditions in which certain community types are found, suggests that marine space-limited communities share properties in common with terrestrial food-limited systems, and furthers the discussion of evolutionary questions in terms appropriate for many marine invertebrates.

Study sites in benthic marine ecology typically are open systems that differ from those studied in terrestrial ecology. A terrestrial population is tacitly viewed as occupying an area containing most of the adults that produce the future recruits. Migration results from propagules that diffuse across its boundaries. In principle, the perimeter of a terrestrial system can be chosen large enough to make the migration rate relative to the reproductive rate within the system arbitrarily small. In contrast, migration into a benthic marine system is not restricted to the boundaries, but can arrive at any point within the system. A benthic marine population can be imagined as an area of substrate immersed in a larval bath, an image quite different from that for a terrestrial population.

A conspicuous feature of benthic marine ecology is the universality with which space on the substrate is the primary limiting resource for populations of sessile organisms. Food is also limiting to some degree. The growth rate of intertidal anemones correlates with the duration of the time for feeding, and that, in turn, with height in the intertidal zone (Sebens 1983). Similarly, depletion of food by neighboring animals can lead to lowered growth rates in bryozoans (Buss 1979b). Nonetheless, there is general consensus that space for attachment or for refuges from predation is the most important limiting resource in benthic marine systems (Chapters 31 and 32; Connell 1961a, Dayton 1971, Menge 1976, Sale 1977, Jackson 1977a, Paine and Levin 1981).

At some sufficiently large scale even marine populations are closed systems. On such a scale a population might usefully be viewed as a collection of space-limited local systems, each coupled to the others by their contributions of larvae to a common larval pool. The larvae in this pool are then redistributed among the local systems according to each system's available vacant space. This view leads to a hierarchical population model, one where local models embodying the processes studied in benthic marine ecology are coupled by processes in the domain of coastal oceanography.

The Open Space-Limited Local Population

Demographic Theory: Premises The demography of an open space-limited population occupying a surface of known area immersed in a larval bath has been modeled by Roughgarden et al. (1985). The processes in the model are settlement onto vacant space, growth of the settled organisms, and mortality of the settled organisms. Both settlement and growth consume vacant space, while mortality releases it.

The main premise of the model is that the total settlement rate into the system is proportional to the amount of unoccupied space (free space) in it. This assumption is supported by the data in Fig. 30.5 for the barnacle *Balanus glandula*. Also, the growth rates and mortality rates may depend on the amount of space available in the system, and if so, these relations must be specified.

It is surprising that the total settlement into a system is proportional to the amount of free space in it. Invertebrate zoologists have discovered that larvae have an interesting behavioral repertoire that leads to substrate specificity, gregariousness, spacing, and sensitivity to microclimate conditions like light, temperature, current velocity, as well as to color and the presence of microflora on the surface (Crisp 1974, Lewis 1978). It is not clear how much such behavioral traits affect settlement *rates*. In both economics and statistical mechanics it is well known that sufficiently large numbers of individuals can be satisfactorily described by simple rules. Fig. 30.5 shows that settlement into a system is a mass-action process. The vacant space in the system must be distributed around existing organisms, however, and not concentrated in a vacant spot of more than, say, 5 cm in diameter. Barnacles are cross-fertilizing hermaphrodites, and landing beyond a certain distance from other adults invites a celibate life.

The theoretical work so far has explored only a density-dependent relation between individual mortality and the amount of free space in the sys-

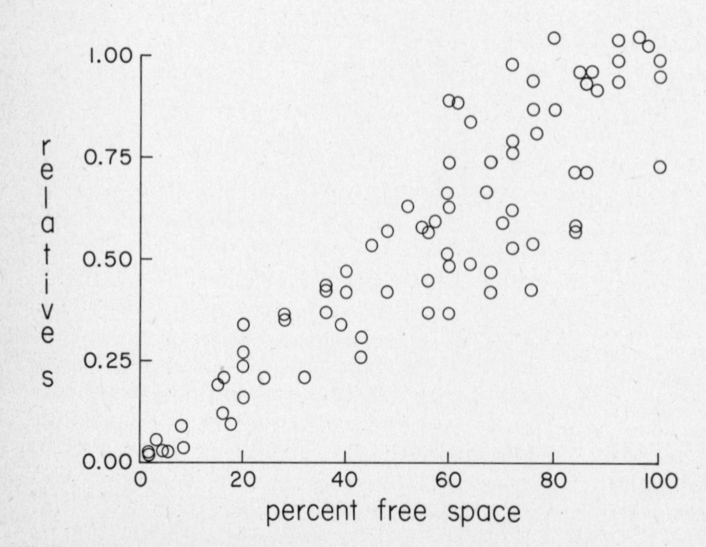

Fig. 30.5 Settlement of *Balanus glandula* larvae as a function of percent of free space in a quadrat. The relationship is approximately a 45° line through the origin, indicating mass-action kinetics. Since settlement is almost exclusively restricted to vacant space, recruitment is space-limited. To obtain these data, quadrats of 36 cm^2 were partitioned into subquadrats. The fraction, in each subquadrat, of the quadrat's total settlement per week was plotted against the percent free space in that subquadrat. (Updated from Roughgarden et al. 1984a).

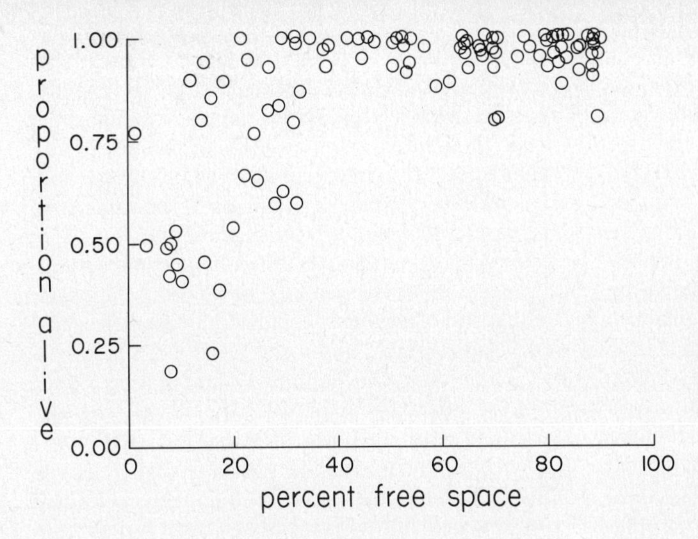

Fig. 30.6 Probability of survival through a period of one week for *Balanus glandula* individual as a function of percent of free space in the quadrat containing that individual. The high mortality that results when vacant space is nearly exhausted represents primarily predation by the starfish *Pisaster ochraceus* and indicates that disturbance is density-dependent. (Updated from Roughgarden et al. 1984a).

tem; density-dependent growth has not been modeled. The density-dependent relation between survivorship and free space has been assumed to be independent of the amount of free space in the system until the free space is nearly exhausted, at which point the survivorship quickly drops. One mechanism that increases mortality when free space is exhausted is ''hummocking,'' that is, barnacles crowding together until they bulge out from the rock surface. Entire hummocks are susceptible to removal by waves (Barnes and Powell 1950, Wethey 1979). Also, as Fig. 30.6 shows crowded barnacles (*Balanus glandula*) are more attractive to predation by starfish (*Pisaster ochraceus*) than barnacles in sparse cover.

Although mechanisms, such as hummocking, for density-dependence in the mortality rate of sessile organisms have been known for over 30 years, their ecological importance is generally overlooked. Since Dayton (1971) mortality has been viewed as an abiotic (density-independent) process generically termed ''disturbance.'' In Dayton's study the mechanism of mortality was frequent damage from the floating logs that litter the coast of Washington State and Oregon as debris from the lumber industry. Clearly, this mechanism of mortality is more abiotic than most. If the mortality is density-dependent, then there is a ''biological targeting'' of the disturbance to spots of high density. In this circumstance it is not correct to conceptualize an intertidal community as the result of abiotic disturbances continually resetting the system to an earlier phase of its dynamics (Paine and Levin 1981). With density-dependent mortality the disturbance becomes part of the system itself (Fig. 30.6).

Although at any one place and time the settlement into a system may be proportional to the vacant space in it, the constant of proportionality (called the settlement parameter, s) varies spatially and temporally both seasonally and stochastically. Variation in the settlement parameter reflects variation in time of exposure of the substrate to the part of the water column where the larvae occur, the physical and chemical character of the substrate's surface, and the effectiveness of factors, like limpets, that cause mortality within the first time interval (typically, one week) after settlement.

Demographic Theory: Predictions There are three main theoretical predictions from the model for the demography of an open, space-limited population.

1. If the settlement parameter s is low, the abundance of animals is directly proportional to s. In this circumstance the system is recruitment-limited. Animals of all sizes and ages intermingle with free space. The population tends to approach a stable age distribution and a dynamic steady-state level of free space resulting from a balance between recruitment and mortality. In practice, the steady state is not one of constant abundance representing a perfect balance between recruitment and mortality. To the contrary, since abundance in the low-settlement limit is directly proportional to the settlement parameter *s*, stochastic fluctuations in *s* have a large impact on abundance, thereby giving the community an air of unpredictability, as lamented in Underwood et al. (1983). Nonetheless, this situation is a stochastic or seasonal equivalent of a steady state in the sense that the fluctuations in abundance are driven by fluctuations in settlement and other vital rates, while the processes involved tend to produce a stable age distribution. Fluctuations in abundance caused by fluctuations in the model's parameters represent a fundamentally different phenomenon from a limit-cycle oscillation caused by the interaction between settlement, growth, and mortality, as discussed below.

2. Regardless of the settlement parameter, if the density-independent component of mortality is high enough relative to the growth rate, then again there is a steady state in which animals of all ages and sizes intermingle with free space. The abundance of animals is, however, not directly proportional to *s* unless *s* is low enough for prediction 1 above to apply.

3. If both the density-independent component of mortality is low enough relative to the growth rate of settled animals *and* the settlement parameter *s* is sufficiently high, then there is no stable steady-state level of free space nor any stable age distribution. Instead, at a local spot there is an oscillation in the amount of cover. At any particular spot, say 100 cm^2 in area, there is a cycle starting with vacant space. This is then quickly filled with recruits. The recruits grow with little mortality and soon all the vacant space is exhausted. But when the vacant space becomes exhausted, additional agents of mortality appear, such as starfish predation or wave damage to barnacle hummocks. Hence, most or all of the animals die simultaneously, leaving the spot nearly vacant again. In this situation animals of all ages and sizes do not intermingle with free space. Instead, because adjacent spots are not synchronized, the animals tend to occur in distinct patches characterized by animals of the same size. In addition, there are gaps or patches of vacant space (Levin and Paine 1974, Paine and Levin 1981). A patch of animals of the same size represents a place where there formerly was a gap; it was then quickly filled by recruits so that it now represents a cohort. The landscape appears as mosaic of cohorts punctuated by discrete gaps of vacant space.

The limit-cycle oscillation of prediction 3 will not, of course, be realized as the exactly regular oscillation of a purely deterministic oscillator. The oscillator is embedded in a stochastic environment, as though a vibrating spring were set out in a hail storm. The realized trajectory would combine the ingredients of the system's oscillatory modes with stochasticity, and time-series analysis is needed to distinguish a realization of prediction 3 from that of prediction 1.

The oscillation of prediction 3 originates in the destabilizing effect of growth. (If the animals do not grow in basal area, then prediction 2 always applies.) Growth in basal area effectively introduces a time lag into the population dynamics because the recruitment depends on the amount of space currently available and not on the amount of space that will be available when the settled animals have grown to their expected area. Hence, the system can "overfill" with more larvae than the space will later support. The additional agents of mortality that enter when space is exhausted tend to stabilize the system and may prevent the oscillations originating from fast growth coupled with high settlement from being expressed.

Application to *Balanus glandula* in Monterey Bay The model above was developed as an aid to understanding the community ecology in the high intertidal zone at Hopkins Marine Station, on the south shore of Monterey Bay in California. The study site consists of granitic rocks whose total area is about 2500 m^2. By inspection one can see that the abundance of *Balanus glandula* varies across the site, with nearly 100% cover of barnacles on rocks with seaward exposure, decreasing to very sparse cover on rocks that lie behind the exposed rocks adjacent to shore.

In one initial hypothesis for this gradation in barnacle abundance rocks adjacent to shore were supposed to be somehow unsuitable for barnacle growth and survival, perhaps as a result of heat stress during spring low tides when the moderating effect of fog is often absent. However, barnacles on the sparsely covered rocks adjacent to shore are approximately as reproductive and attain the same large size as barnacles in other areas. Thus, there is no evidence that heat stress should be considered as a leading factor in determining abundance.

In another initial hypothesis predation was supposed to be responsible for the sparse cover of barnacles on near-shore rocks, much as Connell's (1961b, 1970) suggestion that at Friday Harbor, Washington, differences in barnacle cover between spots were related to accessibility of the spots to predators. Actually, the near-shore rocks at Hopkins are less accessible to starfish than are more exposed rocks, nor are there copious empty tests (from thaid, flat worm, or avian predation) or basal plates (from starfish predation) on near-shore rocks. Nonetheless, variation in the relative magnitude of predation from all sources across the site might be an important factor. In any case, according to the model the importance of predation in affecting abundance should be assessed relative to the settlement and growth rates. Only if the settlement is so high as to saturate free space when it becomes available, should predation be viewed in isolation from these other factors.

Variation in the settlement rate has proved to be the primary cause of variation in the abundance of *Balanus glandula* across the study site at Hopkins (Gaines and Roughgarden 1985). Fig. 30.7 shows that the settlement rate (measured as number of cyprid larvae settling per cm^2 of vacant space per week) is over 20 times higher at seaward rocks (Pete's Rock, Fig. 30.7b) than at near-shore rocks (KLM, Fig. 30.7a). Accordingly, the model predicts that for low settlement the population approaches a stable age distribution and a stochastically sensitive steady-state level of free space; it predicts that for high settlement an oscillation occurs about an average level of free space that is relatively insensitive to the settlement rate.

Fig. 30.7 presents the amount of free space observed at two spots in the Hopkins intertidal zone. At KLM in Fig. 30.7a, a spot with low settlement from a near-shore rock, there is about 75 to 85% free space, and the photographs of quadrats at this site document that animals of all sizes and ages intermingle with this free space. Also, the composition of KLM quadrats is sensitive to stochastic fluctuation in the settlement rate. Notice the drops in the level of free space after a burst of settlement and that the average level of free space each year has directly reflected the intensity of settlement for that year. At Pete's Rock in Fig. 30.7b, a spot with high settlement and accessible to starfish, there is a higher average level of cover (averaged through time), together with large swings in cover. Notice that the swings are not obviously driven by the bursts of settlement. Indeed, the yearly average level of free space is nearly the same in both years even though the settlement was much higher in 1983 than in 1984.

Fig. 30.8 presents the autocorrelation function for the census data from four quadrats, two from KLM and two from Pete's Rock. The autocorrelation is computed for the deviation of the free space from the yearly average. The horizontal lines are the confidence intervals for rejecting the null hypothesis that the fluctuation is a purely random "white noise" fluctuation. The curves show a statistically significant periodicity in the Pete's Rock quadrats, with a period of about 30 weeks. Moreover, inspection of the photographs clearly shows the process of heavy recruitment onto the space left after *Pisaster* predation, followed by rapid growth, followed by the recurrence of starfish predation after the recruits have grown to occupy nearly all the space originally available. The cycle then repeats.

A third type of community found in the high intertidal zone at Hopkins may be termed a "lattice." (R. Paine [personal communication] calls this structure a "sheet.") At spots on Bird Rock, a place of very high settlement, *Balanus glandula* exists in 100% cover and individuals are packed into a lattice, as in a honeycomb. This system

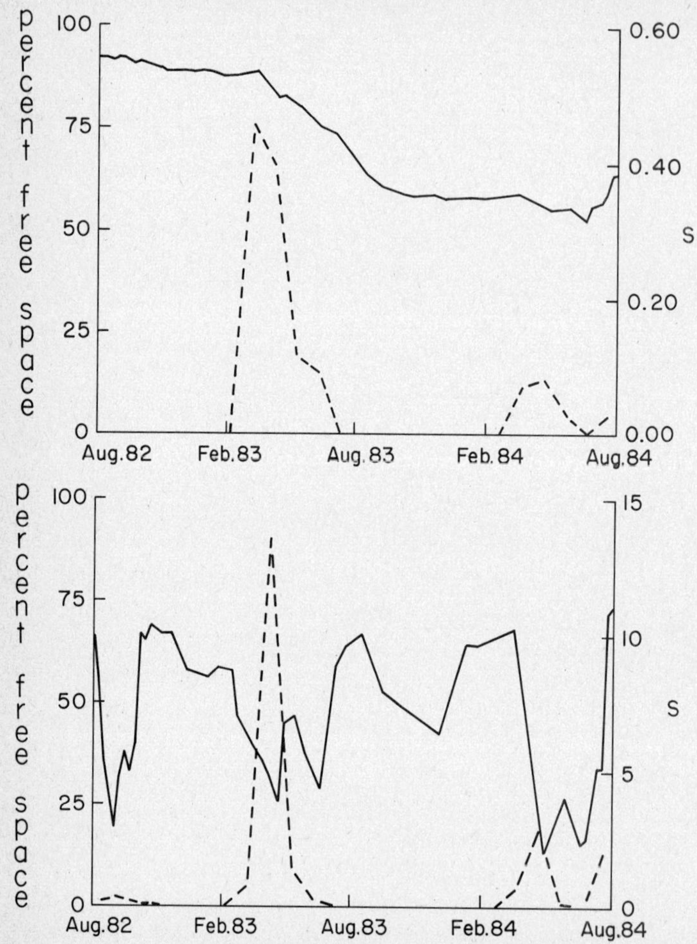

Fig. 30.7 Percent free space (solid line) and settlement rate (dashed line) during two years in two quadrats at Hopkins Marine Station in central California. Settlement (*s*) is in units of number of newly metamorphosed adults per square centimeter of vacant space per week. (Top) A quadrat from the KLM area. Settlement in 1983 was about four times that in 1984. The trajectory of free space in the quadrat is primarily a reflection of the settlement history. (Bottom) A quadrat from the Pete's Rock area. The overall settlement rate is about 20 times higher than in the KLM area. (Compare the right-hand ordinate scale of the two graphs.) The trajectory of free space in the quadrat is essentially independent of the settlement history and primarily reflects the interaction between the kinetics of growth and mortality of adult barnacles.

seems nearly static. Many of the individuals seem very old, with columnar tests 2 to 3 cm long and a basal diameter of 0.5 to 1 cm. I conjecture that this situation forms where the settlement to vacant space is high, but where mortality does not substantially increase when free space becomes exhausted because, for example, the spot is inaccessible to starfish or the wave action is not severe enough to remove hummocks. Here there is density-dependent growth, not density-

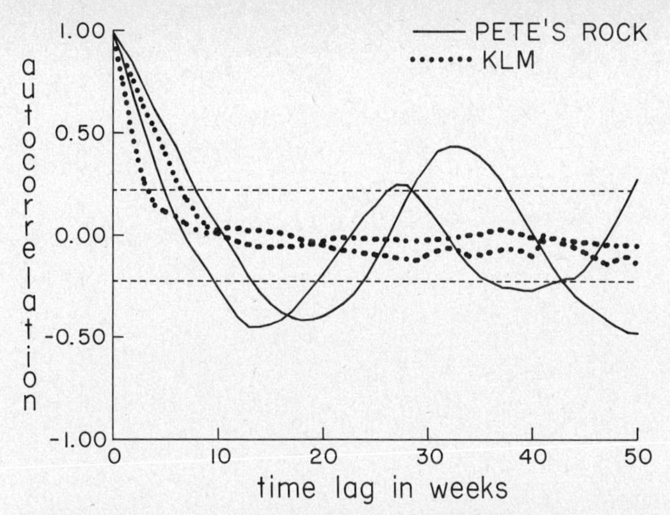

Fig. 30.8 Autocorrelation function for percent free space in four quadrats, two from KLM and two from Pete's Rock at Hopkins Marine Station in central California. The dashed horizontal lines indicate approximately the 95% confidence interval for a null hypothesis of white noise fluctuation. The Pete's Rock autocorrelation functions indicate a statistically significant oscillatory component with a period of about 30 weeks. The autocorrelation functions were computed from the deviation between the free space in a quadrat and the yearly average in the quadrat.

dependent mortality. After the larvae settle, they grow against one another, packing ever more tightly, until all the free space is exhausted. Then further growth stops, yielding a static structure. The structure remains intact until a truly rare and catastrophic event destroys the site, or until the animals die through senescence.

Importance of Settlement The studies at Hopkins Marine Station reveal that the settlement rate plays a role as important as postsettlement processes, including predation and competition, in determining community structure. Although data have been taken on settlement rates in intertidal communities for nearly 50 years (Hatton 1938, de Wolf 1973), the fundamental importance of the settlement rate is often overlooked in contemporary marine ecology. Studies that do stress the importance of settlement include Grosberg (1982) with barnacles, Yoshioka (1982) with bryozoans, and a spirited review by Underwood and Denley (1984). In particular, Connell's classic study (1961a) of zonation between barnacles in Scotland, a study whose findings have been confirmed by later studies at other locations in the North Atlantic (Wethey 1979, 1983), relies on high settlement relative to mortality as a tacit assumption. If the settlement rate is not high, then the barnacles do not come into extensive contact with one another. Without extensive contact *Balanus* does not cause adult *Chthamalus* to remain only in a high intertidal zone. Indeed, as Fig. 30.9 shows, at the KLM site at Hopkins, where the settlement rate is lower than that measured in Scotland (Connell 1961a, Hawkins and Hartnoll 1982), and over much of the central California coast as well (personal observation), *Balanus* and *Chthamalus* do not zone, but have completely overlapping distributions in the intertidal zone. The clue that the settlement rate is not high enough for zonation to result is that there is

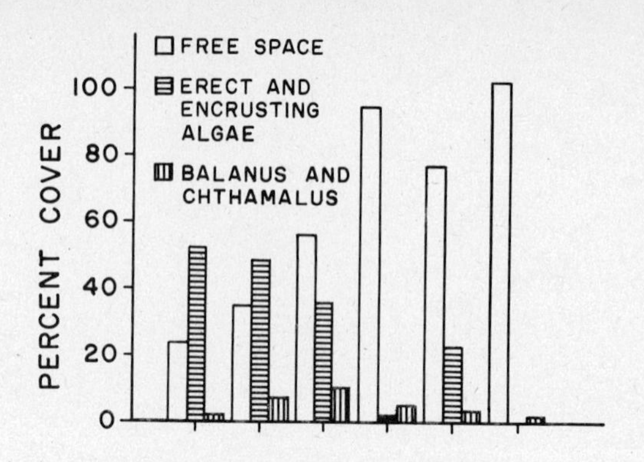

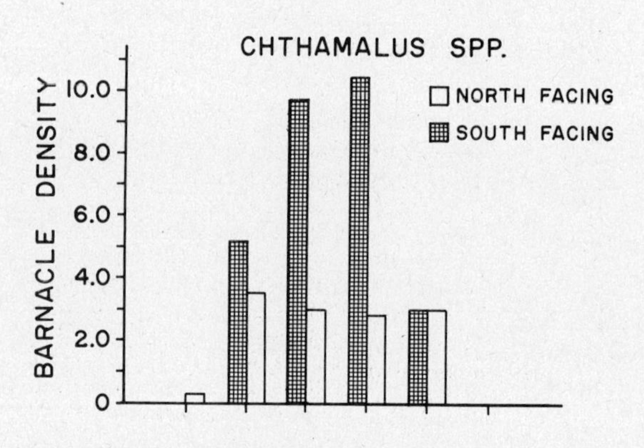

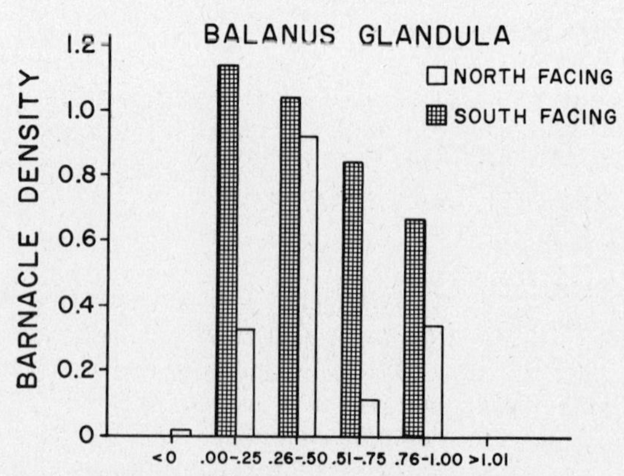

Fig. 30.9 Cover and barnacle density at several heights in the intertidal zone at the KLM area at Hopkins Marine Station in central California during the winter of 1982. (Top) Percent cover of free space, algae, and barnacles. Notice that there is still about 50% free space available at the height with the highest barnacle cover. (Middle) Mean number of barnacles (per 16 cm^2) of *Chthamalus dalli* combined with *C. fissus*. (Bottom) Mean number of barnacles (per 16 cm^2) of *Balanus glandula*. Notice that the distributions of *Balanus* and *Chthamalus* completely overlap.

ample free space where the two species' distributions overlap.

Similarly, the celebrated intermediate disturbance principle of marine ecology (Paine and Vadas 1969, Lubchenco 1978) has a tacit assumption of high settlement. According to this principle, the highest species diversity in a community occurs at an intermediate level of disturbance. But, if the settlement rate is too low for extensive contact to develop among the space-using organisms, then there is no opportunity for a hierarchy of competitive overgrowth to be expressed. In these circumstances diversity and disturbance are probably inversely related. More generally, the intermediate disturbance principle should be replaced by a graph that represents diversity as a function of both disturbance and settlement, as sketched in Fig. 30.10.

Types of Local Communities on Hard Substrate The various processes that affect community structure may be synthesized into a list of community types that are produced depending on the relative strengths of the processes.

1. The high free-space community results from a sufficiently low settlement rate relative to the rate of density-independent mortality. This community is recruitment-limited. Hence, stochastic and spatial fluctuation in the settlement rate greatly affect abundance, and animals of all sizes intermingle with the free space.
2. The low free-space community results from a high settlement rate relative to the rate of density-independent mortality, which itself is high relative to individual growth rates. Settlement is not limiting, and fluctuation in the settlement rate has relatively little effect on abundance. Animals of all sizes intermingle with what little free space there is.
3. The patch-mosaic community results from a combination of low density-independent

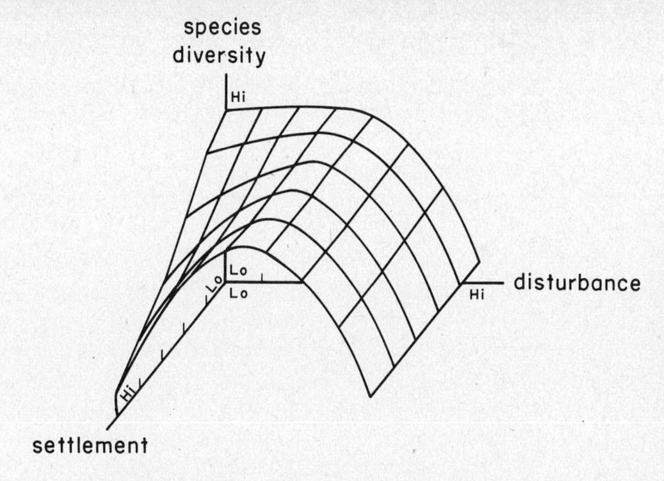

Fig. 30.10 Illustration of how the intermediate disturbance principle may be a special case of a broader codependence of species diversity on disturbance and settlement rate. Species diversity declines with settlement rate at low disturbance, because dominant species can exclude other species, but not at high disturbance. Species diversity peaks at intermediate disturbance only for sufficiently high settlement rates; diversity may decline with disturbance at low settlement rates.

mortality, high density-dependent mortality that enters when space is exhausted, and high settlement. The settlement rate is, as above, assessed relative to the density-independent mortality rate.

4. The lattice community results from a combination of low density-independent mortality, low density-dependent mortality when space is exhausted, growth that stops when space is exhausted, and high settlement relative to density-independent mortality.

The studies at Hopkins suggest that KLM is a high free-space community, Pete's Rock a patch-mosaic community, and Bird's Rock a lattice community. Perhaps subtidal communities such as associations of encrusting sponges, tunicates, anemones, soft corals, or stony corals might also often be lattice communities.

Metapopulation of Space-Limited Local Populations

Since the settlement rate is fundamentally important, what then determines the settlement rate? On a particular substrate the settlement rate reflects the number of larvae in the water column and the time of exposure of the substrate to the water column (de Wolf 1973, Hawkins and Hartnoll 1982, Geraci and Romairone 1982). Thus, the question becomes what determines the number of larvae? For organisms with long-lived pelagic larval dispersal, this question can be answered only by changing from a local to a regional scale.

Theory for the Metapopulation A model for a population on a regional scale may be called a "metapopulation"; it is a species population with open, space-limited local populations. A model for the dynamics of the metapopulation can be developed by supposing that there is a certain number of local systems, say H, all bathed by the same water mass. The model for each local system is exactly that used previously with one modification. The settlement parameter s, which describes the number of larvae settling per unit of vacant space, is itself assumed to be proportional to the number of larvae in the larval pool. Each local system had its own constant of proportionality reflecting its accessibility to the larval pool. Also, each local system has its own regime of mortality and growth rates for the adults in it. Further, the larval pool is constructed by summing the reproductive output over all the local systems, and by accounting for the larvae lost to settlement into the local systems and to a density-independent mortality in the larval pool itself. This model is analyzed in Roughgarden and Iwasa (1985), and a preliminary report appears in Roughgarden et al. (1984a).

Interspecific Competition on the Metapopulation Scale Competitive exclusion is impossible in a local system as a result of interspecific competition between two forms whose larvae settle in vacant space. As long as there are larvae of both species in the water column and vacant space available, then both species land in the local system and are present in it at least for a short while after settlement. Population extinction, from competition or predation, requires a cumulative poor survivorship from all the local systems potentially contributing to the larval pool and is thus inherently a question of regional dynamics.

The metapopulation model has been extended to include several species that compete for space in the local systems without interacting as larvae while in the water column (Iwasa and Roughgarden 1985). The results indicate that the species diversity in a local system is a reflection of specialization (or habitat partitioning) on a regional scale.

CORAL REEF FISH COMMUNITIES

Coral reef fish communities have been approached as though they exemplify marine equivalents of terrestrial food-limited communities and also as though they exemplify the space-limited communities of marine invertebrates on hard substrate. The origin of the approach emphasizing food limitation and resource partitioning as the mechanism for coexistence lies in the studies of Kohn (1959, 1971) that established clear resource partitioning among predatory snails of the genus *Conus*. The early descriptions of coral reef fish communities (e.g., Hiatt and Strasburg 1960, Smith and Tyler 1972) tended to classify fish species according to habits and feeding modes as though classifying distinct niches whose differences provided a mechanism of coexistence. Furthermore, Roughgarden (1974b) pointed out, using Randall's (1967) data on diets of Caribbean reef fish, that coexisting species of groupers differed in body size and correspondingly in prey size and suggested the application of competition theory to the situation.

Sale (1977) offered an important alternative view of coral reef communities, one that in retrospect is similar to that of an open, space-limited community as discussed above for barnacles. Sale emphasized that fish (particularly the Pomacentridae) are territorial and that fish eggs, and often their larvae too, are invariably pelagic. He suggested that their coexistence was somehow explained as a random process involving larval settlement into the vacant space that became available whenever a predator or other agent of disturbance happened to remove an established space-holding adult. Sale suggested that the population dynamics were somehow similar to a lottery, and this approach has come to be known as the lottery hypothesis for the maintenance of coral reef fish diversity.

Sale's paper stimulated attempts to produce models that would clarify exactly what a lottery is, how it works, and whether and how its workings could provide a mechanism of coexistence (Chapter 14; Dale 1978, Chesson and Warner 1981, Sale 1982, Abrams 1984b). These models generally establish that random recruitment to vacant space is not itself a mechanism that promotes regional coexistence, and additional mechanisms are added to obtain such coexistence. However, the focus of Sale's original observations was not on regional patterns of coexistence, but on local coexistence at patch reefs and small sites containing less than 100 individuals.

I suggest that there is sound evidence supporting an analogy of coral reef fish to both terrestrial closed food-limited vertebrate populations and to marine open space-limited invertebrate populations. Here, briefly, are the main facts.

1. Top carnivores like groupers partition prey within local sites with respect to prey size. Herbivorous fish like parrot fish evidence little local resource partitioning (Roughgarden 1974b). Middle-level carnivores like damsel fish and butterfly fish evidence slight resource partitioning (Anderson et al. 1981).
2. The species that are found at one site on a reef are a small subset of the fish that live in the surrounding area. For example, a rotenone collection (Smith 1973) yields a species count that is, say, half of the total count for a surrounding area of a square mile. Moreover, there is species replacement along habitat gradients proceeding, for example, from near-shore reefs to outer exposed reefs (Anderson et al. 1981). Data are not sufficient, however, to conclude that the extent of geographical overlap is inversely related to similarity in feeding habit.
3. Direct evidence is accumulating that coral

reef fish compete for food (Thresher 1983).

4. Many coral reef fish are either territorial or have home ranges (Sale 1977). Furthermore, the provision of additional substrate in the form of artificial reefs made from cement cinder blocks increases fish abundance (Randall 1965). Thus space, especially in the sense of refuges from predation, is limiting in some circumstances.
5. The eggs of almost all coral reef fish and the larvae of most species are pelagic (Sale 1977).
6. The composition of the fish community in a local area may be completely determined by the identity of the larvae that happen to settle out from the plankton into the area (Victor 1983). This observation accords with the view among fishery biologists that stochastic variation in recruitment is the single most important cause of variation in the size of stocks (e.g., Sissenwine 1984).

The first three points tempt one to view coral reef fish, at least for the top carnivores, as a food-limited competition community with local resource partitioning as the mechanism of coexistence. The next three points tempt one to view local coral reef fish communities as an open space-limited community, with disturbance as the mechanism of local coexistence and broad-scale habitat partitioning as the mechanism of regional coexistence. Studies are accumulating that favor one view over the other (Williams 1980, Robertson and Lassig 1980, Sale and Williams 1982). Yet, clearly both views may be correct simultaneously.

What then is the circumstance that should lead to a community having features in common with a closed food-limited terrestrial community and a community assembled with populations that occur in open space-limited communities? As a possible answer, I venture a simple classification that seems to synthesize the various possibilities discussed in this review.

A MODEST PROPOSAL

Table 30.2 classifies competition communities with respect to population structure and to whether the limiting resources are partitionable. The community types are a closed population structure with partitionable resource (CP), an open population structure with partitionable resource (OP), a closed population structure with nonpartitionable resource (CNP), and an open population structure with nonpartitionable resource (ONP).

Coral reef fish share the population structure of many marine organisms including sessile marine invertebrates of the rocky intertidal zone. Yet space is not occupied by fish in the same rather rigid sense as it is by sessile invertebrates. For fish space itself can be partitioned because two or more individuals of different species do share overlapping home ranges when their activities in it are sufficiently different. That resources are partitionable leads automatically to similarities with terrestrial food-limited vertebrate communities like anoles, while the open population structure ensures that recruitment (like settlement) will be the limiting factor in certain local situations, as it can be with sessile invertebrates. In this classification island lizard communities exemplify what might be considered the classical CP properties, intertidal barnacle-dominated communities form an ONP competition community, and coral reef fish have an open population structure with a relatively partitionable resource (OP).

The *Membranipora* bryozoan community on kelp blades was demonstrated by Yoshioka (1982) to be recruitment-limited and appears to be analogous to the barnacle situation (ONP)

Table 30.2 SOME TYPES OF COMPETITION COMMUNITIES

	Resources	
Population structure	**Partitionable**	**Not partitionable**
Closed local population	Island lizards Lacustrine fish Island birds	Subtidal hard substrate
Open local populations	Coral reef fish	Intertidal hard substrate

where space is a nonpartitionable resource. In contrast, marine soft-substrate (infaunal) communities appear to use space in a way that allows partitioning and hence may be comparable to the coral reef fish situation (OP). The picture is complicated, however, by the phenomenon of larvaphagy (Woodin 1976) and by substantial substrate modification that the infaunal inhabitants cause, leading even to facilitation (Gallagher et al. 1983). These facts suggest that it may not be appropriate to consider a soft-substrate community primarily in terms of competition, even though parts of the community, such as suspension-feeding clams, exhibit intraspecific density-dependence and interspecific resource partitioning (Peterson 1982). The evidence is insufficient to judge at this time. Possibly the infaunal community should be viewed in terms of several subcommunities.

Returning briefly to vertebrates, many island bird communities (Chapters 6 and 10), but not the bird communities of small cays: (Chapter 33), and lacustrine fish communities (Chapter 21) seem comparable to island lizard communities in that dispersal is not extensive, so that suitably circumscribed local populations are effectively closed, density-dependence is present, and the resources (primarily food) are partitionable; these are classic CP communities. A nonvertebrate example of a CP community may be provided by bumblebees in North America (Inouye 1977).

Not all vertebrate communities are readily classified in Table 30.2. In small mammal communities interference interactions often predominate, and as Rosenzweig (1985) suggests, it may be useful to distinguish between partitioning with "distinct preferences" and partitioning with "shared preferences." Distinct preferences refer to the use of different resources as a result of possessing specializations for those resources. Shared preferences refer to the use of different resources as a result of a behaviorally dominant species forcing subordinate species to use alternative and presumably marginal resources. Amphibian communities usually occupy different habitats at different phases of their life history (Wilbur 1972, Gill 1978, Hairston 1981, Morin 1983a), and it is not clear which phase to focus on or whether all phases should be studied jointly. Possibly, these amphibian communities are OP communities, but the magnitude of the predation involved makes viewing these communities as competition communities highly suspect. Finally, as Wiens (Chapter 9) has emphasized, long-distance migration also poses difficulties in defining communities for certain continental birds and insects.

The table invites speculation on whether there are closed communities with a nonpartitionable resource (CNP). Perhaps this is the situation that leads to the evolution of colonial space-using forms, like many tunicates, anemones, sponges, and corals (Chapter 31; Porter 1976, Jackson 1977a, Sebens 1983, Olson 1985), many of which are also known to be substrate-limited and to have short dispersal, on the order of meters.

Finally, it is worth adding that many animal associations are not competition communities, either because other types of interspecific interactions are more important or because few interspecific interactions of any kind need be explicitly considered. Hence, not all animal associations, nor, as indicated above, even all competition communities, need be placed in this table. Nonetheless, the table offers a glimpse of how the diverse aspects of competition communities can be seen as part of a general pattern that represents different combinations of population structure with resource characteristics.

ACKNOWLEDGMENTS

I thank T. Case, J. Diamond, S. Gaines, J. Rummel, and J. Wiens for their comments and assistance on this manuscript; K. Naganuma for assistance with the recent anole censuses; the National Science Foundation for support on the studies with *Anolis;* and the Department of Energy for support on the studies of intertidal communities and for research time on a Cray 1 computer that enabled the theoretical studies of faunal buildup on islands.

chapter 31

Competition and Community Organization on Hard Surfaces in the Sea

Leo W. Buss

INTRODUCTION

Investigations of interspecific competition in the sea have flourished at many levels in the past decade. Significant progress has been made in demonstrating that competition chronically occurs on a wide variety of marine hard grounds (Jackson 1977a, 1983); in determining the relative interference competitive abilities of co-occurring species (Lang 1973; Stebbing 1973; Jackson and Buss 1975; Osman 1977; Buss 1979a, 1980, 1981a, 1981b; Buss and Jackson 1979; Jackson 1979b; Karlson 1980; Keen and Neill 1980; Grosberg 1981; Kay and Keough 1981; Rubin 1982; Russ 1982; Sebens 1982; Ayling 1983; Paine 1984); in correlating competitive success (or failure) with environmental conditions or traits of particular organisms (Connell 1961a; Day 1977; Jackson 1979b; Buss 1979b, 1980, 1981a, 1981b; Osman and Haughness 1981; Russ 1982; Rubin 1982); in elucidating the mechanisms by which one species overcomes another (Lang 1973; Buss 1979b, 1981b; Buss et al. 1984); and, finally, in documenting that similar competitive interactions occurred repeatedly throughout the fossil history of groups known to compete actively today (Brasier 1976, Fritz 1977, Stel 1978, Taylor 1979, Lidell and Brett 1982, Lidgard and Jackson 1982).

Coincident with this expanding knowledge of competition on hard surfaces in the sea, the study of predation and physical disturbance has proceeded at a meteoric rate. These studies, particularly in the rocky intertidal zone, have repeatedly demonstrated that predation and disturbance can act to hold the densities of potentially competing species to a level so low that contact-mediated competition between certain structurally or numerically dominant taxa is reduced or even precluded. This suggestion, initially called the predation hypothesis (Paine 1966) and also known, in modified form, as the intermediate disturbance hypothesis (Connell 1978), is based on field experiments repeated several times and in several locations (Paine 1966, 1971, 1974; Paine and Vadas 1969; Harper 1969; Dayton 1971; Menge 1976; Lubchenco 1978, 1980; Lubchenco and Menge 1978; Sousa 1979; and many others).

How might the repeated observations of competition on marine hard substrata be reconciled with the repeated experimental demonstration that predation acts to reduce or preclude the occurrence of competition? It is the object of this chapter to investigate this dilemma. Hypotheses are sought that specify the conditions under which competition will or will not be observed in ecological time.

The first section of this chapter summarizes a specialized subset of the competitive interactions on marine hard grounds, focusing mainly on competition for space. I discuss the occurrence, costs, and mechanisms of competition; the phenomenon of competitive ranking; the pattern of available space and its renewal; and correlates of competitive success such as colony size, population density, and growth morphology. In the second section I briefly review the predation hypothesis and note its limitations. In particular, settlement rates, prey resistance to predators, limitations on predators, refugia, and prey clonality are some factors that permit competition to persist in the face of the predation. In the last section I present a reconciliation of the predation hypothesis with observations of competition. In each section I do not attempt an exhaustive review or enumeration; rather, I present vignettes of particular empirical investigations, chosen largely from my own work.

Both the origin and the subsequent development of the ideas in this paper lie in my collaboration with Jeremy B. C. Jackson. We have discussed at length all of the issues treated here on numerous occasions over the last ten years. While he is not responsible for any failing of this particular rendering of our collaborative efforts, he most certainly shares any success this attempt at provisional synthesis might enjoy.

ON OBSERVING COMPETITION

Occurrence of Competition

Competition between sessile marine invertebrates occupying marine hard substrata may involve interactions for food, light, and space. The study of competition for light almost always requires direct experimental intervention. One must remove a species to remove presumptive shading effects and document the growth or reproductive success of the other species. A number of such studies have been performed in rocky intertidal environments, and several have demonstrated competition (see review by Lubchenco and Gaines 1981).

The study of competition for food is far more difficult. Most sessile invertebrates depend on planktonic food resources. The monitoring of resource levels presents imposing technical problems: plankton populations are never static in distribution, the documentation of the distribution and abundance of plankton is highly labor intensive, and the precise dietary requirements of most invertebrates are largely unknown. Further complicating such analysis is the fact that many invertebrates feed in a low Reynolds number flow regime, in which the actual availability of food to the organism depends not only upon the density of the food in the environment, but also on the detailed fluid mechanics of the particular environmental setting. Despite these rather severe limitations, there have been demonstrations of *in situ* depletion of planktonic food resources (Glynn 1973, Buss and Jackson 1981) and laboratory demonstrations of interference competition for food (Buss 1979b). I know of no case, however, in which competition for food has been documented *in situ* among any sessile marine invertebrates.

Competition for space is by far the easiest form of competition to show, especially between colonial invertebrates. When two such species encounter one another, either of two results may occur: (1) the two colonies cease growth along the shared margin or (2) one or both colonies expands into the space occupied by the other. The latter alternative often occurs as physical overgrowth of one colony by another (Figs. 31.1 and 31.2), frequently resulting in the death of the over grown tissues. Clonal organisms are typically capable of indeterminate growth and can often expand to cover the limits of the available substratum (Jackson 1977a). Since fecundity increases with colony size in an exponential fashion for clonal invertebrates (Jackson 1977a), the reduction in colony size resulting from overgrowth has an unambiguous effect on an important component of colony fitness. Therefore, both the cessation of growth and the occurrence of overgrowth are clearly competitive encounters.

The only exception to the generalization that growth cessation or overgrowth demonstrates the

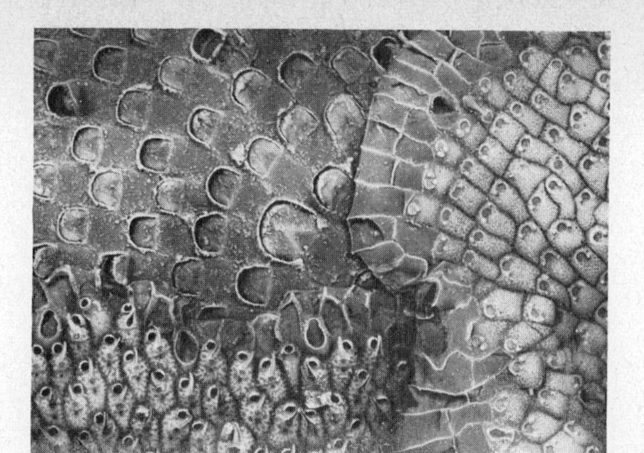

A

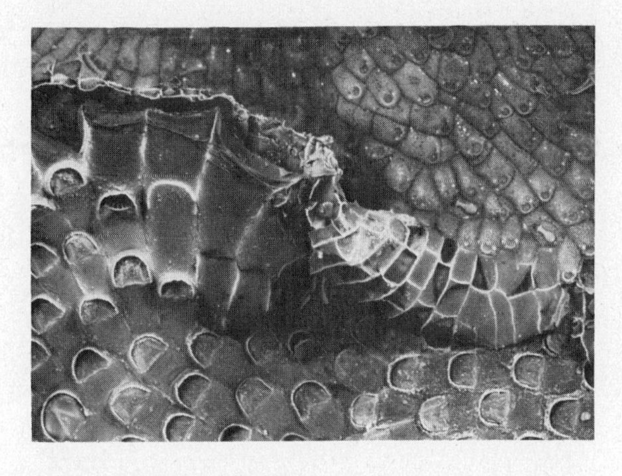

B

C

Fig. 31.1 Examples of overgrowth competition between encrusting bryozoans from Jamaican cryptic coral reefs. (A) *Stylopoma spongities* (right) overgrowing *Reptadeonella plagiopora* (lower left) and *Steganoporella* sp. nov. (upper left). Note the giant buds along the growing margin of *S. spongities*. (B) *Stylopoma spongities* (top) overgrowing *Steganoporella* sp. nov. to the right, while *Steganoporella* sp. nov. has raised its growing edge and prevented overgrowth on the left. (C) *Steganoporella* (top) overgrowing *Stylopoma spongities* (middle right) and *S. spongities* overgrowing *Steganoporella* (bottom). The top *Steganoporella* is also overgrowing the bottom colony of *Steganoporella*. (From Jackson 1979b.)

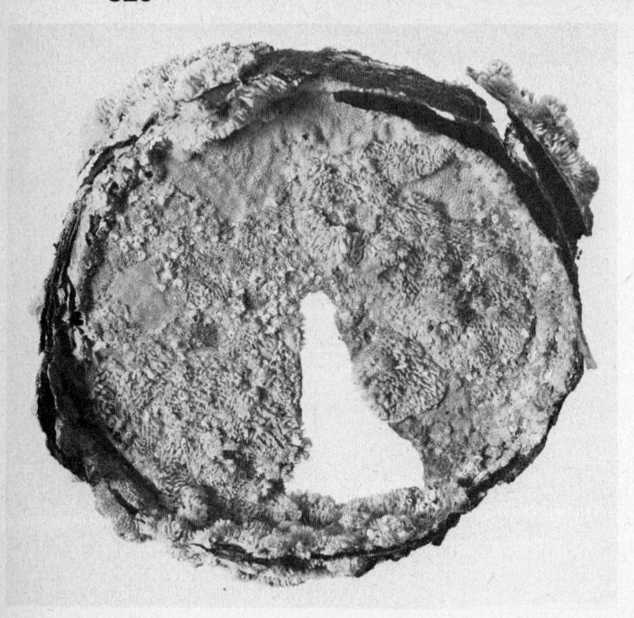

Fig. 31.2 The top of an aluminum beer can collected at 5 m depth near Isla Taboguilla, Panama. Note the almost complete cover by serpulids, barnacles, and bryozoans and the numerous instances of interspecific contact and overgrowth.

existence of competition would be a case in which either (or both) of the two colonies could not survive in the physical environmental regime occupied by the other. The interactive margin would have to lie precisely along a threshold level of some physical environmental factor, such as temperature or flow. Although strong environmental gradients are known to exist in a number of marginal marine environments, such as rocky intertidal shores, these environments are typically dominated by large aclonal organisms (Jackson 1977a, Paine and Suchanek 1983). Clonal groups dominate in the subtidal (Jackson 1977a, 1979b, 1983), where environmental gradients are usually far less pronounced. I know of no evidence for any subtidal species that growth cessation at a colony margin can be attributed to the inability of one organism to persist in the space occupied by the other.

Overgrowth interactions are particularly fruitful for the study of competition. Not only does the observation of overgrowth generally demonstrate that competition occurs, but also the mere inspection of the colonies indicates which individual is dominant in that interaction at that particular time (Fig. 31.1). Overgrowth observations yield only instantaneous data, and any presumption as to the long-term dynamics of an interaction based on such information is necessarily speculative (Stebbing 1973, Buss and Jackson 1979). Nevertheless, overgrowth is one of the simplest interactions in which to study competition, and the bulk of the recent data on competition on marine hard grounds is based on these observations.

The occurrence of overgrowth interactions is not limited to Recent environments. Jackson (1983) reviews examples of overgrowth from the fossil record dating back to early in the Paleozoic (Brasier 1976, Fritz 1977, Stel 1977, Taylor 1979, Liddell and Brett 1982, Lidgard and Jackson 1982). Unfortunately, little attempt has been made to quantify these interactions in order to compare them rigorously either to similar interactions within the same stratigraphic unit or to interactions among extant relatives. Such studies, especially in low-diversity communities living symbiotically on gastropod shells, may turn out to have considerable relevance to problems in ecological and evolutionary theory.

Competitive Rankings

Demonstrating that competition occurs is the first level of analysis of competition. The second level is to determine which species wins the encounter. The relative interference competitive abilities of the abundant species within a community are known for several marine communities (Dayton 1971; Lang 1973; Stebbing 1973; Jackson and

Buss 1975; Osman 1977; Buss 1979a, 1980, 1981a, 1981b; Buss and Jackson 1979; Jackson 1979b; Karlson 1980; Keen and Neill 1980; Kay and Keough 1981; Grosberg 1981; Russ 1982; Rubin 1982; Sebens 1982; Ayling 1983; Paine 1984). Rankings of this sort are available for few other communities and play an important role in discussions of community organization in the sea (Jackson and Buss 1975; Connell 1976, 1978; Buss and Jackson 1979; Karlson and Jackson 1981; Karlson and Buss 1984; Yodzis 1978 and Chapter 29; Paine 1984).

Competitive rankings show great variability both between communities and between species within a community. Many such rankings are largely transitive (Lang 1973, Stebbing 1973, Quinn 1982), that is, if A usually beats B and B usually beats C, then A also beats C. However, different results are found in selected subsets of overgrowth rankings from cryptic coral reef communities in Jamaica, cobble communities in England and Panama, piling communities in Australia, and coralline algal pavements in Washington State (Fig. 31.3). In these more diverse systems competitive rankings are far more complex and form networks of competitive relationships that deviate strongly from transitive patterns. The lack of clear transitivity in these instantaneous data is especially intriguing, given the demonstration that intransitive competitive dynamics can give rise to limit cycle behavior in model communities (Gilpin 1975; May and Leonard 1975; Yodzis 1977a, 1978).

The accumulation of data on competitive rankings within communities has led to the identification of the relative competitive abilities of different taxa (reviewed by Jackson 1983). Aclonal organisms are typically dominated by clonal groups, except in physically stressed environments (Greene and Schoener 1982). Among clonal groups, skeletonized organisms (bryozoans, coralline alga, and corals) are typically overgrown by groups that do not deposit a rigid skeleton (demosponges and ascidians). Among skeletonized organisms hermatypic corals, whose symbiotic dinoflagellates enhance the rate at which skeleton is deposited, are typically dominant over aposymbiotic groups. Patterns in relative competitive ability are known within many of these taxa as well. For example, cheilostome bryozoans with a sheetlike encrusting morphology are typically capable of overgrowing those with a runnerlike encrusting morphology (Buss 1979a, Jackson 1979a). Although clear patterns of this sort have been recognized within and between several taxa in rankings drawn from several communities, the ranking for any community often involves one or more exceptions to these general patterns, leading to cases of highly intransitive within-community competitive rankings.

The Cost of Competition

Although overgrowth rankings provide a simple assessment of winners and losers at a particular time, a third level of analysis of competition—the assessment of the cost of engaging in a competitive encounter—is not revealed by these rankings. To measure the costs of competition, one must assess the risk of death or reduced potential for growth and fecundity of both the overgrowing and the overgrown colony. Such data are rare.

Overgrowth may result in the death of the overgrown colony. For example, colonies of the colonial hydroid *Hydractinia echinata* compete for space on hermit crab shells, and competition typically results in the demise of one of the two competing colonies (Ivker 1972, Buss and Grosberg in preparation). The cost to the victor of engaging in such an encounter in terms of growth and reproductive output varies as a complex function of the relative sizes and competitive abilities of the interacting colonies (Buss and Grosberg in preparation). Far more frequent than actual colony death is the interruption of growth and reduction in colony size by overgrowth. For example, Ayling (1983) monitored overgrowth interactions among encrusting demosponges for a period of nine months and found no cases in which overgrowth led to colony death or even appreciable capture by the overgrowing species of the space occupied by the overgrown species. In contrast to the hydroid example, overgrowth in this system appears to have only a modest effect on survivorship, but rather seems to act to limit the area occupied and hence the fecundity realized by particular colonies.

One clearly cannot assess the cost of competi-

a. Australian Fouling Panels

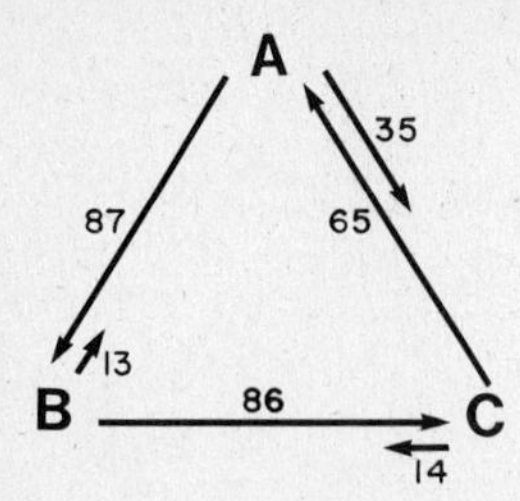

b. Australian Pilings

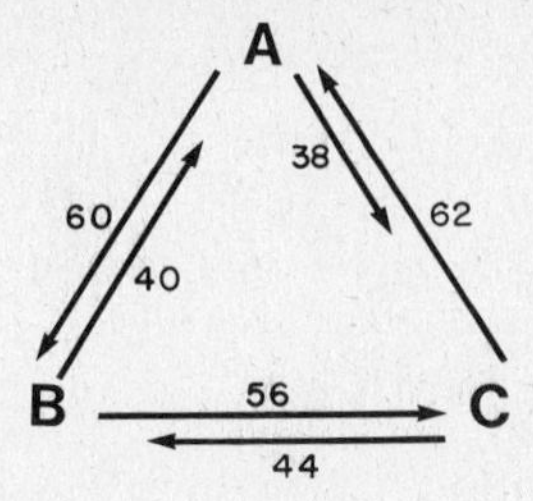

c. Panamanian Cobble

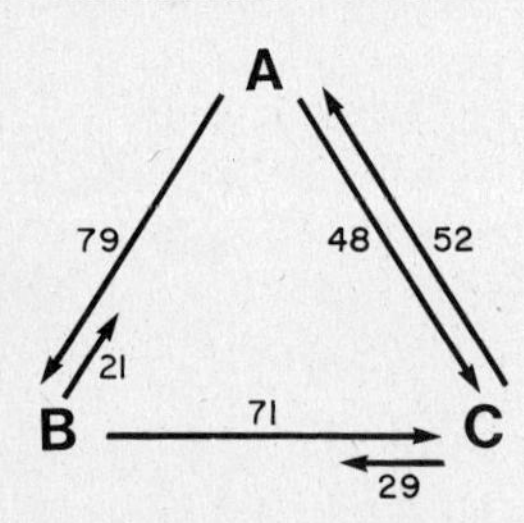

d. British Cobble

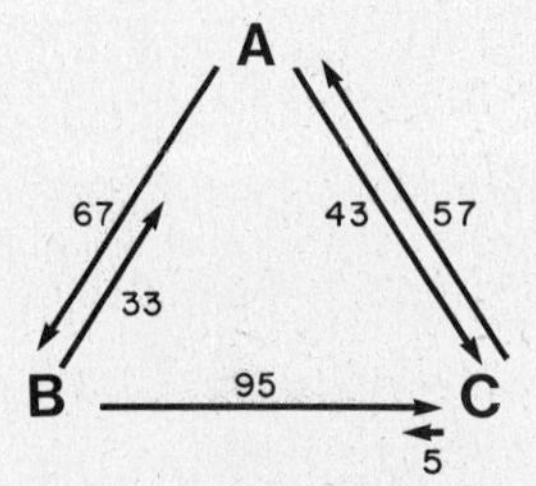

e. Washington State Algal Pavements

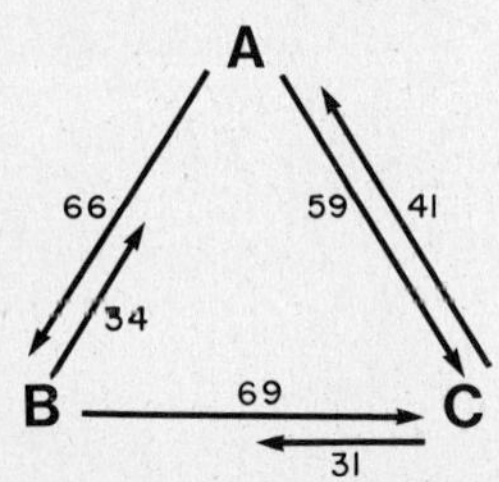

f. Jamaican Cryptic Coral Reefs

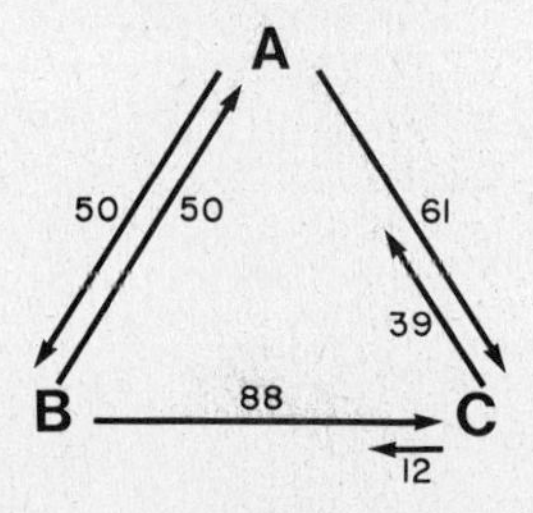

Fig. 31.3 Selected subsets of overgrowth rankings from various marine hard-substratum communities. Arrows point from the dominant competitor to the subordinate. Numbers represent the percentage of total outcomes observed. (a) Australian fouling panels. A = *Distaplia viridus,* B = *Esperiopsis* sp., C = *Botrylloides nigrum.* (Russ 1982.) (b) Australian pilings. A = *Crella* sp., B = *Didemnum* sp., C = unidentified didemnid ascidian. (Kay and Keough 1981.) (c) Panamanian cobble. A = *Antropora tincta,* B = *Onychocella alula,* C = *Neogoniolithum rugulosum.* (Buss 1980.) (d) British cobble. A = *Escharella immersa,* B = *Chorizopora brongniartii,* C = *Callopora rylandi.* (Rubin 1982.) (e) Washington State algal pavements. A = *Pseudolithophyllum lichenare,* B = *Pseudolithophyllum whidbeyense,* C = *Lithothamnium phymatodeum.* (Paine 1984.) (f) Jamaican cryptic coral reefs. A = *Steganoporella magnilabris,* B = *Parasmittina* sp., C = *Reptadeonella violacea.* (Jackson 1979b.)

tion from static observations of overgrowth alone. Overgrowth rankings have two severe limitations. First, the observation of overgrowth at a given time may not accurately reflect the eventual outcome of that interaction. The reversal of overgrowth rankings over time between two individuals is well known, and there is no substitute for long-term observations. Second, some taxa are tolerant of epizooism, that is, a colony may survive being overgrown for an extended period and will recover should some other process remove the successful competitor. The recovery of an overgrown colony may represent the ability of tissues to remain viable while overgrown or, more frequently, may represent the ability of not yet overgrown tissues in the colony to recapture the ground previously yielded to the competitor. Hence, static observation of competition does not allow assessment of either the eventual outcome or the potential severity of a particular interaction.

While the cost of competition in terms of reduced growth, fecundity, and survivorship can be empirically determined, these costs are not the only ones associated with interference competition. Successful competitors are often characterized by morphological or physiological modifications that are central to their ability to overcome others in combat (Case and Gilpin 1974). Such "fixed costs" are often exceedingly difficult to quantify, if these modifications are features that serve roles in addition to their function in competition. There are some structures that are used only in competition (see the later section on mechanisms of competition), and in these cases fixed costs are potentially easier to quantify. In either case, however, a precise understanding of the costs of competition must reveal the compromises between potential for growth and reproduction and capacity for intra- and interspecific competition that these morphological and physiological modifications involve. This ultimately requires an understanding of the genetics of the traits and the manner in which their expression is differentially regulated.

Resource Structure and Renewal

A thorough understanding of the nature of competition requires an appreciation of the structure of the resource and its pattern of renewal. Space for marine hard-substrata organisms occurs as a series of habitable areas that are physically discontinuous and separated by the uninhabitable media of water and soft substrata. Stretches of rocky shore along a coast are separated by stretches of sandy beach, reefal structures are separated by sand channels, and boulders, cobbles, and shell fragments are separated by water, mud, or sand. The sessile nature of the fauna makes each such unit a system closed to communication with other similar units except by exchange of propagules capable of dispersal over considerable distances. On all but the largest substrata (e.g., rocky shores), the number and type of propagules that arrive on a particular unit differ from those arriving on other units, with profound effects on the subsequent development of the fauna on that particular substratum (Sutherland 1974, 1976, 1977, 1978; Menge and Sutherland 1976, Jackson 1977b, Osman 1977, Sutherland and Karlson 1977).

In addition to the resource's structure, one must also consider its generation or renewal, whether as whole units or as patches within a unit. Generation of space has been extensively investigated on rocky shorelines, where patch production is the primary form of resource renewal (Paine and Levin 1981). However, generalizing these results to other marine hard grounds is not entirely appropriate, as space-generating processes operate on different spatial and temporal scales in different environments. For example, in rocky intertidal environments new discrete substrata, that is, new stretches of shore, appear only on geological time scales, whereas patches appear at a far more rapid rate. New coral reef surfaces may arise every 30 or 40 years following a major hurricane, and space is generated both by generation of these substrata and of patches within the substratum. At the opposite extreme, relatively small discrete substrata are renewed even more rapidly through simple introduction of new substrata, such as a new shell or cobble.

Correlates of Competitive Success

Given a knowledge of the frequency at which a species wins or loses a particular interaction, a fifth level of analysis of competition may be profitably explored: determining the ecological correlates associated with competitive success. Considerable progress has been made in this direction, especially in documenting changes in competitive ability as a function of the life history stages involved in the interaction. Three examples follow.

Size In tidal channels on the rocky shore of Punta Paitilla, Panama, small cobbles accumulate. These cobbles are dominated by three encrusting, clonal organisms at midtidal levels, the encrusting cheilostomatous bryozoans *Antropora tincta* and *Onychocella alula* and the coralline alga *Neogoniolithum rugulosum*. Studies of the overgrowth relationships among these three species show that no clear competitive dominant exists (Fig. 31.3c). *A. tincta* wins most of its interactions with *O. alula, O. alula* wins most of its interactions with *N. rugulosum,* and *N. rugulosum* wins slightly over half its interactions with *A. tincta*. All three species are skeletonized organisms without the capacity to lift their growing margins above the substratum that they encrust. As such, thicker colonies generally overgrow thinner ones and thickness is related to colony area (Buss 1980). Thickness increases proportionately with surface area for *A. tincta* and *N. rugolosum,* but is a constant for *O. alula*.

A discriminant function was calculated in an attempt to predict competitive outcome based solely on knowledge of colony area. This analysis demonstrated that no less than 75% of all outcomes between these three species were correctly predicted on the basis of colony size alone. Colony size is an excellent predictor of competitive ability in this and several other interactions (e.g., Connell 1961a, Day 1977, Russ 1982).

Density The cheilostome bryozoans *Bugula turrita* and *Schizoporella errata* are numerically important components of fouling communities along the Atlantic coast of North America. *B. turrita* is a lightly calcified arborescent anascan and is readily pushed over and overgrown by the heavily calcified encrusting ascophoran *S. errata*. Despite this clear dominance of *S. errata* over *B. turrita* in overgrowth encounters, *B. turrita* is an abundant organism, often occurring in dense associations of numerous uprights.

The persistence of dense associations of *B. turrita* uprights in the face of *S. errata* competitive dominance led me to study the mechanism of aggregate formation, the effect of intraspecific competition on the growth of uprights within an aggregate, and the effect of interspecific competition with *S. errata* as a function of *B. turrita* density (Buss 1981a). Results of laboratory habitat selection experiments show that *B. turrita* recruitment is strongly density- and size-dependent; larvae selectively choose habitats in which density of new recruits is high, leading to the establishment of dense monospecific stands (see also Keough 1984). Growth rates of *B. turrita* colonies are strongly influenced by density, with growth stunted at high density, presumably due to intraspecific competition. In contrast to the results of intraspecific competition, growth of *B. turrita* in interspecific competition with *S. errata* is positively associated with *B. turrita* density. Dense populations of *B. turrita* are rarely overgrown by *S. errata,* whereas solitary individuals are uniformly overgrown. In this case the outcome of interspecific competition is strongly density-dependent.

Growth Morphology In temperate soft-bottom environments worldwide, colonies of the athecate hydroid *Hydractinia* live symbiotically on the shells of hermit crabs of the genus *Pagurus*. Colonies of *H. echinata* vary considerably in colony morphology, and this variation has a strong genetic component (McFadden et al. 1984). Some individuals develop as a uniform sheet of ectodermal mat, overlying an extensive gastrovascular canal system. Other colonies develop differing degrees of stolonal extensions (that is, single periderm-covered gastrovascular canals) from the centrally placed mat, with some colonies producing a complex network of anastomosing stolons crisscrossing the substratum (Fig. 31.4).

This variation in growth morphology has important implications for intraspecific combat, because mat and stolonal tissues differ in their morphogenetic potential for producing those tissues required for aggression (Buss et al. 1984). When mat tissue comes into contact with mat tissue, active aggression does not occur; rather, colonies simply cease growth along the interactive

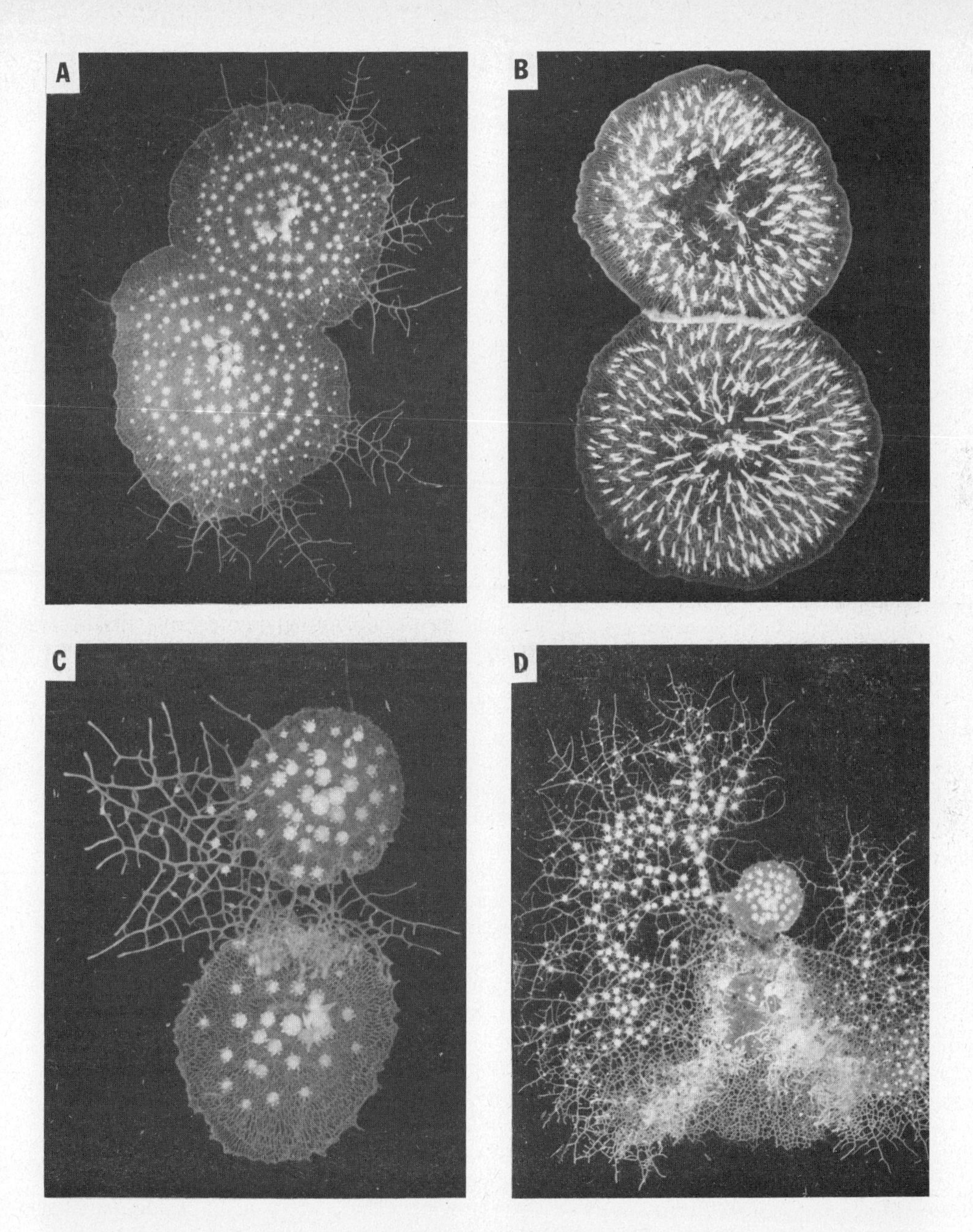

Fig. 31.4 Intraspecific competition between colonies of the colonial, athecate hydroid *Hydractinia echinata*. (A) Fusion between colonies, showing the result of intraspecific contact between histocompatible colonies. (B) Competition without aggression. These stolonless colonies do not fuse; they simply cease growth along the interactive margin. (C) Competition with aggression. The stoloniferous colony (upper) is developing hyperplastic stolons where it contacts the ectodermal mat of the stolonless colony (lower). (D) Competition with aggression. In this interaction between two stoloniferous colonies the upper colony has developed extensive hyperplastic stolons along its zone of contact with the lower colony and has almost completely overgrown it.

margin (Fig. 31.4B). In contrast, when stolons contact foreign tissue, they hypertrophy, lift off the substratum, and effect the destruction of the foreign colony (Figs. 31.4C and 31.4D). The differing morphogenetic potential of the two tissue types results in a strictly deterministic relationship between growth morphology and competitive ability; stoloniferous colonies always overcome less stoloniferous or matty colonies in pairwise symmetrical encounters (Buss and Grosberg in preparation).

Mechanisms of Competition

The final level of analysis of competition is the elucidation of the mechanism of the interaction. That analysis may proceed at a variety of levels. Phenomenological examination of overgrowth interactions may lead to the clear association of certain traits of an organism with different overgrowth results. Examination of such interactions in detail may lead to an understanding of the cellular and subcellular basis of one organism's ability (or inability) to overcome another organism. Studies of the transmission and molecular genetics will ultimately identify the arrangement of competition genes and elucidate their potential for evolutionary modification. Although considerable progress is being made, we are far from any such thorough analysis for any competitive interactions among sessile clonal organisms. Nevertheless, an understanding of mechanism is important if we are to (1) quantify precisely the costs of competition, (2) determine the degree to which ecological circumstance can result in plasticity in competitive ability, (3) determine the extent and control of compromises between competitive ability and growth or reproductive output, and (4) understand the evolutionary potentials for aggression within and between different taxa.

Below I summarize known mechanisms of competition in two groups of common sessile organisms, the encrusting cheilostomatous Bryozoa and the polypoid life stages of Cnidaria. Interactions between individuals within each group are certainly among the best-known examples of interspecific competition among clonal marine invertebrates.

Bryozoa Cheilostomatous bryozoans are skeletonized encrusting organisms that are generally unable to effect the destruction of foreign tissues over a distance. Bryozoans grow as a lineal series or radial arrangement of adjacent lineal series of asexually iterated calcified zooids (see Fig. 31.1). Growth is typically restricted to the margins of a colony. For organisms of this sort, overgrowth mechanisms involve structural features that act to influence either the relative vertical relief of two interacting colonies or the relative rates of lateral and vertical growth along the interactive margin.

A number of encrusting bryozoans have evolved structural modifications on this basic growth plan that allow them to reposition their growing margins in ways that fundamentally alter their competitive abilities (Stebbing 1973; Jackson and Buss 1975; Jackson 1979a, 1979b; Buss 1981a, 1981b; Lidgard and Jackson 1982). Some bryozoans are capable of producing new zooids on top of preexisting ones, a process called frontal budding. This trait allows a colony to reposition the growth margins at considerable distances from the basal layer of zooids. The trait is clearly associated with overgrowth success in competition against species that have no similar capacity to alter the vertical position of their growing margins. Several species of bryozoans are capable of another budding pattern, that of simultaneously producing several immature zooids, or giant buds, along a growing margin. Giant buds allow a colony to achieve a faster rate of lateral growth than a similarly calcified colony that buds zooids one at a time. Giant buds are associated with overgrowth success, especially when combined with a third trait, the capacity to grow marginal zooids unattached to the underlying substrata (Fig. 31.1). The capacity to lift locally off a substratum clearly allows a colony considerable capacity for modifying the vertical relief of its growing surface relative to a substratum-bound competitor. The phyletic distribution of these three traits, along with various others relevant to overgrowth success, suggests that these patterns have evolved convergently in a large number of cheilostome families (Litgard and Jackson 1982).

Species and individual colonies vary considerably in the degree to which these devices are deployed. For example, a colony may grow giant buds only along a portion of the colony, or frontally budded layers may be restricted to certain regions of the colony (Jackson 1979b). It is the variability in the expression of these morphologies at various points along the colony margin that gives rise to the extraordinary variability in overgrowth rankings among bryozoans and the repeated observation that overgrowth results among bryozoans are correlated with encounter conditions (Jackson and Buss 1975; Jackson 1979b; Buss 1980, 1981b; Rubin 1982). Unfortunately, the study of the various budding patterns associated with competitive success among the Bryozoa have been limited almost exclusively to studies at a phenomenological level; little is known of the cellular and subcellular basis for differing growth patterns, and genetic data are entirely lacking.

Cnidaria In contrast to bryozoans, where competition is often mediated by permanent structural features, many cnidarians are capable of the temporary induction of specialized tissues that respond to intra- and interspecific competition and that destroy foreign tissues. An example of such an interaction was introduced earlier, that of encounters between colonies of *Hydractinia echinata* (Fig. 31.4). When stolons of opposing colonies encounter one another, the stolons differentiate into a new structure, called a hyperplastic stolon (Ivker 1972, Buss et al. 1984). These hyperplastic stolons, which appear only when a colony interacts with another hydractiniid hydroid, are formed by the movement of multipotent interstitial cells from the central mat region into the stolons (Figs. 31.5 and 31.6). The interstitial cells differentiate into mature nematocytes, cells containing harpoonlike organelles called nematocysts (Buss et al. 1984). Nematocysts are used in prey capture and are typically armed with potent toxins (Mariscal 1974). Upon contact with the foreign colony the nematocyst batteries of the hyperplastic stolon discharge into the foreign tissue, destroying it (Figs. 31.5 and 31.6).

Interactions of this sort are not limited to hydractiniid hydroids (see Buss et al. 1984, Table 1). Particularly among anthozoans, there are a diversity of inducible structures that inflict damage on neighbors. Scleractinian corals are capable of differentiating sweeper tentacles, elongate tentacles armed with a distinct nematocyst population (Richardson et al. 1979, Wellington 1980, Chornesky 1983). Certain anemones display an analogous phenomenon. Upon contact with neighbors these anemones differentiate catch tentacles armed with a specialized nematocyst population and used to destroy foreign tissues (Williams 1975, Purcell 1977, Watson and Mariscal 1983). Other anemones deploy an entirely different structure. The body columns of these species possess batteries of nematocysts in structures called acrorhagi, which upon contact with neighbors can inflate, reach out, and discharge their nematocysts into foreign tissues (Francis 1973, Williams 1978, Ottaway 1978, Bigger 1980, Brace 1981). It is indeed striking that such a diverse array of specialized, induced structures have evolved to deploy the same nematocyst-based effector system.

The study of competition among cnidarians has revealed an association between the deployment of competitive mechanisms and the capacity for historecognition (see Fig. 31.4A). It is commonly assumed that competition and historecognition are genetically based alternatives in Cnidaria, although genetic data are available only for *Hydractinia echinata* (Hauenschild 1954, 1956; Ivker 1972). The assumption is justified on the basis of the repeated observation that competition occurs only between unrelated individuals or species and that aggression is not observed between related individuals or clone-mates (see review by Buss 1982).

The association of competition among cnidarians with historecognition holds great promise for a more complete description of these interactions on a mechanistic level. If the genetic complexes coding for historecognition in cnidarians are homologous to those coding for historecognition in vertebrates, the opportunities for molecular genetic analysis are enormous. The increasing availability of cDNA clones to various regions of the mouse major histocompatibility complex should allow this study to be pursued. An understanding of cnidarian competition based on data from the community, populational, developmental, cellular, and ultimately molecular-genetic levels may well be possible in a relatively short time.

ON FAILING TO OBSERVE COMPETITION

Just as competition occurs chronically in a number of marine, hard-substratum environments, so

Fig. 31.5 Details of the competitive mechanism of hydractiniid hydroids (A) Scanning electron micrograph showing a hyperplastic stolon arching off the substratum toward a polyp of its competitor (106 ×). (B) Contact between a hyperplastic stolon and the polyp of a competitor (167 ×). Note the concentration of nematocyst threads where the hyperplastic stolon contacts the polyp of the competitor. (C) Discharged basotrichious isorhizal nematocysts extending from a hyperplastic stolon (387 ×). (Buss et al. 1984).

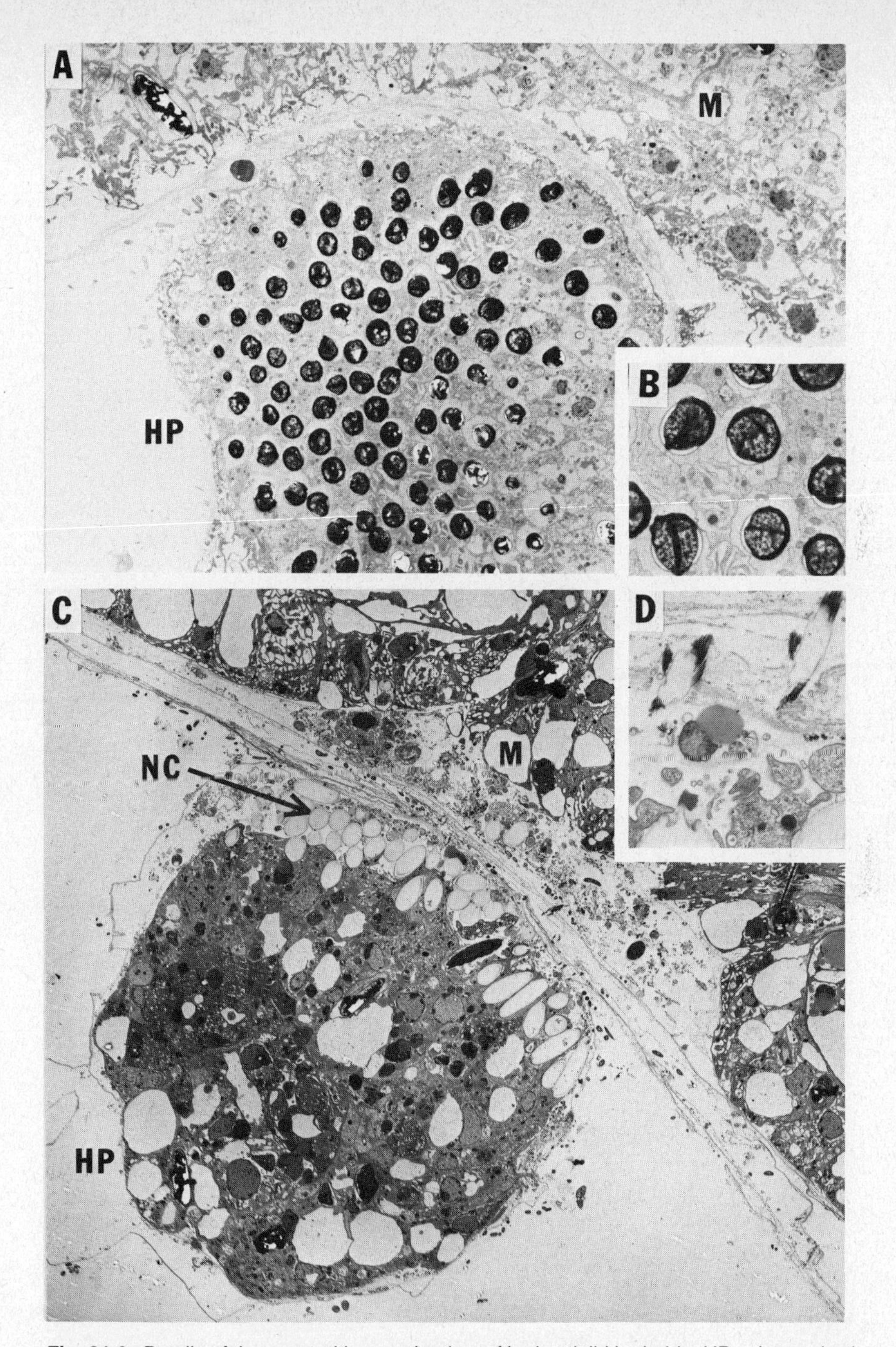

Fig. 31.6 Details of the competitive mechanism of hydractiniid hydroids. HP = hyperplastic stolon; M = ectodermal mat of competitor; NC = nematocyst capsule. (A) Transmission electron micrograph across the tip of a hyperplastic stolon in contact with a competitor (700 ×). Note that almost every cell harbors a nematocyst. (B) Part of section shown in A at greater magnification (2,880 ×). (C) Section across a hyperplastic stolon in contact with a competitor (704 ×). Capsules of discharged nematocysts are concentrated along the margin of the hyperplastic stolon where it is in contact with its competitor's tissue, and there is a zone of necrosis directly underlying this region. These discharged capsules are eventually sloughed off, a new set of nematocytes is differentiated, and the interaction is repeated until one colony has been completely eliminated. (D) The contact zone in C at greater magnification (2,880 ×), showing shafts of the nematocysts embedded in the foreign tissue. (Buss et al. 1984.)

does predation. The manner in which these two processes interact to produce patterns of distribution and abundance is clearly of fundamental importance. Since the publication in 1966 by Robert Paine of the predation hypothesis, many marine ecologists have focused on the suggestion that "local species diversity is directly related to the efficiency with which predators prevent the monopolization of the major environmental requisites by one species" (Paine 1966). The predation hypothesis has motivated an enormous amount of experimental work, mostly concentrated in the marginal marine environments (such as rocky shorelines), and is one of the few ecological suggestions based on solid experimental data, repeated several times, and in several places (e.g., Paine 1966, 1971, 1974; Paine and Vadas 1969; Harper 1969; Dayton 1971; Menge 1976; Lubchenco 1978, 1980; Lubchenco and Menge 1978; Sousa 1979).

The predation hypothesis holds only that "diversity is directly related" to the control of resource monopolization by predation or disturbance. The manner in which it is related has been and continues to be a topic of considerable research interest (see the following section on reconciling the predation hypothesis with observations of competition). One such relationship is that predation acts to preclude the occurrence of interspecific competition by holding structurally or numerically important prey species to densities at which interspecific competition fails to arise (Paine 1966). In this case predation and interspecific competition are mutually exclusive processes [but note that in such a case competition may yet occur between taxa of another morphological scale (Paine personal communication)]. If hypotheses are to be constructed to define the conditions under which competition will or will not be observed in nature, it is necessary to identify clearly the conditions under which predation can preclude the occurrence of competition.

Under what conditions will competition persist in the face of predation? At least five conditions may mitigate against predation precluding competition. Several of these limitations have been noted by others (Chapter 30; Dayton 1971, Connell 1975, Jackson 1977a, Lubchenco and Gaines 1981) and are treated only briefly here. Lubchenco and Gaines (1981) have divided these conditions into two categories: those in which predators and prey will coexist microsympatrically and those in which predator and prey will not coexist. In their classification conditions 1, 2, 3, and 5 (below) would permit coexistence, conditions 3 and 4 preclude coexistence

1. Settlement The relationship between a predator and its dominant prey is sensitive to variation in the density of both predator and prey. For example, Dayton (1971) attempted to repeat Paine's (1966) classic experiments on the effect of the predatory starfish *Pisaster* and failed. This failure was attributed to the failure of the competitively dominant mussel to recruit in sufficiently high densities that year (Dayton 1971, Paine 1974). Abundant evidence suggests that predation will act to preclude competition under only a restricted range of densities (Chapters 29 and 30; Dayton 1971; Yodzis 1978; Denley and Underwood 1979; Underwood et al. 1983; Underwood and Denley 1984; Roughgarden, Iwasa, and Baxter 1985).

2. Predator Resistance Even in the face of high predation pressure, some prey species typically escape predation through chemical, structural, or size-related antipredatory traits. Many such prey species are dependent on the predator, in that they are competitively inferior to species more susceptible to predation (Lubchenco and Gaines 1981, Steneck 1982). If two or more species prove predator-resistant and if recruitment is sufficiently great, predation will not act to preclude competition between the resistant species.

3. Predator Limitation Predators that might otherwise be capable of reducing competition between prey may be prohibited from doing so in some regions by physiological limitations or by reductions in their densities by their own predators (see reviews by Connell 1975, Lubchenco and Gaines 1981).

4. Refugia If predation pressure is sufficiently intense and chronically so, prey species may be forced to compete for refugia (see examples for freshwater fish and herbivorous insects in Chapters 21 and 31; Lubchenco and Gaines

1981, Fig. 4). Imagine two species that in the absence of predation reach high densities and compete for food. If predation is introduced, prey density may drop sufficiently that such interspecific competition no longer occurs. If, however, predation intensity increases yet further, competition may again become important in that prey may be driven to compete for refugia. Predation, if sufficiently intense and chronic, may act to change the resource that is limiting, in this case, from food or space to refugia. As long as the density of prey seeking refugia is greater than the availability of the refugia, competition will occur. Competition for refugia in the face of predation may well be a condition occurring quite frequently among mobile organisms whose foraging habits or breeding habits require that they be active at the same time as predators.

5. Clonality If predation is to preclude competition, it must either reduce the density of potential competitors seeking a limited resource or increase the availability of the resource at a rate faster than the demographic characteristics of the prey allow its exploitation (or both). The act of predation, though, need not result in an immediate reduction in prey density. In fact, predation on established clonal organisms rarely results in the death of the predated individual (Jackson 1977a). Rather, predators only open a patch within the clone which may either be immediately recaptured by regeneration or be captured by recruits. In the former case, the act of partial predation may have no immediate effect on competition. In the latter case, the capacity for clonal growth ensures that the originally predated clone will soon encounter the recruit, hence actually increasing the frequency of competition. Clonal propagation is a demographic character that will frequently allow a clone not only to escape predation, but also rapidly to recapture resources lost to predation. If prey are routinely capable of recapturing the resource freed by the loss of ramets to predators, predators will not necessarily act to preclude any existing competition between prey.

The failure of predation to preclude the occurrence of competition among clonal organisms needs no special demonstration. An encrusting coral colony that has a portion of its center eaten by a sea urchin is no less in competition with a sponge or another coral along its colony margin. The importance of clonality should not be underestimated. The overwhelming majority of the subtidal hard substrata in the sea are dominated by clonal groups (Jackson 1977a, 1983). Furthermore, a large percentage of lower trophic levels in terrestrial communities are clonal, including most grasses and herbs (Harper 1977). In addition, prey need not be clonal for predation to fail to influence density. A tree suffering aphid predation may nonetheless be faced with competition with a liana. As long as the portion of the prey consumed is an iterated unit, which the prey can either regenerate or bud anew (for example, the first several segments of an annelid), predation may also fail to have an immediate impact on prey density (Woodin 1982).

The routine failure of predation to reduce prey density in clonal groups should not be taken to suggest that predators are uniformly unable to reduce prey density. Predators may reduce prey density in clonal groups (1) if they feed primarily on dispersal units (such as in seed predation) or eliminate clones when small, (2) if predator density is sufficiently great, as when the crown-of-thorns starfish devastated acroporid reefs in certain Indo-Pacific islands, or (3) if predation reduces the viability of the clone sufficiently to increase the susceptibility of the colony to further predation or other threats. Nevertheless, these effects are often not sufficient to preclude prey from reaching densities at which they compete, as evidenced by the routine observation of competition among clonal, marine, sessile invertebrates (Jackson 1977a).

These five types of limitations, however, do not necessarily dilute the generality of the predation hypothesis. The predation hypothesis in its original form requires only that predation be "directly related" to prevention of resource monopolization (Paine 1966). For present purposes the hypothesis may be divided into two cases: one in which competition occurs and another in which competition does not occur. The conditions under which predation does not preclude competition are potentially quite frequent ones. The persistence of competition in the face of predation is an area of considerable recent work in marine systems and the subject of the remaining discussion.

RECONCILING THE PREDATION HYPOTHESIS WITH OBSERVATIONS OF COMPETITION

The predation hypothesis holds that predation is directly related to the prevention of resource monopolization. The manner in which it is related has proved to be rather complex. At least two patterns (see Chapter 29 for discussion) are known from experimental removal of predators in the field. (1) Increasing predator intensity may first increase local diversity to a maximum at some intermediate level and may act with further increases to decrease diversity; (2) increasing predation may act to decrease diversity in a monotonic fashion (e.g., Paine 1966, 1971, 1974; Paine and Vadas 1969; Harper 1969; Connell 1978; Lubchenco 1978). A number of authors have suggested that predators can increase diversity only when the predator feeds preferentially on that prey that is the clear competitive dominant (e.g., Paine 1966, 1971; Harper 1969; Hall et al. 1970; MacArthur 1972a; Lubchenco 1978).

Most experimental studies of predation and community organization are based on studies in the rocky intertidal. Unfortunately, data on the relative competitive abilities of intertidal sessile organisms are incomplete. Typically, competitive dominance in the intertidal is treated as synonymous with numerical dominance in the absence of predation (e.g., Paine 1966). While the relative competitive abilities of some subordinate species in these systems have been analyzed experimentally (Connell 1961a, Wethey 1984), most competitive abilities are either unknown or inferred from natural history observations.

Nevertheless, the hypothesis that feeding preferences and competitive rankings of prey in a largely transitive competitive series jointly determine whether diversity falls monotonically with predation or else rises and then falls was tested by Lubchenco (1978). She performed an elegant series of experimental manipulations of the introduced herbivore *Littorina littorea* on various algal prey. In this system the relative competitive abilities of the algal species are apparently reversed in different environmental settings: the perennial red alga *Chrondrus crispus* is dominant on exposed surfaces, whereas the ephemeral alga *Enteromorpha intestinalis* is dominant in tidal pools. Exploiting this competitive reversal in different habitats, Lubchenco manipulated the density of prey and obtained a striking result: diversity decreased monotonically with increasing predator pressure if competitive subordinates were the preferred prey, while diversity showed a maximum at intermediate predation levels if predators preferred the competitive dominant.

In the *Littorina* system a clear competitive dominant was apparent in each habitat type. In a number of other systems, however, overgrowth (dominance) rankings fail to allow the simple identification of a single competitive dominant (Fig. 31.3), raising the question of how predation interacts with competition in these systems. Jackson and Buss (1975) proposed that intransitive competitive relationships would strongly influence the rate at which resources are monopolized by a single species. Specifically, we suggested that the more intransitive the competitive ranking, "the slower will space tend to be occupied by a single competitive dominant, and [hence] the less the amount of external disturbance necessary to maintain a given level of diversity." Simulation studies of this amended hypothesis substantiate this prediction (Karlson and Buss 1984).

The basis for this suggestion is quite simple. Imagine a surface occupied by three encrusting colonial organisms. If the competitive relationships of these three species are purely transitive, the dominant will simply expand to cover the space occupied by its neighbors. If, however, the system is intransitive, each species will lose some ground along one region of its margin and gain ground in others. While it is clear that in the absence of disturbance a single dominant will eventually emerge in such a system (except under the biologically unrealistic condition that overgrowth rates are identical for all species), it is similarly clear that the rate at which the resource is monopolized by a single species will be far slower than for a comparable transitive system (see Buss and Jackson 1979, Figs. 5–7). This suggestion has been further investigated in simulation studies, which confirm that the time required for resource monopolization by a single species is highly sensitive to variation in competitive rankings and in rates of overgrowth (Karlson and Jackson 1981).

How, then, is predation related to diversity in intransitive systems? Given the extraordinary sensitivity of single species resource monopolization rates to relative overgrowth rates and the effect that partial predation on clonal organisms is likely to have on their growth rates, one might expect that predation might interact with competition by modifying overgrowth (or dominance) relationships. I know of only two studies that directly address this question (but see Park 1948, Steneck 1982), and I shall summarize both. One study used Panamanian cobble, and the other used Washington coral algal pavements.

Panamanian Cobble

As described earlier, cobbles in tidal channels at Punta Paitilla, Panama, are commonly dominated by two sessile encrusting cheilostome bryozoa and a coralline alga. Both static and time-course observations of overgrowth relationships among these species are highly intransitive (Fig. 31.2c). The three species occur in midintertidal channels, which I interpret (in the absence of experimental data) to mean that the channels are sandwiched between a higher zone in which thermal stress excludes bryozoans and a lower zone in which anascan bryozoans are largely replaced by ascphorans. These channels are characterized by an enormous diversity of potential predators, including a diverse gastropod guild. Species capable of completely removing the entire sessile encrusting community, such as cowries or urchins, cannot survive in these pools due to thermal stress. In addition to the gastropod predators the cobbles harbor a population of the isopod *Paraleptosphaeroma glynni,* a form known to occur only in association with the two bryozoan species (Buss and Iverson 1981). The isopod preys on the bryozoan species, one zooid at a time. Although the isopod preys on both bryozoans, small isopod individuals are more frequent than large individuals and can prey only on the faster growing species, *Antropora tincta* (Buss and Iverson 1981).

To determine the impact of isopod predation on the competitive relationships in the sessile community and on the maintenance of diversity in this system, I experimentally removed isopods from 25 cobbles and followed the fate of the encrusting community relative to 25 control cobbles. The removal process involved collecting the cobble bimonthly and cleaning the surface with a Water Pik. The isopods blown off in this process were removed and the remaining fauna (including juvenile gastropods) were allowed to repopulate the cobble. Controls were treated in the same fashion, except that isopods were also allowed to repopulate cobbles. The removal process was more than 95% effective. Isopods rarely recruited to the isopod-removal rocks because *P. glynni* lacks swimming appendages.

The results are: (1) The removal of the isopod predators leads to a reduction in diversity from the original three-species condition to a one-species system dominated by the fastest growing species, *Antropora tincta* (Fig. 31.7). (2) The manner in which this domination is achieved is mediated through a change in the dominance (overgrowth) rankings of the sessile community (inset, Fig. 31.7). Shortly after the initial removal the number of cobble on which competitive rankings remained intransitive decreased dramatically. Since the feeding rate of isopods alone is inadequate to account for the dominance of *A. tincta* (compare data in Buss 1981b and Buss and Iverson 1981), the isopods must have acted to influence the competitive ability of the fastest growing species relative to the two slower growing species, maintaining intransitivity, and reducing the rate at which the *A. tincta* achieved dominance. The transition from intransitivity to transitivity in this experiment is not a surprising result. Any multispecies space-limited competitive situation involving encrusting organisms, whether transitive or intransitive, will eventually lead to single species monopolization in the absence of disturbance, except in the biologically unrealistic condition of intransitivity with exactly equal rates of overgrowth (Jackson and Buss 1975, Buss and Jackson 1979, Karlson and Jackson 1981, Karlson and Buss 1984).

Washington State Coralline Algal Pavements

Paine (1984), reporting on collaborative work with Robert Steneck currently in progress, describes a similar pattern emerging in studies of coralline algal pavements on intertidal surfaces in

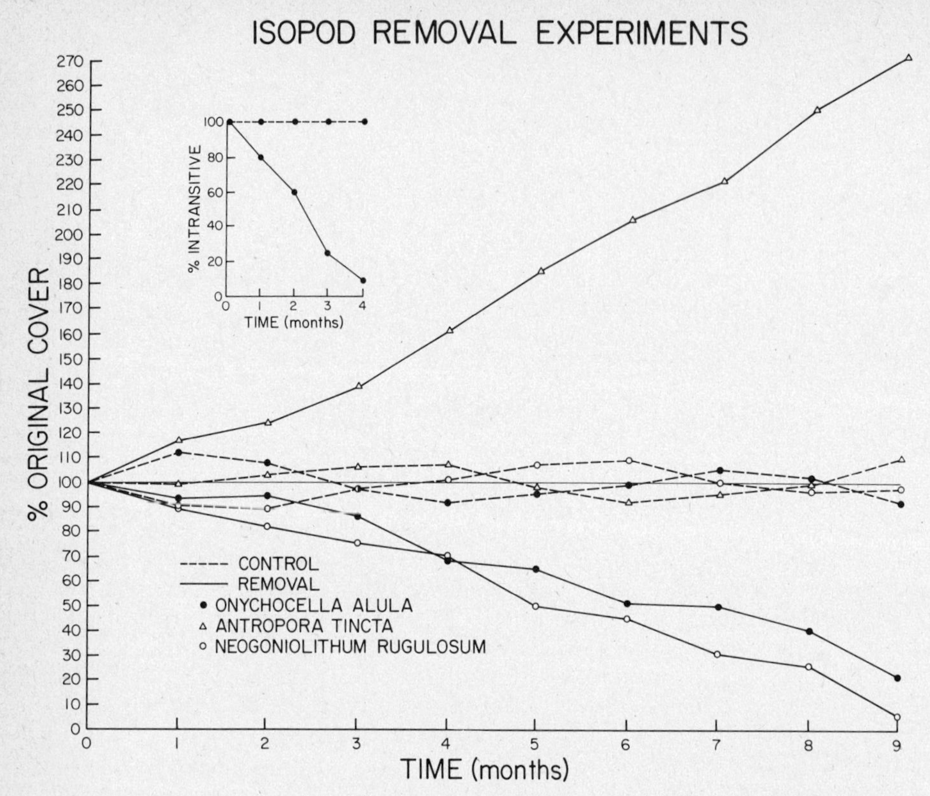

Fig. 31.7 Isopod removal experiments. Data are presented as means of the original cover for 25 control and 25 removal rocks. *Antropora tincta* comes to dominate isopod-removal rocks, whereas control rocks maintain all three species. Within four months of the original removal the competitive ranking on all substrata had shifted from highly intransitive to strictly transitive.

Washington State. Coralline algae are clonal plants that encrust surfaces and compete through overgrowth (Steneck 1982). Static overgrowth interactions are highly intransitive (Fig. 31.8). Corallines are routinely capable of persisting in the face of predation, and these particular species are exposed to activities of numerous herbivores, most commonly the chiton *Katharina tunicata* and the limpets *Acmaea mitra* and *Collisella pelta*.

Paine (1984) and Steneck experimentally reduced the density of predators and observed overgrowth rankings in unmanipulated control sites and predator removal sites (Fig. 31.8). In the absence of predators the overgrowth relationships of 1 of the 5 species treated (2 of 10 possible interactions) changed dramatically. Prior to the removal *Lithothamnium phymatodeum* (species B in Fig. 31.8) overgrew *Pseudolithophylum whidbeyense* (species C in Fig. 31.8) in only 33% of their interactions and overgrew *Lithophyllum impressum* (species D in Fig. 31.8) in 40% of their interactions, whereas subsequent to the removal, it was 98% and 100% successful, respectively, in these same interactions. Both of these shifts in dominance increased the transitivity of the overall dominance rankings in the system. The implication here is identical to that described above: predators modify existing competitive relationships, acting to decrease the incidence of transitivity in dominance relationships.

In addition to these experiments Paine (1984) and Steneck established a series of artificial communities and observed overgrowth relationships in the absence of predators. These artificial communities formed almost perfectly transitive competitive rankings in the absence of predation. The results from artificial communities are rather dif-

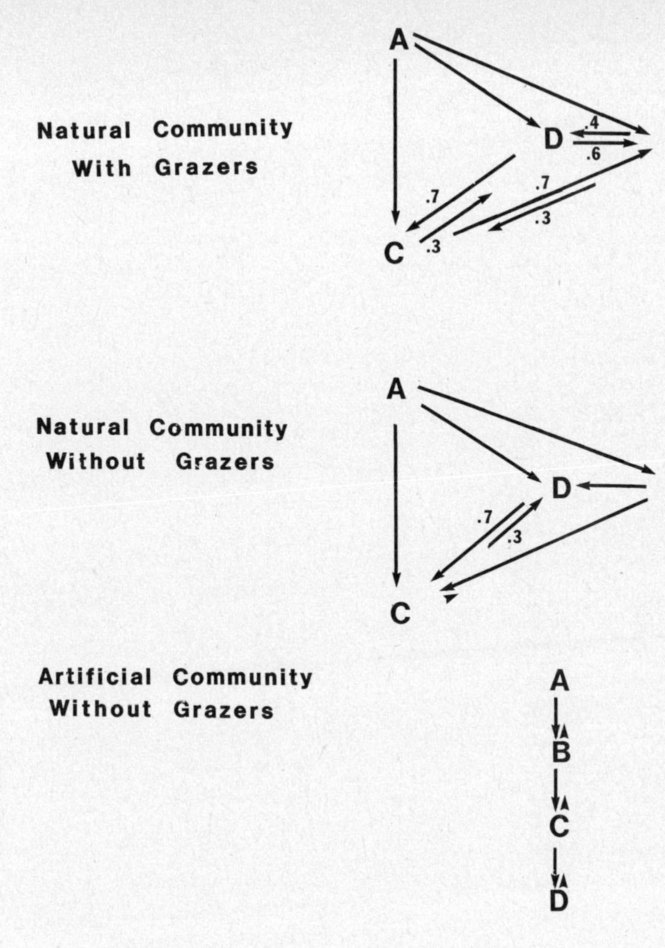

Fig. 31.8 Effect of predator removal on overgrowth relationships within natural and artificial assemblages in coralline algal communities. Arrows point from the winner to the loser. Numbers represent the percentage of total interactions. Competitive dominance was unidirectional, or nearly so, if no number appears. A = *Pseudolithophyllum lichenare;* B = *Lithothamnium phymatodeum;* C = *Pseudolithophyllum whidbeyense;* D = *Lithophyllum impressum*. A fifth species, *Bossiella* sp., is not figured, as it lost all its interactions with all other species in each treatment. (Redrafted from Paine 1984.)

ficult to interpret. These systems were established by chipping off pieces of coralline species and placing them in particular spatial arrangements on the substrata. Although this is a powerful experimental technique, such combats are clearly biased toward symmetrical interactions. On natural substrata thalli contact asymetrically: some thalli are larger than others, some thalli are thicker than others, some thalli are more shaded than others, and so forth. As discussed earlier in this chapter, there is ample evidence that size, encounter conditions, and morphological variability are correlated with competitive ability. Since the establishment of artificial communities removes environmental heterogeneity (Paine 1984), such systems may be unrealistically simple. Without control artificial communities that include grazers (or without detailed knowledge of the asymmetries in naturally occurring interactions) the degree to which the establishment of symmetrical communities biases results toward transitivity is difficult to interpret.

Nevertheless, the Panama and Washington studies are fundamentally similar in conclusion. In both cases predation acts on clonal organisms. In both cases predation does not, at least proximally, result in the death of the predated organism. In both cases predation does not preclude the occurrence of competition; rather, it acts to maintain it. Both experiments are fully consistent with the predation hypothesis (Paine 1966) that "local diversity is directly related to the efficiency at which predators prevent the monopolization of a resource by a single species." Finally, both cases support the original hypothesis that "intransitivity acts to slow the rate of single species resource monopolization and to reduce the level of disturbance required to maintain a given level of diversity'' (Jackson and Buss 1975).

Thus, predation appears to interact with competition in two quite different ways: (1) Predation reduces or precludes the occurrence of competition in some systems, while (2) predation acts to maintain the occurrence of competition in other systems. The limited data available suggest that the former mode is found in systems with transitive competitive relationships (e.g., Paine 1974; Connell 1978; Lubchenco 1978) and the latter mode in systems with intransitive rankings (Paine 1984). This apparent difference in the manner in which predation and competition interact to produce patterns in community organization raises an important question as to ultimate causes underlying these differences in proximate patterns.

Paine (1984 and Fig. 31.8), however, argues that ''the competitive relationships are transitive within this guild of coralline algae,'' since the removal of predators from an intransitive system results in transitivity. While this perspective has the virtue of defining competitive ability as a function of its consequences in the absence of predation, it has difficulties. Predation is only one process affecting competitive ability (Park 1948). Is it not equally arguable that competitive ability be assessed in the absence of surface irregularities, flow conditions, mutualistic species, thermal tolerances, and so forth? The ability of a organism to withstand predation is every bit as ''intrinsic'' to an organism as is the ability to withstand thermal conditions. A second, more important difficulty is that this redefinition focuses attention away from the considerable diversity (and interesting biology) of competitive mechanisms employed by sessile plants and animals. I concur with Miller (1967), who, in reviewing Park's (1948) classic study showing that predation can introduce variation in competitive ability, notes that ''considerable ambiguity and misunderstanding has resulted from defining the competition process solely according to its consequences.'' He further notes that this ambiguity has obscured the more important fact ''that the selective elimination of species . . . requires an extremely long time'' and that ''coexistence even in crowded populations seems to require such a slight alteration in a single factor,'' such as temperature or predation in Park's experiments.

Quite independent of one's choice of definitions, the question remains: What are the ultimate causes selecting for the proximate pattern of competitive intransitivity? Local marine environments typically communicate with one another by swimming larvae. The larvae arriving at any given local environment vary considerably in space and time (Sutherland 1974, 1976, 1977, 1978; Jackson 1977b; Osman 1977; Underwood et al. 1983). Further variation arises in the relative positions of recruits settling on a local substratum. Variation in the relative position of colonies established at settlement results in asymmetrical encounters between colonies (that is, at different sizes, densities, angles, and so forth). Intra- and interspecific competitive ability is known to vary as a function of these encounter asymmetries (size: Connell 1961a, Day 1977, Buss 1980, Russ 1982; encounter angle: Jackson 1979b, Buss 1981b, Rubin 1982; density: Buss 1981a), suggesting that competitive ability may be adapted to exploit particular classes of settlement-induced encounter asymmetries. Competitive intransitivity, thus, may ultimately reflect adaptations of organisms to variation in the timing and location of settlement, such that the various proximate patterns in competitive rankings and community organization result from the interaction of particular settlement patterns with postsettlement processes such as predation.

ACKNOWLEDGMENTS

I thank K. Carle, J. Lubchenco, J. Moore, R. T. Paine, J. Roughgarden, R. S. Steneck, D. Wethey, S. Woodin, P. Yodzis, and P. Yund for discussion or comments on the manuscript and the National Science Foundation (Grant OCE 81-61795) for support.

chapter 32

Relative Importance of Competition and Predation: Early Colonization by Seaweeds in New England

Jane Lubchenco

INTRODUCTION

The relative importance of competition and predation has been a central question in ecologists' continuing attempts to understand the great variety of communities. Under what circumstances does each of these interactions determine observed patterns?

This chapter presents what I believe to be the first experimental, quantitative assessment of this question in any system. The results permit an evaluation of how the relative importance of these two processes depends on physical factors, relative sizes of predator and prey, and life history characteristics of the prey. The most important factors, it turns out, are those characteristics of the organisms and of the environment affecting the ability of predators or physical factors to depress the prey population below the level at which resources are limiting and competition occurs.

Studies of Competition

Although many factors undoubtedly interact to produce community patterns, competition has received the most attention. There has been considerable disagreement over answers to the questions, "How frequent is interspecific competition in nature?" and "How important is it in determining the distribution, abundance, and resource use of species?" (Connell 1975; 1983; Wiens 1977; Schoener 1982, 1983b; Strong 1984c; Strong, Simberloff, Abele, and Thistle 1984). It is clear that there are no simple answers to these questions, and this is undoubtedly due to the variations among organisms (Chapter 28) and variations in results for the same organism at different times or places (Dunham 1980, Smith 1981, Tinkle 1982). Depending on the circumstances, competition may be important, of variable importance, or insignificant. Our challenge is to predict the frequency and intensity of interspecific competition, and to sort out how competition is related to predation or other critical factors.

Four hypotheses have been advanced regarding the types of conditions or organisms for which interspecific competition would be expected to occur and to be important.

1. Large-bodied organisms should be less susceptible to predators and weather than

smaller ones and should therefore be more likely to compete (Chapter 28; Schoener 1974a, 1983b; Connell 1975, 1983).

2. Plants and carnivores are more likely to be regulated by competition than are herbivores (Hairston, Smith, and Slobodkin 1960; Slobodkin et al. 1967). Hereafter, this model is referred to as HSS.
3. Higher trophic levels should be less affected by predators than are lower levels and should therefore be more strongly affected by competition, except where escapes from predators occur at lower trophic levels (Menge and Sutherland 1976). Hereafter, this model is referred to as M&S.
4. The frequency of competition in any particular system may be variable: Competition should occur only during times when resources are scarce (Lack 1947, Svardson 1949, Smith et al. 1978; but see Chapters 10 and 30 for a different view), and competition should be rare for some species in some highly variable environments (Chapters 9 and 15; Wiens 1977, 1983b).

To date, the best and most comprehensive evaluations of these hypotheses have been literature reviews of field studies of competition by Schoener (1983b, 1985c) and Connell (1983). Schoener (1983b) reports competition occurring in 90% of the 164 studies he reviewed and in 76% of the species investigated. Connell (1983) reports competition in most of the 72 studies he reviewed and in more than half of the species studied. The two reviewers agree on some points relative to the four hypotheses, but disagree on others.

1. Both reviewers agree that large organisms show consistently higher frequencies of competition than do smaller ones.
2. Connell concludes that plants, herbivores, and carnivores show similar frequencies of competition in all habitats compared, but that the evidence is still insufficient either to support or reject HSS. Schoener concludes that HSS is strongly supported for terrestrial and freshwater systems, but that no trend exists for marine systems. Schoener (1985c) discusses the reasons for these different conclusions.
3. Connell suggests that the data are inappropriate for an adequate test of M&S, because escapes from predators are not known in most cases
4. Both reviewers cite some studies reporting intermittent competition and other studies reporting continuous competition.

The sample of studies reporting competition in the literature is probably biased toward positive results. As Connell (1983) suggests, few investigators, at least until very recently (Munger and Brown 1981; Pacala and Roughgarden 1982, 1985) design experiments to evaluate competition when none is suspected. Connell (1983) suggests that this bias toward positive results should not grossly affect the comparisons between types of organisms or habitats cited in the previous paragraph. The bias does, however, make it difficult to evaluate the overall importance of competition from such a data base. Thus, while ''interspecific competition does occur in a great variety of natural systems and among many organism types'' (Schoener 1983b), we cannot yet evaluate the frequency of competition or specify the conditions under which it is important.

Studies Investigating Both Competition and Predation

Although the experimental investigations of competition mentioned above shed light on when competition is or is not important, a more complete understanding of the processes governing communities can be gained by evaluating the frequency and intensity of competition in conjunction with other factors affecting community structure. Predation has been the other factor most often investigated. Experimental studies that evaluate competition and predation simultaneously are rare. Some are included in Schoener's (1983b) and Connell's (1983) reviews; others have appeared more recently (Chapter 3; Lubchenco 1982, 1983; Keller 1983; Wilbur et al. 1983). Chapters 21 and 31 and Lubchenco and Gaines (1981) discuss some of the ways in which competition and predation might interact.

The picture emerging from these studies is that interactions between competition and predation are probably complex and vary with characteristics of predators, prey, and environment.

Although many studies support the hypothesis that predators prevent resource monopolization by competitively dominant prey (Chapter 3; Paine 1966, 1980; Harper 1969; Hall et al. 1970; Connell 1975; Menge and Sutherland 1976; Lubchenco 1978; Morin 1981, 1983a; Wilbur et al. 1983; Jara and Moreno 1984), other studies do not support it (Lubchenco 1978; Peterson 1979, 1982; Williams 1981). Predators appear ineffective in many situations in which interspecific competition has been demonstrated (Chapter 31; Menge 1976, Lubchenco and Menge 1978, Williams 1981). Furthermore, the effects of predators vary, from changes in species richness or diversity of prey (Paine 1966, Addicott 1974, Lubchenco 1978) to changes in prey relative abundance without altering richness (Brooks and Dodson 1965, Zaret 1980, Morin 1983a) to simple changes in prey density but not richness (Peterson 1979, 1982).

Various factors have been suggested as affecting how competition and predation interact. These factors include the type, size, abundance, and prey preference of predator; number of predator types and species; competitive abilities, size, and defenses of prey; timing of predation relative to competition and to the life history of the prey; and differential effects of the environment on predators and prey (Chapters 29 and 31; Paine 1966, 1980; DeBenedictis 1974; Connell 1975; Lubchenco 1978; Menge 1978a, 1978b, 1983; Peterson 1979, 1982; Lubchenco and Gaines 1981; Morin 1983a; Hixon and Brostaff 1983; Wilbur et al. 1983). Various indirect interactions also undoubtedly influence the patterns (Chapters 3, 26). Although there is experimental evidence that each of these factors is important in at least one system, a comparative appreciation of their importance in many systems will be necessary for us to make significant progress toward a predictive theory of community organization.

The balance of this chapter is devoted to the initial phase of a systematic, quantitative investigation of the factors influencing the relative importance of competition and herbivory for one system: the seaweeds in the mid zones of rocky shores in New England. This first phase focuses on factors affecting colonization by seaweeds and defers study of factors affecting their persistence. I examine how interspecific competition among seaweeds and effects of grazers vary in importance along a gradient of environmental harshness (wave action) and in different seasons (summer versus winter). Results of a quarter of the experiments have been published (Lubchenco 1983); the rest are summarized here for the first time.

My bias in these studies is that understanding competition between plants depends on understanding the ability of herbivores and abiotic factors to regulate plant abundance. Knowing the impact of herbivores and weather informs us of the potential for competition. The ability of herbivores to control algal abundance may be influenced by (1) features of the plants, such as size, chemistry, life history, and morphology, and (2) features of the environment, such as wave action, that reduce herbivore abundance or activity but not plant growth (Lubchenco and Gaines 1981). The remainder of this chapter considers two questions: How do herbivory and wave action influence competition between algae? What characteristics of the plants and their environment affect herbivory and competition?

SEAWEEDS IN NEW ENGLAND: BACKGROUND

The seaweeds to be discussed live on emergent rock in the mid zone (+1.8 to +0.6 m tidal height) in marine, continuous-rock sites with predominantly horizontal slope along the New England coast north of Cape Cod. Experimental results are reported from three sites that vary in their exposure to wave action (described in J. Menge 1975; B. Menge 1976; Lubchenco 1980, 1983). Canoe Beach Cove on the Nahant peninsula in Massachusetts is the most protected from wave action; Grindstone Neck on the Schoodic peninsula in Maine is intermediate; East Point, Nahant, is exposed. Some observational data (percent cover patterns) are also shown for an additional exposed site: Pemaquid Point, Maine (B. Menge 1976).

Potential Competitors and Their Characteristics

Potential competitors for space during initial colonization events in the mid zone in New England are various species of seaweeds and sessile, filter-feeding animals such as barnacles and mus-

Table 32.1 SOME CHARACTERISTICS OF THE SEAWEEDS STUDIED

	Ephemeral species	Perennial species
Life span	Weeks to months	Years (up to 15–19)*
Adult size	Small (under 20 cm in length)	Large (from 10 cm to 1.5 m)
Growth rate	Fast	Slower
Palatability to generalist grazers†	High	*Fucus*, juvenile: high *Fucus*, adult: low *Ascophyllum*, all stages: low
Successional stage	Early	Middle to late
Juvenile characteristics	Similar to, but smaller than adults	Smaller and more palatable than adults
Dispersal ability	Good	Poor

Common ephemeral species include the genera *Enteromorpha, Rhizoclonium, Spongomorpha, Ulothrix, Ulva* (Chlorophyta); *Bangia, Ceramium, Porphyra* (Rhodophyta); *Petalonia, Scytosiphon* (Phaeophyta); and filamentous diatoms. The common perennial species are *Fucus* spp. and *Ascophyllum nodosum*.

*Individual fronds of *A. nodosum* have been reported to live 15–19 years (David 1943, Keser et al. 1981); holdfasts (which can perennate new fronds) may live much longer (Baardseth 1970).

†Data from Lubchenco 1978, 1983.

sels. Interactions between the filter feeders and the algae will be considered only briefly, as they are complex and incompletely understood. The present focus is on algal-algal competition.

Several characteristics of algae affect their competitive abilities (Table 32.1). Most of the common seaweeds may be categorized as either ephemerals (individual plants that persist in a macroscopic form for less than a year, usually from a few weeks to a few months) or perennials. Relative to adult perennials, ephemerals are generally small, colonize well, grow quickly, and are highly palatable to generalist grazers (Lubchenco 1978, 1980, 1983; see also Littler and Littler 1980, 1984; Steneck and Watling 1982). Young, newly settled individuals are not much different from larger individuals, except in size. Genera of ephemeral algae include *Ulva, Enteromorpha,* and *Porphyra*.

The common perennial species are all fucoid algae. *Fucus vesiculosus* and *F. distichus* ssp. *edentatus* are both middle successional plants, which often colonize when secondary succession is initiated. *Ascophyllum nodosum* (L.) is a late successional species, which usually recruits slowly and in low numbers (see Keser and Larson 1984 for an exception). Although each of these three species can form a dense canopy, the holdfasts (attachment structures) occupy only a fraction of the primary space underneath the canopy and usually leave much primary space unused (Table 32.2). Adult fucoids range in size from approximately 10 cm to 1.5 m. Newly settled *F. vesiculosus* are more palatable than are larger plants to generalist grazers such as *Littorina littorea* (Lubchenco 1983, unpublished data). Although comparable data are not available for *Ascophyllum,* all stages of this plant appear to be unpalatable to most of the common grazers.

Thus, these ephemeral and perennial species differ in their adult size, palatability to generalized grazers, growth rates, colonization abilities, and characteristics of juveniles versus adults. In several respects, as made clear by Tables 28.2 and 28.3, these marine algal communities of large perennial and small ephemeral species resemble the terrestrial chalk grassland communities discussed by Grubb in Chapter 12, with matrix-forming perennial and interstitial short-lived species.

Patterns of Resource Use

Seaweeds can compete for either space or light. Spatial competition involves a place to settle and grow on the rock surface and probably involves preemption of the resource (*sensu* Schoener 1983b). Although ephemeral and perennial species do compete for primary space, that is, space on the rock surface itself (Lubchenco 1983), the interaction is complicated, because ephemeral species can also live as epiphytes attached to pe-

Table 32.2 SEASONAL PATTERNS OF UTILIZATION OF SPACE IN THE MID ZONE AT THREE SITES DIFFERING IN EXPOSURE TO WAVE ACTION

	Winter	Summer	Winter	Summer	Winter	Summer
Canoe Beach Cove (protected)						
Date	4/25/74	8/18/74	3/25/75	8/5/75	2/21/76	7/15/76
Primary						
Bare	9 ± 25	22 ± 31	11 ± 21	14 ± 24	22 ± 25	11 ± 11
Fucoid holdfasts (P)	5 ± 3	2 ± 3	3 ± 3	2 ± 2	t	1 ± 2
Ephemerals	0	0	0	t	0	1 ± 1
Understory						
Fucus vesiculosus (P)	0	0	1 ± 2	2 ± 4	3 ± 5	1 ± 2
Canopy						
Ascophyllum nodosum (P)	87 ± 16	94 ± 9	67 ± 33	86 ± 19	81 ± 19	70 ± 31
F. vesiculosus (P)	2 ± 5	3 ± 6	3 ± 5	1 ± 2	4 ± 6	4 ± 9
F. distichus (P)	0	0	0	0	0	0
Grindstone Neck (intermediate)						
Date	3/9/74	7/22/74	3/2/75	7/14/75	2/15/76	7/12/76
Primary						
Bare	14 ± 10	15 ± 22	10 ± 22	18 ± 27	6 ± 4	10 ± 14
Fucoid holdfasts (P)	4 ± 4	4 ± 12	3 ± 3	2 ± 2	1 ± 1	2 ± 1
Ephemerals	12 ± 25	7 ± 8	10 ± 22	2 ± 4	6 ± 10	t
Understory						
F. vesiculosus (P)	0	0	t	1 ± 3	t	t
Canopy						
A. nodosum (P)	0	0	0	0	1 ± 2	0
F. vesiculosus (P)	82 ± 7	78 ± 25	83 ± 27	89 ± 20	70 ± 24	96 ± 10
F. distichus (P)	0	0	0	0	0	0
Pemaquid Point (exposed)						
Date	3/8/74	7/20/74	2/28/75	7/16/75	2/14/76	7/13/76
Primary						
Bare	t	2 ± 4	0	6 ± 11	5 ± 9	9 ± 7
Fucoid holdfasts (P)	t	t	t	t	t	t
Ephemerals	18 ± 17	8 ± 10	63 ± 32	7 ± 11	27 ± 27	12 ± 17
Understory						
F. distichus (P)	1 ± 1	t	t	t	t	t
Canopy						
A. nodosum (P)	0	0	0	0	0	0
F. vesiculosus (P)	0	0	0	0	0	t
F. distichus (P)	2 ± 4	3 ± 5	3 ± 5	6 ± 7	5 ± 6	12 ± 16

Data are means and standard deviations of percent cover of ten 0.25-m^2 quadrats. Quadrats were placed randomly along a 30-m transect line that ran through the middle of the mid zone parallel to the water's edge (see Menge 1976 for further details). Primary space is defined as space on the rock surface. It may be shaded by canopy species or not. In addition to the primary space occupants shown, mussels, barnacles, and crustose algae are often present (see Lubchenco 1980, Table 1, and Menge 1976, Figs. 2 and 4; for additional information on cover of these species).

t = trace, meaning that the species or group was present, but occupied less than 1% cover.

P = perennial species (see Table 32.1 for a list of common ephemeral species).

rennial species. Thus, preemption of the primary rock surface by perennial species does not totally eliminate ephemeral species from the system.

Competition for light might occur because the distal portions of many seaweeds, especially perennials, are broader than the area occupied by the holdfast. Thus, although space on the rock may be unoccupied, it may be unsuitable for colonization because it is heavily shaded or otherwise affected by one or more adjacent large plants.

The potential types of competitive interactions investigated in this paper are those associated with settlement phenomena: (1) competition for unused space between newly settled propagules of different species, and (2) effects of adult perennials on recruitment of ephemerals or sporeling perennials under the canopy.

How much of the space is used and what species use it vary spatially and temporally (Table 32.2). At protected and intermediate sites perennial seaweeds usurp most of the resources by covering most of the rock surface (averages of 67 to 96% cover) throughout the year. Ephemerals are always rare on the rock surface at protected sites (0 to 1% cover). They are more abundant at intermediate sites (trace to 12% cover) and slightly more common during the winter than the summer. At exposed sites perennial species cover only a fraction of the space (2 to 12% cover), and ephemerals are often abundant (7 to 63% cover), especially during the winter. Crustose algae, which are not considered in this paper, also occupy primary space, most often under the fucoid canopy.

Filter feeders are the other major primary space occupants. They range from rare at protected sites to abundant at exposed sites (Menge 1976, Fig. 2; Lubchenco 1980, Table 1). Since the competitive interactions between filter feeders and seaweeds are beyond the scope of this study, the only habitat of concern at exposed sites is bare rock, a resource created when storms remove sheets of mussels.

Thus, the two general arenas of potential algal-algal competition of interest here are bare rock under canopy (protected and intermediate sites) and bare rock in the open (without canopy; all sites). (''Bare rock'' indicates that no macroscopic organisms, including algal crusts, are present.) Canopy sites provide the opportunity to examine how a dense growth of adults of late successional species affects colonization of potential invaders. Open sites afford the chance to investigate factors affecting competition during early successional stages. In both arenas the potential colonists of interest are the early (ephemeral) and middle (*Fucus* spp.) successional species. (*Ascophyllum nodosum* does not appear until later in the successional sequence). The designation of *Fucus* spp. as middle successional refers to adult plants; juvenile *Fucus* appear early in succession.

Effects of Herbivores on Competitive Interactions

The abundance and activity patterns of gastropod grazers are correlated with degree of wave action. Snails are abundant at protected sites, common at intermediate sites, and rare at exposed sites (Lubchenco 1980). They are active throughout the summer and relatively inactive during the winter (unpublished data). Amphipod grazers are also present, but their abundance, activity, and effects are unknown.

The effects that gastropod grazers (primarily the periwinkle *Littorina littorea*) have on competition between the ephemeral species *Ulva, Enteromorpha,* and *Porphyra* and the perennial *Fucus* species have been experimentally investigated by using snail exclosures and selective removals of the algae (Lubchenco 1983). The results indicate that when bare rock is made available in open habitats, that is, when secondary succession is initiated, at protected and intermediate sites, herbivory normally prevents the competitively superior ephemeral species from becoming abundant and thus allows the perennial *Fucus* species to colonize successfully. Although quite a few juvenile *Fucus* species are eaten by snails, the adults are less vulnerable. A sufficient number of juveniles escape in space or time, grow, and eventually occupy most of the canopy space. If grazers are removed when succession is initiated, ephemeral algae outcompete the perennial species if both colonize (Lubchenco 1983).

Although these experiments indicate that herbivores can prevent competition between ephemeral and perennial species, they do not address the following concerns: (1) How frequently does this happen? (2) Do agents other than grazers, for example, wave action at exposed sites, have similar effects? (3) Can adult perennials inhibit recruitment of ephemeral or perennial spores? The following experiments address these questions.

EXPERIMENTAL DESIGN

Experiments were initiated at the three sites to evaluate the relative importance of herbivory and algal-algal competition in determining initial colonization patterns of seaweeds. The basic experimental design utilized stainless steel mesh cages and was similar to that described in Lubchenco 1983. Each experiment consisted of a series of treatments: (1) unmanipulated control (no cage or roof), (2) roof control (roof without sides; tests for effects of mesh), (3) herbivorous gastropod exclusion (cages), and (4) herbivorous gastropod exclusion combined with removal of potential competitors (cages). Each treatment was performed 1 to 13 times. The data summarized below are from all of the experiments in which the appropriate treatments were included (some other experiments lacked competitor-removal treatments, for example). These 22 experiments were conducted between November 1972 and August 1975 at the three sites. The number of experiments conducted at the different sites and in different microhabitats is indicated in Table 32.3.

Algal competition experiments were designed to focus on interactions between groups of algae, for example, between ephemerals and sporeling perennials and among ephemeral species. Any ephemeral algae found in the competition treatments (treatment 4 above) were first sampled for percent cover and then removed at each sampling date. Perennial species (usually *F. vesiculosus*) were not removed. (There was never more than one perennial algal species in any experiment.) All competition experiments were conducted in the absence of grazers. Competition was deemed to occur and to have a strong influence on algal abundance (percent cover) if herbivore exclusion treatments were significantly different from herbivore-exclusion–plus–algal-removal treatments in any particular experiment. Significance was determined by comparing percent cover of algae

Table 32.3 NUMBER OF EXPERIMENTS AND CONTROL TREATMENTS IN EACH EXPERIMENT IN DIFFERENT MICROHABITATS AT THREE SITES

	Canopy	Open
Protected		
Experiments	3	5*
Control treatments	6	37*
Intermediate		
Experiments	3	8
Control treatments	9	20
Exposed		
Experiments	0	3
Control treatments	0	6
Total		
Experiments	6	16
Control treatments	15	63

Experiments under the canopy test for effects of competition between adult perennials and ephemerals on bare rock and competition between adult perennials and perennial spores on bare rock. Experiments in the open test for effects of herbivory on ephemerals and sporeling perennials, competition between ephemerals on bare rock, and competition between ephemerals and sporeling perennials on bare rock. Control treatments include unmanipulated control and roof control treatments. These two treatments were not statistically different in any experiments (see e.g., Lubchenco 1983).

*Two of these experiments, which included 22 controls, appear in Lubchenco and Cubit 1980, Fig. 3, and Lubchenco 1983, Fig. 1.

in the two treatments throughout the experiment using Hotelling's T^2 with the Bonferroni approximation (see Lubchenco 1983 for details; Morrison 1976). Results were usually dramatic and clear-cut. For example, ephemerals dominated the herbivore exclusion treatments to the exclusion of *F. vesiculosus,* while *F. vesiculosus* became abundant in the herbivore-exclusion–plus–ephemeral-algal removal treatments. The conclusion: ephemeral algae outcompeted sporelings of the perennial *F. vesiculosus* (see Fig. 32.1B and 32.1C for some actual results).

Herbivory was concluded to have a strong influence on algal abundance if there was a significant difference between control treatments and herbivore-exclusion treatments within a particular experiment. Again, significance was assessed with Hotelling's T^2 using the Bonferroni approximation. Percent cover data of individual species were used for analyses.

The effects of herbivory and competition on algal abundance and species composition were thus assessed separately. In any particular experi-

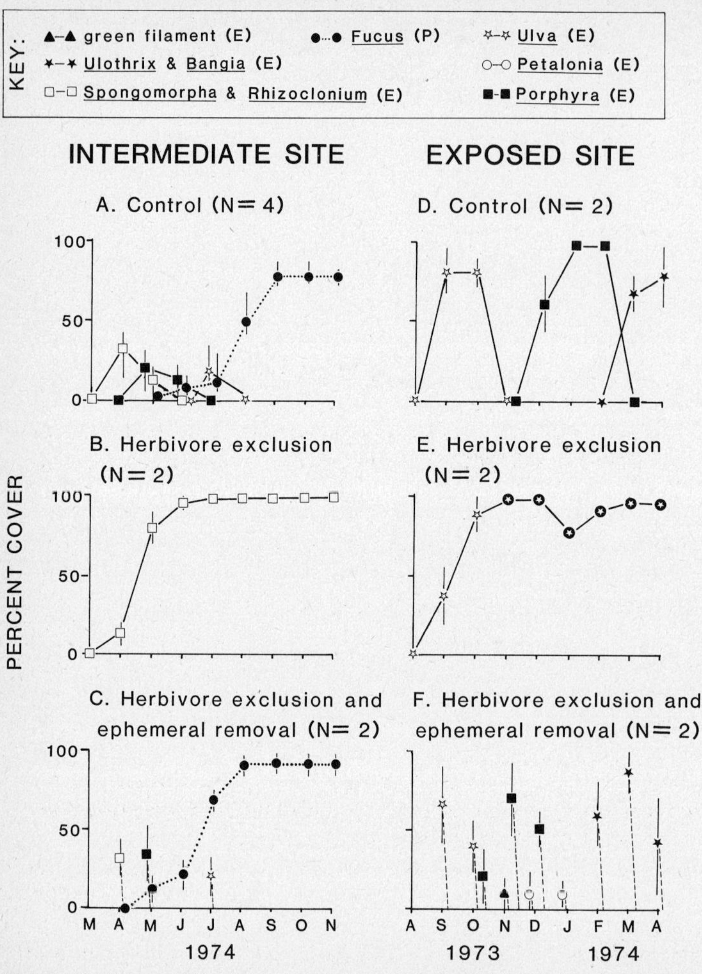

Fig. 32.1 Effects of herbivores and ephemeral algae on percent cover of ephemeral and perennial algae at a site intermediate in exposure to wave action during the summer (A, B, and C) and at a site exposed to wave action during the winter (D, E, and F). Data shown are means and ranges. Number of replicates is indicated in parentheses. In C and F, the ephemeral algae present each month were removed after their cover was measured. Removals are indicated by dashed lines. In E, from November on, *Ulva* was loose in the cages, i.e., no longer attached to the rock. Loose algae are indicated by a circle around the species symbol. Solid lines connect data points for ephemeral species (E in key); dotted lines connect data points for perennial species (P in key).

ment algal abundance could potentially be affected by herbivory and competition, herbivory alone, competition alone, or neither.

Since cages most effectively exclude snails on flat, relatively smooth surfaces, most of the experiments were conducted on these surfaces. Experimental plots did contain small crevices, depressions, and holes, which provide important spatial refuges for *Fucus* species from some snail grazers (Lubchenco 1983), but did not include any tide pools, large crevices, boulders, or other large angularities.

KINDS OF INTERACTIONS

Information about the kinds of interactions operating in this system (e.g., competition versus herbivory) was obtained from manipulations of grazers and competitors. The results are presented in this section, according to habitat. An evaluation of the frequency of each interaction is reserved for the following section.

Open Habitats

Three outcomes were observed on bare rock in noncanopy control treatments. (1) Herbivores prevented competitive exclusion by ephemeral algae, whereupon perennial species became abundant. (2) Herbivores did not control ephemeral algae, and competition occurred among the ephemeral species. (3) Intense wave action periodically removed ephemeral algae, preventing any single species from monopolizing space for more than a few months.

Herbivores Prevent Competitive Exclusion This outcome occurs at protected site experiments in the summer (Lubchenco 1983, Fig. 1) and at some intermediate site experiments in the summer. Although variation occurs between experiments in the identity and amounts of the ephemeral plants, the general sequence of events is uniform and is illustrated by Figs. 32.1A, B, and C. In this experiment various ephemeral species settle in the control plots, but are eaten by the grazers. Eventually the perennial *Fucus vesiculosus* becomes established (Fig. 32.1A). The potential for competitive exclusion by at least one ephemeral species is indicated by the herbivore-removal treatments. The ephemeral species that was initially present monopolized each cage to the exclusion of both other ephemeral species and the perennial *Fucus* (Fig. 32.1B). The effects of the grazers can be mimicked in treatments combining removals of herbivores and ephemeral algae (Fig. 32.1C). A comparison of Figs. 32.1A and 32.1C confirms that if ephemeral species are removed, *Fucus* will colonize. It further suggests that, in the experiments shown, there is a slight competitive effect of intermediate amounts of ephemeral algae on the rates at which *Fucus* appears and becomes abundant. *Fucus* appears sooner and becomes abundant sooner in both of the plots where I removed ephemeral algae (Fig. 32.1C) compared to the four plots where snails removed ephemeral algae less efficiently than I did (Fig. 32.1A).

Among the experiments not illustrated, variation occurred in the identity of the ephemeral alga that monopolized space in the herbivore-exclusion cages and in the degree of competition between ephemeral and sporeling perennial algae. In every herbivore-removal experiment usually one, but occasionally two, ephemeral algae outcompeted the others present and eventually took over all the space. These included *Rhizoclonium, Spongomorpha, Ulva,* and *Porphyra*. As was the case in Fig. 32.1B, the species that eventually monopolized the cage was the one that settled abundantly soon after the experiment was initiated. In most of the other control treatments (22 out of 24) ephemeral algae were less abundant than in Fig. 32.1A. As a result, there was usually no significant difference in recruitment rate or eventual cover of *Fucus* between control and the combination herbivore-exclusion–plus–ephemeral-algal-removal plots. If no such difference was observed, it was concluded that no competition occurred between ephemeral algae and sporeling *Fucus*.

These experiments suggest four conclusions: (1) Herbivores strongly reduce the abundance of ephemeral algae. (2) Herbivores strongly promote the appearance of the perennial *Fucus vesiculosus*. (3) Competition between ephemeral algae does not have a strong effect on their abundance (they are usually not sufficiently abundant to compete). (4) Competition between ephemeral algae and sporeling perennials may occur, but it is not as important as herbivory in determining the abundance or type of algae. In these experiments grazers effectively controlled the abundance of the competitively dominant ephemerals to the extent that either no competition occurred (79% of control plots) or competition had only a slight effect (21% of control plots).

Competition Occurs Between Ephemerals This result occurs where herbivores are either not abundant or not active and where wave action is not severe. These circumstances exist at some intermediate site experiments in the summer, at protected and intermediate sites in the winter, and at exposed sites in the summer. When competition occurs, usually a single ephemeral species becomes abundant, monopolizes the space, and prevents colonization by other ephemeral species or by perennials. In such experiments control treatments are not different from herbivore-exclusion plots. In the herbivore-exclusion–plus–ephemeral-algae-removal treatments at all but exposed sites *Fucus vesiculosus* eventually monopolizes space.

Wave Action Controls Algal Abundance When wave action is severe, it controls the abundance of ephemeral seaweeds. This occurs at exposed sites during the winter (Figs. 32.1D, E, and F). In controls a single ephemeral species often monopolizes space for brief periods (a few months), disappears, and then is replaced by another ephemeral species (Fig. 1D). Algae in control treatments almost always disappeared during storms. Where each of these species is removed monthly, more ephemeral species coexist (Fig. 32.1F). For example, where *Porphyra* is removed, *Petalonia, Bangia,* and *Ulothrix* all colonize (Fig. 32.1F, January and February), but do not appear where *Porphyra* is present (Fig. 32.1D, January and February). Differences between control and herbivore-exclusion cages at exposed sites are due more to effects of cages in ameliorating wave action than to differences in herbivore activity; controls and cages do not differ in herbivore activity, because periwinkle snails are rare to absent at the exposed sites (Lubchenco 1980). The potential for some ephemeral algae to persist for longer than is observed in the control treatments is indicated by the exclusion cages (Fig. 32.1E). *Ulva* persisted for eight months in cages, but for only two months in control treatments. These plants (inside cages) became detached from the rock during the same storm in which *Ulva* disappeared in the control plot. A similar persistence of *Porphyra* in cages, but not controls, was observed in other experiments. Thus, although waves appear to prevent long-term monopolization of space by ephemeral species, single species may be competitively dominant for shorter periods. Therefore, periodic disturbance from waves appears to determine the general pattern of species abundance. Between bouts of disturbance, competition also occurs, but it is of secondary importance.

No perennial species was observed to colonize any of the experimental or control plots at exposed sites. This is probably due to a low abundance of propagules resulting from the paucity of adults of perennial species in the general vicinity (Table 32.2).

These results indicate that competition among ephemeral algae and between ephemerals and sporeling perennial species in the open is mediated by herbivores and wave action. Competition between these groups is most likely when neither herbivory nor wave action is strong.

Canopy Habitats

Experiments placed on bare rock under a canopy of either *Ascophyllum nodosum* or *Fucus vesiculosus* indicate that these perennial species preempt the primary space covered by their canopies. Neither ephemeral nor perennial algae settle under the canopy in these experiments, even in herbivore-removal treatments. This result is reinforced by general observations. Few algal individuals are seen under a dense canopy. The only species observed to colonize beneath the canopy was the climax species *A. nodosum,* which recruited under both *A. nodosum* and *F. vesiculosus* in low numbers in the general habitat but not in experiments. The absence of colonists under the canopy is not caused by a lack of propagules in the water column; both ephemeral and perennial species readily colonize plots that are adjacent to the canopy sites, but lack a canopy. These plots without a canopy include both naturally open and experimentally opened sites (Menge and Lubchenco, unpublished data). Thus, adult perennial plants appear to exhibit strong competitive effects against ephemeral species and against zygotes of *Fucus* species.

FREQUENCY OF INTERACTIONS

The frequency of each interaction within the open or canopy habitats was evaluated in the following manner. Since each habitat is relatively uniform, the frequency of a particular process within the habitat can be estimated by the proportion of control plots affected by the process. The frequency of each interaction within a habitat was determined by tallying the number of control plots in which the interaction had a strong influence on the outcome of the experiment (as judged by experimental treatments) and comparing that number to the total number of control plots in the habitat (Table 32.4). The treatments in which either herbivores or competitors or both were manipulated were thus used only to indicate the interaction occurring; they were not included in frequency calculations. For example, the experiments shown in Figs. 32.1A, B, and C would be scored as follows. During the summer at the intermediate site herbivores had a strong, negative influence on the final abundance of ephemerals in 4 out of 4 times and of perennials in 0 of 4 times. Competition determined the end result in 0 of 4 times between ephemerals and in 0 of 4 times between ephemerals and sporeling perennials. The data in Table 32.4 thus include the control treatments shown in Fig. 32.1 plus those not illustrated. This combination indicates, for example, that during the summer herbivores controlled the abundance of ephemeral algae, thus preventing ephemeral-ephemeral competition and ephemeral–sporeling perennial competition in 28 of the 28 control treatments at the protected site, in 11 of the 18 control plots at the intermediate site, and in 0 of 6 control plots at the exposed site (Table 32.4). The frequencies of each type of interaction will now be discussed by habitat.

Open Habitats

The relative importance of herbivory, competition, and wave action in the open habitat changes with season and exposure to wave action (Table 32.4 and Fig. 32.2). Competition is most frequent where neither herbivores nor waves keep ephemeral and sporeling perennial algae in check. This occurs at protected sites during the winter, at intermediate sites during the winter and sometimes the summer, and at exposed sites during the summer. The ability of herbivores to control ephemeral seaweeds is limited to the summer and decreases along a wave exposure gradient (Fig. 32.3A; G test on number of control treatments showing controlling effects of herbivory versus number not showing such effects at each of three sites, $p < 0.001$). As a consequence, the frequency of summertime ephemeral-ephemeral competition increases with the degree of wave exposure (Fig. 32.3B; G test, $p < 0.001$). Wintertime ephemeral-ephemeral competition occurs more frequently at protected and intermediate sites than at exposed sites (Fig. 32.3B; G test, $p < 0.001$). Although herbivores are inactive or absent at all three sites during the winter, wave action is more intense at the exposed site. At protected sites ephemeral-ephemeral competition occurs more frequently during the winter than during the summer (Fig. 32.3B; G test, $p < 0.001$).

Evaluating the frequency of strong herbivory on sporeling perennials (Table 32.4, row A2) is difficult because the grazers have complex direct and indirect effects on these plants. *Littorina littorea* eats sporeling *Fucus* species (a direct, negative effect). But since substratum heterogeneity created by small cracks and crevices provides spatial refuges for small *Fucus* plants (Lubchenco 1983), the effectiveness of grazing is a function of the local microtopography. Furthermore, snails have an indirect, positive effect on sporeling perennials by eating ephemeral algae that outcompete the perennials (Lubchenco 1983). Table 32.4 shows the number of times that sporeling perennials appeared and persisted in control plots, that is, the number of times that at least some *Fucus* sporelings were neither outcompeted by ephemeral algae nor eaten by grazers. Since the number of sporelings persisting to a stage 3 to 5 cm long (at which point they are more immune to grazing) depends on the spatial heterogeneity of the rock, the numbers shown in row A2 of Table 32.4 represent only what happened in those specific treatments and are less indicative of the entire area than are other results in the table. (Sporeling *Ascophyllum nodosum* were sufficiently rare that their interactions with the grazers are unknown.)

Canopy Habitats

Perennial species preempt the unoccupied space beneath their canopy. Wherever the canopy is dense, neither ephemeral nor perennial species

Table 32.4 PROPORTION OF TIMES THAT HERBIVORY ON/OR COMPETITION BETWEEN SEAWEEDS DETERMINED ALGAL ABUNDANCES, AS A FUNCTION OF PLANT LIFE HISTORY AND AGE, HABITAT, SEASON, AND WAVE EXPOSURE

Strong effect of	Season	Protected site			Intermediate site			Exposed site		
		Number affected	Total number	Proportion	Number affected	Total number	Proportion	Number affected	Total number	Proportion
A. Herbivory on:										
1. Ephemerals (in open)	Summer	28	28	1.0	11	18	0.61	0	6	0
	Winter	0	9	0	0	22	0	0	6	0
2. Sporeling perennials (in open)	Summer	4	26	0.15*	3	11	0.27*			ND
	Winter			ND			ND			ND
B. Competition between:										
1. Ephemerals and ephemerals (in open)	Summer	0	28	0	7	18	0.39	4	6	0.67
	Winter	8	9	0.89	18	20	0.90	1	6	0.17
2. Ephemerals and sporeling perennials (in open)	Summer	0	28	0	7	18	0.39			ND
	Winter			ND			ND			ND
3. Adult perennials and ephemerals (effects of canopy)	Summer	6	6	1.0	9	9	1.0			ND
	Winter	6	6	1.0	9	9	1.0			ND
4. Adult perennials and perennial spores (effects of canopy)	Summer	10	10	1.0†	5	5	1.0			ND
	Winter	10	10	1.0	4	4	1.0			ND

Number affected = number of control treatments in which herbivores directly controlled algal abundance (part A of table), or in which competition was strong (part B of table). Significance was assessed with Hotelling's T^2 using the Bonferroni approximation, with percent cover of individual species used for analyses. Total number = number of control treatments. Proportion = ratio of number affected to total number. Indirect effects of herbivores are not indicated. ND = no data (no experiments addressed this interaction).

Types of experiments used to evaluate the different interactions and total number of experiments and controls are indicated in Table 32.3.

*Since the effects of herbivores on sporeling perennials are determined by substratum heterogeneity (Lubchenco 1983), this proportion will vary, depending on the number of small cracks and depressions present.

†*A. nodosum* will recruit under a canopy of either *F. vesiculosus* or *A. nodosum*, but this did not occur in the experiments. *A. nodosum* recruitment is usually very low.

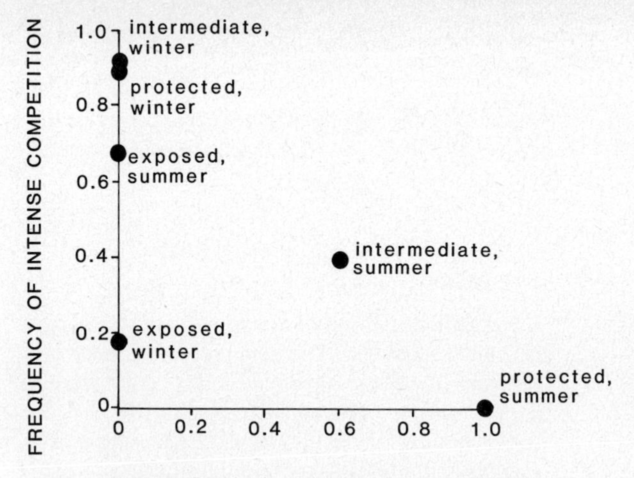

Fig. 32.2 Effect of herbivory on competition between ephemeral seaweeds in open habitats. Data are proportion of times that herbivores controlled algal abundance and proportion of times that strong competition occurred between ephemeral algae (from Table 32.4). Season and degree of wave exposure are indicated for each point.

(with the exception of *A. nodosum,* noted above) colonize underneath (Table 32.4, rows B3 and B4). The mechanisms of this competitive exclusion are not known, but may involve low light levels under the canopy, a physical barrier presented by a thick canopy, whiplash, or other factors.

The harboring of gastropod grazers by the canopy species is not a potential mechanism, since no plants colonized in the grazer-exclusion cages.

Both Habitats Combined

Since my sampling of each site (with control treatments) was not done in proportion to habitat types, calculation of overall interaction frequencies requires weighting each within-habitat frequency by the proportional abundance of canopy and open habitats. These proportional abundances are indicated in Table 32.5; the resulting overall interaction frequencies appear in Table 32.6.

At protected and intermediate sites where fucoids form a thick canopy and occupy most of the space, the dominant interactions are those associated with the fucoid bed: competition between adult perennials and potential colonists under the canopy (Table 32.6). Although herbivores are common under the canopy, they have little effect on colonization on the rock surface, since few spores settle there. They do, however, have strong effects on ephemeral algae colonizing the fucoids (Menge 1975). Because the open habitat is scarce (Table 32.5), herbivores that play a crucial role in determining colonization patterns in this habitat (Table 32.4) are less important when averaged over an entire site (Table 32.6). Thus, the type of habitat strongly affects the relative importance of herbivory and competition, that is, the amount of habitat at different successional stages is of vital importance. If the whole site were cleared of macrophytes and thus turned into open habitat, herbivory would be the dominant interaction instead of competition. Overall, herbivores increase the rate of colonization by perennials (Lubchenco 1983). As succession proceeds and perennials become established, these plants prevent further recruitment by ephemerals under (although not on top of) themselves.

At exposed sites neither competition between seaweeds nor herbivory is particularly important (Tables 32.5 and 32.6). Effects of wave action and competition between seaweeds and mussels determine the overall patterns of distribution and abundance (see Menge 1976 for information on mussel dynamics).

A Key to the Occurrence of Competition in the Mid Zone in New England: a Descriptive Model

A rudimentary understanding of some of the major factors affecting colonization of seaweeds in this community is presented in Table 32.7.

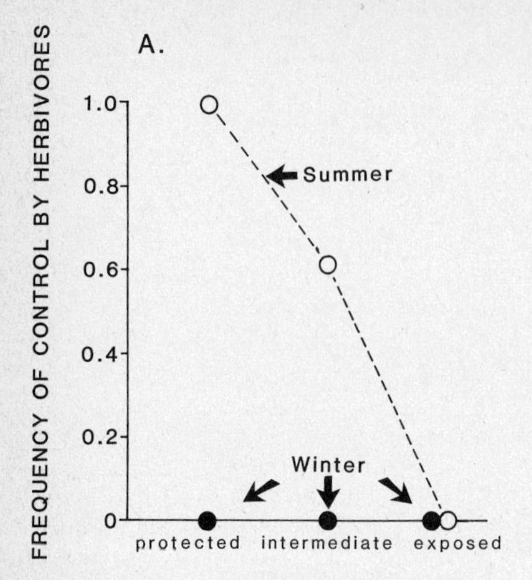

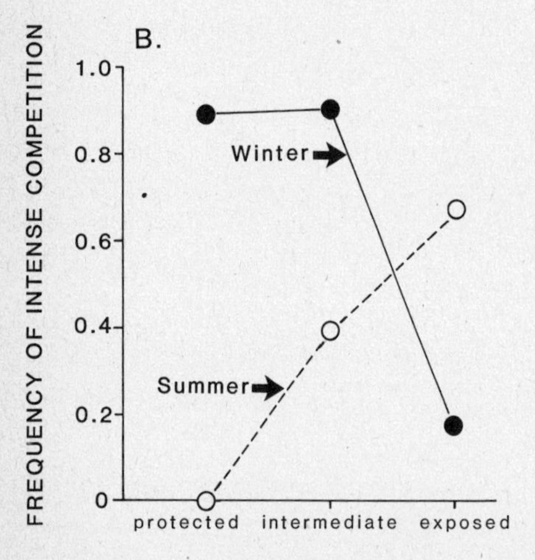

Fig. 32.3 Effect of wave exposure (A) on frequency of herbivore control of ephemeral algae and (B) on frequency of strong competition between ephemeral algae, in summer (unshaded circles) compared to winter (shaded circles) in open habitats. Data are from Table 32.4. The abundance, biomass, and activity of gastropod grazers decrease as wave action increases. (Lubchenco 1980.)

This key is intended to summarize the above discussion and outcomes by suggesting the conditions under which competition will probably occur. None of the events in the table is deterministic. Those with a high probability of occurring are listed. Competition during early succession thus occurs between ephemeral species and between ephemerals and juvenile perennials where neither herbivores nor wave action severely restrict algal abundance. Competition between perennial species often occurs later in the successional sequence, when larger perennial individuals have escaped control by grazers.

DISCUSSION OF NEW ENGLAND SYSTEM

Competition may occur at various stages in a successional sequence. The data presented above are relevant to competition affecting colonization events during (1) early secondary succession (open habitats) and (2) during late succession (canopy habitats). Since any particular site usually represents a mosaic of these different successional stages, the different processes normally co-occur.

Competition During Early Succession

The occurrence of competition between colonists varies with four factors: (1) the degree to which herbivores directly or indirectly control the abundance of the plants, (2) the degree to which abiotic factors such as wave action control the abundance of plants, (3) the life history of the plants (for instance, ephemeral versus perennial), and (4) the size of the plant (for instance, sporeling versus adult *Fucus*).

Herbivory The primary biotic interaction affecting abundance of early colonists in this system is herbivory, the intensity of which sets the stage for competitive interactions. As Fig. 32.2 indicates, intense grazing (high frequency of experiments in which herbivores had a strong effect at intermediate and protected sites during summer) precludes intense competition, because resources are not limiting. Only where grazing effects are minimal is it possible for resources to be limiting and for competition to have a strong effect. Because of wave action, competition does not always occur if grazing effects are minimal, but competition does not occur or is of minor importance if grazers control algal abundance. Both plant characteristics and environmental characteristics affect the ability of herbivores to control algal abundance.

Table 32.5 AVERAGE COVER OF CANOPY AND OPEN HABITATS AT THE THREE SITES

Site	Season	Canopy	Open	Other
Protected site	Summer	86	14	None
	Winter	81	19	None
Intermediate site	Summer	88	12	None
	Winter	78	22	None
Exposed site	Summer	7	15	Mussels
	Winter	3	11	Mussels

Data are average percent cover for three years, 1974 to 1976 (see Table 32.2). Open = noncanopied, bare rock without any macroscopic sessile organisms, and noncanopied rock covered by ephemeral algae.

Plant Characteristics Both algal life history (ephemeral or perennial) and algal size strongly influence the effects of grazers in this system. Plant life history is strongly correlated with the types of antiherbivore defenses. The ephemeral species have no known defenses that allow individual plants to persist in the presence of grazers. Such defenses, termed ''coexistence escapes'' by Menge (1982; see also Lubchenco and Gaines 1981, Connell 1975), include chemical, morphological, or size-related defenses. Lacking coexistence escapes, ephemeral algae persist only where grazers are absent (for example, at very exposed sites) or when grazers are inactive (at protected and intermediate sites during the winter). Their escapes are termed ''noncoexistence escapes'' (*sensu* Menge 1982). Individual plants will sometimes be found at protected and intermediate sites during the summer, but in microhabitats where grazers temporarily miss them. By the end of summer snails have discovered most of these patches, and ephemerals are rare.

In contrast, the perennial fucoid species have coexistence escapes that include chemical (phenolic; Geiselman and McConnell 1981), mechanical, and size-related defenses (Lubchenco 1983). Although small *Fucus* (less than 3 cm long) are readily eaten by *Littorina littorea,* larger individuals are usually avoided if ephemeral algae are available. The size-related escape by *Fucus* species appears to permit the coexistence of herbivores and these perennial species in this community (Lubchenco 1983).

Table 32.6 OVERALL FREQUENCY OF STRONG HERBIVORY AND COMPETITION AS A FUNCTION OF SEAWEED LIFE HISTORY, SEASON, AND DEGREE OF WAVE EXPOSURE

Strong effect of	Season	Protected	Intermediate	Exposed
Herbivory on:				
Ephemerals	Summer	14	7	0
	Winter	0	0	0
Sporeling perennials	Summer	2	3	ND
	Winter	ND	ND	ND
Competition between:				
Ephemerals	Summer	0	5	10
	Winter	17	20	2
Ephemerals and sporeling perennials	Summer	0	5	ND
	Winter	ND	ND	ND
Adult perennials and ephemeral spores	Summer	86	88	ND
	Winter	81	78	ND
Adult perennials and perennial spores	Summer	86	88	ND
	Winter	81	78	ND

ND = no data. Data are overall percent of the time a particular interaction occurs at each site, calculated as the proportion of time the interaction had a strong effect within a habitat (Table 32.4) times the relative abundance of the habitat (Table 32.5).

Table 32.7 KEY TO LIKELIHOOD OF ALGAL COMPETITION FOR SEAWEED COLONISTS IN THE MID ZONE IN NEW ENGLAND

1a. Fucoid canopy is present.	Preemptive competition occurs (perennials prevent recruitment).
1b. Fucoid canopy is absent.	2
2a. Colonist is an ephemeral alga.	3
2b. Colonist is a perennial alga, *Fucus* spp.	6
3a. Season is the winter.	4
3b. Season is the summer.	5
4a. Site is protected from wave action.	Competition will probably occur between ephemeral algae.
4b. Site is exposed to wave action.	Competition will probably occur intermittently between bouts of disturbance by waves during storms; disturbance should have a stronger overall influence than competition.
5a. Site is protected from wave action.	No competition will occur; herbivores should control the abundance of ephemeral algae.
5b. Site is exposed to wave action.	Competition will probably occur between ephemeral algae.
6a. Season is the summer.	No competition will occur between ephemeral algae; herbivores control ephemeral algal abundance. Ability of grazers to prevent establishment of *Fucus* is mediated by spatial, temporal, and size escapes. As a result, *Fucus* usually colonizes.
6b. Season is the winter.	*Fucus* colonists usually outcompeted by ephemeral algae.

Thus, ephemeral and perennial plants escape control by grazers in different ways. Ephemeral species persist where grazers are permanently or temporarily absent or inactive; perennial species persist where grazers are common.

The relationship between the frequencies of competition and herbivory shown in Fig. 32.2 probably holds only for species like the ephemeral algae that lack coexistence escapes. Perennial species like *Fucus* species are affected by both herbivory *and* competition instead of by one *or* the other. For *Fucus vesiculosus*, for example, a significant amount of mortality on juvenile stages is caused by grazing (Lubchenco 1983), but enough individuals often escape being eaten during the more vulnerable stages and compete intensively with conspecifics and individuals of other species (Lubchenco and Menge, unpublished data). Thus, a plot of frequencies of competition and herbivory for species like *Fucus* would include points above a line drawn from the coordinates (0,1.0) to (1.0,0).

Wave Action Wave action has important direct and indirect effects on algal abundance. For example, strong waves intermittently remove algae during winter storms at exposed sites. Waves affect algal abundance indirectly by affecting the abundance and activity patterns of grazers. Periwinkle snails, especially *L. littorea* and *L. obtusata* are rare at exposed sites (Lubchenco 1980), probably because of strong wave action. Exposed sites thus represent a haven for ephemeral algae from grazing, but an intermittently hostile environment because of wave action during storms. Competition between ephemerals occurs in the lull between storm disturbances (Fig. 32.1D and F). This type of control appears analogous to the control by physical factors suggested by Wiens (1977).

Indirect Interactions Indirect interactions (Chapters 3, 26) are obvious and important in this system. The grazer *Littorina littorea* has a strong beneficial effect on colonization of *Fucus* species by removing the competitively superior ephemeral algae. Thus, although snails do graze small *Fucus* individuals (a negative effect), the positive effect of relief from competition with ephemeral seaweeds appears stronger. Once the fucoids are established as adult plants, the periwinkles (here *L. littorea* and *L. obtusata*) provide a different, also indirect, benefit; they remove ephemeral algae that are epiphytes on the perennial fucoids and thereby remove a potential source of mortality for the fucoids. Epiphytes increase drag on their hosts and contribute significantly to loss of the upright portions of plants from wave action during storms. Thus, although the snails have little direct effect on adult fucoids, they have a strong positive effect by negating the negative effect of epiphytes.

Competition During Middle and Late Succession

Two types of competition occur once fucoids become established at protected and intermediate sites: (1) strong intra- and interspecific competition among established perennials (Menge and Lubchenco, unpublished data) and (2) intra- and interspecific competition between established perennials and potential colonists (Table 32.2). The first type occurs in the presence of grazers because the fucoids have effective coexistence escapes from the snails. Thus, although herbivores can affect the abundance of *Fucus* sporelings, enough sporelings usually escape and competition occurs. Hence both herbivory and competition affect patterns of distribution and abundance in *Fucus* species. "Adult-larval competition" (*sensu* Woodin 1976) occurs because middle and late successional perennials preempt space and prevent colonization by early successional ephemerals. However, since most ephemeral algae can settle on top of the perennial species (Menge 1975), they have a "refuge" from this competition for primary space.

COMMUNITY THEORY

The aim of community ecology should be to develop predictive theories about what governs the influence on community structure of biotic interactions such as competition and predation and abiotic factors such as weather. To this end, a number of models have been proposed, such as those of Hairston, Smith, and Slobodkin (1960), Connell (1975), and Menge and Sutherland (1976). We are now at the point of attempting to evaluate and modify these models, as Schoener (1983b and Chapter 28) and Connell (1983) did earlier. The research presented above provides a partial test of portions of Connell's model (1975) and sheds some light on aspects of all three models.

Species Interactions

The seaweed example above and a number of other studies (most recently Chapter 31; Lubchenco 1983, Morin 1983a, Wilbur et al. 1983, Paine 1984) suggest that under certain circumstances predation can control competitive interactions. One key to understanding community structure lies in predicting how important between-trophic-level interactions (predator-prey, etc.) are, since these interactions set the stage for within-trophic-level interactions. Once we predict the impact of predation on a given trophic level, we can better evaluate the possibilities for competition. This does not imply that predation is always more important than competition, or that the two always oppose one another, or that other factors are unimportant. It simply means that whereas heavy predation causes reduced competition, the level of competition indicates nothing about the level of predation (Fig. 32.2). Reduced competition does not cause heavy predation; heavy competition does not cause reduced predation.

Because predation can set the stage for competitive interactions, we need to predict the conditions under which potential competitors escape control by their predators. The seaweed study suggests that a focus on coexistence versus noncoexistence escapes would be profitable. Specific predictions echo and strongly reinforce Connell's (1975) model: (1) For species lacking coexistence defenses, competition should occur only in habitats where or when its natural enemies and weather cannot control its abundance (see Chapters 15 and 33). These organisms should tend to be small in size and early in a successional sequence. (2) For species with coexistence escapes, competition will occur if the species is not controlled by weather, despite the presence of natural enemies. These organisms should be larger and occur later in succession (see Chapters 10 and 33).

The frequency of competition in the seaweed example depends on the kinds of escapes that the plants have from herbivores. These in turn depend on plant, herbivore, and environmental characteristics. Ephemeral algae lack coexistence escapes, are often controlled by herbivores, and compete only when or where physical conditions limit grazers. The perennial algae have effective coexistence escapes and frequently compete, usually in the presence of grazers. Thus, at a single trophic level, competition occurs frequently for some species and only under limited conditions for others. The key to understanding what species compete and when lies in knowing the factors (herbivores and waves) that control abundances of the potential competitors, and knowing the kinds of escapes by which plants prevent this control.

Other Approaches

Settlement Although the above approach emphasizes interactions between species, other phenomena are certainly also important. Roughgarden (Chapter 30), Hawkins and Hartnoll (1983), Underwood and Denley (1984), and others emphasize the importance of settlement to patterns of rocky intertidal structure. In some circumstances settlement has an overwhelming influence on community patterns. For example, the major factor causing variation in the abundance of the barnacles *Balanus glandula* across the study site at Hopkins was variation in settlement (Chapter 30). In a similar fashion, the failure of the fucoid seaweeds to disperse to exposed sites causes major differences between protected and exposed communities.

However, many organisms disperse well and consistently colonize in high enough numbers that variation in settlement densities has no relationship to differences in adult densities among sites. This is certainly true of the ephemeral species discussed above; they colonize abundantly everywhere. The strong effects of herbivory and wave action on their abundance swamp the effect of any initial variation in settlement. Barnacles studies by B. Menge in New England show similar patterns (Menge in preparation): Initial settlement densities have no relationship to adult densities, which are instead determined by predation and competition. Thus, variation in settlement density by itself is not sufficient to explain many patterns of community structure, but the potential importance of settlement variation cannot be ignored. Settlement is one of several factors influencing community patterns. Postcolonization interactions such as those discussed in this chapter are superimposed upon variation in settlement.

Open Versus Closed Systems The model that Roughgarden (Chapter 30) discusses for a local, open system may not be applicable to the ephemeral or perennial species discussed in my chapter, nor to the mid zone community as a whole. In Roughgarden's open model, recruits come from outside the boundaries of the population, so that the number of recruits is not a function of local events (see also Chapter 9), and variation in settlement rate is the primary cause of variation in abundance of adults. Compared to the barnacles that Roughgarden discusses as examples of his open model, ephemeral algae are similar in having broad dispersal (on the order of kilometers), but differ in that postcolonization events are much more important to adult densities than is settlement. Abundance of perennial seaweeds is more influenced by initial settlement than is that of ephemeral algae, but perennial species have limited dispersal (on the order of meters) and are better characterized by a closed model, such as Roughgarden discusses for lizards. Dispersal ability of seaweeds is strongly correlated with life history traits (Table 32.1) and varies considerably from species to species. Thus, one could not apply either a closed or an open model to the mid zone as a whole, since individual species in the community differ greatly in dispersal ability and hence in the relation between local events and number of recruits.

SUMMARY

MacArthur (1972b), Schoener (1972), and Colwell (1984) assert that community ecology can best advance by focusing on models of a limited domain. An understanding of the relative importance of competition and predation in organizing communities will be enhanced by specifying the conditions under which different sorts of interactions are to be expected. This chapter examines the specific conditions under which competition, herbivory, and wave action interact to affect seaweed colonization.

The occurrence, frequency, and importance of competition among the common organisms at a single trophic level were examined. Competition among seaweeds in the New England mid zone varies according to plant life history (ephemeral or perennial), plant size (small or large), season, and degree of wave exposure. Each of these factors affects the degree to which either herbivores or wave action control the abundance of seaweeds, which in turn sets the stage for competitive interactions.

If herbivores are abundant and active or if wave action is severe, ephemeral algae do not become sufficiently abundant to compete intensely. Competition among ephemeral algae occurs primarily where herbivores are ineffective and where wave action is benign. These conditions vary both in space (according to exposure of the site) and in time (between seasons).

In contrast, perennial seaweeds have coexistence escapes from herbivores and compete frequently with each other in the presence of herbivores. Competition is thus a chronic interaction for the perennial algae, but variable for ephemeral seaweeds. It is suggested that an important key to predicting the occurrence of chronic competition lies in understanding the conditions under which coexistence escapes versus noncoexistence escapes from herbivores or other natural enemies occur.

ACKNOWLEDGMENTS

I thank T. Case, M. Dethier, J. Diamond, T. Farrell, P. Kareiva, B. Menge, R. T. Paine, J. Roughgarden, T. W. Schoener, and T. Turner for comments on earlier drafts.

chapter **33**

Patterns in Terrestrial Vertebrate Versus Arthropod Communities: Do Systematic Differences in Regularity Exist?

Thomas W. Schoener

INTRODUCTION

The degree to which ecological communities show regular patterns has long been a matter of controversy.

The Clements Gleason debate about the integrity of plant associations was perhaps the most famous early polarization (McIntosh 1980, Simberloff 1980). Clements' picture of the community as an invariant unit of tightly bound species contrasted markedly with Gleason's ''individualistic'' view of free-spirited species that happen to be in the same place at the same time.

In the 1950s and early 1960s the debate over density-independent versus density-dependent factors dominated much of the ecological literature. The most famous proponents of the density-independent school, Andrewartha and Birch (1954; Andrewartha 1961), argued that biological interactions, operating through negative feedback, are relatively unimportant determinants of population size; rather, mortality directly resulting from climatic vicissitudes predominates. Their view was countered by Nicholson (1957) and Lack (1954), among others, who saw strong self-regulation of population numbers and strong interspecific interactions as characterizing most ecological systems. The density-independent view implies large, possibly erratic variation in population numbers; the density-dependent view implies relative constancy. Although only peripherally directed toward community ecology, the two views necessarily imply differences in that realm as well.

In contemporary times a polarization again seems to have developed (papers in Strong, Simberloff, Abele, and Thistle 1984). The Florida State school, in a series of elaborate analyses, finds little in real communities to distinguish them from randomly constructed ones. While generally not quite advocating that real communities are in fact ''random,'' certain members of this school adopt an individualistic view perhaps somewhat more extreme than that of Gleason's (e.g., Simberloff 1983). In contrast, a large group of investigators, strongly influenced in one way or another by Lack, Hutchinson, and MacArthur, find much pattern in ecological communities and advocate strong interspecific interactions (see Cody 1974a, Pianka 1981).

In all of the above examples, those who find strong regularities propose that interspecific competition is a dominant ecological process. Conversely, those who find much variation frequently question competition's importance. Some critics have gone so far as to claim that competition is a Kuhnian paradigm, suggesting an almost compulsive fervor among competitionists (Strong 1980, Wiens 1983b, but see Schoener 1983a). The debate has been historically so repetitive that one might well ask whether, notwithstanding the disagreement about paradigms, the chronic advocacy of competition's importance may indicate some underlying truth rather than mere historical accident. And the same sort of question might be asked about the chronic insistence that competition is not so important. Moreover, as those finding the most pattern are those advocating competition and vice versa, another underlying truth is suggested—something intrinsic about competition may cause it to produce more regularity than other processes such as predation and the action of physical factors.

A second element of repetition emerges from our historical analysis. In the polarizations involving animal ecologists those advocating weak patterns and strong variability work primarily on terrestrial arthropods, whereas those advocating strong patterns work primarily on terrestrial vertebrates. One must ask, and indeed it is not original to ask, if this segregation is coincidence. A variety of theoretical considerations suggest that it is not. The Hairston, Smith, and Slobodkin (1960) hypothesis proposes for terrestrial systems that animals at the top of food webs should be controlled by competition, whereas the prey of such animals (herbivores) should be controlled by predation. Because vertebrates are much more frequently at the top of terrestrial food webs than arthropods, a statistical difference between the two taxa in the differential importance of competition is suggested. Connell (1975) and Schoener (1974) suggested that small animals should be more vulnerable to predation than large ones; Connell (1975) in addition suggested that small animals should be more affected by physical factors. Because arthropods are much more frequently small than are vertebrates, a differential importance of predation and physical factors versus competition is again suggested. As stated, were predation and physical factors to produce less pattern than competition, the taxonomic break between the two kinds of advocates could find explanation.

A fair amount of data from all ecological systems combined supports the hypotheses of Connell and Schoener and of Hairston, Smith, and Slobodkin, as shown below. Nonetheless, data from terrestrial systems directed toward these questions share, almost without exception, the deficiency that arthropods are studied at one place and time and vertebrates at another place and time. Data directed toward the more general question of the degree of pattern are similarly compromised. Are differences between arthropods and vertebrates, when found, merely the result of differences in type of habitat, locale, or time of year or are they fundamental properties of the two taxa? The observations, in short, lack suitable controls.

This study, which deals with arthropods and vertebrates in the same system, solves this problem of controls and attempts to shed light on possible differences between the taxa in regularity and dominant processes. The system is the Bahamas islands, an archipelago stretching a thousand miles from southern Florida to Hispaniola and Cuba. Those organisms on a single island are considered to belong to the same community (all islands are relatively small). Because arthropods occupy smaller islands than vertebrates, not all communities have both. But, as seen below, this is an advantage, and moreover, many communities do have both: we estimate that 10^4 to 10^5 islands contain at least one vertebrate population (Schoener and Schoener 1983a, 1983b). Thus, the Bahamas archipelago provides a huge number of natural arenas in which we can perform natural experiments or field manipulations (Chapter 1).

Most of the islands we are dealing with are small; the largest rarely exceed 0.5 km^2 (500,000 m^2) and the smallest are mostly 100 to 10,000 m^2 for vertebrates and 50 to 100 m^2 for arthropods (Schoener and Schoener 1983b, Schoener and Toft 1983a). Nearly all the numerically dominant terrestrial vertebrates are birds or lizards. We have observed both taxa intensively and in addition have performed numerous manipulations on the lizards.

A variety of arthropods occur on the study islands, some belonging to groups that are poorly known taxonomically. For practical reasons, as well as because of their obvious numerical importance, we selected diurnal orb-weaving spiders for intensive study. Most genera are treated in recent taxonomic monographs (Levi 1968, 1977a, 1977b, 1978). Orb spiders are conspicuous because of their webs and therefore are easily found. They are usually immobile and therefore are easily counted. Two or three persons have been able to count populations of five species on a hundred islands (the largest of which was less than 10,000 m^2) in three to four weeks. Population estimates are very accurate, as the spiders must build webs in order to feed, that is, to be ecologically active. Orb spiders are also easy to manipulate. Removal of individuals requires little pursuit and handling time. Introduction is logistically simple, and the results are immediately observable; one has in essence an instantaneously germinating seed. Yet orb spiders do change their locations opportunistically (e.g., Olive 1982), so possess a major advantage of mobile organisms—to understand their distribution, one does not have to integrate environmental conditions over long, perhaps unobserved periods of time. Finally, spiders are typically at an intermediate trophic level, but on islands without vertebrates they are at the top of the food web. This allows a comparison of the same kind of organism at different trophic levels, much as has been done in the intertidal and subtidal (e.g., Menge and Sutherland 1976). In particular, we can determine whether spiders at the top of their food webs are more like vertebrates than are spiders in intermediate positions.

The operational definition I assign to patterning includes the classical meaning of the degree to which species' occurrences and abundances in some local area are predictable from the size and diversity of the area, the resources the area contains, and the species that coexist within the area. To the degree that this is true, the implication is that populations grow to carrying capacity and then self-regulate or, sometimes, that they are regulated by predators. To the degree that this is false, the implication is that populations rarely reach carrying capacity and are ultimately controlled by the vagaries of physical factors, again sometimes aided by the action of predators. Also included in my definition of patterning, however, are those properties of communities related to input or immigration; for such patterns, properties of source areas may strongly outweigh properties of recipient areas. One might expect communities predominantly to show one or the other sort of pattern, and, as we shall see, this is indeed often the case.

The remainder of this paper is organized as follows. After methodological preliminaries, I evaluate 11 population or community patterns. For each pattern I first compare birds, lizards, and spiders within the same system, the Bahamian archipelago, with respect to regularity in the pattern. I then compare terrestrial arthropods and vertebrates from other systems for the same pattern; data are gathered from literature surveys or *de novo*.

The sequence of the 11 patterns is organized as follows. The first 9 deal with observational data and the last 2 with experimental data. Of the first 9, the first 7 deal with static observations, that is, observations made at a single time. Such patterns include the classical species-area and species-distance relationships (patterns 1 and 4), as well as analogous patterns for the area and distance-dependence of species-occurrence thresholds (patterns 2 and 5) and of abundance (patterns 3 and 6). The static observations also include patterns of complementarity in distribution and abundance (pattern 7). Static observational patterns often reveal least the underlying processes 9, but can be suggestive, and they are relatively numerous and comprehensive.

The next 2 patterns comprise observations through time of the dynamics of abundances and the dynamics of species lists (patterns 8 and 9). These sorts of observations can confirm underlying mechanisms more than the static observations do, but they depend to some extent on the fortuitous presence of desired natural perturbations and the appropriate controls.

Finally, the last 2 patterns describe results of deliberate manipulations in which populations are introduced into "empty" areas (pattern 10) and populations of hypothetical competitors are manipulated (pattern 11). These are often the most powerful for implicating causes, although of necessity data are relatively limited (Chapter 1; Diamond 1983, Schoener 1983b).

In the last part of the paper I assess overall trends in the differences found, and I speculate on reasons for the differences. In particular, the apparent inability of predation to produce regular patterns among certain terrestrial arthropods is contrasted with the situation in sessile marine and terrestrial organisms.

MATERIALS AND GENERAL METHODS

Study sites in the Bahamas contained six species of diurnal orb spiders, the four most common being *Metepeira datona, Gasteracantha cancriformis, Eustala cazieri,* and *Argiope argentata*. The other two species comprised less than 0.1% of the individuals found. Details of the spiders' natural history are found in Schoener and Toft (1983a, 1983b) and Toft and Schoener (1983).

Six species of lizards were common in one or more study sites. Four of these are anoles: *Anolis sagrei, A. distichus, A. carolinensis,* and *A. angusticeps*. These species typically perch at various places in the vegetation. A fifth species, *Leiocephalus* sp. (one species per island), is a more terrestrial member of the Iguanidae, the same family to which anoles belong. A sixth species, the teid *Ameiva* sp. (one species per island), is a highly terrestrial lizard. All six species consume primarily arthropods, though *Leiocephalus* can take much plant matter. A seventh species, *Cyclura* sp., occurred only on a few islands, is herbivorous, and is much affected by humans because of its edibility. Details of the lizards' natural history are found in Schoener (1968, 1970a, 1974a, 1975), Schoener and Schoener (1978a, 1978b, 1980a, 1980b, 1982a, 1982b) and Schoener et al. (1982).

Birds as a whole are far more diverse than lizards or spiders in our study sites; 33 resident species and numerous migrant species occurred. Because of the varied feeding habits of these birds, I divided them into trophically defined groups, the most important of which are insectivores and herbivores; 14 common species belong to the first group and 6 to the second. Complete lists and details of the birds' natural history are found in Schoener and Schoener (1983a, 1983b), Brudenell-Bruce (1975), and Emlen (1981).

Approximately 600 Bahamian islands were involved in the study, all of which had vertebrates counted at least once, and 106 of which had spiders counted at least once. Almost all islands were smaller than 1,000 m^2, and most were considerably smaller than this. The largest island on which spiders were counted was about 8,600 m^2. Islands on which vertebrates were studied range from the south-central Bahamas (Crooked-Acklins Bank) northward to the northernmost Bahamian island. Details of the locales are given in Schoener and Schoener (1983a, 1983b). Islands on which spiders were studied all are located within a 20-km radius of Staniel Cay, an island in the central Exumas.

The work reported was performed from 1970 through 1983 for lizards and birds and from 1981 through 1983 for spiders. Nearly all data were gathered during April and May, the period immediately before onset of the summer rains. At this time of year activity is at a peak. Birds and lizards are often conspicuous, because of territorial behavior, and spiders are relatively stable in their locations, as little rain falls then. Exact study dates are given in the papers cited above.

We recorded the species of birds and lizards found on each island. For birds, the presence of a species does not necessarily represent successful breeding, but the islands were at least feeding sites. We reasoned that, for islands as close together as ours, it was of little import which island actually contained the nest when a home range could span several islands. For lizards, which scarcely ever cross water (see below), occurrence does imply breeding, and minimal population sizes typically averaged about 50 individuals.

We recorded the number of individual orb spiders found on each island. Species lists were byproducts of these figures. Censuses were performed by two or three people working in tandem, and every plant was inspected for a web. As justified elsewhere (Schoener and Toft 1983a, Toft and Schoener 1983), we believe that we were able to find almost all of the web-building portion of each species' population by this simple (but sometimes lengthy!) visual inspection.

In addition to variables describing the species of interest, we recorded various features of the island, either on the spot or from maps. These included variables describing vegetation structure, area, topography, and distance. Fifty-four such variables were recorded for each island in the surveys limited to vertebrates (list and methods in Schoener and Schoener 1983a), and a smaller number were recorded in the islands surveyed for both spiders and vertebrates (Toft and Schoener 1983).

As explained elsewhere, one of the common spider species (*Eustala cazieri*) is only partly diurnal. This makes inferences about presence more difficult than for the others. We attempted to solve this problem by censusing the species in two ways: counting living spiders in webs, and counting undamaged webs without living spiders plus living spiders in webs. Additional justification is given in Toft and Schoener (1983).

Detailed statistical procedures will be described in the sections to follow. The overall approach is to compute various statistics that summarize the degree of regularity in a pattern for each taxon of interest. Both the direction and magnitude of difference between statistics from vertebrate and arthropod data will be of interest.

THE PATTERNS

I now discuss each pattern according to the following protocol. First, I give the measure or measures of regularity I will use for the pattern, and I briefly discuss what underlying processes might cause any observed regularity in the pattern. Second, I compare orb spiders to vertebrates with respect to the pattern, using data from the Bahamian island system only. Third, I review the general literature for data from other systems with which terrestrial arthropods and vertebrates can be compared with respect to the pattern.

Pattern 1. Increase in Number of Species on an Island with Area

Measures of Regularity One of community ecology's few universal regularities is the species-area relation. With few exceptions plots of number of species on an island versus its area give statistically significant positive correlations under a variety of transformations (reviews in Schoener 1976a, Connor and McCoy 1979, Schoener and Schoener, 1981a). The underlying etiology of so persistent a pattern must surely be varied, and, indeed, a variety of theoretical models predict it. The most famous is the MacArthur-Wilson equilibrium model, which assumes that smaller islands have larger extinction rates because (1) smaller islands have smaller population sizes for the same number of species present and (2) the smaller the population size, the greater the per time extinction probability. Hence for the same immigration rate smaller islands have fewer species at equilibrium. The assumption and the underlying causal mechanisms have been verified in certain systems, although the statistical methodology is tricky (reviews in Toft and Schoener 1983, Schoener 1985d).

A second argument, which can also be made using MacArthur-Wilson machinery, assumes that larger islands have a higher immigration rate because of a larger target size for passively dispersing organisms. The prediction is again more species at equilibrium. The assumption has been demonstrated for marine fouling communities (Osman 1978, Schoener and Schoener 1981a) and for spiders (Toft and Schoener 1983). Moreover, MacArthur-Wilson-like explanations can be cast in terms of habitat availabilities and how these affect extinction via population size, as well as how these affect immigration rates (where immigration is defined as successful colonization and not just arrival). Sometimes correlations of species counts with habitat variables are stronger than those with area (review in Schoener and Schoener 1983b).

Less mechanistic approaches, which view the species-abundance distribution as given and ask what species-area relation is implied, also predict positive relations much like the data (review in May 1975b). Although the MacArthur-Wilson model necessarily implies a positive species-area relation at equilibrium, a positive species-area relation does not necessarily imply equilibrium; not only is this true in theory, but it has been documented experimentally for marine fouling organisms (Schoener and Schoener 1981a). Moreover, stochastic versions of the MacArthur-Wilson model can give very diffuse species-area relationships (Diamond and Gilpin 1980). Various other reasonable models giving a positive species-area relation must also be possible.

Three ways of evaluating the regularity of the species-area relation, each with certain advantages, will be used here. The *log-log transforma-*

tion implies that the species-area relation is a power function; correspondingly, we label the Pearson correlation coefficient for this transformation r_p. It linearizes many species-area plots better than other transformations (Schoener 1976a, Connor and McCoy 1979). A disadvantage is that islands with zero species either must be eliminated or some arbitrary number must be added to each species count. The transformation emphasizes differences between species-poor islands.

The *semi-log transformation* implies an exponential species-area relation; correspondingly, we label the Pearson correlation coefficient for this transformation r_e. This transformation allows zero-species counts to be included directly and also fits many of the data. It does not emphasize differences between species-poor islands.

The *ranking transformation* orders species number and area by increasing magnitude and replaces actual values with rank scores. A convenient measure of regularity for such data is the Spearman correlation coefficient r_s. Zero-species counts can be included directly, and as r_s deals entirely with ranks, extreme points at either end of a plot do not affect its value. This measure also fits naturally into the occurrence-sequence technique of analyzing single species (Schoener and Schoener 1983a, pattern 2 below). Moreover, for small total numbers of species a perfect stepwise increase in species with area (with no ties) gives an r_s much closer to 1 than do the other types of correlation coefficients.

The Bahamas System Species-area relations for lizards within the Bahamas are often extremely regular; nearly perfect, stepwise patterns can exist (Fig. 33.1, top). Relationships for resident birds tend to be somewhat less regular in the extreme but almost as regular overall (Fig. 33.1, bottom). Moreover, birds often show a stronger relationship with habitat variables than with area (Schoener and Schoener 1983b).

Frequency histograms plotting r_e for lizards, all resident birds combined, insectivorous resident birds, and herbivorous resident birds are given in Fig. 33.2 (r_s gives similar histograms). Each is based on separate compilations for 45 archipelagos, some of which are more inclusive groupings that combine islands in several others. (Because of the considerable number of zero-species islands in these data, log-log transformations were not run.) Values of r_e and r_s are fairly similar. They cluster between 0.7 and 0.9 for the first three kinds of vertebrates and somewhat lower than 0.7 for herbivorous birds.

For orb spiders, considering all islands in the central Exumas as a single archipelago, values of r_s computed separately for each year (and for each way of estimating *Eustala*) vary between 0.53 and 0.66 (*N*s range from 40 to 104; detailed statistics available from author upon request). These numbers are lower than most values for all types of vertebrates except herbivorous birds.

Comparable correlations for vertebrates, based on the same region used for spiders, but including fewer small and more large islands, were also computed. Except for herbivorous birds ($r_s = 0.49$), they do not overlap at all with spider values and range from 0.71 to 0.79 ($N = 70$). Values of r_e for spiders are lower than values of r_s, ranging from 0.47 to 0.57. These are substantially lower than typical values for nonherbivorous vertebrates (Fig. 33.2). Values for the exactly comparable island region range from 0.77 to 0.82 for nonherbivorous vertebrates; r_e equals 0.55 for herbivorous birds.

Values of r_p were computed in two ways. When 0.1 was added to each species count, values for spiders ranged from 0.51 to 0.67. The comparable values for vertebrates ranged from 0.73 to 0.76 for nonherbivorous vertebrates and 0.51 for herbivorous birds. When zero-species islands were excluded, values for spiders ranged from 0.17 to 0.31 (*N*s range from 39 to 65). Values for vertebrates in the same island region ranged from 0.85 to 0.88 for nonherbivorous groups; r_p was 0.60 for herbivorous birds (*N*s range from 13 to 28). All values for vertebrates using this second method are substantially higher than those for spiders. Apparently, in contrast to vertebrates, zero-species islands are largely responsible for what positive species-area relation there is for spider counts.

In summary so far, spiders from all islands combined consistently have lower species-area correlations than lizards, insectivorous resident birds and all resident birds, and, less consistently, than herbivorous resident birds. As mentioned, on some islands spiders are at the top of

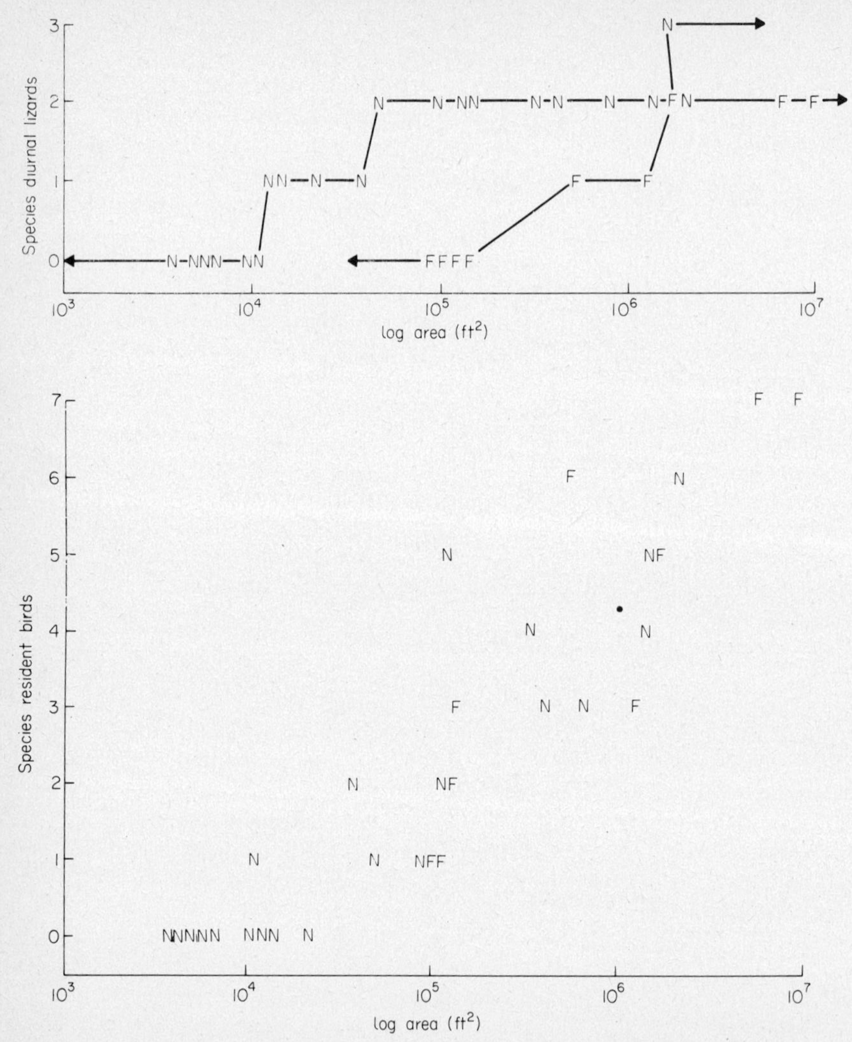

Fig. 33.1 Species-area relation for diurnal lizards (top) and resident birds (bottom) on islands near (N) and far (F) from a presumed colonization source. Near islands, less than 1 mile from the source, are in Lovely Bay, Crooked-Acklins Bank; far islands, between 5 and 10 miles from the source, lie south of Fortune Island, Crooked-Acklins Bank. Notice the strong distance effect for lizards, even over this small range of distances. Over the range of distances represented in these figures, birds show no distance effect. (From Schoener and Schoener 1983a.)

the food web, while on others they co-occur with lizard and sometimes bird populations. Since resident bird populations in the central Exumas almost never occur without lizard populations, we can distinguish two types of spider islands as lizard and no-lizard islands. Species-area statistics were computed for each type separately. With one exception, values of r are higher for no-lizard islands than for all islands combined. With two exceptions, rs are lower for lizard islands than for all islands combined; rs are lower on lizard islands perhaps partly because fewer very small, zero-species islands are included, but for 1982 and 1983 data when *all* zero-species islands are excluded, rs on lizard islands were still much lower than on no-lizard islands. More sensation-

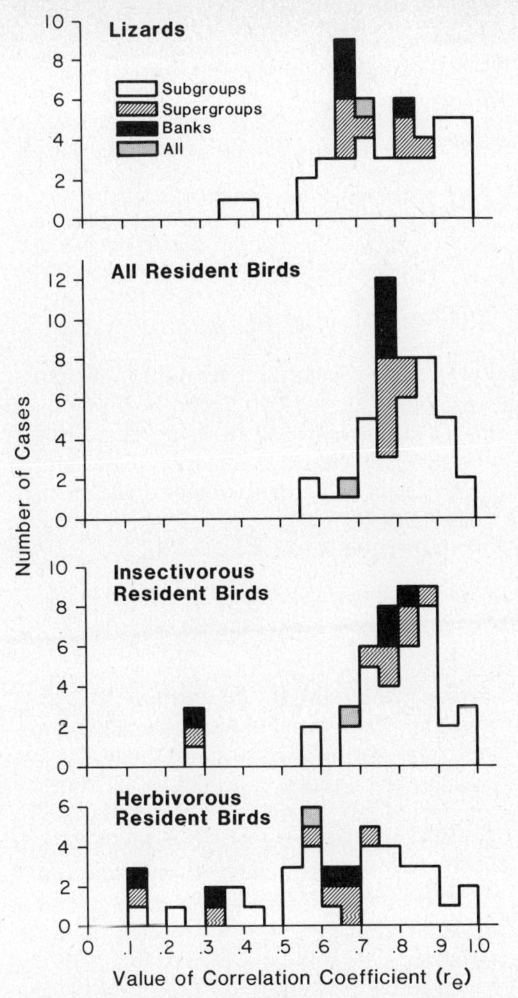

Fig. 33.2 Histograms of number of cases having various values of r_e, the Pearson correlation coefficient between number of species and log (area) of an island. Each entry is an archipelago; the four categories are of increasingly (top to bottom) inclusive archipelagos.

ally, values of r on lizard islands often are not significantly different from zero and rather often are negative! While rs were never significantly negative, these values do indicate the depth to which this nearly universal ecological trend falls for spiders on islands with lizards. Fig. 33.3 gives an example of how the species-area relation collapses on such islands. In this figure, and in the others from 1982–1983, the difference between the top and bottom plots *is not* due to the greater range spanned by the bottom plot; when zero-species islands are excluded, thereby making the magnitude of island-area ranges nearly identical, r_ps are -0.09 (lizard) and 0.61 (no-lizard) (see also figure caption). Because lizard islands have fewer species than no-lizard islands of the same area (Toft and Schoener 1983), the curve is displaced to the right.

Because habitat variables sometimes are better predictors of species counts than area, I also ran species-habitat correlations for spiders, where the habitat variable was maximum vegetation height. This variable is a very competitive predictor of

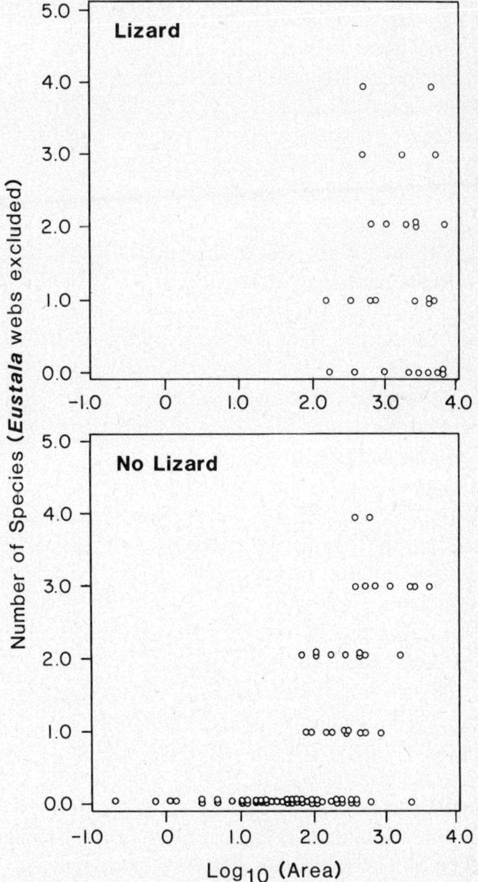

Fig. 33.3 Species-area relations for orb spiders on lizard and no-lizard islands of the central Exumas, 1982 data. Area is in square meters. Pearson correlations for all islands are -0.06 and 0.59, respectively; Pearson correlations for only islands larger than or equal to the smallest island having lizards are -0.06 and 0.48, respectively.

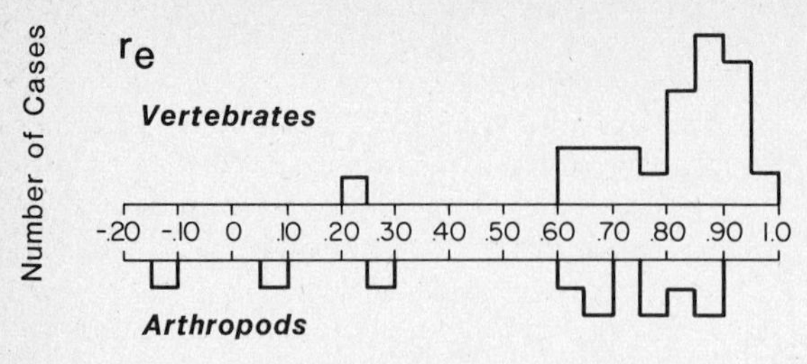

Fig. 33.4 Histogram of number of archipelagos having various values of r_e, the Pearson correlation coefficient between number of species and log (area), for the terrestrial vertebrate and arthropod examples compiled by Connor and McCoy (1979). "Habitat" islands not included. Note that vertebrates have larger coefficients.

species counts as compared to area in multiple regression (Toft and Schoener 1983). For both r_s and r_e, values are similar to those for area, and the same trend between lizard and no-lizard islands holds (statistics available upon request). Habitat variables are at least as good as area in predicting bird species counts and are almost as good for certain lizard species counts (Schoener and Schoener 1983b), so that the overall differences in regularity between spiders and vertebrates found for species-area relations must hold for species-habitat relations.

Other Systems Connor and McCoy (1979) gathered data from the literature on 100 species-area curves and reported r_e and r_p for each. Following the format of Fig. 33.2, I constructed frequency histograms for these correlation coefficients separately for the terrestrial arthropods and terrestrial vertebrates in their tables. I included only "normal" water-surrounded archipelagos, excluding habitat islands and quadrat studies. Results (Fig. 33.4) show exactly the same direction of difference between arthropods and vertebrates as for the Bahamas system. For r_e and r_p, the histograms of vertebrates and arthropods are significantly different (one-tailed $p < 0.05$ by the Kolmogorov-Smirnov two-sample test, which is conservative here [Siegel 1956]). The degree of difference is also fairly comparable; arthropods are a little more regular in general than Bahamian spiders, though this is not statistically significant.

Only one study of which I am aware allows comparison of a taxon's species-area relation on islands with and without predators. In this study (Whitaker 1973) lizards have the tables turned and are now the prey; the predators are Polynesian rats (*Rattus exulans*). Not only is the species-area curve displaced downward in the presence of rats, but the correlation is lowered as well—r_e falls from 0.77 to 0.45. Thus, lizard results with and without predatory rats are similar to spider results with and without predatory lizards (Fig. 33.3).

Pattern 2. Precision of the Threshold in Island Area Separating Islands with and without a Given Species

Measures of Regularity To measure the degree to which an island variable such as area separates islands with and without a given species, I use the occurrence-sequence technique (Schoener and Schoener 1983a). In this method islands are ranked according to the variable of interest (say, from smallest to largest area). Each island is denoted as having or not having the given species. The Mann-Whitney U statistic, which allows a significance level to be assigned to the precision of the threshold between presences and absences, is then computed. For the same number of presences and absences, that is, for a given species in a given archipelago, the statistic also allows different island variables to be compared according to how well each separates presences and absences. The technique can be generalized, using log-linear statistics, to several island variables. The occurrence-sequence method is somewhat similar to Diamond's (1975) incidence functions; the latter are essentially frequency histograms of the number of islands having a species in each of a set of categories defined over an island variable such as area. Incidence functions do not directly allow for a significance test, however.

Occurrence sequences can readily be related to ranking methods of describing the species-area relation. Were each species in an archipelago to show a perfect separation of presences and absences with respect to a given island variable, and were the separations all in the same direction (all absences on small islands, for example), then the sum of all species will show a perfect, stepwise correlation with area. The reverse, of course, is not necessarily true; a perfect, stepwise correlation does not imply that all occurrence sequences show perfect separation.

The Bahamas System Occurrence sequences with respect to area tend to be very regular in Bahamian vertebrates, particularly in lizards (Schoener and Schoener 1983a). The implication is that species-area curves, especially for lizards, are quite regular, and we have seen that this is the case (Fig. 33.1). A small number of possible communities is also implied: All islands having a single species of lizard tend to have the same "solitary" species, all having two species tend to have the first species plus the same second species, and so on. In other words, the communities are nested. This situation is less true for birds when area is considered (more different bird species can alternatively be the solitary species in the same archipelago, for example), but avian communities are still fairly regular. Occurrence sequences for certain habitat variables are very regular also, both for lizards and especially for birds.

In orb spiders occurrence sequences with respect to area are typically less regular than those of vertebrates. In the central Exumas *Anolis sagrei,* that lizard species nearly always occupying a one-species island, shows a more regular occurrence sequence (as measured by the probability associated with the U statistic) than does any species of spider (Schoener and Toft 1983a). If maximum vegetation height, a good predictor relative to area of number of spider species (Toft and Schoener 1983), is used as the ranking variable, Us are slightly higher (less significant) than for area. Thus, the result for spiders is probably not an artifact of using area rather than habitat, although possibly other habitat variables might do better (e.g., Greenstone 1984). Several spider species may commonly occur as the solitary species on an island. For all years combined (*E. cazieri* empty webs excluded), *M. datona* is solitary 13 times, *G. cancriformis* is solitary 6 times, *A. argentatus* is solitary 11 times, and *E. cazieri* is solitary 28 times. In the same time period *A. sagrei* was solitary 75 times, *Leiocephalus* was solitary 3 times, and *A. carolinensis* was solitary 3 times (these figures represent the number of times each year that a species was solitary, added together for the three years; there are actually 26, 27, and 28 lizard islands involved in 1981, 1982, and 1983, respectively). Moreover, this is an unusual amount of variation for Bahamian lizards in general (Schoener in preparation). Finally, the minimal threshold island area for the four common spider species is similar, for example, it ranged from 50 m^2 to 112 m^2 in 1981. Minimal thresholds for Bahamian lizards and birds, incidentally, are much larger, typically falling well above 100 m^2 and often ranging up to 10,000 m^2 (Schoener and Schoener 1983b).

Other Systems As the occurrence-sequence technique has just been published, no comparable studies from other systems are available. Diamond's (1975, Diamond and Marshall 1977) work with birds and mammals from southwest Pacific and Australian regions suggests that occurrence sequences in these vertebrate groups will generally be quite regular. This is at least consistent with the Bahamian results. The existence among southwest Pacific birds of "supertramps," species found on small but not large islands, makes generalization to species-area curves more complex, however (Diamond 1975). Finally, minimal island sizes for reptiles are given by Heatwole (1975), and minimal patch sizes for arthropods are discussed by MacGarvin (1982), Lawton (1978), and Kareiva (1983).

Pattern 3. Increase in Number of Individuals of a Given Species on an Island with Area

Measures of Regularity It is taken as almost axiomatic in theoretical treatments that, assuming no interaction between species, the number of individuals in a species will increase with island area (e.g., Preston 1962, May 1975b, Schoener 1976). If number of competing species increases with area and productivity per area shows no re-

Table 33.1 INDIVIDUALS-AREA STATISTICS FOR BAHAMIAN SPIDERS, 1982

Case	All islands b	All islands r	No-lizard islands b	No-lizard islands r	Lizard islands b	Lizard islands r
	N = 81		N = 53		N = 28	
All spiders	.37	.306	.96	.552	.26	(.216)
M. datona	.28	.233	.83	.492	.21	(.171)
G. cancriformis	.16	.326	.08	(.153)	.18	(.230
A. argentata	.01	(.010)	.41	.325	−.02	(−.057)
E. cazieri (excluding webs)	.15	.269	.28	.407	.04	(.049)
E. cazieri (including webs)	.73	.693	.90	.689	.64	.551

Numbers are similar for 1981 (see Toft and Schoener 1983) and 1983.
b = slope of regression; r = Pearson correlation coefficient (values in parentheses not significant at 0.05 level). All values are for log (number + 1) versus log (area).

lation to island area, average *density* per species must decline with area. However, for all species-area power relations ($S = bA^z$) with z less than 1, the average number of individuals per species will still increase with island area; as almost all real zs are less than 1, this trend should hold in nature.

These calculations, of course, depend on the assumption that changes in niche breadth within species do not prevent whatever species are present from utilizing the entire spectrum of available resources. If groups with more species utilize more resources (or if larger islands are more productive), then the density of individuals of all species combined will increase with island area. Indeed, over certain ranges of island area, this phenomenon seems to hold in birds (Diamond 1970), *Drosophila* (*affinis* group, Jaenike 1978), and herbivorous arthropods on rosebay (MacGarvin 1982), and bracken (Lawton 1978). Where it holds, it reinforces the argument that number of individuals in single species should increase with area, and this is known to be so for certain herbivorous arthropods.

In contrast to most studies, density of individuals of all species combined falls with area for passerine birds on islands near Maine (Morse 1977). Even when the density of individuals of all species combined drops with island area, however, the number of individuals of each species may still increase unless the drop is very precipitous. Such a precipitous drop might be due to a sharp increase in predator efficiency or intensity as island size increases.

The Bahamas System On all islands combined, spider numbers increase with island area at a decreasing rate; the power of the regression of log (number of individuals) on log (area) is always less than 1 (Table 33.1). This is true for all species combined as well as for particular species. In one species (*A. argentata*) correlations are not significant and relationships have zero or close to zero slopes. These low slopes (bs) and correlations (rs) are not the result of including many islands that do not have individuals of the particular species; when only islands with nonzero densities are included, values of b and r usually drop.

Again, additional insight can be gained by dividing islands into those with and without lizards. For the three years combined in 14 of 18 cases bs are smaller for lizard than no-lizard islands, and rs are smaller in 15 of 18 cases. Only 12 cases are completely independent, and conservative ps are not significant except for r (p = 0.02 by a binomial test on the independent cases). In most cases for no-lizard islands rs are not significantly different from zero (Table 33.1). Actual values of slopes for all species active diurnally (that is, not including unoccupied *E. cazieri* webs) range on lizard islands from 0.26 to −0.02 and on no-lizard islands from 0.86 to 0.08. The individuals-area relation for spider species on islands with lizards can be very slight indeed!

Comparable data are unavailable for birds in the Bahamas, but some data exist for lizards. They show a steeper slope of log (number of individuals) on log (area) than do data for spiders. For Acklins Bank *Leiocephalus* (including experimental introductions that have come to equilibrium; Schoener and Schoener 1983c), b = 0.97 and r = 0.87 (N = 5) when only completely censused islands are included; b = 1.34 and r = 0.86 (N = 7) when all available islands are included. For Exumas *Anolis sagrei* (including experimental introductions that have come to equilibrium), $b = 0.60$ and $r = 0.60$ ($N = 6$) for the restricted set, and $b = 0.59$ and $r = 0.60$ ($N = 7$) for all available islands. The *Leiocephalus* slopes are higher than any for spiders, and the *sagrei* values are higher than for all spider species (not including *Eustala* webs) except *Metepeira* on no-lizard islands. These differences between lizards and spiders hold despite the fact that only nonzero densities are included in the analyses for lizards; recall that inclusion of islands with zero densities raises b for spiders.

Other Systems As the numerous studies (review in Wright 1980) on density compensation imply, data surely exist on how number of individuals within particular species increases with island area. In three studies of herbivorous insect species (MacGarvin 1982, Lawton 1978, Cromartie 1975) densities usually rise with increasing patch size, so numbers of individuals must usually (if not always) do so also. I have been unable to find any other data of even this sort, and I would be interested to receive such information.

Pattern 4. Decrease in Number of Species on an Island with Distance from the Source of Colonists

Measures of Regularity That fewer species occur on farther islands is almost as universal a generalization as the species-area relation (e.g., Diamond et al. 1976), and like that relation it has many explanations, only one of which derives from the MacArthur-Wilson equilibrium model. In that model fewer species occur on islands at greater distances from a source area because of declining immigration rates with distance. A declining immigration rate with distance has been directly shown for birds (Jones and Diamond personal communication), mangrove arthropods (Simberloff and Wilson 1969), and Bahamian orb spiders (Toft and Schoener 1983).

However, the Toft and Schoener (1983) study also showed an increasing extinction rate with distance, a factor that also contributed to the effect of distance on species counts. Indeed, in this study population size was lower on distant islands. This phenomenon supports Brown's and Kodric-Brown's (1977) postulate of the "rescue effect," in which the nearer an island is to a source of immigrants, the more population sizes are determined by influx of new individuals from the outside relative to internal processes.

Smaller population sizes on far islands could also result from habitat poverty, and indeed Lack (1976) has proposed that far islands have fewer species primarily for this reason. Habitat poverty has been demonstrated for certain groups of far islands in the Bahamas and shown to exert an effect independently of declining immigration (Schoener and Schoener 1983b and below).

Finally, a distance effect need not imply equilibrium. Far islands may have fewer species simply because they have not had enough time to equilibrate. This argument, which implies that near and far islands differ in their closeness to equilibrium, has never been demonstrated with data, to my knowledge.

The Bahamas System Distance effects have been found for all taxa under consideration.

Combining all Bahamian archipelagos, the number of lizard species is often negatively related to distance from the nearest main island; 18 of 24 Pearson rs are negative, ($p = 0.011$), 9 significantly so, and none is significantly positive. The effect is lower but still present when habitat poverty is taken into account statistically; 16 of 24 partial rs are negative ($p = 0.076$), 5 significantly so, and 2 are significantly positive. The number of resident bird species follows a weaker, though similar trend; 17 and 16, respectively, are negative ($p = 0.032$, $p = 0.076$), 4 and 2 significantly so; 3 and 2 are significantly positive. However, in univariate analyses distance runs very poorly in comparison to area or habitat variables as predictors of species counts, regardless of the vertebrate group considered (Schoener and Schoener 1983b).

The number of species of orb spiders in the central Exumas is strongly negatively related to distance from a large island (Schoener and Toft 1983a, Toft and Schoener 1983). In multiple regressions including distance and log (area) for 1981 and 1982 species counts averaged together, distance coefficients are always significant. While area coefficients are more significant for no-lizard islands, distance coefficients are actually more significant than area coefficients in three or four comparisons involving lizard islands

(Toft and Schoener 1983, Table 9; although only nonzero-species islands were used, this effect could partly result from the still slightly greater range in areas of no-lizard islands). Addition of a third independent variable, maximum vegetation height, reduces distance to the least significant variable for no-lizard islands, but it is the most significant variable for lizard islands. Exactly comparable data are unavailable for vertebrates in the central Exumas. However, the effect of distance requires much greater distances in order to be apparent for vertebrates than for spiders in this and certain other Bahamian regions.

These results may suggest that, were distance taken into account, species-area relationships would be more regular in spiders. In fact, splitting islands into near ($\leq$ 230 m) and far ($>$ 230 m) groups did not systematically increase species-area coefficients except for lizard islands during 1981 (statistics available upon request). In general, correlations became greater for near islands and smaller for far islands. Perhaps this is because of the relatively great importance of immigration in this system as a cause of the species-area effect (see pattern 1). Moreover, for lizard islands during 1982 and 1983, but not 1981, correlation coefficients were much more negative for far islands than for all islands combined. No explanation for this difference between years is apparent to me, and it may be an artifact of somewhat different sets of islands being used for the three years.

In summary, in great contrast to area, distance relationships are apparently more regular for spiders than vertebrates, and distance patterns are especially regular for spiders on islands coinhabited by lizards. A possible reason for the difference between lizard and no-lizard islands is mentioned below in connection with pattern 6. Area is still always important, but it is the importance of distance *relative to* area or habitat diversity that especially characterizes the spiders.

Other Systems Unlike the species-area relation, no systematic survey for the species-distance relation has been published, so comparisons are impossible. One study, that of Abbott (1974) on islands surrounding Antarctica, allows a comparison with the same archipelago. For these 17 islands, area was a poor predictor of both bird and insect species counts, in contrast to most of the other studies summarized in Fig. 33.4. However, distance was a fair predictor for insect species counts, although number of plant species was better. The importance of distance relative to area was somewhat greater for insects than for birds in both univariate and multivariate analyses.

To obtain a preliminary comparison of taxa from disparate archipelagos, I surveyed all the other appropriate studies listed in Connor and McCoy (1979) and a few additional ones. No difference between terrestrial arthropods and vertebrates is obvious, though analyses are varied and explicit comparisons between distance and area spotty. In three arthropod studies (Thornton 1967, Baroni-Urbani 1971, Weissman and Rentz 1976) area performed substantially better than distance as a predictor of species counts. This was also true in certain vertebrate studies (Hamilton et al. 1964 [3 archipelagos], Heatwole 1975, Reed 1984), whereas habitat diversity (especially number of plant species) was dominant to distance in others (Power 1972, Harris 1973, Case 1975, Reed 1974). In three other studies, however, a major effect of some distance variable, comparable or superior to that of area or habitat, was found (Hamilton and Rubinoff 1964, Hamilton and Armstrong 1965, Diamond 1972b). Clearly, whether a distance effect is found depends partly on the scale of distances examined relative to the dispersal capabilities of the taxon. Thus, we argued that lizards, resident birds, and migrant birds should show decreasingly strong effects of distance over the small distances studied in the Bahamas, those distances being much smaller than those studied, for example, in the southwest Pacific (Schoener and Schoener 1983b and Fig. 33.1). Similarly, spiders in the Bahamas could be responding to smaller distances than birds. However, this line of reasoning does not explain differences between spiders and lizards just discussed, nor does it fully explain pattern 6 (below), in which an effect on individuals is found.

Pattern 5. Precision of the Threshold in Distance from a Source Separating Islands With and Without a Given Species

Measures of Regularity As with area (pattern 2), the occurrence-sequence technique can be used to measure how precisely islands ranked according to distance are separated into those having and not having a given species.

The Bahamas System In orb spiders of the central Exumas the relative importance of distance varies for lizard and no-lizard islands. In 5 of 15 comparisons for lizard islands distance is the most important variable (as measured by the *U* statistic, see pattern 2) of the three variables area, maximum vegetation height, and distance. In 1 of these 5 cases, the sequence is statistically significant. In none of the 15 comparisons for no-lizard islands is it the most important variable; this outcome also holds for all islands combined. The relative importance of distance for lizard versus no-lizard islands thus parallels that for number of species (pattern 4), although distance is somewhat less important overall here.

The pattern for spiders on lizard islands differs from patterns for either Bahamian resident birds or Bahamian lizards. In similar three-way comparisons (sample sizes in Schoener and Schoener 1983a, Appendix Tables 1–5), using supergroups, the archipelagal scale most comparable to that of the spider site, distance is most important in 15.5% of cases for all resident bird species combined, and it is the most important in 10.7% of cases for lizards; the bird versus spider frequencies are significantly different at the 0.05 level (chi-square test). As reported elsewhere (Schoener and Schoener 1983a), species counts of migrant birds tend to increase with distance, and distance is slightly more important at the finer archipelagal level of subgroups, but the percents just given are the ones most appropriately compared to the spider data.

Other Systems As with area, comparisons using occurrence sequences do not exist.

Pattern 6. Decrease in Number of Individuals of a Given Species on an Island with Distance from a Source of Colonists

Measures of Regularity As discussed for pattern 4, were island population sizes largely dependent on influx from the outside rather than on internal regulatory processes, farther islands would have smaller populations. Smaller populations on farther islands could also result from a greater habitat poverty there.

The Bahamas System Orb spiders show a striking decrease in number of individuals with increasing distance, both for all species combined and, with scattered exceptions, for individual species (Fig. 33.5). The decline is roughly exponential, suggesting passive dispersal (Schoener and Toft 1983a, Toft and Schoener 1983). In multiple regressions distance coefficients are strong for both lizard and no-lizard islands and, again in great contrast to area, are stronger on the former. Lizard-island coefficients are more negative than no-lizard-island coefficients in over two-thirds of cases (Toft and Schoener 1983, Tables 5–8). This difference suggests that spider populations on remote islands with predatory lizards are "subsidized" by immigrants rather than self-supporting.

I have little information concerning vertebrate population sizes in areas where a strong species-distance relation holds. Within the central Exumas, however, there is no tendency for farther islands to have smaller lizard populations, though completely censused islands are few (see pattern 3). Largely as the result of experiments, we have hypothesized that lizard dispersal is sporadic, occurring nearly always during major storms (Schoener and Schoener 1983c). The implication is that lizard populations are more likely to be controlled by factors internal to their islands than by a steady influx of propagules from the outside. Birds, which in the Bahamas often fly between islands, probably lie in between spiders and lizards.

Other Systems I am unaware of any studies aside from those just discussed that compare number of individuals on near and far islands. This is despite theoretical interest in the topic.

Pattern 7. Complementarities (Negative Associations) Between Species in Distributions or Abundances

Measures of Regularity Complementarities in distribution within an archipelago comprise the most convincing kind of observational evidence for ongoing negative species interactions. When the species involved are similar trophically, com-

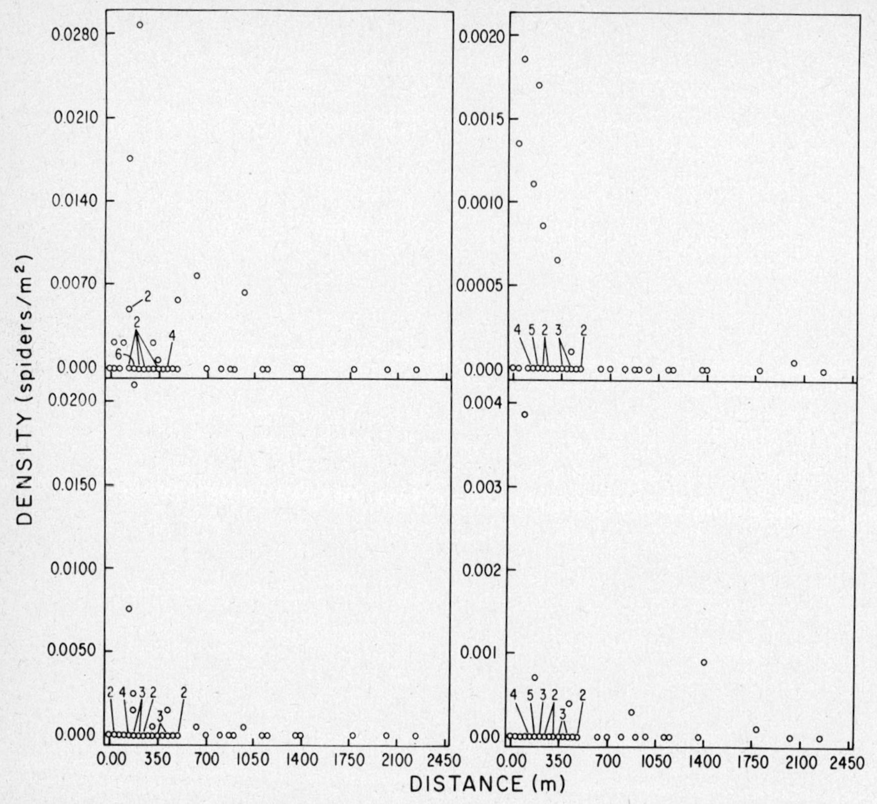

Fig. 33.5 Population densities of individual species of orb spiders versus distance from the nearest source (no-lizard islands exceeding in area the minimum-sized island having spiders, 1981 data.) (Top left) *Metepeira datona.* (Top right) *Gasteracantha cancriformis.* (Bottom left) *Argiope argentata.* (Bottom right) *Eustala cazieri* (occupied webs only). Note that density decreases strikingly with distance. (From Toft and Schoener 1983.)

petition is strongly implicated, especially if islands are identical in habitat and area, so that differences in the kind of island preferred are ruled out. If islands vary in area, competition might still be involved if one species is a good disperser but a poor competitor (and so found only on small islands with a high extinction rate) and the other species is a good competitor but a poor disperser. "Checkerboards" (patterns in which islands having one or the other species are spatially highly interdigitated: Diamond 1975) are even more suggestive of competition, because they rule out the possibility of the two species' having disjunct distributions simply because they invaded the archipelago from different directions. When dispersal rates are high or islands are large, complementarities in occurrence are unlikely, because at least some individuals of each species are likely to be everywhere (Toft et al. 1982). Under such conditions competition, if strong enough, should still produce complementarities in abundances, though complementarities may be difficult to detect if habitat types vary between the islands.

When a high degree of resource partitioning exists between the species (e.g., Chapter 30; Pacala and Roughgarden 1985), complementary island occupancies should be rare relative to species sharing islands. Species must be under strong present-day competition for complementarities to be theoretically likely. Note that a nested community pattern, in which all but one species occurring on n-species islands occur on islands with $n - 1$ species (see pattern 2), is inconsistent with complementary occurrences. It is, however, strongly consistent with a high degree of resource partitioning, and that partitioning may have resulted from past competition, though not necessarily on the archipelago of interest.

Complementarities in occurrence can be measured in several ways. One is to construct contingency tables in which each cell denotes a set of presences and absences for the species of interest (Cohen 1970, Toft et al. 1982). Tables may compare species pairwise or in higher-order combinations. A second way, which considers all species together, is to use the ratio of the variance in total species number in samples to the sum of the variances of the individual species occurrences (Schluter 1984). The expected value of V, the variance ratio statistic, is 1 when no association exists, is less than 1 when a negative association exists, and is more than 1 when a positive association exists. Complementarities in abundances are more difficult to deal with using correlational plots of abundances because of the nonlinear relationships that usually occur. However, the variance-ratio method, with densities rather than species numbers, is an alternative method.

The Bahamas System As reported in connection with pattern 2, lizard species typically form nested communities in small, uniform island groupings, so complementarity occurrences are in general absent. Two lizard species, *Anolis sagrei* and *Leiocephalus* sp., do show a special kind of complementary occurrence in the Bahamas. In the southern and central Bahamas *Leiocephalus* occurs on clumps of islands, within which it occupies both large and small islands; within such clumps, if a single species exists on an island, it is typically *Leiocephalus*. *A. sagrei* is typically the second species in the sequence in such clumps, that is, it occurs on all islands with two or more species. Where *Leiocephalus* does not occur, *A. sagrei* is the first species in the sequence, and it occurs on islands down to the size inhabited solely by *Leiocephalus* within its clumps. Thus there exists a complementarity in distribution but (1) it is restricted to small islands and (2) it has a geographical component, rather than being a checkerboard. Experiments are now underway to test whether the two species are likely to have disjunct distributions on small islands because of competition (see pattern 11).

Obvious complementary distributions for bird species in the Bahamas are rare, though no systematic search has yet been made. One striking example occurs between the northern mockingbird (*Mimus polyglottos*) and the Bahamian mockingbird (*Mimus gundlachi*) on smaller islands near Grand Cay. The two species differ enough in their preferred habitats, however, to make it uncertain that their complementarity is not a simple consequence; the northern mockingbird prefers open, human-altered situations, and the Bahamian mockingbird prefers dense scrub.

I systematically searched for complementary occurrences between spider species in the central Exumas by (1) computing all possible two-by-two contingency tables for presence and absence and doing chi-square tests and (2) using the variance-ratio test on the same sets of data. Data for each year were considered separately, because of high turnover rates (see pattern 9). Only large islands (those exceeding 50 m^2, the minimal island size having spiders in 1981) were used, and islands were grouped into size classes for certain comparisons. As the upper half of Table 33.2 shows, 18.5% of the 162 contingency tables were significant at the 5% or better level. The tables are complexly dependent upon one another (there is continuity between years, island groupings overlap, and all pairwise combinations of species are used), so that it is difficult to know if this percentage is surprisingly high. Whatever the case, in all significant tables the association was positive, not negative! Almost no difference in contingency coefficients (a sample-size–free measure of association) existed between lizard and no-lizard islands. The variance-ratio test (see column headed Occ) gave identical results; all sets of islands gave Vs exceeding 1, indicating positive associations. Moreover, 27 data sets of 36 were significantly positive at $p < 0.05$ or better (lower half of Table 33.2); this is a much more definitive result than with contingency tables. Again, differences between lizard and no-lizard islands were not apparent. These positive associations may be due largely to area or habitat quality covarying among islands, as well as to catastrophic physical factors such as high water

Table 33.2 TESTS FOR COMPLEMENTARITIES IN OCCURRENCES AND ABUNDANCES OF BAHAMIAN SPIDERS

Contingency-Table Tests (Number of Tables)

	Islands below lizard minimal-area threshold (>112–167 m²)*		Islands over lizard minimal-area threshold (>167 m²)						Islands over 112 m²			
	All (and no-lizard)		All		No-lizard		Lizard		All[†]		No-lizard	
	N	N_{sig}	N	N_{sig}	N	N_{sig}	N	N_{sig}	N	N_{sig}	N	N_{sig}
1981	9	0	9	2	9	0	9	0	9	1	9	1
1982	9	0	9	1	9	0	9	1	9	4	9	2
1983	9	0	9	2	9	3	9	2	9	6	9	5

Variance-Ratio Tests (Values of V)

	Islands below lizard minimal-area threshold (>112–167 m²)			Islands over lizard minimal-area threshold (>167 m²)									Islands over 112 m²					
	All (and no-lizard)			All			No-lizard			Lizard			All[†]			No-lizard		
	N	Occ	Den	N	Occ	Den	N	Occ	Den	N	Occ	Den	N	Occ	Den	N	Occ	Den
1981	9	2.00[a]	1.06	55	1.35[a]	0.93	29	1.23	0.86	26	1.55[a]	0.86	64	1.45[b]	0.99	38	1.44[a]	0.97
		1.12	0.99		1.25	0.83		1.41	0.77		1.04	0.82		1.33[a]	0.91		1.54[b]	0.89
1982	7	1.46	0.96	62	1.53[d]	0.98	34	1.54[b]	0.88	28	1.72[b]	0.94	69	1.56[a]	0.97	41	1.60[c]	0.90
		1.71	1.03		1.58[d]	1.03		1.52[a]	0.87		1.90[d]	1.09		1.64[a]	1.02		1.66[c]	0.91
1983	7	1.00	0.96	58	1.79[d]	1.13	31	2.00[d]	1.05	27	1.75[c]	1.02	65	1.98[d]	1.07	38	1.94[d]	1.02
		1.41	1.05		1.73[d]	1.27		1.97[d]	1.14		1.61[b]	1.05		1.77[d]	1.18		2.00[d]	1.11

N = total number of tables. N_{sig} = number of tables significantly associative at the 0.05 level by exact or chi-square tests. V = variance ratio, as defined in text, with significance levels indicated by superscripts (see Schluter 1984 for methods): $a = 0.05$, $b = 0.025$, $c = 0.01$, $d = 0.005$; top number without *Eustala* empty webs, bottom number with *Eustala* empty webs. Values of V greater or less than 1.00 indicate positive or negative associations, respectively. Occ = by occurrence. Den = by density, as explained in text.

*For this case, all islands are too small to have lizards, so All and No-lizard data are identical; 112 m² is the largest minimal-area threshold for any spider species during 1981.

†Lizard islands for this group same as preceding group.

affecting certain islands but not others; mutualisms, e.g., conjoining web structures, are possible, but we see little evidence of these.

In summary so far, spiders show no complementary occurrences, and vertebrates are known to show only a few examples. The most clear-cut case has a geographical component and hence is not a checkerboard.

Let us turn now from complementary distributions to complementary abundances. Data to test for complementary abundances are mostly not available for small-island Bahamian vertebrates, although such abundances do sometimes occur between *Anolis* species on very large islands (Schoener and Schoener 1980a). They also occur in *Leiocephalus* on the Crooked-Acklins Bank. Densities on islands where *Leiocephalus* occurs alone are 0.018 to 0.038 per m^2 ($N = 3$, 1979 data only), whereas where it occurs with other lizard species, densities are 0.002 to 0.023 per m^2 ($N = 9$) (Schoener and Stinson in preparation). These figures hold despite the fact that the smaller islands where *Leiocephalus* is solitary are more habitat-poor (although they may have fewer predators on average, also).

Complementarities in abundances for spiders from our study site are difficult to evaluate with Pearson correlations. While most are negative, none is significantly so, and plots are highly nonlinear (Toft and Schoener 1983). These plots do appear concave, but similar plots could be produced from noninteractive species, each of which does not occur on most of the islands, but has substantial populations on those that it does inhabit (Toft et al. 1982). As many islands are unoccupied by both species, this explanation cannot be ruled out. Variance-ratio tests (see column headed Den in Table 33.2) are equally inconclusive; while slightly fewer ratios exceed than fall below 1, all are close to 1 and none is statistically significant. Because of the ambiguity of distributional evidence here, experiments are underway to see if spider populations are depressed upon introduction of another species (see pattern 11).

Other Systems No surveys of studies using contingency tables to measure complementary occurrences are known to me. Strong (1982) found no such occurrences in his study of hispine beetles on *Heliconia*. Whittam and Siegel-Causey (1981) found only higher-order negative associations in Alaskan seabird colonies; positive two-way interactions were typical. Toft et al. (1982) have the most convincing data for negative occurrences between species. Six species of ducks occupying subarctic ponds showed varying degrees of complementarity depending on niche overlap; those species most similar were most complementary. This study carefully controlled for pond size.

Other scattered examples exist. Diamond (1973, 1975) has reported a variety of checkerboards, though most of the island cases seem to be associated with island-area differences. Another example of complementary distributions, that between the *Emoia* skinks of Arno atoll, also is related to island area (Kiester 1983). One example of complementary distributions apparently entirely unrelated to environmental differences between allopatric localities occurs between two species of *Geospiza* (Schluter and Grant 1982); food supply, vegetation, and substrate are similar on islands having one or the other species (the species are also complementary in altitude on certain large islands).

While vertebrates might show some average difference from arthropods in these comparisons, one is forced to conclude that, in general, complementary occurrences involving many islands or patches are uncommon. More generally, however, a tendency for species similar in morphology, diet, or microhabitat to be allopatric has been found for Galápagos finches (Schluter and Grant 1982) and lizards (Schoener 1970b). Finally, of course, complementarities along habitat gradients are common (review in Schoener 1974a).

Complementarities in abundances (as opposed to complementarities in occurrences) were searched for in some of the above studies. One of 13 species showed significant negative relations in Strong's (1982) hispine beetles, whereas several of Toft's (Toft et al. 1982) duck species did. Complementary abundances are likely to be common even when not specifically tested for (see the density-compensation studies surveyed by Case 1975, Jaenike 1978, Wright 1980). Experimentally, such abundances have been produced many times (review in Schoener 1983b), more often on average for terrestrial vertebrates than terrestrial arthropods (see pattern 11).

In contrast to the above methods, Schluter (1984) has gathered from the literature a variety of studies for which he calculated the variance-ratio test of association (certain censuses were separated temporally rather than spatially). By examining significance levels, he concluded that vertebrates had slightly more negatively associated species than did invertebrates, but differences were not significant. I have extracted from his list variance ratios for vertebrates and arthropods from terrestrial as well as freshwater systems (I included the latter because many of the arthropods have one stage in each type of system). Whether one examines strictly terrestrial species, species with at least one stage that is terrestrial, or terrestrial combined with freshwater species, arthropods are more positively associated than vertebrates. Exact tests for the proportion less and greater than 1 are significant for each vertebrate-arthropod comparison ($p = 0.032$, $p = 0.013$, and $p = 0.017$, respectively). Such tests are not quite suitable because the expected distribution of V is asymmetrical around 1 and depends on sample size; it approaches symmetry as sample size increases (Schluter personal communication). However, probabilities are low enough to suggest real differences.

Pattern 8. Temporal Constancy of Population Size

Measures of Regularity By one definition of stability (e.g., Watt 1968), the more constant the size of a population through time, the more stable it is. Were the underlying deterministic, density-dependent dynamics to be a stable point equilibrium, variation in population size could then be reasonably interpreted as variation extrinsic to the population or populations of interest, e.g., fluctuations in weather. These fluctuations could operate as mortality factors in populations well away from carrying capacities (Ks) set by resource abundance (Chapter 9). Alternatively, populations could track fluctuating Ks caused by a fluctuating resource base (Chapter 10). The resource base, in turn, may be affected directly or indirectly by fluctuating mortality factors associated with weather.

We know from mathematical models that various totally deterministic, density-dependent models with constant Ks can also give oscillatory population sizes, as equilibrium or quasi-equilibrium phenomena; limit cycles, strange attractors, and neutrally stable trajectories are some examples (May 1972, May and Oster 1976, Gilpin 1975, Guckenheimer et al. 1977). However, very long-term data would be required to implicate these more complicated dynamical objects, data beyond the capabilities of most available field studies (but see Hassell et al. 1976). Hence in the absence of data to the contrary, the assumption that constancy of population size for one or more trophically similar species indicates relative independence from external regulation is the parsimonious assumption and can be adopted in an operational spirit.

Connell and Sousa (1983) attempted to survey the literature on variation in population size. They included only studies in which a complete turnover of individuals was known or surmised. For comparability, I have kept a similar convention in my analyses here. For perennial organisms, however, rather than use maximum life span to determine a complete turnover, I used exponential curves fitted to survivorship data and declared a turnover when the product of the per-unit-time survivorships so estimated dropped as low or lower than 5%. This method is less sensitive to "freak" individuals (for additional justification, see Schoener 1985e). Connell and Sousa measured variability as the standard deviation of the log of population size. This measure does not vary with the units in which population size or density is measured and only varies with the base of the logarithms used. A second possible measure, which is dimension-free, is the coefficient of variation of population size. This measure is less sensitive than the first to variation for small populations, because arithmetic rather than logarithmic units are used.

The Bahamas System Data on population size variation satisfying Connell's and Sousa's criteria were available for 11 lizard populations in the Bahamas (including experimentals, once equilibrium was reached). Standard deviations varied from 0.027 to 0.157; the mean was 0.074 ($s = 0.040$). All values fall within Connell's and Sousa's category of least variation (0 to 0.20), the category that included only about 25% of

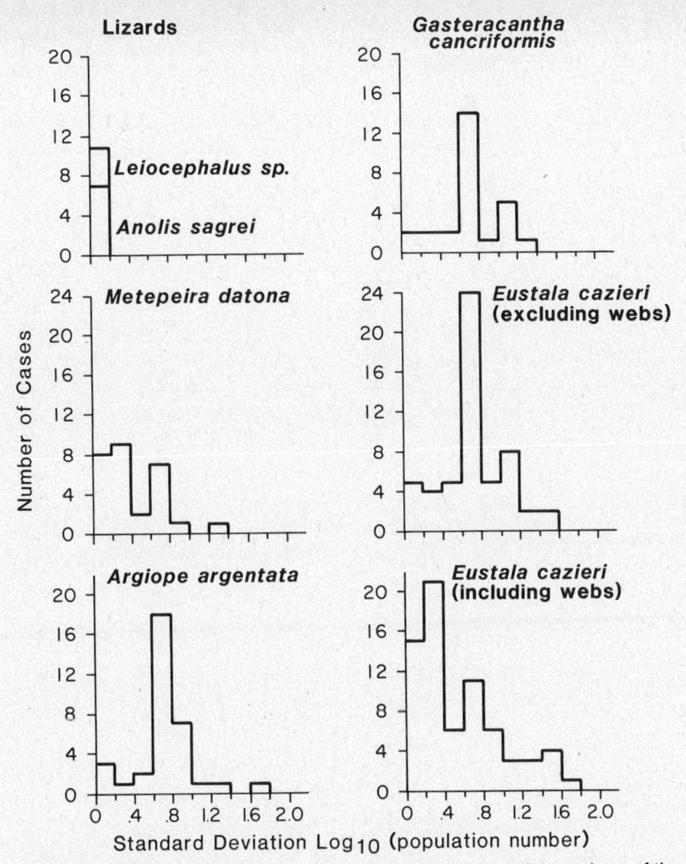

Fig. 33.6 Number histograms of populations having various values of the standard deviation of $\log_{10}$ (population number) through time, for two lizard species (top left graph) and four spider species (other graphs). Categories correspond to those in Connell and Sousa (1983). Note that lizards are far less variable than spiders.

their studies. Bahamian lizard populations are exceptionally stable. Long-term data on population sizes of birds are unavailable for the Bahamas.

To analyze spiders, I assumed that all orb spider species show at least an annual turnover. This is plausible from other studies, some on the same genera (e.g., Horton and Wise 1983, Olive 1981, Robinson et al. 1974, Robinson and Robinson 1970, Spiller 1984, Wise 1983), though I have never followed marked populations through a year. I eliminated all population sequences entirely consisting of zeros. I treated other sequences two ways. First, I added 0.1 to each count, including zero counts, before taking logs. Second, I added 0.1 to each count, but eliminated all sequences with at least one zero. Depending on the species, from 27 to 70 sequences were available per species using the first method; 212 sequences were available in all. With the second method sample sizes were roughly cut in half.

Using the first method, mean standard deviation of log (N) ranged from 0.40 to 0.71, 6 to 10 times that of lizards in the same region. Values ranged as high as 1.69, and on average only 15.1% fell within Connell's and Sousa's class of least variation, as compared to 100% of lizard values (Fig. 33.6).

Using the second method (which is probably less comparable to Connell's and Sousa's), mean

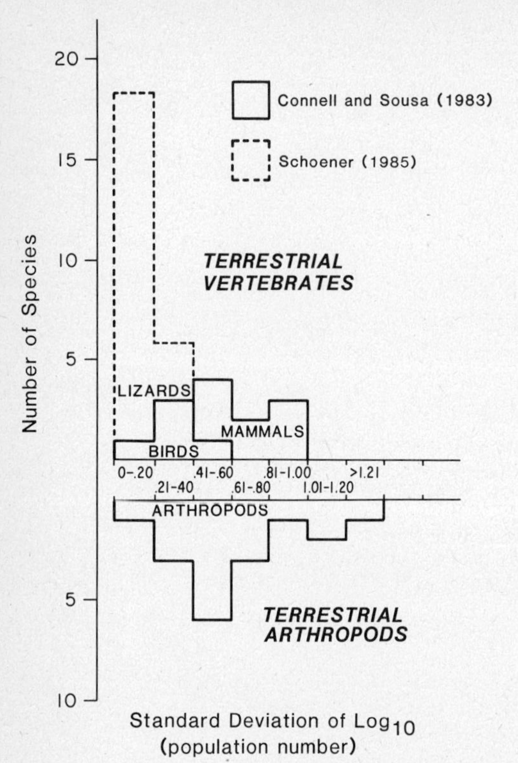

Fig. 33.7 Connell's and Sousa's (1983) published data on species of birds, mammals, and terrestrial arthropods, with newly compiled lizard data from the literature added on. Format is as in Fig. 33.6. Data do not include any in Fig. 33.6. Note that lizards and birds are substantially less variable than arthropods.

standard deviation of log (N) ranged from 0.25 to 0.48. These are lower means than with the first method, but still 3 to 7 times that of lizards. Values ranged as high as 1.06, and on average 34.9% fell within Connell's and Sousa's class of least variation.

Using the second statistic for measuring population size variability, the coefficient of variation, lizard values ranged from 6.1 to 34.5 and averaged 16.9 ($s = 8.9$). Spiders again had much larger coefficients. Using the first method, mean coefficients of variation ranged from 84.5 to 129.7. Using the second method, mean coefficients ranged from 27.1 to 35.9. Because coefficients of variation deemphasize fluctuations in small populations, lizard and spider values are closer than for standard deviation of log (N). Many of the spider populations are quite small, whereas lizard populations minimally have fairly large sizes (Schoener and Schoener 1983c).

Other Systems Connell and Sousa give standard deviation of log (N) for 13 terrestrial vertebrate and 17 terrestrial arthropod species. I have plotted their values in Fig. 33.7, separating bird from mammal species within the vertebrate histogram (the only two classes for which they had values). Although vertebrate populations by species as a whole are less variable than arthropod populations, this tendency is not significant by a Kolmogorov-Smirnov two-sample test (which is conservative for unequal Ns, but $D = 0.176$, chi-square $= 0.918$, one-tailed $p < 0.35$). However, the five bird species and the eight mammal species scarcely overlap in variability, and when birds are tested against arthropods, the former are significantly less variable ($D = 0.565$, chi-square $= 4.978$, one-tailed $p < 0.05$). Unpublished observations of Jones and Diamond for birds on British islands and the British mainland give mean values of coefficients of variation around 20 to 50%; these are larger than those for lizards, but smaller than those for Bahamian spiders.

Unfortunately, Connell's and Sousa's compilation seems to concentrate on very variable kinds of vertebrates; only five bird species are included, and the mammals are mostly the notoriously fluctuating nontropical rodents such as voles and lemmings. No lizard studies are used. I found 29 lizard studies from the literature fitting their criteria (Schoener 1985e). Of these, 25 are in their category of least variability, 3 are in the next least, and 1 is in the third least! These data, which do not include the 11 Bahamian lizard populations reported in the last section, when added to Connell's and Sousa's species graph (Fig. 33.7) totally change its character; now, arthropods win the variability contest going away. Mean variabilities of Bahamian spider species (Fig. 33.6) do fall about at the mean of Connell's and Sousa's arthropod populations, with the exception of *M. datona*, a colonial species with low variability (Schoener and Toft 1983b). A somewhat greater fraction of Bahamian spider populations lies in Connell's and Sousa's class of least variability than for their arthropod populations.

Thus when the Bahamian spider data are added to theirs, conclusions are not too different from those drawn when only their arthropods are included.

Pattern 9. Constancy of Species Lists Through Time

Measures of Regularity The previous section discussed variability in population size through time. The greater such variability, all other things being equal, the greater the extinction probability of the population (e.g., Wright and Hubbell 1983). When an island is at equilibrium, the extinction rate of all species combined equals the turnover rate, which is also then equaled by the immigration rate. Turnover can be calculated away from equilibrium as well. In general, turnover is an inverse measure of the temporal constancy of species lists for an ecological community.

Two types of turnover are distinguished. *Absolute turnover* is calculated as $(I + E)/2T$, where I is the number of new species immigrating during the census interval, E is the number of species going extinct during that interval, and T is the length of time between censuses. *Relative turnover* is defined as $100(I + E)/(S_1 + S_2)T$, where S_1 and S_2 are the number of species on the island at the beginning and end of the census interval, respectively (Diamond 1969b). Relative turnover is a better measure for comparisons between kinds of organisms or systems, whereas absolute turnover is better for testing hypotheses about how turnover is related to island area and distance (Toft and Schoener 1983, Schoener 1985d).

The Bahamas System Turnover for lizards in the central Exumas was low during both intervals for which it was computed; in both 1981 and 1982 only 1 immigration and no extinctions occurred on the 89 islands watched. This gives a mean relative turnover of 1.1% per year. My impression for other Bahamian regions, not systematically observed for turnover, is that this figure is in fact atypically high. Few data are available to calculate turnover among Bahamian birds.

Relative turnover in orb spiders from the central Exumas is much larger than for lizards; in 1981–1982, it was 59% (*Eustala* webs excluded) or 34% (*Eustala* webs included); in 1982–1983, it was 44 or 30%, respectively.

While turnover itself in orb spiders from the central Exumas did not show much relation to island area or distance, absolute extinction (E) and immigration (I) rates did (Toft and Schoener 1983). Absolute extinction during 1981–1982 was negatively related to area and positively related to distance (and to species number). Smaller population sizes at greater distances probably accounted for the distance relation. Absolute immigration was positively related to area and negatively related to distance (and species number). The target effect probably accounted for the area relation (see pattern 1). Among spiders as a whole, population size was strongly negatively associated with extinction; during 1981–1982, for example, the largest population size going extinct was only 9 individuals, and only 3 to 4% of individuals from all populations combined belonged to populations going extinct over the year's interval. Data for 1981–1982 are in Toft and Schoener (1983); data for 1982–1983 are given in Fig. 33.8.

Fig. 33.8 also separates lizard and no-lizard islands. One might expect that a given-sized population of spiders would go extinct more readily on a lizard island, but this was obviously true only for *A. argentata*. As the figure shows, all five populations of this species on lizard islands during 1982 were extinct by 1983. The same was true for 1981–1982, though only one population was involved (these numbers are, of course, minimal, as we monitored populations only at yearly intervals). Possibly other species will show a similar pattern once more years are studied.

One might also expect that immigration rate would be lower on lizard islands, given that we operationally define immigration as the constructing of a web. During the time from arrival to web building, spiders are mostly invisible to us, and they are not counted in our censuses. Nonetheless, they appear from experiment (see pattern 10) to be very apparent to lizards feeding upon them. Thus, the presence of predators might act as does distance in filtering out immigrants.

Despite these arguments, the data show no

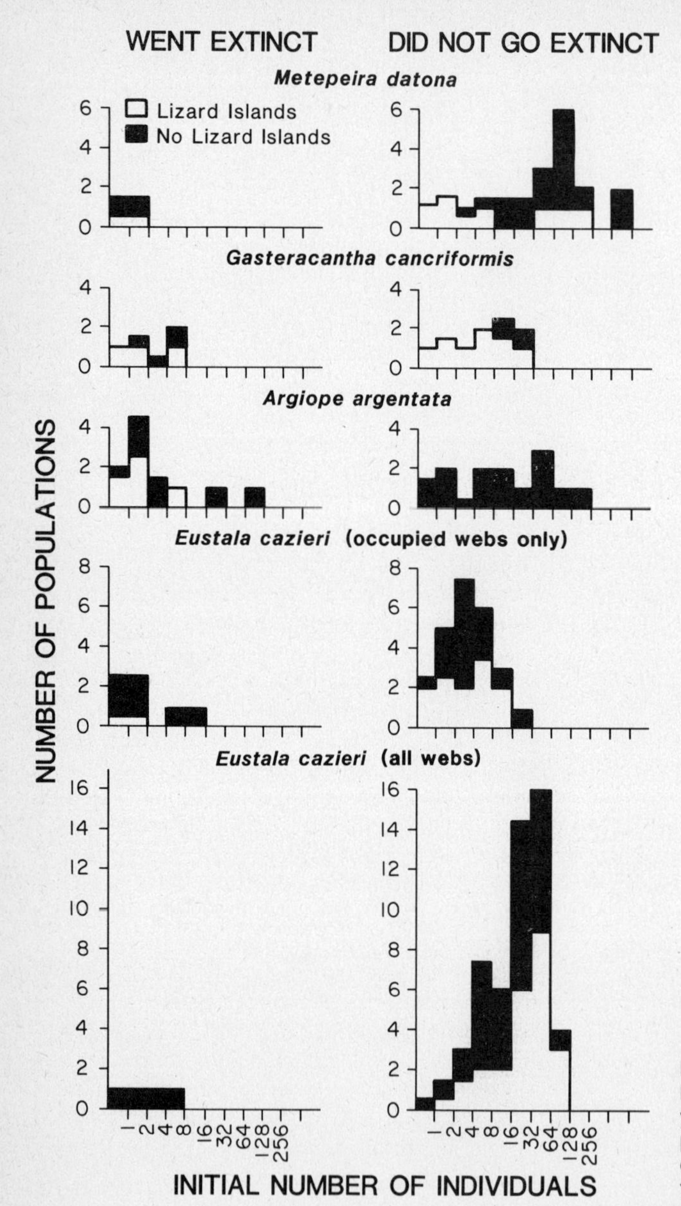

Fig. 33.8 Number histograms for spider populations going extinct and not going extinct over a year's time (1982–1983), categorized by initial population size. Lizard and no-lizard islands are distinguished. Note that populations going extinct are far smaller than those not going extinct.

consistent differences in immigration rates between lizard and no-lizard islands except for *G. cancriformis,* which for all years combined, immigrated *more* frequently on lizard islands (a 10% difference, chi-square = 5.214, 2 df, 2-tailed $p < 0.05$). This species is the only one that never shows a negative effect of lizards on its population size (Toft and Schoener 1983), perhaps because of its thorny integument, or its inaccessibly high webs, or some toxic compound (it has bright orange and black coloring). One must note, however, that we are calculating absolute immigration rates, and that, because lizard islands average much larger than no-lizard islands (Fig. 33.3), immigration per area is much greater for no-lizard islands.

When the regression analyses on absolute rates discussed above are performed for lizard and no-lizard islands separately, only that for immigration works for lizard islands. Whatever is causing extinction on lizard islands seems unrelated to area or distance and may be related to some property of the lizards themselves (Toft and Schoener 1983).

The lack of major differences in absolute immigration or extinction rates between lizard and no-lizard islands implies that the number of species on the two kinds of islands should be similar. In fact, this is roughly so (Fig. 33.3); because lizard islands average so much larger, their number of species *per area* is about half that on no-lizard islands. Moreover, mean population sizes are roughly similar on the two types of islands as well (Fig. 33.8), despite the fact that spider densities on lizard islands average about one-tenth those on no-lizard islands of the same area. Finally, the lack of major differences in immigration, extinction, or species number implies that relative turnover should be fairly similar on lizard and no-lizard islands. In fact, values fall within 2 percentage units (*Eustala* webs excluded) or 20 percentage units (*Eustala* webs included).

Other Systems As my recent review (Schoener 1983c) shows, turnover values for Bahamian spiders and lizards match exactly with turnover data gathered from the literature. Relative turnover decreases from lower to higher organisms, and it varies roughly in inverse proportion to generation time. In particular, arthropods average an order of magnitude greater turnover than do vertebrates. The highest turnover rate for a vertebrate group, averaged over an entire archipelago, is the 16.7% recorded for passerines on small islands off Maine (Morse 1977; this study was inadvertently omitted from my review). Higher rates for birds on single islands are known (Jones and Diamond personal communication).

Pattern 10. High Probability of Colonization by a Small Propagule

Measures of Regularity For a fixed propagule size, the greater the probability of colonization, all other things being equal, the more likely it is that species that can survive in a community when abundant will actually occur there. Put more mathematically, the demographic stochasticity bottleneck is one hazard to successful establishment. Because the natural process of colonization is so unobservable in its initial stages (one has to be in exactly the right place at exactly the right time), information on demographically stochastic extinction is better obtained experimentally. Surprisingly, the only published experiments of this sort for natural island systems are Crowell's (1973) with rodents and ours (Schoener and Schoener 1983c) with lizards. This is despite a rather well-developed theory (MacArthur and Wilson 1967, Goel and Richter-Dyn 1974).

The Bahamas System In 1977 we initiated a series of colonization experiments, using two species of lizards (*Anolis sagrei* and *Leiocephalus* sp.), experiments that are still being monitored. Propagules of 5 or 10 adult individuals with the ratio 3 females to 2 males were introduced onto 30 islands having no lizards naturally. Results reported through 1982 (Schoener and Schoener 1983c) are qualitatively identical to those through 1983 and are briefly as follows.

The time to extinction of colonizing propagules increased monotonically with island area, and many populations "took": Above a certain island area there were no extinctions. Propagules that went extinct never increased in population size, and propagules that successfully colonized increased steadily, often dramatically, to minimum population sizes of 15 to 25 individuals and to maximum sizes of over 400 individuals. Colonization thus was a highly deterministic function of island area (Fig. 33.9).

In 1982 we initiated a similar experiment with the spider *M. datona,* a colonial species with relatively similarly sized sexes and one that is fairly common on our study islands. Because of its coloniality and lack of sexual size dimorphism, we suspected that the species might be especially preadapted for colonization (in fact, we showed subsequently that the presence of established *M. datona* individuals favored establishment of introduced individuals). Propagules of 5 adult individuals (3 females and 2 males) were introduced onto 15 islands. Five islands were smaller than the smallest island on which *M. datona* occurs naturally; 5 islands were larger than the nat-

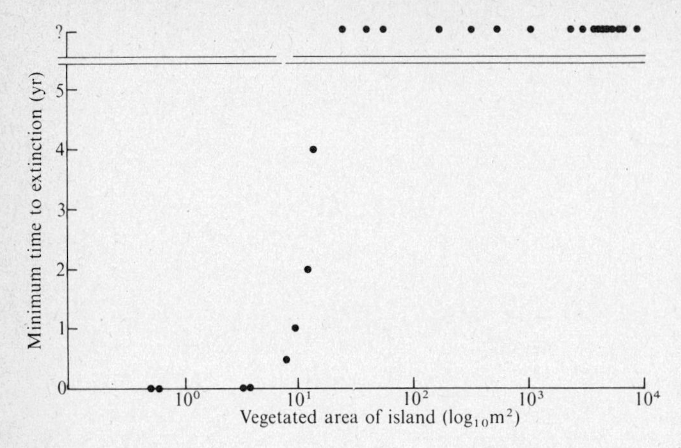

Fig. 33.9 Relation of minimal persistence time to island size for experimentally introduced propagules of *Anolis sagrei,* all sites combined. Minimal persistence time is calculated as: (1) the observation time immediately preceding the time no lizards or only males were found or (2) the first of two consecutive observation times at each of which only nonreproducing females were found. All islands designated by the question mark (?) on the ordinate had populations after four or five years. This relation is statistically significant whether: (1) only the smallest eight islands are used (r_s [Spearman correlation] = 0.939, $p < 0.01$), (2) all islands are used ($r_s = 0.830$, $p < 0.01$), or (3) only the four islands on the ascending part of the curve are used ($r_s = 1.0$, $p < 0.05$). (From Schoener and Schoener 1983c.)

ural minimal size for *M. datona* and did not have populations of that species or populations of lizards; 5 islands were larger than the natural minimal size for *M. datona* and did not have populations of that species, but did have populations of lizards. Spiders were collected from a main island (Staniel Cay) one or two days before introduction; each was placed in a separate vial (they eat one another). The vials were randomized intentionally by shaking them thoroughly in a plastic bag (and unintentionally by dropping them all over the place). All spiders were released within 10 m of one another at sites that we surmised favorable for web construction. We gave the same sort of advantage to our lizard propagules.

The number of spiders found after 4 days, 4 months, and 1 year was recorded for each island. To be ecologically active, a female spider has to build a web. Webs consist of an orb, a barrier web behind the orb, and a "retreat" or kind of shelter fashioned from bits of whatever materials are nearby—leaves, small stones, pieces of shell. As we have said, these webs are very conspicuous, so the number of colonists present can be determined with great ease.

Table 33.3 gives results. After 4 days only 4 (of 15) females survived on lizard islands, as compared to 12 on large no-lizard islands and 10 on small no-lizard islands. No island without lizards lost its entire propagule, whereas 2 with lizards did. Differences between islands with lizards and without lizards are significant, whether only large islands are used ($N = 5,5$; $p < 0.005$; t-test corrected for unequal variances) or all no-lizard islands are used ($N = 10,5$; $p < 0.01$). Differences between small and large no-lizard islands are not significant. The rapidity with which these differences were established suggests that predation on individuals while they are trying to colonize is the major factor involved; healthy spiders will not starve to death in 4 days, and in any event, remnants of their webs would have been found.

After 4 months only one lizard island had a "population" of spiders, and it was a single immature individual, well away from the site of introduction and possibly a natural colonist. Three of each kind of no-lizard island had populations, 5 of which exceeded the initial propagule size. For this interval differences in mean number of spiders present are significant only between all no-lizard islands and lizard islands ($N = 10,5$; $p < 0.05$).

After 1 year, one (a different one than after 4 months) lizard island had a population of 2 adult females, and 2 of each kind of no-lizard island had populations. On 3 of the no-lizard islands, populations had attained rather high numbers (explosions, *sensu* Schoener and Toft 1983a, had been created experimentally). Again, only the difference between all no-lizard and lizard islands is significant ($p < 0.05$).

In conclusion, lizards strongly affected spider establishment, and the effect was exerted in the very short term. A second conclusion is that the time to extinction for spiders showed little rela-

Table 33.3 COLONIZATION EXPERIMENTS ON *METEPEIRA DATONA*

	Number of females released	Number of females after 4 days	Number of spiders after 4 months	Number of spiders after 1 year
Lizard islands				
Ris Cay	3	0	0	0
Cran Cay	3	1	1	0
2nd-highest Cay	3	1	0	0
Cay 313	3	2	0	0
Cay 315	3	0	0	2
Small no-lizard islands				
Cay 405	3	2	13	3
Cay 334	3	1	20	20
Cay 302	3	3	4	0
Cay 316	3	2	0	0
Cay 328	3	2	0	0
Large no-lizard islands				
Bul Cay	3	3	5	35
Yan Cay	3	2	0	0
Longest Cay	3	2	4	27
Cay 314	3	2	0	0
Cay 318	3	3	1	0

tion to island area, in contrast to lizards for which we used the same propagule size and composition. Some seemingly favorable no-lizard islands lost their spiders over 4 to 12 months, inexplicably. This suggests a strong stochastic element in spider colonization, a hypothesis we are testing by repeating the experiment to determine if survival is consistently good or bad on given islands.

Other Systems Only Crowell's (1973) experiments on rodents are comparable to ours. He did not deliberately vary area, but, assuming that the islands he used were fairly similar to one another, his results for *Peromyscus maniculatus* were more like our lizard results (3 of 3 successful introductions using 3 or more pairs), whereas his results for *Clethrionomys gapperi* were intermediate between our lizard and spider results (3 of 5 successful introductions using 3 or more pairs).

Pattern 11. Degree to Which Interspecific Competition Has Been Found in Field Experiments

Measures of Regularity Although a variety of inferences can be made from observation alone concerning the action of interspecific competition, field experiments are an invaluable method for acquiring complementary, supplementary, and often essential information on this process. For example, the nested community sequence discussed above (pattern 2) could exist independently of interspecific competition, but competition may set the threshold island area at which another species is added. If all species-area relations are highly regular, there is no way to detect such competition simply by observation; one must do experiments in which a species is removed on islands near a threshold and another is introduced. Proceeding in exactly this way, Cole (1983) showed for mangrove ant communities that the location of thresholds was often determined by interspecific competition.

The Bahamas System Various competition experiments are presently underway on Bahamian lizards and orb spiders. They have been running for 7 years and 1 year, respectively.

Simultaneously with the introduction of lizards onto empty islands (see pattern 10), we introduced one of the same two species of lizards onto islands inhabited by the other. The empty-island experiments served as "conservative" controls for the invading species, that is, in all respects but occupancy by the putative competitor, nonempty islands were apparently superior for establishment. Normal controls for the resident species were set up as well, using islands matched for area and locale. We invaded two islands each with *A. sagrei* and *Leiocephalus* sp. Results given here are preliminary, as these experiments have yet to resolve themselves completely; population sizes are still changing. In no case did an invasion fail (so far), but in all cases the rate of increase of the invading population was lower than typical rates for the same species invading the larger empty islands. In at least one case the invader (*Leiocephalus*) immediately reduced the resident population; less obvious reductions may have occurred in other cases.

Spider competition experiments are much more inchoate. The introductions of *M. datona* onto islands (see pattern 10) are, in addition to experiments on colonization in this species, experiments on competition for the resident spider species. Controls are potentially all the nonperturbed islands in the central Exumas that are comparable in general characteristics and have the appropriate resident spider species. So far, successful invasions have not increased to nearly their full potential, and effects on resident population sizes are never statistically significant. As no population of *M. datona* has been successfully established on islands with lizards, an evaluation of how predation (and possibly competition from a higher trophic level) affects competition between spider species is presently impossible. Indeed, this will probably never be possible using natural regimes of lizard predation intensity.

Other Systems Elsewhere (Schoener 1983b), I reviewed 164 field-experimental studies of competition in all systems for all types of organisms. In terrestrial systems arthropods have experimentally been shown to compete in a significantly smaller fraction of attempts than have vertebrates. This is true in all three types of comparisons I have made (figures in Schoener 1983b, p. 268, but with the fungus deleted). For example, 19 of 35 arthropod species always or sometimes showed competition, whereas 51 of 67 vertebrate species always or sometimes did. Of the 5 experiments on spiders, 4 failed to find competition (see also Wise 1984), and the one exception—which is very striking—was performed in a salt marsh where spiders are very dense (Spiller 1984). Invertebrates also showed less competition than vertebrates in marine, and sometimes freshwater, comparisons. Connell (1983), in his independent assessment, reached similar conclusions.

SYNTHESIS

Pattern of the Patterns

Table 33.4 summarizes the preceding survey of degree of regularity in 11 patterns. Within the Bahamas vertebrates (birds and lizards) are more regular than arthropods (orb spiders) in most respects. They have stronger species-area (or species-habitat) relationships; occurrences of individual species are more precisely predicted by area (or habitat) thresholds; they show a stronger increase in number of individuals of a given species with area; they show more (but not many more) complementarities in distribution; their population sizes are less variable in time; their species lists are less variable in time; they colonize more predictably and with greater relation to island area; and one pair of species (the only one tested) was shown experimentally to compete.

For other systems where information is available, vertebrates are as regular as or more regular than arthropods in the same set of patterns. They have stronger species-area relationships; they have more negatively associated species; their population sizes are generally less variable with time; their species lists are less variable in time; and they show more experimentally determined competition.

Only in three patterns, all involving distance, are Bahamian spiders as regular as or more regular than vertebrates. Their species counts are as well or sometimes better predicted by distance; compared to island area their occurrence sequences are more regular for distance (but only one sequence is statistically distinguishable from random); and they show a strong tendency for population sizes to be smaller on farther islands. The one study from outside the Bahamas that compares the two taxa on the same archipelago weakly supports the first distance pattern, but studies involving one or the other taxon do not, and even such data are rare. No data from outside

Table 33.4 SUMMARY OF CONCLUSIONS ON COMPARATIVE REGULARITY BETWEEN VERTEBRATES AND ARTHROPODS FOR 11 PATTERNS

Pattern	Bahamas system	Other systems
1. Species-area	Vertebrates more regular	Vertebrates more regular
2. Occurrence sequences for area	Vertebrates more regular	No data
3. Individuals-area	Vertebrates more regular	No comparative data
4. Species-distance	Arthropods somewhat more regular	Arthropods slightly more regular
5. Occurrence sequences for distance	Arthropods more regular (relative to area)	No data
6. Individuals-distance	Arthropods more regular	No data
7. Complementarities	Vertebrates slightly more regular	Vertebrates more regular
8. Population stability	Vertebrates much more regular	Vertebrates more regular
9. Constancy of species lists	Vertebrates much more regular	Vertebrates much more regular
10. Colonization probability determined experimentally	Vertebrates more regular	No comparative data
11. Interspecific competition determined experimentally	Data inconclusive	Vertebrates more regular

the Bahamas exist for either of the other distance patterns.

Interpretation of the Patterns

The general picture that emerges, and a principal message of this chapter, is that spider communities of small islands, and perhaps other terrestrial arthropod communities, are controlled largely by the vicissitudes of dispersal and by predation (and maybe competition) from higher trophic levels. Their greatly fluctuating populations are consistent with the idea that they are affected greatly by physical factors such as wind, high water, and driving rain, but we are unable to present direct evidence here. Vertebrates, on the other hand, are more affected by competition (both inter- and intraspecific), show strong ties to particular types of habitats, and have population sizes that covary positively and sometimes directly with the areas they inhabit. Most of these generalizations are not particularly new, but what is surprising is that they are so decisively supported, not only by a single system where the taxa were deliberately compared, but almost equally so by data at large in the literature.

Arranging all these factors into a causal order for each of the two types of organisms is difficult and possibly not worth the effort; a network rather than a hierarchy is probably more appropriate. But we can highlight some of the connections.

For arthropods such as spiders a high degree of vagility (spiders balloon, for example) implies many populations that are in early stages of establishment, that is, many relatively small populations, at least if small enough isolates are available for colonization. A high degree of indiscriminate predation also implies small population sizes. Small population sizes, in turn, imply a high extinction probability from whatever cause, be it predators, physical factors, or inability to mate (especially likely for a low population *density*). Even large populations, if organisms are small and lack homeostatic mechanisms, can quickly be lowered by physical factors. High vagility and extinction together give high turnover, even though species number can also be relatively high. A large stochastic extinction rate implies relatively little pattern in how species relate to characteristics of the island they occupy, whether such characteristics be area, habitat variables, or co-occurring species. The major patterns shown by such organisms relate to input; dispersal abilities and properties of source areas (their distance, population sizes, and species compositions) are relatively more important than properties of the recipient area in determining community characteristics. Indeed, when input from elsewhere is especially dominant, small, fringing islands may be less interesting for certain purposes than *groups* of islands or patches, distinguished from other such groups by having recruitment come primarily from within.

Because of high maximum rates of increase, recovery or repopulation may be especially rapid, further contributing to great variability in population sizes (see Chapter 15).

Communities of lizards, and to a lesser extent birds, whose populations are safer from stochastic extinction, show more patterns related to characteristics of the places they inhabit; populations expand to limits determined by resources correlated with area or habitat and by interference competition (see Chapter 10). Hence, such species show correlations with area and habitat; they may interact and consequently show negative associations with one another as well (see also chapters 5, 6, and 30). Dispersal is relatively unimportant. Populations fluctuate less in time, and turnover is low, so that community composition is more constant.

When Does Predation Produce Patterns?

The preceding scheme leaves little room for the left-hand portion of the intermediate disturbance curve, or the tendency for species diversity to increase with intensity of predation or physical factors (Paine 1966, Harper 1969, Connell 1975). In fact, on lizard islands counts of spider species are about half those on no-lizard islands for the same area, and this difference is in the opposite direction to that predicted by the intermediate disturbance hypothesis. Thus, over existing ranges of predation intensity in the Bahamas, the hypothesis does not seem to hold. The situation is similar to Lubchenco's (1978) algae on emergent substrata. Moreover, this predation, along with a strong distance effect, apparently contributes to irregularity in patterns such as the species-area relation; correlations for spiders are reduced to zero on lizard islands.

Why should predation reduce diversity and produce irregularity in the spider-vertebrate system, but not in certain other systems? Most systems in which the intermediate disturbance hypothesis works are systems of immobile, space-limited organisms (Quinn 1979, in preparation, Sih et al. 1986). This suggests the following four axes of differentiation as critical for understanding the contrasting systems.

First, for immobile organisms, when space is cleared by disturbance, it cannot be occupied in the near future by adjacent individuals; rather, it is typically recolonized by different life stages dispersing from well outside, whether they be larvae from the water column in marine coastal systems or seeds blowing or brought in from elsewhere in terrestrial fields. This implies that one large apparent community is in fact several rather independent communities, each in its own successional stage (Quinn 1979, in preparation). Diversity in the large community is higher than it would be if component patches were uniform.

Second, organisms that compete for space show relatively little resource partitioning (e.g., Chapters 29 and 30; Harper 1969) and are often in a competitive hierarchy (Chapter 29; Paine 1966, Lubchenco 1978, Quinn 1982, but see Chapter 31). Where predators prefer the competitive dominant, their action increases diversity, as for algae in Lubchenco's (1978) tide pools. Species showing resource partitioning are less likely to show a competitive hierarchy, and only frequency-dependent predation is likely to cause increased diversity (Roughgarden and Feldman 1975). Indiscriminate predation, i.e., with fixed consumption rates (in the units, numbers of prey consumed per prey density per predator), is only occasionally likely to increase prey diversity (Cramer and May 1972).

Third, if predation exerts a very strong effect, population sizes will be so low that stochastic extinction will take over. Natural levels of predation may always be this high, as seems to be the case on Bahamian lizard islands. The state "having lizards" is quantumly different from the state "not having lizards," even for islands having the lowest lizard densities. This in turn results from chronically low dispersal rates of lizards between islands; what dispersal occurs probably does so only during major hurricanes, when "effective" island size is reduced, so that populations that survive this crunch become relatively large. If predation is either nonexistent or always high, the ascending portion of the disturbance curve will be slight or nonexistent.

Fourth, were the species compositions of colonizing organisms sufficiently variable through time, continuously freed space or other resources would imply continuously new communities, even for the same elapsed time since initial colonization. It is a fact, though little understood, that temporal variability in colonizing larvae is extremely high in marine hard-substrate systems (e.g., Sutherland and Karlson 1977, Schoener and Schoener 1981a). This seems obviously much less true in our spider system; I do not know about terrestrial arthropods as a whole.

All four factors may operate to produce differences in how predation affects diversity from one system to another. Where predation keeps most prey populations low, ''resource partitioning'' of shelter sites and other ways in which predation could cause variable prey habits may become obscured by continual stochastic extinction. Thus predation may result in no pattern, rather than either a pattern that mimics competition's action or one that has unique features.

Outlook

I would like to end cheerfully. So far as the system I study is concerned, and very likely many other systems as well, the protagonists on either side of the debate about pattern, population fluctuation, and competition appear both to be largely correct. Even exceptions to vertebrate regularity (and there are many; see, for example, Figs. 33.2 and 33.4), such as shrub-steppe birds (Chapter 9; Wiens 1977, 1984a), may prove some rule. Thus, for example, in certain temperate continental systems where vagility must play a major role at some scale, many species are not well established *in situ* and are subject to local extinction. Increased mortality from major weather disasters may also contribute, making such vertebrates more comparable to arthropods on small islands than to vertebrates there. Carrying the argument farther, I would predict that community studies of arthropods in continental systems such as arid western North America are going to give less regularity than anything we have seen so far, except with respect to immigration phenomena. On the other hand, a considerable number of arthropod populations may resemble the modal vertebrate situation, and such populations may themselves prove a rule. For example, ants and bees, when tested, always show interspecific competition in field experiments (Chapter 3, Schoener 1983b, Davidson 1985); both may be relatively insensitive to predation. As a second example, carabid beetles in a stable forest habitat show substantial niche overdispersion (Loreau in preparation); tiger beetles show food limitation (Pearson and Knisley 1985); both groups are predatory. The distinction between vertebrates and arthropods is far from absolute.

The proposed spectrum of patterned to less patterned communities is, of course, itself a pattern, and I hope I can seduce all but the most inveterate of nihilists into believing at least some of it.

SUMMARY

This paper examines the proposition that terrestrial vertebrates are more regular than arthropods with respect to a set of community and population patterns. Comparisons between birds and lizards on the one hand and orb spiders on the other are made for a single island system, the Bahamas. Published data on other systems are also surveyed.

In most respects vertebrates are more regular.

1. They have stronger species-area (or species-habitat) relationships.
2. Individual species' occurrences are more precisely predicted by area (or habitat) thresholds.
3. They show a stronger increase in number of individuals of a given species with area.
4. They show more complementarities in distribution.
5. Their population sizes are less variable in time.
6. Their species lists are less variable in time.
7. They colonize more predictably and with greater relation to island area.
8. One pair of species (the only one tested) was shown experimentally to compete.

For other systems where information is available, vertebrates are as or more regular in the same set of patterns.

Only in three patterns, all involving distance, are spiders as or more regular than vertebrates.

1. Their species counts are as or sometimes better predicted by distance.
2. Compared to area their occurrence sequences are more regular for distance.
3. They show a strong tendency for population sizes to be smaller on farther islands.

Other systems have little published data with which to compare the two taxa with respect to distance patterns, but those existing are not entirely supportive.

Spiders in the Bahamas are greatly affected by predation where lizards occur, show much extinction, disperse readily, and appear strongly affected by physical factors such as wind and high water. Lizards in the Bahamas are predator-free on small islands, show little extinction or immigration, disperse rarely during normal times, appear close to carrying capacity, and compete. I suggest that, in contrast to competition, predation may often not produce regular patterns, because it may reduce most prey populations to sizes highly vulnerable to stochastic extinction from many factors. In the island system studied, lizard predation on orb spiders is apparently of this sort.

ACKNOWLEDGMENTS

I thank T. Case, J. Diamond, P. Kareiva, D. Schluter, D. Spiller, and C. Toft for reading a previous draft. The research has been supported by National Science Foundation grants DEB 76-02663, 78-22792, and 81-18520.

References

Abbott, I. 1974. Number of plant, insect and land bird species on nineteen remote islands in the Southern Hemisphere. *Biol. J. Linn. Soc.* 6:143–152.

Abbott, I., L. K. Abbott, and P. R. Grant. 1977. Comparative ecology of Galápagos ground finches (*Geospiza gould*): evaluation of the importance of floristic diversity and interspecific competition. *Ecol. Monogr.* 47:151–184.

Abele, L. G. 1976. Comparative species richness in fluctuating and constant environments: coral-associated decapod crustaceans. *Science* 192:461–463.

Abrahamson, W. G., and M. Gadgil. 1973. Growth form and reproductive effort in goldenrods (*Solidago,* Compositae). *Am. Nat.* 107:651–661.

Abrams, P. A. 1977. Density-independent mortality and interspecific competition: a test of Pianka's niche overlap hypothesis. *Am. Nat.* 111:539–552.

Abrams, P. A. 1984a. Variability in resource consumption rates and the coexistence of competing species. *Theor. Popul. Biol.* 25:106–124.

Abrams, P. A. 1984b. Recruitment, lotteries, and coexistence in coral reef fish. *Am. Nat.* 123:44–55.

Abramsky, Z., and C. Sellah. 1982. Competition and the role of habitat selection in *Gerbillus allenbyi* and *Meriones tristrami:* a removal experiment. *Ecology* 63:1242–1247.

Abugov, R. 1982. Species diversity and the phasing of disturbance. *Ecology* 63:289–293.

Acevedo, M. 1981. On Horn's model of forest dynamics with particular reference to tropical forests. *Theor. Popul. Biol.* 19:230–250.

Addicott, J. F. 1974. Predation and prey community structure: an experimental study of the effect of mosquito larvae on the protozoan communities of pitcher plants. *Ecology* 55:475–492.

Addicott, J. F. 1978. The population dynamics of aphids on fireweed: a comparison of local and metapopulations. *Can. J. Zool.* 56:2554–2564.

Addicott, J. F. 1979. A multispecies aphid-ant association: density dependence and species-specific effects. *Can. J. Zool.* 57:558–569.

Addicott, J. F. 1981. Stability properties of 2-species models of mutualism: simulation studies. *Oecologia* 49:42–49.

Addicott, J. F. 1984. Mutualistic interactions in population and community processes. *In* P. W. Price, C. N. Slobodchikoff, and B. S. Guad, eds., *A New Ecology: Novel Approaches to Interactive Systems,* pp. 437–455. Wiley, New York.

Addicott, J. F. 1985. Competition in mutualistic systems. *In* D. H. Boucher, ed., *Mutualism: New Ideas About Nature*. Croom Hlem, London. In press.

Addicott, J. F., and H. I. Freedman. 1984. On the structure and stability of mutualistic systems: analysis of predator-prey and competition models as modified by the action of a slow-growing mutualist. *Theor. Popul. Biol.* 26:320–339.

Adolph, S., and J. Roughgarden. 1983. Foraging by passerine birds and *Anolis* lizards on St. Eustatius (Neth. Antilles): implications for interclass competition and predation. *Oecologia* 56:313–317.

Ahlmann, H. W. 1953. *Glacier Variations and Climatic Fluctuations*. American Geographical Society, New York.

Aker, C. L., and D. Udovic. 1981. Oviposition and pollination behavior of the yucca moth, *Tegeticula maculata* (Lepidoptera: Prodoxidae), and its relation to the reproductive biology of *Yucca whipplei* (Agavacae). *Oecologia* 49:96–101.

Alatalo, R. V. 1982. Evidence for interspecific competition among European tits Parus spp.: a review. *Ann. Zool. Fenn.* 19:309–317.

Alatalo, R. V., and A. Lundberg. 1983. Laboratory experiments on habitat separation and foraging efficiency in Marsh and Willow Tits. *Ornis Scand.* 14:115–122.

Aldrich, J. W. 1982. Rapid evolution in the House Finch (*Carpodacus mexicanus*). *J. Yamashina Inst. Ornithol.* 14:179–186.

Allan, J. D. 1982. The effects of reduction in trout density on the invertebrate community of a mountain stream. *Ecology* 63:1444–1455.

Allen, J. C. 1975. Mathematical models of species interactions in time and space. *Am. Nat.* 109:319–342.

Allen, L. J. S. 1983. Persistence and extinction in Lotka-Volterra reaction-diffusion equations. *Math. Biosci.* 65:1–12.

Alm, G. 1946. Reasons for the occurrence of stunted fish populations. *Inst. Freshwater Res. Drottningholm Rep.* 25:1–146.

Al-Mufti, M., C. Sydes, S. Furness, J. Grime, and S. Band. 1977. A quantitative analysis of shoot phenology and dominance in herbaceous vegetation. *J. Ecol.* 65:759–792.

Alstad, D., and G. Edmunds. 1983. Adaptation, host specificity, and gene flow in the black pineleaf scale. *In* R. Denno and M. McClure, eds., *Variable Plants and Herbivores in Natural and Managed Systems*, pp. 413–426. Academic Press, New York.

Amano, H., and D. A. Chant. 1978. Some factors affecting reproduction and sex ratios in two species of predacious mites, *Phytoseiulus persimilis* (Athias-Henriot) and *Amblyseius andersoni* (Chant) (Acarina: Phytoseiidae). *Can. J. Zool.* 56:1593–1607.

Anderson, G., A. Ehrlich, P. Ehrlich, J. Roughgarden, B. Russell, and F. Talbot. 1981. The community structure of coral reef fishes. *Am. Nat.* 117:476–495.

Anderson, R. C. 1965. *Cerebrospinae nematodiasis* in North American cervids. *North Am. Wildl. Conf. Trans.* 30:156–167.

Anderson, R. M. 1982. *Population Dynamics of Infectious Diseases: Theory and Applications*. Chapman and Hall, London.

Anderson, R. M., and D. M. Gordon. 1982. Processes influencing the distribution of parasite numbers within host populations with special emphasis on parasite-induced host mortalities. *Parasitology* 85:373–398.

Anderson, R. M., and R. M. May. 1978. Regulation and stability of host-parasite population interactions. I. Regulatory processes. *J. Anim. Ecol.* 47:219–247.

Anderson, R. M., and R. M. May. 1979. Population biology of infectious diseases. Part I. *Nature* 280:361–367.

Anderson, R. M., and R. M. May. 1980. Infectious diseases and population cycles in forest insects. *Science* 210:658–661.

Anderson, R. M., and R. M. May. 1985. Helminth infections of humans: mathematical models, population dynamics and control. *Adv. Parasitol.*, in press.

Andrewartha, H. G. 1961. *Introduction to the Study of Animal Populations*. University of Chicago Press, Chicago.

Andrewartha, H. G., and L. C. Birch. 1954. *The Distribution and Abundance of Animals*. University of Chicago Press, Chicago.

Andrusak, H., and T. G. Northcote. 1970. Management implications of spatial distribution and feeding ecology of cutthroat trout and Dolly Varden in coastal British Columbia lakes. *B. C. Fish Wildl. Branch Fish Mgmt.* no. 13.

Andrusak, H., and T. G. Northcote. 1971. Segregation between adult cutthroat trout (*Salmo clarki*) and Dolly Varden (*Salvelinus malma*) in small coastal British Columbia lakes. *J. Fish. Res. Board Can.* 28:1259–1268.

Anon. 1981. Conservation of the genetic resources of fish: problems and recommendations. *FAO Fish. Tech. Pap.* no. 217.

Antonovics, J. 1976. The input from population genetics: "the new ecological genetics." *Syst. Bot.* 1:233–245.

Antonovics, J., A. D. Bradshaw, and R. G. Turner. 1971. Heavy metal tolerance in plants. *Adv. Ecol. Res.* 7:1–85.

Antonovics, J., and D. A. Levin. 1980. The ecological and genetic consequences of density-dependent regulation in plants. *Annu. Rev. Ecol. Syst.* 11:411–452.

Antonovics, J., and R. B. Primack. 1982. Experimental ecological genetics in *Plantago*. VI. The demography of seedlings transplants of *P. lanceolata*. *J. Ecol.* 70:55–75.

Arditi, R. 1983. A unified model of the functional response of predators and parasitoids. *J. Anim. Ecol.* 52:293–303.

Armstrong, R. A., and R. McGehee. 1976. Coexistence of species competing for shared resources. *Theor. Popul. Biol.* 9:317–328.

Armstrong, R. A., and R. McGehee. 1980. Competitive exclusion. *Am. Nat.* 115:151–170.

Arnold, S. J. 1983. Sexual selection: the interface of theory and empiricism. *In* P. Bateson, ed., *Mate Choice,* pp. 67–107. Cambridge University Press, Cambridge.

Askew, R. R., and M. R. Shaw. 1974. An account of the Chalcidoidea (Hymenoptera) parasitising leaf-mining insects of deciduous trees in Britain. *Biol. J. Linn. Soc.* 6:289–335.

Atkinson, I. A. E. 1977. A reassessment of factors, particularly *Rattus rattus* L., that influenced the decline of endemic forest birds in the Hawaiian Islands. *Pac. Sci.* 31:109–133.

Atkinson, I. A. E. 1985. Effect of rodents on islands. *In* P. J. Moors, ed., *Conservation of Island Birds*. International Council for Bird Preservation, Cambridge.

Atkinson, W. D., and B. Shorrocks. 1981. Competition on a divided and ephemeral resource: a simulation model. *J. Anim. Ecol.* 50:461–471.

Ashton, P. S. 1969. Speciation among tropical forest trees: some deductions in light of recent evidence. *Biol. J. Linn. Soc.* 1:155–196.

Aubreville, A. 1938. La forêt coloniale: les forêts de l'Afrique Occidentale Français. *Ann. Acad. Sci. Coloniale* (Paris) 9:1–245.

Aubreville, A. 1971. Regenerative patterns in the closed forest of Ivory Coast. *In* S. R. Eyre, ed., *World Vegetation Types,* pp. 41–55. Columbia University Press, New York.

Augspurger, C. K. 1983. Offspring recruitment around tropical trees: changes in cohort distance with time. *Oikos* 20:189–196.

Axelrod, D. I. 1979. Age and origin of the Sonoran Desert vegetation. *Occ. Pap. Calif. Acad. Sci.* no. 132.

Axelrod, D. I., and P. H. Raven. 1978. Late Cretaceous and Tertiary vegetation history of Africa. *In* M. J. A. Werger, ed., *Biogeography and Ecology of Southern Africa,* pp. 77–130. Junk, Den Haag.

Ayala, F. J. 1967. Dynamics of population. II. Factors controlling population growth and population size in *Drosophila pseudoobscura* and *Drosophila melanogaster*. *Ecology* 48:67–78.

Ayala, F. J. 1969. Genetic polymorphism and interspecific competitive ability in Drosophila. *Genet. Res. Camb.* 14:95–102.

Ayala, F. J., M. E. Gilpin, and J. G. Ehrenfeld. 1973. Competition between species: theoretical models and experimental tests. *Theor. Popul. Biol.* 4:331–356.

Ayling, A. L. 1983. Factors affecting the spatial distribution of thinly encrusting sponges from temperate waters. *Oecologia* 60:412–418.

Ayling, A. M. 1981. The role of biological disturbance in temperate subtidal encrusting communities. *Ecology* 62:830–847.

Baardseth, E. 1970. Synopsis of biological data on knobbed wrack *Ascophyllum nodosum* (Linneaus) Le Jolis. *FAO Fish. Synop.* 38, rev. 1.

Backiel, T., and E. D. LeCren. 1978. Some density relationships for fish population parameters. *In* S. D. Gerking, ed., *Ecology of Freshwater Fish Production,* pp. 279–302. Blackwell, Oxford.

Bagsenal, T. B. 1978. Aspects of fish fecundity. *In* S. D. Gerking, ed., *Ecology of Freshwater Fish Production,* pp. 75–101. Blackwell, Oxford.

Bailey, N. T. J. 1975. *The Mathematical Theory of Infectious Diseases and Its Application,* 2nd ed. Griffin, London.

Bak, R. P. M., and B. E. Luckhurst. 1980. Constancy and change on coral reef habitats along depth gradients at Curacao. *Oecologia* 47:145–155.

Baker, H. G., and G. L. Stebbins. 1965. *The Genetics of Colonizing Species*. Academic Press, New York.

Bakker, K. 1961. An analysis of factors which determine success in competition for food among larvae of *Drosophila melanogaster*. *Arch. Néerl. Zool.* 14:200–281.

Bakker, K. 1964. Backgrounds of controversies about population theories and their terminologies. *Z. Angew. Entomol.* 53:187–208.

Balen, J. H. van. 1980. Population fluctuations of the Great Tit and feeding conditions in winter. *Ardea* 68:143–164.

Ball, R. C., and D. W. Hayne. 1952. Effects of the removal of the fish population on the fish-food organisms of a lake. *Ecology* 33:41–48.

Bambach, R. K. 1983. Ecospace utilization and guilds in marine communities through the Phanerozoic. *In* J. S. Tevesz and P. L. McCall, eds., *Biotic Interactions in Recent and Fossil Benthic Communities,* pp. 719–746. Plenum, New York.

Barbehenn, K. R. 1974. Recent invasions of Micronesia by small mammals. *Micronesica* 10:41–50.

Barber, R. T., and F. P. Chavez. 1983. Biological consequences of El Niño. *Science* 222:1203–1210.

Barker, J. S. F. 1971. Ecological differences and competitive interaction between *Drosophila melanogaster* and *Drosophila simulans* in small laboratory populations. *Oecologia* 8:139–156.

Barker, J. S. F., and W. T. Starmer, eds. 1982. *Ecological Genetics and Evolution: the Cactus-Yeast-Drosophila Model System*. Academic Press, Sydney, Australia.

Barnes, H., and H. Powell. 1950. The development, general morphology and subsequent elimination of barnacle populations of *Balanus crenatus* and *B. balanoides* after a heavy initial settlement. *J. Anim. Ecol.* 19:175–179.

Barnum, C. C. 1930. Rat control in Hawaii. *Hawaii. Plant. Rec.* 34:421–443.

Baroni-Urbani, C. 1971. Studien zur Ameisenfauna Italiens. XI. Die Ameisen des Taskanischen Archipels. Betrachtungen zer Herkunft dur Inselfaunen. *Rev. Suisse Zool.* 78:1037–1067.

Bartholomew, G. A. 1982. Body temperature and energy metabolism. *In* M. S. Gordon, ed., *Animal Physiology: Principles and Adaptations*, pp. 333–406. Macmillan, New York.

Bateman, A. J. 1948. Intra-sexual selection in *Drosophila*. *Heredity* 2:349–368.

Baur, G. N. 1968. *The Ecological Basis of Rainforest Management*. Forestry Commission, New South Wales.

Bawa, K. S. 1979. Breeding systems of trees in a tropical wet forest. *N. Z. J. Bot.* 17:521–524.

Bawa, K. S., and P. Opler. 1975. Dioecism in tropical forest trees. *Evolution* 29:167–179.

Beadle, N. C. W. 1966. Soil phosphate and its role in molding segments of the Australian flora and vegetation, with special reference to zeromorphy and sclerophylly. *Ecology* 47:992–1007.

Beals, E. 1960. Forest bird communities in the Apostle Islands of Wisconsin. *Wilson Bull.* 72:156–181.

Beals, E. W., and J. B. Cope. 1964. Vegetation and soils in an eastern Indiana woods. *Ecology* 45:777–792.

Beard, J. S. 1944. Climax vegetation in tropical America. *Ecology* 25:127–158.

Beard, J. S. 1955. The classification of tropical American vegetation-types. *Ecology* 36:89–100.

Beard, J. S. 1983. Ecological control of the vegetation of Southwestern Australia: moisture versus nutrients. *In* F. J. Kruger, D. T. Mitchell, and J. U. M. Jarvis, eds., *Mediterranean-Type Ecosystems: the Role of Nutrients*, pp. 66–73. Ecological Studies no. 43. Springer-Verlag, Berlin, New York.

Beattie, A. J. 1983. Distribution of ant-dispersed plants. *Sonderb. Naturwiss. Ver. Ham.* 7:249–270.

Beattie, A. J., and D. C. Culver. 1981. The guild of myrmecochores in the herbaceous flora of West Virginia forests. *Ecology* 62:107–115.

Beckman, W. C. 1943. Further studies on the increased growth rate of the rock bass *Ambloplites rupestris* (Rafinesque), following reduction in density of the population. *Trans. Am. Fish. Soc.* 72:72–78.

Beckman, W. C. 1948. Changes in growth rates of fishes following reduction in population densities by winterkill. *Trans. Am. Fish. Soc.* 78:82–90.

Beddington, J. R. 1974. Age distribution and the stability of simple discrete time population models. *J. Theor. Biol.* 47:65–74.

Beddington, J. R., C. A. Free, and J. H. Lawton. 1976. Concepts of stability and resilience in predator-prey models. *J. Anim. Ecol.* 45:791–816.

Behrensmeyer, A. K., and A. P. Hill. 1980. *Fossils in the Making: Vertebrate Taphonomy and Paleoecology*. University of Chicago Press, Chicago.

Behrensmeyer, A. K., and D. E. Schindel. 1983. Resolving time in paleobiology. *Paleobiology* 9:1–8.

Bell, H. L. 1984. A bird community of lowland rainforest in New Guinea. 6. Foraging ecology and community structure of the avifauna. *Emu* 84:142–158.

Bender, E. A., T. J. Case, and M. E. Gilpin. 1984. Perturbation experiments in community ecology: theory and practice. *Ecology* 65:1–13.

Benson, W. W., K. S. Brown, and L. E. Gilbert. 1975. Coevolution of plants and herbivores: passion flower butterflies. *Evolution* 29:659–680.

Bentley, B. L. 1976. Plants bearing extrafloral nectaries and the associated ant community: interhabitat differences in the reduction of herbivore damage. *Ecology* 57:815–820.

Benzing, D. H. 1981. Bark surfaces and the origin and maintenance of diversity among angiosperm epiphytes: a hypothesis. *Selbyana* 5:248–255.

Benzing, D. H. 1983. Vascular epiphytes: a survey with special reference to their interactions with other organisms. *In* S. L. Hutton, T. C. Whitmore, and A. C. Chadwick, eds., *Tropical Rain Forest: Ecology and Management*, pp. 11–24. British Ecological Society, Oxford.

Benzing, D. H. 1984. Epiphytic vegetation: a profile and suggestions for future enquiries. *In* E. Medina, H. A. Mooney, and C. Vasquez-Yanes, eds., *Physiological Ecology of Plants of the Wet Tropics*, pp. 155–172. Junk, Den Haag.

Bequaert, J. 1921. On the dispersal by flies of the spores of certain mosses of the family Splachnaceae. *Bryologist* 24:1–4.

Berger, A. J. 1981. *Hawaiian Birdlife*, 2nd ed. University Press of Hawaii, Honolulu.

Berger, J. 1980. Feeding behaviour of *Didinium nasutum* on *Paramecium bursaria* with normal or apochlorotic zoochlorellae. *J. Gen. Microbiol.* 118:397–404.

Bernabo, J. C. 1981. Quantitative estimates of temperature changes over the last 2700 years in Michigan based on pollen data. *Quat. Res.* 15:143–159.

Bess, H. A., R. van den Bosch, and F. A. Haramoto. 1961. Fruit fly parasites and their activities in Hawaii. *Proc. Hawaii. Entomol. Soc.* 17:376–378.

Betancourt, J. 1984. Late Quaternary plant zonation and climate in southeastern Utah. *Great Basin Nat.* 44:1–35.

Betancourt, J., P. S. Martin, and T. R. Van Devender. 1983. Fossil packrat middens from Chaco Canyon, New Mexico: cultural and ecological significance. *In* S. G. Wells, D. Love, and T. W. Gardner, eds., *Chaco Canyon Country, a Field Guide to the Geomorphology, Quaternary Geology, Paleoecology and Environmental Geology of Northwestern New Mexico,* pp. 207–217. American Geomorphological Field Group, 1983 Field Trip Guidebook.

Betancourt, J., and T. R. Van Devender. 1981. Holocene vegetation of Chaco Canyon, New Mexico. *Science* 214:656–658.

Beverton, R. J. H., and S. J. Holt. 1957. On the dynamics of exploited fish populations. *Fish. Invest. Minist. Agric. Fish. Food (GB) Ser. II.* no. 19.

Biehl, C. 1985. *The distribution of perennial* Eriogonum *shrubs (Polygonaceae) in the eastern Mojave Desert.* Doctoral dissertation, University of California, Los Angeles.

Bierzychudek, P. 1981. Pollinator limitation of plant reproductive effort. *Am. Nat.* 117:838–840.

Bigger, C. H. 1980. Interspecific and intraspecific acrorhagial aggressive behavior among sea anemones: a recognition of self and not-self. *Biol. Bull.* 159:117–134.

Black, F. L. 1975. Infectious diseases in primitive societies. *Science* 187:515–518.

Black, R. 1977. Population regulation in the intertidal limpet *Patelloida alticostata* (Angus 1865). *Oecologia* 30:9–22.

Blair, W. F. 1950. The biotic provinces of Texas. *Tex. J. Sci.* 2:93–117.

Blair, W. F. 1958. Distributional patterns of vertebrates in the southern United States in relation to past and present environments. *In* C. L. Hubbs, ed., *Zoogeography,* pp. 433–468. American Association for the Advancement of Science, Washington, D.C.

Blair, W. F. 1965. Amphibian speciation. *In* H. E. Wright, Jr., and D. G. Frey, eds., *The Quaternary of the United States,* pp. 543–556. Princeton University Press, Princeton, N.J.

Bleakney, J. S. 1972. Ecological implications of annual variation in tidal extremes. *Ecology* 53:933–938.

Boag, P. T. 1983. The heritability of external morphology in Darwin's Ground Finches (*Geospiza*) on Isla Daphne Major, Galápagos. *Evolution* 37:877–894.

Boag, P. T., and P. R. Grant. 1978. Heritability of external morphology in Darwin's Finches. *Nature* 274:793–794.

Boag, P. T., and P. R. Grant. 1981. Intense natural selection in a population of Darwin's Finches (*Geospizinae*) in the Galápagos. *Science* 214:82–85.

Boag, P. T., and P. R. Grant. 1984a. Darwin's Finches (*Geospiza*) on Isla Daphne Major, Galápagos: breeding and feeding ecology in a climatically variable environment. *Ecol. Monogr.* 54:463–489.

Boag, P. T., and P. R. Grant. 1984b. The classical case of character release: Darwin's finches (*Geospiza*) of Isla Daphne Major, Galápagos. *Biol. J. Linn. Soc.* 22:243–287.

Bohlen, C. 1983. *Competition for Space in the Anoles of the Island of St. Maarten (Neth. Antilles).* Master's thesis, Stanford University, Stanford, California.

Bold, H. C. 1970. Some aspects of the taxonomy of soil algae. *In* F. J. Frederick and R. M. Klein, eds., *Phylogenesis and Morphogenesis in the Algae,* pp. 601–616. Annals of New York Academy of Science, New York.

Bond, W. 1983. On alpha-diversity and the richness of the Cape flora: a study in southern Cape fynbos. *In* F. J. Kruger, D. T. Mitchell, and J. U. M. Jarvis, eds., *Mediterranean-Type Ecosystems: the Role of Nutrients,* pp. 337–356. Ecological Studies no. 43. Springer-Verlag, Berlin, New York.

Bonner, J. T. 1965. *Size and Cycle: an Essay on the Structure of Biology.* Princeton University Press, Princeton, N.J.

Bornkamm, R. 1961a. Zur quantitativen Bestimmung von Konkurrenzkraft und Wettberwerbsspannung. *Ber. Dtsch. Bot. Ges.* 74:75–83.

Bornkamm, R. 1961b. Zur Lichtkonkurrenz von Ackerunkraeutern. *Flora,* Jena N. F. 151:126–143.

Botkin, D. B. 1981. Causality and succession. *In* D. C. West, H. H. Shugart, and D. B. Botkin, eds., *Forest Succession Concepts and Application,* pp. 36–55. Springer-Verlag, New York.

Botkin, D. B., J. F. Janak, and J. R. Wallis. 1972. Some ecological consequences of a computer model of forest growth. *J. Ecol.* 60:849–872.

Botkin, D. B., and M. J. Sobel. 1975. Stability in time-varying ecosystems. *Am. Nat.* 109:625–646.

Boucher, D. H., S. James, and K. H. Keeler. 1982.

The ecology of mutualism. *Annu. Rev. Ecol. Syst.* 13:315–347.

Bowers, M. A., and J. H. Brown. 1982. Body size and coexistence in desert rodents: chance or community structure? *Ecology* 63:391–400.

Bowman, R. I. 1961. Morphological differentiation and adaptation in the Galápagos finches. *Univ. Calif. Publ. Zool.* 58:1–302.

Box, E. O. 1981. *Macroclimate and Plant Forms: an Introduction to Predictive Modeling in Phytogeography*. Junk, Den Haag.

Box, G. E. P. 1976. Science and statistics. *J. Amer. Stat. Assoc.* 71:791–799.

Box, T. W. 1961. Relationships between plants and soils of four range plant communities in south Texas. *Ecology* 42:794–810.

Boyden, T. C. 1980. Floral mimicry by *Epidendrum ibaguense* (Orchidaceae) in Panama. *Evolution* 34:135–136.

Boyden, T. C. 1982. The pollination biology of *Calypso bulbosa* var. *americana* (Orchidaceae): initial deception of bumblebee visitors. *Oecologia* 55:178–184.

Brace, R. C. 1981. Intraspecific aggression in the colour morphs of the anemone *Phymactis clematis* from Chile. *Mar. Biol.* 64:85–93.

Bradbury, J. P., B. Leyden, M. Salgado-Labouriau, W. M. Lewis, Jr., C. Schubert, M. W. Binford, D. G. Frey, D. R. Whitehead, and F. H. Weibezahn. 1981. Late Quaternary environmental history of Lake Valencia, Venezuela. *Science* 214:1299–1306.

Bradbury, J. W., and R. M. Gibson. 1983. Leks and mate choice. *In* P. Bateson, ed., *Mate Choice,* pp. 109–138. Cambridge University Press, Cambridge.

Bradley, G. A. 1973. Interference between nest populations of *Formica obscuripes* and *Dolichoderus taschenbergi* (Hymenoptera: Formicidae). *Can. Entomol.* 105:1525–1528.

Bradshaw, A. D. 1965. Evolutionary significance of phenotypic plasticity in plants. *Adv. Genet.* 13:115–155.

Brasier, M. D. 1976. Early Cambrian intergrowths of archaeocyathids, *Renalcis,* and pseudostromatolites from South Australia. *Paleobiology* 19:223–245.

Bray, J. R., and J. T. Curtis. 1957. An ordination of the upland forest communities of southern Wisconsin. *Ecol. Monogr.* 27:325–349.

Bremer, D. 1984. Waipio Christmas bird count. *Elepaio* 44:98–101.

Brenchley, W., and K. Warington. 1958. *The Park Grass Plots at Rothamsted*. Rothamsted Experimental Station, Harpenden, U.K.

Briand, F. 1983. Environmental control of food web structure. *Ecology* 64:253–263.

Briand, F., and J. E. Cohen. 1984. Community food webs have scale-invariant structure. *Nature* 307:264–267.

Bristow, C. M. 1981. *The Structure of a Temperate Zone Mutualism: Ants and Homoptera on New York Ironweed* (Vernonia noveboracensis *L.*). Doctoral dissertation, Princeton University, Princeton, N.J.

Brokaw, N. V. 1982. Treefalls: frequency, timing, and consequences. *In* E. G. Leigh, Jr., A. S. Rand, and D. M. Windsor, eds., *The Ecology of a Tropical Forest: Seasonal Rhythms and Long-term Changes,* pp. 101–108. Smithsonian Institution Press, Washington, D.C.

Brooks, D. R. 1980. Allopatric speciation and non-interactive parasite community structure. *Syst. Zool.* 30:192–203.

Brooks, J. L., and S. I. Dodson. 1965. Predation, body size, and composition of plankton. *Science* 150:28–35.

Brown, J. H. 1971a. Mechanisms of competitive exclusion between two species of chipmunks. *Ecology* 52:305–311.

Brown, J. H. 1971b. Mammals on mountaintops: non-equilibrium insular biogeography. *Am. Nat.* 105:467–478.

Brown, J. H. 1975. Geographical ecology of desert rodents. *In* M. L. Cody and J. M. Diamond, eds., *Ecology and Evolution of Communities,* pp. 315–341. Harvard University Press, Cambridge, Mass.

Brown, J. H. 1978. The theory of insular biogeography and the distribution of boreal birds and mammals. *Great Basin Nat. Mem.* 2:209–227.

Brown, J. H. 1981. Two decades of homage to Santa Rosalia: toward a general theory of diversity. *Am. Zool.* 21:877–888.

Brown, J. H., and D. W. Davidson. 1977. Competition between seed-eating rodents and ants in desert ecosystems. *Science* 196:880–882.

Brown, J. H., D. W. Davidson, and O. J. Reichman. 1979a. An experimental study of competition between seed-eating rodents and ants. *Am. Zool.* 19:1129–1143.

Brown, J. H., and A. C. Gibson. 1983. *Biogeography*. Mosby, St. Louis.

Brown, J. H., J. J. Grover, D. W. Davidson, and G. A. Lieberman. 1975. A preliminary study of seed predation in desert and montane habitats. *Ecology* 56:987–992.

Brown, J. H., and A. Kodric-Brown. 1977. Turnover rates in insular biogeography: effect of immigration on extinction. *Ecology* 58:445–449.

Brown, J. H., and A. Kodric-Brown. 1979. Convergence, competition, and mimicry in a temperate community of hummingbird pollinated flowers. *Ecology* 60:1022–1035.

Brown, J. H., A. Kodric-Brown, T. G. Whitham, and H. W. Bond. 1981. Competition between hummingbirds and insects for the nectar of two species of shrubs. *Southwest. Nat.* 26:133–145.

Brown, J. H., and G. A. Lieberman. 1973. Resource utilization and coexistence of seed-eating rodents in sand dune habitats. *Ecology* 54:788–797.

Brown, J. H., and J. C. Munger. In press. Experimental manipulation of a desert rodent community: food addition and species removal. *Ecology*.

Brown, J. H., O. J. Reichman, and D. W. Davidson. 1979b. Granivory in desert ecosystems. *Annu. Rev. Ecol. Syst.* 54:201–227.

Brown, J. L. 1975. *The Evolution of Behavior*. Norton, New York.

Brown, V. K. 1977. Metamorphosis: a morphometric description. *Int. J. Insect Morphol. Embryol.* 6:221–223.

Brown, W. H. 1919. *Vegetation in the Philippine Islands*. Philippine Bureau of Science, Department of Agriculture and Natural Resources, Manila.

Brown, W. L., Jr., and E. O. Wilson. 1956. Character displacement. *Syst. Zool.* 7:49–64.

Brubaker, L. B. 1981. Long-term forest dynamics. *In* D. C. West, H. H. Shugart, and D. B. Botkin, eds., *Forest Succession Concepts and Application*, pp. 95–106. Springer-Verlag, New York.

Brubaker, L. B., and E. R. Cook. 1983. Tree-ring studies of Holocene environments. *In* H. E. Wright, Jr., ed., *Late Quaternary Environments of the United States. Vol. II. The Holocene*, pp. 222–238. University of Minnesota Press, Minneapolis.

Brudnell-Bruce, P. G. C. 1975. *The Birds of the Bahamas*. Taplinger, New York.

Bryan, E. H., Jr. 1958. *Checklist and Summary of Hawaiian Birds*. Books About Hawaii, Honolulu.

Bryant, J. P., and P. J. Kuropat. 1980. Selection of winter forage by subarctic browsing vertebrates: the role of plant chemistry. *Annu. Rev. Ecol. Syst.* 11:261–285.

Bryant, M., D. J. DePree, S. Dick-Peddie, P. H. Hamilton, and W. G. Whitford. 1976. The impact of seed consumers in a desert ecosystem. *US-IBP Desert Biome Res. Memo.* 76-22:37–46.

Buckley, R. 1983a. A possible mechanism for maintaining diversity in species-rich communities: an addendum to Connell's hypothesis. *Oikos* 40:312.

Buckley, R. 1983b. Interaction between ants and membracid bugs decreases growth and seed set of host plant bearing extrafloral nectaries. *Oecologia* 58:132–136.

Budnik, M., and D. Brncic. 1976. Effects of larval biotic residues on viability in four species of Drosophila. *Evolution* 29:777–780.

Budowski, G. 1961. *Studies on Forest Succession in Costa Rica and Panama*. Doctoral dissertation, Yale University, New Haven, Conn.

Budowski, G. 1970. The distinction between old secondary and climax species in tropical Central American lowland forests. *Trop. Ecol.* 11:44–48.

Bulmer, M. G. 1980. *The Mathematical Theory of Quantitative Genetics*. Oxford University Press (Clarendon Press). Oxford.

Burdon, J. J., and G. A. Chilvers. 1982. Host density as a factor in plant disease ecology. *Annu. Rev. Phytopathol.* 20:143–166.

Burger, W. C. 1981. Why are there so many species of flowering plants? *BioScience* 31:572–581.

Burnet, A. M. R. 1970. Seasonal growth of brown trout in two New Zealand streams. *N. Z. J. Mar. Freshwater Res.* 4:55–62.

Burrows, W. 1959. *Textbook of Microbiology*. Saunders, Philadelphia.

Burton, J. F. 1975. The effects of recent climatic changes on British insects. *Bird Study* 22:203–204.

Bush, A. O., and J. C. Holmes. In preparation. *Intestinal helminths of lesser scaup ducks: patterns of association*.

Bush, A. O., and J. C. Holmes. 1983. Niche separation and the broken-stick model: use with multiple assemblages. *Am. Nat.* 122:849–855.

Bush, G. L. 1969. Mating behavior, host specificity, and the ecological significance of sibling species in frugivorous flies of the genus *Rhagoletis* (Diptera, Tephritidae). *Am. Nat.* 103:669–672.

Bush, G. L. 1974. The mechanism of sympatric host race formation in the true fruit flies (Tephritidae). *In* M. J. D. White, ed., *Genetic Analysis of Speciation Mechanisms*, pp. 3–23. Australia and New Zealand Book Co., Sydney.

Bush, G. L., and S. R. Diehl. 1982. Host shifts, genetic models of sympatric speciation and the origin of parasitic insect species. *In* J. H. Visser and A. K. Minks, eds., *Proceedings Fifth International Symposium on Insect-Plant Relationships*, pp. 297–306. Pudoc, Wangeningen.

Buss, L. W. 1979a. Habitat selection, directional growth, and spatial refuges: why colonial animals have more hiding places. *In* G. Larwood and B. Rosen, eds., *Biology and Systematics of Colonial Organisms*, pp. 459–477. Academic Press, London.

Buss, L. W. 1979b. Bryozoan overgrowth interactions: the interdependence of competition for food and space. *Nature* 281:475–477.

Buss, L. W. 1980. Competitive intransitivity and the size-frequency distributions of interacting populations. *Proc. Natl. Acad. Sci., U.S.A.* 77:5355–5359.

Buss, L. W. 1981a. Group living, competition, and the evolution of cooperation in a sessile invertebrate. *Science* 213:1012–1014.

Buss, L. W. 1981b. Mechanisms of competition between *Onychocella alula* and *Antropora tincta* on an eastern Pacific rocky shoreline. *In* C. Nielson and G. Larwood, eds., *Recent and Fossil Bryozoa,* pp. 39–49. Olsen and Olsen, Denmark.

Buss, L. W. 1982. Somatic cell parasitism and the evolution of somatic tissue compatibility. *Proc. Natl. Acad. Sci., U.S.A.* 79:5337–5341.

Buss, L. W., and E. W. Iverson. 1981. A new genus and species of Sphaeromatidae (Crustacea: Isopoda) with experiments and observations on its reproductive biology, interspecific interactions, and color polymorphisms. *Postilla* 184:1–24.

Buss, L. W., and J. B. C. Jackson. 1979. Competitive networks: nontransitive competitive relationships in cryptic coral reef environments. *Am. Nat.* 113:223–234.

Buss, L. W., and J. B. C. Jackson. 1981. Planktonic food availability and suspension-feeder abundance: evidence of *in situ* depletion. *J. Exp. Mar. Biol. Ecol.* 49:151–161.

Buss, L. W., C. S. McFadden, and D. R. Keene. 1984. Biology of hydractiniid hydroids. 2. Histocompatibility effector system/competitive mechanism mediated by nematocyst discharge. *Biol. Bull.* 167:131–158.

Butler, A. J., and M. J. Keough. 1981. Distribution of *Pinna bicolor* Gmelin (Mollusca: Bivalvia) in South Australia, with observations on recruitment. *Trans. R. Soc. S. Aust.* 105:29–39.

Buzzati-Traverso, A. 1949. A preference of *D. subobscura* for wild yeasts. *Drosoph. Inf. Serv.* 23:88.

Caffey, H. M. 1985. Spatial and temporal variation in settlement and recruitment of an intertidal barnacle. *Ecology,* in press.

Cain, M. L. 1985. Random search by herbivorous insects: a simulation model. *Ecology* 66:876–888.

Cain, M. L., J. Eccleston, and P. M. Kareiva. 1985. The influence of food plant dispersion on caterpillar searching success. *Ecol. Entomol.* 10:1–7.

Campbell, D. T. 1983. The two distinct routes beyond kin selection in ultrasociality: implications for the humanities and social sciences. *In* D. L. Bridgeman, ed., *The Nature of Prosocial Development: Interdisciplinary Theories and Strategies,* pp. 11–41. Academic Press, New York.

Campbell, J. B. 1977. *The Upper Paleolithic of Britain.* Oxford University Press, Oxford.

Campbell, K. E. 1982. Late Pleistocene events along the coastal plain of northwestern South America. *In* G. T. Prance, ed., *Biological Diversification in the Tropics,* pp. 423–440. Columbia University Press, New York.

Campbell, R. S., and J. G. Keller. 1932. Growth and reproduction of *Yucca elata. Ecology* 13:364–374.

Cannon, L. R. G. 1972. Studies on the ecology of papillose allocreadid trematodes of the yellow perch in Algonquin Park, Ontario. *Can. J. Zool.* 50:1231–1239.

Cannon, W. A. 1925. Physiological features of roots, with special reference to the relation of roots to aeration of the soil. *Carnegie Inst. Wash. Publ.* no. 368.

Cappuccino, N., and P. Kareiva. 1985. Coping with a capricious environment: a population study of the rare woodland butterfly, *Pieris virginiensis. Ecology* 66:152–161.

Capurro, H. A., and E. H. Bucher. 1982. Poblaciones de aves granivoras y disponibilidad de semillas en el bosque chaqueño de Chamical. *Ecosur, Argentina* 9:117–131.

Caputa, J. 1948. Untersuchungen ueber die Entwicklung einiger Graeser und Kleearten in Reinsaat und Mischung. *Landwirtsch. Jahrb. der Schweiz* 10:1–127.

Carlander, K. D. 1975. Community relations of bass, large natural lakes. *In* R. H. Stroud and H. Clepper, eds., *Black Bass Biology and Management,* pp. 125–130. Sport Fishing Institute, Washington, D.C.

Carlander, K. D. 1977. *Handbook of Freshwater Fishery Biology.* Iowa State University Press, Ames, Iowa.

Carpenter, F. L. 1979. Competition between hummingbirds and insects for nectar. *Am. Zool.* 19:1105–1114.

Carroll, C. R., and C. A. Hoffman. 1980. Chemical feeding deterrent mobilized in response to insect herbivory and counteradaptation by *Epilachna tredecimnota. Science* 209:414–416.

Carson, H. L. 1978. Speciation and sexual selection in Hawaiian *Drosophila. In* P. F. Brussard, ed., *Ecological Genetics: the Interface,* pp. 93–107. Springer-Verlag, New York.

Case, T. J. 1975. Species numbers, density compensation, and colonizing ability of lizards on islands in the Gulf of California. *Ecology* 56:3–18.

Case, T. J., and E. A. Bender. 1981. Testing for higher order interactions. *Am. Nat.* 118:920–929.

Case, T. J., and R. Casten. 1979. Global stability and multiple domains of attraction in ecological systems. *Am. Nat.* 113:705–714.

Case, T. J., J. Faaborg, and R. Sidell. 1983. The role

of body size in the assembly of West Indian bird communities. *Evolution* 37:1062–1074.
Case, T. J., and M. E. Gilpin. 1974. Interference competition and niche theory. *Proc. Natl. Acad. Sci., U.S.A.* 71:3073–3077.
Case, T. J., M. E. Gilpin, and J. M. Diamond. 1979. Overexploitation, interference competition, and excess density compensation in insular faunas. *Am. Nat.* 113:843–854.
Case, T. J., and R. Sidell. 1983. Pattern and chance in the structure of model and natural communities. *Evolution* 37:832–849.
Casten, R. G., and T. J. Case. 1979. The effect of dispersal on the stability of some model ecological communities. *Math. Biosci.* 43:281–288.
Caswell, H. 1976. Community structure: a neutral model analysis. *Ecol. Monogr.* 46:327–354.
Caswell, H. 1978. Predator-mediated coexistence: a non-equilibrium model. *Am. Nat.* 112:127–154.
Caughley, G., and J. H. Lawton. 1981. Plant-herbivore systems. *In* R. M. May, ed., *Theoretical Ecology: Principles and Applications,* pp. 132–166. Sinauer, Sunderland, Mass.
Cavalloro, R., ed. 1983. *Fruit Flies of Economic Importance.* Balkema, Rotterdam.
Chaloner, W. G. 1968. Paleoecology of fossil spores. *In* E. T. Drake, ed., *Evolution and Environment,* pp. 125–138. Yale University Press, New Haven, Conn.
Chaloner, W. G., and A. Sheerin. 1979. Devonian macrofloras. *Spec. Pap. Palaeontol.* 23:145–161.
Chaney, R. W. 1944. Summary and conclusions. *In* R. W. Chaney, ed., *Pliocene Floras of California and Oregon,* pp. 353–383. Carnegie Inst. Wash. Publ. 553.
Chapin, F. S., III. 1980. The mineral nutrition of wild plants. *Annu. Rev. Ecol. Syst.* 11:233–260.
Chappell, L. H. 1969. Competitive exclusion between two intestinal parasites of the three-spined stickleback, *Gasterosteus aculeatus* L. *J. Parasitol.* 55:775–778.
Charnov, E. L., and W. M. Schaffer. 1973. Life-history consequences of natural selection: Cole's result revisited. *Am. Nat.* 107:791–793.
Chesson, P. L. 1978. Predator-prey theory and variability. *Annu. Rev. Ecol. Syst.* 9:323–347.
Chesson, P. L. 1981. Models for spatially distributed populations: the effect of within-patch variability. *Theor. Popul. Biol.* 19:288–325.
Chesson, P. L. 1982. The stabilizing effect of a random environment. *J. Math. Biol.* 15:1–36.
Chesson, P. L. 1983. Coexistence of competitors in a stochastic environment: the storage effect. *In* H. I. Freedman and C. Strobeck, eds., *Population Biology,* pp. 188–198. Lecture Notes in Biomathematics no. 52. Springer-Verlag, Berlin, New York.
Chesson, P. L. 1984. The storage effect in stochastic population models. *In* S. A. Levin and T. G. Hallam, eds., *Mathematical Ecology: Miramare-Trieste Proceedings,* pp. 76–89. Lecture Notes in Biomathematics no. 54. Springer-Verlag, Berlin, New York.
Chesson, P. L. 1985. Coexistence of competitors in spatially and temporally varying environments: a look at the combined effects of different sorts of variability. *Theor. Popul. Biol.,* in press.
Chesson, P. L., and R. R. Warner. 1981. Environmental variability promotes coexistence in lottery competitive systems. *Am. Nat.* 117:923–943.
Chornesky, E. A. 1983. Induced development of sweeper tentacles on the reef coral *Agaricia agaricities:* a response to direct competition. *Biol. Bull.* 165:569–581.
Christiansen, F. B., and T. M. Fenchel. 1977. *Theories of Population in Biological Communities.* Springer-Verlag, New York.
Christensen, M. S. 1978. Trophic relationships in juveniles of three species of sparid fishes in the South African marine littoral. *Fish. Bull. S. Afr.* 76:389–401.
Clark, D. A. 1982. Foraging behavior of a vertebrate omnivore (*Rattus rattus*): meal structure, sampling and diet breadth. *Ecology* 63:763–772.
Clark, W. C., and C. S. Holling. 1979. Process models, equilibrium structures, and population dynamics: on the formulation and testing of realistic theory in ecology. *Fortschr. Zool.* 25:29–52.
Clausen, C. P. 1958. Biological control of insect pests. *Annu. Rev. Entomol.* 3:291–310.
Clements, F. E. 1916. *Plant Succession.* Carnegie Inst. Wash. Publ. 242.
Clements, F. E. 1936. Nature and the structure of the climax. *J. Ecol.* 24:252–284.
Coblentz, B. E. 1980. Effects of feral goats on the Santa Catalina Island ecosystem. *In* D. M. Power, ed., *The California Islands: Proceedings of a Multidisciplinary Symposium,* pp. 167–170. Santa Barbara Museum of Natural History, Santa Barbara, Calif.
Cockrum, E. L. 1948. The distribution of the hispid cotton rat in Kansas. *Trans. Kans. Acad. Sci.* 51:306–312.
Cody, M. L. 1974a. *Competition and the Structure of Bird Communities.* Princeton University Press, Princeton, N.J.
Cody, M. L. 1974b. Optimization in ecology. *Science* 183:1156–1164.
Cody, M. L. 1975. Towards a theory of continental species diversity: bird distributions over Mediteran-

nean habitat gradients. *In* M. L. Cody and J. M. Diamond, eds., *Ecology and Evolution of Communities,* pp. 214–257. Harvard University Press, Cambridge, Mass.

Cody, M. L. 1978. Habitat selection and interspecific territoriality among the sylviid warblers of England and Sweden. *Ecol. Monogr.* 48:351–396.

Cody, M. L. 1981. Habitat selection in birds: the roles of vegetation structure, competitors, and productivity. *BioScience* 31:107–113.

Cody, M. L. 1983. Continental diversity patterns and convergent evolution in bird communities. *In* F. J. Kruger, D. T. Mitchell, and J. U. M. Jarvis, eds., *Mediterranean-Type Ecosystems: The Role of Nutrients,* pp. 357–402. Ecological Studies no. 43. Springer-Verlag, Berlin, New York.

Cody, M. L. 1984. Branching patterns in columnar cacti. *In* N. Margaris, M. Arianoutsou Faraggitaki, and W. C. Oechel, eds., *Being Alive on Land,* pp. 201–236. Junk, Den Haag.

Cody, M. L. 1985. Spacing patterns in Mojave Desert plant communities. *J. Arid. Envts.* in press.

Cody, M. L. 1986. *Root systems of Mojave Desert shrubs and their relevance to spatial relationships.* Manuscript in preparation.

Cody, M. L., and J. M. Diamond, eds. 1975. *Ecology and Evolution of Communities.* Harvard University Press, Cambridge, Mass.

Cody, M. L., and H. A. Mooney. 1978. Convergence versus nonconvergence in Mediterranean-climate ecosystems. *Annu. Rev. Ecol. Syst.* 9:265–321.

Cody, M. L., R. Moran, and H. J. Thompson. 1984. The plants. *In* T. J. Case and M. L. Cody, eds., *Island Biogeography in the Sea of Cortez,* pp. 49–97. University of California Press, Berkeley.

Cohen, J. E. 1970. A Markov contingency table model for replicated Lotka-Volterra systems near equilibrium. *Am. Nat.* 104:547–559.

Cohen, J. E. 1971. Mathematics as a metaphor. *Science* 172:674–675.

Cohen, J. E. 1978. *Food Webs and Niche Space.* Princeton University Press, Princeton, N.J.

Cole, B. J. 1983. Assembly of mangrove ant communities: patterns of geographic distribution. *J. Anim. Ecol.* 52:339–348.

Cole, K. C., and J. I. Mead. 1981. Late Quaternary animal remains from packrat middens in the eastern Grand Canyon, Arizona. *J. Ariz.-Nev. Acad. Sci.* 16:24–25.

Cole, L. C. 1954. The population consequences of life history phenomena. *Q. Rev. Biol.* 29:103–137.

Cole, L. C. 1960. Competitive exclusion. *Science* 132:348–349.

Coleman, D. C., C. P. P. Reid, and C. V. Vole. 1983. Biological strategies of nutrient cycling in soil systems. *In* A. MacFadyen and E. D. Ford, eds., *Advances in Ecological Research,* pp. 1–44. Academic Press, New York.

Collins, S. L. 1983. Geographic variation in habitat structure of the Black-throated Green Warbler (*Dendroica virens*). *Auk* 100:382–389.

Colwell, R. K. 1973. Competition and coexistence in a simple tropical community. *Am. Nat.* 107:737–760.

Colwell, R. K. 1974. Predictability, constancy, and contingency of periodic phenomena. *Ecology* 55:1148–1153.

Colwell, R. K. 1979. The geographical ecology of hummingbird flower mites in relation to their host plants and carriers. *Recent Adv. in Acarol.* 2:461–468.

Colwell, R. K. 1981. Group selection is implicated in the evolution of female-biased sex ratios. *Nature* 290:401–404.

Colwell, R. K. 1982. Female-biased sex ratios. (Reply to Charlesworth and Toro, Wildish, and Borgia). *Nature* 298:494–496.

Colwell, R. K. 1983. *Rhinoseius colwelli* (Acaro floral del colibri, totolate floral de colibri, hummingbird flower mite). *In* D. H. Janzen, ed., *Costa Rican Natural History,* pp. 767–768. University of Chicago Press, Chicago.

Colwell, R. K. 1984. What's new? Community ecology discovers biology. *In* P. W. Price, C. N. Slobodchikoff, and W. S. Gaud, eds., *A New Ecology: Novel Approaches to Interactive Systems,* pp. 387–396. Wiley, New York.

Colwell, R. K. 1985. Stowaways on the *Hummingbird* Express. Nat. Hist. 94, in press.

Colwell, R. K., B. J. Betts, P. Bunnell, F. L. Carpenter, and P. Feinsinger. 1974. Competition for the nectar of *Centropogon valerii* by the hummingbird *Colibri thalassinus* and the flower-piercer *Diglossa plumbea,* and its evolutionary implications. *Condor* 76:447–452.

Colwell, R. K., and E. R. Fuentes. 1975. Experimental studies of the niche. *Annu. Rev. Ecol. Syst.* 6:281–310.

Colwell, R. K., and S. Naeem. 1979. The first known species of hummingbird flower mite north of Mexico: *Rhinoseius epoecus* n. sp. (Mesostigmata: Ascidae). *Ann. Entomol. Soc. Amer.* 72:485–491.

Colwell, R. K., and D. W. Winkler. 1984. A null model for null models in biogeography. *In* D. R. Strong, D. Simberloff, L. G. Abele, and A. B. Thistle, eds., *Ecological Communities: Conceptual Issues and the Evidence,* pp. 344–359. Princeton University Press, Princeton, N.J.

Comins, H. N., and M. P. Hassell. 1979. The dynamics of optimally foraging predators and parasitoids. *J. Anim. Ecol.* 48:335–351.

Comins, H. N., and I. R. Noble. In press. Dispersal,

variability and transient niches: species coexistence in a uniformly variable environment. *Am. Nat.*

Connell, J. H. 1961a. The influence of interspecific competition and other factors on the distribution of the barnacle *Chthamalus stellatus*. *Ecology* 42:710–723.

Connell, J. H. 1961b. Effects of competition, predation by *Thais lapillus*, and other factors on natural populations of the barnacle *Balanus balanoides*. *Ecol. Monogr.* 31:61–104.

Connell, J. H. 1970. A predator-prey system in the marine intertidal region. I. *Balanus glandula* and several predatory species of *Thais*. *Ecol. Monogr.* 40:49–78.

Connell, J. H. 1971. On the role of natural enemies in preventing competitive exclusion in some marine animals and in rain forest trees. *In* P. J. den Boer and G. R. Gradwell, eds., *Dynamics of Populations*, pp. 298–312. Centre for Agricultural Publishing and Documentation, Wageningen, The Netherlands.

Connell, J. H. 1975. Some mechanisms producing structure in natural communities: a model and evidence from field experiments. *In* M. L. Cody and J. M. Diamond, eds., *Ecology and Evolution of Communities*, pp. 460–490. Harvard University Press, Cambridge, Mass.

Connell, J. H. 1976. Competitive interactions and the species diversity of corals. *In* G. O. Mackie, ed., *Coelenterate Ecology and Behavior*, pp. 51–58. Plenum, New York.

Connell, J. H. 1978. Diversity in tropical rain forests and coral reefs. *Science* 199:1302–1310.

Connell, J. H. 1980. Diversity and the coevolution of competitors, or the ghost of competition past. *Oikos* 35:131–138.

Connell, J. H. 1983. On the prevalence and relative importance of interspecific competition: evidence from field experiments. *Am. Nat.* 122:661–696.

Connell, J. H., and R. O. Slatyer. 1977. Mechanisms of succession in natural communities and their role in community stability and organization. *Am. Nat.* 111:1119–1144.

Connell, J. H., and W. P. Sousa. 1983. On the evidence needed to judge ecological stability or persistence. *Am. Nat.* 121:789–824.

Connor, E. F., and E. D. McCoy. 1979. The statistics and biology of the species-area relationship. *Am. Nat.* 113:791–833.

Connor, E. H., and D. Simberloff. 1979. The assembly of species communities: chance or competition? *Ecology* 60:1132–1140.

Connor, E. F., and D. Simberloff. 1983. Interspecific competition and species co-occurrence patterns on islands: null models and the evaluation of evidence. *Oikos* 41:455–465.

Cooper, D. M. 1960. Food preferences of larval and adult Drosophila. *Evolution* 14:41–55.

Corner, E. J. H. 1954. The evolution of tropical forests. *In* J. Huxley, A. C. Hardy, and E. B. Ford, eds., *Evolution as a Process*, pp. 34–46. Allen and Unwin, London.

Cousens, J. E. 1965. Some reflections on the nature of Malayan lowland rain forest. *Malays. For.* 28:122–128.

Cousminer, H. L. 1961. Palynology, paleofloras, and paleoenvironments. *Micropaleontology* 1:365–368.

Cramer, N. F., and R. M. May. 1972. Interspecific competition, predation and species diversity: a comment. *J. Theor. Biol.* 34:289–293.

Crawley, M. J. 1983. *Herbivory*. University of California Press, Berkeley.

Crepet, W. L. 1974. Investigations of North American cycadeoids: the reproductive biology of *Cycadeoidea*. *Palaeontographica* 148 B:144–169.

Crepet, W. L. 1979. Some aspects of the pollination biology of Middle Eocene angiosperms. *Rev. Palaeobot. Palynol.* 27:213–238.

Crisp, D. J. 1974. Factors influencing the settlement of marine invertebrate larvae. *In* P. T. Grant and A. M. Mackie, eds., *Chemoreception in Marine Organisms*, pp. 177–265. Academic Press, New York.

Crocker, R. L., and B. A. Dickson. 1957. Soil development on the recessional moraines of the Herbert and Mendenhall glaciers of south-eastern Alaska. *J. Ecol.* 45:169–185.

Crocker, R. L., and J. Major. 1955. Soil development in relation to vegetation and surface age at Glacier Bay, Alaska. *J. Ecol.* 43:427–448.

Crofton, H. D. 1971. A quantitative approach to parasitism. *Parasitology* 62:179–193.

Cromartie, W. J. 1975. The effect of stand size and vegetational background on the colonization of cruciferous plants by herbivorous insects. *J. Appl. Ecol.* 12:517–533.

Cromartie, W. J. 1981. The environmental control of insects using crop diversity. *In* D. Pimentel, ed., *Handbook on Pest Management*, pp. 223–251. CRC Publishers, Boca Raton, Fla.

Crook, G., P. Vezina, and Y. Hardy. 1979. Susceptibility of balsam fir to spruce budworm defoliation as affected by thinning. *Can. J. For. Res.* 9:428–435.

Crossman, E. J., and P. A. Larkin. 1959. Yearling liberations and change of food as effecting rainbow trout yield in Paul Lake, British Columbia. *Trans. Am. Fish. Soc.* 88:36–44.

Crowder, L. B., and W. E. Cooper. 1982. Habitat structural complexity and the interaction between bluegills and their prey. *Ecology* 63:1802–1813.

Crowder, L. B., J. J. Magnuson, and S. B. Brandt. 1981. Complementarity in the use of food and thermal habitat by Lake Michigan fishes. *Can. J. Fish. Aquat. Sci.* 38:662–668.

Crowell, K. L. 1973. Experimental zoogeography: introductions of mice to small islands. *Am. Nat.* 107:535–558.

Crowell, K. L. 1983. Islands—insight or artifact? Population dynamics and habitat utilization in insular rodents. *Oikos* 41:442–454.

Crowell, K. L., and S. L. Pimm. 1976. The introduction of small mice onto islands. *Oikos* 27:251–258.

Crowley, P. 1979. Predator-mediated coexistence: an equilibrium interpretation. *J. Theor. Biol.* 80:129–144.

Crush, J. R. 1975. Occurrence of endomycorrhizas in soils of the Mackenzie Basin, Canterbury, New Zealand. *N. Z. J. Agric. Res.* 18:361–364.

Culver, D. C., and A. J. Beattie. 1978. Myrmecochory in *Viola:* dynamics of seed-ant interactions in some West Virginia species. *J. Ecol.* 66:53–72.

Culver, D. C., and A. J. Beattie. 1980. The fate of *Viola* seeds dispersed by ants. *Am. J. Bot.* 67:710–714.

Cunha, A. B. da, T. Dobzhansky, and A. Sokoloff. 1951. On food preferences of sympatric species of Drosophila. *Evolution* 5:97–101.

Cunha, A. B. da, A. M. El-Tabey Shehata, and W. de Oliveira. 1957. A study of the diets and nutritional preferences of tropical species of Drosophila. *Ecology* 38:98–106.

Curtis, J. T. 1959. *The Vegetation of Wisconsin*. University of Wisconsin Press, Madison.

Cushing, D. H. 1982. *Climate and Fisheries*. Academic Press, London.

Cushing, E. J. 1965. Problems in Quaternary phytogeography of the Great Lakes Region. *In* H. E. Wright, Jr., and D. G. Frey, eds., *The Quaternary of the United States,* pp. 403–416. Princeton University Press, Princeton, N.J.

Cushing, E. J. 1967. Late Wisconsinan pollen stratigraphy and the glacial sequence in Minnesota. *In* E. J. Cushing and H. E. Wright, Jr., eds., *Quaternary Paleoecology,* pp. 59–88. Yale University Press, New Haven, Conn.

Dale, G. 1978. Money in the bank model for coral reef fish coexistence. *Environ. Biol. Fishes* 3:103–108.

Dansereau, P. 1957. *Biogeography—An Ecological Perspective*. Ronald Press, New York.

Darlington, P. J. 1957. *Zoogeography: the Geographical Distribution of Animals*. Wiley, New York.

Darwin, C. 1842. *The Structure and Distribution of Coral Reefs. Being the First Part of the Geology of the Voyage of the Beagle*. Smith, Elder, London.

Darwin, C. 1859. *The Origin of Species by Means of Natural Selection*. Murray, London.

Darwin, C. 1871. *The Descent of Man, and Selection in Relation to Sex*. Murray, London.

Daubenmire, R. F. 1936. The Big Woods of Minnesota: its structure and relation to climate, fire and soils. *Ecol. Monogr.* 6:233–268.

David, H. M. 1943. Studies in the autecology of *Ascophyllum nodosum* Le Jol. *J. Ecol.* 31:178–199.

Davidson, D. W. 1977. Foraging ecology and community organization in desert seed-eating ants. *Ecology* 58:725–737.

Davidson, D. W. 1980. Some consequences of diffuse competition in a desert ant community. *Am. Nat.* 116:92–105.

Davidson, D. W. 1985. An experimental study of diffuse competition in desert harvester ants. *Am. Nat.* 125:500–506.

Davidson, D. W., J. H. Brown, and R. S. Inouye. 1980. Competition and the structure of granivore communities. *BioScience* 30:233–238.

Davidson, D. W., R. S. Inouye, and J. H. Brown. 1984. Granivory in a desert ecosystem: experimental evidence for indirect facilitation of ants by rodents. *Ecology* 65:1780–1786.

Davidson, D. W., and S. R. Morton. 1981a. Myrmecochory in some plants (F. Chenopodiaceae) of the Australian arid zone. *Oecologia* 50:357–366.

Davidson, D. W., and S. R. Morton. 1981b. Competition for dispersal in ant-dispersed plants. *Science* 213:1259–1261.

Davidson, D. W., D. A. Samson, and R. S. Inouye. Granivory in the Chihuahuan Desert: interactions within and between trophic levels. *Ecology* 66:486–502.

Davidson, N. C., and P. R. Evans. 1982. Mortality of redshanks and oystercatchers from starvation during severe weather. *Bird Study* 29:183–188.

Davies, S. J. J. F. 1976. Environmental variables and the biology of Australian arid zone birds. *Proc. XVIth Int. Ornithol. Congr., Canberra,* pp. 481–488.

Davis, J. 1973. Habitat preferences and competition of wintering juncos and golden-crowned sparrows. *Ecology* 54:174–180.

Davis, M. B. 1976. Pleistocene biogeography of temperate deciduous forests. *Geoscience and Man* 13:13–26.

Davis, M. B. 1981a. Outbreaks of forest pathogens in Quaternary history. *Proc. IV Int. Palynol. Conf., Lucknow* 3:216–228.

Davis, M. B. 1981b. Quaternary history and the stability of forest communities. *In* D. C. West, H. H. Shugart, and D. B. Botkin, eds., *Forest Succession: Concepts and Application,* pp. 134–153. Springer-Verlag, New York.

Davis, M. B. 1983a. Quaternary history of deciduous

forests of eastern North America and Europe. *Ann. Mo. Bot. Gard.* 70:550–563.

Davis, M. B. 1983b. Holocene vegetational history of the eastern United States. *In* H. E. Wright, Jr., ed., *Late-Quaternary Environments of the United States. Vol. II. The Holocene,* pp. 166–181. University of Minnesota Press, Minneapolis.

Davis, M. B., and D. B. Botkin. 1985. Sensitivity of cool temperate forests and their fossil pollen record to rapid temperature change. *Quat. Res.* 23:327–340.

Davis, M. B., R. W. Spear, and L. C. K. Shane. 1980. Holocene climate of New England. *Quat. Res.* 14:240–250.

Davis, M. B., K. Woods, S. L. Webb, and R. P. Futyma. 1984. Climate versus dispersal as factors limiting the Holocene range extension of beech and hemlock into the Upper Great Lakes Region. *6th Int. Palynological Conf., Calgary,* Abstracts: 29.

Dawood, M. M., and M. W. Strickberger. 1969. The effect of larval interaction on viability in *Drosophila melanogaster*. III. Effects of biotic residues. *Genetics* 63:213–220.

Day, P. R. 1974. *Genetics of Host-parasite Interactions*. Freeman, San Francisco.

Day, R. J. 1972. Stand structure, succession, and the use of southern Alberta's Rocky Mountain forest. *Ecology* 53:472–478.

Day, R. W. 1977. *The Ecology of Settling Organisms on the Coral Reef at Heron Island, Queensland.* Doctoral dissertation, University of Sydney, Sydney.

Dayton, P. K. 1971. Competition, disturbance and community organization: the provision and subsequent utilization of space in a rocky intertidal environment. *Ecol. Monogr.* 41:351–389.

Dayton, P. K. 1973a. Two cases of resource partitioning in an intertidal community: making the right prediction for the wrong reason. *Am. Nat.* 107:662–670.

Dayton, P. K. 1973b. Dispersion, dispersal, and persistence of the annual intertidal alga, *Postelsia palmaeformis* ruprecht. *Ecology* 54:433–438.

Dayton, P. K. 1975a. Experimental evaluation of ecological dominance in a rocky intertidal algal community. *Ecol. Monogr.* 45:137–159.

Dayton, P. K. 1975b. Experimental studies of algal canopy interactions in a sea otter-dominated kelp community at Amchitka Island, Alaska. *Fish. Bull.* 73:230–237.

Dayton, P. K. 1983. Processes structuring some marine communities: are they general? *In* D. R. Strong, D. Simberloff, L. G. Abele, and A. B. Thistle, eds., *Ecological Communities: Conceptual Issues and the Evidence,* pp. 181–197. Princeton University Press, Princeton, N.J.

Dayton, P. K., and M. J. Tegner. 1984. The importance of scale in community ecology: a kelp forest example with terrestrial analogs. *In* P. W. Price, C. N. Slobodchikoff, and W. S. Gaud, eds., *A New Ecology: Novel Approaches to Interactive Systems,* pp. 457–481. Wiley, New York.

Deacon, H. J. 1983. The comparative evolution of Mediterranean-type ecosystems. *In* F. J. Kruger, D. T. Mitchell, and J. U. M. Jarvis, eds., *Mediterranean-Type Ecosystems: the Role of Nutrients,* pp. 3–40. Ecological Studies no. 43. Springer-Verlag, Berlin, New York.

De Angelis, D. L., W. M. Post, and G. Sugihara. 1983. Current trends in food web theory: report on a food web workshop. *Natl. Lab. Tech. Mem.* ORNL-5983.

DeBach, P. 1958. The role of weather and entomophagous species in the natural control of insect populations. *J. Econ. Entomol.* 51:474–484.

DeBach, P. 1966. The competitive displacement and coexistence principles. *Annu. Rev. Entomol.* 11:183–212.

DeBach, P. 1979. *Biological Control by Natural Enemies*. Cambridge University Press, Cambridge.

DeBenedictis, P. A. 1974. Interspecific competition between tadpoles of *Rana pipiens* and *Rana sylvatica:* an experimental field study. *Ecol. Monogr.* 44:129–151.

Decker, J. P. 1952. Tolerance is a good technical term. *J. For.* 50:40–41.

Decker, J. P. 1959. A system for analysis of forest succession. *For. Sci.* 5:154–157.

Degerbøl, M. 1964. Some remarks on late- and postglacial vertebrate fauna and its ecological relations in northern Europe. *J. Anim. Ecol.* 33(supplement):71–85.

Degerbøl, M., and H. Krog. 1951. The European Pond Tortoise (*Emys orbicularis* L.) in Denmark. *Dan. Geol. Unders. Ser. II* no. 78:1–130.

Del Solar, E. 1968. Selection for and against gregariousness in choice of oviposition site by *Drosophila pseudoobscura*. *Genetics* 58:275–282.

Del Solar, E., and H. Palomino. 1966. Choice of oviposition in *Drosophila melanogaster*. *Am. Nat.* 100:127–133.

Dempster, J. P. 1975. *Animal Population Ecology*. Academic Press, New York.

Dempster, J. P. 1983. The natural control of populations of butterflies and moths. *Biol. Rev.* 58:461–481.

Dempster, J. P., and K. H. Lakhani. 1979. A population model for cinnabar moth and its food plant, ragwort. *J. Anim. Ecol.* 48:143–163.

Dempster, J. P., and E. Pollard. 1981. Fluctuations in resource availability and insect populations. *Oecologia* 50:412–416.

Den Boer, P. J. 1970. Stabilization of animal numbers

and the heterogeneity of the environment: the problem of the persistence of sparse populations. *In* P. J. Den Boer and G. Gradwell, eds., *Dynamics of Populations,* pp. 77–97. Proceedings of the Advanced Studies Institute, The Netherlands.

Denley, E. J., and A. J. Underwood. 1979. Experiments on factors influencing settlement, survival and growth of two species of barnacles in New South Wales. *J. Exp. Mar. Biol. Ecol.* 36:269–293.

Denslow, J. S. 1980. Gap partitioning among tropical rainforest trees. *Biotropica* (special issue on tropical succession), pp. 47–55.

Dethier, V. G. 1959. Food-plant distribution and density and larval dispersal as factors affecting insect populations. *Can. Entomol.* 88:581–596.

Dhondt, A. A. 1977. Interspecific competition between great and blue tit. *Nature* 268:521–523.

Dhondt, A. A. 1982. Heritability of blue tit tarsus length from normal and cross-fostered broods. *Evolution* 36:418–419.

Dhondt, A. A., and R. Eyckerman. 1980a. Competition between the Great Tit and the Blue Tit outside the breeding season in field experiments. *Ecology* 61:1291–1296.

Dhondt, A. A., and R. Eyckerman. 1980b. Competition and the regulation of numbers in Great and Blue Tit. *Ardea* 68:121–132.

Diamantouglou, S., and K. Mitrakos. 1981. Leaf longevity in evergreen sclerophylls. *In* N. S. Margaris and H. A. Mooney, eds., *Components of Productivity of Mediterranean-Climate Regions: Basic and Applied Aspects,* pp. 17–19. Junk, Den Haag.

Diamond, J. M. 1969a. Preliminary results of an ornithological exploration of the North Coastal Range, New Guinea. *Am. Mus. Novit.* no. 2362.

Diamond, J. M., 1969b. Avifaunal equilibria and species turnover rates on the Channel Islands of California. *Proc. Natl. Acad. Sci., U.S.A.* 64:57–63.

Diamond, J. M. 1970. Ecological consequences of island colonization by Southwest Pacific birds. II. The effect of species diversity on total population density. *Proc. Natl. Acad. Sci., U.S.A.* 67:1715–1721.

Diamond, J. M. 1972a. *Avifauna of the Eastern Highlands of New Guinea.* Nuttall Ornithological Club, Cambridge, Mass.

Diamond, J. M. 1972b. Biogeographic kinetics: estimation of relaxation times for avifaunas of southwest Pacific islands. *Proc. Natl. Acad. Sci., U.S.A.* 69:3199–3203.

Diamond, J. M. 1973. Distributional ecology of New Guinea birds. *Science* 179:759–769.

Diamond, J. M. 1974. Colonization of exploded volcanic islands by birds: the supertramp strategy. *Science* 184:803–806.

Diamond, J. M. 1975. Assembly of species communities. *In* M. L. Cody and J. M. Diamond, eds., *Ecology and Evolution of Communities,* pp. 342–444. Harvard University Press, Cambridge, Mass.

Diamond, J. M. 1978. Niche shifts and the rediscovery of interspecific competition. *Am. Sci.* 66:322–331.

Diamond, J. M. 1979. Community structure: is it random, or is it shaped by species differences and competition? *In* R. Anderson, B. Turner, and L. Taylor, eds., *Population Dynamics,* Twentieth Symposium of the British Ecological Society, pp. 165–181. Blackwell, Oxford.

Diamond, J. M. 1980. Patchy distributions of tropical birds. *In* M. Soulé and B. Wilcox, eds., *Conservation Biology,* pp. 57–74. Sinauer, Sunderland, Mass.

Diamond, J. M. 1983. Laboratory, field and natural experiments. *Nature* 304:586–587.

Diamond, J. M. 1984a. Historic extinctions: a Rosetta Stone for understanding prehistoric extinctions. *In* P. S. Martin and R. G. Klein, eds., *Quaternary Extinctions: a Prehistoric Revolution,* pp. 824–862. University of Arizona Press, Tucson.

Diamond, J. M. 1984b. Evolution of stable exploiter-victim systems. *Nature* 310:632.

Diamond, J. M. 1985. New distributional records and taxa from the outlying mountain ranges of New Guinea. *Emu* 85, in press.

Diamond, J. M., and M. E. Gilpin. 1980. Turnover noise: contribution to variance in species number and predictions from immigration and extinction curves. *Am. Nat.* 115:884–889.

Diamond, J. M., and M. E. Gilpin. 1982. Examination of the ''null'' model of Connor and Simberloff for species co-occurrences on islands. *Oecologica* 52:64–74.

Diamond, J. M., M. E. Gilpin, and E. Mayr. 1976. Species-distance relation for birds of the Solomon Archipelago, and the paradox of the great speciators. *Proc. Natl. Acad. Sci., U.S.A.* 73:2160–2164.

Diamond, J. M., and M. LeCroy. 1979. Birds of Karkar and Bagabag Islands, New Guinea. *Bull. Amer. Mus. Nat. Hist.* 164:467–531.

Diamond, J. M., and A. G. Marshall. 1976. Origin of the New Hebridean avifauna. *Emu* 76:187–200.

Diamond, J. M., and A. G. Marshall. 1977. Distributional ecology of New Hebridean birds: a species kaleidoscope. *J. Anim. Ecol.* 46:703–727.

Diamond, J. M., and R. M. May. Island biogeography and the design of natural reserves. *In* R. M. May, ed., *Theoretical Ecology,* pp. 228–252. Sinauer, Sunderland, Mass.

Diamond, J. M., and C. R. Veitch. 1981. Extinctions and introductions in the New Zealand avifauna: cause and effect? *Science* 211:499–501.

DiCastri, F., and H. A. Mooney. 1973. *Mediterranean-Type Ecosystems: Origin and Structure*. Ecological Studies no. 7. Springer-Verlag, New York, Heidelberg, Berlin.

Dickie, J. B. 1977. *The Reproduction and Regeneration of Some Common Chalk Grassland Perennials*. Doctoral dissertation, University of Cambridge.

Dickman, M. D., and M. B. Gochnauer. 1978. Impact of sodium chloride on the microbiota of a small stream. *Environ. Pollut.* 17:109–126.

DiMichele, W. A., T. L. Phillips, and R. A. Peppers. In press. The influence of climate and depositional environment on the distribution and evolution of Pennsylvanian coal swamp plants. *In* B. H. Tiffney, ed., *Influences of Physical Environments on Vascular Plant Evolution*. Yale University Press, New Haven, Conn.

Dix, R., and F. Smeins. 1967. The prairie, meadow and marsh vegetation of Nelson County, North Dakota. *Can. J. Bot.* 45:21–58.

Dixon, A. F. G. 1971a. The role of aphids in wood formation I. The effect of the sycamore aphid, *Drepanosiphum platanoides* (Schr.) (Aphididae), on the growth of sycamore, *Acer pseudoplatanus* (L.). *J. Appl. Ecol.* 8:165–179.

Dixon, A. F. G. 1971b. The role of aphids in wood formation II. The effect of the lime aphid, *Eucallipterus tiliae* L. (Aphididae), on the growth of the lime, *Tilia X vulgaris* Hayne. *J. Appl. Ecol.* 8:393–399.

Dixon, A. F. G. 1976. Timing of egg hatch and viability of the sycamore aphid, *Drepanosiphum platanoides* (Schr.), at bud burst of sycamore, *Acer pseudoplatanus* L. *J. Anim. Ecol.* 45:593–603.

Dixon, A. F. G. 1977. Aphid ecology: life cycles, polymorphism, and population regulation. *Annu. Rev. Ecol. Syst.* 8:329–353.

Dobkin, D. S. 1983. *Ecology of Hummingbird Flower Mite (Gamasida: Ascidae) Populations and Their* Heliconia *(Heliconiaceae) Host Plants in Trinidad, West Indies*. Doctoral dissertation, University of California, Berkeley.

Dobkin, D. S. 1984. Flowering patterns of long-lived inflorescences: implications for visiting and resident nectarivores. *Oecologia* 64:245–254.

Dobkin, D. S. 1985. Heterogeneity of tropical floral microclimates and the response of hummingbird flower mites. *Ecology* 66:536–543.

Dobzhansky, T. 1950. Evolution in the tropics. *Am. Sci.* 38:209–221.

Donald, C. M. 1958. The interaction of competition for light and for nutrients. *Aust. J. Agric. Res.* 9:421–435.

Douglas, R. G., and R. F. Betts. 1979. Influenza Virus. *In* G. L. Mandell, R. G. Douglas, and J. E. Bennett, eds., *Principles and Practice of Infectious Diseases*, pp. 1135–1167. Wiley, New York.

Drenner, R. W., W. J. O'Brien, and J. R. Kummert. 1982. Filter-feeding rates of gizzard shad. *Trans. Am. Fish. Soc.* 111:210–215.

Drury, W. H., and I. C. T. Nisbet. 1973. Succession. *J. Arnold Arbor. Harv. Univ.* 54:331–368.

Duff, P. M. D., A. Hallam, and E. K. Walton. 1967. *Cyclic Sedimentation*. Elsevier, Amsterdam.

Duggins, D. O. 1980. Kelp beds and sea otters: an experimental approach. *Ecology* 61:447–453.

Dunham, A. E. 1980. An experimental study of interspecific competition between the iguanid lizards *Sceloporus merriami* and *Urosaurus ornatus*. *Ecol. Monogr.* 50:309–330.

Dunning, J. B., Jr., and J. H. Brown. 1982. Summer rainfall and winter sparrow densities: a test of the food limitation hypothesis. *Auk* 99:123–129.

During, H. J., A. J. Schenkeveld, H. J. Verkaar, and J. H. Willems. 1985. Demography of short-lived forbs in chalk grassland in relation to vegetation structure. *In* J. White and W. G. Beeftink, eds., *The Population Structure of Vegetation* (Handbook of Vegetation Science, Part 3), pp. 341–370. Junk, Den Haag.

Edgar, J. A. 1982. Pyrrolizidine alkaloids sequestered by Solomon Island danaine butterflies. The feeding preferences of the Danainae and Ithomiinae. *J. Zool. (Lond.)* 196:385–399.

Edgar, J. A., C. C. J. Culvenor, and T. E. Pliske. 1974. Coevolution of danaid butterflies and their host plants. *Nature* 250:646–648.

Ehleringer, J., and H. A. Mooney. 1983. Productivity of desert and Mediterranean-climate plants. *In* O. L. Lange, P. S. Nobel, C. B. Ormond, and H. Ziegler, eds., *Encyclopaedia of Plant Physiology*, vol. 12D, pp. 205–231. Springer-Verlag, Berlin, Heidelberg.

Ehrlich, P. R. 1983. Genetics and the extinction of butterfly populations. *In* C. M. Schonewald-Cox, S. M. Chambers, B. MacBryde, and L. Thomas, eds., *Genetics and Conservation: Reference for Managing Wild Animal and Plant Populations*, pp. 152–163. Benjamin/Cummings, Menlo Park, Calif.

Ehrlich, P. R., D. E. Breedlove, P. F. Brussard, and M. A. Sharp. 1972. Weather and the "regulation" of subalpine populations. *Ecology* 53:243–247.

Ehrlich, P. R., and P. H. Raven. 1964. Butterflies and plants: a study in coevolution. *Evolution* 18:586–608.

Ehrlich, P. R., and J. Roughgarden. In press. *Ecology*.

Ehrlich, P. R., R. White, M. Singer, and L. Gilbert. 1975. Checkerspot butterflies: a historical perspective. *Science* 188:221–228.

Ekman, S. 1910. Om manniskans andel i fiskfaunans

spridning till det inre Norrlands vatten. *Ymer* 30:133–140.

Ellenberg, H. 1982. *Vegetation Mitteleuropas mit den Alpen,* 3rd ed. Ulmer, Stuttgart.

Elliot, B. G. 1980. First observation of the European Starling in Hawaii. *Elepaio* 40:100–101.

Elliott, J. M. 1984. Numerical changes and population regulation in young migratory trout *Salmo trutta* in a lake district stream, 1966–83. *J. Anim. Ecol.* 53:327–350.

Ellner, S. P. 1984. Stationary distributions for some difference equation population models. *J. Math. Biol.* 19:169–200.

Elton, C. S. 1927. *Animal Ecology*. Reprinted 1966 by Science Paperbacks and Methuen & Co. Ltd., London.

Elton, C. S. 1949. Population interspersion: an essay on animal community patterns. *J. Ecol.* 37:1–23.

Elton, C. S. 1958. *The Ecology of Invasions by Animals and Plants*. Chapman and Hall, London.

Embree, D. G. 1979. The ecology of colonizing species, with special emphasis on animal invaders. *In* D. J. Horn, G. R. Stairs, and R. D. Mitchell, eds., *Analysis of Ecological Systems,* pp. 51–65. Ohio State University Press, Columbus.

Emerson, E. A. 1960. The evolution of adaptation in population systems. *In* S. Tax, ed., *Evolution After Darwin,* vol. 1, pp. 307–348. University of Chicago Press, Chicago.

Emery, A. R. 1975. Stunted bass: a result of competing cisco and limited crayfish stocks. *In* R. H. Stroud and H. Clepper, eds., *Black Bass Biology and Management,* pp. 154–164. Sport Fishing Institute, Washington, D.C.

Emlen, J. T. 1971. Population densities of birds derived from transect counts. *Auk* 88:323–342.

Emlen, J. T. 1977. Estimating breeding season bird densities from transect counts. *Auk* 94:455–468.

Emlen, J. T. 1981. Divergence in the foraging responses of birds on two Bahama islands. *Ecology* 62:289–295.

Endler, J. A. 1977. *Geographic Variation, Speciation and Clines*. Princeton University Press, Princeton, N.J.

Enright, J. T. 1976. Climate and population regulation: the biogeographer's dilemma. *Oecologia* 24:295–310.

Eriksson, M. O. G. 1979. Competition between freshwater fish and Goldeneyes *Bucephala clangula* (L.) for common prey. *Oecologia* 41:99–107.

Erkamo, V. 1952. On plant biological phenomena accompanying the present climatic change. *Fennia* 75:25–37.

Errington, P. L. 1956. Factors limiting higher vertebrate populations. *Science* 124:304–307.

Esch, G. W. 1971. Impact of ecological succession on the parasite fauna in centrarchids from oligotrophic and eutrophic ecosystems. *Am. Midl. Nat.* 86:160–168.

Etheridge, R. E. 1960. *The Relationships of the Anoles (Reptilia: Sauria: Iguanidae): an Interpretation Based on Skeletal Morphology*. Doctoral dissertation, University of Michigan, University Microfilms, Inc., Ann Arbor.

Etheridge, R. E. 1964. Late Pleistocene lizards from Barbuda, British West Indies. *Bull. Flo. State Mus. Biol. Sci.* 9:43–75.

Evans, E. W. 1982a. Timing of reproduction by predatory stink bugs (Hemiptera: Pentatomidae): patterns and consequences for a generalist and a specialist. *Ecology* 63:147–158.

Evans, E. W. 1982b. Influence of weather on predator/prey relations: stinkbugs and tent caterpillars. *N. Y. Entomol. Soc.* 90:241–246.

Everhart, W. H., and W. O. Young. 1981. *Principles of Fisheries Science*. Comstock, Ithaca, N.Y.

Fager, E. W. 1957. Determination and analysis of recurrent groups. *Ecology* 38:586–595.

Fain, A., K. E. Hyland, and T. H. G. Aitken. 1977. Flower mites of the family Ascidae phoretic in nasal cavities of birds (Acarina: Mesostigmata). *Acta Zool. Pathol. Antverp.* 69:99–154.

Falconer, D. S. 1981. *Introduction to Quantitative Genetics,* 2nd ed. Longman, London, New York.

Fast, A. W., L. H. Bottroff, and R. L. Miller, 1982. Largemouth bass, *Micropterus salmoides,* and bluegill, *Lepomis macrochirus,* growth rates associated with artificial destratification and threadfin shad, *Dorosoma petenense,* introductions at El Capitan Reservoir, California. *Calif. Fish Game* 67:4–20.

Fausch, K. D., and R. J. White. 1981. Competition between brook trout (*Salvelinus fontinalis*) and brown trout (*Salmo trutta*) for positions in a Michigan stream. *Can. J. Fish. Aquat. Sci.* 38:1220–1227.

Fedorov, A. A. 1966. The structure of the tropical rain forest and speciation in the humid tropics. *J. Ecol.* 54:1–11.

Feeny, P. P. 1970. Seasonal changes in oak leaf tannins and nutrients as a cause of spring feeding by winter moth caterpillars. *Ecology* 51:565–581.

Feeny, P. 1976. Plant apparency and chemical defence. *Recent Adv. in Phytochem.* 10:1–40.

Feinsinger, P. 1983. Coevolution and pollination. *In* D. J. Futuyma and M. Slatkin, eds., *Coevolution,* pp. 282–310. Sinauer, Sunderland, Mass.

Feinsinger, P., and R. K. Colwell. 1978. Community

organization among neotropical nectar-feeding birds. *Am. Zool.* 18:779–795.

Feinsinger, P., J. A. Wolfe, and L. A. Swarm. 1982. Island ecology: reduced hummingbird diversity and the pollination biology of plants, Trinidad and Tobago, West Indies. *Ecology* 63:494–506.

Feller, W. 1943. On a general class of "contagious" distributions. *Ann. Math. Stat.* 14:389–400.

Fenchel, T. 1974. Intrinsic rate of natural increase: the relationship with body size. *Oecologia* 14:317–326.

Fenchel, T. 1975. Character displacement and coexistence in mud snails (Hydrobiidae). *Oecologia* 20:19–32.

Fenner, F., and K. Myers. 1978. Myxoma virus and myxomatosis in retrospect: the first quarter century of a new disease. *In* E. Kurstak and K. Maramorosch, eds., *Viruses and Environment,* pp. 539–570. Academic Press, New York.

Fenner, F., and F. N. Ratcliffe. 1966. *Myxomatosis.* Cambridge University Press, Cambridge.

Fernald, M. L. 1950. *Gray's Manual of Botany,* 8th ed. American Book, New York.

Filipsson, O., and C. Svardson. 1976. Principer for fiskevarden i rodingsjoar. *Inf. Inst. Freshwater Res. Drottningholm* 2:1–79.

Finch, S. 1980. Chemical attraction of plant-feeding insects to plants. *Appl. Biol.* 5:67–143.

Findley, J. S. 1973. Phenetic packing as a measure of faunal diversity. *Am. Nat.* 107:580–584.

Findley, J. S. 1976. The structure of bat communities. *Am. Nat.* 110:129–139.

Finger, T. R. 1982. Interactive segregation among three species of sculpins (*Cottus*). *Copeia* 1982:680–694.

Finley, R. B. 1958. The wood rats of Colorado: distribution and ecology. *Univ. Kans. Publ. Mus. Nat. Hist.* 10:213–552.

Fischer, A. G. 1960. Latitudinal variations in organic diversity. *Evolution* 14:64–81.

Fisher, J., N. Simon, and J. Vincent. 1969. *Wildlife in Danger.* Viking, New York.

Fisher, R. A. 1950. *Statistical Methods for Research Workers,* 11th ed. Oliver and Bond, Edinburgh.

Fisher, R. A. 1958. *The Genetical Theory of Natural Selection,* 2nd ed. Dover, New York.

Fleischer, R. C., and R. F. Johnston. 1982. Natural selection on body size and proportions in house sparrows. *Nature* 298:747–749.

Foerster, R. E., and W. E. Ricker. 1941. The effect of reduction of predaceous fish on survival of young sockeye at Cultus Lake. *J. Fish. Res. Board Can.* 5:315–336.

Foley, R. L. 1984. Late Pleistocene (Woodfordian) vertebrates from the driftless area of southwestern Wisconsin, the Moscow Fissure local fauna. *Ill. State Mus. Rep. Invest.* 39:1–52.

Forney, J. L. 1976. Year-class formation in the walleye (*Stizostedion vitreum vitreum*) population of Oneida Lake, New York, 1966–73. *J. Fish. Res. Board Can.* 33:783–792.

Forney, J. L. 1977. Evidence of inter- and intraspecific competition as factors regulating walleye (*Stizostedion vitreum vitreum*) biomass in Oneida Lake, New York. *J. Fish. Res. Board Can.* 34:1812–1820.

Foster, R. B. 1982a. The seasonal rhythm of fruitfall on Barro Colorado Island. *In* E. G. Leigh, A. S. Rand, and D. M. Windsor, eds., *The Ecology of a Tropical Forest: Seasonal Rhythms and Long-Term Changes,* pp. 151–172. Smithsonian Institution Press, Washington, D.C.

Foster, R. B. 1982b. Famine on Barro Colorado Island. *In* E. G. Leigh, A. S. Rand, and D. M. Windsor, eds., *The Ecology of a Tropical Rain Forest: Seasonal Rhythms and Long-Term Changes,* pp. 201–212. Smithsonian Institution Press, Washington, D.C.

Fowler, C. W. 1981. Density dependence as related to life history strategy. *Ecology* 62:602–610.

Fowler, N., and J. Antonovics. 1981. Competition and coexistence in a North Carolina grassland. *J. Ecol.* 69:825–841.

Fowler, S. V., and J. H. Lawton. 1984. Trees don't talk: do they even murmur? *Antenna* 8:69–71.

Fox, J. F. 1979. Intermediate-disturbance hypothesis. *Science* 204:1344–1345.

Fox, L. R., and P. A. Morrow. 1981. Specialization: species property or local phenomenon? *Science* 211:887–893.

Francis, L. 1973. Interspecific aggression and its effects on the distribution of *Anthopleura elegantissima* and some related sea anemones. *Biol. Bull.* 150:361–376.

Frank, P. W. 1965. The biodemography of an intertidal snail population. *Ecology* 46:831–844.

Fraser, J. M. 1978. The effect of competition with yellow perch on the survival and growth of planted brook trout, splake, and rainbow trout in a small Ontario lake. *Trans. Am. Fish. Soc.* 107:505–517.

Fredericksen, N. O. 1980. Middle Eocene to Early Oligocene plant communities of the Gulf Coast. *In* J. Gray, A. J. Boucot, and W. B. N. Berry, eds., *Communities of the Past,* pp. 493–549. Dowden, Hutchinson, and Ross, Stroudsburg, Pa.

Freeland, W. J. 1983. Parasites and the coexistence of animal host species. *Am. Nat.* 121:223–236.

Fretwell, S. D. 1972. *Populations in a Seasonal Environment.* Princeton University Press, Princeton, N.J.

Friedman, J., G. Orshan, and Y. Ziger-Cfir. 1977. Suppression of annuals by *Artemisia herba-alba* in the Negev Desert of Israel. *J. Ecol.* 65:413–426.

Fritz, M. A. 1977. A microbioherm. *In* C. S. Church, ed., *Athlon, Essays on Paleontology in Honor of Loris Shano Russell,* pp. 18–25. Life Science Miscellaneous Publications. Royal Ontario Museum, Toronto.

Fritz, R. S. 1983. Ant protection of a host plant's defoliator: consequence of an ant-membracid mutualism. *Ecology* 64:789–797.

Frye, R. J. 1983. Experimental field evidence of interspecific aggression between two species of kangaroo rat (*Dipodomys*). *Oecologia* 59:74–78.

Futuyma, D. J. 1983. Evolutionary interactions among herbivorous insects and plants. *In* D. J. Futuyma and M. Slatkin, eds., *Coevolution,* pp. 207–231. Sinauer, Sunderland, Mass.

Futuyma, D. J., and G. C. Mayer. 1980. Nonallopatric speciation in animals. *Syst. Zool.* 29:254–271.

Futuyma, D. J., and M. Slatkin, eds. 1983. *Coevolution*. Sinauer, Sunderland, Mass.

Gaines, S., and J. Roughgarden. 1985. *Larval settlement rate: a leading determinant of structure in ecological communities of the marine intertidal zone*. Unpublished manuscript.

Gallagher, E., P. Jumars, and D. Trueblood. 1983. Facilitation of soft-bottom benthic succession by tube builders. *Ecology* 64:1200–1216.

Garcia, E. F. J. 1983. An experimental test of competition for space between Blackcaps *Sylvia atricapilla* and Garden Warblers *Sylvia borin* in the breeding season. *J. Anim. Ecol.* 52:795–805.

Garman, S. 1887. On West Indian reptiles. Iguanidae. *Bull. Essex Inst.* 19:25–50.

Gass, C. L., and K. P. Lertzman. 1980. Capricious mountain weather: a driving variable in hummingbird territorial dynamics. *Can. J. Zool.* 58:1964–1968.

Gates, D. M. 1980. *Biophysical Ecology*. Springer-Verlag, New York.

Gates, D. M., and L. E. Papian. 1971. *Atlas of Energy Budgets of Plant Leaves*. Academic Press, London, New York.

Gause, G. F. 1934. *The Struggle for Existence*. Hafner, New York.

Gause, G. F., and A. A. Witt. 1935. Behaviour of mixed populations and the problem of natural selection. *Am. Nat.* 69:596–609.

Gay, P. E., P. J. Grubb, and H. J. Hudson. 1982. Seasonal changes in the concentrations of nitrogen, phosphorus and potassium, and in the density of mycorrhiza, in biennial and matrix-forming perennial species of closed chalkland turf. *J. Ecol.* 70:571–593.

Gee, A. S., N. J. Milner, and K. J. Hemsworth. 1978. The effect of density on mortality in juvenile Atlantic salmon (*Salmo salar*). *J. Anim. Ecol.* 47:497–505.

Geiselman, J. A., and O. J. McConnell. 1981. Polyphenols in brown algae *Fucus vesiculosus* and *Ascophyllum nodosum*: chemical defenses against the marine herbivorous snail, *Littorina littorea*. *J. Chem. Ecol.* 7:1115–1133.

Gentry, A. H. 1974. Flowering phenology and diversity in tropical Bignoniaceae. *Biotropica* 6:64–68.

George, A. S., A. J. M. Hopkins, and N. G. Marchant. 1979. The heathlands of Western Australia. *In* R. L. Specht, ed., *Heathlands and Related Shrublands of the World, Descriptive Studies,* pp. 211–230. Elsevier, Amsterdam.

Geraci, S., and V. Romairone. 1982. Barnacle larvae and their settlement in Genoa Harbor (North Tyrrhenian Sea). *Mar. Ecol.* 3:225–232.

Gibb, J. A., and J. E. C. Flux. 1973. Mammals. *In* G. R. Williams, ed., *The Natural History of New Zealand,* pp. 334–371. Reed, Wellington.

Gibson, A. C., and K. E. Horak. 1978. Systematic anatomy and phylogeny of Mexican columnar cacti. *Ann. M. Bot. Gard.* 65:999–1057.

Gibson, R. N. 1982. Recent studies on the biology of intertidal fishes. *Oceanogr. Mar. Biol. Annu. Rev.* 20:363–414.

Gilbert, L. E. 1977. The role of insect-plant coevolution in the organization of ecosystems. *In* V. Labyrie, ed., *Comportement des Insectes et Millieu Trophique,* pp. 399–413. C.N.R.S., Paris.

Gilbert, L. E., and P. H. Raven, eds., 1975. *Coevolution of Animals and Plants*. University of Texas Press, Austin.

Gill, D. 1978. The metapopulation ecology of the red-spotted newt, *Notophthalmus viridescens* (Rafinesque). *Ecol. Monogr.* 48:145–166.

Gill, F. B., and L. L. Wolf, 1979. Nectar loss by Golden-winged Sunbirds to competitors. *Auk* 96:448–461.

Gillette, J. B. 1962. Pest pressure, an underestimated factor in evolution. *Syst. Assoc. Publ.* no. 4:37–46.

Gilliam, J. F. 1982. *Habitat Use and Competitive Bottlenecks in Size-structured Fish Populations*. Doctoral dissertation, Michigan State University, Lansing.

Gilliard, E. T. 1969. *Birds of Paradise and Bower Birds*. Natural History Press, Garden City, N.Y.

Gilliard, E. T., and M. LeCroy. 1961. Birds of the Victor Emanuel and Hindenburg Mountains, New Guinea. *Bull. Am. Mus. Nat. Hist.* 123:1–86.

Gilpin, M. E. 1974. Intraspecific competition between Drosophila larvae in serial transfer systems. *Ecology* 55:1154–1159.

Gilpin, M. E. 1975. Limit cycles in competition communities. *Am. Nat.* 109:51–60.

Gilpin, M. E. 1983. Restoration ecology: a note on the theory and practice. *Restoration and Management Notes* 1(4):11–13.

Gilpin, M. E., and T. J. Case. 1976. Multiple domains of attraction in competition communities. *Nature* 261:40–42.

Gilpin, M. E., and J. M. Diamond. 1976. Calculation of immigration and extinction curves for the species-area-distance relation. *Proc. Natl. Acad. Sci., U.S.A.* 73:4130–4134.

Gilpin, M. E., and J. M. Diamond. 1982. Factors contributing to non-randomness in species co-occurrence on islands. *Oecologia* 52:75–84.

Gilpin, M. E., and J. M. Diamond. 1984. Are species co-occurrences on islands non-random, and are null hypotheses useful in community ecology? *In* D. R. Strong, D. Simberloff, L. G. Abele, and A. B. Thistle, eds., *Ecological Communities: Conceptual Issues and the Evidence,* pp. 296–315, 332–341. Princeton University Press, Princeton, N.J.

Gilpin, M. E., and K. E. Justice. 1972. Re-interpretation of the invalidation of the principle of competitive exclusion. *Nature* 236:273–274.

Gilpin, M. E., and G. A. H. McClelland. 1979. Systems analysis of the Yellow Fever Mosquito *Aedes aegypti*. *Fortschr. Zool.* 25(213):355–388.

Givnish, T. J. 1978. On the adaptive significance of compound leaves, with particular reference to tropical trees. *In* P. B. Tomlinson and M. H. Zimmermann, eds., *Tropical Trees as Living Systems,* pp. 351–380. Cambridge University Press, Cambridge.

Givnish, T. J. 1979. On the adaptive significance of leaf form. *In* O. T. Solbrig, S. Jain, G. B. Johnson, and P. H. Raven, eds., *Topics in Plant Population Ecology,* pp. 375–407. Columbia University Press, New York.

Givnish, T. J. 1982. On the adaptive significance of leaf height in forest herbs. *Am. Nat.* 120:353–381.

Givnish, T. J., and G. J. Vermeij. 1976. Sizes and shapes of liane leaves. *Am. Nat.* 110:743–778.

Glasgow, J. P. 1963. *The Distribution and Abundance of Tsetse*. Macmillan, New York.

Glass, G. E., and N. A. Slade. 1980. The effect of *Sigmodon hispidus* on spatial and temporal activity of *Microtus ochrogaster:* evidence for competition. *Ecology* 6:358–370.

Gleason, H. A. 1926. The individualistic concept of plant association. *Bull. Torrey Bot. Club* 53:7–26.

Glesener, R. R., and D. Tilman. 1978. Sexuality and the components of environmental uncertainty: cues from geographic parthenogenesis in terrestrial animals. *Am. Nat.* 117:659–679.

Glynn, P. W. 1973. Ecology of a Caribbean coral reef. The *Porities* reef flat biotope. Part II. Plankton community with evidence of depletion. *Mar. Biol.* 22:1–21.

Glynn, P. W. 1976. Some physical and biological determinants of coral community structure in the eastern Pacific. *Ecol. Monogr.* 46:431–456.

Glynn, P. W. 1983. Crustacean symbionts and the defense of corals: coevolution on the reef? *In* M. H. Nitecki, ed., *Coevolution,* pp. 111–178. University of Chicago Press, Chicago.

Goel, N. S., and N. Richter-Dyn. 1974. *Stochastic Models in Biology*. Academic Press, New York.

Goh, B. S. 1975. Stability, vulnerability and persistence of complex ecosystem. *Ecol. Model.* 1:105–116.

Goh, B. S. 1976. Nonvulnerability of ecosystems in unpredictable environments. *Theor. Popul. Biol.* 10:83–95.

Goh, B. S. 1980. *Management and Analysis of Biological Populations*. Elsevier, Amsterdam.

Goldberg, D. E., and P. A. Werner. 1983. Equivalence of competitors in plant communities: a null hypothesis and a field experimental approach. *Am. J. Bot.* 70:1098–1104.

Goldsmith, F. B. 1978. Interaction (competition) studies as a step towards the synthesis of sea-cliff vegetation. *J. Ecol.* 66:921–931.

Goldwasser, S., D. Gaines, and S. R. Wilbur. 1980. The Least Bell's Vireo in California: a *de facto* endangered race. *Am. Birds* 34:742–745.

Golubic, S., and S. E. Campbell. 1979. Analogous microbial forms in recent subaerial habitats and in Precambrian cherts: *Gloeothece coerulea* Geitler and *Eosynechococcus moorei* Hofmann. *Precambrian Res.* 8:201–217.

Goodman, D. 1974. Natural selection and a cost ceiling on reproductive effort. *Am. Nat.* 108:247–268.

Goodman, D. 1979. Competitive hierarchies in laboratory Drosophila. *Evolution* 33:207–219.

Goodman, D. 1984. Risk spreading as an adaptive strategy in iteroparous life histories. *Theor. Popul. Biol.* 25:1–20.

Goodwin, D. 1983. *Pigeons and Doves of the World*. 3rd ed. British Museum (Natural History), London.

Gorman, G. 1973. The chromosomes of the Reptilia, a cytotaxonomic interpretation. *In* A. Chiarelli and E. Capanna, eds., *Cytotaxonomy and Vertebrate Evolution,* pp. 349–421. Academic Press, London.

Gorman, G., D. Buth, M. Soulé, and S. Yang. 1980a. The relationships of the *Anolis cristatellus* species group: electrophoretic analysis. *J. Herpetol.* 14:269–278.

Gorman, G., D. Buth, and J. Wyles. 1980b. *Anolis* lizards of the eastern Caribbean: a case study in evolution. III. A cladistic analysis of albumin immunological data, and the definition of species groups. *Syst. Zool.* 29:143–158.

Gorman, G., and Y. Kim. 1976. *Anolis* lizards of the eastern Caribbean: a case study in evolution. II. Genetic relationships and genetic variation of the *bimaculatus* group. *Syst. Zool.* 25:62–77.

Gould, F. 1983. Genetics of plant-herbivore systems: interactions between applied and basic study. *In* R. Denno and M. McClure, eds., *Variable Plants and Herbivores in Natural and Managed Systems,* pp. 593–653. Academic Press, New York.

Grace, J. B. 1985. Juvenile versus adult competitive abilities in cattails (*Typha*). *Ecology* in press.

Graham, R. W. 1976. Late Wisconsin mammal faunas and environmental gradients of the eastern United States. *Paleobiology* 2:343–350.

Graham, R. W. 1979. Paleoclimates and late Pleistocene faunal provinces in North America. *In* R. L. Humphrey and D. Stanford, eds., *Pre-Llano Culture of the Americas: Paradoxes and Possibilities,* pp. 49–69. Anthropological Society of Washington, Washington, D.C.

Graham, R. W. In press a. Diversity and community structure of the late Pleistocene mammal fauna of North America. *Acta Zool. Fenn.*

Graham, R. W. In press b. Plant-animal interactions and Pleistocene extinctions. *In* D. K. Elliott, ed., *The Dynamics of Extinction.* Wiley, New York.

Graham, R. W., J. A. Holman, and P. W. Parmalee. 1983. Taphonomy and paleoecology of the Christensen Bog mastodon bone bed, Hancock County, Indiana. *Ill. State Mus. Rep. Invest.* 38:1–29.

Graham, R. W., and E. L. Lundelius, Jr. 1984. Coevolutionary disequilibrium and Pleistocene extinctions. *In* P. S. Martin and R. G. Klein, eds., *Quaternary Extinctions: a Prehistoric Revolution.* University of Arizona Press, Tucson.

Grant, B. R., and P. R. Grant. 1979. Darwin's finches: population variation and sympatric speciation. *Proc. Natl. Acad. Sci., U.S.A.* 76:2359–2363.

Grant, B. R., and P. R. Grant. 1981. Exploitation of *Opuntia* cactus by birds on the Galápagos. *Oecologia* 49:179–187.

Grant, B. R., and P. R. Grant. 1983. Fission and fusion in a population of Darwin's finches: an example of the value of studying individuals in ecology. *Oikos* 41:530–547.

Grant, P. R. 1966a. Ecological compatibility of bird species on islands. *Am. Nat.* 100:451–462.

Grant, P. R. 1966b. Preliminary experiments on the foraging of closely related species of birds. *Ecology* 47:148–151.

Grant, P. R. 1968. Bill size, body size, and the ecological adaptations of bird species to competitive situations on islands. *Syst. Zool.* 17:319–333.

Grant, P. R. 1969a. Experimental studies of competitive interaction in a two-species system. I. *Microtus* and *Clethrionomys* species in enclosures. *Can. J. Zool.* 47:1059–1082.

Grant, P. R. 1969b. Colonization of islands by ecologically dissimilar species of birds. *Can. J. Zool.* 47:41–43.

Grant, P. R. 1972. Interspecific competition among rodents. *Annu. Rev. Ecol. Syst.* 3:79–106.

Grant, P. R. 1975. The classical case of character displacement. *Evol. Biol.* 8:237–337.

Grant, P. R. 1981a. Speciation and the adaptive radiation of Darwin's Finches. *Am. Sci.* 69:653–663.

Grant, P. R. 1981b. The feeding of Darwin's Finches on *Tribulus cistoides* (L.) seeds. *Anim. Behav.* 29:785–793.

Grant, P. R. 1983a. Inheritance of size and shape in a population of Darwin's finches, *Geospiza conirostris*. *Proc. R. Soc. Lond. B* 220:219–236.

Grant, P. R. 1983b. The role of interspecific competition in the adaptive radiation of Darwin's Finches. *In* R. I. Bowman, M. Berson, and A. E. Leviton, eds., *Patterns of Evolution in Galápagos Organisms,* pp. 187–199. Special Publication. American Association for the Advancement of Science, Pacific Division, San Francisco.

Grant, P. R. 1984. Recent research on the evolution of land birds on the Galápagos. *Biol. J. Linn. Soc.* 21:113–136.

Grant, P. R. 1985. Climatic fluctuations on the Galápagos Islands and their influence on Darwin's Finches. *Ornithol. Monogr.* 36.

Grant, P. R., and I. A. Abbott. 1980. Interspecific competition, island biogeography, and null hypotheses. *Evolution* 34:332–341.

Grant, P. R., I. A. Abbott, D. Schluter, R. L. Curry, and L. K. Abbott. 1985. Variation in the size and shape of Darwin's Finches. *Biol. J. Linn. Soc.* 25:1–39.

Grant, P. R., and P. T. Boag. 1980. Rainfall on the Galápagos and the demography of Darwin's Finches. *Auk* 97:227–244.

Grant, P. R., and B. R. Grant. 1980a. The breeding and feeding characteristics of Darwin's Finches on Isla Genovesa, Galápagos. *Ecol. Monogr.* 50:381–410.

Grant, P. R., and B. R. Grant. 1980b. Annual variation in finch numbers, foraging behavior and food supply on Isla Daphne Major, Galápagos. *Oecologia* 46:55–62.

Grant, P. R., B. R. Grant, J. N. M. Smith, I. J. Abbott, and L. K. Abbott. 1976. Darwin's Finches: population variation and natural selection. *Proc. Natl. Acad. Sci., U.S.A.* 73:257–261.

Grant, P. R., and K. T. Grant. 1979. The breeding and

feeding of the Galápagos dove, *Zenaida galapagoensis*. *Condor* 81:397–403.

Grant, P. R., and M. Grant. 1983. The origin of a species. *Nat. Hist.* 92:76–80.

Grant, P. R., and T. D. Price. 1981. Population variation in continuously varying traits as an ecological genetics problem. *Am. Zool.* 21:795–811.

Grant, P. R., and D. Schluter. 1984. Interspecific competition inferred from patterns of guild structure. *In* D. R. Strong, D. Simberloff, L. G. Abele, and A. B. Thistle, eds., *Ecological Communities: Conceptual Issues and the Evidence,* pp. 201–233. Princeton University Press, Princeton, N.J.

Grant, P. R., J. N. M. Smith, B. R. Grant, I. J. Abbott, and L. K. Abbott. 1975. Finch numbers, owl predation and plant dispersal on Isla Daphne Major, Galápagos. *Oecologia* 19:239–257.

Graves, E. R., and N. J. Gotelli. 1983. Neotropical land-bridge avifaunas: new approaches to null hypotheses in biogeography. *Oikos* 41:322–333.

Greene, C. H., and A. Schoener. 1982. Succession on marine hard substrata: a fixed lottery. *Oecologia* 55:289–297.

Greenhill, G. 1881. Determination of the greatest height consistent with stability that a vertical pole or mast can be made, and of the greatest height to which a tree of given proportions can grow. *Proc. Camb. Philos. Soc.* 4:65–73.

Greenstone, M. H. 1984. Determinants of web spider species diversity: vegetation structural diversity vs. prey availability. *Oecologia* 62:299–304.

Greenway, J. C., Jr. 1967. *Extinct and Vanishing Birds of the World*. Dover, New York.

Grigal, D. F., L. M. Chamberlain, H. R. Finney, D. V. Wroblewski, and E. R. Gross. 1974. *Soils of the Cedar Creek Natural History Area*. Miscellaneous Report no. 123. University of Minnesota Agricultural Experimental Station, St. Paul.

Grigg, R. W., and J. R. Maragos. 1974. Recolonization of hermatypic corals on submerged lava flows in Hawaii. *Ecology* 55:387–395.

Grime, J. P. 1977. Evidence for the existence of three primary strategies in plants and its relevance to ecological and evolutionary theory. *Am. Nat.* 111:1169–1194.

Grime, J. P. 1979. *Plant Strategies and Vegetation Processes*. Wiley, New York.

Grime, J. P., and D. W. Jeffrey. 1965. Seedling establishment in vertical gradients of sunlight. *J. Ecol.* 53:621–642.

Grimm, E. C. 1983. Chronology and dynamics of vegetation change in the prairie-woodland region of southern Minnesota, U.S.A. *New Phytol.* 93:311–350.

Grosberg, R. K. 1981. Competitive ability influences habitat choice in marine invertebrates. *Nature* 290:700–702.

Grosberg, R. K. 1982. Intertidal zonation of barnacles: the influence of planktonic zonation of larvae on vertical distribution of adults. *Ecology* 63:894–899.

Grossman, G. D. 1980. Ecological aspects of ontogenetic shifts in prey size utilization in the bay goby (*Pisces:* Gobiidee). *Oecologia* 47:233–238.

Grossman, G. D. 1982. Dynamics and organization of a rocky intertidal fish assemblage: the persistence and resilience of taxocene structure. *Am. Nat.* 119:611–637.

Grossman, G. D., P. B. Moyle, and J. R. Whittaker, Jr. 1982. Stochasticity in structural and functional characteristics of an Indiana stream fish community: a test of community theory. *Am. Nat.* 120:423–455.

Grubb, P. J. 1976. A theoretical background to the conservation of ecologically distinct groups of annuals and biennials in the chalk grassland ecosystem. *Biol. Conserv.* 10:53–76.

Grubb, P. J. 1977. The maintenance of species-richness in plant communities: the importance of the regeneration niche. *Biol. Rev.* 52:107–145.

Grubb, P. J. 1984. Some growth points in investigative plant ecology. *In* J. H. Cooley and F. B. Golley, eds., *Trends in Ecological Research in the 1980's,* pp. 51–74. Plenum, New York.

Grubb, P. J., D. Kelly, and J. Mitchley. 1982. The control of relative abundance in communities of herbaceous plants. *In* E. I. Newman, eds., *The Plant Community as a Working Mechanism,* pp. 79–97. Special Publication Series of the British Ecological Society no. 1. Blackwell, Oxford.

Grubb, P. J., and P. F. Stevens. 1985. *The Forests of the Fatima Basin and Mt. Kerigomna, Papua New Guinea, with a Review of Montane and Subalpine Rain Forests Elsewhere in Papuasia*. Department of Biogeography and Geomorphology Publication BG/5. Australian National University, Canberra.

Guckenheimer, J., G. Oster, and A. Ipaktchi. 1977. The dynamics of density-dependent population models. *J. Math. Biol.* 4:101–147.

Gudmundsson, F. 1951. The effects of the recent climatic changes on the bird life of Iceland. *Proc. 10th Int. Ornith. Congr.* 502:514.

Guilday, J. E., H. W. Hamilton, E. Anderson, and P. W. Parmalee. 1978. The Baker Bluff cave deposit, Tennessee, and the late Pleistocene faunal gradient. *Bull. Carnegie Mus. Nat. Hist.* 11:1–67.

Guilday, J. E., P. S. Martin, and A. D. McGrady. 1964. New Paris No. 4: Pleistocene cave deposit in Bedford County, Pennsylvania. *Bull. Nat. Speleol. Soc.* 26:121–194.

Guilday, J. E., P. W. Parmalee, and H. W. Hamilton.

1977. The Clark's Cave bone deposit and late Pleistocene paleoecology of the central Appalachian Mountains of Virginia. *Bull. Carnegie Mus. Nat. Hist.* 2:1–87.

Gulick, J. T. 1890. Divergent evolution through cumulative segregation. *J. Linn. Soc. Lond. Zool.* 20:189–274.

Gulland, J. A. 1982. Why do fish numbers vary? *J. Theor. Biol.* 97:69–75.

Guthrie, R. D. 1982. Mammals of the mammoth steppe as paleoenvironmental indicators. *In* D. M. Hopkins, J. V. Matthews, Jr., C. E. Schweger, and S. B. Young, eds., *Paleoecology of Beringia,* pp. 307–326. Academic Press, New York.

Haffer, J. 1969. Speciation in Amazonian forest birds. *Science* 165:131–137.

Haila, Y. 1983. Land birds on northern islands: a sampling metaphor for insular colonization. *Oikos* 41:334–351.

Hairston, N. G. 1981. An experimental test of a guild: salamander competition. *Ecology* 62:65–72.

Hairston, N. G. 1985. The interpretation of experiments on interspecific competition. *Am. Nat.* 125:321–325.

Hairston, N. G., J. D. Allan, R. K. Colwell, D. J. Futuyma, J. Howell, M. D. Lubin, J. Mathias, and J. H. Vandermeer. 1968. The relationship between species diversity and stability: an experimental approach with protozoa and bacteria. *Ecology* 49:1091–1101.

Hairston, N. G., F. E. Smith, and L. B. Slobodkin. 1960. Community structure, population control, and competition. *Am. Nat.* 94:421–425.

Haldane, J. B. S. 1940. The conflict between selection and mutation of harmful recessive genes. *Ann. Eugenics* 10:417–422.

Hall, D. J., W. E. Cooper, and E. E. Werner. 1970. An experimental approach to the production dynamics and structure of freshwater animal communities. *Limnol. Oceanogr.* 15:839–928.

Hall, D. J., and T. J. Ehlinger. 1985. Perturbation, planktivory, and pelagic community structure: winterkill in a small lake. *Ecology* (submitted).

Hall, D. J., and E. E. Werner. 1977. Seasonal distribution and abundance of fishes in the littoral zone of a Michigan lake. *Trans. Am. Fish. Soc.* 106:545–555.

Halle, F., and R. A. A. Oldeman. 1970. *Essai sur L'architecture et La Dynamique de Croissance des Arbres Tropicaux.* Maisson, Paris.

Halle, F., R. A. A. Oldeman, and P. B. Tomlinson. 1978. *Tropical Trees and Forests: an Architectural Analysis.* Springer-Verlag, Berlin, New York.

Halliday, T. R. 1983. The study of mate choice. *In* P. Bateson, ed., *Mate Choice,* pp. 3–32. Cambridge University Press, Cambridge.

Halvörsen, O. 1976. Negative interaction amongst parasites. *In* C. R. Kennedy, ed., *Ecological Aspects of Parasitology,* pp. 99–114. North Holland Publishing, Amsterdam.

Hamilton, T. H., and N. E. Armstrong. 1965. Environmental determination of insular variation in bird species abundance in the Gulf of Guinea. *Nature* 207:148–151.

Hamilton, T. H., R. H. Barth, Jr., and I. Rubinoff. 1964. The environmental control of insular variation in bird species abundance. *Proc. Natl. Acad. Sci., U.S.A.* 52:132–140.

Hamilton, T. H., and I. Rubinoff. 1964. On models predicting abundance of species and endemics for the Darwin finches in the Galápagos Archipelago. *Evolution* 18:339–342.

Hanawalt, R. B., and R. H. Whittaker. 1976. Altitudinally coordinated patterns of soils and vegetation in the San Jacinto Mountains, California. *Soil Sci.* 121:114–124.

Handel, S. N. 1978. The competitive relationship of 3 woodland sedges and its bearing on the evolution of ant dispersal of *Carex pedunculata. Evolution* 32:151–163.

Hansen, J., D. Johnson, A. Lacis, S. Lebedeff, P. Lee, D. Rind, and G. Russell. 1981. Climate impact of increasing atmospheric carbon dioxide. *Science* 213:957–966.

Hansen, S. R., and S. P. Hubbell. 1980. Single-nutrient microbial competition: qualitative agreement between experimental and theoretically forecast outcomes. *Science* 207:1491–1493.

Hanski, I. 1981. Coexistence of competitors in patchy environment with and without predation. *Oikos* 37:306–312.

Hanski, I. 1982. Dynamics of regional distribution: the core and satellite species hypothesis. *Oikos* 38:210–221.

Hanski, I. 1983. Coexistence of competitors in patchy environment. *Ecology* 64:493–500.

Hanski, I., and E. Ranta. 1983. Coexistence in a patchy environment: three species of *Daphnia* in rock pools. *J. Anim. Ecol.* 52:263–279.

Hardin, G. 1968. The tragedy of the commons. *Science* 162:1243–1248.

Harger, J. R. E. 1972. Competitive coexistence among intertidal vertebrates. *Am. Sci.* 60:600–607.

Harland, W. B., A. U. Cox, P. G. Llewellyn, C. A. G. Pickton, A. G. Smith, and R. Walters. 1982. *A Geologic Time Scale.* Cambridge University Press, Cambridge.

Harley, J. L. 1970. The importance of micro-organisms to colonizing plants. *Trans. Bot. Soc. Edinb.* 41:65–70.

Harper, J. L. 1961. Approaches to the study of plant competition. *In* F. L. Milthorpe, ed., *Mechanisms in Biological Competition,* pp. 1–39. Symposia of the Society for Experimental Biology no. 15. Cambridge University Press, Cambridge.

Harper, J. L. 1969. The role of predation in vegetational diversity. *Brookhaven Symp. Biol.* 22:48–62.

Harper, J. L. 1977. *Population Biology of Plants.* Academic Press, New York.

Harper, J. L., and I. H. McNaughton. 1962. The comparative biology of closely related species. VII. Interference between individuals in pure and mixed populations of *Papaver* spp. *New Phytol.* 61:175–188.

Harris, A. H. 1970. The Dry Cave mammalian fauna and the late pluvial conditions in southeastern New Mexico. *Tex. J. Sci.* 22:3–27.

Harris, A. H. 1977. Wisconsin age environments in the northern Chihuahuan Desert: evidence from the higher vertebrates. *In* R. H. Wauer and D. H. Riskind, eds., *Trans. Symposium on the Biological Resources of the Chihuahuan Desert Region, United States and Mexico,* pp. 23–52. National Park Service Transactions and Proceedings Series, No. 3, Washington, D.C.

Harris, M. P. 1973. The Galápagos avifauna. *Condor* 75:265–278.

Hartshorn, G. S. 1978. Treefalls and tropical forest dynamics. *In* P. B. Tomlinson and M. H. Zimmermann, eds., *Tropical Trees as Living Systems,* pp. 617–628. Cambridge University Press, Cambridge.

Harvey, P. H., R. K. Colwell, J. V. Silvertown, and R. M. May. 1983. Null models in ecology. *Annu. Rev. Ecol. Syst.* 14:189–211.

Hassell, M. P. 1978. *The Dynamics of Arthropod Predator-Prey Systems.* Princeton University Press, Princeton, N.J.

Hassell, M. P. 1979. Non-random search in predator-prey models. *Fortschr. Zool.* 25:311–330.

Hassell, M. P. 1980. Foraging strategies, population models and biological control: a case study. *J. Anim. Ecol.* 49:603–618.

Hassell, M. P. 1982. Patterns of parasitism by insect parasitoids in patchy environments. *Ecol. Entomol.* 7:365–377.

Hassell, M. P. 1985. Insect natural enemies as regulating factors. *J. Anim. Ecology* 54:323–334.

Hassell, M. P., and H. N. Comins. 1976. Discrete time models for two-species competition. *Theor. Popul. Biol.* 12:202–221.

Hassell, M. P., J. H. Lawton, and R. M. May. 1976. Patterns of dynamical behavior in single-species populations. *J. Anim. Ecol.* 45:471–486.

Hassell, M. P., and R. M. May. 1974. Aggregation in predators and insect parasites and its effect on stability. *J. Anim. Ecol.* 43:567–594.

Hassell, M. P., and R. M. May. 1985. From individual behaviour to population dynamics. *In* R. Sibley and R. Smith, eds., *Behavioral Ecology.* British Ecological Symposium. Blackwell, Oxford.

Hassell, M. P., and J. K. Waage. 1984. Host-parasitoid population interactions. *Annu. Rev. Entomol.* 29: 89–114.

Hastenrath, S., 1984. Predictability of northeast Brazil droughts. *Nature* 307:531–533.

Hastings, A. 1977. Spatial heterogeneity and the stability of predator-prey systems. *Theor. Popul. Biol.* 12:37–48.

Hastings, A. 1978. Spatial heterogeneity and the stability of predator-prey systems: predator-mediated coexistence. *Theor. Popul. Biol.* 14:380–395.

Hastings, A. 1980. Disturbance, coexistence, history, and competition for space. *Theor. Popul. Biol.* 18:363–373.

Hastings, A., and H. Caswell. 1979. Role of environmental variability in the evolution of life history strategies. *Proc. Natl. Acad. Sci., U.S.A.* 76:4700–4703.

Hatton, H. 1938. Essais de bionomie explicative sur quelques especes intercotidales d'algues et d'animaux. *Ann. Inst. Oceanogr.* 17:241–348.

Hauenschild, C. 1954. Genetische und entwicklungphysiologische Untersuchungen über Intersexualität und Gewebeverträglichkeit bei *Hydractinia echinata* Flemm. (Hydroz. Bougainvill.). *Wilhelm. Roux. Arch. Entwicklungsmech. Org.* 147:1–41.

Hauenschild, C. 1956. Über die Vererbung der Gewebeverträglichkeits-Éigenschaft bei dem Hydroidpolypen *Hydractinia echinata. Z. Naturforsch.* 11:132–138.

Haukioja, E. 1980. On the role of plant defenses in the fluctuation of herbivore populations. *Oikos* 35:202–213.

Haukioja, E., and P. Niemalä. 1977. Retarded growth of a geometrical larva after mechanical damage to leaves of its host tree. *Ann. Zool. Fenn.* 14:48–52.

Haukioja, E., and P. Niemelä. 1979. Birch leaves as a resource for herbivores: seasonal occurrence of increased resistance in foliage after mechanical damage of adjacent leaves. *Oecologia* 39:151–159.

Hawboldt, L. S. 1952. Climate and birch "dieback." *Nova Scotia Dept. Lands and Forests. Bull.* no. 6.

Hawkins, S. J., and R. G. Hartnoll. 1982. Settlement patterns of *Semibalanus balanoides* (L.) in the Isle of Man (1977–1981). *J. Exp. Mar. Biol. Ecol.* 62:271–283.

Hawkins, S. J., and R. G. Hartnoll. 1983. Grazing of intertidal algae by marine invertebrates. *Oceanogr. Mar. Biol. Annu. Rev.* 21:195–282.

Hayes, J. L. 1981. The population ecology of a natural population of the Pierid butterfly *Colias alexandra. Oecologia* 49:188–200.

Heads, P., and J. Lawton. 1983. Studies on the natural enemy complex of the holly leaf-miner: the effects of scale on the detection of aggregative responses and the implications for biological control. *Oikos* 40:267–276.

Healey, M. C. 1980. Growth and recruitment in experimentally exploited lake whitefish (*Coregonus clupeaformis*) populations. *Can. J. Fish. Aquat. Sci.* 37:255–267.

Heatwole, H. 1975. Biogeography of reptiles on some of the islands and cays of eastern Papua-New Guinea. *Atoll Res. Bull.* 180.

Heckel, D. G., and J. D. Roughgarden. 1979. A technique for estimating the size of lizard populations. *Ecology* 60:969–975.

Heed, W. B. 1971. Host plant specificity and speciation in Hawaiian *Drosophila. Taxon* 20:115–121.

Heinemann, D. 1984. *Interactions Among Rufous Hummingbirds, Hymenopterans, and a Shared Resource: Exploitative Exclusion of a Vertebrate from a Nectar Source.* Scrophularia montana, *by Insects.* Doctoral Dissertation, University of New Mexico, Albuquerque.

Heinrich, B. 1975. Bee flowers: a hypothesis on flower variety and blooming times. *Evolution* 29:325–334.

Heinrich, B. 1976. Flowering phenologies—bog, woodland, and disturbed habitats. *Ecology* 57:890–899.

Heinselman, M. L. 1973. Fire in the virgin forests of the boundary waters canoe area, Minnesota. *Quat. Res.* 3:329–382.

Helfman, G. S. 1978. Patterns of community structure in fishes: summers and overview. *Env. Biol. Fish.* 3:129–148.

Heller, H. C. 1971. Altitudinal zonation of chipmunks (*Eutamias*): interspecific aggression. *Ecology* 52:312–319.

Hendrickson, J. A., Jr. 1981. Community-wide character displacement reexamined. *Evolution* 35:794–809.

Henshaw, H. W. 1902. *Birds of the Hawaiian Islands.* Thrum, Honolulu.

Hensley, M. M., and J. B. Cope. 1951. Further data on removal and repopulation of the breeding birds in a spruce-fir forest community. *Auk* 68:483–493.

Hepher, B. 1978. Ecological aspects of warm-water fishpond management. *In* S. D. Cerking, ed., *Ecology of Freshwater Fish Production,* pp. 447–468. Blackwell, Oxford.

Hiatt, R. W., and D. W. Strasburgh. 1960. Ecological relationships of the fish fauna on coral reef of the Marshall Islands. *Ecol. Monogr.* 30:65–127.

Hibbard, C. W. 1960. An interpretation of Pliocene and Pleistocene climates in North America. *Annu. Rep. Michigan Academy of Sciences, Arts, and Letters* 62:5–30.

Hibbard, C. W. 1970. Pleistocene mammalian local faunas from the Great Plains and central lowland provinces of the United States. *In* W. Dort, Jr. and J. K. Jones, Jr., eds., *Pleistocene and Recent Environments of the Central Great Plains,* pp. 395–433. University Press of Kansas, Lawrence, Kans.

Hickey, L. J. 1980. Paleocene stratigraphy and flora of the Clark's Fork Basin. *Univ. Mich. Pap. Paleont.* 24:33–49.

Hildén, O. 1965. Habitat selection in birds. *Ann Zool. Fenn.* 2:53–75.

Hixon, M. A. 1980. Competitive interaction between California reef fishes of the genus *Embiotica. Ecology* 61:918–931.

Hixon, M. A., and W. J. Brostoff. 1983. Damselfish as keystone species in reverse: intermediate disturbance and diversity of reef algae. *Science* 220:511–513.

Hoagland, K. D. 1983. Short-term standing crop and diversity of periphytic diatoms in a eutrophic reservoir. *J. Phycol.* 19:30–38.

Hoagland, K. D., S. C. Roemer, and J. R. Rosowski. 1982. Colonization and community structure of two periphyton assemblages, with emphasis on the diatoms (Bacillariophyceae). *Am. J. Bot.* 69:188–213.

Hobbs, R. P. 1980. Interspecific interactions among gastrointestinal helminths in pikas of North America. *Am. Midl. Nat.* 103:15–25.

Hoffman, A. 1979. Community paleoecology as an epiphenomenal science. *Paleobiology* 5:357–379.

Hoffman, R. S., and J. K. Jones, Jr. 1970. Influence of late-glacial and post-glacial events on the distribution of recent mammals on the northern great plains. *In* W. Dort, Jr., and J. K. Jones, Jr., eds., *Pleistocene and Recent Environments of the Central Great Plains,* pp. 355–394. University Press of Kansas, Lawrence, Kans.

Hogg, R. V., and A. T. Craig. 1978. *Introduction to Mathematical Statistics.* 4th ed. Macmillan, New York.

Högstedt, G. 1980. Prediction and test of the effects of interspecific competition. *Nature* 283:64–66.

Hole, F. D. 1976. *Soils of Wisconsin.* University of Wisconsin Press, Madison.

Holling, C. S. 1973. Resilience and stability of ecological systems. *Annu. Rev. Ecol. Syst.* 4:1–23.

Holman, J. A. 1976. Paleoclimatic implications of

''ecologically incompatible'' herpetological species (late Pleistocene: southeastern United States). *Herpetologica* 32:290–294.

Holmes, J. C. 1961. Effects of concurrent infections on *Hymenolepis diminuta* (Cestoda) and *Moniliformis dubius* (Acanthocephala). I. General effects and comparisons with crowding. *J. Parasitol.* 47:209–216.

Holmes, J. C. 1973a. Habitat segregation in sanguinicolid blood flukes (Digenea) of scorpaenid rock fishes (Perciformes) on the Pacific Coast of North America. *J. Fish. Res. Board Can.* 28:903–909.

Holmes, J. C. 1973b. Site selection by parasitic helminths: interspecific interactions, site segregations, and their importance to the development of helminth communities. *Can. J. Zool.* 51:333–347.

Holmes, J. C. 1979. Parasite populations and host community structure. *In* B. B. Nikol, ed., *Host-Parasite Interfaces,* pp. 27–46. Academic Press, New York.

Holmes, J. C. 1982. Impact of infectious disease agents on the population growth and geographical distribution of animals. *In* R. M. Anderson and R. M. May, eds., *Population Biology of Infectious Diseases,* pp. 37–51. Springer-Verlag, Berlin.

Holmes, J. C. 1983. Evolutionary relationships between parasitic helminths and their hosts. *In.* D. J. Futuyma and M. Slatkin, eds., *Coevolution,* pp. 161–185. Sinauer, Sunderland, Mass.

Holmes, J. C., and W. M. Bethel. 1972. Modification of intermediate host behaviour by parasites. *In* E. U. Canning and C. A. Wright, eds., *Behavioural Aspects of Parasite Transmission,* pp. 123–149. *Zool. J. Linn. Soc.* 51 (Suppl. no. 1).

Holmes, J. C., and P. W. Price. 1980. Parasite communities: the roles of phylogeny and ecology. *Syst. Zool.* 29:203–213.

Holmes, J. C., and P. W. Price. 1985. Communities of parasites. *In.* D. J. Anderson and J. Kikkawa, eds., *Community Ecology: Pattern and Process.* In press. Blackwell, Oxford.

Holmes, R., J. Schultz, and P. Nothangle. 1980. Bird predation on forest insects: an enclosure experiment. *Science* 206:462–463.

Holt, R. D. 1977. Predation, apparent competition, and the structure of prey communities. *Theor. Popul. Biol.* 12:197–229.

Holt, R. D., and J. Pickering. 1985. A model for the coexistence of species sharing an infectious disease. *Am. Nat.,* in press.

Hoover, E. E. 1936. Contributions to the life history of the chinook and landlocked salmon in New Hampshire. *Copeia* 1936:193–198.

Hope, G. S. 1976. The vegetational history of Mt. Wilhelm, Papua-New Guinea. *J. Ecol.* 64:627–661.

Hope, J. H. 1973. Mammals of the Bass Strait islands. *Proc. Roy. Soc. Vict.* 85:163–196.

Hope-Simpson, J. F. 1940. Studies of the vegetation of the English chalk. VI. Late stages in succession leading to chalk grassland. *J. Ecol.* 28:386–402.

Hopkins, C. D. 1980. Evolution of electric communication channels of mormyrids. *Behav. Ecol. Sociobiol.* 7:1–13.

Hopkins, C. D., and A. H. Bass. 1981. Temporal coding of species recognition signals in an electric fish. *Science* 212:85–87.

Horn, H. S. 1968. Regulation of animal numbers: a model counter example. *Ecology* 49:776–778.

Horn, H. S. 1971. *Adaptive Geometry of Trees.* Princeton University Press, Princeton, N.J.

Horn, H. S. 1975. Markovian properties of forest succession. *In* M. L. Cody and J. M. Diamond, eds., *Ecology and Evolution of Communities,* pp. 196–211. Harvard University Press, Cambridge, Mass.

Horn, H. S. 1981. Succession. *In* R. M. May, ed., *Theoretical Ecology,* pp. 253–271. Sinauer, Sunderland, Mass.

Horn, H. S., and R. H. MacArthur 1972. Competition among fugitive species in a harlequin environment. *Ecology* 53:749–752.

Horton, C. C., and D. H. Wise. 1983. The experimental analysis of competition among two syntopic species of orb-web spiders. *Ecology* 64:929–944.

Horwitz, R. J. 1978. Temporal variability patterns and the distributional patterns of stream fishes. *Ecol. Monogr.* 48:307–321.

Howe, H. F. 1983. Annual variation of a neotropical seed-dispersal system. *In* S. L. Sutton, T. C. Whitmore, and A. C. Chadwick, eds., *Tropical Rain Forest: Ecology and Management,* pp. 211–228. Blackwell, Oxford.

Howe, H. F. 1985. Constraints on the evolution of mutualism. *Am. Nat.,* in press.

Hoy, M. A. 1977. Inbreeding in the arrhenotokous predator *Metaseiulis occidentalis* (Nesbitt) (Acari: Phytoseiidae). *Int. J. Acarol.* 3:117–121.

Hsu, S. B., S. P. Hubbell, and P. Waltman. 1978. A contribution to the theory of competing predators. *Ecol. Monogr.* 48:337–349.

Hubbell, S. P. 1973. Populations and food webs as energy filters I. One-species systems. *Am. Nat.* 107:94–121.

Hubbell, S. P. 1979. Tree dispersion, abundance, and diversity in a tropical dry forest. *Science* 203:1299–1309.

Hubbell, S. P. 1980. Seed predation and the coexistence of tree species in tropical forests. *Oikos* 35:214–299.

Hubbell, S. P. 1984. Methodologies for the study of

the origin and maintenance of tree diversity in tropical rainforest. *In* G. Maury-Lechon, M. Hadley, and T. Younes, eds., *The Significance of Species Diversity in Tropical Forest Ecosystems,* pp. 8–13. *Biology International,* IUBS Special Issue no. 6.

Hubbell, S. P., and R. B. Foster. 1983. Diversity of canopy trees in a neotropical forest and implications for conservation. *In* S. Sutton, T. C. Whitmore, and A. Chadwick, eds., *Tropical Rain Forest: Ecology and Management,* pp. 25–41. Blackwell, Oxford.

Hubbell, S. P., and R. B. Foster. 1985a. Canopy gaps and the dynamics of a neotropical forest. *In* M. J. Crawley, ed., *Plant Ecology*. Blackwell, Oxford.

Hubbell, S. P., and R. B. Foster. 1985b. La estructura espacial en gran escala de un bosque neotropical. *Revista* de *Biologia Tropical,* in press.

Huffaker, C. B. 1958. Experimental studies on predation: dispersion factors and predator-prey oscillations. *Hilgardia* 27:343–383.

Huffaker, C. B., and C. E. Kennett. 1966. The biological control of *Parlatoria oleae* (Colvee) through the compensatory action of two introduced parasites. *Hilgardia* 37:283–334.

Hughes, N. F. 1976. *Palaeobiology of Angiosperm Origins*. Cambridge University Press, Cambridge.

Hunter, R. D. 1980. Effects of grazing on the quantity and quality of freshwater aufwuchs. *Hydrobiologia* 69:251–259.

Hunter, R. D., and W. D. Russell-Hunter. 1983. Bioenergetic and community changes in intertidal aufwuchs grazed by *Littorina littorea*. *Ecology* 64:761–769.

Huntley, B., and H. J. B. Birks. 1983. *Past and Present Pollen Maps for Europe 0–13,000 years ago*. Cambridge University Press, Cambridge.

Hurd, L. E., and R. M. Eisenberg. 1984. Experimental density manipulation of the predator *Tenodera sinensis* (Orthoptera: Mantidae) in an oldfield community. I. Mortality, development and dispersal of juvenils mantids. *J. Amer. Ecol.* 53:269–281.

Hurlbert, S. H. 1984. Pseudoreplication and the design of ecological field experiments. *Ecol. Monogr.* 54:187–211.

Hurlbert, S. H., W. Loayza, and T. Moreno. 1985. *Fish-Flamingo-Plankton Interactions in the Peruvian Andes*. Unpublished manuscript.

Hustich, I. 1952. The recent climatic fluctuation in Finland and its consequences. *Fennia* 75:1–128.

Huston, M. 1979. A general hypothesis of species diversity. *Am. Nat.* 113:81–101.

Huston, M. 1980. Soil nutrients and tree species richness in Costa Rican forests. *J. Biogeogr.* 7:147–157.

Hutchins, H. E., and R. M. Lanner. 1982. The central role of Clark's nutcracker in the dispersal and establishment of Whitebark Pine. *Oecologia* 55:192–201.

Hutchinson, G. E. 1959. Homage to Santa Rosalia *or* why are there so many kinds of animals? *Am. Nat.* 93:145–159.

Hutchinson, G. E. 1961. The paradox of the plankton. *Am. Nat.* 95:137–145.

Hutchinson, G. E. 1975. Variations on a theme by Robert MacArthur. *In* M. L. Cody and J. M. Diamond, eds., *Ecology and Evolution of Communities,* pp. 492–521. Harvard University Press, Cambridge, Mass.

Hutchinson, G. E. 1978. *An Introduction to Population Ecology*. Yale University Press, New Haven, Conn.

Huxley, C. R. 1982. Ant-epiphytes of Australia. *In* R. C. Buckley, ed., *Ant-Plant Interactions in Australia,* pp. 63–73. Junk, Den Haag.

Imbrie, J., and K. P. Imbrie. 1979. *Ice Ages, Solving the Mystery*. Enslow, Hillside, N.J.

Inouye, D. W. 1977. Species structure of bumblebee communities in North America and Europe. *In* W. Mattson, ed., *The Role of Arthropods in Forest Ecosystems,* pp. 35–40. Springer-Verlag, New York.

Inouye, D. W. 1978. Resource partitioning in bumblebees: experimental studies of foraging behavior. *Ecology* 59:672–678.

Inouye, D. W., and O. R. Taylor, Jr. 1979. A temperate region plant-ant-seed predator system: consequences of extra floral nectar secretion by *Helianthella quinquenervis*. *Ecology* 60:1–7.

Inouye, R. S. 1980. Density dependent germination response by seeds of desert annuals. *Oecologia* 46:235–238.

Inouye, R. S. 1981. Interactions among unrelated species: granivorous rodents, a parasitic fungus, and a shared prey species. *Oecologia* 49:425–427.

Inouye, R. S. 1982. *Population Biology of Desert Plants*. Doctoral dissertation, University of Arizona, Tuscon.

Inouye, R. S., G. S. Byers, and J. H. Brown. 1980. Effects of predation and competition on survivorship, fecundity, and community structure of desert annuals. *Ecology* 61:1344–1351.

Istock, C. A. 1967. The evolution of complex life cycle phenomena: an ecological perspective. *Evolution* 21:592–605.

Ito, Y. 1961. Factors that affect the fluctuations of animal numbers, with special reference to insect outbreaks. *Bull. Natl. Inst. Agric. Sci.* (Tokyo) 13:57–89.

Iversen, J. 1944. *Viscum, Hedera* and *Ilex* as climatic indicators. *Geol. Foren. Stockh. Forh.* 66:463–483.

Ivker, F. B. 1972. A hierarchy of histoincompatibility in *Hydractinia echinata*. *Biol. Bull.* 143:162–174.

Iwasa, Y., and J. Roughgarden. 1985. *Interspecific Competition Among Metapopulations with Space-Limited Subpopulations*. Unpublished manuscript.

Jackson, J. B. C. 1977a. Competition on marine, hard substrata: the adaptive significance of solitary and colonial strategies. *Am. Nat.* 111:743–767.

Jackson, J. B. C. 1977b. Habitat area, colonization, and development of epibenthic community structure. *In* B. F. Keegan, P. O'Ceidigh, and P. J. S. Boaden, eds., *Biology of Benthic Organisms,* pp. 349–358. Pergamon Press, Elmsford, N.Y.

Jackson, J. B. C. 1979a. Morphological strategies of sessile animals. *In* G. Larwood and B. Rosen, eds., *Biology and Systematics of Colonial Organisms,* pp. 499–555. Academic Press, London.

Jackson, J. B. C. 1979b. Overgrowth competition between encrusting cheilostome ectoprocts in a Jamaican cryptic coral reef environment. *J. Anim. Ecol.* 48:805–823.

Jackson, J. B. C. 1981. Interspecific competition and species distributions: the ghosts of theories and data past. *Am. Zool.* 21:889–901.

Jackson, J. B. C. 1983. Biological determinants of present and past sessile animal distributions. *In* M. J. S. Tevesz and P. L. McCall, eds., *Biotic Interactions in Recent and Fossil Benthic Communities,* pp. 39–120. Plenum, New York.

Jackson, J. B. C., and L. Buss. 1975. Allelopathy and spatial competition among coral reef invertebrates. *Proc. Natl. Acad. Sci., U.S.A.* 72:5160–5163.

Jackson, P. B. N. 1961. The impact of predation, especially by the tiger-fish (*Hydrocyon vittatus* Cast.) on African freshwater fishes. *Proc. Zool. Soc. Lond.* 136:603–622.

Jaenike, J. 1978. Effect of island area on *Drosophila* population densities. *Oecologia* 36:327–332.

Jaenike, J. 1982. Environmental modification of oviposition behavior in *Drosophila*. *Am. Nat.* 119:784–802.

Jaksic, F. M. 1981. Abuse and misuse of the term "guild" in ecological studies. *Oikos* 37:397–400.

Jakway, G. E. 1958. Pleistocene Lagamorpha and Rodentia from the San Josecito Cave, Nuevo Leon, Mexico. *Trans. Kans. Acad. Sci.* 61:313–327.

James, F. C. 1983. Environmental component of morphological differentiation in birds. *Science* 221:184–186.

Janos, D. P. 1982. Tropical mycorrhizas, nutrient cycles, and plant growth. *In* S. L. Sutton, T. C. Whitmore, and A. C. Chadwick, eds., *Tropical Rain Forest: Ecology and Management,* pp. 327–345. Blackwell, Oxford.

Janssen, C., J. Ekman, and A. von Brömssen. 1981. Winter mortality and food supply in tits *Parus* spp. *Oikos* 37:313–322.

Janzen, D. H. 1966. Coevolution of mutualism between ants and acacias in Central America. *Evolution* 20:249–274.

Janzen, D. H. 1967. Fire, vegetation structure, and the ant × acacia interaction in Central America. *Ecology* 48:26–35.

Janzen, D. H. 1970. Herbivores and the number of trees in tropical forests. *Am. Nat.* 104:501–528.

Janzen, D. H. 1984. Review of *The Ecology of a Tropical Forest: Seasonal Rhythms and Long-term Changes*. E. G. Leigh, Jr., A. S. Rand, and D. M. Windsor, eds. *Am. Sci.* 72:86.

Janzen, D. H., ed. 1983. *Costa Rican Natural History*. University of Chicago Press, Chicago.

Janzen, D. H., and P. S. Martin. 1982. Neotropical anachronisms: the fruits the gomphotheres ate. *Science* 215:19–27.

Jara, H. F., and C. A. Moreno. 1984. Herbivory and structure in a midlittoral rocky community: a case of Southern Chile. *Ecology* 65:28–38.

Järvinen, O., and S. Ulfstrand. 1980. Species turnover of a continental bird fauna: Northern Europe, 1850–1970. *Oecologia* 46:186–195.

Järvinen, O., and R. A. Väisänen. 1979. Climatic changes, habitat changes, and competition: dynamics of geographic overlap in two pairs of congeneric bird species in Finland. *Oikos* 33:261–271.

Jaquard, P. 1968. Manifestation et nature des relations sociales chez les vegetaux superieurs. *Oecol. Plant.* 3:137–168.

Jenik, J. 1978. Roots and root systems in tropical trees: morphological and ecologic aspects. *In* P. B. Tomlinson and M. H. Zimmermann, eds., *Tropical Trees as Living Systems,* pp. 323–349. Cambridge University Press, Cambridge.

Jenkins, J. M. 1983. *The Native Forest Birds of Guam*. American Ornithologists' Union, Washington, D.C.

Jenny, H. 1980. *Soil Genesis with Ecological Perspectives*. Ecological Studies no. 37. Springer-Verlag, New York.

Johnson, E., and V. T. Holliday. 1980. A Plainview kill/butchering locale on the Llano Estacado—the Lubbock Lake site. *Plains Anthropol.* 25:89–111.

Johnson, F. H. 1977. Responses of walleye (*Stizostedion vitreum vitreum*) and yellow perch (*Perca flavescens*) populations to removal of white sucker

(*Catostomus commersoni*) from a Minnesota lake, 1966. *J. Fish. Res. Board Can.* 34:1633–1642.

Johnson, R. G. 1982. Brunhes-Matuyama magnetic reversal dated at 790,000 years B.P. by marine-astronomical correlation. *Quat. Res.* 17:135–147.

Johnson, W. E. 1965. On mechanisms of self-regulation of population abundance in *Oncorhynchus nerka*. *Mitt. Int. Ver. Limnol.* 13:66–87.

Johnston, M. C. 1977. Brief resume of botanical, including vegetational, features of the Chihuahuan Desert region with special emphasis on their uniqueness. *In* R. H. Wauer and D. H. Riskind, eds., *Trans. Symposium on the Biological Resources of the Chihuahuan Desert Region, United States and Mexico,* pp. 335–362. National Park Service Transactions and Proceedings Series, no. 3, Washington, D.C.

Johnston, R. F., and R. C. Fleischer. 1981. Overwinter mortality and sexual size dimorphism in the house sparrow. *Auk* 98:503–511.

Jones, C. 1983. Phytochemical variation, colonization, and insect communities: the case of bracken fern (*Pteridium aquilinum*). *In* R. Denno and M. McClure, eds., *Variable Plants and Herbivores in Natural and Managed Systems,* pp. 513–558. Academic Press, New York.

Jones, M. S. 1948. Observations and experiments on the biology of *Patella vulgata* at Port St. Mary, Isle of Man. *Proc. Trans. Liverpool Biol. Soc.* 56:60–77.

Jones, P. J., and P. Ward. 1979. A physiological basis for colony desertion by red-billed queleas (*Quelea quelea*). *J. Zool.* (*Lond.*) 189:1–19.

Jong, P. de, L. W. Aarssen, and R. Turkington. 1983. The use of contact sampling in studies of association in vegetation. *J. Ecol.* 71:545–559.

Kalela, O. 1949. Changes in geographic ranges in the avifauna of Northern and Central Europe in relation to recent changes in climate. *Bird Banding* 20:77–103.

Kalela, O. 1952. Changes in the geographic distribution of Finnish birds and mammals in relation to recent changes in climate. *Fennia* 75:38–51.

Kanchan, S., and Jayachandra. 1980. Pollen allelopathy—a new phenomenon. *New Phytol.* 84:739–746.

Kaneshiro, K. Y. 1980. Sexual isolation, speciation, and the direction of evolution. *Evolution* 34:437–444.

Kareiva, P. 1981. *Non-migratory Movement and the Distribution of Herbivorous Insects: Experiments with Plant Spacing and the Application of Diffusion Models to Mark-recapture Data.* Doctoral dissertation, Cornell University, Ithaca, N.Y.

Kareiva, P. 1982a. Experimental and mathematical analyses of herbivore movement: quantifying the influence of plant spacing and quality on foraging discrimination. *Ecol. Monogr.* 52:261–282.

Kareiva, P. 1982b. Exclusion experiments and the competitive release of insects feeding on collards. *Ecology* 63:696–704.

Kareiva, P. 1983a. Influences of vegetation texture on herbivore populations: resource concentration and herbivore movement. *In* R. F. Denno and M. S. McClure, eds., *Variable Plants and Herbivores in Natural and Managed Systems,* pp. 259–290. Academic Press, New York.

Kareiva, P. 1983b. Local movement in herbivorous insects: applying a passive diffusion model to mark-recapture field experiments. *Oecologia* 57:322–327.

Kareiva, P. 1984. Predator-prey dynamics in spatially structured populations: manipulating dispersal in a coccinellid-aphid interaction. *In* S. Levin, ed., *Lecture Notes in BioMathematics,* vol. 54, pp. 368–389. Springer-Verlag, New York.

Kareiva, P. 1985. Flea beetles and their problems finding and staying on host plants: the effects of patch size and surrounding habitat. *Ecology*. In press.

Karlson, R. H. 1980. Alternative competitive strategies in a periodically disturbed habitat. *Bull. Mar. Sci.* 30:894–900.

Karlson, R. H. 1981. A simulation study of growth inhibition and predator resistance in *Hydractinia echinata*. *Ecol. Model.* 13:29–47.

Karlson, R. H., and L. W. Buss. 1984. Competition, disturbance, and local diversity patterns of substratum-bound clonal organisms: a simulation. *Ecol. Model.* 23:13–25.

Karlson, R. H., and J. B. C. Jackson. 1981. Competitive networks and community structure: a simulation study. *Ecology* 62:670–678.

Karr, J. R., and K. E. Freemark. 1983. Habitat selection and environmental gradients: dynamics in the "stable" tropics. *Ecology* 64:1481–1494.

Karr, J. R., and F. C. James. 1975. Eco-morphological configurations and convergent evolution in species and communities. *In* M. L. Cody and J. M. Diamond, eds., *Ecology and Evolution of Communities,* pp. 258–291. Harvard University Press, Cambridge, Mass.

Kato, M., E. Hirai, and Y. Kakinuma. 1963. Further experiments on the interspecific relation in the colony formation among some hydrozoan species. *Sci. Rep. Tôhoku Univ.,* Fourth Ser. (Biol.) 29:317–325.

Kay, A. M., and M. J. Keough. 1981. Occupation of patches in the epifaunal communities on pier pilings and the bivalve *Pinna bicolor* at Edithburgh, South Australia. *Oecologia* 48:123–130.

Keast, A. 1977. Mechanisms expanding niche width

and minimizing intraspecific competition in two centrachid fishes. *In* M. K. Hecht, W. C. Steere, and B. Wallace, eds., *Evolutionary Biology,* vol. 10, pp. 333–395. Plenum, New York.

Keast, A. 1978. Trophic and spatial interrelationships in the fish species of an Ontario temperate lake. *Environ. Biol. Fishes* 3:7–31.

Keeler, K. 1981. A model of selection for facultative nonsymbiotic mutualism. *Am. Nat.* 118:488–498.

Keen, S. L., and W. E. Neil. 1980. Spatial relationships and some structuring processes in benthic intertidal communities. *J. Exp. Mar. Biol. Ecol.* 45:139–155.

Keller, B. D. 1983. Coexistence of sea urchins in seagrass meadows: an experimental analysis of competition and predation. *Ecology* 64:1581–1598.

Kelly, D. 1982. *Demography, Population Control and Stability of Short-lived Plants of Chalk Grassland.* Doctoral dissertation, University of Cambridge, Cambridge.

Kemp, W. P., and G. A. Simmons. 1979. Influence of stand factors on survival of early instar spruce budworm. *Environ. Entomol.* 8:993–996.

Kendall, Robert L. 1969. An ecological history of the Lake Victoria basin. *Ecol. Monogr.* 39:121–176.

Kendeigh, S. C. 1982. Bird populations in east central Illinois: fluctuations, variations, and development over a half-century period. *Ill. Biol. Monogr.* 52.

Keough, M. J. 1983. Patterns of recruitment of sessile invertebrates in two subtidal habitats. *J. Exp. Mar. Biol. Ecol.* 66:213–345.

Keough, M. J. 1984. Kin recognition and the spatial distribution of larvae of the bryozoan *Bugula neritina* (L.). *Evolution* 38:142–147.

Keough, M. J., and A. J. Butler. 1983. Temporal change in species number in an assemblage of sessile marine invertebrates. *J. Biogeogr.* 10:317–330.

Kerr, S. R., and N. V. Martin. 1970. Trophic dynamics of lake trout production systems. *In* J. H. Steele, ed., *Marine Food Chains,* pp. 365–376. Oliver & Boyd, Edinburgh.

Kershaw, A. P. 1976. A late-Pleistocene and Holocene pollen diagram from Lynch's Crater, North-eastern Queensland, Australia. *New Phytol.* 77:469–498.

Kershaw, K. A. 1973. *Quantitative and Dynamic Plant Ecology,* 2nd ed. Edward Arnold, London.

Keser, M., and B. R. Larson. 1984. Colonization and growth of *Ascophyllum nodosum* (Phaeophyta) in Maine. *J. Phycol.* 20:83–87.

Keser, M., R. L. Vadas, and B. R. Larson. 1981. Regrowth of *Ascophyllum nodosum* and *Fucus vesiculosus* under various harvesting regimes in Maine, U.S.A. *Bot. Mar.* 24:29–38.

Kevan, P. G., W. G. Chaloner, and D. B. O. Savile. 1975. Interrelationships of early terrestrial arthropods and plants. *Palaeontology* 18:391–417.

Kiester, A. R. 1979. Conspecifics as cues: a mechanism for habitat selection in the Panamanian grass anole (*Anolis auratus*). *Behav. Ecol. Sociobiol.* 5:323–330.

Kiester, A. R. 1983. Zoogeography of the skinks (Sauria: Scincida) of Arno Atoll, Marshall Islands. *In* A. G. J. Rhodin and K. Miyata, eds., *Advances in Herpetology and Evolutionary Biology,* pp. 359–364. Museum of Comparative Zoology, Cambridge, Mass.

King, J. E., and J. J. Saunders. 1984. Hopwood Farm paleoecology: a *Geochelone*-containing Illinoian-Sangamonian biota from the type region, central Illinois. *AMQUA Abstract,* p. 25.

Kingsolver, J. G. 1983a. Thermoregulation and flight in *Colias* butterflies: elevational patterns and mechanistic limitation. *Ecology* 64:534–545.

Kingsolver, J. G. 1983b. Ecological significance of flight activity in *Colias* butterflies: implications for reproductive strategy and population structure. *Ecology* 64:546–551.

Kira, T., K. Shinozaki, and K. Hozumi. 1969. Structure of forest canopies as related to their primary productivity. *Plant and Cell Physiol.* 10:129–142.

Kirch, P. V. 1982. Transported landscapes. *Nat. Hist.* 91:32–35.

Kirkpatrick, M. 1982. Sexual selection and the evolution of female choice. *Evolution* 36:1–12.

Klein, D. R. 1968. The introduction, increase, and crash of reindeer on St. Matthew island. *J. Wildl. Manage.* 32:350–367.

Klein, R. G. 1971. The Pleistocene prehistory of Siberia. *Quat. Res.* 1:133–161.

Klopfer, P. H. 1973. *Behavioral Aspects of Ecology,* 2nd ed. Prentice-Hall, Englewood, Cliffs, N.J.

Klopfer, P. H., and R. H. MacArthur. 1960. Niche size and faunal diversity. *Am. Nat.* 94:293–300.

Klomp, H. 1962. The influence of climate and weather on the mean density level, the fluctuations and the regulation of animal populations. *Arch. Neerl. Zool.* 15:68–109.

Klomp, H. 1980. Fluctuations and stability in Great Tit populations. *Ardea* 68:205–224.

Knoll, A. H. 1984. Patterns of extinction in fossil record of vascular plants. *In* M. Nitecki, ed., *Extinctions,* pp. 21–67. University of Chicago Press, Chicago.

Knoll, A. H., K. J. Niklas, and B. H. Tiffney. 1979. Phanerozoic land plant diversity in North America. *Science* 206:1400–1402.

Knoll, A. H., K. J. Niklas, P. Gensel, and B. H. Tiffney. 1984. Character diversification and patterns of evolution in early vascular plants. *Paleobiology* 10:134–147.

Knoll, A. H., and G. W. Rothwell. 1981. Paleobotany: perspectives in 1980. *Paleobiology* 7:7–35.

Koch, A. L. 1974. Coexistence resulting from an alternation of density independent and density dependent growth. *J. Theor. Biol.* 44:373–386.

Kohler, C. C., and J. J. Ney. 1981. Consequences of an alewife die-off to fish and zooplankton in a reservoir. *Trans. Am. Fish. Soc.* 110:360–369.

Kohn, A. 1959. The ecology of *Conus* in Hawaii. *Ecol. Monogr.* 29:47–90.

Kohn, A. 1971. Diversity, utilization of resources, and adaptive radiation in shallow-water marine invertebrates of tropical oceanic islands. *Limnol. Oceanogr.* 16:332–348.

Koopman, K. F. 1957. Evolution in the genus *Myzomela* (Aves: Meliphagidac). *Evolution* 74:49–72.

Kostitzin, V. A. 1934. *Symbiose, Parasitisme et Evolution*. Hermann, Paris.

Kotler, B. P. 1984. Predation risk and the structure of desert rodent communities. *Ecology* 65:689–701.

Kramer, R. J. 1971. *Hawaiian Land Mammals*. Tuttle, Tokyo.

Krassilov, A. 1975. *Paleoecology of Terrestrial Plants*. Wiley, New York.

Krebs, C. J., B. Keller, and R. Tamarin. 1969. *Microtus* population biology: demographic changes in fluctuation populations of *M. ochrogaster* and *M. pennsylvanicus* in southern Indiana. *Ecology* 50:587–607.

Krebs, J. R., D. W. Stevens, and W. J. Sutherland. 1983. Perspectives in optimal foraging. *In* A. H. Brush and G. A. Clark, Jr., eds., *Perspectives in Ornithology*, pp. 165–221. Cambridge University Press, Cambridge.

Kruger, F. J. 1979. South African heathlands. *In* R. L. Specht, ed., *Heathlands and Related Shrublands: Descriptive Studies*, pp. 19–80. Ecosystems of the World, vol. 9A. Elsevier, Amsterdam.

Kruger, F. J. 1981. Seasonal growth and flowering rhythms: South African heathlands. *In* R. L. Specht, ed., *Heathlands and Related Shrublands*, pp. 1–4. Ecosystems of the World, vol. 9B. Elsevier, Amsterdam.

Kruger, F. J., D. T. Mitchell, and J. U. M. Jarvis, eds. 1983. *Mediterranean-Type Ecosystems: the Role of Nutrients*. Ecological Studies no. 43. Springer-Verlag, Berlin, New York.

Kruger, F. J., and H. C. Taylor. 1979. Plant species diversity in Cape fynbos: gamma and delta diversity. *Vegetatio* 41:85–93.

Kullman, L., 1979. Change and stability in the altitude of the birch tree-limit in the southern Swedish Scandes 1915–1975. *Acta Phytogeogr. Suec.* no. 65.

Kullman, L. 1981a. Some aspects of the ecology of the Scandinavian subalpine birch forest belt. *Wahlenbergia* 7:99–112.

Kullman, L., 1981b. Recent tree-limit dynamics of Scots pine (*Pinus sylvestris* L.) in the southern Swedish Scandes. *Wahlenbergia* 8.

Kullman, L. 1983. Past and present tree-lines of different species in the Handölan Valley, central Sweden. *In* P. Morisset and S. Payette, eds., *Tree-line Ecology*, pp. 25–45. Collection Nordicana, no. 47. Université Laval, Quebec.

Kurtén, B., and E. Anderson. 1980. *Pleistocene Mammals of North America*. Columbia University Press, New York.

Kutzbach, I. E. 1981. Monsoon climate of the early Holocene: climate experiment with the earth's orbital parameters for 9000 years ago. *Science* 214:59–61.

Kuykendall, G. E., and A. F. Day. 1948. *Hawaii: a History from Polynesian Kingdom to American Commonwealth*. Prentice-Hall, Englewood Cliffs, N.J.

Labeyrie, V. 1971. Trophic relations and sex meetings in insects. *Acta Phytopath. Acad. Sci. Hung.* 6:229–234.

Lack, D. 1933. Habitat selection in birds, with special reference to the effects of afforestation on the Breckland avifauna. *J. Anim. Ecol.* 2:239–262.

Lack, D. 1944. Ecological aspects of species formation in passerine birds. *Ibis* 86:260–286.

Lack, D. 1945. The Galápagos finches (Geospizimae): a study in variation. *Occas. Pap. Calif. Acad. Sci.* 21:1–159.

Lack, D. 1947. *Darwin's Finches*. Cambridge University Press, Cambridge.

Lack, D. 1954. *The Natural Regulation of Animal Numbers*. Oxford University Press, Oxford.

Lack, D. 1968. *Ecological Adaptations for Breeding in Birds*. Methuen, London.

Lack, D. 1971. *Ecological Isolation in Birds*. Harvard University Press, Cambridge, Mass.

Lack, D. 1976. *Island Biology Illustrated by the Land Birds of Jamaica*. Blackwell, Oxford.

Lack, D., and H. N. Southern. 1949. Birds on Tenerife. *Ibis* 91:607–626.

Laidler, D., E. J. Moll, B. M. Campbell, and J. Glyphis. 1981. Phytosociological studies on Table Mountain, S. Africa. 2. The first table. *J. S. Afr. Bot.* 44:291–295.

Laine, K. J., and P. Niemelä. 1980. The influence of ants on the survival of mountain birches during an *Oporinia autumnata* (Lep., Geometridae) outbreak. *Oecologia* 47:39–42.

LaMarche, V. C., Jr. 1973. Holocene climatic fluctuations inferred from treeline fluctuations in White Mountains, California. *Quat. Res.* 3:632–660.

Lamb, H. H. 1977. *Climate—Present, Past and Future. Vol 2. Climatic History and the Future.* Metheun, London.

Land, L. S., E. L. Lundelius, Jr., and S. Valastro. 1980. Isotopic ecology of deer bones. *Palaeogeogr. Palaeoclimatol. Palaeoecol.* 32:143–159.

Landau, H. G. 1951. On dominance relations and the structure of animal societies. I. Effect of inherent characteristics. *Bull. Math. Biophys.* 13:1–19.

Lande, R. 1976. Natural selection and random genetic drift in phenotypic evolution. *Evolution* 30:314–334.

Lande, R. 1979. Quantitative analysis of multivariate evolution applied to brain: body size allometry. *Evolution* 33:234–255.

Lande, R. 1980a. Genetic variation and phenotypic evolution during allopatric speciation. *Am. Nat.* 116:463–479.

Lande, R. 1980b. Sexual dimorphism, sexual selection, and adaptation in polygenic characters. *Evolution* 34:292–305.

Lande, R. 1981. Models of speciation by sexual selection on polygenic traits. *Proc. Natl. Acad. Sci., U.S.A.* 78:3721–3725.

Lande, R. 1982. Rapid origin of sexual isolation and character divergence in a cline. *Evolution* 36:213–223.

Lande, R., and S. J. Arnold. 1983. The measurement of selection on correlated characters. *Evolution* 37:1210–1226.

Lang, G. E., and D. H. Knight. 1983. Tree growth, mortality, recruitment, and canopy gap formation during a 10-year period in a tropical forest. *Ecology* 64:1075–1080.

Lang, J. C. 1973. Interspecific aggression by scleractinian corals. II. Why the race is not only to the swift. *Bull. Mar. Sci.* 23:260–279.

Lanner, R. M., and T. R. Van Devender. 1981. Late Pleistocene piñon pines in the Chihuahuan Desert. *Quat. Res.* 15:278–290.

Larson, R. J. 1980. Competition, habitat selection, and the bathymetric segregation of two rockfish (*Sebastes*) species. *Ecol. Monogr.* 50:221–239.

Laughlin, D. R., and E. E. Werner. 1980. Resource partitioning in two coexisting sunfish: pumpkinseed (*Lepomis sibbosus*) and northern longear sunfish (*Lepomis megalotis peltastes*). *Can. J. Fish. Aqua. Sci.* 37:1411–1420.

Laurence, G. C. 1981. Overview—Modelling—an esoteric or potentially utilitarian approach to understanding larval fish dynamics? *In The Early Life History of Fish: Recent Studies*. Second ICES Symposium. Conseil International pour l'Exploration de la Mer, Copenhagen.

Lawes, J., and J. Gilbert. 1880. Agricultural, botanical and chemical results of experiments on the mixed herbage of permanent grassland, conducted for many years in succession on the same land. I. *Philos. Trans. R. Soc.* 171:189–416.

Lawes, J., J. Gilbert, and M. Masters. 1882. Agricultural, botanical, and chemical results of experiments on the mixed herbage of permanent meadow, conducted for more than twenty years on the same land. II. The botanical results. *Philos. Trans. R. Soc.* 173:1181–1413.

Lawlor, L. R. 1979. Direct and indirect effects of *n*-species competition. *Oecologia* 43:355–364.

Lawton, J. H. 1978. Host plant influences on insect diversity: the effects of space and time. *In* L. A. Mound and N. Waloff, eds., *Diversity of Insect Faunas,* pp. 105–125. Blackwell, Oxford.

Lawton, J. H. 1983. Herbivore community organization: general models and specific tests with phytophagous insects. *In* P. W. Price, C. N. Slobodchikoff, and W. Gaud, eds., *A New Ecology: Novel Approaches to Interactive Systems,* pp. 206–227. Wiley, New York.

Lawton, J. H., and M. P. Hassell. 1981. Asymmetrical competition in insects. *Nature* 289:793–795.

Lawton, J. H., and S. McNeill. 1979. Between the devil and the deep blue sea: on the problem of being a herbivore. *Symp. Br. Ecol. Soc.* 20:223–244.

Lawton, J. H., and D. R. Strong. 1981. Community patterns and competition in folivorous insects. *Am. Nat.* 188:317–338.

Lazell, J. 1972. The *Anolis* (Sauria, Iguanidae) of the Lesser Antilles. *Bull. Mus. Compar. Zool. Harv. Univ.* 143:1–115.

LeCren, E. D. 1973. The population dynamics of young trout (*Salmo trutta*) in relation to density and territorial behaviour. *Rapp. P-V. Reun. Cons. Int. Explor. Mer.* 164:241–246.

Leigh, E. G. 1975. Population fluctuations, community stability and environmental variability. *In* M. L. Cody and J. M. Diamond, eds., *Ecology and Evolution of Communities,* pp. 51–73. Harvard University Press, Cambridge, Mass.

Leigh, E. G., Jr., A. S. Rand, and D. M. Windsor. 1982. *The Ecology of a Tropical Forest: Seasonal Rhythms and Long-Term Changes.* Smithsonian Institution Press, Washington, D.C.

Levi, H. W. 1968. The spider genera *Gea* and *Argiope* in America (Araneae: Araneidae). *Bull. Mus. Comp. Zool. Harv. Univ.* 136:319–352.

Levi, H. W. 1977a. The American orb-weaver genera *Cyclosa, Metazygia* and *Eustala* north of Mexico (Araneae: Araneidae). *Bull. Mus. Comp. Zool.* 148:61–127.

Levi, H. W. 1977b. The orb-weaver genera *Metepeira, Kaira* and *Aculepeira* in America north of Mexico (Araneae: Araneidae). *Bull. Mus. Comp. Zool.* 148:185–238.

Levi, H. W. 1978. The American orb-weaver genera *Colphepeira, Micrathena* and *Gasteracantha* north of Mexico (Araneae, Araneidae). *Bull. Mus. Comp. Zool.* 148:417–442.

Levin, D. A. 1975. Pest pressure and recombination systems in plants. *Am. Nat.* 109:437–452.

Levin, S. A. 1970. Community equilibria and stability, and an extension of the competitive exclusion principle. *Am. Nat.* 104:413–423.

Levin, S. A. 1974. Dispersion and population interactions. *Am. Nat.* 108:207–228.

Levin, S. A. 1976. Spatial patterning and the structure of ecological communities. *In* S. A. Levin, ed., *Some Mathematical Questions in Biology*. Lectures on Mathematics in the Life Sciences, vol. 8. American Mathematical Society, Providence, R.I.

Levin, S. A. 1978. Population models and community structure in heterogeneous environments. *In* S. A. Levin, ed., *Studies in Mathematical Biology. Part II. Populations and Communities,* pp. 439–476. The Mathematical Association of America, Washington, D.C.

Levin, S. A. 1979. Mechanisms for the generation and maintenance of diversity in ecological communities. *In* R. W. Hiorns and D. Cooke, eds., *The Mathematical Theory of the Dynamics of Populations II,* pp. 173–194. Academic Press, London.

Levin, S. A., and L. Segel. 1976. Hypothesis for the origin of plankton patchiness. *Nature* 259:659.

Levin, S. A., and J. D. Udovic. 1977. A mathematical model of coevolving populations. *Am. Nat.* 111:657–675.

Levine, S. 1976. Competitive interactions in ecosystems. *Am. Nat.* 110:903–910.

Levins, R. 1968. *Evolution in Changing Environments*. Princeton University Press, Princeton, N.J.

Levins, R. 1974. Qualitative analysis of partially specified systems. *Ann. N. Y. Acad. Sci.* 231:123–138.

Levins, R. 1975. Evolution in communities near equilibrium. *In* M. L. Cody and J. M. Diamond, eds., *Ecology and Evolution of Communities,* pp. 16–50. Harvard University Press, Cambridge, Mass.

Levins, R. 1979. Coexistence in a variable environment. *Am. Nat.* 114:765–783.

Levins, R., and D. Culver. 1971. Regional coexistence of species and competition between rare species. *Proc. Natl. Acad. Sci., U.S.A.* 68:1246–1248.

Levinton, J. S. 1979. A theory of diversity equilibrium and morphological evolution. *Science* 204:335–336.

Levyns, M. R. 1964. Migrations and origins of the Cape flora. *Trans. R. Soc. S. Afr.* 37:85–107.

Lewin, R. 1983. Santa Rosalia was a goat; predators and hurricanes change ecology. *Science* 221:636–639, 737–740.

Lewin, R. 1984. Parks: how big is big enough? *Science* 225:611–612.

Lewis, A. C. 1984. Plant quality and grasshopper feeding: effects of sunflower condition on preference and performance in *Melanoplus differentialis. Ecology* 65:836–843.

Lewis, C. A. 1978. A review of substratum selection in free-living and symbiotic cirripeds. *In* F. S. Chia and M. E. Rice, eds., *Settlement and Metamorphosis of Marine Invertebrate Larvae,* pp. 207–218. Elsevier, New York.

Lewis, D. H. 1973. The relevance of symbiosis to taxonomy and ecology, with particular reference to mutualistic symbioses and the exploitation of marginal habitats. *In* V. H. Heywood, ed., *Taxonomy and Ecology,* pp. 151–172. Academic Press, London.

Lewontin, R. C. 1965. Comment. *In* H. G. Baker and G. L. Stebbins, eds., *The Genetics of Colonizing Species,* pp. 481–484. Academic Press, New York.

Li, H. W., P. B. Moyle, and R. L. Garrett. 1976. Effect of the introduction of the Mississippi silverside (*Menidia audens*) on the growth of black crappie (*Pomoxis nigromaculatus*) and white crappie (*P. annularis*) in Clear Lake, California. *Trans. Am. Fish. Soc.* 105:404–408.

Licht, P. 1974. Response of *Anolis* lizards to food supplementation in nature. *Copeia* 1974:215–221.

Licht, P., and G. Gorman. 1970. Reproductive and fat cycles in Caribbean *Anolis* lizards. *Univ. Calif. Publ. Zool.* 95:1–52.

Liddell, W. D., and C. E. Brett. 1982. Skeletal overgrowths among epizoans from the Silurian (Wenlockian) Waldron Shale. *Paleobiology* 8:67–78.

Lidgard, S., and J. B. C. Jackson. 1982. How to be an abundant encrusting bryozoan. *Geol. Soc. Am. Abstr.* 14:547.

Lilienfeld, A. M., and D. E. Lilienfield. 1980. *Foundations of Epidemiology,* 2nd ed. Oxford University Press, New York.

Lim, H.-K., and D. Heyneman. 1972. Intramolluscan inter-trematode antagonism: a review of the factors influencing the host-parasite system and its possible role in biological control. *Adv. Parasitol.* 10:191–268.

Lindsey, A. A. 1961. Vegetation of the drainage-aeration classes of northern Indiana soils in 1830. *Ecology* 42:432–436.

Lindsay, S. L. 1958. Food preferences of Drosophila larvae. *Am. Nat.* 92:279–285.

Littlejohn, M. 1981. Reproductive isolation: a critical review. *In* W. R. Atcheley and D. S. Woodruff, eds, *Evolution and Speciation*, pp. 298–334. Cambridge University Press, Cambridge.

Littler, M. M., and D. S. Littler. 1980. The evolution of thallus form and survival strategies in benthic marine macroalgae: field and laboratory tests of a functional form model. *Am. Nat.* 116: 25–44.

Littler, M. M., and D. S. Littler. 1984. Relationships between macroalgal functional form groups and substrata stability in a subtropical rocky intertidal system. *J. Exp. Mar. Biol. Ecol.* 74:13–34.

Livingstone, D. A. 1967. Postglacial vegetation of the Ruwenjori Mountains in equatorial Africa. *Ecol. Monogr.* 37:25–52.

Livingstone, D. A. 1975. Late Quaternary climatic change in Africa. *Annu. Rev. Ecol. Syst.* 6:249–280.

Lomolino, M. V. 1984. Immigrant selection, predation, and the distributions of *Microtus pennsylvanicus* and *Blarina brevicauda* on islands. *Am. Nat.* 123:468–483.

Long, J. 1981. *Introduced Birds of the World*. David and Charles, London.

Loosanoff, V. L. 1966. Time and intensity of settling of the oyster, *Crassostrea virginica*, in Long Island Sound. *Biol. Bull.* 130:211–227.

Lord, F. T., and A. W. MacPhee. 1953. The influence of spray programs on the fauna of apple orchards in Nova Scotia. VI. Low temperatures and the natural control of the oystershell scale, *Lepidosaphes ulmi* (L.) (Homoptera: Coccidae). *Can. Entomol.* 85:282–291.

Loreau, M. Niche differentiation and community organization in forest carabid communities. In preparation.

Losey, G. S., Jr. 1972. The ecological importance of cleaning symbiosis. *Copeia* 1972:820–833.

Losey, G. S., Jr. 1974. Cleaning symbiosis in Puerto Rico with comparison to the tropical Pacific. *Copeia* 1974:960–970.

Losey, G. S., Jr. 1978. The symbiotic behavior of fishes. *In* D. I. Mostofsky, ed., *The Behavior of Fish and Other Aquatic Animals*, pp. 1–31. Academic Press, New York.

Lotka, A. J. 1925. *Elements of Physical Biology*. Williams & Wilkins, Baltimore.

Loucks, O. L. 1970. Evolution of diversity, efficiency, and community stability. *Am. Zool.* 10:17–25.

Louda, S. M. 1982a. Inflorescence spiders: a cost/benefit analysis for the host plant, *Haplopappus venetus* Blake (Asteraceae). *Oecologia* 55:185–191.

Louda, S. M. 1982b. Limitation of the recruitment of the shrub *Haplopappus squarrosus* (Asteraceae) by flower- and seed-feeding insects. *J. Ecol.* 70:43–53.

Lovejoy, T. E., J. M. Rankin, R. O. Bierregaard, K. S. Brown, L. H. Emmons, and M. E. Van der Voort. 1984. Ecosystem decay of Amazon forest remnants. *In* M. H. Nitecki, ed., *Extinctions*, pp. 295–325. University of Chicago Press, Chicago.

Loveless, M. D., and J. L. Hamrick. 1984. Ecological determinants of genetic structure in plant populations. *Annu. Rev. Ecol. Syst.* 15:65–95.

Loveless, M. D., and J. L. Hamrick. 1985. Distribution of genetic variation in tropical tree species. *Rev. Biol. Trop.*, in press.

Lowe-McConnell, R. H. 1969. *Speciation in Tropical Environments*. Academic Press, New York.

Lubchenco, J. 1978. Plant species diversity in a marine intertidal community: importance of herbivore food preference and algal competitive abilities. *Am. Nat.* 112:23–39.

Lubchenco, J. 1980. Algal zonation in the New England rocky intertidal community: an experimental analysis. *Ecology* 61:333–344.

Lubchenco, J. 1982. Effects of grazers and algal competitors on fucoid colonization in tide pools. *J. Phycol.* 18:544–550.

Lubchenco, J. 1983. *Littorina* and *Fucus:* effects of herbivores, substratum heterogeneity, and plant escapes during succession. *Ecology* 64:1116–1123.

Lubchenco, J., and J. Cubit. 1980. Heteromorphic life histories of certain marine algae as adaptations to variations in herbivory. *Ecology* 61:676–687.

Lubchenco, J., and S. D. Gaines. 1981. A unified approach to marine plant-herbivore interactions. I. Populations and communities. *Annu. Rev. Ecol. Syst.* 12:405–437.

Lubchenco, J., and B. A. Menge. 1978. Community development and persistence in a low rocky intertidal zone. *Ecol. Monogr.* 48:67–94.

Luck, R. F., and H. Podoler. 1985. Competitive exclusion of *Aphytis lingnaneasis* by *A. melinus:* potential role of host size. *Ecology* 66:904–913.

Luckinbill, L. S. 1979. Selection and the r/K continuum in experimental populations of protozoa. *Am. Nat.* 113:427–437.

Lundelius, E. L., Jr. 1967. Late Pleistocene and Holocene history of central Texas. *In* P. S. Martin and H. E. Wright, Jr., eds., *Pleistocene Extinctions—the Search for a Cause*, pp. 288–319. Yale University Press, New Haven, Conn.

Lundelius, E. L., Jr. 1983. Climatic implications of late Pleistocene and Holocene faunal associations in Australia. *Alcheringa* 7:125–149.

Lundelius, E. L., Jr., R. W. Graham, E. Anderson, J. Guilday, J. A. Holman, D. Steadman, and S. D.

Webb. 1983. Terrestrial vertebrate faunas. *In* S. C. Porter, ed., *Late-Quaternary Environments of the United States. Vol. 1. The Late Pleistocene,* pp. 311–353. University of Minnesota Press, Minneapolis.

Lyell, C. 1830–1833. *Principles of Geology*. Murray, London.

Lynch, M. 1978. Complex interactions between natural coexploiters—*Daphnia* and *Ceriodaphnia. Ecology* 59:552–564.

Lynch, M. 1979. Predation, competition, and zooplankton community structure: an experimental study. *Limnol. Oceanogr.* 24:253–272.

MacArthur, J. W. 1975. Environmental fluctuations and species diversity. *In* M. L. Cody and J. M. Diamond, eds., *Ecology and Evolution of Communities,* pp. 74–80. Harvard University Press, Cambridge, Mass.

MacArthur, R. H. 1958. Population ecology of some warblers of northeastern coniferous forests. *Ecology* 39:599–619.

MacArthur, R. H. 1972a. *Geographical Ecology*. Harper & Row, New York.

MacArthur, R. H. 1972b. Coexistence of species. *In* J. Behnke, ed., *Challenging Biological Problems,* pp. 253–259. Oxford University Press, New York.

MacArthur, R. H., and R. Levins. 1967. The limiting similarity, convergence and divergence of coexisting species. *Am. Nat.* 101:377–385.

MacArthur, R. H., and E. O. Wilson. 1967. *The Theory of Island Biogeography*. Princeton University Press, Princeton, N.J.

MacGarvin, M. 1982. Species-area relationships of insects on host plants: herbivores on rosebay willow herb. *J. Anim. Ecol.* 51:207–223.

Mack, R. N. 1984. Invaders at home on the range. *Nat. Hist.* 93(2):40–47.

Maclean, G. L. 1976. Arid-zone ornithology in Africa and South America. *Proc. XVIth Internat. Ornithol. Congr.* (Canberra), pp. 468–480.

MacLean, J., and J. J. Magnuson. 1977. Species interactions in period communities. *J. Fish. Res. Board Can.* 34:1941–1951.

MacMahon, J. A. 1979. North American deserts: their floral and faunal components. *In* R. A. Perry and D. W. Goodall, eds., *Arid Land Ecosystems: Structure, Functioning and Management, vol. 1.* Cambridge University Press, Cambridge.

Madison, M. 1977. Vascular epiphytes: their systematic occurrence and salient features. *Selbyana* 2:1–13.

Maelzer, D. A. 1970. The regression of log N_{n+1} on log N as a test of density dependence: an exercise with computer-constructed density-independent populations. *Ecology* 51:810–822.

Magee, J. D. 1965. The breeding distribution of the Stonechat in Britain and the cause of its decline. *Bird Study* 12:83–89.

Magnan, P., and G. J. Fitzgerald. 1982. Resource partitioning between brook trout (*Salvelinus fontinalis* Mitchill) and creek chub (*Semotilus atromaculatus* Mitchill) in selected oligotrophic lakes of southern Quebec. *Can. J. Zool.* 60:1612–1617.

Magnuson, J. J., J. P. Baker, and E. J. Rahel. 1984. A critical assessment of effects of acidification on fisheries in North America. *Phil. Trans. R. Soc. Lond. B* 305:501–516.

Maguire, L. A., and J. W. Porter. 1977. A spatial model of growth and competition strategies in coral communities. *Ecol. Model.* 3:249–271.

Manasse, R. S., and H. F. Howe. 1983. Competition for dispersal agents among tropical trees: influences of neighbors. *Oecologia* 59:185–190.

Mangan, R. L. 1982. Adaptations to competition in cactus breeding *Drosophila*. *In* J. S. F. Barker and W. T. Starmer, eds., *Ecological Genetics and Evolution: the Cactus-Yeast-Drosophila Model System,* pp. 257–272. Academic Press, Sydney.

Margulis, L. 1975. Symbiotic theory of the origin of eukaryotic organelles: criteria for proof. *Symp. Soc. Exp. Biol.* 29:21–38.

Mariscal, R. N. 1974. Nematocysts. *In* L. Muscatine and H. M. Lenhoff, eds., *Coelenterate Biology: Reviews and Perspectives,* pp. 129–178. Academic Press, New York.

Markow, T. A. 1982. Mating systems of cactophilic *Drosophila*. *In* J. S. F. Barker and W. T. Starmer, eds., *Ecological Genetics and Evolution: the Cactus-Yeast-Drosophila Model System,* pp. 273–287. Academic Press, Sydney.

Marks, P. L. 1974. The role of pin cherry (*Prunus pennsylvanica*) in the maintenance of stability in northern hardwood ecosystems. *Ecol. Monogr.* 44:73–88.

Marks, P. L. 1983. On the origin of the field plants of the northeastern United States. *Am. Nat.* 122:210–228.

Marshall, L. G., S. D. Webb, J. J. Sepkoski, and D. M. Raup. 1982. Mammalian evolution and the great American Interchange. *Science* 215:1351–1357.

Martin, E. P. 1956. A population study of the prairie vole (*Microtus ochrogaster*) in northeastern Kansas. *Univ. Kans. Mus. Nat. Hist. Publ.* 8:361–416.

Martin, E. V. 1943. Studies of evaporation and transpiration under controlled conditions. *Carnegie Inst. Wash. Publ.* no. 550.

Martin, N. V. 1966. The significance of food habits in the biology, exploitation, and management of Algonquin Park, Ontario, lake trout. *Trans. Am. Fish. Soc.* 95:415–422.

Martin, P. J. 1982. Digestive and grazing strategies of animals in the arctic steppe. *In* D. M. Hopkins, J. V. Matthews, Jr., C. E. Schweger, and S. B. Young, eds., *Paleoecology of Beringia,* pp. 259–266. Academic Press, New York.

Martin, P. S. 1958. Pleistocene ecology and biogeography of North America. *In* C. L. Hubbs, ed., *Zoogeography,* pp. 375–420. American Association for the Advancement of Science, Washington, D.C.

Martin, P. S. 1984. Prehistoric overkill: the global model. *In* P. S. Martin and R. G. Klein, eds., *Quaternary Extinctions: a Prehistoric Revolution,* pp. 354–403. University of Arizona Press, Tucson.

Martin, P. S., and R. G. Klein, eds. 1984. *Quaternary Extinctions: a Prehistoric Revolution*. University of Arizona Press, Tucson.

May, R. M. 1972. Limit cycles in predator-prey communities. *Science* 177:900–902.

May, R. M. 1973a. *Stability and Complexity in Model Ecosystems*. Princeton University Press, Princeton, N.J.

May, R. M. 1973b. On relationships among various types of population models. *Am. Nat.* 107:46–57.

May, R. M. 1974a. Biological populations with nonoverlapping generations: stable points, stable cycles, and chaos. *Science* 186:645–647.

May, R. M. 1974b. On the theory of niche overlap. *Theor. Popul. Biol.* 5:297–332.

May, R. M. 1975a. Biological populations obeying difference equations: stable points, stable cycles, and chaos. *J. Theor. Biol.* 51:511–524.

May, R. M. 1975b. Patterns of species abundance and diversity. *In* M. L. Cody and J. M. Diamond, eds., *Ecology and Evolution of Communities,* pp. 81–120. Harvard University Press, Cambridge, Mass.

May, R. M. 1978a. Host parasitoid systems in patchy environments: a phenomenological model. *J. Anim. Ecol.* 47:833–843.

May, R. M. 1978b. The dynamics and diversity of insect faunas. *In* L. A. Mound and N. Waloff, eds., *Diversity of Insect Faunas,* pp. 188–204. Blackwell, Oxford.

May, R. M. 1980. Nonlinear phenomena in ecology and epidemiology. *Ann. N. Y. Acad. Sci.* 356:267–281.

May, R. M. 1981a. Models for single populations. *In* R. M. May, ed., *Theoretical Ecology: Principles and Applications,* pp. 5–29. Sinauer, Sunderland, Mass.

May, R. M. 1981b. Models for two interacting populations. *In* R. M. May, ed., *Theoretical Ecology: Principles and Applications,* pp. 78–104. Sinauer, Sunderland, Mass.

May, R. M. 1981c. Modeling recolonization by neotropical migrants in habitats with changing patch structure, with notes on the age structure of populations. *In* R. L. Burgess and D. M. Sharpe, eds., *Forest Island Dynamics in Man-Dominated Landscapes,* pp. 207–213. Springer-Verlag, Berlin.

May, R. M. 1982. Introduction. *In* R. M. Anderson and R. M. May, eds., *Population Biology of Infectious Diseases,* pp. 1–12. Springer-Verlag, Berlin.

May, R. M., and R. M. Anderson. 1978. Regulation and stability of host-parasite population interactions. II. Destabilizing processes. *J. Anim. Ecol.* 47:249–267.

May, R. M., and R. M. Anderson. 1979. Population biology of infectious diseases. Part II. *Nature* 280:455–461.

May, R. M., and R. M. Anderson. 1983a. Epidemiology and genetics in the coevolution of parasites and hosts. *Proc. R. Soc. Lond. B* 219:281–313.

May, R. M., and R. M. Anderson. 1983b. Parasite-host coevolution. *In* D. J. Futuyma and M. Slatkin, eds., *Coevolution,* pp. 186–206. Sinauer, Sunderland, Mass.

May, R. M., J. R. Beddington, C. W. Clark, S. J. Holt, and R. M. Laws. 1979. Management of multispecies fisheries. *Science* 205:267–277.

May, R. M., and W. J. Leonard. 1975. Nonlinear aspects of competition between three species. *SIAM J. Appl. Math.* 29:243–253.

May, R. M., and R. H. MacArthur. 1972. Niche overlap as a function of environmental variability. *Proc. Natl. Acad. Sci., U.S.A.* 69:1109–1113.

May, R. M., and G. F. Oster. 1976. Bifurcations and dynamic complexity in simple ecological models. *Am. Nat.* 110:573–599.

Maynard Smith, J. 1966. Sympatric speciation. *Am. Nat.* 100:637–650.

Maynard Smith, J. 1974. *Models in Ecology*. Cambridge University Press, Cambridge.

Maynard Smith, J. 1978. Optimization theory in evolution. *Annu. Rev. Ecol. Syst.* 9:31–55.

Mayr, E. 1942. *Systematics and the Origin of Species*. Columbia University Press, New York.

Mayr, E. 1944. Wallace's line in the light of recent zoogeographic studies. *Q. Rev. Biol.* 19:1–14.

Mayr, E. 1963. *Animal Species and Evolution*. Harvard University Press, Cambridge, Mass.

Mayr, E. 1966. The nature of colonizations in birds. *In* H. G. Baker and G. L. Stebbins, eds., *Genetics of Colonizing Species,* pp. 29–43. Academic Press, New York.

Mayr, E. 1972. Sexual selection and natural selection. *In* B. Campbell, ed., *Sexual Selection and the Descent of Man,* pp. 87–104. Aldine, Chicago.

Mayr, E. 1982. *The Growth of Biological Thought*. Harvard University Press, Cambridge, Mass.

McCais, R. S. 1980. Effect of sea-run alewives on

rainbow trout and brown trout in reclaimed ponds. *Prog. Fish-Cult.* 42:59–63.

McCauley, E., and F. Briand. 1979. Zooplankton grazing and phytoplankton species richness: field tests of the predation hypothesis. *Limnol. Oceanogr.* 24:243–252.

McComas, S. R., and R. W. Drenner. 1982. Species replacement in a reservoir fish community: silverside feeding mechanics and competition. *Can. J. Fish. Aquat. Sci.* 39:815–821.

McFadden, C. S., M. McFarland, and L. W. Buss. 1984. Biology of hydractiniid hydroids. I. Colony ontogeny in *Hydractinia echinata*. *Biol. Bull.* 166:54–67.

McFadden, J. T. 1969. Dynamics and regulation of salmonid populations in streams. *In* T. G. Northcote, ed., *Symposium on Salmon and Trout in Streams,* pp. 313–329. Institute on Fisheries, University of British Columbia, Vancouver.

McIntosh, R. P. 1980. The background and some current problems of theoretical ecology. *Synthese* 43:195–255.

McIntosh, R. P. 1981. Succession and ecological theory. *In* D. C. West, H. H. Shugart, and D. B. Botkin, eds., *Forest Succession Concepts and Application,* pp. 10–23. Springer-Verlag, New York.

McKeown, S. 1978. *Hawaiian Reptiles and Amphibians*. Oriental, Honolulu.

McMurtrie, R. 1978. Persistence and stability of single-species and prey-predator systems in spatially heterogeneous environments. *Math. Biosci.* 39:11–51.

McNaughton, S. J. 1983a. Serengeti grassland ecology: the role of composite environmental factors and contingency in community organization. *Ecol. Monogr.* 53:291–320.

McNaughton, S. J. 1983b. Compensatory plant growth as a response to herbivory. *Oikos* 40:329–336.

McNaughton, S. J. 1984. Ecology of a grazing ecosystem: the Serengeti. *Ecol. Monogr.,* in press.

McNeill, S., and T. R. E. Southwood. 1978. The role of nitrogen in the development of insect/plant relationships. *In* J. Harborne, ed., *Biochemical Aspects of Plant and Animal Coevolution,* pp. 77–98. Academic Press, New York.

McNeill, W. H. 1976. *Plagues and Peoples*. Doubleday (Anchor Books), Garden City, N.Y.

Mead, J. I., and A. M. Phillips. 1981. The late Pleistocene and Holocene fauna and flora of Vulture Cave, Grand Canyon, Arizona. *Southwest. Nat.* 26:257–288.

Mead, J. I., and T. R. Van Devender. 1981. Late Holocene diet of *Bassariscus astutus* in the Grand Canyon, Arizona. *J. Mammal.* 62:439–442.

Mead, J. I., T. R. Van Devender, and K. L. Cole. 1983. Late Quaternary small mammals from Sonoran Desert packrat middens, Arizona and California. *J. Mammal.* 64:173–180.

Medway, Lord. 1972. Phenology of a tropical rain forest in Malaya. *Biol. J. Linn. Soc.* 4:117–146.

Meents, J. K., J. Rice, B. W. Anderson, and R. D. Ohmart. 1983. Nonlinear relationships between birds and vegetation. *Ecology* 64:1022–1027.

Menge, B. A. 1976. Organization of the New England rocky intertidal community: role of predation, competition, and environmental heterogeneity. *Ecol. Monogr.* 46:355–393.

Menge, B. A. 1978a. Predation intensity in a rocky intertidal community: relation between predator foraging activity and environmental harshness. *Oecologia* 34:1–16.

Menge, B. A. 1978b. Predation intensity in a rocky intertidal community: effect of an algal canopy, wave action and desiccation on predator feeding rates. *Oecologia* 34:17–35.

Menge, B. A. 1982. Effects of feeding on the environment: Asteroidea. *In* M. Jangoux and J. M. Lawrence, eds., *Echinoderm Nutrition,* pp. 521–551. Balkema, Rotterdam.

Menge, B. A. 1983. Components of predation intensity in the low zone of the New England rocky intertidal region. *Oecologia* 58:141–155.

Menge, B. A., and J. Lubchenco. 1981. Community organization in temperate and tropical rocky intertidal habitats: prey refuges in relation to consumer pressure gradients. *Ecol. Monogr.* 51:429–450.

Menge, B. A., and J. P. Sutherland. 1976. Species diversity gradients: synthesis of the roles of predation, competition and temporal heterogeneity. *Am. Nat.* 110:351–369.

Menge, J. L. 1975. *Effects of Herbivores on Community Structure of the New England Rocky Intertidal Region: Distribution, Abundance, and Diversity of Algae*. Doctoral dissertation, Harvard University, Cambridge, Mass.

Merikallio, E. 1951. Der Einfluss der letzten Wärmeperiode (1930–49) auf die Vogelfauna Nordfinnlands. *Proc. 10th Int. Ornithol. Congr.,* pp. 484–493.

Merrell, D. J. 1951. Interspecific competition between *Drosophila funebris* and *Drosophila melanogaster*. *Am. Nat.* 85:159–169.

Messina, F. J. 1978. Mirid fauna associated with old-field goldenrods in Ithaca, N.Y. *New York Entomol. Soc.* 86:137–143.

Messina, F. J. 1981. Plant protection as a consequence of an ant-membracid mutualism: interactions on goldenrod (*Solidago* sp.). *Ecology* 62:1433–1440.

Messina, F. J. 1982. *Food Plant Selection by Golden-*

rod Leaf Beetles: Beetle Foraging in Relation to Plant Quality. Doctoral dissertation, Cornell University, Ithaca, N.Y.

Messina, F. J., and R. B. Root. 1980. The association between leaf beetles and meadow goldenrods (*Solidago* spp.) in central New York. *Ann. Entomol. Soc. Am.* 73:641–646.

Meyer, E. R. 1973. Late-Quaternary paleoecology of the Cuatro Cienegas Basin, Coahuila, Mexico. *Ecology* 54:982–995.

Miller, B. B. 1976. The late Cenozoic molluscan succession in the Meade County, Kansas, area. *In* C. K. Bayne, ed., *Guidebook of the 24th Annual Meeting of the Midwest Friends of the Pleistocene: Stratigraphy and Faunal Sequence—Meade County, Kansas,* pp. 73–85. Kansas Geological Survey, Lawrence.

Miller, P. C., ed. 1981. *Resource Use by Chaparral and Matorral*. Ecological Studies no. 39. Springer-Verlag, New York.

Miller, P. C. 1983. Canopy structure of Mediterranean-type shrubs in relation to heat and moisture. *In* F. J. Kruger, D. T. Mitchell, and J. U. M. Jarvis, eds., *Mediterranean-Type Ecosystems: the Role of Nutrients,* pp. 133–166. Ecological Studies no. 43. Springer-Verlag, Berlin, Heidelberg.

Miller, R. S. 1967. Pattern and process in competition. *Adv. Ecol. Res.* 4:1–74.

Miller, T. E. 1982. Community diversity and interactions between the size and frequency of disturbance. *Am. Nat.* 120:533–536.

Millington, S. J., and P. R. Grant. 1984. The breeding ecology of the cactus finch *Geospiza scandens* on Isla Daphne Major, Galápagos. *Ardea* 72:177–188.

Mills, J. A., and A. F. Mark. 1977. Food preferences of Takahe in Fiordland National Park, New Zealand, and the effect of competition from introduced red deer. *J. Anim. Ecol.* 46:939–958.

Mills, S. 1981. Graveyard of the puffin. *New Sci.* 91:10–13.

Milne, A. 1957. The natural control of insect populations. *Can. Entomol.* 89:193–213.

Milstead, W. W. 1960. Relict species of the Chihuahuan Desert. *Southwest. Nat.* 5:75–88.

Milton, W. 1947. The yield, botanical and chemical composition of natural hill herbage under manuring, controlled grazing and hay conditions. I. Yield and botanical. *J. Ecol.* 35:65–89.

Minot, E. O. 1981. Effects of interspecific competition for food in breeding blue and great tits. *J. Anim. Ecol.* 50:375–385.

Mitchell, J. M., Jr. 1977. The changing climate. *In* Geophysics Study Committee, ed., *Energy and Climate,* pp. 51–58. National Academy of Sciences, Studies in Geophysics, Washington, D.C.

Mitchley, J. 1983. *The Distribution and Control of the Relative Abundance of Perennials in Chalk Grassland*. Doctoral dissertation, University of Cambridge, Cambridge.

Mitrakos, K., and N. Christodoulakis. 1981. Nonconvergence in leaves of Mediterranean scherophylls. *In* N. S. Margaris and H. A. Mooney, eds., *Components of Productivity of Mediterranean-climate Regions: Basic and Applied Aspects,* pp. 21–25. Junk, Den Haag.

Mittelbach, G. G. 1981. Foraging efficiency and body size: a study of optimal diet and habitat use by bluegills. *Ecology* 62:1370–1386.

Mittelbach, G. G. 1983. Optimal foraging and growth in bluegills. *Oecologia* 59:157–162.

Mittelbach, G. G. 1984. Predation and resource partitioning in two sunfishes (Centrarchidae). *Ecology* 65:499–513.

Mitter, C., and D. R. Brooks. 1983. Phylogenetic aspects of coevolution. *In* D. J. Futuyma and M. Slatkin, eds., *Coevolution,* pp. 65–98. Sinauer, Sunderland, Mass.

Mitter, C., and D. Futuyma 1983. An evolutionary-genetic view of host-plant utilization by insects. *In* R. Denno and M. McClure, eds., *Variable Plants and Herbivores in Natural and Managed Systems,* pp. 427–459. Academic Press, New York.

Montgomery, W. M. 1977. Diet and gut morphology in fishes, with special reference to the monkeyface prickleback, *Cebidichthys violaceus* (Stichaeidae: Blennioidei). *Copeia* 1977:178–182.

Moodie, K. B., and T. R. Van Devender. 1979. Extinction and extirpation in the herpetofauna of the southern High Plains with emphasis on *Geochelone wilsoni* (Testudinidae). *Herpetologica* 35:198–206.

Mook, D. H. 1981. Effects of disturbance and initial settlement on fouling community structure. *Ecology* 62:522–526.

Mooney, H. A. 1977. *Convergent Evolution in Chile and California*. Dowden, Hutchinson & Ross, Stroudsburg, Pa.

Mooney, H. A. 1983. Carbon-gaining capacity and allocation patterns of Mediterranean-climate plants. *In* F. J. Kruger, D. T. Mitchell, and J. U. M. Jarvis, eds., *Mediterranean-Type Ecosystems: the Role of Nutrients,* pp. 106–119. Ecological Studies no. 43. Springer-Verlag, Berlin, Heidelberg.

Mooney, H. A., and E. L. Dunn. 1970. Convergent evolution of Mediterranean-climate evergreen sclerophyll shrubs. *Evolution* 24:292–303.

Mooney, H. A., and S. L. Gulmon. 1979. Environmental and evolutionary constraints on the photosynthetic characteristics of higher plants. *In* O. T.

Solbrig, S. Jain, G. W. Johnson, and P. H. Raven, eds., *Topics in Plant Population Biology*, pp. 316–337. Columbia University Press, New York.

Mooney, H. A., S. L. Gulmon, and D. J. Parsons. 1974. Morphological changes within the chaparral vegetation type as related to elevational gradients. *Madrono* 22:281–316.

Moore, J. A. 1952. Competition between *Drosophila melanogaster* and *Drosophila simulans*. I. Population cage experiments. *Evolution* 6:407–420.

Moore, P. D. 1976. Effects of a long hot summer. *Nature* 263:278–279.

Morafka, D. J. 1977. *A Biogeographical Analysis of the Chihuahuan Desert Through Its Herpetofauna*. Junk, Den Haag.

Moran, N. 1981. Intraspecific variability in herbivore performance and host quality: a field study of *Uroleucon caligatum* (Homoptera: Aphididae) and its *Solidago* hosts (Asteraceae). *Ecol. Entomol.* 6:301–306.

Morel, G. J., and M. Y. Morel. 1974. Recherches écologiques sur une savane Sahelienne du Ferlo septentrional, Sénègal: influence de la sécheresse de l'année 1972–1973 sur l'avifaune. *Terre Vie* 28:95–123.

Morgan, A. V., and A. Morgan. 1980. Faunal assemblages and distributional shifts of Coleoptera during the late Pleistocene in Canada and the northern United States. *Can. Entomol.* 112:1105–1128.

Morgan, A. V., A. Morgan, A. C. Ashworth, and J. V. Matthews, Jr. 1983. Late Wisconsin fossil beetles in North America. *In* S. C. Porter, ed., *Late Quaternary Environments of the United States. Vol. I. The Late Pleistocene,* pp. 354–363. University of Minnesota Press, Minneapolis.

Morin, P. J. 1981. Predatory salamanders reverse the outcome of competition among three species of anuran tadpoles. *Science* 212:1284–1286.

Morin, P. J. 1983a. Predation, competition, and the composition of larval anuran guilds. *Ecol. Monogr.* 53:119–138.

Morin, P. J. 1983b. Competitive and predatory interactions in natural and experimental populations of *Notophthalmus viridescens dorsalis* and *Ambystoma tigrinum*. *Copeia* 1983:628–639.

Morris, R. D., and P. R. Grant. 1972. Experimental studies of competitive interaction in a two-species system. IV. *Microtus* and *Clethrionomys* species in a single enclosure. *J. Anim. Ecol.* 41:275–290.

Morrison, D. F. 1976. *Multivariate Statistical Methods*. McGraw-Hill, New York.

Morrison, G., and D. R. Strong. 1980. Spatial variations in host density and the intensity of parasitism: some empirical examples. *Environ. Entomol.* 9:149–152.

Morse, D. H. 1977. The occupation of small islands by passerine birds. *Condor* 79:399–412.

Morton, M. L., and P. W. Sherman. 1978. Effects of spring snowstorm on behavior, reproduction, and survival of Belding's ground squirrels. *Can. J. Zool.* 56:2578–2590.

Moser, J. W. 1972. Dynamics of an uneven-aged forest stand. *For. Sci.* 18:184–191.

Moulton, M. P. 1985. Morphological similarity and coexistence of congeners: an experimental test with introduced Hawaiian birds. *Oikos* in press.

Moulton, M. P., and S. L. Pimm. 1983. The introduced Hawaiian avifauna: biogeographic evidence for competition. *Am. Nat.* 121:669–690.

Moyle, P. B., and H. W. Li. 1979. Community ecology and predator-prey relations in warmwater streams. *In* H. Clepper, ed., *Predator-Prey Systems in Fisheries Management,* pp. 171–180. Sport Fishing Institute, Washington, D.C.

Muller, C. H. 1939. Relations of the vegetation and climatic types in Nuevo Leon, Mexico. *Am. Midl. Nat.* 21:687–729.

Muller, C. H. 1947. Vegetation and climate of Coahuila, Mexico. *Madroño* 9:33–47.

Müller, H. 1877. Ueber den Ursprung der Blumen. *Kosmos* (Leipzig) 1:100–114.

Munger, J. C., and J. H. Brown. 1981. Competition in desert rodents: an experiment with semipermeable enclosures. *Science* 211:510–512.

Munro, G. C. 1944. *Birds of Hawaii*. Tuttle, Rutland, Vt.

Murdie, G., and M. Hassell. 1973. Food distribution, searching success and predator-prey models. *In* R. Hiorns, ed., *The Mathematical Theory of the Dynamics of Biological Populations,* pp. 87–101. Academic Press, London.

Murdoch, W. W. 1966. Population stability and life history phenomena. *Am. Nat.* 100:5–11.

Murdoch, W. W. 1970. Population regulation and population inertia. *Ecology* 51:497–502.

Murdoch, W. W., J. Chesson, and P. L. Chesson. 1985. Biological control in theory and practice. *Am. Nat.* 125:344–366.

Murdoch, W. W., J. Reeve, C. Huffaker, and C. Kennett. 1984. Biological control of olive scale and its relevance to ecological theory. *Am. Nat.* 123:371–392.

Murphy, G. J. 1968. Pattern in life history and the environment. *Am. Nat.* 102:391–403.

Myers, R. H. 1971. *Response Surface Methodology*. Allyn & Bacon, Boston.

Narise, T. 1965. The effect of relative frequency of species in competition. *Evolution* 19:350–354.

Nash, T. H., III. 1975. Influence of effluents from a

zinc factory on lichens. *Ecol. Monogr.* 45:183–198.

Neill, W. E. 1974. The community matrix and interdependence of the competition coefficients. *Am. Nat.* 108:399–408.

Neill, W. E. 1975. Experimental studies of microcrustacean competition, community composition and efficiency of resource utilization. *Ecology* 56:809–826.

Neill, W. E., and A. Peacock. 1980. Breaking the bottleneck: interactions of invertebrate predators and nutrients in oligotrophic lakes. *In* W. C. Kerfoot, ed., *Evolution and Ecology of Zooplankton Communities,* pp. 715–724. University Press of New England, Hanover, N.H.

Newton, I. 1980. The role of food in limiting bird numbers. *Ardea* 68:11–30.

Nichols, J. D., W. Conley, B. Batt, and A. R. Tipton. 1976. Temporally dynamic reproductive strategies and the concept of r- and k-selection. *Am. Nat.* 110:995–1005.

Nicholson, A. J. 1954. An outline of the dynamics of animal populations. *Aust. J. Zool.* 2:9–65.

Nicholson, A. J. 1957. The self-adjustment of populations to change. *Cold Spring Harbor Symp. Quant. Biol.* 22:153–172.

Nicotri, M. E. 1977. Grazing effects of four marine intertidal herbivores on the microflora. *Ecology* 58:1020–1032.

Nielsen, L. A. 1980. Effect of walleye (*Stizostedion vitreum vitreum*) predation on juvenile mortality and recruitment of yellow perch (*Perca flavescens*) in Oneida Lake, New York. *Can. J. Fish. Aquat. Sci.* 37:11–19.

Niklas, K. J., B. H. Tiffney, and A. H. Knoll. 1980. Apparent changes in the diversity of fossil plants: a preliminary assessment. *Evol. Biol.* 12:1–89.

Nilsson, N.-A. 1960. Seasonal fluctuations in the food segregation of trout, char and whitefish in 14 North-Swedish lakes. *Inst. Freshwater Res. Drottningholm Rep.* 41:185–205.

Nilsson, N.-A. 1963. Interaction between trout and char in Scandinavia. *Tran. Am. Fish. Soc.* 92:276–285.

Nilsson, N.-A. 1965. Food segregation between salmonoid species in North Sweden. *Inst. Freshwater Res. Drottningholm Rep.* 46:58–78.

Nilsson, N.-A., and O. Filipsson. 1971. Characteristics of two discrete populations of Arctic char (*Salvelinus alpinus* L.) in a north Swedish lake. *Inst. Freshwater Res. Drottningholm Rep.* 51:90–108.

Nilsson, N.-A., and T. G. Northcote. 1981. Rainbow trout (*Salmo gairdneri*) and cutthroat trout (*S. clarki*) interactions in coastal British Columbia lakes. *Can. J. Fish. Aquat. Sci.* 38:1228–1246.

Nilsson, N.-A., and B. Pejler. 1973. On the relation between fish fauna and zooplankton in North Swedish lakes. *Inst. Freshwater Res. Drottningholm Rep.* 53:51–77.

Nisbet, R. M., and W. S. C. Gurney. 1982. *Modelling Fluctuating Populations.* Wiley, Chicester, New York.

Nobel, P. S. 1979. *Introduction to Biophysical Plant Ecology.* Freeman, San Francisco.

Nobel, P. S. 1980. Morphology, surface temperature, and the northern limits of columnar cacti in the Sonoran Desert. *Ecology* 61:1–7.

Noble, R. L. 1975. Growth of young yellow perch (*Perca flavescens*) in relation to zooplankton populations. *Trans. Am. Fish. Soc.* 104:731–741.

Nolan, V. 1960. Breeding behavior of Bell's Vireo in southern Indiana. *Condor* 62:225–244.

Noon, B. R., D. K. Dawson, D. B. Inkley, C. Robbins, and S. H. Anderson. 1980. Consistency in habitat preference of forest bird species. *Trans. 45th N. Am. Wildl. Nat. Resour. Conf.* pp. 226–244.

Nowak, E. 1971. The range expansion of animals and its causes. *Zesz. Nauk.* 3:1–255.

OConnor, B., S. Naeem, and R. K. Colwell. 1985. Hummingbird flower mites of Trinidad: genera *Rhinoseius* and *Proctolaelaps* (Acari: Ascidae). *Spec. Publ. Museum of Zool., Univ. of Michigan.* In press.

O'Donald, P. 1983. Sexual selection by female choice. *In* P. Bateson, ed., *Mate Choice,* pp. 53–66. Cambridge University Press, Cambridge.

O'Dowd, D. J. 1982. Pearl bodies as ant food: an ecological role for some leaf emergences of tropical plants. *Biotropica* 14:40–49.

O'Dowd, D. J., and E. A. Catchpole. 1983. Ants and extrafloral nectaries: no evidence for plant protection in *Helichrysum* spp.–ant interactions. *Oecologia* 59:191–200.

O'Dowd, D. J., and M. E. Hay. 1980. Mutualism between harvester ants and a desert ephemeral: seed escape from rodents. *Ecology* 61:531–540.

Odum, E. P. 1969. The strategy of ecosystem development. *Science* 164:262–270.

Oechel, W. C., W. T. Lawrence, J. Mustafa, and J. Martinez. 1981. Energy and carbon aquisition. *In* P. C. Miller, ed., *Resource Use by Chaparral and Matorral,* pp. 151–182. Ecological Studies no. 39. Springer-Verlag, New York.

Oldeman, R. A. A. 1979. Architecture and energy exchange of dicotyledonous trees in the forest. *In* P. B. Tomlinson and M. H. Zimmermann, eds., *Tropical Trees as Living Systems,* pp. 535–560. Cambridge University Press, Cambridge.

Olive, C. W. 1981. Optimal phenology and body-size of orb-weaving spiders. *Oecologia* 49:83–87.

Olive, C. W. 1982. Behavioral responses of a sit-and-wait predator to spatial variation in foraging gain. *Ecology* 63:1617–2013.

Oliver, J. H., Jr. 1983. Chromosomes, genetic variance, and reproductive strategies among mites and ticks. *Bull. Entomol. Soc. Amer.* 29:8–17.

Olson, J. S. 1958. Rates of succession and soil changes on southern Lake Michigan sand dunes. *Bot. Gaz.* 119:125–169.

Olson, R. R. 1985. The consequences of short-distance larval dispersal in a sessile marine invertebrate. *Ecology* 66:30–38.

Olson, S. L. 1978. A paleontological perspective of West Indian birds and mammals. *Acad. Nat. Sci. Phil. Special Publ.* 13:99–117.

Olson, S. L., ed. 1982. *Fossil Vertebrates from the Bahamas*. Smithsonian Contrib. Paleobiology no. 48.

Olson, S. L., and H. F. James. 1982a. Fossil birds from the Hawaiian Islands: evidence for wholesale extinctions by man before Western contact. *Science* 217:633–635.

Olson, S. L., and H. F. James. 1982b. Prodromus of the fossil avifauna of the Hawaiian Islands. *Smithsonian Contrib. Zool.* no. 365.

Orians, G. H., and R. T. Paine. 1983. Convergent evolution at the community level. *In* D. J. Futuyma and M. Slatkin, eds., *Coevolution,* pp. 431–458. Sinauer, Sunderland, Mass.

Osman, R. W. 1977. The establishment and development of a marine epifaunal community. *Ecol. Monogr.* 47:37–63.

Osman, R. W. 1978. The influence of seasonality and stability on the species equilibrium. *Ecology* 59:383–399.

Osman, R. W., and J. A. Haughness. 1981. Mutualism among sessile invertebrates: a mediator of competition and predation. *Science* 211:846–848.

Oster, G. 1978. The dynamics of nonlinear models with age structure. *In* S. A. Levin, ed., *Studies in Mathematical Biology. Part II. Populations and Communities,* pp. 411–438. The Mathematical Association of America, Washington, D.C.

Oster, G., and Y. Takahashi. 1974. Models for age-specific interactions in a periodic environment. *Ecol. Monogr.* 44:483–501.

Ostler, W. K., and K. T. Harper. 1978. Floral ecology in relation to plant species diversity in the Wasatch Mountains of Utah and Idaho. *Ecology* 59:848–861.

Ottaway, J. R. 1978. Population ecology of the intertidal anemone *Actinia tenebrosa*. I. Pedal locomotion and intraspecific aggression. *J. Mar. Freshwater Res.* 29:787–802.

Otte, D. 1979. Historical development of sexual selection theory. *In* M. S. Blum and N. A. Blum, eds., *Sexual Selection and Reproductive Competition in Insects,* pp. 1–18. Academic Press, New York.

Overmire, P. G. 1962. Nesting of the Bell's Vireo in Oklahoma. *Condor* 64:75.

Owen, D. F., and R. G. Wiegert. 1976. Do consumers maximize plant fitness? *Oikos* 27:488–492.

Owen, D. F., and R. G. Wiegert. 1981. Mutualism between grasses and grazers: an evolutionary hypothesis. *Oikos* 36:376–378.

Pacala, S., and J. Roughgarden. 1982. Resource partitioning and interspecific competition in two-species insular *Anolis* lizard communities. *Science* 217:444–446.

Pacala, S., and J. Roughgarden. 1984. Control of arthropod abundance by *Anolis* lizards on St. Eustatius (Neth. Antilles). *Oecologia* 64:160–162.

Pacala, S., and J. Roughgarden. 1985. Population experiments with the *Anolis* lizards of St. Maarten and St. Eustatius (Neth. Antilles). *Ecology* 66:129–141.

Pahlavan, A. 1977. Plant geometry in the Mojave Desert. *In* M. L. Cody, ed., *Field Ecology Reports 1977, the New York Mountains*. Unpublished manuscript.

Paine, R. T. 1966. Food web complexity and species diversity. *Am. Nat.* 100:65–75.

Paine, R. T. 1971. A short-term experimental investigation of resource partitioning in a New Zealand rocky intertidal habitat. *Ecology* 52:1096–1106.

Paine, R. T. 1974. Intertidal community structure. Experimental studies on the relationship between a dominant competitor and its principal predator. *Oecologia* 15:93–120.

Paine, R. T. 1980. Food webs: linkage, interaction strength and community infrastructure. *J. Anim. Ecol.* 49:667–685.

Paine, R. T. 1984. Ecological determinism in the competition for space. The first MacArthur lecture. *Ecology* 65:1339–1348.

Paine, R. T., and S. A. Levin. 1981. Intertidal landscapes: disturbance and the dynamics of pattern. *Ecol. Monogr.* 51:145–178.

Paine, R. T., and T. H. Suchanek. 1983. Convergence of ecological processes between independently evolved competitive dominant: a tunicate-mussel comparison. *Evolution* 37:821–831.

Paine, R. T., and R. L. Vadas. 1969. The effect of grazing by sea urchins *Strongylocentrus* spp. on benthic algal populations. *Limnol. Oceanogr.* 14:710–719.

Paloheimo, J. E., and L. M. Dickie. 1966. Food and growth of fishes. II. Effects of food and temperature on the relation between metabolism and body weight. *J. Fish. Res. Board Can.* 23:869–908.

Park, T. 1948. Experimental studies of interspecies competition. I. Competition between populations of the flour beetles, *Tribolium confusum* Duval and *Tribolium castaneum*. *Ecol. Monogr.* 18:265–307.

Park, T. 1962. Beetles, competition, and populations. *Science* 138:1369–1375.

Parker, G. A. 1978. Evolution of competitive mate searching. *Annu. Rev. Entomol.* 23:173–196.

Parker, G. A.1979. Sexual selection and sexual conflict. *In* M. S. Blum and N. A. Blum, eds., *Sexual Selection and Reproductive Competition in Insects,* pp. 123–166. Academic Press, New York.

Parker, L. R. 1977. The paleoecology of the fluvial coal-forming swamps and associated floodplain environments in the Blackhawk Formation (upper Cretaceous) of central Utah. *Brigham Young Univ. Geol. Studies* 22:99–116.

Parker, M. 1984. Local food depletion and the foraging behavior of a specialist grasshopper, *Hesperotettix viridis*. *Ecology* 65:824–835.

Parker, M. 1985. Size-dependent herbivore attack and the demography of an arid grassland shrub. *Ecology* 66:850–860.

Parkhurst, D. F., and O. L. Loucks. 1972. Optimal leaf size in relation to environment. *J. Ecol.* 60:505–537.

Parrish, J. A. P., and F. A. Bazzaz. 1976. Underground niche separation in successional plants. *Ecology* 57:1257–1288.

Parsons, P. A. 1981. Habitat selection in *Drosophila*. *In* W. R. Atcheley and D. S. Woodruff, eds., *Evolution and Speciation,* pp. 219–240. Cambridge University Press, Cambridge.

Pastor, J., J. D. Aber, C. A. McClaughterty, and J. M. Melillo. 1984. Aboveground production and N and P cycling along a nitrogen mineralization gradient on Blackhawk Island, Wisconsin. *Ecology* 65:256–268.

Patten, B. C. 1982. Environs: relativistic elementary particles for ecology. *Am. Nat.* 119:179–219.

Patten, B. C., and G. T. Auble. 1981. System theory and the ecological niche. *Am. Nat.* 118:345–369.

Payette, S. 1983. The forest tundra and present treelines of the northern Quebec-Labrador peninsula. *In* P. Morrisset and S. Payette, eds., *Tree-line Ecology,* pp. 3–23. Collection Nordicana, no. 47. Université Laval, Quebec.

Pearl, R., and S. L. Parker. 1922. On the influence of density of population upon the rate of reproduction in drosophila. *Proc. Natl. Acad. Sci., U.S.A.* 8:212–219.

Pearson, D. L., and C. B. Knisley. 1985. Evidence for food as a limiting resource in the life cycle of tiger beetles (Coleoptera: Cicendelidae). *Oikos* in press.

Pearson, J. C. 1972. A phylogeny of life cycle patterns of the Digenea. *Adv. Parasitol.* 10:153–189.

Peet, R. K., and N. L. Christensen. 1980. Succession: a population process. *Vegetatio* 43:131–140.

Pernetta, J. C., and D. Watling. 1978. The introduced and native terrestrial vertebrates of Fiji. *Pac. Sci.* 32:223–244.

Persson, L. 1985. Effects of reduced interspecific competition on resource utilization of a perch (*Perca fluviatilis*) population. Unpublished manuscript.

Petelle, M. 1980. Aphids and melezitose: a test of Owen's 1978 hypothesis. *Oikos* 35:127–128.

Peterson, C. H. 1979. Predation, competitive exclusion, and diversity in the soft-sediment benthic communities of estuaries and lagoons. *In* R. J. Livingston, ed., *Ecological Processes in Coastal and Marine Systems,* pp. 233–264. Plenum, New York.

Peterson, C. H. 1982. The importance of predation and intra- and interspecific competition in the population biology of two infaunal suspension-feeding bivalves, *Protothaca staminea* and *Chione undatella*. *Ecol. Monogr.* 52:437–475.

Peterson, G. M., T. Webb III, J. E. Kutzbach, T. van der Hammen, T. A. Wijmstra, and F. A. Strut. 1979. The continental record of environmental conditions at 18,000 years B.P.: an initial investigation. *Quat. Res.* 12:47–82.

Peterson, I., and J. S. Wroblewski. 1984. Mortality rate of fishes in the pelagic ecosystem. *Can. J. Fish. Aquat. Sci.* 41:1117–1120.

Petraitis, P. S. 1979. Competitive networks and measures of intransitivity. *Am. Nat.* 114:921–925.

Petraitis, P. S. In press. Laboratory experiments on the effects of a gastropod (*Hydrobia minuta*) on survival of an infaunal deposit-feeding polychaete (*Capitella capitata,* species type II). *Mar. Ecol. Prog. Ser.*

Pfefferkorn, H. W., and M. C. Thomson. 1982. Changes in dominance patterns in upper Carboniferous plant-fossil assemblages. *Geology* 10:641–644.

Phillips, T. L., and W. DiMichele. 1981. Paleoecology of Middle Pennsylvanian age coal swamps. *In* K. J. Niklas, ed., *Paleobotany, Paleoecology, and Evolution,* vol. I, pp. 231–284. Prager, New York.

Phillips, T. L., A. B. Kunz, and D. J. Mickish. 1977. Paleobotany of permineralized peat (coal balls) from the Herrin (No. 6) Coal Member of the Illinois Basin. *In* P. N. Given and A. D. Cohan, eds., *Interdisciplinary Studies of Peat and Coal Origins*, pp. 18–49. *Geological Society of America Microform Publications* 7. Geological Society of America, Boulder, Colo.

Phillips, T. L., R. A. Peppers, M. J. Avcin, and P. F. Laughnan. 1974. Fossil plants and coal: patterns of change in Pennsylvanian coal swamps of the Illinois Basin. *Science* 184:1367–1369.

Pianka, E. R. 1966. Latitudinal gradients in species diversity: a review of concepts. *Am. Nat.* 100:33–46.

Pianka, E. R. 1981. Competition and niche theory. *In* R. M. May, ed., *Theoretical Ecology*. Sinauer, Sunderland, Mass.

Pianka, E. R. 1983. *Evolutionary Ecology*. Harper & Row, New York.

Pickett, S. T. A. 1976. Succession: an evolutionary interpretation. *Am. Nat.* 110:107–119.

Pielou, E. C. 1974. *Population and Community Ecology: Principles and Methods*. Gordon and Breach, New York.

Pigott, C. D., and K. Taylor. 1964. The distribution of some woodland herbs in relation to the supply of nitrogen and phosphorus in the soil. *J. Ecol.* 52 (Suppl):175–185.

Pijl, L. van der. 1969. *Principles of Dispersal in Higher Plants*. Springer-Verlag, New York.

Pimm, S. L. 1978. An experimental approach to the effects of predictability on community structure. *Am. Zool.* 18:797–808.

Pimm, S. L. 1979. Complexity and stability: another look at MacArthur's original hypothesis. *Oikos* 33:351–357.

Pimm, S. L. 1980. Properties of food webs *Ecology* 61:219–225.

Pimm, S. L. 1982. *Food Webs*. Chapman and Hall, New York.

Pimm, S. L. 1984. The complexity and stability of ecosystems. *Nature* 307:321–326.

Pimm, S. L., and J. H. Lawton. 1978. On feeding on more than one trophic level. *Nature* 275:542–543.

Pimm, S. L., and M. L. Rosenzweig. 1981. Competitors and habitat use. *Oikos* 37:1–6.

Pimm., S. L., M. L. Rosenzweig, and W. M. Mitchell. 1985. Competition and food selection: field tests of a theory. *Ecology* 66:798–807.

Pleske, T. 1912. Zur Lösung der Frage, ob *Cyanistes pleskei* Cab. eine Selbständige Art darstellt, oder für einen Bastard von *Cyanistes coeruleus* (Linn.) und *Cyanistes cyanus* (Pall.) angesprochen werden muss. *J. Ornithol.* 60:96–109.

Podoler, H., and D. Rogers. 1975. A new method for the identification of key factors from life-table data. *J. Anim. Ecol.* 44:85–114.

Pollard, E. 1981. Resource limited and equilibrium models of populations. *Oecologia* 49:377–378.

Pomerantz, M. J. 1981. Do higher-order interactions in competition really exist? *Am. Nat.* 117:583–591.

Pomerantz, M. J., W. R. Thomas, and M. E. Gilpin. 1980. Asymmetries in population growth regulated by intraspecific competition: empirical studies and model tests. *Oecologia* 47:311–322.

Pontin, A. J. 1978. The numbers and distribution of subterranean aphids and their exploitation by the ant *Lasius flavus* (Fabr.). *Ecol. Entomol.* 3:203–207.

Poore, M. E. O. 1968. Studies in Malaysian rain forest. I. The forest on Triassic sediments in Jengka Forest Reserve. *J. Ecol.* 56:143–196.

Porter, J. W. 1972. Predation by *Acanthaster* and its effect on coral species diversity. *Am. Nat.* 106:487–492.

Porter, J. W. 1974. Community structure of coral reefs on opposite sides of the Isthmus of Panama. *Science* 186:543–545.

Porter, J. W. 1976. Autotrophy, heterotrophy, and resource partitioning in Caribbean reef-building corals. *Am. Nat.* 110:731–742.

Post, W., and J. W. Wiley. 1977. Reproductive interactions of the Shiny Cowbird and the Yellow-shouldered Blackbird. *Condor* 79:176–184.

Potts, D. C. 1983. Evolutionary disequilibrium among Indo-Pacific corals. *Bull. Mar. Sci.* 33:619–632.

Power, D. M. 1972. Numbers of bird species on the California Islands. *Evolution* 26:451–463.

Pregill, G. K. 1981. An appraisal of the vicariance hypothesis of Caribbean biogeography and its application to West Indian terrestrial vertebrates. *Syst. Zool.* 30:147–155.

Pregill, G. K., and S. L. Olson. 1981. Zoogeography of West Indian vertebrates in relation to Pleistocene climatic cycles. *Annu. Rev. Ecol. Syst.* 12:75–98.

Preston, F. W. 1962. The canonical distribution of commonness and rarity. *Ecology* 43:185–215, 410–432.

Price, M. V., and N. M. Waser. 1979. Pollen dispersal and optimal outcrossing in *Delphinium nelsoni*. *Nature* 277:294–297.

Price, P. W. 1980. *Evolutionary Biology of Parasites*. Princeton University Press, Princeton, N.J.

Price, P. W., C. Bouton, P. Gross, B. McPheron, J. Thompson, and A. Weis. 1980. Interactions among three trophic levels: influence of plants on interactions between insect herbivores and natural enemies. *Annu. Rev. Ecol. Syst.* 11:41–65.

Price, P. W., C. N. Slobodchikoff, and W. S. Gaud. 1984. *A New Ecology: Novel Approaches to Interactive Systems*. Wiley, New York.

Price, T. D., and P. R. Grant. 1984. Life history traits and natural selection for small body size in a population of Darwin's Finches. *Evolution* 38:483–494.

Price, T. D., P. R. Grant, and P. T. Boag. 1984. Genetic changes in the morphological differentiation of Darwin's Ground Finches. *In* K. Wöhrmann and V. Löschcke, eds., *Population Biology and Evolution,* pp. 49–66. Springer-Verlag, Berlin, New York.

Price, T. D., P. R. Grant, H. L. Gibbs, and P. T. Boag. 1984. Recurrent patterns of natural selection in a population of Darwin's Finches. *Nature* 309:787–789.

Prokopy, R. J. 1968. Visual responses of apple maggot flies, *Rhagoletis pomonella* (Diptera: Tephritidae): Orchard studies. *Entomol. Exper. Appl.* 11:403–422.

Prokopy, R. J. 1980. Mating behavior of frugivorous Tephritidae in nature. *In* National Institute of Agricultural Science (Japan), *Proceedings of Symposium on Fruit Fly Problems,* pp. 37–46.

Prokopy, R. J. 1983. Tephritid relationships with plants. *In* R. Cavalloro, ed., *Fruit Flies of Economic Importance,* pp. 230–239. Balkema, Rotterdam.

Prokopy, R. J., A. L. Averill, S. S. Cooley, and C. A. Roitberg. 1982. Associative learning in egglaying site selection by apple maggot flies. *Science* 218:76–77.

Pulliam, H. R. 1975. Coexistence of sparrows: a test of competition theory. *Science* 189:474–476.

Pulliam, H. R. 1983. Ecological community theory and the coexistence of sparrows. *Ecology* 64:45–52.

Pulliam, H. R., and M. R. Brand. 1975. The production and utilization of seeds in the plains grassland of southwestern Arizona. *Ecology* 56:1158–1166.

Pulliam, H. R., and T. H. Parker III. 1979. Population regulation of sparrows. *Fortschr. Zool.* 25:137–147.

Purcell, J. E. 1977. Aggressive function and induced development of catch tentacles in the sea anemone *Metridium senile* (Coelenterata, Actiniaria). *Biol. Bull.* 153:355–368.

Pyle, R. L., and C. J. Ralph. 1981. The eighty-first Audubon Christmas bird count. *Am. Birds* 35:380–381.

Quinlan, R. J., and J. M. Cherret. 1978. Aspects of the symbiosis of the leaf-cutting ant *Acromyrmex octospinosus* (Reich) and its fungus food. *Ecol. Entomol.* 3:221–230.

Quinn, J. F. 1979. *Disturbance, Predation and Diversity in the Rocky Intertidal Zone*. Doctoral dissertation, University of Washington, Seattle.

Quinn, J. F. 1982. Competitive hierarchies in marine benthic communities. *Oecologia* 54:129–135.

Rabinovitch-Vin, A. 1979. *Influence of Parent Rock on Soil Properties and Composition of Vegetation in the Galilee*. Doctoral dissertation, Hebrew University of Jerusalem.

Rabinovitch-Vin, A. 1983. Influence of nutrients on the composition and distribution of plant communities in Mediterranean-type ecosystems of Israel. *In* F. J. Kruger, D. T, Mitchell, and J. U. M. Jarvis, eds., *Mediterranean-Type Ecosystems: the Role of Nutrients,* pp. 74–85. Ecological Studies no. 43. Springer-Verlag, Berlin, New York.

Rabinowitz, D. 1981. Seven forms of rarity. *In* H. Synge, ed., *The Biological Aspects of Rare Plant Conservation,* pp. 205–217. Wiley, Chichester, N.Y.

Rabinowitz, D., and J. K. Rapp. 1981. Dispersal abilities of seven sparse and common grasses from a Missouri prairie. *Am. J. Bot.* 68:616–624.

Rai, B., H. I. Freedman, and J. F. Addicott. 1983. Analysis of three species models of mutualism in predator-prey and competitive systems. *Math. Biosci.* 65:13–50.

Rand, A. S. 1964. Ecological distribution of the anoline lizards of Puerto Rico. *Ecology* 45:745–752.

Rand, A. S. 1967. Ecological distribution of the anoline lizards around Kingston, Jamaica. *Breviora* 22:1–18.

Randall, J. E. 1965. Grazing effect on sea grasses by herbivorous fishes in the West Indies. *Ecology* 46:255–260.

Randall, J. E. 1967. Food habits of reef fishes of the West Indies. *Stud. Tropical Oceanogr.* 5:665–847.

Rathcke, B. J. 1976. Competition and coexistence within a guild of herbivorous insects. *Ecology* 57:76–87.

Raunkiaer, C. 1934. *The Life Forms of Plants and Plant Geography*. Oxford University Press (Clarendon Press), Oxford.

Raup, D. M., and J. J. Sepkoski, Jr. 1984. Periodicity of extinctions in the geologic past. *Proc. Natl. Acad. Sci., U.S.A.* 81:801–805.

Rauscher, M. D. 1983. Conditioning and genetic variation as causes of individual variation in the oviposition behavior of the tortoise beetle, *Deloya guttata*. *Anim. Behav.* 31:743–747.

Raven, P. H. 1977. A suggestion concerning the Cretaceous rise to dominance of the angiosperms. *Evolution* 31:451–452.

Reddingus, J. 1971. Gambling for existence. *Acta Biotheoretica* 20 (Suppl.).

Reddingus, J., and P. J. den Boer. 1970. Simulation experiments illustrating stabilization of animal numbers by spreading of risk. *Oecologia* 5:240–284.

Reed, C. I., and B. P. Reed. 1928. The mechanism of pellet formation in the Great Horned Owl (*Bubo virginianus*). *Science* 68:359–360.

Reed, T. W. 1984. The numbers of landbird species on the Isles of Scilly. *Biol. J. Linn. Soc.* 21:431–437.

Regal, P. J. 1977. Ecology and the evolution of flowering plant dominance. *Science* 196:622–629.

Reichman, O. J. 1979. Desert granivore foraging and its impact on seed densities and distributions. *Ecology* 60:1085–1092.

Reppenning, C. A. 1980. Faunal exchanges between Siberia and North America. *Can. J. Anthropol.* 1:37–44.

Retallack, G. J. 1977. Reconstructing Triassic vegetation of eastern Australasia: a new approach to the biostratigraphy of Gondwanaland. *Alcheringa* 1:247–277.

Rey, J. R., and D. R. Strong. 1983. Immigration and extinction of salt marsh arthropods on islands: an experimental study. *Oikos* 41:396–401.

Reynolds, R. T., J. M. Scott, and R. A. Nussbaum. 1980. A variable circular-plot method for estimating bird numbers. *Condor* 82:309–313.

Rhoades, D. 1985. Offensive-defensive interactions between herbivores and plants: their relevance in herbivore population dynamics and ecological theory. *Am. Nat.* 125:205–238.

Rhodes, R. S., II. 1984. Paleoecology and regional paleoclimatic implications of the Farmdalian Craigmile and Woodfordian Waubonsie mammalian local faunas, southwestern Iowa. *Ill. Sate Mus. Rep. Invest.* 40:1–50.

Richards, P. W. 1952. *The Tropical Rain Forest*. Cambridge University Press, Cambridge.

Richards, R. C., C. R. Goldman, T. C. Frantz, and R. Wickwire. 1975. Where have all the *Daphnia* gone? The decline of a major cladoceran in Lake Tahoe, California-Nevada. *Verh. Int. Verein. Theor. Angew. Limnol.* 19:835–842.

Richardson, C. A., P. Dustan, and J. C. Lang. 1979. Maintenance of the living space by sweeper tentacles of *Montastrea cavernosa*. *Mar. Biol.* 55:181–186.

Richardson, R. H. 1982. Phyletic species packing and the formation of sibling (cryptic) species clusters. *In* J. S. F. Barker and W. T. Starmer, eds., *Ecological Genetics and Evolution: the Cactus-Yeast Drosophila Model System*, pp. 107–123. Academic Press, Sydney.

Ricker, W. E., and R. E. Foerster. 1948. Computation of fish production: a symposium on fish populations. *Bull. Bingham Oceanogr. Collect. Yale Univ.* 11:173–211.

Ricker, W. E., and J. Gottschalk. 1941. An experiment in removing coarse fish from a lake. *Trans. Am. Fish. Soc.* 70:382–390.

Ricklefs, R. E. 1977. Environmental heterogeneity and plant species diversity: a hypothesis. *Am. Nat.* 111:376–381.

Ricklefs, R. E., D. Cochran, and E. R. Pianka. 1981. A morphological analysis of the structure of communities of lizards in desert habitats. *Ecology* 62:1474–1483.

Ricklefs, R. E., and J. Travis. 1980. A morphological approach to the study of avian community organization. *Auk* 97:321–338.

Riley, C. V. 1892. The yucca moth and yucca pollination. *Annu. Rep. Mo. Bot. Gard.* 3:99–159.

Ripley, S. D. 1964. A systematic and ecological study of birds of New Guinea. *Bull. Peabody Mus. Nat. Hist.* 19:1–87.

Risch, S. J., and D. H. Boucher. 1976. What ecologists look for. *Bull. Ecol. Soc. Am.* 57(3):8–9.

Risch, S. J., and C. R. Carroll. 1982. Effect of a keystone predaceous ant, *Solenopsis geminata*, on arthropods in a tropical agroecosystem. *Ecology* 63:1979–1983.

Rising, J. D. 1973. Age and seasonal variation in dimensions of House Sparrows, *Passer domesticus* (L.), from a single population in Kansas. *In* S. C. Kendeigh and J. Pinowski, eds., *Productivity, Population Dynamics and Systematics of Granivorous Birds*, pp. 327–336. PWN-Polish Scientific, Warsaw.

Ritchie, J. C., L. C. Cwynar, and R. W. Spear. 1983. Evidence from north-west Canada for an early Holocene Milankovitch thermal maximum. *Nature* 305:126–128.

Robertson, D. R., and B. Lassig. 1980. Spatial distribution patterns and coexistence of a group of territorial damselfishes from the Great Barrier Reef. *Bull. Mar. Sci.* 30:187–203.

Robinson, M. H., and B. Robinson. 1970. Prey caught by a sample population of the spider *Argiope argentata* (Araneae: Araneidae) in Panama: a year's census data. *Zool. J. Linn. Soc.* 49:345–357.

Robinson, M. H., Y. D. Lubin, and B. Robinson. 1974. Phenology, natural history and species diversity of web-building spiders on three transects at Wau, New Guinea. *Pac. Insects* 16:117–163.

Rock, J. R. 1913. *The Indigenous Trees of the Hawaiian Islands*. Reprinted 1974 by Tuttle, Rutland, Vt.

Roeder, K. D. 1967. *Nerve Cells and Insect Behavior*, rev. ed. Harvard University Press, Cambridge, Mass.

Roff, D. A. 1974. The analysis of a population model demonstrating the importance of dispersal in a heterogeneous environment. *Oecologia* 15:259–275.

Rogers, D. J. 1972. Random search and insect population models. *J. Anim. Ecol.* 41:369–383.

Rogers, D. J. 1979. Tsetse population dynamics and distribution: a new analytical approach. *J. Anim. Ecol.* 48:825–849.

Rogers, R. W., and W. E. Westman. 1979. Niche differentiation and maintenance of genetic identity in cohabiting *Eucalyptus* species. *Aust. J. Ecol.* 4:429–439.

Rohde, K. 1977. A non-competitive mechanism responsible for restricting niches. *Zool. Anz.* 199:164–172.

Rohde, K. 1979. A critical evaluation of intrinsic and extrinsic factors responsible for niche restriction in parasites. *Am. Nat.* 114:648–671.

Roitberg, B. D., and R. J. Prokopy. 1981. Experience required for pheromone recognition by apple maggot fly. *Nature* 292:540–541.

Room, P. M. 1972. The fauna of the mistletoe *Tapinanthus bangwensis* growing on cocoa in Ghana: relationships between fauna and mistletoe. *J. Appl. Ecol.* 41:611–621.

Root, R. B. 1967. The niche exploitation pattern of the blue-gray gnatcatcher. *Ecol. Monogr.* 37:317–350.

Root, R. B., and P. Kareiva. 1984. The search for resources by cabbage butterflies (*Pieris rapae*): ecological consequences and adaptive significance of markovian movement in a patchy environment. *Ecology* 65:147–165.

Rose, B. 1982. Food intake and reproduction in *Anolis acutus*. *Copeia* 1982:322–330.

Rose, E. T., and T. Moen. 1952. The increase in same fish populations in East Okoboji Lake, Iowa, following intensive removal of rough fish. *Trans. Am. Fish. Soc.* 82:104–114.

Rosenzweig, M. L. 1979a. Three probable evolutionary causes for habitat selection. *In* G. P. Patil and M. L. Rosenzweig, eds., *Quantitative Ecology and Related Econometrics,* pp. 49–60. International Co-operative Publishing House, Fairland, Md.

Rosenzweig, M. L. 1979b. Optimal habitat selection in two-species competitive systems. *Fortschr. Zool.* 25:283–293.

Rosenzweig, M. L. 1981. A theory of habitat selection. *Ecology* 62:327–335.

Rosenzweig, M. L. 1985. Habitat selection theory. *In* M. L. Cody, ed., *Habitat Selection in Birds*. Academic Press, New York.

Rosenzweig, M. L., and R. H. MacArthur. 1963. Graphical representation and stability conditions of predator-prey interactions. *Am. Nat.* 97:209–223.

Ross, M. A., and J. L. Harper. 1972. Occupation of biological space during seedling establishment. *J. Ecol.* 60:77–88.

Ross, S. T. 1978. Trophic ontogeny of the leopard searobin, *Prionotus scitulus* (Pisces: Triglidae). *Fish. Bull.* 76:225–234.

Ross, S. T. 1985. Resource partitioning in fish assemblages: a review of field studies. *Copeia* in press.

Rotenberry, J. T. 1980. Dietary relationships among shrubsteppe passerine birds: competition or opportunism in a variable environment? *Ecol. Monogr.* 50:93–110.

Rotenberry, J. T., and J. A. Wiens. 1980a. Habitat structure, patchiness, and avian communities in North American steppe vegetation: a multivariate analysis. *Ecology* 61:1228–1250.

Rotenberry, J. T., and J. A. Wiens. 1980b. Temporal variation in habitat structure and shrubsteppe bird dynamics. *Oecologia* 47:1–9.

Rothschild, M. 1952. A collection of fleas from the bodies of British birds, with notes on their distribution and host preference. *Bull. Br. Mus. (Nat. Hist.) Entomol.* 2:187–232.

Roughgarden, J. 1974a. Niche width: biogeographic patterns among *Anolis* lizard populations. *Am. Nat.* 108:429–442.

Roughgarden, J. 1974b. Species packing and the competition function with illustrations from coral reef fish. *Theor. Popul. Biol.* 5:163–186.

Roughgarden, J. 1975a. Evolution of marine symbiosis—a simple cost-benefit model. *Ecology* 56:1201–1208.

Roughgarden, J. 1975b. A simple model for population dynamics in a stochastic environment. *Am. Nat.* 109:713–736.

Roughgarden, J. 1976. Resource partitioning among competing species: a coevolutionary approach. *Theor. Popul. Biol.* 9:388–424.

Roughgarden, J. 1977. Coevolution in ecological systems: results from "loop analysis" for purely density-dependent coevolution. *In* F. Christiansen and T. Fenchel, eds., *Measuring Selection in Natural Populations,* pp. 499–518. Lecture Notes in Mathematics, vol. 19. Springer-Verlag, New York.

Roughgarden, J. 1979. *Theory of Population Genetics and Evolutionary Ecology: an Introduction*. Macmillan, New York.

Roughgarden, J. 1983. The theory of coevolution. *In* D. J. Futuyma and M. Slatkin, eds., *Coevolution,* pp. 33–64. Sinauer, Sunderland, Mass.

Roughgarden, J., and M. Feldman. 1975. Species packing and predation pressure. *Ecology* 56:489–492.

Roughgarden, J., S. Gaines, and Y. Iwasa. 1984a. In press. Dynamics and evolution of marine populations with pelagic larval dispersal. *In* R. M. May, ed., *Exploitation of Marine Communities*. Dahlem Konferenzen. Springer-Verlag, Berlin, Heidelberg, New York, Tokyo.

Roughgarden, J., D. Heckel, and E. R. Fuentes. 1983a. Coevolutionary theory and the island biogeography of *Anolis*. *In* R. Huey, E. Pianka, and T. Schoener, eds., *Lizard Ecology: Studies on a Model Organism*, pp. 371–410. Harvard University Press, Cambridge, Mass.

Roughgarden, J., and Y. Iwasa. 1985. *Dynamics of a metapopulation with space-limited subpopulations*. Unpublished manuscript.

Roughgarden, J., Y. Iwasa, and C. Baxter. 1985. Demographic theory for an open marine population with space-limited recruitment. *Ecology* 66:54–67.

Roughgarden, J., S. Pacala, and J. Rummel. 1984b. Strong present-day competition between the *Anolis* lizard populations of St. Maarten (Neth. Antilles). *In* B. Shorrocks, ed., *Evolutionary Ecology*, pp. 203–220. Blackwell, Oxford.

Roughgarden, J., J. Rummel, and S. Pacala. 1983b. Experimental evidence of strong present-day competition between the *Anolis* populations of the Anguilla Bank—a preliminary report. *In* A. Rhodin and K. Miyata, eds., *Advances in Herpetology and Evolutionary Biology: Essays in Honor of Ernest Williams*, pp. 499–506. Museum of Comparative Zoology, Cambridge, Mass.

Rouse, I., and L. Allaire. 1978. Caribbean. *In* R. Taylor and C. Meighan, eds., *Chronologies in New World Archeology*, pp. 431–481. Academic Press, New York.

Royama, T. 1971. A comparative study of models for predation and parasitism. *Res. Popul. Ecol.*, Suppl. 1:1–91.

Royama, T. 1984. Population dynamics of the spruce budworm, *Choristoneura fumiferana*. *Ecol. Monogr.* 54:429–462.

Rubin, J. A. 1982. The degree of intransitivity and its measurement in an assemblage of encrusting Bryozoa. *J. Exp. Mar. Biol. Ecol.* 60:119–128.

Rummel, J. D., and J. Roughgarden. 1983. Some differences between invasion-structured and coevolution–structured competitive communities: a preliminary theoretical analysis. *Oikos* 41:477–486.

Rummel, J., and J. Roughgarden. 1985a. Effects of reduced perch-height separation on the competition between two *Anolis* lizards. *Ecology* 66:430–444.

Rummel, J., and J. Roughgarden. 1985b. *A theory of faunal buildup for competition communities*. Unpublished manuscript.

Runkle, J. R. 1981. Gap regeneration in some old-growth forests of the eastern United States. *Ecology* 62:1041–1051.

Russ, G. R. 1982. Overgrowth in a marine epifaunal community: competitive hierarchies and competitive networks. *Oecologia* 53:12–19.

Rutherford, M. C. 1981. Biomass structure and utilization of the natural vegetation in the winter rainfall regions of South Africa. *In* N. S. Margaris and H. A. Mooney, eds., *Components of Productivity of Mediterranean-Climate Regions: Basic and Applied Aspects*, pp. 135–149. Junk, Den Haag.

Sadler, P. M. 1981. Sediment accumulation rates and the completeness of stratigraphic sections. *J. Geol.* 89:569–584.

Sailer, R. I. 1978. Our immigrant insect fauna. *Ecol. Soc. Am. Bull.* 24:3–11.

Sale, P. F. 1977. Maintenance of high diversity in coral reef fish communities. *Am. Nat.* 111:337–359.

Sale, P. F. 1980. The ecology of fishes on coral reefs. *Oceanogr. Mar. Biol. Annu. Rev.* 18:367–421.

Sale, P. F. 1982. Stock-recruit relationships and regional coexistence in a lottery competitive system: a simulation study. *Am. Nat.* 120:139–159.

Sale, P. F. 1984. The structure of communities of fish on coral reefs and the merit of a hypothesis-testing, manipulative approach to ecology. *In* D. R. Strong, D. Simberloff, L. G. Abele, and A. B. Thistle, eds., *Ecological Communities: Conceptual Issues and the Evidence*, pp. 478–490. Princeton University Press, Princeton, N.J.

Sale, P. F., P. J. Doherty, G. J. Eckert, W. A. Douglas, and D. J. Ferrell. 1984. Large scale spatial and temporal variation in recruitment to fish populations on coral reefs. *Oecologia* 64:191–198.

Sale, P. F., and R. Dybdahl. 1975. Determinants of community structure for coral reef fishes in an experimental habitat. *Ecology* 56:1343–1355.

Sale, P. F., and D. M. Williams. 1982. Community structure of coral reef fishes: are the patterns more than those expected by chance? *Am. Nat.* 120:121–127.

Salomonsen, F. 1951. The immigration and breeding of the fieldfare (*Turdus pilaris* L.) in Greenland. *Proc. Xth Int. Ornithol. Cong.* pp. 515–526.

Salt, G. W. 1967. Predation in an experimental protozoa population *(Woodruffia-Paramecium)*. *Ecol. Monogr.* 37:113–144.

Salzburg, M. A. 1984. *Anolis sagrei* and *Anolis cris-*

tatellus in southern Florida: a case study in interspecific competition. *Ecology* 65:14–19.

Sameoto, D. D., and R. S. Miller. 1968. Selection of pupation site by *Drosophila melanogaster* and *Drosophila simulans*. *Ecology* 49:177–180.

Sammarco, P. W. 1982. Effects of grazing by *Diadema antillarum* Philippi (Echinodermata: echinoidae) on algal diversity and community structure. *J. Exp. Mar. Biol. Ecol.* 65:83–105.

Sanders, H. L. 1969. Benthic marine diversity and the stability-time hypothesis. *Brookhaven Symp. Biol.* 22:71–81.

Sang, J. H. 1949. Population growth in Drosophila cultures. *Biol. Rev.* 25:188–219.

Sankurathri, C. S., and J. C. Holmes. 1976. Effects of thermal effluents on parasites and commensals of *Physa gyrina* Say (Mollusca: Gastropoda) and their interactions at Lake Wabamun, Alberta. *Can. J. Zool.* 54:1742–1753.

Sarukhan, J. 1974. Studies of plant demography: *Ranunculus repens* L., *R. bulbosus* L. and *R. acris* L. II. Reproductive strategies and seed population dynamics. *J. Ecol.* 62:151–177.

Savino, J. F., and R. A. Stein. 1982. Predator-prey interaction between largemouth bass and bluegills as influenced by simulated, submerged vegetation. *Trans. Am. Fish. Soc.* 111:255–266.

Schad, G. A. 1966. Immunity, competition, and natural regulation of helminth populations. *Am. Nat.* 100:359–364.

Schaffer, W. M. 1979. The theory of life-history evolution and its application to the Atlantic Salmon. *Symp. Zool. Soc. Lond.* 44:307–326.

Schaffer, W. M. 1981. Ecological abstraction: the consequences of reduced dimensionality in ecological models. *Ecol. Monogr.* 51:383–401.

Schaffer, W. M., and M. D. Gadgil. 1975. Selection for optimal life histories in plants. *In* M. L. Cody and J. M. Diamond, eds., *Ecology and Evolution of Communities,* pp. 142–157. Harvard University Press, Cambridge, Mass.

Scheffer, V. B. 1951. The rise and fall of a reindeer herd. *Sci. Monthly* 73:356–362.

Schenkeveld, A. J. M., and H. J. P. A. Verkaar. 1984. *On the Ecology of Short-Lived Forbs in Chalk Grassland*. Proefschrift, University of Utrecht.

Schimper, A. F. W. 1903. *Plant Geography Upon a Physiological Basis*. Oxford University Press (Clarendon Press), Oxford.

Schindel, D. E. 1980. Microstratigraphic sampling and the limits of paleontological resolution. *Paleobiology* 6:408–426.

Schlesinger, W. H., and D. S. Gill. 1980. Biomass, production, and changes in the availability of light, water and nutrients during the development of pure stands of the chaparral shrub *Ceanothus megacarpus* after fire. *Ecology* 61:781–789.

Schluter, D. 1981. Does the theory of optimal diets apply in complex environments? *Am. Nat.* 118:139–147.

Schluter, D. 1982a. Seed and patch selection by Galápagos ground finches: relation to foraging efficiency and food supply. *Ecology* 63:1106–1120.

Schluter, D. 1982b. Distributions of Galápagos ground finches along an altitudinal gradient: the importance of food supply. *Ecology* 63:1504–1517.

Schluter, D. 1984. A variance test for detecting species associations with some example applications. *Ecology*. 65:998–1005.

Schluter, D., and P. R. Grant. 1982. The distribution of *Geospiza difficilis* in relation to *G. fuliginosa* in the Galápagos Islands: tests of three hypotheses. *Evolution* 36:1213–1226.

Schluter, D., and P. R. Grant. 1984a. Determinants of morphological patterns in communities of Darwin's Finches. *Am. Nat.* 123:175–196.

Schluter, D., and P. R. Grant. 1984b. Ecological correlates of morphological evolution in a Darwin's finch, *Geospiza difficilis*. *Evolution* 38:856–869.

Schmidt, R. H., Jr. 1979. A climatic delineation of the ''real'' Chihuahuan Desert region. *Phytologia* 44:129–133.

Schmitt, R. C. 1977. *Historical Statistics of Hawaii*. University Press of Hawaii, Honolulu.

Schnell, J. H. 1968. The limiting effects of natural predation on experimental cotton rat populations. *J. Wildl. Manage.* 32:698–711.

Schoener, A., and T. W. Schoener. 1981. The dynamics of the species-area relation in marine fouling systems. I. Biological correlates of changes in the species-area slope. *Am. Nat.* 118:339–360.

Schoener, A., and T. W. Schoener. 1984. Experiments on dispersal: short-term flotation of insular anoles, with a review of similar abilities in other terrestrial animals. *Oecologia* 63:289–294.

Schoener, T. W. 1965. The evolution of bill size differences among sympatric species of birds. *Evolution* 19:189–213.

Schoener, T. W. 1968. The *Anolis* lizards of Bimini: resource partitioning in a complex fauna. *Ecology* 49:704–726.

Schoener, T. W. 1969. Size patterns in West Indian *Anolis* lizards. I. Size and species diversity. *Syst. Zool.* 18:386–401.

Schoener, T. W. 1970a. Size patterns in West Indian *Anolis* lizards. II. Correlations with the sizes of particular sympatric species—displacement and convergence. *Am. Nat.* 104:155–174.

Schoener, T. W. 1970b. Nonsynchronous spatial overlap of lizards in patchy habitats. *Ecol.* 51:408–418.

Schoener, T. W. 1972. Mathematical ecology and its place among the sciences. I. The biological domain. *Science* 178:389–391.

Schoener, T. W. 1973. Population growth regulated by intraspecific competition for energy or time: some simple representations. *Theor. Popul. Biol.* 4:56–84.

Schoener, T. W. 1974a. Resource partitioning in ecological communities. *Science* 185:27–39.

Schoener, T. W. 1974b. Competition and the form of habitat shift. *Theor. Popul. Biol.* 6:265–307.

Schoener, T. W. 1975. Presence and absence of habitat shift in some widespread lizard species. *Ecol. Monogr.* 45:233–258.

Schoener, T. W. 1976a. The species-area relation within archipelagos: models and evidence from island land birds. *Proc. 16th Int. Ornith. Congr.* (Canberra), pp. 629–642.

Schoener, T. W. 1976b. Alternatives to Lotka-Volterra competition: models of intermediate complexity. *Theor. Popul. Biol.* 10:309–333.

Schoener, T. W. 1978. Effect of density-restricted food encounter on some single-level competition models. *Theor. Popul. Biol.* 13:365–381.

Schoener, T. W. 1982. The controversy over interspecific competition. *Am. Sci.* 70:586–595.

Schoener, T. W. 1983a. Reply to John Wiens. [Letter to the Editor]. *Am. Sci.* 71:235.

Schoener, T. W. 1983b. Field experiments on interspecific competition. *Am. Nat.* 122:240–285.

Schoener, T. W. 1983c. Rate of species turnover decreases from lower to higher organisms: a review of the data. *Oikos* 41:372–377.

Schoener, T. W. 1984. Size differences among sympatric, bird-eating hawks: a worldwide survey. *In* D. R. Strong, D. Simberloff, L. G. Abele, and A. B. Thistle, eds., *Ecological Communities: Conceptual Issues and the Evidence,* pp. 254–281. Princeton University Press, Princeton, N.J.

Schoener, T. W. 1985a. Mechanistic approaches to community ecology: a new reductionism? *Am. Zool.*

Schoener, T. W. 1985b. Resource partitioning. *In* D. Anderson and J. Kikkawa, eds., *Community Ecology—Pattern and Process.* Blackwell, Oxford.

Schoener, T. W. 1985c. Some comments on Connell's and my reviews of field experiments on interspecific competition. *Am. Nat.* 125:730–740.

Schoener, T. W. 1985d. *On Testing the MacArthur-Wilson Model with Data on Rates.* Manuscript in preparation.

Schoener, T. W. 1985e. Are lizard populations unusually constant through time? *Am. Nat.* in press.

Schoener, T. W., and G. Gorman. 1968. Some niche differences in three Lesser-Antillean lizards of the genus *Anolis*. *Ecology* 49:819–830.

Schoener, T. W., and D. Janzen. 1968. Notes on environmental determinants of tropical versus temperate insect size patterns. *Am. Nat.* 102:207–224.

Schoener, T. W., and A. Schoener. 1978a. Estimating and interpreting body-size growth in some *Anolis* lizards. *Copeia* 1978:390–405.

Schoener, T. W., and A. Schoener. 1978b. Inverse relation of survival of lizards with island size and avifaunal richness. *Nature* 274:685–687.

Schoener, T. W., and A. Schoener. 1980a. Densities, sex ratios and population structure in four species of Bahamian Anolis lizards. *J. Anim. Ecol.* 49:19–53.

Schoener, T. W., and A. Schoener. 1980b. Ecological and demographic correlates of injury rates in some Bahamian *Anolis* lizards. *Copeia* 1980:839–850.

Schoener, T. W., and A. Schoener. 1982a. Intraspecific variation in home-range size in some *Anolis* lizards. *Ecology* 63:809–823.

Schoener, T. W., and A. Schoener. 1982b. The ecological correlates of survival in some Bahamian *Anolis* lizards. *Oikos* 39:1–16.

Schoener, T. W., and A. Schoener. 1983a. Distribution of vertebrates on some very small islands. I. Occurrence sequences of individual species. *J. Anim. Ecol.* 52:209–235.

Schoener, T. W., and A. Schoener. 1983b. Distribution of vertebrates on some very small islands. II. Patterns in species counts. *J. Anim. Ecol.* 52:237–262.

Schoener, T. W., and A. Schoener. 1983c. The time to extinction of a colonizing propagule of lizards increases with island area. *Nature* 302:332–334.

Schoener, T. W., J. Slade, and C. Stinson. 1982. Diet and sexual dimorphism in the very catholic lizard genus *Leiocephalus* of the Bahamas. *Oecologia* 53:160–169.

Schoener, T. W., and C. A.Toft. 1983a. Spider populations: extraordinarily high densities on islands without top predators. *Science* 219:1353–1355.

Schoener, T. W., and C. A. Toft. 1983b. Dispersion of a small-island population of *Metepeira datona* (Araneae: Araneidae) in relation to web-site availability. *Behav. Ecol. Sociobiol.* 12:121–128.

Schreiber, R. W., and E. A. Schreiber. 1984. Central Pacific seabirds and the El Niño southern oscillation: 1892 to 1983 perspectives. *Science* 225:713–716.

Schroder, G. D., and M. L. Rosenzweig. 1975. Perturbation analysis of competition and overlap in habitat utilization between *Dipodomys ordii* and *Dipodomys merriami*. *Oecologia* 19:9–28.

Schultz, J., and I. Baldwin. 1982. Oak leaf quality

declines in response to defoliation by gypsy moth larvae. *Science* 217:149–151.

Schutz, D. C., and T. G. Northcote. 1972. An experimental study of feeding behavior and interaction of coastal cutthroat trout *(Salmo clarki clarki)* and Dolly Varden (*Salvelinus malma*). *J. Fish. Res. Board Can.* 29:555–565.

Scott, A. C. 1977. A review of the ecology of upper Carboniferous plant assemblages with new data from Strathclyde. *Palaeontology* 20:447–473.

Scott, A. C. 1978. Sedimentological and ecological control of Westphalian B plant assemblages from West Yorkshire. *Proc. Yorkshire Geol. Soc.* 41:461–508.

Scott, A. C., and T. L. Taylor. 1983. Plant/animal interactions during the upper Carboniferous. *Bot. Rev.* 49:259–307.

Scott, W. B., and E. J. Crossman. 1973. *Freshwater Fishes of Canada*. Fisheries Research Board of Canada, Ottawa.

Seaburg, K. G., and J. B. Moyle. 1964. Feeding habits, digestive rates, and growth of some Minnesota warmwater fishes. *Trans. Am. Fish. Soc.* 93:269–285.

Sebens, K. 1982. Competition for space: growth rate, reproductive output, and escape in size. *Am. Nat.* 120:189–197.

Sebens, K. 1983. Population dynamics and habitat suitability of the intertidal sea anemones *Anthopleura elegantissima* and *A. xanthogrammica*. *Ecol. Monogr.* 53:405–433.

Seber, G. A. F. 1982. *The Estimation of Animal Abundance and Related Parameters*. Griffin, London.

Segal, L. A., and J. L. Jackson. 1972. Dissipative structure: an explanation and an ecological example. *J. Theor. Biol.* 37:545–559.

Semken, H. A., Jr. 1974. Micromammal distribution and migration during the Holocene. *AMQUA Abstract,* p. 25. University of Wisconsin, Madison.

Semken, H. A., Jr. 1983. Holocene mammalian biogeography and climatic change in the eastern and central United States. *In* H. E. Wright, Jr., ed., *Late-Quaternary Environments of the United States. Vol. II. The Holocene,* pp. 182–207. University of Minnesota Press, Minneapolis.

Sepkoski, J. J., Jr. 1979. A kinetic model of Phanerozoic taxonomic diversity II. Early Phanerozoic families and multiple equilibria. *Paleobiology* 5:222–251.

Sepkoski, J. J., Jr., R. K. Bambach, D. M. Raup, and J. W. Valentine. 1981. Phanerozoic marine diversity and the fossil record. *Nature* 293:435–437.

Serventy, D. 1971. Biology of desert birds. *In* D. S. Farner and J. R. King, eds., *Avian Biology,* pp. 287–339. Academic Press, New York.

Serventy, D. L., and P. J. Curry. 1984. Observations on colony size, breeding success, recruitment, and inter-colony dispersal in a Tasmanian colony of Short-tailed Shearwaters *Puffinus tenuirostris* over a 30-year period. *Emu* 84:71–79.

Service, P. 1984. Genotypic interactions in an aphid-host plant relationship: *Uroleucon rudbeckiae* and *Rudbeckia laciniata*. *Oecologia* 61:271–276.

Shackleton, N. J., and N. D. Opdyke. 1973. Oxygen isotope and paleomagnetic stratigraphy of equatorial Pacific core V28–238: oxygen isotope temperatures and ice volumes on a 10^5 year and 10^6 year scale. *Quat. Res.* 3:39–55.

Shapiro, J., and D. I. Wright. 1984. Lake restoration by biomanipulation: Round Lake, Minnesota, the first two years. *Freshwater Biol.* 14:371–383.

Shaw, R. G. 1983. *Density-Dependence and Demographic Genetics in* Salvia lyrata *L.* Doctoral dissertation, Duke University, Durham, N.C.

Shepherd, J. G., and D. H. Cushing. 1980. A mechanism for density-dependent survival of larval fish as the basis of a stock-recruitment relationship. *J. Cons. Int. Explor. Mer.* 39:160–167.

Sheppard, D. H. 1971. Competition between two chipmunk species (*Eutamias*). *Ecology* 52:320–329.

Shigesada, N., K. Kawasaki, and E. Teramoto. 1979. Spatial segregation of interacting species. *J. Theor. Biol.* 79:83–99.

Shmida, S., and S. P. Ellner. 1984. Coexistence of plant species with similar niches. *Vegetatio* 58:29–55.

Shmida, A., M. Evenari, and I. Noy-Meir. 1985. Hot desert ecosystems: an integrated view. *In* M. Evenari, I. Noy-Meir, and D. W. Goodall, eds., *Hot Desert Ecosystems*. Elsevier, Amsterdam.

Shochat, D. 1976. *Comparative Immunological Study of Albumins of* Anolis *Lizards of the Caribbean Islands*. Doctoral dissertation, Louisiana State University, School of Medicine, New Orleans.

Shugart, H. H., Jr., T. R. Crow, and J. M. Hett. 1973. Forest succession models: a rationale and methodology for modelling forest succession over large regions. *For. Sci.* 19:203–212.

Shugart, H. H., Jr., D. C. West, and W. R. Emmanual. 1981. Patterns and dynamics of forests: an application of simulation models. *In* D. C. West, H. H. Shugart, Jr., and D. B. Botkin, eds., *Forest Succession,* pp. 74–94. Springer-Verlag, New York.

Siegel, S. 1956. *Nonparametric Statistics*. McGraw-Hill, New York.

Sih, A., and J. Dixon. 1983. Tests of some predictions

from the MacArthur-Levins competition models: a critique. *Am. Nat.* 117:550–559.

Sih, A., P. Crowley, M. McPeck, J. Petraaka, and K. Strohmeier. 1986. Predation, competition, and prey communities: a review of field experiments. *Annu. Rev. Ecol. Syst.*, in press.

Silvertown, J. W. 1982. *Introduction to Plant Population Ecology*. Longman, London.

Simberloff, D. 1978. Using island biogeographic distributions to determine if colonization is stochastic. *Am. Nat.* 112:713–726.

Simberloff, D. 1980. A succession of paradigms in ecology: essentialism to materialism and probabilism. *Synthese* 43:3–39.

Simberloff, D. 1981. Community effects of introduced species. *In* M. H. Nitecki, ed., *Biotic Crises in Ecological and Evolutionary Time,* pp. 53–81. Academic Press, New York.

Simberloff, D. 1983. Competition theory, hypothesis testing, and other community-ecological buzzwords. *Am. Nat.* 122:626–635.

Simberloff, D. 1984a. Properties of coexisting bird species in two archipelagoes. *In* D. R. Strong, D. Simberloff, L. G. Abele, and A. B. Thistle, eds., *Ecological Communities: Conceptual Issues and the Evidence,* pp. 234–253. Princeton University Press, Princeton, N.J.

Simberloff, D. 1984b. The great god of competition. *The Sciences* 24:17–22.

Simberloff, D., and W. Bocklen. 1981. Santa Rosalia reconsidered: size ratios and competition. *Evolution* 35:1206–1228.

Simberloff, D., B. J. Brown, and S. Lowric. 1978. Isopod and insect root borers may benefit Florida mangroves. *Science* 201:630–632.

Simberloff, D., and E. F. Connor. 1982. Missing species combinations. *Am. Nat.* 118:215–239.

Simberloff, D., and E. O. Wilson. 1969. Experimental zoogeography of islands. The colonization of empty islands. *Ecology* 50:278–296.

Simpson, G. G. 1969. Species density of North American recent mammals. *Syst. Zool.* 13:57–73.

Simpson, G. G. 1980. *Splendid Isolation*. Yale University Press, New Haven, Conn.

Sinclair, A. R. E. 1975. The resource limitation of trophic levels in tropical grassland ecosystems. *J. Anim. Ecol.* 44:497–520.

Sinclair, A. R. E., and M. Norton-Griffiths. 1982. Does competition or facilitation regulate migrant ungulate populations in the Serengeti? A test of hypotheses. *Oecologia* 53:364–369.

Singer, M. C., and P. R. Ehrlich. 1979. Population dynamics of the checkerspot butterfly *Euphydryas editha*. *Fortschr. Zool.* 25:53–60.

Singh, G., A. P. Kershaw, and R. Clark. 1981. Quaternary vegetation and fire history in Australia. *In* A. M. Gill, R. H. Groves, and F. R. Noble. eds., *Fire and the Australian Biota,* pp. 23–54. Australian Academy of Science, Canberra.

Sissenwine, M. P. 1984. Why do fish populations vary? *In* R. M. May, ed., *Exploitation of Marine Communities*. Dahlem Konferenzen. Springer-Verlag, Berlin, Heidelberg, New York, Tokyo.

Skellam, J. G. 1951. Random dispersal in theoretical populations. *Biometrika* 38:196–218.

Skellam, J. G. 1952. Studies in statistical ecology. I. Spatial pattern. *Biometrika* 39:346–362.

Skinner, G. J., and J. B. Whittaker. 1981. An experimental investigation of inter-relationships between the wood-ant (*Formica rufa*) and some tree-canopy herbivores. *J. Anim. Ecol.* 50:313–326.

Slade, N. A. 1977. Statistical detection of density dependence from a series of sequential censuses. *Ecology* 58:1094–1102.

Slade, N. A., and R. J. Wassersug. 1975. On the evolution of complex life cycles. *Evolution* 29:568–571.

Slagsvold, T. 1978. Competition between the Great Tit *Parus major* and the Pied Flycatcher *Ficedula hypoleuca:* an experiment. *Ornis Scand.* 9:46–50.

Slatkin, M. 1974. Competition and regional coexistence. *Ecology* 55:128–134.

Slatkin, M. 1980. Ecological character displacement. *Ecology* 61:163–177.

Slaughter, B. H. 1975. Ecological interpretations of Brown Sand Wedge local fauna. *In* F. Wendorf and J. J. Hester, eds., *Late Pleistocene Environments of the Southern High Plains,* pp. 179–192. Fort Burgwin Research Center, Taos, N.Mex.

Slobodkin, L. B. 1964. Experimental populations of Hydrida. *J. Anim. Ecol.* 33(Suppl.):131–148.

Slobodkin, L. B., F. E. Smith, and N. G. Hairston. 1967. Regulation in terrestrial ecosystems, and the implied balance of nature. *Am. Nat.* 101:109–124.

Smale, S. 1976. On the differential equations of species in competition. *J. Math. Biol.* 3:5–7.

Smart, J., and N. F. Hughes. 1972. The insect and the plant: progressive palaeoecological integration. *Symp. R. Entomol. Soc. Lond.* 6:3–30.

Smith, C. L. 1973. Small rotenone stations: a tool for studying coral reef fish communities. *Am. Mus. Novit.* no. 2512:1–21.

Smith, C. L., and A. J. Tyler. 1972. Space resource sharing in a coral reef fish community. *Bull. Nat. Hist. Los Angeles County* 14:115–170.

Smith, D. C. 1981. Competitive interaction of the striped plateau lizard (*Sceloporus virgatus*) and the tree lizard (*Urosaurus ornatus*). *Ecology* 62:679–687.

Smith, J. N. M., and A. A. Dhondt. 1980. Experimental confirmation of heritable morphological variation in a natural population of song sparrows. *Evolution* 34:1155–1158.

Smith, J. N. M., P. R. Grant, B. R. Grant, I. J. Abbott, and L. K. Abbott. 1978. Seasonal variation in feeding habits of Darwin's ground finches. *Ecology* 59:1137–1150.

Smith, J. N. M., and R. Zach. 1979. Heritability of some morphological characters in a song sparrow population. *Evolution* 33:460–467.

Smith, K. G. 1982. Drought-induced changes in avian community structure along a montane sere. *Ecology* 63:952–961.

Smith, N. G. 1968. The advantage of being parasitized. *Nature* 219:690–694.

Smith, N. G. 1979. Alternate responses by hosts to parasites which may be helpful or harmful. *In* B. B. Nickol, ed., *Host Parasite Interfaces,* pp. 7–15. Academic Press, New York.

Smith, S. H. 1968. Species succession and fishery exploitation in the Great Lakes. *J. Fish. Res. Board Can.* 25:667–693.

Smith, V. H. 1983. Low nitrogen to phosphorus ratios favor dominance by blue-green algae in lake phytoplankton. *Science* 221:669–671.

Snaydon, R. W. 1962. Micro-distribution of *Trifolium repens* L. and its relation to soil factors. *J. Ecol.* 50:133–143.

Solomon, M. E. 1957. Dynamics of insect populations. *Annu. Rev. Entomol.* 2:121–142.

Sousa, W. P. 1979. Disturbance in a marine intertidal boulder field: the non-equilibrium maintenance of species diversity. *Ecology* 60:1225–1239.

Southward, A. J., and D. J. Crisp. 1954. Recent changes in the distribution of the intertidal barnacles *Chthamalus stellatus* Poli and *Balanus balanoides* L. in the British Isles. *J. Anim. Ecol.* 23:163–177.

Southwood, T. R. E. 1977. Habitat, the templet for ecological strategies? *J. Anim. Ecol.* 46:337–365.

Southwood, T. R. E. 1978. *Ecological Methods.* Chapman and Hall, London.

Southwood, T. R. E., and H. N. Comins. 1976. A synoptic population model. *J. Anim. Ecol.* 45:949–965.

Southwood, T. R. E., and P. M. Reader. 1976. Population census data and key factor analysis for the viburnum whitefly *Aleurotrachelus jelinikii* (fravenf.) on three bushes. *J. Anim. Ecol.* 45:313–325.

Souza, H. L. de, A. B. da Cunha, and E. P. dos Santos. 1968. Adaptive polymorphism of behavior developed in laboratory populations of *Drosophila willistoni*. *Am. Nat.* 102:583–586.

Spaulding, W. G., E. B. Leopold, and T. R. Van Devender. 1983. Late Wisconsin paleoecology of the American Southwest. *In* S. C. Porter, ed., *Late-Quaternary Environments of the United States*. The Late Pleistocene, pp. 259–293. University of Minnesota Press, Minneapolis.

Spear, R. W. 1985. Vegetational history of the alpine and subalpine zones of the White Mountains of New Hampshire. *Ecol. Monogr.,* 1985.

Specht, R. L. 1963. Dark Island heath (Ninety Mile Plain, South Australia). VII. The effect of fertilizers on composition and growth, 1950–60. *Aust. J. Bot.* 11:62–66.

Specht, R. L. 1973. Structure and functional response of ecosystems in the Mediterranean climate of Australia. *In* F. diCastri and H. A. Mooney, eds., *Mediterranean-Type Ecosystems: Origins and Structure,* pp. 113–120. Ecological Studies no. 7. Springer-Verlag, Berlin, New York.

Specht, R. L., and E. J. Moll. 1983. Mediterranean-type heathlands and sclerophyllous shrublands of the world: an overview. *In* F. J. Kruger, D. T. Mitchell, and J. U. M. Jarvis, eds., *Mediterranean-Type Ecosystems: the Role of Nutrients,* pp. 41–65. Ecological Studies no. 43. Springer-Verlag, Berlin, New York.

Speight, T. 1975. On a snail's chances of becoming a year old. *Oikos* 16:9–14.

Spicer, R. A., and C. R. Hill. 1979. Principal components and correspondence analysis of quantitative data from a Jurassic plant bed. *Rev. Palaeobot. Palynol.* 28:273–299.

Spieth, H. T. 1974. Courtship behavior in *Drosophila*. *Annu. Rev. Entomol.* 19:385–405.

Spiller, D. A. 1984. Competition between two spider species: an experimental field study. *Ecology* 65:905–919.

Springett, B. P. 1968. Aspects of the relationship between burying beetles, *Necrophorus* spp., and the mite, *Poecilochirus necrophori* Vitz. *J. Anim. Ecol.* 37:417–424.

St. Amant, J. L. S. 1970. The detection of regulation in animal populations. *Ecology* 51:823–828.

Stamps, J. A. 1976. Egg retention, rainfall, and egg-laying in a tropical lizard *Anolis aeneus*. *Copeia* 1976:759–764.

Stamps, J. A. 1977. Rainfall, moisture, and dry season growth rates in *Anolis aeneus*. *Copeia* 1977:415–419.

Stanton, M. 1983. Spatial patterns in the plant community and their effects upon insect search. *In* S. Ahmad, ed., *Herbivorous Insects: Host-seeking Behavior and Mechanisms,* pp. 125–157. Academic Press, New York.

Starks, K. J., R. Muniappan, and R. Eikenbary. 1972. Interaction between plant resistance and parasitism against greenbug or barley and sorghum. *Ann. Entomol. Soc. Am.* 65:650–655.

Steadman, D. W., and P. S. Martin. 1984. Extinction of birds in the late Pleistocene of North America. *In* P. S. Martin and R. G. Klein, eds., *Quaternary Extinctions: a Prehistoric Revolution,* pp. 466–477. University of Arizona Press, Tucson.

Steadman, D. W., G. K. Pregill, and S. L. Olson. 1984. Fossil vertebrates from Antigua, Lesser Antilles: evidence for late Holocene human-caused extinctions in the West Indies. *Proc. Natl. Acad. Sci. U.S.A.* 81:4448–4451.

Stebbing, A. R. D. 1973.Competition for space between the epiphytes of *Fucus serratus* L. *J. Mar. Biol. Assoc. U.K.* 53:247–261.

Stebbins, G. L., and G. J. C. Hill. 1980. Did multicellular plants invade the land? *Am. Nat.* 115:342–353.

Steele, J. H., and E. W. Henderson. 1984. Modeling long-term fluctuations in fish stocks. *Science* 224:985–987.

Steenis, C. G. C. J. van. 1969. Plant speciation in Malesia with reference to the theory of nonadaptive saltatory evolution. *Biol. Linn. Soc.* 1:97–133.

Stein, G. H. W. 1936. Ornithologische Ergebnisse der Expedition Stein 1931-1932. V. Beiträge zur Biologie papuanischer Vögel. *J. Ornithol.* 84:21–57.

Stein, G. H. W. 1951. Populationsanalytische Untersuchungen am europäischen Maulwurf. *Zool. Jahrb. (Syst.)* 79:567–590.

Stel, J. H. 1978. *Studies on the Paleobiology of Favositids.* Stabol/All-Round, Groningen.

Steneck, R. S. 1982. A limpet-coralline alga association: adaptations and defenses between a selective herbivore and its prey. *Ecology* 63:507–522.

Steneck, R. S., and L. Watling. 1982. Feeding capabilities and limitation of herbivorous molluscs: a functional group approach. *Mar. Biol.* 68:299–319.

Stenson, J. A. E., T. Bohlin, L. Henrikson, B. T. Nilsson, H. G. Nyman, H. G. Oscarson, and P. Larsson. 1978. Effects of fish removal from a small lake. *Verh. Int. Verein. Theor. Angew. Limnol.* 20:794–801.

Stephens, G. R., and P. E. Waggoner. 1970. The forests anticipated from 40 years of natural transitions in mixed hardwoods. *Bull. Conn. Agric. Exp. Station* (New Haven) 707:1–58.

Stephenson, A. G. 1981. Flower and fruit abortion: proximate causes and ultimate functions. *Annu. Rev. Ecol. Syst.* 12:253–279.

Stephenson, A. G. 1982. The role of extrafloral nectaries of *Catalpa speciosa* in limiting herbivory and increasing fruit production. *Ecology* 63:663–669.

Stewart, D. J., J. F. Kitchell, and L. B. Crowder. 1981. Forage fishes and their salmonid predators in Lake Michigan. *Trans. Am. Fish. Soc.* 110:751–763.

Stiling, P. D., and D. R. Strong. 1982. Egg density and the intensity of parasitism in *Prokelisia marginata* (Homoptera: Delphacidae). *Ecology* 63:1630–1635.

Stiven, A. E. 1971. The spread of Hydramoeba infections in mixed hydra species systems. *Oecologia* 6:118–132.

Stoner, A. W. 1980. Feeding ecology of *Lagodon rhomboides* (Pisces: Sparidae): variation and functional responses. *Fish. Bull.* 78:337–352.

Stoner, A. W., and R. J. Livingston. 1984. Ontogenetic patterns in diet and feeding morphology in sympatric sparid fishes from seagrass meadows. *Copeia* 1984:174–187.

Strong, D. R. 1977. Epiphyte loads, treefalls, and perennial forest disruption: a mechanism for maintaining higher tree species richness in the tropics without animals. *J. Biogeogr.* 4:215–218.

Strong, D. R. 1980. Null hypotheses in ecology. *Synthese* 43:271–285.

Strong, D. R. 1982. Harmonious coexistence of hispine beetles in *Heliconia* in experimental and natural communities. *Ecology* 63:1039–1049.

Strong, D. R. 1983. Natural variability and the manifold mechanisms of ecological communities. *Am. Nat.* 122:636–660.

Strong, D. R. 1984a. Exorcising the ghost of competition past: phytophagous insects. *In* D. R. Strong, D. Simberloff, L. G. Abele, and A. B. Thistle, eds., *Ecological Communities: Conceptual Issues and the Evidence,* pp. 28–41. Princeton University Press, Princeton, N.J.

Strong, D. R. 1984b. Density-vague ecology and liberal population regulation in insects. *In* P. W. Price, C. N. Slobodchikoff, and W. S. Gaud, eds., *A New Ecology: Novel Approaches to Interactive Systems,* pp. 313–327. Wiley, New York.

Strong, D. R., J. H. Lawton, and T. R. E. Southwood. 1984. *Insects on Plants: Community Patterns and Mechanisms.* Blackwell, Oxford.

Strong, D. R., D. Simberloff, L. G. Abele, and A. B. Thistle, eds. 1984. *Ecological Communities: Conceptual Issues and the Evidence.* Princeton University Press, Princeton, N. J.

Strong, D. R., L. A. Szyska, and D. Simberloff. 1979. Tests of community-wide character displacement against null hypotheses. *Evolution* 33:897–913.

Stuart, A. J. 1979. Pleistocene occurrences of the European pond tortoise (*Emys orbicularis* L.) in Britain. *Boreas* 8:359–371.

Stuart, A. J. 1982. *Pleistocene Vertebrates in the British Isles*. Longman, New York.

Suchanek, T. H. 1985a. *Mutualism in a Species-rich Mussel Bed (Mytilus californianus): Enhancement of Community Stability*. Unpublished manuscript.

Sugihara, G. 1983. Holes in niche space: a derived assembly rule and its relation to intervality. *In* D. L. DeAngelis, W. M. Post, and G. Sugihara, eds., *Current Trends in Food Web Theory,* pp. 25–35. Oak Ridge National Laboratory, Oak Ridge, Tenn.

Suppe, F. 1977. *The Structure of Scientific Theories,* 2nd ed. University of Illinois Press, Urbana.

Sutherland, J. P. 1974. Multiple stable points in natural communities. *Am. Nat.* 108:859–873.

Sutherland, J. P. 1976. Life histories and the dynamics of fouling communities. *In* J. D. Costlow, ed., *The Ecology of Fouling Communities,* pp. 137–153. Duke University Marine Laboratory, Beaufort, N. C.

Sutherland, J. P. 1977. Effects of *Schizoporella* removal on the fouling community at Beaufort, N. C. *In* B. C. Coull, ed., *Ecology of Marine Benthos,* pp. 155–189. University of South Carolina Press, Columbia.

Sutherland, J. P. 1978. Functional roles of *Schizoporella* and *Styela* in the fouling community at Beaufort, N.C. *Ecology* 59:257–264.

Sutherland, J. P., and R. H. Karlson. 1977. Development and stability of the fouling community at Beaufort, North Carolina. *Ecol. Monogr.* 47:425–446.

Svärdson, G. 1949. Competition and habitat selection in birds. *Oikos* 1:157–174.

Svärdson, G. 1976. Interspecific population dominance in fish communities of Scandinavian lakes. *Inst. Freshwater Res. Drottningholm Rep.* 55:144–171.

Swarbrick, S. L. 1984. *Disturbance, Recruitment, and Competition in a Marine Invertebrate Community*. Doctoral dissertation, University of California, Santa Barbara.

Taggart, R. E., A. T. Cross, and L. Satchell. 1982. Effects of periodic volcanism on Miocene vegetation distribution in eastern Oregon and western Idaho. *In* B. Mamet and M. J. Copeland, eds., *North American Paleontological Convention* (Third), pp. 535–540. Business and Economic Service, Toronto.

Tansley, A. G. 1935. Use and abuse of vegetation concepts and terms. *Ecology* 16:284–307.

Tansley, A. G., and R. S. Adamson. 1925. Studies of the vegetation of the English chalk. III. The chalk grasslands of the Hampshire-Sussex border. *J. Ecol.* 13:177–223.

Taylor, D. L. 1973. Some ecological implications of forest fire control in Yellowstone National Park, Wyoming. *Ecology* 54:1394–1396.

Taylor, H. C., and F. van der Meulen. 1981. Structural and floristic classifications of Cape Mountain fynbos on Rooiberg, southern Cape. *Bothalia* 13:557–567.

Taylor, L. R. 1965. A natural law for the spatial disposition of insects. *Proc. XII Int. Congr. Entomol.* pp. 396–397.

Taylor, L. R. 1971. Aggregation as a species characteristic. *In* G. Patil, E. Pielou, and W. Waters, eds., *Statistical Ecology,* vol. 1, pp. 357–377. Pennsylvania State University Press, Philadelphia.

Taylor, P. L. 1979. Paleoecology of the encrusting epifauna of some British Jurassic bivalves. *Paleogeogr. Paleoclimatol. Paleoecol.* 28:241–262.

Taylor, S. E. 1975. Optimal leaf form. *In* D. M. Gates and R. B. Schmerl, eds., *Perspectives in Biophysical Ecology,* pp. 73–86. Springer-Verlag, New York.

Taylor, T. L., and M. A. Millay. 1979. Pollination biology and reproduction in early seed plants. *Rev. Palaeobot. Palynol.* 27:329–355.

Temple, S. A. 1977. Plant-animal mutualism: coevolution with Dodo leads to near extinction of plants. *Science* 197:885–886.

Templeton, A. R. 1980. Theory of speciation via the founder principle. *Genetics* 94:1011–1038.

Templeton, A. R., and L. E. Gilbert. 1985. Population genetics and the coevolution of mutualism. *In* D. H. Boucher, ed., *Mutualism: New Ideas About Nature*. Croom Helm, London.

Terborgh, J. 1971. Distribution on environmental gradients: theory and a preliminary interpretation of distributional patterns in the avifauna of the Cordillera Vilcabamba, Peru. *Ecology* 52:23–40.

Terborgh, J. 1983. *Five New World Primates*. Princeton University Press, Princeton, N.J.

Terborgh, J., and J. S. Weske. 1975. The role of competition in the distribution of Andean birds. *Ecology* 56:562–576.

Terborgh, J., and B. Winter. 1980. Some causes of extinction. *In* M. E. Soulé and B. A. Wilcox, eds., *Conservation Biology,* pp. 119–134. Sinauer, Sunderland, Mass.

Thomas, A. G., and H. M. Dale. 1976. Cohabitation of three *Hieracium* species in relation to the spatial heterogeneity in an old pasture. *Can. J. Bot.* 54:2517–2529.

Thomas, W. R., M. J. Pomerantz, and M. E. Gilpin. 1980. Chaos, asymmetric growth and group selection for dynamical stability. *Ecology* 61:1312–1320.

Thompson, J. N. 1982. *Interaction and Coevolution*. Wiley, New York.

Thompson, J. N. 1984. Variation among individual seed masses in *Lomatium grayi* (Umbelliferae) under controlled conditions: magnitude and partitioning of the variance. *Ecology* 65:626–631.

Thompson, P. H., and J. H. Lawton. 1983. Seed size diversity, bird species diversity and interspecific competition. *Ornis Scand.* 14:327–336.

Thompson, W. R. 1956. The fundamental theory of natural and biological control. *Annu. Rev. Entomol.* 1:379–402.

Thomson, G. M. 1922. *The Naturalization of Animals and Plants in New Zealand*. Cambridge University Press, Cambridge.

Thomson, J. D. 1980. Implications of different sorts of evidence for competition. *Am. Nat.* 116:719–726.

Thomson, J. D., B. J. Andrews, and R. C. Plowright. 1981. The effect of a foreign pollen on ovule development in *Diervilla lonicera* (Caprifoliaceae). *New Phytol.* 90:777–783.

Thornhill, R., and J. Alcock. 1983. *The Evolution of Insect Mating Systems*. Harvard University Press, Cambridge, Mass.

Thorton, I. W. B. 1967. The measurement of isolation on archipelagos, and its relation to insular faunal size and endemism. *Evolution* 21:842–849.

Thorpe, J. H., and E. A. Bergey. 1981. Field experiments on responses of a freshwater, benthic macroinvertebrate community to vertebrate predators. *Ecology* 62:365–375.

Thresher, R. 1983. Habitat effects on reproductive success in the coral reef fish, *Acanthochromis polyacanthus* (Pomacentridae). *Ecology* 64:1184–1199.

Thrower, N. J. W., and D. E. Bradbury, eds. 1977. *Chile-California Mediterranean Scrub Atlas*. Dowden, Hutchinson and Ross, Stroudsburg, Pa.

Thurston, J. 1969. The effect of liming and fertilizers on the botanical composition of permanent grassland, and on the yield of hay. *In* I. Rorison, ed., *Ecological Aspects of the Mineral Nutrition of Plants,* pp. 3–10. Blackwell, Oxford.

Tiffney, B. H. 1977. Fossil angiosperm fruits and seeds. *J. Seed Technol.* 2:54–71.

Tiffney, B. H. 1981. Diversity and major events in the evolution of land plants. *In* K. J. Niklas, ed., *Paleobotany, Paleoecology, and Evolution*. vol. 2, pp. 193–230. Praeger, New York.

Tilles, D. A., and D. L. Wood. 1982. The influence of carpenter ant *(Camponotus modoc)* (Hymenoptera: Formicidae) attendance on the development and survival of aphids (*Cinara* spp.) (Homoptera: Aphididae) in a giant sequoia forest. *Can. Entomol.* 114:1133–1142.

Tilman, D. 1977. Resource competition between planktonic algae: an experimental and theoretical approach. *Ecology* 58:338–348.

Tilman, D. 1978. Cherries, ants and tent caterpillars: timing of nectar production in relation to susceptibility of caterpillars to ant predation. *Ecology* 59:686–692.

Tilman, D. 1980. Resources: a graphical-mechanistic approach to competition and predation. *Am. Nat.* 116:362–393.

Tilman, D. 1982. *Resource Competition and Community Structure*. Princeton University Press, Princeton, N. J.

Tilman, D. 1983. Plant succession and gopher disturbance along an experimental gradient. *Oecologia* 60:285–292.

Tilman, D. 1984. Plant dominance along an experimental nutrient gradient. *Ecology* 65:1445–1453.

Tilman, D. 1985. The resource ratio hypothesis of succession. *Am. Nat.* 125, in press.

Tinkle, D. W. 1982. Results of experimental density manipulation in an Arizona lizard community. *Ecology* 63:135–146.

Toft, C. A. 1984. Activity budgets in two species of bee flies (*Lordotus:* Bombyliidae, Diptera): a comparison of species and sexes. *Behav. Ecol. Sociobiol.* 14:287–296.

Toft, C. A. 1985. Resource partitioning in amphibians and reptiles. *Copeia* 1985:1–21.

Toft, C. A., and T. W. Schoener. 1983. Abundance and diversity of orb spiders on 106 Bahamian islands: biogeography at an intermediate trophic level. *Oikos* 41:411–426.

Toft, C. A., and P. J. Shea. 1983. Detecting community-wide patterns: estimating power strengthens statistical inference. *Am. Nat.* 122:618–625.

Toft, C. A., D. L. Trauger, and H. W. Murdy. 1982. Tests for species interactions: breeding phenology and habitat use in subarctic ducks. *Am. Nat.* 120:586–613.

Tomanek, G. W., and G. K. Hulett. 1970. Effects of historical droughts on grassland vegetation in the central Great Plains. *In* W. Dort, Jr., and J. K. Jones, Jr., eds., *Pleistocene and Recent Environments of the Central Great Plains,* pp. 203–210.

University of Kansas Special Publications no. 3. University Press of Kansas, Lawrence, Kans.

Tomback, D. F. 1982. Dispersal of Whitebark Pine seeds by Clark's Nutcracker: a mutualism hypothesis. *J. Anim. Ecol.* 51:451–467.

Tomblin, J. F. 1975. The Lesser Antilles and Aves Ridge. *In* A. E. M. Nairn and F. G. Stehli, eds., *Ocean Basins and Margins. Vol. 3. The Gulf of Mexico and the Caribbean,* p. 467. Plenum, New York.

Tomich, P. Q. 1969. Mammals in Hawaii—a synopsis and notational bibliography. *Spec. Publ. Bernice P. Bishop Mus.* no. 57:1–238.

Tomkins, D. J., and W. F. Grant. 1977. Effects of herbicides on species diversity of two plant communities. *Ecology* 58:398–406.

Tomlinson, P. B., and M. H. Zimmermann, eds. 1978. *Tropical Trees as Living Systems*. Cambridge University Press, Cambridge.

Tonn, W. M. 1985. Density compensation in *Umbra-Perca* fish assemblages of northern Wisconsin lakes. *Ecology* 66:415–429.

Tonn, W. M., and J. J. Magnuson. 1982. Patterns in the species composition and richness of fish assemblages in northern Wisconsin lakes. *Ecology* 63:1149–1166.

Traub, R., and H. Starcke. 1980. *Fleas*. Proceedings of the International Conference on Fleas. Balkema, Rotterdam.

Travis, J. 1982. A method for the statistical analysis of time energy budgets. *Ecology* 63:19–25.

Trivers, R. L. 1972. Parental investment and sexual selection. *In* B. Campbell, ed., *Sexual Selection and the Descent of Man,* pp. 136–179. Aldine, Chicago.

Troyer, W. A. 1960. The Roosevelt elk on Afognak island, Alaska. *J. Wildl. Manage.* 24:15–21.

Tschumy, W. O. 1982. Competition between juveniles and adults in age-structured populations. *Theor. Popul. Biol.* 21:255–268.

Tsubaki, Y., and Y. Shiotsu. 1982. Group feeding as a strategy for exploiting food resources in the burnet moth *Pryeria sinica. Oecologia* 55:12–20.

Turbott, E. G. 1963. Three Kings Islands, New Zealand. *In* J. L. Gressitt, ed., *Pacific Basin Biogeography,* pp. 485–498. Bishop Museum Press, Honolulu.

Turelli, M. 1978. Does environmental variability limit niche overlap? *Proc. Natl. Acad. Sci. U.S.A.* 75:5085–5089.

Turelli, M. 1981. Niche overlap and invasion of competitors in random environments I. Models without demographic stochasticity. *Theor. Popul. Biol.* 20:1–56.

Turelli, M., and J. H. Gillespie. 1980. Conditions for the existence of stationary densities for some two-dimensional diffusion processes with applications in population biology. *Theor. Popul. Biol.* 17:167–189.

Turing, A. M. 1952. On the chemical basis of morphogenesis. *Philos. Trans. R. Soc. Lond. B.* 237:27–72.

Turkington, R., M. A. Cahn, A. Vardy, and J. L. Harper. 1979. The growth, distribution, and neighbor relationships of *Trifolium repens* in a permanent pasture. III. The establishment and growth of *Trifolium repens* in natural and perturbed sites. *J. Ecol.* 67:231–243.

Turkington, R., and J. L. Harper. 1979. The growth, distribution, and neighbor relationships of *Trifolium repens* in a permanent pasture. IV. Fine-scale biotic differentiation. *J. Ecol.* 67:245–254.

Turnbull, C. L., and D. C. Culver. 1983. The timing of seed dispersal in *Viola nuttallii:* attraction of dispersers and avoidance of predators. *Oecologia* 59:360–365.

Tyler, A. V., and R. S. Dunn. 1976. Ration, growth, measures of somatic and organ condition in relation to meal frequency in white flounder. *Pseudopleuronectes americanus,* with a hypothesis on population homeostasis. *J. Fish. Res. Board Can.* 33:63–75.

Udovic, D. 1981. Determinants of fruit set in *Yucca whipplei:* reproductive expenditure vs. pollinator availability. *Oecologia* 48:389–399.

Udovic, D., and C. Aker. 1981. Fruit abortion and the regulation of fruit number in *Yucca whipplei. Oecologia* 49:245–248.

Underwood, A. J. 1978. An experimental evaluation of competition between three species of intertidal prosobranch gastropods. *Oecologia* 33:185–202.

Underwood, A. J., and E. L. Denley. 1984. Paradigms, explanations, and generalizations in models for the structure of intertidal communities on rocky shores. *In* D. R. Strong, D. Simberloff, L. G. Abele, and A. B. Thistle, eds., *Ecological Communities: Conceptual Issues and the Evidence,* pp. 151–180. Princeton University Press, Princeton, N. J.

Underwood, A. J., E. J. Denley, and M. J. Moran. 1983. Experimental analyses of the structure and dynamics of mid-shore rocky intertidal communities in New South Wales. *Oecologia* 56:202–219.

Underwood, G. 1959. Anoles of the eastern Caribbean. Part III. Revisionary notes. *Bull. Mus. Comp. Zool. Harv. Univ.* 121:191–226.

Utida, S. 1957. Population fluctuation, an experimental and theoretical approach. *Cold Spring Harbor Symp. Quant. Biol.* 22:139–151.

Uyenoyama, M., and M. W. Feldman. 1980. Theories of kin and group selection: a population genetics perspective. *Theor. Popul. Biol.* 17:380–414.

Vaartaga, O. 1959. Evidence of photoperiodic ecotypes in trees. *Ecol. Monogr.* 29:91–111.

Vance, R. R. 1978a. Predation and resource partitioning in one predator–two prey model communities. *Am. Nat.* 112:797–813.

Vance, R. R. 1978b. A mutualistic interaction between a sessile marine clam and its epibionts. *Ecology* 59:679–685.

Vance, R. R. 1979. Effects of grazing by the sea urchin, *Centrostephanus coronatus,* on prey community composition. *Ecology* 60:537–546.

Vandermeer, J. H. 1969. The competitive structure of communities: an experimental approach with protozoa. *Ecology* 50:362–372.

Vandermeer, J. H. 1980. Indirect mutualism: variations on a theme by Stephen Levine. *Am. Nat.* 116:441–448.

Vandermeer, J. H., and D. H. Boucher. 1978. Varieties of mutualistic interaction in population models. *J. Theor. Biol.* 74:549–558.

Vander Wall, S. B., and R. P. Balda. 1977. Coadaptations of the Clark's Nutcracker and the piñon pine for efficient seed harvest and dispersal. *Ecol. Monogr.* 47:89–111.

Van Devender, T. R. 1977. Holocene woodlands in the southwestern deserts. *Science* 198:189–192.

Van Devender, E. R. In press. Pleistocene climates and endemism in the Chihuahuan Desert flora. *In* J. C. Barlow, B. N. Timmerman, and A. M. Powell, eds., *Proceedings of the Second Chihuahuan Desert Symposium.* Alpine, Tex.

Van Devender, T. R., J. L. Betancourt, and M. L. Wimberly. 1984. Biogeographical implications of a packrat midden sequence from the Sacramento Mountains, south-central New Mexico. *Quat. Res.* 22:344–360.

Van Devender, T. R., and T. L. Burgess. In press. Late Pleistocene woodlands in the Bolson de Mapimi: a refugium for the Chihuahuan Desert biota? *Quat. Res.*

Van Devender, T. R., and J. I. Mead. 1978. Early Holocene and late Pleistocene amphibians and reptiles in Sonoran Desert packrat middens. *Copeia* 1978:464–475.

Van Devender, T. R., K. B. Moodie, and A. H. Harris. 1976. The desert tortoise *(Gopherus agassizi)* in the Pleistocene of the northern Chihuahuan Desert. *Herpetologica* 32:298–304.

Van Devender, T. R., A. M. Phillips, and J. I. Mead. 1977. Late Pleistocene reptiles and small mammals from the lower Grand Canyon of Arizona. *Southwest. Nat.* 22:49–66.

Van Devender, T. R., and W. G. Spaulding. 1979. Development of vegetation and climate in the southwestern United States. *Science* 204:701–710.

Van Devender, T. R., and L. J. Toolin. 1983. Late Quaternary vegetation of the San Andres Mountains, Sierra County, New Mexico. *In* P. L. Eidenbach, ed., *The Pre-history of Rhodes Canyon: Survey and Mitigation,* pp. 33–54. Report to Holloman Air Force Base, Tularosa, N. Mex.

Van Tyne, J. 1951. The distribution of the Kirtland warbler (*Dendroica kirtlandii*). *Xth Int. Ornithol. Cong. Proc.* pp. 537–544.

Varley, G. C., G. R. Gradwell, and M. P. Hassell. 1973. *Insect Population Ecology: an Analytical Approach.* Blackwell, Oxford.

Vaurie, C. 1957. Systematic notes on palearctic birds, no. 26. Paridae: the *Parus caeruleus* complex. *Am. Mus. Novit.* no. 1833.

Vereshchagin, M. K., and G. F. Baryshnikov. 1984. Quaternary mammalian extinctions in northern Eurasia. 1984. *In* P. S. Martin and R. G. Klein, eds., *Quaternary Extinctions: a Prehistoric Revolution,* pp. 483–516. University of Arizona Press, Tucson.

Vermeij, G. J. 1983. Intimate associations and coevolution in the sea. *In* D. J. Futuyma and M. Slatkin, eds., *Coevolution,* pp. 311–327. Sinauer, Sunderland, Mass.

Verosoglou, D. S., and A. H. Fitter. 1984. Spatial and temporal patterns of growth and nutrient uptake of five co-existing grasses. *J. Ecol.* 72:259–272.

Via, S. 1984. The quantitative genetics of polyphagy in an insect herbivore. I. Genotype-environment interaction in larval performance on different host plant species. *Evolution* 38:881–895.

Victor, B. 1983. Recruitment and population dynamics of a coral reef fish. *Science* 219:419–420.

Virnstein, R. W. 1977. The importance of predation by crabs and fishes on benthic infauna in Chesapeake Bay. *Ecology* 58:1199–1218.

Volterra, F. 1926. Variations and fluctuations of the number of individuals in animal species living together. *J. Cons. Perm. Int. Ent. Mer.* 3:3–51.

Vuilleumier, B. S. 1971. Pleistocene changes in the fauna and flora of South America. *Science* 173:771–780.

Waage, J. K. 1979. The evolution of insect/vertebrate associations. *Biol. J. Linn. Soc.* 12:187–224.

Waage, J. K., and C. R. Davies. *Host-Mediated Competition in a Bloodsucking Insect Community.* Unpublished manuscript.

Waddington, K. D. 1983. Pollen flow and optimal outcrossing distance. *Am. Nat.* 122:147–151.

Wahl, E. W., and T. L. Lawson. 1970. The climate of the mid-nineteenth century United States compared to the current normals. *Monthly Weather Rev.* 98:259–265.

Wainhouse, D., and R. S. Howell. 1983. Intraspecific variation in beech scale populations and in susceptibility of their host *Fagus sylvatica. Ecol. Entomol.* 8:351–359.

Walker, D. 1970. The changing vegetation of the montane tropics. *Search* 1:217–221.

Walker, J., and R. K. Peet. 1984. Composition and species diversity of pine-wiregrass savannas of the Green Swamp, North Carolina. *Vegetatio* 55:163–179.

Walker, J., C. H. Thompson, I. F. Fergus, and B. R. Tunstall. 1981. Plant succession and soil development in coastal sand dunes of subtropical eastern Australia. *In* D. C. West, H. H. Shugart, and D. B. Botkin, eds., *Forest Succession Concepts and Application,* pp. 107–131. Springer-Verlag, New York.

Walkinshaw, L. H. 1983. *Kirtland's Warbler*. Cranbrook Institute of Science, Bloomfield Hills, Mich.

Wallace, A. R. 1876. *The Geographical Distribution of Animals*. Macmillan, London.

Wankowski, J. W. J., and J. E. Thorpe. 1979. The role of food particle size in the growth of juvenile Atlantic salmon (*Salmo salar* L.). *J. Fish. Biol.* 14:351–370.

Ware, D. M. 1975. Relation between egg size, growth, and natural mortality of larval fish. *J. Fish. Res. Board Can.* 32:2503–2512.

Warner, A., and B. Mosse. 1982. Factors affecting the spread of vesicular mycorrhizal fungi in soil. *New Phytol.* 90:529–536.

Warner, R. E. 1968. The role of introduced diseases in the extinction of the endemic Hawaiian avifauna. *Condor* 70:101–120.

Warner, R. R., and P. L. Chesson. 1985. Coexistence mediated by recruitment fluctuations: a field guide to the storage effect. *Am. Nat.*

Waser, N. M. 1983. Competition for pollination and floral character differences among sympatric plant species: a review of evidence. *In* C. E. Jones and R. J. Little, eds., *Handbook of Experimental Pollination Ecology,* pp. 277–293. Van Nostrand Reinhold, New York.

Waser, N. M., and M. V. Price. 1981. Effects of grazing on diversity of annual plants in the Sonoran Desert. *Oecologia* 50:407–411.

Wassersug, R. J., and K. Hoff. 1982. Developmental changes in the orientation of the anuran jaw suspension: a preliminary exploration into the evolution of anuran metamorphosis. *Evol. Biol.* 15:223–246.

Watson, G. M., and R. N. Mariscal. 1983. The development of a sea anemone tentacle specialized for aggression: morphogenesis and regression of the catch tentacle of *Haliplanella luciae* (Cnidaria, Anthozoa). *Biol. Bull.* 164:506–517.

Watt, A. S. 1947. Pattern and process in the plant community. *J. Ecol.* 35:1–22.

Watt, A. S. 1957. The effects of excluding rabbits from a grassland B (Mesobrometum) in Breckland. *J. Ecol.* 45:861–878.

Watt, A. S. 1960. The effects of excluding rabbits from acidiphilous grassland in Breckland. *J. Ecol.* 48:601–604.

Watt, K. E. F. 1968. *Ecology and Resource Management*. McGraw-Hill, New York.

Watt, W. B., P. C. Hoch, and S. G. Mills. 1974. Nectar resource use by *Colias* butterflies. *Oecologia* 14:353–374.

Watts, W. A., and J. P. Bradbury. 1982. Paleoecological studies at Lake Patzcuaro on the west-central Mexican Plateau and at Chalco in the Basin of Mexico. *Quat. Res.* 17:56–70.

Way, M. J. 1963. Mutualism between ants and honeydew-producing homoptera. *Annu. Rev. Entomol.* 8:307–344.

Way, M. J., and C. J. Banks. 1967. Intra-specific mechanisms in relation to the natural regulation of numbers of *Aphis fabae*. *Ann. Appl. Biol.* 59:189–205.

Weatherhead, P. J., and R. H. Robertson. 1979. Offspring quality and the polygyny threshold: "the sexy son hypothesis." *Am. Nat.* 113:201–208.

Weatherly, A. H. 1972. *Growth and Ecology of Fish Populations*. Academic Press, New York.

Weaver, J. E. 1965. *Native Vegetation of Nebraska*. University of Nebraska Press, Lincoln.

Weaver, J. E., and F. E. Clements. 1938. *Plant Ecology*. McGraw-Hill, New York.

Webb, L. J. 1958. Cyclones as an ecological factor in tropical lowland rainforest, North Queensland. *Aust. J. Bot.* 6:220–228.

Webb, S. D. 1974. *Pleistocene Mammals of Florida*. The University Presses of Florida, Gainesville.

Webb, T. W., III. 1982. Temporal resolution in Holocene pollen data. *Proc. 3rd N. Am. Paleontol. Conv.* 2:569–572.

Webb, T. W., III. 1984. Criteria for inferring climatic equilibrium in the temporal and spatial variation of plant taxa. *6th Int. Palynological Conf. Calgary,* Abstracts.

Webb, T. W., III, E. J. Cushing, and H. E. Wright, Jr. 1983. Holocene changes in the vegetation of the midwest. *In* H. E. Wright, Jr., ed., *Late-Quaternary Environments of the United States. Vol. II. The Holocene,* pp. 142–165. University of Minnesota Press, Minneapolis.

Weisbrot, D. R. 1966. Genotype interactions among competing strains and species of Drosophila. *Genetics* 53:427–435.

Weissman, D. B., and D. E. Rentz. 1976. Zoogeography of the grasshoppers and their relatives (Orthoptera) on the California Channel Islands. *J. Biogeogr.* 3:105–114.

Wellington, G. M. 1980. Reversal of digestive interactions between Pacific reef corals: mediation by sweeper tentacles. *Oecologia* 47:340–343.

Wells, L. 1969. Effects of alewife predation on zooplankton populations in Lake Michigan. *Limnol. Oceanogr.* 14:556–565.

Wells, P. V. 1966. Late Pleistocene vegetation and degree of pluvial climatic change in the Chihuahuan Desert. *Science* 153:970–975.

Wells, P. V. 1983. Paleobiogeography of montane islands in the Great Basin since the last glaciopluvial. *Ecol. Monogr.* 53:341–382.

Wells, P. V., and J. H. Hunziker. 1976. Origin of the creosote bush *(Larrea)* deserts of southwestern North America. *Ann. Mo. Bot. Gard.* 63:843–861.

Werner, E. E. 1977. Species packing and niche complementarity in three sunfishes. *Am. Nat.* 111:553–578.

Werner, E. E. 1984. The mechanisms of species interactions and community organization in fish. *In* D. R. Strong. D. Simberloff, L. G. Abele, and A. B. Thistle, eds., *Ecological Communities: Conceptual Issues and the Evidence,* pp. 360–382. Princeton University Press, Princeton, N. J.

Werner, E. E., and J. F. Gilliam. 1984. The ontogenetic niche and species interactions in size-structured populations. *Annu. Rev. Ecol. Syst.* 15:393–425.

Werner, E. E., J. F. Gilliam, D. J. Hall, and G. G. Mittelbach. 1983a. An experimental test of the effects of predation risk on habitat use in fish. *Ecology* 64:1540–1548.

Werner, E. E., and D. J. Hall. 1976. Niche shifts in sunfishes: experimental evidence and significance. *Science* 191:404–406.

Werner, E. E., and D. J. Hall. 1977. Competition and habitat shift in two sunfishes (Centrarchidae). *Ecology* 58:869–876.

Werner, E. E., and D. J. Hall. 1979. Foraging efficiency and habitat switching in competing sunfishes. *Ecology* 60:256–264.

Werner, E. E., D. J. Hall, D. R. Laughlin, D. J. Wagner, L. A. Wilsmann, and F. C. Funk. 1977. Habitat partitioning in a freshwater fish community. *J. Fish. Res. Board Can.* 34:360–370.

Werner, E. E., G. G. Mittelbach, D. J. Hall, and J. F. Gilliam. 1983b. Experimental tests of optimal habitat use in fish: the role of relative habitat profitability. *Ecology* 64:1525–1539.

Werner, P. A. 1979. Competition and coexistence of similar species. *In* O. T. Solbrig, S. Jain, G. B. Johnson, and P. H. Raven, eds., *Topics in Plant Population Biology,* pp. 287–310. Columbia University Press, New York.

Werner, P. A., and W. J. Platt. 1976. Ecological relationship of co-occurring goldenrods (*Solidago:* Compositae). *Am. Nat.* 110:959–971.

West, D. G. 1970. Pleistocene history of the British flora. *In* D. Walker and R. G. West, eds., *Studies in the Vegetational History of the British Isles,* pp. 1–11. Cambridge University Press, Cambridge.

West, R. G. 1964. Inter-relations of ecology and Quaternary paleobotany. *J. Anim. Ecol.* 33(suppl.):47–57.

West, R. G. 1980. Pleistocene forest history of East Anglia. *New Phytol.* 85:571–622.

Wethey, D. 1979. *Demographic Variation in Intertidal Barnacles.* Doctoral dissertation, University of Michigan, Ann Arbor, University Microfilms No. 80-07857.

Wethey, D. 1983. Geographic limits and local zonation: the barnacles *Semibalanus* (*Balanus*) and *Chthamalus* in New England. *Biol. Bull.* 165:330–341.

Wethey, D. S. 1984. Sun and shade mediate competition in the barnacles *Chthamalus* and *Semibalanus*. A field experiment. *Biol. Bull.* 167:176–185.

Whitaker, A. H. 1973. Lizard populations on islands with and without Polynesian rats, *Rattus exulans* (Peale). *Proc. N. Z. Ecol. Soc.* 20:121–130.

Whitcomb, R. F., J. F. Lynch, M. K. Klinkiewicz, C. S. Robbins, B. L. Whitcomb, and D. Bystrack. 1981. Effects of forest fragmentation on avifauna of the eastern deciduous forest. *In* R. L. Burgess and D. M. Sharpe, eds., *Forest Island Dynamics in Man-Dominated Landscapes,* pp. 125–292. Springer-Verlag, New York.

White, T. C. R. 1969. An index to measure weather induced stress of trees associated with outbreaks on psyllids in Australia. *Ecology* 50:905–909.

Whitehead, D. R., S. T. Jackson, M. C. Sheehan, and B. W. Leyden. 1982. Late-glacial vegetation associated with caribou and mastodon in central Indiana. *Quat. Res.* 17:241–257.

Whitham, T. 1983. Host manipulation of parasites: within-plant variation as a defense against rapidly evolving pests. *In* R. Denno and M. McClure, eds., *Variable Plants and Herbivores in Natural and Managed Systems,* pp. 15–42. Academic Press, New York.

Whitmore, T. C. 1974. Change with time and the role of cyclones in tropical rainforest on Kolombangara, Solomon Islands. *Commonw. For. Inst. Pap.* no. 46. Holywell, Oxford.

Whitmore, T. C. 1975. *Tropical Rain Forests of the Far East.* Oxford University Press (Clarendon Press), Oxford.

Whitmore, T. C. 1978. Gaps in the forest canopy. *In*

P. B. Tomlinson and M. H. Zimmerman, eds., *Tropical Trees as Living Systems*, pp. 639–655. Cambridge University Press, Cambridge.

Whittaker, R. H. 1951. A criticism of the plant association and climatic climax concepts. *Northwest Sci.* 25:17–31.

Whittaker, R. H. 1956. Vegetation of the Great Smoky Mountains. *Ecol. Monogr.* 26:1–80.

Whittaker, R. H. 1967. Gradient analysis of vegetation. *Biol. Rev.* 42:207–264.

Whittaker, R. H. 1972. Evolution and measurement of species diversity. *Taxon* 21:213–251.

Whittaker, R. H. 1975. *Communities and Ecosystems*, 2nd ed. Macmillan, New York.

Whittaker, R. H. 1977. Evolution of species diversity in land communities. *Evol. Biol.* 10:1–67.

Whittaker, R. H., and W. A. Niering. 1965. Vegetation of the Santa Catalina Mountains, Arizona. II. A gradient analysis of the south slope. *Ecology* 46:429–452.

Whittaker, R. H., and W. A. Niering. 1975. Vegetation of the Santa Catalina Mountains, Arizona. V. Biomass, production, and diversity along the elevation gradient. *Ecology* 56:771–790.

Whittam, T. S., and D. Siegel-Causey. 1981. Species interactions and community structure in Alaskan seabird colonies. *Ecology* 62:1515–1524.

Wiener, J. G., and W. R. Hanneman. 1982. Growth and condition of bluegills in Wisconsin lakes: effects of population density and lake pH. *Trans. Am. Fish. Soc.* 111:761–767.

Wiens, J. A. 1969. An approach to the study of ecological relationships among grassland birds. *Ornithol. Monogr.* 8:1–93.

Wiens, J. A. 1973a. Pattern and process in grassland bird communities. *Ecol. Monogr.* 43:237–270.

Wiens, J. A. 1973b. Interterritorial habitat variation in Grasshopper and Savannah sparrows. *Ecology* 54:877–884.

Wiens, J. A. 1974. Climatic instability and the "ecological saturation" of bird communities in North American grasslands. *Condor* 76:385–400.

Wiens, J. A. 1976. Population responses to patchy environments. *Annu. Rev. Ecol. Syst.* 7:81–120.

Wiens, J. A. 1977. On competition and variable environments. *Am. Sci.* 65:590–597.

Wiens, J. A. 1981. Scale problems in avian censusing. *Stud. Avian Biol.* 6:513–521.

Wiens, J. A. 1982. On size ratios and sequences in ecological communities: are there no rules? *Ann. Zool. Fenn.* 19:297–308.

Wiens, J. A. 1983a. Avian community ecology: an iconoclastic view. *In* A. H. Brush and G. A. Clark, Jr., eds., *Perspectives in Ornithology*, pp. 355–403. Cambridge University Press, Cambridge.

Wiens, J. A. 1983b. Interspecific competition [letter to the editor]. *Am. Sci.* 71:234–235.

Wiens, J. A. 1984a. On understanding a non-equilibrium world: myth and reality in community patterns and processes. *In* D. R. Strong, D. Simberloff, L. G. Abele, and A. B. Thistle, eds., *Ecological Communities: Conceptual Issues and the Evidence*, pp. 439–457. Princeton University Press, Princeton, N. J.

Wiens, J. A. 1984b. Resource systems, populations, and communities. *In* P. W. Price, C. N. Slobodchikoff, and W. S. Gaud, eds., *A New Ecology: Novel Approaches to Interactive Systems*, pp. 397–436. Wiley, New York.

Wiens, J. A. 1985a. Habitat selection in variable environments: shrubsteppe birds. *In* M. L. Cody, ed., *Habitat Selection in Birds*, pp. 227–251. Academic Press, New York.

Wiens, J. A. 1985b. Vertebrate responses to environmental patchiness in arid and semi-arid ecosystems. *In* S. T. A. Pickett and P. S. White, eds., *Natural Disturbance: the Patch Dynamics Perspective*, pp. 169–193. Academic Press, New York.

Wiens, J. A., and J. T. Rotenberry. 1979. Diet niche relationships among North American grassland and shrubsteppe birds. *Oecologia* 42:253–292.

Wiens, J. A., and J. T. Rotenberry. 1980a. Patterns of morphology and ecology in grassland and shrubsteppe bird populations. *Ecol. Monogr.* 50:287–308.

Wiens, J. A., and J. T. Rotenberry. 1980b. Bird community structure in cold shrub deserts: competition or chaos? *Proc. XVIIth Int. Ornithol. Congr., Berlin*, pp. 1063–1070.

Wiens, J. A., and J. T. Rotenberry. 1981a. Habitat associations and community structure of birds in shrubsteppe environments. *Ecol. Monogr.* 51:21–41.

Wiens, J. A., and J. T. Rotenberry. 1981b. Morphological size ratios and competition in ecological communities. *Am. Nat.* 117:592–599.

Wiens, J. A., J. T. Rotenberry, and B. Van Horne. Submitted. The response of shrubsteppe birds to an experimental habitat alteration: a lesson in the limitations of field experiments. *Ecology*.

Wilbur, H. M. 1972. Competition, predation, and the structure of the *Ambystoma-Rana sylvatica* community. *Ecology* 53:3–20.

Wilbur, H. M. 1980. Complex life cycles. *Annu. Rev. Ecol. Syst.* 11:67–93.

Wilbur, H. M. 1984. Complex life cycles and community organization in amphibians. *In* P. W. Price, C. N. Slobodchikoff, and W. S. Gaud, eds., *A New Ecology: Novel Approaches to Interactive Systems*, pp. 195–224. Wiley, New York.

Wilbur, H. M., P. J. Morin, and R. N. Harris. 1983.

Salamander predation and the structure of experimental communities: anuran responses. *Ecology* 64:1423–1429.

Wilbur, H. M., and J. Travis. 1984. An experimental approach to understanding pattern in natural communities. *In* D. R. Strong, D. Simberloff, L. G. Abele, and A. B. Thistle, eds., *Ecological Communities: Conceptual Issues and the Evidence*, pp. 113–122. Princeton University Press, Princeton, N. J.

Wilgen, B. W. van. 1982. Some effects of post-fire age on the above-ground plant biomass of fynbos (macchia) vegetation in South Africa. *J. Ecol.* 70:217–225.

Willems, J. H. 1983. Phytosociological and geographical survey of Mesobromion communities. *Vegetatio* 48:227–240.

Williams, A. H. 1981. Analyses of competitive interactions in a patchy back-reef environment. *Ecology* 62:1107–1120.

Williams, C. B. 1964. *Patterns in the Balance of Nature*. Academic Press, New York.

Williams, D. 1980. Dynamics of the pomacentrid community on small patch reefs in One Tree Island lagoon (Great Barrier Reef). *Bull. Mar. Sci.* 30:159–170.

Williams, E. E. 1972. The origin of faunas: evolution of lizard congeners in a complex island fauna—a trial analysis. *Evol. Biol.* 6:47–89.

Williams, E. E. 1976. West Indian anoles: a taxonomic and evolutionary summary. I. Introduction and a species list. *Breviora* 440:1–21.

Williams, E. E. 1983. Ecomorphs, faunas, island size, and diverse endpoints in island radiations of *Anolis*. *In* R. B. Huey, E. R. Pianka, and T. W. Schoener, eds., *Lizard Ecology: Studies of a Model Organism*, pp. 326–370. Harvard University Press, Cambridge, Mass.

Williams, G. C. 1966. *Adaptation and Natural Selection*. Princeton University Press, Princeton, N. J.

Williams, I. J. M. 1972. A revision of the genus *Leucadendron* (Proteaceae). *Contrib. Bolus Herb.* no. 3:1–425. (University of Cape Town, Rondebosch, South Africa.)

Williams, J. B., and G. O. Batzli. 1979. Competition among bark-foraging birds in central Illinois: experimental evidence. *Condor* 81:122–132.

Williams, R. B. 1975. Catch-tentacles in anemones: occurrence in *Haliplanella luciae* (Verill) and a review of current knowledge. *J. Nat. Hist.* 9:241–248.

Williams, R. B. 1978. Some recent observations on the acrorhagi of sea anemones. *J. Mar. Biol. Assoc. U.K.* 58:787–788.

Williamson, K. 1975. Birds and climatic change. *Bird Study* 22:143–164.

Williamson, M. 1983. The land-bird community of Skokholm: ordination and turnovers. *Oikos* 41:378–384.

Willis, A., and E. Yemm. 1961. Braunton Burrows: mineral nutrient status of the dune soils. *J. Ecol.* 49:377–390.

Willis, E. O. 1980. Species reduction in remanescent woodlots in southern Brazil. *Proc. 19th Int. Ornithol. Congr.* pp. 783–786.

Willis, E. O., and E. Eisenmann. 1979. A revised list of the birds of Barro Colorado Island, Panama. *Smithsonian Contrib. Zool.* no. 291.

Wilson, D. S. 1975. The adequacy of body size as a niche difference. *Am. Nat.* 109:769–784.

Wilson, D. S. 1976. Evolution of the level of communities. *Science* 192:1358–1360.

Wilson, D. S. 1980. *The Natural Selection of Populations and Communities*. Benjamin-Cummings, Menlo Park, Calif.

Wilson, D. S. 1982. Genetic polymorphism for carrier preference in a phoretic mite. *Ann. Entomol. Soc. Am.* 75:293–296.

Wilson, D. S. 1983a. The effect of population structure on the evolution of mutualism: a field test involving burying beetles and their phoretic mites. *Am. Nat.* 121:851–870.

Wilson, D. S. 1983b. The group selection controversy: history and current status. *Annu. Rev. Ecol. Syst.* 14:159–187.

Wilson, D. S., and R. K. Colwell. 1981. Evolution of sex ratio in structure demes. *Evolution* 35:882–897.

Wilson, D. S., and J. Fudge. 1984. Burying beetles: intraspecific interactions and reproductive success in the field. *Ecol. Entomol.* 9:195–203.

Wilson, D. S., W. G. Knollenberg, and J. Fudge. 1984. Species packing and temperature dependent competition among burying beetles (Silphidae, *Nicrophorus*). *Ecol. Entomol.* 9:205–216.

Wilson, E. O. 1961. The nature of the taxon cycle in the Melanesian ant fauna. *Am. Nat.* 95:169–193.

Wilson, E. O., and D. S. Simberloff. 1969. Experimental zoogeography of islands: deformation and monitoring techniques. *Ecology* 50:267–278.

Wilson, W. H. 1983. The role of density dependence in a marine infaunal community. *Ecology* 64:295–306.

Windberg, L. A., and L. B. Keith. 1976. Snowshoe hare population response to artificial high densities. *J. Mamm.* 57:523–553.

Wing, S. L., and L. J. Hickey. 1982. Time scales of megafloral assemblages (abstract). *J. Paleontol.* 56(Suppl. no. 2):30.

Wingate, D. 1965. Terrestrial herpetofauna of Bermuda. *Herpetology* 21:202–213.

Winstanley, D., R. Spencer, and K. Williamson.

1974. Where have all the Whitethroats gone? *Bird Study* 21:1–14.

Winston, P. H., and B. K. P. Horn. 1984. *LISP*, 2nd ed. Addison-Wesley, Reading, Mass.

Wirth, N. 1976. *Algorithms + Data Structures = Programs*. Prentice-Hall, Englewood Cliffs, N.J.

Wirth, N. 1984. History and goals of Modula-2. *Byte* 9:145–152.

Wise, D H. 1983. Competitive mechanisms in a food-limited species: relative importance of interference and exploitative interactions among labyrinth spiders. *Oecologia* 58:1–9.

Wise, D. H. 1984. The role of competition in spider communities: insights from field experiments with a model organism. *In* D. R. Strong, D. Simberloff, L. Abele, and A. B. Thistle, eds., *Ecological Communities: Conceptual Issues and the Evidence*, pp. 42–53. Princeton University Press, Princeton, N. J.

Wit, C. T. de. 1960. On competition. *Versl. Landbouwk. Onderz*. 66:1–82.

Witman, J. D., and T. H. Suchanek. 1984. Mussels in flow: drag and dislodgement by epizonas. *Mar. Ecol. Prog. Ser*. 16:259–268.

Wolf, P. de. 1973. Ecological observations on the mechanisms of dispersal of barnacle larvae during planktonic life and settling. *Neth. J. Sea Res*. 6:1–129.

Wolfe, J. A., and E. S. Barghoorn. 1960. Generic change in tertiary floras in relation to age. *Am. J. Sci*. 258-A:388–399.

Wolin, C. L., and L. R. Lawlor. 1985. Models of facultative mutualism: density effects. *Am. Nat*. in press.

Wollkind, D., A. Hastings, and J. Logan. 1982. Age structure in predator-prey systems. II. Functional response and stability and the paradox of enrichment. *Theor. Popul. Biol*. 21:57–68.

Wood, T. K. 1982. Ant-attended nymphal aggregations in the *Enchenopa binotata* complex (Homoptera: Membracidae). *Ann. Entomol. Soc. Am*. 75:649–653.

Woodin, S. A. 1976. Adult-larval interactions in dense infaunal assemblages: patterns of abundance. *J. Mar. Res*. 34:25–41.

Woodin, S. A. 1982. Browsing: important in marine sediments? Spionid polychaete examples. *J. Exp. Mar. Biol. Ecol*. 60:35–45.

Woodin, S A., and J. A. Yorke. 1975. Disturbance, fluctuating rates of resource recruitment, and increased diversity. *In* S. A. Levin, ed., *Ecosystem Analysis and Prediction*, pp. 38–41. Proceedings SIAM-SIMS Conference, Alta, Utah.

Woodring, W. P. 1954. Caribbean land and sea through the ages. *Bull. Geol. Soc. Am*. 65:719–732.

Woods, K. D. 1979. Reciprocal replacement and the maintenance of codominance in a beech-maple forest. *Oikos* 33:31–39.

Woods, K., and M. B. Davis. 1982. Sensitivity of Michigan pollen diagrams to Little Ice Age climatic change. *AMQUA Prog. Abstracts*, 1982, p. 181.

Wootten, R. 1973. The metazoan parasite-fauna of fish from Hanningfield Reservoir, Essex, in relation to features of the habitat and host populations. *J. Zool. Lond*. 171:323–331.

Wright, H. E., Jr. 1981. Vegetation east of the Rocky Mountains 18,000 years ago. *Quat. Res*. 15:113–125.

Wright, S. J. 1980. Density compensation in island avifaunas. *Oecologia* 45:385–389.

Wright, S. J. 1982. Character change, speciation and the higher taxa. *Evolution* 36:427–443.

Wright, S. J., and S. P. Hubbell. 1983. Stochastic extinction and reserve size: a focal species approach. *Oikos* 41:466–476.

Wu, L. S.-Y., and D. B. Botkin. 1980. Of elephants and men: a discrete, stochastic model for long-lived species with complex life histories. *Am. Nat*. 116:831–849.

Wydoski, R. S., and D. H. Bennett. 1981. Forage species in lakes and reservoirs of the western United States. *Trans. Am. Fish. Soc*. 110:764–771.

Wyles, J., and G. Gorman. 1980. The classification of *Anolis*: conflict between genetic and osteological interpretation as exemplified by *Anolis cybotes*. *J. Herpetol*. 14:149–153.

Yang, S., M. Soulé, and G. Gorman. 1974. *Anolis* lizards of the eastern Caribbean: a case study in evolution. I. Genetic relationships, phylogeny, and colonization sequence of the *roquet* group. *Syst. Zool*. 23:387–399.

Yapp, C. J., and S. Epstein. 1977. Climatic implications of D/H ratios of meteoric water over North America, 9500–22,000 B.P., as inferred from ancient wood and cellulose. *Earth Planetary Science Letters* 34:333–350.

Yarranton, G. A. 1967. Organismal and individualistic concepts and the choice of methods of vegetation analysis. *Vegetatio* 15:113–116.

Yeaton, R. I., and M. L. Cody. 1974. Competitive release in island Song Sparrow populations. *Theor. Popul. Biol*. 5:42–58.

Yeaton, R. I., and M. L. Cody. 1976. Competition and spacing in plant communities: the northern Mojave Desert. *J. Ecol*. 64:689–696.

Yeaton, R. I., and M. L. Cody. 1979. Distribution of cacti along environmental gradients in the Sonoran and Mojave deserts. *J. Ecol*. 67:529–541.

Yeaton, R. I., J. Travis, and E. Gilinsky. 1977. Competition and spacing in plant communities: the Arizona upland association. *J. Ecol.* 65:587–595.

Yeaton, R. I., R. W. Yeaton, J. P. Waggoner III, and J. E. Horenstein. 1985. The ecology of *Yucca* (Agavaceae) over an environmental gradient in the Mohave Desert: distribution and interspecific interactions. *J. Arid Envts.* 8:33–44.

Yodzis, P. 1976a. Species richness and stability of space-limited communities. *Nature* 264:540–541.

Yodzis, P. 1976b. The effects of harvesting on competitive systems. *Bull. Math. Biol.* 38:97–109.

Yodzis, P. 1977a. Limit cycles in space-limited communities. *Math. Biosci.* 37:19–22.

Yodzis, P. 1977b. Harvesting and limiting similarity. *Am. Nat.* 111:833–843.

Yodzis, P. 1978. *Competition for Space and the Structure of Ecological Communities.* Springer-Verlag, Berlin.

Yodzis, P. 1980. The connectance of real ecosystems. *Nature* 284:544–545.

Yodzis, P. 1982. The compartmentation of real and assembled ecosystems. *Am. Nat.* 120:551–570.

Yoshioka, P. M. 1982. Role of planktonic and benthic factors in the population dynamics of the bryozoan *Membranipora membranacea. Ecology* 63:457–468.

Zaret, T. M. 1980. *Predation and Freshwater Communities.* Yale University Press, New Haven, Conn.

Zaret, T. M., and R. T. Paine. 1973. Species introduction in a tropical lake. *Science* 182:449–455.

Zedler, J., and P. Zedler. 1969. Association of species and their relationship to microtopography within old fields. *Ecology* 50:432–442.

Zimmerman, M. 1982. The effect of nectar production on neighborhood size. *Oecologia* 52:104–108.

Zinsser, H. 1935. *Rats, Lice, and History.* Little, Brown, Boston.

Zwölfer, H. 1961. A comparative analysis of the parasitic complexes of the European fir budworm, *Choristoneura muritana* (Hb.) and the North American spruce budworm, *C. fumiferana* (Clem.). *Tech. Bull. Commonw. Inst. Biol. Control* 1:1–162.

Zwölfer, H. 1969. *Urophora siruna-seva* Hg. (Diptera: Trypetidae), a potential insect for the biological control of *Centaurea solstitialis* L. in California. *Tech. Bull. Commonw. Inst. Biol. Control* 11:105–155.

Zwölfer, H. 1974. Das Treffpunkt-Prinzip als Kommunikationsstrategie und Isolationsmechanismus bei Bohrfliegen (Diptera: Trypetidae). *Entomol. Ger.* 1:11–20.

Zwölfer, H. 1983. Life systems and strategies of resource exploitation in tephritids. *In* R. Cavalloro, ed., *Fruit Flies of Economic Importance,* pp. 16–30. Balkema, Rotterdam.

Index